AA006614

International Conference "Applied Mathematics, Computational Science and Mechanics: Current Problems" (AMSCM 2017)

Journal of Physics: Conference Series Volume 973

Voronezh, Russia
18 – 20 December 2017

ISBN: 978-1-5108-6041-4
ISSN: 1742-6588

Printed from e-media with permission by:

Curran Associates, Inc.
57 Morehouse Lane
Red Hook, NY 12571

Some format issues inherent in the e-media version may also appear in this print version.

Printed by Curran Associates, Inc. (2018)

For permission requests, please contact the Institute of Physics at the address below.

Institute of Physics
Dirac House, Temple Back
Bristol BS1 6BE UK

Phone: 44 1 17 929 7481
Fax: 44 1 17 920 0979

techtracking@iop.org

Additional copies of this publication are available from:

Curran Associates, Inc.
57 Morehouse Lane
Red Hook, NY 12571 USA
Phone: 845-758-0400
Fax: 845-758-2633
Email: curran@proceedings.com
Web: www.proceedings.com

Table of contents

Volume 973

International Conference "Applied Mathematics, Computational Science and Mechanics: Current Problems"

18–20 December 2017, Voronezh, Russian Federation

Accepted papers received: 16 February 2018
Published online: 19 March 2018

Preface

Papers

Mechanics

Differential equations

System analysis

Mathematical and computer software

Computer modelling

IOP Conf. Series: Journal of Physics: Conf. Series **973** (2018) 011001 doi:10.1088/1742-6596/973/1/011001

Preface

This conference keeps the tradition of having regular scientific and technical conferences on the urgent issues of applied mathematics, computer science and mechanics at the Voronezh State University, which has been developed over the past 20 years. Every year, well-known authorities, who represent various scientific areas, educational organizations, industrial enterprises and IT companies, gather at this scientific forum. In all these years, this conference has been supported by the Russian Foundation for Scientific Research (RFBR).

Within the framework of the conference, thorough analyses of modern scientific advances in fundamental and applied areas are made, original reports on the current trends in the development of modern mathematics, computer science and mechanics are made. In general, the work of the conference is aimed at identifying and systematizing the issues and trends in the development of the scientific areas of modern mathematics, computer science and mechanics; discussion of the results of solving scientific problems and projects; development of scientific cooperation; involving students and graduate students in scientific work.

The main scientific lines of orientation of the conference are the following:

development of essential foundations, mathematical models and numerical methods for solving scientific and applied problems of a wide range,

development and application of methods for system analysis of complex systems and processes including issues of optimization, control and decision-making,

detection of laws of deformation, damage and destruction of materials with complex properties,

study of problems of mathematical hydrodynamics,

creation of technologies and software for distributed and high-performance computing systems,

development of mathematical basis of modern information technologies.

Aleksandr Shashkin

Dean of the Faculty of Applied Mathematics, Informatics and Mechanics, Voronezh State University, Russian Federation

IOP Conf. Series: Journal of Physics: Conf. Series **973** (2018) 011001 doi:10.1088/1742-6596/973/1/011001

Organizing Institutions

The conference was organized by the Faculty of Applied Mathematics, Informatics and Mechanics, Voronezh State University, Russian Federation.

Faculty of Applied Mathematics, Informatics and Mechanics	**www.amm.vsu.ru**
Voronezh State University	**www.vsu.ru**

Supporting Institutions

The conference was organized with the support of the Ministry of Education and Science of the Russian Federation and Russian Foundation for Basic Research.

Ministry of Education and Science of the Russian Federation	https://минобрнауки.рф
Russian Foundation for Basic Research	www.rfbr.ru

IOP Conf. Series: Journal of Physics: Conf. Series **973** (2018) 011001 doi:10.1088/1742-6596/973/1/011001

Organizing Committee

Prof. Shashkin A.I. (Voronezh)
Prof. Jorg Becker (Münster, Germany)
Prof. Semyon Levitsky (Israel)
Assoc. Prof. Andrew Schumann (Rzeszów, Poland)
Prof. Algazinov E.K. (Voronezh)
Assoc. Prof. Aristova E.M. (Voronezh)
Prof. Baev A.D. (Voronezh)
Prof., Corresponding Member, Russian Academy of Sciences Burenin A.A. (Komsomolsk-on-Amur)
Prof., Corresponding Member, Russian Academy of Sciences Voevodin V.V. (Moscow)
Prof. Vervejko N.D. (Voronezh)
Prof. Zhabko A.P. (Saint Petersburg)
Prof. Zadorozhnij V.G. (Voronezh)
Prof., Academician of Russian Academy of Sciences Ievlev V.M. (Voronezh)
Prof., Academician of Academy of Sciences of the Republic of Tajikistan Ilolov Mamadsho (Tajikistan)
Assoc. Prof. Kaplieva N.A. (Voronezh)
Prof. Kovalev A.V. (Voronezh)
Prof. Kovalev V.A. (Moscow)
Assoc. Prof. Korolkov O. G. (Voronezh)
Assoc. Prof. Kuznecov A.V. (Voronezh)
Prof. Ledeneva T.M. (Voronezh)
Prof., Corresponding Member, Russian Academy of Sciences Lomakin E.V. (Moscow)
Prof. Markin A.A. (Tula)
Prof. Matveev M.G. (Voronezh)
Prof. Matchenko N.M. (Tula)
Prof. Mahortov S.D. (Voronezh)
Assoc. Prof. Medvedeva O.A. (Voronezh)
Assoc. Prof. Medvedev S.N. (Voronezh)
Prof. Mirsalimov V.M. (Azerbaijan)
Prof., Academician of Russian Academy of Sciences Moiseev E.I. (Moscow)
Prof., Academician of Russian Academy of Sciences Morozov N.F. (Saint Petersburg)
Prof. Petrosyan L.A. (Saint Petersburg)
Prof. Radaev Yu.N. (Moscow)
Prof. Radchenko V.P. (Samara)
Prof. Revuzhenko A.F. (Novosibirsk)
Prof. Sirota A.A. (Voronezh)
Prof. Sporyhin A.N. (Voronezh)
Prof. Timoshenko Yu.K. (Voronezh)
Prof. Chigarev A.V. (Minsk, Belarus)

Conference Paper Review Process

All papers published in this volume of *IOP Journal of Physics: Conference Series* have been peer reviewed through processes administered by the Editors. Reviews were conducted by organizing committee to the professional and scientific standards expected of a proceedings journal published by IOP Publishing. Conference organizers would like to thank all the reviewers for their time and effort.

IOP Conf. Series: Journal of Physics: Conf. Series **973** (2018) 011002 doi:10.1088/1742-6596/973/1/011002

Peer review statement

All papers published in this volume of *Journal of Physics: Conference Series* have been peer reviewed through processes administered by the proceedings Editors. Reviews were conducted by expert referees to the professional and scientific standards expected of a proceedings journal published by IOP Publishing.

Determination of the critical plane and durability estimation for a multiaxial cyclic loading

N G Burago[1], A D Nikitin[2,3], I S Nikitin[2,3,*], V L Yakushev[2]

[1]Ishlinsky Institute for Problems in Mechanics of RAS, Pr. Vernadskogo 101-1, 119536 Moscow, Russia
[2]Institute of Computer Aided Design of RAS, 2-ya Brestskaya ul. 19/18, 123056 Moscow, Russia
[3]Moscow Aviation Institute (National Research University), Volokolamskoe shosse 4, 125993 Moscow, Russia

E-mail: *i_nikitin@list.ru

Abstract. An analytical procedure is proposed to determine the critical plane orientation according to the Findley criterion for the multiaxial cyclic loading. The cases of in-phase and anti-phase cyclic loading are considered. Calculations of the stress state are carried out for the system of the gas turbine engine compressor disk and blades for flight loading cycles. The formulas obtained are used for estimations of the fatigue durability of this essential element of structure.

1. Introduction

Modern approaches to the construction of multi-axial fatigue fracture criteria often use the notion of the critical plane [1-6]. The first criterion with determination of critical plane for classical low-cyclic (LCF, conditionally $10^3 < N < 10^5$) and high-cyclic (HCF, conditionally $10^6 < N < 10^8$) fatigue failure was proposed by Findley [7]. According to this criterion, the destruction in the cyclic loading process occurs along so called critical plane where the maximum of the combination reaches a certain critical value. To determine service life of a sample up to its fatigue failure under the uniaxial cyclic loading, there is the Baskin relation [8], which is an analytical representation of the fatigue curve for various coefficients of the asymmetry of the cycle. The generalization of the Baskin relation in the case of the multiaxial stress state for the Findley criterion has the form:

$$(\Delta\tau_n / 2 + \alpha_F \sigma_n)_{MAX_n} = S_F + A_F N^{\beta_F}, \quad \sigma_n = \mathbf{n}\cdot\boldsymbol{\sigma}\cdot\mathbf{n}, \quad \tau_n = \left|\boldsymbol{\sigma}\cdot\mathbf{n} - (\mathbf{n}\cdot\boldsymbol{\sigma}\cdot\mathbf{n})\mathbf{n}\right|$$

Here $\boldsymbol{\sigma}$ is the stress tensor, which determines the stress state in the point of the sample, parameters $\beta_F < 0$, α_F, S_F, A_F are defined experimentally, N is the number of cycles to failure, $\Delta\tau_n$ is the range of the tangential stress per cycle, σ_n is the normal stress on critical plane, $\mathbf{n}$ is the unit normal vector to the critical plane.

Calculation of durability by the Findlay criterion requires the determination of the orientation ($\mathbf{n}$) of the critical plane in a given material point. It means to find the plane ($\mathbf{n}$) where during cycle

$t \in [0,T]$ the Findlay's function $F = \Delta\tau_n + 2\alpha_F \max_{t\in[0,T]} \sigma_n$ is reached its maximum. For a multiaxial stressed state this is not easy task, which, as a rule, is to be solved numerically. Below we propose analytical solution of this problem for in-phase and anti-phase cyclic loading.

2. Determination of the critical plane for in-phase multiaxial stressed state

Let us consider a case of triaxial cyclic loading in a coordinate system connected with the principal axes of the stress tensor. We assume that these axes are not changed during the cycle and the principal values of the stress tensor vary according to a harmonic law without a phase shift relative to each other:

$$\sigma_1(t) = \sigma_{1m} + \sigma_{1a}\sin\omega t\,,\ \sigma_2(t) = \sigma_{2m} + \sigma_{2a}\sin\omega t\,,\ \sigma_3(t) = \sigma_{3m} + \sigma_{3a}\sin\omega t\,,\ \sigma_{1,2,3a} \geq 0$$

where the additional index "m" denotes the average values of the stresses per cycle, and the index "a" denotes their amplitudes.

The ranges of the principal stresses in the cycle are

$$\Delta\sigma_1 = 2\sigma_{1a} \geq 0\,,\ \Delta\sigma_2 = 2\sigma_{2a} \geq 0\,,\ \Delta\sigma_3 = 2\sigma_{3a} \geq 0\,.$$

We choose the principal axes so that the maxima of the principal stresses satisfy the inequalities

$$\Sigma_1 \geq \Sigma_2 \geq \Sigma_3\,, \text{ where } \Sigma_1 = \sigma_{1m} + \sigma_{1a}\,,\ \Sigma_2 = \sigma_{2m} + \sigma_{2a}\,,\ \Sigma_3 = \sigma_{3m} + \sigma_{3a}\,.$$

We use the following notation:

$$\Sigma_{12} = \sigma_1 - \sigma_2\,,\ \Sigma_{13} = \sigma_1 - \sigma_3\,,\ \Sigma_{23} = \sigma_2 - \sigma_3\,,$$
$$\Delta\sigma_{12} = \Delta\sigma_1 - \Delta\sigma_2\,,\ \Delta\sigma_{13} = \Delta\sigma_1 - \Delta\sigma_3\,,\ \Delta\sigma_{23} = \Delta\sigma_2 - \Delta\sigma_3\,.$$

In the coordinate system associated with the principal axes of the stress tensor we have the following formulas for the normal and tangential stresses on the plane with normal n_k (summation over the repeated indices k and l):

$$\sigma_n = \sigma_{kl} n_k n_l = \sigma_k n_k^2\,,\quad \tau_i = \sigma_{ik} n_k - \sigma_n n_i = (\sigma_i - \sigma_n) n_i$$

For ranges of the tangential stress components and their modulus, we have:

$$\Delta\tau_i = (\Delta\sigma_i - \Delta\sigma_n) n_i\,,\ \Delta\tau = \sqrt{\sum_{k=1}^{3} (\Delta\sigma_k - \Delta\sigma_n)^2 n_k^2}\,,\ \Delta\sigma_n = \Delta\sigma_k n_k^2$$

Finally

$$\Delta\tau = \sqrt{\Delta\sigma_k^2 n_k^2 - (\Delta\sigma_k n_k^2)^2}$$

This formula can be transformed to the form:

$$\Delta\tau = \sqrt{\Delta\sigma_k^2 n_k^2 - (\Delta\sigma_k n_k^2)^2} = \sqrt{(\Delta\sigma_1 - \Delta\sigma_2)^2 n_1^2 n_2^2 + (\Delta\sigma_1 - \Delta\sigma_3)^2 n_1^2 n_3^2 + (\Delta\sigma_2 - \Delta\sigma_3)^2 n_2^2 n_3^2}$$

In the notation adopted, we obtain the compact formula

$$\Delta\tau_n = \sqrt{(\Delta\sigma_{12})^2 n_1^2 n_2^2 + (\Delta\sigma_{13})^2 n_1^2 n_3^2 + (\Delta\sigma_{23})^2 n_2^2 n_3^2}$$

Excluding n_1^2 we obtain the formula used to calculate the maximum of the function F:

$$\Delta\tau = \sqrt{\Delta\sigma_{12}^2 n_2^2 + \Delta\sigma_{13}^2 n_3^2 - \left(\Delta\sigma_{12} n_2^2 + \Delta\sigma_{13} n_3^2\right)^2}$$

Taking these formulas into account, we define the orientation of the critical plane by the components of the normal:

$$x_1 = n_1^2 \geq 0,\ x_2 = n_2^2 \geq 0,\ x_3 = n_3^2 \geq 0,\ x_1 + x_2 + x_3 = 1.$$

It can be shown that the problem of determining the critical plane for a triaxial stressed state reduces to determining the maximum of the function

$$F(x_2,x_3) = \sqrt{\Delta\sigma_{12}^2 x_2 + \Delta\sigma_{13}^2 x_3 - \left(\Delta\sigma_{12} x_2 + \Delta\sigma_{13} x_3\right)^2} + 2\alpha_F\sigma_1 - 2\alpha_F\left(\Sigma_{12}\, x_2 + \Sigma_{13}\, x_3\right)$$

under constraints $0 \leq x_2 + x_3 \leq 1$, $x_2 \geq 0$, $x_3 \geq 0$.

Below we present the results of solving this problem for all possible values of the maxima and variations of the principal stresses.

Firstly, we define the extremum of a function inside the zone of the constraints

$$0 < x_2 + x_3 < 1,\ x_2 > 0,\ x_3 > 0:\ \partial F(x_2,x_3)/\partial x_2 = 0,\ \partial F(x_2,x_3)/\partial x_3 = 0$$

Depending on the magnitude of the principal stresses, we have the following cases.

2.1. Case 1

If $\Sigma_{12}/\Delta\sigma_{12} \neq \Sigma_{13}/\Delta\sigma_{13}$, $\Delta\sigma_{12} \neq 0$, $\Delta\sigma_{13} \neq 0$ then

I-a. For the case $\Delta\sigma_{12} \neq \Delta\sigma_{13}$ conditions of extremum for $F(x_2,x_3)$:

$$x_2 = \Delta_2/(\Delta\sigma_{12}\Delta\sigma_{23})\quad x_3 = \Delta_3/(\Delta\sigma_{13}\Delta\sigma_{23})$$

I-b. For the case $\Delta\sigma_{12} = \Delta\sigma_{13} \neq 0$ conditions of extremum for $F(x_2,x_3)$:

$$x_2 + x_3 = \Delta_S/(\Delta\sigma_{12}\Delta)$$

where

$$\beta_{12} = 4\alpha_F\,\Sigma_{12}/\Delta\sigma_{12},\ \beta_{13} = 4\alpha_F\,\Sigma_{13}/\Delta\sigma_{13},\ \Delta = 2(\beta_{13} - \beta_{12}) \neq 0$$
$$\Delta_S = \Delta\sigma_{12}\beta_{13} - \Delta\sigma_{13}\beta_{12},\ \Delta_T = 2(\Delta\sigma_{13} - \Delta\sigma_{12})$$
$$\Delta_2 = \Delta\sigma_{13}\,\Delta_S/\Delta - \left(\Delta_S^2 + \Delta_T^2\right)/\Delta^2,\ \Delta_3 = \left(\Delta_S^2 + \Delta_T^2\right)/\Delta^2 - \Delta\sigma_{12}\,\Delta_S/\Delta$$

2.2. Case 2

If $\Sigma_{12}/\Delta\sigma_{12} = \Sigma_{13}/\Delta\sigma_{13}$ then $\Delta\sigma_{12} = \Delta\sigma_{13} \neq 0$, $\Sigma_{12} = \Sigma_{13}$.

Conditions of extremum $F(x_2,x_3)$: $x_2 + x_3 = \left(1 - \beta_{12}/\sqrt{4 + \beta_{12}^2}\right)/2$

Values (x_2, x_3) should satisfy inequalities: $x_2 > 0$, $x_3 > 0$, $x_2 + x_3 < 1$

and conditions of maximum $F(x_2,x_3)$: $\Delta\sigma_{12}^2 + 4\Delta\sigma_{13}\Delta\sigma_{23}x_3 > 0$, $\Delta\sigma_{13}^2 - 4\Delta\sigma_{12}\Delta\sigma_{23}x_2 > 0$

The inequalities follow from the condition

$$\frac{\partial^2 F}{\partial x_2^2}(dx_2)^2 + 2\frac{\partial^2 F}{\partial x_2 \partial x_3}dx_2 dx_3 + \frac{\partial^2 F}{\partial x_3^2}(dx_3)^2 < 0$$

If such values of principal stresses do not exist, then it needs to seek maximum $F(x_2,x_3)$ on boundaries: $x_2 = 0$, or $x_3 = 0$, or $x_2 + x_3 = 1$.

2.3. *Case 3*

In this case maximum $F(x_2, x_3)$ is reached for values:

$$x_2 = 0,\ x_3 = \left(1 - \beta_{13}/\sqrt{4 + \beta_{13}^2}\right)/2$$

$$x_3 = 0,\ x_2 = \left(1 - \beta_{12}/\sqrt{4 + \beta_{12}^2}\right)/2$$

$$x_2 + x_3 = 1,\ x_2 = \left(1 + \beta_{23}/\sqrt{4 + \beta_{23}^2}\right)/2,\ x_3 = \left(1 - \beta_{23}/\sqrt{4 + \beta_{23}^2}\right)/2$$

From these three pairs (x_2, x_3) we choose one that provides maximum value of function $F(x_2, x_3)$.

So, for the case of multiaxial stressed state the critical plane is found analytically. Knowing components of normal to the critical plane it is possible to calculate Findley's function and corresponding number of cycles to fracture N [9].

3. Determination of the critical plane for anti-phase multiaxial stressed state

In the coordinate system connected with the principal axes of the stress tensor, an anti-phase harmonic cyclic loading can always be represented in the form:

$$\sigma_1(t) = \sigma_{1m} + \sigma_{1a} \sin \omega t,\ \sigma_2(t) = \sigma_{2m} + \sigma_{2a} \sin \omega t,$$

$$\sigma_3(t) = \sigma_{3m} + \sigma_{3a} \sin(\omega t + \pi) = \sigma_{3m} - \sigma_{3a} \sin \omega t,\ \sigma_{1,2,3a} \geq 0$$

In this case

$$\Delta\sigma_1 = 2\sigma_{1a} \geq 0,\ \Delta\sigma_2 = 2\sigma_{2a} \geq 0,\ \Delta\sigma_3 = -2\sigma_{3a} \leq 0$$

and

$$\Delta\tau_n = \sqrt{(\Delta\sigma_{12})^2 n_1^2 n_2^2 + (\Delta\sigma_{13})^2 n_1^2 n_3^2 + (\Delta\sigma_{23})^2 n_2^2 n_3^2}\ .$$

The normal stress on the plane with the normal **n** depends on time:

$$\sigma_n(t) = (\sigma_{1m} n_1^2 + \sigma_{2m} n_2^2 + \sigma_{3m} n_3^2) + (\sigma_{1a} n_1^2 + \sigma_{2a} n_2^2 - \sigma_{3a} n_3^2) \sin \omega t$$

Therefore to determine $\max_{t \in [0,T]} \sigma_n$ we consider two cases.

3.1. *Case 1*

If $\sigma_{1a} n_1^2 + \sigma_{2a} n_2^2 - \sigma_{3a} n_3^2 \geq 0$, then $\max_{t \in [0,T]} \sigma_n = \Sigma_1 n_1^2 + \Sigma_2 n_2^2 + \Sigma_3 n_3^2$ for $\sin \omega t = 1$,

where $\Sigma_1 = \sigma_{1m} + \sigma_{1a}$, $\Sigma_2 = \sigma_{2m} + \sigma_{2a}$, $\Sigma_3 = \sigma_{3m} - \sigma_{3a}$.

We renumber the values Σ_k, $k = 1,2,3$ so that they satisfy the inequalities $\Sigma_1 \geq \Sigma_2 \geq \Sigma_3$. After that the same analysis and calculation of the orientation of the critical plane are carried out for the renumbered values Σ_k, $\Delta\sigma_k$, $k = 1,2,3$ as for in-phase cycle loading.

The resulting values of the normal components are verified to satisfy condition $\sigma_{1a} n_1^2 + \sigma_{2a} n_2^2 - \sigma_{3a} n_3^2 \geq 0$. If this condition is not satisfied we consider the following alternative case.

3.2. *Case 2*

If $\sigma_{1a} n_1^2 + \sigma_{2a} n_2^2 - \sigma_{3a} n_3^2 \leq 0$, then $\max_{t \in [0,T]} \sigma_n = \Sigma_1 n_1^2 + \Sigma_2 n_2^2 + \Sigma_3 n_3^2$ for $\sin \omega t = -1$,

where $\Sigma_1 = \sigma_{1m} - \sigma_{1a}$, $\Sigma_2 = \sigma_{2m} - \sigma_{2a}$, $\Sigma_3 = \sigma_{3m} + \sigma_{3a}$.

In this case we also renumber the values Σ_k, $k=1,2,3$ so that they satisfy the inequalities $\Sigma_1 \geq \Sigma_2 \geq \Sigma_3$. After that the critical plane is defined by renumbered values Σ_k, $\Delta\sigma_k$, $k=1,2,3$ in the same manner as for in-phase cycle loading. The resulting values of the normal components are verified to satisfy condition $\sigma_{1a}n_1^2 + \sigma_{2a}n_2^2 - \sigma_{3a}n_3^2 \leq 0$.

If both conditions are satisfied the orientation of the chosen unit normal should provide the maximal value of the Findley function on the plane.

4. Durability estimates for gas turbine engine compreessor disc in flight cycles

As an example let us consider fatigue durability calculation for typical compressor disk during flight cycles (low cycle fatigue mode). It is assumed that during loading cycle, the maximum loads are achieved at the cruising speed of the aircraft and the corresponding rotational speeds of the compressor disk. A multiaxial Findley's criterion for LCF regime is used to calculate the safe operation time of the disk (number of flight cycles N to fracture).

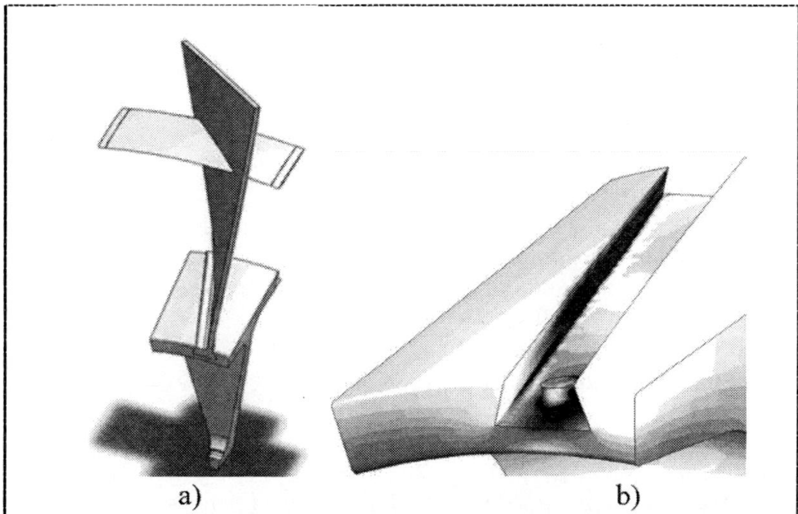

Figure 1. Disk-blade sector under action of centrifugal forces (a), maximum principal stress in groove of disk-blade connection (b).

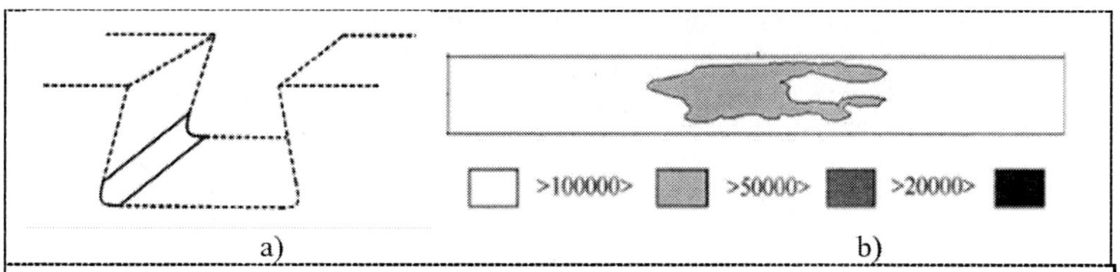

Figure 2. Selected zone of maximum stress concentration (a) and durability distribution in selected zone (b).

Using to the procedure proposed in [10] the parameters of the fatigue criterion were determined from the results of two uniaxial fatigue tests with the coefficients of asymmetry of the cycle $R=-1$, $R=0$. For this purpose calculation of the stress-strain state of the disc with blades under action of centrifugal forces was performed.

IOP Conf. Series: Journal of Physics: Conf. Series **973** (2018) 012001 doi:10.1088/1742-6596/973/1/012001

The following input data for loading were used: angular rotation velocity $\omega = 314\,\text{rad/s}$ (that corresponds to $3000\,\text{RPM}$), velocity head at infinity $\rho v_\infty^2/2 = 26000\,\text{N/m}^2$ (that corresponds to cruising speed $200\,\text{m/s}$ and air density $1.3\,\text{kg/m}^3$).

Elastic material properties were taken as: $E = 116\,\text{MPa}$, $\nu = 0.32$, $\rho = 4370\,\text{kg/m}^3$ for disk (titanium alloy), $E = 69\,\text{MPa}$, $\nu = 0.33$, $\rho = 2700\,\text{kg/m}^3$ for blades (aluminum alloy) and $E = 207$ MPa, $\nu = 0.27$, $\rho = 7860\,\text{kg/m}^3$ for fixing pins (steel).

Firstly, stress-strain state (SSS) was calculated for the whole disk and 32 blades. Secondly, SSS was recalculated using refined grid for typical sector of the disk with blade that is shown in Figure 1-a.

Calculations showed that safe operation life reaches its minimum near the contact zone between the disk and the blades (in a dovetail-type connection). Figure 1-b shows the zone of concentration of the maximum tensile stresses in the left (rounded) corner of the groove of contact connection. So the lowest durability was obtained for left corner of the groove.

The left corner of the groove is shown in Figure 2-a using fat continuous lines. Figure 2-b shows the calculated values N (the number of flight cycles to failure) for the chosen criterion for multiaxial fatigue failure. In Figure 2-b the dimensionless coordinate of the curvature of the left corner of the connection groove is plotted along the horizontal axis, and the dimensionless coordinate along the groove depth is plotted along the vertical axis. The smallest value was ~ 30000 cycles. If we take that the average flight cycle time is equal to 2 hours then in real time the durability of the safe exploitation for disk-blade structural element will be 60 000 hours.

5. Conclusions

The critical plane orientation in the Findley's criterion is found analytically for the cases of in-phase and anti-phase cyclic multiaxial loading. This solution is used to estimate the durability of safe operation life for gas turbine engine compressor disk under cyclic flight loading.

Acknowledgments

The research was partially supported under the project RSF No. 16-19-10376.

References

[1] Matake T 1977 *Bull. JSME* **20** 257
[2] McDiarmid D 1999 *Fatigue Fract. Eng. Mater. Struct.* **17** 1475
[3] Morel F 2000 *Int. J. of Fatigue* **22** 101
[4] Papadopoulos I, Davoli P and Gorla C 1997 *Int. J. of Fatigue* **19** 219
[5] Papadopoulos I 2001 *Int. J. of Fatigue* **23** 839
[6] Carpinteri A, Karolczuk A, Macha E and Vantadori S 2002 *Int. J. of Fatigue* **24** 87
[7] Fatemi A and Socie D 1988 *Fatigue Fract. Eng. Mater. Struct.* **11** 149
[8] Findley W 1959 *J. of Eng. for Indust.* 301
[9] Collins J 1993 *Failure of Materials in Mechanical Design: Analysis, Prediction, Prevention.* (New York: John Wiley) p 654
[10] Bourago N, Zhuravlev A and Nikitin I 2011 *Mechanics of Solids* **46** 828

IOP Conf. Series: Journal of Physics: Conf. Series **973** (2018) 012002 doi:10.1088/1742-6596/973/1/012002

Analysis of the stress field in a wedge using the fast expansions with pointwise determined coefficients

A D Chernyshov[1], **V V Goryainov**[2,3] **and A A Danshin**[4]

[1] Voronezh State University of Engineering Technologies, Revolution av., 19, Voronezh, Russia
[2] Voronezh State Technical University, 20 let Oktyabrya st., 84, Voronezh, Russia
[3] Voronezh State University, University sq., 1, Voronezh, Russia
[4] PJSC Rostelecom, Revolution av., 35, Voronezh, Russia

E-mail: chernyshovad@mail.ru, gorvit77@mail.ru, hb79@pisem.net

Abstract. The stress problem for the elastic wedge-shaped cutter of finite dimensions with mixed boundary conditions is considered. The differential problem is reduced to the system of linear algebraic equations by applying twice the fast expansions with respect to the angular and radial coordinate. In order to determine the unknown coefficients of fast expansions, the pointwise method is utilized. The problem solution derived has explicit analytical form and it's valid for the entire domain including its boundary. The computed profiles of the displacements and stresses in a cross-section of the cutter are provided. The stress field is investigated for various values of opening angle and cusp's radius.

1. Introduction

A lot of research is devoted to consideration of elastic infinite wedge-shaped domains. The stress field for the antiplane deformation of elastic wedge is examined in [1-3]. The plane deformation of the wedge is discussed in [4-6]. Some particular results are obtained by applying Mellin transforms [1,5,6] and using functions of a complex variable [2,3,7]. The exact solutions to the problems of plane deformation of the elastic wedge under zero loading on its lateral edges can be derived using Wiener-Hopf method [4]. The wedge has straight cracks on its axis of symmetry. An intrusion of the wedge into the plastic half-space has been investigated in the monograph [8]. The loaded wedge with smooth edges was considered in [9]. The particular solutions for the truncated circular sector are given in [10]. Some studies are devoted to three-dimensional problems for an elastic wedge [11,12]. In [11] the explicit matrix algorithm for three-dimensional wedge problem solution is proposed. In [12] one of the wedge's surfaces is reinforced with a coating of Winkler type. On the other surface some arbitrary boundary conditions are set. In this case, the methods of nonlinear boundary integral equations and of successive approximations are used.

In this work the new analytical method of fast expansions [13] will be applied. It allows to obtain the high accuracy solution to the problem under study in an explicit analytical form. The method of fast expansions is applicable for solution to the problems associated with partial differential [14], integro-differential [13] and ordinary differential [15] equations.

2. Materials and methods

The equilibrium equations in terms of displacements in cylindrical coordinates for the wedge-shaped domain Ω ($0<r_0\le r\le R,\ 0\le\theta\le\theta_0$) have form

$$\begin{aligned}&\frac{2}{1-\nu}\frac{\partial}{\partial r}\left(\frac{\partial U}{\partial r}+\frac{U}{r}\right)+\frac{1+\nu}{(1-\nu)r}\frac{\partial^2 V}{\partial r\partial\theta}-\frac{3-\nu}{(1-\nu)r^2}\frac{\partial V}{\partial\theta}+\frac{1}{r^2}\frac{\partial^2 U}{\partial\theta^2}=0,\\&\frac{1+\nu}{(1-\nu)r}\frac{\partial^2 U}{\partial r\partial\theta}+\frac{3-\nu}{(1-\nu)r^2}\frac{\partial U}{\partial\theta}+\frac{\partial}{\partial r}\left(\frac{\partial V}{\partial r}+\frac{V}{r}\right)+\frac{2}{1-\nu}\frac{1}{r^2}\frac{\partial^2 V}{\partial\theta^2}=0.\end{aligned}\tag{1}$$

The displacements U and V are defined at the rear edge $r=R$:

$$U\big|_{r=R}=U_R(\theta),\ V\big|_{r=R}=V_R(\theta)\tag{2}$$

The rest edges are loaded by external forces

$$\begin{aligned}&\left.\left(\frac{1}{r}\frac{\partial U}{\partial\theta}+\frac{\partial V}{\partial r}-\frac{V}{r}\right)\right|_{\theta=0}=\Phi_1(r),\ \left.\left[\left(\frac{U}{r}+\frac{1}{r}\frac{\partial V}{\partial\theta}\right)+\nu\frac{\partial U}{\partial r}\right]\right|_{\theta=0}=\Phi_3(r),\\&\left.\left(\frac{1}{r}\frac{\partial U}{\partial\theta}+\frac{\partial V}{\partial r}-\frac{V}{r}\right)\right|_{\theta=\theta_0}=\Phi_2(r),\ \left.\left[\left(\frac{U}{r}+\frac{1}{r}\frac{\partial V}{\partial\theta}\right)+\nu\frac{\partial U}{\partial r}\right]\right|_{\theta=\theta_0}=\Phi_4(r),\end{aligned}\tag{3}$$

$$\left.\left(\frac{\partial U}{\partial r}+\nu\left(\frac{1}{r}\frac{\partial V}{\partial\theta}+\frac{U}{r}\right)\right)\right|_{r=r_0}=F_1(\theta),\ \left.\left(\frac{1}{r}\frac{\partial U}{\partial\theta}+\frac{\partial V}{\partial r}-\frac{V}{r}\right)\right|_{r=r_0}=F_2(\theta).\tag{4}$$

Relying on a physical meaning of the problem, the stresses and strains should be bounded throughout the wedge-shaped domain Ω (otherwise the cutter would be broken). For this reason, the additional conditions must be imposed

$$(U,V)\in\left\{C^{(6)}(0\le\theta\le\theta_0),\ C^{(6)}(r_0\le r\le R)\right\},\ |U,V|<\infty.\tag{5}$$

The formulation of the elastic problem (1)-(5) is given in a classical form, but it is not complete. The requirement to the functions U and V of being smooth and bounded is essential, since it imposes specific restrictions on the boundary conditions formulation. For the case of mixed boundary conditions (2)-(4), the relations (5) lead to the necessity of the "consistency" conditions formulation. If these conditions aren't satisfied, a discontinuity will occur at the angular points $(r=R,\ \theta=0)$, $(r=R,\ \theta=\theta_0)$, $(r=r_0,\ \theta=0)$, $(r=r_0,\ \theta=\theta_0)$, that is, the conditions (5) will be broken. Hence, the accuracy of the solution will be essentially affected.

The boundary conditions (2) and (4) at the arcs $r=r_0$, $r=R$ are to be conformed with the conditions (3) at the rays $\theta=0,\ \theta=\theta_0$. For this purpose the predetermined functions $U_R(\theta),V_R(\theta)$ at the arc $r=R$ are replaced by some new functions $U_R^*(\theta),V_R^*(\theta)$ and the boundary conditions (2) are rewritten as:

$$\forall\theta\in[0,\theta_0],\ U\big|_{r=R}=U_R^*(\theta),\ V\big|_{r=R}=V_R^*(\theta)\tag{6}$$

The functions $U_R^*(\theta),V_R^*(\theta)$ are supposed to coincide with the original functions $U_R(\theta),V_R(\theta)$ outside the $\varepsilon-$neighborhoods of the angular points. And inside the $\varepsilon-$neighborhoods these functions are such that they take on some unknown values at points $\theta=0$ and $\theta=\theta_0$:

$$\forall\theta\in[\varepsilon,\theta_0-\varepsilon],\ U_R^*(\theta)=U_R(\theta),\ V_R^*(\theta)=V_R(\theta),$$
$$U_R^*(0)=U_{R1}^*,\ V_R^*(0)=V_{R1}^*,\ U_R^*(\theta_0)=U_{R2}^*,\ V_R^*(\theta_0)=V_{R2}^*, \tag{7}$$

where the constants U_{R1}^*, V_{R1}^*, U_{R2}^*, V_{R2}^* are assumed to be unknown in advance and they are to be determined throughout the solution process.

Further, the boundary functions of sixth order [13] will be used to determine the displacements U and V, so the functions $U_R^*(\theta), V_R^*(\theta)$ will be smoothly conjugated with $U_R(\theta), V_R(\theta)$ from (2) at the points (R,ε) and $(R,\theta_0-\varepsilon)$ up to sixth order derivatives as it is required by the condition (5).

Assuming fixed condition at the cutter's rear edge $r=R$, the relations (7) reduce

$$\forall\theta\in[\varepsilon,\theta_0-\varepsilon],\ U_R^*(\theta)=0,\ V_R^*(\theta)=0,\ U_R^*(0)=U_{R1}^*,\ V_R^*(0)=V_{R1}^*,\ U_R^*(\theta_0)=U_{R2}^*,\ V_R^*(\theta_0)=V_{R2}^*. \tag{8}$$

Thus, under the "fixed boundary" we will imply the equality of the displacement components to zero only for $\varepsilon\le\theta\le\theta_0-\varepsilon$, and inside the $\varepsilon-$neighborhoods of the angular points the displacements are determined by some additional relations $U_R^*(\theta), V_R^*(\theta)$ satisfying the conditions (5) and (8). Assuming the smallness of $\varepsilon-$neighborhoods the concrete relations for $U_R^*(\theta), V_R^*(\theta)$ are not given. This is sufficient to derive the solution to the problem in a finite form.

At the cusp of the cutter at points $(r_0,0), (r_0,\theta_0)$ some "consistency" conditions are also to be satisfied. In order to derive these conditions we will make use of the following considerations. For the boundary conditions defined in (2)-(4) the unique solution to the boundary problem exists. However, discontinuities of the stresses might occur at the angular points if this circumstance isn't envisaged in advance when the boundary conditions are being specified. That is, the stresses will depend on the direction of approach to the angular points for arbitrarily chosen functions $F_1(\theta)$ and $F_2(\theta)$. This drawback can be avoided by specifying two "consistency" conditions at every point $(r_0,0), (r_0,\theta_0)$. The first two conditions can be obtained using the symmetry property of the stress tensor $\sigma_{r\theta}=\sigma_{\theta r}$. This leads to the following relations:

$$\Phi_1(r_0)=F_2(0),\ \Phi_2(r_0)=F_2(\theta_0) \tag{9}$$

These conditions impose the restrictions on a choice of the function $F_2(\theta)$.

The remaining two conditions are derived by formally substituting the radial loading $F_1(\theta)$ at the cusp from relations (4) by some new function $F_1^*(\theta)$. This function is supposed to coincide with the original function $F_1(\theta)$ inside the interval $\delta\le\theta\le\theta_0-\delta$, and it takes on some unknown values $F_1^*(0)=F_{11}^*$ and $F_1^*(\theta_0)=F_{12}^*$ at the angular points

$$\forall\theta\in[\delta,\theta_0-\delta],\ F_1^*(\theta)=F_1(\theta),\ F_1^*(0)=F_{11}^*,\ F_1^*(\theta_0)=F_{12}^*, \tag{10}$$

where the constants F_{11}^*, F_{12}^* are assumed to be determined throughout the solution process.

Following the condition (5), the function $F_1^*(\theta)$ should be smoothly conjugated with $F_1(\theta)$ at the points (r_0,δ) and $(r_0,\theta_0-\delta)$ up to sixth order derivatives. Assuming the smallness of $\delta-$ neighborhoods the concrete relation for $F_1^*(\theta)$ is not given to simplify the construction of the solution. It is only sufficient to suppose that the function $F_1^*(\theta)$ satisfies the conditions (5) and (10).

Further, we need to define $\Phi_i(r)$ in the boundary conditions (3) to solve the problem. As an example, the simplest rapidly decreasing functions were chosen. The main loading throughout the sample processing is applied to the upper edge ($\theta=\theta_0$) of the cutter

$$\Phi_2(r)=\Phi_{02}\left(1-\frac{r}{R}\right),\ \Phi_4(r)=\Phi_{04}\left(1-\frac{r}{R}\right) \tag{11}$$

The lower edge ($\theta=0$) is assumed to be zero loaded, that is $\Phi_1(r)=\Phi_3(r)=0$.

Under the conditions (9)-(11), the loadings $F_1(\theta)$, $F_2(\theta)$ on the cutter's cusp are expressed in the simplest linear form

$$\forall\theta\in[\delta,\theta_0-\delta]\Rightarrow F_1^*(\theta)=F_1(\theta)=F_{01}\frac{\theta}{\theta_0},\ F_1^*(0)=F_{11}^*,\ F_1^*(\theta_0)=F_{12}^*,\ F_2(\theta)=\Phi_{02}\left(1-\frac{r_0}{R}\right)\frac{\theta}{\theta_0}. \tag{12}$$

The constant coefficients $F_{01},\Phi_{02},\Phi_{04}$ in relations (11), (12) depend on pressing force applied to the sample being processed and they are determined experimentally.

From the previous considerations, it follows that instead of relations (4), the consistent boundary conditions should be used

$$\left.\left(\frac{\partial U}{\partial r}+\nu\left(\frac{1}{r}\frac{\partial V}{\partial\theta}+\frac{U}{r}\right)\right)\right|_{r=r_0}=F_1^*(\theta),\ \left.\left(\frac{1}{r}\frac{\partial U}{\partial\theta}+\frac{\partial V}{\partial r}-\frac{V}{r}\right)\right|_{r=r_0}=F_2(\theta), \tag{13}$$

where the function $F_1^*(\theta)$ must satisfy the conditions (12). The form of function $F_2(\theta)$ is not changed as it satisfies the consistency conditions (9) due to relations (11).

Thus, the original boundary value problem (1)-(4) reduces to its altered formulation involving consistent boundary conditions (3), (8), (13) with 6 additional unknown constants

$$U_{R1}^*,\ U_{R2}^*,\ V_{R1}^*,\ V_{R2}^*,\ F_{11}^*,\ F_{12}^* \tag{14}$$

to be determined throughout the solution process.

According with the method of fast expansions [13] the displacements $U(r,\theta)$ and $V(r,\theta)$ can be expressed as the superposition of boundary function and truncated Fourier sine series

$$U=M_6^U(r,\theta)+\sum_{m=1}^{N_1}u_m(r)\sin m\pi\frac{\theta}{\theta_0},\quad V=M_6^V(r,\theta)+\sum_{m=1}^{N_1}\upsilon_m(r)\sin m\pi\frac{\theta}{\theta_0}, \tag{15}$$

where N_1 is the number of Fourier series terms retained, $M_6^U(r,\theta)$ and $M_6^V(r,\theta)$ are the boundary functions of sixth order

$$\begin{aligned}M_6^U(r,\theta)=&A_1(r)\left(1-\frac{\theta}{\theta_0}\right)+A_2(r)\frac{\theta}{\theta_0}+A_3(r)\left(\frac{\theta^2}{2}-\frac{\theta^3}{6\theta_0}-\frac{\theta_0\theta}{3}\right)+A_4(r)\left(\frac{\theta^3}{6\theta_0}-\frac{\theta_0\theta}{6}\right)\\&+A_5(r)\left(\frac{\theta^4}{24}-\frac{\theta^5}{120\theta_0}-\frac{\theta_0\theta^3}{18}+\frac{\theta_0^3\theta}{45}\right)+A_6(r)\left(\frac{\theta^5}{120\theta_0}-\frac{\theta_0\theta^3}{36}+\frac{7\theta_0^3\theta}{360}\right)\\&+A_7(r)\left(\frac{\theta^6}{720}-\frac{\theta^7}{5040\theta_0}-\frac{\theta_0\theta^5}{360}+\frac{\theta_0^3\theta^3}{270}-\frac{2\theta_0^5\theta}{945}\right)\\&+A_8(r)\left(\frac{\theta^7}{5040\theta_0}-\frac{\theta_0\theta^5}{720}+\frac{7\theta_0^3\theta^3}{2160}-\frac{31\theta_0^5\theta}{15120}\right),\end{aligned} \tag{16}$$

$$\begin{aligned}
M_6^V(r,\theta) &= B_1(r)\left(1-\frac{\theta}{\theta_0}\right)+B_2(r)\frac{\theta}{\theta_0}+B_3(r)\left(\frac{\theta^2}{2}-\frac{\theta^3}{6\theta_0}-\frac{\theta_0\theta}{3}\right)+B_4(r)\left(\frac{\theta^3}{6\theta_0}-\frac{\theta_0\theta}{6}\right)\\
&+B_5(r)\left(\frac{\theta^4}{24}-\frac{\theta^5}{120\theta_0}-\frac{\theta_0\theta^3}{18}+\frac{\theta_0^3\theta}{45}\right)+B_6(r)\left(\frac{\theta^5}{120\theta_0}-\frac{\theta_0\theta^3}{36}+\frac{7\theta_0^3\theta}{360}\right)\\
&+B_7(r)\left(\frac{\theta^6}{720}-\frac{\theta^7}{5040\theta_0}-\frac{\theta_0\theta^5}{360}+\frac{\theta_0^3\theta^3}{270}-\frac{2\theta_0^5\theta}{945}\right)\\
&+B_8(r)\left(\frac{\theta^7}{5040\theta_0}-\frac{\theta_0\theta^5}{720}+\frac{7\theta_0^3\theta^3}{2160}-\frac{31\theta_0^5\theta}{15120}\right).
\end{aligned} \tag{17}$$

This representation of the displacement components involves $16+2N_1$ unknown functions

$$A_1(r)\div A_8(r),\ B_1(r)\div B_8(r),\ u_m(r),\ \upsilon_m(r),\ m=1\div N_1, \tag{18}$$

depending only on the radial coordinate r. Following the same approach the unknown functions (18) can also be represented in a fast expansion form:

$$\begin{aligned}
A_j(r) &= M_6^{A(j)}(r)+\sum_{n=1}^{N_2}a_{n+8}^{(j)}\sin n\pi z,\quad z=\frac{r-r_0}{R-r_0}\in[0,1],\quad j=1\div 8,\quad M_6^{A(j)}(r)=a_1^{(j)}(1-z)+a_2^{(j)}z\\
&+a_3^{(j)}\left(\frac{z^2}{2}-\frac{z^3}{6}-\frac{z}{3}\right)+a_4^{(j)}\left(\frac{z^3}{6}-\frac{z}{6}\right)+a_5^{(j)}\left(\frac{z^4}{24}-\frac{z^5}{120}-\frac{z^3}{18}+\frac{z}{45}\right)\\
&+a_6^{(j)}\left(\frac{z^5}{120}-\frac{z^3}{36}+\frac{7z}{360}\right)+a_7^{(j)}\left(\frac{z^6}{720}-\frac{z^7}{5040}-\frac{z^5}{360}+\frac{z^3}{270}-\frac{2z}{945}\right)\\
&+a_8^{(j)}\left(\frac{z^7}{5040}-\frac{z^5}{720}+\frac{7z^3}{2160}-\frac{31z}{15120}\right),
\end{aligned} \tag{19}$$

$$\begin{aligned}
B_j(r) &= M_6^{B(j)}(r)+\sum_{n=1}^{N_2}b_{n+8}^{(j)}\sin n\pi z,\quad j=1\div 8,\quad M_6^{B(j)}(r)=b_1^{(j)}(1-z)+b_2^{(j)}z+b_3^{(j)}\left(\frac{z^2}{2}-\frac{z^3}{6}-\frac{z}{3}\right)\\
&+b_4^{(j)}\left(\frac{z^3}{6}-\frac{z}{6}\right)+b_5^{(j)}\left(\frac{z^4}{24}-\frac{z^5}{120}-\frac{z^3}{18}+\frac{z}{45}\right)+b_6^{(j)}\left(\frac{z^5}{120}-\frac{z^3}{36}+\frac{7z}{360}\right)\\
&+b_7^{(j)}\left(\frac{z^6}{720}-\frac{z^7}{5040}-\frac{z^5}{360}+\frac{z^3}{270}-\frac{2z}{945}\right)+b_8^{(j)}\left(\frac{z^7}{5040}-\frac{z^5}{720}+\frac{7z^3}{2160}-\frac{31z}{15120}\right),
\end{aligned} \tag{20}$$

$$\begin{aligned}
u_m(r) &= M_6^{u(m)}(r)+\sum_{n=1}^{N_2}u_{n+8}^{(m)}\sin n\pi z,\ m=1\div N_1,\ M_6^{u(m)}(r)=u_1^{(m)}(1-z)+u_2^{(m)}z+u_3^{(m)}\left(\frac{z^2}{2}-\frac{z^3}{6}-\frac{z}{3}\right)\\
&+u_4^{(m)}\left(\frac{z^3}{6}-\frac{z}{6}\right)+u_5^{(m)}\left(\frac{z^4}{24}-\frac{z^5}{120}-\frac{z^3}{18}+\frac{z}{45}\right)+u_6^{(m)}\left(\frac{z^5}{120}-\frac{z^3}{36}+\frac{7z}{360}\right)\\
&+u_7^{(m)}\left(\frac{z^6}{720}-\frac{z^7}{5040}-\frac{z^5}{360}+\frac{z^3}{270}-\frac{2z}{945}\right)+u_8^{(m)}\left(\frac{z^7}{5040}-\frac{z^5}{720}+\frac{7z^3}{2160}-\frac{31z}{15120}\right),
\end{aligned} \tag{21}$$

$$\upsilon_m(r) = M_6^{\upsilon(m)}(r) + \sum_{n=1}^{N_2} \upsilon_{n+8}^{(m)} \sin n\pi z,\ m = 1 \div N_1,\ M_6^{\upsilon(m)}(r) = \upsilon_1^{(m)}(1-z) + \upsilon_2^{(m)} z + \upsilon_3^{(m)}\left(\frac{z^2}{2} - \frac{z^3}{6} - \frac{z}{3}\right)$$
$$+\upsilon_4^{(m)}\left(\frac{z^3}{6} - \frac{z}{6}\right) + \upsilon_5^{(m)}\left(\frac{z^4}{24} - \frac{z^5}{120} - \frac{z^3}{18} + \frac{z}{45}\right) + \upsilon_6^{(m)}\left(\frac{z^5}{120} - \frac{z^3}{36} + \frac{7z}{360}\right) \tag{22}$$
$$+\upsilon_7^{(m)}\left(\frac{z^6}{720} - \frac{z^7}{5040} - \frac{z^5}{360} + \frac{z^3}{270} - \frac{2z}{945}\right) + \upsilon_8^{(m)}\left(\frac{z^7}{5040} - \frac{z^5}{720} + \frac{7z^3}{2160} - \frac{31z}{15120}\right).$$

Substituting the relations (16), (17) and (19)-(22) into the differential equations (1) and into the boundary conditions (3), (8), (13) and differentiating with respect to r and θ gives the representation of the original boundary value problem containing $2(N_1+8)(N_2+8)$ unknown constant coefficients

$$a_n^{(j)},\ b_n^{(j)},\ u_n^{(m)},\ \upsilon_n^{(m)},\ j = 1 \div 8,\ n = 1 \div N_2 + 8,\ m = 1 \div N_1. \tag{23}$$

As the resultant expressions are quite cumbersome, they are not given here and further referred to as *double fast expansion* form of the original problem.

In order to determine the unknown constant coefficients (23), the pointwise method, which was developed and tested in [15] will be employed. Following this technique it is necessary to discretize the wedge-shaped domain Ω in a uniform grid with $(N_1+6)(N_2+6)$ nodes (r_s, θ_k) so that the interval $[0, \theta_0]$ is divided by N_1+6 inner points $\theta_k = k\theta_0/(N_1+5)$, $k = 0,1,..,N_1+5$ and the interval $[r_0, R]$ - by N_2+6 inner points $r_s = r_0 + s(R-r_0)/(N_2+5)$, $s = 0,1,..,N_2+5$. Substituting the values of each node (r_s, θ_k) into the double fast expansion of governing differential equations (1) gives $2(N_1+6)(N_2+6)$ linear algebraic equations in terms of unknown coefficients (23).

Further, it is required to discretize the boundary of the domain Ω in the same way. Dividing the interval $[0, \theta_0]$ by N_1+7 inner computational points $\theta_k = k\theta_0/(N_1+6)$, $k = 0,1,..,N_1+6$ and substituting these values into the relations (8) and (13) gives $4(N_1+7)$ linear equations. Similarly, dividing the interval $[r_0, R]$ by N_2+7 inner computational points $r_s = r_0 + s(R-r_0)/(N_2+6)$, $s = 0,1,..,N_2+6$ and substituting these values into the relations (3) gives $4(N_2+7)$ more linear equations. Thus, from the boundary conditions (3), (8) and (13) we obtain $4(N_1+7)+4(N_2+7)$ linear equations in terms of unknown coefficients (23).

All in all, we derive the consistent system of $2(N_1+8)(N_2+8)$ linear algebraic equations in the unknowns (23) and 6 additional equations for the constants U_{R1}^*, U_{R2}^*, V_{R1}^*, V_{R2}^*, F_{11}^*, F_{12}^* following from the consistency conditions. This system then has been solved numerically in Maple software.

3. Results and discussion

As an example the numerical results for solution to the boundary value problem (1), (3), (8), (13) for the cutter made of instrumental quick-cutting steel of grade R18 are provided. This steel has the following values of elastic moduli [16, 17]

$$\sigma_{0.2} = 5.1 \cdot 10^8 \text{ Pa},\ \nu = 0{,}33,\ E = 2.28 \cdot 10^{11} \text{ Pa},\ \lambda = 1.66 \cdot 10^{11},\ \mu = 8.57 \cdot 10^{10},$$

where $\sigma_{0.2}$ is the 0.2% offset yield stress.

Since the solution determined by the expressions (15)-(17), (19)-(22) approximately satisfies the differential equations (1) and the boundary conditions (3), (8), (13), then we will use a relative residual $\delta = \delta(r,\theta)/\max\left|\{f_i(r,\theta)\}\right|$ to estimate an accuracy of the solution derived, where $f_i(r,\theta)$ denotes

IOP Conf. Series: Journal of Physics: Conf. Series **973** (2018) 012002 doi:10.1088/1742-6596/973/1/012002

the i-th term of the equations (1) or boundary conditions (3), (8), (13).

The distributions of the relative residual δ_D of the approximate solution (15) to the equations (1) retaining only three terms $(N_1 = 3)$ in first expansion, and thirty terms $(N_2 = 30)$ in the second one are shown in figure 1. These results were computed for the following values of loadings and wedge dimensions

$$\Phi_{02} = 10^{-8}, \quad \Phi_{04} = 10^{-8}, \quad F_{01} = 10^{-8}, \quad r_0 = 10^{-5}\text{m}, \quad R = 10^{-1}\text{m}, \quad \theta_0 = 5\pi/180 \tag{24}$$

It can be seen (refer to figure 1) that the maximal values of relative residual δ_D are concentrated nearby wedge's cusp and they don't exceed $1.6 \cdot 10^{-6}$ and $2.5 \cdot 10^{-8}$ for the first and second differential equations, respectively.

The distribution of relative residual δ_G along the region's boundary as well as residual's magnitudes for all boundary conditions (3), (8), (13) are essentially the same. The typical behavior of relative residual δ_G at the edges $\theta = \theta_0$ and $r = r_0$ of the domain Ω is shown in figure 2.

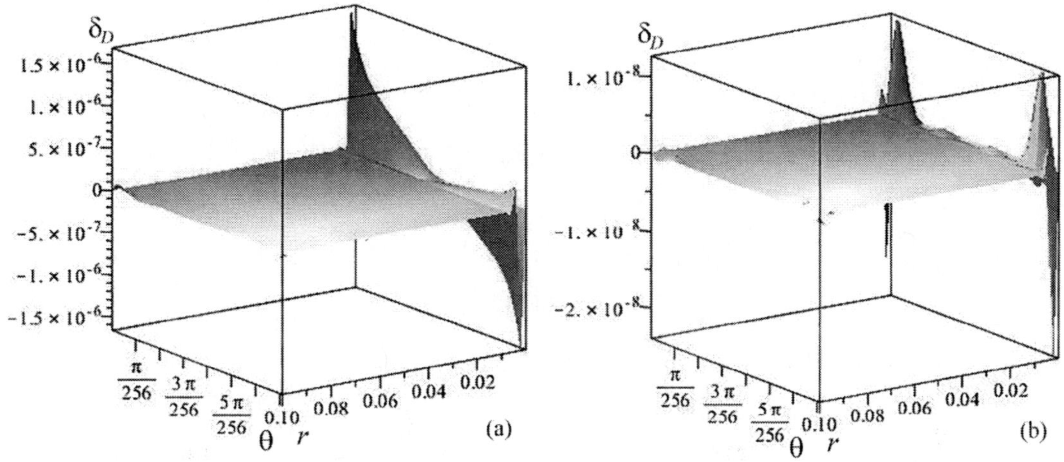

Figure 1. The relative residual of the differential equations: (a) $(1)_1$; (b) $(1)_2$.

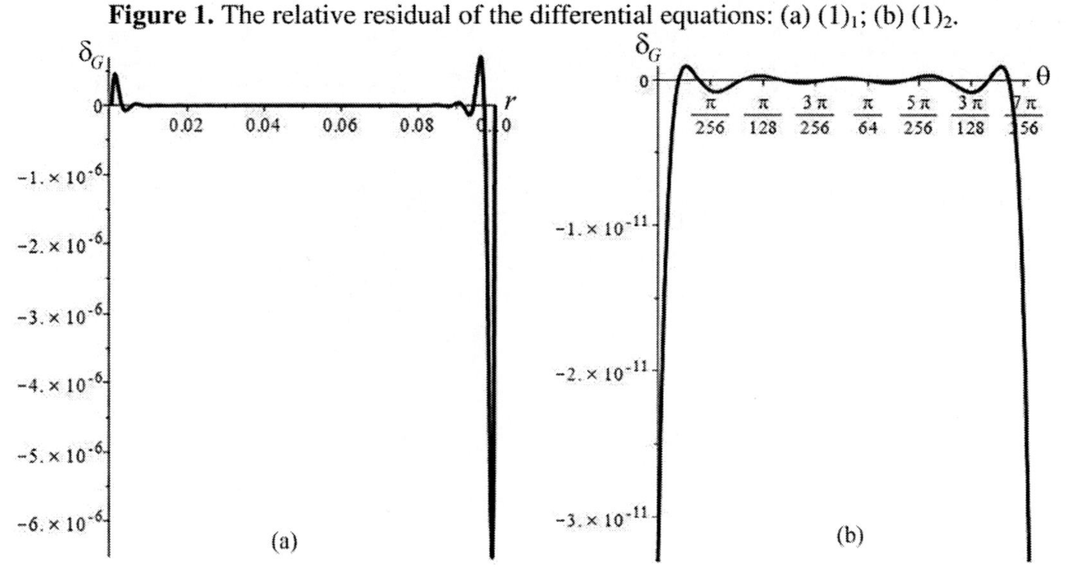

Figure 2. The relative residual of the boundary conditions: (a) $\delta_G\big|_{\theta=\theta_0}$; (b) $\delta_G\big|_{r=r_0}$.

It is apparent that the accuracy achieved (refer to figures 1, 2) is acceptable for the most technical purposes. It should be noted that the similar high accuracy approximate solutions were accomplished in [13-15] where the method of fast expansions has also been involved.

The computed profiles of the displacements $U(r,\theta)$, $V(r,\theta)$ and stress components σ_r, σ_θ, $\sigma_{r\theta}$ in a cross-section Ω of the cutter are shown in figure 3 and figure 4, respectively.

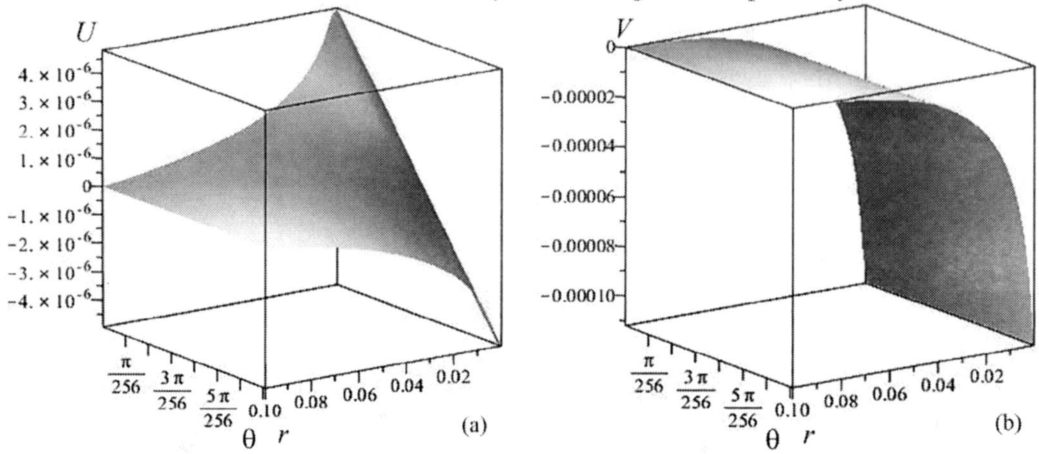

Figure 3. The displacement components: (a) $U(r,\theta)$; (b) $V(r,\theta)$.

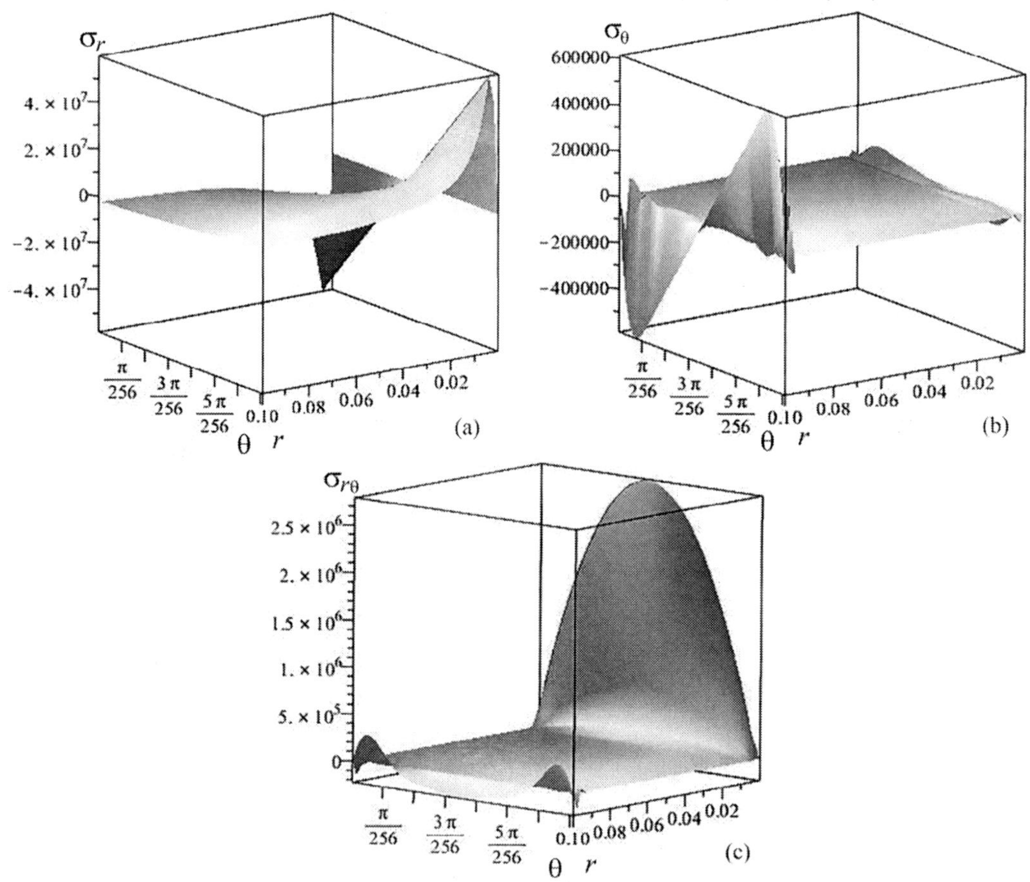

Figure 4. The stress components: (a) σ_r; (b) σ_θ; (c) $\sigma_{r\theta}$.

The numerical results corresponding to the parameters' values (24) satisfy the yield condition [18]

$$\sigma_{0,2} = 5.1\cdot 10^8 \text{ Pa} \geq \tilde{\sigma}, \quad \tilde{\sigma} = \sqrt{\frac{(\sigma_r^*)^2 + (\sigma_\theta^*)^2 + (\sigma_z^*)^2 + 2(\sigma_{r\theta})^2}{2}} = 3.5\cdot 10^7 \text{ Pa},$$

where $\sigma_r^* = \sigma_r - \frac{1}{3}(\sigma_r + \sigma_\theta)$, $\sigma_\theta^* = \sigma_\theta - \frac{1}{3}(\sigma_r + \sigma_\theta)$, $\sigma_z^* = \sigma_z - \frac{1}{3}(\sigma_r + \sigma_\theta)$, $\sigma_z = 0$.

The distribution of $\tilde{\sigma}$ in the cutter's cross-section is shown in figure 5. It is obvious that the maximal stress $\tilde{\sigma}$ in a cutter occurs at its edge $\theta = \theta_0$ in the vicinity of point $r = 0.0035$ m.

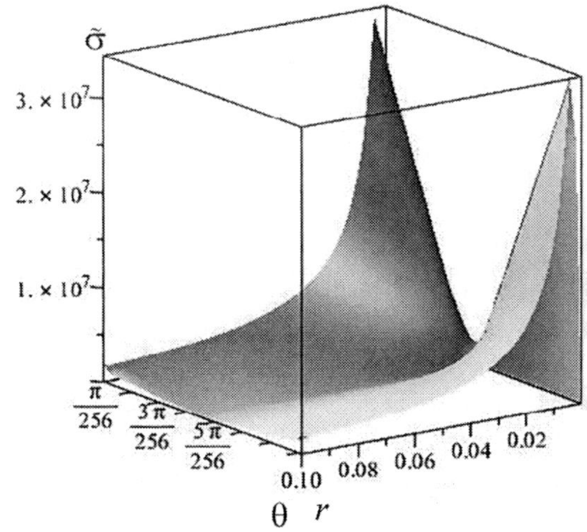

Figure 5. The distribution of $\tilde{\sigma}$.

The computed values of $\tilde{\sigma}$ for various opening angles θ_0 and radii r_0 are presented in table 1. Examining these data it can be concluded that the displacements and stresses decline with an increase of opening angle θ_0 and they raise with a decrease of the cusp's radius r_0. These results are consistent with the general elasticity and strength of materials theories.

Table 1. The values of $\tilde{\sigma}$, Pa.

θ_0	r_0, m		
	$1\cdot 10^{-4}$	$1\cdot 10^{-5}$	$1\cdot 10^{-6}$
$\pi/180$	$1.959\cdot 10^8$	$2.043\cdot 10^8$	$2.527\cdot 10^8$
$2\pi/180$	$5.207\cdot 10^7$	$5.432\cdot 10^7$	$5.918\cdot 10^7$
$3\pi/180$	$2.634\cdot 10^7$	$2.831\cdot 10^7$	$2.984\cdot 10^7$
$4\pi/180$	$2.052\cdot 10^7$	$2.143\cdot 10^7$	$2.167\cdot 10^7$

4. Conclusion

Thus, the application of the method of fast expansions allowed to obtain the solution to the problem in an explicit analytical form which is valid for the entire domain Ω including its boundary. The displacement and stress profiles derived can be utilized in the analysis of the stress state arising in wedge-shaped cutters.

References

[1] Shahani A R and Adibnazari S 2000 Analysis of perfectly bonded wedges and bonded wedges

with an interfacial crack under antiplane shear loading *Int. J. of Solids and Structures* vol 37 is 19 pp 2639-50

[2] Shahani A R 1999 Analysis of an anisotropic finite wedge under antiplane deformation *J. of Elasticity* vol 56 is 1 pp 17-32

[3] Shahani A R and Ghadiri M 2010 Analysis of anisotropic sector with a radial crack under anti-plane shear loading *Int. J. of Solids and Structures* vol 47 is 7-8 pp 1030-39

[4] Nekislykh E M and Ostrik V I 2010 Problems on elastic equilibrium of a wedge with cracks on the axis of symmetry *Mechanics of Solids* vol 45 is 5 pp 743-756

[5] Stampouloglou I H and Theotokoglou E E 2003 The plane elasticity problem of an isotropic wedge under normal and shear distributed loading - application in the case of a multi-material problem *Int. J. of Solids and Structures* vol 40 is 21 pp 5839-60

[6] Linkov A and Rybarska-Rusinek L 2010 Plane elasticity problem for a multi-wedge system with a thin wedge *Int. J. of Solids and Structures* vol 47 is 24 pp 3297–3304

[7] Chue C H and Liu Ch I 2001 A general solution on stress singularities in an anisotropic wedge *Int. J. of Solids and Structures* vol 38 is 38–39 pp 6889-6906

[8] Bykovtsev G I and Ivlev D D 1998 *Theory of plasticity* (Vladivostok: Dal'nauka) p 528

[9] Chernyshov A D 1975 On deformation of continuous media in a wedge-like region with smooth faces *J. of Applied Mathematics and Mechanics* vol 39 is 6 pp 1047-53

[10] Timoshenko S P and Goodier J N 1970 *Theory of elasticity* (McGraw-Hill, Inc., 3rd edn) p 567

[11] Guo L, Zhang Z M, Wang W and Wong P L 2017 An explicit matrix algorithm for solving three-dimensional elastic wedge under surface loads *Int. J. of Solids and Structures* vol 128 pp 231-242

[12] Aleksandrov V M and Pozharskii D A 2008 Three-dimensional contact problems for an elastic wedge with a coating *J. of Applied Mathematics and Mechanics* vol 72 is 1 pp 62-65

[13] Chernyshov A D 2014 Method of fast expansions for solving nonlinear differential equations *Computational Mathematics and Mathematical Physics* vol 54 is 1 pp 11-21

[14] Chernyshov A D, Goryainov V V and Marchenko A N 2016 Study of temperature fields in a rectangular plate with a temperature-dependent internal source with the aid of fast expansions *Thermophysics and Aeromechanics* vol 23 is 2 pp 237-245

[15] Chernyshov A D, Goryainov V V and Chernyshov O A 2015 Application of the fast expansion method for spacecraft trajectory calculation *Russian Aeronautics* vol 58 is 2 pp 180-186

[16] http://metallicheckiy-portal.ru/marki_metallov/sti/R18

[17] http://docs.cntd.ru/document/1200093817

[18] Ishlinskiy A Yu and Ivlev D D 2001 *The mathematical theory of plasticity* (Moscow: Fizmatlit) p 704

IOP Conf. Series: Journal of Physics: Conf. Series **973** (2018) 012003 doi:10.1088/1742-6596/973/1/012003

Model of the discrete destruction process of a solid body

V V Glagolev[1], A A Markin[1]

[1]Tula State University, prospect Lenina 92, Tula 300012, Russia

E–mail: vadim@tsu.tula.ru

Abstract. Destruction is considered as a discrete thermomechanical process, in which the deformation of a solid body is achieved by changing the boundary stresses acting on the part of the volume being destroyed with the external load unchanged. On the basis of the proposed concept, a model for adhesive stratification of a composite material is constructed. When adhesive stratification is used, the stress state of one or two boundaries of the adhesive layer changes to zero if the bonds with the joined body are broken. As a result of the stratification, the interaction between the part of the composite, which may include an adhesive layer and the rest of the body stops. When solving the elastoplastic problem of cohesive stratification, the region in which the destruction criterion is achieved is identified. With the help of a repeated solution of the problem of subcritical deformation with the known law of motion of the boundary of the region, the distribution of the load (nodal forces) acting from the region to the body is located. The next step considers the change in the stress–strain state of the body in the process of destruction of the selected area. The elastoplastic problem is solved with a simple unloading of the formed surface of the body and preservation of the external load corresponding to the beginning of the process of destruction.

1. Introduction

Modeling formation of new material surfaces involves the description of two stages of the process of loading the body: subcritical and postcritical. The subcritical stage can be described in the framework of the known equations of mechanics of a deformed solid body, both without taking into account nonlinear effects, and in a nonlinear (elastoplastic) formulation. The main problem is the formulation of criteria of the transition to the postcritical stage. The general formulation of the problem of postcritical behavior in elastoplastic deformation is far from being complete. This is due, firstly, to the choice of conditions for the termination of the interaction between surfaces, and secondly to the necessity of formulating boundary conditions on newly formed surfaces. When performing this or that criterion of destruction, the phase of formation of new material surfaces begins. The principal moment of this stage is the discreteness of the increment of the discontinuity surfaces [1] and the localization of the destruction process. In the modeling of destruction within the material surface of measure zero, cohesive elements are used [2,3]. The widely used bilinear law of the distribution of interaction in the cohesive zone from the opening of a crack with a falling section was considered in the works [4–5]. The main disadvantage of this approach is that the cohesive elements must be located on the trajectory of destruction, which, in general, depends on the new material surfaces. In addition, the various laws of cohesive interaction [6–8] and the material characteristics of cohesive elements significantly affect the distribution of stress–strain state [9] and require experimental confirmation.

Complexity of describing the destruction of the material volume when using the falling section of the tensile diagram is related to the construction of the determining relations of the unstable Drucker deformation and their confirmation in experiments [10]. The main approach to finite element modeling of the destruction process is the "kill element" procedure, described in detail in [11]. In this case, when the criterion characteristic of destruction is reached in the final element, the isolated material volume is excluded from consideration by multiplying the local stiffness matrix of the element by a number close to zero. Note that this procedure is correct for elastic deformation, when the load and unloading are determined by the same modules in the determining ratios. In the case of elastoplastic deformation of a body with the element excluded in this way, it is necessary to take into account possible unloading.

The discretization of the problem by the finite element method implies the description of the interaction among the particles of the body by means of nodal forces. When the material volume, in the general case interacting with several finite elements, is destroyed, we assume that the node interaction forces will be unloaded to zero within a simple process, with the unchanged external load reached at the time of destruction. Thus, the unstable Drucker deformation is excluded from consideration. Note that the determination of the nodal forces of interaction with the excluded material volume, at the internal points of the deformed body, is a separate problem. In the article [12], a procedure is proposed for the repeated loading of a body without a destructible volume, but with a displacement field found along its boundary, to find the corresponding nodal forces. The finding of forces of interaction of material volumes, in such a way, does not require additional information on the characteristics of the connection of these volumes except the determining relationships of the material.

In this paper, when modeling the process of destruction of a composite along the adhesive layer, along with destruction along the adhesive array, we considered the procedure for stratification of the adhesive layer from the adjacent material. In this case, the determination of the interaction forces of the separated materials with the repeated loading method is considered on the surface of the measure zero, and the use of the averaged characteristics of the stress–strain state of the adhesive layer allows us to avoid the singularity at the blind point of the mathematical cut being formed.

2. Materials and methods

Figure 1 shows the excluded volume of the adhesive layer 3 with the size $\delta \times \delta$ in the state of pre–destruction that connects the bodies 1 and 2. Its effect on the body is compensated by the external nodal load $\mathbf{F}^{+(i)}$, $\mathbf{F}^{-(j)}$ on the part of the element.

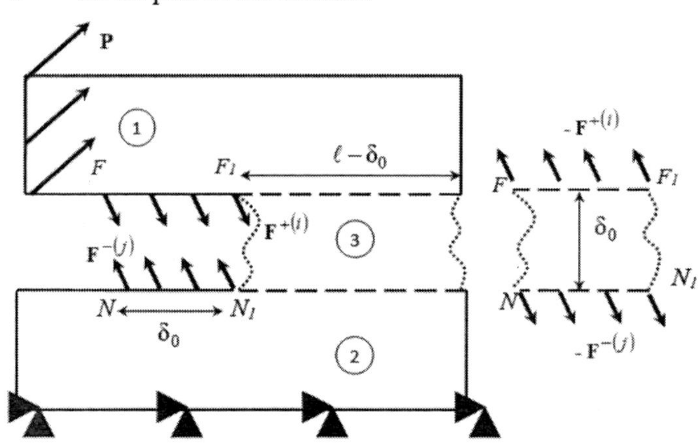

Figure 1. Equivalent loads at the time of pre–destruction of a material volume.

We assume that during the interval Δt δ–element will be completely destroyed if the load acting on the part of the layer element on the body becomes zero. In this case, during the time interval Δt, with the external load unchanged $\mathbf{P}$, it is necessary to unload the new material surfaces. We assume that the unloading process is simple, depending on one parameter Δt . This corresponds to the specification of the velocities of the external nodal loads:

$$\dot{\mathbf{F}}^{FF_1} = -\mathbf{F}^{+(i)}/\Delta t\,,\ \dot{\mathbf{F}}^{NN_1} = -\mathbf{F}^{-(j)}/\Delta t\ . \tag{1}$$

Thus, the description of the destruction process reduces to solving the problem with boundary conditions (1) at zero external–load velocity $\dot{\mathbf{P}} = 0$. Note that the determination of the nodal forces of interaction with the excluded material volume, at the internal points of the deformed body, is a separate problem. To determine the nodal forces $\mathbf{F}^{+(i)}$, $\mathbf{F}^{-(j)}$ it is proposed to use the procedure of repeated loading, which consists in replacing the action of the destructible element by setting the process of change with the "time" of nodal displacements of the element when the law of external action on the body is repeated, according to figure 2.

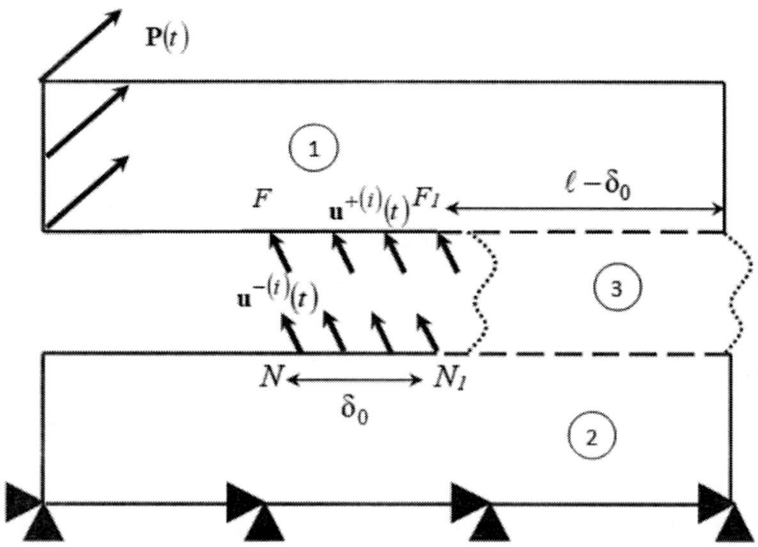

Figure 2. The scheme of repeated loading.

When the adhesive layer is delaminated the surface with the material adjacent to it, the state of pre–destruction is shown in figure 3. We will mentally divide the layer and the adjacent material along the length δ_0, and apply oppositely directed nodal forces $\mathbf{F}^{-(j)}$ corresponding to the stressed state at the boundary to the corresponding new surfaces. Without loss of generality, we assume that the separation occurs along the lower boundary of the layer. We assume that during the interval Δt there will be a stratification in the δ –part of the surface, if the load acting from the layer to the body becomes zero. In this case, during the time interval Δt, with the external load $\mathbf{P}$ unchanged, the unloading of new material surfaces will correspond to the specification of the velocities of external nodal loads:

$$\dot{\mathbf{F}}^{N'N_1} = \mathbf{F}^{-(j)}/\Delta t\,,\ \dot{\mathbf{F}}^{NN_1} = -\mathbf{F}^{-(j)}/\Delta t\ . \tag{2}$$

As an example of the use of the proposed procedure, the problem of stratification of a composite material along the adhesive layer is considered. In this case, destruction is possible both over the adhesive array and along the boundary of the adhesive layer with the material adjacent to it. In the first case, we mean a cohesive character of the stratification, and in the second case, we mean an adhesional character of the stratification.

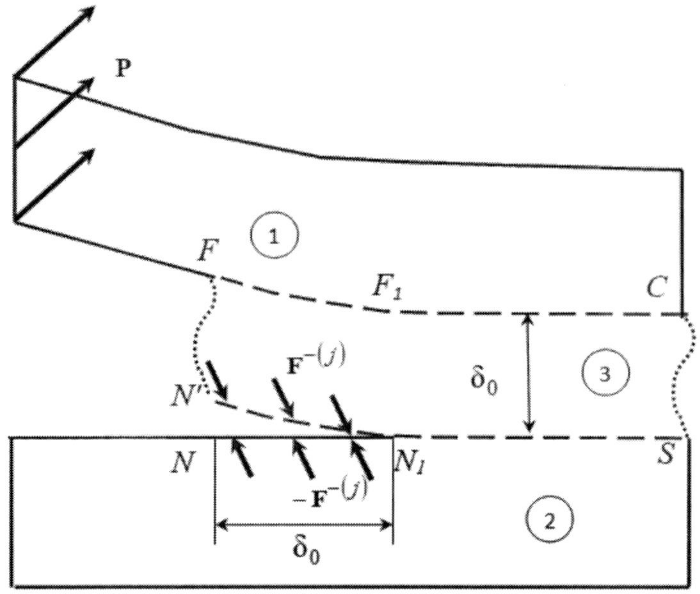

Figure 3. Stratification the surface.

The equilibrium flow conditions of the process with small deformation and rotations of material fibers was taken in the form [13]:

$$\iint_S \dot{\boldsymbol{\sigma}} \cdot\cdot \delta\mathbf{w} \mathrm{d}s = \int_L \dot{\mathbf{P}} \cdot \delta\mathbf{v} \mathrm{d}l\,, \tag{3}$$

where $\mathbf{v}$ is the velocity field; $\mathbf{w} = 0.5(\nabla\mathbf{v} + \mathbf{v}\nabla)$ is the velocity strain tensor; $\nabla = \mathbf{e}_i \frac{\mathrm{d}}{\mathrm{d}x_i}$; $\dot{\mathbf{P}} = \frac{\mathrm{d}\mathbf{P}}{\mathrm{d}t}$ is the velocity of external load on the circuit; $\dot{\boldsymbol{\sigma}} = \frac{\mathrm{d}\boldsymbol{\sigma}}{\mathrm{d}t}$ is the velocity of the stress tensor; t is the time–like parameter; S is the inner region of the complex body; δ is the variation.

In the adhesive layer, the stress–strain state is considered on the basis of the average stresses along the thickness δ_0 and the associated boundary stress velocities [12]:

$\dot{\bar{\sigma}}_{21}(x_1) = \frac{1}{\delta_0}\int_{-\delta_0/2}^{\delta_0/2} \dot{\sigma}_{21}(x_1, x_2)\mathrm{d}x_2$, $\dot{\bar{\sigma}}_{12}(x_1) = \frac{1}{\delta_0}\int_{-\delta_0/2}^{\delta_0/2} \dot{\sigma}_{12}(x_1, x_2)\mathrm{d}x_2$, $\dot{\bar{\sigma}}_{22}(x_1) = \frac{1}{\delta_0}\int_{-\delta_0/2}^{\delta_0/2} \dot{\sigma}_{22}(x_1, x_2)\mathrm{d}x_2$, $\dot{\bar{\sigma}}_{11}(x_1) = \frac{1}{\delta_0}\int_{-\delta_0/2}^{\delta_0/2} \dot{\sigma}_{11}(x_1, x_2)\mathrm{d}x_2$, and the average velocities, and deformation velocities through their boundary values:

$$\mathrm{w}_{22}(x_1) = \left(\frac{\mathrm{v}_2^+(x_1) - \mathrm{v}_2^-(x_1)}{\delta_0}\right), \quad \mathrm{w}_{11}(x_1) = 0.5\left(\frac{\mathrm{d}\mathrm{v}_1^+(x_1)}{\mathrm{d}x_1} + \frac{\mathrm{d}\mathrm{v}_1^-(x_1)}{\mathrm{d}x_1}\right), \tag{4}$$

$$\frac{\mathrm{d}\bar{\mathrm{v}}_1(x_1)}{\mathrm{d}x_2} = \frac{\mathrm{v}_1^+(x_1) - \mathrm{v}_1^-(x_1)}{\delta_0}, \quad \frac{\mathrm{d}\bar{\mathrm{v}}_2(x_1)}{\mathrm{d}x_1} = 0.5\left(\frac{\mathrm{d}\mathrm{v}_2^+(x_1)}{\mathrm{d}x_1} + \frac{\mathrm{d}\mathrm{v}_2^-(x_1)}{\mathrm{d}x_1}\right), \tag{5}$$

$$\bar{\mathrm{v}}_1(x_1) = 0.5(\mathrm{v}_1^+(x_1) + \mathrm{v}_1^-(x_1)), \quad \bar{\mathrm{v}}_2(x_1) = 0.5(\mathrm{v}_2^+(x_1) + \mathrm{v}_2^-(x_1)), \tag{6}$$

where $\mathbf{v}^+$, $\mathbf{v}^-$ is the velocity vector of the upper and lower boundaries of the region 3.

From the expressions (5) we arrive at the representation of the average shear strain velocity along the considered region:

$$w_{21}(x_1)=0.5\left(\frac{d\bar{v}_2(x_1)}{dx_1}+\frac{d\bar{v}_1(x_1)}{dx_2}\right)=0.5\left(\frac{v_1^+(x_1)-v_1^-(x_1)}{\delta_0}+0.5\left(\frac{dv_2^+(x_1)}{dx_1}+\frac{dv_2^-(x_1)}{dx_1}\right)\right). \quad (7)$$

We assume that the stress velocity vectors on the conjugate boundaries of the adhesive layer are equal and opposite to the stress velocity vectors of the conjugate boundaries of the body. In addition, rigid adhesion among the boundaries of the layer and regions 1, 2 is postulated (see figure 1). Considering the velocities of the boundary stress vectors of the layer as boundary conditions for regions adjacent to it [14], we arrive at a simultaneous solution of the variational equilibrium equations for the body 1:

$$\int_{S_1}\dot{\sigma}\cdot\cdot\delta w ds+\int_{\ell}\dot{\bar{\sigma}}_{22}\delta v_2 dx_1+\int_{\ell}\dot{\bar{\sigma}}_{21}\delta v_1 dx_1+0.5\delta_0\left(\int_{\ell}\dot{\bar{\sigma}}_{11}\frac{d\delta v_1}{dx_1}dx_1+\int_{\ell}\dot{\bar{\sigma}}_{21}\frac{d\delta v_2}{dx_1}dx_1\right)=\int_{L^1}\dot{\mathbf{P}}\cdot\delta v dl \quad (8)$$

and body 2:

$$\int_{S_2}\dot{\sigma}\cdot\cdot\delta w ds-\int_{\ell}\dot{\bar{\sigma}}_{22}\delta v_2 dx_1-\int_{\ell}\dot{\bar{\sigma}}_{21}\delta v_1 dx_1+0.5\delta_0\left(\int_{\ell}\dot{\bar{\sigma}}_{11}\frac{d\delta v_1}{dx_1}dx_1+\int_{\ell}\dot{\bar{\sigma}}_{21}\frac{d\delta v_2}{dx_1}dx_1\right)=\int_{L^2}\dot{\mathbf{P}}\cdot\delta v dl. \quad (9)$$

The behavior of the material of bodies 1 and 2 under active loading ($\sigma\cdot\cdot\dot{\sigma}>0$) is determined by the following physical relationships:

$$\dot{\tilde{\sigma}}=2G^{(i)}\tilde{w}\ ,\dot{p}=3K^{(i)}\dot{\theta}, \quad (10)$$

where $\dot{\tilde{\sigma}}$ is the velocity of the deviator of the stress tensor; $i=1,2$; $\tilde{w}$ is the deviator component of the strain velocity tensor ; $\dot{p}=\dot{\sigma}\cdot\cdot\mathbf{E}$; $K^{(i)}$ is the volume compression module; $G^{(i)}$ is the shear modulus $G^{(i)}=G_e^{(i)}$ at, $T\le T_p^{(i)}$, $G^{(i)}=G_p^{(i)}$ at $T>T_p^{(i)}$; $T_p^{(i)}$ is the yield limit of the corresponding material; T is the intensity of tangential stresses.

In the unloading state ($\sigma\cdot\cdot\dot{\sigma}\le 0$), the defining relations are written as:

$$\dot{\tilde{\sigma}}=2G_e^{(i)}\tilde{w}\ ,\dot{p}=3K^{(i)}\dot{\theta}. \quad (11)$$

In the material of the layer, the defining relationships are assumed to be valid for the medium–thick layer of the velocity characteristics of the stress–strain state:

$$\dot{\bar{\tilde{\sigma}}}=2G^{(3)}\bar{\tilde{w}},\ \dot{p}=3K^{(3)}\dot{\theta},\quad \sigma\cdot\cdot\dot{\sigma}>0\ ; \quad (12)$$

$$\dot{\bar{\tilde{\sigma}}}=2G_e^{(3)}\bar{\tilde{w}},\ \dot{p}=3K^{(3)}\dot{\theta},\quad \sigma\cdot\cdot\dot{\sigma}\le 0; \quad (13)$$

where $K^{(3)}$ is the volume compression module of the layer material; $G^{(3)}$ is the shear modulus of the layer material $G^{(3)}=G_e^{(3)}$ at $T\le T_p^{(3)}, G^{(3)}=G_p^{(3)}$ at $T>T_p^{(i)}$; $T_p^{(3)}$ is the yield limit of the layer material.

As a result of substituting expressions in the defining relations (12) – (13) the component of the average velocity deformation, the average stress velocities are determined through the boundary velocities and their derivatives. Thus, the solution of the system (8) – (13) reduces to determining the velocity field $v(x_1,x_2)$ in bodies 1 and 2. In this case, boundary velocities of the body 2 adjacent to the layer (see) will be present in equation (8), and in equation (9) boundary velocities of the body 1, which adjacent to the layer. We note that the use of the average–thickness adhesive layer characteristics of the stress–strain state makes it possible to exclude the dependence on the form of the end of the adhesive layer in the formulation of problem (8) – (13). Therefore, in figure 1–3, the left and right boundaries of the layer are shown as a wavy dotted line. Fields of displacement and deformation are determined with the help of integration according to the parameter t of the found velocity fields.

In stratification along the surface (see figure 3), we will take into account that along the lower boundary of the layer and the body 2, on the exfoliated region, there is no rigid adherence condition:

$\mathbf{v}^- \neq \mathbf{v}(x_1, -\delta_0/2);\ x_1 \in [N, N_1]$ while along the upper boundary of the layer and body 1, there is a rigid adherence of the boundaries: $\mathbf{v}^+ = \mathbf{v}(x_1, \delta_0/2);\ x_1 \in [F, F_1]$. In this case, due to the small thickness of the layer compared with the thicknesses of bodies 1 and 2, we consider the hypothesis of the membrane theory of the homogeneity of velocity fields over the layer thickness to be valid. Thus, in the exfoliated part of the layer, the deformation velocities (4) and (7) will be equal to:

$$\mathrm{w}_{11}(x_1) = \frac{\mathrm{d}v_1^+(x_1)}{\mathrm{d}x_1},\quad \mathrm{w}_{22}(x_1) = 0,\quad \mathrm{w}_{21}(x_1) = 0.5\frac{\mathrm{d}v_2^+(x_1)}{\mathrm{d}x_1}.$$

3. Results and discussion

Calculations carried out for a composite material consisting of two elements in a state of a plane deformation, with material characteristics close to the aluminum alloy: $G_e = 2.8 \cdot 10^{10}$ Pa; $G_p = 5.2 \cdot 10^8$ Pa; $K = 6 \cdot 10^{10}$ Pa; $T_p = 3 \cdot 10^8$ Pa is the elasticity limit; $\sigma_k = 4.2 \cdot 10^8$ is the strength limit bonded epoxy resin with the following properties: $G_e = 1.3 \cdot 10^9$ Pa; $K = 1.7 \cdot 10^9$ Pa; $\sigma_k = 9 \cdot 10^7$ Pa. For adhesive, a variant of adhesion strength for peeling and shearing along the boundary between the material and the layer also considered. The geometric characteristics of the composite shown in figure 4 were taken as follows: $AD = 5 \cdot 10^{-3}$ m; $\delta_0 = FN = 10^{-3}$ m; $MQ = 2 \cdot 10^{-2}$ m; $AB = 10^{-1}$ m; $DF = 5 \cdot 10^{-2}$ m. The velocity of the external load $\dot{\mathbf{P}} = 1$ Pa/s is directed at an angle $\pi/4$ to the axis $0X_1$.

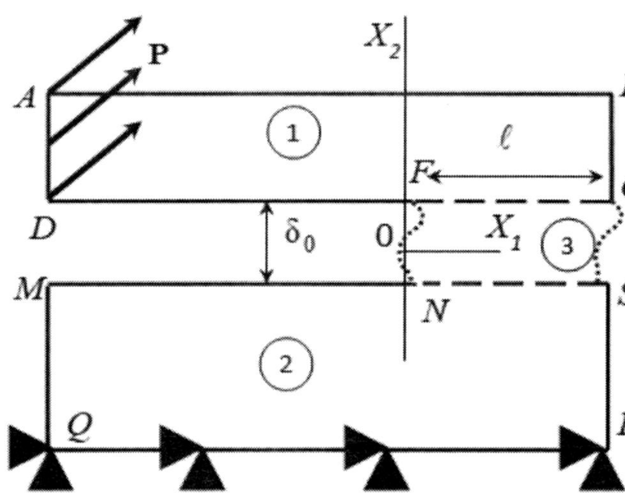

Figure 4. Scheme of composite loading.

As a test, the problem of destruction of an element of a layer of the size $\delta \times \delta$ in an elastic setting is considered. Let the critical state in the δ–element be determined by the external load $\mathbf{P} = 1$ Pa. The computational convergence of the solution establishes a partition of the boundary region conjugate to the δ–element into four finite elements [15]. Taking into account the quadratic law of distribution of the velocity field on a finite element, nine nodal forces along the upper and lower boundaries of the destructible δ–element will determine the equilibrium of the composite at the pre–destruction moment. Without loss of generality, the minimal module of projections of the found nodal forces can be considered as a local unloading parameter Δt for determining the boundary conditions (1). Graph 2 in figure 5 a) shows the distribution of dimensionless vertical displacement along the boundary of body 1 and layer 3 toward the end of subcritical deformation, graph 3 shows the distribution of displacements due to the process of destruction of the structural element of the layer, graph 1 shows the superposition of displacement fields. Displacements related to the value of displacement $u_2^+(0,\delta_0/2)$ at the beginning of the process of destruction. The dimensionless coordinate in the direction of the abscissa axis is defined as $x_1^b = x_1/\delta_0$. Graph 1 in figure 5 b) determines the corresponding vertical displacements when solving a problem without the first structural element of the layer at the end of the interval. This solution simulates the "kill element" approach used in finite element modeling [10] procedures, where the local stiffness matrix of the excluded element is multiplied by a number close to zero. Graph 2 in figure 5 b) repeats curve 1 in figure 5 a). As we can see from figure 5 b), the coincidence of graphs 1 and 2 shows the adequacy of the results of the proposed procedure of unloading new surfaces during the destruction process of the "kill element" approach within elastic deformation.

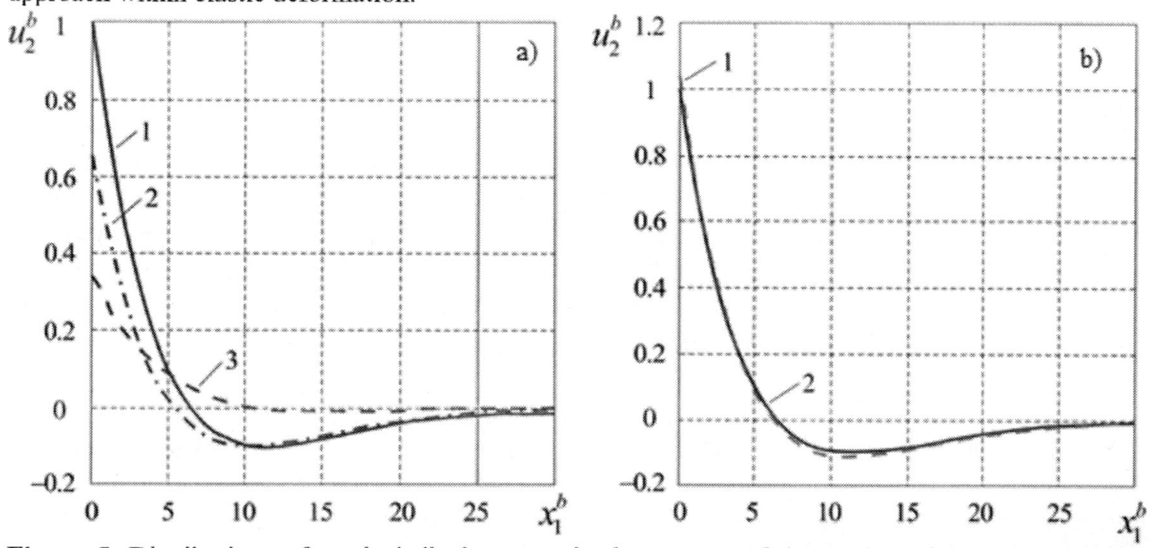

Figure 5. Distributions of vertical displacement in the process of destruction of the first structural element of the layer. In Fig. a) Graph 2 describes subcritical deformation, Graph 3 describes process of destruction of the structural element, Graph 1 describes a superposition of solutions. In Fig. b) Graph 1 repeats the corresponding line Fig. a), Graph 2 describes subcritical deformation without the first structural element.

Figure 6 shows a comparison of vertical displacement along the boundary of body 1 and layer 3 with total destruction of the layer material and when the layer is exfoliated along the boundary with body 2. Calculations made for the elastic deformation of the composite. Graph 1 in figure 6 a) shows the distribution of dimensionless vertical displacement u_2^b along the boundary of body 1 and layer 3 with destruction of the first structural element of the layer with the size $\delta_0 \times \delta_0$, graph 2 shows the distribution of displacements when the layer element is exfoliated along the length δ_0. Displacements are referred to the value of the displacement $u_2^+(0,\delta_0/2)$ toward the end of the process of destruction of the structural element of the layer. Graph 1 in figure 6 b) shows the distribution of dimensionless vertical displacements u_2^b along the boundary of body 1 and layer 3 toward the end when the five structural elements of the layer are destroyed, and Graph 2 shows the distribution of displacements when the layer element is exfoliated along the length $5\delta_0$. From figure 6 we can see that the material of the layer creates additional rigidity due to tensile–compressive and shear stresses, and its account can play an important role in the stratification of the material along the length exceeding the thickness of the layer. For a stratification along the length equal to the size δ_0, the results of calculation in comparison with the complete destruction of the structural element are practically identical.

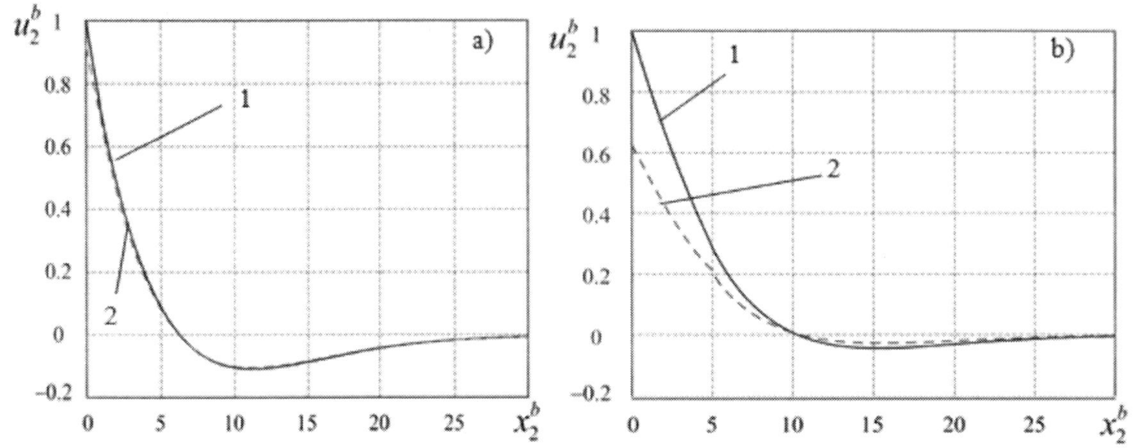

Figure 6. Distributions of vertical displacements in the process of destruction and stratification of the adhesive layer. In Fig. a) the solid line (Graph 1) shows the distribution of displacements during the cohesive destruction of the layer along the length δ_0, and the dotted line (Graph 2) shows the distribution of displacements during adhesive delamination along the length δ_0. In Fig. b) the solid line (Graph 1) shows the distribution of displacements during the cohesive destruction of the layer along the length $5\delta_0$, and the dotted line (Graph 2) shows the distribution of displacements during adhesive delamination along the length $5\delta_0$.

Let us consider the case of cohesive destruction of the adhesive layer of a composite, taking into account the elastoplastic properties of the materials. Figure 7 shows the area of plastic deformation in the state of pre–destruction. The value of the external load is denoted as $\mathbf{P}_k$. The maximum principal stress in the adhesive layer on the characteristic element is equal to the strength limit, and the strength limit is not reached outside the layer. Consequently, destruction in the layer will occur faster than in the materials conjugate with it. It was assumed that the adhesive bond – resin–alloy is sufficiently strong, and the destruction will take place along the array of the adhesive component.

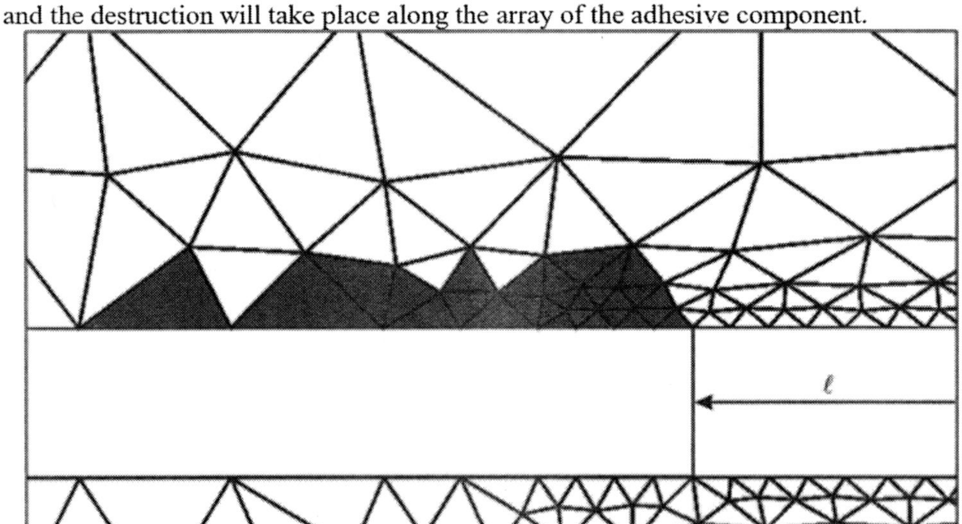

Figure 7. Zone of plasticity in the state of pre–destruction. The area is highlighted

in color.

Destruction of the first element leads to redistribution of the plasticity zone and unloading of a number of elements of the composite. In this case, the result of consideration of destruction as a thermomechanical process is different from the result received with "kill element" procedure. In figure 8, the zone of plastic loading is marked with a dark filling, and the lighter are the elements where elastic unloading from the plastic region occurred, after the destruction of the first δ –element.

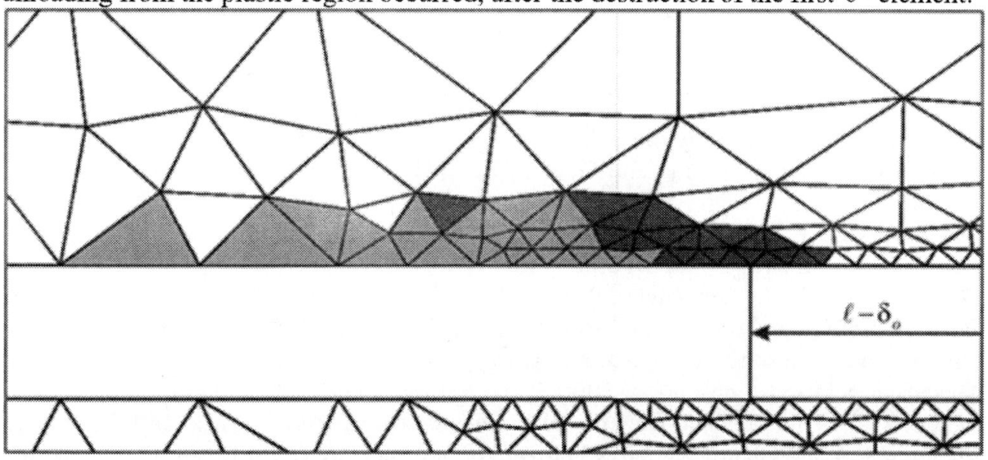

Figure 8. Evolution of the plasticity zone and unloading in the process of destruction of the first element. The unloading region is highlighted in gray, and the elastoplastic region is black.

Solving the problem of loading a composite with a critical load without the first structural element of the layer, we arrive at the distribution of the plastic region shown in figure 9. Comparing figure 9 and figure 8 we see that unloading from the plastic region is not taken into account in the "kill element" approach, the plasticity region is slightly larger than the combination of the unloading regions from the plastic region and the additional plastic load shown in figure 8.

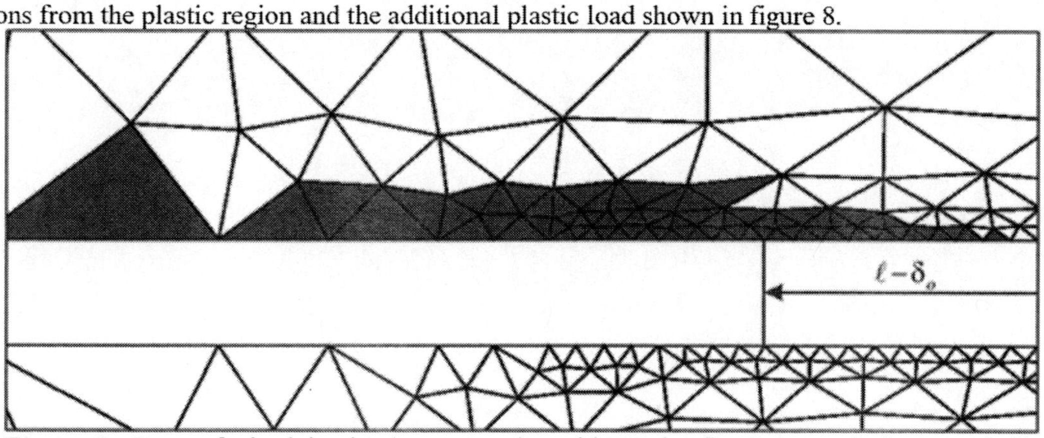

Figure 9. Zone of plasticity in the composite without the first structural element of the adhesive layer. The area is highlighted in color.

Destruction of the first element leads to an excess of the strength limit on the second element of the layer, which means its destruction at a fixed external load.

4. Conclusion

The results of the calculations according to the proposed model within the elastic behavior of the material do not contradict the known method of calculation – the "kill element". In the case with an elastoplastic material, the stress state obtained by simulating the destruction process with the local unloading method may differ significantly from the state determined by the kill element method. The proposed approach allows taking into account the redistribution of plastic zones and the possibility of forming new zones of destruction as a result of local unloading.

Acknowledgments
The results of the research project are published with the financial support of Tula State University within the framework of the scientific project № 2017–36.

References

[1] Petrov Y V 1996 *Physics of the Solid State* **38** pp 1846–50
[2] Xiangting Su, Zhenjun Yang and Guohua Liu 2010 *Acta Mechanica Solida Sinica* **23** pp 271–282
[3] Sua X T, Yang Z J and Liu G H 2010 *Int. J. Solids Struct.* **47** pp 2336–45
[4] Dávila C G, Camanho P P and Turon A 2008 *Journal of Aircraft* **42** pp 663–672
[5] De Moura MFSF and Gonçalves JPM 2014 *Int. J. Solids Struct.* **51** pp 1123–31
[6] Jain S, Na S R, Liechti K M and Bonnecaze R T 2017 *Int. J. Solids Struct.* **129** pp 167–176
[7] Olsson E and Larsson P–L 2013 *J. Mech. Phys. Solids* **61** pp 1185–1201
[8] Mesarovic S D and Johnson K L 2000 *J. Mech. Phys. Solids* **48** 2009–33
[9] Panettieri E, Fanteria D and Danzi F 2016 *Composite Structures* **137** pp 140–7
[10] Lebedev A A and Chausov N G 1983 *Strength of Materials* **15** pp 155–160
[11] ANSYS 2006 User's Guide, Release 11.0 (Pennsylvania USA: ANSYS Inc.)
[12] Glagolev V V, Markin A A and Fursaev A A 2017 *PNRPU Mechanics Bulletin* pp 45–59
[13] Markin A A and Sokolova M Y 2014 *Thermomechanics of Elastoplastic Deformation* (Cambridge International Science Publishing) p 380
[14] Glagolev V V, Glagolev L V and Markin A A 2015 *Acta Mechanica Solida Sinica* **28** pp 375–383
[15] Glagolev V V, Markin A A and Fursaev A A 2016 *PNRP Mechanic Bulletin* pp 34–44

IOP Conf. Series: Journal of Physics: Conf. Series **973** (2018) 012004 doi:10.1088/1742-6596/973/1/012004

Stress–strain state in an elastoplastic pipe taking into account the temperature and compressibility of the material

K K Gornostaev, A V Kovalev, Y V Malygina
Voronezh State University, 1 Universitetskaya pl., Voronezh, 394018, Russia
E-mail: ymkahavren@gmail.com

Abstract. In the article the authors have considered the problem of determining the stress–strain state of the elastoplastic pipe with the Mises' condition in case of plane strain for the compressible material taking into account the temperature.The task was solved using the method of the small parameter. The expressions for the fields of stresses and displacements were received as well as the ratio of the radius of the elastoplastic boundary in the zero and first approximations.

1. Introduction

Many of the works are devoted to the problem of determining the stress strain–state in the compressible elastoplastic bodies, among which there are some works like [1–5]. In the works [1], [2] it was presented an approach that allows to get the solution of the problem of determining the stress–strain state for the case of the plane strain material compressible by elastic deformation (the case of small elastic compressibility). An algorithm for solving problems of determining the stress–strain state for the case of plane strain and plane stress state, which allows to remove the assumption of a low elastic compressibility of the material was proposed in works [3], [4]. The plasticity condition is piecewise linear. The work [5] considers the effect of isotropic hardening and compressibility on the solution of plane elastoplastic problems. Also there was shown the strong influence of elastic compressibility of the material and the small effect of hardening on the stress–strain state.

The accounting of the temperature effects in the solution of these tasks leads to a significant complication of the mathematical calculations. The research of this question is the subject of many works such as [6–8]. The stress–strain state in the viscoelastoplastic hardening steel with respect to temperature is determined in work [6]. The material of the pipe was considered incompressible. The problems of determining the stress–strain state in an elastic space weakened by cylindrical cavity with respect to temperature and elastoplastic pipe with account for compressibility of the material were solved using perturbation method in [7, 8].

Many of the tasks that today's mathematicians, physicists, and engineers face are not amenable to exact solution. Among the reasons that hinder to search for exact solutions, it is possible to specify, for example, the nonlinear equations of motion, variable coefficients and nonlinear boundary conditions at known or unknown boundaries of complex form. In this situation, the researcher is forced to use various kinds of approximations and here it is the most appropriate to use approximate analytical approaches. One of such approaches is the method of the small parameter, or the perturbation method that allows you to find the solution close to the already known exact one. At the same time both the body shape and boundary conditions [12] can be subjected to the perturbation.

In this article, we consider the problem of determining the stress–strain state of the elastic–plastic pipe including the temperature and compressibility of the material. In the first approximation, the stress and displacement fields are defined in the elastic and plastic zones, and also a relation for the radius of the elastoplastic boundary is found.

2. Materials and methods

The solution of our problem will be sought by the method of a small parameter [2], [12]. It was first used to solve the practical problems of mechanics in the works of Poincare [13], Van–Dyck [14] and Nayfe [15, 16]. This method is based on introduction of small quantities that more or less "disturbing" certain original solution in comparison some data. Due to the fact that the small quantities are used as "disturbing" ones, in many works the perturbation method is called the small parameter method. The small parameter in the theory of plasticity was introduced in different ways. In [2] a small parameter characterizes the difference between a planar and an axisymmetric state. In [17], as a small parameter, it's used the reverse of the volume compression modulus, also the normal and tangential stresses were investigated for pure bending of the girder beyond the elastic limit. For the solution of elastoplastic problems by the method of small parameter, uses various schemes. D D Ivlev and L V Ershov [2] considered the case when a plastic zone develops from a certain boundary and completely covers it. In this approach, a number of two–dimensional and three–dimensional problems were solved [18–22]. B D Annin and G P Cherepanov [23] gave a solution to the problem of comprehensively compressing a plane with a hole. In this case, in contrast to the Ivlev–Ershov scheme, the solution in the elastic zone was determined by the methods of a function of the complex variable. It was shown that for the plate with elliptic hole the proposed scheme and the Ivlev–Ershov scheme lead to the same result. M A Artemov received [24–27] a number of approximate solutions based on the Ivlev–Ershov scheme for the problems of the tension plane of hardening elastic–plastic material with a circular holeas well as about the eccentric pipe exposed to internal pressure. Also, a large number of elastic–plastic tasks considered in the works of A N Sporykhin and his disciples [28–38]. The modeling of the thermoelastic behavior of solids was considered in [39-48]

Assume that the desired solution depends on some parameter δ $(1 \ll \delta)$. Linearization of the small parameter is the decomposition of all the source relations: the equations of equilibrium, equations of the relationship of strain and stress, boundary conditions, etc. in the ranks in this parameter. Then members of the decomposition are discriminated with the same degrees of this parameter, which define the system of equations, which allows to develop the method of successive approximations, if the solution when $\delta = 0$ is known [12].

Consider the elastic–plastic state of thick–walled round pipe, the cross section of which is limited by circles of radiuses a and b $(a < b)$ of compressible material under the action of uniform internal pressure p. The temperature of the pipe $T = T(r)$ is the known solution of the heat equation, and modified by logarithmic law [10, 11], the yield strength is temperature independent. The problem can be solved in a cylindrical coordinate system (r, θ, z).

It used the following system of equations:

– the equation of equilibrium

$$\frac{\partial \sigma_r}{\partial r} = \frac{\sigma_\theta - \sigma_r}{r}, \tag{1}$$

where σ_r, σ_θ — the components of the stress tensor;

– the ratio of Cauchy

$$e_r = \frac{du}{dr}, e_\theta = \frac{u}{r}, \tag{2}$$

where e_r, e_θ — the components of the tensor of total deformation, u — component of radial movement;

– the ratio of Hooke's law with respect to temperature connecting stress and elastic strain

$$\begin{aligned} e_r^e &= \frac{1}{E}[\sigma_r - \mu(\sigma_\theta + \sigma_z)] + (\alpha T), \\ e_\theta^e &= \frac{1}{E}[\sigma_\theta - \mu(\sigma_r + \sigma_z)] + (\alpha T), \\ e_z^e &= \frac{1}{E}[\sigma_z - \mu(\sigma_r + \sigma_\theta)] + (\alpha T), \\ e_{r\theta}^e &= \frac{1+\mu}{E}\tau_{r\theta}, \end{aligned} \tag{3}$$

where $e_z, e_{r\theta}$ — the components of the tensor of total deformation, E — elastic modulus, $T = T(r)$ — temperature that is a known solution of the heat equation (see, for example, [10, 11]), α — coefficient of linear thermal expansion, μ — Poisson's ratio, upper index e — indicates belonging to the elastic region;

– the condition of plasticity

$$(\sigma_\theta - \sigma_r)^2 + (\sigma_\theta - \sigma_z)^2 + (\sigma_r - \sigma_z)^2 + 6\tau_{r\theta}^2 = 6k^2, \tag{4}$$

where $\sigma_z, \tau_{r\theta}$ — the components of the stress tensor, $k = k(T)$ — yield stress [10];

– the ratio of the associative law of plastic flow

$$\begin{aligned} de_r^p &= \frac{d\lambda}{3}(2\sigma_r - \sigma_\theta - \sigma_z), \\ de_\theta^p &= \frac{d\lambda}{3}(2\sigma_\theta - \sigma_r - \sigma_z), \\ de_z^p &= \frac{d\lambda}{3}(2\sigma_z - \sigma_r - \sigma_\theta), \\ de_{r\theta}^p &= d\lambda\tau_{r\theta}, \end{aligned} \tag{5}$$

where $d\lambda$ — positive scalar multiplier, upper index p — indicates belonging to the plastic region;

– boundary conditions

$$\sigma_r|_{r=a} = -p, \sigma_r|_{r=b} = 0, \tag{6}$$

– the continuity conditions of stresses and displacements for an elasto–plastic boundary

$$[\sigma_r] = [\sigma_\theta] = [u] = 0. \tag{7}$$

Square brackets here and below denote the difference between the values of expressions enclosed in parentheses, corresponding to the elastic and plastic regions.

Full deformations of the body in the plastic zone consist of two parts: elastic and plastic

$$e_r = e_r^e + e_r^p, e_\theta = e_\theta^e + e_\theta^p, e_z = e_z^e + e_z^p, e_{r\theta} = e_{r\theta}^e + e_{r\theta}^p. \tag{8}$$

The elastic deformations are connected with the stress by Hooke's law with respect to temperature, and the increment of plastic deformation – with associated law of plastic flow. In the case of plane strain $e_z = 0$. Then from (3), (4), (5) and (8) we get:

IOP Conf. Series: Journal of Physics: Conf. Series **973** (2018) 012004 doi:10.1088/1742-6596/973/1/012004

$$
\begin{aligned}
de_r &= \frac{1}{E}[d\sigma_r - \mu(d\sigma_\theta + d\sigma_z)] + d(\alpha T) + \frac{d\lambda}{3}(2\sigma_r - \sigma_\theta - \sigma_z),\\
de_\theta &= \frac{1}{E}[d\sigma_\theta - \mu(d\sigma_r + d\sigma_z)] + d(\alpha T) + \frac{d\lambda}{3}(2\sigma_\theta - \sigma_r - \sigma_z),\\
0 &= \frac{1}{E}[d\sigma_z - \mu(d\sigma_r + d\sigma_\theta)] + d(\alpha T) + \frac{d\lambda}{3}(2\sigma_z - \sigma_r - \sigma_\theta),\\
de_{r\theta} &= \frac{1+\mu}{E}d\tau_{r\theta} + d\lambda\tau_{r\theta}.
\end{aligned} \tag{9}
$$

Imagine the ratio in the form of series in the parameter δ according to the method of small parameter. In this work, we restrict ourselves to the first two members. Thus, the solution will be sought in the form:

$$
\begin{aligned}
e_{ij} = e_{ij}^{(0)} + \delta e_{ij}^{(1)}, \sigma_{ij} = \sigma_{ij}^{(0)} + \delta\sigma_{ij}^{(1)},\\
T = T^{(0)} + \delta T^{(1)}, \lambda = \lambda^{(0)} + \delta\lambda^{(1)},\\
\alpha = \alpha_0 + \delta\alpha^{(1)}, \mu = \mu_0 + \delta\mu^{(1)},
\end{aligned} \tag{10}
$$

where δ — small parameter, $\mu_0 = 0.5$, $\mu^{(1)}$ — known constant, upper index (0), (1) specifies the zero and first approximations, respectively.

Substituting these expansions into the equations (1), (2), (3), (4) and (7) and equating expressions at identical degrees δ, in each approximation, we get the system of differential equations.

There is a flat deformation of elasto–plastic incompressible material in a zero approximation, and, therefore, it will be just the condition of incompressibility

$$e_r + e_\theta = 0. \tag{11}$$

The solution of this problem has the form [2]

$$\sigma_r^{(p)(0)} = -p + 2kln\left(\frac{r}{a}\right), \sigma_\theta^{(p)(0)} = -p + 2kln\left(\frac{r}{a}\right) + 2k, \tag{12}$$

$$\sigma_z^{(p)(0)} = -p + 2kln\left(\frac{r}{a}\right) + k, \tag{13}$$

$$u^{(p)(0)} = u^{(e)(0)} = \frac{3kr_s^{(0)^2}}{2Er}, \tag{14}$$

$$\sigma_r^{(e)(0)} = kr_s^{(0)2}\left[\frac{1}{b^2} - \frac{1}{r^2}\right], \sigma_\theta^{(e)(0)} = kr_s^{(0)2}\left[\frac{1}{b^2} + \frac{1}{r^2}\right], \tag{15}$$

$$\sigma_z^{(e)(0)} = \frac{kr_s^{(0)^2}}{b^2}. \tag{16}$$

Consider first approximation.

The condition of plasticity (3) at the first approximation will be

$$2(\sigma_\theta^{(0)} - \sigma_r^{(0)})(\sigma_\theta^{(1)} - \sigma_r^{(1)}) + 2(\sigma_\theta^{(0)} - \sigma_z^{(0)})(\sigma_\theta^{(1)} - \sigma_z^{(1)}) + 2(\sigma_r^{(0)} - \sigma_z^{(0)})(\sigma_r^{(1)} - \sigma_z^{(1)}) = 0. \tag{17}$$

The relations (7) in the first approximation will take the form of

$$
\begin{gathered}
de_r^{(1)} = \frac{1}{E}[d\sigma_r^{(1)} - \mu_0(d\sigma_\theta^{(1)} + d\sigma_z^{(1)}) - \mu^{(1)}(d\sigma_\theta^{(0)} + d\sigma_z^{(0)})] + \\
+ d(\alpha_0 T^{(1)}) + d(\alpha^1 T^{(0)}) + \frac{d\lambda^{(0)}}{3}(2\sigma_r^{(1)} - \sigma_\theta^{(1)} - \sigma_z^{(1)}) + \frac{d\lambda^{(1)}}{3}(2\sigma_r^{(0)} - \sigma_\theta^{(0)} - \sigma_z^{(0)}), \\
de_\theta^{(1)} = \frac{1}{E}[d\sigma_\theta^{(1)} - \mu_0(d\sigma_r^{(1)} + d\sigma_z^{(1)}) - \mu^{(1)}(d\sigma_r^{(0)} + d\sigma_z^{(0)})] + \\
+ d(\alpha_0 T^{(1)}) + d(\alpha^{(1)} T^{(0)}) + \frac{d\lambda^{(0)}}{3}(2\sigma_\theta^{(1)} - \sigma_r^{(1)} - \sigma_z^{(1)}) + \frac{d\lambda^{(1)}}{3}(2\sigma_\theta^{(0)} - \sigma_r^{(0)} - \sigma_z^{(0)}), \\
0 = \frac{1}{E}[d\sigma_z^{(1)} - \mu_0(d\sigma_r^{(1)} + d\sigma_\theta^{(1)}) - \mu^{(1)}(d\sigma_r^{(0)} + d\sigma_\theta^{(0)})] + \\
+ d(\alpha_0 T^{(1)}) + d(\alpha^{(1)} T^{(0)}) + \frac{d\lambda^{(0)}}{3}(2\sigma_z^{(1)} - \sigma_r^{(1)} - \sigma_\theta^{(1)}) + \frac{d\lambda^{(1)}}{3}(2\sigma_z^{(0)} - \sigma_r^{(0)} - \sigma_\theta^{(0)}).
\end{gathered} \tag{18}
$$

The boundary conditions (4) in the first approximation take the form

$$
\sigma_r^{(1)}|_{r=a} = 0, \sigma_r^{(1)}|_{r=b} = 0. \tag{19}
$$

The relations of the equations of continuity of stresses and displacements on the boundary (5) in the first approximation have the form [2]

$$
\left[\sigma_r^{(1)} + \frac{d\sigma_r^{(0)}}{dr} r_s^{(1)}\right] = \left[\sigma_\theta^{(1)} + \frac{d\sigma_\theta^{(0)}}{dr} r_s^{(1)}\right] = \left[u^{(1)} + \frac{du^{(0)}}{dr} r_s^{(1)}\right] = 0. \tag{20}
$$

Converting the third equation of (16), we obtain

$$
\frac{1}{2}d(2\sigma_z^{(1)} - \sigma_r^{(1)} - \sigma_\theta^{(1)}) + \frac{E}{3}d\lambda^{(0)}(2\sigma_z^{(1)} - \sigma_r^{(1)} - \sigma_\theta^{(1)}) = \mu^{(1)}(d\sigma_r^{(0)} + d\sigma_\theta^{(0)}) - \alpha^{(1)} E dT^{(0)}. \tag{21}
$$

Let's introduce the designation $\chi^{(1)} = 2\sigma_z^{(1)} - \sigma_r^{(1)} - \sigma_\theta^{(1)}$. Then the previous equation can be represented in the form

$$
\frac{d\chi^{(1)}}{d\lambda^{(0)}} + \frac{2E}{3}\chi^{(1)} = 4\mu^{(1)}\frac{d\sigma_z^{(0)}}{d\lambda^{(0)}} - 2\alpha^{(1)} E \frac{dT^{(0)}}{d\lambda^{(0)}}. \tag{22}
$$

After solving the equation (20), we obtain

$$
\begin{gathered}
\chi^{(1)} = \exp(-\frac{2}{3}E\lambda^{(0)}) \int \left[4\mu^{(1)}\frac{d\sigma_z^{(0)}}{d\lambda^{(0)}} - 2\alpha^{(1)} E \frac{dT^{(0)}}{d\lambda^{(0)}}\right] \exp(\frac{2}{3}E\lambda^{(0)}) d\lambda^{(0)} + \chi_e^{(1)}, \\
\chi_e^{(1)} = 4\mu^{(1)}\sigma_z^{(0)} - 2\alpha^{(1)} E T^{(0)}.
\end{gathered} \tag{23}
$$

According to the introduced notation, we can determine the component voltage $\sigma_z^{(1)}$

$$
\sigma_z^{(1)} = \frac{1}{2}(\sigma_r^{(1)} + \sigma_\theta^{(1)} + \chi^{(1)}). \tag{24}
$$

From relation (16), taking into consideration the solution in the zero approximation and the expression (22), we define $e_r^{(1)}$ and $e_\theta^{(1)}$ in the elastic zone

$$
e_r^{(1)} + e_\theta^{(1)} = -\frac{6\mu^{(1)}}{E}\sigma_z^{(0)} + 3\alpha^{(1)} T^{(0)}, \tag{25}
$$

$$e_r^{(1)} - e_\theta^{(1)} = \frac{1}{2E}\left(3(\sigma_r^{(1)} - \sigma_\theta^{(1)}) + 2\mu^{(1)}(\sigma_r^{(0)} - \sigma_\theta^{(0)})\right). \tag{26}$$

In the plastic zone on the basis of plasticity condition (15) given (21), (22) and known zero approximation, we get

$$\sigma_\theta^{(1)} - \sigma_r^{(1)} = 0. \tag{27}$$

Solving these equations, we determine expressions for the fields of stresses and displacements in the first approximation in the form

$$\sigma_r^{(p)(1)} = 0, \sigma_\theta^{(p)(1)} = 0, \tag{28}$$

$$\sigma_z^{(p)(1)} = \frac{1}{2}\exp\left(1 - \frac{r_s^{(0)2}}{r^2}\right)\left[\int (4\mu^{(1)} + 2\alpha^{(1)}EQ)\exp\left(\frac{r_s^{(0)2}}{r^2} - 1\right)\frac{dr}{r} + \chi_e^{(1)}\right], \tag{29}$$

where

$$\chi_e^{(1)} = 4\mu^{(1)}\left(-p + 2kln\left(\frac{r}{a}\right) + k\right) - 2\alpha^{(1)}EP + 2\alpha^{(1)}EQln(r). \tag{30}$$

$$\begin{aligned} u^{(p)(1)} &= \frac{k}{E}\mu^{(1)}\left[-3\frac{r_s^{(0)2}}{b^2}r - 6rln\left(\frac{r}{r_s^{(0)}}\right) + 3r - 2\frac{r_s^{(0)2}}{r}\right] + \\ &+\delta\alpha^{(1)}\left[\frac{3}{2}Pr - \frac{3}{2}Qrln(r) + \frac{3}{4}Qr - \frac{3}{2}Q\frac{1}{r}\frac{r_s^{(0)2}b^2}{r_s^{(0)2} - b^2}ln\left(\frac{r_s^{(0)}}{b}\right)\right], \end{aligned} \tag{31}$$

$$\sigma_r^{(e)(1)} = \alpha^{(1)}EQ\left[ln\left(\frac{r}{b}\right) - \frac{r_s^{(0)2}}{r^2}\frac{r^2 - b^2}{r_s^{(0)2} - b^2}ln\left(\frac{r_s^{(0)}}{b}\right)\right], \tag{32}$$

$$\sigma_\theta^{(e)(1)} = \alpha^{(1)}EQ\left[ln\left(\frac{r}{b}\right) - \frac{r_s^{(0)2}}{r_s^{(0)2} - b^2}ln\left(\frac{r_s^{(0)}}{b}\right) + 1\right], \tag{33}$$

$$\sigma_z^{(e)(1)} = \alpha^{(1)}EQ\left[ln\left(\frac{r}{b}\right) + \frac{1}{2} - \frac{1}{2}\frac{r_s^{(0)2}}{r_s^{(0)2} - b^2}ln\left(\frac{r_s^{(0)}}{b}\right)\left(2 - \frac{b^2}{r^2}\right)\right] + \frac{1}{2}\chi_e^{(1)}, \tag{34}$$

$$\begin{aligned} u^{(e)(1)} &= \frac{k}{E}\mu^{(1)}\left[-3\frac{r_s^{(0)2}}{b^2}r + \frac{r_s^{(0)2}}{r}\right] + \\ &+\delta\alpha^{(1)}\left[\frac{3}{2}Pr - \frac{3}{2}Qrln(r) + \frac{3}{4}Qr - \frac{3}{2}Q\frac{1}{r}\frac{r_s^{(0)2}b^2}{r_s^{(0)2} - b^2}ln\left(\frac{r_s^{(0)}}{b}\right)\right]. \end{aligned} \tag{35}$$

3. Results and discussion

Thus, having the necessary expressions in the zero and first approximations, and using the decomposition (8), we can obtain the ratio for the fields of stresses and displacements

$$\sigma_r^{(p)} = -p + 2kln\left(\frac{r}{a}\right), \tag{36}$$

$$\sigma_\theta^{(p)} = -p + 2kln\left(\frac{r}{a}\right) + 2k, \tag{37}$$

$$\sigma_z^{(p)} = -p + 2kln\left(\frac{r}{a}\right) + k+$$
$$+\delta\frac{1}{2}\exp\left(1 - \frac{r_s^{(0)2}}{r^2}\right)\left[\int(4\mu^{(1)} + 2\alpha^{(1)}EQ)\exp\left(\frac{r_s^{(0)2}}{r^2} - 1\right)\frac{dr}{r} + \chi_e^{(1)}\right], \tag{38}$$

$$u^{(p)} = \frac{3kr_s^{(0)2}}{2Er} + \delta\frac{k}{E}\mu^{(1)}\left[-3\frac{r_s^{(0)2}}{b^2}r - 6rln\left(\frac{r}{r_s^{(0)}}\right) + 3r - 2\frac{r_s^{(0)2}}{r}\right] +$$
$$+\delta\alpha^{(1)}\left[\frac{3}{2}Pr - \frac{3}{2}Qrln(r) + \frac{3}{4}Qr - \frac{3}{2}Q\frac{1}{r}\frac{r_s^{(0)2}b^2}{r_s^{(0)2} - b^2}ln\left(\frac{r_s^{(0)}}{b}\right)\right], \tag{39}$$

$$\sigma_r^{(e)} = kr_s^{(0)2}\left[\frac{1}{b^2} - \frac{1}{r^2}\right] + \delta\alpha^{(1)}EQ\left[ln\left(\frac{r}{b}\right) - \frac{r_s^{(0)2}}{r^2}\frac{r^2 - b^2}{r_s^{(0)2} - b^2}ln\left(\frac{r_s^{(0)}}{b}\right)\right], \tag{40}$$

$$\sigma_\theta^{(e)} = kr_s^{(0)2}\left[\frac{1}{b^2} + \frac{1}{r^2}\right] + \delta\alpha^{(1)}EQ\left[ln\left(\frac{r}{b}\right) - \frac{r_s^{(0)2}}{r_s^{(0)2} - b^2}ln\left(\frac{r_s^{(0)}}{b}\right) + 1\right], \tag{41}$$

$$\sigma_z^{(e)} = \frac{kr_s^{(0)2}}{b^2} + \delta\alpha^{(1)}EQ\left[ln\left(\frac{r}{b}\right) + \frac{1}{2} - \frac{1}{2}\frac{r_s^{(0)2}}{r_s^{(0)2} - b^2}ln\left(\frac{r_s^{(0)}}{b}\right)\left(2 - \frac{b^2}{r^2}\right)\right] + \frac{1}{2}\chi_e^{(1)}, \tag{42}$$

$$u^{(e)} = \frac{3kr_s^{(0)2}}{2Er} + \delta\frac{k}{E}\mu^{(1)}\left[-3\frac{r_s^{(0)2}}{b^2}r + \frac{r_s^{(0)2}}{r}\right] +$$
$$+\delta\alpha^{(1)}\left[\frac{3}{2}Pr - \frac{3}{2}Qrln(r) + \frac{3}{4}Qr - \frac{3}{2}Q\frac{1}{r}\frac{r_s^{(0)2}b^2}{r_s^{(0)2} - b^2}ln\left(\frac{r_s^{(0)}}{b}\right)\right]. \tag{43}$$

The ratio for determination of the radius r_s of elastic–plastic border has the form

$$r_s = r_s^{(0)} + \delta\frac{\alpha^{(1)}EQ}{4k}\left[r_s^{(0)}ln\left(\frac{r_s^{(0)}}{b}\right) + r_s^{(0)} - r_s^{(0)}\frac{r_s^{(0)2} + b^2}{r_s^{(0)2} - b^2}ln\left(\frac{r_s^{(0)}}{a}\right)\right], \tag{44}$$

where $r_s^{(0)}$ is determined from the equation

$$r_s^{(0)2} = b^2 + 2b^2ln\left(\frac{r_s^{(0)}}{a}\right) - p. \tag{45}$$

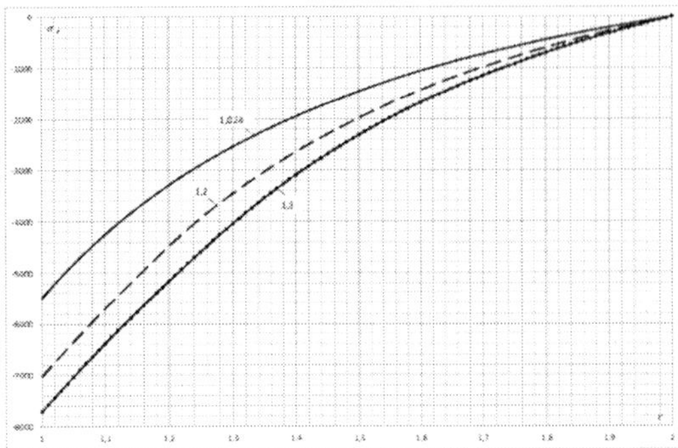

Figure 1. The dependence of the stress component σ_r on radius r.

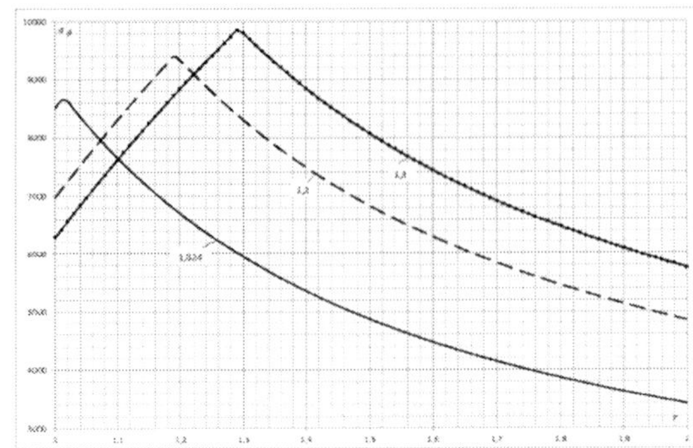

Figure 2. The dependence of the stress component σ_θ on radius r.

Let's assume the following values of constant [10]: $E = 2.1 \cdot 10^6$ H; $\alpha = 0.000011/°C$; $a = 1$ cm; $b = 2$ cm; $T = 100 - 115\ \ln(r)$ °C; $k = 7 \cdot 10^4$ H.

In figure 1–3 are graphs of the component of stresses and radial displacements depending on the radius r for different values of the elasto–plastic boundary, given the adopted values of constants. In figure 1 by the solid line it is plotted stress component σ_r of radius r when the radius of elastic–plastic boundary $r_s = 1,024$ cm. The dotted line is plotted stress component σ_r of radius r when the radius of elastic–plastic boundary $r_s = 1,2$ cm. Line with marker shows stress component σ_r of radius r when the radius of elastic–plastic border $r_s = 1,3$ cm. In figure 2 by the solid line it is plotted stress component σ_θ of radius r when the radius of elastic–plastic boundary $r_s = 1,024$ cm. The dotted line is plotted stress component σ_θ of radius r when the radius of elastic–plastic boundary $r_s = 1,2$ cm. Line with marker shows stress component σ_θ of radius r when the radius of elastic–plastic border $r_s = 1,3$ cm. In figure 3 by the solid line it is plotted radial displacements u of radius r when the radius of elastic–plastic boundary $r_s = 1,024$ cm. The dotted line is plotted radial displacements u of radius r when the radius of elastic–plastic boundary $r_s = 1,2$ cm. Line with marker shows radial displacements u of radius r when the radius of elastic–plastic border $r_s = 1,3$ cm.

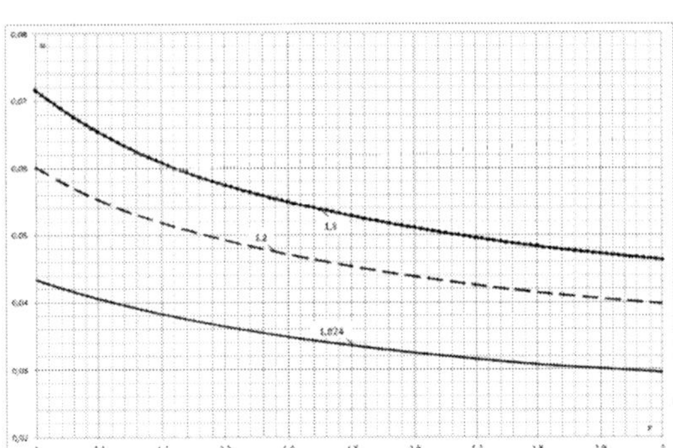

Figure 3. The dependence of the radial displacements u on radius r.

Comparison of the results obtained solving this problem with results of works [1, 6], shows good agreement between these results.

4. Conclusion

The article presents the solution of the problem of determining the stress–strain state of an elasto–plastic pipe, taking into consideration temperature and compressibility of the material.

IOP Conf. Series: Journal of Physics: Conf. Series **973** (2018) 012004 doi:10.1088/1742-6596/973/1/012004

In the first approximation, stress and displacement fields are defined in the elastic and plastic regions, and a ratio for the radius of the elasto–plastic boundary is found. The plots of the stress component σ_r and σ_θ and radial displacements u of radius r are presented.

References

[1] Ivlev D D, Makarov Ye V and Marushkey Yu M 1978 *Journal of Russian Academy of Sciences: Mechanics of Solids* **4** 80–87
[2] Ivlev D D and Yershov L V 1978 *The Perturbation Method in the Theory of an Elastoplastic Body* (Moscow: Nauka)
[3] Artemov M A and Larin I A 2009 *Journal Proceedings of Voronezh State Technical University* **5 (7)** 39–42
[4] Artemov M A and Potapov N S 2009 *Journal Proceedings of Voronezh State Technical University* **5 (8)** 25–29
[5] Kilikovskaya O A and Ovchinikova N V 2012 *Journal of Russian Academy of Sciences: Mechanics of Solids* **1** 75–91
[6] Gornostaev K K and Kovalev A V 2015 *Journal Proceedings of Voronezh State University. Series: Physics. Mathematics* **1** 135–140
[7] Kovalev A V and Khvostov I G 2014 *Bulletin of the Yakovlev Chuvash State Pedagogical University. Series: Mechanics of Limit State* **2(20)** 29–35
[8] Kovalev A V 2013 *Bulletin of the Yakovlev Chuvash State Pedagogical University. Series: Mechanics of Limit State* **1(15)** 64–69
[9] Sporykhin A N, Kovalev A V and Shcheglova Yu D 2004 *Nonuniform Problems of Elastic-viscoplasticity with an Unknown Boundary* (Voronezh: Voronezh State University)
[10] Danilovskaya V I 1965 *Applied Mechanics* **1(6)** 8–13
[11] Parkus G. 1963 *Unsteady Temperature Stresses* (Moscow: Fizmatlit)
[12] Kovalev A V 2006 *The Problems of Determining the Elastoplastic State of Complex and Hardening Media (Doctoral dissertation)* Voronezh
[13] Puankare A 1971 *Selected Works. New Methods of Celestial Mechanics* (Moscow: Nauka)
[14] Van-Dayk M 1967 *Perturbation Methods in Fluid Mechanics* (Moscow: Mir)
[15] Nayfe A Kh 1976 *Perturbation Methods* (Moscow: Mir)
[16] Nayfe A Kh 1984 *Introduction to Perturbation Methods* (Moscow: Mir)
[17] Ilyushin A.A. 1954 *Engineering Collection* **19** 3–12
[18] Vulman S A 1969 *Proceedings of the USSR Academy of Sciences. Mechanics of Solids* **3** 164–169
[19] Kovalev A V and Sporykhin A N 1998 *Bulletin of the Faculty of Applied Mathematics and Mechanics* **1** 85–90
[20] Marushkey Yu M 1976 *Applied Mechanics* **12(2)** 126–130
[21] Marushkey Yu M 1975 *Proceedings of Higher Educational Institutions. Mchine Building* **12** 25–30
[22] Semykina T D 1963 *Proceedings of the USSR Academy of Sciences. Mechanics and mechanical engineering* **1** 17–21
[23] Annin B D and Cherepanov G P 1983 *The Elastic-plastic Problem* (Novosibirsk: Nauka)
[24] Artemov M A 1985 *Journal of Applied Mechanics and Technical Physics* **6** 158–163
[25] Artemov M.A. 1986 *Current Problems of Continuum Mechanics* (Cheboksary: Yakovlev Chuvash State Pedagogical University) pp 8–13
[26] Artemov M.A. 1988 *Applied Problems of Continuum Mechanics* (Voronezh: Voronezh State University) pp 51–53
[27] Artemov M.A. 1988 *Current Problems in the Theory of Boundary Value Problems and Their Ppplications* (Cheboksary: Yakovlev Chuvash State Pedagogical University) pp 4–8
[28] Kovalev A V, Sporykhin A N and Yakovlev A Yu 2004 *Modern Problems of Mechanics and Applied Mathematics: Collection of Works of the International School-Seminar* (Voronezh, Voronezh State University) pp 280–282
[29] Artemov M A and Kovalev A V 1992 *Modern methods in the theory of boundary value problems: Proceedings of the School-Conference* (Voronezh: Voronezh State University) pp 7–27
[30] Kovalev A V, Sporykhin A N and Yakovlev A Yu 1999 *Mechanics-99: Proceedings of the II Byelorussian Congress on Theoretical and Applied Mechanics* (Belarus, Minsk) pp 85-86
[31] Andreeva Yu V, Vnukov A N and Kovalev A V 2015 *The Mechanics of Limit State and Related Questions: Materials of the all-Russian scientific school-conference, dedicated to the 85th anniversary of Professor D.D. Ivleva* (Cheboksary: Yakovlev Chuvash State Pedagogical University) **1** pp 167–172
[32] Andreeva Yu V, Vnukov A N and Kovalev A V 2016 *Mechanics of a Deformable Solid: Proceedings of the IX All-Russian Conference* (Voronezh: Voronezh State University) pp 153–155

[33] Kovalev A V, Shcheglova Yu D and Sviridov I E. 2014 *Materials of the VIII All-Russian Conference on Mechanics of Deformable Solids* (Cheboksary: Yakovlev Chuvash State Pedagogical University) **1** pp 193–199
[34] Kovalev A V, Shcheglova Yu D and Sviridov I E. 2015 *Applied Mathematics, Computational Science and Mechanics: Proceedings of the International Conference* (Voronezh: Voronezh State University) pp 68–72
[35] Kovalev A V, Shcheglova Yu D and Sviridov I E 2015 *The Mechanics of Limit State and Related Questions: Materials of the all-Russian scientific school-conference, dedicated to the 85th anniversary of Professor D.D. Ivleva* (Cheboksary: Yakovlev Chuvash State Pedagogical University) **2** pp 113–117
[36] Kovalev A V, Shcheglova Yu D and Sviridov I E 2016 *Mechanics of a Deformable Solid: Proceedings of the IX All-Russian Conference* (Voronezh: Voronezh State University) pp 104–108
[37] Kovalev A V, Shcheglova Yu D and Sviridov I E 2016 *Bulletin of the Yakovlev Chuvash State Pedagogical University. Series: Mechanics of Limit State* **4(30)** 42–54
[38] Kovalev A V, Rusina Ye Yu and Yakovlev A Yu 2015 *The Mechanics of Limit State and Related Questions: Materials of the all-Russian scientific school-conference, dedicated to the 85th anniversary of Professor D.D. Ivleva* (Cheboksary: Yakovlev Chuvash State Pedagogical University) **2** pp 107–115
[39] Kovalev V A and Radayev Yu N 2010 *Wave Problems of the Field Theory and Thermomechanics* (Saratov: Saratov University Press)
[40] Kovalev V A and Radayev Yu N 2012 *Journal Izvestiya of Saratov University. New Series. Series: Mathematics. Mechanics. Informatics* **12(3)** 72–82
[41] Murashkin E V and Radayev Yu N 2016 *Mechanics 2016: Works of the International School-Conference of Young Scientists* (Erevan, National University of Architecture and Construction of Armenia) pp 160–163
[42] Kovalev V A, Murashkin E V and Radayev Yu N 2017 *Journal of Physics: Conference Series* **788(1)** 012043
[43] Murashkin E V and Radayev Yu N 2017 *Advanced Structured Materials* **46** 363–376
[44] Ziegler H 1958 *Zeitschrift fur Angewandte Mathematik and Physik ZAMP* **9(5)** 748–763
[45] Green A E and Naghdi P M 1992 *Journal Thermal Stresses* **15(2)** 253–264
[46] Green A E and Naghdi P M 1993 *Journal Elasticity* **31(3)** 189–208
[47] Green A E and Naghdi P M 1998 *The Mathematical Theory of Plasticity* (Oxford: Oxford university press)
[48] Radayev Yu N 2004 *Three dimensional problems of mathematical theory of plasticity* (Samara: Samara university press)

IOP Conf. Series: Journal of Physics: Conf. Series **973** (2018) 012005 doi:10.1088/1742-6596/973/1/012005

Equilibrium stability of a cylindrical body subject to the internal structure of the material and inelastic behaviour of the completely compressed matrix

D V Gotsev[1,2], N S Perunov[1] and E N Sviridova[2]

[1] Department of Applied Mathematics, Informatics and Mechanics, Voronezh State University, Voronezh 394006, Russia

[2] Department of Mathematics, Military Educational and Scientific Center of the Air Force "N.E. Zhukovsky and Y.A. Gagarin Air Force Academy" (Voronezh), Voronezh 394064, Russia

E-mail: rbgotsev@mail.ru, perunovnikolays@gmail.com, sviridova.e.n@mail.ru

Abstract. The mathematical model describing the stress-strain state of a cylindrical body under the uniform radial compression effect is constructed. The model of the material is the porous medium model. The compressed skeleton of the porous medium possesses hardening elastic-plastic properties. Deforming of the porous medium under the specified compressive loads is divided into two stages: elastic deforming of the porous medium and further elastic-plastic deforming of the material with completely compressed matrix. The analytical relations that define the fields of stress and displacement at each stage of the deforming are obtained. The influence of the porosity and other physical, mechanical and geometric parameters of the construction on the size of the plastic zone is evaluated. The question of the ground state equilibrium instability is investigated within the framework of the three-dimensional linearized relationships of the stability theory of deformed bodies.

1. Introduction

At present, the construction and protection of underground structures of various kinds, including underground cylindrical cavities, remain the topical issue. As well, there is a number of related problems that need to be considered, such as rock burst control, underground rock blasting, pollution prevention, seismic problems, etc. The underground structural analysis includes (a) the definition of the fields of stress and displacement that arise in the structural components; (b) the definition of the conditions of structural strength and stability. According to the results of the structural analysis, the rational designs of the support constructions and the optimum dimensions of their sections are selected to ensure reliable operation of structures at the minimal cost. Ensuring the stability of shallow excavations with sufficiently strong housing rocks is generally not very difficult. At present, however, there is a steady increase in volume of excavations at great depths and at complex geological conditions (permafrost, high seismicity, neotectonic phenomena, etc.), partly due to the active development of Arctic space. When the condition of a favorable combination of depth and strength of materials is violated, the stability of the excavations and their support constructions becomes a complex engineering and scientific problem [1-11]. To solve all these problems, it is necessary to view the destruction of rock massifs and the stability of the support constructions. The destruction of the underground structure support construction can happen as a result of following two situations: (1) The stress-strain state reaches

IOP Conf. Series: Journal of Physics: Conf. Series **973** (2018) 012005 doi:10.1088/1742-6596/973/1/012005

the strength limits; (2) The stress-strain state reaches the critical values corresponding to the loss of stability (failure) of the support construction. The solution of the first problem is based on the comparison of the stress-strain state, which can be found in analytical or numerical form, and the material ultimate strength [1]. In the second case, the initial phase of the stability problem solution is to find the ground stress-strain state of the structure in the analytical view [2-8]. As a result, it is a matter of urgency to obtain analytical relations describing the subcritical stress-strain state in the analytical form. Cylindrical constructions are widely used in the study of rock mechanics, building mechanics and engineering. Therefore, the question of the definition and analysis of stress-strain states and the study of its stability, taking into account the different physical, mechanical and geometric parameters for cylindrical structures, is the subject of considerable recent and ongoing research.

2. Material model

In the paper under consideration, the deforming of the porous medium in the inelastic behavior of the completely compressed skeleton is described within the framework of the model analysis proposed in the paper [4].

The deforming of the porous material with a characteristic value ε_0, determined by the specific volume of pores, is divided into two stages: deforming of the material *before* and *after* the pores are compressed at a given point in space. The approach used to determine the stress-strain state in this paper differs from the one used in work [5]. (In [5] some analytical relations that use the values of applied loads and allow to determine the existence of partly compressed pores at a given point in space were obtained. To study the process of pore compression for the part of the medium with completely compressed pores it was necessary to introduce an additional term in the rheological relationships).

In this paper, as well as in [5], we obtain the conditions of complete pore compression at a given point in space. As opposed to [5], to determine the stress-strain state in the completely compressed area, we have to consider two independent problems: (1) obtaining the history of deforming and deformations before the pores are compressed (the first stage); (2) investigation of the further deforming and obtaining the deformations of the compressed skeleton (the second stage). In the first stage, we consider elastic deforming of the porous medium under the pressure of the loads at which complete pore compression occurs (these loads will be further defined). In the second stage, we consider inelastic deforming of the compressed skeleton under the effects of the initial loads without the portion of loads that caused the complete pore compression.

Thus, the stress-strain state of the first stage is assumed to be the initial state of the body for the second stage of the deforming. The resulting stress-strain state is obtained by composition of solutions from the first and the second stages.

The obtained results have made it possible to move forward with the stability of the ground stress-strain state in the framework of three-dimensional linearized stability theory of the deformed bodies.

3. Main relationships used for model-based analysis of the stress-strain state of the porous body in the inelastic behaviour of the completely compressed skeleton

The resulting stress-strain state is obtained by composition of solutions from each stage. We use the following formulas for displacement, deformation and stress, respectively

$$u_i = u_i^{(1)} + u_i^{(2)}; \qquad \varepsilon_{ij} = \varepsilon_{ij}^1 + \varepsilon_{ij}^2 + \varepsilon_{ij}^1 \varepsilon_{ij}^2; \qquad \sigma_{ij} = \sigma_{ij}^1 + \sigma_{ij}^2, \tag{1}$$

where the values with (1)-superscript refer to the first stage, the ones with (2)-superscript refer to the second stage.

The relation between stresses and deformations at the first stage of deforming is taken from Hooke's law for the compressible body. In the second stage, the elastic deformation of the compressed skeleton is subjected to the Hooke's law for the incompressible body. In the plastic deforming zone of the compressed skeleton, we use the incompressible hardening elastic-plastic body model [12] with the load surface

$$F = \left(S_j^\beta - c \overset{p}{\varepsilon}{}_j^\beta \right)\left(S_\beta^j - c \overset{p}{\varepsilon}{}_\beta^j \right) - k^2 , \tag{2}$$

where S_j^β are the components of the deviator stress tensor; $\overset{p}{\varepsilon}{}_j^\beta$ are the components of plastic strain tensor; c is the strengthening factor of the material with the completely compressed matrix; k is the fluidity limit of the material with the completely compressed matrix.

In the inelastic deforming zone that occurs in a region of completely compressed pores, the total deformation is composed of the elastic and plastic parts

$$\varepsilon_j^\beta = \overset{e}{\varepsilon}{}_j^\beta + \overset{p}{\varepsilon}{}_j^\beta . \tag{3}$$

The plastic and elastic components of the volumetric deformation respectively satisfy the conditions of incompressibility

$$\overset{p}{\varepsilon}{}_n^n = 0 , \quad \overset{e}{\varepsilon}{}_n^n = -\varepsilon_0 , \tag{4}$$

where ε_j^β and $\overset{e}{\varepsilon}{}_j^\beta$ are the components of the total deformation tensor and elastic deformation tensor, respectively, ε_0 is the value determined by the specific volume of pores.

4. Mathematical modelling of the stress-strain state of the porous cylindrical body under uniform radial compression

Consider the stress-strain problem for a cylindrical heavy-wall body with outer and inner radii b and a, respectively (figure 1). The outer and inner surfaces are under uniformly distributed compressive loads with the intensities q_b and q_a, respectively.

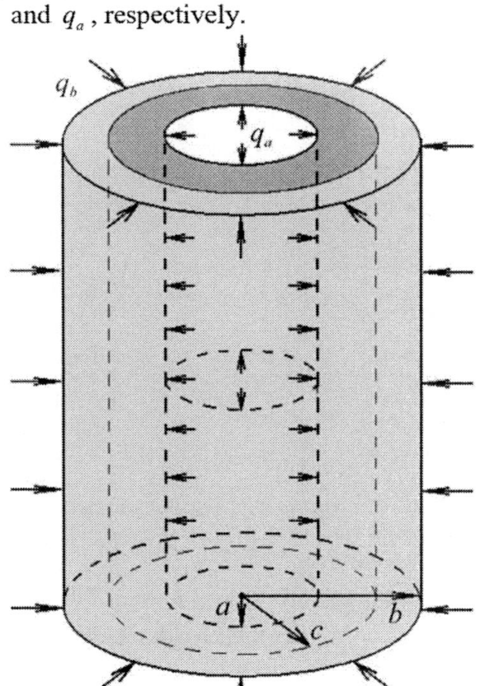

Figure 1. The cylindrical body under uniform radial compression

The axisymmetric stress-strain state is evaluated at the first stage of deforming within the framework of plane deformation, in the cylindrical coordinate system (r, θ, z). We will use the following relations of geometric linear theory:

the equilibrium equation

$$\frac{d\sigma_r}{dr}+\frac{\sigma_r-\sigma_\theta}{r}=0; \tag{5}$$

the Cauchy relations

$$\varepsilon_r=\frac{du}{dr},\ \varepsilon_\theta=\frac{u}{r},\varepsilon_z=0; \tag{6}$$

the Hooke's law for the elastic compressible body

$$\sigma_r=(\lambda_1+2\mu_1)\varepsilon_r+\lambda_1\varepsilon_\theta,\ \sigma_\theta=\lambda_1\varepsilon_r+(\lambda_1+2\mu_1)\varepsilon_\theta,\ \sigma_z=\lambda_1(\varepsilon_r+\varepsilon_\theta). \tag{7}$$

Hereinafter, u is the radial component of the displacement vector, λ_1 and μ_1 are Lamé parameters for the compressible body.

The boundary conditions

$$\sigma_r|_{r=b}=-q_b,\ \sigma_r|_{r=a}=-q_a. \tag{8}$$

From (5)–(7), the volumetric deformation can be written as $\varepsilon_r+\varepsilon_\theta=2C$, where C is an integration constant.

Hence, the volumetric deformation is independent on a coordinate, being the same at each point of the body under consideration. Therefore, complete pore compression occurs in the entire body when the volumetric deformation reaches some given value.

Let the limit value of the volumetric deformation be equal to $-\varepsilon_0$ ($\varepsilon_0>0$ is the value defining by the ratio of the total specific volume of pores to the volume of the medium). Then the condition of the existence of uncompressed pores in the cylindrical body is

$$-2C<\varepsilon_0. \tag{9}$$

Subject to the boundary conditions (8), inequality (9) takes the form

$$\frac{q_b b^2-q_a a^2}{(\lambda_1+\mu_1)(b^2-a^2)}<\varepsilon_0. \tag{10}$$

4.1. Displacement, deformation and stress at the elastic deforming stage of the porous medium

If inequality (10) takes place, the body has uncompressed pores and the stress-strain state is described by the following relations

$$\begin{gathered}u=-\frac{\varepsilon_0}{2}\cdot r+\frac{(q_a-\varepsilon_0(\lambda_1+1))}{2}\cdot\frac{a^2}{r};\\ \varepsilon_r=-\frac{\varepsilon_0}{2}-\frac{(q_a-\varepsilon_0(\lambda_1+1))}{2}\frac{a^2}{r^2},\ \varepsilon_\theta=-\frac{\varepsilon_0}{2}+\frac{(q_a-\varepsilon_0(\lambda_1+1))}{2}\frac{a^2}{r^2};\\ \sigma_r=-\varepsilon_0(\lambda_1+1)-\frac{(q_a-\varepsilon_0(\lambda_1+1))a^2}{r^2},\ \sigma_\theta=-\varepsilon_0(\lambda_1+1)+\frac{(q_a-\varepsilon_0(\lambda_1+1))a^2}{r^2},\ \sigma_z=-\lambda_1\varepsilon_0.\end{gathered} \tag{11}$$

In (11) all values are used without the superscripts since the second stage of the deforming doesn't occur due to the inequality (10). Hence, equalities (11) give the definitive result for the stress-strain state.

Hereinafter, all relations are written in the non-dimensional form. All values of the stress dimension are referred to the parameter μ_1. All values of the length dimension are referred to the parameter b.

If intensities $q_b=q_b^*$ and $q_a=q_a^*$ satisfy the condition

$$q_b^*=\varepsilon_0(\lambda_1+1)(1-a^2)+q_a^*a^2, \tag{12}$$

then all pores are completely compressed in the cylindrical body.

Subject to (12), the stress-strain state at the time of complete pore compression has the form

$$u^{(1)} = -\frac{\varepsilon_0}{2}\cdot r + \frac{\left(q_a^* \cdot f(\varepsilon_0) - \varepsilon_0(\lambda_1+1)\right)}{2}\cdot\frac{a^2}{r};$$

$$\varepsilon_r^{(1)} = -\frac{\varepsilon_0}{2} - \frac{\left(q_a^* \cdot f(\varepsilon_0) - \varepsilon_0(\lambda_1+1)\right)}{2}\frac{a^2}{r^2},\ \varepsilon_\theta^{(1)} = -\frac{\varepsilon_0}{2} + \frac{\left(q_a^* \cdot f(\varepsilon_0) - \varepsilon_0(\lambda_1+1)\right)}{2}\frac{a^2}{r^2};$$

$$\sigma_r^{(1)} = -\varepsilon_0(\lambda_1+1) - \frac{\left(q_a^* \cdot f(\varepsilon_0) - \varepsilon_0(\lambda_1+1)\right)a^2}{r^2}, \tag{13}$$

$$\sigma_\theta^{(1)} = -\varepsilon_0(\lambda_1+1) + \frac{\left(q_a^* \cdot f(\varepsilon_0) - \varepsilon_0(\lambda_1+1)\right)a^2}{r^2},\quad \sigma_z^{(1)} = -\lambda_1\varepsilon_0,$$

where $f(\varepsilon_0) = \begin{cases} 1, \text{ if } \varepsilon_0 \neq 0 \\ 0, \text{ if } \varepsilon_0 = 0 \end{cases}$ ensures the possibility of limiting process when $\varepsilon_0 \to 0$.

4.2. The stress-strain state at the stage of inelastic deforming of the material with the completely compressed matrix

The stage of deforming a material with the completely compressed matrix for the problem under consideration occurs when the condition $q_b > \varepsilon_0(\lambda_1+1)(1-a^2) + q_a a^2$ holds. If at the same time the plasticity condition (2) holds, an inelastic deforming zone forms and starts growing near the inner surface of the cylindrical body. Material hardening holds back this expansion, i.e. compressed skeleton deforms like a hardening incompressible elastic-plastic material with parameters $\mu = 1 + \mu_0$, k, c, where μ_0 is the displacement modulus of the element of the incompressible material. In this case, the continuous medium divides into two zones – elastic and plastic.

At the second stage of deforming, equations (3)–(6) and the following rheological relations describe the stress-strain state:

in the elastic zone

$$s_r = 2\mu\varepsilon_r,\ s_\theta = 2\mu\varepsilon_\theta,\ s_z = 2\mu\varepsilon_z; \tag{14}$$

in the plastic zone

$$s_r = 2\mu\varepsilon_r^e,\ s_\theta = 2\mu\varepsilon_\theta^e,\ s_z = 2\mu\varepsilon_z^e; \tag{15}$$

$$\left(s_r - c\varepsilon_r^p\right)^2 + \left(s_\theta - c\varepsilon_\theta^p\right)^2 + \left(s_z - c\varepsilon_z^p\right)^2 = 2k^2. \tag{16}$$

The boundary conditions at current stage have the form

$$\sigma_r\big|_{r=a} = -\left(q_a - q_a^*\right),\ \sigma_r\big|_{r=b} = -\left(q_b - q_b^*\right). \tag{17}$$

The consistency conditions at the edge of the elastic and plastic zones have the form

$$\sigma_r^p\big|_{r=\gamma} - \sigma_r^e\big|_{r=\gamma} = 0,\ \varepsilon_r^p\big|_{r=\gamma} = 0. \tag{18}$$

The stress-strain state of the cylindrical body at the stage of inelastic deforming of the material with the completely compressed matrix is found from the solution of the boundary value problem (5), (6), (13)–(18) in the following way:

in the elastic area ($\gamma < r < 1$)

$$\sigma_r^{e(2)} = -\left(q_b - q_b^*\right) + 2\mu D\left(1 - \frac{1}{r^2}\right),\ \sigma_\theta^{e(2)} = -\left(q_b - q_b^*\right) + 2\mu D\left(1 + \frac{1}{r^2}\right); \tag{19}$$

in the plastic area ($a < r < \gamma$)

$$\varepsilon_r^{p(2)} = -\frac{1}{2\mu + c}\left(\frac{2D\mu}{r^2} + \chi\sqrt{k^2 - \mu^2\varepsilon_0^2}\right); \tag{20}$$

$$\sigma_r^{p(2)} = -\left(q_a - q_a^*\right) - 4\mu\left(\frac{Dc}{2(2\mu+c)}\left(\frac{1}{r^2}-\frac{1}{a^2}\right)+\left(\frac{\chi\sqrt{k^2-\mu^2\varepsilon_0^2}}{2\mu+c}+\varepsilon_0\right)\ln\frac{r}{a}\right),$$
$$\sigma_\theta^{p(2)} = -\left(q_a - q_a^*\right) + 4\mu\left(\frac{Dc}{2(2\mu+c)}\left(\frac{1}{a^2}+\frac{1}{r^2}\right)-\left(\frac{\chi\sqrt{k^2-\mu^2\varepsilon_0^2}}{2\mu+c}-\varepsilon_0\right)\left(\ln\frac{r}{a}-1\right)\right). \quad (21)$$

The displacements and total deformations in the elastic and plastic regions are determined by the relations

$$u = \frac{D}{r} - \frac{\varepsilon_0}{2}r, \quad \varepsilon_r = -\frac{D}{r^2} - \frac{\varepsilon_0}{2}, \quad \varepsilon_\theta = \frac{D}{r^2} - \frac{\varepsilon_0}{2}. \quad (22)$$

The equation for defining the radius γ, which separates the elastic and plastic deforming zones of the cylindrical body with the completely compressed skeleton, has the form

$$\left(q_b - q_b^*\right) - \left(q_a - q_a^*\right) + 4\mu\left(\frac{D}{2}\left(\frac{1}{a^2}-1\right)+\frac{D\mu}{2\mu+c}\left(\frac{1}{\gamma^2}-\frac{1}{a^2}\right)-\left(\frac{\chi\sqrt{k^2-\mu^2\varepsilon_0^2}}{(2\mu+c)}+\varepsilon_0\right)\ln\frac{\gamma}{a}\right)=0. \quad (23)$$

Here in (19)–(23)

$$\chi = sign\left(\left(q_b - q_b^*\right) - \left(q_a - q_a^*\right)\right), \quad D = \frac{-\chi\gamma^2\sqrt{k^2-\mu^2\varepsilon_0^2}}{2\mu}.$$

Thus, the resulting stress-strain state of the cylindrical body under consideration within the framework of the assumed material model is determined by the relations (1). Besides, the components with (1)-superscript are defined by relations (13) and the components with (2)-superscript are defined by relations (19)–(23).

We note, that for the obtained solutions (13), (19)–(23) it is possible to pass to the limit when $\varepsilon_0 \to 0$. In this case, we obtain the results presented in the paper [7].

5. The results of a numerical simulation of the ground stress-strain state of the cylindrical body subject to the initial internal material structure

A numerical experiment is the form of the obtained solutions describing the stress-strain state of the cylindrical body during the stage of the inelastic material deforming. The results of the numerical experiment are presented in figures 2, 3.

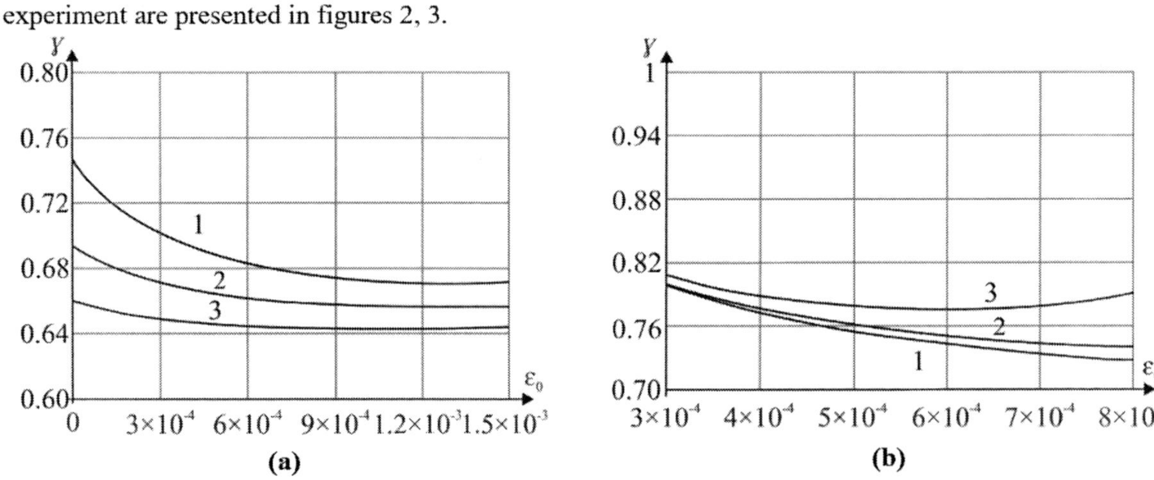

Figure 2. Graph of radius of the elastic-plastic interface γ against value of the initial pore size ε_0 (a) at varying values of the material fluidity limit k (curve 1: $k = 0.007$; curve 2: $k = 0.008$; curve 3: $k = 0.009$); (b) at varying values of the parameter μ_0 (curve 1: $\mu_0 = 2$; curve 2: $\mu_0 = 3$; curve 3: $\mu_0 = 5$).

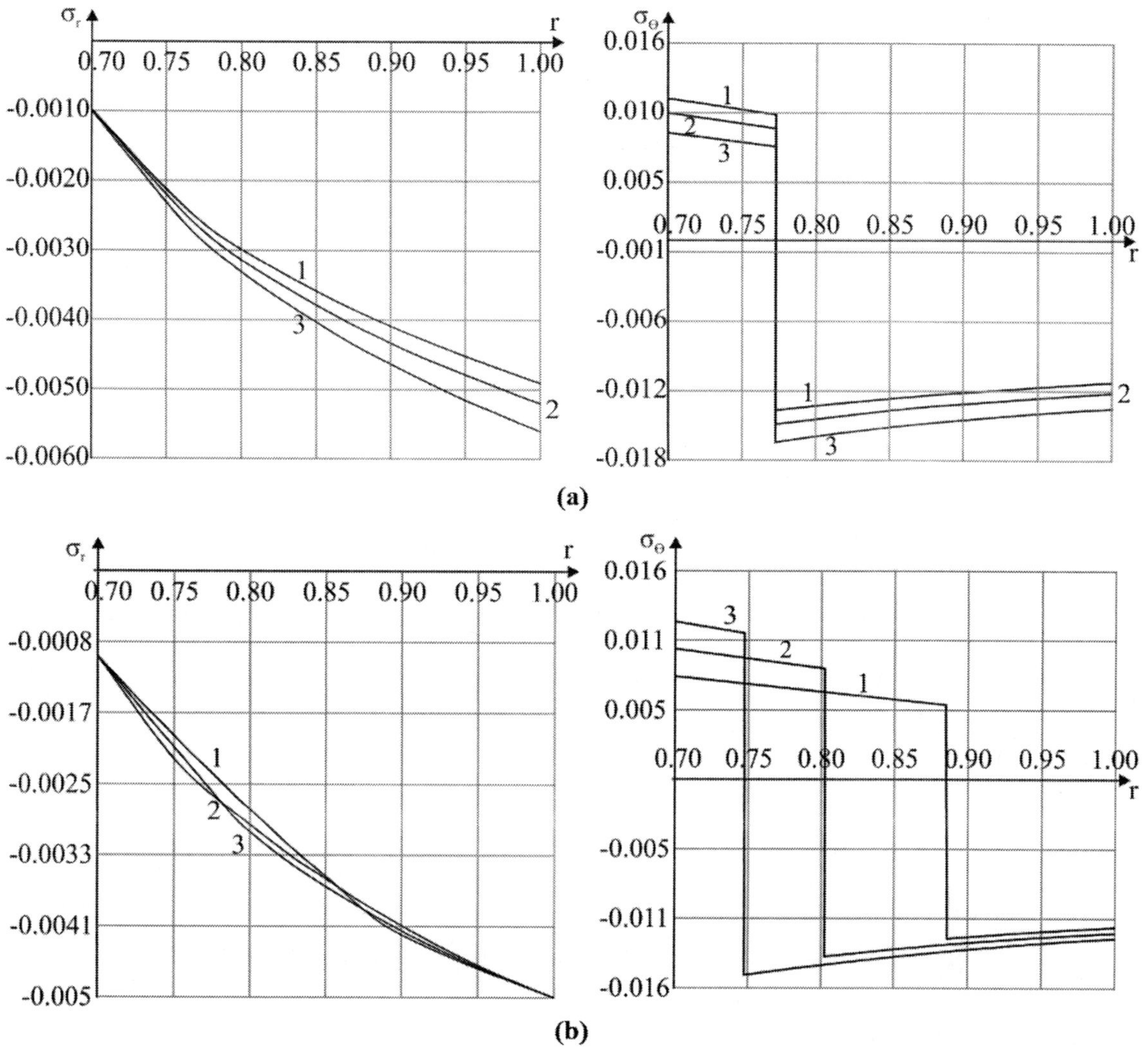

Figure 3. Graph of principal stresses σ_r, σ_θ against the radius r
(a) at varying values of the parameter λ (curve 1: $\lambda = 1.5$; curve 2: $\lambda = 3$; curve 3: $\lambda = 5$);
(b) at varying values of the material fluidity limit k
(curve 1: $k = 0.0035$; curve 2: $k = 0.0045$; curve 3: $k = 0.0055$).

The values of other physical, mechanical and geometric parameters are as follows: $q_a = 0.001$, $q_b = 0.005$, $q_a^* = 0$, $a = 0.7$, $b = 1$, $\varepsilon_0 = 0.0004$, $k = 0.005$, $c = 0.005$, $\mu_0 = 2$, $\lambda = 2$, $\mu_1 = 1$.

The obtained analytical solutions describing the subcritical stress-strain state of the cylindrical construction in question can be used for studying stability of this state.

6. Basic relationships that model the equilibrium stability of the bodies in case of non-homogeneous subcritical states

Investigation of stability of the systems under elastic-plastic deformations is conducted within the framework of the perturbation method assuming the concept of continuing loading [8]. According to this method, the stability of the ground state of such systems is specified by the behaviour of small perturbations of the corresponding linearized problem in the stable unloading areas forming in subcritical state.

IOP Conf. Series: Journal of Physics: Conf. Series **973** (2018) 012005 doi:10.1088/1742-6596/973/1/012005

The perturbed state of the nonconservative systems in question is described by the relationships of three-dimensional linearized stability theory of deformed bodies [7, 8].

The equilibrium equations for each area of elastic and plastic deforming of the material with the completely compressed matrix have the form

$$\nabla_i\left(\sigma_j^i + \overset{0}{\sigma}{}_\alpha^i \nabla^\alpha u_j\right) = 0, \tag{24}$$

where ∇ is the symbol of covariant differentiation; values without "0"-superscript correspond to the components of perturbed state; values with "0"-superscript correspond to the components of the ground unperturbed state defined by the relationships (19)–(23).

The boundary conditions on the part of body surface with given stresses have the form

$$N_i\left(\sigma_j^i + \overset{0}{\sigma}{}_\alpha^i \nabla^\alpha u_j\right) = 0, \tag{25}$$

where N_i are the components of the surface unit normal vector.

The matching conditions of stresses and displacements on the elastic-plastic interface γ have the form

$$\left[N_i\left(\sigma_j^i + \overset{0}{\sigma}{}_\alpha^i \nabla^\alpha u_j\right)\right] = 0, \; \left[u_j\right] = 0, \tag{26}$$

where the brackets indicate the difference of the corresponding values belonging to the elastic and plastic areas.

The equations of state for the material with the completely complex elastic-plastic matrix have the form

$$\sigma_j^\beta = \left(x_{\beta\alpha} g^{\alpha\alpha} \nabla_\alpha u_\alpha + p\right) g_j^\beta + \left(1 - g_j^\beta\right) g^{\beta\beta} \mu\left(\nabla_\beta u_j + \nabla_j u_\beta\right), \mathrm{X}_{j,\beta}, \Sigma_\alpha, \tag{27}$$

where

$$x_{\beta\alpha} = 2\mu g_{\beta\alpha} - \upsilon f_{\alpha\alpha} f_{\beta\beta}, \; f_{ij} = \overset{0}{S}_{ij} - c \overset{0p}{\varepsilon}_{ij}, \; \upsilon = \begin{cases} \dfrac{4\mu^2}{k^2\left(2\mu + c\right)}, & \text{in the plastic area;} \\ 0, & \text{in the elastic area.} \end{cases} \tag{28}$$

The incompressibility condition in perturbations due to linearity can be rewritten as

$$\nabla^\alpha u_\alpha = 0 \tag{29}$$

(for each area of elastic and plastic deforming).

7. Spatial equilibrium stability of the cylindrical body under uniform radial compression

Stability analysis of the ground state (1), (13), (19)–(23) is based on investigating the closed system of equations (24)–(29). It is a partial differential equation system in terms of the perturbations of the components of the displacement vector (*u*, *v*, *w*) and the hydrostatic pressure *p*. The nontrivial solution of this problem corresponds to the instability of the ground state.

We will seek a double trigonometric series solution in elastic and plastic deforming areas

$$\begin{gathered} u = \sum_n^\infty \sum_m^\infty A_{nm}(r)\cos(m\theta)\cos(nz), \; v = \sum_n^\infty \sum_m^\infty B_{nm}(r)\sin(m\theta)\cos(nz), \\ w = \sum_n^\infty \sum_m^\infty C_{nm}(r)\cos(m\theta)\sin(nz), \; p = \sum_n^\infty \sum_m^\infty D_{nm}(r)\cos(m\theta)\cos(nz), \end{gathered} \tag{30}$$

where *n* and *m* are the wave formation parameters.

The system (24)–(29) is a linear homogeneous system, hence, it can be written with the same values of the wave formation parameters for each member. Later on we don't write *m* and *n* indices for notational convenience.

Taking into account (27)–(30), the boundary value problem (24)–(26) in terms of the perturbations of displacements and stresses after a number of transformations is rearranged to the following boundary value problem in terms of the functions $A(r)$, $B(r)$, $D(r)$.

Equations of equilibrium

$$\begin{gathered}
\overset{0}{\xi}_1(r)A(r)+\overset{0}{\xi}_2(r)A'(r)+\overset{0}{\xi}_3(r)A''(r)+\overset{0}{\xi}_4(r)B(r)+\overset{0}{\xi}_5(r)B'(r)+r\cdot D'(r)=0,\\
\overset{0}{\xi}_6(r)A(r)+\overset{0}{\xi}_7(r)A'(r)+\overset{0}{\xi}_8(r)B(r)+\overset{0}{\xi}_9(r)B'(r)+\overset{0}{\xi}_{10}(r)B''(r)+m\cdot D(r)=0,\\
\overset{0}{\xi}_{11}(r)A(r)+\overset{0}{\xi}_{12}(r)A'(r)+\overset{0}{\xi}_{13}(r)A''(r)+\overset{0}{\xi}_{14}(r)r\cdot A'''(r)+\overset{0}{\xi}_{11}(r)mB(r)+\\
+\overset{0}{\xi}_{15}(r)B'(r)+\overset{0}{\xi}_{14}(r)m\cdot B''(r)+D(r)\cdot r^3n^2=0,
\end{gathered} \tag{31}$$

where

$$\overset{0}{\xi}_1(r)=\frac{1}{r}\left(r^2\rho\omega^2-\left(\overset{0}{\sigma}_\theta+\mu\right)\left(1+m^2\right)+\upsilon\cdot\overset{0}{\psi}_2-r^2\left(\overset{0}{\sigma}_z+\mu\right)n^2\right),\ \overset{0}{\xi}_2(r)=\overset{0}{\sigma}_r+r\overset{0}{\sigma}_{r,r}-\upsilon\cdot\overset{0}{\psi}_1+\mu,$$

$$\overset{0}{\xi}_3(r)=r\left(-\upsilon\cdot\overset{0}{\psi}_1+\mu+\overset{0}{\sigma}_r\right),\ \overset{0}{\xi}_4(r)=\frac{m}{r}\left(-2\overset{0}{\sigma}_\theta-2\mu+\upsilon\cdot\overset{0}{\psi}_2\right),\ \overset{0}{\xi}_5(r)=-m\upsilon\cdot\overset{0}{\psi}_3,$$

$$\overset{0}{\xi}_6(r)=-\frac{m}{r}\left(2\overset{0}{\sigma}_\theta-\upsilon\cdot\overset{0}{\psi}_2+4\mu\right),\ \overset{0}{\xi}_7(r)=m\left(\upsilon\cdot\overset{0}{\psi}_3-2\mu\right),$$

$$\overset{0}{\xi}_8(r)=\frac{1}{r}\left(r^2\rho\omega^2-\overset{0}{\sigma}_\theta-r^2\left(\mu+\overset{0}{\sigma}_z\right)n^2-\left(3\mu-\upsilon\cdot\overset{0}{\psi}_2+\overset{0}{\sigma}_\theta\right)m^2-\mu\right),\ \overset{0}{\xi}_9(r)=\overset{0}{\sigma}_r+r\overset{0}{\sigma}_{r,r}+\mu,$$

$$\overset{0}{\xi}_{10}(r)=r\left(\overset{0}{\sigma}_r+\mu\right),\ \overset{0}{\xi}_{11}(r)=\left(r^2\left(\rho\omega^2-n^2\left(2\mu+\overset{0}{\sigma}_z\right)\right)-m^2\left(\mu+\overset{0}{\sigma}_\theta\right)+\mu n^2r^2+\mu+\overset{0}{\sigma}_r-r\overset{0}{\sigma}_{r,r}\right),$$

$$\overset{0}{\xi}_{12}(r)=r\left(r^2\left(\rho\omega^2-n^2\left(2\mu+\overset{0}{\sigma}_z\right)\right)-m^2\left(\mu+\overset{0}{\sigma}_\theta\right)+\mu n^2r^2-\mu-\overset{0}{\sigma}_r+r\overset{0}{\sigma}_{r,r}\right),$$

$$\overset{0}{\xi}_{13}(r)=r^2\left(2\left(\mu+\overset{0}{\sigma}_r\right)+r\overset{0}{\sigma}_{r,r}\right),\ \overset{0}{\xi}_{14}(r)=r^2\left(\mu+\overset{0}{\sigma}_r\right),\ \overset{0}{\xi}_{15}(r)=-rm\left(\mu+\overset{0}{\sigma}_r-r\overset{0}{\sigma}_{r,r}\right),$$

$$\overset{0}{\psi}_1=\left(\frac{\mu_1\varepsilon_0}{3}+\chi\sqrt{k^2-\frac{1}{3}\mu_1^2\varepsilon_0^2}\right)^2,\ \overset{0}{\psi}_2=\left(\frac{\mu_1\varepsilon_0}{3}-\chi\sqrt{k^2-\frac{1}{3}\mu_1^2\varepsilon_0^2}\right)^2,\ \overset{0}{\psi}_3=\left(\frac{2\mu_1\varepsilon_0}{3}\right)^2-k^2.$$

The equilibrium equations (31) are valid for the plastic zone if we add "p"-superscript to all variables in (31). If we put in (31) $\upsilon=0$ and place "e"- superscript to all variables, the equilibrium equations (31) are valid for the elastic zone.

The boundary conditions on the inner ($r=a$) and outer ($r=b$) surfaces of the body, respectively, are written as

$$\begin{gathered}
\left(A^p(a)+B^p(a)m\right)\overset{0}{\varphi}_1+\overset{0}{\varphi}_2A'^p(a)+D^p(a)=0,\\
m\mu A^p(a)+\mu B^p(a)+\overset{0}{\varphi}_3B'^p(a)=0,\\
\overset{0}{\varphi}_4A^p(a)+\left(A'^p(a)a+A''^p(a)a^2-B^p(a)m+B'^p(a)am\right)\overset{0}{\varphi}_5=0;
\end{gathered} \tag{32}$$

$$\begin{gathered}
A'^e(1)\overset{0}{\varphi}_6+D^e(1)=0,\\
m\mu A^e(1)+\mu B^e(1)-\overset{0}{\varphi}_5B'^e(1)=0,\\
A^e(1)\overset{0}{\varphi}_7+\left(A'^e(1)+A''^e(1)-B^e(1)m+B'^e(1)m\right)\overset{0}{\varphi}_5=0,
\end{gathered} \tag{33}$$

where

IOP Conf. Series: Journal of Physics: Conf. Series **973** (2018) 012005 doi:10.1088/1742-6596/973/1/012005

$$\overset{0}{\varphi}_1 = -\frac{\upsilon \cdot \overset{0}{\psi}_3}{a}, \; \overset{0}{\varphi}_2 = -q_a - \upsilon \cdot \overset{0}{\psi}_1 + 2\mu, \; \overset{0}{\varphi}_3 = -a\left(\mu - q_a\right), \; \overset{0}{\varphi}_4 = n^2 a^2 \mu - \mu - q_a, \; \overset{0}{\varphi}_5 = \mu - q_a,$$

$$\overset{0}{\varphi}_6 = 2\mu - q_b, \; \overset{0}{\varphi}_7 = n^2 \mu - \mu - q_a.$$

The matching conditions on the elastic-plastic interface γ have the form

$$\begin{gathered} \overset{0}{\varphi}_8 \left[A'\right] + \left[D\right] + \overset{0}{\varphi}_9 \left(A^P + mB^P\right) - A'^P \upsilon \cdot \overset{0}{\psi}_1 = 0, \\ m\mu\left[A\right] + \mu\left[B\right] + \overset{0}{\varphi}_{10}\left[B'\right] = 0, \\ \overset{0}{\varphi}_{11}\left[A\right] - \gamma\left[A'\right] - \gamma^2\left[A''\right] + m\left[B\right] - \gamma m\left[B'\right] = 0, \end{gathered} \tag{34}$$

where

$$\overset{0}{\varphi}_8 = \overset{0}{\sigma}{}_r^e(\gamma) + 2\mu, \; \overset{0}{\varphi}_9 = -\frac{\upsilon \cdot \overset{0}{\psi}_3}{\gamma}, \; \overset{0}{\varphi}_{10} = -\gamma\left(\overset{0}{\sigma}{}_r^e(\gamma) + \mu\right), \; \overset{0}{\varphi}_{11} = 1 - \frac{\mu n^2 \gamma^2}{\overset{0}{\sigma}{}_r^e(\gamma) + \mu}.$$

It is not possible to obtain an exact analytical solution of the boundary value problem (31)–(34). The approximate solution can be obtained by the use of the finite-difference method [13]. As a result we have the infinite system of homogeneous algebraic equations linear with respect to parameters A_{nm}, B_{nm}, D_{nm}. Hence, sizing the critical internal load q_a corresponding to instability of the cylindrical body under uniform radial compression for the material with a completely compressed skeleton resolves into a matrix equation solvability. In calculating the algebraic system determinant we must take into account the ground stress-strain state for each of the areas of elastic and plastic deforming (1), (13), (19)–(22) and the equation (23). The latter determines the position of the elastic-plastic interface γ entirely covering the internal boundary of the cylindrical body under consideration. The minimization is for the following parameters: the difference interval, the wave formation parameters n and m, the material and structure parameters λ_j. Thus, we have a multidimensional q_a- optimization problem depending on parameters n and m subject to the requirement that the resulting algebraic system determinant equal to zero.

8. Conclusions

The results are summarized as follows.

1. Mathematical models describing the stress-strain states of the cylindrical body under radial compression have been constructed within the framework of the assumed material model at the stages of the elastic deforming of the compressible porous medium and the elastic-plastic deforming of the body with the completely compressed matrix.
2. The analytical model-based calculation of the non-homogeneous subcritical stress-strain state of the cylindrical body in question has been performed within the framework of plane deformation of the body under static uniformly distributed radial loads.
3. The influence of the physical and mechanical parameters of the material (characteristic value determined by the specific volume of pores, the material fluidity limit, the strengthening factor, etc.), the external loads and the geometric parameters of the cylindrical body on its stress-strain state has been evaluated.

In particular, it has been found that

- the value of elastic-plastic interface γ decreases when the Lamé parameter μ_0 of the compressible porous material becomes large;
- increase of the characteristic value ε_0 determined by the specific volume of pores leads to enlarging the region of the compressed matrix inelastic deformations;
- the value of elastic-plastic interface γ decreases when the values of material fluidity limit k and

IOP Conf. Series: Journal of Physics: Conf. Series **973** (2018) 012005 doi:10.1088/1742-6596/973/1/012005

strengthening factor c of the material with the completely compressed matrix increase.

4. The question of the nonhomogeneous ground state equilibrium instability of the cylindrical body under uniform radial compression has been investigated within the framework of three-dimensional linearized stability theory of the deformed bodies. The question has been considered within the bounds of model that takes into account the porous structure of the material, which compressed skeleton has elastic and plastic properties simultaneously.

References

[1] Tsytovich N A 1983 *Soil mechanics* (Moscow: Vysshaya shkola) (in Russ.)

[2] Gotsev D V, Yenenko I A and Sporykhin A N 2005 Local instability of the horizontal excavations of the elliptic shape in elasto-viscous-plastic massifs *Journal of Applied Mechanics and Technical Physics SBRAS* **46**(2) 141–150 (in Russ.)

[3] Gotsev D V, Yenenko I A and Sporykhin A N 2007 Local instability of the horizontal excavations of the polygonal shape in elasto-viscous-plastic massifs *Mechanics of Solids RAS* **2** 183–192 (in Russ.)

[4] Gotsev D V and Sporykhin A N 2010 *Perturbation method in stability problems of supported excavations* (Voronezh: Publishing House of Voronezh State University) (in Russ.)

[5] Gotsev D V, Perunov N S and Buntov A E 2016 Mathematical model of deformation process of the support excavation taking into account the initial porosity of the material and elasto-viscous-plastic properties of the compressed skeleton *The Problems of Strength and Plasticity Journal* **78**(3) 289–299 (in Russ.)

[6] Gotsev D V and Perunov N S 2017 Mathematical model of the stress-strain state of the elastic cylindrical body with porous filler *Tomsk State University Journal of Mathematics and Mechanics* **47** 43–50 (in Russ.)

[7] Sporykhin A N and Shashkin A I 2004 *Equilibrium stability of the space bodies and rock mechanic problems* (Moscow: FIZMATLIT) (in Russ.)

[8] Guz' A N 1977 *Foundation of the stability theory of the mine excavations* (Kiev: NaukDumka) (in Russ.)

[9] Jefferies M G and Shuttle D A 2005 Norsand: calibration and use. *Prediction, analysis and design in geomechanical applications (The 11th Conf. of IACMAG* vol 1*)* 345–352

[10] Tien Y M and Kuo M C 2001 A failure criterion for transversely isotropic rocks *Int. J Rock Mech Min* **38** 399–412

[11] Vervoort A, Min K, Konietzkyc H, Cho J W, Debecker B, Dinh Q, Frühwirt T and Tavallali A 2014 Failure of transversely isotropic rock under Brazilian test conditions *Int. J Rock Mech Min* **70** 343–352

[12] Ivlev D D 2002 *Plastic continuum mechanics* (Moscow: FIZMATLIT) (in Russ.)

[13] Causon D M and Mingham C G 2010 *Introductory Finite Difference Methods for PDEs* (Ventus Publishing ApS).

IOP Conf. Series: Journal of Physics: Conf. Series **973** (2018) 012006 doi:10.1088/1742-6596/973/1/012006

Parameter determination of hereditary models of deformation of composite materials based on identification method

R A Kayumov[1], I Z Muhamedova[1], B F Tazyukov[2], F R Shakirzjanov[1]

[1] Kazan State University of Architecture and Engineering, 1 Zelenaya Street, Kazan, 420043, Russian Federation
[2] Kazan (Volga region) Federal University, 18 Kremlyovskaya Street, Kazan, 420008, Russian Federation

E-mail: muhamedova-inzilija@mail.ru

Abstract. In this paper, based on the analysis of some experimental data, a study and selection of hereditary models of deformation of reinforced polymeric composite materials, such as organic plastic, carbon plastic and a matrix of film-fabric composite, was pursued. On the basis of an analysis of a series of experiments it has been established that organo-plastic samples behave like viscoelastic bodies. It is shown that for sufficiently large load levels, the behavior of the material in question should be described by the relations of the nonlinear theory of heredity. An attempt to describe the process of deformation by means of linear relations of the theory of heredity leads to large discrepancies between the experimental and calculated deformation values. The use of the theory of accumulation of micro-damages leads to much better description of the experimental results. With the help of the hierarchical approach, a good approximation of the experimental values was successful only in the first three sections of loading.

1. Introduction

Polymeric and composite materials belong to the class of materials with a long memory. This means that the stresses at this point in time depend not only on the current values of deformations, temperature and other determining parameters, but also on the values of these parameters at all preceding points in time – from the history of the deformation process. Dependence on the history of the process is particularly shown in the fact that in the simplest experiments regarding pure tension such phenomena as creep and relaxation take place [1]-[4].

Defining equations of hereditary type, written in integral form, take into account the influence of load history, temperature history and other certain factors. In connection with this circumstance, they can be used to describe the mechanical behavior under various loading regimes, in particular, creep, relaxation, and other regimes of time-variant loading. The material described by such equations can be considered as a dynamic system determined by a number of interrelated material functions, such as long-term moduli, long-term compliances, creep and relaxation functions, complex moduli, creep and relaxation spectrum [5]-[8].

In this paper, based on the analysis of some experimental data, a study and selection of hereditary models of deformation of reinforced polymeric composite materials, such as organic plastic, carbon plastic and a matrix of film-fabric composite, was pursued [9].

IOP Conf. Series: Journal of Physics: Conf. Series **973** (2018) 012006 doi:10.1088/1742-6596/973/1/012006

2. Solutions

When constructing the model of creeping of polymeric composite materials under constant loading, they are usually based on the linear theory of heredity:

$$\varepsilon = e + \varepsilon^c, \quad \text{where} \quad e = \sigma / E, \ \varepsilon^c = \int_0^t H(t-\tau)\sigma(\tau)d\tau . \tag{1}$$

Here E is an elastic module of the studied material, H is a creep kernel, ε is a total deformation, ε^c is a creep deformation, e is an elastic part of deformation.

One of the simplest cores is Abel core, which well describes creep of the composite materials, at least the initial sections. It has the form:

$$H(t-\tau) = \frac{C}{(t-\tau)^\alpha}, \quad C \geq 0, \quad 0 < \alpha < 1. \tag{2}$$

Substituting (2) into relation (1), when $\sigma = const$ we can obtain:

$$\varepsilon^c = \varepsilon^c(t) = \sigma \frac{C}{1-\alpha} t^{1-\alpha}. \tag{3}$$

First, the experimental data of the organic plastic tests were processed in one-stage and two-stage loading with unloading.

To find the constants C, α we used the experimental values ε_j^c and σ_j at the different points in time t_j (j=1….m). Substituting them into the expression (3) gives a system of nonlinear algebraic equations with respect to C, α which can be written in the following form:

$$\{\varepsilon^{\exp}\} = \{f(C,\alpha)\}, \quad \text{where} \quad \{\varepsilon^{\exp}\} = \{\varepsilon_1^c, \varepsilon_2^c, ..., \varepsilon_m^c\}^T . \tag{4}$$

Here an index "T" means a transposition operation, an index "exp" means a experimental data, $\{..\}$ is a sign of vector.

To ensure the conditions (2), the required values were replaced:

$$C = s^2, \qquad 1 - \alpha = \frac{1}{1+\theta^2}.$$

To determine the unknowns s, θ a quadratic discrepancy of the equation system was composed (4):

$$\rho^2 = (\{f(s,\theta)\} - \{\varepsilon^{\exp}\})^T (\{f(s,\theta)\} - \{\varepsilon^{\exp}\}), \tag{5}$$

and its minimization by standard gradient methods in the application package Mathematica was done.

Figure 1 shows a plot of deformation ε versus time t, obtained from the formula (1), the parameters of which were found by the above method in the case of one-step loading. The received values are: $C = 0.049$ $(MPa \cdot s^{1-\alpha})^{-1}$; $\alpha = 0.69$. The dots on the plot represent the experimental data.

The results of the calculations showed that for different load levels under a single loading the constants C, α differed greatly. This suggests that it is necessary to use the nonlinear theory of heredity. It can be represented in the following form:

$$\varepsilon = \frac{1}{E}[\sigma + \int_0^t H[t-\tau, \sigma(\tau)]\sigma(\tau)d\tau]. \tag{6}$$

As a non-linear kernel, a modification of the Abel kernel was used, in which it was assumed that $C = C(\sigma), \alpha = \alpha(\sigma)$.

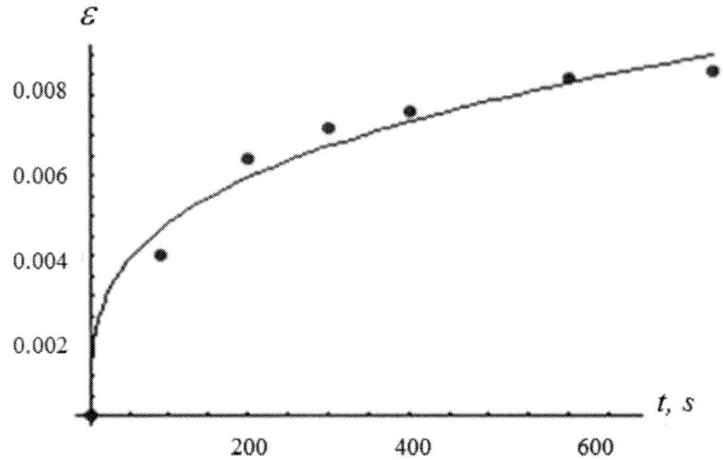

Figure 1. A plot of deformation ε versus time t.

To ensure the relations (2) we used the replacement of the unknown values like this:

$$\begin{cases} C = (s_0 + s_1\sigma)^2, \\ 1 - \alpha = \dfrac{1}{\sqrt{1 + (\alpha_0 + \alpha_1\sigma + \alpha_2\sigma^2)^2}}. \end{cases} \quad (7)$$

Using the forth going procedure a system of equations similar to (4) is constructed. By minimizing the quadratic discrepancy (5), the unknowns were sought s_0, s_1, α_0, α_1, α_2.

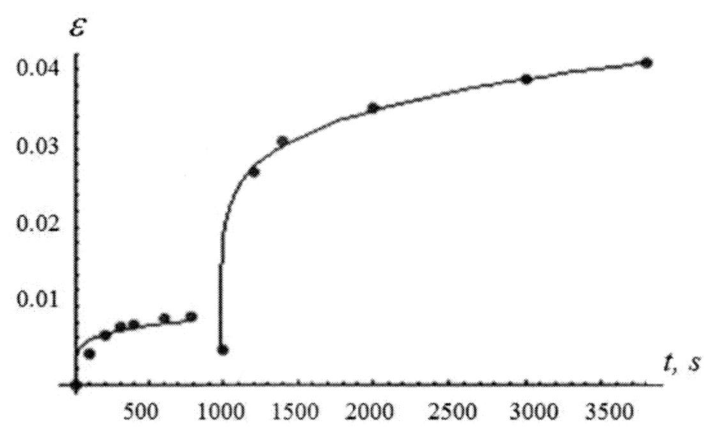

Figure 2. The deformation ε dependence on time t for the case of two-stage loading.

Figure 2 shows the deformation dependence ε on time t for the case of two-stage loading. The dots represent the experimental data, and the curves are constructed on the basis of the mathematical model (6), (2), (7) with the help of the values obtained as a result of minimization ρ^2:

$$s_0 = -0.526(\sqrt{MPa \cdot s^{1-\alpha}})^{-1};\quad s_1 = 0.39(MPa \cdot \sqrt{MPa \cdot s^{1-\alpha}})^{-1};$$
$$\alpha_0 = 0.013;\quad \alpha_1 = 0.393;\quad \alpha_2 = -0.005.$$

Furthermore, a mathematical model of behavior of carbon plastic was constructed with due account for the process of accumulation of microdamages since an attempt to describe its deformation with the help of (1) used to lead to large discrepancies between the calculated and experimental values of deformations in some experiments [10]. That was observed when the shell was brought to ruin.

In the experiment, a five-step loading with the same tension was considered for all five sections with unloadings between them [9]. Two approaches were considered.

a) Hierarchical approach

In this case, when writing the creep equations, the possibility is considered that a damage degree of the material (an accumulation degree of microcracks) affects the process of creep. The modernized Abel kernel was used in the form:

$$H(t-\tau) = \frac{C(\omega)}{(t-\tau)^{\alpha(\omega)}},\quad C(\omega) \geq 0,\quad 0 < \alpha(\omega) < 1. \tag{8}$$

where ω is a damage parameter of the material. The kinetic equation for the damage parameter was taken in the form:

$$\dot{\omega} = f(\sigma,\omega),\quad f(\sigma,\omega) = \frac{B\sigma^{m^2}}{(1-\omega)^{n^2}},\quad \omega(0) = 0, \tag{9}$$

where B, m, n are some constants.

The simplified strength condition according to the works of Rabotnova Yu.N. was accepted in the form:

$$\omega = 1,\quad t = t^* = t_{\max}. \tag{10}$$

When $\sigma = const$ after integration (9) and an exception $B\sigma^{m^2}$ the expression for the damage parameter ω takes the following form:

$$\omega = (1 - [1 - \frac{t}{t^*}]^{\frac{1}{n^2+1}}),\quad \text{where}\quad t^* = t_{\max}. \tag{11}$$

Just as before, to ensure the relations (2) we used the replacement of the required values, but with ω:

$$\begin{cases} C = (s_0 + s_1\omega)^2, \\ 1 - \alpha = \dfrac{1}{1 + (\alpha_0 + \alpha_1\omega)^2}. \end{cases} \tag{12}$$

To find out C, α, which are already functions $C = C(\omega),\ \alpha = \alpha(\omega)$, we used the minimization of the quadratic discrepancy ρ^2 according to the procedure mentioned above. As a result, the required constants are obtained:

$$s_0 = -0.00206(\sqrt{MPa \cdot s^{1-\alpha}})^{-1},\ s_1 = 0.00267(\sqrt{MPa \cdot s^{1-\alpha}})^{-1},$$
$$\alpha_0 = 2.253,\ \alpha_1 = -2.6825,\ n = 0.0000339.$$

Figure 3 shows a graph where the dots represent the experimental data, and the lines represent the calculated curves.

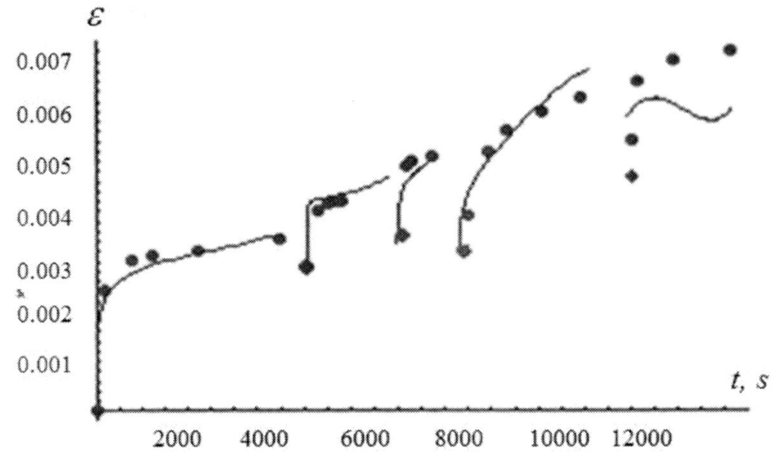

Figure 3. A graph where the dots represent the experimental data, and the lines represent the calculated curves with using the Hierarchical approach.

b) Approach to Kachanov's hypothesis

In this approach, the assumption is made that the damage parameter is not included in the creep equation, that is, the mechanisms of creep and fracture processes are generally different. This hypothesis for our case can be written in the following form:

$$\varepsilon = e + \varepsilon^*, \quad \text{where} \quad \varepsilon^* = \varepsilon^c + \varepsilon^\omega, \tag{13}$$

here ε^c is a deformation, responsible for creep, ε^ω is a deformation, responsible for accumulation of microdamages. It is assumed that the simplest expression of the dependence ε^ω on ω has this form:

$$\varepsilon^\omega = s\omega^{k^2}. \tag{14}$$

For ε^c, as before, the expression (1) is accepted. Following the procedure mentioned above, by means of minimization ρ^2 the required constants are obtained s_1, α_1, n, s, k. On Figure 4 the dots also represent the experimental data, and the lines represent the calculated curves.

Furthermore, the results of processing the experimental data on the creep of a film coating of PVC (polyvinyl chloride) were processed. To describe the creep process, the Abel kernel was used in the form (2), where C, α are functions of stresses: $C = C(\sigma), \alpha = \alpha(\sigma)$.

To ensure the conditions (2), the following replacement of the required values was made:

$$\begin{cases} C = (s_0 + s_1\sigma)^2, \\ \alpha = 1 - \dfrac{1}{\sqrt{1 + (\alpha_0 + \alpha_1\sigma)^2}}. \end{cases} \tag{15}$$

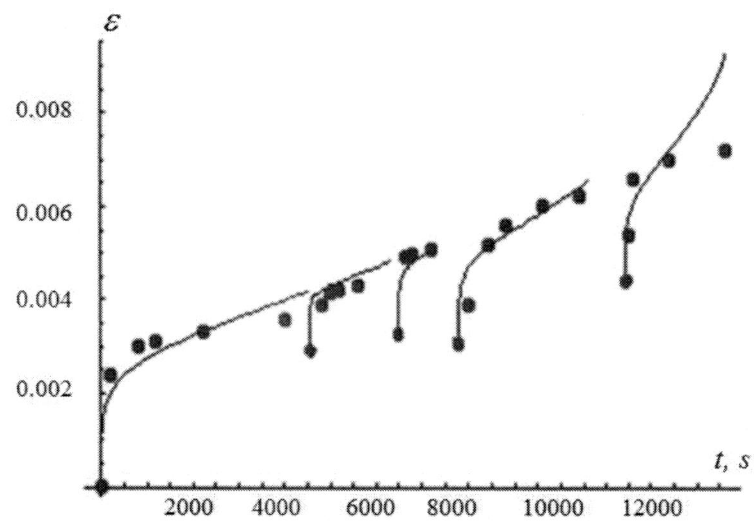

Figure 4. A graph where the dots represent the experimental data, and the lines represent the calculated curves with using the Kachanov's hypothesis.

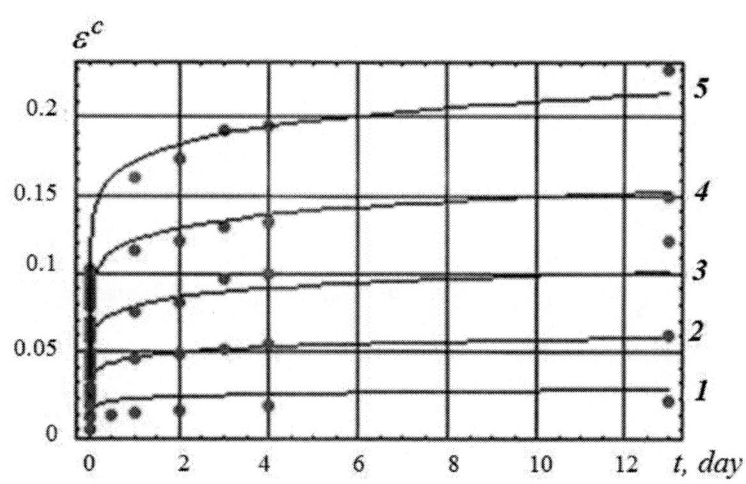

Figure 5. The dependence of creep deformation ε^c on time for PVC at different load levels, where the points represent the experimental data, and the solid line represents the calculated curves.

By minimizing the quadratic discrepancy ρ^2 according to the process mentioned above, the required values were obtained:

$$
\begin{aligned}
s_0 &= 0.09(\sqrt{MPa \cdot s^{1-\alpha}}\,)^{-1}, \\
s_1 &= 0.03(MPa\sqrt{MPa \cdot s^{1-\alpha}}\,)^{-1}, \\
\alpha_0 &= 9.14,\ \alpha_1 = 2.4 / MPa.
\end{aligned} \tag{16}
$$

Figure 5 shows the dependence of creep deformation ε^c on time for PVC at different load levels, where the points represent the experimental data, and the solid line represents the calculated curves.

3. Results and discussion

On the basis of an analysis of a series of experiments it has been established that organo-plastic samples behave like viscoelastic bodies. It is shown that for sufficiently large load levels, the behavior of the material in question should be described by the relations of the nonlinear theory of heredity. An attempt to describe the process of deformation by means of linear relations of the theory of heredity leads to large discrepancies between the experimental and calculated deformation values. The use of the theory of accumulation of micro-damages leads to much better description of the experimental results. With the help of the hierarchical approach, a good approximation of the experimental values was successful only in the first three sections of loading. According to Kachanov's hypothesis, the calculated curves are closer to the experimental points in all five sections of loading.

Acknowledgments

This work was done with support of the Russian Foundation for Basic Research (projects No. 15-08-06018, No. 16-38-00736 mol_a) and the state task of the Ministry of Education and Science of the Russian Federation (project No. 9.1395.2017/PCh)

References

[1] Abdrakhmanova A I and Sultanov L U 2016 *Materials Physics and Mechanics* **26 (1)** 30–32
[2] Davydov R L and Sultanov L U 2015 *Journal of Engineering Physics and Thermophysics* **88 (5)** 1280–1288
[3] Heß H, Roth Y C and Himmel N 2007 *Composites Science and Technology* **67 (6)** 1081–1095
[4] Nagy-Györgya T, Sasb G, Dăescua A C , Stoian V 2012 *Engineering Structures* **33 (44)** 291–303
[5] Kayumov R A, Suleimanov A M, Mukhamedova I Z 2005 *Mechanics of composite materials and structures* **11 (4)** 519-530
[6] Kayumov R A 2004 *Izv. RAS. Mechanics of solid body* **2** 94-103
[7] Kayumov R A, Kupriyanov V N, Mukhamedov I Z, Suleimanov A M, Shakirova A M 2007 *Mechanics of composite materials and constructions* **13 (2)** 165-173
[8] Hicks S J, Lawson R M, Lan D 2006 *Composite Construction in Steel and Cocrete V. American Society of Civil Engineers* 190–201
[9] Alekseev K P, Teregulov I.G 1999 *Mechanics of Composite Materials and Constructions* **5 (2)** 89-100
[10] Obrazcov I F, Saveljev L M and Hazanov H S 1985 *The finite element method in problems of aircraft structural mechanic* 329–364

IOP Conf. Series: Journal of Physics: Conf. Series **973** (2018) 012007 doi:10.1088/1742-6596/973/1/012007

Dynamics of unstable sound waves in a non-equilibrium medium at the nonlinear stage

Sergey Khrapov*, **Alexander Khoperskov**

Mathematics and Information Technology Institute, Volgograd State University, Volgograd, 400062, Russia

E-mail: *khrapov@volsu.ru

Abstract.
A new dispersion equation is obtained for a non-equilibrium medium with an exponential relaxation model of a vibrationally excited gas. We have researched the dependencies of the pump source and the heat removal on the medium thermodynamic parameters. The boundaries of sound waves stability regions in a non-equilibrium gas have been determined. The nonlinear stage of sound waves instability development in a vibrationally excited gas has been investigated within CSPH-TVD and MUSCL numerical schemes using parallel technologies OpenMP-CUDA. We have obtained a good agreement of numerical simulation results with the linear perturbations dynamics at the initial stage of the sound waves growth caused by instability. At the nonlinear stage, the sound waves amplitude reaches the maximum value that leads to the formation of shock waves system.

1. Introduction

The dynamics of sound waves in thermodynamically nonequilibrium media significantly differs from the equilibrium media acoustics [1, 2, 3, 4, 5]. Examples of such media are the vibrationally excited gas, the nonisothermal plasma, the chemically active mixtures, etc. In the non-equilibrium media the second (bulk) viscosity and the acoustic dispersion may turn negative. The negative acoustic dispersion denotes that the equilibrium (low-frequency) sound velocity c_0 becomes greater than the "frozen" (high-frequency) sound velocity c_∞. The medium is acoustically active when the negative second viscosity gives the main contribution to the total viscosity value. Various aspects of this problem have been discussed in Ref.[6]. In such gas the increase in the sound waves amplitude due to the instability may lead to a shock waves system formation [7].

Conditions for sound instability caused by a special viscosity may arise in accretion disks around relativistic compact objects (white dwarfs, neutron stars and black holes) or in circumstellar gas-dust disks. The aim of such astrophysical studies is an attempt to construct an accretion model of rotating gas using turbulence phenomenological models [8, 9, 10].

In Ref. [7] the mechanisms of nonequilibrium gas medium influence on the structure of small amplitude shock waves caused by new dispersion-viscosity properties were considered. It was also shown there that a finite area perturbation in a nonequilibrium medium decays with the formation of autowave pulses and periodic autowave.

It should be noted that research of essentially nonlinear stage of sound waves and strong shock waves dynamics in nonequilibrium media requires the numerical gasdynamic methods of high accuracy. In current paper, we have numerically simulated the nonlinear stage of unstable sound waves development in non-equilibrium media, in which gas molecules are excited by vibrational degrees of freedom stipulated by the energy of the external pump source.

As a rule, the numerical schemes for the hydrodynamics equations are being developed and tested for gas in the thermodynamic equilibrium state. Therefore, studying the nonlinear structures (shock waves and rarefaction waves) in a nonequilibrium gas we used the numerical algorithms based on various approaches of continuous medium description (Lagrangian and Eulerian). Since the Euler and Lagrangian numerical schemes have different dispersion properties, our combined approach allows to identify unphysical numerical solutions that may appear in non-equilibrium medium models.

2. Materials and Methods

2.1. Mathematical model

A nonequilibrium (acoustically active) medium may be simulated using the vibrationally excited gas equations along with the exponential relaxation model [7]:

$$\frac{\partial \varrho}{\partial t} + \mathrm{div}(\varrho \mathbf{v}) = 0\,, \tag{1}$$

$$\frac{\partial \varrho \mathbf{v}}{\partial t} + \mathrm{div}(\varrho \mathbf{v}\mathbf{v}) = -\nabla p + \varrho\, \mathbf{f}\,, \tag{2}$$

$$\frac{\partial E}{\partial t} + \mathrm{div}[(E+p)\mathbf{v}] = \varrho\,(Q - I + \mathbf{v}\mathbf{f})\,, \tag{3}$$

$$\frac{\partial E_\nu}{\partial t} + \mathrm{div}(E_\nu \mathbf{v}) = \varrho\left(\frac{E_{\nu e} - E_\nu}{\tau} + Q\right)\,, \tag{4}$$

where ϱ is the gas density, p is the gas pressure, $\mathbf{v}$ is the velocity vector, $\mathbf{f}$ is the specific external force, Q, I is the pump and heat sink power, respectively, E_ν is the energy of the vibrational degrees of freedom of the excited gas, $E_{\nu e}$ is the equilibrium value of the vibrational energy, τ is the vibrational relaxation time. The expression of total energy for a vibrationally excited gas can be represented as follows

$$E = \frac{p}{\gamma - 1} + \frac{\varrho|\mathbf{v}|^2}{2} + E_\nu\,, \tag{5}$$

where γ is the adiabatic exponent.

In general the quantities of Q, I and τ depend on the thermodynamic parameters of the medium (the density ϱ and temperature T). The functions $Q(\varrho, T)$, $I(\varrho, T)$ and $\tau(\varrho, T)$ in the vicinity of the local equilibrium values of $\tau_0, Q_0, I_0, \varrho_0, T_0$ we represent as a power series expansion:

$$\tau = \tau_0 \left(\frac{\varrho}{\varrho_0}\right)^{\tau_\varrho} \left(\frac{T}{T_0}\right)^{\tau_T}, \quad Q = Q_0 \left(\frac{\varrho}{\varrho_0}\right)^{Q_\varrho} \left(\frac{T}{T_0}\right)^{Q_T}, \quad I = I_0 \left(\frac{\varrho}{\varrho_0}\right)^{I_\varrho} \left(\frac{T}{T_0}\right)^{I_T}. \tag{6}$$

Dependences (6) allow to investigate the dynamics of sound waves in a variety of acoustically active media by selection of appropriate parameters $\tau_\varrho, \tau_T, Q_\varrho, Q_T, I_\varrho, I_T$.

The equation of state is applied to close the system of equations (1)–(6)

$$p = R\varrho T\,, \tag{7}$$

where R is the gas constant.

2.2. Linear stability analysis

Let us conduct the standard procedure of linearizing the system of equations (1)–(4) in the one-dimensional case and represent all the quantities in the form $f = f_0 + \widetilde{f}$. Using the WKB approximation and substituting a solution in the form of plane harmonic waves $\widetilde{f}(t,x) = \widehat{f}(\omega,k)\exp\{-i\omega t + ikx\}$ (where ω is the natural frequency, k is the wave number) into the obtained linearized equations, we obtain a system of linear algebraic equations (SLAE). Equating the determinant of SLAE with zero and taking into account Eqs. (5)–(7), we get the dispersion equation for an acoustically active medium:

$$\omega^2\left[(A - i\delta) + i\frac{(1-i\delta)\gamma_1}{\delta R}(I_0 I_T - Q_0 Q_T)\right] =$$

$$k^2 c_0^2\left[(B - i\delta) + i\frac{(1-i\delta)\gamma_1}{\omega\gamma T_0 R}(I_0(I_T - I_\varrho) - Q_0(Q_T - Q_\varrho))\right], \qquad (8)$$

where $\delta = \omega\tau_0$, $\gamma_1 = \gamma - 1$, $c_0 = \sqrt{\gamma R T_0}$ is the equilibrium low-frequency sound velocity in the gas, $A = 1 + (C_k - S(\tau_T + Q_T))\gamma_1$, $B = 1 + (C_k - S(\tau_T - \tau_\varrho + Q_T - Q_\varrho))\gamma_1/\gamma$, $C_k = \dfrac{\partial E_{ve}}{R\partial T}$, $S = \dfrac{\tau_0 Q_0}{RT_0}$ is a degree of non-equilibrium of the medium.

The dispersion relation (8) is a generalization of the equation obtained in Ref.[7] to the case $Q_\varrho \neq 0, Q_T \neq 0, I_\varrho \neq 0, I_T \neq 0$. For certain values of the parameters $\tau_\varrho, \tau_T, Q_\varrho, Q_T, I_\varrho, I_T$ and degree of medium non-equilibrium, S, corresponding to the negative second (bulk) viscosity and acoustic dispersion, solutions with acoustic increment ($\mathrm{Im}k < 0$) appear in Eq. (8).

For acoustically active medium described by Eq. (8), four characteristic nonequilibrium regions can be distinguished depending on the degree of nonequilibrium S [7]:

(i) An area with positive acoustic dispersion and second viscosity ($0 < S < S_1$). In this region sound waves are damped like in the equilibrium medium with viscosity.

(ii) An area with negative acoustic dispersion and second viscosity ($S_1 \leq S < S_2$). In this region unstable sound waves with $\mathrm{Im}k < 0$ exist. For $S \to S_2$ the equilibrium sound velocity is substantially higher than the high-frequency ("frozen") velocity.

(iii) An area with positive dispersion and negative bulk viscosity ($S_2 \leq S < S_3$). In this area a low-frequency sound cannot propagate due to $c_0^2/c_\infty^2 < 0$.

(iv) An area with positive dispersion and negative bulk viscosity ($S_3 \leq S$). In this region, along with acoustic instability, a thermal instability is possible.

In accordance with (8), the values of S_1, S_2, S_3 have the form:

$$S_1 = -\frac{\gamma_1 C_k}{\tau_\varrho + \gamma_1 \tau_T}, \quad S_2 = -\frac{\gamma_1 C_k}{(\tau_\varrho + I_\varrho) + \gamma_1(\tau_T + I_T)},$$

$$S_3 = -\frac{I_T - Q_T + (I_\varrho - Q_\varrho)(C_k + 1/\gamma_1)}{\tau_T(I_\varrho - Q_\varrho) - \tau_\varrho(I_T - Q_T) + Q_T I_\varrho - Q_\varrho I_T}.$$

We adopt following values of the parameters: $\gamma = 1.4$, $C_k = 0.1$, $S = 0.1$, $\delta_\varrho = -1$, $\delta_T = -2$, $Q_\varrho = Q_T = I_\varrho = I_T = 0$ as the basic parameters corresponding to a typical laser CO_2–containing medium under normal conditions.

The analysis of solutions of the dispersion equation (8) indicates an increase in the acoustic increment with decreasing of $\tau_\varrho, \tau_T, I_\varrho, I_T$ and increasing of Q_ϱ, Q_T.

2.3. Numerical nonlinear model

To construct the numerical model, we use the standard procedure for discretization of a continuous medium on a space-time grid defined by the grid points $(x_j = x_{j-1} + h, y_k = y_{k-1} + h, t_n = t_{n-1} + \Delta t)$ and for each function we have $f(x, y, t) \to f(x_j, y_k, t_n,) = f^n_{j,k}$.

We use two numerical methods: CSPH-TVD (Combined Smoothed Particle Hydrodynamics – Total Variation Diminishing, [11, 12]) and MUSCL (Monotone Upwind Scheme for Conservation Laws, [13, 14]), which are based on different approaches to the continuous medium description. Let's consider each of them in more detail below.

The CSPH-TVD method contains two main stages. At the first Lagrangian stage, a modified SPH algorithm is used [15, 16]. The explicit method of Godunov type [17] and the TVD principle underlie the second Eulerian stage. For the non-equilibrium gas on the basis of equations (1)–(4), the general calculation scheme is:

$$\mathbf{U}^{n+1}_{j,k} = \widetilde{\mathbf{U}}^{n+1}_{j,k} - \frac{\Delta t}{h}\left(\mathbf{F}^{n+1/2}_{j+1/2,k} - \mathbf{F}^{n+1/2}_{i-1/2,k} + \mathbf{G}^{n+1/2}_{j,k+1/2} - \mathbf{G}^{n+1/2}_{j,k-1/2}\right), \tag{9}$$

where

$$\mathbf{U} = \begin{pmatrix} \varrho \\ \varrho u \\ \varrho v \\ E \\ E_\nu \end{pmatrix}, \quad \mathbf{F} = \begin{pmatrix} \varrho u \\ \varrho u^2 \\ \varrho uv \\ uE \\ uE_\nu \end{pmatrix}, \quad \mathbf{G} = \begin{pmatrix} \varrho v \\ \varrho vu \\ \varrho v^2 \\ vE \\ vE_\nu \end{pmatrix}.$$

In equation (9), the quantity $\widetilde{\mathbf{U}}^{n+1}_{j,k}$ is calculated on the Lagrangian stage using a two-step time recalculation (predictor-corrector), providing the second order of accuracy in time:

Predictor

$$\widetilde{\mathbf{U}}^{*}_{j,k} = \mathbf{U}^{n}_{j,k} + \Delta t\, \mathbf{Y}_{j,k}\left(\mathbf{U}^n, \mathbf{r}^n\right), \qquad \mathbf{r}^{*}_{j,k} = \mathbf{r}^{n}_{j,k} + \Delta t \frac{\mathbf{v}^n_{j,k} + \mathbf{v}^*_{j,k}}{2}, \tag{10}$$

Corrector

$$\widetilde{\mathbf{U}}^{n+1}_{j,k} = \frac{\mathbf{U}^{n}_{j,k} + \widetilde{\mathbf{U}}^{*}_{j,k}}{2} + \frac{\Delta t}{2}\mathbf{Y}_{j,k}\left(\mathbf{U}^*, \mathbf{r}^*\right), \qquad \mathbf{r}^{n+1}_{j,k} = \frac{\mathbf{r}^n_{j,k} + \mathbf{r}^*_{j,k}}{2} + \frac{\Delta t}{2}\frac{\mathbf{v}^n_{j,k} + \mathbf{v}^{n+1}_{j,k}}{2}, \tag{11}$$

where $\mathbf{r}$ is the radius vector of the position of the "liquid" SPH-particles, which simulates a continuous medium at the Lagrangian stage. At each time point t_n, the particles are in the center of the Euler cells before the start of the predictor-corrector procedure. The quantity $\mathbf{Y}_{j,k}$ entering into equations (10)–(11) is equal to

$$\mathbf{Y}_{j,k} = -\left(\begin{array}{c} 0 \\ \varphi_{j,k} \sum\limits_{j'=j-1}^{j+1} \sum\limits_{k'=k-1}^{k+1} \varphi_{j',k'} \nabla W_{j,k} + \varrho_{j,k}\mathbf{f}_{j,k} \\ \varphi_{j,k} \sum\limits_{j'=j-1}^{j+1} \sum\limits_{k'=k-1}^{k+1} \varphi_{j',k'} \dfrac{\mathbf{v}_{j,k} + \mathbf{v}_{j',k'}}{2} \nabla W_{j,k} + \varrho_{j,k}\left(Q_{j,k} - I_{j,k} + \mathbf{v}_{j,k}\mathbf{f}_{j,k}\right) \\ \varrho_{j,k}\left(\dfrac{E_{\nu e} - E_{\nu\, j,k}}{\tau_{j,k}} + Q_{j,k}\right) \end{array}\right), \tag{12}$$

where $\varphi = \sqrt{2p}$, $W_{j,k} = W(|\mathbf{r}_{j,k} - \mathbf{r}_{j',k'}|, h)$ is the smoothing kernel used to approximate the spatial derivatives in equations (1)–(4) in the SPH approach.

The values of the fluxes $\mathbf{F}^{n+1/2}_{j\pm1/2,k}$ and $\mathbf{G}^{n+1/2}_{j,k\pm1/2}$ are computed on the cell boundaries using solutions of the Riemann problem at time $t_{n+1/2}$. Approximate methods of Lax-Friedrichs (LF), Harten-Lax-van Lier (HLL), and the modified Harten-Lax-van Lier method (HLLC) [18] are most often used to solve the Riemann problem. In this paper, we use the HLL method for the flux $\mathbf{F}^{n+1/2}_{j+1/2,k}$

$$\mathbf{F}^{n+1/2}_{j+1/2,k} = \begin{cases} \mathbf{F}_L, & 0 < S_L \\ \dfrac{S_R\mathbf{F}_L - S_L\mathbf{F}_R + S_L S_R(\mathbf{U}_R - \mathbf{U}_L)}{S_R - S_L}, & S_L \le 0 \le S_R \\ \mathbf{F}_R, & 0 > S_R \end{cases}, \tag{13}$$

where $S_L = \min(u_L - c_L, u_R - c_R)$, $S_R = \max(u_L + c_L, u_R + c_R)$, and the indices L, R determine the flow parameters to the left and right of the boundary $j + 1/2$, respectively.

The construction of the second-order accuracy scheme together with the time predictor-corrector algorithm requires the use of piecewise-linear approximation of $\mathbf{U}$ values inside cells. In the CSPH-TVD method, the values of the flux parameters on the left (L) and on the right (R) from the boundary $j + 1/2$ have the following form:

$$\begin{aligned} \mathbf{U}^L &= \mathbf{U}^{n+1/2}_{j,k} + \frac{1}{2}\left(h - \xi^{n+1}_{j,k}\right)\mathbf{\Theta}^n_{j,k}, \\ \mathbf{U}^R &= \mathbf{U}^{n+1/2}_{j+1,k} - \frac{1}{2}\left(h + \xi^{n+1}_{j+1,k}\right)\mathbf{\Theta}^n_{j+1,k}, \end{aligned} \tag{14}$$

where $\xi^{n+1}_{j,k}$ is the displacement of the "liquid" particle relative to the cell center (j, k) at time t_{n+1}. Since the slopes $\mathbf{\Theta}$ of the piecewise linear dependence (14) must satisfy the TVD-condition, it is necessary to apply the limiter functions.

The MUSCL scheme [13, 14] also uses the two-step predictor-corrector method, the piecewise linear approximation with TVD-limiters and the HLL method for solving the Riemann problem. For the MUSCL method, equation (9) takes the form:

$$\mathbf{U}^{n+1}_{j,k} = \mathbf{U}^n_{j,k} - \frac{\Delta t}{h}\left(\mathbf{F}^{n+1/2}_{j+1/2,k} - \mathbf{F}^{n+1/2}_{i-1/2,k} + \mathbf{G}^{n+1/2}_{j,k+1/2} - \mathbf{G}^{n+1/2}_{j,k-1/2}\right) + \Delta t\,\mathbf{Y}^{n+1/2}_{j,k}, \tag{15}$$

Since the expressions for the fluxes and source terms in equation (15) differ from the corresponding expressions for CSPH-TVD, we write them explicitly as

$$\mathbf{F} = \begin{pmatrix} \varrho u \\ \varrho u^2 + p \\ \varrho uv \\ u(E+p) \\ uE_\nu \end{pmatrix}, \quad \mathbf{G} = \begin{pmatrix} \varrho v \\ \varrho vu \\ \varrho v^2 + p \\ v(E+p) \\ vE_\nu \end{pmatrix}, \quad \mathrm{Y} = \begin{pmatrix} 0 \\ \varrho f_x \\ \varrho f_x \\ \varrho(\mathbf{vf} + Q - I) \\ \varrho((E_{\nu e} - E_\nu)/\tau + Q) \end{pmatrix}.$$

For the MUSCL method, we set $\xi = 0$ in equation (14) for constructing piecewise linear dependencies.

New combined methods based both on the Euler (grid method) approach and on the Lagrangian (particle motion) approach, began to develop in recent years. The combined TVD + SPH method demonstrated high efficiency for problems of the numerical simulation of long waves runup on a shore [19]. The hybrid code SPH-TVD was used in astrophysical gas dynamics to simulate the stellar disruption by a massive black hole [20].

In general, the stability condition for the explicit numerical schemes CSPH-TVD and MUSCL is:

$$\Delta t < \min_{j,k} \left(\frac{h}{2\,|\mathbf{v}_{j,k}|}, \frac{h}{|\mathbf{v}_{j,k}| + c_{j,k}} \right), \tag{16}$$

To increase the speed of gas-dynamic simulation on high-resolution grids, we parallelized the numerical algorithms CSPH-TVD and MUSCL using OpenMP/CUDA technologies and performed calculations on a Computing Clusters with NVIDIA Tesla K40 and K80 coprocessors.

We normalize all physical quantities using ϱ_0, p_0, τ_0 and $\sqrt{p_0/\varrho_0}$ as units of density, pressure, time and velocity, respectively. And then we will use dimensionless quantities only.

2.4. Testing numerical model

Let's consider the test of the collision of two shock waves [21]. The initial state consists of three regions filled with an ideal gas with an adiabatic index $\gamma = 1.4$ and the computational domain is bounded by solid walls. The initial density is homogeneous ($\rho(x) = 1$, $u(x) = 0$, $x \in [0, 1]$), but we set two pressure discontinuities:

$$p(x) = \begin{cases} 1000, & 0 \le x < 0.1; \\ 0.01, & 0.1 \le x \le 0.9; \\ 100, & 0.9 < x \le 1. \end{cases}$$

As a result, two strong convergent shock waves, two contact density discontinuities and two diverging discharge waves are formed. Figure 1 shows the dynamics of interaction of two shock waves for $N = 10^3$. At $t_1 = 0.028$ the shock waves collide with the formation of a characteristic peak at the collision position $x = 0.7$ and as a result two divergent shock waves are formed (Figure 2).

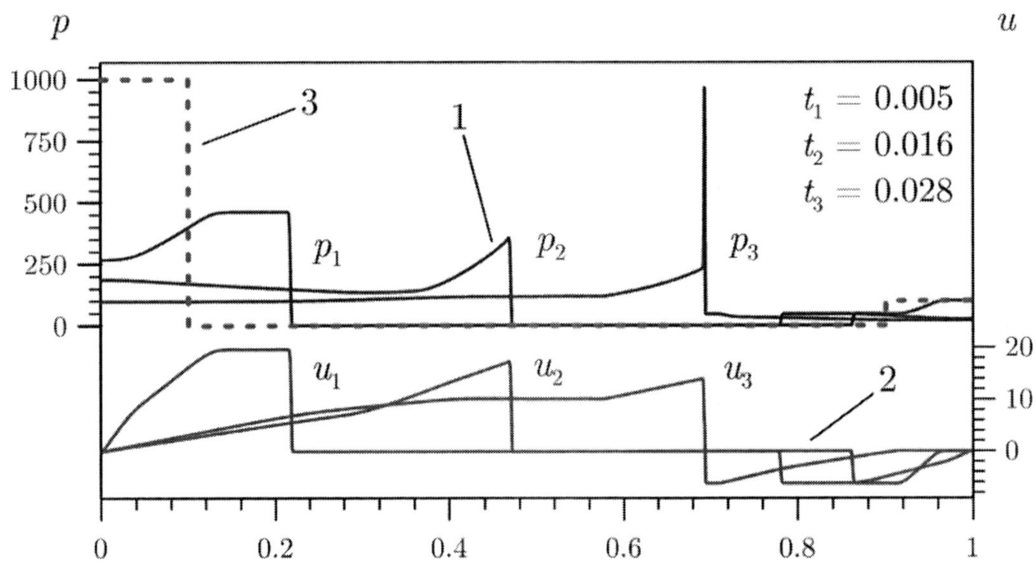

Figure 1. The interaction of two shock waves is shown at different times. The solid blue lines on the upper panel are the pressure distributions $p(x)$ and the red lines are the velocities $u(x)$ on the bottom panel at different times. The initial pressure profile $p(x, t = 0)$ is represented by the green dotted line.

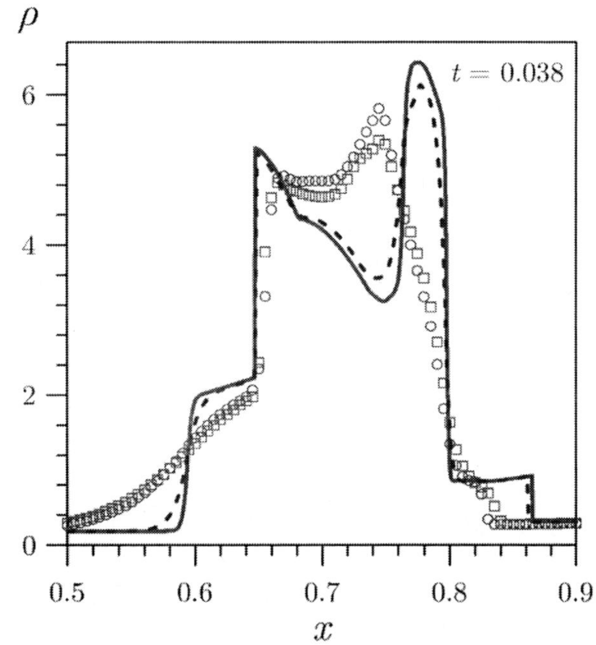

Figure 2. The result of a collision of two shock waves. The blue circles are the density after the collision ($t = 0.038$) for the CSPH-TVD scheme with $N = 200$ and the dotted blue line is the model with $N = 1600$; the red squares are the result for MUSCL ($N = 200$). The exact solution is shown by the solid line.

Figure 2 shows the results obtained by the CSPH-TVD and MUSCL schemes at the time $t = 0.038$ after the collision of shock waves. Both methods demonstrate a good resolution of the contact discontinuity formed as a result of the collision of convergent shock waves, and the width of the fronts for the generated divergent shock waves are approximately equal. The numerical solution on a very detailed grid for $N = 10^5$ is shown as an exact solution. Convergence of our numerical solutions is clearly noticeable with increasing spatial resolution in discretization.

3. Results and Discussion

For numeric simulation of the acoustic waves dynamics in a nonequilibrium medium, we specify forced oscillations from a source with initial conditions $u = 0$, $\varrho = 1$, $p = 1$. To calculate the vibrational relaxation time, we define the values of parameters $\tau_\varrho = -1$, $\tau_T = -1.5$, $Q_\varrho = Q_T = I_\varrho = I_T = 0$. The wavelength of the generated perturbations is $\lambda = 0.1$. We set the value of the degree of medium nonequilibrium to be $S = 0.5$.

Consider the initial linear stage of sound waves propagation. Within the linear analysis, the perturbation amplitude grows according to the law $\sim \exp \frac{-\mathrm{Im}k}{\lambda} x$. For the environmental parameters given above Figure 3a shows the results of numerical gasdynamic simulations with acoustic increment $\mathrm{Im}k = -0.43$. As Figure 3a demonstrates in the linear stage the exponential growth of the perturbation amplitude corresponding to a linear stability analysis (solid red line) is well tracked.

We set $S = 1$, $\mathrm{Im}k = -1.034$ to study the dynamics of sound waves at the nonlinear stage. For given medium parameters Figure 3b shows the distributions of velocity perturbation in the sound wave. At time $t = 0.8$, the initial harmonic perturbations turn into a system of strong

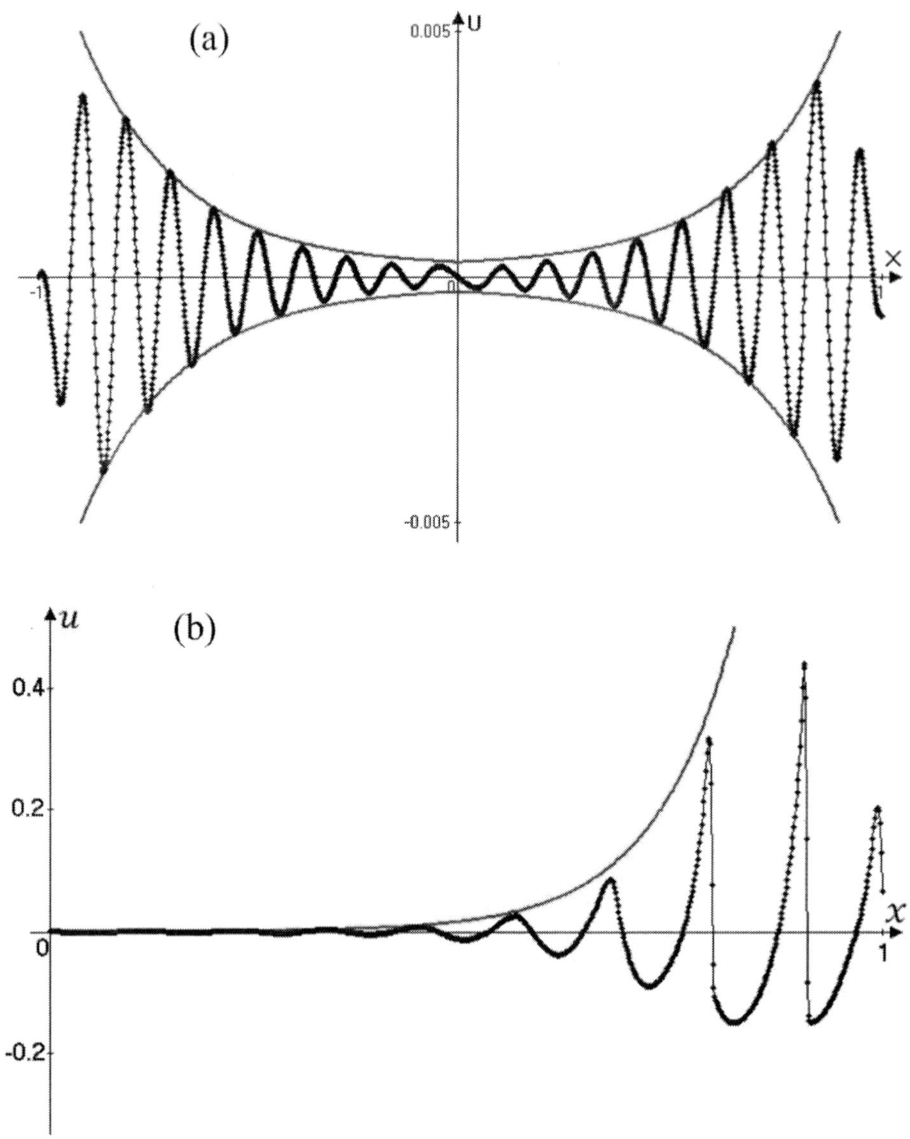

Figure 3. The distribution of the velocity perturbation: a) at the linear stage at time $t = 0.8$ and $N = 800$; b) at the nonlinear stage at time $t = 0.8$ and $N = 1600$ (the section of the calculation region $0 \leq x \leq 1$ is shown.

shock waves. The amplitude, velocity and shape of such disturbances are determined only by the nonequilibrium medium parameters.

Consider the problem of simulation of the dynamics of acoustic waves in two-dimensional space. The active medium is set in some part of the computational domain, at a triangle with an angle $\alpha = 30°$ (see Figure 4). The non-percolation condition (a solid wall) is set at the boundary G_1, and the free flow conditions are specified at the boundaries G_2, G_3, G_4. According to such problem formulation the source begins to generate perturbations diverging in all directions, but the acoustic waves amplitude increases only in the sector where the non-equilibrium medium is determined (see Figure 5).

At time $t = 1$, the wave reaches the boundary G_1 and is reflected (see Figure 5d). The calculation continues that demonstrates a good stability and balance of the numerical scheme for reflection problems of a strong shock waves system.

Thus, the results obtained above indicate a significant effect of stationary non-equilibrium on the sound waves evolution. In the active media sound waves are unstable, so a strong increase of the acoustic disturbance amplitude is observed. In this case, the wave packet passes through all the stages of the instability development: linear and nonlinear stages in the formation of strong shock waves system.

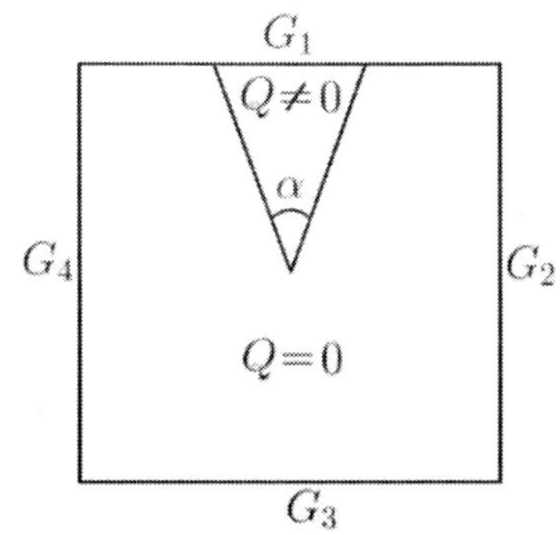

Figure 4. The computational domain scheme (the active medium is given by the sector with the angle $\alpha = 30°$).

4. Conclusion

In the article we have researched the sound waves steadyness in a nonequilibrium acoustically active medium. Using a linear steadyness analysis, we have obtained a new dispersion equation accounting for the following functional dependences: $\tau \propto \varrho^{\tau_\varrho} T^{\tau_T}$, $Q \propto \varrho^{Q_\varrho} T^{Q_T}$ $I \propto \varrho^{I_\varrho} T^{I_T}$ for the vibrational relaxation time, pump power and heat sink, respectively. The influence of parameters τ_ϱ, τ_T, Q_ϱ, Q_T, I_ϱ, I_T on the acoustic increment has been studied in detail. The boundaries of the steadyness regions in a nonequilibrium vibrationally excited gas have been determined as well.

Numerical algorithms using the CSPH-TVD and MUSCL methods have been developed to study the nonlinear stage of unstable acoustic waves development in a nonequilibrium vibrationally excited gas. The parallel computing technologies OpenMP-CUDA have been implemented to accelerate the calculations.

At the initial stage of sound waves propagation in a nonequilibrium medium, the results of numerical gas dynamic modeling are in good agreement with the linear steadyness analysis, which indicates the adequacy of the developed algorithms. At the nonlinear stage of development unstable sound waves form a shock waves system which maximum amplitude depends on the degree of the medium nonequilibrium S and model parameters τ_ϱ, τ_T, Q_ϱ, Q_T, I_ϱ, I_T.

Applying the numerical gas dynamic methods of simulation, the dynamics of cylindrical sound waves which propagates simultaneously in acoustically active and inactive media has been investigated. Due to the presence of interface between two media, the shock waves formed at the nonlinear stage of the acoustic instability development have an inhomogeneous structure along the azimuthal coordinate.

Acknowledgments
The second author has been supported by the Ministry of Education and Science of the Russian Federation (government task No. 2.852.2017/4.6). The authors are thankful to the RFBR (grants

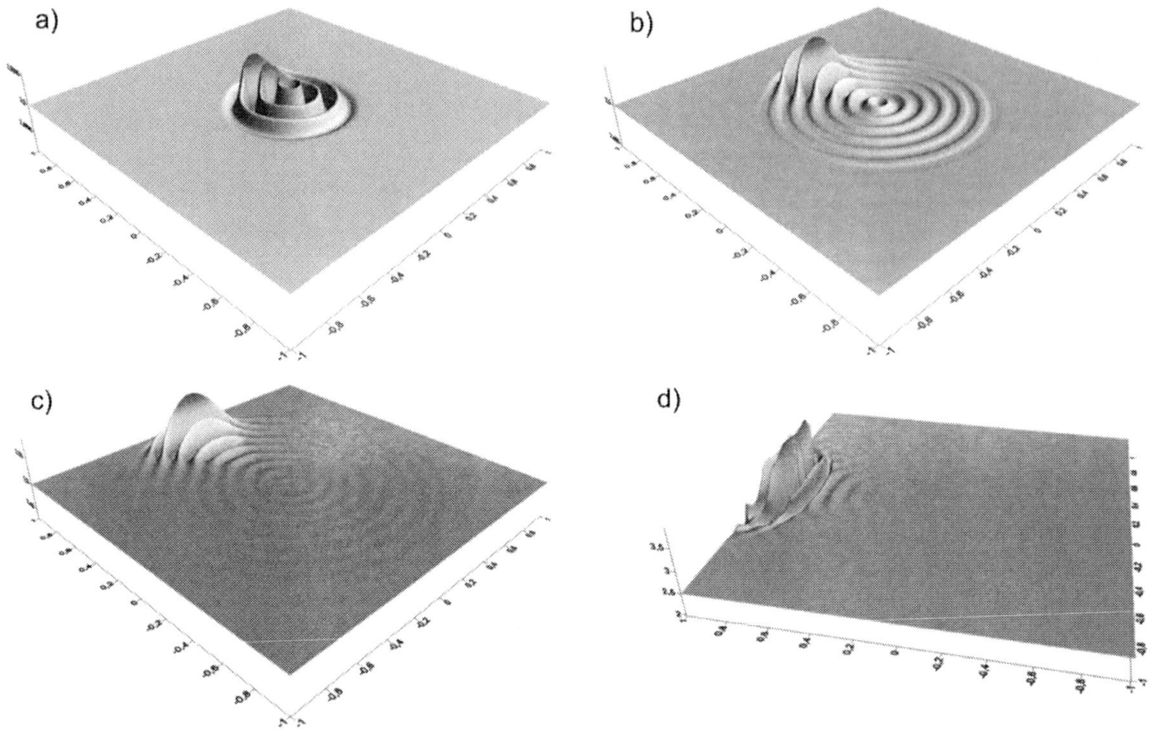

Figure 5. The density perturbation distribution at different moments of time: a) $t = 0.25$, b) $t = 0.5$, c) $t = 0.75$, d) $t = 1$.

16-07-01037, 15-02-06204). The research is carried out using the equipment of the shared research facilities of HPC computing resources at Lomonosov Moscow State University.

References

[1] Bauer H J and Bass H E 1973 *Phys. Fluid* **16** 988–996
[2] Kogan E Ia and Molevich N E 1985 *Soviet Physics Technical Physics* **30** 447–449
[3] Kogan E Ia and Molevich N E 1986 *Soviet Physics Technical Physics* **29** 547–551
[4] Molevich N E The Journal of the Acoustical Society of America **123** 3691
[5] Zavershinskiy D I and Molevich N E 2014 *Computer Optics* **38** 619–622
[6] Galimov R N, Makaryan V G and Molevich N E 2006 *Technical Physics* **51** 1644–1646
[7] Makaryan V G and Molevich N E 2005 *Technical Physics* **50** 685–691
[8] Khoperskov A V and Khrapov S S 1999 *Astronomy and Astrophysics* **345** 307–314
[9] Khoperskov A V, Khrapov S S and Nedugova E A 2003 *Astronomy Letters* **29** 246–257
[10] Yao Z R and Li X Q 2007 *Journal of Plasma Physics* **73** 367–375
[11] Khrapov S S, Khoperskov A V, Kuz'min N M, Pisarev A V and Kobelev I A 2011 *Vychisl. Metody Programm* **12** 282–297
[12] Khrapov S S, Pisarev A V, Kobelev I A, Zhumaliev A G, Agafonnikova E O, Losev A G and Khoperskov A V 2013 *Advances in Mechanical Engineering* **5** 787016
[13] Khoperskov S A, Vasiliev E O, Ladeyschikov D A, Sobolev A M and Khoperskov A V 2016 *MNRAS* **455** 1782–1795
[14] Khoperskov S A, Vasiliev E O, Sobolev A M and Khoperskov A V 2013 *MNRAS* **428** 2311–2320
[15] Monaghan J J 2005 *Report on Progress in Physics* **68** 1703–1759
[16] Khrapov S S and Khoperskov A V 2017 *Communications in Computer and Information Science* **793** 266–277
[17] Toro E F 2012 Godunov Methods: Theory and Applications *Verlag: Springer* 1077
[18] Toro E F 1999 Riemann solvers and numerical methods for fluid dynamics *Verlag: Springer* 624

[19] Shokin Y, Rychkov A, Khakimzyanov G, et al. 2016 *Russian Journal of Numerical Analysis and Mathematical Modelling* **31** 217–227
[20] Lee H M and Kim S S 1996 *Jour. of Korean Astron. Soc.* **29** 195–205
[21] Woodward P and Colella P 1984 *J. Comput. Phys.* **54** 115–173

IOP Conf. Series: Journal of Physics: Conf. Series **973** (2018) 012008 doi:10.1088/1742-6596/973/1/012008

Discrete element method modification for the transition to a linearly elastic body model

S V Klishin, S V Lavrikov[1], O A Mikenina and A F Revuzhenko

Chinakal Institute of Mining, Siberian Branch, Russian Academy of Sciences, 54 Krasny Prospect, Novosibirsk, 6300091 Russia

E-mail: [1]lvk64@mail.ru

Abstract. The article addresses the issue of description of lateral pressure in the discrete element method. The authors formulate a one-dimensional problem on elastic deformation of a particle under the action of the uncompensated normal force and torque. The elastic change in the profile of the particle is calculated and is then included in problem solving by DEM and a specifically developed algorithm. The calculations of compression applied to a regular packing of discrete particles with regard to lateral pressure are exemplified. The developed procedure is the DEM modification enabling transition to modeling a linearly elastic body.

1. Introduction

Currently, the calculations of deformation and failure of solids use mostly the discrete element method (DEM), or the methods based on continuum models (finite elements, boundary integral equations, gridless methods, etc). In the discrete element method, a real medium is represented by the packing of discrete particles. Then, the laws of interaction between particles are assigned, and the problem of deformation of a body is replaced by the problem of deformation of the set packing of discrete particles. Such approach is faced with no fundamental difficulties in description of physical nonlinearity or even any large deformation. Moreover, it readily takes into account dynamic effects. On the other hand, you know, when you win something, you lose something. In the case discussed, we lose in the degree of adequacy of the obtained results. Indeed, after setting a synthesis algorithm for a stochastic particle packing, as well as properties of the particles and their interaction laws, and then having solved the problem of deformation of the packing, it will be difficult to answer which macroproperties of the packing the solution of the problem corresponds to [1–4]. Furthermore, a problem connected with the consistency arises. The classical approach to deformation problems involves the models of elasticity, plasticity, creep, etc. Naturally, when one and the same object is analyzed using different methods, the results should be consistent. For example, transition to a linearly elastic body should be fulfilled in the limit case. The method of discrete elements [5–8] dissatisfies this condition since the forces at the contact of particles are connected with the relative overlaps and displacements of particles only at that contact. Therefore, it is simple to set such a regular packing of particles that no lateral pressure arises in the packing under the uniaxial compression. It is noteworthy that the attempts to modify DEM with a view to describing the lateral pressure phenomenon are currently undertaken [9, 10].

This article describes a DEM modification that allows avoiding such paradoxes. To this effect, it is required to set more complex conditions at the contacts of particles. Now, the forces on a contact of particles depend on the overlaps at the given contact and at the other contacts of these particles.

2. Particle deformation under uncompensated forces

Let us consider dynamics of a particle packing. It is intended to take into account deformation of particles. In principle, this problem can be formulated sufficiently strictly. It is required to write dynamic equations inside each particle and to take into account contact conditions between the particles and between the particles and a loading device. At the same time, it is necessary to include conditions at the free boundary of particles and all initial conditions. Evidently, such formulation makes the problem immeasurable both in terms of solving and analyzing results even if reachable—there is too much information.

Thus, it is of no sense to deal with a strict problem formulation. Some simplifications should be made. For this reason, we neglect wave dynamics inside particles and account for the dynamics of a particle as a whole. This is a major premiss.

Let us discuss the simplest and sufficiently adequate variant. In the discrete element method, as a rule, we deal with an unbalanced system of forces applied to a particle. The particle accelerates, interacts with the neighbor particles and, finally, attains equilibrium. Based on that, first, we analyze a problem on movement of a particle under an unbalanced force. The particle is assumed linearly elastic and rounded.

We limit ourselves to planar deformation. It is convenient to demonstrate the key ideas using a particle shaped as a rod with a length L (figure 1):

$$\frac{\partial \sigma_x}{\partial x} = \rho \frac{\partial^2 u_x}{\partial t^2}, \quad \frac{\partial u_x}{\partial x} = \frac{1}{E}\sigma_x. \tag{1}$$

Here, σ_x, u_x — stresses and displacements, respectively; ρ — density, E — elasticity modulus. Let the right-hand end $x = L$ of the rod be free of stress and the left-hand end $x = 0$ be subject to a constant compression stress:

$$\sigma_x(0) = -p < 0, \quad \sigma_x(L) = 0. \tag{2}$$

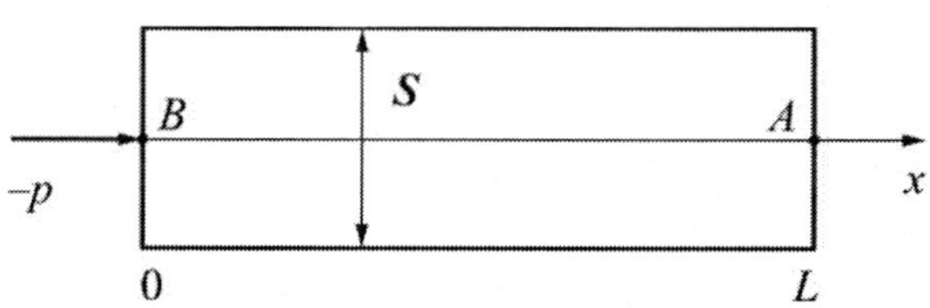

Figure 1. The simplest one-dimensional problem on uncompensated force.

To put it simpler, the rod is pushed rightwards in the line of the Ox axis. Let us also assume that the wave process quickly attenuates in the rod (the causes of which are beyond the scope of the discussion). At the steady-state stage, the rod has a constant acceleration a given by:

$$a = \frac{p}{\rho L}. \tag{3}$$

This follows from the equation of dynamics: $ma = F, m = \rho LS, F = pS,$ where S — cross section area of the rod, and m — mass of the rod. Consequently,

$$\frac{\partial^2 u_x}{\partial t^2} = a = const, \\ u_x = \frac{at^2}{2} + C_1(x)t + C_2(x), \tag{4}$$

$C_1(x), C_2(x)$ — arbitrary functions. Then, the first equation in the system (1) with the constant acceleration a

$$\frac{\partial \sigma_x}{\partial x} = \rho a, \text{ or } \frac{\partial \sigma_x}{\partial x} - \rho a = 0, \tag{5}$$

with regard to the boundary conditions (2), yields a solution:

$$\sigma_x = -p\left(1-\frac{x}{L}\right). \tag{6}$$

Regarding the second equation in the system (1):

$$\frac{\partial u_x}{\partial x} = \frac{-p}{E}\left(1-\frac{x}{L}\right), \tag{7}$$

its solution, considering (4), is given by:

$$u_x(x,t) = \frac{-p}{E}\left(1-\frac{x}{2L}\right)x + C_1 t + C_2 + \frac{at^2}{2}. \tag{8}$$

Let at $t=0$

$$u_x(0,0)=0,\ \frac{\partial u_x(0,0)}{\partial t}=0, \tag{9}$$

then

$$u_x(x,t) = \frac{-p}{E}\left(1-\frac{x}{2L}\right)x + \frac{at^2}{2}. \tag{10}$$

Based on the problem discussed, some conclusions are drawn:

1. The initially dynamic problem can be reduced to a static problem with the mass forces equal to the inertia forces with an opposite sign (Eq. (5)) — d'Alembert principle.
2. The force application point, i.e. $x=0$, moves by the law (10).
3. Including lateral distortion in the solution of (10), we determine the shape of the particle after its deformation.

In this case, the numerical solution of a dynamic problem on motion and deformation of an elastic particle under an uncompensated force follows the steps below:

1. Presetting the value of the uncompensated force, using the formula (3) and finding the acceleration a.
2. Setting the boundary condition $u_x(0)=0$ which "automatically" provides the solution with the stress $\sigma_x(0)$ equal to the preset value. Then, solving the static problem in terms of displacements.
3. Adding the static solution with the dynamic summand $\frac{at^2}{2}$.

Regarding the issue of moments and rotations, we start from the simplest problem as well. Let a particle be shaped as a cylinder and rotate around an immobile center. The angle of rotation is denoted by $\varphi(t)$ (figure 2). It is obvious that the element of the medium has the coordinates given by:

$$\begin{aligned} x(r,t) &= r\cos(\varphi(t)), \\ y(r,t) &= r\sin(\varphi(t)), \end{aligned} \tag{11}$$

whence it follows that:

$$\begin{aligned} \ddot{x}(r,t) &= -r\ddot{\varphi}\sin\varphi - r(\dot{\varphi})^2\cos\varphi, \\ \ddot{y}(r,t) &= r\ddot{\varphi}\cos\varphi - r(\dot{\varphi})^2\sin\varphi. \end{aligned} \tag{12}$$

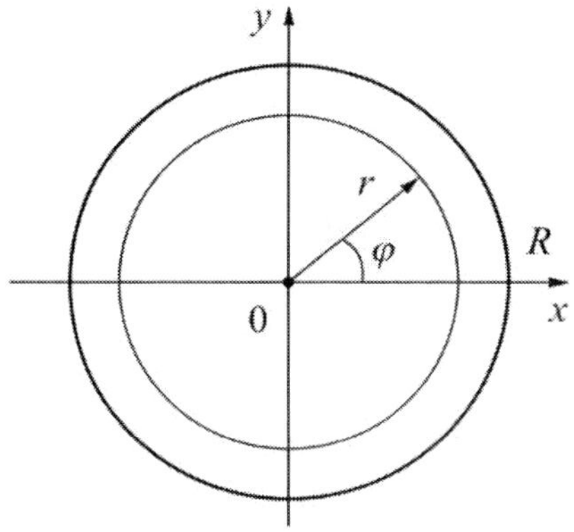

Figure 2. The simplest one-dimensional problem with the uncompensated torque.

The rotation is governed by the uncompensated portion of the force moment applied at the boundary $r = R$. The uncompensated moment is only connected with the rotational acceleration $\ddot{\varphi}$. The centripetal forces are connected with the rotation speed $\dot{\varphi}^2$. Above all, they are internally compensated, i.e. are in no way connected with the boundary forces. Additionally, in the case under discussion, the particles are in contact, and the values of turns and rotational speeds $\dot{\varphi}$ are small. For this reason, the centripetal forces are neglected. Finally, we have:

$$\begin{aligned} \ddot{x}(r,t) &= -\ddot{\varphi} y, \\ \ddot{y}(r,t) &= \ddot{\varphi} x. \end{aligned} \tag{13}$$

Considering that $\ddot{\varphi} = M / J$, where M — uncompensated torque applied at the boundary of a particle and J — inertia moment, it can be written that:

$$\begin{aligned} \rho \ddot{x} &= -\rho \frac{M}{J} y, \\ \rho \ddot{y} &= \rho \frac{M}{J} x. \end{aligned} \tag{14}$$

The forces (14) contribute to the particle deformation.

These constructs allow formulating a general algorithm of constitutive equations for an elastic particle in the method of discrete elements:

1. At a current time t the position and configuration of a particle are known. The velocity of the particle center of inertia and the rotational speed are known, too. These values are the initial conditions of the problem.

2. At the time t at the known boundary points, the particle is in contact with the other particles. By the values of overlaps, the forces applied to the particle by the neighbor particles are determined: $T_x(A_1), T_y(A_1), \ldots, T_x(A_k), T_y(A_k)$ (figure 3). The system of forces is self-unbalanced, i.e., in a general case:

IOP Conf. Series: Journal of Physics: Conf. Series **973** (2018) 012008 doi:10.1088/1742-6596/973/1/012008

$$F_x = \sum_{i=1}^{k} T_x(A_i) \neq 0,$$
$$F_y = \sum_{i=1}^{k} T_y(A_i) \neq 0, \quad (15)$$
$$M = \sum_{i=1}^{k} \left(-T_x(A_i)\sin\alpha_i + T_y(A_i)\cos\alpha_i\right) r_i \neq 0.$$

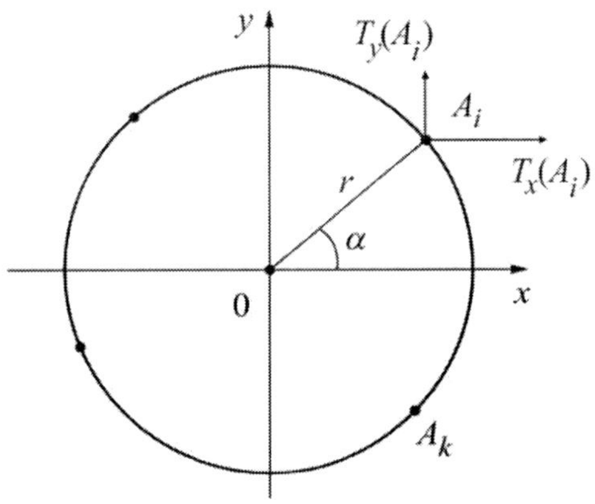

Figure 3. Uncompensated force system acting on the particle.

Under solution is an elastic problem for a particle subjected to the mentioned forces and to the bulk forces. The bulk and the surface forces are mutually balanced. The equilibrium equation takes the form:

$$\frac{\partial \sigma_{xx}}{\partial x} + \frac{\partial \sigma_{xy}}{\partial y} - \frac{F_x}{S} + \rho \frac{M}{J} y = 0,$$
$$\frac{\partial \sigma_{xy}}{\partial x} + \frac{\partial \sigma_{yy}}{\partial y} - \frac{F_y}{S} - \rho \frac{M}{J} x = 0. \quad (16)$$

3. The static problem solution is connected with the initial position of the particle and acts as the **constitutive relations** later on. As a result, we find:

$$u_x(A_1) = \Phi_{x1}\left(T_x(A_1),\ldots,T_y(A_k)\right),$$
$$\ldots \quad (17)$$
$$u_y(A_k) = \Phi_{yk}\left(T_x(A_1),\ldots,T_y(A_k)\right).$$

Moreover, the shape of the particle is determined: $R = R(\alpha)$.

4. The time increment τ is selected. From the equations of motion

$$m\frac{d^2 v_x}{dt^2} = F_x, \quad m\frac{d^2 v_y}{dt^2} = F_y, \quad J\frac{d^2 \Omega}{dt^2} = M \quad (18)$$

and the initial conditions, new average values of the displacements and rotation are determined:

$$v_x(t, t+\tau),\ v_y(t, t+\tau),\ \Omega(t, t+\tau). \quad (19)$$

The resultant displacements of points at the particle contact are determined:

$$u_x(A_i) = v_x - \Omega r_i \sin\alpha_i + u_x^0(A_i),$$
$$u_y(A_i) = v_y + \Omega r_i \cos\alpha_i + u_y^0(A_i). \quad (20)$$

Accordingly, the new position of the particle at the time $t+\tau$ is calculated. Then, the new system of forces is determined, etc.

3. Example of calculation of an elastic particle profile

Below in this article an example of calculating a plane static problem with the preset bulk forces is given. Let a round particle at a boundary point A be subjected to an unbalanced surface force F and an unbalanced torque M (figure 4). In accord with the above algorithm, we replace the surface forces by the bulk forces, i.e. analyze the equilibrium equation (16). The boundary conditions are taken as:

$$\begin{aligned} &u_x(A)=0,\ u_y(A)=0,\\ &\sigma_n(\Gamma \setminus A)=0,\ \tau_n(\Gamma \setminus A)=0. \end{aligned} \quad (21)$$

Such conditions ensure the required preset load at the point A.

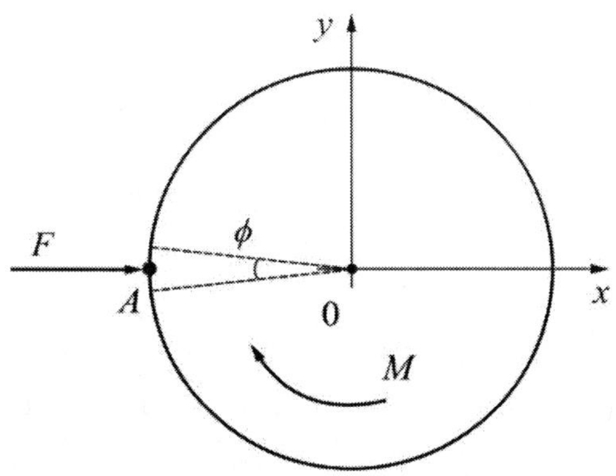

Figure 4. Formulation of the static problem.

The solution of (16), (21) was implemented in the elastic formulation using the finite element method. The elastic characteristics of the particle were chosen as the average values of the elasticity modulus $E=2\cdot10^3$ MPa and Poisson's ratio $\nu=0.25$. It should be mentioned that the described theoretical formulation assumes that the conditions (21) are preset at a single infinitely small point A. Regarding DEM, the normal forces and, accordingly, the shearing forces due to the particle interactions are applied through a finite size area (governed by the overlap of the particles). For this reason, in the numerical calculation, the point A was replaced by the finite area with a size characterized by the polar angle ϕ (figure 4). The calculations assumed $\phi=10^\circ$.

The problem solution was the displacement profile of the particle boundary. Figure 5 shows a normalized profile ($\max(|u|)=1$) of the boundary point displacement only under the preset normal force, i.e. $F=const, M=0$ (figure 5a — component u_x, figure 5b — 10 times magnified component u_y). Figure 6 depicts the calculation results in case when $F=0, M=const$ (components u_x and u_y in figure 6a and figure 6b, respectively). It is seen in the figures that the particle exposed to the unbalanced load undergoes the change in the profile of its boundary. Compression of the particle induces lateral pressure (refer to figure 5b).

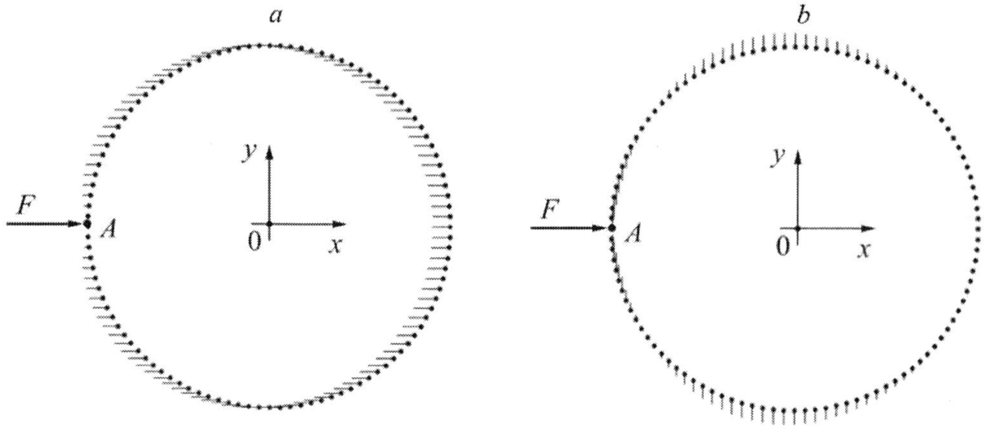

Figure 5. Estimated profile of displacement at $F = const, M = 0$: (a) component u_x; (b) component u_y magnified by 10 times for the convenience of visualization.

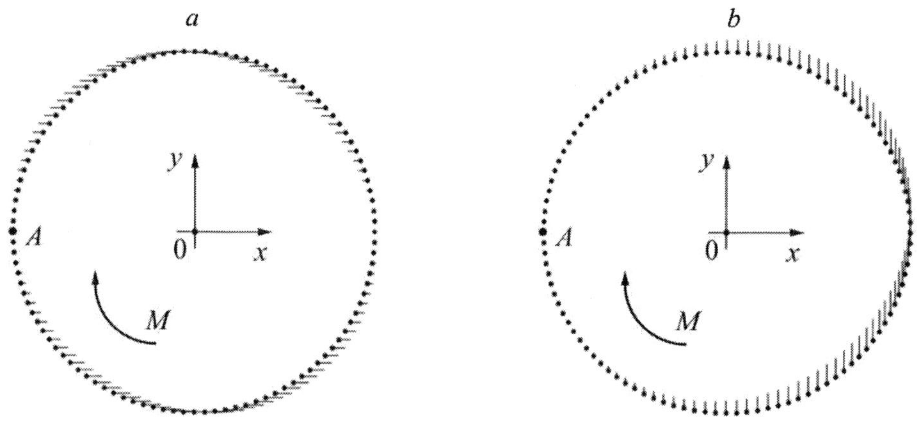

Figure 6. Estimate displacement profile when $F = 0, M = const$: (a) component u_x; (b) component u_y.

As a result, the obtained field can be used in the method of discrete element by the algorithm described in Section 2. The profile of an elastic particle during deformation is only calculated once. Later on, the calculated profile is a parameter of the problem solving by DEM.

4. Application of the modified discrete element method

In the framework of DEM, we present an example of calculating compression of a flat regular packing of particles centers of which lie at the points of a square lattice in the initial state (figure 7). It is obvious that in the problem of compression of a solid elastic body with a square cross section, the coefficient of lateral pressure will coincide with Poisson's ratio.

In the classical method of discrete elements [5–8], movement of particles is calculated using equations of translational and rotational motion (18). The computation of forces and moments of forces at the contacts of particles and at the contacts of particles and loading plates involves Kelvin–Voigt viscoelastic model with regard to the normal and shear components. The normal force is found using the Hertz law from the contact mechanics [11], and the shear force is determined based on the Mindlin–Deresiewicz concept [12, 13].

When setting loading conditions, all boundary walls are assumed perfectly smooth, the side and bottom boundary are rigidly fixed, and the top boundary is moved downward at a constant velocity v_n (figure 7). Denote by F_n the total normal resistance force exerted to the top boundary by the particles when pushed downward and by F_t the total normal force applied by the particles to each side boundary. The calculation aims to find the lateral pressure coefficient $\lambda = F_t / F_n$ versus time in the course of loading. Evidently, the classical DEM approach (neglecting lateral pressure) under given conditions in the regular particle packing yields $\lambda(t) = 0$.

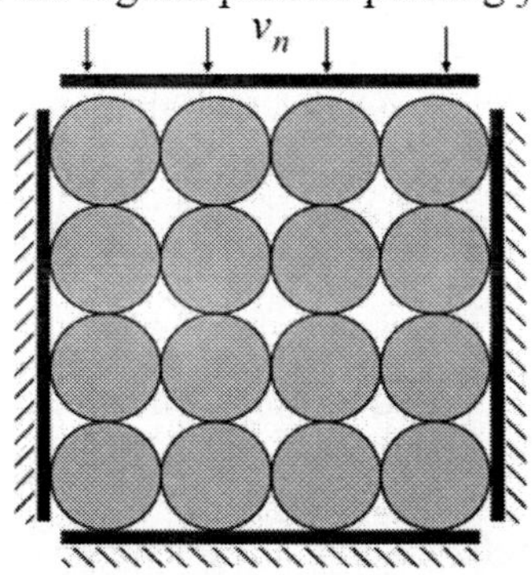

Figure 7. Compression of a regular packing of particles (DEM-based calculation).

In order to include lateral pressure induced by elastic interaction between particles, we take the algorithm from Section 2 and the calculated elastic change in the particle profile from Section 3. It appears sufficient to consider the profile change only under the action of the uncompensated normal force without regard to the torque: $F = const, M = 0$ (figure 5). The DEM-based solution of the problem produces the result $\lambda(t)$ illustrated in figure 8a. When averaged, $\lambda \approx 0.06$.

The resultant value $\lambda \approx 0.06$ is considerably lower than in the solution of the problem of compression of a solid elastic body. An obvious explanation is that we analyze compression of a regular packing composed of round elastic particles a relatively small areas for contact interaction.

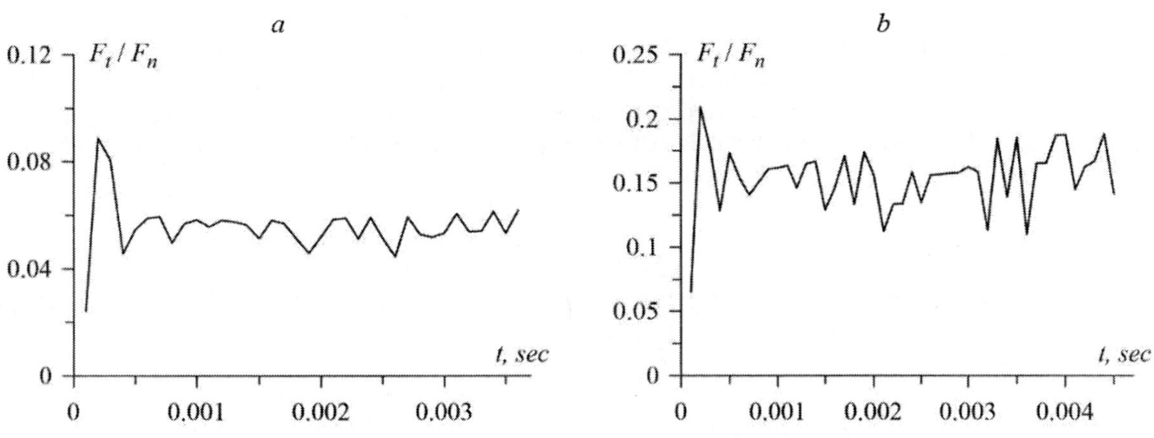

Figure 8. The calculated value of lateral pressure coefficient versus time $\lambda(t)$: (a) averaged elasticity characteristics of a particle; (b) specifically set characteristics of elasticity of a particle.

Let us vary elastic properties of particles by increasing compressive stiffness and ductility orthogonally to the compression orientation. Assume $E = 2 \cdot 10^9$ MPa and $\nu = 0.45$. Increase the dimension of the contact areas for the force interaction between the neighbor particles, and set $\phi = 20°$. Given such parameters, an elastic problem is again calculated using the finite element method (Section 3) and a new profile of a particle is determined. Inclusion of the latter in the problem of the particle packing compression in the framework of the discrete element method results in $\lambda(t)$ depicted in figure 8b; in this case, the average value is $\lambda \approx 0.15$.

In this way, by varying elastic characteristics of a particle and sizes of contact areas between particles, it is possible to obtain sufficiency realistic values of lateral pressure. Consequently, the presented procedure of numerical calculation is the discrete element method modification which permits transition to linear elasticity.

5. Conclusions

The presented approach allows taking into account elastic properties of particles and enables description of lateral pressure using the method of discrete elements.

The proposed procedure is the modification of the discrete element method for the transition to modeling a linearly elastic body.

This study was supported by the Russian Science Foundation, Project No. 16-17-10121.

References

[1] Revuzhenko A F and Klishin S V 2012 Numerical Method for Constructing a Continual Deformation Model Equivalent to a Specified Discrete Element Model *Phys. Mesomech.* **16**(2) 152–161

[2] Murray P J, Edwards C M, Tindall M J, Maini P K 2009 From a discrete to a continuum model of cell dynamics in one dimension *Phys Rev E* **80** 031912

[3] Mühlhaus H-B, Sakaguchi H, Moresi L, Fahey M 2001 Discrete and continuum modelling of granular materials *Continuous and Discontinuous Modelling of Cohesive-Frictional Materials. Lecture Notes in Physics* ed Vermeer P A, Herrmann H J, Luding S, Ehlers W, Diebels S, Ramm E (Springer, Berlin, Heidelberg) **568** 185–204

[4] Coetzee C J, Basson A H and Vermeer P A 2006 Discrete and Continuum Modelling of Silo Discharge *R&D Journal* **22**(2) 26–38

[5] Cundall P A and Strack O D L 1979 A distinct element model for granular assemblies *Géotechnique* **29**(1) 47–65

[6] Zhu H P, Zhou Z Y, Yang R Y and Yu A B 2008 Discrete particle simulation of particulate systems: A review of major applications and findings *Chem. Eng. Sci.* **63**(23) 5728–5770

[7] Lavrikov S V and Revuzhenko A F 2016 Numerical modeling of elastic energy accumulation and release in structurally heterogeneous geomaterials *J. Min. Sci.* **52**(4) 632–637

[8] Klishin S V and Revuzhenko A F 2017 Spatial discrete element analysis of problem on floor buckling in underground openings *IOP Conference Series: Earth and Environmental Science* **53**(1) 012012

[9] Brodu N, Dijksman J A and Behringer R P 2015 Multiple-contact discrete-element model for simulating dense granular media *Phys. Rev. E.* **91** 032201

[10] Gonzalez M and Cuitiño A M 2012 A nonlocal contact formulation for confined granular systems *Journal of the Mechanics and Physics of Solids* **60**(2) 333–350

[11] Johnson K L 1985 *Contact mechanics* (Cambridge: Cambridge University Press) p 452

[12] Mindlin R D 1949 Compliance of elastic bodies in contact *J. Appl. Mech.* **16** 259–268

[13] Mindlin R D and Deresiewicz H 1953 Elastic spheres in contact under varying oblique forces *J. Appl. Mech.*, Trans. ASME **20** 327–344

IOP Conf. Series: Journal of Physics: Conf. Series **973** (2018) 012009 doi:10.1088/1742-6596/973/1/012009

An Interpretation of the Laminar-Turbulent Transition Startup against the Consideration of the Transverse Viscosity Factor

V N Kolodezhnov

Air Force Academy, 54A, Star. Bolshevikov Street, Voronezh, 394064, Russia

E-mail : kvn117@mail.ru

Abstract. This paper proposes a rheological model of a fluid having the Newtonian model applicability limit and a potential for further "addition" of the transverse viscosity factor. The dynamic equations for a fluid that has such rheological model are discussed, the analysis of which demonstrates the possibility of "generating" the cross stream velocity components. The transition to the dimensionless notation introduces four dimensionless complexes of local characterization for the transition conditions in the neighborhood of the flow region point in question. Based on such dimensionless complexes and using the known experimental data, the empiric conditions of "generating" the cross stream velocity components and starting the laminar-turbulent transition are proposed.

1. Introduction

The key approaches to modeling the laminar-turbulent transition suggest superimposition of the pressure and velocity disturbance pulse background on an initially laminar flow that obeys the Navier-Stokes equations to further define the conditions under which such disturbance amplitudes or kinetic energies will rise indefinitely over time [1-5].

However, along with the conventional flow disturbance factors which result in an instability of such flow and, ultimately, in turbulence, other circumstances that impel the startup of transition and could be related to some specific "intrinsic" properties of fluids may as well be in place there. In the early 1970-s, Academician L I Sedov supposed the following in his discussion of the new physical models: "It is clear from practice that such rheological properties of fluids may manifest themselves in turbulent flows that are hidden and insignificant from the laminar liquid motion standpoint" [6]. Therefore, such hypothesis allows for possible consideration of the rheological factor impact on the startup of the laminar-turbulent transition as well. Quite naturally, such an assumption will demand respective adjustments of the rheological model of fluids in order to have not only their conventional properties taken into account, but also such properties that "initiate" the transition on certain conditions.

This paper is an attempt to study this sort of such rheological model and propose, on its basis, an interpretation of the startup of the laminar-turbulent transition.

Note that the preceding local characterizations of the laminar flow form transition into the turbulent form often involved various "local" dimensionless complexes similar to the Reynolds number which included, among other things, the "differential" properties of the velocity field. A fairly simple example of such complex in relation to the boundary layer flow was discussed in [7]. A little later, a more complex parameter was introduced for the cylindrical channel flow in [8]. The result was developed in [9], which suggested another example of a dimensionless complex

$$K = \frac{1}{2} \cdot \rho \cdot \frac{|grad(\vec{v} \cdot \vec{v})|}{|\vec{F} - grad(p)|},$$

distributed across the flow region that characterizes the start of the transition locally. Here: ρ is density; $\vec{v}$ is fluid velocity vector; $\vec{F}$ is the body force density factor; and p is pressure.

Various other examples of such dimensionless complexes were suggested later at different times [10-14]. Despite that those complexes were different in form, their common trait was that they shaped up a basis for the following hypothesis. An excess of the maximum value of such complex, at a certain spatial point within the flow region area, of the respective threshold value triggers the transition of the laminar flow form to the turbulent flow form. Then the neighborhood of that point acts as a kind of "zone of initial instability" [7].

2. Rheological model.

Assume that there is some boundary to the deployment of the Newtonian incompressible fluid model beyond which the threshold transverse viscosity factor "joins in".

It is well known, modeling of channel flows for fluids whose rheology involves the consideration of transverse viscosity, leads to solutions which show the existence of secondary flows [15]. These flows are characterized by cross stream velocity components in relation to the original flow lines (obtained by flow modeling without the consideration of transverse viscosity). In this respect, note that a similar phenomenon (in terms of "generation" of cross stream velocity components in relation to the original laminar flow lines) exists in the laminar-turbulent transition, too. Indeed, the "surge" of not previously existing cross stream components can be always seen on the initial stage of transition.

Given the relation between the transverse viscosity and secondary flows characterized by the cross stream velocity components, let us introduce the following rheological model with the key relationships represented in the Cartesian system

$$\tau_{ij} = -P \cdot \delta_{ij} + 2 \cdot \mu \cdot \varepsilon_{ij} + 4 \cdot \eta_c(I_2) \cdot \xi_{ij}, \tag{1}$$

$$I_2 = \varepsilon_{11} \cdot \varepsilon_{22} + \varepsilon_{22} \cdot \varepsilon_{33} + \varepsilon_{33} \cdot \varepsilon_{11} - \varepsilon_{12}^2 - \varepsilon_{23}^2 - \varepsilon_{31}^2, \qquad \varepsilon_{ij} = \frac{1}{2} \cdot \left(\frac{\partial u_i}{\partial x_j} + \frac{\partial u_j}{\partial x_i} \right),$$

$$\xi_{ij} = \sum_{k=1}^{3} \varepsilon_{ik} \cdot \varepsilon_{kj}, \quad i, j = 1, 2, 3, \quad \eta_c(I_2) = \begin{cases} 0, & |I_2| < I_{2\eta} \\ \eta_q \cdot \left(|I_2| - I_{2\eta}\right)^q, & |I_2| \ge I_{2\eta} \end{cases},$$

where τ_{ij}, ε_{ij} are components of stress tensors and strain rates; P is pressure; δ_{ij} is the Kronecker symbol; $\mu = const$ is dynamic viscosity; u_i are velocity vector components; $\eta(I_2)$ is transverse viscosity as a function of the second strain rate tensor invariant I_2; $I_{2\eta}$ is some critical value of $|I_2|$ above which level the transverse viscosity factor "joins in" and, consequently, cross stream velocity components begin "generating" η_q, q are the empiric constants of the model.

A similar rheological model, yet with a constant transverse viscosity value ($q = 0$), was studied in [16]. A model with a linear law ($q = 1$) of the dependence of the transverse viscosity on the second invariant was proposed in [17].

The analysis of (1) reveals the following. If, at some spatial point of the flow region, the condition $|I_2| < I_{2\eta}$ is satisfied, the behavior of fluids satisfies the classical Newtonian model in the minor neighborhood of such point. Otherwise, if $|I_2| > I_{2\eta}$, the transverse viscosity factor "joins in" to create conditions for "generating" the cross stream velocity components. Such condition is suggested as a necessary, yet insufficient, condition for triggering the laminar-turbulent transition.

3. Dynamic equations

Let us consider the following problem. Assume that the respective velocity and pressure regions, which exactly satisfy the Navier-Stokes equations and continuity condition, have been formed in a set liquid flow region having the rheological model (1) by a certain moment of time conventionally set as the initial moment of time ($t = 0$). With such initial conditions, let us, however, admit that the following condition is satisfied at a certain point and its minor neighborhood at the initial moment of time

$$|I_2| > I_{2\eta} \quad . \tag{2}$$

Condition (2) means that the transverse viscosity factor manifests itself here right after the initial moment of time.

Let us introduce the Cartesian system and place its origin at that point in question. One of the axes, e.g. Ox_1, runs tangentially to the current line at this flow region point. Now we have the equations that describe the evolution of velocity component and pressure distribution in the neighborhood of that point over time

$$u_i = u_i(x_1, x_2, x_3, t), \quad i = 1, 2, 3, \qquad P = P(x_1, x_2, x_3, t). \tag{3}$$

For characterizing the velocity and pressure fields (3), let us introduce the two following vectors

$$\vec{E} = \text{grad}\left\{P + \frac{\rho \cdot u^2}{2}\right\}, \quad \vec{D} = \text{grad}\left\{2 \cdot \mu \cdot \sqrt{|I_2|}\right\}, \quad u = \sqrt{u_1^2 + u_2^2 + u_3^2}\,.$$

In this case, vector $\vec{E}$ characterizes the maximum increase direction and "speed" (along the space axes) of the full mechanical liquid flow region energy at the flow region point in question. Vector $\vec{D}$, in its turn, is again a characteristic of the maximum increase direction and speed, but it relates to viscous dissipation at the flow region point in question.

Let us introduce, along with scale velocity u_s, the following parameters also taken as scale values for measuring length, time, pressure and the second invariant of the strain rate tensor, respectively

$$L_s = \frac{\mu \cdot E_s}{\rho \cdot u_s \cdot D_s}, \quad t_s = \frac{L_s}{u_s}, \quad P_s = \frac{\rho^3 \cdot u_s^5}{\mu \cdot E_s}, \quad I_s = \left(\frac{E_s}{\rho \cdot u_s}\right)^2, \tag{4}$$

where E_s, D_s are vector $\vec{E}$, $\vec{D}$ modules, respectively.

Here and hereafter, subscript s shows that the respective functions have been calculated at the spatial point in questions (zero point) for $t = 0$.

Given the scale values (4) introduced above, the incompressible fluid equations having rheological model (1) for special case $q = 1$ with (2) having the dimensionless form, acquire the form of

$$\frac{\partial u_i'}{\partial t'} = -K_3 \cdot \frac{\partial P'}{\partial x_i'} + \sum_{j=1}^{3}\left\{\frac{2}{K_1} \cdot \frac{\partial \varepsilon_{ij}'}{\partial x_j'} - u_j' \cdot \frac{\partial u_i'}{\partial x_j'} - 4 \cdot K_4 \cdot \frac{\partial}{\partial x_j'}\left[\left(I_2' + K_{2g}\right) \cdot \sum_{k=1}^{3} \varepsilon_{ik}' \cdot \varepsilon_{kj}'\right]\right\}, \tag{5}$$

$$I_2' = K_5^2 \cdot (\varepsilon_{11}' \cdot \varepsilon_{22}' + \varepsilon_{22}' \cdot \varepsilon_{33}' + \varepsilon_{33}' \cdot \varepsilon_{11}' - \varepsilon_{12}'^2 - \varepsilon_{23}'^2 - \varepsilon_{31}'^2), \quad \varepsilon_{ij}' = \frac{1}{2} \cdot \left(\frac{\partial u_i'}{\partial x_j'} + \frac{\partial u_j'}{\partial x_i'}\right), \quad i, j = 1, 2, 3,$$

$$K_1 = \frac{E_s}{D_s}, \; |I_2'| = \left|\frac{I_2}{I_{2s}}\right| = K_2 > K_{2g}, \; K_{2g} = \frac{I_{2\eta}}{I_{2s}}, \; K_3 = \frac{\rho^2 \cdot u_s^3}{\mu \cdot E_s}, \; K_4 = \frac{\eta_1 \cdot D_s^2}{\rho \cdot \mu^2}, \; K_5 = \frac{K_3}{K_1}, \quad (6)$$

$$t' = \frac{t}{t_s}, \quad P' = \frac{P}{P_s}, \quad x_i' = \frac{x_i}{L_s}, \quad u_i' = \frac{u_i}{u_s},$$

where K_1, K_{2g}, K_3, K_4 are the individual dimensionless complexes built of invariant values and not connected directly with the specific linear dimensions of the flow region.

Note that, in terms of its meaning and "location" in equations (5), the dimensionless array K_1 is similar to the Reynolds number. Given that, the inequality in (6) results from condition (2) of "generating" the cross stream velocity components via the transition to the dimensionless form.

4. Modeling the development of the initial stage of transition

Let us consider an approach to building the first approximation of the problem about the initial (with $t = 0$) velocity distribution development in the minor neighborhood of a certain spatial point where condition (2) is satisfied.

Assume that the initial velocity and pressure distributions

$$t = 0, \quad u_i = U^{(0,i)}(x_1,x_2,x_3), \; i = 1,2.3, \; P = P^{(0)}(x_1,x_2,x_3), \quad (7)$$

exactly satisfy the Navier-Stokes equations and flow continuity condition.

In (7) and further on in this section, the strokes over the dimensionless values have been omitted.

The solution of equation system (5) is suggested in the form of

$$u_i(x_1,x_2,x_3,t) = U^{(0,i)}(x_1,x_2,x_3) + u^{(t,i)}(x_1,x_2,x_3,t), \qquad i = 1,2,3, \quad (8)$$

$$P(x_1,x_2,x_3,t) = P^{(0)}(x_1,x_2,x_3) + p^{(t)}(x_1,x_2,x_3,t),$$

where $u^{(t,i)}$; $p^{(t)}$ are the non-stationary components of the pressure and velocity fields which "originate" in the minor neighborhood of the point in question on the initial stage of flow development from original condition (7) and satisfy the following conditions

$$u^{(t,i)}(x_1,x_2,x_3,0) = 0, \quad i = 1,2.3, \quad p^{(t)}(x_1,x_2,x_3,0) = 0. \quad (9)$$

By linearizing equations (5) with reference to (8) and conventional, for such procedures, assumptions, we arrive at the following equation system [17]

$$\frac{\partial u^{(t,i)}}{\partial t} = -K_3 \cdot \frac{\partial p^{(t)}}{\partial x_i} + \sum_{j=1}^{3} \left\{ \frac{2}{K_1} \cdot \frac{\partial \varepsilon_{ij}^{(t)}}{\partial x_j} - u^{(t,j)} \cdot \frac{\partial U^{(0,i)}}{\partial x_j} - U^{(0,j)} \cdot \frac{\partial u^{(t,i)}}{\partial x_j} \right.$$

$$- 4 \cdot K_4 \cdot \frac{\partial}{\partial x_j} \sum_{k=1}^{3} \left[\left(I_2^{(t)} + I_2^{(0)} + K_{2g} \right) \cdot \varepsilon_{ik}^{(0)} \cdot \varepsilon_{kj}^{(0)} \right.$$

$$\left. \left. + \left(I_2^{(0)} + K_{2g} \right) \cdot \left(\varepsilon_{ik}^{(0)} \cdot \varepsilon_{kj}^{(t)} + \varepsilon_{ik}^{(t)} \cdot \varepsilon_{kj}^{(0)} \right) \right] \right\}, \; i = 1,2,3. \quad (10)$$

In developing (10), it was assumed that

$$\varepsilon_{ij} = \varepsilon_{ij}^{(0)} + \varepsilon_{ij}^{(t)}, \qquad I_2 = I_2^{(0)} + I_2^{(t)},$$

$$\varepsilon_{ij}^{(t)}=\frac{1}{2}\cdot\left(\frac{\partial u^{(t,i)}}{\partial x_j}+\frac{\partial u^{(t,j)}}{\partial x_i}\right),\quad \varepsilon_{ij}^{(0)}=\frac{1}{2}\cdot\left(\frac{\partial U^{(0,i)}}{\partial x_j}+\frac{\partial U^{(0,j)}}{\partial x_i}\right),\quad i,j=1,2,3\ ,$$

$$I_2^{(0)}=K_5^2\cdot\left\{\varepsilon_{11}^{(0)}\cdot\varepsilon_{22}^{(0)}+\varepsilon_{22}^{(0)}\cdot\varepsilon_{33}^{(0)}+\varepsilon_{33}^{(0)}\cdot\varepsilon_{11}^{(0)}-(\varepsilon_{12}^{(0)})^2-(\varepsilon_{23}^{(0)})^2-(\varepsilon_{31}^{(0)})^2\right\},$$

$$I_2^{(t)}=K_5^2\cdot\left\{(\varepsilon_{22}^{(0)}+\varepsilon_{33}^{(0)})\cdot\varepsilon_{11}^{(t)}+(\varepsilon_{33}^{(0)}+\varepsilon_{11}^{(0)})\cdot\varepsilon_{22}^{(t)}+(\varepsilon_{11}^{(0)}+\varepsilon_{22}^{(0)})\cdot\varepsilon_{33}^{(t)}-2\cdot\varepsilon_{12}^{(0)}\cdot\varepsilon_{12}^{(t)}-2\cdot\varepsilon_{23}^{(0)}\cdot\varepsilon_{23}^{(t)}-2\cdot\varepsilon_{31}^{(0)}\cdot\varepsilon_{31}^{(t)}\right\}.$$

Let us represent the initial velocity component distributions and non-stationary components for the pressure and velocity components in the minor neighborhood of the point in question (zero point) through space coordinate power expansion

$$U^{(0,i)}(x_1,x_2,x_3)=\sum_{k=0}^{\infty}\sum_{m=0}^{\infty}\sum_{n=0}^{\infty}U_{kmn}^{(0,i)}\cdot x_1^k\cdot x_2^m\cdot x_3^n\ ,$$

$$p^{(t)}(x_1,x_2,x_3,t)=\sum_{k=0}^{\infty}\sum_{m=0}^{\infty}\sum_{n=0}^{\infty}p_{kmn}^{(t)}(t)\cdot x_1^k\cdot x_2^m\cdot x_3^n\ ,$$

$$u^{(t,i)}(x_1,x_2,x_3,t)=\sum_{k=0}^{\infty}\sum_{m=0}^{\infty}\sum_{n=0}^{\infty}u_{kmn}^{(t,i)}(t)\cdot x_1^k\cdot x_2^m\cdot x_3^n\ ,\qquad i=1,2,3. \tag{11}$$

Here, $U_{kmn}^{(0,i)}$ are the known expansion coordinates; $u_{kmn}^{(t,i)}(t)$, $p_{kmn}^{(t)}(t)$ are the expansion factors representing the yet unknown time functions which, with reference to (9), satisfy the following initial conditions

$$u_{kmn}^{(t,i)}(0)=0\ ,\qquad p_{kmn}^{(t)}(0)=0\ ,\qquad i=1,2,3\ ,\qquad k,m,n=0,1,2.... \tag{12}$$

First of all, let us estimate the tendency for the consequential velocity component evolution over time at the spatial point in question at the initial moment of time. Let us plug (11) into (10). Assuming that $x_1=0$; $x_2=0$; $x_3=0$, in the resulting equations and taking into account initial conditions (12), we deduce that

$$\left.\frac{du_{000}^{(t,i)}}{dt}\right|_{t=0}=W^{(t,i)}=-4\cdot K_4\cdot\sum_{j=1}^{3}\left\{\left(I_2^{(0,0)}+K_{2g}\right)\cdot\xi_{ij}^{(0,j)}+I_2^{(0,j)}\cdot\xi_{ij}^{(0,0)}\right\},\qquad i=1,2,3. \tag{13}$$

In deriving the latter relationships and subsequent transformations, the following representations, with reference to (11), of the respective functions in (10) were taken into account through their space-wise expansion into power series to further restrict them with the first four terms

$$\varepsilon_{ij}^{(0)}(x_1,x_2,x_3)\approx\varepsilon_{i,j}^{(0,0)}+\varepsilon_{i,j}^{(0,1)}\cdot x_1+\varepsilon_{i,j}^{(0,2)}\cdot x_2+\varepsilon_{i,j}^{(0,3)}\cdot x_3\ ,$$

$$\xi_{ij}^{(0)}(x_1,x_2,x_3)=\varepsilon_{i1}^{(0)}\cdot\varepsilon_{1j}^{(0)}+\varepsilon_{i2}^{(0)}\cdot\varepsilon_{2j}^{(0)}+\varepsilon_{i3}^{(0)}\cdot\varepsilon_{3j}^{(0)}\approx\xi_{ij}^{(0,0)}+\xi_{ij}^{(0,1)}\cdot x_1+\xi_{ij}^{(0,2)}\cdot x_2+\xi_{ij}^{(0,3)}\cdot x_3\ ,$$

$$I_2^{(0)}(x_1,x_2,x_3)\approx I_2^{(0,0)}+I_2^{(0,1)}\cdot x_1+I_2^{(0,2)}\cdot x_2+I_2^{(0,3)}\cdot x_3\ .$$

All coefficients in the latter relationships are expressed in the known manner for $t=0$ via the set initial velocity distribution parameters $U_{kmn}^{(0,i)}$.

Formula (13) produces the following result. If condition (2) is satisfied at the point in question (zero point) at the initial moment of time, time derivatives $W^{(t,i)}$ for the cross stream velocity

IOP Conf. Series: Journal of Physics: Conf. Series **973** (2018) 012009 doi:10.1088/1742-6596/973/1/012009

components, that were not here before, may not be equal to zero ($W^{(t,i)} \neq 0$). This means that the cross stream velocity components can "originate" at this point right after the initial moment of time. Such "generation" of the cross stream velocity components will result in changing the velocity field in the minor neighborhood of the spatial point in question that was formed around the initial moment of time. A similar result was achieved in [16] for the plane problem statement with an example of a rheological model similar to (1), having a constant transverse viscosity value ($q = 0$).

Let us further restrict ourselves just to a concept of the simplest approximation to a solution of the problem stated above. Let us plug (11) into (10) and, after the respective transformations, equate the multiplier coefficients in the form of $x_1^k \cdot x_2^m \cdot x_3^n$ to the same exponent combination k,m,n for the respective coordinates in the left and right sides of the equation. In such case, each of the three equations (10) may become a basis for an infinite ordinary differential equation system in relation to expansion coefficients $u_{kmn}^{(t,i)}(t)$, $p_{kmn}^{(t)}(t)$. As the first approximation, let us restrict ourselves only to the first equation in each such infinite system built of the coefficients before complex $x_1^k \cdot x_2^m \cdot x_3^n$ with zero exponents $k = 0$; $m = 0$; $n = 0$. As a result, we arrive at the following system of three ordinary differential first-order equations which may be structurally represented as

$$\frac{du_{000}^{(t,1)}}{dt} = -K_3 \cdot p_{100}^{(t)} + W^{(t,1)} + \sum_{\substack{k=0 \\ k+m+n\le 2}}^{2} \sum_{m=0}^{2} \sum_{n=0}^{2} \left\{ \sum_{j=1}^{3} \alpha_{kmn}^{(1,j)} \cdot u_{kmn}^{(t,j)} \right\},$$

$$\frac{du_{000}^{(t,2)}}{dt} = -K_3 \cdot p_{010}^{(t)} + W^{(t,2)} + \sum_{\substack{k=0 \\ k+m+n\le 2}}^{2} \sum_{m=0}^{2} \sum_{n=0}^{2} \left\{ \sum_{j=1}^{3} \alpha_{kmn}^{(2,j)} \cdot u_{kmn}^{(t,j)} \right\}, \quad (14)$$

$$\frac{du_{000}^{(t,3)}}{dt} = -K_3 \cdot p_{001}^{(t)} + W^{(t,3)} + \sum_{\substack{k=0 \\ k+m+n\le 2}}^{2} \sum_{m=0}^{2} \sum_{n=0}^{2} \left\{ \sum_{j=1}^{3} \alpha_{kmn}^{(3,j)} \cdot u_{kmn}^{(t,j)} \right\}.$$

The set of coefficients $\alpha_{kmn}^{(i,j)}$; $i,j = 1, 2, 3$; $k,m,n = 0,1,2$; $k+m+n \le 2$ is assumed as assigned in this case and may be represented in the known manner through velocity component initial distribution (11) coefficients $U_{kmn}^{(0,i)}$ or dimensionless complexes K_1, K_{2g} и K_4.

Equation system (14) is not locked as, for the approximation in question, it should contain 34 unknown time functions: $p_{000}^{(t)}$, $p_{100}^{(t)}$, $p_{010}^{(t)}$, $p_{001}^{(t)}$, $u_{000}^{(t,j)}$, $u_{100}^{(t,j)}$, $u_{010}^{(t,j)}$, $u_{001}^{(t,j)}$, $u_{200}^{(t,j)}$, $u_{020}^{(t,j)}$, $u_{002}^{(t,j)}$, $u_{110}^{(t,j)}$, $u_{011}^{(t,j)}$, $u_{101}^{(t,j)}$, $j = 1,2,3$.

First of all, in order to lock the three equation system (14), the continuity condition should be used which provides for four additional equations. The other 27 equations should result from meeting the respective "boundary" conditions which must be placed at eight spatial nodal points $M_\beta(x_{1\beta}, x_{2\beta}, x_{3\beta})$, $\beta = 1,2,...,8$; three velocity components, and in three nodal points $M_\gamma(x_{1\gamma}, x_{2\gamma}, x_{3p\gamma})$, $\gamma = 1, 2, 3$, for pressure. Assuming that such nodal points M_β may act as vertexes of a "hexagon", let us proceed to building the respective finite element. Note that such elements with six, generally speaking, "curvy" faces and eight vertexes should "cover" the entire flow region in the general case

As a result, the system of three differential equations (14) and 31 linear algebraic equation for each such finite elements may, after all respective transformations, be simplified to a simply translated equation, such as

$$\frac{d^2 u_{000}^{(t,2)}}{dt^2} + a_1 \cdot \frac{du_{000}^{(t,2)}}{dt} + a_0 \cdot u_{000}^{(t,2)} = H(t). \tag{15}$$

Coefficients a_0, a_1 of equation (15) and its constituent function $H(t)$ are determined by the totality of initial process parameters with reference to the relationships mentioned above (including the "boundary" conditions on the finite element vertexes).

Any further evolution scenarios for the cross stream velocity components are determined by a specific type of solving equation (15) with reference to initial conditions (12), (13), and may be drastically different in meaning. Given certain additional conditions (overlaid on the value and signs of coefficients a_0, a_1 and, consequently, dimensionless complexes K_1, K_{2g}, K_3, K_4), the "generated" velocity components may decay. There may also be a situation where the originating cross stream velocity components, which generate the secondary flow, will "convert" the existing laminar flow pattern into another still laminar pattern. Finally, in the event of infinite growth, such components may become a basis for instigating the initial stage of the laminar transition to turbulent flow. In the latter case, it may concern a continuous stage cascade where the cross stream velocity components generate that are "new" in relation to the recently (at the preceding moments of time) existing "old" cross stream velocity components.

An analysis of the above problem solution scheme regarding the development of the initial flow stage in the minor neighborhood of a certain spatial point where condition (2) is satisfied shows that obvious analytical conditions for the startup of the laminar-turbulent transition are quite difficult to build even for the simplest approximation cited above due to cumbersomeness of the aforementioned relationships. In this respect, it is suggested that the anticipated consequences of the approach discussed be represented in the form of empiric conditions based on the known experimental data.

5. Empiric "generation" conditions

The class of studies associated with the "generation" of cross stream velocity components should also include quite spectacular experiments related to the study of secondary flows (in the form of interspersing, toroidal, vortex structures) in a gap between the two coaxial cylinders [18-26]. Such experimental data may be admitted as a basis for an empiric estimate of the startup conditions for the "generation" of cross stream velocity components.

Let us restrict ourselves to a well-known and fairly simple special case of such flows where the internal cylinder with the radius R_1 rotates with a constant angular speed ω_1, and the external cylinder whose radius is R_2 remains motionless.

In a gap between the cylinders and against a fairly low angular speed values for the internal cylinder rotation, the distribution of velocity and second invariant of the strain rate tensor is determined by the following relationships

$$u(r) = \frac{\omega_1 \cdot R_1^2}{r} \cdot \left(\frac{R_2^2 - r^2}{R_2^2 - R_1^2} \right), \quad I_2(r) = -\varepsilon_{r\varphi}^2 = -\frac{\omega_1^2 \cdot R_1^4 \cdot R_2^4}{(R_2^2 - R_1^2)^2 \cdot r^4}, \tag{16}$$

where r is the radial coordinate measured from the cylinders' axis.

Let us interpret the startup of the toroidal structures formation from the standpoint of the initial "generation" of cross stream velocity components due to "joining in" the transverse viscosity factor in the respective flow region areas. Let us also assume that condition $|I_2| = I_{2\eta}$ satisfies the startup of this process satisfies, with reference to the fulfillment of (2).

Taking into consideration (16), we arrive at the expressions that determine the distribution of primary dimensionless complexes (6) in the form of

$$K_1(r')=\frac{Re\cdot(R_2'^2-r'^2)\cdot r'^2}{2\cdot R_2'^2\cdot(R_2'^2-1)\cdot(R_2'-1)},\quad K_2(r')=\frac{1}{4}\cdot\left(\frac{R_2'}{r'}\right)^4,$$

$$K_3(r')=\frac{Re\cdot(R_2'^2-r'^2)^2}{2\cdot r'^2\cdot(R_2'^2-1)\cdot(R_2'-1)},\quad Re=\frac{\rho\cdot\omega_1\cdot R_1\cdot(R_2-R_1)}{\mu},\quad r'=\frac{r}{R_1},\quad R_2'=\frac{R_2}{R_1}. \tag{17}$$

It follows from (16) that the second invariant module is a monotonic decreasing radial coordinate function. That is why its maximum value can be achieved on the internal cylinder surface. This means that, as the angular speed increases and against its critical value of $\omega_1=\omega_{1\eta}$, threshold level $|I_2|=I_{2\eta}$ may be primarily achieved on the surface of the second rotating cylinder. In this respect, we will further associate the critical value of $\omega_{1\eta}$ with such angular speed that instigates the origination of toroidal structures determined by the "generation" of the cross stream velocity components. Then primary dimensionless arrays (17) in such situation will be defined as follows

$$K_{1g}=K_1(1)=\frac{Re_g}{2\cdot R_2'^2\cdot(R_2'-1)},\quad K_{2g}=K_2(1)=\frac{R_2'^4}{4},\quad K_{3g}=K_3(1)=\frac{Re_g\cdot(R_2'+1)}{2}, \tag{18}$$

$$Re_g=\frac{\rho\cdot\omega_{1\eta}\cdot R_1\cdot(R_2-R_1)}{\mu}.$$

These critical values (18) of dimensionless complexes calculated on the basis of the known experimental data taken from various authors are shown in the first 17 lines of Table 1. The results from paper [26] have been assumed by reference in [27], and the results of [19] – by reference in monograph [1].

These experimental data may be amplified with the results related to the Couette flow in a plane channel with the following velocity distribution $u(y)=V_w\cdot y/h$. Here V_w is the velocity of the channel's mobile wall; h is the channel width; y is the coordinate measured from the motionless channel wall. Quite often such flow pattern is treated as a specific flow case in a gap between the coaxial cylinders in the extreme case $R_2'\to 1$. At this point, we can add that paper [28] sets forth experimental evidence of vortex structures associated with the originating cross stream velocity components for the plane Couette flows, and offers a current line pattern for such vortexes.

For such flow pattern, complexes (6) may be represented in the form of

$$K_1\to\infty,\qquad K_2=\frac{1}{4},\qquad K_3=Re\cdot\left(\frac{y}{h}\right)^2,\qquad Re=\frac{\rho\cdot h\cdot V_w}{\mu}. \tag{19}$$

Using known experimental data [29-33] related to the laminar-turbulent transition startup and assuming a priori that this process is concomitant with the "generation" of the cross stream velocity components in the vicinity of the movable channel wall (for $y=h$), we arrive at the critical values of dimensionless arrays (19) shown in the last four lines of Table 1.

Table 1. Relationships between the primary dimensionless complexes on the "generation" stage obtained by processing the known experimental data.

Item	References	R_2'	Re_g	K_{1g}	K_{2g}	K_{3g}
1	[18]	1.345	94.1	75.37	0.818	110
2	[22]	1.250	94.4	120.8	0.610	106
3	[19]	1.190	94.6	175.3	0.502	104
4	[21]	1.177	106	216.1	0.480	116
5	[20]	1.144	114	302.4	0.428	122
6	[21]	1.143	117	313.0	0.426	125
7	[23]	1.142	128	345.3	0.425	137
8	[18]	1.137	122	344.8	0.417	130
9	[24]	1.133	120	352.8	0.410	128
10	[21]	1.111	130	475.1	0.381	138
11	[21]	1.081	150	793.2	0.341	156
12	[18]	1.062	170	1217	0.318	175
13	[21]	1.053	183	1571	0.307	188
14	[22]	1.043	217	2335	0.295	221
15	[21]	1.039	212	2521	0.291	216
16	[25]	1.018	316	8711	0.268	319
17	[26]	1.010	350	16890	0.260	352
18	[29]	1.000	280	∞	0.250	280
19	[33]	1.000	325	∞	0.250	325
20	[31]	1.000	360	∞	0.250	360
21	[30, 32]	1.000	370	∞	0.250	370

The data from Table 1 demonstrate that complex K_{2g} introduced earlier is not a constant value. With this in mind, Figure 1 shows the experimental points of dependence of K_{2g} on K_{3g}.

Analyzing the behavior of the data produced, it may be assumed that the approximating function is a hyperbolic curve having the form of

$$K_{2g} = F_g(K_{3g}) = k_0 + \frac{k_1}{K_{3g} - k_2}, \quad k_0 = 0.221,\ k_1 = 8.572,\ k_2 = 84.015. \tag{20}$$

Processing of the experimental data did not include the results of the first, third and seventh lines of Table 1 since they demonstrate an evident disruption of monotony in varying the values of complex K_{3g} against the variations of complex values K_{2g}. Note that the data excluded from processing largely reflect the general dependence (20) trend detected. Thus the correlating points are shown in figure 1.

The above results allow for the following interpretation. If condition (6) is satisfied for any flow point for complex K_2 representable, with regard to (20), in the form of

$$K_2 > k_0 + \frac{k_1}{K_3 - k_2}, \tag{21}$$

that means that the "generation" of cross stream velocity components starts in the neighborhood of such point.

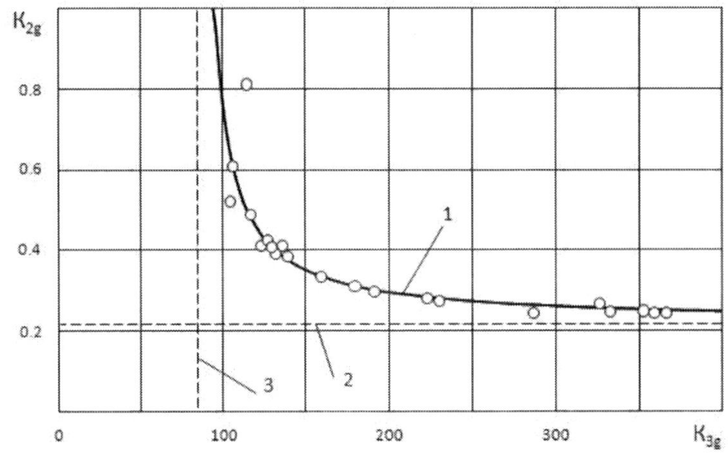

Figure 1. The nature of dependence of K_{2g} on K_{3g} built on the basis of known experimental data [18-26, 29-33]; 1 is the approximating function (20); graph. 2, 3 is the horizontal asymptote $K_{2g} = k_0$ and vertical asymptote $K_{3g} = k_2$, respectively.

Quite naturally, general case "generation" condition (21) for the cross stream velocity components may be satisfied not for a separate spatial point, but in a certain flow region area.

6. Empiric condition of the transition startup

The empiric relationship (21) of the cross stream velocity component generation may be seen as a necessary yet insufficient startup condition for the laminar-turbulent transition. The sufficient startup condition may be sought empirically with reference to the known experimental data in the form of restricted dimensionless complex K_1 (a Reynolds number analog) in the following form

$$K_{1max} > K_{1c} = F_c(K_{3c}), \tag{22}$$

because this is exactly the complex that characterizes the relationship between the full energy density increase factor for the spatial point in question and viscous dissipation factor. In condition (22), K_{1c} is the critical threshold for complex K_1, and K_{1max} is the maximum value of dimensionless array K_1 in that flow region area where the necessary transition startup (21) is satisfied, i.e. the condition of "generating" the cross stream velocity components. Complex K_{3c} should be defined in the same spatial point as where $K_1 = K_{1max}$ is achieved. In that case, it is assumed that the spatial point at which conditions (21) and (22) are satisfied at the same time starts to act as an "initiator" of the laminar-turbulent transition.

With the exemplary transition startup for the rotational Couette flow, such approach may be interpreted, with reference to (17), as follows.

Due to the monotonic decrease of function $K_2(r')$, the flow region area where necessary transition condition (21) may be satisfied is essentially a ring with an internal radius $r' = 1$. Formally, function $K_1(r')$, with reference to (17), achieves the maximum type extremum at the point $r' = r'_{MAX} = R'_2 / \sqrt{2}$. However, the condition $r'_{MAX} < 1$ is satisfied for values R'_2 set forth in Table 1. In such situation, the maximum value of function $K_1(r')$ is achieved on the internal cylinder surface, i.e. $K_{1max} = K_1(1)$. Using the known experimental data for the critical Reynolds number Re_c in relation to the transition startup [19, 20, 23, 34-38], the respective critical values of dimensionless complexes K_{1c} and K_{3c} set forth in Table 2 have been calculated. The experimental data from [38] were assumed by reference in [39].

Table 2. Relationships between the primary dimensionless arrays on the transition-to-turbulence stage obtained by processing the known experimental data.

Item	References	Re_c	K_{3c}	K_{1c}
1	[36]	3357	3551	11620
2	[37]	3076	3296	8232
3	[20]	2877	3057	9094
4	[34]	2879	3024	11760
5	[38]	2687	2821	11100
6	[23, 35]	2608	2791	7168
7	[37]	2603	2789	6967
8	[19]	1892	2072	3516
9	[40-43]	2653	1021	1021
10	[44-46]	3344	967	965
11	[47, 48]	1743	380	168

Similarly, the calculations have been performed for some other flow patterns.

The most well-known are the transition-to-turbulence results for the Poiseuille flow in the cylindrical channel with the radius R. It can be proven that, in such case, the maximum value of complex K_1 is achieved at a point with the radial coordinate of $r = R/\sqrt{3}$. Given the scatter of the known experimental data with various authors, both classical transition results [40] and recent year results [41-43] have been averaged. The resulting values are shown in line 9, Table 2. The average channel flow velocity has been conventionally assumed as the characteristic velocity for building the Reynolds number used in the calculation.

Similarly, the results of experimental data processing [44-46] for the startup transition with the Poiseuille flow in a plane channel, width $2 \cdot h$ have been obtained and produced in line 10, Table 2.

In line 11, Table 2, also averaged, the experimental data [47, 48] processing results are produced for the laminar-turbulent transition with the Poiseuille flow in a gap between the two coaxial cylinders. Here, the double gap width $2 \cdot (R_2 - R_1)$ between the external and internal cylinders has been used as the characteristic linear dimension for the Reynolds number, and the average velocity for the annular cross-section has been taken as the characteristic velocity.

After processing these known experimental results, the data from Table 2 are represented as an assembly of points in Figure 2.

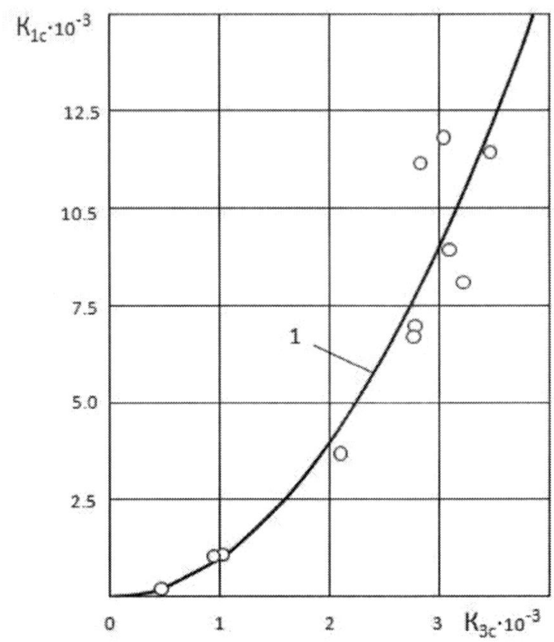

Figure. 2. The nature of dependence of K_{1c} on K_{3c} built on the basis of known experimental data [19, 20, 23, 34-48]; 1 is the approximating function (23) graph.

For such point assembly, the approximating function has been obtained in the form of the following power law

$$K_{1c} = F_c(K_{3c}) = q_0 \cdot (K_{3c})^{q_1}, \qquad q_0 = 9.382 \cdot 10^{-4}, \qquad q_1 = 2.008. \tag{23}$$

Thus, if condition (21) related to the "generation" of cross stream velocity components is satisfied at a certain flow region area, and, simultaneously, condition (22) is satisfied with reference to (23) at a certain point of that area where the maximum value of dimensionless complex K_1 is obtained, we suggest that such situation be interpreted as a startup of the laminar-turbulent transition.

Note that the laminar-to-turbulent transition startup hypothesis used in this section and based on the excess of a critical threshold level by the maximum value of a dimensionless complex (similar to the Reynolds number) was assumed as a startup condition in a series of well-known works [7-14]. Quite naturally, those works discussed the dimensionless complexes that were different from K_1 that was introduced in accordance with relationship (6).

7. Conclusion.

The proposed rheological model, which, within a certain variation range of the second invariant of the strain rate tensor, perfectly coincides with the classical Newtonian fluid model, and takes into account the transverse viscosity factor beyond such range, allowed us to arrive at the respective notation for the fluid dynamics equations of such kind. In translating the dimensionless forms of such equations, all scale values were determined using the invariant local characteristics of the velocity and pressure field. As a result, this introduced four dimensionless complexes that are not directly associated with macrocharacteristics of the set flow region. The analysis of the fluid dynamics equations demonstrates that a process of "generating" the cross stream velocity components starts in the neighborhood of a spatial point where "joining in" of the transverse viscosity factor is satisfied. In the event that the additional condition of infinite increase for such "generated" cross stream velocity components is satisfied, we suggest that the latter circumstance be interpreted as a startup of the laminar-turbulent transition. Based on the analysis of the known experimental data related, among other things, to the secondary flow origination due to the generation of cross stream velocity components, certain (necessary and sufficient) empiric conditions overlaid on the local dimensionless complexes are proposed for the startup of laminar-turbulent transition.

References

[1] Schlichting H 1955 *Boundary-layer Theory* (New York: McGraw-Hill) p 535
[2] Joseph D D 1976 *Stability of Fluid Motions, Vol . I , II* (Berlin-Heidelberg-New York: Springer-Verlag) Vol. I: p 282, Vol. II: p 274
[3] Goldschtick M A and Stern V N 1977 *Hydrodynamic Stability and Turbulence* (Novosibirsk: Nauka Publishers) p 367
[4] Zhigulev V N and Tumin A N 1987 *Origination of Turbulence*. (Novosibirsk: Nauka Publishers) p 283
[5] Boiko A V, Grek G R, Dovgal A V and Kozlov V V 1999 *Origination of Turbulence in Wall-adjacent Flows* (Novosibirsk: Nauka Publishers) p 328
[6] Sedov L I 1980 Invention of Physical Models In: Sedov L I *Thoughts on Science and Scientists* (Moscow: Nauka Publishers) pp 119–123
[7] Rose H 1946 *Elementary Mechanics of Fluids* (New York: Dover Publications) p 376
[8] Ryan N W and Johnson M 1959 Transition from laminar to turbulent flow in pipes *AIChE J.* **5** No 4 pp 433–435
[9] Hanks R W 1963 The Laminar-Turbulent Transition for flow in Pipes, Concentric Annuli, and Parallel Plates *AIChE J.* **9** (1) pp 45–48
[10] Artyushkov L S 1974 Transition from Laminar to Turbulent Flow for Pure Viscous Power Law Non-Newtonian Fluids *Proc. of the Leningrad shipbuilding Institute* (Leningrad: Leningrad shipbuilding Institute Press) **89** pp 19–24
[11] Artyushkov L S 1979 *Dynamics of the Non-Newtonian Fluid* (Leningrad: Leningrad shipbuilding Institute Press) p 228
[12] Kolodezhnov V N. 2005 On a Dimensionless Complex for Modeling Viscous Non-Newtonian Fluid Flow *The Modern Issues of Mechanics and Applied Mathematics. In two parts. Part 1* (Voronezh: Voronezh State University Press) pp 166-168
[13] Dou H S 2006 Mechanism of flow instability and transition to turbulence Int.. J. Non-Linear Mechanics **41**(4) pp 512–517
[14] Tao J J, Chen S Y and Su W D 2013 Local Reynolds number and thresholds of transition in shear flows *Science China Physics, Mechanics & Astronomy* **56** (2) pp 263–269
[15] Litvinov V G 1982 *Motion of Non-linear Viscous Fluid (* Moscow: Nauka Publishers) p 376
[16] Kolodezhnov V N 2012 On a Possible Model of the Initial Stage of the Laminar-turbulent Transition *Proc. of Voronezh State Technical University* (Voronezh: Voronezh State Technical University Press) **8** (5) pp 25–30
[17] Kolodezhnov V N 2016 Modeling the initial stage of the flow of non-Newtonian fluid, taking into account the threshold threshold "connection" of the transverse viscosity factor *Actual problems of applied mathematics, Informatics and mechanics* (Voronezh: Voronezh State University Press) pp 338–340
[18] Taylor G I 1923 Stability of a Viscous Liquid contained between Two Rotating Cylinders *Phil. Trans. Royal Society,* A **223** pp 289–343
[19] Schultz-Grunow F and Hein H 1956 Beitrag zur Couettestromung *Z. Flugwiss* No 4 pp 28–30
[20] Coles D 1965 Transition in circular Couette flow *J. of Fluid Mechanics* **21** (03) pp 385–425
[21] Donnelly R J and Schwarz K W 1965 Experiments on the stability of Viscous Flow Between Rotating Cylinders. VI Finite-Amplitude Experiments *Proc. of the Royal Society* A **283** pp 531–556

[22] Snyder H A 1968 Stability of rotating Couette flow. II. Comparison with numerical results *The Physics of Fluids* **11** (8) pp 1599–1605
[23] Gollab J P and Swinney H L 1975 Onset of turbulence in a rotating fluid *Phisical Review Letters* **36** (14) pp 927–930
[24] Andereck C D, Liu S S and Swinney H L 1986 Flow regimes in circular Couette system with independently rotating cylinders *J. of Fluid Mechanics* **164** pp 155–183
[25] Prigent A and Dauchot O 2002 "Barder pole turbulence" in large aspect ratio Taylor-Couette flow *Phisical Review Letters* **89** (1) p 014501
[26] Hinko K A 2003 Transition in the Small Gap Limit of Taylor – Couette Flow *The Ohio State University Physics Summer Institute, REU Summer 2003*; Advisor: Dr. C.D. Andereck, Department of Physics, The Ohio State University
[27] Dou H S, Khoo B C and Yeo K S 2008 Instability of Taylor – Couette Flow between Concentric Rotating Cylinders *Inter. J. of Termal Science* **47** (11) pp 1422–35
[28] Bottin S, Dauchot O, Daviaud F and Manneville P. 1998 Experimantal evidence of streamwise vortices as finite amplitude solutions in transitional plane Couette flow *Physics of Fluids* **10** (10) pp 2597–607
[29] Aydin M and Leutheusser H J 1979 Novel experimental facility for the study of plane Couette flow *Review of Scientific Instruments* **50** (11) pp 1362–66
[30] Daviaud F, Hegseth J and Berge P 1992 Subcritical Transition to Turbulence in Plane Flow *Physical Review Letters* **69** (17) pp 2511–14
[31] Tillmark N and Alfredsson P H 1992 Experiments on transition in plane Couette flow *J. Fluid Mech.* **235** pp 89–102
[32] Malerud S, Maloy K J and Goldburg W I 1995 Measurements of turbulent velocity fluctuations in a planar Couette cell Physics of Fluids **7** (8) pp 1949–55
[33] Dauchot O and, Daviaud F 1995 Finite amplitude perturbation and growth mechanism in plane Couette flow *Physics of Fluids* **7** (2) pp 335–343
[34] Barcilon A, Brindley J, Leesen M and Modds F R 1979 Marginal instability in Taylor-Couette flows at very high Taylor number *J. of Fluid Mechanics* **94**(3) pp 453–463
[35] Fenstermacher P R, Swinney H L and Gollab J P 1979 Dinamical instabilities and the transition to chaotic Taylor vortex flow *J. of Fluid Mechanics* **94** (1) pp 103–128
[36] Koschmieder E L 1979 Turbulent Taylor vortex flow *J. of Fluid Mechanics* **93** (3) pp 515–527
[37] Welden R W and Donnelly R J 1979 Reemergent order of chaotic circular Couette flow *Phisical Review Letters* **42** (5) p 301–304
[38] Bouabdallah A and Cognet G 1980 Laminar-turbulent transition in Taylor-Couette flow *Laminar-Turbulent Transition, (IUTAM Conference*) pp 368–377
[39] Di Prima R S and Sweeney H L. 1984 Instabilities and Flow Transition between Concentric Rotating Cylinders *In: Hydrodynamic Instabilities and Transition to Turbulence* (Moscow: Mir Publishers) pp 169–217
[40] Nikuradse J 1933 Stromungsgesetze in rauhen Rohren *VDI. Forschungsheft* No 361 pp 1–22
[41] Novopashin S and Muriel A 2002 The critical Reynolds number universal *J. of Experimental and Theoretical Physics* **122** Issue 2 (8) pp 306–309
[42] Swanson C J, Julian B, Ihas G G and Donnelly R J 2002 Pipe flow measurements over a wide range of Reynolds numbers using liquid helium and various gases *J. of Fluid Mechanics* **461** pp 51–60
[43] Paveliev A A, Reshmin A I, Teplovodsky S H and Fedoseev S G 2003 On the lower Critical Reynolds Number for Flow in Circular Pipe *Fluid Dynamics* **38** (4) pp 545–551
[44] Karnitz M A, Potter M C and Smith M C 1974 An experimental investigation of transition of a plane Poiseulle flow *Trans. ASME. I.* **96** (4) pp 384–388
[45] Carlson D R., Widnall S E and Peeters M F 1982 A flow-visualization study of transition in plane Poiseuille flow *J. of Fluid Mechanics* **121** pp 487–505
[46] Lemoult G, Gumowski K, Aider J and Wesfreid J 2014 Turbulent spots in channel: An

experimental study *The Euroean Physical Journal E* **37** (25) p 11
[47] Walker J E, Whan G A and Rothfus RR 1957 Fluid Friction in Noncircular Ducts. *AIChE. J.* **3** (4) pp 484–489
[48] Hanks R.W and Bonner W F 1971 Transition flow phenomena in concentric annuli *Industrial& Engineering Chemistry Fundamentals* **10** (1) pp 105–113

IOP Conf. Series: Journal of Physics: Conf. Series **973** (2018) 012010 doi:10.1088/1742-6596/973/1/012010

On the determination of stress fields and displacements in a thin elastoplastic plate containing elastic inclusion - a shim

A V Kovalev[1], E Y Rusina[1] and A Y Yakovlev[1]

[1] Department of Applied Mathematics, Informatics and Mechanics, Voronezh State University, University pl, 1, Voronezh, 394018, RU

E-mail: Elena_RS@ibbox.ru

Abstract. The paper is devoted to the determination of the stress-strain state of a mechanical structure, which consists of a thin infinite elastoplastic plate with a hole and a continuous fine elastic inclusion. The complexity of this problem lies in the fact that the shape of the boundary between the elastic and plastic zones in the plate is not known in advance. The small parameter method is used as the solution method, while the small parameter characterizes the deviation of the shape of the contour from the circle and the perturbation of external static boundary conditions. As the zero solution, the axisymmetric elastoplastic state of the plate with inclusion is chosen. Two variants of inclusion fixation in a plate are considered: inclusion is enclosed with tension or soldered. As a result of solving the problem within the framework of ideal plasticity, the distribution of the stress and displacement fields and the shape of the elastoplastic boundary are obtained. To illustrate the case of a plane-stressed state, the results of a numerical experiment on the mathematical model obtained are presented.

1. Introduction

Stressed and deformed states in planar elastoplastic problems have been studied in a number of papers by various authors [1–13, 16–18, 20–23]. In this paper within the framework of a plane stress state the distribution of the stress and displacement fields within a thin plate with an hole close in shape to the regular n-gon is determined by the small parameter method. In this case, a slightly larger continuous inclusion with a similarly shaped contour is pressed into the hole with tension or soldered. The solution in the plastic region is determined according to the condition of plasticity of Tresca-Saint-Venant [6].

At infinity the plate is stretched by mutually perpendicular forces with intensities P_1 P_2. It is assumed that the loading scheme is such that the inclusion is in the resilient state, and a plastic deformation zone of the material appears within the plate, which completely encloses the contour of the hole in the plate.

The curve of the boundary between the plate and the inclusion is in the form of a regular smooth polygon. The number of smooth angles is denoted by n, when $n = 2$ we have the elliptical shape of the hole in the plate and in the outer boundary of the inclusion.

The solution is in cylindrical coordinates (ρ, θ, z), while the axis $0z$ is perpendicular to the plane of the plate. We select the origin in the center of the hole.

IOP Conf. Series: Journal of Physics: Conf. Series **973** (2018) 012010 doi:10.1088/1742-6596/973/1/012010

2. Problem solution

We introduce a small parameter characterizing the deviation of the contour of the hole from the circle, as well as the perturbation of the static boundary conditions. According to what has been said, the solution of the problem is sought in the form

$$\sigma_\rho = \sigma_\rho^{(0)} + \delta\sigma_\rho^{(1)},$$
$$\sigma_\theta = \sigma_\theta^{(0)} + \delta\sigma_\theta^{(1)},$$
$$\tau_{\rho\theta} = \tau_{\rho\theta}^{(0)} + \delta\tau_{\rho\theta}^{(1)},$$
$$\sigma_z = 0, \tag{1}$$
$$u_\rho = u_\rho^{(0)} + \delta u_\rho^{(1)},$$
$$u_\theta = u_\theta^{(0)} + \delta u_\theta^{(1)},$$
$$r_s = 1 + \delta r_s^{(1)},$$

where the superscript indicates the approximation number, δ is a small parameter, σ_ρ, σ_θ, σ_z, $\tau_{\rho\theta}$, u_ρ, u_θ are the components of the stress tensor and the displacement vector, r_s is the radius of the elastoplastic boundary in the plate.

In the plane perpendicular to the axis $0z$, according to [14, 19], we write the equation of the contour that bounds the hole in the plate before deformation

$$\rho = \alpha(1 + \delta d_1 \cos n\theta - \ldots), \tag{2}$$

equation of the contour that limits inclusion prior before deformation

$$\rho = \alpha_1(1 + \delta d_1 \cos n\theta - \ldots), \tag{3}$$

where $\alpha_1 > \alpha$, α, α_1 — are the radii of the circular contours in the zeroth approximation, respectively: the holes in the plate, the outer outline of the inclusion, δ — the small parameter, d_1 — the dimensionless constants.

In view of the smallness of the quantity ε, we take the external boundary of the inclusion [14, 19] beyond the contact line of the plate and the inclusion, which, in decomposition, is represented in the form

$$\rho_{kon} = R^{(0)} + \delta R^{(1)}, \tag{4}$$

where $R^{(0)} = \alpha_1$, $R^{(1)} = \alpha_1 d_1 \cos n\theta$.

The calculation is carried out with values that has been written in a dimensionless form. So the quantities having the dimensionality of the stresses are referred to $2k$ — doubled value of yield strength for the shear of the plate material. The displacements and geometric characteristics are related to the radius of the elastoplastic boundary in the plate r_{s0} in the zero approximation. To denote dimensionless quantities, we use their previous notation.

IOP Conf. Series: Journal of Physics: Conf. Series **973** (2018) 012010 doi:10.1088/1742-6596/973/1/012010

3. Zero approximation

For the zero approximation, we choose the axisymmetric state of a thin plate with a circular hole of radius α, and a thin resilient continuous inclusion with a circular contour of radius α_1. At infinity this construction is stretched by mutually perpendicular forces with intensities $P = (P_1 + P_2)/4k$.

We present the results of solving the problem in the zero approximation separately for the plate and the inclusion.

3.1. The solution in the zero approximation within the plate

Following [6], for the zero approximation, we have the solution in the resilient region of the plate

$$\sigma_\rho^{e(0)} = P - \frac{(q+1)\alpha}{2\rho^2},$$

$$\sigma_\theta^{e(0)} = P + \frac{(q+1)\alpha}{2\rho^2},$$

$$\tau_{\rho\theta}^{e(0)} = 0, \tag{5}$$

$$u_\rho^{e(0)} = \frac{1}{E_1}\left[(2-(q+1)\alpha)\rho + \frac{3(q+1)\alpha}{\rho}\right],$$

$$u_\theta^{e(0)} = 0.$$

Within the plastic zone of the plate we have

$$\sigma_\rho^{p(0)} = 1 - \frac{(q+1)\alpha}{\rho},$$

$$\sigma_\theta^{p(0)} = 1,$$

$$\tau_{\rho\theta}^{p(0)} = 0, \tag{6}$$

$$u_\rho^{p(0)} = \frac{1}{2E_1}\left(\rho + (q+1)\alpha(1 - 2\ln\rho)\right),$$

$$u_\theta^{p(0)} = 0,$$

where E_1 is the Young's modulus of the plate material, $q = q/2k$ is the normal pressure at the interface between the plate and the inclusion.

3.2. The solution in the zero approximation in the inclusion

In the elastic inclusion, in the zero approximation, the distribution of the stress and displacement field has the form [15]

$$\sigma_{B\rho}^{e(0)} = -q,$$

$$\sigma_{B\theta}^{e(0)} = -q,$$

$$\tau_{B\rho\theta}^{e(0)} = 0, \tag{7}$$

$$u_{B\rho}^{e(0)} = -\frac{qk}{E_2}\rho,$$

$$u_{B\theta}^{e(0)} = 0,$$

where E_2 is the Young's modulus of the inclusion material.

3.3. Determination of the contact and radius of the elastoplastic boundary in the plate in the zero approximation

From the conditions for the compatibility of deformations of the plate and the inclusion along the contact line, it follows that

$$u_{\rho=\alpha}^{p(0)} = u_{B\rho=\alpha_1}^{e(0)} + \varepsilon, \text{ where } \varepsilon = \alpha_1 - \alpha \tag{8}$$

and from the interfacing conditions on the elastoplastic boundary in the plate

$$\sigma_\theta^{p(0)} = \sigma_\theta^{e(0)}, \quad \rho = 1 \tag{9}$$

we have the following system of equations for determining the contact pressure q and the radius of the elastoplastic boundary in the plate in the zero approximation r_{s0}

$$r_{s0}\left(1 - 2\ln\alpha + 2\ln r_{s0} + \frac{\alpha_1 E_1}{\alpha E_2}\right) + \frac{\alpha k E_2 - \alpha_1 k E_1 - E_1 E_2 (\alpha_1 - \alpha)}{2kE_2\left(1 - \frac{P}{2k}\right)} = 0,$$

$$q = \frac{2}{\alpha}\left(1 - \frac{P}{2k}\right) - 1. \tag{10}$$

4. First approximation

We now turn to the definition of the first approximation.

The boundary conditions at infinity have the form [14, 19]

$$\sigma_\rho^\infty = P - \delta d_2 \cos 2\theta,$$

$$\tau_{\rho\theta}^\infty = \delta d_2 \sin 2\theta, \tag{11}$$

where $P = \frac{P_1+P_2}{4k}$, $\delta d_2 = \frac{P_1-P_2}{4k}$, where d_2 is a dimensionless constant.

Linearized interfacing conditions on the elastoplastic boundary in the plate can be presented

$$\left.\left[\sigma_{ij}^{(1)} + \frac{d\sigma_{ij}^{(0)}}{d\rho} r_s^{(1)}\right]\right|_{\rho=1} = 0, \tag{12}$$

where $\sigma_{ij}^{(0)}$, $\sigma_{ij}^{(1)}$ are the components of the stress tensor for the zero and first approximations, $r_s^{(1)}$ is the ratio describing the contour of the contact boundary in the first approximation.

Along the contact line of the plate and the inclusion, according to [6, 15], we write

1) inclusion is enclosed with tension in the plate

$$\sigma_\rho^{p(1)} + \frac{d\sigma_\rho^{p(0)}}{d\rho} R^{(1)} = \sigma_{B\rho}^{e(1)} + \frac{d\sigma_{B\rho}^{e(0)}}{d\rho} R^{(1)},$$

$$\tau_{B\rho\theta}^{e(1)} - \left(\sigma_{B\theta}^{e(0)} - \sigma_{B\rho}^{e(0)}\right) \dot{s}_1 = 0,$$

$$\tau_{\rho\theta}^{p(1)} - \left(\sigma_{\theta}^{p(0)} - \sigma_{\rho}^{p(0)}\right)\dot{s_1} = 0, \tag{13}$$

$$u_{\rho}^{p(1)} + \frac{du_{\rho}^{p(0)}}{d\rho}R^{(1)} = u_{B\rho}^{e(1)} + \frac{du_{B\rho}^{e(0)}}{d\rho}R^{(1)}, \text{ with } \rho = R^{(0)}$$

2) the inclusion is soldered into the plate

$$\sigma_{\rho}^{p(1)} + \frac{d\sigma_{\rho}^{p(0)}}{d\rho}R^{(1)} = \sigma_{B\rho}^{e(1)} + \frac{d\sigma_{B\rho}^{e(0)}}{d\rho}R^{(1)},$$

$$\tau_{B\rho\theta}^{e(1)} - \left(\sigma_{B\theta}^{e(0)} - \sigma_{B\rho}^{e(0)}\right)\dot{s_1} = \tau_{\rho\theta}^{p(1)} - \left(\sigma_{\theta}^{p(0)} - \sigma_{\rho}^{p(0)}\right)\dot{s_1},$$

$$u_{\rho}^{p(1)} + \frac{du_{\rho}^{p(0)}}{d\rho}R^{(1)} = u_{B\rho}^{e(1)} + \frac{du_{B\rho}^{e(0)}}{d\rho}R^{(1)}, \tag{14}$$

$$u_{\theta}^{p(1)} + u_{\rho}^{p(0)}\dot{s_1} = u_{B\theta}^{e(1)} + u_{B\theta}^{e(0)}\dot{s_1}, \text{ when } \rho = R^{(0)},$$

where $R^{(1)} = \alpha_1 d_1 \cos n\theta$, $s_1 = R^{(1)}/R^{(0)} \Rightarrow \dot{s_1} = -nd_1 \sin n\theta$.

We give the solution of the problem in the first approximation for the resilient and plastic zones of the plate and for a resilient inclusion.

4.1. The solution with the first approximation within the plate

According to [6], taking into account the boundary conditions (11), we have the solution with the first approximation in the resilient region the plates for $\infty > \rho > 1$

$$\sigma_{\rho}^{e(1)} = \left[\left(1 - \frac{4}{\rho^2} + \frac{3}{\rho^4}\right)d_2 + \left(-\frac{1}{\rho^4} + \frac{2}{\rho^2}\right)a_{21} + \left(\frac{2}{\rho^4} - \frac{2}{\rho^2}\right)a_{22}\right]\cos 2\theta +$$
$$+\frac{1}{2}\left[\left(-\frac{n}{\rho^{n+2}} + \frac{n+2}{\rho^n}\right)a_{n1} + \left(\frac{n+2}{\rho^{n+2}} - \frac{n+2}{\rho^n}\right)a_{n2}\right]\cos n\theta,$$

$$\sigma_{\theta}^{e(1)} = \left[-\left(1 + \frac{3}{\rho^4}\right)d_2 + \frac{1}{\rho^4}a_{21} - \frac{2}{\rho^4}a_{22}\right]\cos 2\theta +$$
$$+\frac{1}{2}\left[\left(\frac{n}{\rho^{n+2}} - \frac{n-2}{\rho^n}\right)a_{n1} + \left(-\frac{n+2}{\rho^{n+2}} + \frac{n-2}{\rho^n}\right)a_{n2}\right]\cos n\theta,$$

$$\tau_{\rho\theta}^{e(1)} = \left[\left(-1 - \frac{2}{\rho^2} + \frac{3}{\rho^4}\right)d_2 + \left(-\frac{1}{\rho^4} + \frac{1}{\rho^2}\right)a_{21} + \left(\frac{2}{\rho^4} - \frac{1}{\rho^2}\right)a_{22}\right]\sin 2\theta +$$
$$+\frac{1}{2}\left[\left(-\frac{n}{\rho^{n+2}} + \frac{n}{\rho^n}\right)a_{n1} + \left(-\frac{n+2}{\rho^{n+2}} - \frac{n}{\rho^n}\right)a_{n2}\right]\sin n\theta, \tag{15}$$

$$u_{\rho}^{e(1)} = \frac{\alpha}{E_1}\left[\left(\frac{3}{2}\left(\rho - \frac{1}{\rho^3}\right) + \frac{4}{\rho}\right)d_2 + \left(\frac{1}{2\rho^3} - \frac{2}{\rho}\right)a_{21} + \left(-\frac{1}{\rho^3} + \frac{2}{\rho}\right)a_{22}\right]\cos 2\theta +$$
$$+\frac{\alpha}{E_1}\left[\left(\frac{3n}{4(n+1)}\frac{1}{\rho^{n+1}} - \frac{3n+2}{4(n-1)}\frac{1}{\rho^{n-1}}\right)a_{n1} +\right.$$
$$\left. + \left(-\frac{3(n+2)}{4(n+1)}\frac{1}{\rho^{n+1}} + \frac{3n+2}{4(n-1)}\frac{1}{\rho^{n-1}}\right)a_{n2}\right]\cos n\theta,$$

$$u_\theta^{e(1)} = \frac{\alpha}{E_1}\left[\left(-\frac{3}{2}\left(\rho+\frac{1}{\rho^3}\right)+\frac{1}{\rho}\right)d_2+\frac{1}{2}\left(\frac{1}{\rho^3}+\frac{1}{\rho}\right)a_{21}+\left(-\frac{1}{\rho^3}-\frac{1}{2\rho}\right)a_{22}\right]\sin 2\theta +$$

$$+\frac{\alpha}{E_1}\left[\left(\frac{3n}{4(n+1)}\frac{1}{\rho^{n+1}}-\frac{3n-8}{4(n-1)}\frac{1}{\rho^{n-1}}\right)a_{n1}+\right.$$

$$\left.+\left(-\frac{3(n+2)}{4(n-1)}\frac{1}{\rho^{n+1}}+\frac{3n-8}{4(n-1)}\frac{1}{\rho^{n-1}}\right)a_{n2}\right]\sin n\theta,$$

where a_{21}, a_{22}, a_{n1}, a_{n2} are constants determined from the interfacing conditions (13) or (14) at the contact boundary between the plate and the inclusion.

In the plastic region of the plate, according to [6, 19], we derive for stresses and displacements

$$\sigma_\rho^{p(1)} = \left[\frac{a_{21}-2a_{22}}{\rho}+\frac{2a_{22}}{\rho^2}\right]\cos 2\theta + \left[\frac{a_{n1}+na_{n2}}{\rho}-\frac{na_{n2}}{\rho^2}\right]\cos n\theta,$$

$$\sigma_\theta^{p(1)} = 0,$$

$$\tau_{\rho\theta}^{p(1)} = \frac{2a_{22}}{\rho^2}\sin 2\theta - \frac{na_{n2}}{\rho^2}\sin n\theta, \tag{16}$$

$$u_\rho^{p(1)} = \left[\frac{a_{21}-2a_{22}}{E_1}\ln\rho+\frac{a_{21}-2a_{22}}{E_1}-\frac{2a_{22}}{\rho E_1}+\right.$$

$$\left.+4d_2\alpha - a_{21}\left(\frac{3}{2}\alpha+1\right)+a_{22}(\alpha+4)\right]\cos 2\theta+$$

$$+\left[\frac{a_{n1}+na_{n2}}{E_1}\ln\rho+\frac{a_{n1}+na_{n2}}{E_1}+\frac{na_{n2}}{\rho E_1}-a_{n1}\left(\frac{4n+1}{2(n^2-1)}\alpha+1\right)+\right.$$

$$\left.+a_{n2}\left(\frac{3n^2+4n-2}{2(n^2-1)}\alpha-2n\right)\right]\cos n\theta,$$

$$u_\theta^{p(1)} = \left[\frac{2a_{22}}{E_1}\ln\rho+\frac{5a_{22}}{2E_1}-2\left(4d_2\alpha-a_{21}\left(\frac{3}{2}\alpha+1\right)+a_{22}(\alpha+4)\right)+\right.$$

$$\left.+\rho\left(6d_2\alpha-2a_{21}(\alpha+1)+a_{22}\left(\frac{1}{2}\alpha+7\right)\right)\right]\sin 2\theta+$$

$$+\left[\frac{a_{n2}}{E_1}\ln\rho-\frac{a_{n2}}{E_1}+2a_{n1}\left(\frac{4n+1}{2(n^2-1)}\alpha+1\right)-a_{n2}\left(\frac{3n^2+4n-2}{(n^2-1)}\alpha-n\right)+\right.$$

$$\left.+\rho\left(-a_{n1}\left(\frac{3n+3}{2(n^2-1)}\alpha+1\right)+3a_{n2}\left(\frac{n^2-n-3}{2(n^2-1)}\alpha-n\right)\right)\right]\sin n\theta,$$

4.2. The solution with the first approximation within the inclusion

According to [8], we obtain the form of the components of the stress tensor and the displacement vector in the first approximation for the resilient inclusion

$$\sigma_{B\rho}^{e(1)} = -2C_1\cos 2\theta + \left(-n(n-1)C_{1n}\rho^{n-2} - (n-2)(n+1)C_{2n}\rho^n\right)\cos n\theta,$$

$$\sigma_{B\theta}^{e(1)} = 2\left(C_1 + 6C_2\rho^2\right)\cos 2\theta + \left(n(n-1)C_{1n}\rho^{n-2} + (n+1)(n+2)C_{2n}\rho^n\right)\cos n\theta,$$

$$\tau_{B\rho\theta}^{e(1)} = 2\left(C_1 + 3C_2\rho^2\right)\sin 2\theta + \left(n(n-1)C_{1n}\rho^{n-2} + n(n+1)C_{2n}\rho^n\right)\sin n\theta, \qquad (17)$$

$$Eu_{B\rho}^{e(1)} = 2\left(-3C_1\rho + 4C_2\rho^3\right)\cos 2\theta + \left(-3nC_{1n}\rho^{n-1} - (3n+2)C_{2n}\rho^{n+1}\right)\cos n\theta,$$

$$Eu_{B\theta}^{e(1)} = 2\left(3C_1\rho + 5C_2\rho^3\right)\sin 2\theta + \left(3nC_{1n}\rho^{n-1} + (3n+4)C_{2n}\rho^{n+1}\right)\sin n\theta,$$

where C_1, C_2, C_{1n}, C_{2n} are constants determined from the ratios on the interface of the plate and the inclusion.

4.3. Systems of equations for unknown constants determination

The ratio (13) at the contact boundary, in the case when the inclusion is inserted with tension has the form

$$\frac{a_{21}}{\alpha_1} + 2a_{22}\left(\frac{1}{\alpha_1^2} - \frac{1}{\alpha_1}\right) + 2C_1 + a_{n1}\frac{1}{\alpha_1} + a_{n2}\left(\frac{n}{\alpha_1} - \frac{n}{\alpha_1^2}\right) + n(n-1)C_{1n}\alpha_1^{n-2} +$$

$$+(n-2)(n+1)C_{2n}\alpha_1^n = -\frac{(q+1)\alpha d_1}{\alpha_1},$$

$$2C_1 + 6C_2\alpha_1^2 + n(n-1)C_{1n}\alpha_1^{n-2} + n(n+1)C_{2n}\alpha_1^n = 0,$$

$$\frac{2a_{22}}{\alpha_1^2} + \frac{na_{n2}}{\alpha_1^2} = \frac{(q+1)\alpha n d_1}{\alpha_1}, \qquad (18)$$

$$a_{21}\left(\frac{\ln\alpha_1}{E_1} + \frac{1}{E_1} - \left(\frac{3}{2}\alpha + 1\right)\right) + a_{22}\left(-\frac{2\ln\alpha_1}{E_1} - \frac{2}{E_1} - \frac{2}{\alpha_1 E_1} + (\alpha+4)\right) +$$

$$+a_{n1}\left(\frac{\ln\alpha_1}{E_1} + \frac{1}{E_1} - \frac{4n+1}{2(n^2-1)}\alpha + 1\right) + a_{n2}\left(\frac{n\ln\alpha_1}{E_1} + \frac{n}{E_1} + \frac{n}{\alpha_1 E_1} + \frac{3n^2+4n-2}{2(n^2-1)}\alpha - 2n\right) +$$

$$+C_1\frac{6\alpha_1}{E_2} - C_2\frac{8\alpha_1^3}{E_2} + 3nC_{1n}\alpha_1^{n-1} + (3n+2)C_{2n}\alpha_1^{n+1} =$$

$$= -4d_2\alpha - \frac{1}{2E_1}\left(\alpha_1 - 2(q+1)\alpha\right)d_1 - \frac{qk\alpha_1 d_1}{E_2}.$$

For the soldered inclusion case (14), the system for determining unknown constants has the form

$$\frac{a_{21}}{\alpha_1} + 2a_{22}\left(\frac{1}{\alpha_1^2} - \frac{1}{\alpha_1}\right) + 2C_1 + a_{n1}\frac{1}{\alpha_1} + a_{n2}\left(\frac{n}{\alpha_1} - \frac{n}{\alpha_1^2}\right) + n(n-1)C_{1n}\alpha_1^{n-2} +$$

IOP Conf. Series: Journal of Physics: Conf. Series **973** (2018) 012010 doi:10.1088/1742-6596/973/1/012010

$$+(n-2)(n+1)C_{2n}\alpha_1^n = -\frac{(q+1)\alpha d_1}{\alpha_1},$$

$$-\frac{2a_{22}}{\alpha_1^2} + 2C_1 + 6C_2\alpha_1^2 - \frac{na_{n2}}{\alpha_1^2} + n(n-1)C_{1n}\alpha_1^{n-2} + n(n+1)C_{2n}\alpha_1^n = \frac{(q+1)\alpha n d_1}{\alpha_1},$$

$$a_{21}\left(\frac{\ln\alpha_1}{E_1} + \frac{1}{E_1} - \left(\frac{3}{2}\alpha + 1\right)\right) + a_{22}\left(-\frac{2\ln\alpha_1}{E_1} - \frac{2}{E_1} - \frac{2}{\alpha_1 E_1} + (\alpha+4)\right) +$$
$$+a_{n1}\left(\frac{\ln\alpha_1}{E_1} + \frac{1}{E_1} - \frac{4n+1}{2(n^2-1)}\alpha + 1\right) +$$
$$+a_{n2}\left(\frac{n\ln\alpha_1}{E_1} + \frac{n}{E_1} + \frac{n}{\alpha_1 E_1} + \frac{3n^2+4n-2}{2(n^2-1)}\alpha - 2n\right) +$$
$$+C_1\frac{6\alpha_1}{E_2} - C_2\frac{8\alpha_1^3}{E_2} + 3nC_{1n}\alpha_1^{n-1} + (3n+2)C_{2n}\alpha_1^{n+1} =$$
$$-4d_2\alpha - \frac{1}{2E_1}\left(\alpha_1 - 2(q+1)\alpha\right)d_1 - \frac{qk\alpha_1 d_1}{E_2}, \tag{19}$$

$$a_{21}\left(3\alpha + 2 - 2\alpha_1(\alpha+1)\right) + a_{22}\left(\frac{2\ln\alpha_1}{E_1} + \frac{5}{2E_1} - 2\alpha - 8 + \alpha_1\left(\frac{1}{2}\alpha + 7\right)\right) +$$
$$+a_{n1}\left(\frac{4n+1}{n^2-1}\alpha + 2 - \alpha_1\left(\frac{3n+3}{2(n^2-1)}\alpha + 1\right)\right) +$$
$$+a_{n2}\left(\frac{\ln\alpha_1 - 1}{E1} - \frac{3n^2+4n-2}{n^2-1}\alpha - n + \alpha_1\left(\frac{3n^2-3n-9}{2(n^2-1)}\alpha - 3n\right)\right) -$$
$$-6C_1\alpha_1 - 10C_2\alpha_1^3 - 3nC_{1n}\alpha_1^{n-1} - (3n+4)C_{2n}\alpha_1^{n-1} =$$
$$= 8d_2\alpha - 6d_2\alpha\alpha_1 + \left(1 - \frac{(q+1)\alpha}{\alpha_1} + \frac{qk\alpha_1}{E_2}\right)nd_1.$$

To determine the type of constants a_{21}, a_{22}, a_{n1}, a_{n2}, C_1, C_2, C_{1n}, C_{2n} we use the relations at the interface between the plate and the inclusion (18) or (19). Thus, the obtained relations (15) – (19) completely determine the stressed and strained states in the plate and the inclusion for the first approximation.

4.4. Finding the radius of the elastoplastic boundary within the plate

The form of the elastoplastic boundary of the first approximation $r_s^{(1)}$ is determined by the linearized condition [6, 14]

$$r_s^{(1)} = -\left[\sigma_\theta^{(1)}\right]\left[\frac{\partial\sigma_\theta^{(0)}}{\partial\rho}\right]^{-1}\Bigg|_{\rho=1}. \tag{20}$$

Using (15) and (16), expression (20) gives the radius of the elastoplastic boundary in the plate, which in this case has the form

$$r_s^{(1)} = \frac{(4d_2 + a_{21} - 2a_{22})}{(q+1)\alpha}\cos 2\theta + \frac{a_{n1} + 2a_{n2}}{(q+1)\alpha}\cos n\theta, \tag{21}$$

where q is the contact pressure at the interface between the plate and the connection.

IOP Conf. Series: Journal of Physics: Conf. Series **973** (2018) 012010 doi:10.1088/1742-6596/973/1/012010

5. Results

Let $\delta = 0.01$, $\alpha = 0.2$ m, $\alpha_1 = 0.201$ m, $E_1 = 810$ MH/m^2, $E_2 = 1100$ MH/m^2, $d_1 = 2.79$, $d_2 = 2.9$, $k = 12/\sqrt{3}$ MH/m^2,$P_1 = 9.5$ MH/m^2, $P_2 = 9.0$ MH/m^2, $n = 6$.

In figure 1, curve 1 reflects the dependence r_s on the angle θ, i.e. curve 1 is the shape of the elastoplastic boundary in the plate. The contour 2 corresponds to the contour of the hole in the plate.

The condition on the contact boundary for the pattern corresponds to the condition that the inclusion is enclosed with tension into the plate (13).

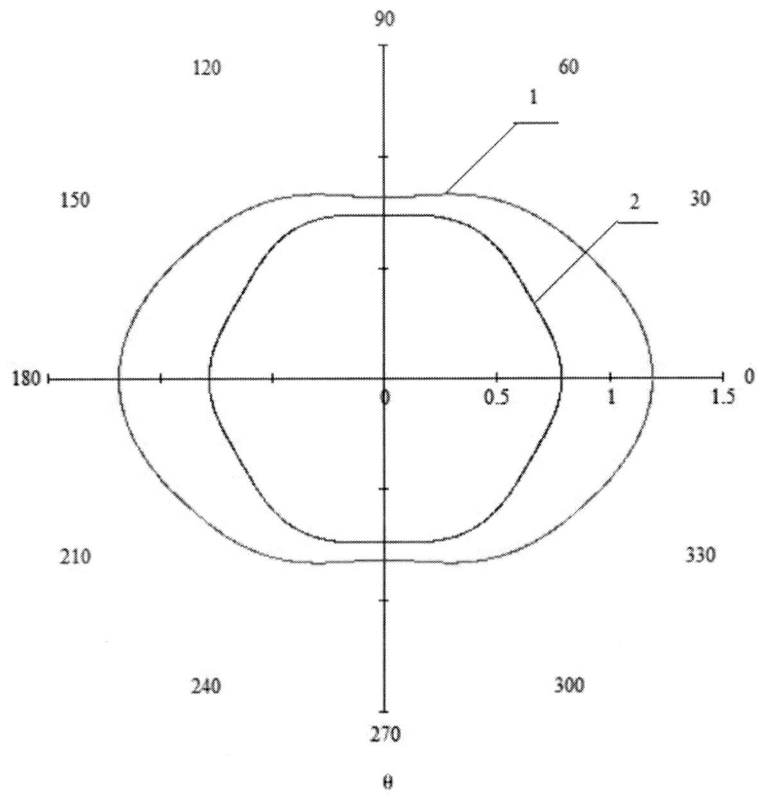

Figure 1. Form of elastoplastic boundary

6. Conclusion

In the paper of the stress and displacement has been in a thin elastoplastic plate and in elastic inclusion. The radius of the elastoplastic boundary in the plate is found. An example is considered and given when the inclusion is enclosed with tension in the plate.

7. References

[1] Artemov M A 1985 On the biaxial stretching of a thick plate with a circular hole of a reinforced elastoplastic material *AMTF.* **6** 158–63

[2] Barber J R 2008 Shakedown in frictional contact problems for the continuum *C R Mecanique* vol 336 **1–2** 34–41

[3] Biceno K B and Grammel R 1952 *Technical Dynamics. Trans. from Ger. Y I Perelman and E P Popova* vol 2 ed A I Lurie (Moscow: Leningrad)

[4] Dollar A 1983 Influence of metal inhomogeneity from the shape of non-circular sections of thick-walled cylinders in the state of complete plasticity and the stage of destruction *Rozpz. Inz.* vol 31 **2** 241–57
[5] Gornostaev K K and Kovalev A V 2015 On the elastoplastic state of a thick-walled tube with temperature for a complex environment model *Vestnik of the Voronezh State University. Series: Phys. Math.* **1** 135–40
[6] Ivlev D D and L V Ershov 1978 *Perturbation method in the theory of an elastoplastic body* (Moscow: Science) p 208
[7] Johnson K L 1985 *Contact mechanics* (Cambridge University Press) p 452
[8] Kovalev A V 2013 On the account of the associated compressibility of elastoplastic bodies in the case of plane deformation *Vestnik of the Chuvash State Pedagogical University name I. Yakovleva. Series: The Mechanics of the Ultimate State* **1(15)** 65–9
[9] Kovalev A V and Khvostov I G 2014 On the determination of stresses and displacements in elastic space weakened by a spherical cavity, taking temperature into account *Vestnik Chelyabinsk State Pedagogical University name I Y Yakovleva. Series: The Mechanics of the Ultimate State* **2(20)** 29–35
[10] Kovalev A V, Rusina E Y and Yakovlev A Y 2015 On the study of the mechanical interaction of elements of a thin elastoplastic design *In the collection: The Mechanics of the Ultimate State and Related Questions. Proceedings of the All-Russian Scientific School-Conference, dedicated to the 85th anniversary of Professor D D Ivlev* 107–14
[11] Kovalev A V, Sporykhin A N and Yakovlev A Y 2000 Biaxial expansion of the elastoplastic space with prismatic inclusions *Applied Mechanics* vol 36 **6** 114–20
[12] Kovalev A V, Sporykhin A N and Yakovlev A Y 2004 Investigation of the mechanical interaction of elements of various shapes and properties in a system of compacted bodies *In the collection: Modern Problems of Mechanics and Applied Mathematics, a collection of works of the international school-seminar, ed A D Chernyshov* 280–1
[13] Mackenzie D, Boyle J T and Hamilton R 2000 The elastic compensation method for limit and shakedown analysis: a review *The Journal of Strain Analysis for Engineering Design* vol 35 **3** 171–88
[14] Maroushkei Y M 1975 Biaxial stretching of an elastoplastic space with inclusion *Izv. Universities. Mechanical engineering* **12** 25–30
[15] Muskhelishvili N I 1966 *Some basic problems of the mathematical theory of elasticity* (Moscow: Science) p 707
[16] Spenser A J M 1961 Perturbation methods in plasticity. I. Plane strain of non-homogeneity plastic solids *J. Mech. and Phys. Solid* vol 9 **4** 279–88
[17] Spenser A J M 1962 Perturbation methods in plasticity. II. Plane strain of slightly irregular bodies *J.Mech. and Phys. Solid* vol 10 **1** 17–26
[18] Spenser A J M 1962 Perturbation methods in plasticity. III. Plane strain solids with body forces *J.Mech. and Phys. Solid* vol 10 **1** 165–77
[19] Sporykhin A N, Kovalev A V and Shcheglova Y D 2004 *Nonuniform problems of elastic-viscoplasticity with an unknown boundary* (Voronezh: Izd-vo VSU) p 219
[20] Stepanova L V and Yakovleva E M 2016 *Stress-strain state near the crack tip under mixed-mode loafing: Asymptotic approach and numerical solutions of nonlinear eigenvalue problems* (AIP Conference Proceedings) **1785** 30030
[21] Yakovlev A Y 2014 On the influence of the shape of the external contour of inclusion on the stress-strain state of the elastoplastic space *In the collection: Prospects for the development of science and education, a collection of scientific papers on the materials of the International Scientific and Practical Conf. Tambov* 160–2
[22] Zadorozhny V G, Kovalev A V and Sporykhin A N 2008 On the analyticity of the solution of a plane elastoplastic problem *Izvestiya Rossiiskoi Akademii Nauk. Mechanics of a solid body* **1** 138–46
[23] Zimmermann M, Klemenz M and Schulze V 2010 Literature review on shot peening simulation *Int/ J. og Computational Materials Science and Surface Engineering* vol 3 **4** 289–310

IOP Conf. Series: Journal of Physics: Conf. Series **973** (2018) 012011 doi:10.1088/1742-6596/973/1/012011

Determination of the stress state of anisotropic hollow bar with eccentricity for the elastoplastic torsion problem

A V Kovalev, I E Sviridov and Ju D Scheglova

Department of Mechanics and Computer Simulation, Voronezh State University, 394006, Voronezh, University square, 1, Russia

E-mail: synettt@gmail.com

Abstract. The article is devoted to the problem of elastoplastic torsion of anisotropic hollow bar with eccentricity. The influence of mass forces is not taken into account. The bar is twisted about the symmetry axis by equal and opposite pairs of forces. The value of the moment has such a value that in some parts of the cross section the material passes into a plastic state and plastic zones are formed. The propagation of plastic flow comes from the outer contour to the inside. It's assumed that the value of the moment is such that the plastic region entirely covers the outer contour of the cross section and an elastoplastic boundary is formed. It is assumed that the bar material is anisotropic and in special cases there are the kinematic properties of the anisotropy and anisotropy according to Hill. With using perturbation method, stress state and elastoplastic boundary at first approximate is defined.

1. Introduction

The work [1] uses wellknown formulation of the torsion problem in complex variable theory and used to obtain an expression for shear stresses, angle of twist and cross-sectional warping. The obtained mapping is used to achieve a solution for torsion problem of bars with specially shaped hollow cross-section. Calculated results are compared with the finite-element method solution. By using the meshless regularized integral equation method (MRIEM), the solution of elastic torsion problem of a uniform bar with arbitrary cross-section is presented by the first kind Fredholm integral equation on an artificial circle, which just encloses the bar's cross-section [2]. The paper considers [3] a parabolic partial differential equation involving the infinity Laplace operator and a Leray-Lions operator with no coercitive assumption. The existence and uniqueness of the corresponding approached problem is proved and shows that the limit the solution solves the parabolic variational inequality arising in the elastoplastic torsion problem. For the first time, the problem of determining the stress-strain state of an eccentric hollow bar under the influence of internal pressure for an elastic perfectly plastic material was solved in [4] with using perturbation method. In the work [5] with application of the perturbation method, a stress state is determined in a cylindrical hollow bar subjected to external and internal pressures, with boundaries of the cross section close to circular. The question of the existence and uniqueness of the solution of the problem is considered on the basis of the implicit function theorem. The work [6] is devoted to the investigation of the stress-strain state of a hollow bar of a non-circular cross section.

2. Model

In the presented work with application of the perturbation method [4], [7] the stress state and elastoplastic boundary is determined for elastoplastic torsion [8] of hollow bar with eccentricity including anisotropy properties of material.

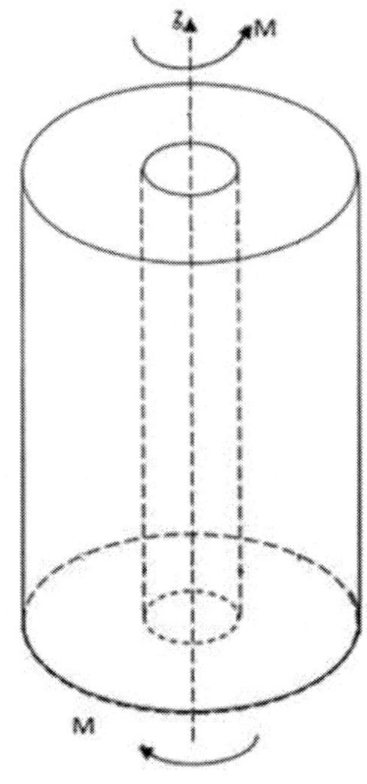

Figure 1. cylindrical hollow bar with small eccentricity.

We consider a cylindrical hollow bar (figure 1). a and b radius of outer and inner circular cylinders respectively. Suppose that the lateral surface of the bar is free of loads. Eccentricity is equal e. It is assumed that the plastic zone covers completely the outer contour of the pipe. The case of small eccentricity is considered. The solution uses a cylindrical coordinate system. The origin of coordinate system is aligned with the center of the inner contour. In this case, the equation of the outer contour of the hollow bar has the form

$$(x-e)^2+y^2=b^2.$$

Passing to the polar coordinate system, we have

$$\rho^2-2\rho\delta\cos\theta+\delta^2-\beta^2=0, \qquad (1)$$

where $\rho=\rho/r_s$, $\delta=e/r_s$, $\beta=b/r_s$, where r_s radius of the elastoplastic boundary in the zero approximation. From equation (1) we find

$$\rho=\delta\cos\theta+\sqrt{\beta^2-\delta^2\sin^2\theta}.$$

Expanding this expression in powers of the parameter δ for the case of small eccentricity $\delta<<1$, we obtain the equation of the outer contour linearized with respect to the small parameter.

$$\rho=\beta+\delta\cos\theta-\frac{\delta^2}{2\beta}\sin^2\theta+\dots. \qquad (2)$$

IOP Conf. Series: Journal of Physics: Conf. Series **973** (2018) 012011 doi:10.1088/1742-6596/973/1/012011

The yield criteria for the torsion problem can be written in the form [9]

$$A(\tau_{xz}-k_1)^2+B(\tau_{yz}-k_2)^2=k_0^2 \tag{3}$$

where A, B, k_1, k_2 is anisotropy constant; where k_0 is yield strength.

Next, we pass to the cylindrical coordinate system ρ, θ, z, where the axis z coincides with the axis of the hollow bar, and we use dimensionless quantities. Divide quantities having the dimensionality of the stresses by the yield strength k_0 and the quantities having the dimensionality of length divide into the radius of the elastoplastic boundary in the zero approximation ρ_0.

Taking into account the formulas for the transition of stress components from Cartesian coordinates to cylindrical coordinates,

$$\tau_{xz}=\tau_{\rho z}\cos\theta-\tau_{\theta z}\sin\theta,$$

$$\tau_{yz}=\tau_{\rho z}\sin\theta+\tau_{\theta z}\cos\theta,$$

We write down the yield criterion (3) in the form.

$$A(\tau_{\rho z}\cos\theta-\tau_{\theta z}\sin\theta-k_1)^2+B(\tau_{\rho z}\sin\theta+\tau_{\theta z}\cos\theta-k_2)^2=k_0^2 \tag{4}$$

The equation of equilibrium reduces to a single differential equation

$$\frac{\partial\tau_{\rho z}}{\partial\rho}+\frac{1}{\rho}\frac{\partial\tau_{\theta z}}{\partial\theta}+\frac{\tau_{\rho z}}{\rho}=0. \tag{5}$$

In the elastic region, the Cauchy relationships

$$\varepsilon_{\rho z}^{e}=\frac{\omega}{2}\frac{\partial w^{e}}{\partial\rho},\ \varepsilon_{\theta z}^{e}=\frac{\omega}{2}\left(\frac{1}{\rho}\frac{\partial w^{e}}{\partial\theta}+\rho\right). \tag{6}$$

The Cauchy relations connecting components of the strain tensor with the function w^e. w^e characterizes the cross-sectional warping, where ω is twist.
Hooke's Law relationships

$$\tau_{\rho z}^{e}=2G\varepsilon_{\rho z}^{e},\ \tau_{\theta z}^{e}=2G\varepsilon_{\theta z}^{e}, \tag{7}$$

where G is shear modulus.

Since the side surfaces of the hollow bar are free from the load, the boundary conditions are satisfied with the outer and inner contours of the cross section

$$\left(\tau_{\rho z}^{p}n_\rho+\tau_{\theta z}^{p}n_\theta\right)\Big|_L=0, \tag{8}$$

$$\tau_{\rho z}^{e}\Big|_{\rho=\alpha}=0,$$

where L is outer contour, and n_θ, n_ρ is the components of the unit normal. The components of the unit normal are determined by the formulas

$$n_\theta=\frac{1}{\rho}\frac{\partial\Phi}{\partial\theta}\left(\left(\frac{\partial\Phi}{\partial\rho}\right)^2+\left(\frac{1}{\rho}\frac{\partial\Phi}{\partial\theta}\right)^2\right)^{-\frac{1}{2}}, \tag{9}$$

$$n_\rho=\frac{\partial\Phi}{\partial\rho}\left(\left(\frac{\partial\Phi}{\partial\rho}\right)^2+\left(\frac{1}{\rho}\frac{\partial\Phi}{\partial\theta}\right)^2\right)^{-\frac{1}{2}}.$$

where $\Phi(\rho,\theta)=0$ is equation of the outer contour of the hollow bar.
On the elastoplastic boundary the stress-continuity condition

$$[\tau_{\rho z}]_{L_s}=[\tau_{\theta z}]_{L_s}=0, \tag{10}$$

where L_S is elastoplastic boundary.

In the elastic region from the equilibrium equation (5), taking equations (6) and (7) into account, we obtain

$$\Delta w^e=0, \tag{11}$$

where $\Delta=\frac{\partial^2}{\partial\rho^2}+\frac{1}{\rho^2}\frac{\partial^2}{\partial\theta^2}+\frac{1}{\rho}\frac{\partial}{\partial\rho}$ is Laplace operator.
On the elastoplastic boundary L_s the displacement continuity condition is fulfilled.

$$[w]|_{L_s}=0. \tag{12}$$

We seek solutions in the form of the decomposition with respect to a small dimensionless parameter δ $(\delta<<1)$. The translational anisotropy and anisotropy constants according to Hill [10], stresses, cross-sectional warping and elastoplastic boundary are representable in the form

$$A=1+\delta a_1, k_1=\delta k_1^{(1)}, k_2=\delta k_2^{(1)}, B=1+\delta b_1,\ \tau_{ij}=\tau_{ij}^{(0)}+\delta\tau_{ij}^{(1)}+\ldots,$$

$$w^e=w^{e(0)}+\delta w^{e(1)}+\ldots,\ \rho_s=\rho_0+\delta\rho_1+\ldots, \tag{13}$$

$$\Phi(\rho,\theta)=\rho-\left(\beta+\delta\cos\theta-\frac{\delta^2}{2\beta}\sin^2\theta+\ldots\right), \tag{14}$$

where ρ_s is elastoplastic boundary.
Substitute decomposition (13) into relationships (4) – (12) and equate the terms at the same powers. We obtain the system of equations for each approximation.

In the zero approximation, the known problem of elastoplastic torsion of a homogeneous hollow rod with a ring cross-section takes place. The solution of this problem has the form

$$\tau_{\theta z}^{p(0)}=1,\ \tau_{\rho z}^{p(0)}=0,\ w^{e(0)}=0,\ \tau_{\theta z}^{e(0)}=G\omega\rho,\ \tau_{\rho z}^{e(0)}=0,\ \rho_0=1. \tag{15}$$

Consider only the solution of the problem in the first approximation. Since the relationships (5) and (7) are linear in δ, they preserve their form for each approximation.

Substituting decomposition (13) into equation (4) in the plastic region, we obtain the yield criteria in the first approximation and express from it the stress component $\tau_{\theta z}^{p(1)}$ as a coordinate function θ containing the anisotropy constants. This expression takes the form

$$\tau_{\theta z}^{p(1)}=k_2^{(1)}\cos\theta-k_1^{(1)}\sin\theta-\frac{1}{2}(a_1^{(1)}\sin^2\theta+b_1^{(1)}\cos^2\theta). \tag{16}$$

To determine the stress component $\tau_{\rho z}^{p(1)}$, we obtain a system by substituting solution (15) into the equilibrium equation (5) and the boundary condition on the outer contour (8), taking into account the decomposition (13)

$$\begin{cases} \dfrac{\partial}{\partial\rho}\left(\rho\,\tau_{\rho z}^{p(1)}\right) = -\dfrac{\partial\tau_{\theta z}^{p(1)}}{\partial\theta}, \\ \tau_{\rho z}^{p(1)}\Big|_{\rho=\beta} = -\dfrac{\sin\theta}{\beta}. \end{cases} \tag{17}$$

Solving system of equations (17), we obtain the component $\tau_{\rho z}^{p(1)}$ as a function of the coordinates, the anisotropy constants, and the radius of the outer contour in the zero approximation. Write this in the form

$$\tau_{\rho z}^{p(1)} = \left(1-\frac{\beta}{\rho}\right)\left(k_2^{(1)}\sin\theta + k_1^{(1)}\cos\theta - \frac{b_1^{(1)}-a_1^{(1)}}{2}\sin 2\theta\right) - \frac{\beta}{\rho}\sin\theta . \tag{18}$$

From the relationships (6) and (7) in the elastic region, taking into account the decomposition (13), we obtain the stress components in the first approximation

$$\tau_{\rho z}^{e(1)} = G\omega\frac{\partial w^{e(1)}}{\partial\rho},\ \tau_{\theta z}^{e(1)} = G\omega\frac{1}{\rho}\frac{\partial w^{e(1)}}{\partial\theta}. \tag{19}$$

The stress-continuity condition on the elastoplastic boundary (10), taking into account the decomposition (13), in the first approximation will take the form

$$\left(\frac{\partial\tau_{\rho z}^{p(0)}}{\partial\rho}\rho_1 + \tau_{\rho z}^{p(1)}\right)_{\rho=1} = \left(\frac{\partial\tau_{\rho z}^{e(0)}}{\partial\rho}\rho_1 + \tau_{\rho z}^{e(1)}\right)_{\rho=1}, \tag{20}$$

$$\left(\frac{\partial\tau_{\theta z}^{p(0)}}{\partial\rho}\rho_1 + \tau_{\rho z}^{p(1)}\right)_{\rho=1} = \left(\frac{\partial\tau_{\theta z}^{e(0)}}{\partial\rho}\rho_1 + \tau_{\rho z}^{e(1)}\right)_{\rho=1}. \tag{21}$$

After the substitution of solution (15) and relationships (19) the relationship (2) is transformed into

$$\left.\frac{\partial w^{e(1)}}{\partial\rho}\right|_{\rho=1} = \frac{1}{G\omega}\tau_{\rho z}^{p(1)}\Big|_{\rho=1}. \tag{22}$$

For the displacement function $w^{e(1)}$ in the elastic region, combining boundary conditions (8), equation (11), relationships (19) and relationship (22), we obtain the Neumann problem for the ring. The necessary and sufficient condition of this problem for the existence of a solution is satisfied.

$$\begin{cases} \Delta w^{e(1)} = 0, \alpha<\rho<1, \theta\in[0;2\pi], \\ \left.\dfrac{\partial w^{e(1)}}{\partial\rho}\right|_{\rho=1} = \dfrac{1}{G\omega}\tau_{\rho z}^{p(1)}\Big|_{\rho=1}, \\ \left(\dfrac{\partial w^{e(1)}}{\partial\rho}\right)_{\rho=\alpha} = 0. \end{cases} \tag{23}$$

3. Results and discussion

Solving the problem, we obtain $w^{e(1)}$ in the elastic region as a function of the coordinates containing the anisotropy constants and the radius of inner and outer contours in the zero approximation. We represent this in the form.

IOP Conf. Series: Journal of Physics: Conf. Series **973** (2018) 012011 doi:10.1088/1742-6596/973/1/012011

$$w^{e(1)} = \left(k_2^{(1)} \sin\theta + k_1^{(1)} \cos\theta\right) \frac{(\beta-1)(\rho^2+\alpha^2)}{\rho G\omega(\alpha^2-1)} + \sin\theta \frac{\rho^2+\alpha^2}{\rho G\omega(\alpha^2-1)} - \frac{1}{4}\sin 2\theta \frac{\left(b_1^{(1)} - a_1^{(1)}\right)(\beta-1)(\rho^4+\alpha^4)}{\rho^2 G\omega(\alpha^4-1)}. \tag{24}$$

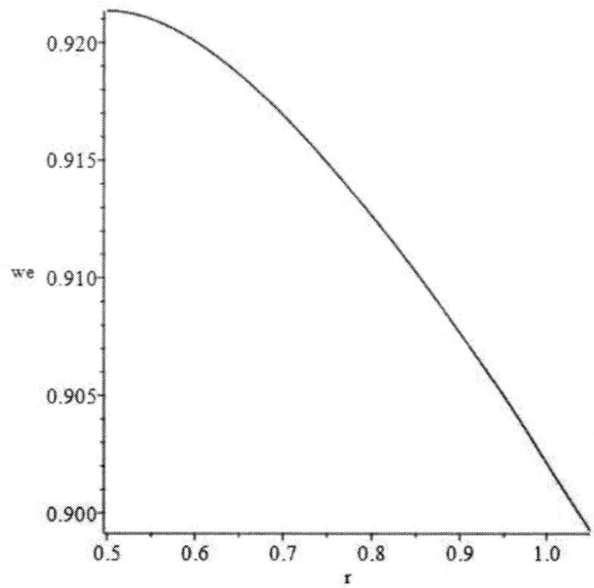

Figure 2. Graph of the dependence of displacement function in elastic region on the radius of hollow bar.

This graph (figure 2) indicates the dependence of the displacement function in elastic region, characterizes the cross-sectional warping on the radius of hollow bar. The graph is constructed for a fixed value of $\theta = \frac{\pi}{4}$. It can be seen from the graph that the maximum displacement occurs near the outer contour.

In the elastic region, the stress state in the first approximation is determined by substituting relationship (24) in relationships (19).

$$\tau_{\theta z}^{e(1)} = \left(k_2^{(1)} \cos\theta - k_1^{(1)} \sin\theta\right) \frac{(\beta-1)(\rho^2+\alpha^2)}{\rho^2(\alpha^2-1)} + \cos\theta \frac{\rho^2+\alpha^2}{\rho^2(\alpha^2-1)} \tag{25}$$

$$-\frac{1}{2}\cos 2\theta \frac{\left(b_1^{(1)} - a_1^{(1)}\right)(\beta-1)(\rho^4+\alpha^4)}{\rho^3(\alpha^4-1)}$$

$$\tau_{\rho z}^{e(1)} = \left(k_2^{(1)} \sin\theta + k_1^{(1)} \cos\theta\right) \frac{(\beta-1)(\rho^2-\alpha^2)}{\rho^2(\alpha^2-1)} + \sin\theta \frac{\rho^2-\alpha^2}{\rho(\alpha^2-1)} - \frac{1}{2}\sin 2\theta \frac{\left(b_1^{(1)} - a_1^{(1)}\right)(\beta-1)(\rho^4-\alpha^4)}{\rho^3(\alpha^4-1)} \tag{26}$$

Define the elastoplastic boundary in the first approximation using condition (21) with allowance for solution (15)

$$\rho_1 = \left(\tau_{\theta z}^{p(1)} - \tau_{\theta z}^{e(1)}\right)\Big|_{\rho=1}. \tag{27}$$

Substituting relationship (25) and relationship (16) into equation (27), we obtain the required expression for an elastoplastic boundary in the first approximation.

$$\rho^{(1)} = \frac{1}{G\omega}\left(\left(k_2^{(1)}\cos\theta - k_1^{(1)}\sin\theta\right)\frac{\alpha^2-1-(\beta-1)(1+\alpha^2)}{\alpha^2-1} - \frac{1}{2}(a_1^{(1)}\sin^2\theta + b_1^{(1)}\cos^2\theta) - \frac{1+\alpha^2}{\alpha^2-1}\cos\theta \right. \quad (28)$$
$$\left. + \frac{1}{2}\frac{(b_1^{(1)}-a_1^{(1)})(\beta-1)(1+\alpha^4)}{\alpha^4-1}\cos 2\theta\right)$$

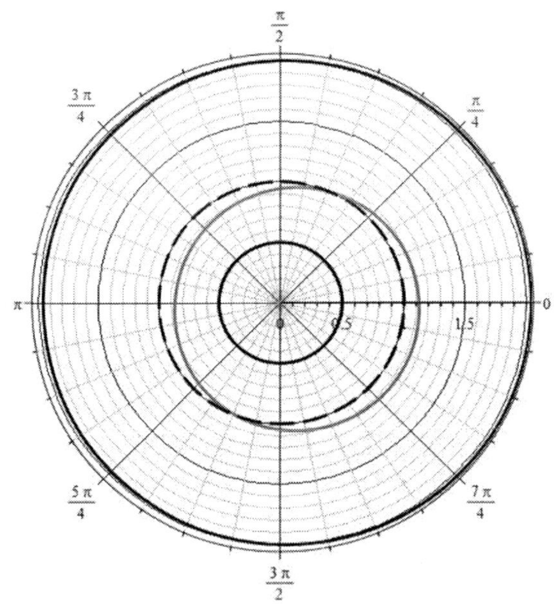

Figure 3. Elastoplastic boundary of the elastoplastic torsion problem of an anisotropy elastoplastic cylindrical hollow bar with eccentricity.

The results of the numerical experiment are presented in the following figures, where $\delta = 0.025$, $k_1 = 0.5$, $k_2 = 0.6$, $a_1 = 0.3$, $b_1 = 0.4$, $G\omega = 1$, $G = 323.67$. In the figure 3 the gray represents the elastoplastic boundary. The dash line is an elastoplastic boundary in the zero approximation. For analyzing the obtained solution and estimating the influence of the anisotropy constants on the expression for the elastoplastic boundary, it is necessary to compare the obtained relationship (28) with the solution of the elastoplastic torsion problem of an elastic perfectly plastic cylindrical hollow bar with eccentricity. The elastoplastic boundary in the first approximation of the elastoplastic torsion problem of an elastic perfectly plastic cylindrical hollow bar with eccentricity has the form

$$\rho^{(1)} = \frac{1+\alpha^2}{G\omega(1-\alpha^2)}\cos\theta. \quad (29)$$

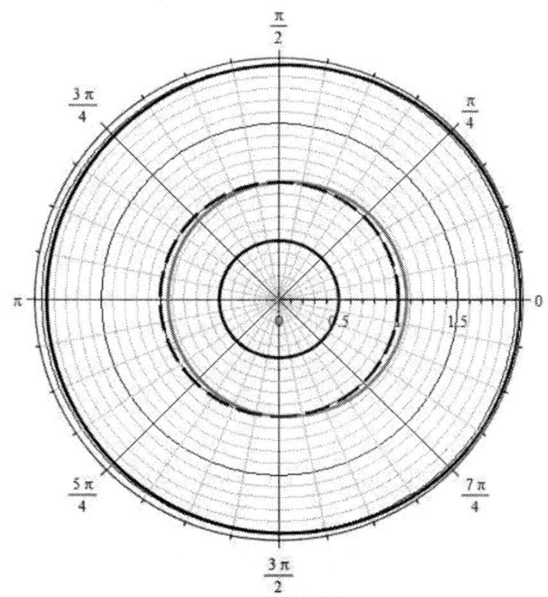

Figure 4. Elastoplastic boundary of the elastoplastic torsion problem of an elastic perfectly plastic cylindrical hollow bar with eccentricity.

This graph (figure 4) indicates the dependence of the elastoplastic boundary (gray), on the components of the polar coordinate system of hollow bar. The dash line is an elastoplastic boundary in the zero approximation. The graph shows that the deviation between the elastoplastic boundary and the elastoplastic boundary in the zero approximation (dash line) is substantially less than the deviation between the elastoplastic boundary for the problem with allowance for the anisotropy and the elastoplastic boundary in the zero approximation (figure 3). The zero approximation corresponds to the problem of elastoplastic torsion of a homogeneous hollow rod with a ring cross-section.

Then we calculate the maximum difference between relationship (29) and relationship (28) in absolute value and plot the dependence of the given deviation on $k_1^{(1)}$, running through the values from 0..0.1, fixing the remaining parameters of the anisotropy.

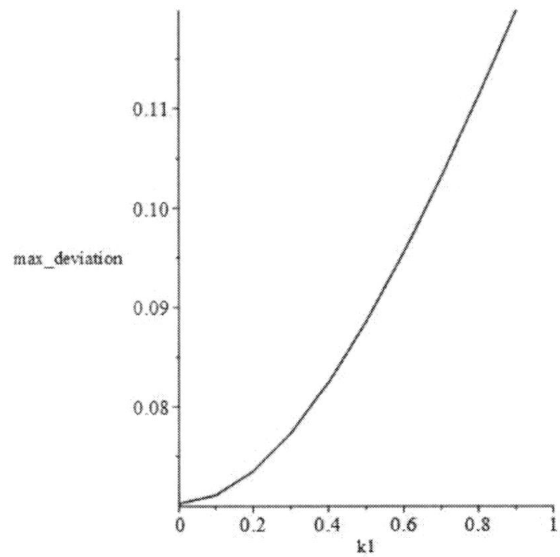

Figure 5. Graph of the dependence of the maximum deviation on the anisotropy constant $k_1^{(1)}$.

This graph (figure 5) indicates the character of the dependence of the anisotropy constant on the maximum deviation between the solutions of the problems of elastoplastic torsion of an eccentric anisotropic hollow bar and the elastoplastic torsion of an eccentric hollow bar with elastic perfectly plastic material. Similar relationships can also be created for other anisotropy constants $a_1^{(1)}$, $b_1^{(1)}$, $k_2^{(1)}$.

4. Conclusion

In the course of solving the problem posed, the dependencies of the stress components in the elastic and plastic regions, the displacement function in the elastic region, characterizing the cross-section warping in the elastic region and the expressions for the elastoplastic boundary have been obtained. The dependence of the radius of the elastoplastic boundary on the anisotropy constant has been analyzed. The plot for the maximum deviation of the radius of the elastoplastic boundary for the solution of the problem of elastoplastic torsion of an eccentric anisotropic hollow bar from elastoplastic torsion of an eccentric hollow bar of an elastic perfectly plastic material has been created.

References

[1] Bazehhour BG and Rezaeepazhand J 2014 *Mathematics and mechanics of solids* **19** pp 260–276
[2] Chein–Shan Liu 2007 *Computers, Materials and Continua* **5** pp 31–42
[3] Messelmi F 2017 *Analysis and Mathematical Physics* **7** pp 437–447
[4] Ivlev D D and Ershov L V 1978 *Perturbing approximation in theory of elastoplastic body* (Moscow: Science) p 208
[5] Zadorozhniy V G, Kovalev A V and Sporykhin A N 2008 *Russian Academy of Sciences: Mechanics of solid body* **1** pp 138–146
[6] Kovalev A V, Shcheglova Yu D and Sviridov I E 2016 *Bulletin of the Yakovlev Chuvash State Pedagogical University: Mechanics of Limit State* **4(30)** pp 42–54
[7] Sporykhin A N, Kovalev A V and Shcheglova Yu D 2004 *Nonuniform elastic-viscoplasticity problems with an unknown boundary* (Voronezh: Voronezh State University Press) p 219
[8] Kachanov L M 1969 *Basics of the theory of plasticity* (Moscow: Science) p 420
[9] Fominyh S O 2013 *Bulletin of the Yakovlev Chuvash State Pedagogical University: Mechanics*

of Limit State **2(16)** pp 150–153
[10] Hill R 1950 *The Mathematical theory of plasticity* (Oxford: The Clarendon Press) p 407
[11] Ivlev D D and Mironov B G 2010 *Bulletin of the Yakovlev Chuvash State Pedagogical University: Mechanics of Limit State* **3** pp 560–600 (in Russian)
[12] Mironov B G and Mitrofanova T V 2011 *Bulletin of the Yakovlev Chuvash State Pedagogical University: Mechanics of Limit State* **1(9)** pp 150–155

IOP Conf. Series: Journal of Physics: Conf. Series **973** (2018) 012012 doi:10.1088/1742-6596/973/1/012012

Stress-strain state of the elastic strip with nearly rectangular cross section

N V Minaeva

Professor, Department of Mechanics and Computer Modeling, Voronezh State University, Universitetskaya pl.1, Voronezh, 364018, Russia

E-mail: nminaeva@yandex.ru

Abstract. The article considers behavior of elastic strip in the framework of plane strain under compression. Conditions are formulated in integral form on the boundaries where the forces are imposed. All the boundary conditions are imposed on the boundary of the body in the strained state, which is necessary for investigating the continuous dependence of the solution to the corresponding task on the functions describing the difference between the shape of cross-section of the strip and a rectangle. The study of the analyticity of the problem solution with respect to small near zero parameters is carried out. The solution is found by perturbation method up to the first order of terms.

1. Introduction

When studying the behavior of elastic bodies, it is often assumed that the displacements and strains are small. Within the framework of made assumptions, this approach yields useful results in different problems [1-3]. Deformation of strips and plates by various external forces under diverse types of support is considered in [4-6].

In [7] it is shown that taking into account the strain of the boundary surface allows one to solve the problems of stability of the elastic bodies by the methods of the mathematical theory of elasticity. Using as an example the compression of a strip in the direction of an infinite extension, it is shown that the results obtained on the basis of the methods of material resistance are limiting for critical loads determined by the mathematical theory of elasticity. This approach was implemented in [8], where the components of rotation were taken into account not only in the boundary conditions, but also in the equilibrium equations.

Since it is necessary to take into account the peculiarities of a real construction in the most complete and accurate way, further development of the methods for calculating and analyzing the solutions obtained is still an urgent task.

In this paper, we consider the problem of analyzing the continuous dependence of the solution describing the stress-strain state of an elastic strip under compression on the functions that determine the contours of the cross section. It is assumed that these characteristics are known up to small parameters.

In such cases, to find the solution the perturbation asymptotic methods are usually used [9-11]. In some problems it is possible to estimate the expansion error [12]. The accuracy of the found solution is demonstrated in [13] by the comparison with the accurate known solution. In some special cases, this question has been analyzed. In the works of H Wendt, and even earlier of M V Keldysh and F I Frankl [14], the convergence of the series obtained for purely subsonic flow was shown.

IOP Conf. Series: Journal of Physics: Conf. Series **973** (2018) 012012 doi:10.1088/1742-6596/973/1/012012

Poincare [15] demonstrated that for the second-order equation describing periodic oscillations, when a small parameter enters the right-hand side under the condition of analyticity of the right-hand side, the small parameter method is a convergent one.

For quasilinear systems with non-analytic characteristics of nonlinearity by S N Shimanov and I G Malkin [16] the method for constructing periodic solutions by means of successive approximations in the case of multiple degrees of freedom has been developed, the convergence of these approximations to the periodic desired solution has been proved, and the conditions for the existence of such a solution in the case of resonance has been shown.

Solving the problems the authors have restricted themselves, as a rule, with only one parameter. The idea of finding the solution by expansion with respect to two or more small parameters has been used in the study of oscillations of quasilinear systems [17].

In [18] (for some solution spaces and initial data [19-21]) the conditions are stated when the solution is analytic one with respect to small parameters in the vicinity of zero.

2. Continuous dependence of the solution characterizing the stress-strain state of the strip on the initial data

On the basis of the results obtained in [7, 8], various mathematical models of boundary conditions in stresses are presented in [22]. As a result of the performed analysis, it has been found that in the study of the continuous dependence of the solution on the initial data (functions, parameters entering the mathematical model) the static boundary conditions are required to be stated on the boundary of the real body in the strain state. In [22], special cases of differential operators used in solving problems of quasi-static deformation of solids are considered. For some Banach spaces with the corresponding norms, usually used in solving the problems of continuum mechanics (Helder spaces, Hilbert space, $C^2([a,b],\mathbf{R}^m)$, $C^4([a,b],\mathbf{R}^m)$, etc.), in which Frechet derivative is an isomorphism [20, 21], the conditions are obtained for the continuous dependence of the solution on the initial data.

Let us consider the deformation of a compressed elastic strip under the conditions of a plane strain. The strip is subject to the forces under which accurate boundary conditions for normal stresses cannot be set. Similar situations arise, for example, if the load is a concentrated or distributed force applied along the line running along the strip. Let them be reduced to the principal vector equal modulo $2ph$. In the loaded state, the upper and lower edges are characterized by the functions $g_i(x)$, and lateral edges by $q_i(y)$ $(i=1,2)$. In the unloaded state, by f_j $(j=1,..,4)$, respectively (see figure 1).

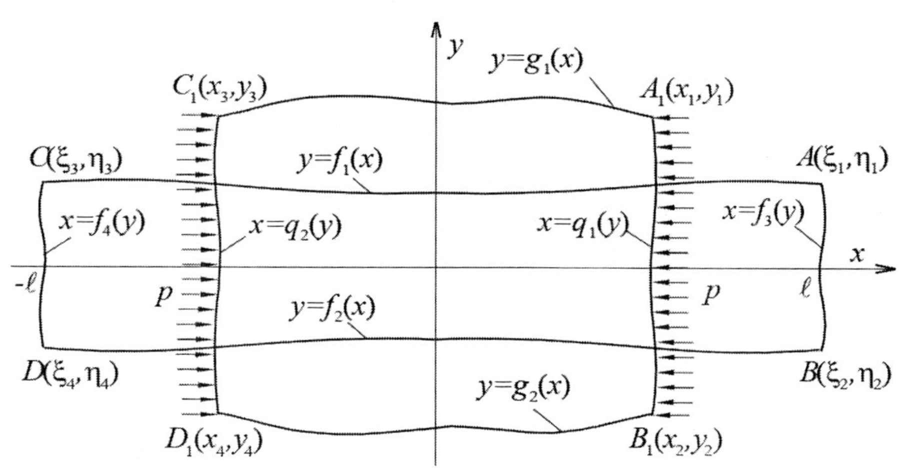

Figure 1. The cross section of the strip in unloaded and loaded conditions

The behavior of the plate will be described by the solution to the system of equations [7]:

$$\frac{\partial \sigma_x}{\partial x} + \frac{\partial \tau}{\partial y} = 0, \qquad \frac{\partial \sigma_y}{\partial y} + \frac{\partial \tau}{\partial x} = 0,$$
$$\sigma_x = \lambda\theta + 2\mu \frac{\partial u}{\partial x}, \quad \sigma_y = \lambda\theta + 2\mu \frac{\partial v}{\partial y}, \tag{1}$$
$$\tau = \mu\left(\frac{\partial v}{\partial x} + \frac{\partial u}{\partial y}\right), \quad \theta = \frac{\partial u}{\partial x} + \frac{\partial v}{\partial y}.$$

We state the boundary conditions in the following form

$$P_x\big|_{y=g_1} = P_x\big|_{y=g_2} = 0,$$
$$P_y\big|_{y=g_1} = P_y\big|_{y=g_2} = 0. \tag{2}$$

$$\int_{A_1B_1} P_x\big|_{x=q_1} \mathrm{d}s = \int_{C_1D_1} P_x\big|_{x=q_2} \mathrm{d}s = -2ph,$$
$$\int_{A_1B_1} P_x\big|_{x=q_1} y\mathrm{d}s = \int_{C_1D_1} P_x\big|_{x=q_2} y\mathrm{d}s = 0, \tag{3}$$
$$\int_{AB} v\big|_{x=f_3} \mathrm{d}s = \int_{CD} v\big|_{x=f_4} \mathrm{d}s = 0.$$

Since in (3) $\mathrm{d}s = \sqrt{1+(q_i')^2}\,\mathrm{d}y \quad (i=1,2)$, (2), (3) will take the form [7, 22]

$$\left(\sigma_x g_1' - \tau\right)_{y=g_1} = \left(\sigma_x g_2' - \tau\right)_{y=g_2} = 0,$$
$$\left(\sigma_y - \tau g_1'\right)_{y=g_1} = \left(\sigma_y - \tau g_2'\right)_{y=g_2} = 0. \tag{4}$$

$$\int_{y_2}^{y_1} \left(\sigma_x - \tau q_1'\right)_{x=q_1} \mathrm{d}y = \int_{y_4}^{y_3} \left(\sigma_x - \tau q_2'\right)_{x=q_2} \mathrm{d}y = -2ph,$$
$$\int_{y_2}^{y_1} \left(\sigma_x - \tau q_1'\right)_{x=q_1} y\mathrm{d}y = \int_{y_4}^{y_3} \left(\sigma_x - \tau q_2'\right)_{x=q_2} y\mathrm{d}y = 0, \tag{5}$$
$$\int_{\eta_2}^{\eta_1} v\big|_{x=f_3} \mathrm{d}y = \int_{\eta_4}^{\eta_3} v\big|_{x=f_4} \mathrm{d}y = 0.$$

$$\eta_1 = f_1(\xi_1), \quad \xi_1 = f_3(\eta_1), \quad \eta_2 = f_2(\xi_2); \quad \xi_2 = f_3(\eta_2),$$
$$\eta_3 = f_1(\xi_3), \quad \xi_3 = f_4(\eta_3), \quad \eta_4 = f_2(\xi_4); \quad \xi_4 = f_4(\eta_4),$$
$$y_1 = \eta_1 + v(\xi_1, \eta_1), \qquad y_2 = -\eta_1 + v(\xi_2, -\eta_1), \tag{6}$$
$$y_3 = \eta_3 + v(-\xi_1, \eta_3), \quad y_4 = -\eta_1 + v(-\xi_2, -\eta_3).$$

Let us consider a special case when the cross section in an unloaded state differs only slightly from a rectangular one

$$f_1(x)=h+h^{(1)}(x),\quad f_2(x)=-f_1(x),$$
$$f_3(y)=\ell+\ell^{(1)}(y),\quad f_4(y)=-f_3(y),\tag{7}$$
$$h=\text{const},\quad \ell=\text{const}$$

If we neglect this difference, i.e. $h^{(1)}(x)\equiv 0$, $\ell^{(1)}(y)\equiv 0$, then problem (1)-(7) admits solution [4,8]:

$$v=v^{(0)}=\varepsilon_y^0 y,\quad u=u^{(0)}=-\frac{2\mu+\lambda}{\lambda}\varepsilon_y^0 x,$$
$$\sigma_x=\sigma_x^{(0)}=-\frac{4\mu(\mu+\lambda)}{\lambda}\varepsilon_y^0,\quad \sigma_y=\sigma_y^{(0)}=0,\quad \tau=\tau^{(0)}=0,\tag{8}$$

where the constant ε_y^0 is determined from (5) by the following relation

$$\varepsilon_y^0=-\frac{1}{2}+\sqrt{\frac{1}{4}+\frac{\lambda}{4\mu(\lambda+\mu)}p}.\tag{9}$$

In order to find out for what values of the external load parameter p the solution (8) to the problem (1)–(7) for $h^{(1)}(x)<<h$, $\ell^{(1)}(y)<<\ell$ could be taken as an approximate one, we must consider the question of the continuous dependence of the solution to the problem (1) - (7) on the functions $h^{(1)}(x)$, $\ell^{(1)}(y)$ for $h^{(1)}(x)\equiv 0$, $\ell^{(1)}(y)\equiv 0$. For this purpose, as it follows from [22], we consider the auxiliary problem with respect to the functions ζ_i, obtained on the basis of (1), (4), (5) ($\sigma_x=\sigma_x^{(0)}+\zeta_1;\ \sigma_y=\sigma_y^{(0)}+\zeta_2;\dots\ v=v^{(0)}+\zeta_5$):

$$\frac{\partial\zeta_1}{\partial x}+\frac{\partial\zeta_3}{\partial y}=0,\qquad \frac{\partial\zeta_2}{\partial y}+\frac{\partial\zeta_3}{\partial x}=0,$$
$$\zeta_1=\lambda\zeta_6+2\mu\frac{\partial\zeta_4}{\partial x},\quad \zeta_2=\lambda\zeta_6+2\mu\frac{\partial\zeta_5}{\partial y},\tag{10}$$
$$\zeta_3=\mu\left(\frac{\partial\zeta_5}{\partial x}+\frac{\partial\zeta_4}{\partial y}\right),\quad \zeta_6=\frac{\partial\zeta_4}{\partial x}+\frac{\partial\zeta_5}{\partial y}.$$

$$\left[\left(\sigma_x^{(0)}+\zeta_1\right)\varphi_1'-\zeta_3\right]_{y=\varphi_1}=\left[\left(\sigma_x^{(0)}+\zeta_1\right)\varphi_2'-\zeta_3\right]_{y=\varphi_2}=0,$$
$$\left[\zeta_2-\zeta_3\varphi_1'\right]_{y=\varphi_1}=\left[\zeta_2-\zeta_3\varphi_2'\right]_{y=\varphi_2}=0.\tag{11}$$

$$\int_{y_2}^{y_1}\left(\sigma_x^{(0)}+\zeta_1-\zeta_3\psi_1'\right)_{x=\psi_1}\mathrm{d}y=\int_{y_4}^{y_3}\left(\sigma_x^{(0)}+\zeta_1-\zeta_3\psi_2'\right)_{x=\psi_2}\mathrm{d}y=-2ph,$$
$$\int_{y_2}^{y_1}\left(\sigma_x^{(0)}+\zeta_1-\zeta_3\psi_1'\right)_{x=\psi_1}y\mathrm{d}y=\int_{y_4}^{y_3}\left(\sigma_x^{(0)}+\zeta_1-\zeta_3\psi_2'\right)_{x=\psi_2}y\mathrm{d}y=0,\tag{12}$$
$$\int_{-h}^{h}\left(v^{(0)}+\zeta_5\right)_{x=\psi_1}\mathrm{d}y=\int_{-h}^{h}\left(v^{(0)}+\zeta_5\right)_{x=\psi_2}\mathrm{d}y=0,$$

where

$$y_1=h+v^{(0)}(\ell,h)+\zeta_5(\ell,h),\qquad y_2=-h+v^{(0)}(\ell,-h)+\zeta_5(\ell,-h),$$
$$y_3=h+v^{(0)}(-\ell,h)+\zeta_5(-\ell,h),\quad y_4=-h+v^{(0)}(-\ell,-h)+\zeta_5(-\ell,-h).\tag{13}$$

The function $\varphi_1(x)$ is determined by the following parametric form

$$\begin{aligned} x &= t + u^{(0)}(t,h) + \zeta_4(t,h), \\ y &= h + v^{(0)}(t,h) + \zeta_5(t,h), \end{aligned} \tag{14}$$

the function $\varphi_2(x)$:

$$\begin{aligned} x &= t + u^{(0)}(t,-h) + \zeta_4(t,-h), \\ y &= h + v^{(0)}(t,-h) + \zeta_5(t,-h), \end{aligned} \tag{15}$$

the function $\psi_1(x)$:

$$\begin{aligned} x &= \ell + u^{(0)}(\ell,t) + \zeta_4(\ell,t), \\ y &= t + v^{(0)}(\ell,t) + \zeta_5(\ell,t), \end{aligned} \tag{16}$$

and the function $\psi_2(x)$:

$$\begin{aligned} x &= \ell + u^{(0)}(-\ell,t) + \zeta_4(-\ell,t), \\ y &= t + v^{(0)}(-\ell,t) + \zeta_5(-\ell,t). \end{aligned} \tag{17}$$

Following [22], it is necessary to perform linearization of the problem (10)–(17) with respect to ζ_i. The system of equations (10) is already linear. Functions $\varphi_i(x)$ and $\psi_i(x)$ linearized with respect to ζ_i take the form

$$\varphi_1 = \left(1+\varepsilon_y^0\right)h + \zeta_5\left(\frac{x}{1+\varepsilon_x^0}, h\right), \quad \varphi_2 = -\left(1+\varepsilon_y^0\right)h + \zeta_5\left(\frac{x}{1+\varepsilon_x^0}, -h\right),$$

$$\psi_1 = \left(1+\varepsilon_x^0\right)\ell + \zeta_4\left(\ell, \frac{y}{1+\varepsilon_y^0}\right), \quad \psi_2 = -\left(1+\varepsilon_x^0\right)\ell + \zeta_4\left(-\ell, \frac{y}{1+\varepsilon_y^0}\right),$$

$$\varepsilon_x^0 = -\frac{2\mu+\lambda}{\lambda}\varepsilon_y^0.$$

The boundary conditions (11) will take the form

$$\begin{aligned} &\left[\zeta_3 - \sigma_x^{(0)} \frac{d\zeta_5\left(\frac{x}{1+\varepsilon_x^0}, h\right)}{dx}\right]_{y=\left(1+\varepsilon_y^0\right)h} = 0; \quad \zeta_2(x,y)\Big|_{y=\left(1+\varepsilon_y^0\right)h} = 0, \\ &\left[\zeta_3 - \sigma_x^{(0)} \frac{d\zeta_5\left(\frac{x}{1+\varepsilon_x^0}, -h\right)}{dx}\right]_{y=-\left(1+\varepsilon_y^0\right)h} = 0; \quad \zeta_2(x,y)\Big|_{y=-\left(1+\varepsilon_y^0\right)h} = 0. \end{aligned} \tag{18}$$

And conditions (12) are as follows

$$\int_{-y_1^0}^{y_1^0} \zeta_1\Big|_{x=\pm\left(1+\varepsilon_x^{(0)}\right)\ell} \mathrm{d}y + \sigma_x^0\left[\zeta_5(\pm\ell,h)-\zeta_5(\pm\ell,-h)\right]=0,$$
$$\int_{-y_1^0}^{y_1^0} \zeta_1\Big|_{x=\pm\left(1+\varepsilon_x^{(0)}\right)\ell} y\mathrm{d}y + \sigma_x^0\left[h\zeta_5(\pm\ell,h)+h\zeta_5(\pm\ell,-h)\right]=0, \quad (19)$$
$$\int_{-h}^{h} \zeta_5\Big|_{x=\pm\ell} \mathrm{d}y = 0,$$

where $y_1^0 = \left(1+\varepsilon_y^0\right)h$.

In order the solution to the initial problem depends continuously on the functions $h^{(1)}(x)$, $\ell^{(1)}(y)$ it is necessary that the problem (10) and (12) has only a trivial solution [20, 22]. It is rather difficult to carry out the study in the general case, therefore we will further consider the case, which is often encountered in practice, when $\frac{p}{\mu} << 1$. Then, according to (9), the strain is also small

$$\varepsilon_y^0 << 1, \qquad \varepsilon_x^0 << 1.$$

In this case, the boundary conditions (18) and (19) can be replaced by the following approximate ones

$$\zeta_3(x,\pm h) - \sigma_x^{(0)}\frac{\partial\zeta_5(x,\pm h)}{\partial x} = 0,$$
$$\zeta_2(x,\pm h) = 0. \quad (20)$$

$$\int_{-h}^{h}\zeta_1(\pm\ell,y)\mathrm{d}y + \sigma_x^0\left[\zeta_5(\pm\ell,h)-\zeta_5(\pm\ell,-h)\right]=0,$$
$$\int_{-h}^{h}\zeta_1(\pm\ell,y)y\mathrm{d}y + \sigma_x^0\left[h\zeta_5(\pm\ell,h)+h\zeta_5(\pm\ell,-h)\right]=0, \quad (21)$$
$$\int_{-h}^{h}\zeta_5(\pm\ell,y)\mathrm{d}y = 0.$$

The boundary condition (20) coincides with the results in [7, 8].

Following [7, 8] we seek the solution to the problem (10), (20), (21) in the form

$$\zeta_4 = -\left[A\mathrm{sh}\alpha y + D\left(\frac{\lambda+3\mu}{\lambda+\mu}\mathrm{sh}\alpha y + \alpha y\mathrm{ch}\alpha y\right)\right]\sin\alpha x,$$
$$\zeta_5 = \left[A\mathrm{ch}\alpha y + D\alpha y\mathrm{sh}\alpha y\right]\cos\alpha x, \qquad \alpha = \frac{\pi}{2\ell}. \quad (22)$$

Since, on the basis of (22) and (10), ζ_1 and ζ_5 for $x = \pm\ell$ turn to zero, the conditions (21) are satisfied. We obtain that the problem (10), (20), (21) has a nontrivial solution for

$$p = p^* = 2\mu \frac{\text{ch}\beta\text{sh}\beta - \beta}{\frac{\mu}{\lambda+\mu}\text{ch}\beta\text{sh}\beta + \beta}, \quad (23)$$

where $\beta = \frac{\pi h}{2\ell}$. This result coincides, to within the notations, with the value of the critical force in [7].

Figure 2 shows the graph (23) ($\frac{p}{\mu} << 1$) for $\lambda = 117.68\text{GPa}$ и $\mu = 78.45\text{GPa}$. If we compare p^* with the critical load under the loss of stability of a compressed rectangular plate [4] for $\beta = 0.04$, then we obtain insignificant discrepancies: $p^* = 119.544\text{MPa}$, but $p_{cr} = 119.543\text{MPa}$. For the values $\beta > 0.1$ the condition $\frac{p}{\mu} << 1$ is still fulfilled, but the stresses already exceed the elastic limit of real materials, therefore the graph is not given.

Thus, we obtain that for $p < p^*$ the solution to the problem (1)-(7) depends continuously on $h^{(1)}(x)$, $\ell^{(1)}(y)$ for $h^{(1)}(x) \equiv 0$, $\ell^{(1)}(y) \equiv 0$, and the stress-strain state (under small strain) of the strip will be close to the homogeneous one, described by (8).

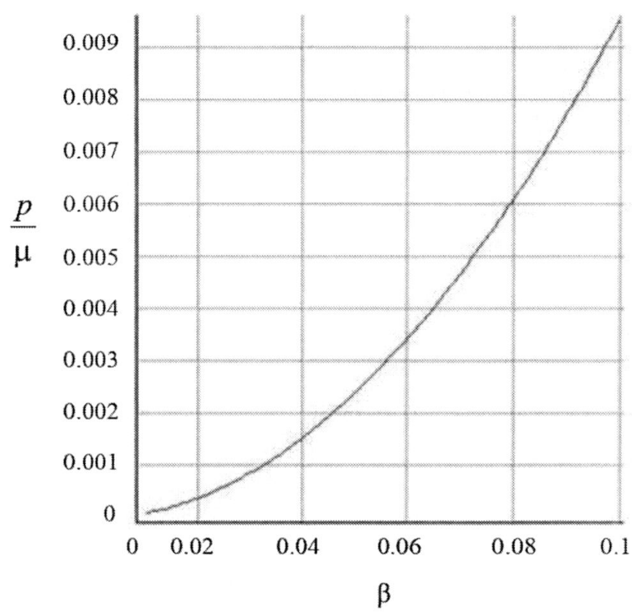

Figure 2. The graph of the function $p(\beta)$.

3. Application of the perturbation method

Let in (1) - (3) the functions characterizing the deviation of the unloaded cross-section edges from straight lines be known to within small parameters

$$h^{(i)}(x) = \varepsilon_i f^{(1)}(x), \quad l^{(i)}(y) = \varepsilon_3 f^{(2)}(y) \qquad (i=1,2), \ (|\varepsilon_j| < 1 \ j = 1,2,3).$$

Then, according to [18, 19], for $p < p^*$ the solution to problem (1) - (3) are analytic one with respect to the parameters $\varepsilon_i, \varepsilon_3$ in the vicinity of the point $\varepsilon_i = \varepsilon_3 = 0$.

We seek it in the form of power series, and for zero approximation we use (8):

$$u=\sum_{k,l,m=0}^{\infty}\varepsilon_1^k\varepsilon_2^l\varepsilon_3^m u^{klm}(x,y),\qquad v=\sum_{k,l,m=0}^{\infty}\varepsilon_1^k\varepsilon_2^l\varepsilon_3^m v^{klm}(x,y),$$
$$\sigma_x=\sum_{k,l,m=0}^{\infty}\varepsilon_1^k\varepsilon_2^l\varepsilon_3^m \sigma_x^{klm}(x,y),\quad \sigma_y=\sum_{k,l,m=0}^{\infty}\varepsilon_1^k\varepsilon_2^l\varepsilon_3^m \sigma_y^{klm}(x,y),\quad \tau=\sum_{k,l,m=0}^{\infty}\varepsilon_1^k\varepsilon_2^l\varepsilon_3^m \tau^{klm}(x,y). \tag{24}$$

As a result of substitution (24) into (1) - (3), according to the small parameter method, we obtained the relations for finding the corresponding approximations.

For example, for the components "100" they have the form

$$\left(\frac{\partial\sigma_x^{100}}{\partial x}+\frac{\partial\tau^{100}}{\partial y}\right)=0,\qquad \left(\frac{\partial\sigma_y^{100}}{\partial y}+\frac{\partial\tau^{100}}{\partial x}\right)=0,$$
$$\sigma_x^{100}=\lambda\theta^{100}+2\mu\frac{\partial u^{100}}{\partial x},\qquad \sigma_y^{100}=\lambda\theta^{100}+2\mu\frac{\partial u^{100}}{\partial y}, \tag{25}$$
$$\tau^{100}=\mu\left(\frac{\partial v^{100}}{\partial x}+\frac{\partial u^{100}}{\partial y}\right),\qquad \theta^{100}=\left(\frac{\partial u^{100}}{\partial x}+\frac{\partial v^{100}}{\partial y}\right)$$

with the boundary conditions

$$\tau^{100}(x,h)-\sigma_x^0\frac{\partial v^{100}(x,h)}{\partial x}=\sigma_x^0\frac{\mathrm{d}f^{(1)}}{\mathrm{d}x}, \tag{26}$$
$$\tau^{100}(x,-h)-\sigma_x^0\frac{\partial v^{100}(x,-h)}{\partial x}=\sigma_y^{100}(x,\pm h)=0.$$

$$\int_{-h}^{h}\sigma_x^{100}(\pm l,y)\mathrm{d}y+\sigma_x^0\left[v^{100}(\pm l,h)-v^{100}(\pm l,-h)\right]=0,$$
$$\int_{-h}^{h}\sigma_x^{100}(\pm l,y)y\mathrm{d}y+\sigma_x^0 h\left[v^{100}(\pm l,h)-v^{100}(\pm l,-h)\right]=0, \tag{27}$$
$$\int_{-h}^{h}v^{100}(\pm l,h)\mathrm{d}y=0.$$

For the remaining components of the first approximation, the system of differential equations are the same as (25).

The conditions (27) are analogous for the approximation "010", and (26) take the form:

$$\tau^{100}(x,h)-\sigma_x^0\frac{\partial v^{100}(x,h)}{\partial x}=\sigma_y^{100}(x,\pm h)=0,$$
$$\tau^{100}(x,-h)-\sigma_x^0\frac{\partial v^{100}(x,-h)}{\partial x}=\sigma_x^0\frac{\mathrm{d}f^{(1)}}{\mathrm{d}x}.$$

For the components "001" all boundary conditions are trivial one.

Let $f^{(1)}(x)=\sin ax$, then we seek the solution to the problem (25)-(27), following [7], in the form

$$u^{100}=g^{100}(ay)\sin ax,\qquad v^{100}=f^{100}(ay)\sin ax. \tag{28}$$

As a result of the substitution of (28) into (25) we obtained the system of equations

$$\mu\ddot{g}^{100} - (\lambda + 2\mu)g^{100} + (\lambda + \mu)f^{100} = 0,$$
$$(\lambda + \mu)\ddot{f}^{100} - \mu f^{100} + (\lambda + \mu)\dot{g}^{100} = 0,$$

where the dot denotes the finding of the derivative.

Its solution is as follows

$$g^{100}(\eta) = A^{100}\text{ch}\eta + B^{100}\text{sh}\eta + C^{100}\eta\text{ch}\eta + D^{100}\eta\text{sh}\eta, \tag{29}$$
$$f^{100}(\eta) = A_1^{100}\text{ch}\eta + B_1^{100}\text{sh}\eta + C_1^{100}\eta\text{ch}\eta + D_1^{100}\eta\text{sh}\eta,$$

where the arbitrary constants are connected by the relations:

$$A_1^{100} = B^{100} - \frac{\lambda + 3\mu}{\lambda + \mu}C^{100}, \quad B_1^{100} = A^{100} - \frac{\lambda + 3\mu}{\lambda + \mu}D^{100},$$
$$C_1^{100} = D^{100}, \qquad D_1^{100} = C^{100}.$$

The boundary conditions (27) for $a = n\frac{\pi}{l}$ are satisfied. Substituting (29) into (26) we will obtain the system for determining the constants

$$\begin{aligned}
\alpha_{11}A^{100} + \alpha_{12}B^{100} + \alpha_{13}C^{100} + \alpha_{14}D^{100} &= -p, \\
-\alpha_{11}A^{100} + \alpha_{12}B^{100} + \alpha_{13}C^{100} - \alpha_{14}D^{100} &= 0, \\
\alpha_{31}A^{100} + \alpha_{32}B^{100} + \alpha_{33}C^{100} + \alpha_{34}D^{100} &= 0, \\
\alpha_{31}A^{100} - \alpha_{32}B^{100} - \alpha_{33}C^{100} + \alpha_{34}D^{100} &= 0,
\end{aligned} \tag{30}$$

where

$$\alpha_{11} = (2\mu + p)\text{sh}\beta, \quad \alpha_{12} = (2\mu + p)\text{ch}\beta,$$
$$\alpha_{13} = (2\mu + p)(\beta\text{sh}\beta - \frac{\mu}{\lambda + \mu}\text{ch}\beta) - p\frac{\lambda + 2\mu}{\lambda + \mu}\text{ch}\beta,$$
$$\alpha_{14} = (2\mu + p)(\beta\text{ch}\beta - \frac{\mu}{\lambda + \mu}\text{sh}\beta) - p\frac{\lambda + 2\mu}{\lambda + \mu}\text{sh}\beta,$$
$$\alpha_{31} = 2\mu\text{ch}\beta, \quad \alpha_{32} = 2\mu\text{sh}\beta, \quad \alpha_{33} = 2\mu(\beta\text{ch}\beta - \frac{\lambda + 2\mu}{\lambda + \mu}\text{sh}\beta),$$
$$\alpha_{34} = 2\mu(\beta\text{sh}\beta - \frac{\lambda + 2\mu}{\lambda + \mu}\text{ch}\beta), \qquad \beta = ah.$$

As a result of the transformations, (30) is divided into two systems, from which all the unknowns are determined:

$$\alpha_{11}A^{100} + \alpha_{14}D^{100} = -\frac{1}{2}p, \qquad \alpha_{12}B^{100} + \alpha_{13}C^{100} = -\frac{1}{2}p,$$
$$\alpha_{31}A^{100} + \alpha_{34}D^{100} = 0. \qquad \alpha_{32}B^{100} + \alpha_{33}C^{100} = 0.$$

To find u^{010} and v^{010} in the form which is analogous to (28), the systems for arbitrary constants take the form

$$\alpha_{11}A^{010} + \alpha_{14}D^{010} = 0, \qquad \alpha_{12}B^{010} + \alpha_{13}C^{010} = 0,$$

$$\alpha_{31}A^{010} + \alpha_{34}D^{010} = -\frac{1}{2}p. \qquad \alpha_{32}B^{010} + \alpha_{33}C^{010} = -\frac{1}{2}p.$$

For the functions $g^{001}(ay)$ and $f^{001}(ay)$ the system of equations is homogeneous one for finding arbitrary constants. Since its determinant for $p < p^*$ is not equal to zero, all arbitrary constants, and consequently, the components themselves, will be equal to zero.

Thus, the functions

$$\sigma_x = \sigma_x^0 + \varepsilon_1\sigma_x^{100} + \varepsilon_2\sigma_x^{010}, \ldots, \quad v = v^0 + \varepsilon_1 v^{100} + \varepsilon_2 v^{010},$$

according to Saint-Venant's principle, at a sufficient distance (because of the integral form of the boundary conditions) from the edges of the strip for $x = \pm l$, describe the stress-strain state of the strip for $p < p^*$ to within the first-order quantities of smallness, since the remaining terms of the series (24) are the second-order quantities of smallness.

4. Conclusion

The limitation on the compressive forces parameter is obtained for a strip with a cross-sectional shape close to a rectangle. If the loads are such that they do not exceed the limit found, then the similarity of the cross-sectional shape is also retained during deformation. If the value found is exceeded, the stress-strain state will no longer be homogeneous one, corresponding to (8). For such values, one needs either consider another solution to the nonlinear problem posed, or change the mathematical model, for example, use equilibrium equations in the deformed state, nonlinear physical relationships, etc.

If the contours of the cross section in an unloaded state are known up to small parameters, then for compressive forces satisfying the condition obtained above, one can find the solution by the perturbation method with the required accuracy. In this case the parameters are independent. To estimate the error of the solution obtained, one can use any of the estimates of the Taylor series. For a particular case of deviation of the cross section contour, the solution describing the stress-strain state of the strip has been obtained, up to first order terms. The nonideality of the unloaded edges does not influence significantly (is characterized by not more than second order values) the state of the strip.

References

[1] Papkovich P F 1940 *Dokl. Akad. Nauk SSSR* On one form of solution of the plane problem of the theory of elasticity for a rectangular strip **27** 359
[2] Gotsev D V, Kovalev A V and Sporykhin A N 2001 *J. of Appl. Mech. and Tech. Phys.* Local instability of plates with pressed-in annular incusions at the elastoplastic behavior of materials **42(3)** 505
[3] Sheidakov D N 2007 *J. of Appl. Mech. and Tech. Phys.* Stability of a recgtangular plate under biaxial tension **48(4)** 547
[4] Vol'mir A S 1963 *Stability of elastic systems* (Moscow: Fizmatgiz) p 880 [in Russian]
[5] Pflüger A 1950 *Stabilitätsprobleme der Elastostatik* (Springer-Verlag)
[6] Green A E and Zerna W 1968 *Theoretical Elasticity* (Oxford, London: Univ. Press) p 457
[7] Ishlinskii A Y 1954 *Ukr. Math. J.* Consideration of problems of stability of equilibrium of elastic bodies from the viewpoint of the mathematical theory of elasticity **6(2)** 140
[8] Ershov LV and Ivlev D D 1961 *Dokl. Akad. Nauk SSSR* On stability of a strip in compression **138(5)** 1047
[9] Nayfeh A H 2000 *Perturbation Methods* (New York: John Wiley and Sons) p 456
[10] Ivlev D D and Ershov L V 1978 *The Perturbation Method in the Theory of Elasto-Plastic Body* (Moscow: Nauka) p 208 [in Russian]

IOP Conf. Series: Journal of Physics: Conf. Series **973** (2018) 012012 doi:10.1088/1742-6596/973/1/012012

[11] Guz'A N and Nemish Y N 1989 *The Method of Perturbation of the Boundary Form in Continuum Mechanics* (Kiev: Vishcha Shkola) p 352 [in Russian]

[12] Il'in A M and Danilin A P 2009 *Asymptotic Methods in Analysis* (Moscow: Fizmatlit) p 248 [in Russian]

[13] Cherepanov G P 1963 *J. of Appl. Math. and Mech.* On a method of solving the elasto-plastic problem **27(3)** 644

[14] Keldysh M V and Frankl F I 1934 *Izv. Akad. Nauk SSSR* The exterior problem for nonlinear elliptic differential equations with application to the theory of a wing in a compressible gas

[15] Poincare H 1971 *Selected Works* (in Russian) (*in 3 volumes)* vol 1, 2 (Moscow: Nauka)

[16] Malkin I G 2004 *Lyapunov and Poincare Methods in the Theory of Nonlinear Oscillations* (Moscow: Nauka) p 234 [in Russian]

[17] Proskuryakov A P 1977 *Poincare Method in the Theory of Nonlinear Oscillations* (Moscow: Nauka) p 256 [in Russian].

[18] Minaeva N V 2008 *Mechanics of Solids* Application of the perturbation method in mechanics of deformable solids **43(1)** 31

[19] Krasnosel'skii M A and Zabreiko P P 1975 *Geometric Methods of Nonlinear Analysis* (Moscow: Nauka) p 512 [in Russian]

[20] Zachepa V R and Sapronov Y I 2002 *Local Analysis of Fredholm Equations* (Voronezh: Voronezh State Univ.) p 185 [in Russian]

[21] Darinskii B M, Sapronov Y I and Tsarev S L 2007 *J. Math. Sci.* **145** 5311

[22] Minaeva N V 2006 *The Adequacy of Mathematical Models of Deformable bodies* (Moscow: Nauchnaya kniga) p 236 [in Russian]

IOP Conf. Series: Journal of Physics: Conf. Series **973** (2018) 012013 doi:10.1088/1742-6596/973/1/012013

Simulation of surface hardening in the deep rolling process by means of an axial symmetric nodal averaged finite element

P G Morrev and V A Gordon

Oryol State University, 29a Naugorskoe Road, Oryol, Russia

E-mail: paulorel@mail.ru gordon1312@mail.ru

Abstract. Surface hardening by deep rolling can be considered as the axial symmetric problem in some special events (namely, when large R and small r radii of the deforming roller meet the requirement $R \gg r$). An axisymmetric nodal averaged stabilized finite element is formulated. The formulation is based on a variational principle with a penalty (stabilizing) item in order to involve large elastic-plastic strain and near to incompressible materials. The deep rolling process for a steel rod is analyzed. Axial residual stress, yield stress, and Odkvist's parameter are calculated. The residual stress is compared with the data obtained by other authors using a three-dimensional statement of the problem. The results obtained demonstrate essential advantages of the newly developed finite element.

1. Introducton

Plastic deformation is used not only in material forming processes. This physical property of metals and alloys is also applied for improving the surface behaviour under mechanical and chemical influences. As is known, surface affects such characteristics of metallic components as resistance to plastic deformation (surface hardness), resistance to corrosion, fatigue fracture, fretting fatigue, and wear, because most cases of material failure originate from the exterior layers of the workpiece [1]. When surface experiences mechanical contact loads, the role of those characteristics grows dramatically. Mechanical working is one of the ways to improve the surface behavior. Such working has been employed for a long time and the three most popular techniques are: shot peening, burnishing, and deep rolling. Each of them has its own advantages and disadvantages but the deep rolling stands out against a background of other (both mechanical and many non-mechanical) techniques owing to large values of hardness, residual stress, and hardening depth reached. By combining deep rolling and burnishing processes one can elevate the surface hardness by several times for some materials [2, 3]. This combined process is shown in Figure 1 and Figure 2 where the burnishing roller follows the deforming one and all the process can be realized by means of a turning lathe.

The present work is dedicated to the construction of a simple mathematical model such that the model can be studied by means of a personal computer. Mechanical surface treatment has attracted considerable scientific interest in the last few decades due to both spread of modern experimental equipment for microstructure investigations and accelerated possibilities of computers. Note that the process of multiple local deforming is the most complicated one in solid mechanics because large gradients of tensor fields superimpose on material non-uniformities induced by plastic flowing. Classical finite elements fail in this case because they are not consistent with the incompressibility requirement. Many improved finite element schemes were developed but in spite of essential

IOP Conf. Series: Journal of Physics: Conf. Series **973** (2018) 012013 doi:10.1088/1742-6596/973/1/012013

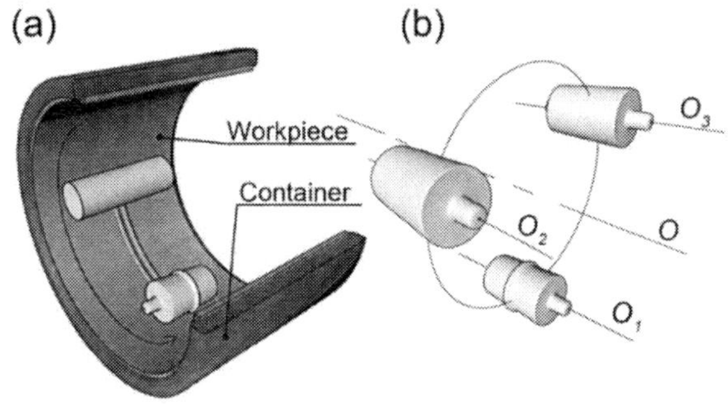

Figure 1. Schematic illustration of deep rolling-burnishing (DRB) process

Figure 2. Experimental setup: (1) Container, (2) Workpiece, and (3) DRB tool

progress most of them suffered from volume locking, bending locking, and spurious low-energy modes. Efforts to eliminate these drawbacks have resulted in several new approaches and the simplest of them is the stabilized nodal averaging method [4-6]. In the present work, simplicity is the main criterion of choice of an appropriate modern approach. Due to many reasons, finite elements of low order prove to be preferable for non-linear problems and large plastic deformation. Besides, mesh generators produce triangles and tetrahedrons as a rule; more complex figures especially for irregular meshes cause serious problems. The procedure of nodal averaging was first formulated by Dohrmann et al. [4] and further analyzed by Bonet et al. [5]. As it turns out, nodal averaging provides weakened constraints such that coarse mesh accuracy can often be achieved for near incompressible materials. It was also pointed out that the formulation was prone to spurious low-energy modes. In order to suppress them, Pusso and Solberg [6] proposed to enter stabilizing terms into a discrete form of a variational equation. A number of different formulations have been developed to rectify poor performance of low-order elements under incompressible and near incompressible conditions but the approach by Dohrmann et al. [4] alleviates both the volumetric locking and locking due to bending. However, some modern methods give the same result as in Pusso and Solberg [6] – see for example Puso et al. [7]. But, stabilized nodal averaging is the simplest among them owing to a trivial theoretical basis and a short program code.

One more problem is that periodic local deformation requires three-dimensional modelling despite the fact that a workpiece looks like an axisymmetric body. However, 3D simulation is too time-consuming therefore the task has to be reduced to a simpler one (see Section 3). In contrast to this approach, the Authors propose to consider the deep rolling problem as the axial symmetric one in some special events (namely, when large R and small r radii of the deforming roller meet the requirement $R >> r$). Thus, if an effective axial symmetric finite element will be developed then modelling of the process can be simplified considerably. In this work, an appropriate nodal (i.e. attached to a mesh node) stabilized finite element is formulated. An approximation scheme based on the simplest triangle element and the averaging procedure over triangles of a nodal complex is proposed and the methodology [6] is extended on axial symmetric problems.

2. Materials and methods

Let an arbitrary triangulation (i.e. N nodes $A_1,...,A_N$ and M triangles $T_1,...,T_M$) of an arbitrary plane region is given. Let a nodal complex C_n be a set of triangles sharing node A_n . Denote by $<n>$ a set of triangle numbers in complex C_n; by $<np>$ – a set of triangle numbers in intersection of complexes C_n and C_p (i.e. $<np>=<n>\cap<p>$); by $[n]$ – a set of vertex numbers in complex C_n; by $]m[$ – a set of node numbers of triangular T_m . Figure 3 illustrates these denotations.

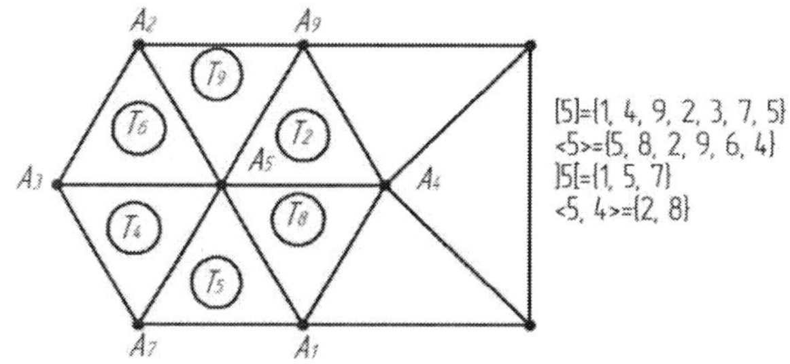

Figure 3. Nodal complexes.

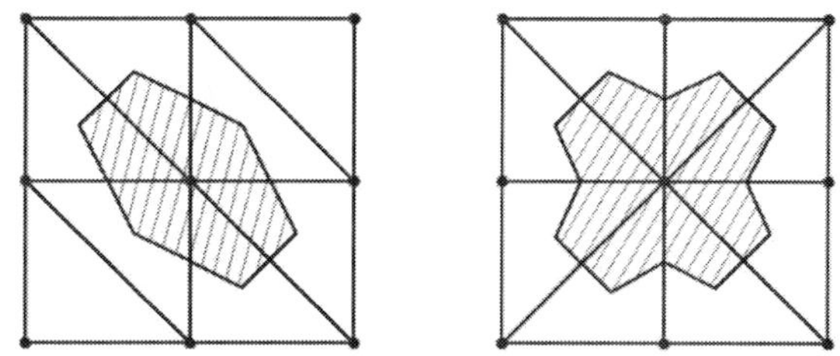

Figure 4. Nodal finite elements with 7 and 9 nodes.

The idea of stabilized nodal averaging can be demonstrated by the simplest example – plane strain problem with small elastic strain. Let $\mathbf{u}=(u^1,u^2)$ be a FEM-approximation of the displacement field. By $\boldsymbol{\varepsilon}_n$ denote the average strain at node A_n and by $\bar{S}_n$ denote the average area at that node, i.e.

$$\boldsymbol{\varepsilon}_n = \sum_{i\in<n>} w_i \boldsymbol{\varepsilon}_i \text{ and } \bar{S}_n = S_n/3$$

where $\boldsymbol{\varepsilon}_i$ – deformation on triangle T_i ($\boldsymbol{\varepsilon}_i \equiv const$ because T_i is a linear finite element); $w_i = s_i/S_n$ – weight coefficient; s_i – area of T_i; S_n – area of C_n. The discretized variational equation is of the form

$$\sum_{n=1}^{N} \bar{S}_n \delta\boldsymbol{\varepsilon}_n \cdot\cdot(\mathbf{E}-\hat{\mathbf{E}})\cdot\cdot\boldsymbol{\varepsilon}_n + \sum_{m=1}^{M} s_m \delta\boldsymbol{\varepsilon}_m \cdot\cdot\hat{\mathbf{E}}\cdot\cdot\boldsymbol{\varepsilon}_m = \mathbf{f}\cdot\delta\mathbf{u} \tag{1}$$

where the first sum corresponds with nodes while the second one – with triangles; **f** is an external load; **E** is the isotropic elasticity tensor and $\hat{\mathbf{E}}$ is the stabilization tensor (on optimal choice of $\hat{\mathbf{E}}$ see Pusso and Solberg [6]).

It is clear that the influence of the stabilization tensor drops as a finite element mesh is refined but for a coarse mesh tensor $\hat{\mathbf{C}}$ becomes to be necessary. Formally, one can consider that the tessellation procedure follows the meshing one and new finite elements appear while each of them is a one-third of a nodal complex – see Figure 2 where one-thirds of triangles (each of triangles is dissected into three parts of equal area) form a new (nodal averaged) finite element. Note that all the nodes except for the central one are located outside the new finite element.

As for the axisymmetric event, a volume formula for a ring of triangular profile is not so trivial as for triangular or tetrahedron. This results in too cumbersome expressions for both nodal force and stiffness matrix. In the present work, an approximate formula of volume for such a ring is exploited in order to simplify a program code and to accelerate calculations. Let us proceed to the axisymmetric formulation itself. In cylindrical coordinates $y^1 = r$, $y^2 = h$, $y^3 = \varphi$, the Cauchy stress tensor and the small strain one take the form

$$\boldsymbol{\sigma} = \begin{bmatrix} \sigma_r & \sigma_{rh} & 0 \\ \sigma_{rh} & \sigma_h & 0 \\ 0 & 0 & \sigma_\varphi \end{bmatrix}, \ \boldsymbol{\varepsilon} = \frac{1}{2}\begin{bmatrix} 2u_{r,r} & u_{r,h}+u_{h,r} & 0 \\ u_{r,h}+u_{h,r} & 2u_{h,h} & 0 \\ 0 & 0 & 2u_r/r \end{bmatrix}$$

It is necessary to rewrite the variational principle (1) taking into account large irreversible (plastic) strain (the Euler approach is adopted throughout the work i.e. all tensor fields refer to the current configuration rather the initial one). Contracting by a pair of indexes, we obtain

$$\sum_{n=1}^{N} \bar{V}_n \boldsymbol{\sigma}_n \cdot \delta\boldsymbol{\varepsilon}_n + \sum_{m=1}^{M} v_m \hat{\boldsymbol{\sigma}}_m \cdot \delta\boldsymbol{\varepsilon}_m = \mathbf{f}\cdot\delta\mathbf{u} \tag{2}$$

where:

$\boldsymbol{\sigma}_n = \sum_{m\in<n>} w_m(\boldsymbol{\sigma}_m - \hat{\boldsymbol{\sigma}}_m)$ – average stabilized stress at node A_n;

$\hat{\boldsymbol{\sigma}}_m$ - stabilizing stress on triangle T_m;

$\boldsymbol{\varepsilon}_n = \sum_{i\in<n>} w_i\boldsymbol{\varepsilon}_i$ – average strain at node A_n;

$w_i = v_i / \bar{V}_n$ – weight coefficient of T_i in nodal complex C_n;

$\bar{V}_n = V_n / 3$ – averaged nodal volume;

$V_n = \sum_{m\in<n>} v_m$ – volume of nodal complex C_n;

$v_m = 2\pi r_m s_m$ – approximate volume for a ring of triangular profile T_m;

$r_m = (r_1 + r_2 + r_3)/3$ – average radius of a ring of triangular profile T_m.

Equation (2) has to be supplemented by a constitutive equation. As far as metals and alloys refer to near rigid-plastic materials, it is convenient to write that equation by means of any corotational rate "r" (for example, Yaumann's rate "j"):

$$^r\boldsymbol{\sigma} = (\mathbf{C} - \hat{\mathbf{C}})\cdot\cdot\,\mathbf{d}\,,\ ^r\hat{\boldsymbol{\sigma}} = \hat{\mathbf{C}}\cdot\cdot\,\mathbf{d}$$

where $\mathbf{d}$ is the deformation rate tensor. Therefore, two stress tensors take part in step-by-step numerical analysis: real stress $\boldsymbol{\sigma}$ and stabilizing stress $\hat{\boldsymbol{\sigma}}$. Introduce a FEM-approximation for the displacement field $\mathbf{u}$, its gradient $\partial u^i / \partial y^j$ $(1 \le i, j \le 2)$, and circular deformation ε_{33}:

$$u^i = u_n^i N_n(y^1, y^2);\ u^i_{,j} = u_n^i N_{n,j}(y^1, y^2)\ (1 \le i, j \le 2);\ \varepsilon_{33} = u_r / r = u_n^1 N_n(y^1, y^2) / y^1$$

where the summation over repeated indexes is implied ; the differentiation sign is separated by a comma; $N_n(y^1, y^2)$ – shape function corresponding to node A_n; (u_n^1, u_n^2) – nodal displacement of A_n. In the sequel, a shape function will be numerated by a pair of indexes: $N_n^m(y^1, y^2)$ where n corresponds to node A_n and m – to triangle T_m. Denote by $\bar{u}^{(n)i}_{,j}$ the average nodal gradient, i.e.

$$\bar{u}^{(n)i}_{,j} = \sum_{m\in<n>} w_m^n u^{(m)i}_{,j}\ (1 \le i, j \le 2)$$

where $\bar{u}^{(n)i}_{,j}$ and $u^{(m)i}_{,j}$ correspond to node A_n and triangle T_m respectively. The nodal gradient FEM-approximation yields

$$\bar{u}^{(n)i}_{,j} = \sum_{m\in<n>} w_m^n \sum_{p\in]m[} N^m_{p,j} u^i_p = \sum_{p\in[n]} u^i_p \sum_{m\in<pn>} w_m^n N^m_{p,j} = \sum_{p\in[n]} u^i_p B^n_{pj}\ \ (1 \le i, j \le 2)$$

while the summation in the last but one sum runs over the intersection of nodal complexes C_p and C_n (this intersection can be empty, consists of one or two triangles or coincides with the entire nodal complex if $n = p$); matrix $\mathbf{B}^{(n)}$ is of the form

$$B^n_{pj} = \sum_{m\in<pn>} w_m^n N^m_{p,j}\ . \tag{3}$$

Circular deformation ε_{33} is calculated according to

$$\varepsilon_{33} = \sum_{p\in[n]} u^1_p \tilde{B}^n_p$$

where vector $\tilde{\mathbf{B}}^{(n)}$ is of the form

$$\tilde{B}^n_p = \sum_{m\in<pn>} w^n_m N^m_p (y^1, y^2)/y^1 \ . \qquad (4)$$

Due to the approximate formula for volume v_m, expression (3) yields

$$B^n_{pj} = \sum_{m\in<pn>} w^n_m N^m_{p,j} = \sum_{m\in<pn>} \frac{v_m}{V_n}\frac{b^m_{pj}}{D_m} = \frac{2\pi}{V_n}\sum_{m\in<pn>} r_m s_m \frac{b^m_{pj}}{D_m} = \frac{\pi}{V_n}\bar{b}^n_{pj} \quad (1 \le i, j \le 2) \qquad (5)$$

where $D_m = 2s_m$; $\bar{b}^n_{pj} = \sum_{m\in<pn>} r_m b^m_{pj}$; $\mathbf{b}^{(m)}$ is the well-known in FEM matrix of coordinate differences of triangle vertices (three nodes are numerated by 1,2,3 counter clockwise and index m is omitted):

$$\mathbf{b} = \begin{bmatrix} y_2 - y_3 & x_3 - x_2 \\ y_3 - y_1 & x_1 - x_3 \\ y_1 - y_2 & x_2 - x_1 \end{bmatrix}.$$

As for expression (4), taking into account the one-point integration scheme for the classical linear triangle element, function $N^m_p(y^1, y^2)/y^1$ should be evaluated at the center of triangle T_m where this function equals to $1/(3r_m)$ such that

$$\tilde{B}^n_p = \frac{1}{3}\sum_{m\in<pn>} \frac{v_m}{V_n r_m} = \frac{2\pi}{3V_n}\sum_{m\in<pn>} s_m = \frac{2\pi}{3V_n}\tilde{b}^n_p \qquad (6)$$

where

$$\tilde{b}^n_p = \tilde{b}^p_n = \sum_{m\in<pn>} s_m \, .$$

Substituting the displacement gradient approximation in variational principle (2), find the nodal force at node A_p:

$$f^i_p = \sum_{n\in[p]} \bar{V}_n B^n_{pj}\sigma^i_n + \delta^{i1}\sum_{n\in[p]} \bar{V}_n \tilde{B}^n_p \sigma^{33}_n + \sum_{m\in<p>} v_m N^m_{p,j}\bar{\sigma}^i_n + \delta^{i1}\sum_{m\in<p>} v_m \bar{\sigma}^{33}_n N^n_p / y^1 \quad (1 \le i, j \le 2) \qquad (7)$$

where δ^{ij} is Kronecker's delta-symbol. Substituting (5) and (6) into (7) and taking into account $\bar{V}_n = V_n/3$, obtain

$$f^i_p = \frac{\pi}{3}\sum_{n\in[p]} \bar{b}^n_{pj}\sigma^i_n + \delta^{i1}\frac{2\pi}{9}\sum_{n\in[p]} \tilde{b}^n_p \sigma^{33}_n + \frac{\pi}{3}\sum_{m\in<p>} r_m b^m_{pj}\bar{\sigma}^{ij}_n + \delta^{i1}\frac{2\pi}{9}\sum_{m\in<p>} s_m \bar{\sigma}^{33}_n \quad (1 \le i, j \le 2) \, . \qquad (8)$$

Note that this expression is simpler than (7) because all volumes (both nodal and of rings of triangular profile) disappeared due to the successful choice of the approximate formula for volume v_m (see the elucidations to equation (2)). Finally, the stiffness matrix is obtained by means of the nodal force variation. In order to deduce a proper expression, find the stress tensor variation from the constitutive equation with Yaumann's rate:

$$\mathrm{d}\boldsymbol{\sigma}/\mathrm{d}t = {}^J\boldsymbol{\sigma} - \boldsymbol{\sigma}\cdot\boldsymbol{\omega} + \boldsymbol{\omega}\cdot\boldsymbol{\sigma} = (\mathbf{C} - \hat{\mathbf{C}})\cdot\cdot\mathbf{d} - \boldsymbol{\sigma}\cdot\boldsymbol{\omega} + \boldsymbol{\omega}\cdot\boldsymbol{\sigma}\,;$$

$$d\bar{\sigma}/dt={}^{J}\bar{\sigma}-\bar{\sigma}\cdot\omega+\omega\cdot\bar{\sigma}=\hat{\mathbf{C}}\cdot\cdot\mathbf{d}-\bar{\sigma}\cdot\omega+\omega\cdot\bar{\sigma}$$

where $\omega=(\nabla\mathbf{v}-\nabla\mathbf{v}^{T})/2$ is the vortex tensor. Then the nodal stress variation at node A_n is

$$\delta\sigma_n^{ij}=(C_\pi^{ijkl}-\hat{C}_\pi^{ijkl})\delta\varepsilon_{kl}^{\pi}+\frac{1}{2}\sum_{q\in[n]}\delta u_q^k(-\sigma_\pi^{kj}B_{qi}^{\pi}-\sigma_\pi^{ik}B_{qj}^{\pi}+\delta_k^i\sigma_\pi^{lj}B_{ql}^{\pi}+\delta_k^j\sigma_\pi^{li}B_{ql}^{\pi})\ (1\le i,j,k,l\le 2)$$

where π denotes "no summation" over index n. Besides, $\mathbf{\bar{b}}$, $\mathbf{\tilde{b}}$, r_m, and s_m are also dependent variables and subjected to variation. Therefore, the total expression for the nodal force variation is too cumbersome and consists of two items: symmetrical and non-symmetrical. In practice of metal forming analysis, a non-symmetrical item can be rejected without the loss of convergence. The symmetrical positive-definite item of the stiffness matrix of the nodal finite element is of the form (p and q are node numbers of complex C_n)

$${}^{(n)}K_{qk}^{pi}=\frac{\pi^2}{3V_n}\bar{b}_{pj}^{\pi}(C_\pi^{ijkl}-\hat{C}_\pi^{ijkl})\bar{b}_{ql}^{\pi}\ (1\le i,j,k,l\le 2)$$

while if $i=1$ or $k=1$ then some special items should be added to the stiffness matrix respectively:

$$\Delta^{(n)}K_{qk}^{p1}=\frac{2\pi^2}{9V_n}\bar{b}_{ql}^{\pi}(C_\pi^{33kl}-\hat{C}_\pi^{33kl})\tilde{b}_p^{\pi}\ \text{or}\ \Delta^{(n)}K_{q1}^{pi}=\frac{2\pi^2}{9V_n}\bar{b}_{pj}^{\pi}(C_\pi^{ij33}-\hat{C}_\pi^{ij33})\tilde{b}_q^{\pi}\quad(1\le i,j\le 2)$$

and if $i=k=1$ then one more item should be added:

$$\Delta^{(n)}K_{q1}^{p1}=\frac{4\pi^2}{27V_n}\tilde{b}_p^{\pi}(C_\pi^{3333}-\hat{C}_\pi^{3333})\tilde{b}_q^{\pi}.$$

When assembling the global stiffness matrix, the stabilizing matrices are also taking into account (they are calculated according to classical expressions [8]). This completes the nodal averaged axisymmetric finite element formulation.

3. Results and discussion

The problem of simulation for multiple local deforming (including the deep rolling treatment) is very difficult from the computational point of view due to the fact that this process may be accompanied not only by high gradient values of the tensor fields but also by essential transformations of the material microstructure leading to high material non-homogeneity. So, the superposition of the material non-homogeneity with high gradients of tensor's components leads to an extremely complicated isoline pattern for stress, strain and other variables. As a rule, such problems are considered in a three-dimensional statement (though a workpiece looks like an axisymmetric body) when a small "representative" part is cut out the entire workpiece and some "reasonable" boundary conditions are imposed at the cut surface. The developed axial symmetric finite element reduces the total computational time radically and allows considering the whole workpiece rather than its small "representative" part (this may be necessary in some cases, for example in the event of deep penetration of the residual stress). Further, consider a numerical example. A steel rod of radius $\rho=20$ mm and of length $l=100$ mm is rolled by two rollers of transverse radius $r=0.908$ mm and of longitudinal one $R>>r$ while the rollers move in opposite directions (Figure 5). Note that a concrete value of R does not matter and if $R\approx r$ then the problem cannot be considered as the axisimmetric one. Rolling length $l=10$ mm, rolling depth $h=0.15$ mm; rolling step d=0.5 mm; number of passes is equal to 1. Material is a steel AISI 4340; the hardening law is $\sigma=\sigma_0+b\varepsilon^n$ where $\sigma_0=792$ MPa;

IOP Conf. Series: Journal of Physics: Conf. Series **973** (2018) 012013 doi:10.1088/1742-6596/973/1/012013

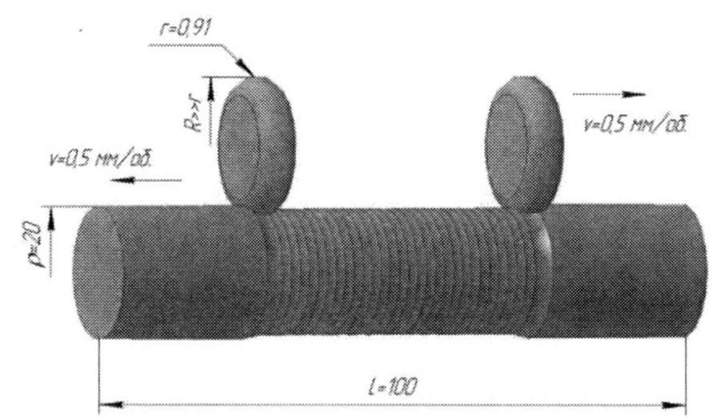

Figure 5. Symmetric deep rolling.

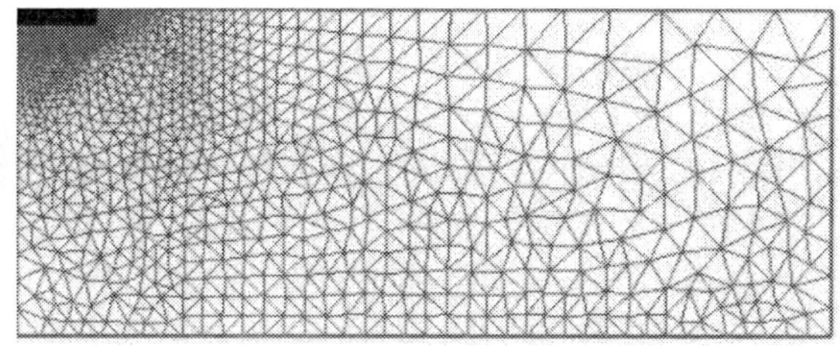

Figure 6. Finite element mesh.

$b = 510$ MPa; $n = 0.26$; Young's modulus $E = 210$ GPa; Poisson's ratio $\mu = 0.324$. The process shown in Figure 5 is not realistic but this process possesses the bilateral symmetry therefore the total number of nodes may reduced twice. In practice, the working is carried out by a single roller or by two-three rollers rotating in a common plane. The finite element mesh is presented in Figure 6 where the finest discretisation is marked out by black; the size of this region is 5×1mm and the splitting numbers here are 50×10. A contact problem statement conforms to [9, 10]. A similar problem in 3D statement was investigated in [11] with the only difference: longitudinal roller's radius was equal to $R = 2.5$ mm. The results of both works (the present and [11]) are compared in Figure 7 (the finite elements meshes correspond to each other), where the axial component of residual stresses is presented (two other components are small).

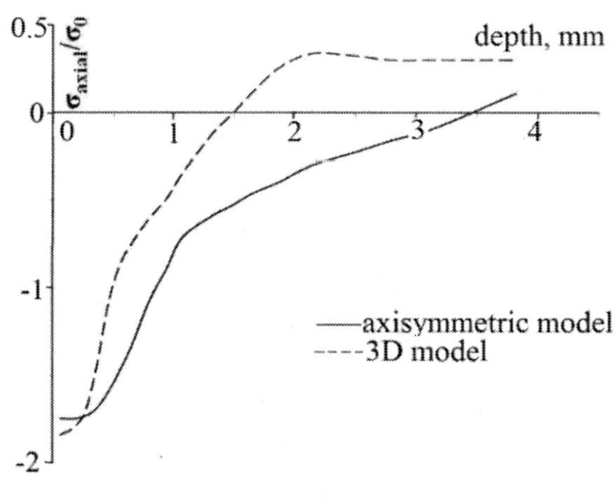

Figure 7. Axial residual stress up to depth 4mm.

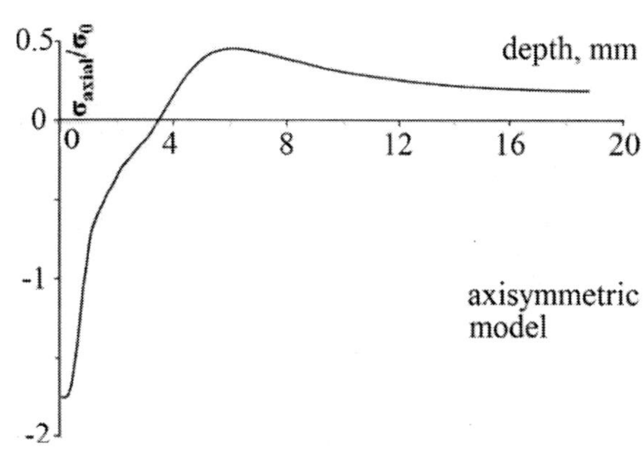

Figure 8. Axial residual stress along the entire depth

The negative sign demonstrates the compressive character of this stress. The peak values are almost equal. The difference between the two plots is explained by different values of radius R. It is obvious that if $R >> r$ then the contact zone is of the form of a very oblong oval and is much larger than in case $R = 2.5$ mm. Correspondingly, the hardening depth is larger too when $R >> r$. It should be pointed out that the three-dimensional model [11] was calculated up to depth 3 mm only and in the present work an extrapolation of those results up to depth 4 mm is given. The developed finite element permits to calculate the model without any restriction in depth and the proper result is presented in Figure 8. It is seen that at depth 3.5 mm the axial component changes its sign and reaches a maximum at depth 6mm. The fractional increase of yield stress (here, it is a measure of hardening) in different cross

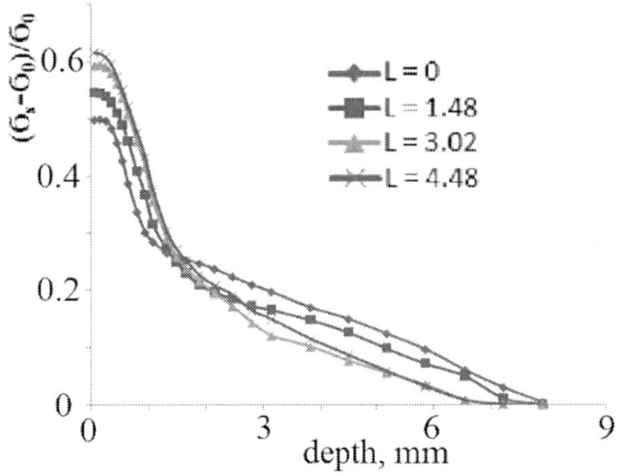

Figure 9. Relative hardening in different cross sections.

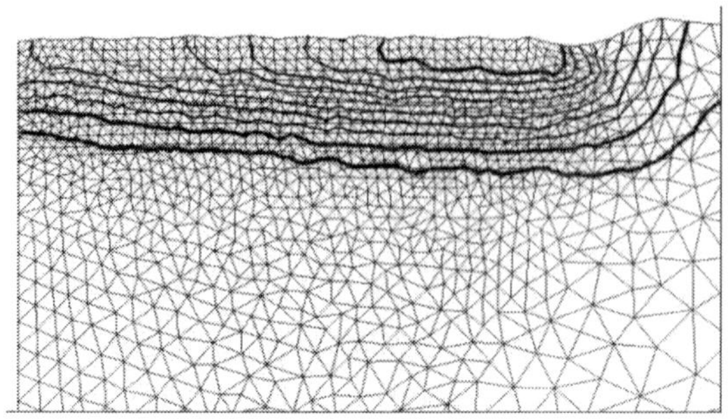

Figure 10. Odkvist's parameter isolines.

sections of the rod is shown in Figure 9, where L is the distance from the rod middle. It is seen that the hardening profiles are not identical hence the deep rolling process is not stabilized. The reason is surface buckling in front of the roller, and this local distortion grows during the process of working. Finally, the isolines of Odkvist's parameter are presented in Figure 10 in the range 0.078-0.78. It is seen, they are not parallel to the surface. This confirms that the process is not stabilized. Thus, it is necessary to model a whole workpiece in order to obtain an adequate result but 3D modelling of periodic local deformation is too time-consuming for modern computers. In contrast to 3D approach, it is proposed to consider the deep rolling problem as the axisymmetric one in some special events (the restriction in radii $R >> r$ is required) and to simulate the behaviour of the whole workpiece.

In this work, only the axial component of the residual stress tensor is analyzed. As for the other components, one can turn to the results [11, 12] where all the components were calculated. Those results indicate that the stress component in the radial direction is not as significant and may be ignored altogether. However, the circular stress component in the direction which coincides with rolling direction is considerable but still is significantly lower than the stress component in the axial direction which is perpendicular to the direction of rolling. Therefore, only the axial stress component is considered for evaluation of the effects of the rolling parameters on the residual stress distribution induced by deep rolling.

Discuss some practical aspects of the work. To fulfil the deep rolling treatment, it is necessary to design an appropriate technological process (numerical modelling is an essential stage of this design) and to manufacture a proper rig (see for example Figure 1 and Figure 2). Therefore, the total cost of a produced detail will depend on the sum of the design cost and the rig manufacturing one. At first sight, a roller of small radius is preferable because its cost is lower in comparison to a roller of large radius. On the other hand, the design will require a three dimensional numerical model in such case because a two dimensional model leads to great errors. The results on ball burnishing modelling [13] indicate that the difference between calculated and measured residual stress values vary by 40% for 2D simulation and only by 7% for 3D one. According to that work, the plane strain assumption for 2D model prohibits the material from flowing along the out-of-plane tangential direction, and thus, compared to 3D simulations, the plastic strain decreases clearly. Consequently, the maximum residual stress introduced is affected. As for deep rolling simulation, the results [13] should be corrected. Indeed, the contact zone for the deforming roller in the deep rolling process is of the oval form and plastic flowing in tangential direction is far less than in axial one if this oval is strongly oblong. Hence, the error of axial stress 2D calculations may be not as large as 40% for a sufficiently large roller. One can conclude from this resume: 3D model is necessary if a small roller is employed in the treatment. But, this requires a supercomputer application. In turn, each supercomputer has its own operating system and its own programming peculiarities. Therefore, the program debugging cost is added to the detail one. Note that a 3D model is too time-consuming and that supercomputer time cost is much greater than the time cost of a personal computer. A simplified 3D simulation may bring essential mistakes as it was shown in this Section hence the initial model has to be considered and the application of supercomputers appears to be necessary. On the other hand, the cost of a big roller is greater but all the design of the technological process can be carried out by means of a personal computer because the axisymmetric model proves to be sufficient. The comparison of these two variants (small roller or large roller) is the task to be resolved at the initial stage of the process design. It is obvious that if the material of which the roller will be made is cheap then large roller's size is preferable. And if the material is of a high price then the opportunity of compound roller may be considered when the roller consists of a work part (material of a high price) and an inside (a cheap material).

One more issue to be considered is how to determine a concrete value of roll's radius R. Strictly speaking; a series of three dimensional numerical experiments has to be performed for increasing values of R as $R \rightarrow \infty$. Taking into account a great cost of such investigation, it may be reasonable to be confined to initial stages of the treatment. An appropriate approximation (that is adequate for a given concrete event) by the axial symmetric model will give the optimal value of radius R. In this case, both the design and the control of the process will be an easy task. If it is not the case (for example if $R \approx r$) then serious problems may occur.

4. Conclusion

A stabilized nodal averaged axisymmetric finite element was developed. The formulation is based on the virtual work principle for large irreversible strain together with a nodal averaged approximation and an approximate formula for a ring of triangular profile. An additional stabilizing stress tensor appears to be necessary for incremental numerical analysis. Simple expressions for nodal force and stiffness matrix permit to obtain fast algorithms and short program codes. The finite element can be exploited successfully for numerical simulation of the most complicated task in solid mechanics – multiple periodic local deformations as it takes place in the deep rolling process. It becomes possible to analyze a whole detail under working rather than its small three-dimensional part. The necessary condition is that longitudinal roller's radius must be sufficiently large. If this is not the case then the task cannot be considered as the axisymmetric one hence both a mathematical simulation and a process control may be too problematic.

A steel rod under deep rolling treatment is simulated. Profiles of residual stress (its axial component) and the yield stress are obtained. The residual stress is calculated in all the cross section rather than in a thin near-surface layer as this occurs in three dimensional models. The Odkvist's parameter isolines are not parallel to the surface wrought hence the process is not stabilized. A bulged surface forms in front of the roller while this bulging grows in the process of working. Taking into account that deep rolling may require many passes, three dimensional simulations of such processes seem to be ineffective.

Acknowledgements
Special thanks to E. Yu. Beregovaya for her assistance in program code debugging and for developing of pre- and post-processors. The work was carried out within the bounds of the Russian Federation government target 1.5265.2017/БЧ №1.5265.2017/8.9.

References

[1] Bagherifard S, Ghelichi R and Guagliano M 2010 A Numerical Model of Severe Shot Peening (SSP) to predict the Generation of a Nanostructured Surface Layer of Material *Surface and Coatings Technology* **204** 4081
[2] Radchenko S Yu, Dorokhov D O and Gryadunov I M 2015 The volumetric surface hardening of hollow axisymmetric parts by roll stamping method *J. of Chem. Tech. and Metallurgy*

50 104

[3] Gryadunov I M, Radchenko S Yu, Dorokhov D O and Morrev P G 2015 Deep Hardening of Inner Cylindrical Surface by Periodic Deep Rolling - Burnishing Process *Modern Applied Science* **9** 251

[4] Dohrmann C R, Heinstein M W, Jung J, Key S W and Witkowski W R 2000 Node-based uniform strain elements for three-node triangular and four-node tetrahedral meshes *Int. J. for Num. Meth. in Eng.* **47** 1549

[5] Bonet J, Marriot M and Hassan O 2001 An averaged nodal deformation gradient linear tetrahedral element for large strain explicit dynamic applications *Com. in Num. Meth. in Eng.* **17** 551

[6] Puso M A and Solberg J 2006 A formulation and analysis of a stabilized nodally integrated tetrahedral *Int. J. for Num. Meth. in Eng.* **67** 841

[7] Puso M A, Chen J S, Zywicz E and Elmer W 2008 Meshfree and finite element nodal integration methods *Int. J. for Num. Meth. in Eng.* **74** 416

[8] Zienkiewicz O C 1971 *The Finite Element Method in Engineering Science* (London: McGraw-Hill) p 521

[9] Morev P G 2004 A version of the finite element method for frictional contact problems *Mechanics of Solids* **42** 640

[10] Morrev P G 2011 A variational statement of quasistatic “rigid-deformable” contact problems at large strain involving generalized forces and friction *Acta Mechanica* **222** 115

[11] Majzoobi G H, Teimoorial Motlagh S and Amiri A 2010 Numerical simulation of residual stress induced by roll-peening *Trans. of The Indian Inst. of Metals* **63** 499

[12] Majzoobi G H, Azadikhah K and Nemati J 2009 The Effects of Deep Rolling and Shot Peening on Fretting Fatigue Resistance of Aluminum *Materials Sc. and Eng. A* **516** 235

[13] Sayahi M, Sghaier S, and Belhadjsalah H 2013 Finite Element Analysis of Ball Burnishing Process: Comparisons Between Numerical Results and Experiments *The Int. J. of Adv. Manuf. Tech.* **67**

Dynamics of layered reinforced concrete beam on visco-elastic foundation with different resistances of concrete and reinforcement to tension and compression

Y V Nemirovsky[1], **S V Tikhonov**[2]

[1]Physics of Fast Processes Laboratory, Khristianovich Institute of Theoretical and Applied Mechanics Siberian Branch of the Russian Academy of Sciences, Institutskaya str., 4/1, Novosibirsk, 630090, Russia

[2]Faculty of Information and Computer Systems, I. Ulianov Chuvash State University, 15, Moskovskiy pr., Cheboksary, Chuvash Republic, 428015, Russia

E-mail: nemiryury@mail.ru

Abstract. Originally, fundamentals of the theory of limit equilibrium and dynamic deformation of building metal and reinforced concrete structures were created by A. A. Gvozdev [1] and developed by his followers [4, 5, 6, 7, 11, 12]. Forming the basis for the calculation, the model of an ideal rigid-plastic material has enabled to determine in many cases the ultimate load bearing capacity and upper (kinematically possible) or lower (statically valid) values for a wide class of different structures with quite simple methods. At the same time, applied to concrete structures the most important property of concrete to significantly differently resist tension and compression was not taken into account [10]. This circumstance was considered in [3] for reinforced concrete beams under conditions of quasistatic loading. The deformation is often accompanied by resistance of the environment in construction practice [8, 9]. In [2], the dynamics of multi-layered concrete beams on visco-elastic foundation under the loadings of explosive type is considered. In this work we consider the case which is often encountered in practical applications when the loadings weakly change in time.

1. Introduction

In the scientific literature the calculation of reinforced concrete structures is often limited to the case of the simplest forms of the rod cross-section and simplest conditions of loading and fixing. Modern technological capabilities to create flexible sets of hybrid laminated reinforced structures, where various grades of concrete, reinforcing elements and reinforcement structures can be implemented in layers of cross-section, are not taken into account. In this paper, the problem of limit equilibrium and dynamic deformation of arbitrary reinforced rods based on concrete is examined by the ideas of the model of perfectly plastic deals.

2. Methods

In this paper we used a classical model of ideal rigid-plastic body for all materials (concrete and reinforcement) in accordance with the diagram shown in Figure 1.

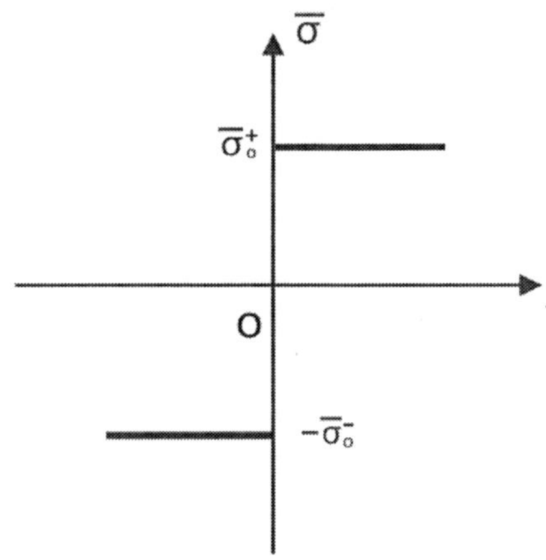

Figure 1. The diagram of concrete and reinforcement deformation, where the $\overline{\sigma}_0^+$ is a tensile yield strength, the $\overline{\sigma}_0^-$ is the yield strength in compression

We assume that the structure of the reinforcement in layers 1 and 3 of the beam cross section varies according to the intensity and properties of the phase materials. For all considered sections we assume the classical kinematical hypotheses of Kirchgoff-Lyav to be valid, according to which the deformation will have the expression

$$\overline{\varepsilon}(\overline{x}, \overline{z}) = \overline{e}_0(\overline{x}) + \overline{z}\kappa(\overline{x}), \quad \overline{e}_0(\overline{x}) = \frac{d\overline{u}_0}{d\overline{x}}, \qquad \kappa(\overline{x}) = -\frac{d^2\overline{w}}{d\overline{x}^2},$$

where the $u_0(x)$ is movement along the reporting axis $\overline{x}$ of the beam, $\overline{w}(\overline{x})$ is a de-flection.

3. Results

Consider an ideal-plastic section of the beam, for which

$$\overline{e}_0(\overline{x}) < 0, \quad \overline{e}_0 + \overline{h}_3\kappa(\overline{x}) > 0.$$

Both inequalities will be satisfied if

$$\overline{e}_0(\overline{x}) < 0, \quad \kappa(\overline{x}) > 0.$$

The state of stress will correspond to figure 2.

Then, for the considered part of the beam the longitudinal force $\overline{N}$ is defined by the equality

$$\begin{gathered}\overline{N} = -2\overline{\sigma}_{01}^- \int_0^{\overline{h}_1} \overline{b}_1(\overline{z})d\overline{z} - 2\overline{\sigma}_{02}^- \int_{\overline{h}_1}^{\overline{z}_1} \overline{b}_2^0 d\overline{z} + 2\overline{\sigma}_{02}^+ \int_{\overline{z}_1}^{h_2} \overline{b}_2^0 d\overline{z} + \\ +2\overline{\sigma}_{03}^+ \int_{\overline{h}_2}^{\overline{h}_3} \overline{b}_3(\overline{z})d\overline{z} = -2\overline{b}_2^0\left(\overline{\sigma}_{02}^- + \overline{\sigma}_{02}^+\right)\overline{z}_1 + 2\overline{b}_2^0\left(\overline{\sigma}_{02}^-\overline{h}_2 + \overline{\sigma}_{02}^+\overline{h}_1\right) - \\ -2\overline{\sigma}_{01}^- \int_0^{\overline{h}_1} \overline{b}_1(\overline{z})d\overline{z} + 2\overline{\sigma}_{03}^+ \int_{\overline{h}_2}^{\overline{h}_3} \overline{b}_3(\overline{z})d\overline{z}.\end{gathered}$$

The expression for the bending moment $\overline{M}$ will be

$$\begin{gathered}\overline{M} = -2\overline{\sigma}_{01}^- \int_0^{\overline{h}_1} \overline{b}_1(\overline{z})\overline{z}d\overline{z} - 2\overline{\sigma}_{02}^- \int_{\overline{h}_1}^{\overline{z}_1} \overline{b}_2^0\overline{z}d\overline{z} + 2\overline{\sigma}_{02}^+ \int_{\overline{z}_1}^{h_2} \overline{b}_2^0\overline{z}d\overline{z} + \\ +2\overline{\sigma}_{03}^+ \int_{\overline{h}_2}^{\overline{h}_3} \overline{b}_3(\overline{z})\overline{z}d\overline{z} = -\overline{b}_2^0\left(\overline{\sigma}_{02}^- + \overline{\sigma}_{02}^+\right)\overline{z}_1^2 + \overline{b}_2^0\left(\overline{\sigma}_{02}^-\overline{h}_1^2 + \overline{\sigma}_{02}^+\overline{h}_2^2\right) - \\ -2\overline{\sigma}_{01}^- \int_0^{\overline{h}_1} \overline{b}_1(\overline{z})\overline{z}d\overline{z} + 2\overline{\sigma}_{03}^+ \int_{\overline{h}_2}^{\overline{h}_3} \overline{b}_3(\overline{z})\overline{z}d\overline{z}.\end{gathered} \qquad (1)$$

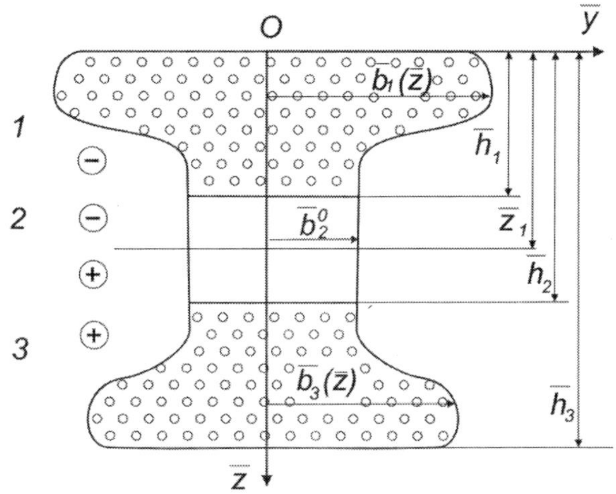

Figure 2. The beam cross section. Numbers 1, 2, 3 refer to the layers of the beam.

In the case of transverse bending ($\overline{N} = 0$), we have

$$\bar z_1 = \frac{\bar b_2^0\left(\bar\sigma_{02}^-\bar h_2 + \bar\sigma_{02}^+\bar h_1\right) - \bar\sigma_{01}^-\int_0^{\bar h_1}\bar b_1(\bar z)d\bar z + \bar\sigma_{03}^+\int_{\bar h_2}^{\bar h_3}\bar b_3(\bar z)d\bar z}{\bar b_2^0(\bar\sigma_{02}^- + \bar\sigma_{02}^+)},$$

and for $\bar z_1$ the ratio $\bar h_1 \le \bar z_1 \le \bar h_2$ must be satisfied.

Substituting $\bar z_1$ in (1), we obtain the expression for the limit bending moment

$$\overline{M_0^+} = \frac{\left(\bar b_2^0(\bar\sigma_{02}^-\bar h_2 + \bar\sigma_{02}^+\bar h_1) - \bar\sigma_{01}^-\int_0^{\bar h_1}\bar b_1(\bar z)d\bar z + \bar\sigma_{03}^+\int_{\bar h_2}^{\bar h_3}\bar b_3(\bar z)d\bar z\right)^2}{\bar b_2^0(\bar\sigma_{02}^- + \bar\sigma_{02}^+)} +$$
$$+\bar b_2^0\left(\bar\sigma_{02}^-\bar h_1^2 + \bar\sigma_{02}^+\bar h_2^2\right) - 2\bar\sigma_{01}^-\int_0^{\bar h_1}\bar b_1(\bar z)\bar z d\bar z + 2\bar\sigma_{03}^+\int_{\bar h_2}^{\bar h_3}\bar b_3(\bar z)\bar z d\bar z.$$

For an ideal-plastic section of the beam with the deformed state

$$\bar e_0(\bar x) > 0, \quad \bar e_0 + \bar h_3 \ \ \kappa(\bar x) < 0, \quad \bar e_0(\bar x) > 0, \quad \kappa(\bar x) < 0.$$

Similarly, it is possible to determine the limiting bending moment $\overline{M_0^-}$

$$\overline{M_0^-} = \frac{\left(\bar b_2^0(\bar\sigma_{02}^+\bar h_1 + \bar\sigma_{02}^-\bar h_2) - \bar\sigma_{01}^+\int_0^{\bar h_1}\bar b_1(\bar z)d\bar z - \bar\sigma_{03}^-\int_{\bar h_2}^{\bar h_3}\bar b_3(\bar z)d\bar z\right)^2}{\bar b_2^0(\bar\sigma_{02}^+ + \bar\sigma_{02}^-)} -$$
$$-\bar b_2^0\left(\bar\sigma_{02}^+\bar h_1^2 + \bar\sigma_{02}^-\bar h_2^2\right) + 2\bar\sigma_{01}^+\int_0^{\bar h_1}\bar b_1(\bar z)\bar z d\bar z - 2\bar\sigma_{03}^-\int_{\bar h_2}^{\bar h_3}\bar b_3(\bar z)\bar z d\bar z.$$

Thus, in the general case the beam is divided into sections corresponding to the deformed states $\bar e_0 < 0$, $\kappa < 0$ or $\bar e_0 > 0$, $\kappa > 0$ and $\bar e_0 = \kappa = 0$. The last correspond to the rigid undeformed state. The number of rigid sections and the plastic hinges separating them will depend on the conditions of fastening of the beams, the type of distributed and concentrated loads and the law of resistance of the supporting environment.

Consider a reinforced concrete beam of length $\bar l$, lying on visco-elastic foundation and cantilever-fixed on the left edge of $\bar x = 0$.

IOP Conf. Series: Journal of Physics: Conf. Series **973** (2018) 012014 doi:10.1088/1742-6596/973/1/012014

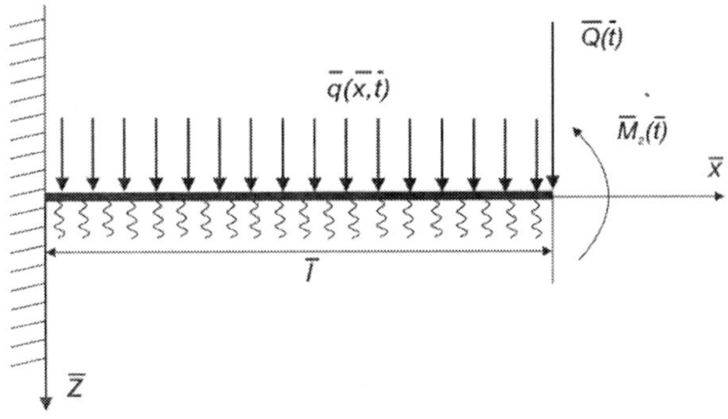

Figure 3. A multilayer beam cantilever-fixed on the left edge.

The beam is loaded by a distributed transverse load $\overline{q(x,t)}$, its own weight $\overline{q}_s = \overline{mg}$, concentrated load $\overline{Q}(\overline{t})$ and moment $\overline{M}_2(\overline{t})$ at the end of $\overline{x} = \overline{l}$. Here, $\overline{m}$ is the mass of length unit of the beam. For a given beam, the mass will be determined by

$$\overline{m} = \int_0^{\overline{h_1}} \left(\overline{\rho}_1^a \mu_1 + \overline{\rho}_1^c (1-\mu_1)\right) \overline{b}_1(\overline{z}) d\overline{z} + 2\overline{b}_2^0 \overline{\rho}_2^c (\overline{h}_2 - \overline{h}_1) +$$
$$+ \int_{\overline{h}_2}^{\overline{h_3}} \left(\overline{\rho}_3^a \mu_3 + \overline{\rho}_3^c (1-\mu_3)\right) \overline{b}_3(\overline{z}) d\overline{z}$$

where $\rho_1^a, \rho_3^a, \rho_1^c, \rho_2^c, \rho_3^c$ are the density of reinforcement materials and binding concrete, μ_1, μ_3 are the coefficients of reinforcement.

Represent the current external loads in the form of

$$\overline{q}(\overline{x},\overline{t}) = \overline{q}_1 \overline{\varphi}(\overline{x}) \overline{\psi}_1(\overline{t}), \qquad \overline{Q}(\overline{t}) = \overline{Q}_1 \overline{\psi}_2(\overline{t}), \quad \overline{M}_2(\overline{t}) = \overline{M}_{12} \overline{\psi}_3(\overline{t}).$$

A model of an ideal resisting rigid-plastic material will be used for all phase materials of the reinforced concrete beam. Then we have

$$\overline{\sigma}_{01}^{\pm} = \mu_1 \left(\overline{\sigma}_{01}^{\pm}\right)^a + (1-\mu_1) \left(\overline{\sigma}_{01}^{\pm}\right)^c,$$
$$\overline{\sigma}_{03}^{\pm} = \mu_3 \left(\overline{\sigma}_{03}^{\pm}\right)^a + (1-\mu_3) \left(\overline{\sigma}_{03}^{\pm}\right)^c.$$

For beams of constant cross section, loaded with an uniformly distributed load ($\varphi(x) = 1$), if $M_2(\overline{t}) \geq -\overline{M_0^-}$, the movement will be provided, as a rigid rod, with the formation of plastic hinge in the left edge $x = 0$. Then for the deflection $\overline{w}$ we will have the expression

$$\overline{w}(\overline{x},\overline{t}) = \overline{w}_0(\overline{t}) \frac{\overline{x}}{\overline{l}}, \quad 0 \leq \overline{x} \leq \overline{l}$$

and taking into account that according to the conditions of dynamic equilibrium the main point regarding the support $\overline{x} = 0$ of all active forces, including inertia forces must be equal to zero, we obtain the equation

$$\overline{M}_0^+ = -\overline{M}_2(\overline{t}) + \overline{l}\overline{Q}(\overline{t}) + \int_0^{\overline{l}} \left[\overline{q}(\overline{x},\overline{t}) + \overline{q}_s - \overline{q}_r - \overline{m}\overline{x}\ddot{\overline{\theta}}\right] \overline{x} d\overline{x}, \tag{2}$$

where $\overline{\theta} = \frac{\partial \overline{w}}{\partial \overline{x}} = \frac{\overline{w}_0(\overline{t})}{\overline{l}}$, $\overline{q}_r = \overline{b}_0 \overline{k}_1 \overline{w} + \overline{b}_0 \overline{k}_2 \dot{\overline{w}}$, $\overline{k}_1, \overline{k}_2$ are the coefficients of the elastic and viscous resistance of the base, and then we assume that $\overline{k}_1 > 0, \overline{k}_2 > 0$, $\overline{b}_0 = \overline{b}_3(\overline{h}_3)$ are the length of the base of the beam cross section. A point denotes a partial derivative in time $\overline{t}$.

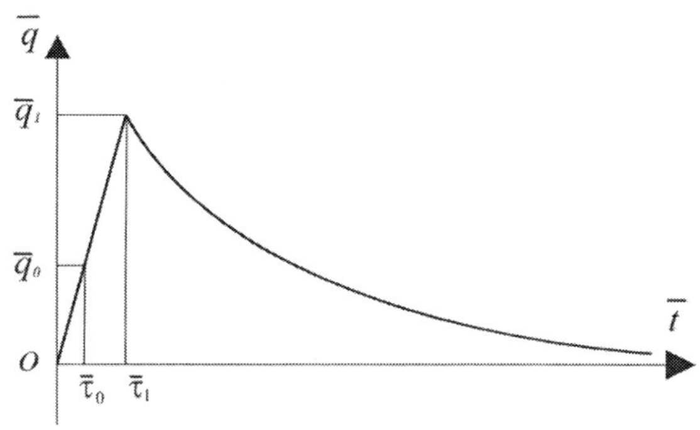

Figure 4. The law of change $\overline{q}(\overline{x},\overline{t})$

From (2), we get the equation to determine the deflection

$$\ddot{\overline{w}}_0(\overline{t}) + \frac{\overline{b}_0\overline{k}_2}{\overline{m}}\dot{\overline{w}}_0(\overline{t}) + \frac{\overline{b}_0\overline{k}_1}{\overline{m}}\overline{w}_0(\overline{t}) = \gamma(\overline{t}), \quad (3)$$

where

$$\gamma(\overline{t}) = \frac{3}{2\overline{m}}\overline{q}_1\psi_1(\overline{t}) + \frac{3\overline{g}}{2} + \frac{3}{\overline{m}\overline{l}^2}(-\overline{M}_0^+ + \overline{l}\overline{Q}_1\psi_2(\overline{t}) - \overline{M}_{12}\psi_3(\overline{t})). \quad (4)$$

Discarding in (3) the inertial member $\ddot{\overline{w}}_0(\overline{t})$, we obtain the equation of quasi-static change of the beam overlimited deflection

$$\frac{\overline{b}_0\overline{k}_2}{\overline{m}}\dot{\overline{w}}_0(\overline{t}) + \frac{\overline{b}_0\overline{k}_1}{\overline{m}}\overline{w}_0(\overline{t}) = \gamma(\overline{t}), \quad (5)$$

where $\gamma(\overline{t})$ is defined by equation (4).

Taking the value $\overline{w}_0(\overline{t}) = 0$, we obtain the equation for the first limiting amplitudes of the considered beam

$$\frac{3}{2\overline{m}}\overline{q}_1\psi_1(\overline{t}) + \frac{3\overline{g}}{2} + \frac{3}{\overline{m}\overline{l}^2}(-\overline{M}_0^+ + \overline{l}\overline{Q}_1\psi_2(\overline{t}) - \overline{M}_{12}\psi_3(\overline{t})) = 0. \quad (6)$$

From relation (6) we determine the point in time $\overline{\tau}_0$ when the loads reach the limit value and the plastic hinge occurs at the pinch point of the beam.

If at $\overline{t} \geq \overline{\tau}_0$ the loads monotonically quasistatically temporarily increase, then integrating the equation (5) under the initial conditions

$$\overline{w}_0(\overline{\tau}_0) = 0. \quad (7)$$

for the deflection we get the expression

$$\overline{w}_0(\overline{t}) = \frac{\overline{m}}{\overline{b}_0\overline{k}_2}e^{-\frac{\overline{k}_1\overline{t}}{\overline{k}_2}}\int\limits_{\overline{\tau}_0}^{\overline{t}}\gamma(\overline{z})e^{\frac{\overline{k}_1\overline{z}}{\overline{k}_2}}\,d\overline{z}.$$

If we accept that the law of pressure variation on the beam $\overline{q}(\overline{x},\overline{t}) = \overline{q}_1\overline{\psi}_1(\overline{t})$ changes according to the law as shown in figure 4

$$\psi_1(\overline{t}) = \begin{cases} \frac{\overline{t}}{\overline{\tau}_1}, & \text{if } \overline{\tau}_0 \leq \overline{t} \leq \overline{\tau}_1, \\ e^{-(\overline{t}-\overline{\tau}_1)}, & \text{, if } \overline{\tau}_1 \leq \overline{t} \leq \infty, \end{cases} \tag{8}$$

and the acting external moment and concentrated force at $\overline{x} = \overline{l}$ are absent

$$\overline{M}_2(\overline{t}) = 0, \quad \overline{Q}(\overline{t}) = 0.$$

Then before reaching the first load limit $\overline{q}_0$ at the time $0 \leq \overline{t} < \overline{\tau}_0$ the beam will be rigid, and for the time moment of formation of the plastic hinge and the value of the first limit load it is true that

$$\overline{\tau}_0 = \frac{2\overline{m}}{3\overline{q}_1}\left(\frac{3\overline{M}_0^+}{\overline{m}\overline{l}^2} - \frac{3\overline{g}}{2}\right)\overline{\tau}_1, \quad \overline{q}_0 = \frac{2\overline{m}}{3}\left(\frac{3\overline{M}_0^+}{\overline{m}\overline{l}^2} - \frac{3\overline{g}}{2}\right). \tag{9}$$

From relations (7) and assumptions $0 < \overline{\tau}_0 < \overline{\tau}_1$ the following inequalities have the form of

$$0 < \frac{3\overline{M}_0^+}{\overline{m}\overline{l}^2} - \frac{3\overline{g}}{2} < 1.$$

In case of load increase according to the law (8) after reaching the first limit load, for the time interval $\overline{\tau}_0 \leq \overline{t} \leq \overline{\tau}_1$ the relations (4), (5) take the form of

$$f_1\dot{\overline{w}}_0(\overline{t}) + f_2\overline{w}_0(\overline{t}) = f_4\overline{t} + f_3 \tag{10}$$

where

$$f_1 = \frac{\overline{b}_0\overline{k}_2}{\overline{m}}, \quad f_2 = \frac{\overline{b}_0\overline{k}_1}{\overline{m}}, \quad f_3 = \frac{3\overline{g}}{2} - \frac{3}{\overline{m}\overline{l}^2}\overline{M}_0^+, \quad f_4 = \frac{3}{2\overline{m}}\frac{\overline{q}_1}{\overline{\tau}_1}.$$

Integrating equation (10) under the initial conditions (7) we obtain an expression for deflection for the time interval $\overline{\tau}_0 \leq \overline{t} \leq \overline{\tau}_1$

$$\overline{w}(\overline{t}) = g_1 e^{g_2\overline{t}} + g_3\overline{t} + g_4, \tag{11}$$

where

$$g_1 = \left(-\frac{f_3\overline{\tau}_0 + f_4}{f_2} + \frac{f_3 f_1}{f_2^2}\right)e^{\frac{f_2}{f_1}\overline{\tau}_0}, \quad g_2 = -\frac{f_2}{f_1}, \quad g_3 = \frac{f_3}{f_2}, \quad g_4 = \frac{f_4}{f_2} - \frac{f_3 f_1}{f_2^2}.$$

The deflection rate for the indicated interval will be equal to

$$\dot{\overline{w}}(\overline{t}) = g_1 g_2 e^{g_2\overline{t}} + g_3.$$

We denote the deflection by $\overline{w}_{0\tau}$ at time $\overline{t} = \overline{\tau}_1$

$$\overline{w}(\overline{\tau}_1) = g_1 e^{g_2\overline{\tau}_1} + g_3\overline{\tau}_1 + g_4 = \overline{w}_{0\tau}. \tag{12}$$

For time interval $\overline{\tau}_1 \leq \overline{t} \leq \infty$ the relations (4), (5) take the form

$$f_1\dot{\overline{w}}(\overline{t}) + f_2\overline{w}_0(\overline{t}) = j_4 e^{-\overline{t}} + j_3, \tag{13}$$

where

$$j_3 = \frac{3\overline{g}}{2} - \frac{3}{\overline{ml}^2}\overline{M}_0^+, \quad j_4 = \frac{3\overline{q}_1}{2\overline{m}}e^{\overline{\tau}_1}.$$

The solution to (13) with conditions (12) has the form

$$\overline{w}(\overline{t}) = v_1 e^{v_2\overline{t}} + v_3 e^{-\overline{t}} + v_4, \tag{14}$$

where

$$v_1 = \left(\overline{w}_{0\tau} - \frac{j_3}{f_2} + \frac{j_4}{f_1 - f_2}e^{-\tau_1}\right)e^{\frac{f_2}{f_1}\overline{\tau}_1}, \quad v_2 = -\frac{f_2}{f_1}, \quad v_3 = -\frac{j_4}{f_1 - f_2}, \quad v_4 = \frac{j_3}{f_2}.$$

If $0 \le \overline{t} < \overline{\tau}_0$, the beam remains rigid, then the equation to determine the moment has the form

$$\frac{\partial^2 \overline{M}}{\partial \overline{x}^2} = \overline{q}(\overline{x}, \overline{t}) + \overline{q}_s, \tag{15}$$

integrating which with the initial conditions

$$\overline{M}(\overline{l}, \overline{t}) = 0, \quad \frac{\partial \overline{M}}{\partial x}(\overline{l}, \overline{t}) = 0, \tag{16}$$

we obtain an expression for the moment that is valid when $0 \le \overline{t} < \overline{\tau}_0$

$$\overline{M}(\overline{x}) = \frac{\overline{q}_1\overline{t} + \overline{mg}\overline{\tau}_1}{2\overline{\tau}_1}\overline{x}^2 - \frac{\overline{q}_1\overline{t}\overline{l} + \overline{mg}\overline{l}\overline{\tau}_1}{\overline{\tau}_1}\overline{x} - \frac{3(\overline{q}_1\overline{t} + \overline{mg}\overline{\tau}_1)\overline{l}^2}{2\overline{\tau}_1}.$$

The equation of beam deflection in the case of plastic hinge in the place of pinch of a beam has the form

$$\frac{\partial^2 \overline{M}}{\partial \overline{x}^2} = \overline{q}(\overline{x}, \overline{t}) + \overline{q}_s - \frac{\overline{k}_1\overline{b}_0\overline{x}\overline{w}_0(\overline{t})}{\overline{l}} - \frac{\overline{k}_2\overline{b}_0\overline{x}\dot{\overline{w}}_0(\overline{t})}{\overline{l}}, \tag{17}$$

adding to which the expression for deflection (10), we obtain

$$\frac{\partial^2 \overline{M}}{\partial \overline{x}^2} = \overline{x}\left(g_5 e^{g_2\overline{t}} + g_6\overline{t} + g_7\right) + g_8\overline{t} + g_9, \tag{18}$$

where

$$g_5 = -\frac{\overline{k}_1\overline{b}_0}{\overline{l}}g_1 - \frac{\overline{k}_2\overline{b}_0}{\overline{l}}g_1g_2, \quad g_6 = -\frac{\overline{k}_1\overline{b}_0}{\overline{l}}g_3, \quad g_7 = -\frac{\overline{k}_1\overline{b}_0}{\overline{l}}g_4 - \frac{\overline{k}_2\overline{b}_0}{\overline{l}}g_3, \quad g_8 = \frac{\overline{q}_1}{\overline{\tau}_1}, \quad g_9 = \overline{mg}.$$

Integrating (18) under the initial conditions (16), we obtain the expression for the moment that is valid for $\overline{\tau}_0 \le \overline{t} < \overline{\tau}_1$

$$\overline{M} = \overline{x}^3\left(\tfrac{g_5}{6}e^{g_2\overline{t}} + \tfrac{g_6}{6}\overline{t} + \tfrac{g_7}{6}\right) + \overline{x}^2\left(\tfrac{g_8}{2}\overline{t} + \tfrac{g_9}{2}\right) - \overline{x}\left(\tfrac{g_5\overline{l}^2}{2}e^{g_2\overline{t}} + \left(\tfrac{g_6\overline{l}^2}{2} + g_8\overline{l}\right)\overline{t} + \tfrac{g_7\overline{l}^2}{2} + g_9\overline{l}\right) +$$
$$+ \tfrac{g_5\overline{l}^3}{3}e^{g_2\overline{t}} + \left(\tfrac{g_6\overline{l}^3}{3} + \tfrac{g_8}{2}\overline{l}^2\right)\overline{t} + \tfrac{g_7\overline{l}^3}{3} + \tfrac{g_9}{2}\overline{l}^2.$$

Substituting in (17) the expression for the deflection (13), we obtain

$$\frac{\partial^2 \overline{M}}{\partial \overline{x}^2} = \overline{x}\left(v_5 e^{v_2\overline{t}} + v_6 e^{-\overline{t}} + v_7\right) + v_8 e^{-\overline{t}} + v_9, \tag{19}$$

where

$$v_5 = -\frac{\overline{k}_1\overline{b}_0}{\overline{l}}v_1 - \frac{\overline{k}_2\overline{b}_0}{\overline{l}}v_1v_2, \quad v_6 = -\frac{\overline{k}_1\overline{b}_0}{\overline{l}}v_3 + \frac{\overline{k}_2\overline{b}_0}{\overline{l}}v_3, \quad v_7 = -\frac{\overline{k}_1\overline{b}_0}{\overline{l}}v_4, \quad v_8 = q_1e^{\overline{\tau}_1}, \quad v_9 = \overline{mg}.$$

Integrating (19) under the initial conditions (16), we obtain the expression for the moment that is valid for $\overline{\tau}_1 \leq \overline{t} < \infty$

$$\overline{M} = \overline{x}^3\left(\tfrac{v_5}{6}e^{v_2\overline{t}} + \tfrac{v_6}{6}e^{-\overline{t}} + \tfrac{v_7}{6}\right) + \overline{x}^2\left(\tfrac{v_8}{2}e^{-\overline{t}} + \tfrac{v_9}{2}\right) -$$
$$-\overline{x}\left(\tfrac{v_5\overline{l}^2}{2}e^{v_2\overline{t}} + \left(\tfrac{v_6\overline{l}^2}{2} + v_8\overline{l}\right)e^{-\overline{t}} + \tfrac{v_7\overline{l}^2}{2} + v_9\overline{l}\right) + \tfrac{v_5\overline{l}^3}{3}e^{v_2\overline{t}} + \left(\tfrac{v_6\overline{l}^3}{3} + \tfrac{v_8}{2}\overline{l}^2\right)e^{-\overline{t}} + \tfrac{v_7\overline{l}^3}{3} + \tfrac{v_9}{2}\overline{l}^2.$$

4. Conclusions

Thus, the obtained expressions are determined for the deflections of a beam and the expression of the moments. This solution is true for the entire length of the beam $0 \leq \overline{x} \leq \overline{l}$ if the bending moment $\overline{M}(\overline{x}, t)$ is within $-\overline{M}_0^- \leq \overline{M}(\overline{x}, t) \leq \overline{M}_0^+$.

This work is carried out with the partial financial support of RFBR grants (projects 17-41-210272, 15-01-00825, 16-31-00511).

References

[1] Gvozdev A To calculation of designs on action of a blast wave 1940 *Building industry.* **1** 18–21
[2] Nemirovski Y and Tikhonov S Dynamics of reinforced concrete beams on the visco-elastic foundation 2017 *Vestnik I. Yakovlev Chuvach State Pedagogical University. Series: Mechanics of a limit state.* **2** 45–64
[3] Nemirovski Y and Tikhonov S Limit conditions of reinforced concrete beams 2016 *Vestnik I. Yakovlev Chuvach State Pedagogical University. Series: Mechanics of a limit state.* **3** 134–158
[4] Erkhov M 1978 *Theory of Perfectly Plastic Solids and Structures* (Moscow: Nauka) p 352
[5] Shamin V 1989 *Calculation of elements of constructions on action of explosive loadings* (Moscow: Stroyizdat) p 72
[6] Komarov K and Nemirovski Y 1984 *Dynamics of rigid and plastic elements of designs* (Novosibirsk: Nauka) p 236
[7] Rzhanicin 1954 *Calculation of constructions taking into account plastic properties of materials* (Moscow: Stroyizdat) p 288
[8] Savelyev N Calculation of beams of variable section on the elastic base 1956 *Calculations on durability.* **2**
[9] Gorbunov-Posadov M 1949 *Beams and plates on the elastic foundation* (Moscow: Mashstroyizdat)
[10] Nemirovski Y and Boltayev A Nonliner deformation of concrete elements at a longitudinally cross bend 2015 *News of higher education institution. Construction.* **6** 125–129
[11] Cox A and Morland W Dynamic plastic deformation simple supported square plates 1959 *J. Mech. Solids.* **7** 229–241
[12] Johns N, Uran T and Tekin S The dynamic plastic behavior of fully clamped rectangular plates 1970 *Int. J. Solid Struct.* **12** 1499–1512

IOP Conf. Series: Journal of Physics: Conf. Series **973** (2018) 012015 doi:10.1088/1742-6596/973/1/012015

An algorithm for full parametric solution of problems on the statics of orthotropic plates by the method of boundary states with perturbations

V B Penkov[1], D A Ivanychev[1], O S Novikova[1] and L V Levina[2]

[1] Department of General mechanics, Lipetsk State Technical University, Moskovskaya 30, Russia
[2] Department of Applied Mathematics, Lipetsk State Technical University, Moskovskaya 30, Russia

E-mail: vbpenkov@mail.ru

Abstract. The article substantiates the possibility of building full parametric analytical solutions of mathematical physics problems in arbitrary regions by means of computer systems. The suggested effective means for such solutions is the method of boundary states with perturbations, which aptly incorporates all parameters of an orthotropic medium in a general solution. We performed check calculations of elastic fields of an anisotropic rectangular region (test and calculation problems) for a generalized plane stress state.

1. Introduction

Historically, there are three phases in the methodology for solving problems that aim to determine the strength of linear-elastic bodies.

Classical (pre-computer) *method*. It is characterized by an analytical approach to the formulation and solution of problems. The results are presented in a strict (closed or quadrature) or approximate (series-based) form and contain all the parameters of the problem (FPS: analytical full parametric solution). The obvious features and advantages of this approach are: wide use and concomitant development of the mathematical apparatus (special functions expanded into series, functions of a complex variable, integral transforms, ...); simplicity of computational procedures used in engineering and design work (including for optimization of solutions as regards geometric and physical parameters). The main disadvantages are: a relative narrowness of the classes available for the analysis of bodies in relation to their "geometry" (unlimited 3D or 2D space, half-space, layers, cylindrical bodies, cones (wedges), etc.); limited options for types of boundary conditions (BC; first main, second main, main combined, contact problems); and rareness of problems focusing on the interaction of homogeneous elastic bodies of different materials.

Computational (computer) *method*. The widespread use of computer technology has practically removed restrictions with regard to classes of geometric configurations of bodies considered. It became possible to find precise solutions for problems with numerous boundary conditions and problems involving piecewise-homogeneous and inhomogeneous bodies. However, the advantages inherent in the classical approach turned into disadvantages: a change of any parameter (outside the P–theorem [1]) requires time-consuming and resource-intensive re-computation; this spurred the need to compare the results of the application of various computational techniques; the energy intensity of optimization problems dramatically increased. Nevertheless, sophisticated numerical computing systems, especially their finite-element and boundary-element options, are now the ultimate analysis tool used by structural engineers.

Computer algebraic (numerical analytical) *method.* The development of modern computing systems supporting "computer algebra" essentially allows combining the advantages of the classical and computational approaches. The (quite solvable) bottleneck here is the plurality of geometric parameters of bodies and physical constants of media. The methodology of FPS for isotropic media has been by and large carefully thought-out by now [2].

The *purpose* of this paper is to develop a method for structuring FPSs for homogeneous anisotropic bodies. It is to be achieved through a system of interconnected procedures: correct problem formulation, non-dimensionization (P-theorem), choice of solution methods, and incorporation of parameters of boundary conditions and physical parameters of the medium in the explicit solution.

Some of the above tasks require clarification. An effective means of building FPSs is the modern power-operated method of boundary states (MBS) [3], initially created to address computerized algebra issues. Its further enhancement to incorporate the perturbation method (MBSP) [3] allows to effectively address particular physical properties of media.

Below we provide a methodology for structuring solutions taking into account medium parameters, using a two-dimensional problem seeking to determine the anisotropic elasticity value as an example.

The most complete presentation of materials on the state of two-dimensional bodies (isotropic, orthotropic) is contained in the fundamental work of S. Timoshenko and S. Woinowsky-Krieger [4] and S. Lehnitsky [5].

2. Reduction of the model of weakly orthotropic environment by the perturbation method

The state of the environment is governed by the generalized Hooke's law (the coordinate axes coinciding with the axes of orthotropy), which expresses the stresses $\sigma_{xx}, \sigma_{yy}, \sigma_{xy}$ with their respective deformations ε_{xx}, ε_{yy}, ε_{xy} ($\theta = \varepsilon_{xx} + \varepsilon_{yy}$):

$$\begin{aligned} \sigma_{xx} &= \lambda_1 \theta + 2\mu_1 \varepsilon_{xx}; \\ \sigma_{yy} &= \lambda_1 \theta + 2\mu_2 \varepsilon_{yy}; \\ \sigma_{xy} &= 2\mu_0 \varepsilon_{xy}, \end{aligned} \tag{2.1}$$

where λ_1, μ_1, μ_2, μ_0 are the modulus of volume expansion and the shift moduli, respectively. In the case of a generalized plane stress state it is convenient to use the following isotropic medium parameters: $\mu = (E_1 + E_2)/(4(1+\nu))$, $\lambda = (E_1 + E_2)\nu/(2(1+\nu)(1-\nu))$, where E_1, E_2, ν are the moduli of elasticity and Poisson's ratio for an orthotropic medium, respectively.

Such an assignment allows us to describe real elastic constants using small parameters α, β, δ, with the parameter α used twice: $\mu_1 = \mu(1+\beta)$, $\mu_2 = \mu(1+\alpha)$, $\lambda_1 = \lambda(1+\alpha)$, $\mu_0 = \mu(1+\delta)$. Now Hooke's law (2.1) is rewritten in a convenient form:

$$\begin{aligned} \sigma_{xx} &= \lambda(1+\alpha)\theta + 2\mu(1+\beta)\varepsilon_{xx}; \\ \sigma_{yy} &= \lambda(1+\alpha)\theta + 2\mu(1+\alpha)\varepsilon_{yy}; \\ \sigma_{xy} &= 2\mu(1+\delta)\varepsilon_{xy}. \end{aligned} \tag{2.2}$$

Zero values of the parameters correspond to isotropy.

Detailed summary of features on the asymptotic perturbation method is given in the monograph of Nayfeh A. [6] and Minaeva N. [7].

For brevity we introduce the designation $\sum\limits_{m\,n\,k} \ldots = \sum\limits_{m=0}^{\infty} \sum\limits_{n=0}^{\infty} \sum\limits_{k=0}^{\infty} \ldots$

Below in asymptotic series the upper indices, which correspond to the powers of the small parameters, identify particular elements of the expansion:

$$\{\sigma_{ij},\varepsilon_{ij},u_i,\theta\}=\sum_{mnk}\alpha^m\beta^n\delta^k\{\sigma_{ij}^{mnk},\varepsilon_{ij}^{mnk},u_i^{mnk},\theta^{mnk}\},\quad \theta^{mnk}=\varepsilon_x^{mnk}+\varepsilon_y^{mnk}.$$

Expansion of the values incorporated into Hooke's law (2.2) after replacement of the summation and postulation variables with zero values for any formally non-existent element of expansion having at least one index with a negative value ($a^{mnk}=0, m<0\vee n<0\vee k<0$) leads to the following consequence:

$$\sigma_{xx}^{mnk}=\lambda\theta^{mnk}+2\mu\,\varepsilon_{xx}^{mnk}+\tilde{\sigma}_{xx}^{mnk},\quad \tilde{\sigma}_{xx}^{mnk}=\lambda\theta^{m-1\,n\,k}+2\mu\,\varepsilon_{xx}^{m\,n-1\,k},$$
$$\sigma_{yy}^{mnk}=\lambda\theta^{mnk}+2\mu\,\varepsilon_{yy}^{mnk}+\tilde{\sigma}_{yy}^{mnk},\quad \tilde{\sigma}_{yy}^{mnk}=\lambda\theta^{m-1\,n\,k}+2\mu\,\varepsilon_{yy}^{m-1\,n\,k},$$
$$\sigma_{xy}^{mnk}=2\mu\,\varepsilon_{xy}^{mnk}+\tilde{\sigma}_{xy}^{mnk},\quad \tilde{\sigma}_{xy}^{mnk}=2\mu\,\varepsilon_{xy}^{m\,n\,k-1}.$$

After a change of notation (tensorial and indicial notation combined) [8]

$$s_{ij}^{mnk}=\sigma_{ij}^{mnk}-\tilde{\sigma}_{ij}^{mnk} \tag{2.3}$$

we present the elements of the expansion in the form conventional for the generalized Hooke's law for isotropic bodies:

$$s_{ij}^{mnk}=\lambda\theta^{mnk}+2\mu\,\varepsilon_{ij}^{mnk} \tag{2.4}$$

The Cauchy formula is put in the same notation:

$$\varepsilon_{ij}^{mnk}=\frac{1}{2}\left(u_{i,j}^{mnk}+u_{j,i}^{mnk}\right) \tag{2.5}$$

Using X_i^0 to denote volume forces and assuming the expansion

$$X_i^0=\sum_{mnk}\alpha^m\beta^n\delta^k X_i^{0\,mnk}$$

as known, we rewrite the equilibrium equations in the following form:

$$s_{ij,j}^{mnk}+X_i^{mnk}=0,\quad X_i^{mnk}=X_i^{0\,mnk}+\tilde{\sigma}_{ij,j}^{mnk}. \tag{2.6}$$

Equations (2.4) – (2.6) formally describe the plane deformation state of isotropic elastic bodies (or generalized plane stress, in which case, though, the modulus of volumetric strain will have a different value [9]). Assuming the elastic field formed based on the fictitious body forces X_i^{mnk}, we define the state $\{u_i,\ \varepsilon_{ij}^{mnk},\ s_{ij}^{mnk}\}$ through the biharmonic equation for the Airy stress function [9], for which the Kolosov–Muskhelishvili general representation formula is obtained. This allows to obtain a basis of space of internal states by the MBS method [3]. Since the original basis is independent from the parameters α,β,δ, an orthonormal basis ("body in the sense of MBS") is obtained only once with high precision and then used in each iteration.

3. The basics of the method of boundary states with perturbations

The MBS is based on the isomorphism of spaces Ξ and Γ, Ξ being the space of internal states and Γ being the space of boundary states.

IOP Conf. Series: Journal of Physics: Conf. Series **973** (2018) 012015 doi:10.1088/1742-6596/973/1/012015

The main provisions of the perturbation method as applied to the medium state concept are as follows. Operator A governs the constitutive relations of a medium. By making operator A act on the internal state ξ , we obtain the following right-hand members of constitutive relations f :

$$A(\beta)\xi = f(\beta),\ x \in V, \tag{3.1}$$

where β is a small parameter. Nonlinear operator L generates a set of boundary conditions φ from attributes of the boundary state γ :

$$L(\beta)\gamma = \varphi(\beta),\ x \in \partial V. \tag{3.2}$$

We will analyze a class of operators A representable by series in terms of powers β involving linear operators A_k :

$$A = \sum_{k=0}^{\infty} \beta^k A_k .$$

Let the right-hand member (2.6) be presented by the following series:

$$f(\beta) = \sum_{s=0}^{\infty} \beta^s f_s ,$$

then the operator equation (3.1) takes the following form:

$$\sum_{k=0}^{\infty} \beta^k A_k \xi = \sum_{s=0}^{\infty} \beta^s f_s . \tag{3.3}$$

The internal state is treated as a series in terms of the small parameter β :

$$\xi = \sum_{j=0}^{\infty} \beta^j \xi^j , \tag{3.4}$$

where ξ^j is a state of j – order in the method of perturbations. Ratios (3.3) adjusted for equation (3.4) deliver the following:

$$\sum_{k=0}^{\infty} \beta^k A_k \sum_{j=0}^{\infty} \beta^j \xi^j = \sum_{s=0}^{\infty} \beta^s f_s$$

which is later converted as follows:

$$\sum_{k=0}^{\infty}\sum_{j=0}^{\infty} \beta^{k+j} A_k \xi^j = \sum_{s=0}^{\infty} \beta^s f_s . \tag{3.5}$$

After bringing (3.5) to a common index s we obtain the following:

$$\sum_{s=0}^{\infty}\sum_{k=0}^{s} \beta^s A_k \xi^{s-k} = \sum_{s=0}^{\infty} \beta^s f_s .$$

For approximation s we have:

$$\sum_{k=0}^{s} A_k \xi^{s-k} = f_s ,$$

which brings us to the operator equation:

$$A_0\xi^s = f_s - \sum_{k=1}^{s} A_k \xi^{s-k} . \tag{3.6}$$

These calculations are also valid for operator L, which is applied to boundary state γ. The resultant formula of the boundary state is:

$$L_0\gamma^s = \varphi^s - \sum_{k=1}^{s} L_k \gamma^{s-k} . \tag{3.7}$$

The result of the decomposition of the original nonlinear boundary value problem (3.1), (3.2) is reduced to a sequence of linear problems (3.6), (3.7).

4. The method of boundary states with perturbations for a multiple-parameter physical medium

Any internal state of a linear isotropic elastostatic medium is a set of displacements, strains, and stresses that correlate with each other through constitutive relations: $\xi = \{u_i, \varepsilon_{ij}, \sigma_{ij}\} \in \Xi$. Their trace at the boundary ∂V of the region V with an outer unit normal n_j contains information about the displacements and forces at work along the boundary $\gamma = \{u_i, p_i\} \in \Gamma$, $p_i = \sigma_{ij} n_j$ and corresponds to the boundary state. The spaces of possible internal and boundary states are isomorphic: $\Xi \leftrightarrow \Gamma$; moreover, this is a Hilbert isomorphism [3]:

$$a_1\xi^{(1)} + a_2\xi^{(2)} \leftrightarrow a_1\gamma^{(1)} + a_2\gamma^{(2)}, \quad (\xi^{(1)}, \xi^{(2)})_\Xi = (\gamma^{(1)}, \gamma^{(2)})_\Gamma .$$

Any well-defined problem is reduced to an infinite system of linear algebraic equations (ISE) [3]

$$Q\mathbf{c} = \mathbf{q} \tag{4.1}$$

relative to the vector of Fourier coefficients $\mathbf{c}$, the unknown state being expanded into a series with an orthonormal basis

$$\xi = \sum_l c_l \xi^{(l)} , \tag{4.2}$$

where all the coefficients c_l are related to any component of the element basis. The matrix Q ("the skeleton of the problem") is determined structurally only by the type of boundary conditions and numerically through an orthonormal basis (with identity matrices used in basic problems). The vector of the right-hand members includes the information about the specific boundary conditions.

When using the MBSP, it is convenient to organize the enumeration of exponents of small parameters by assigning the consolidated index $m+n+k=p$ in ascending order:

$$p = 0,1,2,3\ldots: \quad m = 0,\ldots,p: \quad n = 0,\ldots,p-m: \quad k = 0,\ldots,p-m-n$$

and solving specific problems for each combination of m, n, k.

An ISE is formulated at each step (4.1) in accordance with the boundary condition of the iteration. In practice, it suffices to take the real boundary conditions into account only when $p=0$, solve only the basic problem with $Q \equiv E$ in subsequent iterations, and take account of the corrections brought about by the new fictitious volume forces (which are not actually potential forces, but are polynomial) in the right-hand members. A general method for searching for internal states for a class of such forces has been recently described [10].

There are certain steps to be taken before the iterations: the bases of the spaces Ξ and Γ are formed using the general solution of Lamé's equations and the basis of functions harmonic in

IOP Conf. Series: Journal of Physics: Conf. Series **973** (2018) 012015 doi:10.1088/1742-6596/973/1/012015

$V \cup \partial V$; isomorphic orthonormal bases are made; and members $X_i^{0\,mnk}$ are set for expansion series for X_i^0 .

At step $p=0$ $(m=n=k=0)$: state $\tilde{\xi}^{000}$ caused by volume forces $X_i^{0\,000}$ is searched for; a correction corresponding to this state is made in the real boundary conditions; an ISE (4.1) $Q\mathbf{c}^{000}=\mathbf{q}$ is formed; its solution and a linear combination (4.2) lead to the internal state value ξ^{000}; this value added to that of the state caused by volume forces prepares an initial approximation for ξ: $\xi=\xi^{000}+\tilde{\xi}^{000}$. Based on the previous (2.3) formulas, the tensor $\tilde{\sigma}_{ij}^{000}$ is set.

At step $m+n+k=p>0$: the tensor $s_{ij}^{mnk}=\sigma_{ij}^{mnk}-\tilde{\sigma}_{ij}^{mnk}$ and vector X_i^{mnk} are drawn; the state $\tilde{\xi}^{mnk}$ caused by volume forces X_i^{mnk} is searched for in accordance with formula (2.6); the boundary conditions are amended accordingly; the first basic problem for a system of Lamé's equations (2.4) – (2.6) is solved; the resultant state value is added to that of the state caused by fictitious volume forces; the stress field is adjusted in accordance with (2.3) $\sigma_{ij}^{mnk}=s_{ij}^{mnk}+\tilde{\sigma}_{ij}^{mnk}$; and the new value is added to the resultant state being accumulated with an index of $\alpha^m\beta^n\delta^k$.

After performing a sufficient number of approximations it is necessary to make a final substitution $\beta=\mu_1/\mu-1$, $\alpha=\lambda_1/\lambda-1$, $\delta=\mu_0/\mu-1$ and turn to dimensional values. The result is an FPS.

5. Full parametric solution of problems with rectangular plates

The FPS methodology described above is tested on a fairly simple first basic problem featuring an aluminium plate occupying the area $D=\{(x,y)\|x|\le 2R,|y|\le R\}$. Following a non-dimensionalization (scales: on stresses – η^*, linear – R), we formulated and solved a test problem and a calculation problem (figure 1).

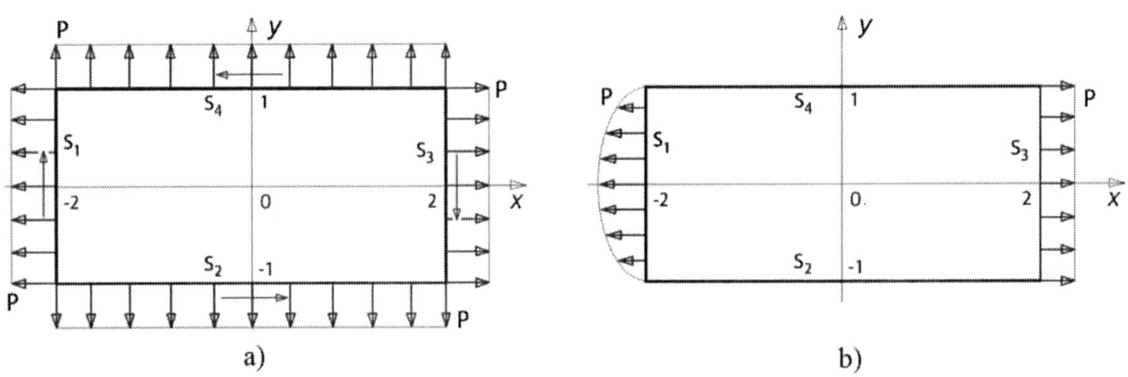

Figure 1. Orthotropic plate: a) test; b) calculation.

Test problem. The lateral faces of the plate are subjected to forces (figure 1,a):

$$\{p_x,p_y\}=\begin{cases}\{-1,\,1\},(x,y)\in S_1\cup S_4\\ \{1,-1\},(x,y)\in S_2\cup S_3\end{cases}.$$

There are no volume forces at work: $X_i^0=0$.

The solution of the homogeneous problem is plain: $\sigma_{xx}=1$, $\sigma_{yy}=1$, $\tau_{xy}=-1$. For aluminium ($\eta^*=10^9\ Pa$; dimensionless elasticity parameters are taken as $\lambda_1=19.96$, $\mu_1=24.46$, $\mu_2=31.61$, $\nu=0.29$) the deformation values are: ε_{xx} = 0.0118, ε_{yy} = 0.00917, ε_{xy} = −0.01724.

The MBSP applied allows to analyze an isotropic medium with a dimensionless Young's modulus of $E=70.75$ and a Poisson's ratio of $\nu=0.24$ (hence $\mu=28.53$, $\lambda=18.018$), which correspond to the small parameters of $\alpha=0.108$, $\beta=-0.142$, and $\delta=0.0165$, at each iteration. The third-order solution delivers the following components of the displacement vector $\mathbf{u}=\{u_x,u_y\}$ (with the highest degrees of the product of small parameters greater than the third order being dropped):

$$u_x \approx (10.742x-17.526y-2.079x\alpha+0.4x\alpha^3-8.663x\beta+1.676x\alpha^2\beta+1.676x\alpha\beta^2+$$
$$+6.986x\beta^3-17.526y\delta-17.526y\delta^3)\cdot 10^{-3};$$
$$u_y \approx (-17.526x+10.742y-12.821y\alpha+13.223y\alpha^3+2.079y\beta-0.402y\alpha^2\beta-$$
$$-1.676y\beta^3+17.526x\delta-17.526x\delta^3)\cdot 10^{-3}.$$

Using the above real values of α, β, δ, we obtain: $\sigma_{xx}=1$, $\sigma_{yy}=1$, $\tau_{xy}=-1$, ε_{xx} = 0.01173, ε_{yy} = 0.0091, ε_{xy} = −0.0172, $u_x=0.01173x-0.01723y$, $u_y=-0.01723x+0.0091y$. The relative asymptotic errors for $\varepsilon_{xx},\varepsilon_{yy},\varepsilon_{xy}$ are 0.59, 0.79, and 0.23%, respectively.

Calculation problem. As an example of how the proposed method for taking account of physical parameters for media is to be refined, we are going to solve the first basic problem for an orthotropic plate (fig.1,b) in its heterogeneous state, it being understood that its orthotropic parameters are known. The non-dimensionalized load parameters are matched by the following equilibrium:

$$\{p_x,p_y\}=\begin{cases}\{-1+y^2,0\},\ (x,y)\in S_1,\\ \{0,0\},\ (x,y)\in S_2\cup S_4,\\ \{2/3,0\},\ (x,y)\in S_3.\end{cases}$$

We propose a second-order solution of the problem. To get around the awkwardness of the formulas involved, we will provide the components of the displacement vector in a truncated form. Their structures are:

$$u_x \approx (9.568x-0.238x^2+0.197x^3-0.094x^4+\mathrm{K}+0.157y^2-0.68xy^2+0.98x^2y^2+\mathrm{K}+$$
$$+0.634y^4\alpha^2\beta+\mathrm{K}+73.635xy^4\alpha\beta^2-\mathrm{K}-5.983xy^2\beta\delta^2-\mathrm{K}-2.291x^4y^6\alpha\delta^2+\mathrm{K})\cdot 10^{-3};$$
$$u_y \approx (-2.45y+0.576xy-0.537x^2y+0.207x^3y+\mathrm{K}+0.038x^2y^5\alpha-\mathrm{K}-17.465x^6y\alpha\beta^2-\mathrm{K}-$$
$$-1.31x^2y^7\alpha^2\delta-\mathrm{K}-4.07x^8y^5\delta^3+\mathrm{K})\cdot 10^{-3}.$$

Reverse parameterization delivers the following:

$$u_x \approx R(9.568x-0.238x^2+0.197x^3-0.094x^4+\mathrm{K}+0.157y^2-0.68xy^2+0.98x^2y^2+\mathrm{K}+$$
$$+0.634y^4(\lambda_1/\lambda-1)^2(\mu_1/\mu-1)+\mathrm{K}+73.635xy^4(\lambda_1/\lambda-1)(\mu_1/\mu-1)^2-\mathrm{K}-$$
$$-5.983xy^2(\mu_1/\mu-1)(\mu_0/\mu-1)^2-\mathrm{K}-2.291x^4y^6(\lambda_1/\lambda-1)(\mu_0/\mu-1)^2+\mathrm{K})\cdot 10^{-3};$$

$$u_y \approx R(-2.45y+0.576xy-0.537x^2y+0.207x^3y+\mathrm{K}+0.038x^2y^5(\lambda_1/\lambda-1)-\mathrm{K}-$$
$$-17.465x^6y(\lambda_1/\lambda-1)(\mu_1/\mu-1)^2-\mathrm{K}-1.31x^2y^7(\lambda_1/\lambda-1)^2(\mu_0/\mu-1)-\mathrm{K}-$$
$$-4.07x^8y^5(\mu_0/\mu-1)^3+\mathrm{K})\cdot 10^{-3}.$$

We note here that x, y are to be understood as their actual values assigned to R.

Displacements, corresponding basis element of the space of internal states have the form (given 10 elements, table 1):

Table 1. The basic elements of displacements.

n	u_x	u_y
1	$0.021x$	$0.021y$
2	$0.004x^2 - 0.074y^2$	$0.078x\,y$
3	$-0.035x$	$0.035y$
4	$-0.078x\,y$	$0.074x^2 - 0.004y^2$
5	$0.035y$	$0.035x$
6	$-0.014x^3 - 0.17x\,y^2$	$0.17x^2\,y + 0.014y^3$
7	$-0.053x^2 + 0.053y^2$	$0.11x\,y$
8	$-0.064x^2 y + 0.092y^3$	$-0.064xy^2 + 0.092x^3$
9	$0.11x\,y$	$0.053x^2 - 0.053y^2$
10	$-0.031x^4 - 0.23x^2\,y^2 + 0.11y^4$	$0.3x^3 y - 0.016x\,y^3$

The accuracy of the approximate parametric solution of the problem mainly depends on the accuracy of the solution in the zero approximation in which given boundary conditions are set on the boundary of the body. We present several explicit and implicit characteristics of the convergence of the problem in the zero approximation (table 2).

Table 2. Nonzero Fourier coefficients.

k	c_k	k	c_k	k	c_k	k	c_k	k	c_k	k	c_k	k	c_k
1	0.138	6	0.012	11	-0.014	18	-0.012	23	0.005	30	0.002	35	0.003
2	-0.006	7	0.005	14	0.015	19	-0.017	26	-0.004	31	-0.005	46	0.001
3	-0.177	10	-0.015	15	0.021	22	0.007	27	0.003	34	-0.001	50	-0.001

Figure 2 is a graph illustrating the saturation of the Bessel sum (left part of the inequality Bessel). It does not prove anything, but convinces us of the correctness of the received solution.

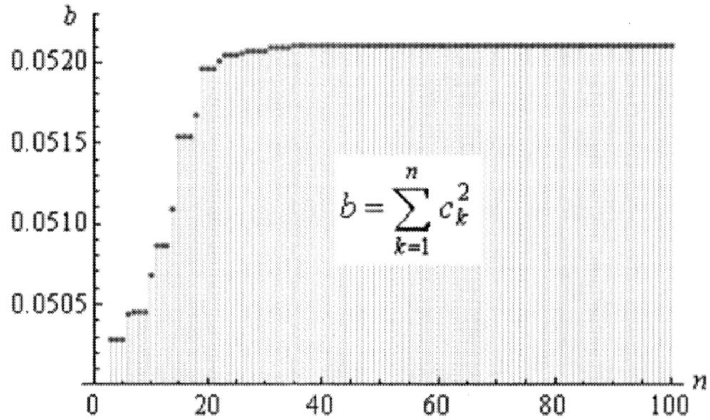

Figure 2. Saturation of the Bessel sum.

Figures 3, 4 and 5 show comparison of boundary conditions (bar) with the same value from the boundary state, obtained by solving problem (solid line). In the drawings there has been a convergence with the given calculated values with a sufficient degree of accuracy.

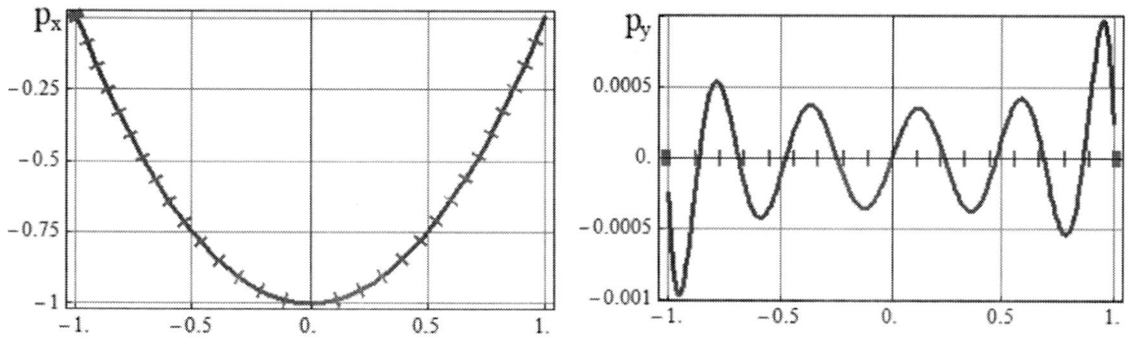

Figure 3. Verification of the boundary conditions with calculated values on the boundary S_1.

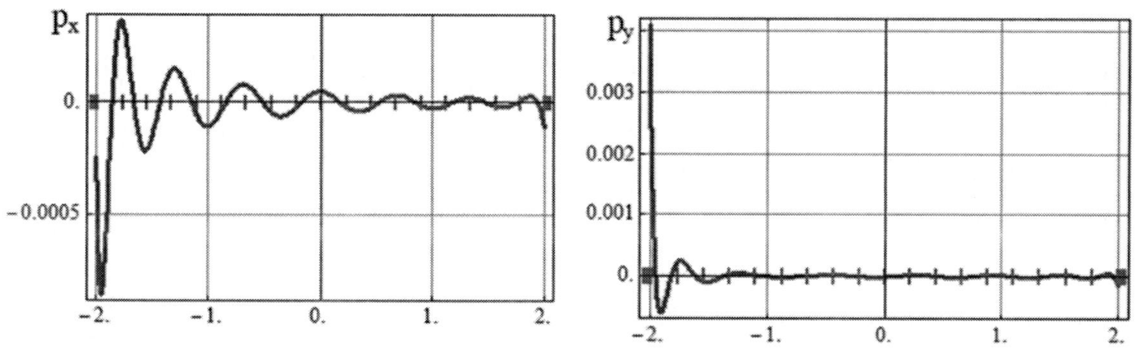

Figure 4. Verification of the boundary conditions with calculated values on the boundary S_2.

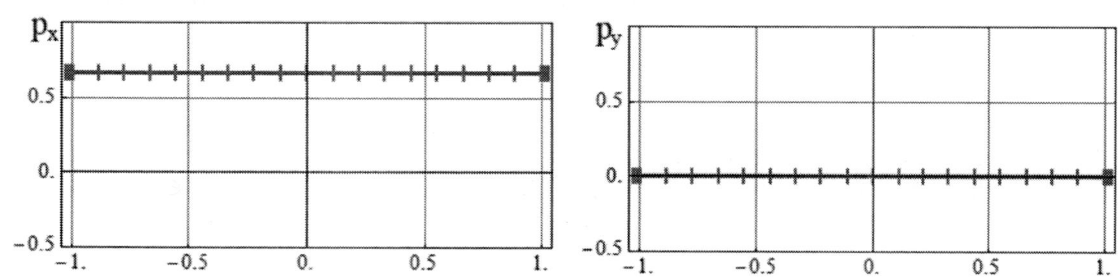

Figure 5. Verification of the boundary conditions with calculated values on the boundary S_3.

Figure 6 shows isolines of the components of the stress tensor σ_{xx}, σ_{yy}, σ_{xy} and shear stress Tmax, drawn based on the actual parameters of the material.

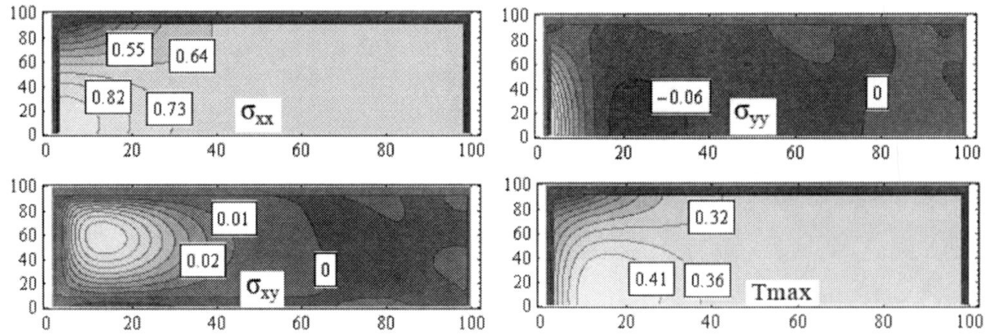

Figure 6. Isolines of components of the stress tensor.

As we can see from the drawings, the greatest tensile stress σ_{xx} along the axis of symmetry is focused along the left edge in the center of the plate, whereas compressive stress is localized at its edges. Stress σ_{yy} has approximately the same character. The nonzero shear stress σ_{xy} is concentrated near the left side and each side approximately the same distance (20% of the width of the plate). The character of the maximum shear stress Tmax differs little from the character of the axial tensile stresses, except that its highest concentration is not adjacent to the left side.

The foregoing analysis leads to the conclusion that MBSP proves to be an effective means of explicit full parametric solution of mechanics problems containing a finite number of physical parameters of the medium.

The research was supported by RFFR with grant No. 16-41-480729 "r_a"

References

[1] Sedov L I 1977 *Similarity methods and dimensional analysis in mechanics* (Moscow: Nauka) p 440
[2] Levina L V, Novikova O S and Penkov V B 2016 Full parametric solution of problems of the theory of elasticity of simply connected finite bodies *Bulletin of Lipetsk State Technical University* **2 (28)** pp 16-24
[3] Penkov V B and Satalkina L V 2012 *Method of boundary states with the perturbation: the inhomogeneous and nonlinear problems in the theory of elasticity and thermoelasticity* (Germany: LAP LAMBERT Academic Publishing GmbH & Co.) p 108
[4] Timoshenko S and Woinowsky-Krieger S 1959 *Theory of plates and shells* (USA: Mcgraw-Hill Book Company,Inc) p 580
[5] Lehnitsky S G 1957 *Anisotropic plate* Editor Feldman G I (Moscow: State publishing house of technical-theoretical literature) p 463
[6] Nayfeh A H 1993 *Introduction to perturbation techniques* (USA: A Wiley-interscience publication. John Wiley & Sons, Inc) p 519
[7] Minaeva N I 2002 *The perturbation method in mechanics of deformable bodies* (Moscow: Scientific book) p 156
[8] Mase G E 1970 *Theory and problems of continuum mechanics* (New York, St. Louis, San Francisco, London, Sydney, Toronto, Mexico and Panama: Mcgraw-Hill Book Company,Inc) p 221
[9] Muskhelishvili N I 1966 *Some basic problems of the mathematical theory of elasticity* (Moscow: Nauka) p 70
[10] Penkov V B, Levina L V, Levin M Y and Kuzmenko N V 2016 A new method for analyzing the effect of body forces induced by nanodispersed magnetic fluids on states of elastic solids *Science in the Central Russia* (journal on science and production technology) **2(20)** pp 12-16

An algorithm for analytical solution of basic problems featuring elastostatic bodies with cavities and surface flaws

V B Penkov[1], L V Levina[2], O S Novikova[1] and A S Shulmin[3]

[1] Department of General Mechanics, Lipetsk State Technical University, Moskovskaya 30, Russia
[2] Department of Applied Mathematics, Lipetsk State Technical University, Moskovskaya 30, Russia
[3] Department of Mathematics, Lipetsk State Technical University, Moskovskaya 30, Russia

E-mail: vbpenkov@mail.ru

Abstract. Herein we propose a methodology for structuring a full parametric analytical solution to problems featuring elastostatic media based on state-of-the-art computing facilities that support computerized algebra. The methodology includes: direct and reverse application of P-Theorem; methods of accounting for physical properties of media; accounting for variable geometrical parameters of bodies, parameters of boundary states, independent parameters of volume forces, and remote stress factors. An efficient tool to address the task is the sustainable method of boundary states originally designed for the purposes of computerized algebra and based on the isomorphism of Hilbertian spaces of internal states and boundary states of bodies. We performed full parametric solutions of basic problems featuring a ball with a nonconcentric spherical cavity, a ball with a near-surface flaw, and an unlimited medium with two spherical cavities.

1. Introduction

The classical approach to boundary value problems in mathematical physics suggests that a solution should contain all parameters of a problem (full parametric solution, or FPS). Computer technology has made away with the traditional approach to analytical presentation of solutions. The present-day technology that makes computerized algebra possible allows to organize the presentation of solutions in the classical form. A verifiably efficient method to achieve this is the method of boundary states (MBS).

This study purports to generalize the methods for structuring full parametric analytical solutions to elastostatics problems featuring multiply connected restricted bodies and unrestricted spaces with multiple cavities and boundaries that may also contain flaws.

2. A current approach to structuring full parametric solutions

The methodology for structuring full parametric solutions is synoptically yet non-exhaustively described in study [1]. The FPS procedure includes the following stages, each of which corresponds to particular parameters to be introduced:

1) formulation of a well-defined mechanical problem;

Published under licence by IOP Publishing Ltd

IOP Conf. Series: Journal of Physics: Conf. Series **973** (2018) 012016 doi:10.1088/1742-6596/973/1/012016

2) non-dimensionalization of relations and parameters contained in the problem so as to come to a minimum set of parameters that allows to present the solution analytically;
3) making provisions for including the parameters of boundary states (non-dimensionalized constants);
4) if the problem features an unrestricted space, the explicit parameters must be extended to include symbols to denote remote stress states;
5) making provisions for parameters of elasticity. High-power computers of today use fixed values of medium parameters. Traditional computational approaches do not allow for their direct application in FPS;
6) accounting for geometrical parameters of the bodies involved;
7) accounting for parameters of various volume forces at work;
8) reversal to dimensional values in accordance with P-Theorem [2].

These points also apply to full parametric solutions to problems featuring multiply connected restricted bodies. A key difference of solutions addressing multiply connected bodies as opposed to simple-connected ones is that the basis of space of internal states of the medium in the former case must include a set of states describing each cavity involved [3]. The experience gained in the area so far suggests that the expansion of the basis of space by introducing sets of states to describe surface flaws significantly reduces the time needed for solution even in the case of problems featuring restricted bodies parts whose boundaries are negative curvatures.

A crucial choice to be made further is that of the solution method. As was shown in [3], the MBS is an efficient state-of-the-art method to address our needs. Its "up-to-dateness" resides in the use of computer facilities with capacity for computerized algebra. The method relies on the concept of an internal state of a body $\xi = \{u_i, \varepsilon_{ij}, \sigma_{ij}\}$ whose attributes identically match constitutive equations of the medium: Cauchy equations, generalized Hooke's law, and balance equations [4]:

$$\varepsilon_{ij} = \frac{1}{2}\left(u_{i,j} + u_{j,i}\right), \tag{1}$$

$$\sigma_{ij} = \lambda\theta\delta_{ij} + 2\mu\,\varepsilon_{ij}, \quad \theta = \varepsilon_{ii}, \tag{2}$$

$$\sigma_{ij,j} + \mathrm{X}_i = 0, \tag{3}$$

where ε_{ij}, σ_{ij} are components of the deformation and strain tensors, u_i is a component of the displacement vector, θ is the volume strain, δ_{ij} is the Kronecker delta, X_i is a component of the vector of volume forces, $\lambda = 2\nu\mu/(1-2\nu)$ is the Lamé parameter, μ is the shear modulus, and ν is the Poisson ratio.

Simultaneous equations (1)–(3) are equivalent to three Lamé's equations [4]

$$\mu u_{i,jj} + (\lambda + \mu) u_{j,ji} + X_i = 0,$$

general solutions to which have been separately and independently performed by H. Neuber [5] and P.F. Papkovich [6]:

$$u_i^k = 4(1-\nu_0)B_i - (x_j B_j + B_0)_{,i} + u_i^P,$$

where B_i is a component of the arbitrary harmonic vector $\mathbf{B}$, B_0 is a harmonic scalar, and u_i^P corresponds to the displacements caused by volume forces.

To fill the basis of the Hilbertian space of internal states, it's most convenient to use the Arzhanykh-Slobodyansky solution [4] as applied to the interior of a restricted body (volume forces being left out):

$$u_i = 4(1-\nu_0)B_i + x_j B_{i,j} - x_i B_{j,j}, \tag{4}$$

or as applied to the exterior of a cavity in unrestricted space:

$$u_i = 4(1-\nu_0)B_i - (x_j B_j)_{,i}, \tag{5}$$

where B_i is a component of the arbitrary harmonic vector

$$\mathbf{B} \in \left(\begin{pmatrix} \varphi \\ 0 \\ 0 \end{pmatrix}, \begin{pmatrix} 0 \\ \varphi \\ 0 \end{pmatrix}, \begin{pmatrix} 0 \\ 0 \\ \varphi \end{pmatrix} \right).$$

For the interior of a restricted space φ is an element of a set of linearly independent harmonic polynomials, i.e.,

$$\varphi \in \{x, y, z, yz, xz, xy, x^2 - z^2, y^2 - z^2, \dots\},$$

whereas for the exterior of a cavity it's a harmonic of the following type [3]:

$$\varphi \in \left\{1/r, x/r^3, y/r^3, z/r^3, yz/r^5, xz/r^5, xy/r^5, \dots \right\},$$

where the radius vector is reckoned from the center inside the cavity.

Below we provide some templates of state bases if the Poisson ratio ν equals 0.25 in case of the interior of the restricted body:

$$\mathbf{u}^{(1)} = \begin{pmatrix} 3x \\ -y \\ -z \end{pmatrix};\ \hat{\boldsymbol{\varepsilon}}^{(1)} = \begin{pmatrix} 3 & 0 & 0 \\ 0 & -1 & 0 \\ 0 & 0 & -1 \end{pmatrix}, \hat{\boldsymbol{\sigma}}^{(1)} = \begin{pmatrix} 7 & 0 & 0 \\ 0 & -1 & 0 \\ 0 & 0 & -1 \end{pmatrix},$$

$$\mathbf{u}^{(2)} = \begin{pmatrix} 0 \\ 4x \\ 0 \end{pmatrix};\ \hat{\boldsymbol{\varepsilon}}^{(2)} = \begin{pmatrix} 0 & 2 & 0 \\ 2 & 0 & 0 \\ 0 & 0 & 0 \end{pmatrix}, \hat{\boldsymbol{\sigma}}^{(2)} = \begin{pmatrix} 0 & 4 & 0 \\ 4 & 0 & 0 \\ 0 & 0 & 0 \end{pmatrix}, \dots$$

and now in case of the exterior of a restricted cavity:

$$\mathbf{u}^{(1)} = r^{-3/2} \begin{pmatrix} 3x^2 + 2(y^2 + z^2) \\ xy \\ xz \end{pmatrix},$$

$$\hat{\boldsymbol{\varepsilon}}^{(1)} = r^{-5/2} \begin{pmatrix} -3x^3 & -y(7x^2 + y^2 + z^2) & -z(7x^2 + y^2 + z^2) \\ -y(7x^2 + y^2 + z^2) & x(x^2 - 2y^2 + z^2) & -3xyz \\ -z(7x^2 + y^2 + z^2) & -3xyz & x(x^2 + y^2 - 2z^2) \end{pmatrix},$$

$$\hat{\boldsymbol{\sigma}}^{(1)} = r^{-5/2} \begin{pmatrix} -x(7x^2 + y^2 + z^2) & -y(7x^2 + y^2 + z^2) & -z(7x^2 + y^2 + z^2) \\ -y(7x^2 + y^2 + z^2) & x(x^2 - 5y^2 + z^2) & -6xyz \\ -z(7x^2 + y^2 + z^2) & -6xyz & x(x^2 + y^2 - 5z^2) \end{pmatrix},$$

$$\mathbf{u}^{(2)} = r^{-3/2}\begin{pmatrix} xy \\ 2x^2+3y^2+2z^2 \\ yz \end{pmatrix},$$

$$\hat{\boldsymbol{\varepsilon}}^{(2)} = r^{-5/2}\begin{pmatrix} y(-2x^2+y^2+z^2) & -x\left(x^2+7y^2+z^2\right) & -3xyz \\ -x\left(x^2+7y^2+z^2\right) & -3y^3 & -z\left(x^2+7y^2+z^2\right) \\ -3xyz & -z\left(x^2+7y^2+z^2\right) & y\left(x^2+y^2-2z^2\right) \end{pmatrix},$$

$$\hat{\boldsymbol{\sigma}}^{(2)} = r^{-5/2}\begin{pmatrix} y\left(-5x^2+y^2+z^2\right) & -x\left(x^2+7y^2+z^2\right) & -6xyz \\ -x\left(x^2+7y^2+z^2\right) & -y\left(x^2+7y^2+z^2\right) & -z\left(x^2+7y^2+z^2\right) \\ -6xyz & -z\left(x^2+7y^2+z^2\right) & y\left(x^2+y^2-5z^2\right) \end{pmatrix},\ldots$$

Relations (1) and (2) enable us to formulate a k-th internal state for each displacement vector $\mathbf{u}^{(k)}$: $\xi^{(k)} = \{u_i^{(k)}, \varepsilon_{ij}^{(k)}, \sigma_{ij}^{(k)}\}$. After removing linearly dependent elements from the list $\left\{\xi^{(k)}\right\}$, $k \in \{1,\ldots,\infty\}$ we come to the basis of space of internal states of an isotropic elastic body V. The basis of internal states $\Xi = \left\{\xi^{(k)}\right\}_\infty$ engenders at boundary ∂V a basis of boundary states $\Gamma = \left\{\gamma^{(k)}\right\}_\infty$, $\gamma^{(k)} = \{u_i^{(k)}, p_i^{(k)}\}$, $p_i^{(k)} = \sigma_{ij}^{(k)} n_j$ isomorphic to Ξ, with n_j being a component of the unit vector of the external normal to the boundary. The isomorphism of the spaces of internal and boundary states is conditioned by the Somigliana theorem [7] and Betti's reciprocity principle [2].

The next step is the orthogonalization of the basis in accordance with the scalar product

$$(\xi^{(1)}, \xi^{(2)})_\Xi = \int_V \sigma_{ij}^{(2)} \varepsilon_{ij}^{(1)} \mathrm{d}V = \int_{\partial V} p_i^{(2)} u_i^{(1)} \mathrm{d}S = (\gamma^{(1)}, \gamma^{(2)})_\Gamma,$$

followed by the formulation and solution of the infinite set of equations (ISE)

$$Q\mathbf{c} = \mathbf{q},$$

where the matrix of coefficients Q is determined only by the type of boundary conditions (BCs) and is based only on elements of the basis of space of boundary states [3], it being understood that this matrix as featured in the first and second general elasticity problems is a unit matrix and the vector $\mathbf{q}$ uses both basic elements and BCs. In the case of the first general problem we come to the following:

$$\bar{q}_k = \int_{\partial V} (p_x u_x^{(k)} + p_y u_y^{(k)} + p_z u_z^{(k)})\mathrm{d}S. \tag{6}$$

The unknown internal state ξ and boundary state γ are determined via the following Fourier series:

$$u_i = \sum_k c_k u_i^{(k)}, \quad \sigma_{ij} = \sum_k c_k \sigma_{ij}^{(k)}, \quad \varepsilon_{ij} = \sum_k c_k \varepsilon_{ij}^{(k)},$$

$$u_i\big|_{\partial V} = \sum_k c_k u_i^{(k)}\big|_{\partial V}, \quad p_i = \sum_k c_k p_i^{(k)}.$$

The general advantages of the MBS reside in two facts. Firstly, the equations used to describe the medium are matched identically. Secondly, the degree of accuracy of the solution performed is easy to estimate by looking at the disparity between the resultant boundary state and the boundary conditions, after which the interval on the original basis of states may be adjusted as necessary. It removes the need for comparing test problems with ones solved by other methods (sustainability of the MBS).

3. Physical parameters of media, geometrical parameters of domains, and remote load

Study [1] describes the methodology for expanding the FPS to contain, among scale factors, the parameters of surface forces in their linear combinations as boundary conditions of the first basic problem, and the Poisson ratio. In order to obtain ν, the analytical solution was structured so as to include a Lagrange interpolation, which pre-determined the need for multiple modelling of the original basis of space Ξ and multiple orthogonalization attempts. It suffices to perform interpolation for the displacement vector only, as the deformation and strain tensors will further follow from the Cauchy formulas and the generalized Hooke's law.

An approach that appears more rational is that based on the Poincaré perturbation method, which is quite well described in various works [8]. This approach requires to perform orthogonalization just once, which significantly reduces the energy intensity of the solution process. However, the approach brings a new difficulty: each new approximation triggers non-potential fictitious volume forces that require modelling of specific states of the elastic body. It would be easy to do if the forces were potential ones [4]. To address non-potential forces at work, we have discovered an efficient approach to modelling fields [9].

The methodology for modelling an internal state with a finite number of restricted cavities is described in study [3]. When applied to specific sets of values of the Poisson ratio ν_l and specific sets of geometric parameters of the body h_m, the methodology fundamentally allows us to model each of the internal states $\xi_{l\,m}$. Interpolation of the solutions taken as a whole brings us to the following description of the internal state of the medium: $\xi(\nu,h)$. This notation explicitly accounts for both physical and geometric properties of the body. A similar approach may be applied if a problem contains multiple geometric parameters of a body. The resultant state can be safely considered to contain all non-dimensionalization parameters.

To account for a homogenous stressed state at infinity (it being enough to maintain the principal stresses $X^\infty, Y^\infty, Z^\infty$ only, although the set can be extended to contain the shear stress), we need to have reference states ξ_X, ξ_Y, ξ_Z that correspond to each of $X^\infty, Y^\infty, Z^\infty$ equaling 1 in alternative variants. The actual character representation of the cumulative state is determined by the following linear combination:

$$\xi = X^\infty \xi_X + Y^\infty \xi_Y + Z^\infty \xi_Z.$$

This equation is easy to generalize in case we wish to account for remote shear stress.

4. A ball with a non-concentric spherical cavity

In its dimensionless form (linear scale R is the radius of the ball; stress scale coincides with the shear modulus μ), the problem reads as follows: a ball with its center at point $O_1(0,0,1/8)$ contains a load-free spherical cavity with a radius of 1/2. The surface of the ball is loaded with hydrostatic pressure p. A full parametric solution is required.

Bases of spaces of internal states were assigned alternately to each of the Poisson ratio values $\nu \in [0.2, 0.25, 0.3]$ and were formed based on general Arzhanykh-Slobodyansky solutions [5] for the interior of a restricted body (4) and for the exterior of a restricted cavity (5).

Let's present a fragment of the vector of the right-hand members of the ISE (rounded to the thousandth place) for a problem with a Poisson ratio $\nu = 0.25$. The vector fragment is calculated in accordance with (6) as follows:

$$\mathbf{q} = \{0,\ 0,\ -0.051,\ -0.458,\ 0,\ 0,\ -0.692,\ 0,\ -1.477,\ -0.493,\ \ldots\}$$

Non-zero Fourier coefficients (rounded to the thousandth place) for a problem with a Poisson ratio $\nu = 0.25$ are presented in table 1.

IOP Conf. Series: Journal of Physics: Conf. Series **973** (2018) 012016 doi:10.1088/1742-6596/973/1/012016

Table 1. Non-zero Fourier coefficients in the problem featuring a ball with a cavity.

k	c_k	k	c_k	k	c_k	k	c_k	k	c_k	k	c_k
3	-0.051	10	-0.493	24	0.008	39	0.002	52	0.010	66	0.010
4	-0.458	14	-0.359	28	-0.033	43	-0.005	56	-0.003	70	0.001
7	-0.692	18	-0.284	32	-0.066	47	-0.004	59	0.001	84	0.001
9	-1.477	21	-0.015	36	-0.006	49	-0.002	63	-0.110		

In each instance the solution is based on the reference external pressure value of 1 unit. The real stress field differs from 1 by a factor of p [3]. A Lagrange interpolation of the actual states in all of the problems after a reversal to dimensional values brings us to the following FPS (below we provide the initial intervals only due to the immensity of the full solution):

$$u_x \approx \frac{p_0 x}{\mu}[-0.554-0.041z/R+0.008x^2z/R^3-0.021z^2/R^2-0.039z^3/R^3-0.001\,z^3/(R^3A^{7/2})-$$
$$-0.001/A^{5/2}-0.036x^2/(R^2A^{5/2})-0.036y^2/(R^2A^{5/2})+0.009z/(RA^{5/2})-0.037z^2/(R^2A^{5/2})+\ldots+$$
$$+(1.53-0.005\,x^2/R^2-0.002y^2/R^2+0.018z/R-0.063x^2z/R^3+0.042\,z^2/R^2+0.308z^3\,/\,R^3-$$
$$-0.001x^2z/(R^3A^{7/2})-0.001(2x^2+y^2)/(R^2A^{5/2})+\ldots)\nu+(-0.872+0.001x^2/R^2-0.004y^2/R^2+$$
$$+0.143z/R+0.123x^2z/R^3-0.027z^2/R^2-0.610z^3/R^3+0.001y^2z^2/(R^4A^{9/2})+0.001z^4/(R^4A^{9/2})+$$
$$+0.002x^2z/(R^3A^{7/2})-0.001z^3/(R^3A^{7/2})+0.001(2x^2+y^2-z^2)/(R^2A^{5/2})+\ldots)\nu^2],$$

where $A=x^2/R^2+y^2/R^2+(-1/8+z/R)^2$.

The remaining characteristics of the internal state may be obtained as usual.

Since calculations have shown that the difference between the various states corresponding to different values of ν within the limits stated above are almost imperceptible to the naked eye, we have drawn up table 3 to show charts reflecting the difference between stress strain state fields at real values of ν and $\nu_0 = 1/4$.

5. Elastic ball with a spherical surface flaw

The problem is defined in a form similar to the one above: a ball with a radius of 1 contains a spherical surface flaw (the center of the sphere being $(0,0,\sqrt{3}/2)$) with a radius of $1/2$ with the following boundary conditions: $\mathbf{p}=\begin{cases}\{-Cos\theta\ Cos\varphi,\ -Cos\theta\ Sin\varphi,\ 0\},\ r\in S_1\\ \{0,\ 0,\ 0\},\ r\in S_2\end{cases}$. A full parametric solution is required.

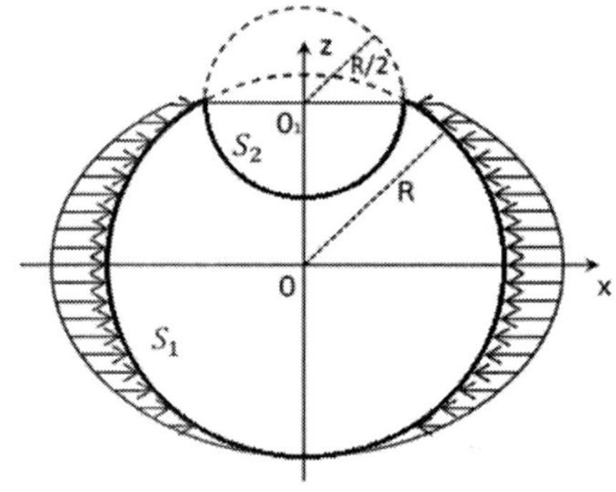

Figure 1. Load configuration for a ball with a surface flaw.

Let's present a fragment of the vector of the right-hand members of the ISE for a problem with a Poisson ratio of $\nu = 0.25$, rounded to a thousandth place:

$$\mathbf{q} = \{0,\ 0,\ 0.378,\ -0.828,\ 0,\ 0,\ -1.304,\ 0,\ -0.360,\ -0.325,\ \ldots\}.$$

Non-zero Fourier coefficients (rounded to the thousandth place) for a problem with a Poisson ratio $\nu = 0.25$ are presented in table 2.

Table 2. Non-zero Fourier coefficients in the problem featuring a ball with a surface flaw.

k	c_k	k	c_k	k	c_k	k	c_k	k	c_k
3	0.378	36	0.082	73	0.046	116	0.017	154	-0.006
4	-0.828	39	-0.084	77	-0.051	120	0.019	167	-0.037
7	-1.304	43	-0.021	80	0.073	127	-0.027	170	-0.006
9	-0.360	47	-0.026	84	-0.011	130	0.025	174	0.006
10	-0.325	49	-0.096	87	0.020	135	0.019	177	0.007
14	-0.310	52	0.035	93	0.054	138	0.012	178	-0.022
18	-0.205	56	-0.119	100	-0.014	140	-0.012	182	-0.022
21	0.176	59	0.084	103	0.025	143	0.029	186	-0.004
24	-0.130	63	-0.003	108	0.070	145	-0.024	187	-0.010
28	-0.043	66	0.005	111	0.024	149	-0.024	190	-0.009
32	-0.088	70	-0.088	113	-0.013	153	-0.007	200	-0.029

A Lagrange interpolation of the actual states in all of the problems after the reversal to dimensional values brings us to the following FPS (below we provide the initial intervals only due to the immensity of the full solution):

$$\begin{aligned}
u_x \approx \frac{p_0 x}{\mu}[&-0.573 + 0.047x^2/R^2 + 0.049y^2/R^2 - 0.020y^4/R^4 - 0.077z/R - 0.010/A^{13/2} + \\
&+ 0.015x^2/(R^2 A^{13/2}) + 0.026x^4/(R^4 A^{13/2}) - 0.019y^2/(R^2 A^{13/2}) + 0.046x^2y^2/(R^4 A^{13/2}) + \\
&+ 0.056/A^{11/2} - 0.153/A^{9/2} + 0.326/A^{7/2} - 0.554/A^{5/2} + 0.435/A^{3/2} + \ldots + \\
&+ (1.19 - 0.192x^2/R^2 + 0.047x^4/R^4 - 0.202y^2/R^2 - 0.019x^2y^2/R^4 + 0.145\,y^4/R^4 - \\
&- 0.072z/R + 0.089/A^{13/2} - 0.229x^2/(R^2 A^{13/2}) + 0.884x^2/(R^2 A^{11/2}) + \ldots)\,\nu + \\
&+ (-1.66 + 0.557x^2/R^2 \text{-} 0.122x^4/R^4 + 0.577\,y^2/R^2 \text{-} 0.0176\,x^2y^2)/R^4 \text{-} 0.318\,y^4/R^4 + \\
&+ 0.099z/R \text{-}\ 0.435x^2z/R^3 - 0.205/A^{13/2} + 0.538x^2/(R^2 A^{13/2}) + \ldots)\,\nu^2],
\end{aligned}$$

where $A = x^2/R^2 + y^2/R^2 + (-\sqrt{3}/2 + z/R)^2$.

A careful analysis of the images (table 3) shows that:

1) qualities of the radial stresses are changeable: tensile stresses in looser media are concentrated in smaller thicknesses;
2) circumferential stresses in loose media operate in a directly opposite way: they create pulling in denser regions and compression otherwise;
3) in less loose media axial stresses concentrate compression load around the cavity, whereas in looser media the opposite is true;
4) depending on the Poisson ratio, the concentration of shear stresses operates in a directly opposite fashion with respect to the surfaces that restrict the body;
5) surface flaws concentrate all stresses in the neighborhood of the geometric singularities of the boundary.

IOP Conf. Series: Journal of Physics: Conf. Series **973** (2018) 012016 doi:10.1088/1742-6596/973/1/012016

Table 3. Variations in the internal state characteristics depending on varying Poisson ratios.

	Spherical layer		Ball with a surface flaw	
	$\xi(0.3)-\xi(0.25)$	$\xi(0.25)-\xi(0.2)$	$\xi(0.3)-\xi(0.25)$	$\xi(0.25)-\xi(0.2)$
σ_{xx}	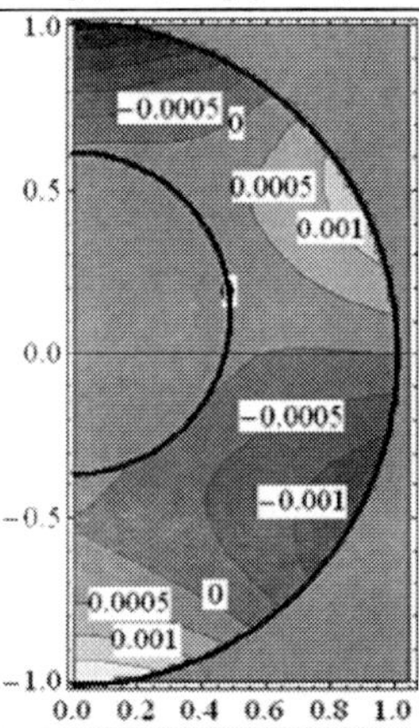	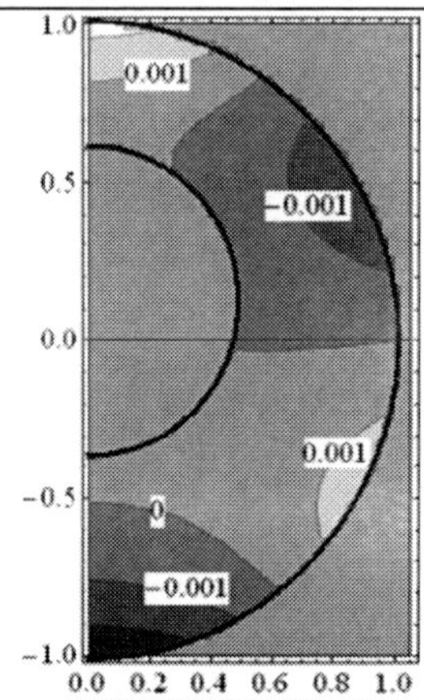	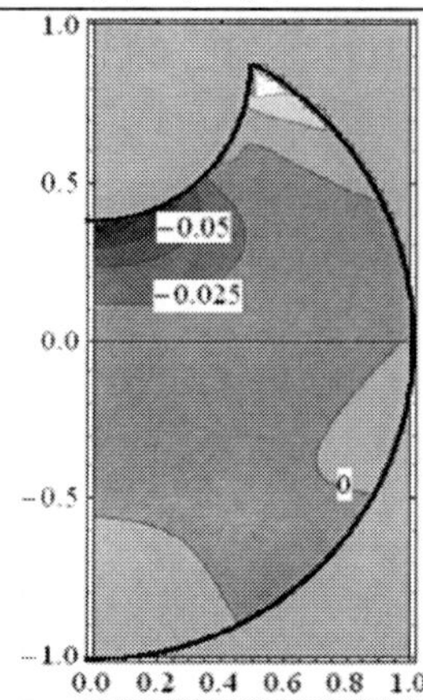	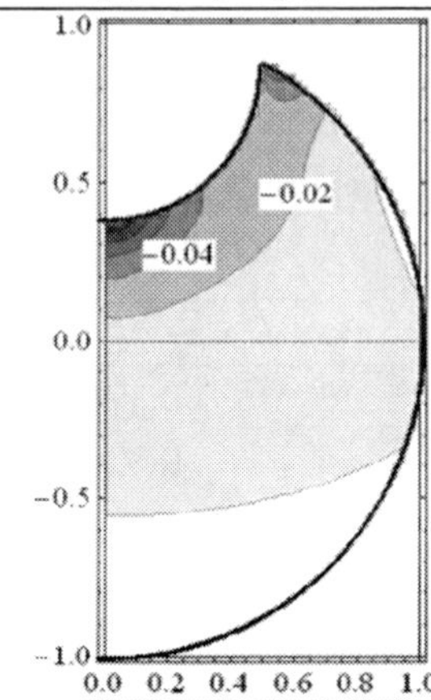
σ_{yy}	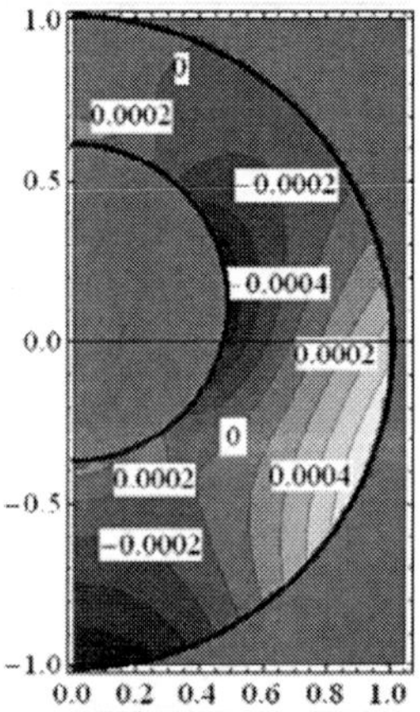	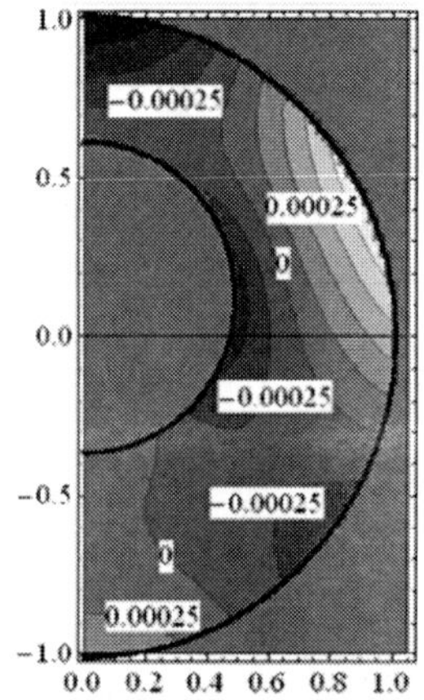	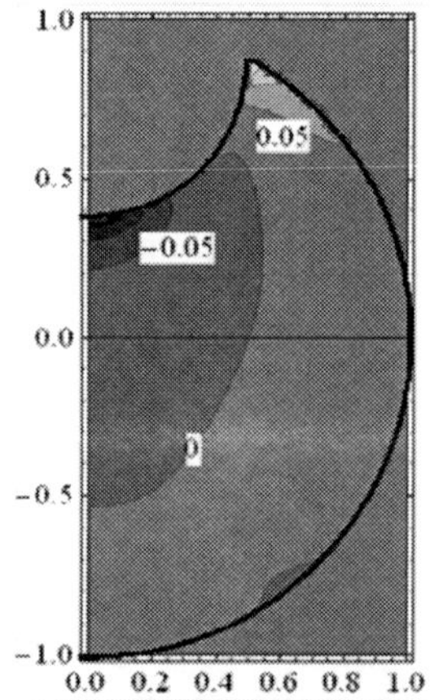	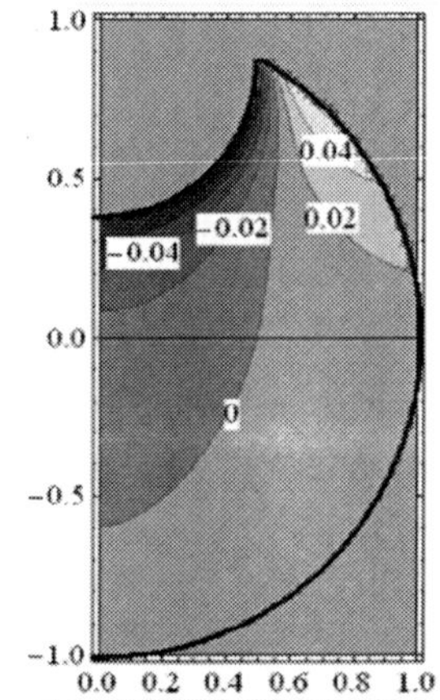
σ_{zz}	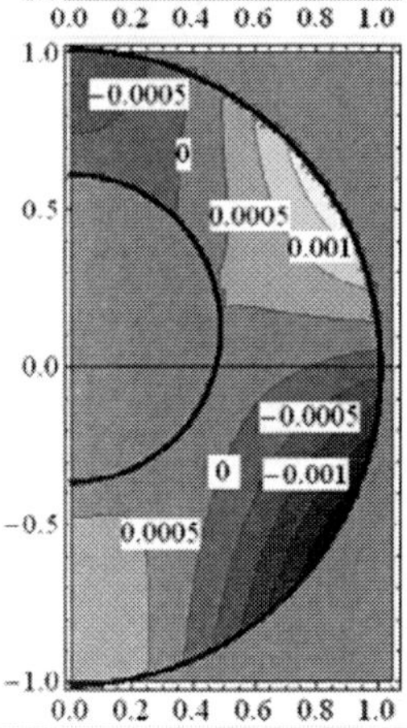	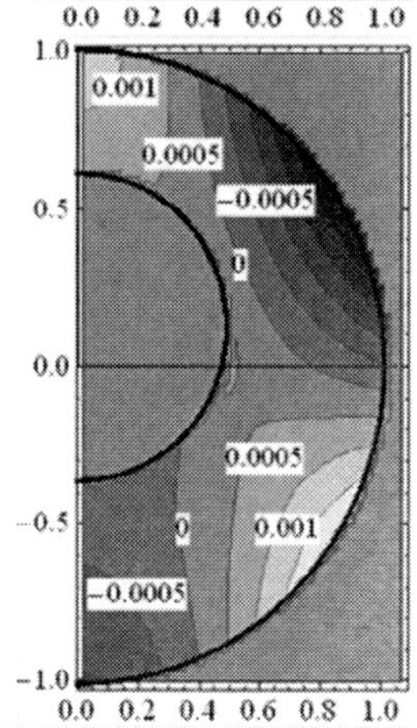	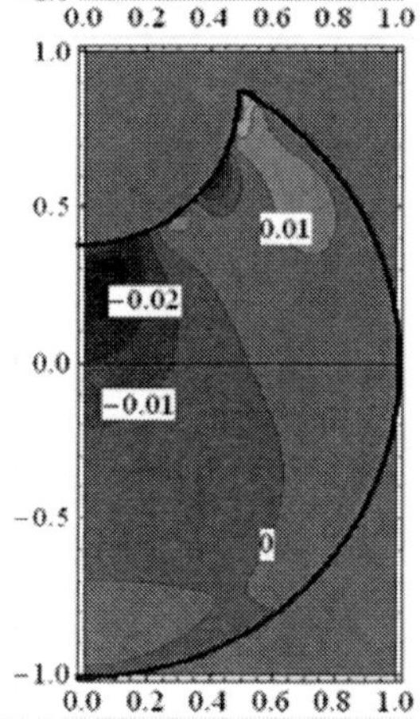	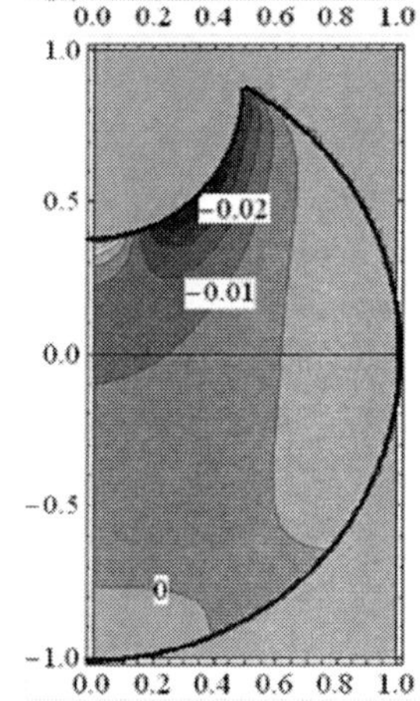

	Spherical layer		Ball with a surface flaw	
	$\xi(0.3)-\xi(0.25)$	$\xi(0.25)-\xi(0.2)$	$\xi(0.3)-\xi(0.25)$	$\xi(0.25)-\xi(0.2)$
σ_{xz}	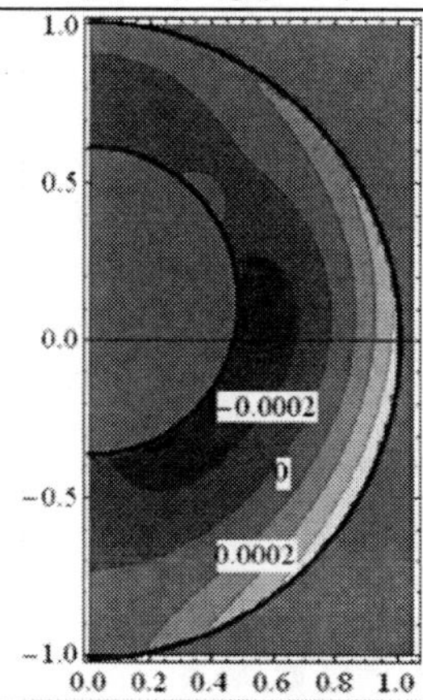	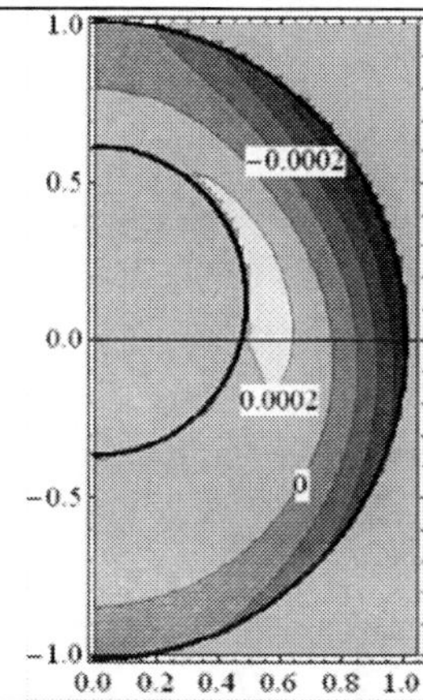	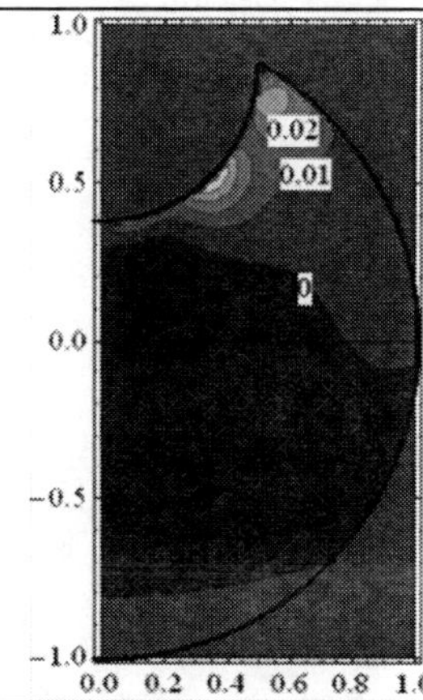	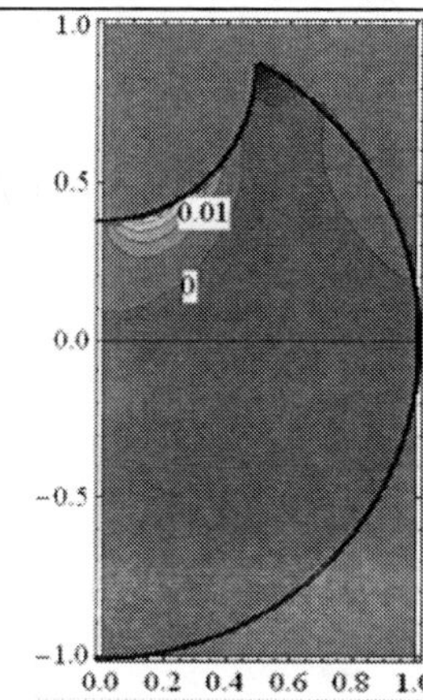

6. Single-axis loading of an unrestricted medium with two cavities

As an example, let us consider a problem about single-axis tensioning of an unrestricted medium that contains two equal-sized spherical cavities, the Poisson ratio and the distance between the cavities being understood as arbitrary values. It is possible to use the Southwell's solution [10] combined with an iterative process of Schwartz for building the solution, but using MBS is more brief approach. Remote stress is directed along axis z, which is perpendicular to the common equatorial plane of the cavities. Once non-dimensionalization is performed, the problem is found to contain three variable parameters: ν, h, and Z^{∞}. Reference solutions (at $Z^{\infty}=1$) were performed for possible combinations of the values $\nu\in\{0.2, 0.25, 0.3\}$ and $h\in\{0.05, 0.4, 12.6\}$ (linear scale being equal to R and stress scale being equal to shear modulus μ). Non-zero Fourier coefficients (rounded to the thousandth place) for a problem with $\nu=0.2$, $h=0.05$ are presented in table 4 below:

Table 4. Non-zero Fourier coefficients in the problem featuring a loaded medium with two cavities.

k	c_k	k	c_k	k	c_k	k	c_k	k	c_k	k	c_k
1	0.047	25	-0.019	62	-0.166	90	-0.344	124	-0.014	163	0.008
4	-0.085	32	-0.035	65	0.207	93	0.210	131	-0.018	176	0.018
7	0.197	36	0.005	69	-0.345	97	0.015	135	0.023	180	-0.001
11	0.049	40	0.019	72	0.208	104	0.018	138	-0.009	184	0.010
15	0.555	47	0.036	76	-0.279	108	-0.023	139	0.002	192	0.020
16	0.197	51	-0.004	79	0.117	111	0.009	142	-0.028	196	0.008
20	0.049	55	-0.284	83	-0.172	112	-0.002	151	0.010	209	0.018
24	0.558	58	0.120	86	0.203	115	0.027	159	0.020	213	-0.001

Figure 2 shows stress fields $\sigma_{xx},\sigma_{yy},\sigma_{zz},\sigma_{xz}$ at section $y=0$ for $\nu=0.2, h=0.05, R=0.7$. There is little difference in the stress fields in reference solutions at sections $y=0$, $x>0$, and $z>0$.

Let's present a fragment of the vector of the right-hand members of the ISE for a problem with a Poisson ratio of $\nu=0.2$, rounded to a thousandth place:

$$\mathbf{q}=\{0.047,\ 0,\ 0,\ -0.085,\ 0,\ 0,\ 0.197,\ 0,\ 0,\ 0,\ \ldots\}.$$

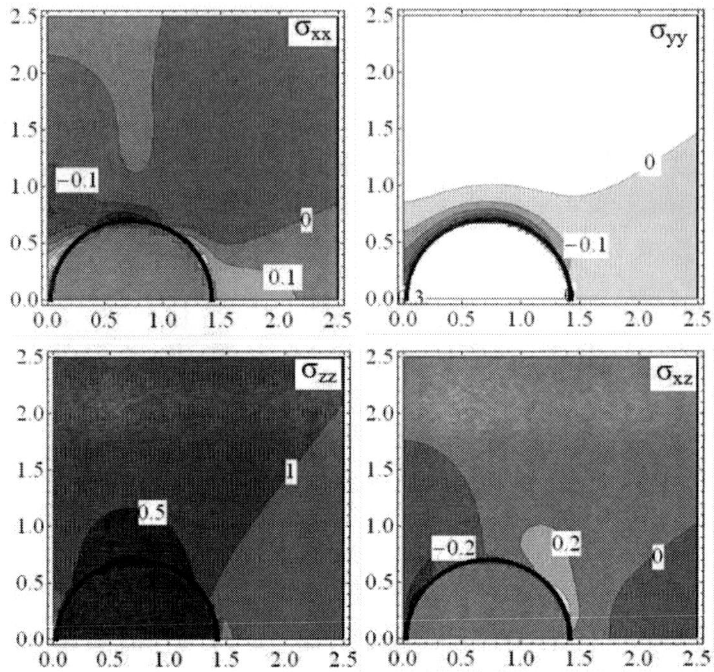

Figure 2. Stress level lines in the problem featuring two cavities.

Interpolation of the set of reference solutions by the Lagrangian method, incorporation of the actual remote stress and further reversal to dimension values produced an explicit solution. The entire solution being cumbersome, below we provide only fragments of the results obtained for displacement values:

$$u_x \approx Z^{\infty} x / \mu[-0.004 + 0.029/A^{9/2} - 0.008x/(R\,A^{9/2}) + 0.013x^2z^2/(R^4B^{11/2}) +$$
$$+ 0.197x^5/(R^5C^{15/2}) - 0.449\,v - 0.134\,x^4v/(R^4B^{11/2}) + 3.123xy^4v^2/(R^5\,C^{11/2}) + \ldots +$$
$$+ \left(-0.641/A^{9/2} + 0.117x^2y^2/(R^4B^{9/2}) + 0.242y^2/(R^2C^{5/2}) + 7.634xv/(RB^{7/2}) + \ldots\right)h +$$
$$+ \left(1.425/A^{9/2} + 0.015y^2/(RxB^{11/2}) + 0.847\,v/A^{9/2} + 0.082x^2y^4v^2/(R^6C^{15/2}) + \ldots\right)h^2],$$

$$u_y \approx Z^{\infty} y / \mu[-0.004 + 0.006/A^{9/2} - 0.003x/(R\,A^{9/2}) + 0.043x^2z^2/(R^4B^{11/2}) -$$
$$- 0.212x^5/(R^5C^{15/2}) - 0.449\,v - 0.023z^2v/(R^2B^{5/2}) + 0.141x^3y^2v^2/(R^5C^{13/2}) + \ldots +$$
$$+ \left(-0.128/A^{9/2} + 0.456z^2v/(R^2B^{5/2}) - 0.182x^2z^2v^2/(R^4C^{13/2}) + \ldots\right)h +$$
$$+ \left(0.285/A^{9/2} - 0.054x^3y^2/(R^5C^{11/2}) + 0.003xz^2v/(R^3A^{9/2}) + 0.183z^4v^2/(R^4C^{15/2}) + \ldots\right)h^2],$$

$$u_z \approx Z^{\infty} z / \mu[0.496 + 0.017/A^{9/2} - 0.010x/(R\,A^{9/2}) + 0.018x^2z^2/(R^4B^{11/2}) +$$
$$+ 1.232x^5/(R^5C^{15/2}) - 0.449\,v - 0.152\,x^4v/(R^4B^{11/2}) + 2.924xy^4v^2/(R^5\,C^{11/2}) + \ldots +$$
$$+ \left(-0.385/A^{9/2} + 0.003z^4/(R^4B^{11/2}) + 2.399x^7v/(R^7C^{15/2}) + 0.028v^2/A^{5/2} + \ldots\right)h +$$
$$+ \left(0.855/A^{9/2} - 0.013xz^2/(R^3\,C^{9/2}) + 0.043xv/(R\,B^{9/2}) + 0.005z^2v^2/(R^2A^{9/2}) + \ldots\right)h^2],$$

where $A = (-7 + x/R)^2 + y^2/R^2 + z^2/R^2$, $B = (-0.9 + x/R)^2 + y^2/R^2 + z^2/R^2$,
$C = (-0.725 + x/R)^2 + y^2/R^2 + z^2/R^2$.

7. Conclusions and prospects

1. The above methodology makes it fundamentally possible to perform FPSs for restricted and unrestricted bodies with a finite number of restricted cavities.
2. The MBS proves an efficient method for formulating analytical solutions.
3. We see prospects of applying the MBS with perturbations to FPSs. This approach holds a promise of a more rational use of computer technology.

The study was supported by RFFR with grant No. 16-41-480729 “r_a”.

References

[1] Levina L V, Novikova O S and Penkov V B 2016 Full parametric solution of problems of the theory of elasticity of simply connected finite bodies *Bulletin of Lipetsk State Technical University* **2 (28)** pp 16-24
[2] Sedov L I 1977 *Similarity methods and dimensional analysis in mechanics* (Moscow: Nauka) p 440
[3] Penkov V B, Satalkina L V and Shulmin A S 2014 The use of the method of boundary states to analyse an elastic medium with cavities and inclusions *J of Applied Mathematics and Mechanics* vol 78 **4** pp 384-394
[4] Lurie A I 1955 *Spatial problems of the theory of elasticity* (Moscow: Gostekhizdat) p 492
[5] Neuber H 1934 *Ein neuer Ansatz zur Lözung räumlicher Probleme der Elastizitätstheorie* Z Angev. Math. Mech. **1** pp 203-212
[6] Papkovich P F 1932 Expression of the General integral of the fundamental equations of elasticity theory using harmonic functions *Proceedings of the Academy of Sciences of the USSR series math. and natural sciences* **10** pp 1425-35
[7] Kupradze V D, Gegelia T G, Basheleishvili M O and Burchuladze T V 1976 *Three dimensional problems of mathematical theory of elasticity and thermoelasticity ed V D Kupradze* (Moscow: Nauka) p 664
[8] Nayfeh A H 1993 *Introduction to perturbation techniques* (USA: A Wiley-interscience publication. John Wiley & Sons, Inc) p 519
[9] Penkov V B, Levina L V, Levin M Y and Kuzmenko N V 2016 A new method for analyzing the effect of body forces induced by nanodispersed magnetic fluids on states of elastic solids *Science in the Central Russia* (journal on science and production technology) **2(20)** pp 12-16
[10] Southwell R V and Gough H J 1926 On the concentration of stress in the neighborhood of a small spherical flaw *Philos. Mag.* series 7 vol 1 pp 71-97

IOP Conf. Series: Journal of Physics: Conf. Series **973** (2018) 012017 doi:10.1088/1742-6596/973/1/012017

Fracture of functionally graded thermal barrier coating on a homogeneous substrate: models, methods, analysis

V E Petrova[1,2] **and S Schmauder**[1]

[1] IMWF, University of Stuttgart, Pfaffenwaldring 32, D-70569 Stuttgart, Germany
[2] Voronezh State University, University Sq.1, Voronezh, 394006, Russia

[1] e-mail: veraep@gmail.com; Siegfried.Schmauder@imwf.uni-stuttgart.de

Abstract. Fracture of functionally graded thermal barrier coatings on a homogeneous semi-infinite substrate (FGC/H) is studied under the influence of thermal loadings. A mathematical model for the FGC/H with pre-existing systems of cracks is presented in point of view of the formulation of the boundary conditions and assumptions, and afterwards the integral equations. Methods for solving the problem are described, and then demonstrated for some special cases of the arrangement of cracks. The effect of geometry (the thickness of the coating and location – orientation of multiple cracks) and inhomogeneity parameters of FGCs on the main fracture characteristics of the material is analyzed. The study can be used to improve the thermal fracture resistance of FGC/H systems.

1. Introduction

Thermal barrier coatings are widely used in industry, for example, in aerospace and airplane construction, and in power engineering. The effectiveness of rocket equipment, power turbines and aircraft gas turbine engines depends on the operating temperature. High temperatures and abrupt temperature changes lead to high-temperature cracking and melting of metal parts. Thermal barrier coatings create a layer with low thermal conductivity on the surface of the workpiece, which reduces the negative impact of high temperatures. For this purpose, various ceramics are used. Ceramics have a low thermal conductivity, but they are brittle. In addition, as a result of the difference in coefficients of thermal expansion of the coating and the substrate, high thermal stresses at the bond interface occur, which leads to cracking and debonding of the coatings. To overcome this problem, new materials are being sought, and new coatings are designed, namely, layered coatings and coatings from functionally gradient materials [1]. Functionally graded materials (FGMs) are a special class of composites with properties that vary along one direction (one spatial coordinate). Generally, the change in the properties of FGMs is connected with a corresponding variation in the chemical composition and/or physical structure of the material along this direction.

The study of the strength and fracture toughness of such FGMs is of great importance, especially under the combined effect of temperature, mechanical stresses and aggressive environment. Nowadays, a lot of studies has been devoted to these problems, reviews can be found in [1, 2]. However, the interaction of cracks and defects in FGMs under thermal loading is not sufficiently developed.

Investigation of the fracture of the functionally graded coating on a homogeneous substrate (FGC/H) under the influence of thermal loads includes the following tasks:

- Modeling of inhomogeneous physical and mechanical properties of FGMs and the determination of the stress-strain state in FGC/H structures.
- The problem of edge, internal and interface cracks and their interaction.
- The influence of material heterogeneity on the interaction of cracks, as well as the mutual effect of thermal and mechanical loads on cracks and on the interaction of cracks.

In the present paper, the problem of the interaction of systems of cracks (edge and internal) in a functionally graded coating on a homogeneous semi-infinite substrate subjected to thermal loads is considered. The applied methods are similar to those used in reference [3] for an infinite bimaterial formed from functionally gradient and homogeneous materials, and in [4] for a FGM coating on a semi-infinite substrate with a periodic system of edge cracks in the coating.

The paper is organized as follows. The statement of the problem is presented in Section 2, where the geometry of the problem and the loading are described, and also the boundary conditions and assumptions are formulated. Then, in Section 3 models for functionally graded materials (FGMs) and for cracks in FGMs are described. In Section 4 the formulation of integral equations is followed by methods for solving these equations. Afterwards, in Section 5 the main fracture characteristics are calculated, namely, the stress intensity factors, as well as the fracture angles and the critical loads. In Section 6 a few illustrative examples are presented and parametric analysis is performed. Some concluding remarks are made in the final Section 6.

2. Statement of the problem

2.1. *Geometry of the problem*

The general case of the geometry of the problem is depicted in figure 1a. The upper layer of thickness h is made from a functionally graded material (FGM), and the semi-infinite substrate consists of a homogeneous material. The functionally graded coating (FGC) and the substrate are perfectly bonded with the exception of an interface crack disposed at the interface between the two materials. The FGC contains arbitrary located cracks of length $2a_k$ (k=1,...,N), which can be internal and/or edge cracks. The coordinate systems are chosen as follows: the global coordinates (x, y) with the x-axis located on the surface of the FGC/H structure and the local coordinates (x_k, y_k) connected with cracks and with centers in the midpoint of the k-th crack. The crack positions are determined explicitly by the midpoint coordinates z_k^0 and the inclination angles α_k to the x-axis (or β_k for edge cracks, $\beta_k = -\alpha_k$) (figure 1b).

2.2. *Thermal and mechanical loads*

The following loads are considered: a steady state thermal flux applied normal to the surface of the FGC/H structure (figure 1a), a tension parallel to this surface (in the x-direction), and cooling by ΔT. In the case of cooling of the FGC/H structure tensile residual stresses are observed as shown in the experimental investigations [5]. Since the problem is linear, the results for each load can be superimposed. In the case of FGC/H under a heat flux the problem is solved in two steps, first, the thermal problem for the FGC/H structure with cracks, and then the thermo-elastic problem for the same geometry.

2.3. *Assumptions*

The following assumptions are used for this problem:

- The uncoupled, quasi-static thermo-elasticity theory is applied. In this theory the temperature distribution is independent of the mechanical field, and the solution consists of the determination of the temperature field and then the determination of the thermal stresses.
- The thermal and mechanical properties of an FGC are continuous functions of the thickness coordinate y.
- The non-homogeneity of the functionally graded material is revealed in the form of the corresponding inhomogeneous stress distributions on the surfaces of the cracks [3, 4, 6]. In this case, the properties of the FGM should vary slightly with the depth of the layer.

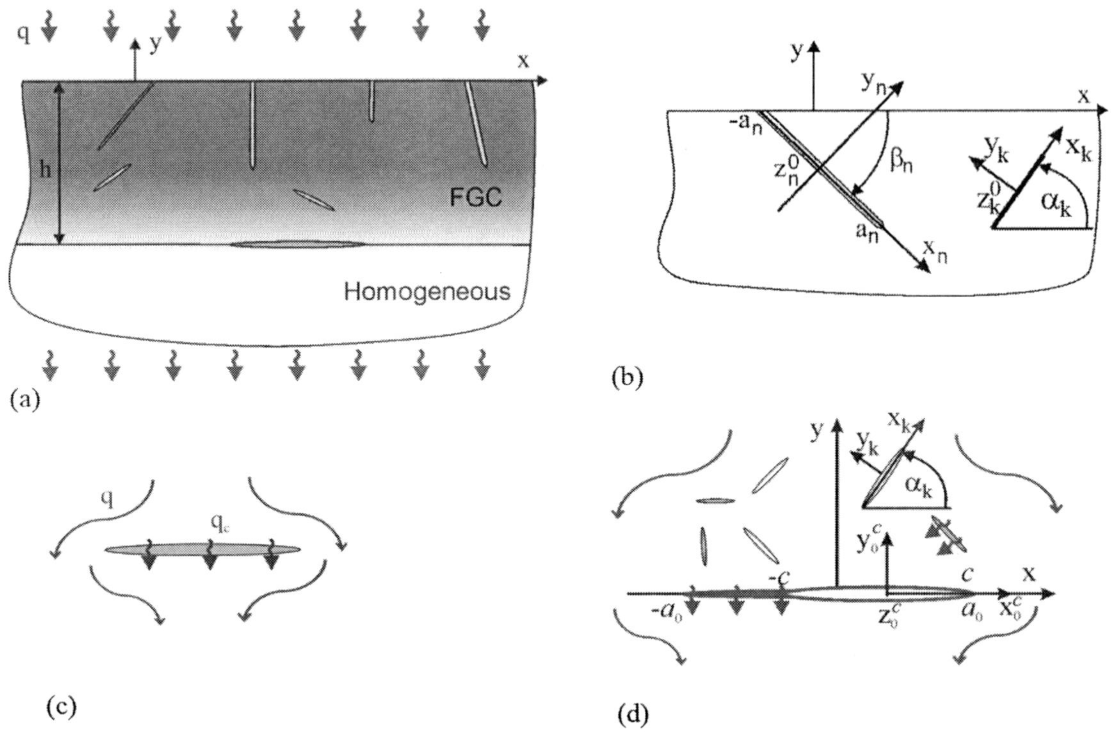

Figure 1. (a) An FGC/H structure with a system of cracks. (b) Coordinate systems connected with cracks. (c) A partially thermal conducting crack. (d) Partially closed crack with an open zone of length 2c and a system of open or closed cracks.

2.4. Boundary conditions

The thermal problem for a system of thermally insulated cracks in an infinite bimaterial consisting of a functionally graded material and a homogeneous substrate has been formulated in [3]. In the present problem the partially thermal insulated cracks in the FGC are considered with the insulation coefficient η_k $(0 \le \eta_k \le 1)$. The description of the model is given below in Section 3. Cracks are free of stresses, unless other conditions are mentioned. The FGC and the substrate are perfectly bonded, i.e. ideal thermal and mechanical conditions are fulfilled outside an interface crack, namely, the temperatures and the thermal fluxes are equal at the interface, and the stresses and displacements are also equal.

2.5. Methods

For the formulation of the problem the method of complex variables is used, and the method of superposition. Due to the superposition principle the common problem is decomposed into sub-problems, each of which contains one crack. Besides, the loads (thermal and mechanical) at infinity are reduced to the corresponding loads on the crack faces. The solution of each sub-problem can be obtained in explicit form (and can be analyzed separately) or in an integral form, and this integral form is used for constructing the equations of the complete problem. The methods for constructing integral equations for systems of cracks in a homogeneous material can be found in [7].

3. Modeling of functionally graded materials and cracks in FGMs

3.1. Models for functionally graded materials

The thermal and mechanical properties of an FGC are continuous functions of the thickness coordinate y. As in the previous works [3, 8, 9] the exponential form of these properties is used:

$$f(y) = f_1 \exp(\zeta_n (y + h)), \quad -h \le y \le 0 \qquad (1)$$

here

$$f = \{k,\ \alpha_t,\ E\},\ f_1 = \{k_1,\ \alpha_{t1},\ E_1\},\ \zeta_n = \{\delta,\ \varepsilon,\ \omega\},$$

k is thermal conductivity, α_t – thermal expansion coefficient and E – the Young's modulus with non-homogeneity parameters δ, ε and ω, respectively. f_1 are thermal and mechanical properties of the homogeneous substrate.

The Poisson's ratio is assumed to be constant and is equal to the value of the homogeneous substrate. The previous studies show that the effect of Poison's ratio on the stress intensity factors is negligibly small (see, for example, [10]).

The values of the dimensionless graded parameters $\zeta_n h$ (h is the thickness of the FGC) are obtained from Eq. (1) as

$$\zeta_n h = \ln(f_2 / f_1), \quad f_2 = f(y)|_{y=0} \quad f_1 = f(y)|_{y=-h}$$

An example of ceramic/metal FGC/H is (ZrO_2/Ti-6Al-4V)/Ti-6Al-4V with $f_2 / f_1 = \{2/6.7,\ 10/8.6,\ 200/114\}$ and, respectively, with the inhomogeneity parameters $\{\delta h,\ \varepsilon h,\ \omega h\} = \{-1.2,\ 0.15,\ 0.56\}$ (see [8, 11]). The variation of non-dimensional properties $f / f_1 = \exp(\zeta_n h(y/h + 1))$ with dimensionless coordinate y/h is displayed in figure 2a. The coefficient of thermal expansion and the Young's modulus increase toward the upper part of the FGC/H structure, while the coefficient of thermal conductivity decreases (figure 2a).

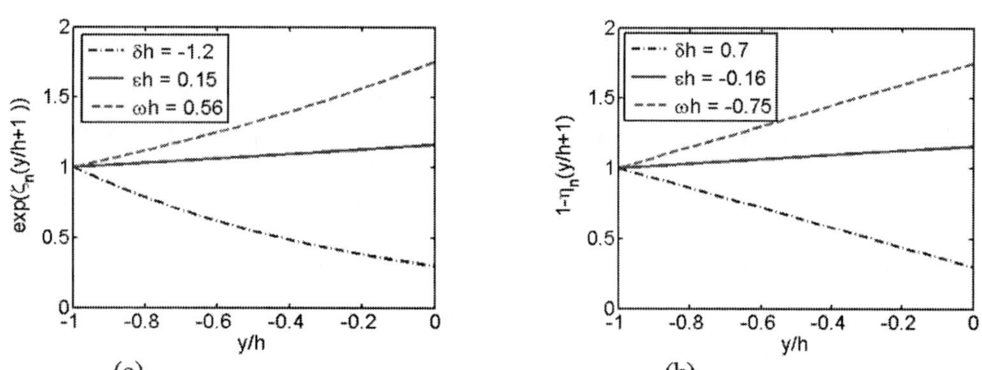

Figure 2. The variation of the exponential function $f / f_1 = \exp(\zeta_n h(y/h + 1))$ (a) and the linear function $f(y) = f_1(1 - \eta_n h(y/h + 1))$ (b) with the coating depth y/h for the FGC/H structure.

The exponential form of properties (1) is convenient for the mathematical formulation of equations. Other types of variation laws can be a linear function, a power law or other continuous functions [12]. An example of the linear law for our case is written as

$$f(y) = f_1(1 - \eta_n (y + h)) \text{ with } \eta_n = \{\delta,\ \varepsilon,\ \omega\}. \qquad (2)$$

The inhomogeneity parameters are defined as $\eta_n h = 1 - f_2 / f_1$, and for the above mentioned ceramic/metal FGC/H structure are the following: $\eta_n h = \{\delta h,\ \varepsilon h,\ \omega h\} = \{0.7, -0.16, -0.75\}$. Figure 2b shows the variation of this linear function with coordinate y/h.

For practical application, if the spatial composition profile is known, it can be useful to apply a formula based on the rule of mixtures for determining the effective properties of composites. Theoretical mixing laws have been considered extensively for different types of composites, and have been applied to FGMs [13, 14, 15]. For example, the thermal conductivity is defined by

$$k(y) = k_m \left[1 + \frac{V_i(y)(k_i - k_m)}{k_m + (k_i - k_m)(1 - V_i(y))/3} \right],$$

where the subscripts i and m stand for the inclusion and matrix properties, respectively. The volume fraction of the inclusion phase $V_i(y)$ is assumed in the form of a power-law function with a non-homogeneity parameter p

$$V_i(y) = ((y + h)/h)^p,$$

and the matrix phase is calculated as $V_m = 1 - V_i$.

3.2. *Models for cracks*

Partially thermally permeable cracks. A model of partially thermally permeable (insulated) cracks is used, figure 1c. The simplest model of partially insulated interfacial crack was used, for example, in [9, 16]. The insulation coefficient η_k $(0 \le \eta_k \le 1)$ is introduced, here the limiting value $\eta_k = 0$ represents the fully insulating crack and $\eta_k = 1$ – the fully conducting crack. In common case the coefficient $\eta_k(x)$ can be the function of the crack coordinate.

Cracks with contact zones. In the present work possible crack closure and contact of the crack surfaces is taken into account in the model (figure 1d), and its contribution is assessed. Previously, the model of a crack with contact zones was considered in the problem of the interaction between an interface crack and internal cracks in FGM/homogeneous bimaterials under the influence of thermal flux in [17] and under the combination of Mode II load and thermal flux in [18]. It was supposed that the closure of the crack does not affect the temperature distribution, and therefore only the thermoelastic problem was reformulated and solved.

Cracks with plastic zones. The plasticity is considered through a special mode, e.g. Dugdale model [19]. In [20] this model was applied for the problem of the interaction between a system of microcracks and a main crack with plastic zones. This plasticity model allows to formulate the equations within the framework of linear fracture mechanics.

4. Formulation of the integral equations of the problem and solution

4.1. *System of singular integral equations*

As it was mentioned in Section 2, assuming that the inhomogeneity of the FGM is revealed in the form of the corresponding inhomogeneous stress distributions on the surfaces of cracks, the solution of the boundary value problem of elasticity is reduced to the solution of the system of singular integral equations [3, 6, 7]:

$$\int_{-a_n}^{a_n} \frac{g_n'(t)\mathrm{d}t}{t - x} + \sum_{\substack{k=1 \\ k \ne n}}^{N} \int_{-a_k}^{a_k} [g_k'(t)R_{nk}(t,x) + \overline{g_k'(t)}S_{nk}(t,x)]\mathrm{d}t = \pi p_n(x), \quad |x| < a_n, \tag{3}$$

$$(n = 1, 2, \ldots, N).$$

Here the unknown functions $g_n'(x)$ are the derivatives of displacement jumps on the crack lines

$$g_n'(x) = \frac{2\mu}{i(\kappa+1)} \frac{\partial}{\partial x}\left([u_n] + i[v_n]\right),$$

where $[u_n]$ and $[v_n]$ are shear and vertical displacement jumps, respectively, on the n-th crack line, $\mu = E/2(1+\nu)$ is the shear modulus, E - Young's modulus, ν - Poisson's ratio, $\kappa = 3-4\nu$ for the plane strain state and $\kappa = (3-\nu)/(1+\nu)$ for the plane stress state. In equations (3) an overbar $\overline{(...)}$ denotes the complex conjugate.

The regular kernels R_{nk} and S_{nk} in equations (3) determine the geometry of the problem, and on the right side of equations (3) the functions p_n are known functions determined by the load on the crack lines.

For internal cracks, the condition of displacement continuity at crack tips has to be taken into account. The system of equations for the heat conduction problem has a similar form, where the unknowns are the derivatives of temperature jumps on the crack lines, and on the right side of the equations – the known functions of the heat fluxes on the cracks.

4.2. Solution of the equations

Many results concerning the interaction of cracks were obtained using methods of integral equations [3, 4, 6-9]. Various numerical methods such as series expansions, boundary element methods and others can be used to obtain an approximate solution. The solutions presented below are obtained for FGM/H structures.

4.2.1. Approximate analytical method. In the previous works [3, 17, 18] results for an infinite bimaterial consisting of an FGM and a homogeneous material with an interface crack and a system of internal cracks in the FGM were obtained by the small parameter method. The solution of equations (3) was presented in explicit approximate form and was written for the interface crack as

$$g_0'(x) = g_{00}'(x) + \lambda^2 g_{02}'(x),$$

and the stress intensity factors (SIFs) at the interface crack tips were presented as

$$k_I^{\pm} - ik_{II}^{\pm} = k_0\left\{1 + \lambda^2 \sum_{n=1}^{N} g_n(\alpha_n, z_n^0/a_0, \delta a_0, \varepsilon a_0)\right\}.$$

The stress intensity factors are found by the formula [4, 7]

$$k_{nI}^{\pm} - ik_{nII}^{\pm} = \mp \lim_{x_n \to \pm a_n} \sqrt{(a_n^2 - x_n^2)/a_n}\, g_n'(x_n), \tag{4}$$

where the upper part of the "±" or "∓" signs concerns to the right tip and the lower to the left tip of the crack.

It was assumed that an interface crack is significantly larger in size than internal cracks in the FGM. For this special case the small parameter λ is equal to the ratio of the size of small internal cracks to the interface crack size, i.e. $\lambda = a/a_0$ with $a = a_n$. The solution takes into account the interaction of the interface crack and each of the internal cracks (the interaction between small cracks is not accounted). These SIF functions contain parameters of the geometry of the problem (the midpoint coordinates z_n^0 of cracks, the orientation angles α_n of cracks and the crack sizes) and the parameters of the non-homogeneity of the FGM (δ and ε). It was shown that the non-homogeneity parameters δ and ε of thermo-conductivity and of thermal expansion coefficients, correspondingly, notably affect the SIFs of the interface crack. The SIFs can be amplified or shielded by the system of microcracks.

4.2.2. Numerical solution. The solution of the equations is obtained by the method of mechanical quadratures [7, 21]. Applying quadrature formulas based on Chebyshev polynomials, the integral equations (3) reduce to systems of algebraic equations. In [8], a scheme for solving this method is given. After determining the unknowns, the main characteristics of the fracture mechanics are calculated, namely, the stress intensity factors at the crack tips (4) or crack tip opening displacements.

5. Fracture criteria

Experimental and theoretical studies show that with a mixed-mode type of loading, the crack propagation deviates from its original one. In FGMs, a mixed stress-strain state can also arise due to the heterogeneity of the material. Applying a fracture criterion, it is possible to determine the angles of the deflection of the propagation of cracks from their original direction and the critical loads at which this propagation occurs. For example, according to the criterion of maximum normal stresses [22] (the formulation of this criterion and the method of application can be also found in [7]), the crack deflection angle (or the fracture angle) is calculated by the formula

$$\phi_0 = 2\mathrm{arctg}\left[\left(k_I - \sqrt{k_I^2 + 8k_{II}^2}\right) / 4k_{II}\right], \quad (5)$$

where k_I and k_{II} are the stress intensity factors.

The critical stresses are calculated from the expression

$$\cos^3(\phi/2)(k_I - 3k_{II}\tan(\phi/2)) = K_{ICtip} / \sqrt{\pi} \,.$$

Here K_{ICtip} is the fracture toughness of the material near the crack tip. First, it is obtained the angle of the crack propagation using the results for the calculation of SIFs. Then, the local fracture stability is evaluated. This criterion is the simplest one to employ.

In the minimum strain energy density criterion [23] the strain energy density factor (SDF) is introduced so that it is a quadratic form of SIFs (with coefficients in this form presented as functions of fracture angles). Consequently, it is postulated that the crack will grow in the direction where the SDF is minimal and when this minimum reaches a critical value. This criterion depends on Poisson's ratio.

The maximum strain energy release rate (SERR) criterion [24] posts that the crack will grow along the direction for which the strain energy release rate reaches a critical value. Note, this critical value is material dependent.

As a result (by using these criteria) the fracture angles are obtained as functions of the geometry of the problem and of the non-homogeneity parameters of the FGC with additional parameters due to thermal loadings. Then, the critical stresses and critical heat fluxes are obtained near cracks (for different crack locations).

In [25], the results for these three criteria were compared for the problem for an infinite FGM/H structure and it was shown that they are nearly the same. In the present work the criterion of maximum normal stresses [22] is applied.

6. Results and analysis

The problem contains geometric parameters, such as the size of the cracks, the coordinates of their centers and the inclination angles, and the thickness of the FGM layer, and the parameters of material inhomogeneity, determined by formulas (1) or (2). The presented model allows analyzing the influence of these parameters on the main characteristics of the problem.

Fig. 3 schematically shows three edge cracks and their deviation angles ϕ, in Fig. 3a the cracks are of the same size, and in Fig. 3b the length of the first crack is twice the length of the second and third ones. The following parameters were used for the calculation: $h/a = 4$, $d/a = 2, a = \max a_k$, $a = 1mm$, $\varepsilon a = -1$ and $\omega a = 0$. The distance between the cracks is equal to the length of the large crack (d/a=2), $\beta = 90^0$. At $\varepsilon a = -1$ the coefficient of thermal expansion increases with the depth of the layer. The

FGC/H structure is cooled by ΔT, in this case tensile residual stresses parallel to the boundary ($y = 0$) arise. With this load, for one edge crack the value of the angle is equal to $\phi = 0°$. The change in the angles ϕ due to the interaction of the cracks is shown in Fig. 3. It should be noted that the thermal expansion inhomogeneity coefficient does not greatly affect the deflection angles ϕ. Earlier it was shown that the coefficient of inhomogeneity of thermal conductivity significantly affects the interaction of cracks [3, 17].

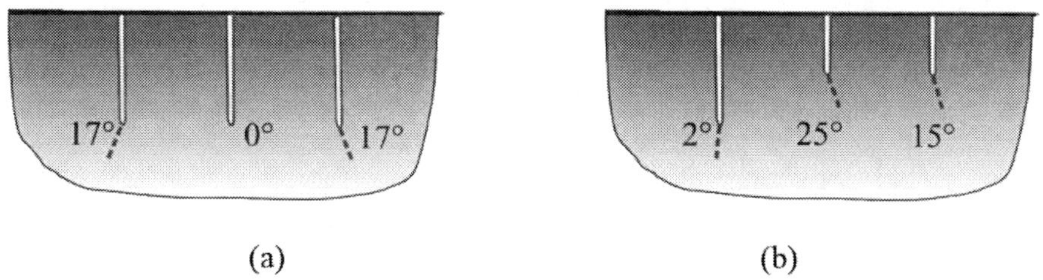

Figure 3. Angles of deflection of the propagation of cracks under thermal loading. (a) Three edge cracks of equal length. (b) Edge cracks of different lengths $a=a_2=a_3=0.5a_1$. The distance between the cracks is equal to the length of the large crack $d/a=2$, $\beta = 90^0$.

The influence of the remaining parameters, geometrical and physical, will be considered later separately for the corresponding problems for thermal and mechanical loads.

7. Conclusions

A general theoretical formulation of the model for the thermal fracture analysis of functionally graded coatings on a homogeneous substrate (FGC/H) has been performed by means of integral equations. The FGC/H structure with pre-existing systems of interacting cracks in the FGC is subjected to thermal and mechanical loadings. The FGM properties are modeled by continuous functions of the spatial coordinate (the thickness of the coating). This approximate model allows taking into account some special crack models such as the partially thermally permeable cracks and the cracks with contact and plastic zones. The solution of equations is obtained numerically, using the special quadrature formulae for the singular and regular integrals in the integral equations. Besides, an approximate analytical solution previously obtained for a special case, when an interface crack is significantly larger in size than internal cracks in the functionally graded material for an infinite FGM/H structures, is given. In the present study the main emphasis is done on the fracture with multiple crack interactions. At first, the stress intensity factors are calculated, then, using the fracture criterion of maximum normal stresses [22], the crack deflection angles (or the fracture angles) are obtained. An illustrative example is presented to show the influence of the parameters of the problem on the edge crack interaction and their deviation from the initial propagation. Other crack models in the FGC/H structures will be considered in future works. The study of fracture of the FGC/H structures is important for a better understanding of the thermal fracture of graded coatings and for improving the fracture resistance of these systems.

Acknowledgments

The authors would like to acknowledge the financial support of the German Research Foundation under Grant Schm 746/139-2.

References

[1] Miyamoto Y, Kaysser W A, Rabin B H, Kawasaki A and Ford R G 1999 *Functionally Graded Materials: Design, Processing and Applications* (Dordrecht: Kluwer Academic)
[2] Birman V and Byrd L W 2007 Modeling and analysis of functionally graded materials and structures *ASME Appl. Mech. Rev.* **60** 195–216

[3] Petrova V and Schmauder S 2011 Thermal fracture of a functionally graded/homogeneous bimaterial with a system of cracks *Theor. Appl. Fract. Mech.* **55** 148 –157

[4] Afsar A M and Sekine H 2000 Crack spacing effect on the brittle fracture characteristics of semi-infinite functionally graded materials with periodic edge cracks *Int. J. Fract.* **102** L61–L66.

[5] Gilbert A, Kokini K and Sankarasubramanian S 2008 Thermal fracture of zirconia-mullite composite thermal barrier coatings under thermal shock: An experimental study *Surf. Coat. Technol.* **202** 2152–61

[6] Petrova V and Sadowski T 2014 Theoretical modeling and analysis of thermal fracture of semi-infinite functionally graded materials with edge cracks *Meccanica* **49** 2603–15

[7] Panasyuk V, Savruk M and Datsyshin A 1976 *Stress Distribution near Cracks in Plates and Shells* (Kiev: Naukova Dumka) (in Russian)

[8] Petrova V and Schmauder S 2017 Modeling of thermo-mechanical fracture of FGMs with respect to multiple cracks interaction *Phys. Mesomech.* **20** 241–249

[9] El-Borgi S, Hidri L and Abdelmoula R 2006 An embedded crack in a graded coating bonded to a homogeneous substrate under thermo-mechanical loading *J. Therm. Stresses* **29** 439–466

[10] Delale F and Erdogan F 1983 The crack problem for a nonhomogeneous plane *ASME J. Appl. Mech.* **50** 609–614

[11] Shackelford J F and Alexander W 2001 *CRC Materials Science and Engineering Handbook* (Boca Raton: CRC Press)

[12] Sevcik M, Hutar P, Nahlik L and Knesl Z 2009 An evaluation of the stress intensity factor in functionally graded materials *Appl. Comput. Mech.* **3** 401–410

[13] Tilbrook M, Moon R and Hoffman M 2005 Crack propagation in graded composites *Composites Science and Technology* **65** 201-220

[14] Fujimoto T and Noda N 2001 Influence of the Compositional Profile of Functionally Graded Material on the Crack Path under Thermal Shock *J. Am. Ceram. Soc.* **84** 1480–86

[15] Burlayenko V N, Altenbach H, Sadowski T, Dimitrova S D and Bhaskar A 2017 Modelling functionally graded materials in heat transfer and thermal stress analysis by means of graded finite elements *Appl. Math. Model.* **45** 422–438

[16] Lee K Y and Park S J 1995 Thermal stress intensity factors for partially insulated interface crack under uniform heat flow *Eng. Fract. Mech.* **50** 475–482

[17] Petrova V and Schmauder S 2012 Mathematical modelling and thermal stress intensity factors evaluation for an interface crack in the presence of a system of cracks in functionally graded *Comput. Mater. Sci.* **52** 171–177

[18] Petrova V and Schmauder S 2015 Crack closure effects in thermal fracture of functionally graded/homogeneous bimaterials with systems of cracks *ZAMM* **195** 1027–36

[19] Dugdale D S 1960 Yielding in steel sheets containing slits *J. Mech. Phys. Solid.* **8** 100–104

[20] Tamuzs V, Petrova V and Tarasovs S 2004 Interaction of micro-cracks with a macro-crack yielded in a narrow strip *Theor. Appl. Fract. Mech.* **41** 291–299

[21] Erdogan F and Gupta G 1972 On the numerical solution of singular integral equations *Quart. Appl. Math.* **29** 525–534

[22] Erdogan F and Sih G C 1963 On the crack extension in plates under plane loading and transverse shear *J. Basic. Eng.* **85** 519–527

[23] Sih G C 1991 *Mechanics of Fracture Initiation and Propagation: Surface and Volume Energy Density Applied as Failure Criterion* (Dordrecht, The Netherlands: Kluwer Academic Publishers)

[24] Hussain M A, Pu S L and Underwood J 1974 Strain energy release rate for a crack under combined mode I and mode II *ASTM STP* **560** 2–28

[25] Petrova V and Schmauder S 2014 FGM/homogeneous bimaterials with systems of cracks under thermo-mechanical loading: Analysis by fracture criteria *Eng. Fract. Mech.* **130** 12–20

IOP Conf. Series: Journal of Physics: Conf. Series **973** (2018) 012018 doi:10.1088/1742-6596/973/1/012018

Nonlinear constitutive relations for anisotropic elastic materials

Marina Sokolova[1], **Dmitrii Khristich**[2]

Department of Computational Mechanics and Mathematics, Tula State University, 92, Prospect Lenina, Tula 300012, Russia

E-mail: [1] m.u.sokolova@gmail.com, [2] dmitrykhristich@rambler.ru

Abstract. A general approach to constructing of nonlinear variants of connection between stresses and strains in anisotropic materials with different types of symmetry of properties is considered. This approach is based on the concept of elastic proper subspaces of anisotropic materials introduced in the mechanics of solids by J. Rychlewski and on the particular postulate of isotropy proposed by A. A. Il'yushin.

The generalization of the particular postulate on the case of nonlinear anisotropic materials is formulated. Systems of invariants of deformations as lengths of projections of the strain vector into proper subspaces are developed.

Some variants of nonlinear constitutive relations for anisotropic materials are offered. The analysis of these relations from the point of view of their satisfaction to general and limit forms of generalization of partial isotropy postulate on anisotropic materials is performed. The relations for particular cases of anisotropy are written.

1. Introduction

Anisotropic materials characterized by nonlinear behavior in the process of deformation are increasingly becoming the subject of the mechanics of deformable solids. Such materials can include various kinds of tissues of various types, blood vessels and bones of living organisms as well as composite materials widely used in engineering. Models of elastic and inelastic behavior of such materials were represented in the works [1, 2, 3, 4, 5, 6, 7, 8]. Finite deformations of anisotropic materials were considered in the issues [2, 8, 9, 10, 11].

The main problem of constructing of relations which determine the behavior of anisotropic material is the choice of a system of invariants of a strain tensor which characterize material symmetry properties. For a linearly elastic material this problem is solved in works by J. E. Green [12], C. Truesdell [13], A. J. M. Spencer [14], V. V. Novozhilov [15] and many other authors. Systems of invariants used in constitutive relations are also given in the papers [2, 11, 16].

In the present article we use the system of invariants connected with the concept of elastic proper subspaces of materials introduced by J. Rychlewski in [17]. For different materials the number of invariants is different: for the isotropic material the system includes two invariants, for the transversally-isotropic (hexagonal) — four and for orthotropic (rhombic) — six invariants. Functions which determine the behavior of the anisotropic material can depend on all invariants forming the system for the material under consideration. This fact significantly complicates the model of the material. For this reason it is appropriate to introduce additional hypotheses

IOP Conf. Series: Journal of Physics: Conf. Series **973** (2018) 012018 doi:10.1088/1742-6596/973/1/012018

concerning the behavior of the nonlinear anisotropic material. In our investigation we use A. A. Il'yushin's particular postulate as such hypothesis [18].

We also propose variants of nonlinear constitutive relations for anisotropic materials constructed within the frameworks of the generalization of the A. A. Il'yushin's particular postulate to the case of anisotropic materials [8, 19, 20]. Adopting the particular postulate as a hypothesis reduces the number of arguments of material functions which are included into nonlinear constitutive relations and consequently to reduce the number of experiments which are necessary for concretization of these relations.

2. Elastic proper subspaces

Let us establish the compliance between the process of deforming $\varepsilon(t)$ and its image in the six-dimensional space which represents the trajectory of deforming i.e. the hodograph of the six-dimensional vector $\vec{e}(t)$ with the stress vector $\vec{\sigma}(t)$ built at each point. The relation between the coordinates of six-dimensional vectors $\vec{e}$, $\vec{\sigma}$ and the components of the strain tensor ε and the stress tensor $\mathbf{S}$ is given in the work [20] and is used further:

$$\vec{e} = e_0\vec{\imath}_0 + e_1\vec{\imath}_1 + e_2\vec{\imath}_2 + e_3\vec{\imath}_3 + e_4\vec{\imath}_4 + e_5\vec{\imath}_5,$$

where $\vec{\imath}_\alpha$, $\alpha = 0, ..., 5$ are basis vectors of the six-dimensional space of deformations,

$$\begin{gathered} e_0 = \frac{1}{\sqrt{3}}\left(\varepsilon_{11} + \varepsilon_{22} + \varepsilon_{33}\right), \quad e_1 = \frac{1}{\sqrt{6}}\left(2\varepsilon_{33} - \varepsilon_{11} - \varepsilon_{22}\right), \\ e_2 = \frac{1}{\sqrt{2}}\left(\varepsilon_{11} - \varepsilon_{22}\right), \quad e_3 = \frac{1}{\sqrt{2}}\left(\varepsilon_{12} + \varepsilon_{21}\right), \\ e_4 = \frac{1}{\sqrt{2}}\left(\varepsilon_{23} + \varepsilon_{32}\right), \quad e_5 = \frac{1}{\sqrt{2}}\left(\varepsilon_{31} + \varepsilon_{13}\right). \end{gathered} \tag{1}$$

The inverse relations are of the form

$$\varepsilon_{11} = \frac{1}{\sqrt{3}}e_0 - \frac{1}{\sqrt{6}}e_1 + \frac{1}{\sqrt{2}}e_2, \quad \varepsilon_{22} = \frac{1}{\sqrt{3}}e_0 - \frac{1}{\sqrt{6}}e_1 - \frac{1}{\sqrt{2}}e_2,$$

$$\varepsilon_3 = \frac{1}{\sqrt{3}}e_0 + \sqrt{\frac{2}{3}}e_1, \quad \varepsilon_{12} = \frac{e_3}{\sqrt{2}}, \quad \varepsilon_{23} = \frac{e_4}{\sqrt{2}}, \quad \varepsilon_{31} = \frac{e_5}{\sqrt{2}}.$$

The requirement that components of the strain tensor ε_{ij} and the stress tensor S_{ij} must be determined in the Cartesian coordinate system with axes directed along main axes of anisotropy is important in the case of anisotropic materials.

Let us write the Hooke's law in the six-dimensional space in the form

$$\vec{\sigma} = \mathbf{n} \cdot \vec{e},$$

where $\mathbf{n}$ is the image of the elasticity tensor of the material in the six-dimensional space. We define eigenvalues λ_α and eigenvectors $\vec{\omega}_\alpha$ of the tensor $\mathbf{n}$:

$$\mathbf{n} \cdot \vec{\omega}_\alpha = \lambda_\alpha \vec{\omega}_\alpha.$$

For a second-order tensor in the six-dimensional space the maximum number of different eigenvalues is 6. It is possible to put into compliance a one-dimensional proper subspace with the basis tensor $\mathbf{\Omega}_\alpha = \vec{\omega}_\alpha \vec{\omega}_\alpha$ and each single root λ_α. We can also correlate a k-dimensional proper subspace with the basis tensor

$$\mathbf{\Omega}_\alpha = \sum_{\beta=\alpha}^{\beta=\alpha+k-1} \vec{\omega}_\beta \vec{\omega}_\beta \tag{2}$$

with each root λ_α of the multiplicity k.

In view of the invariance of eigenvalues and eigenvectors, the basis tensors $\mathbf{\Omega}_\alpha$ are invariant with respect to the group of eigen orthogonal transformations.

Elastic proper subspaces of a material in the conception of J. Rychlewski [17] are strain tensors ε_α, which correspond to six-dimensional vectors $\vec{\omega}_\alpha$. Elastic proper subspaces for materials belonging to different crystallographic systems are defined by N. I. Ostrosablin in [21]. In the issues [8, 19, 20] this problem is solved in the six-dimensional space under consideration. In particular it is determined [8, 19, 20] that multidimensional proper subspaces are inherent not in all anisotropic materials, but only in uniaxial crystals (tetragonal, trigonal and hexagonal) as well as in cubic and isotropic media. The elastic properties of a transversally-isotropic material coincide with those of a hexagonal crystal.

We present the expressions for the basis tensors $\mathbf{\Omega}_\alpha$ which characterize anisotropic materials of different types. We assign vectors and tensors of the six-dimensional space to the basis $\vec{i}_\alpha$.

Isotropic materials have their two proper subspaces: the one-dimensional subspace with the basis tensor $\mathbf{\Omega}_1 = \vec{i}_0 \vec{i}_0$ and the five-dimensional one with the basis tensor $\mathbf{\Omega}_2 = \vec{i}_1 \vec{i}_1 + \vec{i}_2 \vec{i}_2 + \vec{i}_3 \vec{i}_3 + \vec{i}_4 \vec{i}_4 + \vec{i}_5 \vec{i}_5$. The elasticity tensor in the six-dimensional space is represented as

$$\mathbf{n}^{(is)} = \lambda_1 \mathbf{\Omega}_1 + \lambda_2 \mathbf{\Omega}_2, \tag{3}$$

where $\lambda_1 = 3K$, $\lambda_2 = 2G$ are the elastic modules.

Cubic materials have three proper subspaces: the one-dimensional subspace with the basis tensor $\mathbf{\Omega}_1 = \vec{i}_0 \vec{i}_0$, the two-dimensional subspace with the basis tensor $\mathbf{\Omega}_2 = \vec{i}_1 \vec{i}_1 + \vec{i}_2 \vec{i}_2$ and the three-dimensional one with the basis tensor $\mathbf{\Omega}_3 = \vec{i}_3 \vec{i}_3 + \vec{i}_4 \vec{i}_4 + \vec{i}_5 \vec{i}_5$.

Hexagonal and trigonal materials have four proper subspaces: two two-dimensional subspaces with the basis tensors $\mathbf{\Omega}_1 = \vec{\omega}_1 \vec{\omega}_1$ and $\mathbf{\Omega}_2 = \vec{\omega}_2 \vec{\omega}_2$, where the eigenvectors are associated with the basic vectors by the relations

$$\vec{\omega}_1 = \vec{i}_0 \cos\varphi + \vec{i}_1 \sin\varphi, \quad \vec{\omega}_2 = -\vec{i}_0 \sin\varphi + \vec{i}_1 \cos\varphi,$$

and the angle φ is expressed in terms of components $n_{\alpha\beta} = \vec{i}_\alpha \cdot \mathbf{n} \cdot \vec{i}_\beta$:

$$\tan\varphi = \frac{2n_{01}}{n_{00} - n_{11} + \sqrt{(n_{00} - n_{11})^2 + 4n_{01}^2}}$$

and two two-dimensional proper subspaces with the basis tensors $\mathbf{\Omega}_3 = \vec{i}_2 \vec{i}_2 + \vec{i}_3 \vec{i}_3$, $\mathbf{\Omega}_4 = \vec{i}_4 \vec{i}_4 + \vec{i}_5 \vec{i}_5$ for hexagonal materials and with basis tensors $\mathbf{\Omega}_3 = \vec{\omega}_3 \vec{\omega}_3 + \vec{\omega}_5 \vec{\omega}_5$, $\mathbf{\Omega}_4 = \vec{\omega}_4 \vec{\omega}_4 + \vec{\omega}_6 \vec{\omega}_6$ for trigonal materials. In the latter case the orientation of the eigenvectors with respect to the basis $\vec{i}_\alpha$ is defined by the following relations:

$$\vec{\omega}_3 = \vec{i}_2 \cos\psi + \vec{i}_4 \sin\psi, \quad \vec{\omega}_4 = -\vec{i}_2 \sin\psi + \vec{i}_4 \cos\psi,$$

$$\vec{\omega}_5 = \vec{i}_3 \cos\psi + \vec{i}_5 \sin\psi, \quad \vec{\omega}_6 = -\vec{i}_3 \sin\psi + \vec{i}_5 \cos\psi,$$

and the angle ψ is expressed in terms of components $n_{\alpha\beta}$:

$$\tan\psi = \frac{2n_{24}}{n_{22} - n_{44} + \sqrt{(n_{22} - n_{44})^2 + 4n_{24}^2}}.$$

Tetragonal materials have five proper subspaces: four one-dimensional subspaces with basis tensors $\mathbf{\Omega}_1 = \vec{\omega}_1\vec{\omega}_1$, $\mathbf{\Omega}_2 = \vec{\omega}_2\vec{\omega}_2$, $\mathbf{\Omega}_3 = \vec{i}_2\vec{i}_2$, $\mathbf{\Omega}_4 = \vec{i}_3\vec{i}_3$, where the eigenvectors $\vec{\omega}_1$ and $\vec{\omega}_2$ are related to the basis vectors by the same relations as in the case of hexagonal or trigonal materials, and only one two-dimensional proper subspace with the basis tensor $\mathbf{\Omega}_5 = \vec{i}_4\vec{i}_4 + \vec{i}_5\vec{i}_5$.

From the expressions for eigen basis tensors we obtain the invariance of the operators $\mathbf{n}$ relative to the eigen transformations. For example, the operator of the isotropic material (3) is invariant with respect to rotations and reflections of the six-dimensional basis with respect to the vector $\vec{i}_0$. This transformation can be interpreted as the rotation of the basis $\vec{i}_\alpha$ ($\alpha = 0, ..., 5$) around the axis $\vec{i}_0$, orthogonal to the deviatoric plane $\vec{i}_1$, $\vec{i}_2$. Herewith the angle of the type of the deformed state, which is proportional to the angle of rotation, changes, but the change of this angle does not affect the linear relationship between stress and strain, described by the Hooke's law.

Thus, the operator $\mathbf{n}$ for the linearly elastic materials satisfies the invariance conditions for the subgroup of eigen orthogonal transformations. This condition is a consequence of the linearity of the relationship between stress and strain tensors and is determined by the uniqueness of this relationship.

3. Variants of constitutive relations

According to A. A. Il'yushin's particular postulate, the six-dimensional image of the deformation process of the initially isotropic body is invariant not only in relation to the group of orthogonal transformations, associated with the choice of the initial frame of reference, but also with respect to arbitrary transformations of rotation and reflection in the five-dimensional deviatoric subspace. Since under such transformations there are changes in the third invariants of strain and stress tensors, the particular postulate requires that the invariants of this type do not appear explicitly in the number of functions for which the scalar functionals, defining the properties of materials, are given. The particular isotropy postulate is a hypothesis that allows to formulate quasi-linear constitutive equations, with the linear relations of the Hooke's law satisfying the particular postulate.

The particular postulate of isotropy has found numerous experimental confirmation. In the works by V. S. Lenskiy and I. D. Mashkov, R. A. Vasin and R. I. Shirov, Dao Zuy Bik, L. S. Andreev, R. G. Terekhov and Yu. N. Shevchenko, A. M. Zhukov, V. P. Degtyarev, V. G. Zubchaninov, S. A. Elsuf'ev programs of experiments were developed for the implementation of complex loading for small strains and the results of experiments that confirm the validity of the postulate of isotropy for most materials at small strains were presented. The review and the analysis of these articles is performed in [20].

In his book A. A. Il'yushin [18] mentioned the possibility of generalization of the isotropy postulate for initially anisotropic materials. The idea of this generalization is that it is formulated in the space of generalized stresses, for example, obtained by multiplying the stresses with anisotropic basic tensors. In this space it is required to define a class of invariant transformations preserving the form of the connection between stresses and strains.

We proceed from the fact that for linearly elastic anisotropic materials the stress and strain vectors in each proper subspace are collinear. This relationship is preserved under any orthogonal transformation of the deformation path in each proper space, which is not one-dimensional. In particular, for an isotropic material this condition is satisfied in the five-dimensional deviatoric subspace. In the case of the cubic crystal system the condition of isotropy of images is fulfilled in the plane of the vectors $\vec{\omega}_2$, $\vec{\omega}_3$ and in a three-dimensional subspace with the basis $\vec{\omega}_4$, $\vec{\omega}_5$, $\vec{\omega}_6$, in the case of the hexagonal (trigonal) material — in planes $\vec{\omega}_3$, $\vec{\omega}_4$ ($\vec{\omega}_3$, $\vec{\omega}_5$) and $\vec{\omega}_5$, $\vec{\omega}_6$ ($\vec{\omega}_4$, $\vec{\omega}_6$) and in the case of the tetragonal material — in the plane $\vec{\omega}_5$, $\vec{\omega}_6$. We extend this property to all stages of the deformation process.

As a result, we obtain the following generalization of the particular postulate for the nonlinear anisotropic materials: the image of the thermomechanical process with a deformation path disposed in the proper subspace of the tensor of initial elasticity is invariant under the group of intrinsic orthogonal transformations.

From this generalization it follows that the process of loading in each multidimensional proper subspace is defined only by the inner deformation path in this subspace and is not dependent on the orientation of the trajectory relative to the basis vectors of this subspace. This formulation allows the deviation of the stress vector from the strain vector located in the proper subspace. Moreover, this deviation is the effect of the second order with respect to deformation since asymptotically the strain and stress vectors are collinear. According to the postulate the deviation should not change at intrinsic orthogonal transformations. An example of this behavior is the appearance of a hydrostatic stress in shear deformation of an isotropic material.

The limit form of the generalization of the particular postulate is the assumption that the image of the stress process and the strain trajectory lie in the same proper subspace. In view of this assumption the variant of nonlinear relations for anisotropic materials is written [19, 20] in the form

$$\vec{\sigma}_{(\alpha)} = \sum_{i=1}^{i=k} A^i_{(\alpha)} \vec{r}^{(\alpha)}_i,$$

where $A^i_{(\alpha)} \left[s_{(\alpha)}(\tau), \chi^1_{(\alpha)}(\tau), ..., \chi^{m-1}_{(\alpha)}(\tau) \right]_{\tau=t_0}^{\tau=t}$ is the functional of the deformation process $\vec{e}_{(\alpha)}(t)$; $\vec{r}^{(\alpha)}_i$ is the basis of a k-dimensional proper subspace. This basis is covariant with respect to eigen orthogonal transformations. $s_{(\alpha)}$ and $\chi^i_{(\alpha)}$ are the arc length and curvatures of the eigen trajectory respectively. The Frenet basis as the basis $\vec{r}^{(\alpha)}_i$ is convenient to accept as connected with the trajectory $\vec{e}_{(\alpha)}(t)$ of the process in the case of smooth eigen trajectories.

If an anisotropic material has p single eigenvalues and q eigenvalues with multiplicities k_1, k_2,..., k_q respectively the connection between stress and strain vectors has the form

$$\vec{\sigma} = \sum_{i=1}^{i=p} \sigma_{(i)} \vec{\omega}_i + \sum_{\alpha=p+1}^{\alpha=q+p+1} \sum_{j=1}^{j=k} A^j_{(\alpha)} \vec{r}^{(\alpha)}_j, \tag{4}$$

where $\sigma_{(i)} = \sigma_{(i)} \left[s_i(\tau) \right]_{\tau=t_0}^{\tau=t}$ is the functional of the deformation process in the one-dimensional proper subspace with the basis vector $\vec{\omega}_i$.

The relations (4) are significantly simplified if the deformation process is simple and the strain trajectory is a ray:

$$\vec{\sigma} = \sum_{i=1}^{i=p} \sigma_{(i)}(e_{(i)}) \vec{\omega}_i + \sum_{\alpha=p+1}^{\alpha=q+p+1} \tilde{A}_{(\alpha)}(s_{(\alpha)}) \vec{e}_{(\alpha)}, \tag{5}$$

where $e_{(i)} = \vec{e} \cdot \vec{\omega}_i$ are the linear invariants of deformations; $\vec{e}_{(\alpha)} = \vec{e} \cdot \mathbf{\Omega}_\alpha$ are the strain eigenvectors, i.e. the projections of the strain vector into multidimensional proper subspace with basis tensors $\mathbf{\Omega}_\alpha$ (2); $s_{(\alpha)} = \sqrt{\vec{e}_{(\alpha)} \cdot \vec{e}_{(\alpha)}}$ are the quadratic invariants of deformations.

For anisotropic materials of different types and for the isotropic material linear and quadratic invariants of deformations are defined in the table 1. The invariants given in the table 1 for different anisotropic materials are based on the invariant bases. They coincide with algebraic invariants which were brought in the article of K. F. Chernykh [16].

The orthogonal system of eigenvectors $\vec{\omega}_i$, $i = 1, ..., p$ which match the single eigenvalues is connected with the basis vectors $\vec{i}_0$, $\vec{i}_1$,..., $\vec{i}_{p-1}$ of the six-dimensional space by the rotation

Table 1. Invariants of the six-dimensional vector $\vec{e}$ and of the symmetrical second-order tensor $\boldsymbol{\varepsilon}$.

Linear invariants of the vector $\vec{e}$	Linear invariants of the tensor $\boldsymbol{\varepsilon}$	Quadratic invariants of the vector $\vec{e}$	Quadratic invariants of the tensor $\boldsymbol{\varepsilon}$
		triclinic material	
$e_0, e_1, e_2,$ e_3, e_4, e_5	$\varepsilon_{11}, \varepsilon_{22}, \varepsilon_{33},$ $\varepsilon_{12}, \varepsilon_{23}, \varepsilon_{31}$		
		monoclinic material	
e_0, e_1, e_2, e_3	$\varepsilon_{11}, \varepsilon_{22}, \varepsilon_{33}, \varepsilon_{12}$	e_4^2, e_5^2, e_4e_5	$\varepsilon_{23}^2, \varepsilon_{31}^2, \varepsilon_{23}\varepsilon_{31}$
		rhombic material	
e_0, e_1, e_2	$\varepsilon_{11}, \varepsilon_{22}, \varepsilon_{33}$	e_3^2, e_4^2, e_5^2	$\varepsilon_{12}^2, \varepsilon_{23}^2, \varepsilon_{31}^2$
		tetragonal material	
e_0, e_1	$\varepsilon_{11} + \varepsilon_{22}, \varepsilon_{33}$	$e_2^2, e_3^2, e_4^2 + e_5^2$	$(\varepsilon_{11} - \varepsilon_{22})^2, \varepsilon_{12}^2, \varepsilon_{23}^2 + \varepsilon_{31}^2$
		trigonal material	
e_0, e_1	$\varepsilon_{11} + \varepsilon_{22}, \varepsilon_{33}$	$e_2^2 + e_3^2, e_4^2 + e_5^2$ $e_2e_4 + e_3e_5$	$(\varepsilon_{11} - \varepsilon_{22})^2 + 4\varepsilon_{12}^2, \varepsilon_{23}^2 + \varepsilon_{31}^2,$ $(\varepsilon_{11} - \varepsilon_{22})\varepsilon_{23} + 2\varepsilon_{12}\varepsilon_{31}$
		hexagonal material	
e_0, e_1	$\varepsilon_{11} + \varepsilon_{22}, \varepsilon_{33}$	$e_2^2 + e_3^2, e_4^2 + e_5^2,$	$(\varepsilon_{11} - \varepsilon_{22})^2 + 4\varepsilon_{12}^2, \varepsilon_{23}^2 + \varepsilon_{31}^2$
		cubic material	
e_0	$\varepsilon_{11} + \varepsilon_{22} + \varepsilon_{33}$	$e_1^2 + e_2^2, e_3^2 + e_4^2 + e_5^2$	$(\varepsilon_{11} - \varepsilon_{22})^2 + (\varepsilon_{22} - \varepsilon_{33})^2+$ $+(\varepsilon_{33} - \varepsilon_{11})^2, \varepsilon_{12}^2 + \varepsilon_{23}^2 + \varepsilon_{31}^2$
		isotropic material	
e_0	$\varepsilon_{11} + \varepsilon_{22} + \varepsilon_{33}$	$e_1^2 + e_2^2 + e_3^2 + e_4^2 + e_5^2$	$(\varepsilon_{11} - \varepsilon_{22})^2 + (\varepsilon_{22} - \varepsilon_{33})^2+$ $+(\varepsilon_{33} - \varepsilon_{11})^2 + 6(\varepsilon_{12}^2 + \varepsilon_{23}^2 + \varepsilon_{31}^2)$

tensor. Therefore the first term in the formulae (4), (5) can be rewritten in the form

$$\sum_{i=1}^{i=p} \sigma_{(i)}(e_{(i)})\vec{\omega}_i = \sum_{\beta=0}^{\beta=p-1} \sigma_\beta(e_\beta)\vec{i}_\beta,$$

where $\sigma_\beta = \vec{\sigma} \cdot \vec{i}_\beta$, $e_\beta = \vec{e} \cdot \vec{i}_\beta$ are the projections of the stress and strain vectors on coordinate axes.

Let us represent this term in the form

$$\sum_{\beta=0}^{\beta=p-1} \sigma_\beta(e_\beta)\vec{i}_\beta = \sum_{\alpha=0}^{\alpha=p-1} \sum_{\beta=0}^{\beta=p-1} N_{\alpha\beta}(e_0, e_1, ..., e_{p-1})e_\alpha\vec{i}_\beta,$$

where $\mathbf{N} = N_{\alpha\beta}\vec{i}_\alpha\vec{i}_\beta$ is the symmetric second-order tensor in the subspace with the basis $\vec{i}_0$, $\vec{i}_1$,..., $\vec{i}_{p-1}$. Then the constitutive relations (5) take the form

$$\vec{\sigma} = \sum_{\alpha=0}^{\alpha=p-1} \sum_{\beta=0}^{\beta=p-1} N_{\alpha\beta}(e_0, e_1, ..., e_{p-1})e_\alpha\vec{i}_\beta + \sum_{\alpha=p+1}^{\alpha=q+p+1} \tilde{A}_{(\alpha)}(s_{(\alpha)})\vec{e}_{(\alpha)}. \tag{6}$$

If the functions $N_{\alpha\beta}(e_0, e_1, ..., e_{p-1})$ and $\tilde{A}_{(\alpha)}(s_{(\alpha)})$ are assumed to be constant the expression

(6) becomes an expression for the Hooke's law

$$\vec{\sigma} = \sum_{\beta=0}^{\beta=p-1} N_{\alpha\beta} e_\alpha \vec{i}_\beta + \sum_{\alpha=p}^{\alpha=q+p} G_\alpha \vec{e}_{(\alpha)},$$

where $N_{\alpha\beta}$, G_α are the components of the elasticity tensor of a material.

Nonlinear constitutive relations (6) satisfy the limit form of the particular postulate for anisotropic materials, since the functions $N_{\alpha\beta}(e_0, e_1, ..., e_{p-1})$ depend only on the linear invariants of the strain tensor, and the functions $\tilde{A}_{(\alpha)}(s_{(\alpha)})$ depend only on the quadratic invariant of the same name.

A variant of relations taking into account the mutual influence of processes in different proper subspaces and satisfying the generalization of the particular postulate on anisotropic materials can be written in the form

$$\vec{\sigma} = \sum_{i=1}^{i=p} \sigma_{(i)}(e_{(i)}, s_{(p+1)}, s_{(p+2)}, ..., s_{(q+p+1)})\vec{\omega}_i +$$

$$+ \sum_{\alpha=p+1}^{\alpha=q+p+1} \tilde{A}_{(\alpha)}(e_{(1)}, e_{(2)}, ..., e_{(p)}, s_{(\alpha)})\vec{e}_{(\alpha)}, \tag{7}$$

where the material functions $\sigma_{(i)}$, $\tilde{A}_{(\alpha)}$ depend on linear and quadratic invariants of the strain tensor simultaneously.

If one transforms the first term in the same manner as in the relations (5) a variant of relations will be obtained from (7) in the form

$$\vec{\sigma} = \sum_{\alpha=0}^{\alpha=p-1} \sum_{\beta=0}^{\beta=p-1} N_{\alpha\beta}(e_0, e_1, ..., e_{p-1}, s_{(p+1)}, s_{(p+2)}, ..., s_{(q+p+1)}) e_\alpha \vec{i}_\beta +$$

$$+ \sum_{\alpha=p+1}^{\alpha=q+p+1} \tilde{A}_{(\alpha)}(e_{(1)}, e_{(2)}, ..., _{(p)}, s_{(\alpha)})\vec{e}_{(\alpha)}. \tag{8}$$

In the relations (8) the influence of processes in different proper subspaces affects only on scalar material properties. If some process lies entirely in a multidimensional proper subspace i.e. $\vec{e} = \vec{e}_{(\alpha)}$ the corresponding stress vector is also located in this subspace i.e. $\vec{\sigma} = \vec{\sigma}_{(\alpha)}$.

Let us represent functions $\sigma_{(i)}$ in the relations (7) in the form

$$\sigma_{(i)} = K_i(e_{(i)})e_{(i)} + \sum_{\alpha=p+1}^{\alpha=q+p+1} G_{\alpha i}(s_{(\alpha)})s_{(\alpha)}^2,$$

then these relations will be written as

$$\vec{\sigma} = \sum_{i=1}^{i=p} \left(K_i(e_{(i)})e_{(i)} + \sum_{\alpha=p+1}^{\alpha=q+p+1} G_{\alpha i}(s_{(\alpha)})s_{(\alpha)}^2 \right) \vec{\omega}_i +$$

$$+ \sum_{\alpha=p+1}^{\alpha=q+p+1} \left(\tilde{A}_{(\alpha)}(e_{(1)}, e_{(2)}, ..., e_{(p)}, s_{(\alpha)}) + \sum_{i=1}^{i=p} 2G_{\alpha i}(s_{(\alpha)})s_{(\alpha)} \right) \vec{e}_{(\alpha)}. \tag{9}$$

The relations (9) also satisfy the generalization of the particular postulate on anisotropic materials but allow to describe a deviation of the stress vector from the strain eigen vector: when $\vec{e} = \vec{e}_{(\alpha)}$ the stress vector

$$\vec{\sigma} = \sum_{i=1}^{i=p} G_{\alpha i} s^2_{(\alpha)} \vec{\omega}_i + \left(\tilde{A}_{(\alpha)} + \sum_{i=1}^{i=p} 2G_{\alpha i} s_{(\alpha)} \right) \vec{e}_{(\alpha)}.$$

Besides, this deviation has the second order of infinitesimals. Therefore in the relations (9) the influence of processes in different proper subspaces affects both scalar and vector material properties.

4. Nonlinear relations for the hexagonal material

The hexagonal material has two one-dimensional proper subspaces and two two-dimensional subspaces with basis tensors given above. The linear and quadratic invariants of deformations for such material are brought in the table 1.

The spectral decomposition of the strain vector in this case has the form

$$\vec{e} = e_{(1)} \vec{\omega}_{(1)} + e_{(2)} \vec{\omega}_{(2)} + \vec{e}_{(3)} + \vec{e}_{(4)}, \tag{10}$$

where $e_{(1)} = \vec{e} \cdot \vec{\omega}_1$, $e_{(2)} = \vec{e} \cdot \vec{\omega}_2$ are the linear invariants of deformations; $\vec{e}_{(3)}$, $\vec{e}_{(4)}$ are the projections of the strain vector into multidimensional proper subspaces.

The vectors $\vec{\omega}_1$, $\vec{\omega}_2$ are connected with the basis vectors $\vec{i}_0$, $\vec{i}_1$ by a matrix of rotation on the angle which is independent of the strain process and is defined only by the initial elastic properties of the material. Therefore it is possible to choose $e_0 = \vec{e} \cdot \vec{i}_0$ and $e_1 = \vec{e} \cdot \vec{i}_1$ as the linear invariants of deformations defined by the expressions (1).

The quadratic invariants of deformations are defined by the relations

$$s_{(3)} = \sqrt{\vec{e}_{(3)} \cdot \vec{e}_{(3)}}, \quad s_{(4)} = \sqrt{\vec{e}_{(4)} \cdot \vec{e}_{(4)}}. \tag{11}$$

The spectral decomposition of the strain tensor ε

$$\varepsilon = \varepsilon_{(1)} + \varepsilon_{(2)} + \varepsilon_{(3)} + \varepsilon_{(4)}, \tag{12}$$

where $\varepsilon_{(3)} = \varepsilon_{11}(\vec{e}_1\vec{e}_1 - \vec{e}_2\vec{e}_2) + \varepsilon_{12}(\vec{e}_1\vec{e}_2 + \vec{e}_2\vec{e}_1)$, $\varepsilon_{(4)} = \varepsilon_{23}(\vec{e}_2\vec{e}_3 + \vec{e}_3\vec{e}_2) + +\varepsilon_{31}(\vec{e}_1\vec{e}_3 + \vec{e}_3\vec{e}_1)$ and expressions for the quadratic invariants in the form

$$s_{(3)} = \frac{1}{2}\sqrt{\varepsilon_{11}^2 + \varepsilon_{12}^2}, \quad s_{(4)} = \frac{1}{2}\sqrt{\varepsilon_{23}^2 + \varepsilon_{31}^2}$$

correspond to the relations (10) and (11).

Let us write the suggested variants of the constitutive relations for the hexagonal material. The relations (6) take the form

$$\vec{\sigma} = (N_{00}(e_0, e_1)e_0 + N_{01}(e_0, e_1)e_1)\,\vec{i}_0 + (N_{01}(e_0, e_1)e_0 + N_{11}(e_0, e_1)e_1)\,\vec{i}_1 +$$
$$+\tilde{A}_{(3)}(s_{(3)})\vec{e}_{(3)} + \tilde{A}_{(4)}(s_{(4)})\vec{e}_{(4)}. \tag{13}$$

If the material functions in the relations (13) are assumed to be constants: $N_{00}(e_0, e_1) = n_{00}$, $N_{01}(e_0, e_1) = n_{01}$, $N_{11}(e_0, e_1) = n_{11}$, $\tilde{A}_{(3)}(s_{(3)}) = n_{22}$, $\tilde{A}_{(4)}(s_{(4)}) = n_{44}$ the expression of the Hooke's law for the hexagonal material is obtained

$$\vec{\sigma} = (n_{00}e_0 + n_{01}e_1)\,\vec{i}_0 + (n_{01}e_0 + n_{11}e_1)\,\vec{i}_1 + n_{22}\left(e_2\vec{i}_2 + e_3\vec{i}_3\right) + n_{44}\left(e_4\vec{i}_4 + e_5\vec{i}_5\right),$$

IOP Conf. Series: Journal of Physics: Conf. Series **973** (2018) 012018 doi:10.1088/1742-6596/973/1/012018

where $\mathbf{n} = n_{\alpha\beta}\vec{i}_\alpha\vec{i}_\beta$ is the elasticity tensor.

The relation (8) for the hexagonal material has the form

$$\begin{aligned}\vec{\sigma} = &\left(N_{00}(e_0, e_1, s_{(3)}, s_{(4)})e_0 + N_{01}(e_0, e_1, s_{(3)}, s_{(4)})e_1\right)\vec{i}_0 + \\ &+\left(N_{01}(e_0, e_1, s_{(3)}, s_{(4)})e_0 + N_{11}(e_0, e_1, s_{(3)}, s_{(4)})e_1\right)\vec{i}_1 + \\ &+\tilde{A}_{(3)}\left(e_0, e_1, s_{(3)}\right)\vec{e}_{(3)} + \tilde{A}_{(4)}\left(e_0, e_1, s_{(4)}\right)\vec{e}_{(4)}\end{aligned} \tag{14}$$

and requires to define five material functions for its concretization.

5. Nonlinear relations for the cubic and isotropic materials

The cubic material has one one-dimensional proper subspace, one two-dimensional subspace and one three-dimensional one. The basis tensors for each of three subspaces are mentioned above in the Section 2. The linear and quadratic invariants of deformations for such material are given in the table 1.

The spectral decomposition of the strain vector in the case of the cubic material has the following form:

$$\vec{e} = e_{(1)}\vec{\omega}_{(1)} + \vec{e}_{(2)} + \vec{e}_{(3)}, \tag{15}$$

where $e_{(1)} = \vec{e}\cdot\vec{\omega}_1$ is the linear invariant of deformations; $\vec{e}_{(2)}$, $\vec{e}_{(3)}$ are the projections of the strain vector into multidimensional proper subspaces. Since $\vec{\omega}_1 = \vec{i}_0$ for the cubic material its linear invariant $e_{(1)}$ takes the form $e_{(1)} = \vec{e}\cdot\vec{i}_0 = e_0$.

For the cubic material the eigenvectors $\vec{\omega}_\beta$ coincide with the basis vectors $\vec{i}_\alpha$ in the multidimensional proper subspaces, namely: $\vec{\omega}_2 = \vec{i}_1$, $\vec{\omega}_3 = \vec{i}_2$ in the two-dimensional subspace, $\vec{\omega}_4 = \vec{i}_3$, $\vec{\omega}_5 = \vec{i}_4$, $\vec{\omega}_6 = \vec{i}_5$ in the three-dimensional subspace.

Therefore the basis tensors in the multidimensional proper subspaces of the cubic material can be determined as $\mathbf{\Omega}_2 = \vec{\omega}_2\vec{\omega}_2 + \vec{\omega}_3\vec{\omega}_3 = \vec{i}_1\vec{i}_1 + \vec{i}_2\vec{i}_2$ and $\mathbf{\Omega}_3 = \vec{\omega}_4\vec{\omega}_4 + \vec{\omega}_5\vec{\omega}_5 + \vec{\omega}_6\vec{\omega}_6 = \vec{i}_3\vec{i}_3 + \vec{i}_4\vec{i}_4 + \vec{i}_5\vec{i}_5$.

Let us write the suggested variants of the constitutive relations for the cubic material. The relations (6) take the following form:

$$\vec{\sigma} = N_{00}(e_0)e_0\vec{i}_0 + \tilde{A}_{(2)}(s_{(2)})\vec{e}_{(2)} + \tilde{A}_{(3)}(s_{(3)})\vec{e}_{(3)}. \tag{16}$$

If the material functions in the relations (16) are assumed to be constants: $N_{00}(e_0) = n_{00}$, $\tilde{A}_{(2)}(s_{(2)}) = n_{11}$, $\tilde{A}_{(3)}(s_{(3)}) = n_{33}$ the expression of the Hooke's law for the cubic material is obtained

$$\vec{\sigma} = n_{00}e_0\vec{i}_0 + n_{11}\left(e_1\vec{i}_1 + e_2\vec{i}_2\right) + n_{33}\left(e_3\vec{i}_3 + e_4\vec{i}_4 + e_5\vec{i}_5\right),$$

where $\mathbf{n} = n_{\alpha\beta}\vec{i}_\alpha\vec{i}_\beta$ is the elasticity tensor.

The relation (8) for the cubic material has the form

$$\vec{\sigma} = N_{00}(e_0, s_{(2)}, s_{(3)})e_0\vec{i}_0 + \tilde{A}_{(2)}\left(e_0, s_{(2)}\right)\vec{e}_{(2)} + \tilde{A}_{(3)}\left(e_0, s_{(3)}\right)\vec{e}_{(3)}. \tag{17}$$

It is necessary to define three material functions for the concretization of the last relation.

On the basis of results given in the sections 2 and 3 the relations (6) for the isotropic material take the form

$$\vec{\sigma} = N_{00}(e_0)e_0\vec{i}_0 + \tilde{A}_{(1)}\left(s_{(1)}\right)\vec{e}_{(1)}, \tag{18}$$

where $e_0 = \vec{e}\cdot\vec{i}_0$, $s_{(1)} = \sqrt{e_1^2 + e_2^2 + e_3^2 + e_4^2 + e_5^2}$ are the invariants of deformations for the isotropic material which characterize the change of material volume and the intensity of forming respectively in the case of small deformations.

A more general variant of relations follows from (8):

$$\vec{\sigma} = N_{00}\left(e_0, s_{(1)}\right)e_0\vec{i}_0 + \tilde{A}_{(1)}\left(e_0, s_{(1)}\right)\vec{e}_{(1)}. \tag{19}$$

The mutual influence of processes in different proper subspaces of the isotropic material is taken into account in this relation.

6. Results and discussion

The obtained constitutive relations (6) are concretized for materials with different types of symmetry of properties (13), (16) and (18). All these relations satisfy the limit form of the generalization of A. A. Il'yushin's postulate on anisotropic materials. In the particular case of the isotropic material (18) this hypothesis means that the volume change e_0 does not depend on the forming $s_{(1)}$ and the process of forming defined by the vector $\vec{e}_{(1)}$ is independent of the volume change.

The relations (8) and their particular variants for the hexagonal (14), cubic (17) and isotropic (19) materials are constructed on the basis of the general form of generalization of the particular postulate on anisotropic materials. In the case of the isotropic material it means that the material functions $N_{00}\left(e_0, s_{(1)}\right)$ and $\tilde{A}_{(1)}\left(e_0, s_{(1)}\right)$ depend on both the material volume change and forming. This fact significantly complicates the process of experimental concretization of the material functions.

7. Conclusion

Systems of invariants of deformations are defined as lengths of projections of the strain vector into elastic proper subspaces. The systems are written for anisotropic materials of different types. In the article several variants of nonlinear constitutive relations for anisotropic materials are offered. For the hexagonal, cubic and isotropic materials the analysis of satisfaction of the constitutive relations to the general and limit forms of the generalization of A. A. Il'yushin's particular postulate is fulfilled.

Acknowledgments

The results of the research project are published with the financial support of Tula State University within the framework of the scientific project No 2017-40.

References

[1] Cooper R C, Lee C, Marianetti C A, Wei X, Hone J and Kysar J W 2013 *Phys. Rev.* B **87** 035423
[2] Holzapfel G A and Ogden R W *Math. Mech. Solids* **14** 474–489
[3] Vergori L, Destrade M, McGarry P and Ogden R W 2013 *Comput. Mech.* **52(5)** 1185–97
[4] Wagner D R and Lotz J C 2004 *J. Orthop. Res.* **22** 901–909
[5] Lomakin E V and Fedulov B N 2015 *Meccanica* **50** 1527–35
[6] Bogetti T A, Hoppel C P R, Harik V M, Newill J F and Burns B P 2004 *Compos. Sci. Technol.* **64(3-4)** 329–342
[7] Markin A A and Sokolova M Yu 2003 *J. Appl. Mech. Techn. Phys.* **44(1)** 141–145
[8] Markin A A and Sokolova M Yu 2007 *J. Appl. Math. Mech.* **71** 536–542
[9] ten Thije R H W, Akkerman R and Huétink J 2007 *Comput. Methods Appl. Mech. Engrg.* **196** 3141–50
[10] Swendsen B 2001 *Int. J. Solids Struct.* **38** 9579–99
[11] Olive M, Kolev B and Auffray N 2017 *Arch. Rational Mech. Anal.* **226(1)** 1–31
[12] Adkins A E and Green J E 1960 *Large Elastic Deformations, and Non-Linear Continuum Mechanics* (Oxford: Oxford Clarenden Press)
[13] Truesdell C 1977 *A First Course in Rational Continuum Mechanics* (San Diego, CA: Academic Press)
[14] Spencer A J M 1971 *Theory of invariants* (Continuum Physics vol 1) ed A Eringen (New York, NY: Academic Press) pp 239–353
[15] Novozhilov V V 1961 *Theory of Elasticity* (New York, NY: Pergamon Press)
[16] Chernykh K F 2004 *Mech. Solids* **39(1)** 1–13
[17] Rychlewski J 1984 *J. Appl. Math. Mech.* **48** 303–314
[18] Il'yushin A A 1963 *Plasticity. Basics of General Mathematical Theory* (Moscow: Academy of Sciences of the USSR Publishing)
[19] Markin A A, Sokolova M Yu and Khristich D V 2011 *Mechanics of Solids* **46(1)** 30–35
[20] Markin A A and Sokolova M Yu 2015 *Thermomechanics of Elastoplastic Deformation* (Cambridge: Cambridge International Science Publishing)
[21] Ostrosablin N I 1986 *Dynamics of elasto-plastic systems* **75** 113–125

IOP Conf. Series: Journal of Physics: Conf. Series **973** (2018) 012019 doi:10.1088/1742-6596/973/1/012019

Nonlinear dynamics in problems of stability of complex media

A I Sumin, A A Boger and V A Sumin

Military Educational Scientific Center of the Military – Air Forces "Military Air Academy named after Professor N E Zhukovsky and Yu A Gagarin", Staryih Bolshevikov 54 «A», Voronezh 394064, Russia

E-mail: sumin_ai@mail.ru

Abstract. The problem of stability of nonlinear viscoelastic bodies with respect to finite perturbations is considered in this article. The analysis of the basic process of deformation of a viscoelastic medium is reduced to solution of a nonlinear boundary value problem with variable coefficients. Solutions for perturbations of displacements are in the form of series in eigenfunctions. Using the principle of possible displacements the question on stability of the ground state of a variational nonlinear problem is reduced to investigation of stability of the zero solution of an infinite system of ordinary differential equations with constant coefficients. For the resulting system of equations we construct a function, which under certain restrictions on the initial perturbations is a Lyapunov function. The dimension of the strange attractor of the dynamical system which allows to limit the number of terms in a Bubnov-Galerkin set is found.

1. Introduction

The theory of the stability of deformable systems, the beginning of which was laid by Euler's work, has now turned into a highly ramified branch of mechanics that has numerous applications that have created their methods and approaches. In particular, there is no branch of industry and construction where the methods and results of the theory of stability of deformable media are not used. There are a lot of monographs and articles devoted to the theory of stability of elastic and inelastic systems [1], [2]. From the analysis of these works it follows that almost all the problems of stability of elastic and inelastic systems are fulfilled with two basic approaches: static and dynamic with the appropriate stability criteria and assumptions depending on the body model and, as a rule, reduce to the determination of a single critical parameter, in which together with the initial state, there can be a contiguous state of the system. As a rule, according to the statement of the problem, perturbations were taken as arbitrarily small, which made it possible to linearize the initial equations for the perturbations.

Further development of engineering activities related to the use and creation of new materials, the problems of precision in mechanical engineering, the problems of geomechanics, especially in the areas with increased seismicity, and the internal needs of the mechanics of deformable bodies have necessitated the development of a theory of stability under finite perturbations.

Solving the problem of stability of elements and objects with respect to finite disturbances is important not only for the purpose of reducing the weight of structures and objects, but also for ensuring the reliability by estimating the permissible boundary of the region with respect to finite perturbations for specified loading parameters and constructions. Thus, the development of the theory of elastic stability at finite perturbations makes it possible to trace the connection between this theory and the general theory of stability of motion on the one hand, as well as with the theory of bifurcations on the other hand, and evaluate the accuracy of the three-dimensional linearized stability theory for small and finite deformations simultaneously. Moreover, the constructed finite sequences of bifurcation points show that, in contrast to the classical stability theories, which determine a unique bifurcation point, there always exists a hierarchy of stable equilibrium states. These sequences make it possible to find the dimensions of a strange attractor for nonlinear systems, and, consequently, the number of terms in the Bubnov-Galerkin set, which allows to present the solutions to the equations in perturbations.

2. Materials and methods

We consider three states of the body [1]. The first state is a natural state when the stresses and deformations are absent in a continuous medium and the temperature θ_0 is independent on the time and coordinates. The second state is obtained from the first state when the mechanical and thermal loads are constant over time after a time interval sufficient for the creep and relaxation processes to completely end in the body. This state corresponds to a displacement vector $\overset{0}{u}$, a strain tensor $\overset{0}{E}$, a temperature $\overset{0}{\theta}$, and these characteristics do not depend on the time *t*, i.e., they correspond to a certain equilibrium state. The third state of motion is achieved by thermomechanical perturbation of the second state. All the values of this state will be marked with a stroke and represented as a sum of quantities related to the second state and perturbations of the corresponding quantities. The perturbations will be considered as a finite and we will not mark them with any index [2], [3].

The third state is characterized by a displacement vector, a stress tensor, a temperature, etc., i.e.

$$u' = \overset{0}{u} + u, \quad E' = \overset{0}{E} + E, \quad S' = \overset{0}{S} + S, \quad \theta' = \overset{0}{\theta} + \theta, \quad \ldots .$$

We represent the relations of thermomechanics in the Cartesian coordinates of the natural state $\xi = \{\xi_1, \xi_2, \xi_3\}$, which coincide with the material coordinates at the initial instant. The final deformation corresponding to the displacement $\overset{0}{u} = \left\{\overset{0}{u}_1, \overset{0}{u}_2, \overset{0}{u}_3\right\}$ is determined by the equations in [1], in which it is necessary to put the zero-marks over all quantities. The same values are satisfied by the quantities with primes. Taking this into account it is possible to obtain the relations for the perturbations assuming that the quantities with zero-marks are known [4].

We consider a nonlinearly viscoelastic medium described in monographs [1], [5]. The problem of solution of such a system is very difficult, in this way some simplifications are necessary. The influence of connectivity of mechanical and thermal fields in the propagation of transient disturbances is important at relatively high rates of temperature increasing. Therefore, in order to simplify the problem we consider the dynamical unrelated problems of thermomechanics that are fall in to two distinct problems:

- the problem of thermal conductivity;
- the problem of thermoviscoelastics, in which the mechanical properties of the material are considered as dependent on the temperature and the initial deformation. Within the framework of these ideas the energy equation can be obtained from the disturbances in [4], [6] by neglecting the dissipation function and can be written as:

$$\rho_R \, \dot{\theta} \dot{\eta} + h_{Ri,i} = 0. \tag{1}$$

Here dots determine the time derivative of the corresponding quantity.

The equations of motion and boundary conditions have the same form. The corresponding relations [1], [5] can be written as:

$$S=S^{\infty}+\int_{-\infty}^{t}\left[G^{(1)}(0,\xi-\xi';\overset{0}{E},\overset{0}{\theta})\frac{\partial E(\tau')}{\partial\tau'}+G^{(2)}(0,\xi-\xi';\overset{0}{E},\overset{0}{\theta})\frac{\partial\theta(\tau')}{\partial\tau'}\right]\mathrm{d}\tau'; \tag{2}$$

$$\eta=\eta^{\infty}+\int_{-\infty}^{t}G^{(2)}(0,\xi-\xi';\overset{0}{E},\overset{0}{\theta})\frac{\partial\theta(\tau')}{\partial\tau'}\mathrm{d}\tau'. \tag{3}$$

For the heat flow we have:

$$-\overset{0}{h}_{\mathrm{R}}=M(\overset{0}{E},\overset{0}{\theta})g_{\mathrm{R}}. \tag{4}$$

Relations for the perturbation of stress tensor can be rewritten as:

$$S^{\infty}=S^{\infty}(1)+S^{\infty}(2)+\cdots+S_{\mathrm{H}}. \tag{5}$$

The number of terms in (5) is specified by the form of elastic potential, and the ratio S_{H} takes the form:

$$S_{\mathrm{H}}=S^{\infty}+\int_{-\infty}^{t}\left[G^{(1)}(0,q-q';\overset{0}{E},\overset{0}{\theta})\frac{\partial E(\tau')}{\partial\tau'}+G^{(2)}(0,q-q';\overset{0}{E},\overset{0}{\theta})\frac{\partial\theta(\tau')}{\partial\tau'}\right]\mathrm{d}\tau'. \tag{6}$$

Let us suppose that the temperature distribution has the form:

$$\theta=\theta(t,\overset{0}{E},\overset{0}{\theta},\xi_k). \tag{7}$$

We construct the variational equations in the frame of the Bubnov-Galerkin method that corresponds to nonlinear boundary value problem. Assuming that the left sides of the equations of motion are taken with the opposite sign of some components of the volume forces, and the left sides of the equations of equilibrium contain some components of the surface forces we can construct the conditions that take into account the fact that the work of these forces on the possible displacements δu_m is zero.

$$\int_V\left\{\frac{\partial}{\partial\xi_j}\left[\left(\delta_{ik}+\frac{\partial\overset{0}{u}_i}{\partial\xi_k}\right)T_{*kj}+\overset{0}{T}_{*kj}\frac{\partial u_i}{\partial\xi_k}+T_{*kj}\frac{\partial u_i}{\partial\xi_k}\right]+\rho_{\mathrm{R}}\mathrm{X}_i-\rho_{\mathrm{R}}\ddot{u}_i\right\}\delta u_i\mathrm{d}V$$

$$+\int_S\left\{\left[\left(\delta_{ik}+\frac{\partial\overset{0}{u}_i}{\partial\xi_k}\right)T_{*kj}+\overset{0}{T}_{*kj}\frac{\partial u_i}{\partial\xi_k}+T_{*kj}\frac{\partial u_i}{\partial\xi_k}\right]n_{\mathrm{R}j}-P_i\right\}\delta u_i\mathrm{d}S=0. \tag{8}$$

Without loss of generality, let us assume that there are no mass and surface forces in the body. Then, the variational equation can be written as:

$$\int_V\left\{\frac{\partial}{\partial\xi_j}\left[\left(\delta_{ik}+\frac{\partial\overset{0}{u}_i}{\partial\xi_k}\right)T_{*kj}+\overset{0}{T}_{*kj}\frac{\partial u_i}{\partial\xi_k}+T_{*kj}\frac{\partial u_i}{\partial\xi_k}\right]-\rho_{\mathrm{R}}\ddot{u}_i\right\}\delta u_i\mathrm{d}V$$

$$+\int_S\left\{\left[\left(\delta_{ik}+\frac{\partial\overset{0}{u}_i}{\partial\xi_k}\right)T_{*kj}+\overset{0}{T}_{*kj}\frac{\partial u_i}{\partial\xi_k}+T_{*kj}\frac{\partial u_i}{\partial\xi_k}\right]n_{\mathrm{R}j}\right\}\delta u_i\mathrm{d}S=0. \tag{9}$$

We transform the first integral in (9) by the Gauss-Ostrogradsky theorem

$$-\int_V \left\{ \frac{\partial}{\partial \xi_j} \left[\left(\delta_{ik} + \frac{\partial \overset{0}{u}_i}{\partial \xi_k} \right) T_{*kj} + \overset{0}{T}_{*kj} \frac{\partial u_i}{\partial \xi_k} + T_{*kj} \frac{\partial u_i}{\partial \xi_k} \right] \right\} \delta u_i \mathrm{d}V$$

$$= -\int_S \left\{ \left[\left(\delta_{ik} + \frac{\partial \overset{0}{u}_i}{\partial \xi_k} \right) T_{*kj} + \overset{0}{T}_{*kj} \frac{\partial u_i}{\partial \xi_k} + T_{*kj} \frac{\partial u_i}{\partial \xi_k} \right] n_{\mathrm{R}j} \right\} \delta u_i \mathrm{d}S$$

$$+\int_V \left\{ \frac{\partial}{\partial \xi_j} \left[\left(\delta_{ik} + \frac{\partial \overset{0}{u}_i}{\partial \xi_k} \right) T_{*kj} + \overset{0}{T}_{*kj} \frac{\partial u_i}{\partial \xi_k} + T_{*kj} \frac{\partial u_i}{\partial \xi_k} \right] \right\} \delta u_{i,j} \mathrm{d}V. \tag{10}$$

Substituting (10) in to (9) we obtain:

$$\int_V \left[\left(\delta_{ik} + \overset{0}{u}_{i,k} \right) T_{*kj} + \overset{0}{T}_{*kj}\, u_{i,k} + T_{*kj} u_{i,k} \right] \delta u_{i,j} \mathrm{d}V + \rho_{\mathrm{R}} \int_V \ddot{u}_i \delta u_i \mathrm{d}V = 0. \tag{11}$$

Let us search the solution of nonlinear boundary value problem (11) in the form of a series

$$u_i(\xi_k, t) = \sum_{n,m} f_{nm}(t) \cdot \varphi_{inm}(\xi_k), \tag{12}$$

$$i = 1, 2, 3;\ n, m = 1, 2, \dots \infty.$$

Here $f_{nm}(t)$ are the undetermined coefficients depending on the time, $\varphi_{inm}(\xi_k)$ are the functions satisfying the geometric boundary conditions for u_i. As a variation δu_i we consider the expressions

$$\delta u_i = \sum_{n,m} \varphi_{inm}(\xi_k) \cdot \delta f_{nm}(t). \tag{13}$$

As a function φ_{inm} we take the known forms of troughs that correspond to linearized problem.

Substituting (12) and (13) in to (11) we obtain

$$\int_V \left[\left(\delta_{ik} + \overset{0}{u}_{i,k} \right) T_{*kj} + \overset{0}{T}_{*kj}\, u_{i,k} + T_{*kj} u_{i,k} \right] \varphi_{inm,j} \delta f_{nm} \mathrm{d}V + \rho_{\mathrm{R}} \int_V \varphi_{inm} \delta f_{nm} \mathrm{d}V = 0. \tag{14}$$

From (14) we obtain the equation (for each time moment, taking into account the arbitrariness of the variations $\delta f_{nm}(t)$):

$$\int_V \left[\left(\delta_{ik} + \overset{0}{u}_{i,k} \right) T_{*kj} + \overset{0}{T}_{*kj}\, u_{i,k} + T_{*kj} u_{i,k} \right] \varphi_{inm,j} \mathrm{d}V + \rho_{\mathrm{R}} \int_V \varphi_{inm} \mathrm{d}V = 0. \tag{15}$$

It is obvious, that when specifying the perturbation potential the components of stress tensor are expressed in terms of the perturbation of the displacement, and the relation (5) can be represented in the form:

$$S_{ij}^{\infty}(1) = f_{nm} S_{ijnm}; \quad S_{ij}^{\infty}(2) = f_{nm} f_{kl} S_{ijnmkl}; \quad \cdots. \tag{16}$$

Here we assume that

$$T_* = \rho_{\mathrm{R}} S. \tag{17}$$

We introduce the notation:

$$f_{nm} = f^{(1)}; \quad f_{kl} = f^{(2)}; \quad \cdots \quad S_{ijnm} = S^{(1)}; \qquad S_{ijnmkl} = S^{(2)}; \quad \cdots. \tag{18}$$

Then, (16) can be rewritten as:

$$S^{\infty}(1) = f^{(1)}S^{(1)}; \quad S^{\infty}(2) = f^{(1)}f^{(2)}S^{(2)}; \quad S^{\infty}(3) = f^{(1)}f^{(2)}f^{(3)}S^{(3)}. \tag{19}$$

In these notations the relation (12) can be written as:

$$u_i(X_k, t) = f^{(1)}(t)\varphi^{(1)}(\xi_k). \tag{20}$$

The components of the strain tensor perturbations in (8) with (12) and (20) can be written as:

$$E_{mn}(1) = \frac{1}{2} f_{ij}\left(\varphi_{mij,n} + \varphi_{nij,m} + \overset{0}{u}_{k,m}\varphi_{kij,n} + \varphi_{kij,m}\overset{0}{u}_{k,m}\right); \quad E_{mn}(2) = \frac{1}{2} f_{ij} f_{kl}\varphi_{rij,m}\varphi_{rij,n}. \tag{21}$$

The time derivative of the strain tensor components will have the form:

$$\dot{E}_{mn}(1) = \frac{1}{2}\dot{f}_{ij}\left(\varphi_{mij,n} + \varphi_{nij,m}\right). \tag{22}$$

The expression for the temperature perturbation (7) can also be presented in the form similar to (20):

$$\theta\left(t, \overset{0}{E}, \overset{0}{\theta}, \xi_k\right) = \theta^{(1)}(t)\psi^{(1)}\left(\overset{0}{E}, \overset{0}{\theta}, \xi_k\right). \tag{23}$$

We have:

$$\dot{\theta}\left(t, \overset{0}{E}, \overset{0}{\theta}, \xi_k\right) = \dot{\theta}^{(1)}(t)\psi^{(1)}\left(\overset{0}{E}, \overset{0}{\theta}, \xi_k\right). \tag{24}$$

Substituting (22) and (24) in to (6) we obtain

$$S_H = \int_{-\infty}^{t}\left[G^{(1)}\dot{f}^{(1)}E^{(0)} + G^{(2)}\dot{\theta}^{(1)}\psi^{(1)}\right]\mathrm{d}\tau', \tag{25}$$

where we use the following notations:

$$E^{(0)}_{ijmn} = \frac{1}{2}\left(\varphi_{mij,n} + \varphi_{nij,m}\right);$$

$$E^{(1)}_{ijmn} = \frac{1}{2}\left(\varphi_{mij,n} + \varphi_{nij,m} + \overset{0}{u}_{k,m}\varphi_{kij,n} + \varphi_{kij,m}\overset{0}{u}_{k,n}\right);$$

$$E^{(2)}_{ijmnkl} = \frac{1}{2}\varphi_{rij,m}\varphi_{nkl,r}. \tag{26}$$

Substituting (19), (25) and (16) in to (15) we obtain

$$\rho_k\int_V\left[\left(I + \overset{0}{H}\right)\left(f^{(1)}S^{(1)} + f^{(1)}f^{(2)}S^{(12)} + \cdots + S_{\mathrm{H}}\right) + \left(f^{(1)}S^{(1)} + f^{(1)}f^{(2)}S^{(12)} + \cdots + S_{\mathrm{H}}\right)f^{(1)}H^{(1)}\right]H^{(1)}\mathrm{d}V$$

$$+\rho_k\int_V\dot{f}^{(1)}\varphi^{(1)}\varphi^{(1)\mathrm{T}}\mathrm{d}V = 0. \tag{27}$$

Finally, substituting the expression (25) in (27), we obtain the system of equations

$$A\dot{f}^{(1)}+B\left(\dot{f}^{(1)}\right)+C\left(\dot{f}^{(1)}\right)f^{(1)}+D\left(\dot{f}^{(1)}\right)+L\left(\dot{\theta}^{(1)}\right)f^{(1)}$$

$$+K_1 f^{(1)}+K_2 f^{(1)}f^{(2)}+K_3 f^{(1)}f^{(2)}f^{(3)}+\cdots=0, \tag{28}$$

where the coefficients are as follows:

$$A=\rho_k\int_V \dot{f}^{(1)}\varphi^{(1)}\varphi^{(1)\mathrm{T}}\mathrm{d}V;$$

$$B\left(\dot{f}^{(1)}\right)=\int_{-\infty}^{t}\dot{f}^{(1)}(\tau)\int_V\left[\left(I+\overset{0}{H}\right)E^{(0)}H^{(1)}\right]\mathrm{d}VG^{(1)}\mathrm{d}\tau;$$

$$D\left(\dot{\theta}^{(1)}\right)=\int_{-\infty}^{t}\dot{\theta}^{(1)}(\tau)\int_V\left[\left(I+\overset{0}{H}\right)\psi^{(1)}H^{(1)}\right]\mathrm{d}VG^{(2)}\mathrm{d}\tau;$$

$$C\left(\dot{f}^{(1)}\right)=\int_{-\infty}^{t}\dot{f}^{(1)}(\tau)\int_V\left[E^{(0)}H^{(1)}H^{(1)}\right]\mathrm{d}VG^{(2)}\mathrm{d}\tau; \tag{29}$$

$$L\left(\dot{\theta}^{(1)}\right)=\int_{-\infty}^{t}\dot{\theta}^{(1)}(\tau)\int_V\left[\psi^{(1)}H^{(1)}\right]\mathrm{d}VG^{(2)}\mathrm{d}\tau;$$

$$K_1=\int_V\left[\left(I+\overset{0}{H}\right)S^{(1)}+\overset{0}{S}H^{(1)}\right]H^{(1)}\mathrm{d}V;$$

$$K_2=\int_V\left[\left(I+\overset{0}{H}\right)S^{(2)}+S^{(1)}H^{(1)}\right]H^{(1)}\mathrm{d}V;$$

$$K_3=\int_V\left[\left(I+\overset{0}{H}\right)S^{(3)}+S^{(2)}H^{(1)}\right]H^{(1)}\mathrm{d}V;$$

etc.

The number of terms in the system of equations (28) and the kind of coefficients are specified by the form of elastic potential.

Multiplying the system of equations (28) to the $\dot{f}$ (where $f=\{f^{(1)},f^{(2)},\cdots,f^{(N)}\}$) we obtain:

$$A\ddot{f}\dot{f}+B\left(\dot{f}\right)\dot{f}+C\left(\dot{f}\right)f\dot{f}+D\left(\dot{f}\right)\dot{f}+L\left(\dot{\theta}\right)f\dot{f}+K_1 f\dot{f}+K_2 ff\dot{f}+K_3 fff\dot{f}+\cdots=0. \tag{30}$$

The obtained relation can be rewritten in the form:

$$\frac{\mathrm{d}}{\mathrm{d}t}\left[\frac{1}{2}A\dot{f}\dot{f}+\frac{1}{2}K_1 ff+\frac{1}{3}K_2 fff+\frac{1}{4}K_3 ffff+\cdots+\int_{-\infty}^{t}B\left(\dot{f}\right)\dot{f}\,\mathrm{d}\tau\right.$$

$$\left.+\int_{-\infty}^{t}C\left(\dot{f}\right)\dot{f}\,\mathrm{d}\tau+\int_{-\infty}^{t}D\left(\dot{\theta}\right)\dot{f}\,\mathrm{d}\tau+\int_{-\infty}^{t}L\left(\dot{\theta}\right)f\dot{f}\,\mathrm{d}\tau\right]=0. \tag{31}$$

We introduce the function

$$P=\frac{1}{2}A\dot{f}\dot{f}+\frac{1}{2}K_1ff+\frac{1}{3}K_2fff+\frac{1}{4}K_3ffff+\cdots+\int_{-\infty}^{t}B\left(\dot{f}\right)\dot{f}\,\mathrm{d}\tau$$

$$+\int_{-\infty}^{t}C\left(\dot{f}\right)\dot{f}\,\mathrm{d}\tau+\int_{-\infty}^{t}D\left(\dot{\theta}\right)\dot{f}\,\mathrm{d}\tau+\int_{-\infty}^{t}L\left(\dot{\theta}\right)f\dot{f}\,\mathrm{d}\tau. \quad (32)$$

If the function (32) is positive defined for the values of the initial coordinates and velocities that do not exceed the values found from the relations

$$\left(\frac{\partial P}{\partial \dot{f}}\right)_0=0;\quad \left(\frac{\partial P}{\partial f}\right)_0=0;\quad \left(\frac{\partial P}{\partial \dot{\theta}}\right)_0=0, \quad (33)$$

then, according to the first Lyapunov stability theorem the homogeneous solution of the system (28) will be stable in this area, because the derivative of the function P by the time is non-positive.

Let us consider the structure of the function (32). It can be divided into several terms

$$P=P_1+P_2+P_3+P_4+P_5, \quad (34)$$

where

$$P_1=\frac{1}{2}A\dot{f}\dot{f}+\frac{1}{2}K_1ff+\frac{1}{3}K_2fff+\frac{1}{4}K_3ffff+\cdots; \quad (35)$$

$$P_2=\int_{-\infty}^{t}B\left(\dot{f}\right)\dot{f}\,\mathrm{d}\tau=\int_{-\infty}^{t}\int_{-\infty}^{t}\int_{V}\left[\left(I+\overset{0}{H}\right)E^{(0)}H^{(1)}\right]\mathrm{d}VG^{(1)}\dot{f}(\tau_1)\dot{f}(\tau_2)\,\mathrm{d}\tau_1\mathrm{d}\tau_2; \quad (36)$$

$$P_3=\int_{-\infty}^{t}C\left(\dot{f}\right)\dot{f}\,\mathrm{d}\tau=\int_{-\infty}^{t}\int_{-\infty}^{t}\int_{V}\left[\left(I+\overset{0}{H}\right)E^{(1)}H^{(1)}\right]\mathrm{d}VG^{(1)}\dot{f}(\tau_1)\dot{f}(\tau_2)\,\mathrm{d}\tau_1\mathrm{d}\tau_2; \quad (37)$$

$$P_4=\int_{-\infty}^{t}D\left(\dot{\theta}\right)\dot{f}\,\mathrm{d}\tau=\int_{-\infty}^{t}\int_{-\infty}^{t}\int_{V}\left[\left(I+\overset{0}{H}\right)\psi^{(1)}H^{(1)}\right]\mathrm{d}VG^{(2)}\dot{\theta}(\tau_1)\dot{f}(\tau_2)\,\mathrm{d}\tau_1\mathrm{d}\tau_2; \quad (38)$$

$$P_5=\int_{-\infty}^{t}L\left(\dot{\theta}\right)\dot{f}\,\mathrm{d}\tau=\int_{-\infty}^{t}\int_{-\infty}^{t}\int_{V}\psi^{(1)}H^{(1)}H^{(1)}\mathrm{d}V\,\dot{\theta}(\tau_1)f(\tau_1)\dot{f}(\tau_2)\,\mathrm{d}\tau_1\mathrm{d}\tau_2. \quad (39)$$

The function P_1 is similar to the corresponding function obtained in [3], [4]. The functions $P_i\,(i=2,3,4,5)$ are the kernel multilinear forms, and the question of their positive definiteness is solved within the positive definiteness of the kernels of these forms. The specific form of the kernels in the (36)–(39) is determined from the experiments. The expressions (36) and (38) are the kernel bilinear forms. If we take the linear dependence relative to perturbations, the relations P_3 and P_5 in (37) and (39) can be considered equal to zero.

The equations (33) with respect to the initial perturbation amplitudes are the following:

$$\left(\frac{\partial P}{\partial \dot{f}}\right)_0=A\dot{f}(0)+B\left(\dot{f}(0)\right)\dot{f}(0)+C\left(\dot{f}(0)\right)f(0)\dot{f}(0)$$

$$+D\left(\dot{\theta}(0)\right)f(0)\dot{f}(0)+L\left(\dot{\theta}(0)\right)f(0)\dot{f}(0)+\cdots=0; \quad (40)$$

$$\left(\frac{\partial P}{\partial f}\right)_0 = K_1 f(0) + K_2 f(0) f(0) + K_3 f(0) f(0) f(0) + K_4 f(0) f(0) f(0) f(0) + \cdots = 0. \quad (41)$$

These relations represent a system of algebraic equations for the values of the initial perturbations and their velocities. The solution of (40), (41) gives the area of initial perturbations and velocities, wherein the homogeneous solution of the system (28) is stable.

3. Results and discussion

Thus, we see that the initial perturbation region, for which the zero solution of the system (28) is stable, consistent with the same output for a nonlinear elastic body [3]. For nonlinear elastic body the obtained velocities of initial perturbation equal to zero. For nonlinear viscoelastic body from the equation (40) it can be obtained that there is a region of initial velocities of perturbation, wherein the homogeneous solution is stable (except the limiting case). The system of equations (40), (41) can be solved numerically by setting the initial deformation and fixed strain rate. The difference between the presented approach and the linearized theory is that for given values of initial deformations and fixed strain rates (load history) a nonlinear viscoelastic body can always lose stability if the amplitudes of initial perturbations exceed a certain value. Outside this region the time derivative of the function P is negative definite and according to Lyapunov's first theorem on instability, the zero solution of system (28) will be unstable.

It is known that in the phase space of nonlinear dynamical systems the strange attractors arise. The strange attractor is an attracting region characterized by the regime of steady-state non-periodic oscillations [7]. The conditions for the appearance of a strange attractor are a combination of a global compression with a local instability. The regime of the strange attractor is realized only in the dissipative systems and is characterized by the presence of positive indicators in the spectrum of Liapunov's characteristic exponents. In this case, the attractor is in some region of the phase space and includes the Cantor's set of hypersurfaces. The limit set of trajectories that correspond to a strange attractor is not a manifold. The set of trajectories corresponding to a strange attractor is characterized by Lyapunov instability, but Poisson stability.

The fractal dimension D_F of an arbitrary limit set G in N-dimensional space can be calculated by the expression [7]:

$$D_F = \lim_{\varepsilon \to 0} \left\{ \ln M(\varepsilon) \cdot \left[\ln\left(\varepsilon^{-1}\right) \right]^{-1} \right\}, \quad (42)$$

where $M(\varepsilon)$ is the minimal number of N-dimensional cubes with a side ε needed to cover all the elements of the set.

If the spectrum of the Lyapunov exponents $\lambda_1 > \lambda_2 > \lambda_3 > \cdots > \lambda_N$ is known, then the concept of Lyapunov dimension can by introduced, namely:

$$D_L = j + \sum_{i=1}^{j} \lambda_1 \left| \lambda_{j+1}^{-1} \right|,$$

where j is the largest number satisfying the condition

$$L = \lambda_1 + \lambda_2 + \lambda_3 + \cdots + \lambda_j > 0. \quad (43)$$

As it follows from the works of Takens [8], the dimension of a strange attractor allows us to quantify the number of phase variables involved in the motion. If this dimension is finite and relatively small, then, in the distributed systems the simulation of the processes is possible within a finite number of ordinary differential equations the number of which corresponds to the dimension of the phase space the strange attractor is embedded in to.

If the sequence of experimental values for the variables is known, then on the basis of this sequence we construct the spaces of dimensions $m = 2,3,\ldots$, for which the correlation dimension can be calculated [7], [9], [10]:

$$\gamma_m = d_k^m = \lim_{\varepsilon \to 0} \frac{\ln C_m(\varepsilon)}{\ln \varepsilon}, \tag{44}$$

where

$$C_m(\varepsilon) = \lim_{N\to\infty} \frac{1}{N^2} \sum_{i,j=1}^{N} \Theta\left(\varepsilon - \left|\xi(t_1) - \xi(t_2)\right|\right),$$

where Θ is the Heaviside step-function

In the case when $\varepsilon \to 0,\ m \to \infty$ we have:

$$C_m(\varepsilon) \approx \varepsilon^{d_k} \exp(-m\tau L). \tag{45}$$

The dimension of a strange attractor is defined as follows:

$$L = \frac{1}{\tau} \ln \frac{C_m}{C_{m+1}}, \tag{46}$$

where τ is the time interval between the measurements of neighbor bifurcation points on the trajectory of the dynamical system.

The process of calculation of the correlation dimension continues until the value of $d_k^{*(m=m*)}$ stabilizes for some $m = m^*$.

According to [8], the dynamic stochasticity can develop after a finite sequence of bifurcations that ensure the achievement of a chaotic regime. For a viscoelastic body many bifurcation points can be obtained, which can be taken as traces of the solution of a strange attractor that has fallen into the region of attraction. Therefore, instead of the sequence of experimental values of T_j it is proposed to take a sequence of bifurcation values of $\{f_{nm}\}$, for which it is possible to carry out the calculation of the correlation dimension of a strange attractor. This will calculate the dimension of the phase space of the dynamical system, which simulates the processes occurring in the original system. Knowing the dimension of the strange attractor we can confine ourselves to the series of the expansion of displacements by coordinate functions by the number of terms equal to the dimension of the space the strange attractor is embedded in to.

From the expression (46) it is possible to obtain the value of the predictability interval, namely:

$$T = \frac{1}{L}.$$

This interval corresponds to the time necessary for the solution of the system of differential equations describing the process of loss of stability of nonlinear elastic and nonlinear viscoelastic media through a bounded sequence of bifurcations to be entered in the regime of a strange attractor.

For the obtained dynamical system using the approach presented in [6], [8] we can obtain the dimension of the strange attractor – γ_m, which allows to limit the number of terms in the Bubnov-Galerkin series (12), in which the perturbations of displacements can be decomposed.

As a result, we can obtain the graphical representations of dependencies between the disturbance module and load parameters.

4. Conclusions

IOP Conf. Series: Journal of Physics: Conf. Series **973** (2018) 012019 doi:10.1088/1742-6596/973/1/012019

Using the approach described above, the stability problems for the concrete structures [4], [6] have been solved. Namely, we obtained the following facts:

- a decrease in the ratio of the dimensions of the structures reduces the stability region both with respect to the initial perturbations and relative to the elongations; the replacement of the compressible material by an incompressible material leads to the same results;
- for all considered cases, the dimension of the strange attractor decreases with increasing of the load parameter.

We also emphasize that the dimension of the strange attractor, found for all the considered problems makes it possible to recommend the number of terms in the Bubnov-Galerkin set in the case when the structure operates in different ranges of variation of the initial stresses and strains.

The use of the stability criterion with respect to finite perturbations allows to obtain a limited sequence of admissible initial perturbations (for a specific value of the load parameter) in which the basic process of deformation will be stable and a hierarchy of stable equilibrium states is observed.

Setting the maximum allowable value of the initial perturbation leads to finding the region of the load parameter variation in which the main process of deformation will be stable.

References

[1] Karnaukhov V G and Gumenyuk B P 1990 *Thermomechanics of previously deformed viscoelastic bodies* (Kiev: Naukova Dumka) p 304

[2] Green A E, Rivlin R S and Shield R T 1952 General theory of small elastic deformation superposed on finite elastic deformation. *Proc. Roy. Soc.* London A **211** 128–154

[3] Green A E, Rivlin R S and Spenser A J M 1959 The mechanics of non-linear materials with memory. Part II *Arch. Rational Mech. Anal.* **3** 82–90

[4] Sporykhin A N and Sumin A I 1999 *Hierarchy of stable states in the mechanics of nonlinear media* (Voronezh: Voronezh State University) p 210

[5] Ilyushin A A 1978 *Continuum mechanics* (Moscow: Moskcow State University Press) p 288

[6] Sumin A I and Sporykhin A N 2014 *Nonlinear dynamics in problems of stability of mechanics of complex media* (LAP LAMBERT Academic Publishing) p 173

[7] Ruelle D and Takens F 1981 *On the nature of turbulence. Strange attractors*, ed Ya G Sinaya and L P Shilnikova (Moscow: Mir) pp 117–151

[8] Takens F 1981 Detecting strange attractors in turbulence *Dynamic systems and turbulence* (Heidelberg: Springer-Verlag) pp 366–381

[9] Sinay Ya G 1979 *Stohastichnost dinamicheskih sistem. Nelineynyie volnyi*, ed A V Gaponova (Moscow: Nauka) pp 192–211

[10] Farmer J D, Ott E and Yorke J A 1983 The Dimension of Chaotic Attractors *Physica D* **7** 153–180

A solution to the problem of elastic half-plane with a cohesive edge crack

Le Thi Thanh, L A Belaya and I M Lavit

Department of Computational Mechanics and Mathematics, Tula State University, 300012, 92 Lenin Avenue, Tula, Russia

E-mail: IgorLavit@yandex.ru

Abstract. This paper considers the problem of extension of an elastic half-plane slackened by a rectilinear edge crack. The opposite edges of the crack are attracted to each other. The intensity of attracting forces – the forces of cohesion – depends on displacements of the edges; this dependence is nonlinear in the general case. External load and cohesive forces are related to each other by the condition of finite stresses at the crack tip. The authors apply Picard's method of successive approximation. In each iteration, Irwin's method is used to solve the problem of a half-plane with a crack, the edges of which are subjected to irregularly distributed load. The solution of the resulting integral equation is found by Galerkin's method. The paper includes examples of calculations and their results. Some of them are compared with the data of previous studies.

1. Introduction

Cracks usually appear on the surface of the material in the stress concentration zone. Therefore, studying the conditions of surface crack growth is of undoubted interest. This work considers this problem in a plane formulation, as it is customary in fracture mechanics. The calculation model is an elastic half-plane with a rectilinear crack perpendicular to the half-plane boundary (Figure 1).

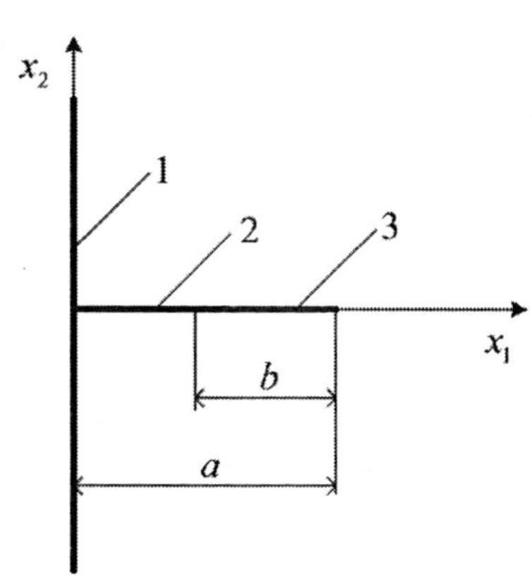

Figure 1. A half-plane with an edge cohesive crack: 1 – the half-plane boundary; 2 – a crack with the length a ; 3 – a cohesive zone with the length b .

Self-balanced distributed normal load is applied to the edges of the crack. Thus, the crack under consideration belongs to the cracks in mode I. The problem is symmetric relative to the abscissa axis; so, it is reasonable to consider the deformation of only the upper half of the half-plane. The load applied to the upper edge of the crack is represented as

$$p_1 = 0; \quad p_2(x_1) = p(x) = p_0(x) - g(x); \quad p_0, g \geq 0; \quad x = x_1 \tag{1}$$

where $p_0(x)$ is external load and $g(x)$ is cohesive forces. Cohesive forces are non-zero only in the interval $x \in [a-b, a]$ (in the cohesive zone). The equation of state for them is usually written as

$$g = g[v(x)] \tag{2}$$

where $v(x) = u_2(x_1, 0)$ is displacement of the upper edge of the crack along the ordinate axis.

If the length of the cohesive zone is much smaller than the length of the crack, the solution of a fracture mechanics problem is not dependent on the function form (2) [1]. However, in the general case considered in the present study, it is not true.

It is possible to distinguish three types of relationships (2). The first type is the case when the function (2) is reduced to the constant $g = \text{const}$. This problem for a half-plane with an edge crack was solved by Howard and Otter [2]. The second type is a linear ratio:

$$g = \begin{cases} g_M(1 - v/v_M), \, v \leq v_M \\ 0, \, v > v_M \end{cases} \tag{3}$$

where g_M, v_M are certain constants. A solution to the case with the allocation (3) was found by Wang and Dempsey [3]. The third type is a non-linear relationship, i.e. the general case of the formula (2). A solution to this problem is presented in this paper.

2. Mathematical model

2.1. *Iterative process*

A similar but simpler problem of a cohesive crack in a plane with nonlinear defining relationship (1) was considered in other works [4, 5]. To solve the nonlinear problem, Huang [4] applied Newton's method. Ungsuwarungsri and Knauss [5] showed that Picard's method of successive approximation is even more effective and much simpler in this case. This paper follows the recommendations of the work [5]. In each iteration of the solution process of the nonlinear problem, we should solve the problem of the linear elasticity for the model shown in Figure 1. The difference between the iterations lies in different distribution of forces $p(x)$. In this paper, the solution to this linear problem is obtained by Irwin's method [6, 7]. Its main point is that the initial problem is represented as a composition of two basic problems: the problem of a crack with length $2a$ in a plane (problem 1) and the problem of a half-plane loaded along the boundary contour (problem 2). These problems are connected by introducing additional fictitious forces that ensure the fulfillment of boundary conditions.

2.2. *Solution to the problem of a crack in a half-plane*

We consider an elastic half-plane with an edge rectilinear crack of length a (Figure 2); a distributed normal load of intensity $p(x_1)$ is symmetrically applied to the edges of the crack. It is necessary to calculate the stress intensity factor K_I and the stresses at interior points of the half-plane. The procedure of this problem solution will be as follows [6, 7]. Let us consider a plane with two intersecting cracks (Figure 3): a crack 1 with length $2a$ and an unloaded crack 2 with length $2b$. At

$b \to \infty$, the crack 2 is transformed into boundary contours of two half-planes; the design model shown in Figure 3 becomes equivalent to the design model shown in Figure 2. Thus, the initial problem is represented [6, 7] as a composition of two problems, whose solutions are known [8]. The first of these problems (hereinafter, Problem 1) is about a crack of length $2a$ in a plane and the second one (Problem 2) is about the right half-plane loaded along the boundary contour.

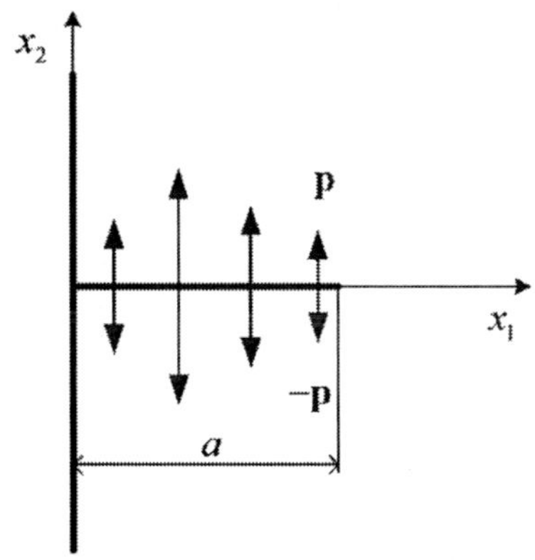

Figure 2. A half-plane with an edge crack: a – the length of the crack. Equal oppositely directed loads are applied to the edges of the crack, $p_1 = 0,\, p_2 = p(x_1)$.

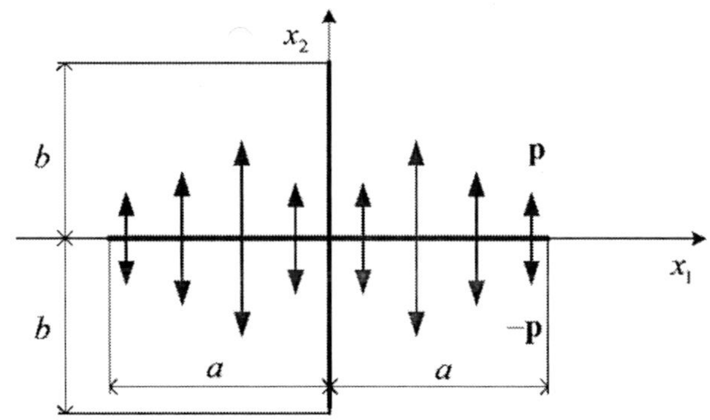

Figure 3. A plane with two intersecting cracks. The load intensity $p(x_1)$ is an even function.

The stresses in the plane problem of the elasticity theory are determined by the formulas [3]:

$$\begin{cases} \sigma_{11} + \sigma_{22} = 2\left[\Phi(z) + \overline{\Phi(z)}\right] \\ \sigma_{22} - \sigma_{11} + 2\mathrm{i}\sigma_{12} = 2\left[\bar{z}\Phi'(z) + \Psi(z)\right] \end{cases} \tag{4}$$

where σ_{km} are stresses in Cartesian coordinates x_k; $z = x_1 + \mathrm{i}x_2$ is a complex variable (i is an imaginary unit); $\Phi(z)$, $\Psi(z)$ are holomorphic functions determined from the boundary conditions of the problem; the bar over the symbol denotes complex conjugation, the prime denotes a derivative. Due to the linearity of the problem, these functions are represented as

$$\Phi(z) = \Phi_{(1)}(z) + \Phi_{(2)}(z); \quad \Psi(z) = \Psi_{(1)}(z) + \Psi_{(2)}(z) \tag{5}$$

where the subscript in parentheses indicates the problem number.

The first problem, taking into account its symmetry (Figure 3), has the following form [8]:

$$\begin{aligned} \Phi_{(1)}(z) &= -\frac{z}{\pi\left(z^2 - a^2\right)^{1/2}} \int_0^a \frac{\left(a^2 - t^2\right)^{1/2} p_{(1)}(t)\,\mathrm{d}t}{t^2 - z^2}; \\ \Psi_{(1)}(z) &= -z\Phi'_{(1)}(z); \quad p_{(1)}(t) = p(t) + q(t) \end{aligned} \tag{6}$$

where $q(t)$ is unknown additional load compensating for the impact of the cut along the ordinate axis.

In view of symmetry, the solution to the second problem is represented as [8]:

$$\Phi_{(2)}(z) = -\frac{z}{\pi}\int_0^\infty \frac{p_{(2)}(\eta)\,\mathrm{d}\eta}{\eta^2 + z^2}; \quad \Psi_{(2)}(z) = z\Phi'_{(2)}(z) \tag{7}$$

where the load $p_{(2)}(\eta)$, being applied to the boundary contour of the half-plane in the direction of the abscissa axis, compensates for the impact of the cut (crack).

According to formulas (6) and (4), the load coming from the left half-plane affects the right half-plane; this load is equal to $-\sigma_{(1)11}(0,y)$, where $y = x_2$. It is compensated by the distributed force $p_{(2)}(y)$. It follows from relations (6) and (4) that

$$p_{(2)}(y) = \sigma_{(1)11}(0,y) = -\frac{2}{\pi}\frac{\mathrm{d}}{\mathrm{d}y}\left[\frac{y^2}{\left(y^2+a^2\right)^{1/2}}\int_0^a \frac{\left(a^2-t^2\right)^{1/2} p_{(1)}(t)\,\mathrm{d}t}{t^2+y^2}\right] \tag{8}$$

Solution (7) and formulas (4) suggest that the upper edge of the crack has a load coming from the lower edge. This load is equal to $-\sigma_{(2)22}(x,0)$, where $x = x_1$. It is compensated by the load $q(x)$. From formulas (7) and (4), we have:

$$q(x) = \sigma_{(2)22}(x,0) = -\frac{4x}{\pi}\int_0^\infty \frac{\eta^2 p_{(2)}(\eta)\,\mathrm{d}\eta}{\left(\eta^2+x^2\right)^2}; \quad x\in[0,a] \tag{9}$$

The system of integral equations (8) and (9) with respect to the functions $q(x)$, $p_{(2)}(y)$ is identical to the system obtained by Irwin [6, 7] with the help of integral transformations.

Let us substitute the function $p_{(2)}(y)$, which is defined by expression (8), into equation (9). In the obtained expression, we perform integration by parts for the variable η and then change the order of integration. The resulting equation is:

$$q(x) = \frac{16x}{\pi^2}\int_0^a \left(a^2-t^2\right)^{1/2} p_{(1)}(t)\,\mathrm{d}t \int_0^\infty \frac{\eta^3\left(\eta^2-x^2\right)\mathrm{d}\eta}{\left(\eta^2+x^2\right)^3\left(\eta^2+t^2\right)\left(\eta^2+a^2\right)^{1/2}} \tag{10}$$

The inner integral can be expressed in terms of elementary functions by the substitution $\eta = \left(\xi^2-a^2\right)^{1/2}$:

$$\begin{gathered} I(x,t) = \int_0^\infty \frac{\eta^3\left(\eta^2-x^2\right)\mathrm{d}\eta}{\left(\eta^2+x^2\right)^3\left(\eta^2+t^2\right)\left(\eta^2+a^2\right)^{1/2}} = \int_a^\infty \frac{\left(\xi^2-a^2\right)\left[\xi^2-\left(a^2+x^2\right)\right]\mathrm{d}\xi}{\left[\xi^2-\left(a^2-x^2\right)\right]^3\left[\xi^2-\left(a^2-t^2\right)\right]} = \\ = \frac{t^2\left(t^2+x^2\right)}{\left(t^2-x^2\right)^3}\left[F_1(x)-F_1(t)\right] - \frac{x^2\left(3t^2-x^2\right)}{\left(t^2-x^2\right)^2}F_2(x) + \frac{2x^4}{t^2-x^2}F_3(x) \end{gathered} \tag{11}$$

where

$$F_1(x)=\frac{1}{\left(a^2-x^2\right)^{1/2}}\ln\left[\frac{a+\left(a^2-x^2\right)^{1/2}}{x}\right];\quad F_2(x)=\frac{1}{2\left(a^2-x^2\right)}\left[\frac{a}{x^2}-F_1(x)\right];$$
$$F_3(x)=\frac{1}{4\left(a^2-x^2\right)}\left[\frac{a}{x^4}-3F_2(x)\right] \tag{12}$$

Taking into account expressions (6), equation (10) can be represented in the standard form:

$$q(x)-\lambda\int_0^a K(x,t)q(t)\,\mathrm{d}t=f(t);\quad \lambda=16/\pi^2;$$
$$K(x,t)=x\left(a^2-t^2\right)^{1/2}I(x,t);\quad f(t)=\lambda\int_0^a K(x,t)p(t)\,\mathrm{d}t \tag{13}$$

Equation (13) will become a Fredholm equation under the following condition:

$$\int_0^a\int_0^a\left[K(x,t)\right]^2\mathrm{d}x\,\mathrm{d}t<M \tag{14}$$

where M is an arbitrarily large real number. It follows from the relations (11) and (12) that the kernel $K(x,t)$ is continuous for all possible arguments, except for the case when they both tend to zero. Let us find an asymptotic expression for the kernel at $x,t\to 0$, resorting to polar coordinates

$$x=r\cos\varphi;\quad t=r\sin\varphi;\quad \varphi\in[0,\pi/2] \tag{15}$$

and letting r tend to zero. Formulas (11) and (12) allow making an asymptotic expression for the kernel $K(x,t)$:

$$-\frac{\cos\varphi\sin^2\varphi}{r\cos^2 2\varphi}\left(\frac{\ln\tan\varphi}{\cos 2\varphi}+1\right) \tag{16}$$

Using L'Hospital rule, it is easy to show that at $r\neq 0$, this expression is finite at $\varphi=0$, as well as at $\varphi=\pi/4$ or $\varphi=\pi/2$. Because of the singularity at the point $r=0$, the kernel does not satisfy inequality (14). However, it follows from expression (16) that the integral

$$\int_0^a\int_0^a K(x,t)\psi(x,t)\,\mathrm{d}x\,\mathrm{d}t \tag{17}$$

where $\psi(x,t)$ is a continuous function, converges. This gives grounds to use Galerkin's method [9] for the solution of equation (13). The solution can be found by the finite number of terms in the series of coordinate functions [9]:

$$q(\zeta)=\sum_{k=0}^{N-1}c_k P_k(\zeta) \tag{18}$$

where c_k are desired expansion coefficients. In this study, the coordinate functions $P_k(\zeta)$ are Legendre polynomials [10], which are orthogonal in the interval $[-1,1]$ and form a complete system of functions [10]. Here, Galerkin's method gives a solution that tends to be exact at $N\to\infty$ [9]. To convert the integration interval in equation (13) to the interval $[-1,1]$, we introduce new variables ζ, τ:

$$x = 0.5a(1+\zeta); \quad t = 0.5a(1+\tau); \quad \zeta, \tau \in [-1,1] \tag{19}$$

Multiplying equation (13) by $P_j(\zeta)$ and integrating over the interval $[-1,1]$, we obtain a system of N linear algebraic equations for N of indeterminate c_k:

$$\begin{gathered}\sum_{k=0}^{N-1} \mu_{jk} c_k = b_j; \quad j = 0,1,\ldots,N-1; \\ \mu_{jk} = \int_{-1}^{1} P_k(\zeta) P_j(\zeta) \mathrm{d}\zeta - 0.5\lambda \int_{-1}^{1}\int_{-1}^{1} K(\zeta,\tau) P_k(\tau) P_j(\zeta) \mathrm{d}\zeta \, \mathrm{d}\tau \\ b_j = 0.5\lambda \int_{-1}^{1}\int_{-1}^{1} K(\zeta,\tau) p(\tau) P_j(\zeta) \mathrm{d}\zeta \, \mathrm{d}\tau\end{gathered} \tag{20}$$

The formula can be simplified for μ_{jk}, considering the property of Legendre polynomials [10]:

$$\int_{-1}^{1} P_k(\zeta) P_j(\zeta) \mathrm{d}\zeta = \frac{2\delta_{jk}}{2j+1} \tag{21}$$

where δ_{jk} is Kronecker delta.

The integrals in the formulas for the equation system coefficients can be determined numerically. In this study we use Gaussian quadrature. The next steps after solving the system of equations are calculation of the stress intensity factor

$$K_I = 2\left(\frac{a}{\pi}\right)^{1/2} \int_0^a \frac{p_{(1)}(t)\mathrm{d}t}{\left(a^2 - t^2\right)^{1/2}} \tag{22}$$

and determination of the stress – by the formulas (4) – (8).

The case of constant load calculation can serve as an example. The dimensionless stress intensity factor for this problem is determined by Wigglesworth's formula [11]:

$$K_I^* = \frac{K_I}{p\sqrt{\pi a}} = \exp\left[-\frac{1}{\pi} \int_0^{\infty} \frac{\ln\left(1 - \dfrac{\eta^2}{\sinh^2(0.5\pi\eta)}\right)}{1+\eta^2} \mathrm{d}\eta \right] \tag{23}$$

Equation (23) determines K_I^* as a value, which is accurate to six significant digits and is equal to 1.12152. Table 1 shows the values of K_I^* depending on the number of coordinate functions N; they were obtained with the developed method. Table 1 reflects acceptable accuracy and sufficiently rapid convergence of the solution with an increase in the number N.

Table 1. Dependence of the dimensionless stress intensity factor on the number of coordinate functions

N	2	20	200
K_I^*	1.07785	1.12102	1.12152

2.3. *The algorithm for solving the cohesive crack problem*

The distribution of cohesive forces and the actual load are related by the condition of absent stress singularity at the tip of the crack [1]:

$$K_I = 0 \tag{24}$$

When constructing an iterative process algorithm, it is convenient to represent the external load in the following form:

$$p_0(x) = Cq_0(x) \tag{25}$$

where $q_0(x)$ is a given function, C is a constant varying from iteration to iteration. For the given distribution of cohesive forces, equation (24) is a linear equation for determining C.

To calculate the cohesive forces, it is necessary to find the displacement of the upper edge of the crack along the ordinate axis according to formula (1). Since the crack is a great discontinuity of displacement field and the displacements in the basic problem 2 are continuous, the above displacement of the upper edge of the crack is completely determined by the solution to the basic problem 1 – the problem of a crack in a plane. Thus, it is defined (for the plain strain case) by the following formula [8]:

$$v(x) = \frac{4(1-\nu^2)}{E}\operatorname{Im}\left[\varphi_{(1)}(x)\right] \tag{26}$$

where E is Young's modulus, ν is Poisson's ratio, $\varphi_{(1)}(z)$ is the complex potential [8] related to the function $\Phi_{(1)}(z)$ through the formula $\Phi_{(1)}(z) = \varphi'_{(1)}(z)$. Let us transform the expression for the function $\Phi_{(1)}(z)$:

$$\Phi_{(1)}(z) = \frac{z}{\pi}\left[\frac{1}{\left(z^2-a^2\right)^{1/2}}\int_0^a \frac{p_{(1)}(t)\mathrm{d}t}{\left(a^2-t^2\right)^{1/2}} + \left(z^2-a^2\right)^{1/2}\int_0^a \frac{p_{(1)}(t)\mathrm{d}t}{\left(t^2-z^2\right)\left(a^2-t^2\right)^{1/2}}\right] \tag{27}$$

It follows from (22) and (24) that the first summand is zero. Integration of the second summand in (27) with respect to z and simple transformations result in formula (26) in the following form:

$$v(x) = \frac{2(1-\nu^2)}{\pi E}\int_0^a \ln\left|\frac{\left(a^2-t^2\right)^{1/2} + \left(a^2-x^2\right)^{1/2}}{\left(a^2-t^2\right)^{1/2} - \left(a^2-x^2\right)^{1/2}}\right| p_{(1)}(t)\mathrm{d}t \tag{28}$$

It is natural to assume that the cohesive forces along the crack length change continuously. Since they are non-zero only in the cohesive zone, expression (1) takes the form

$$g = \begin{cases} g_M f(v/v_M), & v \le v_M \\ 0, & v > v_M \end{cases}; \quad f(0) = 1; \quad f(1) = 0 \tag{29}$$

From equation (29) and the problem formulation, it follows that $v\big|_{x=a-b} = v_M$.

The iterative process is made as follows. By the beginning of the k-th iteration, we know the displacement of the upper edge of the crack $v_{k-1}(x)$ and, consequently, the boundary of the cohesive zone b_{k-1}. Formula (29) defines the distribution of the cohesive forces along the edge of the crack $g_{k-1}(x)$. The next step is to solve the linear problem of a half-plane with an edge crack and to define

IOP Conf. Series: Journal of Physics: Conf. Series **973** (2018) 012020 doi:10.1088/1742-6596/973/1/012020

the constant C_{k-1} from equation (24). Next, we find the displacements $v_k(x)$ by formula (28) and, basing on the condition $v|_{x=a-b} = v_M$, a new value for the boundary of the cohesive zone b_k. The iterative process is repeated until convergence, which is determined by the criterion

$$|b_k - b_{k-1}|/a < \varepsilon \tag{30}$$

where $\varepsilon > 0$ is a small given number.

To solve the problem by Picard's method of successive approximation, it is necessary to find the initial approximation $v_0(x)$, which should be sufficiently close to the solution. If the initial approximation is unsuccessful, the iterative process converges to some infinite set of trivial solutions – solutions for a fully cohesive crack, which are characterized by the condition

$$b = a; \quad v(0) \in [0, v_M) \tag{31}$$

In this case, the second constraint (29) on the function $f(v/v_M)$ is not satisfied.

In accordance with earlier recommendations [5], the initial approximation is chosen as:

$$v_0(x) = v_M \left(\frac{a-x}{b_0} \right)^{1.5} \tag{32}$$

Changing the value b_0, it is possible to find an initial approximation that ensures convergence of the iterative process to a nontrivial solution.

The effect of the cohesive zone extent is manifested in the magnitude of the J -integral [12]. Let us introduce a cohesive force potential:

$$W(v) = \int_0^v g(\eta)\,\mathrm{d}\eta \tag{33}$$

Using Rice's formula [12], we obtain:

$$J = -2\int_{a-b}^{a} g(x)\frac{\mathrm{d}v}{\mathrm{d}x}\mathrm{d}x = 2W(v_M) \tag{34}$$

This value can naturally be compared with the value of the J -integral for a non-cohesive crack:

$$J_* = (1-\nu^2)K_{I*}^2/E \tag{35}$$

where K_{I*} is the stress intensity factor calculated by formula (22) provided that $p(x) = p_0(x)$, i.e. in the absence of cohesive forces.

3. Calculation results

In the examples discussed below, the external load was assumed constant along the length of the crack, i.e. $q_0 = 1$. The chosen number of coordinate functions of Galerkin's method was 60. We consequently obtained $K_I^* = 1.12147$, which differs from the exact solution by 0.00446 %. The permissible relative error of the solution was assumed equal to $\varepsilon = 10^{-7}$; the Poisson's ratio was $\nu = 0.3$.

Let us compare the results of the calculations via the developed method with the results of the work [3] obtained for the linear law (3). Figure 4 shows the graphs of external load intensity. They indicate that the results of the present paper and work [3] become significantly different when the cohesive zone extends almost over the entire crack, i.e. in the most important applied case of short cracks. This difference is caused by imperfection of the method used in the work [3], where trivial but meaningful for physics solutions are out of the problem consideration – when the cohesive zone covers the entire crack, but the crack opening is not equal to zero.

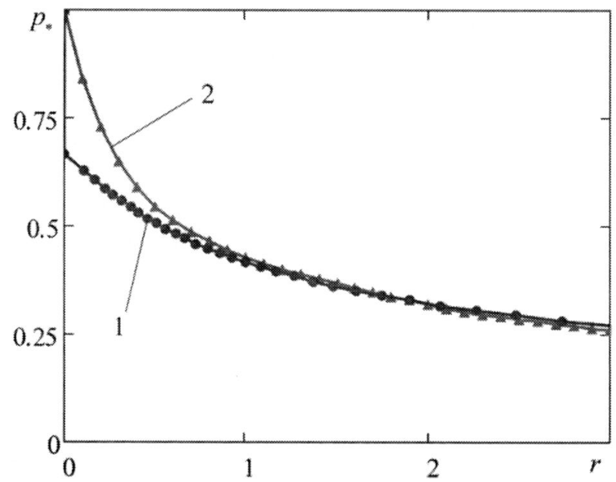

Figure 4. Comparison of the calculation results with the data of the work [3]: 1 – results of calculation by the developed method; 2 – the results of Wang and Dempsey's calculation [3]; $p_* = p_0/g_M$; $r = (1-\nu^2) g_M (a-b)(E v_M)^{-1}$.

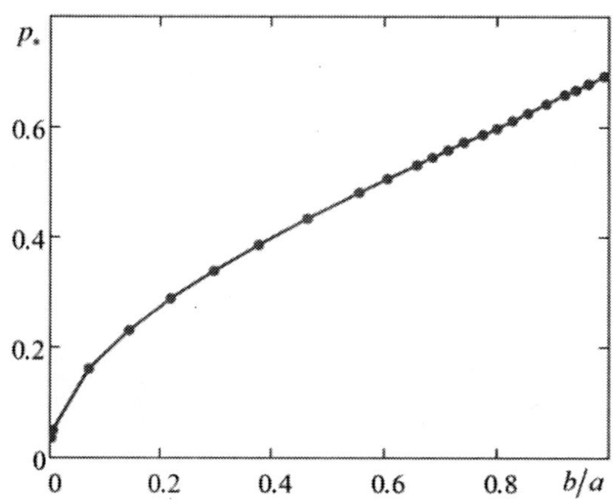

Figure 5. External load intensity depending on cohesive zone length.

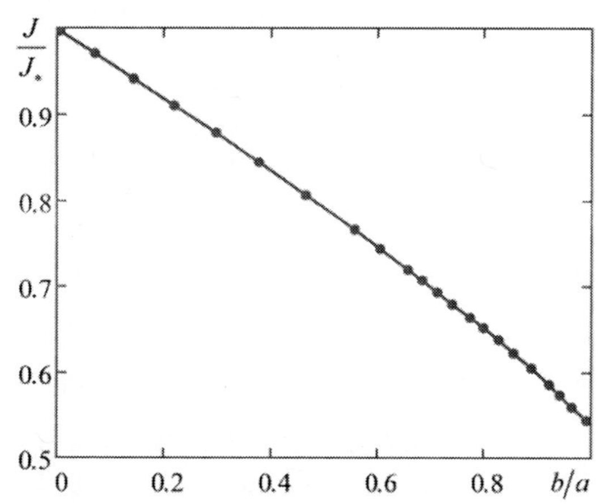

Figure 6. Correlation of J-integral values depending on cohesive zone length.

Now we proceed to the problem solution for nonlinear defining relation (29). The function $f(v/v_M)$ is continuous; according to the Weierstrass theorem, it can be arbitrarily precisely approximated by a polynomial. It is natural to consider it not only continuous, but also smooth. The simplest function of this kind is as follows:

$$f(v/v_M) = \frac{g}{g_M} = \begin{cases} 1 - 3(v/v_M)^2 + 2(v/v_M)^3, v \le v_M; \\ 0, v > v_M \end{cases} \qquad (36)$$

$$f(0) = 1; \quad f(1) = 0; \quad f'(0) = 0; \quad f'(1) = 0$$

Some calculation results are represented in Figure 5 and 6. Figure 5 shows the dependence of external load on cohesive zone length, Figure 6 – the dependence of the ratio of the J-integral values calculated by formulas (34) and (35).

4. Conclusion

Formulation and solution of problems on a crack with a cohesive zone commensurable with the crack are associated with attempts to simulate the anomalous properties of short cracks [13-15]. Within the framework of the model considered in this paper, a short crack differs from a long one only in the fact that the cohesive zone length in a long crack is negligible compared to the length of the crack. Therefore, all Barenblatt postulates are valid for a long crack; its growth is determined by the magnitude of the stress intensity factor from external load. A short break should be justified by other laws; it is evidently presented in Figure 6. Their construction should be based on experimental data (for example, see [13-15]). Most likely, all those laws will be non-linear. Thus, when creating methods for solving specific problems, it is necessary to provide for the possibility to use nonlinear defining relations for cohesive forces. This study develops a numerical method for solving such a problem – the problem of a cohesive crack in a half-plane. It is based on the progressive approximation method combined with Irwin's method of solving the elastic problem of a crack in a half-plane. Numerical examples demonstrate the effectiveness of the method.

Acknowledgments

The results of the research project are published with the financial support of Tula State University within the framework of the scientific project № 2017-60.

References

[1] Barenblatt G I 1962 *Adv. Appl. Mech.* **7** 55
[2] Howard I C and Otter N R 1975 *J. Mech. Phys. Solids* **23** 139
[3] Wang S and Dempsey J P 2011 *Eng. Fract. Mech.* **78** 1353
[4] Huang N C 1985 *J. Appl. Mech.* **52** 490
[5] Ungsuwarungsri T and Knauss W G 1988 *J. Appl. Mech.* **55** 44
[6] Irwin G R 1958 *U. S. Naval Research Lab.* Rept. 5120.
[7] Hartranft R J and Sih G C 1973 *Methods of Analysis and Solutions of Crack Problems* ed G C Sih (Leyden: Noordhoff Int. Publ.) p 179
[8] Muskhelishvili N I 1953 *Some Basic Problems of the Mathematical Theory of Elasticity* (Groningen: Noordhoff Int. Publ.)
[9] Mikhlin S G 1964 *Variational Methods in Mathematical Physics* (Oxford: Pergamon Press)
[10] Bateman H and Erdelyi A 1953 *Higer Transcendental Functions* vol 1 (New York: Mc Graw – Hill Book Company)
[11] Wigglesworth L A 1957 *Mathematika* **4** 76.
[12] Rice J R 1968 *Fracture* vol 2, ed H Liebowitz (New York: Academic Press) p 191
[13] Suresh S 2006 *Fatigue of Materials* (New York: Cambridge University Press)
[14] Proudhon H, Li J, Wang F, Roos A, Chiaruttini V and Forest S 2016 *Int. J. Fatigue* **82** 238
[15] Mayén J, Abúndez A, Pereyra I, Colín J, Blanco A and Serna S 2017 *Eng. Fract. Mech.* **177** 203

IOP Conf. Series: Journal of Physics: Conf. Series **973** (2018) 012021 doi:10.1088/1742-6596/973/1/012021

On impact by a hard cone on elasto-viscoplastic material, leading to the generation of a conical crack

N D Verveiko[1,3], A I Shashkin[1,4] and S E Krupenko[2]

[1] Voronezh State University, University Sq.1, Voronezh, 394018, Russia

[2] Joint-stock company Voronezh pilot plant of software products, (VOZPP) Tekstilshchikov St., Building 10a, Voronezh, 394026, Russia

E-mail: [3] verveyko2017@yandex.ru and [4] dean@amm.vsu.ru

Abstract. The destruction of solid physical objects is a complex process in which mechanical, chemical, thermobaric and other matter transformations take place. Under mechanical destruction is understood the violation of the integrity of the object due to the occurrence of cracks. High-speed impact of a solid body on deformable materials is accompanied by the spread of cracks and is of a wave nature. This article presents an analysis of the dynamic stress-strain state in an elastoviscoplastic (EVP) material near the leading edge of a moving crack, approximated by a zone of continuous deformation. An analysis of the distribution of the intensity of tangential stresses and plastic deformations that occur behind the front of the longitudinal and shear head waves of a spherical shape generated by the impact of the vertex of the solid cone is carried out on the model EVP of the medium by the ray method. It is shown that the presence of a maximum of the jump of the tangential velocity component on the shear wave leads to a development with time of a jump in the displacements of the tangents to the front of the shear wave. This can be interpreted as the moment of initiation of the head part of a crack running along with the front of the elastic wave with the velocity of shear waves.

1. Introduction

A significant number of publications are devoted to the problems of formation and development of cracks. In work [1], interaction of edge cracks and factors affecting the further formation of cracks are considered. A number of illustrative examples is presented.

There are numerous experimental results [2], which indicate that a thermal shock on the ceramic surface causes cracks. It is shown that the failure pattern depends strongly on the microstructure of the material and the type of load.

Questions of the ultimate state of elastic materials with cracks are considered in detail in [3, 4, 5] from the standpoint of the static stress-strain state of an elastic medium and the use of criteria for the ultimate state of fracture or shear cracks.

The problems of the dynamics of the initiation of cracks and their propagation can not in principle be formulated in the framework of this approach. Modeling the dynamics of crack propagation requires the use in the mathematical model of the laws of elastic, viscous and plastic behavior of the material.

One of the approaches to studying the propagation of the front edges of cracks is the approximation of cracks by thin layers of continuous plastic deformation [6, 7]. In this paper, we analyze the dynamic stress-strain state in the EVP material near the front edge of the moving crack, approximated by the zone of continuous deformation.

The initial moment of a dynamic impact by a cone on a deformable material is of a wave nature [8]. For a short time Δt, the region of the material perturbed by impact will be concentrated in a neighborhood of the radius $c\Delta t$ (c is the propagation velocity of the wave) of the vertex of the cone. With a small velocity v_0 of the cone, the produced perturbation is reversible. However, at a sufficiently high impact speed v_0, irreversible plastic deformation of the material takes place.

2. Materials and methods

Let's consider the deformed state of the material in the neighborhood of the cone vertex, bounded by the weak waves Σ_1 and Σ_2 with a singularity on their front.

For convenience, we approximate the velocity of the diffracted wave Σ at its front by its linear approximation

$$\tilde{v}_n\big|_{\Sigma} \approx v_n(\delta) + \delta \frac{\partial v_n(\delta)}{\partial n}, \tag{1}$$

using the expansion of the function $v_n(n)$ in a Taylor series in the neighborhood of the point $n = \delta$

$$v_n(n) = \sum_{k=0}^{\infty} \frac{1}{k!} \frac{\partial v_n(\delta)}{\partial n} (n-\delta)^k . \tag{2}$$

This approximation of (1) allows us to consider the diffracted wave Σ as a strong wave [8, 9] with the intensity $v_n(\delta)$ generated by the impact of the cone. The vertex of the cone is represented by a sphere of radius δ, which excludes the singularity at $n = 0$ (figure 1).

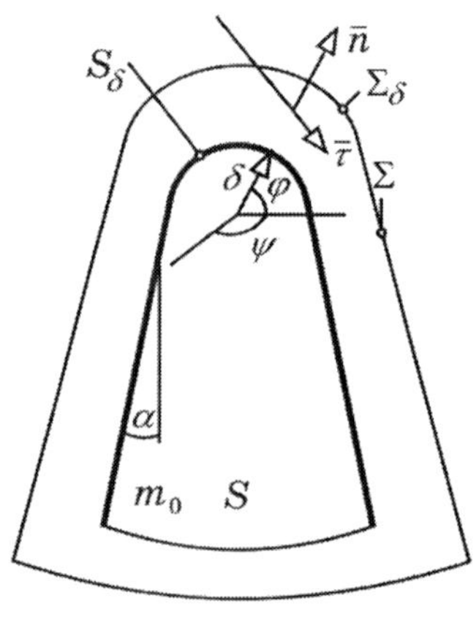

Figure 1. The image of the surface of a cone S, the surface of a sphere S_δ (δ is its radius) and, generated by impact, diffracted (weak) wave Σ_δ and shock (strong) wave Σ

The velocity v_n of the material behind a weak sigular diffracted wave Σ_1 satisfies the conditions: $v_n\big|_{\Sigma_1} = v_n(0) = 0$; $\left.\frac{\partial v_n}{\partial n}\right|_{n\to 0} \to \infty$; $\frac{\partial^2 v_n}{\partial n^2} \neq 0$.

Further on let's confine by the piecewise-constant approximation of stresses and velocities in the zones behind the fronts of the diffracted waves Σ_1 and Σ_2. Let's consider the EVP material as a model of the dynamic deformation of the medium. On its shock fronts, the following conditions are satisfied for jumps of stresses, strains and velocities [9, 10]:

$$[\sigma_{ij}] = \lambda [\varepsilon_{kk}] \delta_{ij} + 2\mu [\varepsilon_{ij}]; \quad (i, j = 1, 2, 3);$$

$$[\varepsilon_{ij}] = \frac{1}{2}\left([u_{i,j}] + [u_{j,i}]\right); \quad [u_{i,j}] = -[v_i]\frac{n_j}{c}. \tag{3}$$

Here: σ_{ij} are the components of the stress tensor; ε_{ij} are the components of the strain tensor; δ_{ij} is the Kronecker symbol; u_i are the displacements; v_i are the speeds; λ, μ are the Lame elastic parameters; c is the velocity of the front of the shock wave; a comma between the lower indices i, j means the differentiation of the component u_i along the coordinate x_j $\left(\dfrac{\partial u_i}{\partial x_j} = u_{i,j}\right)$; the square brackets at the functions mean the difference between their values on the right and on the left on Σ $\left([f] = f\big|_{\Sigma+} - f\big|_{\Sigma-}\right)$.

The initial values of the velocity of the medium behind the front Σ of the shock waves Σ_1 or Σ_2 $\left(v_i\big|_{\Sigma-}\right)$ are determined at the initial moment of impact from the conditions on the surface of a sphere of radius δ approximating the vertex of the cone (figure 2).

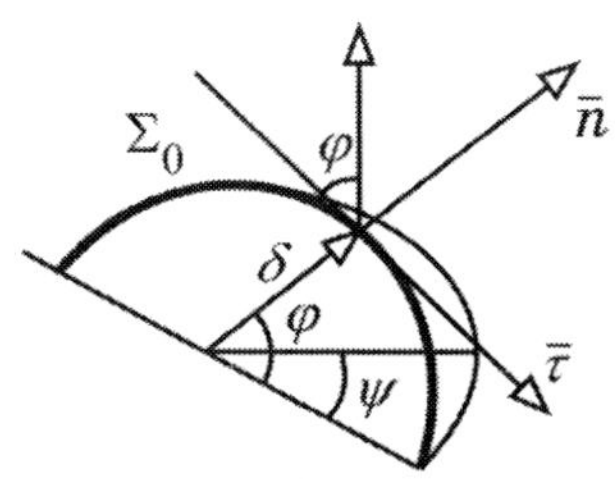

Figure 2. The image of the initial moment of the shock wave Σ coinciding with the surface of the cone (its singular point approximated by a sphere of radius δ)

Let's consider the stress state of the EVP material in the regions behind the diffracted waves Σ_1 and Σ_2, which are generated by the impact of the cone with the angle 2α at the vertex with velocity v_0 (figure 1). This figure shows the vertex of the cone, approximated by a sphere S_δ of small radius δ, and the diffracted waves generated by the impact of the vertex of the cone, at a position that appeared after a period of time Δt after the start of the impact. At the time of impact, the EVP material at the cone boundary S and S_δ will acquire a velocity $\overline{v}_s$ that depends on the impact velocity $\overline{v}_0$ and the surface S properties of the cone, which causes either complete adherence of the material, or slip with friction, or complete slippage. On the surface of the sphere S_δ the following relations takes place:

$$\overline{v}_s = v_n\overline{n} + v_\tau\overline{\tau}, \quad \text{i.e.} \quad v_{si} = v_n n_i + v_\tau \tau_i, \tag{4}$$

where $v_\tau = f v_0 \cos\varphi$; $v_n = v_0 \sin\varphi$;

$$\overline{n} = (\cos\varphi\cos\psi;\ \cos\varphi\sin\psi;\ \sin\varphi);$$
$$\overline{\tau} = (\sin\varphi\cos\psi;\ \sin\varphi\sin\psi;\ -\cos\varphi). \tag{5}$$

On the surface S of the cone $\varphi = \dfrac{\pi}{2} - \alpha$.

Note that the sticking coefficient $f \in [0; 1]$. The value $f = 1$ corresponds to the complete adhesion of the material to the surface S of the cone, and $f = 0$ corresponds to slippage. At $f \in (0; 1)$ different degrees of adhesion are realized.

In the case of complete slippage, the impact of a cone generates one front Σ_1 of the longitudinal wave, in the case of different degrees of adhesion, a second Σ_2 (shear) wave appears, which lags in its motion from the longitudinal wave Σ_1.

To determine the stresses σ^I_{ij} and σ^{II}_{ij} in the zones behind Σ_1 and Σ_2 we use expressions (3) for stress jumps and strain rates in the regions of a piecewise constant approximation of stresses and velocities.

On Σ_1 the conditions (3) are fulfilled

$$[\sigma^I_{ij}] = \sigma^0_{ij} - \sigma^I_{ij} = -[v_n]\frac{(\lambda\delta_{ij} + 2\mu n_i n_j)}{c_1}, \tag{6}$$

where $[v_n] = -v_0 \sin\varphi$.

On Σ_2 conditions (3) take the form

$$[\sigma^{II}_{ij}] = \sigma^I_{ij} - \sigma^{II}_{ij} = -[v_\tau]\frac{\mu}{c_2}\tau_i n_j, \tag{7}$$

where $[v_\tau] = -f v_0 \cos\varphi$.

From (6)–(7), one can obtain expressions for the components of the stress tensors σ^I_{ij} and σ^{II}_{ij} as functions of the velocity v_0 of the cone and directions of the vectors $\overline{n}$, $\overline{\tau}$:

$$\sigma^I_{ij} = \sigma^0_{ij} + \frac{v_0}{c_1}(\lambda\delta_{ij} + 2\mu u_i u_j)\sin\varphi; \tag{8}$$

$$\sigma^{II}_{ij} = \sigma^0_{ij} + \frac{v_0}{c_1}(\lambda\delta_{ij} + 2\mu u_i u_j)\sin\varphi + f\left(\frac{v_0}{c_2}\right)\mu(\tau_i n_j + \tau_j n_i)\cos\varphi. \tag{9}$$

The first invariants $J_1 = \sigma_{kk}$ of the stress tensor in the zones behind Σ_1 and Σ_2 coincide

$$J^I_1 = J^{II}_1 = \sigma_{kk} = \sigma^0_{kk} + \frac{v_0}{c_1}(3\lambda + 2\mu)\sin\varphi. \tag{10}$$

The deviator components of the stress tensors in the zones behind Σ_1 and Σ_2 are determined from the relations (8)–(9) as follows:

$$\sigma'^I_{ij} = \sigma^I_{ij} - \frac{1}{3}\sigma^I_{kk}\delta_{ij} = \sigma'^0_{ij} + \frac{v_0}{c_1}\sin\varphi\left(-\frac{2}{3}\mu\delta_{ij} + 2\mu n_i n_j\right); \tag{11}$$

$$\sigma'^{II}_{ij} = \sigma'^I_{ij} + f\frac{v_0}{c_2}\cos\varphi\mu(\tau_i n_j + \tau_j n_i). \tag{12}$$

It follows from (12) that the stresses in zone II differ from the stresses in zone I only in the case of friction ($f \neq 0$) on the surface of the cone.

The second invariants of the deviator of the stress tensor in the zones behind Σ_1 and Σ_2, which characterize the intensity of the tangential stresses, are calculated as follows:

$$J^I_2 = \sigma'^I_{ij}\tau'^I_{ij} = J^0_2 + 4\frac{v_0}{c_1}\mu\sigma'^0_{ij}n_i n_j\sin\varphi + \frac{16}{3}\left(\frac{v_0}{c_1}\mu\sin\varphi\right)^2, \tag{13}$$

where $\sigma'^0_{ij}n_i n_j = \sigma'_{nn}$;

$$J^{II}_2 = J^I_2 + 2f\frac{v_0}{c_1}\mu\sigma'^0_{ij}n_i\tau_j\cos\varphi + 2\left(f\frac{v_0}{c_2}\mu\cos\varphi\right)^2, \tag{14}$$

here $\sigma'^0_{ij}n_i\tau_j = \sigma'^0_{n\tau}$.

Let's estimate the influence of the static stress state σ^0_{ij} for the axisymmetric case on the intensity of tangential stresses J_2.

$$\sigma^0_{11} = \sigma_1; \quad \sigma^0_{22} = \sigma_1; \quad \sigma^0_{33} = \sigma_3; \quad \sigma^0_{ij} = 0 \quad \text{for} \quad i \neq j;$$

$$\sigma_{ij}^{\prime 0} = \frac{1}{3}\begin{pmatrix} \sigma_1 - \sigma_3 & 0 & 0 \\ 0 & \sigma_1 - \sigma_3 & 0 \\ 0 & 0 & 2(\sigma_3 - \sigma_1) \end{pmatrix}. \tag{15}$$

Knowing the components of the vectors $\overline{n}$ and $\overline{\tau}$ (5), we define the contribution of $\sigma_{ij}^{\prime 0}$ to I_2^I and I_2^{II}.

$$\sigma_{ij}^{\prime 0} n_i n_j = \overline{\sigma}\left(\cos^2\varphi\cos^2\psi + \cos^2\varphi\sin^2\psi - 2\sin^2\varphi\right) = \overline{\sigma}\left(\cos^2\varphi - 2\sin^2\varphi\right); \tag{16}$$

$$\sigma_{ij}^{\prime 0} n_i \tau_j = \overline{\sigma}\left(\sin\varphi\cos\varphi\cos^2\psi + \sin\varphi\cos\varphi\sin^2\psi + 2\sin\varphi\cos\varphi\right) = \sigma\sin\varphi\cos\varphi = \frac{\sigma}{2}\sin 2\varphi, \tag{17}$$

here $\sigma = \sigma_1 - \sigma_3$; $\overline{\sigma} = \dfrac{\sigma}{3}$.

Let's express the second invariant J_2^I of the deviator of the stress tensor $\sigma_{ij}^{\prime I}$ in the first zone behind the diffracted longitudinal wave Σ_δ, taking into account the stresses $\sigma_{ij}^{\prime 0}$ in the unperturbed zone before Σ_δ (16)–(17). We get

$$J_2^I = J_2^0 + \frac{16}{3}\left(\frac{v_0}{c_1}\mu\sin\varphi\right)^2 + \frac{4}{3}\frac{v_0}{c_1}\mu\left(\sigma_1 - \sigma_3\right)\sin\varphi\left(1 - 3\sin^2\varphi\right), \tag{18}$$

where

$$J_2^0 = \sigma'_{ij}\sigma_{ij}^{\prime 0} = \left(\frac{2}{3}\left(\sigma_3 - \sigma_1\right)\right)^2.$$

In the second zone, the account of the static stress state leads the expression (14) for J_2^{II} to the form

$$J_2^{II} = J_2^I + 2\left(f\mu\frac{v_0}{c_2}\cos\varphi\right)^2 + f\mu\frac{v_0}{c_2}\left(\sigma_1 - \sigma_3\right)\cos\varphi\sin 2\varphi. \tag{19}$$

Here

$$J_2^0 = \sigma_{ij}^{\prime 0}\sigma_{ij}^{\prime 0} = \frac{4}{9}\left(\sigma_1 - \sigma_3\right)^3. \tag{20}$$

3. Results and analysis

Let's consider the following cases.

3.1. Analysis of the variation of the tangential stress in a non-stressed material

We analyze the variation in the intensity of the tangential stresses J_2 behind the waves Σ_1 and Σ_2, which propagate in the unstressed material. In this case $\sigma_1^0 = \sigma_2^0 = \sigma_3^0 = 0$.

3.1.1. In the case of impact by an ideal smooth cone at $f = 0$, the shear wave Σ_2 does not arise, and behind the wave Σ_1, the shear stress intensity is given by

$$\Lambda_1 = \frac{\sqrt{J_2^I}}{\mu\sqrt{\dfrac{16}{3}}}\frac{v_0}{c_1} = \sin\varphi, \tag{21}$$

so that the maximum J_2^I is attained at $\varphi = \dfrac{\pi}{2}$, i.e. in the direction of impact (see figure 3).

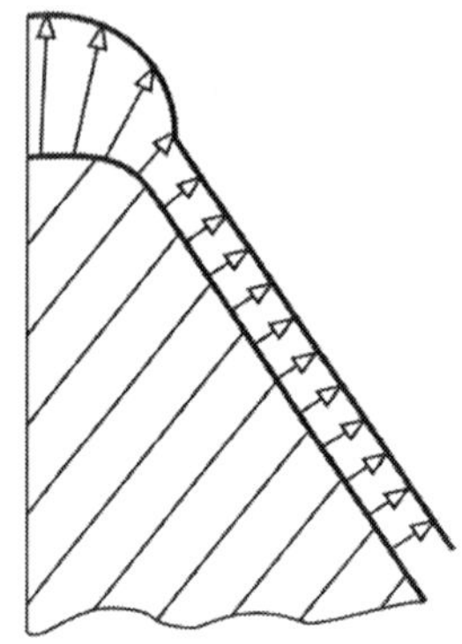

Figure 3. An image of the stress J_2^I intensity on the surface of the cone at the moment of impact as a function of the direction φ

3.1.2. In the case of a rough cone impact at $f \neq 1$, two waves arise: a longitudinal strain wave Σ_1 and a shear deformation wave Σ_2. The expression for the intensity of the tangential stresses J_2^{II} behind Σ_2 takes the form

$$\Lambda_2 = \frac{\sqrt{J_2^{II}}}{\mu\sqrt{\dfrac{16}{3}}}\frac{v_0}{c_1} = \sqrt{\sin^2\varphi + \left(\frac{c_1}{c_2} f \cos\varphi\right)^2}. \tag{22}$$

Λ_2 takes the maximum value in the directions $\varphi = 0$ and $\varphi = \dfrac{\pi}{2}$.

3.2. The influence of the static stress state of a material on the intensity of tangential stresses
The static stress state of a material influences its behavior behind Σ_1 and Σ_2 and, as follows from formulas (18)-(20), the intensity of tangential stresses J_2 behind Σ_1 and Σ_2 increases by a static value $J_2^0 = J_2^{\prime 0} = \left(\dfrac{2}{3}(\sigma_1 - \sigma_3)\right)^2$ in any direction φ.

Depending on the direction φ the additives to J_2 have the following form:

$$\lambda_2^I = \frac{\Delta J_2^I(\varphi)}{\dfrac{4}{3}\dfrac{v_0}{c_1}\mu(\sigma_1 - \sigma_3)} = \sin\varphi\left(1 - 3\sin^2\varphi\right); \tag{23}$$

$$\lambda_2^{II} = \frac{\Delta J_2^{II}(\varphi)}{f\dfrac{v_0}{c_2}\mu(\sigma_1 - \sigma_3)} = \cos\varphi \sin 2\varphi, \tag{24}$$

where $\Delta J_2^i(\varphi) = J_2^i(\varphi) - J_2^0$; $i = I, II$.

The quantities λ_2^I and λ_2^{II} (figure 4) take the maximum values in the directions $\varphi_1^* = \sqrt{\dfrac{1}{6}} \approx 24°$ and $\varphi_2^* = \sqrt{\dfrac{1}{3}} \approx 36°$.

3.3. Distribution of residual plastic deformations after impact by a cone
Let's investigate the distribution of residual plastic deformations remaining in the material after passage of the waves of plastic loading Σ_1 and Σ_2 for the case of a short (small Δt) impact by

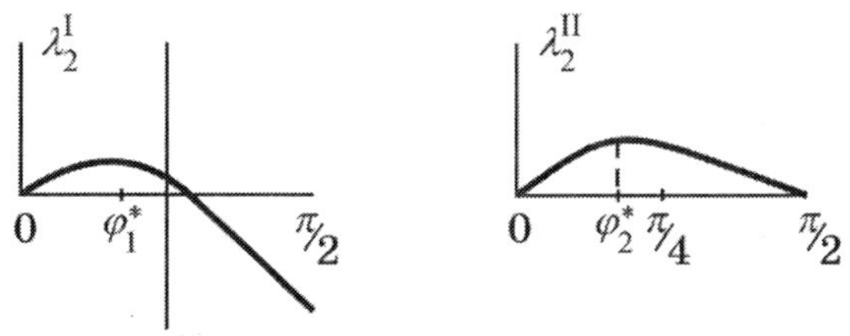

Figure 4. Graphs of the relative change in the intensity λ_2 of tangential stresses due to the static stress axisymmetric state $(\sigma_1^0, \sigma_1^0, \sigma_3^0)$ as a function of direction φ

a cone. Directly behind Σ plastic deformations are absent, but the rates of plastic deformations $\dot{\varepsilon}^p_{nn}\big|_{\Sigma_1^-}$ and $\dot{\varepsilon}^p_{n\tau}\big|_{\Sigma_2^-}$ are not equal to zero. In the linear approximation, plastic deformations $\varepsilon^p_{nn} = \dot{\varepsilon}^p_{nn}\big|_{\Sigma_1^-}\Delta t$ and $\varepsilon^p_{n\tau} = \dot{\varepsilon}^p_{n\tau}\big|_{\Sigma_2^-}\Delta t$ develop in the material (behind the wave Σ) over the time Δt at a distance $\Delta n = c\Delta t$ behind the wave Σ before the arrival of the waves of plastic unloading.

The rate of plastic deformation $\dot{\varepsilon}^p_{ij}$ is determined through the stress σ_{ij} [10]

$$\dot{\varepsilon}^p_{ij} = \dot{\varepsilon}'^p_{ij} = \frac{\psi}{1+\eta\psi}\sigma'_{ij}\,, \tag{25}$$

where $\psi = \dfrac{1}{\eta}\left(\dfrac{\sqrt{J_2}}{k\sqrt{2}} - 1\right)$.

The second invariant of the deviator of the tensor of the rates of plastic deformation $\dot{I}_2 = \dot{\varepsilon}^p_{ij}\dot{\varepsilon}^p_{ij}$ is expressed in terms of the second invariant of the deviator of the stress tensor as follows:

$$\dot{I}_2 = \frac{2k^2}{\eta^2}\left(\frac{\sqrt{J_2}}{k\sqrt{2}} - 1\right). \tag{26}$$

As a measure of plastic deformation, we take a value $\Lambda_2 = \sqrt{\dot{I}_2}\Delta t\dfrac{\eta}{k\sqrt{2}}$, that can be represented in terms of the intensity J_2 of the deviator of the stress tensor as follows

$$\Lambda_2 = \frac{\sqrt{J_2}}{k\sqrt{2}} - 1. \tag{27}$$

Plastic deformation accumulated over time Δt on the surface of the cone, by the time of the termination of the impact process, is transported by waves Σ_1^p or Σ_2^p of plastic unloading, damped by geometric expansion of spherical waves and viscous friction [11]. On diffracted waves Σ_1 and Σ_2 we have

$$\Lambda_2 = \Lambda_2^0(\varphi)\,\frac{\delta}{\delta+ct}\,\frac{\alpha}{\mu}\exp\left(-\,\alpha t\right); \quad \varphi \in \left[\alpha; \frac{\pi}{2}\right], \tag{28}$$

where $\alpha = \dfrac{4\mu^2}{3\eta\rho c^2}$; $\rho c^2\big|_{\Sigma_1} = \lambda + 2\mu$; $\rho c\big|_{\Sigma_2} = \mu$.

On the waves generated by the impact of the lateral surface of the cone we have

$$\Lambda_2 = \Lambda_2^0(\alpha)\sqrt{\frac{z\sin\alpha}{z\sin\alpha+ct}}\,\frac{\alpha}{\mu}\exp\left(-\,\alpha t\right), \tag{29}$$

where z is counted from the vertex of the cone along its axis.

3.4. A note on the possibility of formation of thin zones of plastic deformation leading to the formation of cracks as discontinuities of displacements

In the direction $\varphi = \varphi_2^*$ behind the shear wave Σ_2 front, the maximum value of the residual deformation measure Λ_2 takes place and, consequently, the maximum value of the intensity J_2 of the second invariant of the deviator of the stress tensor and the maximum value of the jump of the tangential velocity $[v_\tau]_\Sigma$ takes place also, since it follows from (3)

$$J_2 = \left[\sigma'_{ij}\right]\left[\sigma'_{ij}\right] = \left(\frac{\mu}{c_2}\left[v_\tau\right]\right)^2. \tag{30}$$

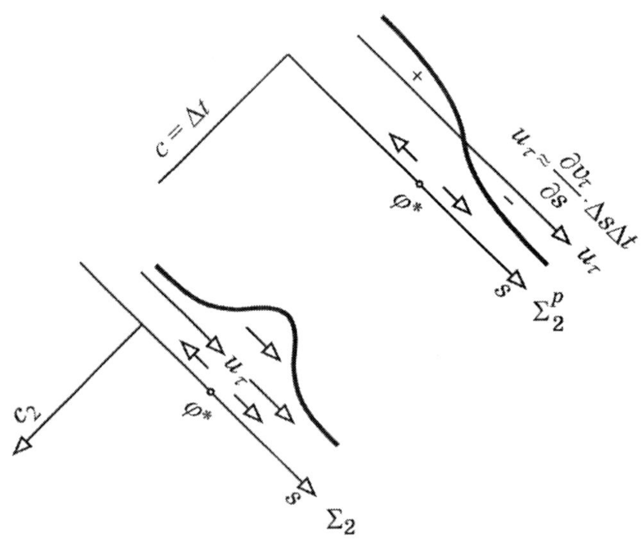

Figure 5. Schematic representation of the shear wave Σ_2 of the load and the following it with the time shift Δt of the shear wave Σ_2^p of unloading at a distance $c_2\Delta t$

During the time Δt in the neighborhood of the direction $\varphi = \varphi_2^*$ on the wave Σ_2^p of plastic unloading, the displacements of the opposite direction will occur (figure 5)

$$\Delta u_\tau = \Delta s \left.\frac{\partial v_\tau}{\partial s}\right|_{\overleftarrow{\varphi}_2^*} \Delta t - \Delta s \left.\frac{\partial v_\tau}{\partial s}\right|_{\overrightarrow{\varphi}_2^*} \Delta t, \tag{31}$$

which lead to disruption of the continuity of the material and development of the crack, the head part of which moves along with the shearing discharge wave Σ_2^p.

In the neighborhood of the direction φ^*, the velocity v_τ tangent to Σ_2 has a maximum, its gradients on the left and on the right from φ^* have different signs $\left.\frac{\partial v_\tau}{\partial s}\right|_{\overleftarrow{\varphi}} > 0$; $\left.\frac{\partial v_\tau}{\partial s}\right|_{\overrightarrow{\varphi}} < 0$ (figure 6), which in time Δt leads to tangential displacements u_τ of different signs to the left and to the right from φ^* on Σ^p

$$\overleftarrow{u}_\tau \approx \left.\frac{\partial u_\tau}{\partial s}\right|_{\overleftarrow{\varphi}^*} \Delta s \Delta t > 0; \quad \overrightarrow{u}_\tau \approx \left.\frac{\partial u_\tau}{\partial s}\right|_{\overrightarrow{\varphi}^*} \Delta s \Delta t < 0,$$

so that a disruption of displacements $[u_\tau]_{\Sigma^p}$ is expected after the unloading wave.

A crack that appeared at the time $t = \Delta t$ on the surface Σ_2^p will in time be a conical surface with an angle $2\left(\frac{\pi}{2} - \varphi_2^*\right)$. Its propagation along the generatrix of the cone will cease upon reaching the plasticity limit at the length $L = c_2 t_{limit}$

$$J_2 = 2k^2 \frac{\delta}{\delta + L} \exp\left(-\frac{\alpha L}{c_2}\right). \tag{32}$$

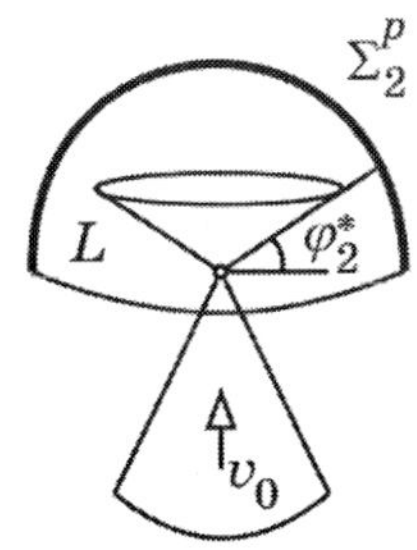

Figure 6. An image of a crack in the form of a part of a conical surface of length L along the generatrix. The head part of the crack is born on Σ_2^p and damps at the time moment t_{limit} after which its propagation stops

The above presented result on the propagation velocity of the head part of the crack agrees with the known experimental data [5, 12].

4. Conclusion

The occurrence of longitudinal and shear waves in an unstressed material is investigated depending on the properties of the surface of the cone: in the case of complete slippage, a shock generates one front Σ_1 of a longitudinal wave, in the case of different degrees of adhesion, a second (shear) wave Σ_2 appears that collapses in its motion from the longitudinal wave Σ_1.

It is shown that a static stress state σ_{ij}^0 increases the intensity of tangential stresses J_2 by Σ_1 and Σ_2 and for a static value $J_2^0 = J_2'^0 = \left(\frac{2}{3}(\sigma_1 - \sigma_3)\right)^2$ in any direction φ.

The distribution of residual plastic deformations remaining in the material after passing through the waves of plastic loading Σ_1 and Σ_2 for the case of a short-time impact by a cone is investigated.

The possibility of formation of thin zones of plastic deformation leading to the formation of cracks as discontinuities of displacements is estimated. It should be noted that the possibility of the development of a jump in the velocities in the discontinuity jump was expressed in [7].

The patterns of propagation of the front edge of cracks are local in nature and are determined by the initial perturbation and the stressed state in front of the edge.

Acknowledgments

The authors are sincerely grateful to N. G. Bilchenko for translating this article from the Russian.

References

[1] Petrova V, Schmauder S and Shashkin A 2016 Modeling of edge cracks interaction *Frattura ed integrità strutturale (Fracture and structural integrity)* **36** 8–26 DOI:10.3221/IGF-ESIS.36.02

[2] Rangaraj S and Kokini K 2003 Multiple surface cracking and its effect on interface cracks in functionally graded thermal barrier coatings under thermal shock *Trans. ASME J. Appl. Mech.* **70** 234–245 DOI: 10.1115/1.1533809

[3] Freund L B and Rice J R 1974 On the determination of elastodynamic crack tip stress fields *Int. J. of solids and struct.* **10** 411–417

[4] Rice J R 1968 *Mathematical Analysis in the Mechanics of Fracture Chapter 3 of Fracture: An Advanced Treatise (Vol 2, Mathematical Fundamentals)* ed H Liebowitz (New York: Academic Press) pp 191–311

[5] Anderson T L 2005 *Fracture Mechanics: Fundamentals and Applications, third ed.* (Boca Raton: CRC Press, Taylor & Francis Group) xviii+610 ISBN-13: 978-1-4200-5821-5

[6] Parton V Z and Borisovskiy V G 1988 *Dynamics of brittle fracture* (Moscow: Mashinostroenie) 240

[7] Kovalev V A and Radayev Yu N 2010 *Wave problems of field theory and thermomechanics* (Saratov: Publishing House of the Saratov University 328

[8] Thomas T Y 1961 *Plastic flow and fracture in solids* (New York-London: Academic Press) vi+267

[9] Babicheva L A, Bykovtsev G I and Verveiko N D 1973 Ray method of solving dynamic problems in elastic-viscoplastic media *Journal of Applied Mathematics and Mechanics* **37:1** 132–141 URL: https://doi.org/10.1016/0021-8928(73)90142-1

[10] Verveiko N D 1997 *Ray theory of elastoviscoplastic waves and hydrodynamic waves* (Voronezh: VSU Publishers) 204

[11] Verveiko N D, Krupenko S E and Naoman S A 2016 Mathematical modeling of dynamic wedging of elastoviscoplastic material *"Actual Problems of Applied Mathematics, Informatics and Mechanics" Proc. Int. Scientific and Technical Conf.* (Voronezh: Research Publishers) pp 314–317

[12] Parton V Z and Morozov E M 1974 *Mechanics of elastoplastic fracture* (Moscow: Nauka) 416

The application of the large particles method of numerical modeling of the process of carbonic nanostructures synthesis in plasma

G V Abramov[1] **and A N Gavrilov**[2]

[1]Voronezh State University, 1, University Sq., Voronezh, 394018, Russia
[2] Voronezh State University of Engineering Technologies, 19, Revolution Ave., Voronezh, 394036, Russia

E-mail: ganinvrn@yandex.ru

Abstract. The article deals with the numerical solution of the mathematical model of the particles motion and interaction in multicomponent plasma by the example of electric arc synthesis of carbon nanostructures. The high order of the particles and the number of their interactions requires a significant input of machine resources and time for calculations. Application of the large particles method makes it possible to reduce the amount of computation and the requirements for hardware resources without affecting the accuracy of numerical calculations. The use of technology of GPGPU parallel computing using the Nvidia CUDA technology allows organizing all General purpose computation on the basis of the graphical processor graphics card. The comparative analysis of different approaches to parallelization of computations to speed up calculations with the choice of the algorithm in which to calculate the accuracy of the solution shared memory is used. Numerical study of the influence of particles density in the macro particle on the motion parameters and the total number of particle collisions in the plasma for different modes of synthesis has been carried out. The rational range of the coherence coefficient of particle in the macro particle is computed.

1. Introduction

The study of synthesis processes characteristics of carbon nanostructures (CNS) which have unique mechanical and electrical properties is one of the modern science development promising directions. Having a set of unique physicochemical properties CNS (fullerenes, nanotubes, nanofibers) are considered a promising reinforcing material for polymer composites. Uniform introduction of CNS in the resin matrix only in the amount of 0.01 ÷ 5% by mass makes possible to have a promising composite materials with enhanced or improved set of properties [1]. Endowed with unique electrical, mechanical, barrier and other properties these nanostructural formated materials are of great interest for various industries. However their industrial growth is constrained by low efficiency and high cost of existing technologies of the filler synthesis. Insufficient knowledge of the mechanism of nucleation and growth of carbon clusters forming the bulk of the nanostructure prevents the increase of the productivity and the quality of the synthesized product and reduction of its cost.

Thus the solution of modeling the process of carbon with various communication types clusters formation problem is an urgent task in terms of development and improvement modern technologies of new materials obtaining. Knowledge of the mechanism of the CNS formation will enable researchers to purposefully create and vary the methods and conditions for obtaining various types of carbon

IOP Conf. Series: Journal of Physics: Conf. Series **973** (2018) 012022 doi:10.1088/1742-6596/973/1/012022

nanostructures and their derivatives which will significantly improve the efficiency and performance of existing synthesis technologies.

Mainly used in industry the technologies of CNS synthesis involve the application of various methods of thermal evaporation of atomic graphite by energy flux (laser, plasma, electron ion beam, etc.) or decomposition of carbon containing gases with subsequent deposition of the formed structures on the cooling surface. In all technologies there is a thermal destruction of the raw material structure with the subsequent formation and growth of polyatomic carbon clusters with different types of bonds which form the bulk of CNS.

The application of various forms of catalysts in the synthesis allows to increase significantly the output and to modify the characteristics of the finished product, but results in additional defects i.e. heterogeneity, contamination by the catalyst nanomaterial that requires further purification which significantly affects the yield, quality and cost of the finished product.

One of the most common methods for the CNS synthesis is the method of graphite thermal evaporation by plasma of arc discharge in the buffer gas environment. It allows to involve in the technological process large amounts of raw materials, to synthesize under different conditions the various types of CNS (single layer, multilayer nanotubes, nanofibers, fullerenes) and is characterized by the possibility of obtaining high-quality material.

The synthesis proceeds under the influence of low-temperature plasma (temperature of about 4500 ÷ 5500 K) in the environment of a buffer gas (usually He, Ar) followed by the deposition of material on the cooling surface. The process is characterized by a high concentration of energy, fast operation, a large number of different effects between the particle phase and structural transformations, by the influence of the large number of factors. To obtain the necessary carbon structures with the maximum output stable optimal technological parameters of the synthesis in real-time are required.

The study and search of optimum conditions of synthesis of the CNS by experimental methods are ineffective and difficult, therefore it is more productive to use methods of mathematical modeling.

Modern approaches to simulation of processes in plasma can be divided conventionally into classical, semiclassical and quantum-mechanical. To solve this class of problems in which the registration of the collective phenomena in plasma, as well as the complexity of computational calculation are characteristic is to use kinetic description which fits the task in the best way [2].

To study the kinetics of the movement and interaction of particles in a multicomponent plasma, taking into account the moving boundaries of the electrodes and condensation of supersaturated carbon vapor on the cooled cathode as well as the influence of process parameters and starting material on the formation process of the CNS a comprehensive mathematical model has been developed [3].

The Boltzmann equations (1) describing each species of particles in the plasma (electrons, carbon ions, buffer gas) and supplemented by the terms elastic and inelastic binary collisions between particles are the basis for the model describing the movement and interaction of particles in plasma [4].

$$\frac{\partial f_\alpha}{\partial t}+\vec{\vartheta}\frac{\partial f_\alpha}{\partial \vec{r}}-\frac{q_\alpha}{m_\alpha}(\vec{E}+\frac{1}{C}(\vec{\vartheta}\times\vec{B}))\frac{\partial f_\alpha}{\partial \vec{\vartheta}}=\sum_{\beta=e,c,h}\iint_V(f'_\alpha f'_\beta - f_\alpha f_\beta)|\vec{\vartheta}-\vec{\vartheta}'|d\sigma d\vec{\vartheta}', \alpha=e,h,c \quad (1)$$

where f_α, f'_α – the distribution function of the component of the plasma before and after the collision (α: e – electron, h – ion-buffer gas (He), c – ion carbon); $\vec{\vartheta}, \vec{\vartheta}'$ - the velocities of the particles before and after collision; $d\sigma = 4R_1R_2\cos\theta d\Omega$ – the differential effective scattering cross section of particles R_1 and R_2 in the solid angle $d\Omega$; θ – the angle between the velocity of the colliding particles and the line of movement; $\vec{r}$ – the coordinates of the particles; q_α, m_α – the charge and mass of particles.

To find the parameters of the electromagnetic field the equation system (1) is supplemented by the Maxwell's equations system (2) describing the self-consistent electric field.

$$\begin{cases} rot\,\vec{H} = \dfrac{4\pi\vec{j}}{C} + \dfrac{1}{C}\dfrac{\partial\vec{D}}{\partial t}, \\ rot\,\vec{E} = -\dfrac{1}{C}\dfrac{\partial\vec{B}}{\partial t}, \\ div\,\vec{B} = 0, \\ div\,\vec{D} = 4\pi\rho, \\ \rho = \dfrac{q_e}{m_e}\int\limits_V (f_c + f_h - f_e)\mathrm{d}\vec{\vartheta}, \\ \vec{j} = \dfrac{q_e}{m_e}\int\limits_V (f_c + f_h - f_e)\vec{\vartheta}\mathrm{d}\vec{\vartheta}, \end{cases} \tag{2}$$

where $\vec{E},\vec{H}$ – the strength of the electric and magnetic fields; $\vec{D},\vec{B}$ – electric and magnetic induction; ρ – charge density; $\vec{j}$ – the density of current; C – the speed of light.

The use of distribution functions of various plasma components in the process model allows to predict on the basis of the probabilistic approach the behavior of the carbon particles in the synthesis process through the consideration of the collective phenomena of plasma: plasma oscillations, fluctuations of various characteristics, concentration and particle flows.

The presence in the right side vector of the equation (1) collision integral causes considerable difficulties in the explicit solution of this problem. Therefore using the representation of the collision integral as the Fokker-Planck equation it is possible to move from vector algebra in each point of the space to tensor analysis and to reduce the model to a dimensionless form in three-dimensional coordinate system [5].

The solution of the proposed model is possible only using a numerical method by splitting into three subsidiary problems to be solved consecutively: the formation of the initial position of the particle, the calculation of their displacement and the calculation of collisions of charged particles in plasma.

The availability of a large variety of particles present simultaneously in the plasma of the interelectrode space, make the calculations very resources demanding. Therefore to obtain physically reasonable results of numerical simulation it is necessary to use the method of large particles (MCC) [6]. This approach allows to reduce the amount of computation and accordingly the requirements for computer resources by reducing the number of similar particles in the calculation by grouping them to a reasonable level in larger (macro particles) having the same ratio of charge to mass as the original ones and the averaged speed and direction of movement. However, when combining the particles in the macro particles, it is necessary to determine the rational density of their placement, which directly determines the size of a large particles and affects the calculation accuracy.

2. Statement of the problem

The aim of this paper is to study the influence of parameters of the large particles method on the results of numerical simulation of processes in the formation of the CNS, by the method of thermal sublimation with graphite plasma and to develop efficient numerical methods for processing large volumes of data necessary for parameters calculation of these processes. It is necessary to calculate parameters such as the cell geometry of the computational domain, the number of particles forming the large particle of a certain type, the radius of macro particles and to develop the algorithms for calculation of coordinates of the centers of large particles in cells, their charge, mass, average velocity and direction of movement. The charge weighing and the current at the nodes of the computational grid should be done which are required to determine the new coordinates of the macro particles in each time step.

To perform calculations by the model in question it is necessary to develop an efficient algorithm for the numerical solution which allows to perform all calculations by a personal computer based on distributed parallel processing of large data amounts.

3. Solution method

To construct the algorithm of numerical solution of the model all the phase space computational domain is broken down into individual cubic cells the sides of which are parallel to the axes of the Cartesian coordinate system. In each cell instead of actual number of particles of each type in accordance with the distribution function by a macro particle with the total charge and mass of all particles of this sort contained in one cell is generated. To do this, the coordinates of the nodes of the computational grid and the coordinates of all particles of each species falling within the pitch range of each cell are calculated [7].

The radius and the charge of a large particle and consequently the effect of the electromagnetic field under the action of which charged particles move depends on the size of the macro particles, i.e. on the number of micro particles combined in the macro particle. The main parameter influencing the properties of large particles is the density of particles in the macro particle.

Assuming that the initial particle shape in the model is spherical the radius of the formed large particles $R_{\alpha_{lp}}$ will be calculated by the formula:

$$R_{\alpha_{lp}} = (\chi \cdot N_\alpha)^{\frac{1}{3}} \cdot R_\alpha{}^3, \tag{3}$$

where R_α – the radius of particles of class α; χ – the factor bases on the assigned density of the particles in the macro particle;$\cdot N_\alpha$ – the number of particles in a macro particle.

The radius of the macro particles must be large enough to eliminate the interaction of particles at short range distances and to provide an approximation of self-consistent field. The coordinates of the centers of large particles in *i* cells are calculated by finding the center of their masses.

The general algorithm for calculating the movement and interaction of particles in a multicomponent plasma is given in figure 1. It includes the consistent implementation of the 11 key tasks.

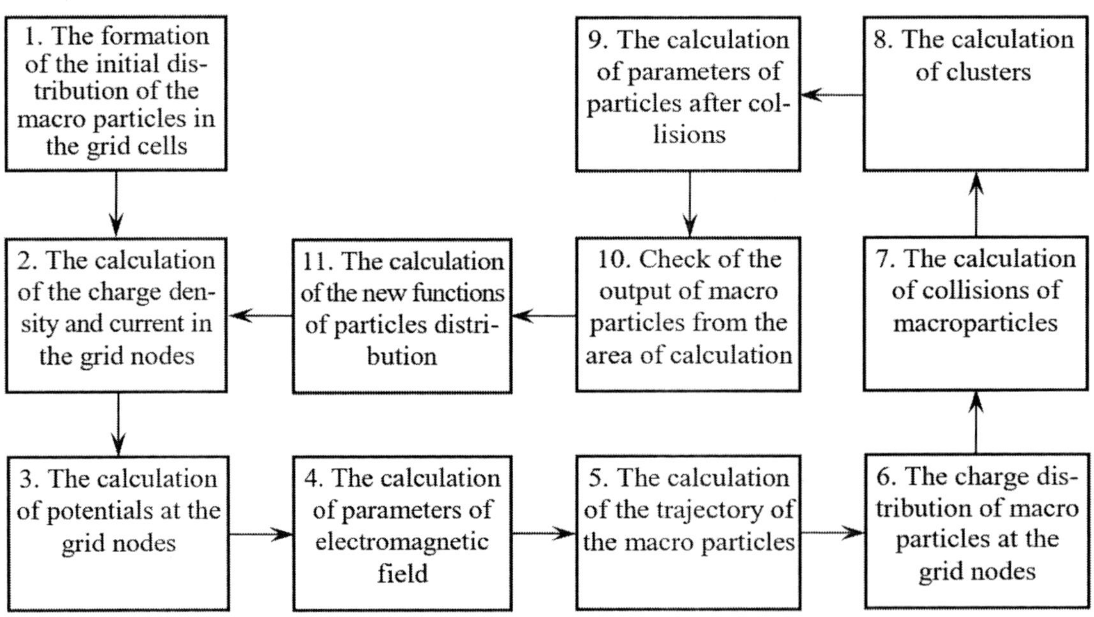

Figure 1. Calculation scheme of the method of large particles

IOP Conf. Series: Journal of Physics: Conf. Series **973** (2018) 012022 doi:10.1088/1742-6596/973/1/012022

After calculating the trajectory of a large particle under the action of electromagnetic field and its new position in phase space at later point of time, based on the current distribution functions is the dispersion of the particle charge over the nodes of the phase space is made and the process is repeated. To do this, we use the method of weighing the cloud in a cell for three-dimensional space [8].

The distribution of charge and current at the nodes occurs using the method of "cloud in a cell". The particle contributes only to the nodes that are vertices of the cell in which it has been inserted. Charge and therefore the current are split between them according to a linear law. Figure 2 shows the weighting for the three-dimensional case. The weighting is proportional to cell volume V_i.

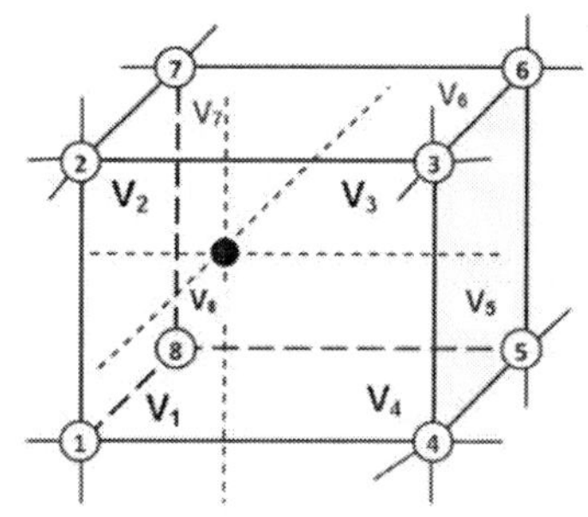

Figure 2. Weighing on a three-dimensional grid

To find the values of the potential and the forces that act on the particles in the cells of the computational grid the Poisson equation describing the electric field is solved which enables us to define subsequent coordinates of the macro particles from the equations of their motion.

Under the influence the electromagnetic field due to the collective interactions taking into account elastic and inelastic collisions with other particles each macro particle changes its speed and trajectory. The search of their collision was to verify the fact of the pairwise intersections of the particle trajectories in space. A special algorithm for the search of collisions consisting of two stages was developed. In the first phase, the information about the starting and ending position of particles in the grid cells is used as input data. Each calculated flow based on the considered data determines the number of cells where the particle had been, where the particle moved and generates a cell list in which an interaction could have taken place. In the next step according to the received list of cells to process each pair of particles to determine the coordinates of intersection of their trajectories is processed.

To develop effective methods and algorithms of calculation for the resource consuming tasks a number of additional studies were made. The main characteristic of the algorithm for the calculation that has a direct impact on the accuracy, CPU time and computational complexity is the dimension of grid of the modeled area. To obtain the necessary precision of calculations the grid of 100x100x100 dimension was specified. A further increase in dimensionality of the computational grid has practically no effect on the final results of the calculation, but significantly increases the time and resource-consuming computing. To improve computational efficiency of numerical solution of the model under consideration the analysis of the CPU time required to solve each problem indicated in figure 1 was made. The results of the time costs are shown in figure 3.

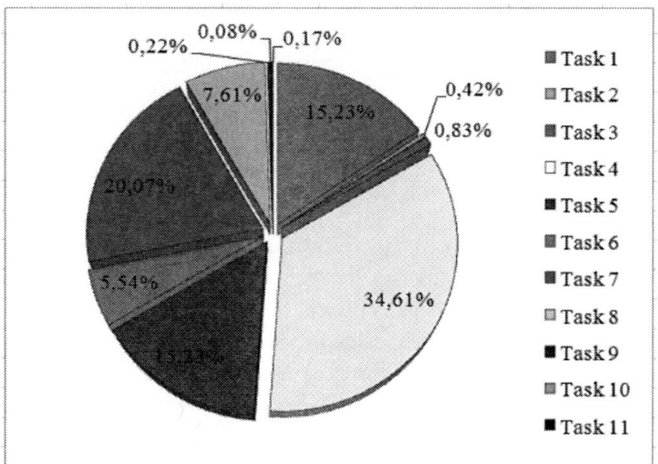

Figure 3. The time spent on the solution of the basic task model

The analysis shows that the main CPU time (85.14%) in the process of numerical calculation by the model is 1, 4, 5 and 7. Moreover the longest is problem 4 – calculation of electromagnetic fields parameters by the establishment method. To reduce the overall continuous CPU time required for the calculation by the model, the first task was carried out separately by pre-generating the array of initial parameters of the particles taking into account the boundary and initial conditions of the simulation. The reduction in CPU time for the solution of other tasks was achieved through the organization of parallel calculations. As a paralleling technology GPGPU (General-purpose graphics processing units) technology was used. The technology of parallel programming allows a computer to achieve a high enough level of parallelism, while avoiding the considerable time loss in the data transfer between compute nodes and the synchronization of the computations. The GPGPU technology allows applying high-performance PC graphics card for computing which greatly increases the efficiency of processing large volumes of data without using supercomputers or large computing clusters. All calculations were performed on hardware-software complex CUDA (Compute Unified Device Architecture) by technology Nvidia CUDA which makes possible programming of the GPU [9].

A characteristic feature of the CUDA architecture is block-grid organization, implying that all threads are grouped in blocks and the hardware independently allocates resources of the computing device between them. All threads executing a particular area are grouped together in blocks, and the blocks in the grid. To identify the threads a two-dimensional indexes were used.

Using the GPGPU technology one part of the computing threads run in parallel at instant will finish the process execution not at the same time. Consequently we solved the problem of synchronization of parallel sections calculations. For each instant when the calculations of all variables were made by graphical accelerator the control was transmitted to the CPU where the results were processed, and only after that the transition passed to the next iteration of the algorithm.

The efficiency of parallel computing based on Nvidia CUDA technology besides the CPU cores performance depends on startup configuration. Therefore with the goal of minimizing the CPU time different versions of the organization of the parallelization algorithms were considered. Table 1 shows the comparative analysis of the application of various versions of parallel calculations organization with the sequential execution of iterations (sequential algorithm). All calculations have been performed for finding the parameters of the electric field by the establishment method (task 4) with the use of computational grids of $25{\times}25{\times}25$ and $50{\times}50{\times}50$ dimensions on a personal computer equipped with a Intel® Core™2 QuadQ 8200 2.3 GHz processor and the Nvidia GeForce GTX 660 Ti graphics card.

Table 1. Comparison of the parallelization algorithms

Version of the algorithm	Calculation time, sec.	
	Grid 25×25×25	Grid 50×50×50
Consequent	5594.49	55927.15
Algorithm 1	420.48	1956.72
Algorithm 2	150.51	1140.28
Algorithm 3	26.08	158.31

In algorithm 1 the error calculation only in serial mode was used. Solution of initializing the source data representing the grid in the previous and current time step is executed in single-threaded mode, and the calculation of the potential at the grid points and the accuracy of the solution in parallel mode. In algorithm 2, each block of parallel threads has been synchronized and calculates the accuracy of their nodes and obtaining the total value of the accuracy is performed in single-threaded mode. In algorithm 3 shared memory was used to calculate the precision of the solution. This implementation of the algorithm appeared to be the most time-efficient calculation and was later used in the calculations of the model. A distinctive feature of the use of shared memory is that it is addressed the in same way for all tasks within a single block, so it can be used to exchange data between the threads of that block. This memory has a high access speed which helps to maximize the number of operations performed in parallel mode, i.e. before the synchronization process, and to minimize the number of cycles within the thread, thereby to reduce the computational complexity of thread.

4. Results and discussion

The adequacy of the constructed model is confirmed by a number of experimental studies of various CNS production by the device of electric arc synthesis with AISU [10].

The research by the model to determine the effect of the coherence coefficient χ on the results of numerical calculation was performed for the two main modes of synthesis: obtaining the cathode deposit with the maximum output of multi-walled nanotubes ("Nanotubes", the power of the arc current is 150 A) and the mode with maximum input range of fullerenes $C_{60} \div C_{70}$ deposited on the chamber walls ("Fullerenes", current 350 A). The rest calculation parameters: interelectrode distance of 1 mm, the diameter of graphite electrodes of 12 mm, the voltage between the electrodes of 25 V, the pressure of the buffer gas (He) in the cell synthesis chamber of 400 Torr.

In figures 4a, 4b the influence of coefficient χ on the average speed change of carbon macro particles movement in plasma of the interelectrode space without considering collisions.

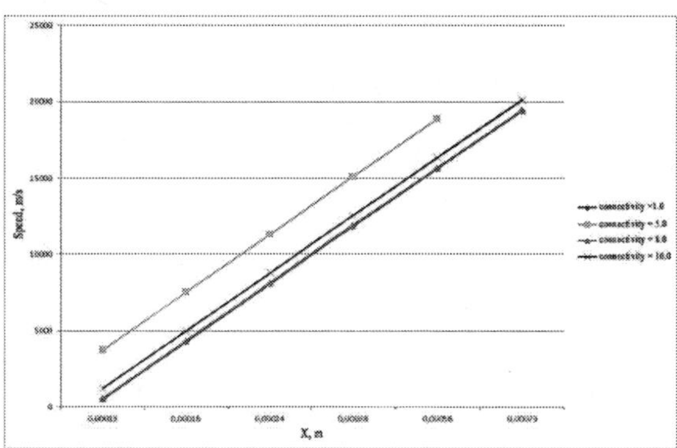

(a)

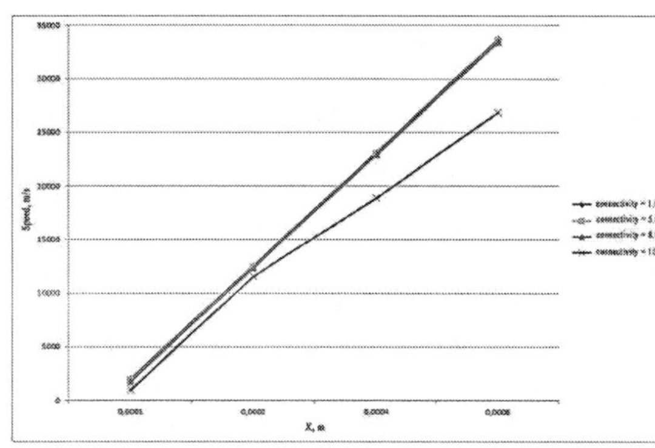

(b)

Figure 4. The change of the average velocity of the macro particles in the plasma for different coefficients of cohesion: (a) - mode «Nanotubes», (b) - mode «Buckyballs»

The Analysis of calculation results shows that the size of the macro particles in different modes of synthesis for $\chi = 1 \div 10$ affect its acceleration by electromagnetic field in the plasma.

The influence of the cohesion coefficient on the total number of collisions of particles of carbon in the plasma for the considered modes of synthesis is shown in figures 5a, 5b. The calculation time is 360 ns, with a time step of 10 ns.

The number of interactions of carbon particles depends on the size of macro particles. When increasing the cohesion coefficient the number of collisions increases respectively.

In different modes of electric arc synthesis of CNS the interaction of carbon particles is different. This is because the speed of carbon ions and their interaction energy depends on the magnitude of the current in the discharge. With the increase in the current of 150 A to 350 A the plasma temperature, magnetic field intensity increases, the average velocity of the particles without regard to collisions increases as well.

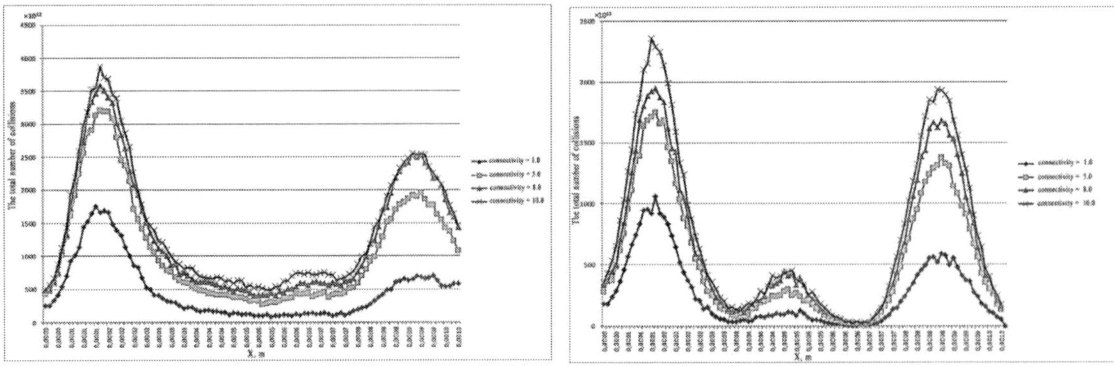

(a) (b)

Figure 5. The change in the total number of collisions of carbon particles in the plasma along the length of the inter-electrode distance for various cohesion coefficients: (a) -mode «Nanotubes», (b) - mode «Buckyballs»

The presence of peaks in collisions in figures 5a, 5b can be explained by the fact that in the initial moment of time the particles are generated in the model at random with regard to the profile of the evaporation of the anode on the basis of the Maxwell distribution. At the initial stage there is the largest number of particles with a relatively low speed. A large number of particles gives a large number of collisions, some of which result in the formation of linkages and agglomeration of particles. The number of particles is reduced, decreasing their concentration, which leads to a decrease in the total number of collisions and collisions with formation of bonds. Further, the larger particles are accelerated by electromagnetic field that leads to an increase in the total number of collisions in the cathode region, and, consequently, the number of generated links.

The presence of intermediate extrema in the «Buckyballs» mode is due to a more rapid increase in the speed of the particles, which formerly leads to an increase in collisions, then the number of particles decreases and the number of collisions reduces. This results in intermediate maximum point on the plot.

The increase of mass of deposit sediment on the cathode containing carbon tubes, in time for different of cohesion in comparison with experimental values is shown in figure 6.

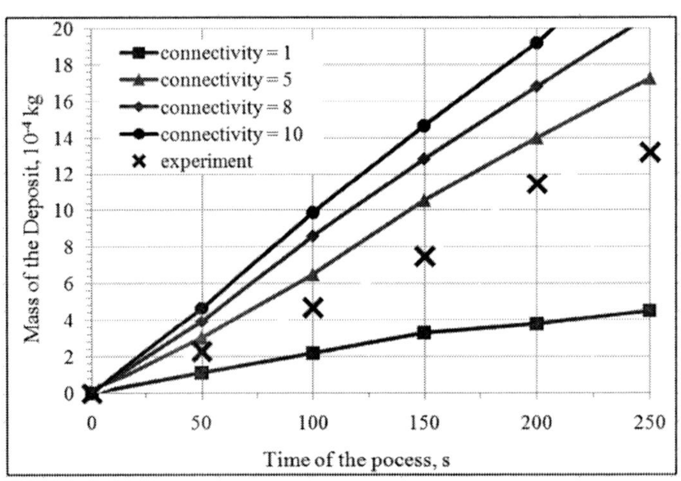

Figure 6. The change of mass deposit sediment on the cathode in time (mode «Nanotubes»)

The analysis of the numerical calculation results presented in figure 6 considering the received data for the other modes of synthesis has led to the conclusion that to correspond the proposed model to the physical process of CNS coherence coefficient synthesis of particles in the large particle method for different conditions of synthesis must be in the range $\chi = 3 \div 5$.

5. Conclusion

In this article we propose numerical solution of complex resource-consuming tasks of modeling of CNS synthesis based on the method of large particles which enables one to reduce the volume and the total CPU time computation with the given accuracy. This solutions based on the analysis of possible algorithms of calculation and the use of technology of parallel computing and GPGPU with application Nvidia CUDA technology are offered. A method which allows to weigh a charge and current at the nodes of the computational grid required to calculate the new coordinates of the macro particles in each time step is also offered. Dependences for definition of radius, coordinates of the centers of masses, charges, velocities and directions of motion of large particles of a certain type in the cells are given. Based on numerical studies by the model the range of the cohesion coefficient between particles ($\chi = 3 \div 5$) is made, which allows to describe adequately the real physical process for the synthesis of CNS.

References

[1] Yershova O V, Ivanovskaya S K, Chuprova L V and Bagaeva A N 2015 *International journal of applied and fundamental research* **4** 14
[2] Tsvetkov I V 2007 *The application of numerical methods for simulation of processes in plasma* (Moscow: Moscow engineering physics Institute) p 84
[3] Abramov G V, Gavrilov A N and Tatarkin E S 2012 *Modeling of kinetics of the ions in the electric arc synthesis of carbon nanostructures* (Saratov: Saratov state technical University) **1** 70
[4] Cercignani K 1978 *Theory and applications of the Boltzmann equation* (Moscow: Mir) p 495
[5] Abramov G V and Gavrilov A N 2012 *Mathematical modelling of the motion of interacting particles based on the distribution functions in the plasma of electric arc synthesis of CNS* (Voronezh: Voronezh state University of engineering technologies) **2** 71
[6] Dekaluk A S and Kuratov S E 2017 *Mathematical modeling* **29** 33
[7] Abramov G V, Gavrilov A N and Tolstova I S 2014 *The formation of the initial distribution of plasma components on the phase plane in the method of large particles in the electric arc synthesis of CNS* (Voronezh: Voronezh state University of engineering technologies) **3** 67
[8] Decyk V K and Singh T V 2014 *Computer Physics Communications* **185** 708
[9] Cook S 2013 *CUDA programming: A Developer's Guide to Parallel Computing with GPUs* (San Francisco: Morgan Kaufmann) p 576
[10] Abramov G V and Gavrilov A N 2016 *Automation. Modern technology* **3** 10

IOP Conf. Series: Journal of Physics: Conf. Series **973** (2018) 012023 doi:10.1088/1742-6596/973/1/012023

On solving the equation of the firm size distribution

I N Shchepina[1], I P Polovinkin[1], M V Polovinkina[2] and S A Rabeeakh[1]

[1] Voronezh State University, 1 Universitetskaya pl., Voronezh, Russia
[2] Voronezh State University of Engineering Technologies, Revolution Av., 19, Voronezh, Russia

E-mail: polovinkin@yandex.ru

The work is carried out with the support of the Russian Foundation for Basic Research No. 16-06-00535

Abstract. We build an explicit solution to the equation of the firm size distribution in the case of quadratic firm size dependence on the speed of its changes.

1. Introduction

The modern economic science includes research in area of business demography dynamics of new small innovative enterprises in a region, in a particular market, or in an industry. This part of research can use a population-based approach, which considers a set of homogeneous objects, whose dynamics obeys certain General regularities. We are interested in estimates of business demography dynamics with respect to the change in the dimensions of small innovative enterprises. The size distribution of economic entities allows to explore the structure of employment and to use this structure in the institutional sphere, in particular, to develop the system of state support of innovation activities. The firm size can be interpreted as the number of employees, revenue, profit, total assets. There are many papers devoted to firm size distribution. One of the first works on the firm size distribution dynamics is the paper by R. Gibrat [1]. It was established in that paper that if increments of the firm size are identically distributed independent random variables then the firm size distribution is coming to obey the lognormal distribution law. That fact is called the Gibrat law or the law of proportionate growth. The law contains three elements. Those are

1) growth rate of each company during certain period does not depend on its size;
2) growth rate depends on individual firm's characteristics;
3) growth rates of the firm in two successive periods are independent variables.

Gibrat's research resulted in a number of works devoted to applying of the law of proportional growth for various markets, industries and business sectors.

D. Evans, B. Hall, T. Dunne, M. J. Roberts, L. Samuelson (see the papers [2]–[5]) analyzed the conditional density of probability distribution $p(g|x)$ of growth rates g for firms with a characteristics vector. The density of the surviving firms' growth rate $h(g|x)$ and the density of the measured growth rates $f(g|x)$ were considered separately. D. Evans and B. Hall applied a technique of simultaneous assessing growth regression and mortality regression using the maximum likelihood method. T. Dunne, M. J. Roberts and L. Samuelson proposed a method of grouping of enterprises by intervals in accordance with their size and age. Important statistical regularities were identified in those papers. Those are

1) the probability of survival increases with the company size increasing;
2) the growth rate of surviving firms decreases with the company size increasing;
3) for any given firm size, the growth rate becomes smaller with increasing the size, but the probability of survival becomes greater.

Papers on the survival of enterprises deal with the firm survival probability to be considered as a function. This function depends on enterprise characteristics, industry characteristics, characteristics of the environment functioning and ways of organizing the production process. D. Evans studied the relationships among firm growth, firm size, and firm age for manufacturing firms. Firm growth is found to decrease with firm age and firm size.

D. Audretsch [6] found out that the firm survival probability depends on the technological regime and the inter-industry characteristics such as economies of scale and capital capacity. The results show that the minimum efficient level of output increasing in this sector decreases the probability of new-firm survival and the ability to achieve average branch size.

K. V. Arshakuni [7, 8] fulfilled a multifactor research on genesis and development of small innovative enterprises. The author estimated two joint dynamics models: a model of employment dynamics, and a lognormal survival model. These models were estimated using contemporary simulation-based techniques. The author infers that both financial and human capitals are the main factors that condition firm post-entry performance. The results indicate that the public subsidies and tax exemptions have a significant impact not only on the firm dynamics but also on the initial capital and on the bank loan provision.

Pin'koveckaya [9] analyzed the size distribution of small enterprises. It is proved that the number of employees is the most constructive characteristics among the various characteristics of small firms used in practice because it reflects the specificity of small enterprises and how they differ from other types of firms. The author considered mathematical models of firm size distribution for small enterprises. These models were tested to estimate small enterprises' characteristics in the Russian Federation and in particular Russian regions in 2007. The results indicate that the size distribution of small enterprises obeys a normal distribution with the density $f(x)=\left(2.3\sqrt{2\pi}\right)^{-1}\exp\left((x-8.66)^2/10.54\right)$.

The authors of the paper [10] studied the effect of the firm size distribution on the relationship between employment and output. They constructed a theoretical model, which predicts that changes in demand for industry output have larger effects on employment in industries characterised by a distribution that is more skewed towards smaller firms. Industry-specific shape parameters of the firm size distributions were estimated using firm-level data from Germany, Sweden and the UK, and used to augment a relationship between industry-level employment and output. Their empirical results align with the predictions of the theory and confirm that the size distribution of firms is an important determinant of the relationship between changes in output and employment. We would like to draw attention to an interesting publications review on the topic in that paper.

The authors of the paper [11] studied empirically and analytically growth and fluctuation of firm size distribution. An empirical analysis was carried out on a US data set on firm size, with emphasis on one-time distribution as well as growth-rate probability distribution. Both Pareto's law and Gibrat's law are often used to study firm size distribution. Their theoretical relationship was discussed, and it was shown how they are complementable with a bimodal distribution of firm size. They introduce economic mechanisms that suggest a bimodal distribution of firm size in the long run in the context of a dynamic decision problem of the firm. The authors presented different types of tests on US data on firm size which indicate a bimodal distribution of firm size.

A. Kantarbayeva and A. Mustafin [12] formulated a mathematical model of population development of firms. This model is based on kinetic equation of transfer for the firm size distribution function. The authors analyzed the stationary version of the model. The stationary distributions reveal a prevalent statistical weight of small firms.

In this article we investigate theoretical and algorithmic aspects of methods of mathematical modeling to assess the business demography dynamics of small innovative enterprises in a region. The study is based on the paper [12], where the authors considered linear and exponential growth of the firm size. We consider an intermediate case.

2. The equation of the firm size distribution and a scheme of its integration

Consider a small innovative enterprises population. The population life is carried out in a region within a particular industrial sector. The existence of the population is accompanied by the processes of creation, growth and closure of its objects [13]. The authors of the paper [12] established a linear partial differential equation of the first order as the balance ratio (the differential conservation law) modeling the dynamics of the firm size distribution. The equation has the following form

$$\frac{\partial n}{\partial t}+r\frac{\partial n}{\partial x}+pn=Q. \tag{1}$$

Here x is a firm size i.e. the number of employees, $n=n(x,t)\mathrm{d}x$ is the number of objects whose size lies in the span $[x, x+\mathrm{d}x]$, $r=r(x,t)=\frac{\mathrm{d}x}{\mathrm{d}t}$ is the speed of the size objects changing, $Q=Q(x,t)$ is the number of new enterprises size x created at the moment of time t, $D=D(x,t)$ is the relative frequency of objects closing, $p=\frac{\partial r}{\partial x}+D$.

For the function $n(x,t)$ the authors impose additional requirements

$$n(x,t)_{|x=0}=0, \tag{2}$$

$$n(x,t)_{|x\to+\infty}=0. \tag{3}$$

Requirements (2) and (3) were justified by the authors of [12] as formalization of the fact that there are no firms with zero or infinite size. We don't discuss this thesis. We just consider the question about the integration of equation (1) in a certain case.

We suppose that $r=r(x)$, $D=D(x)$ doesn't depend on x. Let's put into our consideration a new unknown function $u(x,t)$, suggesting that

$$n(x,t)=\mathrm{e}^{-\Phi(x)}u(x,t). \tag{4}$$

Here $\Phi(x)$ must be taken in a special form to make the coefficient of the function $u(x,t)$ in a new equation vanishing.

Substituting (4) in equation (1) we obtain

$$\mathrm{e}^{-\Phi}\left(\frac{\partial u}{\partial t}+r\frac{\partial u}{\partial x}+(p-\phi')u)\right)=Q. \tag{5}$$

It is clear from equation (5) that, for the achievement of our aim it will be enough to define

$$\Phi'=p=r'+D. \tag{6}$$

We can guarantee equality (6), putting

$$\Phi=\int_{x_0}^{x}r'(s)\mathrm{d}s+\int_{x_0}^{x}D(s)\mathrm{d}s+const \tag{7}$$

or

IOP Conf. Series: Journal of Physics: Conf. Series **973** (2018) 012023 doi:10.1088/1742-6596/973/1/012023

$$\Phi(x) = r(x) - r(x_0) + \int_{x_0}^{x} D(s)\mathrm{d}s + const. \qquad (7')$$

If $D = const$, we get

$$\Phi(x) = r(x) - r(x_0) + D(x - x_0) + const. \qquad (8)$$

If we make replacement (4) with the function Φ, defined by formula (7), we will get the equation (1) in the following form

$$\frac{\partial u}{\partial t} + r\frac{\partial u}{\partial x} = \mathrm{e}^{\Phi}Q. \qquad (9)$$

Now we make a replacement of the variable x by implementing a new variable $\xi = \xi(x)$, where $\xi(x)$ is continuously differentiable function. In this replacement we will consider ξ as a new variable and x as an old variable. We express the partial derivative $\partial u / \partial x$ with respect to the old variable x using the partial derivative with respect to the new variable ξ. As a result we get

$$\frac{\partial u}{\partial x} = \frac{\partial u}{\partial \xi} \cdot \frac{\mathrm{d}\xi}{\mathrm{d}x}. \qquad (10)$$

Substituting (10) in (9), we obtain

$$\frac{\partial u}{\partial t} + r\frac{\mathrm{d}\xi}{\mathrm{d}x} \cdot \frac{\partial u}{\partial \xi} = \mathrm{e}^{\Phi}Q \qquad (11)$$

Assuming

$$\frac{\mathrm{d}\xi}{\mathrm{d}x} = \frac{1}{r} \qquad (12)$$

we get from (11)

$$\frac{\partial u}{\partial t} + \frac{\partial u}{\partial \xi} = \mathrm{e}^{\Phi}Q. \qquad (13)$$

Here

$$\phi = r(x(\xi)) - r(x_0) + \int_{x_0}^{x(\xi)} D(s)\mathrm{d}s. \qquad (14)$$

And now we make another new replacement. Let new variables τ, μ be expressed in terms of the old variables t and ξ by formulas

$$\mu = \xi - t, \qquad (15)$$

$$\tau = t. \qquad (16)$$

We can see that the reversed replacement is defined by formulas

$$\tau = t, \qquad (17)$$

$$\xi = \mu + \tau. \qquad (18)$$

By the rules of computing the partial derivatives of complex functions, we have:

IOP Conf. Series: Journal of Physics: Conf. Series **973** (2018) 012023 doi:10.1088/1742-6596/973/1/012023

$$\frac{\partial u}{\partial t}=\frac{\partial u}{\partial \tau}\cdot\frac{\partial \tau}{\partial t}+\frac{\partial u}{\partial \mu}\cdot\frac{\partial \mu}{\partial t}=\frac{\partial u}{\partial \tau}-\frac{\partial u}{\partial \mu}, \tag{19}$$

$$\frac{\partial u}{\partial \xi}=\frac{\partial u}{\partial \tau}\cdot\frac{\partial \tau}{\partial \xi}+\frac{\partial u}{\partial \mu}\cdot\frac{\partial \mu}{\partial \xi}=\frac{\partial u}{\partial \mu}. \tag{20}$$

Substituting (19) and (20) in (13), we get the equation

$$\frac{\partial u}{\partial \tau}=\mathrm{e}^{\Phi}Q. \tag{21}$$

Here

$$\Phi=\Phi(x(\mu+\tau)),\ Q=Q(x(\mu+\tau),\tau).$$

The variable μ is included in equation (21) as a parameter because the partial derivatives with respect to μ are not included in the equation. Therefore, equation (21) is an ordinary differential equation with respect to τ for the function $u(x(\mu+\tau),\tau)$. This differential equation can be solved by integrating both sides.

3. The solution to the equation of the firm size distribution

The authors of the paper [12] considered two kinds of speed *r(x)* for the stationary equation (1) ($\partial n/\partial t=0$). Those are

$$\text{Linear growth, i.e. } r(x)=\alpha. \tag{22}$$

$$\text{Exponential growth, i.e. } r(x)=\beta x. \tag{23}$$

In each of these situations the authors constructed an explicit solution for equation (1) under stationarity assumption.

Cases (22) and (23) can be considered as boundary situations for values of parameter $\gamma\in[0,1]$ when the speed *r(x)* is a power function

$$r(x)=\beta x^{\gamma},\quad \gamma\in[0,1]. \tag{24}$$

We consider a case when $D=const$,

$$r(x)=\beta\sqrt{x} \tag{25}$$

i.e. $\gamma=1/2$ in (24). Equality (25) can be written in the form

$$x=\beta^{-2}(\mathrm{d}x/\mathrm{d}t)^{2}. \tag{25'}$$

Thus we assume the quadratic firm size dependence on the speed of its changes. The function $\Phi(x)$ which is defined by (7) or (14) could be taken in the form

$$\Phi(x)=\beta\sqrt{x}+Dx. \tag{26}$$

Equality (12) will take the form

$$\frac{\mathrm{d}\xi}{\mathrm{d}x}=\frac{1}{\beta\sqrt{x}}.$$

Then

IOP Conf. Series: Journal of Physics: Conf. Series **973** (2018) 012023 doi:10.1088/1742-6596/973/1/012023

$$\xi = \frac{2\sqrt{x}}{\beta}, \quad x = \left(\frac{\beta\xi}{2}\right)^2.$$

Let

$$0 < D < \lambda^{-1}.$$

Then $(1-\lambda D)\lambda^{-1} > 0$. Let's denote

$$Q(x) = Q_0\lambda^{-1} \cdot \exp(-x/\lambda) \tag{27}$$

(see [12]),

$$\alpha = \lambda^{1/2}(1-\lambda D)^{-1/2}, \quad u_0 = 2\alpha Q_0(\beta\lambda)^{-1}\exp(\alpha^2\beta^2/4).$$

Then we can find the general solution of equation (21) calculating an integral

$$u = \frac{Q_0}{\lambda}\int_{\alpha^2-\mu}^{\tau} \exp\left(-\beta^2/(4\alpha^2)(\mu+s)^2 + \beta^2/2(\mu+s)\right)ds + G(-\mu).$$

We transform the power in (31) as the following

$$-\beta^2/(4\alpha^2)\left((\mu+s)^2 - 2\alpha^2(\mu+s) + \alpha^4 - \alpha^4\right) =$$
$$= -\beta^2/(4\alpha^2)(\mu+s-\alpha^2)^2 + \alpha^2\beta^2/4.$$

Let's make replacement of the variable in integral (31)

$$z = \beta(2\alpha)^{-1}\left(s+\mu-\alpha^2\right).$$

Let $F(\eta)$ be defined by the formula

$$F(\eta) = \int_0^{\eta} \exp(-z^2)\,dz.$$

Then

$$u = u_0F(\beta(2\alpha)^{-1}(\tau+\mu-\alpha^2)) + G(-\mu) =$$
$$= u_0F(\beta(2\alpha)^{-1}(\xi-\alpha^2)) + G(t-\xi).$$

Condition (2) leads to

$$0 = u\big|_{\xi=0} = u_0F(-\alpha\beta/2) + G(t) = -u_0F(\alpha\beta/2) + G(t).$$

Hence

$$G(t) = u_0F(\alpha\beta/2)$$

and we have

$$u = u_0\left(F(\beta(2\alpha)^{-1}(\tau+\mu-\alpha^2) + F(\alpha\beta/2)\right) =$$
$$= u_0\left(F(\beta(2\alpha)^{-1}(\xi-\alpha^2) + F(\alpha\beta/2)\right).$$

Finally from (4), (15) – (18), (25) and (26) we obtain

$$n(x) = u_0 \exp\left(-\beta\sqrt{x} - Dx\right)\left(F\left(\frac{\beta}{2}\sqrt{\frac{1-\lambda D}{\lambda}}\left(\frac{2\sqrt{x}}{\beta} - \frac{\lambda}{1-\lambda D}\right)\right) + F\left(\frac{\beta}{2}\sqrt{\frac{\lambda}{1-\lambda D}}\right)\right). \qquad (28)$$

Note that for any constants $A > 0, B$

$$\lim_{\xi\to+\infty} F(A\xi + B) = \lim_{\eta\to+\infty} F(\eta) = \int_0^{+\infty} \exp(-z^2)\,\mathrm{d}z = \sqrt{\pi}/2 .$$

Hence

$$\lim_{x\to+\infty} n(x) = 0 .$$

Thus condition (3) holds after holding condition (2). A possible graph of the function $n(x)$ is shown in figure 1 $(\beta = 1, \lambda = 1, D = 1/2, u_0 = 5\sqrt{\pi})$.

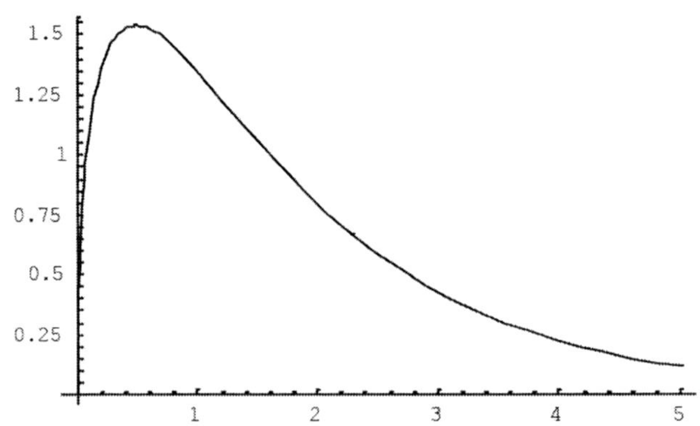

Figure 1. A typical form of a graph of the function $n(x)$ **.**

After we found the function $n(x)$ by formula (28) we can find the normalized firm size distribution.

$$\nu(x) = \frac{n(x)}{N} . \qquad (29).$$

Here

$$N = \int_0^{\infty} n(x)\mathrm{d}x \qquad (30)$$

is the total number of objects in the economic sector (in the population). Function (29) satisfies an equation (see the paper [12])

$$\frac{\mathrm{d}\nu}{\mathrm{d}x} + \left(\frac{\mathrm{d}r}{\mathrm{d}x} + D\right)\frac{\nu}{r} = \frac{Q}{Nr} \qquad (31)$$

and conditions

$$\nu(0)=\nu(\infty)=0\,, \quad \int_0^\infty \nu(x)\mathrm{d}x = 1\,. \tag{32}$$

We could consider equalities (32) as a kind of additional terms for equation (30) as like as the authors of the paper [12] did. However, this approach would lead us to a new question about the correctness of overdetermined problem (31)–(32). To our mind we'd rather firstly find the solution problem (1)–(2), then find number (30), and then we could find $\nu(x)$. We have demonstrated the above scheme for $\gamma = 1/2$. Numerical calculations were carried out for other values of parameter γ (see the paper [14]). Analysis of numerical calculations shows that the increase of the parameter γ shifts the function $\nu(x)$ to a scope of small sizes.

References

[1] Gibrat R 1931 *Les inégalités économiques; applications : aux inégalités des richesses, à la concentration des entreprises, aux populations des villes, aux statistiques des familles, etc., d'une loi nouvelle, la loi de l'effet proportionnel* (Paris: Librairie du Recueil Sirey)

[2] Evans D S 1987 Tests of Alternative Theories of Firm Growth *Journal of Political Economic* **95** 4 657-74

[3] Hall B 1987 The Relationship between Firm Size and Firm Growth in the U.S. Manufacturing Sector *Journal of Industrial Economics* **35** Jun 583-605

[4] Dunne T Roberts M J and Samuelson L 1988 Patterns of Firm Entry and Exit in U.S. Manufacturing Industries *Rand Journal of Economics* **19** 4 495-515

[5] Dunne T Roberts M J and Samuelson L 1989 The growth and Failure of U.S. Manufacturing Plants *Quarterly Journal of Economics* **104** 4 671-98

[6] Audretsch D B 1991 New-Firm survival and Technological Regime *Review of Economics and Statistics* **73** 3 441-50

[7] Arshakuni K V 2005 Processes of formation of new small enterprises. An econometric study of the initial conditions of Genesis on the basis of individual data. (in Russian) *HSE Economic journal* **9** 1 17-50

[8] Arshakuni K V 2005 Dynamics of new small enterprises and endogenous initial conditions. The econometric approach based on the simulated likelihood. (in Russian) *HSE economic journal* **9** 3 291-324

[9] Pin'kovetskaya Yu S 2009 The Number of small enterprises: an analysis (in Russian) *Journal "Economy of a region"* Ekaterinburg: Institute of Economics UB RAS –**2** 18 224-30

[10] Görg H et al. 2016 Firm Size Distribution and Employment Fluctuations: Theory and Evidence *IZA Discussion Papers* (November 2016)10371

[11] Aoyama H Grüne L Semmler W Fujiwara Y and Souma W 2011 Fluctuation of Firm Size in the Long-Run and Bimodal Distribution *Advances in Operations Research* (2011) ID 269239

[12] Mustafin A Kantarbayeva A 2000 On the firm size distribution(in Russian) *Economics and the Mathematical Methods* **36** 3 105–12

[13] Baeva N B 2006 Forecasting of employment in a regional economic system on the normative-statistical approach basis (in Russian) *Economic forecasting: models and methods: proceedings of the International scientific-practical conference* Voronezh state University 1 180-6

[14] Azarnova T V Bondarenko J V Kashirina I L Shchepina I N Gogoleva T N 2016 Mathematical modeling of dinamics business demography of new small innovative region enterprises (in Russian) *Modern Economics: problems and solutions* 10 82 117 - 31

IOP Conf. Series: Journal of Physics: Conf. Series **973** (2018) 012024 doi:10.1088/1742-6596/973/1/012024

Development of dynamic Bayesian models for web application test management

T V Azarnova, P V Polukhin, Yu V Bondarenko and I L Kashirina

Department of Applied Mathematics, Informatics and Mechanics, Voronezh State University, 1 Universitetskaya pl., Voronezh, 394006, Russia

E-mail: ivdas92@mail.ru

Abstract. The mathematical apparatus of dynamic Bayesian networks is an effective and technically proven tool that can be used to model complex stochastic dynamic processes. According to the results of the research, mathematical models and methods of dynamic Bayesian networks provide a high coverage of stochastic tasks associated with error testing in multiuser software products operated in a dynamically changing environment. Formalized representation of the discrete test process as a dynamic Bayesian model allows us to organize the logical connection between individual test assets for multiple time slices. This approach gives an opportunity to present testing as a discrete process with set structural components responsible for the generation of test assets. Dynamic Bayesian network-based models allow us to combine in one management area individual units and testing components with different functionalities and a direct influence on each other in the process of comprehensive testing of various groups of computer bugs. The application of the proposed models provides an opportunity to use a consistent approach to formalize test principles and procedures, methods used to treat situational error signs, and methods used to produce analytical conclusions based on test results.

1. Introduction

Modern development of information systems tends to be aimed at integrating technological processes and hardware by means of high-speed communication channels, especially the Internet. This perspective gives an opportunity to solve a wide variety of problems and provides dynamic cluster scalability. Such architecture is built on web applications that simplify structuring information and transmission process organization, and data processing and storage. According to the classical interpretation, web applications are a type of a client-server architecture of information systems with a number of formal characteristics. Cross-platform technologies used for creation and maintenance of web applications provide conditions for their continuous renewal and improvement. Meanwhile specific character of the dynamically changing environment and the tools of application development, which is favorable for development, might give way to stability errors. Testing aimed at early error detection and localization is an inseparable function used at all stages of the architectural design, software implementation, and operation in the context of the information system. The increasing complexity of the internal structure of web applications and the external infrastructure of their operation requires new approaches to testing and modeling of test management processes. Test toolkits based on classical approaches usually require significant resources and are time consuming. At the same time, their capability to detect new errors steadily decreases with time. The fuzzing test method is a flexible, ever-evolving technology with currently insufficiently developed mechanisms for automated intelligent control. The fuzzing test method is stochastic in nature. To model tools that can be used to manage such processes it is necessary to use models and methods that can reflect the specific character of stochastic processes. Among mathematical modeling methods that can reproduce the specific character of stochastic testing processes by means of the fuzzing method are models based on dynamic Bayesian networks, automata theory, Petri nets, fuzzy and neuron systems, which have a wide range of tools to describe random test variables and parameters and simulation tools to initialize and predict the behavior of the analyzed test process. Different aspects associated with the adaptation of such tools for testing are considered in papers by the following researchers: T.I. Buldakova [1], P.D. Zegzhda [2], I.V. Kotenko [3], A.I. Mischenko [4], P.V. Tumoyan, D.A. Kavchuk [5], etc. For our research, we choose mathematical fuzzing models based on dynamic Bayesian networks due to the high quality of reflection of the test managing tasks in the scope of the formalized possibilities used to describe dynamic processes, to transmit probabilistic information between time slices, and to provide probabilistic inference by dynamic Bayesian networks.

A dynamic Bayesian network is one of the modeling tools used for intellectual data analysis, intended for the deep investigation of complex static and dynamic processes under risk and indeterminateness. In the scientific research, the use of dynamic Bayesian networks to model the process of web application test management will allow us to form a single computational structure that reproduces the functional model of the testing process, describes the logic and probabilistic structure of information and the methodological base of vulnerabilities detection, and implements stochastic intelligent mechanisms of learning, inference, and forecasting in the test procedures. Primary test management findings and results represent results of tasks of filtering, prediction, and smoothing for dynamic Bayesian networks. This paper uses approximate stochastic algorithms and their resource optimization methods to solve the tasks of filtering, prediction, and smoothing.

2. Materials and methods

Testing strategy for specific groups of web application errors is quite specific. It requires a specific setting of the test mechanisms. This article considers the use of dynamic Bayesian networks to test vulnerabilities such as cross-site scripting (XSS) by means of the fuzzing method [9]. The nature of these vulnerabilities is closely related with the client code running in the browser when the web page is loading. In particular, an additional code can be added to the page body without proper authorization and validation. Cross-site scripting covers a wide range of intrusion techniques which can be divided by the code insertion method into the following categories: stored, reflected, and dom [12]. Document object model injection category is associated with the page view in the form of a structured model document and is common for interactive web applications that update data at runtime [13].

The research proposes a conceptual approach to testing that combines methods of random testing and dynamic Bayesian models of testing processes. The random testing by means of the fuzzing method is used to generate samples to be used in: learning structure and probability distributions of individual vertices of a slice, and the formation of evidence for different time slices of the dynamic Bayesian network. Inherently, the testing control process algorithms use a number of formal signs that allow us to determine the structure of such process, as well as to simulate connection handling managerial influence. The following principles can be distinguished: disconnection and feedback. The principle of disconnection is aimed at isolating control signals from external influences, which allows us to block the impact of external factors on the algorithm of the control module. The principle of feedback offers the possibility to correct control action with the help of a correcting signal that signals the lack of effectiveness of a specific functional module that needs adjusting its testing algorithms. The approaches of

IOP Conf. Series: Journal of Physics: Conf. Series **973** (2018) 012024 doi:10.1088/1742-6596/973/1/012024

self-tuning control models are used to construct test management systems. Such models allow us to get an adaptive and intelligent control system that is able to adjust itself to test a wide range of software errors.

Bayesian models represent directed graphs without cycles, which nodes are modeled as random variables X, $i = 1,..,n$ with the domains of values $D_n, i \in D_n$. The arcs leading from the vertex X to the vertex Y, mean that $X \in Parents(Y)$, i.e. X is the parent node for Y. The vertex X_i corresponds to the conditional probability distribution $P(X_i|Parents(X_i)$. The full joint probability distribution for Bayesian networks is as follows [7, 10]:

$$P(X_1, X_2, ..., X_n) = \prod_{i=1}^{n} P(X_i|Parents(X_i)) \tag{1}$$

Dynamic Bayesian networks represent sequences of Bayesian networks taken in chronological order, connected by logical and probabilistic connections. The vertices of dynamic networks describe the spatial-temporal state of the process under investigation. The set of all vertices of the network $Z_t = \{X_t, E_t, Y_t\}$ can be distributed into three subsets: Y_t — hidden variables, X_t — observation variables, and E_t — evidence variables.
The mathematical apparatus of static Bayesian networks is aimed at solving the inference problem of the computation of the posteriori probabilities $P(X|E)$ for the set of query variables in the occurrence of some events (evidence) $e \in E$. The following formula is used to calculate the posteriori probability distribution of a Bayesian network [6].

$$P(X|E) = \alpha \sum\nolimits_Z S_N \omega(X, Z, E), \tag{2}$$

where α is the normalizing constant.

Interface, brute force, and clustering algorithms are used as basic algorithms to compute $P(XE)$. However, in case of the Bayesian networks that model test processes with a high connectivity and a complex structure, these algorithms are inefficient as they are time consuming and their computational cost is quite high. Randomized probabilistic algorithms based on the Monte Carlo method can be used to solve the above-mentioned problem. This family of algorithms is based on the formation and processing of specific samples. The accuracy of the algorithms depends on the method of organization and the dimensions of N samples, which allows us to find the optimal ratio in terms of adequacy, accuracy, and time. The classical interpretation of the Monte Carlo algorithm is a likelihood weighing sampling algorithm. The main feature of this algorithm is its ability to adjust weight in the process flow of the evidence E and to reject events inconsistent with the evidence $e_i \in E_m$. This means that the values of the variables E are fixed, and the process of formation of the samples is carried out only for the variables X and Z ($S_N = N(X, Z, E)$ — the total number of generated probability samples). Weights likelihood is computed through recalculation of conditional probabilities for each node of the Bayesian network at the incoming event time e_i in case of presence of the parent nodes $Parents(e_i)$ and can be described by the following expression:

$$\omega(X, Z, E) = \prod_{i=1}^{m} P(e_i|Parents(e_i)) \tag{3}$$

The posterior probability distribution in case of all evidences E is calculated by the formula:

$$P(X|E) = \alpha \sum\nolimits_Z S_N \omega(X, Z, E) \tag{4}$$

The analysis of the adaptation of the algorithm for testing tasks in the context of dynamic Bayesian networks shows that it is appropriate to combine Monte Carlo and Markov chains, which makes it possible to tune the dynamical probabilistic inference in temporal models. The mathematical apparatus of Markov chains is used to describe the transitive relations. The transition process is modeled as a first order Markov process [11]. The dynamic Bayesian network can be described as a set of Bayesian networks, arranged in chronological order on the time interval $[0, t]$. Let us denote $X_{t-k:t}$, $E_{t-k:t}$ is respectively the set of query and evidence variables for time slices during the transition from the time slice $t-k$ to t. The mathematical model of the dynamic Bayesian network imposes several limitations. The first limitation is associated with conditional probability tables for each of the presented time slices. It means that the table data can be specified once for the 0-th slice, and they do not change when moving from slice to slice. The second limitation is that transitive connections only exist between two neighboring time slices in compliance with the assumption of Markov Rule $P(X_t|X_{0:t-1}) = P(X_{t-1}|X_t)$. The probability distribution $P(X_{t-1}|X_t)$ describes the transition between the slices and is called a transition model. The state of evidence variable at the time t is described by the model $P(E_t|X_{0:t}, E_{0:t-1}) = P(X_t|E_t)$ (it only depends on the current state) [6]. The full joint conditional probability distribution for dynamic Bayesian network is as follows:

$$P(X_n, E_n) = P(X_0) \prod_{i=1}^{n} P(X_{t-1}|X_t) P(E_t|X_t), \tag{5}$$

where $P(X_0)$ is the initial probability distribution for zero time.

In case of dynamic Bayesian model for testing of cross-site scripting vulnerabilities, algorithmic components that are used in the error detection process and a set of advanced features to bypass filtering mechanisms and validation parameters can be described as the network nodes. Arcs between vertices in the network determine the probabilistic relationships between these components, both within the slice and the transition between the slices, and determine cause and effect relationships between the algorithms in the context of testing of specific cross-site scripting error groups. Designations for the dynamic Bayesian model for testing cross-site scripting vulnerabilities are given in table 1.

Table 1. Description of bayesian network nodes

Node	Description
XSS_TYPE	Varieties of XSS to embed JavaScript code on a web page.
ENCODER	XSS encoding algorithms to overcome software filters.
EVASION	Algorithms for XSS obfuscation to overcome software filters.
XSS_PLD	Variation of the XSS payload (html tag, event handlers).
KEYLOG	Algorithms used to save keyboard pressed keys.
SPY_EYE	Mechanism for obtaining web page screenshots using XSS.
DDOS	Mechanism to use web browser as denial of service attack element.
PORT_SCAN	Algorithm for scanning ports on the computer.
NET_SCAN	Scanning the local network, building a network topology.
NAT_PIN	Algorithms to bypass NAT traversal rules.
DR_BY_DW	Redirect the user to resources that distribute malware and viruses.
BROW_FINPT	Overview the components installed in the user's browser.
AUTHEN,AUTHOR, INTEG,CONF	Authentication, authorization, integrity, confidentiality mechanisms.

IOP Conf. Series: Journal of Physics: Conf. Series **973** (2018) 012024 doi:10.1088/1742-6596/973/1/012024

Figure 1 shows a fragment of a dynamic Bayesian network (two time slices) of the cross-site scripting vulnerabilities testing process. It is important to note that the nodes in the first time slice do not have external relations, and the distribution of conditional probabilities is only determined by the probability $P(X_0)$. If there is a conditional relation between the variables of two time slices $P(X_{t-1})$ and $P(X_t)$ for two time moments, such a relationship will be constant, otherwise — temporal. Slices of the dynamic Bayesian network describe the state of the test environment and characterize a certain finished step for testing a particular group of applications. Each step is characterized by the processes of calculating parameters of a dynamic Bayesian network (observation and evidence variables) based on the representative data resulting from testing of each error type. Consideration of the testing process in time is, first of all, due to the accumulation of statistical information for each time slice, which offers the possibility to adapt the network to detect anomalous behavior and perform network configuration to identify a certain group of errors [6]. The formation of software error groups is based on the concept of the homogeneity criterion that defines the degree of membership for a particular error by a set of formal characteristics. This enables to generalize some error detection algorithms and optimize the search procedure for the optimal set of test sequences used within the experiment. Obtaining initial distribution, transition, and sensor model also requires initial testing by means of the black box with the subsequent learning of network parameters.

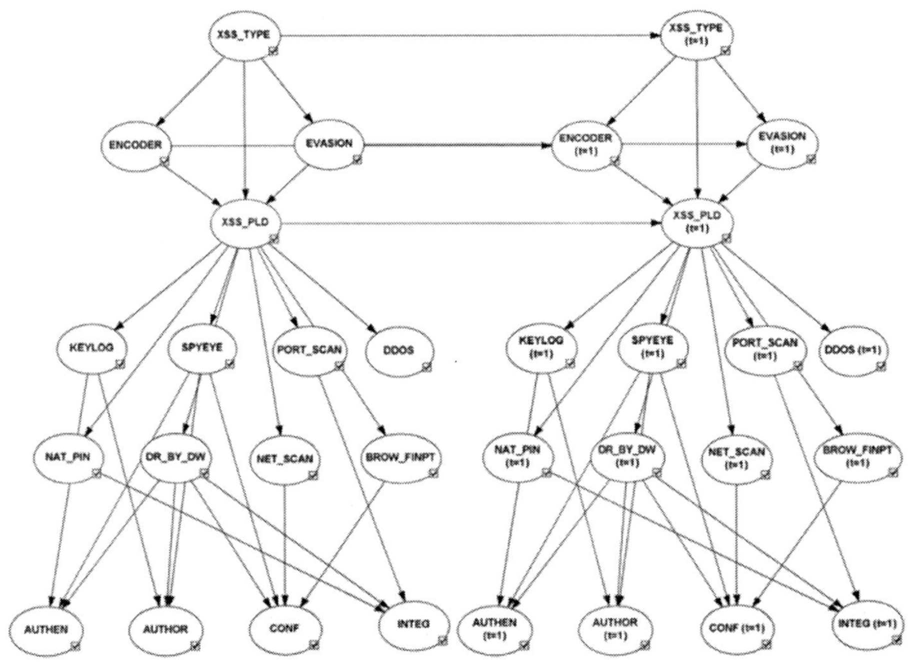

Figure 1. The part of dynamic Bayesian network for testing cross site scripting errors.

A fragment of a dynamic Bayesian network shown in figure 1 consists of two layers (slices) of a Bayesian network for the time instants t and $t+1$ with defined topological relations between the slices that are modeled by means of conditional probabilities between the nodes of the slices.

Naive Bayesian classifier is used as a criterion of learning. The mathematical apparatus for this classifier is based on the maximum posteriori decision rule (MAP) with the following form:

$$\theta^G_{MAP} = \arg\max_{\theta} \log P(Y|\theta^G) + \log P(\theta^G), \tag{6}$$

where $\log P(Y|\theta^G) = \sum_{i=1}^{n}\sum_{j=1}^{q_i}\sum_{k=1}^{r_i} N_{i,j,k} \ln \theta^G_{i,j,k}$ is the likelihood log, $Y = y^1_{1:T}, ..., y^N_{1:T}$ is the training sample, G is the structure of the Bayesian network, and θ^G is the set of the network parameters.

The Lagrange function is used as the basis for the solution of conditional optimization problem and is represented by the formula:

$$Lagr(G, \theta^G, D) = \sum_{i=1}^{n}\sum_{j=1}^{q_i}\sum_{k=1}^{r_i} N_{i,j,k} \ln \theta^G_{i,j,k} + \sum_{i=1}^{n}\sum_{j=1}^{q_i} u_{i,j} \sum_{k=1}^{r_i} \theta^G_{i,j,k} \tag{7}$$

where $u_{i,j}$ are the Lagrange coefficients.

The optimization problem can be interpreted as a problem of convex programming, and the necessary and sufficient condition for the maximization of the likelihood log is equality to zero of partial derivatives of the Lagrange function:

$$\begin{gathered} \frac{\partial}{\partial \theta^G_{i,j,k}} Lagr(G, \theta^G, D) = \frac{N_{i,j,k}}{\theta^G_{i,j,k}} + u_{i,j} = 0 \Rightarrow \frac{N_{i,j,k}}{\theta^G_{i,j,k}} = -u_{i,j}, \\ \forall k \Rightarrow \sum_{k=1}^{r_i} = -u_{i,j} \sum_{k=1}^{u_i} \theta^G_{i,j,k} \Rightarrow N_{i,j} := \sum_{k=1}^{r_i} N_{i,j,k} = -u_{i,j} \end{gathered} \tag{8}$$

where $N_{i,j,k} = \sum_{k=1}^{r_i} N_{i,j,k} = \{l \in \{1..N\} | Parents(X_i) = j\}$

The parameters of the probabilistic distributions of the Bayesian network are as follows:

$$\frac{N_{i,j,k}}{\theta^G_{i,j,k}} = N_{i,j}, \forall k \Rightarrow \theta^G_{i,j,k} = \frac{N_{i,j,k}}{N_{i,j}} \tag{9}$$

The network training procedure is used to fill up the conditional probabilities tables and to calculate the initial probability distribution $P(X_0)$, the transition $P(X_{t+1}|X_t)$, and the sensor $P(X_t|E_t)$ models.

Adaptation of dynamic Bayesian networks to the testing process allows us to solve a number of tasks aimed at detecting anomalous behavior and predicting occurrence probability of error groups by solving the main tasks of probabilistic inference: filtering $P(X_t|E_{1:t})$, predictions $P(X_{t+k}|E_{1:t})$, and smoothing $P(X_k|E_{1:t})$. To address these tasks the research uses a variation of an algorithm for probabilistic inference based on likelihood weighing, i.e. Rao-Blackwell particle filtering algorithm. This algorithm involves the formation of samples in the process of network deployment with their subsequent likelihood weighting. In classical particle filtering algorithm N samples are formed from the initial distribution $P(X_0)$, then samples are updated with the transition model $P(X_{t-1}|X_t)$ during the transition step from the state $t-1$ to t. The total number of samples for the state X_t is determined by the following [14]:

$$\begin{gathered} N(X_{t+1}|E_{1:t}) = \sum_{X_t} P(X_{t+1}|X_t) N(X_t|E_{1:t}), \\ N(X_{t+1}|E_{1:t+1})/N = \alpha W(X_{t+1}|E_{1:t+1}) = P(X_{t+1}|E_{1:t+1}), \end{gathered} \tag{10}$$

where $W(X_{t+1}|E_{1:t+1}) = P(E_{t+1}|X_{t+1})N(X_{t+1}|E_{1:t})$ - total weight of the samples in X_{t+1}.

The structure of the particle filtering algorithm is as follows:

IOP Conf. Series: Journal of Physics: Conf. Series **973** (2018) 012024 doi:10.1088/1742-6596/973/1/012024

Algorithm 1 Particle filtering

Input e: incoming evidence
N: samples number
dbn : dynamic bayesian network with $P(X_0)$, transition $P(X_t|X_{t+1})$ and sensor $P(E_{t+1}|X_{t+1})$ models

1: **Static**: S, samples vector of size N constructed from $P(X_0)$
2: **Local Variable**: W, vector of size N
3: **for** $i = 1, \ldots, N$ **do**
4: $S[i] \leftarrow samples\_from\ P(X_{t+1}|X_t) = S[i]$
5: $W[i] \leftarrow P(e|X_{t+1} = S[i])$
6: **end for**
7: $S \leftarrow weighted\_sample\_with\_replace(N, S, W)$

Return S

To reduce the total number of samples in the optimization process of the particle filtration algorithm and to increase the resource efficiency of the algorithm as a whole we use the Rao-Blackwell theorem [8]. This theorem allows us to adapt the theory of sufficient statistics to improve estimates of distribution parameters in terms of mean-square deviation. The application of the Rao-Blackwell theorem is aimed at the separation of the set of hidden variables Z_t into the subsets X_t and R_t, which provides the opportunity to estimate each sample generated for each subset. According to the previously introduced designations the transition model can be transformed into the following form:

$$P(Z_{t+1}|Z_t) = P(X_{t+1}|R_{t:t+1}, X_t)P(R_{t+1}|R_t) \tag{11}$$

The calculation of the conditional probability distributions$P(X_{0:t}|Y_{1:t}, R_{0:t})$ is performed analytically. Estimation of $P(R_{0:t}|Y_{0:t})$ is based on the following expression:

$$P(R_{0:t}|Y_{0:t}) = \frac{P(Y_t|Y_{1:t-1}, R_{0:t})P(R_t|Rt-1)P(R_{0:t-1}|Y_{1:t-1})}{P(Y_t|Y_{1:t-1})} \tag{12}$$

Adaptation of the Rao-Blackwell theorem to the topology of the dynamic Bayesian network is based on the comparison criterion of marks (weights) of N samples generated in the process of basic particle filtration algorithm. The mathematical representation of this comparison is shaped by the following inequality [8]

$$var_{Q(R_{0:t}|E_{1:t})}(\omega(R_{0:t})) \leq var_{Q(R_{0:t},X_{0:t}|E_{1:t})}(\omega(R_{0:t}, X_{0:t})) \tag{13}$$

3. Results and discussion

The research involves a computational experiment in testing cross-site scripting errors. For the purpose of the experiment, a special software was developed and a test environment was created. The test environment represents a system that consists of the following components: a web server with an installed set of web applications with different criticality levels of software errors; the database server PostgreSQL for testing a stored XSS; and the browser factories: Firefox, Chrome, Internet Explorer, Edge. Structurally, the experiment process can be divided into four main phases. The first phase involves a primary testing of applications by means of random fuzzing (white and black box methods). The next phase uses the results of the previous stage to build the Bayesian network and to implement parameter learning procedures: the initial distribution $P(X_0)$, the transition model $P(X_{t-1}|X_t)$,the sensor model $P(X_t|E_t)$, and the time moments $t-1$ and t. The third phase of the experiment involves probabilistic inference procedure aimed at obtaining samples that characterize the most typical test assets that can be used to test new applications and detect anomalous behavior within web applications. The final phase is dedicated to the evaluation of the quality of the experiment and the adequacy of test assets to error detection criteria when testing a number of new applications from the homogeneous group.

The proposed mathematical algorithms and models are used to develop the general model of testing cross-site scripting errors, to optimize Rao-Blackwell particle filtering algorithm by parallelizing individual blocks of the algorithm, aimed at the creation of samples and calculating sufficient statistics for the variables of a dynamic Bayesian network. A generalized block diagram of this algorithm is shown in figure 2.

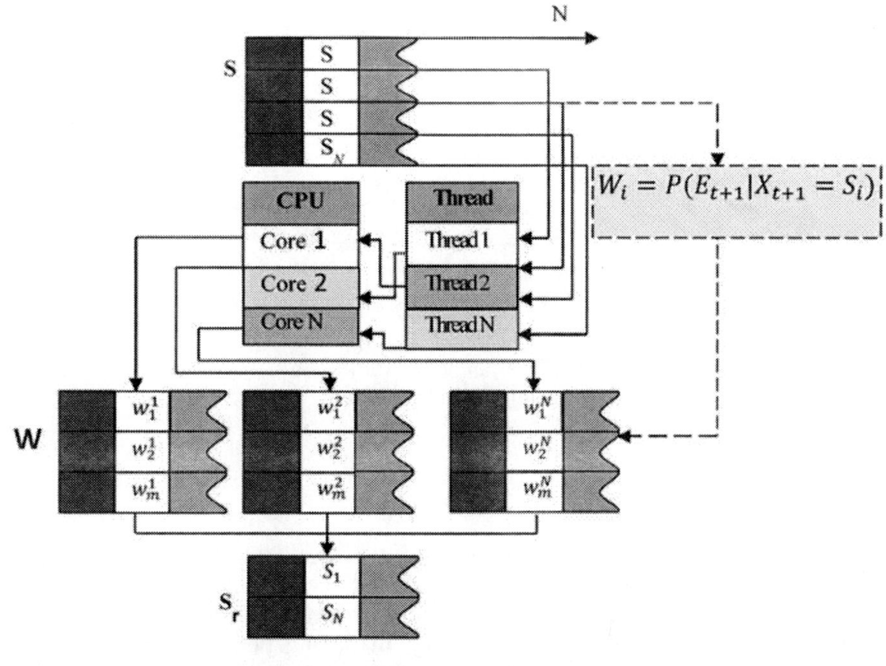

Figure 2. Parallel Rao-Blackwell particle filtering algorithm.

Evaluation of the effectiveness of the proposed algorithm involves a comparison of the developed algorithm and the classical particle filtering algorithm. The comparison is shown in figure 3. It is based on the analysis of two different algorithms within the same computing environment.

The use of parallel operations together with the Rao-Blackwell theorem allows us to optimize the algorithmic base of probabilistic inference as applicable to Bayesian networks with a complex structure without losing precision and effectiveness. At the same time, the testing process becomes more sustainable and adaptive.

The results of the research, supported by computational experiments allow us to speak about the effectiveness of using in testing procedures testing tools that were created as a result of combining of random testing and the mathematical techniques of dynamic Bayesian networks, which are capable of self-training and predicting abnormal errors. The application of this approach to building complex test systems contributes to the effectiveness of solving test management tasks by formalizing relations between the individual elements of testing and by building a clear approach to the timely correction based on the analytical data.

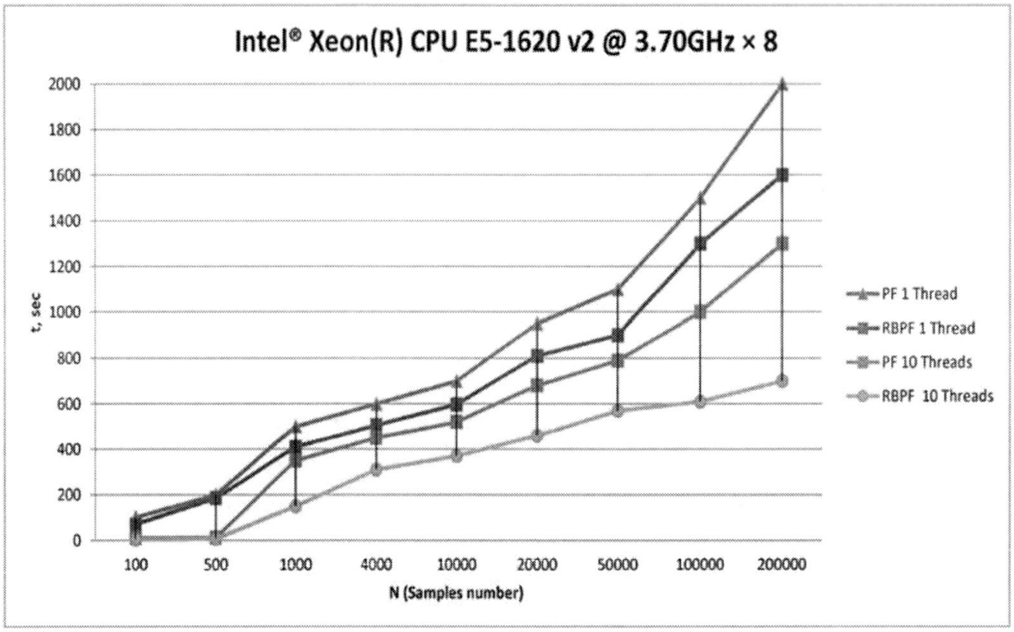

Figure 3. Comparison of algorithms: particle filtering and Rao-Blackwell particle filtering algorithm (**PF** – particle filtering algorithm, **RBPF** – Rao-Blackwell particle filtering algorithm).

4. Conclusion

The paper considers issues of modeling testing processes and their management based on dynamic Bayesian models. It describes dynamic Bayesian models for the testing of cross-site scripting vulnerabilities. The mathematical apparatus of Bayesian networks allowed us to perform step-by-step modeling, to implement all testing phases, to build structural links between the individual test blocks, and to optimize the existing approaches used in testing. In order to improve the resource efficiency, it was proposed to adapt the Rao-Blackwell theorem to the particle filtering algorithm and to use the parallelizing of individual blocks of the algorithm responsible for the formation of randomized samples. By solving the problems of filtering, prediction and smoothing, the strategy of adaptive static and dynamic web application analysis was successfully implemented to optimize the detection of software errors in testing web applications on different time slices. Comparative analysis with the classical approaches demonstrates the effectiveness of the proposed algorithm and software. The detailed modeling of test management processes by the fuzzing method allows us: to make timely correction of data without direct adjustment of the functional principles of the individual testing modules and to normalize the process of conversion of test chains in logical inference procedures with simultaneous correction of the individual units responsible for test data generation.

5. References

[1] Buldakova T I and Djalalov A H 2013 *J. Science And Innovation* B **11** 1
[2] Zegzhda P D, Kort S S and Suprun A F 2015 *J. Automatic Control and Computer Sciences* B **49** 820
[3] Kotenko D I, Kotenko I V and Saenko I B 2012 *J. SPIIRAS Proceedings* B **2** 5

IOP Conf. Series: Journal of Physics: Conf. Series **973** (2018) 012024 doi:10.1088/1742-6596/973/1/012024

[4] Mischenko A I 2013 *J. Systems and Means of Informatics* B **1** 27
[5] Tumoyan E P and Kavchuk D A 2013 *J. Information Technologies Security* B **1** 25
[6] Azarnova T V, Barkalov S A and Polukhin P V 2017 *J. Of Ural State University. Computer technologies, control and radio electronics* B **2**(17) 51
[7] Tulupyev A L 2009 *Bayesian belief networks: logical-probabilistic inference in acyclic direct graphs* (Saint-Petersburg:Saint-Petersburg State University Press)
[8] Suhov Y and Kelbert M 2005 *Probability and Stochastic by Example. Volume 1: Basic Probability and Statistic* (New York: Cambridge University Press)
[9] Beizer B 1995 *Black-Box Testing* (New York: Wiley Publishing)
[10] Murphy K P 2012 *Machine Learning. A Probabilistic Perspective* (Massachusetts: The MIT Press)
[11] Perl J 2009 *Causality: Models, Reasoning and Inference* (New York: Cambridge University Press)
[12] Zalewsky M 2012 *A guide to securing web Applications* (San Francisco: No Starch Press)
[13] Baloch R 2015 *Ethnical Hacking and Penetration Testing Guide* (Boca Raton: CRC Press)
[14] Russel S and Norvig P 2010 *Artificial Intelligence A Modern Approach* (New Jersey: Prentice Hall)

IOP Conf. Series: Journal of Physics: Conf. Series **973** (2018) 012025 doi:10.1088/1742-6596/973/1/012025

The algorithmic details of polynomials application in the problems of heat and mass transfer control on the hypersonic aircraft permeable surfaces

G G Bilchenko[1,2] **and N G Bilchenko**[1,3]

[1] Research Laboratory #16, Department of Heat Engineering and Power Engineering Machinery, Kazan National Research Technical University (KNRTU-KAI) named after A N Tupolev, 10, K Marx St., Kazan, Tatarstan 420111, Russia

E-mail: [2] ggbil2@gmail.com and [3] bilchnat@gmail.com

Abstract. The hypersonic aircraft permeable surfaces heat and mass transfer effective control mathematical modeling problems are considered. The analysis of the control (the blowing) constructive and gasdynamical restrictions is carried out for the porous and perforated surfaces. The functions classes allowing realize the controls taking into account the arising types of restrictions are suggested. Estimates of the computational complexity of the W. G. Horner scheme application in the case of using the C. Hermite interpolation polynomial are given.

1. Introduction

To solve the hypersonic aircraft permeable surfaces heat and mass transfer effective control (with the use of blowing into boundary layer) mathematical modeling problems it is necessary to take into account some constructive and gasdynamical restrictions [1, 2].

To simulate the controls tending to discontinuous the use of glued functions [3, 4] is suggested. The right choice of polynomials arithmetics algorithms is very important from the point of view of computational complexity for the multinode meshes and the high degree polynomials [5, 6, 7] realizing the glued functions elements.

2. Special blowing laws for permeable porous surfaces

In the case of *porous* area for some problems with special blowing laws taking into account the *constructive* restrictions in the stagnation point neighborhood [1]

$$m(x) = \begin{cases} m^* & \text{for} \quad x \in [0\,;\, x^*]\,; \\ m^c_{opt}(x) & \text{for} \quad x \in [x^*\,;\, 1]\,, \end{cases} \tag{1}$$

where the blowing system power [8] satisfies the condition $N(m) \le N_c$ for $N_c = N\left(m^c_{\mathrm{IV}}\right)$, the following variants are considered:

the law 1 : $m\,(0) = m_1\,, \quad m_1 \le m^{analyt}_{opt}\,(0)$;
the law 2 : $m\,(0) = 0\,$;
the law 3 : according to [9, 10]: $m^* = 0\,$,
and displayed on the left half of fig. 1.

Taking into account the necessity to guarantee the determined "level" of *weak blowing* let's introduce an additional *gazdynamical* restriction into laws 1-3:

$$m(x) \le m_{weak}^{\lim}, \tag{2}$$

where

$$m_{\mathrm{IV}}^{c} \le m_{weak}^{\lim},$$

and obtain laws 4-6 displayed on the right half of fig. 1.

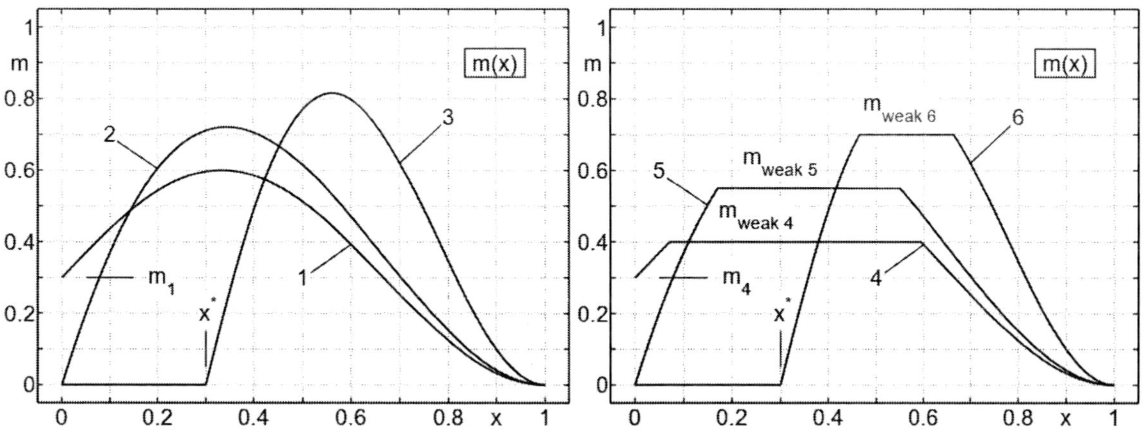

Figure 1. The special blowing laws

The numerical solutions obtained for strengthened conditions:

$$m^{*} \to 0 \qquad \text{and/or} \qquad x^{*} \to 1 \qquad \text{and/or} \qquad m_{weak}^{\lim} \to m_{\mathrm{IV}}^{c}$$

tend to discontinuous solutions, which can not be realized physically for porous blowing.

3. Special blowing laws for permeable perforated surfaces

Let's consider some problems with special blowing laws simulating blowing through holes.

The blowing law variant ("decreasing triangle") similar to "simple" law 1 from [8] and realized for surface perforated with the ratio $d/h = 1/5$ (here d is the hole diameter and h is the distance between holes) is displayed on fig. 2 as an example. The ratio 1/5 is recommended in [11].

The necessity to simulate the controls corresponding to blowing through holes [2] comes as a variant to the use of rectangular "impulses" (the "jets"):

$$m(x) = \begin{cases} 0 & \text{for} \quad x \in [x_0\,;\, x_1)\,; \\ m_k & \text{for} \quad x \in (x_{2k-1}\,;\, x_{2k})\,; \\ 0 & \text{for} \quad x \in (x_{2k}\,;\, x_{2k+1})\,, \end{cases} \tag{3}$$

having for $m_k > 0$ discontinuous character in the joint points x_{2k-1}, x_{2k}.

4. The boundary layer on hypersonic aircraft permeable surfaces effective control problems solution classes

Let's describe the functions [12] similar to the glued functions introduced in [3, 4].

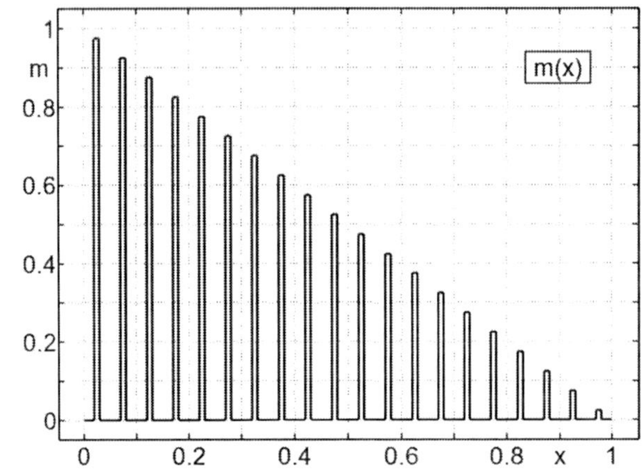

Figure 2. The law "decreasing triangle" for surface with perforation ratio "1/5"

Let the segmentation be fixed for $X = [0; 1]$:

$$x_0 = 0 < x_1 < \ldots < x_{n-1} < x_n = 1 \,. \tag{4}$$

Each segment $[x_{j-1}; x_j]$, $j = 1, \ldots, n$ is associated with an "element" $m_j\,(x)$ of function $m\,(x)$:

$$m\,(x) = m_j\,(x) \quad \text{for} \quad x \in [x_{j-1}; x_j] \,. \tag{5}$$

The following opportunities were added to the computation experiment algorithm scheme for the fixed integer number $\nu \geq 0$.

4.1. Restrictions on segments

Let for the continuity intervals $[x_{j-1}; x_j]$ of the elements m_j (here $j = 1, \ldots, n$) and for $k = 0, \ldots, \nu$ the continuous functions $b_{j,k}(x)$ and $t_{j,k}(x)$ be preset given as:

$$b_{j,k}(x) \leq t_{j,k}(x) \quad \text{for} \quad x \in [x_{j-1}; x_j] \,. \tag{6}$$

Let's the elements $m_j(x)$ of unknown glued function satisfy the conditions

$$m_j^{(k)}(x) \in I_{j,k}(x) \quad \text{for} \quad x \in [x_{j-1}; x_j] \,, \tag{7}$$

where $I_{j,k}(x) = [b_{j,k}(x); t_{j,k}(x)]$.

4.2. Restrictions for jumps

For the joint points x_j of the segments $[x_{j-1}; x_j]$ and $[x_j; x_{j+1}]$ (here $j = 1, \ldots, n-1$) and for $k = 0, \ldots, \nu$ let's define the jumps

$$\Delta m_{j,k} = m_{j+1}^{(k)}(x_j) - m_j^{(k)}(x_j) = m^{(k)}(x_j + 0) - m^{(k)}(x_j - 0). \tag{8}$$

Let the restrictions $\underline{\Delta m_{j,k}}$, $\overline{\Delta m_{j,k}}$ be given as follows:

$$\underline{\Delta m_{j,k}} \leq \overline{\Delta m_{j,k}} \,. \tag{9}$$

Let's the jumps $\Delta m_{j,k}$ of unknown glued function satisfy the conditions

$$\underline{\Delta m_{j,k}} \le \Delta m_{j,k} \le \overline{\Delta m_{j,k}}\,. \tag{10}$$

This allows choose one of the following alternatives.
"The jump of fixed size":

$$\underline{\Delta m_{j,k}} = \overline{\Delta m_{j,k}} \neq 0\,; \tag{11}$$

"the necessity of "+jump"":

$$0 < \underline{\Delta m_{j,k}} < \overline{\Delta m_{j,k}}\,; \tag{12}$$

"the necessity of "–jump"":

$$\underline{\Delta m_{j,k}} < \overline{\Delta m_{j,k}} < 0\,; \tag{13}$$

"the jump prohibition":

$$\underline{\Delta m_{j,k}} = 0 = \overline{\Delta m_{j,k}}\,. \tag{14}$$

For

$$0 < \overline{\Delta m_{j,k}} \tag{15}$$

and/or for

$$\underline{\Delta m_{j,k}} < 0 \tag{16}$$

the jump is **possible**, but not obligatory.

4.3. Notes
1) In the case, when $m_j(x)$ are polynomials, the glued function is a generalization of spline.
2) It is preferable for the functions $b_{j,k}(x)$ and $t_{j,k}(x)$ to be simple enough, for example, they may be the polynomials of not high degree.
3) The values of

$$b_{j-1,k}(x_j-0)\,, \qquad b_{j,k}(x_j+0)\,, \qquad t_{j-1,k}(x_j-0)\,, \qquad t_{j,k}(x_j+0)$$

must satisfy the conditions (6) only.
4) The interval conditions (7) and the joint conditions (10) may require mutual coordination.
5) **If** on a certain proper subinterval $[x_*;x^*] \subset [x_{j-1};x_j]$ for a certain k^*, $0 \le k^* \le \nu$ the restrictions

$$b_{j,k^*}(x) = t_{j,k^*}(x) \quad \text{for} \quad x \in [x_*;x^*], \tag{17}$$

coincide; besides, the values x_*, x^* are *exact*, i.e. for them such value $\delta_{\mathrm{isol}} > 0$ exists that for $x \in [x_* - \delta_{\mathrm{isol}}; x^* + \delta_{\mathrm{isol}}] \cap [x_{j-1};x_j]$, but $x \notin [x_*;x^*]$ it follows that $b_{j,k^*}(x) < t_{j,k^*}(x)$, **then** such points x_*, x^* must be included in the nodes system (4).

The procedure for including nodes of this type in the system (4) should be **executed** until such subintervals finish, **or** the size $(x^* - x_*)$ of each remaining subinterval becomes less than the preset given threshold value $\delta_{\min} > 0$.

For $k^* = 0$ the unknown function $m(x)$ may be obtained on $[x_*;x^*]$ with the use of (17) and (7). For $k^* > 0$ the function $m(x)$ may be obtained with the accuracy upon k^* integration constants.
6) **If** in a certain interior point x^* of interval $[x_{j-1};x_j]$ for a certain k^*, $0 \le k^* \le \nu$ the restrictions

$$b_{j,k^*}(x^*) = t_{j,k^*}(x^*)\,, \tag{18}$$

coincide; besides, x^* is an *isolated* point, i.e. for it such value $\delta_{\mathrm{isol}} > 0$ exists that for $x \in [x^* - \delta_{\mathrm{isol}}; x^* + \delta_{\mathrm{isol}}] \cap [x_{j-1};x_j]$, but $x \neq x^*$ it follows that $b_{j,k^*}(x) < t_{j,k^*}(x)$, **then** such point x^* must be included in the nodes system (4).

The procedure for including nodes of this type in the system (4) should be **executed** until such nodes finish, **or** the including of any of such remaining nodes into system (4) gives the interval which length is shorter than the preset given threshold value $\delta_{\min} > 0$.

The unknown function $m(x)$ must satisfy the point condition (18) and (7). The jump prohibition (14) (with the order k^*) must be prescribed for the point x^*, which was included into the system of nodes.

5. The polynomials application algorithmic details

Among functions being representable on computer one of the simpliest variant allowing satisfy the conditions (7) is the case of the *polynomials*. Taking into account the necessity of the conditions (7) control on the intervals ends

$$m_j^{(k)}(x_{j-1}+0) \in [b_{j,k}(x_{j-1}+0)\,;\, t_{j,k}(x_{j-1}+0)]\,, \tag{19}$$

$$m_j^{(k)}(x_j-0) \in [b_{j,k}(x_j-0)\,;\, t_{j,k}(x_j-0)] \tag{20}$$

and the fulfillment of conditions (10), it is convenient to use the boundaries x_{j-1}, x_j of segments (4) as the points, where the polynomials parameters are given. Because the numbers of conditions are $(\nu+1)$ both for the left and the right ends of the segment $[x_{j-1}; x_j]$, then a special case of C. Hermit interpolation polynomial (with two nodes) having degree $2\mu-1 \geq 2\nu+1$ may be used as polynomial.

5.1. *Complete construction of interpolation polynomial by means of divided differences*

Let for $a < b$ and $\mu \geq 1$ on the segment $[a; b]$ the values

$$\left(f^{(p-1)}(a)\right)_{p=1,\ldots,\mu}, \qquad \left(f^{(p-1)}(b)\right)_{p=1,\ldots,\mu} \tag{21}$$

of derivatives in the points a and b having the order from 0 to $\mu-1$ be given as the control $f(x)$ initial approximation. To obtain the polynomial let's make a table of a special form. Let's place the values

$$f_{p,0} = f(\underbrace{a,\ldots,a}_{p}) = \frac{f^{(p-1)}(a)}{(p-1)!}, \quad f_{0,q} = f(\underbrace{b,\ldots,b}_{q}) = \frac{f^{(q-1)}(b)}{(q-1)!} \quad \text{for} \quad p,q = 1,\ldots,\mu \tag{22}$$

in its left and upper headings. Let's denote $\Delta = b-a$. Then, from left to right and from top to bottom let's compute the divided differences

$$f_{p,q} = f(\underbrace{a,\ldots,a}_{p},\underbrace{b,\ldots,b}_{q}) = \frac{f_{p-1,q}-f_{p,q-1}}{\Delta} \quad \text{for} \quad p,q = 1,\ldots,\mu\,. \tag{23}$$

In table 1 and further the arrow to the right means the construction "by a, then by b", and the arrow to the left means the construction "by b, then by a". The C. Hermit interpolation polynomial constructed for the nodes a and b can be written in the following forms:

$$P_{\mu,\mu}(x;a,b) = L_\mu^{\rightarrow}(x;a,b) + H_\mu^{\rightarrow}(x;a,b) = \tag{24}$$

$$= L_\mu^{\leftarrow}(x;a,b) + H_\mu^{\leftarrow}(x;a,b). \tag{25}$$

Hereinafter L is the *lower* part of P and H is the *higher* part of P:

$$L_\mu^{\rightarrow}(x;a,b) = \sum_{k=0}^{\mu-1} \lambda_k^{\rightarrow}\cdot(x-a)^k, \qquad H_\mu^{\rightarrow}(x;a,b) = (x-a)^\mu \cdot \sum_{k=0}^{\mu-1} \eta_k^{\rightarrow}\cdot(x-b)^k, \tag{26}$$

Table 1.

	$\lambda_0^{\leftarrow} = f_{0,1} = f(b)$	$f_{0,2} = f'(b)$	$\dots$	$f_{0,\mu} = \dfrac{f^{(\mu-1)}(b)}{(\mu-1)!}$
$\lambda_0^{\rightarrow} = f_{1,0} = f(a)$	$f_{1,1}$	$f_{1,2}$	$\dots$	$\eta_0^{\leftarrow} = f_{1,\mu}$
$f_{2,0} = f'(a)$	$f_{2,1}$	$f_{2,2} = \dfrac{f_{1,2} - f_{2,1}}{\Delta}$	$\dots$	$f_{2,\mu}$
$\vdots$	$\vdots$	$\vdots$	$\ddots$	$\vdots$
$f_{\mu,0} = \dfrac{f^{(\mu-1)}(a)}{(\mu-1)!}$	$\eta_0^{\rightarrow} = f_{\mu,1}$	$f_{\mu,2}$	$\dots$	$f_{\mu,\mu}$

$$L_\mu^{\leftarrow}(x;a,b) = \sum_{k=0}^{\mu-1} \lambda_k^{\leftarrow} \cdot (x-b)^k, \qquad H_\mu^{\leftarrow}(x;a,b) = (x-b)^\mu \cdot \sum_{k=0}^{\mu-1} \eta_k^{\leftarrow} \cdot (x-a)^k, \tag{27}$$

$$\lambda_k^{\rightarrow} = f_{k+1,0}\,, \quad \eta_k^{\rightarrow} = f_{\mu,k+1}\,, \qquad \lambda_k^{\leftarrow} = f_{0,k+1}\,, \quad \eta_k^{\leftarrow} = f_{k+1,\mu}\,, \qquad k = 0,\dots,\mu-1\,. \tag{28}$$

To compute the divided differences (23) it is necessary to accomplish μ^2 divisions (further on for the complexity estimates the operations with floating-point numbers are considered everywhere) and 2μ divisions (22) by factorials. It takes 2μ memory cells to compute one of the forms (24) or (25) and requires $(4\mu - 1)$ cells for two forms.

5.2. Interpolation polynomials construction by means of basis polynomials

The interpolation polynomials may be constructed in a different way: let's prepare the "basis" polynomials beforehand [13], i.e. for the preset given values of a, b, μ let's construct the system

$$F = (F_{p,0}\,, F_{0,p})_{p=1,\dots,\mu}$$

of such polynomials, that

$$F_{j,0}^{(k-1)}(x) = \begin{cases} \dfrac{1}{k!} \cdot \delta_j^k & \text{for} \quad x = a; \\ 0 & \text{for} \quad x = b, \end{cases} \qquad F_{0,j}^{(k-1)}(x) = \begin{cases} 0 & \text{for} \quad x = a; \\ \dfrac{1}{k!} \cdot \delta_j^k & \text{for} \quad x = b \end{cases} \tag{29}$$

for $j,k = 1,\dots,\mu$, where the L. Kronecker's delta is $\delta_j^k = \begin{cases} 0 & \text{for} \quad j \neq k\,; \\ 1 & \text{for} \quad j = k\,. \end{cases}$ In this case

$$P_{\mu,\mu}(x;a,b) = \sum_{k=0}^{\mu-1} F_{k,0}(x;a,b) \cdot f^{(k-1)}(a) + \sum_{k=0}^{\mu-1} F_{0,k}(x;a,b) \cdot f^{(k-1)}(b)\,. \tag{30}$$

In particular for $\mu = 2$:

$F_{1,0}$	0	0
1	$\dfrac{-1}{\Delta^1}$	$\dfrac{+1}{\Delta^2}$
0	$\dfrac{-1}{\Delta^2}$	$\dfrac{+2}{\Delta^3}$

$F_{2,0}$	0	0
0	0	0
1	$\dfrac{-1}{\Delta^1}$	$\dfrac{+1}{\Delta^2}$

IOP Conf. Series: Journal of Physics: Conf. Series **973** (2018) 012025 doi:10.1088/1742-6596/973/1/012025

$$F_{1,0}(x;a,b) = \left[1 + 0\cdot(x-a)^1\right] + (x-a)^2\cdot\left[\frac{-1}{\Delta^2} + \frac{+2}{\Delta^3}\cdot(x-b)^1\right] = \quad (31)$$

$$= \left[0 + 0\cdot(x-b)^1\right] + (x-b)^2\cdot\left[\frac{+1}{\Delta^2} + \frac{+2}{\Delta^3}\cdot(x-a)^1\right] = \quad (32)$$

$$= 1 + 0\cdot(x-a)^1 + \frac{-3}{\Delta^2}(x-a)^2 + \frac{+2}{\Delta^3}\cdot(x-a)^3 = \quad (33)$$

$$= 0 + 0\cdot(x-b)^1 + \frac{+3}{\Delta^2}(x-b)^2 + \frac{+2}{\Delta^3}\cdot(x-b)^3, \quad (34)$$

$$F_{2,0}(x;a,b) = \left[0 + 1\cdot(x-a)^1\right] + (x-a)^2\cdot\left[\frac{-1}{\Delta^1} + \frac{+1}{\Delta^2}\cdot(x-b)^1\right] = \quad (35)$$

$$= \left[0 + 0\cdot(x-b)^1\right] + (x-b)^2\cdot\left[0 + \frac{+1}{\Delta^2}\cdot(x-a)^1\right] = \quad (36)$$

$$= 0 + 1\cdot(x-a)^1 + \frac{-2}{\Delta^1}\cdot(x-a)^2 + \frac{+1}{\Delta^2}\cdot(x-a)^3 = \quad (37)$$

$$= 0 + 0\cdot(x-b)^1 + \frac{+1}{\Delta^1}\cdot(x-b)^2 + \frac{+1}{\Delta^2}\cdot(x-b)^3. \quad (38)$$

The formulas for $F_{0,1}(x;a,b)$ and $F_{0,2}(x;a,b)$ may be constructed by analogy. The first two forms: (31), (35) and (32), (36) are two-point, constructed with the use of the tables of divided differences. The next two forms: (33), (37) and (34), (38) are one-point, obtained by the substitution $(x-a) = (x-b) + \Delta$.

Generally, table 2 of divided differences for $F_{1,0}$ is a set of coefficients at monomials $a^p b^q$ in expansion $-\left(\frac{a-b}{\Delta}\right)^{p+q}$. It is the union of two fragments of the B. Pascal triangle with signs changing from column to column and with degrees of Δ increasing in denominators.

Table 2.

	$\lambda_0^{\leftarrow} = F_{1,0}^{(0)}(b) = 0$	$F_{1,0}^{(1)}(b) = 0$	$\cdots$	$F_{1,0}^{(\mu-1)}(b) = 0$
$\lambda_0^{\rightarrow} = F_{1,0}^{(0)}(a) = 1$	$\frac{-1}{\Delta^1}$	$\frac{+1}{\Delta^{1+2-1}}$	$\cdots$	$\eta_0^{\leftarrow} = \frac{(-1)^\mu}{\Delta^\mu}\cdot\binom{\mu-1}{\mu-1}$
$F_{1,0}^{(1)}(a) = 0$	$\frac{-1}{\Delta^{2+1-1}}$	$\frac{(-1)^2}{\Delta^{2+2-1}}\cdot\binom{2}{1}$	$\cdots$	$\frac{(-1)^\mu}{\Delta^{2+\mu-1}}\cdot\binom{\mu}{\mu-1}$
$\vdots$	$\vdots$	$\vdots$	$\ddots$	$\vdots$
$F_{1,0}^{(\mu-1)}(a) = 0$	$\eta_0^{\rightarrow} = \frac{-1}{\Delta^\mu}\cdot\binom{\mu-1}{0}$	$\frac{+1}{\Delta^{\mu+2-1}}\cdot\binom{\mu}{1}$	$\cdots$	$\frac{(-1)^\mu}{\Delta^{2\mu-1}}\cdot\binom{2\mu-2}{\mu-1}$

The tables for $F_{p,0}$ may be obtained from $F_{1,0}$ by means of shift down for $(p-1)$ lines (the top lines becoming empty must be filled in by zeros) and by means of the result division by $(p-1)!$.

The table for $F_{0,1}$ is a copy of the one for $F_{1,0}$: only a sign replacement and the headers correction are necessary. The tables for $F_{0,p}$ may be obtained from $F_{0,1}$ by means of shift to the right (with the use filling in of columns by zeros) and division by $(p-1)!$.

IOP Conf. Series: Journal of Physics: Conf. Series **973** (2018) 012025 doi:10.1088/1742-6596/973/1/012025

For $p = 1, \ldots, \mu$ the form "by a, then by b" can be written as

$$F_{p,0}^{\rightarrow}(x;a,b) = \frac{(x-a)^{p-1}}{(p-1)!} + \frac{(x-a)^{\mu}}{(p-1)!} \cdot \left[\sum_{k=0}^{\mu-1} \frac{(-1)^{k+1}}{\Delta^{\mu+k-p+1}} \cdot \left(\begin{array}{c} \mu+k-p \\ k \end{array} \right) \cdot (x-b)^k \right], \tag{39}$$

$$F_{0,p}^{\rightarrow}(x;a,b) = \frac{(x-a)^{\mu}}{(p-1)!} \cdot \left[\sum_{k=p-1}^{\mu-1} \frac{(-1)^{k+p-1}}{\Delta^{\mu+k-p+1}} \cdot \left(\begin{array}{c} \mu+k-p \\ k-p+1 \end{array} \right) \cdot (x-b)^k \right]. \tag{40}$$

It is reasonably to compute only once (by means of integer additions) and to save $(\mu+1)\cdot\mu/2$ binomial coefficients.

To compute $P_{\mu,\mu}$ in (30) it is necessary to add up 2μ basis polynomials of the same form multiplied by the corresponding 2μ preset given coefficients. Formally it requires to execute $4\mu^2$ multiplications. However, in two-point form ("by a, then by b") for the lower part of $F_{p,0}^{\rightarrow}(x;a,b)$ only one coefficient is not equal to zero; for $F_{0,p}^{\rightarrow}(x;a,b)$ the lower part entirely is equal to zero, and in the high part the coefficients are equal to zero at the items $(x-a)^{\mu}\cdot(x-b)^k$ for $k < p-1$. In fact, it requires to accomplish $(\mu^2+\mu)\cdot 3/2$ multiplications and to prepare μ negative degrees of Δ. The first one-point form ("by a") almost always has μ non-zero coefficients in the higher parts $F_{k,0}(x;a,b)$ because of decomposition $(x-a)^k = \sum\limits_{p=0}^{k}(x-b)^p \cdot \Delta^{k-p} \cdot \left(\begin{array}{c} k \\ p \end{array} \right)$.

Unlike of subsection 5.1 the method of basis polynomials is better because (when the basis polynomials and $P_{\mu,\mu}(x;a,b)$ are already constructed) it allows replace k values in the set (21) (for example, for variation) by $O(2k\mu)$ operations. But if all 2μ parameters vary, then the subsection 5.1 method is preferable.

5.3. On $P_{\mu,\mu}$ computation and reconstructions

Statement 1. *For $c \in [a;b]$ the computation of value $P_{\mu,\mu}(c;a,b)$ requires 2μ multiplications.*

Statement 2. *The presentation $P_{\mu,\mu}(x;a,b) = T_{2\mu}(x;a)$ (one-point – in the point a), where*

$$T_{2\mu}(x;a) = \sum_{k=0}^{2\mu-1} t_k(a)\cdot(x-a)^k, \qquad t_k(a) = \frac{T_{2\mu}^{(k)}(a;a)}{k!}, \quad k = 0,\ldots,2\mu-1 \tag{41}$$

may be obtained from (24) by $O(\mu)$ multiplications/divisions.

Really, the coefficients $(\lambda_k^{\rightarrow})_{k=0,\ldots,\mu-1}$ coincide with $t_k(a)$ for $k = 0,\ldots,\mu-1$. Then, for $(\eta_k^{\rightarrow})_{k=0,\ldots,\mu-1}$ it is necessary to find the coefficients $\left(A_{\mu+k}^{\rightarrow}\right)_{k=0,\ldots,\mu-1}$ in the multiplier

$$\sum_{k=0}^{\mu-1} \eta_k^{\rightarrow}\cdot(x-b)^k = \sum_{k=0}^{\mu-1} A_{\mu+k}^{\rightarrow}\cdot(x-a)^k \tag{42}$$

of the polynomial $P_{\mu,\mu}(x;a,b)$ higher part $H_{\mu}^{\rightarrow}(x;a,b)$. It requires $O(\mu)$ multiplications/divisions. At last, $t_k(a) = A_k^{\rightarrow}$ for $k = \mu,\ldots,2\mu-1$.

Statement 3. *If the coefficients of $T_{2\mu}(x;a)$ are known, then for $c \in (a;b]$ the presentation (one-point – in the point c) $P_{\mu,\mu}(x;a,b) = T_{2\mu}(x;c)$, where*

$$T_{2\mu}(x;c) = \sum_{k=0}^{2\mu-1} t_k(c)\cdot(x-c)^k, \qquad t_k(c) = \frac{T_{2\mu}^{(k)}(c;c)}{k!}, \quad k = 0,\ldots,2\mu-1 \tag{43}$$

may be obtained by $O(\mu)$ multiplications/divisions.

IOP Conf. Series: Journal of Physics: Conf. Series **973** (2018) 012025 doi:10.1088/1742-6596/973/1/012025

Really, using the coefficients $(t_k(a))_{k=0,\ldots,2m-1}$ we can find the coefficients $(t_k(c))_{k=0,\ldots,2m-1}$ in the presentation

$$\sum_{k=0}^{2\mu-1} t_k(a)\cdot(x-a)^k = \sum_{k=0}^{2\mu-1} t_k(c)\cdot(x-c)^k \tag{44}$$

of the polynomial $P_{\mu,\mu}(x;a,b)$, that requires $O(2\mu)$ multiplications/divisions.

Statement 4. *If the coefficients of $P_{\mu,\mu}(x;a,b)$ are known, then for $[\tilde{a};\tilde{b}]\subset[a;b]$ the presentation (two-point "by $\tilde{a}$, then by $\tilde{b}$")*

$$P_{\mu,\mu}\left(x;\tilde{a},\tilde{b}\right) = L_\mu^{\rightarrow}\left(x;\tilde{a},\tilde{b}\right) + H_\mu^{\rightarrow}\left(x;\tilde{a},\tilde{b}\right), \tag{45}$$

$$L_\mu^{\rightarrow}(x;\tilde{a},\tilde{b}) = \sum_{k=0}^{\mu-1}\tilde{\lambda}_k^{\rightarrow}\cdot(x-\tilde{a})^k, \qquad H_\mu^{\rightarrow}(x;\tilde{a},\tilde{b}) = (x-\tilde{a})^\mu\cdot\sum_{k=0}^{\mu-1}\tilde{\eta}_k^{\rightarrow}\cdot(x-\tilde{b})^k \tag{46}$$

may be obtained by $O(3\mu)$ multiplications/divisions.

Really, after $O(2\mu)$ multiplications/divisions, according to Statements 2 or 3 we can find

$$T_{2\mu}(x;\tilde{a}) = \sum_{k=0}^{2\mu-1} t_k(\tilde{a})\cdot(x-\tilde{a})^k \tag{47}$$

and with the use of the found $\tilde{\lambda}_k^{\rightarrow} = t_k(\tilde{a})$ for $k=0,\ldots,\mu-1$ we can determinate $L_\mu^{\rightarrow}(x;\tilde{a},\tilde{b})$. The pass to $H_\mu^{\rightarrow}$ requires additionally $O(\mu)$ multiplications/divisions:

$$(x-\tilde{a})^\mu\cdot\sum_{k=0}^{\mu-1} t_{k+\mu}(\tilde{a})\cdot(x-\tilde{a})^k = (x-\tilde{a})^\mu\cdot\sum_{k=0}^{\mu-1}\tilde{\eta}_k^{\rightarrow}\cdot\left(x-\tilde{b}\right)^k = H_\mu^{\rightarrow}\left(x;\tilde{a},\tilde{b}\right). \tag{48}$$

5.4. Notes

1) In Statement 1 the variant of the W. G. Horner scheme [14] is applied.
2) In Statements 2-4 the M. Shaw and J. F. Traub variant [14] of the W. G. Horner scheme is applied. This variant has the complexity $O(\mu)$ multiplications/divisions. Besides, this variant contains double cycle. So it has the complexity $O(\mu^2)$ additions/subtractions.
3) The pass from $P_{\mu,\mu}(x;a,b)$ to $P_{\mu,\mu}\left(x;\tilde{a},\tilde{b}\right)$ is applied for example when it is necessary to localize on subintervals [6] initial approximations being preset given on continuity intervals. In particular, it applies for addition of points in (4), when the condition (17) or (18) is satisfied. Further, the individual polynomials will be used on the subintervals (for the controls variation for the functionals [8, 15, 16] optimization).
4) From the presented estimates it follows that the use of explicit expressions as the algorithmic base for computation of derivatives $P_{\mu,\mu}^{(k)}(c;a,b)$ for $k=0,\ldots,\mu-1$ at $c\in[a;b]$

$$P_{\mu,\mu}^{(k)}(c;a,b) = \sum_{p=k}^{\mu-1} f_{p+1,0}\cdot\frac{p!\cdot(c-a)^{p-k}}{(p-k)!} +$$
$$+\left[\sum_{q=0}^{k}\sum_{s=0}^{q} + \sum_{q=k+1}^{\mu-1}\sum_{s=0}^{k}\right] f_{\mu,q+1}\cdot\frac{k!\cdot\mu!\cdot q!\cdot(c-a)^{\mu-(k-s)}\cdot(c-b)^{q-s}}{s!\cdot(k-s)!\cdot(\mu-(k-s))!\cdot(q-s)!} \tag{49}$$

at high μ is worse than the use of $T_{2\mu}(x;c)$, where $P_{\mu,\mu}^{(k)}(c;a,b) = k!\cdot t_k(c)$.

IOP Conf. Series: Journal of Physics: Conf. Series **973** (2018) 012025 doi:10.1088/1742-6596/973/1/012025

Acknowledgments
This work has been supported financially by the Ministry of Education and Science of the Russian Federation (the unique identifier of the agreement – RFMEF157715X0195).

References
[1] Bilchenko G G and Bilchenko N G 2015 Comparative analysis of some special blowing laws application in the problems of heat and mass transfer optimal control on the permeable porous surfaces of hypersonic aircraft *"Actual Problems of Applied Mathematics, Informatics and Mechanics" Proc. Int. Scientific and Technical Conf.* (Russia, Voronezh: "Research Publishers") pp 137-139 (in Russian)
[2] Bilchenko G G and Bilchenko N G 2015 Comparative analysis of some special blowing laws application in the problems of heat and mass transfer optimal control on the permeable perforated surfaces of hypersonic aircraft *"Actual Problems of Applied Mathematics, Informatics and Mechanics" Proc. Int. Scientific and Technical Conf.* (Russia, Voronezh: "Research Publishers") pp 134-136 (in Russian)
[3] Bilchenko G G 2011 On valency of Cisotti's generalized integrals *"Modern Problems of Applied Mathematics, Control Theory and Mathematical Modeling" Proc. IV Int. Scientific Conf.* (Russia, Voronezh: VSU Publishing Center) pp 31-33 (in Russian)
[4] Bilchenko G G 2011 Starlike modification of strictly close-to-convex Cisotti's integral *"Actual Problems of Applied Mathematics, Informatics and Mechanics" Proc. Int. Scientific Conf.* (Russia, Voronezh: VSU Publishing Center) pp 69-71 (in Russian)
[5] Bilchenko G G and Bilchenko N G 2016 On the possibility of control restoration in some inverse problems of heat and mass transfer *11th Int. Conf. "Mesh methods for boundary-value problems and applications" IOP Conf. Series: Mater. Sci. Eng.* **158:1** 012020 pp 1-6 (Russia, Kazan: IOP Publishing) DOI: 10.1088/1757-899X/158/1/012020 URL: http://iopscience.iop.org/article/10.1088/1757-899X/158/1/012020/pdf
[6] Bilchenko G G and Bilchenko N G 2016 On computational experiments in some inverse problems of heat and mass transfer *11th Int. Conf. "Mesh methods for boundary-value problems and Applications" IOP Conf. Series: Mater. Sci. Eng.* **158:1** 012021 pp 1-5 (Russia, Kazan: IOP Publishing) DOI: 10.1088/1757-899X/158/1/012021 URL: http://iopscience.iop.org/article/10.1088/1757-899X/158/1/012021/pdf
[7] Bilchenko G G and Bilchenko N G 2017 Inverse problems of heat and mass transfer on hypersonic aircrafts permeable surfaces III. On the statement of two dimensional problems and the ranges of allowed values "heat-friction" *Proc. Voronezh State Univ. Ser. System Analysis and Inform. Technologies* **1** 18-25 (in Russian) URL: http://www.vestnik.vsu.ru/pdf/analiz/2017/01/2017-01-03.pdf
[8] Bilchenko N G 2015 Permeable surfaces hypersonic aircraft optimal heat protection mathematical modeling *Proc. 2015 Int. Conf. "Stability and Control Processes" SCP-2015 (in Memory of V I Zubov)* ed L A Petrosyan and A P Zhabko (Russia, Saint-Petersburg: IEEE) pp 310-313 DOI: 10.1109/SCP.2015.7342145 URL: http://ieeexplore.ieee.org/document/7342145/
[9] Polyakov A F and Reviznikov D L 1999 Singularities of thermal protection of the front edge under conditions of combination porous penetration and convection-conduction cooling *High Temperature* **37:6** 895-898
[10] Varaksin A Yu, Polyakov A F, Reviznikov D L, Strat'ev V K and Tret'yakov A F 2002 Investigation of flow in the vicinity of porous material on a cylindrical body subjected to flow *High Temperature* **40:6** 843-849 URL: https://doi.org/10.1023/A:1021473116156
[11] Polezhaev Yu V and Yurevich F B 1976 *Thermal Protection* (Russia, Moscow: "Energiya") 392 (in Russian)
[12] Bilchenko G G and Bilchenko N G 2016 The classes of solutions of the problems of boundary layer optimal control on the permeable surfaces of hypersonic aircrafts *Proc. Int. Conf. "Voronezh Winter Math. School of S G Krein"* ed V A Kostin (Russia, Voronezh: "Nauchnaya kniga" Publishers) pp 82-86 (in Russian)
[13] Samarskiy A A and Gulin A V 1989 *Numerical Methods* (Russia, Moscow: "Nauka") 432 (in Russian) ISBN 5-02-013996-3
[14] Knuth D E 1997 *The Art of Computer Programming. vol 2. Seminumerical Algorithms. – 3rd ed.* (Reading, Massachusetts: Addison-Wesley) xiv+762 ISBN 0-201-89684-2
[15] Bilchenko G G and Bilchenko N G 2016 Construction of range of extreme values of functionals *Diff. Equations and Control Processes* **2:2** "Herzen Readings 2016. Some Actual Probl. of Modern Math. and Math. Educ." ed V F Zaytsev, V D Budaev and A V Flegontov (Russia, Saint-Petersburg) pp 56-61 (in Russian) URL: http://www.math.spbu.ru/diffjournal/pdf/herzen2016.pdf
[16] Bilchenko G G and Bilchenko N G 2016 Functional extreme values range *Proc. 2016 Int. Conf. "Stability and Oscillations of Nonlinear Control Systems" (Pyatnitskiy's Conf.)* ed V N Tkhai (Russia, Moscow: IEEE) pp 1-2 DOI: 10.1109/STAB.2016.7541166 URL: http://ieeexplore.ieee.org/document/7541166/

IOP Conf. Series: Journal of Physics: Conf. Series **973** (2018) 012026 doi:10.1088/1742-6596/973/1/012026

Motion of a carrier with a mobile load along a rough inclined plane

G G Bilchenko

Department of Theoretical and Applied Mechanics and Mathematics, Kazan National Research Technical University (KNRTU-KAI) named after A N Tupolev, 10, K Marx St., Kazan, Tatarstan 420111, Russia

E-mail: ggbil40@gmail.com

Abstract. The mechanical system consisting of a carrier and a load is considered. The load can move respectively the carrier according to the preset given motion law. The carrier motion from rest caused the load motion is investigated. The carrier can move translationally along rectilinear trajectory along rough inclined plane. The trajectory is the line of the greatest descent. The axis of rectilinear channel along which the load moves is situated in vertical plane containing the carrier trajectory. The Coulomb model is taken to describe the friction forces on sloped plane. Differential equations of motion of carrier with load are obtained. The sufficient condition of the carrier motion without detachment from inclined plane is given. For two special cases of the channel installation angle and the plane inclination angle combination the motion types are described. The computation experiments results are presented: the carrier motions in the special cases are illustrated, the phase portraits for some types of motions are constructed.

1. Introduction

The vibration-driven systems are a new type of movable dynamical systems whose motion is carried out by periodic motion of internal masses. These systems application in disaster rescue, pipeline inspection and cardiovascular surgery is preferable in comparison with ordinary legged and wheeled robots because of their simplicity, controllability and miniaturization potential.

Such systems have been studied from many points of view. The laws of motion of the load, providing the required carrier motion, were found in [1, 2, 3, 4, 5]. The inverse problem of finding the carrier motion for the *initial conditions*

$$x(t_0) = 0 \qquad \text{and} \qquad \dot{x}(t_0) = 0$$

at $t_0 = 0$ for a preassigned law of load motion $x_2(t) = \ell \cdot \sin(\omega t)$, where $\ell = Const$, $\omega = Const$, when the carrier moves along the *horizontal* plane, is posed and solved in [6, 7, 8]. The problem of definition of load motion influence on carrier dynamics is considered in [9] for the other load motion law.

In this paper the motion of carrier with mobile load along *inclined* rough plane is investigated. The author has no information on any research on motion of vibration-driven systems on inclined rough plane.

2. The carrier motion differential equations

The carrier motion dynamics is described exactly, if the carrier motion initial conditions are defined correctly, that allow avoid the situations like the P. Painlevé paradoxes [10]. It means that the carrier must be in rest on inclined plane before the load starts to move.

Let's consider the vertical plane Q containing the carrier trajectory and the load motion channel axis as the system material symmetry plane. Then all the forces acting on the system: $\vec{P}_C$ – the carrier weight, $\vec{P}_L$ – the load weight, $\vec{N}$ – the sloped inclined plane normal reaction and $\vec{F}_{fr}$ – the sliding friction force in rest are situated in the plane Q. Since $\vec{P}_C \uparrow\uparrow \vec{P}_L$ and, hence, $\left(\vec{P}_C, \vec{P}_L\right) \sim \vec{P}$, where $\vec{P} = \vec{P}_C + \vec{P}_L$, then the carrier will be in rest on the inclined plane if the system of forces acting on it is balanced, i.e. $\left(\vec{P}, \vec{N}, \vec{F}_{fr}\right) \sim 0$. As the considered system is a system of convergent forces on plane, then writing its balance conditions we obtain (see fig. 1)

$$\sum_{s=1}^{3} F_{s_x} \equiv -(P_C + P_L) \cdot \sin\alpha + F_{fr} = 0\,, \tag{1}$$

$$\sum_{s=1}^{3} F_{s_z} \equiv -(P_C + P_L) \cdot \cos\alpha + N = 0\,. \tag{2}$$

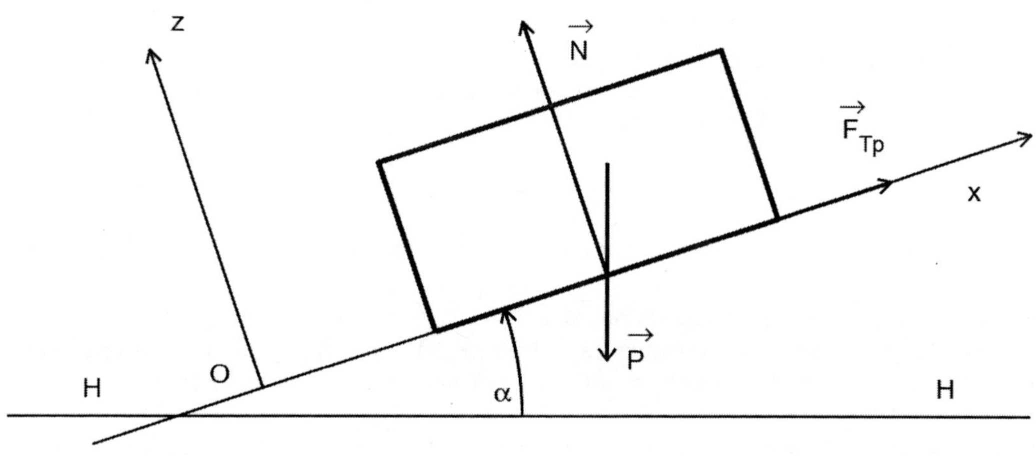

Figure 1. The carrier position on inclined plane scheme

Let's add to (1), (2) the Coulomb-Amonton law

$$F_{fr} \le F_{fr_{\max}}\,, \tag{3}$$

where $F_{fr_{\max}} = f \cdot N$ and f is the coefficient of sliding friction in rest for the pair of materials "carrier-inclined plane". The conditions (1), (2) are the balance equations. Solving them we can find F_{fr} and N. The substitution of F_{fr} and N into (3) allows obtain the condition for the plane inclination angle

$$\operatorname{tg}\alpha \le f\,.$$

Further on the condition is presented as follows

$$|\alpha| \le \operatorname{arctg} f\,. \tag{4}$$

Note. On fig. 1 HH is the line of intersection of the horizontal plane and the vertical plane Q; the plane inclination angle $\alpha > 0$.

According to [6] let's introduce:

$Oxyz$ is the motionless coordinates system being an inertial counting out system, where Oxy is the plane coinciding with the inclined plane; the axis Ox is directed along the line of the greatest descent (to ascent for $\alpha > 0$ or to descent for $\alpha < 0$), and axis Oz is directed normally to inclined plane;

$O_1x_1y_1z_1$ is the mobile coordinates system fastened together with the carrier, besides, $O_1x_1z_1$ coincides with the vertical plane, and $O_1x_1 \uparrow\uparrow Ox$;

$O_2x_2y_2z_2$ is the mobile coordinates system fastened together with the channel, besides, $O_2x_2z_2$ coincides with the vertical plane, and φ is the channel installation angle – the angle between axes O_1x_1 and O_2x_2 counting up from inclined plane for $\varphi > 0$.

If the load motion in channel law is given as $x_2(t) = \ell \cdot \sin(\omega t)$, where $\ell = Const$, $\omega = Const$, and the carrier motion medium resistance forces are the Coulomb friction type forces, then the *carrier motion differential equations* (CMDE) for the case inclined plane can be obtained analogously [6] and written as follows:

$$\ddot{x} = \beta\,(\cos\varphi + f \cdot \sin\varphi) \cdot \sin(\omega t) - g\,(f \cdot \cos\alpha + \sin\alpha) \qquad \text{for} \qquad \dot{x} > 0\,, \tag{5}$$

$$\ddot{x} = \beta\,(\cos\varphi - f \cdot \sin\varphi) \cdot \sin(\omega t) + g\,(f \cdot \cos\alpha - \sin\alpha) \qquad \text{for} \qquad \dot{x} < 0\,, \tag{6}$$

$$\ddot{x} = 0 \qquad \text{for} \qquad \dot{x} = 0\,, \tag{7}$$

where x is the carrier coordinate in $Oxyz$; $\beta = \dfrac{m}{M+m} \cdot \ell \cdot \omega^2$; M is the carrier mass; m is the load mass; g is the free fall acceleration; f is the sliding friction in motion coefficient, which is equal to the sliding friction in rest coefficient for the pair "carrier-inclined plane".

Analogously [6] the following inequality is obtained

$$\beta \cdot |\sin\varphi| < g \cdot \cos\alpha\,. \tag{8}$$

Its fulfillment guarantees the carrier motion without detachment from inclined plane.

3. The carrier with mobile load dynamics

Let's the system "carrier-load" parameters values m, M, ℓ, ω, f, φ, α be given, and the parameters are such, that the conditions (4) and (8) are fulfilled. Let's determine the *carrier motion* (CM) from rest.

The **necessary condition** of carrier motion from rest in the axis Ox **positive direction** is

$$\beta\,(\cos\varphi + f \cdot \sin\varphi) > g\,(f \cdot \cos\alpha + \sin\alpha)\,. \tag{9}$$

Then from

$$\beta\,(\cos\varphi + f \cdot \sin\varphi) \cdot \sin(\omega\tau_+) = g\,(f \cdot \cos\alpha + \sin\alpha)$$

we can find the carrier start moment for positive direction

$$\tau_+ = \frac{1}{\omega} \arcsin\left(\frac{g}{\beta} \cdot \frac{f \cdot \cos\alpha + \sin\alpha}{\cos\varphi + f \cdot \sin\varphi}\right). \tag{10}$$

The **necessary condition** of carrier motion from rest in the axis Ox **negative direction** is

$$\beta\,(\cos\varphi - f \cdot \sin\varphi) > g\,(f \cdot \cos\alpha - \sin\alpha)\,. \tag{11}$$

Then from

$$\beta\,(\cos\varphi - f\cdot\sin\varphi)\cdot\sin\left(\omega\left(\frac{T}{2}+\tau_-\right)\right) = -\,g\,(f\cdot\cos\alpha-\sin\alpha)\,,$$

where $T = \dfrac{2\pi}{\omega}$, we can find the carrier start moment for negative direction

$$\tau_- = \frac{1}{\omega}\arcsin\left(\frac{g}{\beta}\cdot\frac{f\cdot\cos\alpha-\sin\alpha}{\cos\varphi - f\cdot\sin\varphi}\right). \tag{12}$$

If the (9) and (11) take place, then the carrier motion from rest along inclined plane is possible both in the axis Ox positive and negative directions. Let's consider a *special case* of carrier motion when

$$\tau_+ = \tau_- = \tau\,. \tag{13}$$

Taking α as preset given from ratio

$$\frac{f\cdot\cos\alpha+\sin\alpha}{\cos\varphi_* + f\cdot\sin\varphi_*} = \frac{f\cdot\cos\alpha-\sin\alpha}{\cos\varphi_* - f\cdot\sin\varphi_*}$$

we can find

$$\operatorname{tg}\varphi_* = \frac{1}{f^2}\cdot\operatorname{tg}\alpha\,. \tag{14}$$

Hence, from (14) we obtain that α and φ_* must have the same sign.

Let's $\alpha > 0$, then $\varphi_* > 0$. Since $\operatorname{tg}\alpha \le f$, then

$$\operatorname{tg}\varphi_* \le \frac{1}{f}\,.$$

Hence, with respect to (4) in the considered case of carrier motion the channel installation angle along which the load moves is such, that

$$\varphi_* < \operatorname{arctg}\left(\frac{1}{f}\right). \tag{15}$$

Taking into account (14), let's find

$$\cos\alpha = \frac{1}{\sqrt{1+f^4\cdot\operatorname{tg}^2\varphi_*}} \qquad \text{and} \qquad \sin\alpha = \frac{f^2\cdot\operatorname{tg}\varphi_*}{\sqrt{1+f^4\cdot\operatorname{tg}^2\varphi_*}}\,. \tag{16}$$

Then the CMDE for the considered special case are the following

$$\ddot{x} = (\cos\varphi_* + f\cdot\sin\varphi_*)\cdot(\beta\sin(\omega t) - \widetilde{\gamma}) \qquad \text{for} \qquad \dot{x} > 0\,, \tag{17}$$

$$\ddot{x} = (\cos\varphi_* - f\cdot\sin\varphi_*)\cdot(\beta\sin(\omega t) + \widetilde{\gamma}) \qquad \text{for} \qquad \dot{x} < 0\,, \tag{18}$$

$$\ddot{x} = 0 \qquad \text{for} \qquad \dot{x} = 0\,, \tag{19}$$

where

$$\widetilde{\gamma} = \frac{\gamma}{\sqrt{\cos^2\varphi_* + f^4\cdot\sin^2\varphi_*}}\,,$$

and

$$\gamma = g\cdot f\,.$$

For the CMDE (17)-(19) there is

$$\tau_+ = \tau_- = \tau = \frac{1}{\omega} \arcsin \left(\frac{\gamma}{\beta} \frac{1}{\sqrt{\cos^2 \varphi_* + f^4 \cdot \sin^2 \varphi_*}} \right). \tag{20}$$

Let's find $\beta_1 > \widetilde{\gamma}$ such, for that

$$\int\limits_{\tau_1}^{\frac{T}{2}+\tau_1} \left(\beta_1 \cdot \sin(\omega t) - \widetilde{\gamma} \right) dt = 0\,, \tag{21}$$

where

$$\tau_1 = \frac{1}{\omega} \arcsin \left(\frac{\widetilde{\gamma}}{\beta_1} \right).$$

From (21) we can find that

$$\beta_1 = \widetilde{\gamma} \cdot \frac{\sqrt{\pi^2+4}}{2} = \frac{\gamma}{2} \frac{\sqrt{\pi^2+4}}{\sqrt{\cos^2 \varphi_* + f^4 \cdot \sin^2 \varphi_*}}. \tag{22}$$

Analogously we can find $\beta_2 > \widetilde{\gamma}$ such, for that

$$\int\limits_{\frac{T}{2}+\tau_2}^{T+\tau_2} \left(\beta_2 \cdot \sin(\omega t) + \widetilde{\gamma} \right) dt = 0\,, \tag{23}$$

where

$$\tau_2 = \frac{1}{\omega} \arcsin \left(\frac{\widetilde{\gamma}}{\beta_2} \right).$$

From (23) we can obtain

$$\beta_2 = \widetilde{\gamma} \cdot \frac{\sqrt{\pi^2+4}}{2} = \frac{\gamma}{2} \frac{\sqrt{\pi^2+4}}{\sqrt{\cos^2 \varphi_* + f^4 \cdot \sin^2 \varphi_*}}. \tag{24}$$

So, in the considered special case of carrier motion

$$\beta_1 = \beta_2 = \frac{\gamma}{2} \frac{\sqrt{\pi^2+4}}{\sqrt{\cos^2 \varphi_* + f^4 \cdot \sin^2 \varphi_*}} = \frac{\beta_0}{\sqrt{\cos^2 \varphi_* + f^4 \cdot \sin^2 \varphi_*}}, \tag{25}$$

where $\beta_0 = \frac{\gamma}{2} \cdot \sqrt{\pi^2+4}$.

Therefore, if in the studying system "carrier-load" the parameter

$$\beta < \frac{\beta_0}{\sqrt{\cos^2 \varphi_* + f^4 \cdot \sin^2 \varphi_*}},$$

then the CM belongs to the $R2$ type [6] for which the following regular time consequence of the CMDE (17)-(19) alternations takes place:

$$t_0 \xrightarrow{19} \underbrace{\mathbf{t}_1 \xrightarrow{17} t_2 \xrightarrow{19} t'_2 \xrightarrow{18} t_3 \xrightarrow{19} \mathbf{t}'_3} \overbrace{\xrightarrow{17} t_4 \xrightarrow{19} t'_4 \xrightarrow{18} t_5 \xrightarrow{19} \mathbf{t}'_5} \xrightarrow{17} \dots$$

Here $t_0 = 0\,;\, t_1 = \tau\,;\, \dot{x}\,(t_2) = 0\,;\, t'_i = \dfrac{T}{2}\cdot(i-1)+\tau$, where $i = 2, 3, \ldots;\, t_k = \dfrac{T}{2}\cdot(k-2)+t_2$, where $k = 3, 4, \ldots$

If on a level with the system (17), (18) we consider a similar system for which $\beta_3 > \widetilde{\gamma}$ and

$$\int\limits_{\tau_3}^{\frac{T}{2}+\theta} (\beta_3 \cdot \sin(\omega t) - \widetilde{\gamma}) \cdot dt = 0 \tag{26}$$

and

$$\int\limits_{\frac{T}{2}+\theta}^{T+\tau_3} (\beta_3 \cdot \sin(\omega t) + \widetilde{\gamma}) \cdot dt = 0\,, \tag{27}$$

where

$$\tau_3 = \frac{1}{\omega}\arcsin\left(\frac{\widetilde{\gamma}}{\beta_3}\right),$$

then we obtain from (26)

$$\beta_3 \cdot \cos(\omega\theta) + \sqrt{\beta_3^2 - \widetilde{\gamma}^2} = \widetilde{\gamma}\,[\pi + \omega\theta - \omega\tau_3]\,, \tag{28}$$

and find from (27)

$$\beta_3 \cdot \cos(\omega\theta) + \sqrt{\beta_3^2 - \widetilde{\gamma}^2} = \widetilde{\gamma}\,[\pi + \omega\tau_3 - \omega\theta]\,. \tag{29}$$

Comparing (28) and (29) we ascertain that $\omega\theta = \omega\tau_3$, and then $2\sqrt{\beta_3^2 - \widetilde{\gamma}^2} = \widetilde{\gamma}\pi$, so

$$\beta_3 = \widetilde{\gamma}\cdot\frac{\sqrt{\pi^2+4}}{2}\,, \tag{30}$$

i.e. it is clarified, that

$$\beta_3 = \beta_2 = \beta_1 = \frac{\beta_0}{\sqrt{\cos^2\varphi_* + f^4\cdot\sin^2\varphi_*}}\,.$$

Hence, if in the investigated system "carrier-load" for the considered case the parameter

$$\beta = \beta_3 = \frac{\beta_0}{\sqrt{\cos^2\varphi_* + f^4\cdot\sin^2\varphi_*}}\,,$$

then the CM belongs to the $R5$ type for which the following regular time consequence of the CMDE (17)-(19) alternations takes place:

$$t_0 \xrightarrow{19} \underbrace{t_1 \xrightarrow{17} t_2 \xrightarrow{18} t_3}\overbrace{\xrightarrow{17} t_4 \xrightarrow{18} t_5}\xrightarrow{17} \ldots$$

Here $t_0 = 0\,;\, t_i = \dfrac{T}{2}\cdot(i-1)+\tau$, where $i = 1, 2, \ldots$

If the parameter $\beta > \beta_3$, then the CM realizes according to the NR type with diverse moments of switching in each half-period of the load motion.

Note. If $\alpha < 0$ and hence $\varphi_* < 0$, then in this case all the above established (for $\alpha > 0$) statements on the carrier motion are correct.

4. The computation experiments

For two cases of CM of R type ($R2$ and $R5$) and for one case of CM of NR type the mathematical modeling results are presented on figures 2-10. The left halves of fig. 2-10 correspond to the the case $\alpha > 0$ and the rights ones correspond to the case $\alpha < 0$. In all the computation experiments the following parameter values are used: $f = 0.204$, $\omega = 6\,[1/\mathrm{s}]$ and $\alpha \cong \pm 0.0285$. Then $\gamma = gf \cong 2\,[\mathrm{m/s^2}]$ and $\varphi_* = \pm 0.6$. Let's find

$$\widetilde{\gamma} = \frac{\gamma}{\sqrt{\cos^2\varphi_* + f^4\cdot\sin^2\varphi_*}} \cong 2.42\,[\mathrm{m/s^2}] \quad \text{and} \quad \widetilde{\beta} = \frac{\gamma}{2}\frac{\sqrt{\pi^2+4}}{\sqrt{\cos^2\varphi_* + f^4\cdot\sin^2\varphi_*}} \cong 4.51\,[\mathrm{m/s^2}]\,.$$

So,

if $\beta = 4.0$, then $\beta < \widetilde{\beta}$ and the CM realizes according the $R2$ type (see fig. 2-4);
if $\beta = 4.51$, then $\beta = \widetilde{\beta}$ and the CM realizes according the $R5$ type (see fig. 5-7);
if $\beta = 5.0$, then $\beta > \widetilde{\beta}$ and the CM realizes according the NR type (see fig. 8-10).

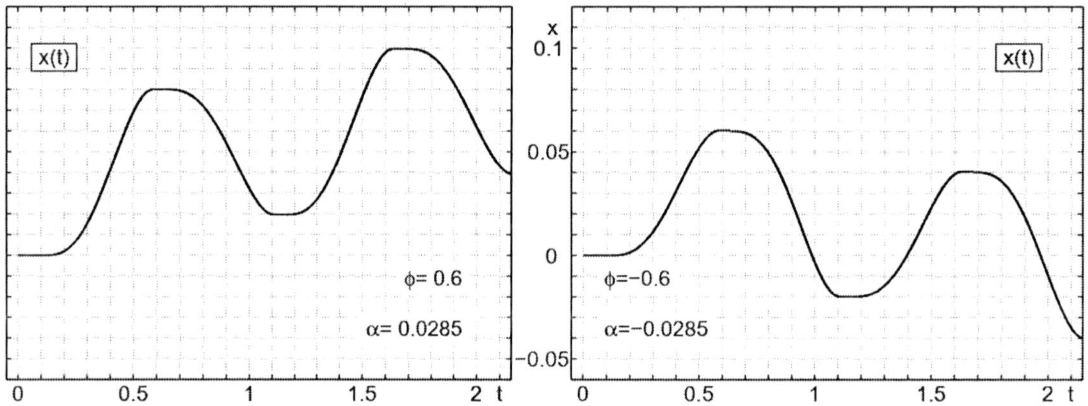

Figure 2. The x dependence on t for CM $R2$ type

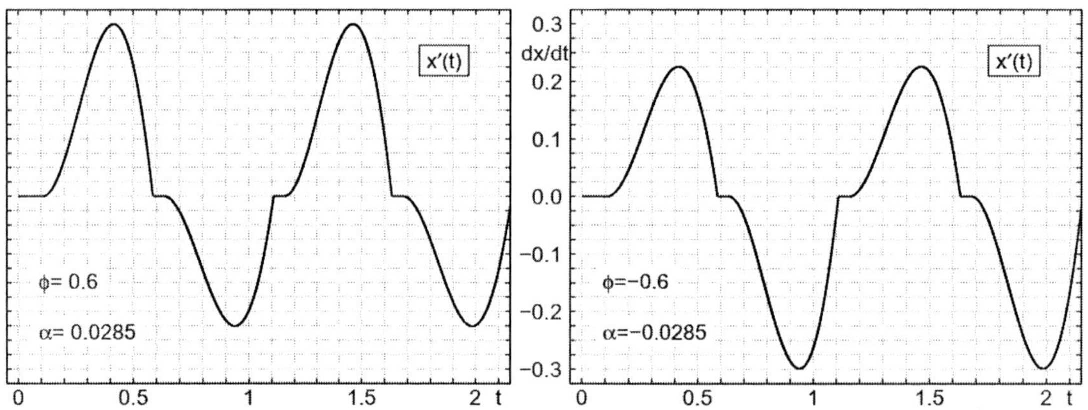

Figure 3. The $\dot{x}$ dependence on t for CM $R2$ type

IOP Conf. Series: Journal of Physics: Conf. Series **973** (2018) 012026 doi:10.1088/1742-6596/973/1/012026

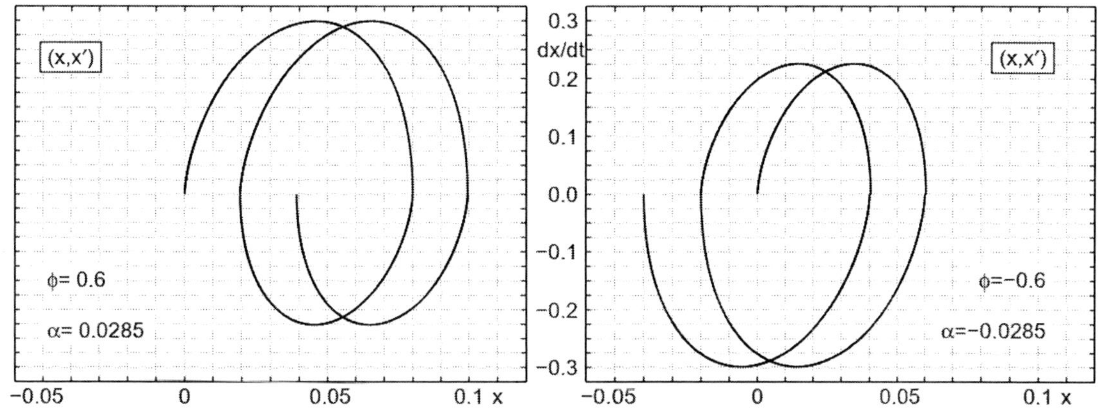

Figure 4. The CM $R2$ type phase portrait

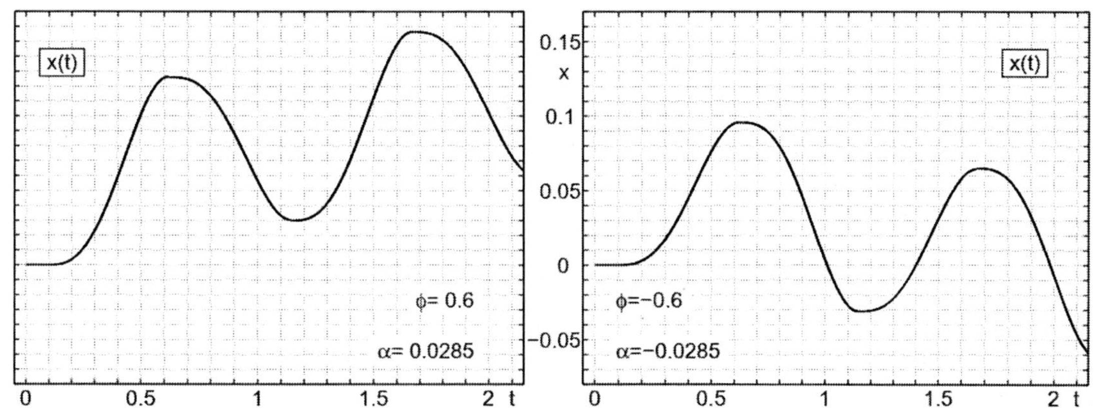

Figure 5. The x dependence on t for CM $R5$ type

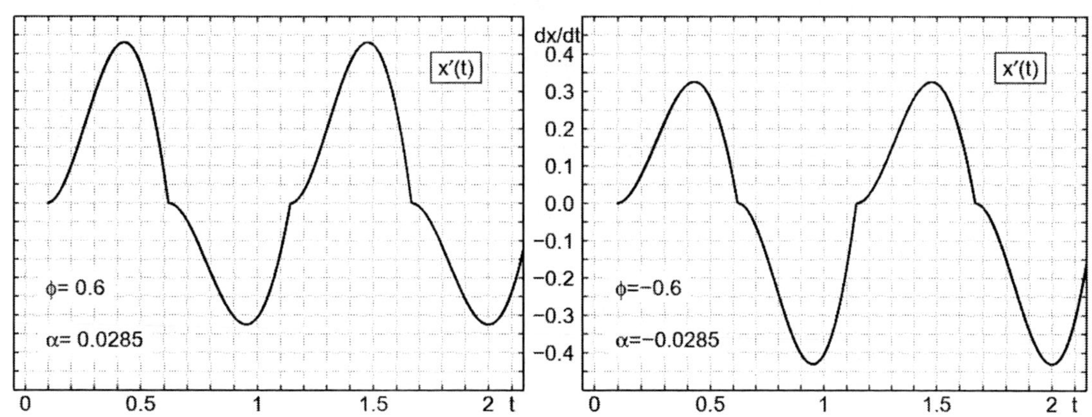

Figure 6. The $\dot{x}$ dependence on t for CM $R5$ type

IOP Conf. Series: Journal of Physics: Conf. Series **973** (2018) 012026 doi:10.1088/1742-6596/973/1/012026

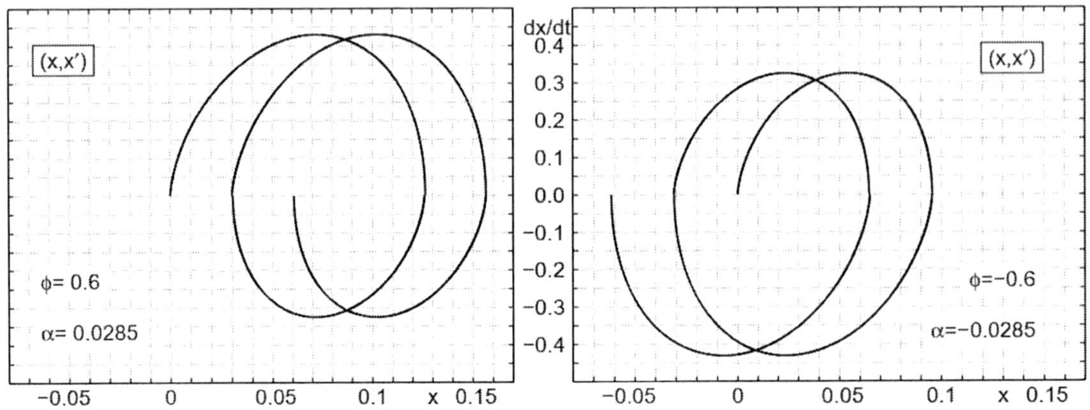

Figure 7. The CM $R5$ type phase portrait

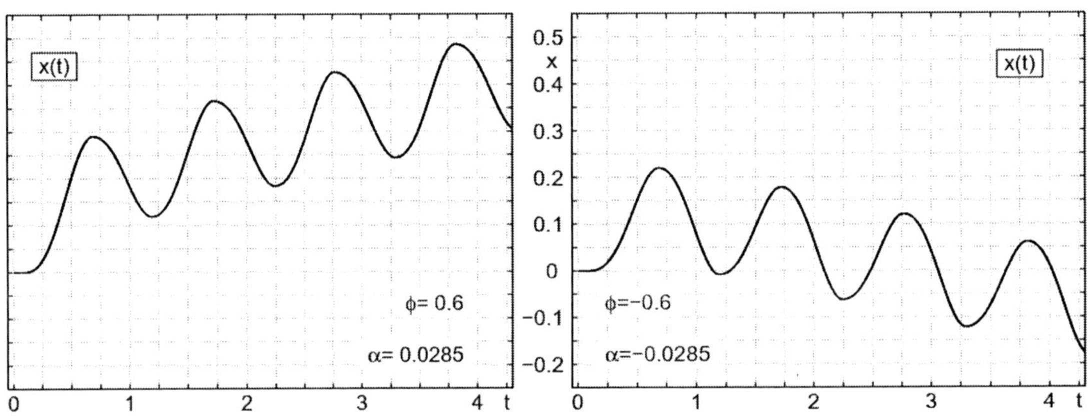

Figure 8. The x dependence on t for CM NR type

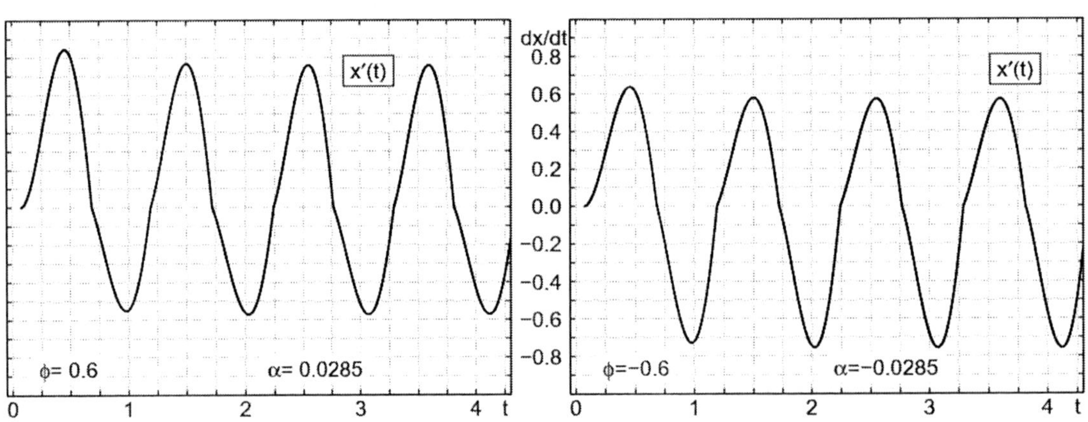

Figure 9. The $\dot{x}$ dependence on t for CM NR type

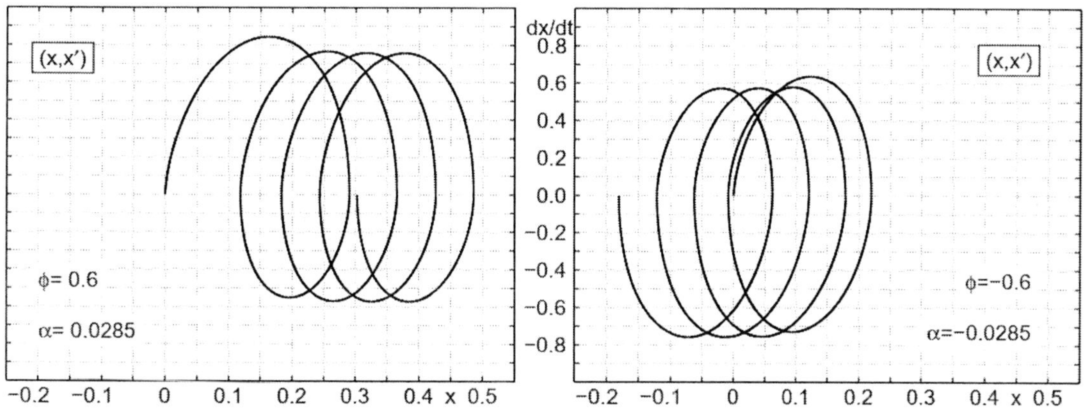

Figure 10. The CM NR type phase portrait

Acknowledgments
This work has been financially supported by the Ministry of Education and Science of the Russian Federation (the unique identifier of the agreement – RFMEF157715X0195).

References

[1] Chernous'ko F L 2005 On the motion of a body containing a movable internal mass *Dokl. Phys.* **50:11** 593-597 DOI: 10.1134/1.2137795

[2] Chernous'ko F L 2006 Analysis and optimization of a body motion controlled by a movable internal mass *J. Appl. Math. Mech.* **70:6** 819-842 DOI: 10.1016/j.jappmathmech.2007.01.003 URL: http://www.sciencedirect.com/science/article/pii/S0021892807000111

[3] Bolotnik N N and Figurina T Yu 2008 Optimal control of rectilinear motion of a rigid body on a rough surface by means of displacement of two internal masses *J. Appl. Math. Mech.* **72:2** 126-135 DOI: 10.1016/j.jappmathmech.2008.04.013 URL: http://www.sciencedirect.com/science/article/pii/S0021892808000452

[4] Chernous'ko F L and Bolotnik N N 2010 Mobile robots controlled by the motion of internal bodies *Trudy Instituta Matematiki i Mekhaniki UrO RAN* **16:5** 213-222 (in Russian) URL: http://www.mathnet.ru/links/60f4a8d9a407770ed2b2e2f0856bba4a/timm624.pdf

[5] Bolotnik N N, Zeidis I M, Zimmerman K and Yatsun S F 2006 Dynamics of Controlled Motion of Vibration-Driven Systems *J. Comput. Syst. Sci. Int.*, **45:5** 831-840 DOI: 10.1134/S1064230706050145

[6] Bilchenko G G 2017 The influence of mobile load on the carrier dynamics *Proc. 2017 Int. Conf. "Constructive Nonsmooth Analysis and Related Topics" CNSA-2017 (dedicated to the memory of V F Demyanov)* ed V N Malozemov, G Sh Tamasyan and M V Udot (Russia, Saint-Petersburg: IEEE) pp 1-4 DOI: 10.1109/CNSA.2017.7973939 URL: http://ieeexplore.ieee.org/document/7973939/

[7] Bilchenko G G 2017 The influence of mobile load on the carrier motion *11th Int. Chetaev Conf. "Analytical Mechanics, Stability and Control"* vol 1, Section 1 "Analitical dynamics" ed S N Vassilyev, V V Kozlov et al (Russia, Kazan: KNRTU-KAI Publishers) pp 37-44 (in Russian)

[8] Bilchenko G G 2017 On the carrier with mobile load motions on the horizontal plane with friction *XII Int. Conf. "High Technology Control System" HTSC-2017* vol 1 ed D A Novikov, O Čokorilo et al (Russia, Lipetsk: Lipetsk State Technical Univ. Publishers) pp 28-32 (in Russian) URL: http://htcs2017.stu.lipetsk.ru/files/HTCS_2017_vol_1.pdf

[9] Bardin B S and Panev A S 2015 On periodic motions of the body with movable internal mass over a horizontal surface *Trudy MAI* **84** (in Russian) URL: http://www.mai.ru/science/trudy/published.php?ID=62995

[10] Rozenblat G M 2009 On the settings of problems in dynamics of a rigid body with constraints and Painlevé paradoxes *The Bulletin of Udmurt Univ. Mathematics. Mechanics. Computer Science* **2** 75-88 (in Russian) DOI: 10.20537/vm090208 URL: http://vst.ics.org.ru/journal/article/1527/

IOP Conf. Series: Journal of Physics: Conf. Series **973** (2018) 012027 doi:10.1088/1742-6596/973/1/012027

On two special values of temperature factor in hypersonic flow stagnation point

G G Bilchenko[1,2] **and N G Bilchenko**[1,3]

[1] Research Laboratory #16, Department of Heat Engineering and Power Engineering Machinery, Kazan National Research Technical University (KNRTU-KAI) named after A N Tupolev, 10, K Marx St., Kazan, Tatarstan 420111, Russia

E-mail: [2] ggbil2@gmail.com and [3] bilchnat@gmail.com

Abstract. The hypersonic aircraft permeable cylindrical and spherical surfaces laminar boundary layer heat and mass transfer control mathematical model properties are investigated. The nonlinear algebraic equations systems are obtained for two special values of temperature factor in the hypersonic flow stagnation point. The mappings bijectivity between heat and mass transfer local parameters and controls is established. The computation experiments results are presented: the domains of allowed values "heat-friction" are obtained.

1. Introduction

Satisfaction of the requirements of strength, lightness, efficiency and reliability of structural elements is very important for the design of hypersonic aircraft intended for movement in dense layers of the atmosphere. Various physical and chemical processes (the oscillations excitation, the dissociation, the ionization) characterize the aircraft flight in hypersonic flows [1]. The simulation of all real conditions of hypersonic flight in test bench is currently impossible. Hence, the elaboration of adequate mathematical models describing the phenomena and taking into account their particularities is necessary. With the use of A. A. Dorodnitcyn [2] generalized integral relations method the PDE systems describing laminar boundary layers can be reduced to the approximating ODE systems. This approach is very popular in engineering for compressible gas aerodynamic characteristics computation [3, 4, 5, 6].

The approximating ODE systems describing laminar boundary layers on the control area for hypersonic aircraft permeable cylindrical and spherical surfaces are obtained in [7, 8]. The nonlinear algebraic equations systems [8] are the approximating ODE systems corollaries and allow obtain the hypersonic laminar boundary layers characteristics in stagnation point neighborhood. The nonlinear algebraic systems solutions properties investigation has a great practical significance because it allows answer the hypersonic aircraft permeable surfaces heat and mass control inverse problems [9] solvability questions.

In this paper continuing the research of hypersonic aircraft permeable cylindrical and spherical surfaces laminar boundary layer control mathematical model [8, 10, 11, 12] the following results are discussed:

1) the nonlinear algebraic equations systems allowing determine the boundary layer parameters in the stagnation point neighborhood are obtained for two special temperature factor values $\tau_0 = 0$ and $\tau_0 = 1$;

2) the statements describing the boundary layer parameters and heat and mass transfer local parameters dependences on control parameters for $\tau_0 \in T_1^c = [0.75; 1.25]$ are established;

3) the computation experiments results are presented in the form of allowed values "heat-friction" domains;

4) the comparison with the results [10, 12] is carried out.

2. The algebraic systems

To obtain the boundary layer mathematical model parameters $\overline{\theta}_0(m_0, \tau_0, s_0)$, $\overline{\theta}_1(\ldots)$, $\overline{\omega}_0(\ldots)$, $\overline{\omega}_1(\ldots)$ being the functions of values $m_0 = m(x_0)$, $\tau_0 = \tau_w(x_0)$, $s_0 = \sigma B_0^2$ according to the functions $\theta_0(x; m, \tau_w, s)$, $\theta_1(\ldots)$, $\omega_0(\ldots)$, $\omega_1(\ldots)$ presentations (14), (30) from [8] in the stagnation point neighborhood let's rewrite the algebraic systems (15), (31) from [8] together

$$18m_0 - k_0\overline{\theta}_0 - 7\overline{\theta}_1 + 8\overline{\omega}_0 + 10\overline{\omega}_1 + \frac{34b_0(0)C}{\overline{\theta}_0} - \frac{32b_1(0)C}{\overline{\theta}_1} + \\ + \frac{1}{5}C_0B_0^2 \cdot \left(5\overline{\theta}_0 + 10\overline{\theta}_1 - 4\overline{\omega}_0 - 11\overline{\omega}_1\right) = 0\,; \tag{1}$$

$$12m_0 - 4\overline{\theta}_0 - k_1\overline{\theta}_1 - 4\overline{\omega}_0 + 8\overline{\omega}_1 + \frac{20b_0(0)C}{\overline{\theta}_0} - \frac{16b_1(0)C}{\overline{\theta}_1} + \\ + \frac{2}{5}C_0B_0^2 \cdot \left(5\overline{\theta}_1 - \overline{\omega}_0 - 4\overline{\omega}_1\right) = 0\,; \tag{2}$$

$$(1 - \tau_0) \cdot \overline{\theta}_0 - \overline{\omega}_0 = 0\,; \tag{3}$$

$$6m_0\frac{\overline{\omega}_0}{\overline{\theta}_0} - \overline{\omega}_0 - k_2\overline{\omega}_1 + \frac{\overline{\omega}_0^2}{\overline{\theta}_0} + 4\frac{\overline{\omega}_1^2}{\overline{\theta}_1} + 6\frac{\overline{\omega}_0 b_0(0)C}{\overline{\theta}_0^2} + \\ + C\left(\frac{1}{\Pr} + 1\right)\left[\frac{b_0(0)}{\overline{\theta}_0}\left(-3\frac{\overline{\omega}_0}{\overline{\theta}_0} + 4\frac{\overline{\omega}_1}{\overline{\theta}_1}\right) - 4\frac{b_1(0)}{\overline{\theta}_1}\frac{\overline{\omega}_0}{\overline{\theta}_0}\right] - \\ - \frac{6Cb_0(0)}{\Pr\,\overline{\theta}_0}\left(-3\frac{\overline{\omega}_0}{\overline{\theta}_0} + 4\frac{\overline{\omega}_1}{\overline{\theta}_1}\right) + \frac{1}{10}B_0^2C_0\left(10\overline{\omega}_1 - \frac{\overline{\omega}_0^2}{\overline{\theta}_0} - 8\frac{\overline{\omega}_1^2}{\overline{\theta}_1}\right) = 0\,, \tag{4}$$

where $(k_0; k_1; k_2) = (10; 7; 4)$ is applied for side surface of circular cylinder and $(k_0; k_1; k_2) = (11; 8; 5)$ is applied for spherical nose surface; control $m(x)$ is the *blowing* into the laminar boundary layer; control $s(x) = \sigma B_0^2(x)$ is the *magnetic field*; control $\tau(x) = T_w(x)/T_{e_0}$ is the *temperature factor*; $T_w(x)$ is the wall temperature and T_{e_0} is the temperature in the flow stagnation point $x_0 = 0$; $x \in X = [0; 1]$ (the axis x is directed along the body contour).

3. The case $\tau_0 = 0$

It follows from (3) that

$$\overline{\omega}_0 = \overline{\theta}_0\,. \tag{5}$$

The equations (1), (2), (4) taking into account (5) can be rewritten in the form

$$18m_0 + (8 - k_0)\,\overline{\theta}_0 - 7\overline{\theta}_1 + 10\overline{\omega}_1 + \frac{34b_0(0)C}{\overline{\theta}_0} - \frac{32b_1(0)C}{\overline{\theta}_1} + \frac{1}{5}C_0B_0^2 \cdot \left(\overline{\theta}_0 + 10\overline{\theta}_1 - 11\overline{\omega}_1\right) = 0\,; \tag{6}$$

$$12m_0 - 8\overline{\theta}_0 - k_1\overline{\theta}_1 + 8\overline{\omega}_1 + \frac{20b_0(0)C}{\overline{\theta}_0} - \frac{16b_1(0)C}{\overline{\theta}_1} + \frac{2}{5}C_0B_0^2 \cdot \left(-\overline{\theta}_0 + 5\overline{\theta}_1 - 4\overline{\omega}_1\right) = 0\,; \tag{7}$$

$$6m_0 - k_2\overline{\omega}_1 + 4\frac{\overline{\omega}_1^2}{\overline{\theta}_1} + 6\frac{b_0(0)C}{\overline{\theta}_0} + C\left(\frac{1}{\Pr} + 1\right)\left[\frac{b_0(0)}{\overline{\theta}_0}\left(4\frac{\overline{\omega}_1}{\overline{\theta}_1} - 3\right) - 4\frac{b_1(0)}{\overline{\theta}_1}\right] - \\ - \frac{6Cb_0(0)}{\Pr\,\overline{\theta}_0}\left(4\frac{\overline{\omega}_1}{\overline{\theta}_1} - 3\right) + \frac{1}{10}B_0^2C_0\left(-\overline{\theta}_0 + 10\overline{\omega}_1 - 8\frac{\overline{\omega}_1^2}{\overline{\theta}_1}\right) = 0\,. \tag{8}$$

4. The case $\tau_0 = 1$

As far as $\overline{\theta}_0 > 0$ after substitution $\tau_0 = 1$ in (3) we obtain

$$\overline{\omega}_0 = 0\,. \tag{9}$$

Since $\overline{\theta}_1 > 0$ then taking into account (9) it follows from equation (4) that

$$\overline{\omega}_1 \left[\left(5 - B_0^2 C_0\right) \frac{\overline{\omega}_1}{5} + \left(B_0^2 C_0 - k_2\right) \frac{\overline{\theta}_1}{4} + \frac{\Pr - 5}{\Pr} \cdot \frac{C b_0(0)}{\overline{\theta}_0} \right] = 0. \tag{10}$$

At $\tau_0 = 1$, i.e. for $T(x_0) = T_{e_0}$, the local heat flow $q_0 = q\,(x_0) = 0$ and for two multipliers in the left part of (10) the physically realizing case is

$$\overline{\omega}_1 = 0\,. \tag{11}$$

The equations (1) and (2) for the case $\tau_0 = 1$ with respect to (9), (11) can be rewritten in the form of "shortened" system [13]

$$18 m_0 - k_0 \overline{\theta}_0 - 7\overline{\theta}_1 + \frac{34 b_0(0) C}{\overline{\theta}_0} - \frac{32 b_1(0) C}{\overline{\theta}_1} + C_0 B_0^2 \cdot \left(\overline{\theta}_0 + 2\overline{\theta}_1\right) = 0\,; \tag{12}$$

$$12 m_0 - 4\overline{\theta}_0 - k_1 \overline{\theta}_1 + \frac{20 b_0(0) C}{\overline{\theta}_0} - \frac{16 b_1(0) C}{\overline{\theta}_1} + 2 C_0 B_0^2 \overline{\theta}_1 = 0\,. \tag{13}$$

Let's denote $T_1^c = [0.75; 1.25]$. Analogously to Statements 1-3 [12] let's describe the boundary layer parameters and heat and mass transfer local parameters dependences on control parameters for $\tau \in T_1^c$. Let the unchangeable parameters values be fixed:

$$\text{the Mach number} \quad M_\infty \in [10; 40]\,, \tag{14}$$

$$\text{the flight altitude} \quad H \in [10; 30]\ [\text{km}]\,, \tag{15}$$

$$\text{the body radius} \quad R \in [0.1; 1]\ [\text{m}]\,. \tag{16}$$

Let the controlling parameters changing ranges in stagnation point be restricted as follows:

$$m_0 = m(0) \in [0; 1]\,, \tag{17}$$

$$s_0 = \sigma B_0^2(0) \in [0; 5 \cdot 10^4]\ [\text{T/Ohm} \cdot \text{m}]\,. \tag{18}$$

Further on the control parameter s_0 dimension $[\text{T/Ohm} \cdot \text{m}]$ is omitted. For a small value of the parameter $\varepsilon > 0$ let's denote $T_\varepsilon^< = [0.75; 1 - \varepsilon]$, $T_\varepsilon^> = [1 + \varepsilon; 1.25]$.

Statement 1. *For any $\varepsilon > 0$ in the conditions (14)-(18) there exist:*

$$\min_{\tau_0 \in T_1^c} \overline{\theta}_k > 0\,, \quad \min_{\tau_0 \in T_\varepsilon^<} \overline{\omega}_k > 0\,, \quad \max_{\tau_0 \in T_\varepsilon^>} \overline{\omega}_k < 0\,, \quad \min_{\tau_0 \in T_\varepsilon^<} q_0 > 0\,, \quad \max_{\tau_0 \in T_\varepsilon^>} q_0 < 0\,, \quad \min_{\tau_0 \in T_1^c} \overline{f}_0 > 0\,,$$

and the dependences $\overline{\theta}_k$, $\overline{\omega}_k$, q_0, $\overline{f}_0$ on m_0, τ_0, s_0 satisfy the conditions (everywhere $k = 0, 1$)

$$\frac{\partial \overline{\theta}_k}{\partial m_0} > 0, \ \frac{\partial \overline{\theta}_k}{\partial \tau_0} < 0, \ \frac{\partial \overline{\theta}_k}{\partial s_0} > 0, \ \frac{\partial \overline{\omega}_k}{\partial \tau_0} < 0, \ \frac{\partial q_0}{\partial \tau_0} < 0, \ \frac{\partial \overline{f}_0}{\partial m_0} < 0, \ \frac{\partial \overline{f}_0}{\partial \tau_0} > 0, \ \frac{\partial \overline{f}_0}{\partial s_0} < 0 \ \textit{for} \ \tau_0 \in T_1^c\,,$$

$$\frac{\partial \overline{\omega}_k}{\partial m_0} > 0\,, \ \frac{\partial \overline{\omega}_k}{\partial s_0} > 0\,, \ \frac{\partial q_0}{\partial m_0} < 0\,, \ \frac{\partial q_0}{\partial s_0} < 0 \quad \textit{for} \quad \tau_0 \in T_\varepsilon^<\,,$$

$$\frac{\partial \overline{\omega}_k}{\partial m_0} < 0\,, \ \frac{\partial \overline{\omega}_k}{\partial s_0} < 0\,, \ \frac{\partial q_0}{\partial m_0} > 0\,, \ \frac{\partial q_0}{\partial s_0} > 0 \quad \textit{for} \quad \tau_0 \in T_\varepsilon^>\,.$$

For the case τ_0=1 let's give a variant of Statements 4-6 [12] and expand Statement 1.

Statement 2. *In the conditions (14)-(18) and $\tau_0 \in T_1^c$ the dependences $q_0 = q(0; m_0, \tau_0, s_0)$ on m_0, τ_0, s_0 for $\tau_0 = 1$ take the form*

$$q_0 = 0, \quad \frac{\partial q_0}{\partial m_0} = 0, \quad \frac{\partial q_0}{\partial \tau_0} < 0, \quad \frac{\partial q_0}{\partial s_0} = 0.$$

Besides,

$$\frac{D\left(q_0, \overline{f}_0\right)}{D\left(m_0, s_0\right)} = \det \frac{\partial\left(q_0, \overline{f}_0\right)}{\partial\left(m_0, s_0\right)} = 0,$$

$$\frac{D\left(q_0, \overline{f}_0\right)}{D\left(m_0, \tau_0\right)} = \det \frac{\partial\left(q_0, \overline{f}_0\right)}{\partial\left(m_0, \tau_0\right)} < 0, \quad \frac{D\left(q_0, \overline{f}_0\right)}{D\left(\tau_0, s_0\right)} = \det \frac{\partial\left(q_0, \overline{f}_0\right)}{\partial\left(\tau_0, s_0\right)} > 0.$$

5. The computation experiments for $\tau_0 \in [0; 1)$

The computation experiments results (accomplished for $M_\infty = 10$, $H = 10$ [km], $R = 0.1$ [m] to compare with the results [10, 12]) are presented on figures 1, 2. The cases of circular cylinder side surface are displayed on the left halves of the pictures and the cases of spherical nose are displayed on the right ones. The systems (1)-(4) solutions series are presented on fig. 1 in terms of pairs "heat-friction" $(q_0, \overline{f}_0)$. The marked points A, B, C, D correspond to the parameters combinations which are written in round brackets:

$$A: \ (m_0 = 0, \ \tau_0 = 0, \ s_0 = 0), \quad C: \ (m_0 = 0, \ \tau_0 = 0.999, \ s_0 = 0),$$
$$B: \ (m_0 = 1, \ \tau_0 = 0, \ s_0 = 0), \quad D: \ (m_0 = 0, \ \tau_0 = 0, \ s_0 = 5 \cdot 10^4).$$

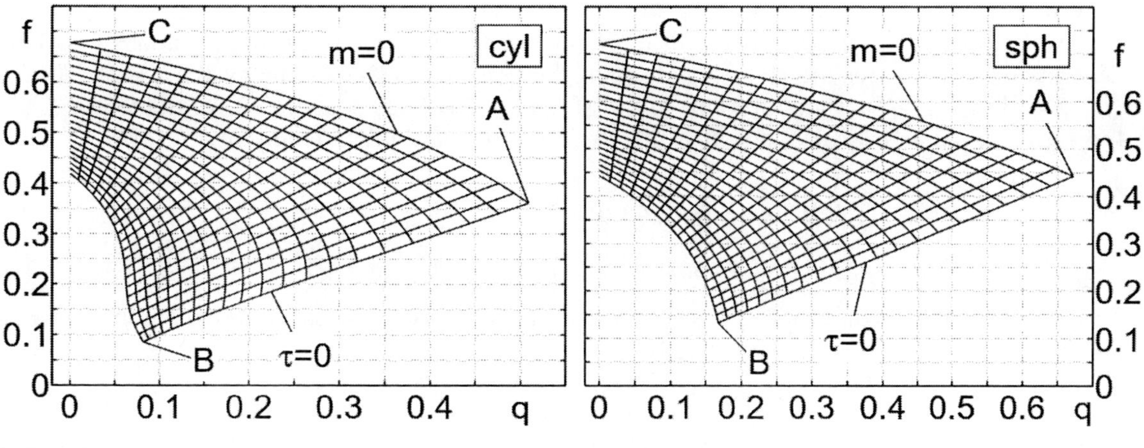

Figure 1. The domains $\Omega = \{(q_0, \overline{f}_0)\}$ for $\tau_0 \in T^c$ at $s_0 = 0$

On fig. 1 magnetic field is absent; the grid is formed by set of lines

$$\left\{\{(m_0, \tau_0, s_0) \mid m_0 \in M^c\} \mid \tau_0 \in T_{05}^d, s_0 = 0\right\},$$

$$\left\{\{(m_0, \tau_0, s_0) \mid \tau_0 \in T^c\} \mid m_0 \in M_{05}^d, s_0 = 0\right\};$$

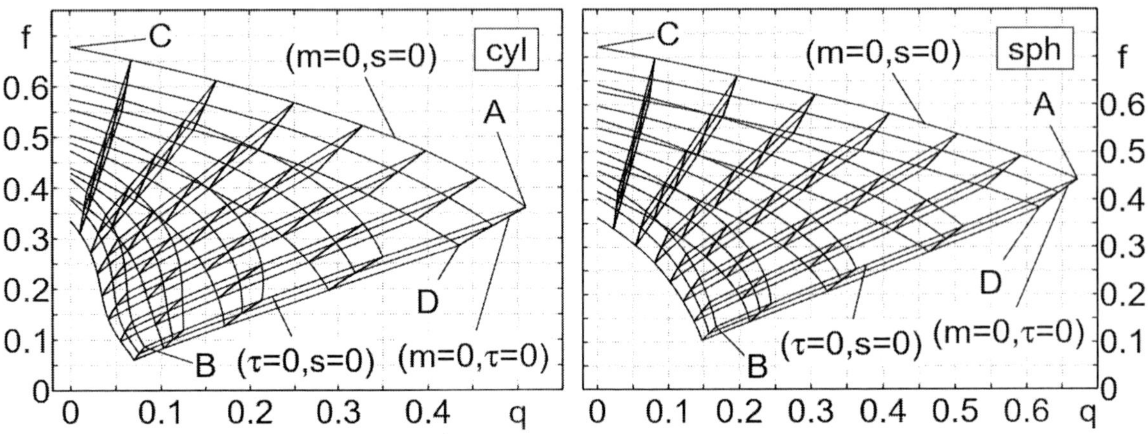

Figure 2. Three sets of lines: M, T, S

three sets of lines corresponding to the below mentioned variable parameters combinations are shown on fig. 2:

$$M = \left\{\{(m_0, \tau_0, s_0) \mid m_0 \in M^c\} \mid \tau_0 \in T^d_{15},\ s_0 \in S^d_{25}\right\},$$

$$T = \left\{\{(m_0, \tau_0, s_0) \mid \tau_0 \in T^c\} \mid m_0 \in M^d_{25},\ s_0 \in S^d_{25}\right\},$$

$$S = \left\{\{(m_0, \tau_0, s_0) \mid s_0 \in S^c\} \mid m_0 \in M^d_{25},\ \tau_0 \in T^d_{15}\right\},$$

where $M^c = [0;1]$, $T^c = [0;0.999]$, $S^c = [0;5\cdot 10^4]$, $M^d_{05} = \{0;0.05;\ldots;1\}$, $M^d_{25} = \{0;0.25;\ldots;1\}$, $T^d_{05} = \{0;0.05;\ldots;0.95\}$, $T^d_{15} = \{0;0.15;\ldots;0.9\}$, $S^d_{25} = \{0;25\cdot 10^3;50\cdot 10^3\}$. The line AB characterizes the case $\tau_0 = 0$.

6. The computation experiments for $\tau_0 \in [0.75;1.25]$

The computation experiments results (accomplished for $M_\infty = 10$, $H = 10$ [km], $R = 0.1$ [m] to compare with the results [10, 12]) are presented on figures 3-11. The systems (12), (13) for $\tau_0 = 1$ with $\overline{\omega}_0 = \overline{\omega}_1 = 0$ and systems (1)-(4) for $\tau_0 \neq 1$ solutions series are presented on fig. 3, 4 in terms of pairs "heat-friction" $(q_0, \overline{f}_0)$. On fig. 3 magnetic field is absent; the grid is formed by set of lines

$$\left\{\{(m_0, \tau_0, s_0) \mid m_0 \in M^c\} \mid \tau_0 \in T^d_{05,1}, s_0 = 0\right\},$$

$$\left\{\{(m_0, \tau_0, s_0) \mid \tau_0 \in T^c_1\} \mid m_0 \in M^d_{05}, s_0 = 0\right\}.$$

Three sets of lines corresponding to the below mentioned variable parameters combinations are shown on fig. 4:

$$M_1 = \left\{\{(m_0, \tau_0, s_0) \mid m_0 \in M^c\} \mid \tau_0 \in T^d_{15;10},\ s_0 \in S^d_{25}\right\},$$

$$T_1 = \left\{\{(m_0, \tau_0, s_0) \mid \tau_0 \in T^c_1\} \mid m_0 \in M^d_{25},\ s_0 \in S^d_{25}\right\},$$

$$S_1 = \left\{\{(m_0, \tau_0, s_0) \mid s_0 \in S^c\} \mid m_0 \in M^d_{25},\ \tau_0 \in T^d_{15;10}\right\},$$

where $T^d_{15;10} = \{0.75;0.90;1.00;1.10;1.25\}$, $T^d_{05,1} = \{0.75;0.80;\ldots;1.25\}$. The marked points A_1, B_1, C_1, D_1 correspond to the parameters combinations which are written in round brackets:

$$A_1: (m_0 = 0,\ \tau_0 = 0.75,\ s_0 = 0), \qquad B_1: (m_0 = 1,\ \tau_0 = 0.75,\ s_0 = 0),$$
$$C_1: (m_0 = 0,\ \tau_0 = 1.25,\ s_0 = 0), \qquad D_1: (m_0 = 0,\ \tau_0 = 0.75,\ s_0 = 5\cdot 10^4).$$

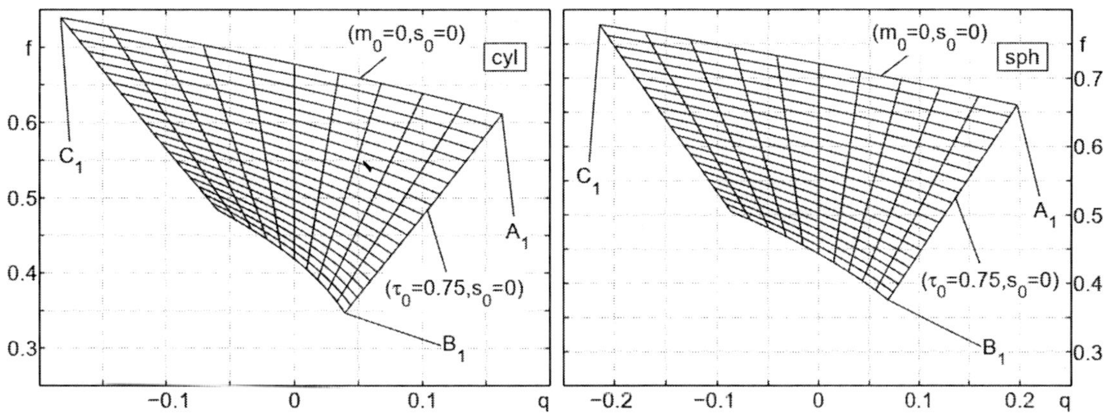

Figure 3. The domains $\Omega = \{(q_0, \overline{f}_0)\}$ for $\tau_0 \in T_1^c$ at $s_0 = 0$

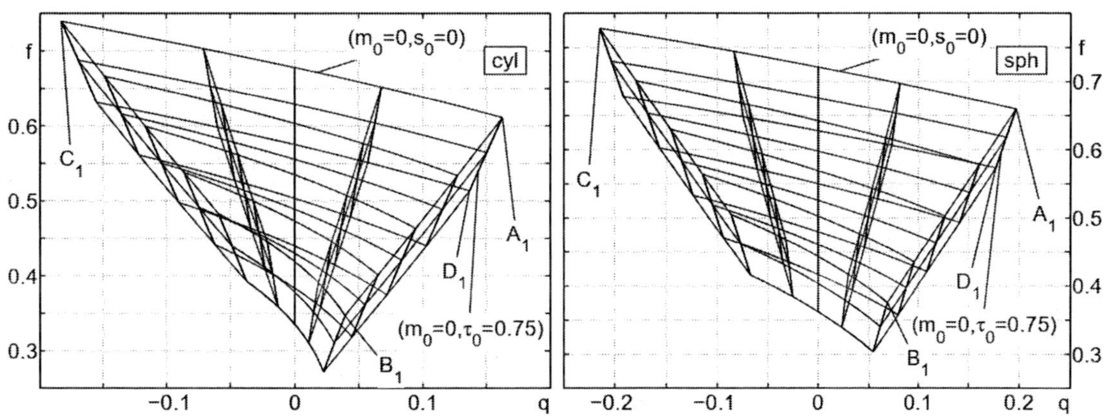

Figure 4. Three sets of lines: M_1, T_1, S_1

Fig. 5 demonstrates the influence of s_0: the lines of sets M and T are displayed.
Fig. 6 demonstrates the influence of τ_0: the lines of sets M and S are displayed.
Fig. 7 demonstrates the influence of m_0: the lines of sets T and S are displayed.

On fig. 5-7 the domains $\Omega = \{(q_0, \overline{f}_0)\}$ in the case $\tau_0 = 1$ degenerate into the vertical lines with $q_0 = 0$. It is possible to note the difference in elementary areas (dm_0, ds_0) orientations for $\tau_0 < 1$ and $\tau_0 > 1$.

In case $\tau_0 = 1$ fig. 6 must be supplemented with the graphs of dependences $\overline{\theta}_0$, $\overline{\theta}_1$ and $\overline{f}_0$ on $m_0 \in M^c$ at $s_0 \in S_{25}^d$ (fig. 8, 9) and on $s_0 \in S^c$ at $m_0 \in M_{25}^d$ (fig. 10, 11).

The lower curve on fig. 8 and the upper curve on fig. 9 correspond to $s_0 = 0$.

The lower curve on fig. 10 and the upper curve on fig. 11 correspond to $m_0 = 0$.

The established in Statements 1 and 2 mappings bijectivity between heat and mass transfer parameters and controls make possible to construct the initial part of inverse problem [9] solution in interpolation statement for local heat flow in stagnation point "test" value $q^{\vee}(0) = 0$ corresponding to $\tau_0 = 1$ both for one-dimensional and two-dimensional cases [12].

IOP Conf. Series: Journal of Physics: Conf. Series **973** (2018) 012027 doi:10.1088/1742-6596/973/1/012027

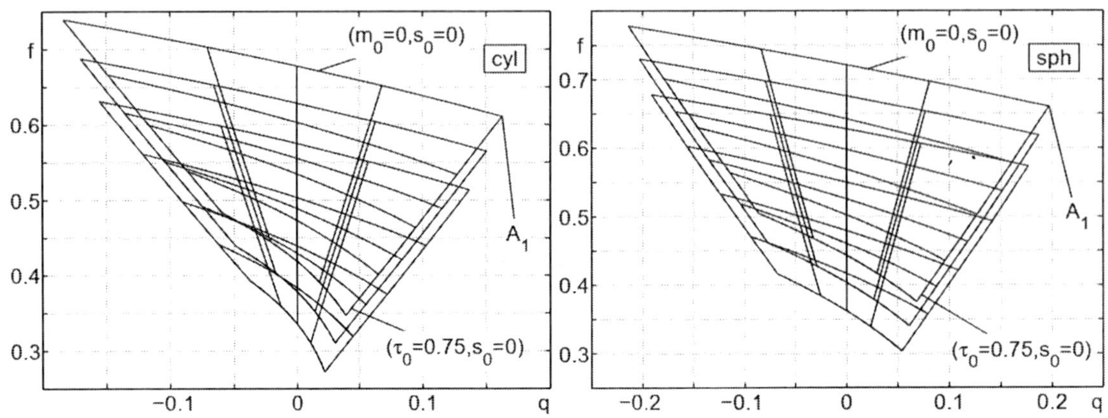

Figure 5. The domains $\Omega = \{(q_0, \overline{f}_0)\}$ for $s_0 \in S^d_{25}$

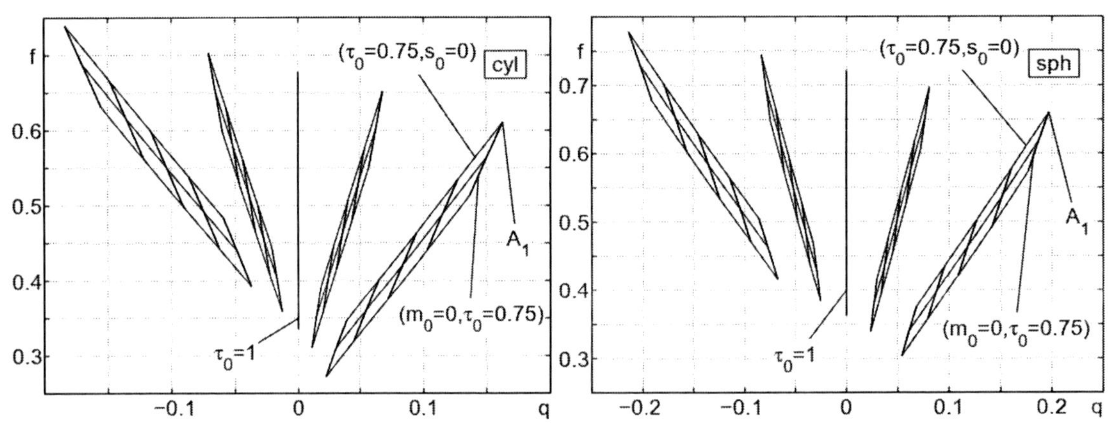

Figure 6. The domains $\Omega = \{(q_0, \overline{f}_0)\}$ for $\tau_0 \in T^d_{15;10}$

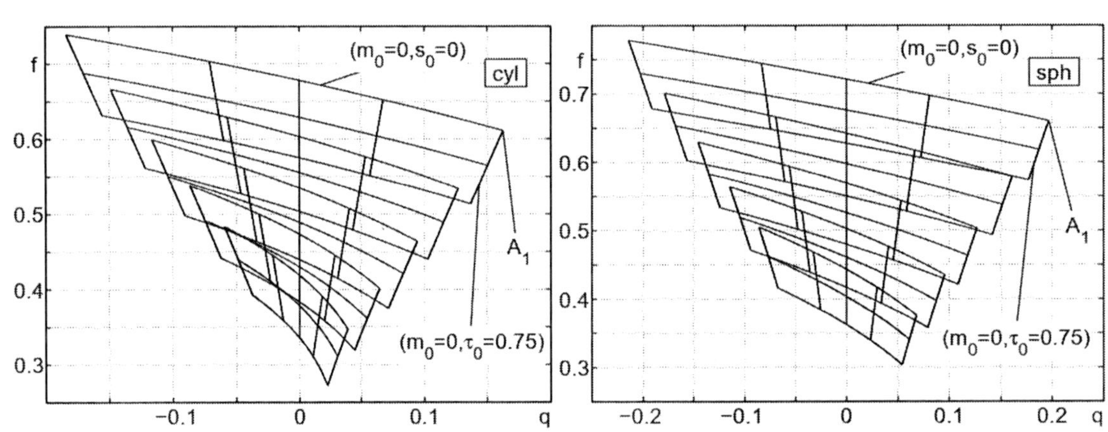

Figure 7. The domains $\Omega = \{(q_0, \overline{f}_0)\}$ for $m_0 \in M^d_{25}$

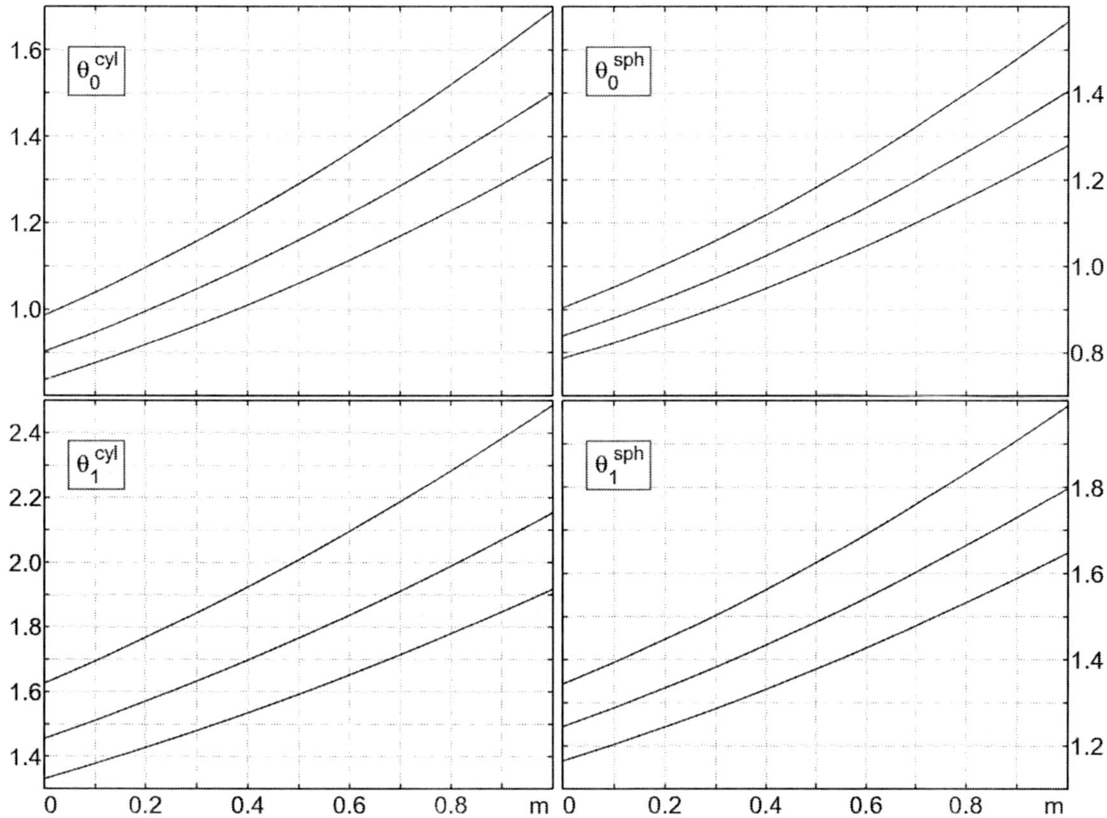

Figure 8. The dependences $\overline{\theta}_0$, $\overline{\theta}_1$ on $m_0 \in M^c$ at $\tau_0 = 1$ for $s_0 \in S_{25}^d$

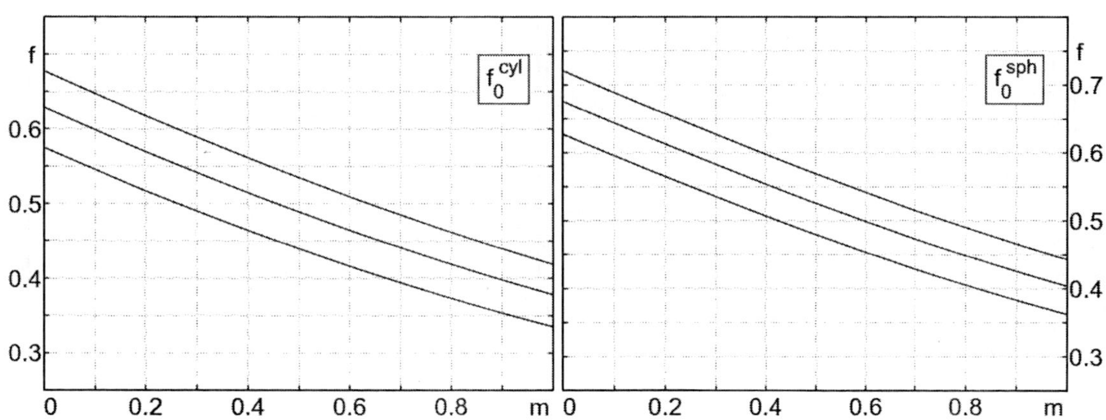

Figure 9. The dependences $\overline{f}_0$ on $m_0 \in M^c$ at $\tau_0 = 1$ for $s_0 \in S_{25}^d$

Acknowledgments
This work has been supported financially by the Ministry of Education and Science of the Russian Federation (the unique identifier of the agreement – RFMEF157715X0195).

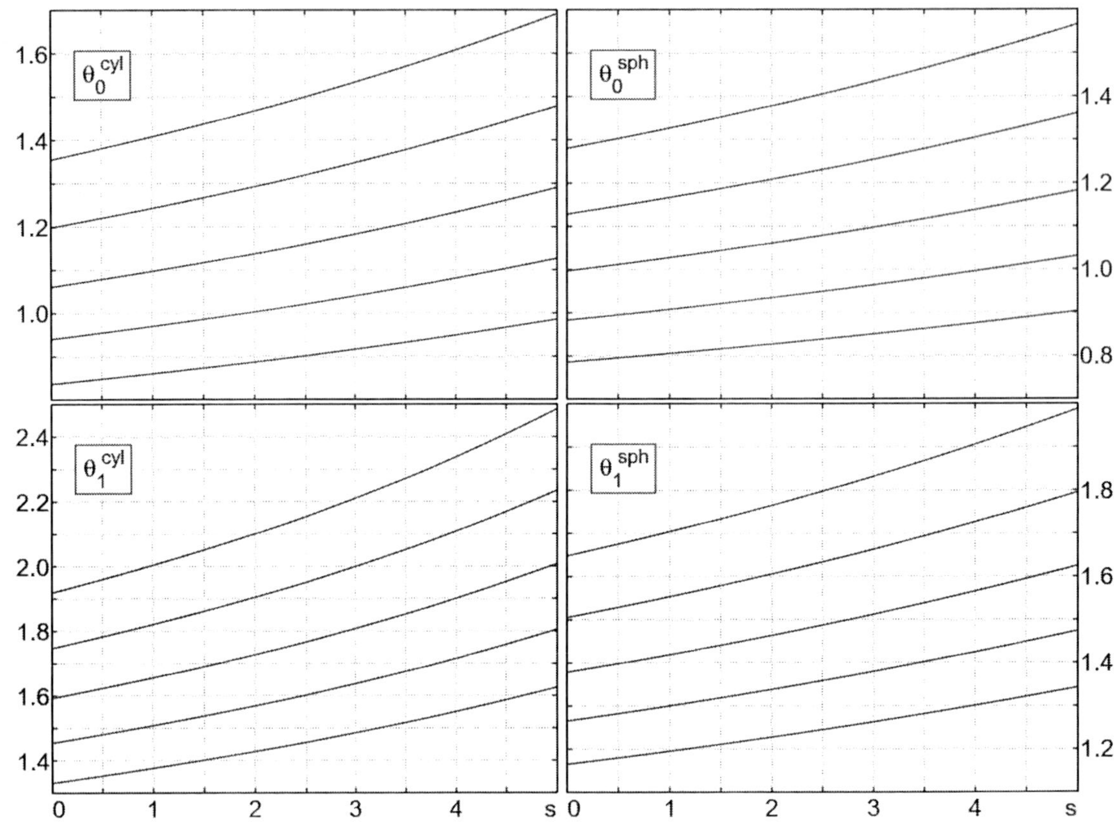

Figure 10. The dependences $\overline{\theta}_0$, $\overline{\theta}_1$ on $s_0 \in S^c$ at $\tau_0 = 1$ for $m_0 \in M^d_{25}$

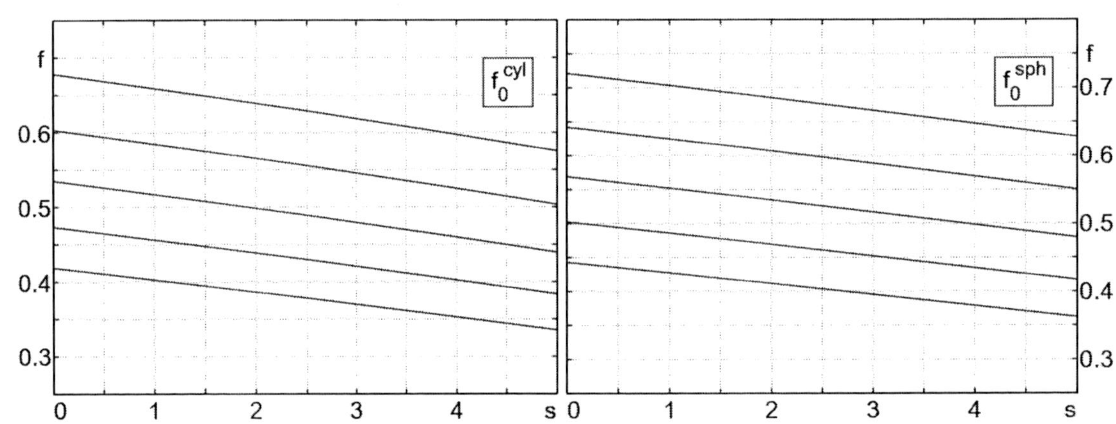

Figure 11. The dependences $\overline{f}_0$ on $s_0 \in S^c$ at $\tau_0 = 1$ for $m_0 \in M^d_{25}$

References

[1] Dorrance W H 1962 *Viscous Hypersonic Flow* (New York: McGraw-Hill) x+334
[2] Dorodnitsyn A A 1960 On a method of laminar boundary layer equations solution *Applied mathematics and technical physics* **3** 111-118 (in Russian) URL: http://sibran.ru/upload/iblock/54a/

54a6619425ad4650191e6c716470588b.pdf

[3] Pavlovsky Yu N 1963 The numerical calculation of the laminar boundary layer in a compressible gas *USSR Computational Mathematics and Mathematical Physics* **2:5** 1021-1042 URL: https://doi.org/10.1016/0041-5553(63)90509-3

[4] Lyu Shen'-Tsyuan' 1963 Calculation of the laminar boundary layer in a compressible gas in the presence of suction or blowing *USSR Computational Mathematics and Mathematical Physics* **2:5** 1000-1020 URL: https://doi.org/10.1016/0041-5553(63)90508-1

[5] Garaev K G 1988 Optimal control of heat and mass transfer in a laminar compressible-gas boundary layer on permeable surfaces *Fluid Dynamics* **23:3** 399-406 URL: https://doi.org/10.1007/BF01054748

[6] Bashkin V A 1968 Calculation of the equations of a laminar boundary space layer by the method of integral relations *USSR Computational Mathematics and Mathematical Physics* **8:6** 125-138 URL: https://doi.org/10.1016/0041-5553(68)90097-9

[7] Bilchenko N G and Garaev K G 1994 On optimum control of laminar boundary layer of electroconductive gas by supersonic flow conditions *Proc. 12th National Heat Transfer Conf. (UIT)* (Italy, L'Aquila) pp 213-224

[8] Bilchenko N G 2016 A A Dorodnitsyn's method in the problems of permeable surfaces in electroconductive gas laminar boundary layer heat and mass transfer optimal control *Proc. Voronezh State Univ. Ser. System Analysis and Inform. Technologies* **1** 5-14 (in Russian) URL: http://www.vestnik.vsu.ru/pdf/analiz/2016/01/2016-01-01.pdf

[9] Bilchenko G G and Bilchenko N G 2017 The one-dimensional and two-dimensional inverse problems of heat and mass transfer on hypersonic aircraft permeable surfaces *Proc. 2017 Int. Conf. "Constructive Nonsmooth Analysis and Related Topics" CNSA-2017 (dedicated to the memory of V F Demyanov)* ed V N Malozemov, G Sh Tamasyan and M V Udot (Russia, Saint-Petersburg: IEEE) pp 1-4 DOI: 10.1109/CNSA.2017.7973940 URL: http://ieeexplore.ieee.org/document/7973940/

[10] Bilchenko N G 2015 Permeable surfaces hypersonic aircraft optimal heat protection mathematical modeling *Proc. 2015 Int. Conf. "Stability and Control Processes" SCP-2015 (in Memory of V I Zubov)* ed L A Petrosyan and A P Zhabko (Russia, Saint-Petersburg: IEEE) pp 310-313 DOI: 10.1109/SCP.2015.7342145 URL: http://ieeexplore.ieee.org/document/7342145/

[11] Bilchenko N G 2016 Computational experiments in the problems of heat and mass transfer optimal control on permeable surfaces in electro-conductive gas laminar boundary layer *Proc. Voronezh State Univ. Ser. System Analysis and Inform. Technologies* **3** 5-11 (in Russian) URL: http://www.vestnik.vsu.ru/pdf/analiz/2016/03/2016-03-01.pdf

[12] Bilchenko G G and Bilchenko N G 2017 On the bijectivity of controls pairs and pairs of heat and mass transfer local parameters in the hypersonic flow stagnation point *Proc. 2017 Int. Conf. "Constructive Nonsmooth Analysis and Related Topics" CNSA-2017 (dedicated to the memory of V F Demyanov)* ed V N Malozemov, G Sh Tamasyan and M V Udot (Russia, Saint-Petersburg: IEEE) pp 1-3 DOI: 10.1109/CNSA.2017.7973941 URL: http://ieeexplore.ieee.org/document/7973941/

[13] Bilchenko G G and Bilchenko N G 2017 On a special case of temperature factor value in hypersonic flow stagnation point *Proc. Int. Conf. "Voronezh Spring Math. School "Modern Methods in Theory of Boundary Value Problems" – "Pontryagin Readings-XXVIII" "* (Russia, Voronezh: VSU Publishers) pp 38-41 (in Russian)

IOP Conf. Series: Journal of Physics: Conf. Series **973** (2018) 012028 doi:10.1088/1742-6596/973/1/012028

Ultrametric properties of the attractor spaces for random iterated linear function systems

A G Buchovets and P V Moskalev

Voronezh State Agricultural University, 1 Michurin street, Voronezh, 394087, Russia

E-mail: abuchovets@mail.ru, moskaleff@mail.ru

Abstract. We investigate attractors of random iterated linear function systems as independent spaces embedded in the ordinary Euclidean space. The introduction on the set of attractor points of a metric that satisfies the strengthened triangle inequality makes this space ultrametric. Then inherent in ultrametric spaces the properties of disconnectedness and hierarchical self-similarity make it possible to define an attractor as a fractal. We note that a rigorous proof of these properties in the case of an ordinary Euclidean space is very difficult.

1. Introduction

Expanding the practical use of ultrametric spaces is an important problem in mathematical modeling. The classical solution of this problem consists in mapping the model into p-adic spaces, as it is done in mathematical physics and cognitive sciences [9, 7], in astrophysics and genome research [6, 2], in number theory and complex systems [1, 8] and so on. In this case, the problems of mapping a model into an ultrametric space are solved in various ways. For example, in [11] the author constructs an ultrametric space from R^n by stochastic generating of new points and increasing the dimension of the original space. Another approach is to determine the parameter that characterizes the mapping of the Euclidean metric in the ultrametric. In this paper, we use cluster analysis to determine the ultrametric classification space associated with the original characteristic space by randomized procedures.

2. Materials and methods

Iterated function system (IFS) in the most general form represent a certain set of functions $\{f_j\}_{j=1}^k$ that are executed in a given sequence. As a result of the implementation of this IFS, there is some compact set, which, as a rule (but not necessarily), has fractal properties [3]. However, for computer implementation it is more convenient to execute on each iteration only one of the functions f_i in accordance with a given probability distribution $\{f_j|p_j\}_{j=1}^k$, where $p_j > 0$, $\sum_{j=1}^k p_j = 1$ [10]. The use of a random order in the execution of functions is called randomization, and the system obtained in this way is called a random iterated function system (RIFS) [5, 4].

2.1. Random iterated linear function systems

The simplest form for functions $\{f_j|p_j\}_{j=1}^k$ from the IFS is linear, and under additional constraints the function is a convex combination:

$$x_{i+1} = \xi x_i + (1-\xi)z_{i,j}, \tag{1}$$

where ξ — parameter; $Z = \{z_j\}_{j=1}^k$ — parameters defining the form of linear functions, $Z \subset R^p$; $i = 1, 2, \ldots, n$ — iteration number.

The iterative process begins at an arbitrary point $x_0 \in R^p$, not necessarily, but preferably $x_0 \in \text{conv}(Z)$. After n iterations $x_{i+1} = f_j(x_i)$, where $i = 1, 2, \ldots, n$, $j \in [1, k]$, a set of points $X = \{x_i\}_{i=1}^n$, called protofractals, will be obtained. The limit set, called the RIFS attractor, was shown in [4] to be compact, open-closed, has zero Lebesgue measure and fractional fractal dimension [9]. In the general case, the attractor RIFS is a homomorphic mapping of the Cantor set. This way of constructing the RIFS attractor will be denoted by F1.

This procedure was implemented by the authors in the RIFS package [10] with the free programming language and software environment for statistical computing and graphics R. The following listing shows the source code that implements the procedure F1:

```
> R2ngon <- function(n1=3, n2=1, r=1, o=c(0,0), cycle=FALSE) {
+   phi <- seq(0, 2*pi, length=n1+1)
+   x <- approx(o[1] + r*cos(phi), n=n1*n2+1)$y
+   y <- approx(o[2] + r*sin(phi), n=n1*n2+1)$y
+   if (cycle) return(cbind(x = x, y = y))
+   else return(cbind(x = x[-1], y = y[-1])) }
> preRIFS <- function(n=10000, Z=R2ngon(),
+                      P=rep(1/nrow(Z),nrow(Z)), M=rep(1,nrow(Z))) {
+   z <- sample.int(nrow(Z), size=n, prob=P, replace=TRUE)
+   X <- array(0, dim=c(n,ncol(Z)))
+   X[1,] <- colMeans(Z)
+   for (i in 2:n) X[i,] <- (X[i-1,] + M[z[i]]*Z[z[i],])/(1+M[z[i]])
+   return(list(pre=X, proto=Z, distr=cbind(p=P, mu=M), index=z)) }
```

Examples of using this code for fractal sets simulation are shown in Fig. 1.

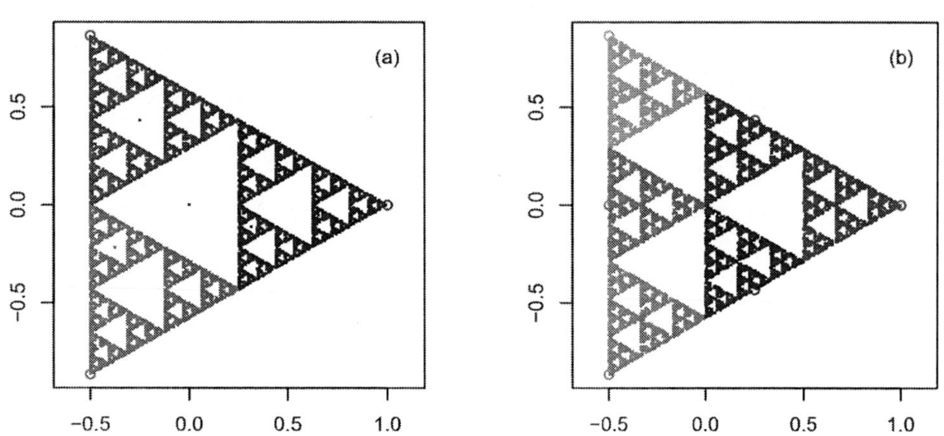

Figure 1. Fractal sets simulation using the procedure F1.

IOP Conf. Series: Journal of Physics: Conf. Series **973** (2018) 012028 doi:10.1088/1742-6596/973/1/012028

Figure 1 (a) shows the implementation of a fractal set known as the Sierpinski triangle. The parameter ξ is equal to one for it, and the set Z corresponds to an equilateral triangle with equiprobable vertices $Z_{31} = \{(-\frac{1}{2}, -\frac{\sqrt{3}}{2}),\ (-\frac{1}{2}, \frac{\sqrt{3}}{2}),\ (1, 0)\}$. Figure 1 (b) shows the implementation of a fractal set, which can be called the Sierpinski triangle of the second order. The parameter ξ is equal to two for it, and the set Z corresponds to the union of the vertices of an equilateral triangle with the bases of its medians $Z_{32} = Z_{31} \cup \{(-\frac{1}{2}, 0),\ (\frac{1}{4}, \frac{\sqrt{3}}{4}),\ (\frac{1}{4}, -\frac{\sqrt{3}}{4})\}$. All points of the set Z_{32} are also equiprobable.

When implementing the iterative process, we can obtain the following representation for x_n value [3]:

$$x_n = \beta_{n-1}\xi + \beta_{n-2}\xi^2 + \ldots + \beta_0\xi^n, \tag{2}$$

where $\beta_i = (1 - \xi)z_{i,j}$ for $j = 1, 2, \ldots, k$, and the limiting value of x^* after grouping elements with respect to z_j takes the form

$$x^* = \Big(1 - \xi\Big)\Big(z_1 \sum_s \xi^s + z_2 \sum_t \xi^t + \ldots + z_k \sum_u \xi^u\Big) = \sum_{j=1}^{k} a_j z_j,$$

where $a_j = (1 - \xi)\sum_t \xi^t$; $a_j > 0$; $\sum_{j=1}^{k} a_j = 1$.

Consequently, the value of x^* is finite and represents a convex combination of protofractal points $Z = \{z_j\}_{j=1}^{k}$. We can say that the set of coefficients a_j is the barycentric coordinates of the point x^* in the affine coordinate system.

Relation (2) shows that the initial value of x_0 entering the term with the highest degree of ξ, when iteration is repeated, gets less and less weight in the limiting value x^*. In other words, the limiting value of x^* is practically independent of the initial value x_0. This fact, already noted in publications (see, for example, [5, p. 99]), allows us to move from procedure F1 to another randomized procedure, which we will denote as F2.

In the case where the parameter ξ is constant and the same for all functions, we can form the RIFS attractor using some analog of the urn scheme. To this end, we introduce the series

$$\mu \sum_{t=1}^{\infty} \xi^t = 1, \tag{3}$$

where μ is the valuation constant associated with the parameter ξ by the relation $\mu = (1 - \xi)/\xi$. Elements of the series (3) are distributed over the k classes in accordance with the probability distribution: $\{p_j\}_{j=1}^{k}$, where $p_j > 0$, $\sum_{j=1}^{k} p_j = 1$.

The sums of the series (3) obtained at each step are written as rows $A_r = (a_{r1}, a_{r2}, \ldots, a_{rk})$ for the matrix A of size $n \times k$. In this case, the result of the procedure will be represented as the product of two matrices

$$X = AZ. \tag{4}$$

Such a method of constructing the RIFS attractor, as mentioned above, will be denoted by F2. The proposed F2 procedure allows the calculation of new matrices X, changing the matrix Z, and preserving the matrix A, which gives an effective method of forming sets X different from each other.

This procedure was also implemented by the authors in the RIFS package [10] with the free programming language and software environment for statistical computing and graphics R. The following listing shows the source code that implements the procedure F2:

```
> preRSum0 <- function(n=10000, mu=1, eps=1e-09,
+                      Z=R2ngon(), P=rep(1/nrow(Z),nrow(Z))) {
```

```
+    k <- nrow(Z)
+    S <- array(0, dim=c(n,k))
+    m <- 1 - log(eps*mu)/log(1+mu)
+    X <- mu/(mu+1)^seq(m)
+    for (i in seq(n)) {
+        l <- sample.int(k, size=m, prob=P, replace=TRUE)
+        for (j in seq(k)) S[i,j] <- sum(X[l==j]) }
+    return(list(pre=S%*%Z, proto=Z, distr=cbind(p=P, mu=rep(mu,k)))) }
```

Fig. 2 shows the commutative diagram for obtaining the RIFS attractor. By commutativity in this case we mean that there is a nonzero probability that equality $X_{i+1} = A_{i+1}Z$ will be satisfied.

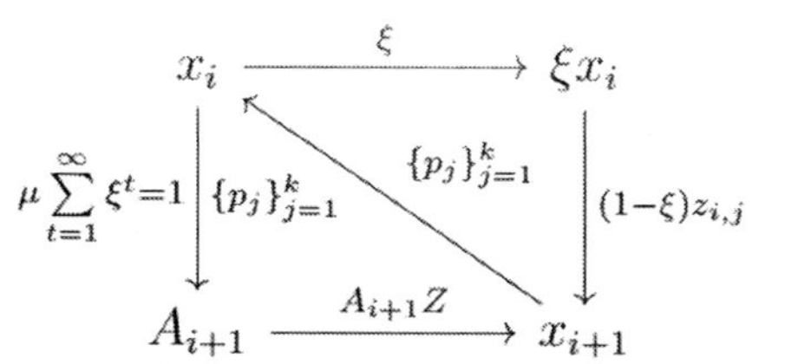

Figure 2. Commutative diagram of RIFS implementation.

Note that the space formed by row vectors of the input data matrix is usually called a characteristic space, and the space formed by F2 will be called the classification space [4]. It is obvious that the set of rows from the matrix A does not depend on the values of X, but characterizes the structural features of the matrix X, and in this sense is an invariant of the classification problem.

2.2. Spaces of random partitions

We consider the collection of different subsets of the series formed by the procedure F2

$$\mu\sum_{t=1}^{\infty}\xi^t = \mu(\xi + \xi^2 + \ldots + \xi^t + \ldots) = 1 \tag{5}$$

as the set of numbers defined on these subsets by the standard rules of addition and multiplication, that is, numbers of the form

$$\alpha = \mu\sum_{t'=1}^{\infty}\xi^{t'}, \tag{6}$$

where t' are the indices of randomly chosen elements of the series (5). Note that if in the fraction $\xi = 1/(1+\mu)$, the denominator $(1+\mu) \equiv p$ has properties analogous to the properties of prime numbers, then we can assert that these numbers are, in some sense, p-adic norms of numbers $(1+\mu)$.

Now consider the set of subsets of the elements of the series (5), which will correspond to the set of numbers formed from the elements of the sequence (5). Obviously, the collection of elements of the sequence (5) is a countable set, while the set of all its subsets has the cardinality of the continuum.

The numbers (6) introduced above can be bijectively mapped into a binary set of digits $\{0, 1\}$. Each such number can be associated with a sequence of digits 0 and 1, depending on whether the element of the series (5) was excluded or included when forming the number (6).

IOP Conf. Series: Journal of Physics: Conf. Series **973** (2018) 012028 doi:10.1088/1742-6596/973/1/012028

The previously introduced set of all subsets of the elements of the series (5) is combined with the limiting values 0, 1 and denoted by $A_\xi = \{\{\alpha\} \cup 0 \cup 1\}$. This set, as noted above, has the cardinality of the continuum.

For two arbitrary elements $\alpha, \beta \in A_\xi$ we define operation $\alpha \otimes \beta = \alpha \wedge \beta$, where α and β are represented by the corresponding binary code. Then A_ξ is regarded as an commutative multiplicative group with generator ξ and neutral element 1.

In addition, on this set for the binary representations of the numbers $\alpha, \beta \in A_\xi$ we define operation $\alpha \oplus \beta = \alpha + \beta - (\alpha \wedge \beta)$, where $\alpha + \beta = \alpha \vee \beta$, sign "$-$" corresponds to the opposite element, such that $\alpha + (-\alpha) = 0$. Then A_ξ is regarded as an associative group with neutral element 0.

With the operations introduced above, the set A_ξ will represent a ring without zero divisors.

We define a norm on the set A_ξ as follows

$$|\alpha|_\xi = \left|\mu\left(\xi^s + \xi^{s'} + \xi^{s''} + \ldots\right)\right|_\xi = \xi^s. \tag{7}$$

The properties of symmetry and homogeneity of this norm are easily verified by direct transformations. For example, if $s < t$ then

$$|\alpha \oplus \beta|_\xi = \left|\mu\left(\xi^s + \xi^{s'} + \xi^{s''} + \ldots\right) + \mu\left(\xi^t + \xi^{t'} + \xi^{t''} + \ldots\right)\right|_\xi =$$
$$= \left|\mu\left(1 + \xi^{s'-s} + \xi^{s''-s} + \ldots\right) + \mu\left(\xi^{t-s} + \xi^{t'-s} + \xi^{t''-s} + \ldots\right)\right|_\xi =$$
$$= \xi^s \leqslant \max\{\xi^s, \xi^t\} = \max\{|\alpha|_\xi, |\beta|_\xi\} \leqslant |\alpha|_\xi + |\beta|_\xi.$$

Then for any α, β, $\delta \in A_\xi$, we can define an ultrametric

$$d_\xi(\alpha, \beta) = |\alpha - \beta|_\xi \leqslant \max\{d_\xi(\alpha, \delta), d_\xi(\delta, \beta)\}.$$

Indeed,

$$d_\xi(\alpha, \beta) = |\alpha - \beta|_\xi = |(\alpha - \delta) - (\beta - \delta)|_\xi \leqslant$$
$$\leqslant \max\{|\alpha - \delta|_\xi, |\beta - \delta|_\xi\} = \max\{d_\xi(\alpha, \delta), d_\xi(\delta, \beta)\} < d_\xi(\alpha, \delta) + d_\xi(\delta, \beta).$$

We define the set A_ξ^k as the Cartesian product $A_\xi^k = \overbrace{A_\xi \times A_\xi \times \cdots \times A_\xi}^{k \text{ times}}$, and B as the subset of A_ξ^k consisting of the objects A_b

$$B = \left\{ A_b = \{a_{b1}, a_{b2}, \ldots, a_{bk}\} \mid a_{bj} \geqslant 0, \sum_{j=1}^{k} a_{bj} = 1 \right\}.$$

We define a norm on the set B as follows

$$||B_i||_\xi = \max_{1 \leqslant j \leqslant k} |a_{ij}|_\xi,$$

where $a_j \in A_\xi$, and $B_i \in A_\xi^k$. For this norm the triangle inequality is satisfied

$$||B_s + B_t||_\xi = \max_{1 \leqslant j \leqslant k}\left(a_{sj} \oplus a_{tj}\right) \leqslant \max_{1 \leqslant j \leqslant k}\left(\max\left\{|a_{sj}|_\xi, |a_{tj}|_\xi\right\}\right) =$$
$$= \max\left\{\max_{1 \leqslant j \leqslant k}\left(|a_{sj}|_\xi, |a_{tj}|_\xi\right)\right\} = \max\left\{\max_{1 \leqslant j \leqslant k}\left(|a_{sj}|_\xi\right), \max_{1 \leqslant j \leqslant k}\left(|a_{tj}|_\xi\right)\right\} =$$
$$= \max\left\{||B_s||_\xi, ||B_t||_\xi\right\} \leqslant ||B_s||_\xi + ||B_t||_\xi.$$

Then the natural metric in the space B satisfies the properties of homogeneity and symmetry. The third property of the metric is related to the triangle inequality

$$d_\xi(B_s, B_t) = ||B_s - B_t||_\xi = ||(B_s - B_q) - (B_t - B_q)||_\xi \leqslant \\ \leqslant \max\{||B_s - B_q||_\xi, ||B_t - B_q||_\xi\} = \max\{d_\xi(B_s, B_q), d_\xi(B_t, B_q)\} < \\ < d_\xi(B_s, B_q) + d_\xi(B_t, B_q).$$

As you can see, in this case not only the triangle inequality is satisfied, but also the stronger inequality — the ultrametric.

$$d_\xi(B_s, B_t) \leqslant \max\{d_\xi(B_s, B_q), d_\xi(B_t, B_q)\},$$

where B_s, B_t, and $B_q \in A_\xi^k$.

3. Results and discussion

The above norm defines an ultrametric on the set B, which turns the space A_ξ^k into an ultrametric one. In practice, this leads to the fact that the geometric properties of many objects from this space will be counterintuitive to the properties of objects from a finite-dimensional Euclidean space. Indeed, many “strange” properties of fractal sets can be explained precisely by the properties of ultrametric space.

As an example, we compare the distances between pairs of points of a fractal set in the characteristic Euclidean space R^2 and in the classification space $A_{1/2}^3$.

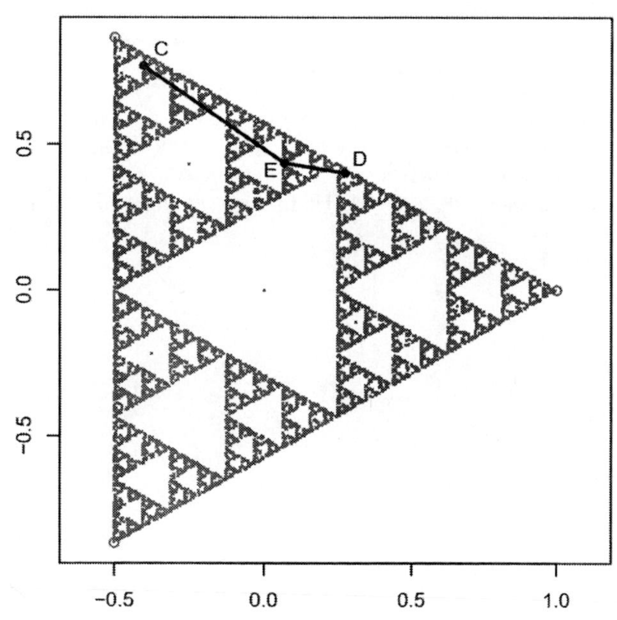

Figure 3. Fractal set simulation using the procedure F2.

The distance between the points shown in Fig. 3, in the Euclidean metric is equal to $\rho(CE) = 0.58$ and $\rho(DE) = 0.21$. Obviously, $\rho(CE) > \rho(DE)$, although the points C and E belong to the same cluster, and the point D belongs to another cluster.

In the classification ultrametric space, the above distances will be equal to $d_{1/2}(CE) = 0.22$ and $d_{1/2}(DE) = 0.25$. In this case, already $d_{1/2}(CE) < d_{1/2}(DE)$, which better reflects the location of the points.

The following properties of ultrametric spaces are known from the literature (see, for example, [7, 9]): 1) the ultrametric spaces are non-Archimedean and completely disconnected; 2) in any ultrametric space all triangles are isosceles; 3) any ball in an ultrametric space is an open-closed set, and any point inside the ball is its center; 4) any two balls in the ultrametric space either do not intersect, or one is inside the other; 5) any points C and D in an ultrametric space will not be ϵ-connected if $d_\xi(C, D) > \epsilon$ is true.

4. Conclusion

Interpretation of the properties of the RIFS attractor makes it possible to approximate data from the multidimensional classification problem by using a set of tuples from the space A_ξ^k that has ultrametric properties. Note that the classification ultrametric space A_ξ^k has all properties (1-5), whereas it is very difficult to find out their presence in a characteristic Euclidean space.

References

[1] Avetisov V A, Bikulov A H and Osipov V A 2003 *J. Phys. A: Math. Gen.* **36** 4239 (DOI: 10.1088/0305-4470/36/15/301)
[2] Barnaby N, Biswas T and Cline J M 2007 *J. High Energy Phys.* JHEP04(2007)056 (DOI: 10.1088/1126-6708/2007/04/056)
[3] Bukhovets A G and Bukhovets E A 2012 *Autom. Remote Control.* **73** 381–385 (DOI: 10.1134/S0005117912020154)
[4] Bukhovets A G and Borucinska T Y 2016 *Proceedings of VSU. Series: Systems analysis and information technologies.* **2** 5–10 (URL: http://www.vestnik.vsu.ru/pdf/analiz/2016/02/2016-02-01.pdf)
[5] Crownover R M 1995 *Introduction to fractals and chaos* (Jones & Bartlett Pub.)
[6] Dragovich B. and Dragovich A 2007 *Comput. J.* **53** 432–442 (DOI: 10.1093/comjnl/bxm083)
[7] Dragovich B, Khrennikov A Y, Kozyrev S V, Volovich I V and Zelenov E I 2017 *p-Adic Num. Ultrametr. Anal. Appl.* **9** 87–121 (DOI: 10.1134/S2070046617020017)
[8] Greinecker F 2017 *J. Fractal Geom.* **4** 105–126 (DOI: 10.4171/JFG/46)
[9] Khrennikov A Y 2004 *Russ. J. Math. Phys.* **11** 45–70 (arXiv:nlin/0402042)
[10] Moskalev P V, Bukhovets A G and Biruchnskay T Ya 2012 *RIFS: Random iterated function system* (URL: https://cran.r-project.org/package=RIFS)
[11] Zubarev A P 2013 *Vestnik SamGUPS.* **2** 55–64 (URL: https://elibrary.ru/item.asp?id=20245733)

IOP Conf. Series: Journal of Physics: Conf. Series **973** (2018) 012029 doi:10.1088/1742-6596/973/1/012029

Asymptotics of eigenvalues and eigenvectors of some infinite three-diagonal matrices

V E Chernov[1], **K I Suvorov**[1], **M Yu Knyazev**[1] **and A I Zetkina**[2]
[1] Department of Physics, Voronezh State University, 394018 Voronezh, Russia
[2] Department of Economics, P. G. Demidov Yaroslavl' University, 150003 Yaroslavl', Russia

E-mail: chernov@niif.vsu.ru

Abstract. We consider the eigenvalue problem for an infinite three-diagonal matrix whose main diagonal is a sequence tending to infinity. We derive asymptotic expansions for the eigenvalues and eigenvectors. The small parameter of the asymptotics is determined by quotient of second-diagonal elements to the differences between the main-diagonal elements. Some applications to molecular spectroscopy are considered.

1. Introduction

1.1. Motivation: molecular physics

The Schrödinger equation is one of the most important concepts in modern physics and chemistry, as well to quantum computing. At the same time there exists only a small set of potentials (as a rule, one-dimensional) which allow exact solutions of the corresponding Schrödinger equations [1]. The set of exactly solvable two- and three-dimensional non-central potentials is much smaller [2]. Therefore searching for exactly solvable Schrödinger equations with non-central potentials is an important problem.

One of exactly solvable three-dimensional Schrödinger equations with non-central potential describes one-electron model of a highly excited (Rydberg) molecule [3–5] or dipole-bound anion [6, 7]. In this model implies that the molecular is a source of central potential, $U(r)$ and non-central potential, $V(r,\vartheta)$, where r,ϑ,φ are the polar coordinates characterizing the position of electron with respect to the center of charge of the molecular core, local z axis being directed along the core dipole moment vector, $\vec{\mu}$. Hence the Schrödinger equation for the Rydberg electron has the following form (atomic units are used henceforward):

$$H\Psi = E\Psi, \qquad H = T_{\text{el}} + U(r) + V(r,\vartheta), \tag{1}$$

where the kinetic energy operator

$$T_{\text{el}} = -\frac{\Delta}{2} = -\frac{1}{2r^2}\frac{\partial}{\partial r}\left(r^2\frac{\partial}{\partial r}\right) + \frac{\vec{l}^2}{2r^2} \tag{2}$$

is expressed in terms of the electron's squared orbital momentum operator, $\vec{l}^2$, which, in turn, is proportional to the Legendrian (angular part of the Laplacian) operator:

$$\vec{l}^2 = -\Delta_{\vartheta,\varphi} = -\frac{1}{\sin\vartheta}\frac{\partial}{\partial\vartheta}\left(\sin\vartheta\frac{\partial}{\partial\vartheta}\right) - \frac{1}{\sin^2\vartheta}\frac{\partial^2}{\partial\varphi^2}. \tag{3}$$

For the point dipole potential

$$V(r,\vartheta) = -\frac{(\vec{\mu}\cdot\vec{r})}{r^3} = -\frac{\mu\cos\vartheta}{r^2}, \tag{4}$$

one can separate the variables in (1). Indeed, assuming $\Psi(r,\vartheta,\varphi) = R(r)\mathcal{Z}(\vartheta,\varphi)$ we obtain the following eigenvalue problem for the angular function $\mathcal{Z}$:

$$\left(\vec{l}^2 - 2\mu\cos\vartheta\right)\mathcal{Z} = \eta\mathcal{Z}; \qquad \mathcal{Z}(\vartheta,\varphi+2\pi) = \mathcal{Z}(\vartheta,\varphi), \quad |\mathcal{Z}(\vartheta,\varphi)| < \infty \quad \text{when } |\cos\vartheta| = 1. \tag{5}$$

The eigenfunction of the problem (5), so called dipole-spherical functions, play an important role in molecular physics. It seems that the dipolespherical angular functions mentioned above were used for the first time by Debye [8] in the analysis of the problem of the Stark effect for a symmetric rotator. Zon [3] and Watson [9] used thes these functions for the description of the Rydberg states of polar molecules. Dipole-spherical functions appear in the theory of electron photodetachment from atomic anions in a strong field [10] as well as in theoretical models of dipole-bound anion photodetachment [6] and collisions [7]. Dipole-spherical functions are convenient basis for calculation of dynamic polarizability of polar molecules [4].

At $\mu = 0$ the dipole-spherical functions are reduced to the well-known spherical functions $Y_{\ell\lambda}(\vartheta,\varphi)$ and the eigenvalues $\eta_\ell = \ell(\ell+1)$. The orbital quantum number $\ell = 0,1,2\ldots$ and the magnetic quantum number $\lambda = -\ell, -\ell+1, \ldots, \ell$ are integers.

At $\mu \neq 0$ it is natural to expand the dipole-spherical functions (5) over the spherical harmonics:

$$\mathcal{Z}(\vartheta,\varphi) = \mathcal{Z}_{\ell\lambda}(\vartheta,\varphi) = \sum_{l=|\lambda|}^{\infty} x_{\ell\lambda,l} Y_{l\lambda}(\vartheta,\varphi). \tag{6}$$

After substituting Eq. (6) into the angular Schrödinger equation (5) we obtain a recurrence relation (or, mathematically, a difference equation) of the type (7a) below for the coefficients $x_{\ell\lambda,l}[\mu]$ which are responsible for all the properties of the dipole-spherical functions.

Recurrence relations of the type (6) occur, *e.g.* in analyzing the spectrum of asymmetric top [11, § 18] or a symmetric top in an external field [12]. In the latter case the problem, the three-diagonal matrix of the recurrence relation (7a) has constant (*i.e.* independent of the representation index $l = l$) elements, while in the former case the matrix elements are weakly dependent on l, that corresponds to the semiclassical approximation in l-representation.

1.2. Mathematical formulation of the problem

Consider a second-order difference equation

$$-2\mu\beta_l x_{l-1} + \gamma_l x_l - 2\mu\alpha_{l+1}x_{l+1} = \eta x_l, \qquad l = l_0, l_0+1, \ldots, \tag{7a}$$

with the boundary conditions

$$x_{l_0-1} = 0, \quad \sum_{l=l_0}^{+\infty} |x_l|^2 < \infty. \tag{7b}$$

Let γ_l be mutually distinct and grow rapidly enough:

$$l \neq l' \Rightarrow \gamma_l \neq \gamma_{l'}; \qquad \gamma_l \to \infty, \ \frac{\beta_{l+1}}{\gamma_l} = O(\gamma_l), \ \frac{\alpha_{l-1}}{\gamma_l} = O(\gamma_l) \quad \text{at } l \to \infty. \tag{8}$$

We will treat Eqs. (7) as an eigenvalue, η, and eigenvector, $\vec{x} = \{x_l\}$, problem for the operator

$$(\mathcal{Q}\vec{x})_l = (\mathcal{Q}[\mu]\vec{x})_l = \sum_{l'=l_0}^{\infty} q_{l,l'}x_{l'}, \qquad l = l_0, l_0+1, \ldots, \tag{9}$$

defined by the infinite three-diagonal matrix Q with the elements

$$q_{l,l'} = -2\mu\beta_l\delta_{l-1,l'} + \gamma_l\delta_{l,l'} - 2\mu\alpha_l\delta_{l+1,l'}, \qquad l = l_0, l_0+1, \dots \tag{10}$$

The Kronecker delta $\delta_{i,j} = \begin{cases} 1, & i = j \\ 0, & i \neq j \end{cases}$ is used henceforward.

To the authors' knowledge [13, Eq. (1.4.14), p. 14], all orthogonal polynomials $x_l(\eta)$ of a continuous variable η satisfy some difference equation of the type (7a) treated as a recurrence relation without initial/boundary conditions in the discrete variable l. The classical orthogonal polynomials of discrete variable l [13, tables 2.1–2.3] satisfy a recurrence relation (7a), provided $\alpha_l, \beta_l, \gamma_l = O(l^2)$ at $l \to \infty$, that does not match the last requirement (8). The case $\gamma_l = \text{const} \cdot l^2$ is close to the recurrence relation for the Fourier coefficients of the Mathieu functions [14, sec. 16.2].

This work studies the dependence of eigenvalues and eigenvectors of infinite three-diagonal matrices (10) on the parameter μ. For the eigenvalues and eigenvectors we obtain simple approximate analytic expressions which are their asymptotic estimates at $\mu \to 0$. We also prove some symmetry properties of eigenvalues and eigenvectors related to their behaviour with respect to the substitution $\mu \to -\mu$. A continuous fraction representation is given for the eigenvectors. The efficiency of the proposed approximation is demonstrated on the example of Fourier coefficients of the dipole-spherical functions which has wide application in molecular spectroscopy.

2. Methods

2.1. Properties of the operator $\mathcal{Q}$

According to the boundary conditions, we shall treat the operator $\mathcal{Q}$ to act in the $l_2 = l_2([l_0, \infty))$ space and to have the domain of definition $D(\mathcal{Q}) = \{x : \mathcal{Q}x \in l_2\}$.

Proposition 1. *The domain of the operator $\mathcal{Q}[\mu]$, which is defined by the formula (9), consists of the sequences $x \in l_2$, such that $\sum_{l=l_0}^{\infty} |\gamma_l|^2 \cdot |x_l|^2 < \infty$.*

Proposition 2. *The operator $\mathcal{Q}[\mu]$, defined by the formula (9), is closed.*

Proof. Let $\vec{x}_k = (x_{k,l_0}, x_{k,l_0+1}, \dots,) \in D(\mathcal{Q})$, $\mathcal{Q}\vec{x}_k = \vec{y}_k = (y_{k,l_0}, y_{k,l_0+1}, \dots,)$ and $\vec{x}_k$ converges in norm to $\vec{x}_* = (x_{*,l_0}, x_{*,l_0+1}, \dots,) \in l_2$, and $\vec{y}_k$ converges in norm to $\vec{y}_* = (y_{*,l_0}, y_{*,l_0+1}, \dots,) \in l_2$. Let us show that $\vec{x}_* \in D(\mathcal{Q})$ and $\mathcal{Q}\vec{x}_* = \vec{y}_*$.

The convergence of $\vec{x}_k$ to $\vec{x}_*$ and $\vec{y}_k$ to $\vec{y}_*$ in norm implies the pointwise convergence *i.e.* $x_{k,l} \to x_{*,l}$ when $k \to \infty$ for all $l = l_0, l_0+1, \dots$ and, similarly, for $\vec{y}_k$. Then one can pass to the limit in the following equations:

$$q_{l,l-1}\, x_{k,l-1} + q_{l,l}\, x_{k,l} + q_{l,l+1}\, x_{k,l+1} = y_{k,l}, \qquad l = l_0, l_0+1, \dots,$$

that results in

$$q_{l,l-1}\, x_{*,l-1} + q_{l,l}\, x_{*,l} + q_{l,l+1}\, x_{*,l+1} = y_{*,l}, \qquad l = l_0, l_0+1, \dots,$$

This means that $\vec{x}_* \in D(\mathcal{Q})$ and $\mathcal{Q}\vec{x}_* = \vec{y}_*$. □

It is easy to see that the family $\mathcal{Q}[\mu]$, $\mu \in \mathbb{C}$ is holomorphic in a sense of [15, ch. VII, § 1.2]. The operator $\mathcal{Q}[0]$ has compact resolvent because of $\gamma_l \to \infty$ at $l \to \infty$. Therefore [15, ch. VII, theor. 2.4], the resolvent of the operator $\mathcal{Q}[\mu]$ is compact for all $\mu \in \mathbb{C}$.

Theorem 3. *The spectrum of the operator $\mathcal{Q}[\mu]$ consists entirely of countable number of isolated eigenvalues with finite multiplicities for all $\mu \in \mathbb{C}$.*

Proof. Follows from [15, ch. III, theor. 6.29]. □

Clearly, the spectrum of the operator $\mathcal{Q}[\mu]$ consists of the in-diagonal elements, γ_n, of Q matrix. Due to the condition $\gamma_l \neq \gamma_{l'}$ at $l \neq l'$, all the eigenspaces of the operator $\mathcal{Q}[0]$ have dimension 1. Let the points of the spectrum of the operator $\mathcal{Q}[0]$ and the corresponding eigen-sequences are enumerated in the natural way:

$$\eta_n[0] = \gamma_n, \qquad x_{n,l}[0] = \delta_{n,l}. \tag{11}$$

We emphasize that these eigen-sequences $\vec{x}_n[0]$ have norm 1.

Theorem 4. *For all $n \in \mathbb{N}$ there exists $\varepsilon_n > 0$ such that the eigenvalue $\eta_n[\mu]$ remains isolated at $|\mu| < \varepsilon$ and the eigenvalues $\eta_n[\mu]$ and the corresponding eigenprojections depend on μ holomorphically.*

Proof. Follows from [15, ch. VII, theors. 1.7 and 1.8]. □

2.2. The behavior of the eigenvalues and eigenvectors under the replacement $\mu \to -\mu$

Theorem 5. *The replacement $\mu \to -\mu$ does not change the spectrum of the operator $\mathcal{Q}[\mu]$, while the components of its eigen-sequences are changed according to $x_l[-\mu] = (-1)^l x_l[\mu]$.*

Proof. Consider the operator $\mathcal{T} : l_2[(l_0, \infty)] \to l_2[(l_0, \infty)]$, which acts according to $(\mathcal{T}\vec{x})_l = (-1)^l x_l$; its matrix is clearly consist of the elements $t_{l,l'} = (-1)^l \delta_{l,l'}$. Consider now the operator $\mathcal{Q}' = \mathcal{T}^{-1}\mathcal{Q}\mathcal{T}$, which is similar to the operator $\mathcal{Q}$. The operator $\mathcal{Q}'$ is obviously orthogonal and is defined by the matrix Q' whose elements are

$$q'_{l,l'} = \left(T^{-1}QT\right)_{l,l'} = (-1)^l \delta_{l,i} q_{i,j} (-1)^j \delta_{j,l'} = (-1)^{l-l'} q_{l,l'}. \tag{12}$$

The standard summation over repeated indices is assumed in (12). It is easy to see that the transition from Q to Q' is brought to the change of signs in the off-diagonal elements, that is equivalent to $\mu \to -\mu$ replacement. Therefore, $Q[-\mu] = Q'[\mu]$. This fact, together with the invariance of the spectrum under the similarity transformations, result in the identity of the spectra of the operators $\mathcal{Q}[-\mu]$ and $\mathcal{Q}[\mu]$. To finish the proof, we note that it follows from $\mathcal{Q}'\mathcal{T}^{-1} = \mathcal{T}^{-1}\mathcal{Q}$ that the eigen-sequences, $\vec{x}_n$ and $\vec{x}'_n$, of the operators $\mathcal{Q}$ and $\mathcal{Q}'$ are related by $x'_{n,l} = (-1)^l x_{n,l}[\mu]$. □

Remark 1. We are focused here on the dependence of the eigen-sequences, $\vec{x}_n[\mu]$, on the parameter μ, but the eigen-sequences are defined up to an arbitrary scale factor. It is convenient to choose this factor in such a way that the dependence on μ would be continuous in norm and that we would have the eigen-sequence (11) at $\mu = 0$ holding the identity $x_l[-\mu] = (-1)^l x_l[\mu]$ from Theorem 5. These conditions imply the following relationship:

$$x_{n,l}[-\mu] = (-1)^{n-l} x_{n,l}[\mu]. \tag{13}$$

2.3. Continuous fraction representation of the eigenvalues

Theorem 6. *If all the off-diagonal elements α_l, β_l are nonzero, then the eigenvalues $\eta = \eta_n$ of the operator $\mathcal{Q}$ satisfy the following equation:*

$$\gamma_{l_0} - \eta = \cfrac{(2\mu)^2 \alpha_{l_0+1}\beta_{l_0+1}}{\gamma_{l_0+1} - \eta - \cfrac{(2\mu)^2 \alpha_{l_0+2}\beta_{l_0+2}}{\gamma_{l_0+2} - \eta - \ddots \, \gamma_{l-1} - \eta - \cfrac{(2\mu)^2 \alpha_l \beta_l}{\gamma_l - \eta - \cdots}}}$$

or, in a short form
$$\mathop{\mathrm{K}}_{l=l_0}^{\infty} \left(-(2\mu)^2 \alpha_{l+1}\beta_{l+1} \middle/ \gamma_l - \eta\right) = 0. \tag{14}$$

Proof. Denoting the n-th eigen-sequence as $\vec{x}_n$, we rewrite Eq. (7a) in the following form:

$$\begin{aligned} \frac{x_{n,l}}{x_{n,l+1}} &= \frac{2\mu\alpha_{l+1}}{\gamma_l - \eta_n - 2\mu\beta_l \dfrac{x_{n,l-1}}{x_{n,l}}}, \\ \frac{x_{n,l}}{x_{n,l-1}} = &= \frac{2\mu\beta_l}{\gamma_l - \eta_n - 2\mu\alpha_{l+1} \dfrac{x_{n,l+1}}{x_{n,l}}}. \end{aligned} \tag{15}$$

For the initial $l = l_0$ in (15) one should assume $x_{n,l_0-1} = 0$ and/or $\beta_l = 0$:

$$\frac{x_{n,l_0}}{x_{n,l_0+1}} = \frac{2\mu\alpha_{l_0+1}}{\gamma_{l_0} - \eta_n} \quad \Rightarrow \quad \gamma_{l_0} - \eta_n = 2\mu\alpha_{l_0+1} \frac{x_{n,l_0+1}}{x_{n,l_0}}. \tag{16}$$

Applying (15) recursively, we obtain (16) in the continuous fraction form (14) □

The continuous-fraction expansion (14) can be treated as an implicit equation to determine the eigenvalues $\eta[\mu]$. The parameter μ appears in (14) squared only, that ensures that the eigenvalues $\eta[\mu]$ are even functions of μ.

2.4. Perturbation theory

If we write formal perturbation series for the eigenvectors $\vec{x}_n$ at $\mu \to 0$, then it is easy to see that, for $k > 0$, the first non-vanishing contribution to $x_{n,l}$ has the order of μ^k and arise at $l = n \pm k$. It can also be seen that

$$\begin{aligned} \frac{x_{n,l+1}}{x_{n,l}} &= = o(\mu), \quad l \geq \max(n, l_0 + 1), \qquad \text{and} \\ \frac{x_{n,l-1}}{x_{n,l}} &= = o(\mu), \quad l_0 \leq l \leq n. \end{aligned} \tag{17}$$

By (17) one can neglect the last terms in the denominators of the expressions (15) and write them approximately (we denote this approximation by the PT mark):

$$x^{(\mathrm{PT})}_{n,l} = \begin{cases} \dfrac{2\mu\alpha_{l+1}}{\gamma_l - \eta_n} x^{(\mathrm{PT})}_{n,l+1}, & l_0 \leq l \leq n, \\ \dfrac{2\mu\beta_l}{\gamma_l - \eta_n} x^{(\mathrm{PT})}_{n,l-1}, & l \geq \max(n, l_0 + 1). \end{cases} \tag{18}$$

Some standard calculations result in the following statement.

Theorem 7. *The eigenvectors in PT-approximation have the following form ($k \geq 0$):*

$$x^{(PT)}_{n,l\leq n} = x^{(PT)}_{n,n-k} = \frac{x^{(PT)}_{n,n}(2\mu)^k \prod\limits_{s=n-k+1}^{n} \alpha_s}{\prod\limits_{s=n-k}^{n-1} (\gamma_s - \eta_n)} = \frac{x^{(PT)}_{n,n}(2\mu)^{n-l} \prod\limits_{s=l+1}^{n} \alpha_s}{\prod\limits_{s=l}^{n-1} (\gamma_s - \eta_n)}, \tag{19a}$$

$$x^{(PT)}_{n,l\geq n} = x^{(PT)}_{n,n+k} = \frac{x^{(PT)}_{n,n}(2\mu)^k \prod\limits_{s=n+1}^{n+k} \beta_s}{\prod\limits_{s=n+1}^{n+k} (\gamma_s - \eta_n)} = \frac{x^{(PT)}_{n,n}(2\mu)^{l-n} \prod\limits_{s=n+1}^{l} \beta_s}{\prod\limits_{s=n+1}^{l} (\gamma_s - \eta_n)}. \tag{19b}$$

As usual, all the products in (19) are assumed to equal to unity when the lower bound of the product is greater than its upper bound.

Theorem 8. *PT-approximation is the first non-vanishing term of the order of* $\mu^{|l-n|}$ *in the asymptotics of the eigen-sequences:*

$$x_{n,l} = x_{n,l}^{(PT)} + o\left(\mu^{|l-n|}\right), \tag{20}$$

Note that the expressions (19) satisfy the transformation law (13). It is convenient to chose the component $x_{n,n}^{(\mathrm{PT})}$ as normalizing factor of the vectors (19).

Proposition 9. *If*

$$x_{n,n}^{(PT)} = 1, \tag{21}$$

then the vectors with the components calculated according to (19), satisfy approximately the orthogonality conditions

$$\sum_l x_{n',l}^{(PT)} x_{n'',l}^{(PT)*} = \delta_{n',n''} \cdot O(1) + (1 - \delta_{n',n''}) \cdot O(\mu^{|n'-n''+2|}). \tag{22}$$

Now we proceed to the analogous perturbation series for the eigenvalues η_n. Multiplying the difference equation (7a) by $x_{n,l}^*$ and summing it over l, using the conditions (22) and (21), we have in PT-approximation (20):

$$\eta_n = \gamma_n + S_-^{(\mathrm{PT},k)}(\eta_n) + S_+^{(\mathrm{PT},k)}(\eta_n) + o(\mu^k), \tag{23}$$

$$S_-^{(\mathrm{PT},k)}(\eta_n) = (2\mu)^2 \sum_{l=\max(l_0+1,n-k)}^{n} \frac{\alpha_l\beta_l\left|x_{n,l}^{(\mathrm{PT})}\right|^2}{\eta_n - \gamma_{l-1}}, \tag{24}$$

$$S_+^{(\mathrm{PT},k)}(\eta_n) = (2\mu)^2 \sum_{l=n}^{n+k} \frac{\alpha_{l+1}\beta_{l+1}\left|x_{n,l}^{(\mathrm{PT})}\right|^2}{\eta_n - \gamma_{l+1}}. \tag{25}$$

As usual, the sum (24) turns to zero when the lower bound of summation is greater than the lower bound of summation. In particular, $\forall k > 0 \;\; S_-^{(\mathrm{PT},k)}(\eta_{l_0}) = 0$.

The relationship (23) has a form of an implicit equation for η_n. Given that $S_\pm^{(\mathrm{PT},k)}(\eta) = O(\mu^{2k+2})$, it would be reasonable to get the asymptotics (of the same order) of this equation. This can be done by different ways which are characterized by different estimates of the residual term, $o(\mu^{2k+2})$, of the asymptotics. One of such ways (we denote it as PT0-approximation) consists in using the non-perturbed values $\eta_n = \gamma_n$ in formulas (19) and (23), *i. e.* $S_\pm^{(\mathrm{PT0},k)} = S_\pm^{(\mathrm{PT},k)}(\eta_n = \gamma_n)$. The second way (we denote it as PT-approximation, in analogy with that for the eigen-sequences (19)) consists in using the PT0-estimate for $\eta_n = \eta_n^{(\mathrm{PT0},k)}$ in formulas (19) and (23) .

Theorem 10. *(PT0)- and (PT)-approximations for the eigenvalues*

$$\eta_n^{(PT,k)} = \gamma_n + S_-^{(PT,k)}\left(\eta_n^{(PT0)}\right) + S_+^{(PT,k)}\left(\eta_n^{(PT0)}\right), \tag{26a}$$

$$\eta_n^{(PT0,k)} = \gamma_n + S_-^{(PT0,k)} + S_+^{(PT0,k)} \tag{26b}$$

are their asymptotic estimates of the order of $2k+2$*:*

$$\eta_n = \eta_n^{(PT,k)} + o(\mu^{2k+2}). \tag{26c}$$

IOP Conf. Series: Journal of Physics: Conf. Series **973** (2018) 012029 doi:10.1088/1742-6596/973/1/012029

3. Numerical results and discussion

A disadvantage of the theory presented above is a lack of explicit estimates for accuracy of the asymptotics (20), (22), (26). To illustrate their accuracy, we consider a particular example appearing in molecular spectroscopy.

The Fourier coefficients of dipole-spherical functions (used in theory of Rydberg states in polar molecules [3] and dipole-bound anions [6]), satisfy the recurrence relation (7a) with the following matrix elements (10):

$$\alpha_l = \beta_l = \left[\frac{l^2-\lambda^2}{4l^2-1}\right]^{1/2}, \quad \gamma_l = l(l+1); \qquad n=\ell. \tag{27}$$

The parameter μ is the dipole moment of the molecular core (in atomic units); $n=\ell$ is an integer quantum number which enumerates the eigenvalues in order of growth (at constant λ). In the unperturbed case (11), which physically corresponds to the central potential, ℓ is a conserved quantum number satisfying the standard relationship from the angular momentum theory: $\eta_\ell[0]=\ell(\ell+1)$, $\ell\in\mathbb{N}_0$. For $\mu\neq 0$ the orbital angular momentum is not a conserved quantum number, so ℓ is merely an integer numerator of the eigenvalue series with the same projection, λ, of the orbital angular momentum on the dipole moment direction. This projection is conserved even if $\mu\neq 0$. From (27) it is natural to assume $l_0=|\lambda|$. We emphasize that the notation ℓ for the eigenvalue enumerator should not be mixed with the discrete variable l (in quantum mechanics, l is referred to as the index of the representation, while $n=\ell$ as the index of the state).

Introducing

$$P_{\ell\lambda,l} = \frac{(l+\lambda)!(l-\lambda)!(2\ell-1)!!(2\ell+1)!!}{(\ell+\lambda)!(\ell-\lambda)!(2l-1)!!(2l+1)!!}, \tag{28}$$

and inserting the matrix elements (27) into eqns. (19) we obtain for the eigenvectors in PT0-approximation:

$$x^{(\mathrm{PT0})}_{\ell\lambda,l}[\mu] = \begin{cases} (2\mu)^{l-\ell}\left[P_{\ell\lambda,l}\right]^{\frac{1}{2}}\dfrac{(2\ell+1)!}{(l-\ell)!(\ell+l+1)!}, & l\geq\ell, \\ (-2\mu)^{\ell-l}\left[P_{\ell\lambda,l}\right]^{-\frac{1}{2}}\dfrac{(\ell+l)!}{(\ell-l)!(2\ell)!}, & |\lambda|\leq l\leq\ell; \end{cases} \tag{29}$$

Using the PT0-approximation (29) in (23) we can make summation in (24,25) up to $k=\infty$ and express the result in terms of generalized hypergeometric functions:

$$\begin{aligned} \eta^{(\mathrm{PT0})}_{\ell\lambda} &= \ell(\ell+1) \\ &+ \frac{\left[\lambda^2-(\ell+1)^2\right](2\mu)^2}{(2\ell+1)(2\ell+2)(2\ell+3)}\,{}_2F_5\left(\begin{matrix}2+\ell-\lambda,\ \ 2+\ell+\lambda\\ \frac{3}{2}+\ell,\ \frac{5}{2}+\ell,\ 2,\ 2\ell+3,\ 2\ell+2\end{matrix}\middle|\mu^2\right) \\ &+ \frac{(1-\delta_{\ell,0})(\ell^2-\lambda^2)(2\mu)^2}{(2\ell-1)(2\ell)(2\ell+1)}\,{}_2F_5\left(\begin{matrix}1-\ell-\lambda,\ \ 1-\ell+\lambda\\ \frac{3}{2}-\ell,\ \frac{1}{2}-\ell,\ 2,\ 1-2\ell,\ -2\ell\end{matrix}\middle|\mu^2\right). \end{aligned} \tag{30}$$

Here the factor $(1-\delta_{\ell,0})$ is zero for $\ell=0$ in accord with (24).

In the first non-vanishing approximation in powers of μ, the hypergeometric functions in (30) turn to unities, that yields the simple expression for the eigenvalues [9]:

$$\eta_{\ell\lambda} = \begin{cases} \ell(\ell+1)+\dfrac{2\mu^2\left[\ell(\ell+1)-3\lambda^2\right]}{\ell(\ell+1)(2\ell-1)(2\ell+3)}+O(\mu^4), & \ell>0 \\ -\dfrac{2}{3}\mu^2+O(\mu^4), & \ell=0 \end{cases} \tag{31}$$

The analogous results in (PT)-approximation can also be obtained. To this end, we define so called quasiorbital momentum $\tilde{\ell}$ (in (PT0)-approximation) according to $\tilde{\ell}(\tilde{\ell}+1) = \eta_n^{(\mathrm{PT0},k)}$, and express the the eigenvectors in PT0-approximation:

$$x^{(\mathrm{PT})}_{\ell\lambda,l}[\mu] = \begin{cases} (2\mu)^{l-\ell}\left[P_{\ell\lambda,l}\right]^{\frac{1}{2}}\dfrac{\Gamma(1+\ell-\tilde{\ell})\Gamma(2+\ell+\tilde{\ell})}{\Gamma(1+l-\tilde{\ell})\Gamma(2+l+\tilde{\ell})}, & l \geq \ell, \\ (-2\mu)^{\ell-l}\left[P_{\ell\lambda,l}\right]^{-\frac{1}{2}}\dfrac{\Gamma(1-\ell+\tilde{\ell})\Gamma(1+l+\tilde{\ell})}{\Gamma(1-l+\tilde{\ell})\Gamma(1+\ell+\tilde{\ell})}, & |\lambda| \leq l \leq \ell. \end{cases} \tag{32}$$

Using the PT-approximation (32) in (23) we again make summation in (24,25) up to $k = \infty$ and express the result in terms of generalized hypergeometric functions:

$$\begin{aligned} \eta^{(\mathrm{PT})}_{\ell\lambda} &= \ell(\ell+1) + \frac{\left[\lambda^2 - (\ell+1)^2\right](2\mu)^2}{(2\ell+1)(2\ell+3)(\ell-\tilde{\ell}+1)(\tilde{\ell}+\ell+2)} \\ &\times\ {}_3F_6\left(\left.\begin{matrix} 1,\quad 2+\ell-\lambda,\quad 2+\ell+\lambda \\ \frac{3}{2}+\ell,\ \frac{5}{2}+\ell,\ 2-\tilde{\ell}+\ell,\ 1-\tilde{\ell}+\ell,\ 2+\tilde{\ell}+\ell,\ 3+\tilde{\ell}+\ell \end{matrix}\right| \mu^2\right) \\ &+ \frac{(\ell^2-\lambda^2)(2\mu)^2}{(2\ell-1)(2\ell+1)(\tilde{\ell}-\ell+1)(\tilde{\ell}+\ell)} \\ &\times\ {}_3F_6\left(\left.\begin{matrix} 1,\quad 1-\ell-\lambda,\quad 1-\ell+\lambda \\ \frac{3}{2}-\ell,\ \frac{1}{2}-\ell,\ 2+\tilde{\ell}-\ell,\ 1-\tilde{\ell}-\ell,\ 1+\tilde{\ell}-\ell,\ -\tilde{\ell}-\ell \end{matrix}\right| \mu^2\right) \end{aligned} \tag{33}$$

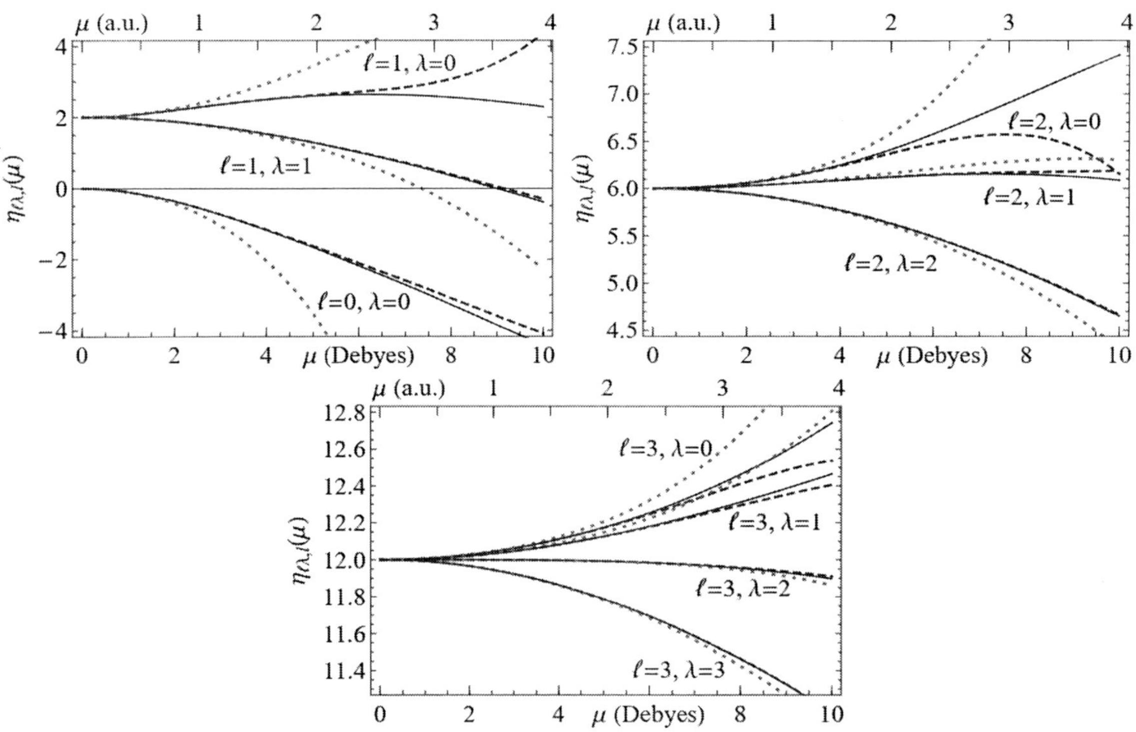

Figure 1. Eigenvalues $\eta_\ell[\mu]$ for $\ell = 0..3$. Dotted lines (······) correspond to PT0-approximation (29); dashed lines (- - - -) correspond to PT-approximation (32); solid lines (——) correspond to the numerical solution of the eigenvector problem (7)

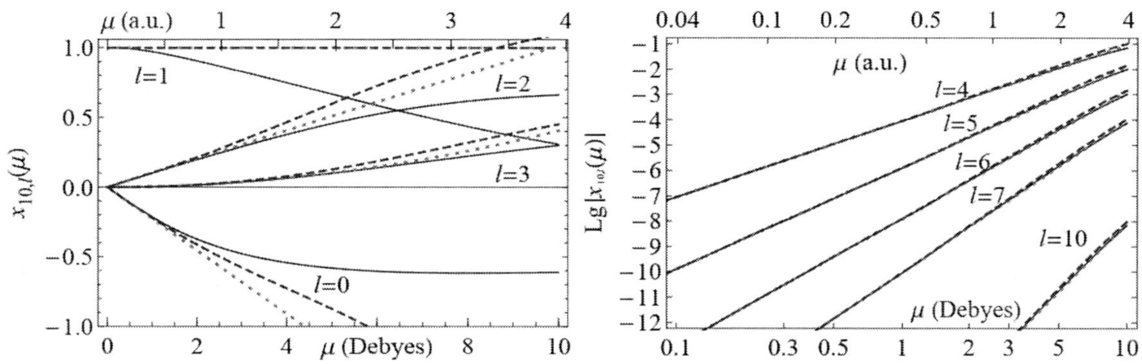

Figure 2. Eigenvector components $x_{n,l}[\mu]$ at $\ell = 1$ $(\lambda = 0)$. Dotted lines $(\cdots\cdots)$ correspond to PT0-approximation (30); dashed lines (- - - -) correspond to PT-approximation (33); solid lines (——) correspond to the numerical solution of the eigenvalue problem (7)

Some plots of eigenvalues, $\eta_{n=\{\ell\lambda\}}$, and eigenvectors, $x_{\ell,l}$, (so called l-mixing coefficients) as functions of the dipole moment, μ, and index of the representation, l are given in figures 1 and 2 for different ℓ and λ.

It can bee seen from the figures that, at fixed l and ℓ, λ, the accuracy of our analytical approximations is the better the closer to zero are the values of μ. At fixed μ the accuracy is improved with increase of l and ℓ, however this improvement depends non-monotonically on the ratio between ℓ and λ). The plots demonstrate good agreement of the above analytical approximations with the numerical results up to $\mu = 7..10$ Debye units that surpass the dipole moments of most of known molecules.

4. Conclusion

This work studies the dependence of eigenvalues and eigenvectors of infinite three-diagonal matrices depending on the parameter μ. For the eigenvalues and eigenvectors we obtain simple approximate analytic expressions which are their asymptotic estimates at $\mu \to 0$. We also prove some symmetry properties of eigenvalues and eigenvectors related to their behaviour with respect to the substitution $\mu \to -\mu$. A continuous fraction representation is given for the eigenvectors. The efficiency of the proposed approximation is demonstrated on the example of Fourier coefficients of the dipole-spherical functions which has wide application in molecular spectroscopy.

Acknowledgments

The authors acknowledge the support of Russian Ministry of Education and Science (State order No. 3.1761.2017/4.6).

References

[1] Abdalla M S and Eleuch H 2014 *J. Appl. Phys.* **115** 234906
[2] Chabab M, El Batoul A and Oulne M 2015 *J. Math. Phys.* **56** 062111
[3] Zon B A 1992 *J. Exp. Theor. Phys.* **75**(1) 36–46
[4] Akindinova E V, Chernov V E, Kretinin I Y and Zon B A 2010 *Phys. Rev. A* **81** 042517
[5] Elfimov S V, Dorofeev D L and Zon B A 2014 *Phys. Rev. A* **89**(2) 022507
[6] Chernov V E, Dolgikh A V and Zon B A 2005 *Phys. Rev. A* **72** 052701

IOP Conf. Series: Journal of Physics: Conf. Series **973** (2018) 012029 doi:10.1088/1742-6596/973/1/012029

[7] Chernov V E, Danilyan A V and Zon B A 2009 *Phys. Rev. A* **80** 022702

[8] Debye P J W 1929 *Polar Molecules* (New York: The Chemical Catalog Co., Inc.)

[9] Watson J K G 1994 *Mol. Phys.* **81** 277–289

[10] Chernov V E, Kiyan I Y, Helm H and Zon B A 2005 *Phys. Rev. A* **71**(3, B) 033410

[11] Braun P A and Kiselev A A 1983 *Introduction to the theory of molecular spectra* (Leningrad: Leningr.Univ.) in Russian

[12] Zon B A and Katsnel'son B G 1976 *J. Exp. Theor. Phys.* **42** 595–601

[13] Nikiforov A F, Uvarov V B and Suslov S K 1985 *Classical Orthogonal Polynomials of a Discrete Variable* Springer Series in Computational Physics (Berlin, Heidelberg: Springer)

[14] Bateman H and Erdélyi A 1955 *Higher Transcendental Functions* Bateman Manuscript Project (New York: McGraw-Hill)

[15] Kato T 1995 *Perturbation theory for linear operators* Classics in Mathematics (Berlin: Springer–Verlag) reprint of the 1980 edition

Analysis of methods of processing of expert information by optimization of administrative decisions

D Y Churakov[1], E G Tsarkova[1], N D Marchenko[2], E V Grechishnikov[1]

[1]ET Department, Voronezh Institute Federal Penitentiary Service of Russia, 1a Irkutskaya Street, Voronezh, RU
[2]IT Department, Research institute of information technologies Federal Penitentiary Service of Russia, 17 Vagzhanova Street, Tver, RU

E-mail: university69@mail.ru

Abstract. In the real operation the measure definition methodology in case of expert estimation of quality and reliability of application-oriented software products is offered. In operation methods of aggregation of expert estimates on the example of a collective choice of an instrumental control projects in case of software development of a special purpose for needs of institutions are described. Results of operation of dialogue decision making support system are given an algorithm of the decision of the task of a choice on the basis of a method of the analysis of hierarchies and also. The developed algorithm can be applied by development of expert systems to the solution of a wide class of the tasks anyway connected to a multicriteria choice.

1. Introduction

The efficiency of application of the software in the official activities of employees in many respects is defined by quality and reliability of the used automated means [1]. In the real operation the measure definition methodology in case of expert estimation of quality and reliability of application-oriented software products is offered. Results of operation of dialogue decision making support system are given an algorithm of the decision of the task of a choice on the basis of a method of the analysis of hierarchies and also.

Often by development of software of a special purpose for needs of penal correction systems institutions the situation when several employees are engaged in development of one general task meets.

At the same time the only form of control over the implementation of goals are meetings and plenary sessions. These circumstances result in need of use of specialized flexible and convenient software in use which would allow all members of the team to watch efficiency of performance of work, to reduce labor costs and to successfully finish the project in time.

Control systems of projects can be such effective tool. Now in the world the wide class of similar systems functions. In this regard before the project manager there is a problem of the choice of the most suitable means.

The choice of the platform of the automated system is defined by the criteria which are based on completeness, functionality, continuous development of system, integrability of decisions, availability of technical support, a technological basis.

Objective of this research is development of methods of support of decision-making during definition of the most suitable software.

2. Theoretical part

2.1. Comparative analysis of platforms

Identification of the criteria causing suitability of system requires a detailed research of advantages and shortcomings of concrete platforms [1]. During the research the following systems have been considered: YouTrack, Redmine, JIRA, Planiro, Wrike, Basecamp. In table 1a, b the main attention is paid to the main features of the considered control systems of projects.

Table 1a. Features of project management systems.

Opportunity/system	YouTrack	Redmine	Jira	Planiro	Wrike	Basecamp
Schedule	-	+	+	-	+	+
Reports	+	+	+	+	+	+
Priorities	+	-	+	+	+	+
Cloudy storage	-	unlimited	10Gb	unlimited	50Gb	500Gb
Comments to tasks	+	+	+	+	+	-
Notices	+	+	+	+	+	+
Investment of files	+	+	+	+	+	+
Delegation of tasks	+	+	+	+	+	-
Access control	+	+	+	+	+	+
Templates of projects	+	+	-	+	+	+
Deshborda	+	-	+	+	+	
Frequent questions	+	-	+	-	-	-
Management of appointments	+	-	+	+	+	-
Tracking of time	+	-	+	+	+	-
Management of tasks	+	-	+	+	+	-
Management of access	+	-	+	-	+	-
Discussions	+	-	+	+	+	-
Statistics	+	-	+	+	+	-
Management Agile-project	+	-	+	-	-	-
Management of sprints	+	-	+	+	-	-
Management of a backlog	+	-	+	+	-	-
Assessment conditions of the project	+	-	+	+	-	+
Assessment of load of participants	+	-	-	+	-	-

Table 1b. Features of project management systems.

Opportunity/system	YouTrack	Redmine	Jira	Planiro	Wrike	Basecamp
Chart of combustion of tasks	+	-	+	+	-	-
Management of time	+	-	+	+	-	+
Android application	+	-	+	+	+	+
IOS application	+	-	+	+	+	+
Phone Windows application	-	-	+	-	-	-
Russian localization	+	+	+	+	+	-
Cloud	+	-	+	+	+	-
Server	+	+	+	-	-	-
Realization language	Java	Ruby on Rails	Java	Ruby, Swift	Java	Ruby
License	proprietary software	proprietary software	proprietary software	proprietary software	proprietary software	proprietary software
Free of charge for open-source of porekt	+	-	+	-	-	+
Access under the HTTPS protocol	+	-	+	+	+	+

Proceeding from the revealed features of platforms during consultations with the staff of departments of administration and development, the possibility of determination of the most significant criteria for assessment of systems (table 2) appeared.

Table 2. The main criteria for evaluating systems.

Criterion	Description
Existence of the user's guide	Meeting of frequently asked questions and answers to them
Dashboard	Practicality of the user interface for the organization and information representations by a method, simple for perception
Convenience of the graphic user interface	Attention is paid to clearness, convenience of useand friendliness of the interface of the user
Adjusted fields	Possibility of adding of additional information to some data types
Possibility of free use of the platform	Completely or open-source free software
Accumulation of functionality	Possibility of customization for expansion of standard opportunities of system
Management of access rights	Setting up the rules regulating conditions of access for the user to objects of an information system
File storage	The cloudy storage providing a possibility of file transfer
Task management	Convenience of planning, statement and control of tasks

1. ***YouTrack*** – the service developed especially for programmers and teams of development by the JetBrains company. The system provides fast and convenient search by means of search queries,

maintains command syntax for editing several tasks at the same time, possesses flexible control of system under working processes of any complexity. Supports*Scrum and Kanban*, is localized into Russian, gives an opportunity of creation of a set of reports and carries out the accounting of time.

2. ***Redmine*** – the flexible web application for project management including the Gantt charts, the calendar, forums, setup of roles and the email notification.

3. ***JIRA*** – system which is suitable for tracking of mistakes and project management in the company of any size. This tool can be used by all employees in team and project managers.

4. ***Planiro*** – the flexible and scalable web application intended for convenient project management and collaboration. Developers realized building block system that provides to users basic functionality and modules under its increase.

5. ***Wrike*** – control system of projects, including for creation of tasks, gathering of team, connection to already familiar services of collaboration, integration of mail.

6. ***Basecamp*** – one of the first platforms in the market for full project management created by the company 37 signals. Developers have provided work only with English. Differences between subscriptions to this service consist only in the volume of storage and number of active projects.

The method of the analysis of hierarchies can be applied to the solution of a problem of the optimum choice of system [2-14].

2.2. Model on the basis of a set of the studied parameters

We will enter into consideration of a set:

1) $Q_i = \{Q_1, Q_2, ..., Q_i, ..., Q_I\}$ – alternatives of the chosen control systems of projects;

2) $k = \{1, 2, ..., K\}$ – identifying numbers of criteria;

3) $n = \{n_1, n_2, ..., n_K\}$ – set of coefficients of the relative importance (scales) of the entered criteria;

4) $F_k = \{F_{1k}, F_{2k}, ..., F_{ik}, ..., F_{Ik}\}$ – set of coefficients (scales) of alternatives at the choice of system on the basis of compliance to a certain criterion;

5) $F = \{F_1, F_2, ..., F_i, ..., F_I\}$ – total set of coefficients of decisions (alternatives).

2.3. The decision algorithm based on analytic hierarchy process

1. In case of a choice of a project management system we install a finite set of criteria: {user guide; task management; Dashboard; convenience of the graphic user interface; possibility of a tuning of fields; possibility of free use of the platform; possibility of increase in functionality; control of access rights, file storage}.

Table 3. Matrix of pair comparisons of criteria.

		1	2	3	4	5	6	7	8	9
User's guide	1	1	1/5	1/3	1/3	1	1/7	1	1/3	1/3
Task management	2	5	1	3	3	5	1/5	3	3	3
Dashboard	3	3	1/3	1	1	3	1/5	3	1	1
Convenience of the graphic user interface	4	3	1/3	1	1	3	1/7	3	1	1
Configurationfield	5	1	1/5	1/3	1/3	1	1/5	1	1/3	1/3
Free use of the platform	6	7	5	5	5	5	1	3	3	3
Increase in functionality	7	1	1/3	1/3	1/3	1	1/3	1	1/3	1/3
Management of access rights	8	3	1/3	1	1	3	1/3	3	1	1
File storage	9	3	1/3	1	1	3	1/3	3	1	1

IOP Conf. Series: Journal of Physics: Conf. Series **973** (2018) 012030 doi:10.1088/1742-6596/973/1/012030

2. For formation of estimates of the relative importance of criteria we form a matrix of paired comparisons of criterion $N_{k,r}$, $k,r=\overline{1,K}$.

We choose intuitively reasonable scale of preferences of a look: 1, 3, 5, 7, 9, where:

a) 1 – criteria are equivalent;
b) 3 – insignificant importance of criterion k over r;
c) 5 – existence of considerable importance of criterion k over r;
d) 7 – criterion k it is obvious more important than criterion r;
e) 9 – the absolute importance of criterion k in comparison r.

At the choice of this scale the following values of estimates are possible: $1/9$, $1/7$, $1/5$, $1/3$, 1, 3, 5, 7, 9, at the same time $N_{k,r}=1/N_{r,k}$, $k,r=\overline{1,K}$.

We form matrixes of pair comparisons on the basis of the analysis of the possibilities of the considered systems and also results of interviewing of experts.

3. We create matrixes of paired comparisons of preference of alternatives of the chosen systems depending on the chosen criterion $T_{i,z,k}$. For this purpose, we compare quality of the chosen systems *i* and *z* by various criteria *k*, $k=\overline{1,I}$, $T_{z,i,k}=1/T_{i,z,k}$, $i,z=\overline{1,K}$.

Paired comparisons are made according to a scale:

a) 1 – equivalence of alternatives;
b) 3 – existence of insignificant preference of an alternative *i* over *z*;
c) 5 – considerable preference of an alternative *i* in comparison with *z*;
d) 7 – alternative of *i* it is obvious more important than an alternative *z*;
e) 9 – absolute preference of an alternative of *i* in comparison *z*.

Results are given in tables 4-12.

Table 4. Matrix of pair comparisons of systems by criterion «the user's guide» ($T_{z,i,1}$).

		1	2	3	4	5	6
YouTrack	1	1	3	1/3	5	3	3
Redmine	2	3	1	1/5	5	1	1
Jira	3	3	5	1	5	3	3
Planiro	4	1/5	1/5	1/5	1	1/3	1/3
Wrike	5	1/3	1	1/3	3	1	1
Basecamp	6	1/3	1	1/3	3	1	1

Table 5. Matrix of pair comparisons by criterion «management of tasks» ($T_{z,i,2}$).

		1	2	3	4	5	6
YouTrack	1	1	5	1	3	3	5
Redmine	2	1/5	1	1/5	1/3	1/3	1
Jira	3	1	5	1	3	3	5
Planiro	4	1/3	3	1/3	1	1	3
Wrike	5	1/3	3	1/3	1/3	1	3
Basecamp	6	1/5	1	1/3	1/3	1/3	1

IOP Conf. Series: Journal of Physics: Conf. Series **973** (2018) 012030 doi:10.1088/1742-6596/973/1/012030

Table 6. Matrix of pair comparisons by criterion «Dashboard» ($T_{z,i,3}$).

		1	2	3	4	5	6
YouTrack	1	1	5	1	5	1	7
Redmine	2	1/5	1	1/3	1	1/3	3
Jira	3	1	3	1	5	1	7
Planiro	4	1/5	1	1/5	1	1/5	3
Wrike	5	1	3	1	5	1	5
Basecamp	6	1/7	1/3	1/7	1/3	1/5	1

Table 7. Matrix of paired comparisons by criterion «convenience of the graphic user interface» ($T_{z,i,4}$).

		1	2	3	4	5	6
YouTrack	1	1	1	1/5	1/3	1/5	1/3
Redmine	2	1	1	1/5	1/3	1/5	1/3
Jira	3	5	5	1	3	1	3
Planiro	4	3	3	1/3	1	1/3	1
Wrike	5	5	5	1	3	1	3
Basecamp	6	3	3	1/3	1	1/3	1

Table 8. Matrix of pair comparisons by criterion «a configurability of fields» ($T_{z,i,5}$).

		1	2	3	4	5	6
YouTrack	1	1	3	3	5	5	5
Redmine	2	1/3	1	1	3	3	3
Jira	3	1/3	1	1	3	3	3
Planiro	4	1/5	1/3	1/3	1	1	1
Wrike	5	1/5	1/3	1/3	1	1	1
Basecamp	6	1/5	1/3	1/3	1	1	1

Table 9. Matrix of pair comparisons by criterion «cost» ($T_{z,i,6}$).

		1	2	3	4	5	6
YouTrack	1	1	1	3	5	5	3
Redmine	2	1	1	3	5	5	3
Jira	3	1/3	1/3	1	3	3	1
Planiro	4	1/5	1/5	1/3	1	1	1/3
Wrike	5	1/5	1/5	1/3	1	1	1/3
Basecamp	6	1/3	1/3	1	3	3	1

Table 10. Matrix of pair comparisons by criterion «increase in functionality» ($T_{z,i,7}$).

		1	2	3	4	5	6
YouTrack	1	1	1	1	3	3	3
Redmine	2	1	1	1	3	3	3
Jira	3	1	1	1	3	3	3
Planiro	4	1/3	1/3	1/3	1	1	1
Wrike	5	1/3	1/3	1/3	1	1	1
Basecamp	6	1/3	1/3	1/3	1	1	1

Table 11. Matrix of pair comparisons by criterion «management of access» ($T_{z,i,8}$).

		1	2	3	4	5	6
YouTrack	1	1	3	3	3	3	3
Redmine	2	1/3	1	1	1	1	1
Jira	3	1/3	1	1	1	1	1
Planiro	4	1/3	1	1	1	1	1
Wrike	5	1/3	1	1	1	1	1
Basecamp	6	1/3	1	1	1	1	1

Table 12. Matrix of pair comparisons by criterion «file storage» ($T_{z,i,9}$).

		1	2	3	4	5	6
YouTrack	1	1	1/7	1/5	1/7	1/3	1/5
Redmine	2	7	1	3	1	5	3
Jira	3	5	1/3	1	1/3	3	1
Planiro	4	7	1	3	1	5	3
Wrike	5	3	1/5	1/3	1/5	1	1/3
Basecamp	6	5	1/3	1	1/3	3	1

4. For a matrix of pair comparisons of criteria we find the vector $\overline{n_{\max}}$ equal to the maximum own value: $\overline{n_{\max}} = \left(n_{\max 1}, n_{\max 2}, ..., n_{\max K}\right)$ $\lambda_{\max K}$ – maximum own value of a matrix of paired comparisons of criteria.

5. We carry out transformation of the received vector according to the rule:

$$n_{\max k} = n_{\max k} \Big/ \sum_k n_{\max k} \,. \tag{1}$$

6. For all matrixes of paired comparisons of preference of alternatives of the chosen control system of projects we calculate a vector $\overline{T_{\max k}}$ which corresponds to the maximum own values of matrixes: $\overline{T_{\max k}} = \left(T_{\max 1k}, T_{\max 2k}, ..., T_{\max Ik}\right)$ where $\lambda_{\max ik}$ – maximum ownvalues of matrixes of paired comparisons of preference of alternatives of the chosen control system of projects depending on the chosen criterion.

7. We carry out transformation of elements of the received vector $\overline{T_{\max k}}$ by the rule:

$$T_{ik} = T_{\max ik} \Big/ \sum_i T_{\max ik} \,. \tag{2}$$

8. We calculate coefficients of decisions (alternatives) on a formula:

$$F_i = \sum_k T_{ik} \cdot n_k . \tag{3}$$

9. We order versions of decisions F_i. The greatest value of coefficient corresponds to the most preferable version of the decision. Sort the solutions variants F_i.

10. End of an algorithm.

This method is the basis for freely distributed software of *Mpriority* representing dialogue decision making support system in the most different spheres of human activities [8-14].

We will apply the possibilities of this system for the decision of the considered task. In the *Mpriority* interface the researched model has the appearance provided in a figure 1.

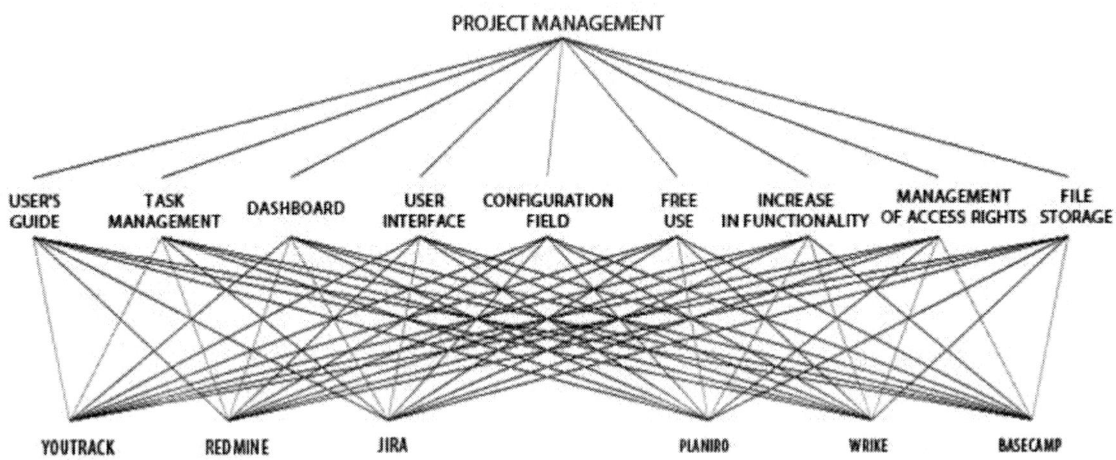

Figure 1. The scheme of a problem of the choice of a control system of projects in *Mpriority*.

The matrix of paired comparisons of criteria is entered into system (see the figure 2).

The matrixes of paired comparisons of alternatives presented in drawings are entered 3-11.

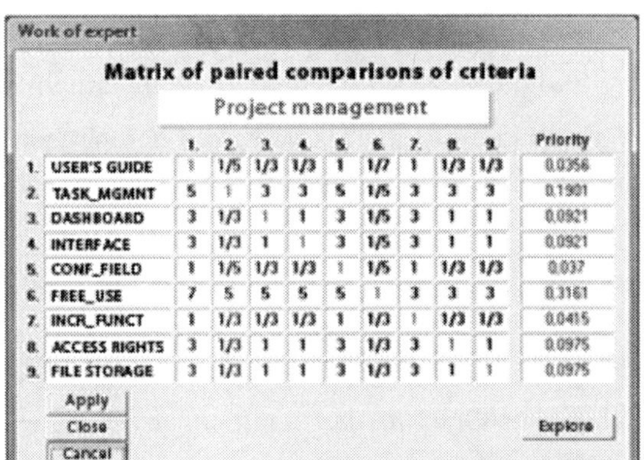

Work of expert

Matrix of paired comparisons of criteria

Project management

		1.	2.	3.	4.	5.	6.	7.	8.	9.	Priority
1.	USER'S GUIDE	1	1/5	1/3	1/3	1	1/7	1	1/3	1/3	0.0356
2.	TASK_MGMNT	5	1	3	3	5	1/5	3	3	3	0.1901
3.	DASHBOARD	3	1/3	1	1	3	1/5	3	1	1	0.0921
4.	INTERFACE	3	1/3	1	1	3	1/5	3	1	1	0.0921
5.	CONF_FIELD	1	1/5	1/3	1/3	1	1/5	1	1/3	1/3	0.037
6.	FREE_USE	7	5	5	5	5	1	3	3	3	0.3161
7.	INCR_FUNCT	1	1/3	1/3	1/3	1	1/3	1	1/3	1/3	0.0415
8.	ACCESS RIGHTS	3	1/3	1	1	3	1/3	3	1	1	0.0975
9.	FILE STORAGE	3	1/3	1	1	3	1/3	3	1	1	0.0975

Apply Close Cancel Explore

Figure 2. Matrix $N_{k,r}$.

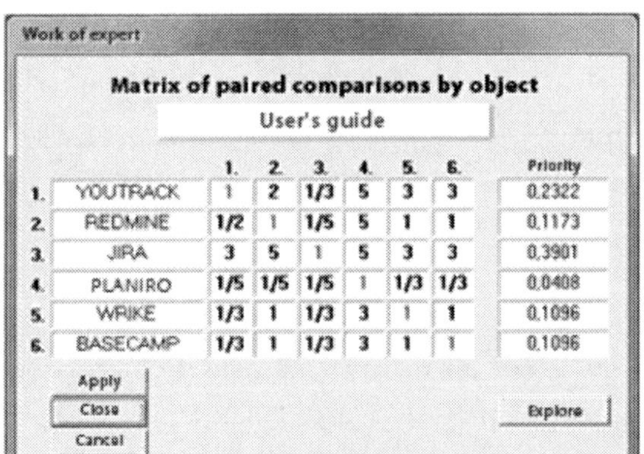

Work of expert

Matrix of paired comparisons by object

User's guide

		1.	2.	3.	4.	5.	6.	Priority
1.	YOUTRACK	1	2	1/3	5	3	3	0.2322
2.	REDMINE	1/2	1	1/5	5	1	1	0.1173
3.	JIRA	3	5	1	5	3	3	0.3901
4.	PLANIRO	1/5	1/5	1/5	1	1/3	1/3	0.0408
5.	WRIKE	1/3	1	1/3	3	1	1	0.1096
6.	BASECAMP	1/3	1	1/3	3	1	1	0.1096

Apply Close Cancel Explore

Figure 3. Matrix $T_{z,i,1}$.

IOP Conf. Series: Journal of Physics: Conf. Series **973** (2018) 012030 doi:10.1088/1742-6596/973/1/012030

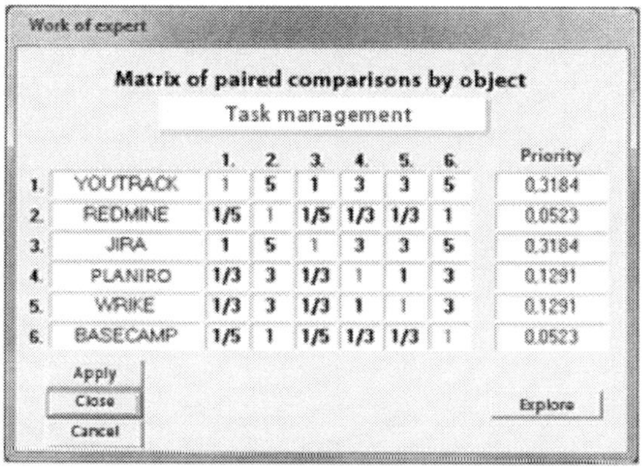

		1.	2.	3.	4.	5.	6.	Priority
1.	YOUTRACK	1	5	1	3	3	5	0,3184
2.	REDMINE	1/5	1	1/5	1/3	1/3	1	0,0523
3.	JIRA	1	5	1	3	3	5	0,3184
4.	PLANIRO	1/3	3	1/3	1	1	3	0,1291
5.	WRIKE	1/3	3	1/3	1	1	3	0,1291
6.	BASECAMP	1/5	1	1/5	1/3	1/3	1	0,0523

Figure 4. Matrix $T_{z,i,2}$.

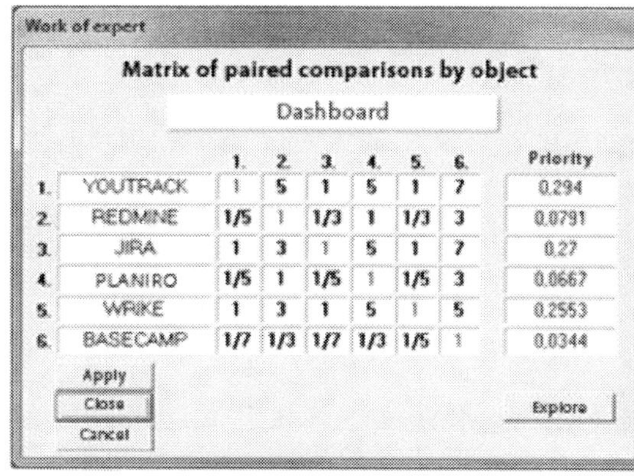

		1.	2.	3.	4.	5.	6.	Priority
1.	YOUTRACK	1	5	1	5	1	7	0,294
2.	REDMINE	1/5	1	1/3	1	1/3	3	0,0791
3.	JIRA	1	3	1	5	1	7	0,27
4.	PLANIRO	1/5	1	1/5	1	1/5	3	0,0667
5.	WRIKE	1	3	1	5	1	5	0,2553
6.	BASECAMP	1/7	1/3	1/7	1/3	1/5	1	0,0344

Figure 5. Matrix $T_{z,i,3}$.

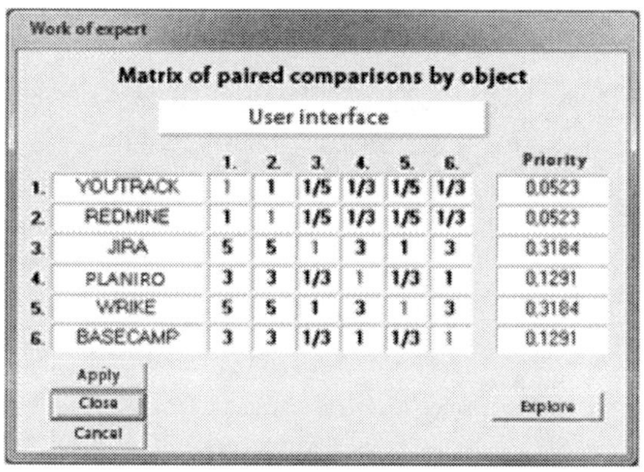

		1.	2.	3.	4.	5.	6.	Priority
1.	YOUTRACK	1	1	1/5	1/3	1/5	1/3	0,0523
2.	REDMINE	1	1	1/5	1/3	1/5	1/3	0,0523
3.	JIRA	5	5	1	3	1	3	0,3184
4.	PLANIRO	3	3	1/3	1	1/3	1	0,1291
5.	WRIKE	5	5	1	3	1	3	0,3184
6.	BASECAMP	3	3	1/3	1	1/3	1	0,1291

Figure 6. Matrix $T_{z,i,4}$.

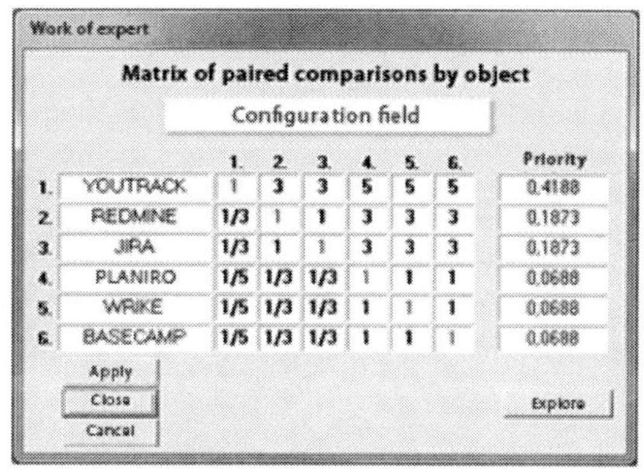

		1.	2.	3.	4.	5.	6.	Priority
1.	YOUTRACK	1	3	3	5	5	5	0,4188
2.	REDMINE	1/3	1	1	3	3	3	0,1873
3.	JIRA	1/3	1	1	3	3	3	0,1873
4.	PLANIRO	1/5	1/3	1/3	1	1	1	0,0688
5.	WRIKE	1/5	1/3	1/3	1	1	1	0,0688
6.	BASECAMP	1/5	1/3	1/3	1	1	1	0,0688

Figure 7. Matrix $T_{z,i,5}$.

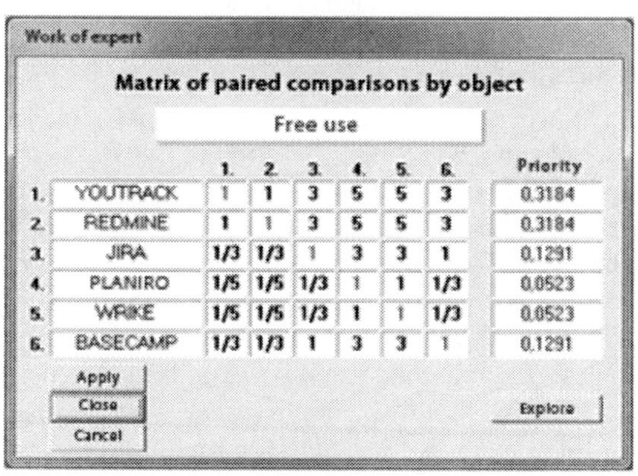

		1.	2.	3.	4.	5.	6.	Priority
1.	YOUTRACK	1	1	3	5	5	3	0,3184
2.	REDMINE	1	1	3	5	5	3	0,3184
3.	JIRA	1/3	1/3	1	3	3	1	0,1291
4.	PLANIRO	1/5	1/5	1/3	1	1	1/3	0,0523
5.	WRIKE	1/5	1/5	1/3	1	1	1/3	0,0523
6.	BASECAMP	1/3	1/3	1	3	3	1	0,1291

Figure 8. Matrix $T_{z,i,6}$.

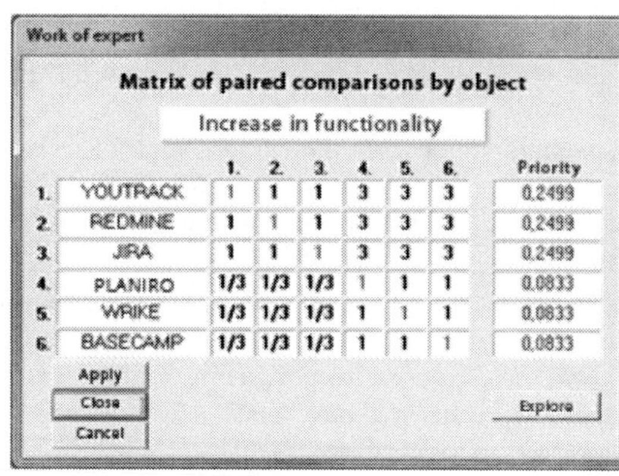

		1.	2.	3.	4.	5.	6.	Priority
1.	YOUTRACK	1	1	1	3	3	3	0,2499
2.	REDMINE	1	1	1	3	3	3	0,2499
3.	JIRA	1	1	1	3	3	3	0,2499
4.	PLANIRO	1/3	1/3	1/3	1	1	1	0,0833
5.	WRIKE	1/3	1/3	1/3	1	1	1	0,0833
6.	BASECAMP	1/3	1/3	1/3	1	1	1	0,0833

Figure 9. Matrix $T_{z,i,7}$.

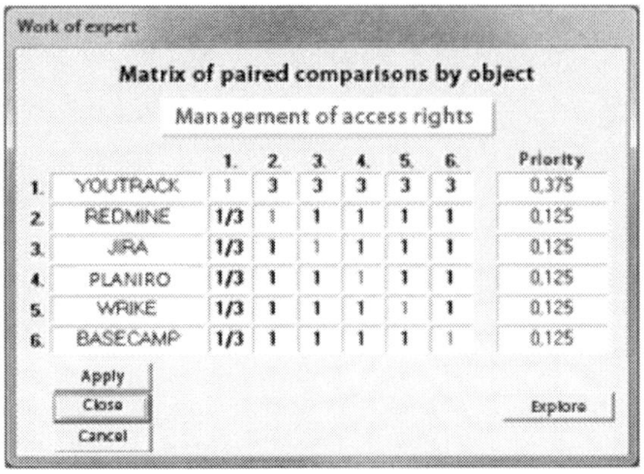

		1.	2.	3.	4.	5.	6.	Priority
1.	YOUTRACK	1	3	3	3	3	3	0,375
2.	REDMINE	1/3	1	1	1	1	1	0,125
3.	JIRA	1/3	1	1	1	1	1	0,125
4.	PLANIRO	1/3	1	1	1	1	1	0,125
5.	WRIKE	1/3	1	1	1	1	1	0,125
6.	BASECAMP	1/3	1	1	1	1	1	0,125

Figure 10. Matrix $T_{z,i,8}$.

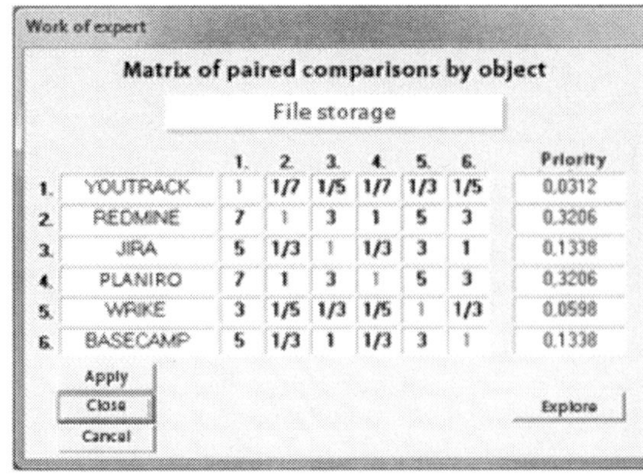

		1.	2.	3.	4.	5.	6.	Priority
1.	YOUTRACK	1	1/7	1/5	1/7	1/3	1/5	0,0312
2.	REDMINE	7	1	3	1	5	3	0,3206
3.	JIRA	5	1/3	1	1/3	3	1	0,1338
4.	PLANIRO	7	1	3	1	5	3	0,3206
5.	WRIKE	3	1/5	1/3	1/5	1	1/3	0,0598
6.	BASECAMP	5	1/3	1	1/3	3	1	0,1338

Figure 11. Matrix $T_{z,i,9}$.

The result of the solution of a task in the *Mpriority* system is presented in the figure 12.

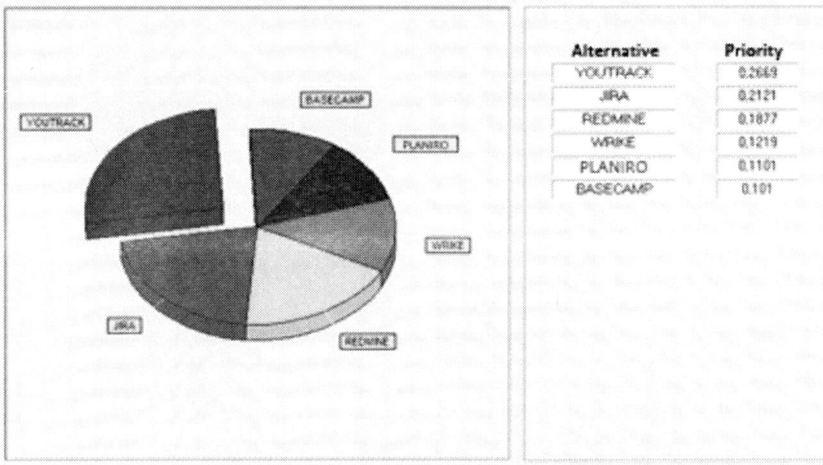

Figure 12. Result of calculations.

Taking into account the entered values as a result of work of *Mpriotity* of the most preferable there is a control system of projects of ***YouTrack***. It should be noted that this software for several years is actively used in office activity of staff of the center of information technology development Research Institute of Information Technologies, in particular, at team development of an information system of the personified account and personnel office-work in institutions and bodies of a penal correction system (IS PCS). At the same time, considering insignificant lag of the received values of the priorities corresponding to alternatives of ***Jira*** and ***Redmine*** from ***YouTrack*** value it is possible to draw a conclusion on expediency of possible use of any of this software. Thus, *Mpriotity* appears the effective tool of hierarchy of alternatives realizing the algorithm given above. However, apparent defect of this system is restriction for amount of the criteria entered for processing (no more than 9 names, see the figure 13).

These restrictions essentially influence a possibility of use of this software in activity of employees Federal state institution "Research Institute of Information Technologies of the Federal Penitentiary Service", as in practice at adoption of administrative decisions the amount of criteria or alternatives is considerably big. Therefore, development of the specialized software realizing the developed algorithm of a method of the analysis of hierarchies is expedient.

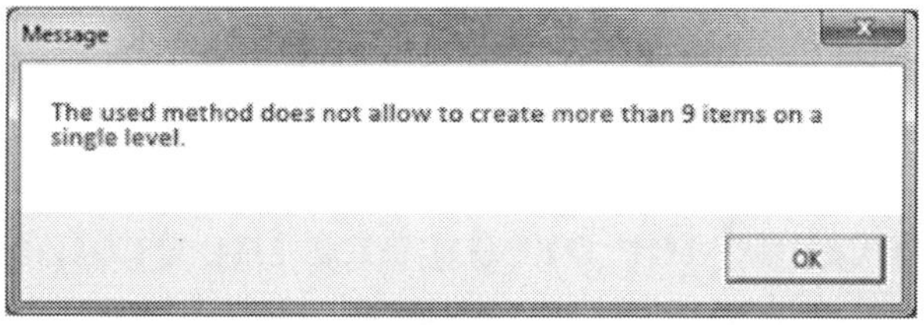

Figure 13. The message about restriction of amount of criteria in *Mpriotity*.

Thus, during the research the algorithm of selection of an optimum control facility projects based on a method of the analysis of hierarchies Saati is constructed. It is shown that in office activity of employees Research Institute of Information Technologies the most effective is use of the ***YouTrack*** system also perhaps application of ***Redmine*** or ***Jira***. The developed algorithm can be applied when developing expert systems to the solution of a wide class of the tasks anyway connected with the multicriteria choice (for example, at staff recruitment, software, drawing up ratings, etc.).

References

[1] Final report on ROC-21-2015 2015 *Development of the software for data collection and processing on the goods and production capacities made by institutions and the penal correction systems enterprises* (Tver: RIIT FPS of Russia) p 121

[2] Ridchenko A I 2014 *Metod of the analysis of hierarchies as the effective tool of the systems concept for acceptance of administrative decisions* (Omsk: Institute of Radiotronics, Service and Diagnostics) p 80

[3] Dushkin A V, Kravchenko A S, Novoseltsev V I, Smolentseva T E and Sumin V I 2014 *Mathematical models and information processes of management of a difficult object* (Voronezh: Scientific book) p 125

[4] Anosov A A 2001 *Selection term of the DBMS during creation of information systems* (Moscow: KRONE) p 124

[5] Trakhtengerts E A 1997 Generation, assessment and a choice of the scenario in decision making support systems *AiT* p 38-43

[6] Trakhtengerts E A 1998 *Computer support of decision-making* (Moscow: SINTEG) p 143

[7] Schneiderman M V 1988 Procedures of collective expert inquiry and their pilot study *AiT* pp 3-16

[8] Saati T L 2015 About measurement of impalpable. Approach to the relative measurements on the basis of the principal latent vector of a matrix of paired comparisons *Cloud of Science* **1** pp 5-39

[9] Gerald V 2001 Comparison of database management systems *Journal of Computer Information Systems* **3** pp 47-53

[10] Saati T 1993 *Making decisions. Method for analyzing hierarchies* (Moscow: Radio and communication) p 278

[11] Danshin F A, Dushkin A V and Shchekin V A 2015 Experimental research of the information structure of the process of preparation and decision-making in the automated system of personnel resource management *Fundamental research* **5/2** pp 284-288

[12] Dushkin A V, Novoseltsev V I and Sumin V I 2017 *Introduction to systems analysis* (Moscow: Hot line – Telecom) p 234

[13] Dushkin A V, Novoseltsev V I and Sumin V I 2017 *Introduction to system engineering of intellectual knowledge bases* (Moscow: Hot line – Telecom) p 108

[14] Dushkin A V, Dmitriev E V, Irkhin V P and Rossikhina L V 2017 Substantiation of the choice of the formalized model of information process of knowledge base management *Bulletin of the Voronezh Institute of the FPS of Russia* **1** pp 45-52

An improved method for predicting the evolution of the characteristic parameters of an information system

A V Dushkin, T I Kasatkina, V I Novoseltsev, S V Ivanov

ET Department, Voronezh Institute Federal Penitentiary Service of Russia, 1a Irkutskaya Street, Voronezh, RU

E-mail: kasatkinatatian@googlemail.com

Abstract. The article proposes a forecasting method that allows, based on the given values of entropy and error level of the first and second kind, to determine the allowable time for forecasting the development of the characteristic parameters of a complex information system. The main feature of the method under consideration is the determination of changes in the characteristic parameters of the development of the information system in the form of the magnitude of the increment in the ratios of its entropy. When a predetermined value of the prediction error ratio is reached, that is, the entropy of the system, the characteristic parameters of the system and the depth of the prediction in time are estimated. The resulting values of the characteristics and will be optimal, since at that moment the system possessed the best ratio of entropy as a measure of the degree of organization and orderliness of the structure of the system. To construct a method for estimating the depth of prediction, it is expedient to use the maximum principle of the value of entropy.

1. Introduction

Considering the process of improvement as one of the main ways of the evolutionary development of any information system (IS) with the purpose of developing a new one, one must know the optimal perspective characteristics of this new system at a specific time. As such, a measure of evolutionary development of the system, information entropy can be used. The longer the prediction time interval, the greater the uncertainty and, correspondingly, the lower reliability and reliability of the characteristics obtained with such a forecast. That is, we can assume that there is a so-called optimal forecasting time, at which it is possible to determine with sufficient accuracy the characteristics of the system's behavior and its parameters. A lot of mathematical models and methods that can be used to predict the behavior of the characteristic parameters of the system in a certain time interval [1-3]. Nevertheless, it is precisely the methods and the developed approaches for estimating the depth of the forecast for the development of the system that there is little.

The aim of the work is the development of a forecasting method that allows, based on the given values of entropy and the level of errors of the first and second kind, to determine the permissible time for forecasting the development of the characteristic parameters of a complex information system.

The main feature of the method under consideration is the determination of changes in the characteristic parameters of the development of the information system in the form of the magnitude of the increment in the ratios of its entropy. When a predetermined value of the prediction error ratio is reached, that is, the entropy of the system, the characteristic parameters of the system and the depth of the prediction in time are estimated. The resulting values of the characteristics and will be optimal, since at that moment the system possessed the best ratio of entropy as a measure of the degree of organization and orderliness of the structure of the system. The great value of the increment of entropy indicates not the development and the emergence of a new system, but its modernization. With a small amount of increment of entropy from the previous system, there is a danger that the new version of the system will not be adapted, that is, it is not adapted to the existing environment and circumstances. In other words, the new system has a small number of hereditary characteristics, has a small continuity from the previous system-progenitor. To construct a method for estimating the depth of prediction, it is expedient to use the maximum principle of the value of entropy.

Similar questions have already been of interest to domestic and foreign scientists, for example [5-8], but they focused on the theory of decision-making as the basis for choice, from the totality of existing systems, the best system. As a result of the application of the information entropy method, the predicted characteristic parameters of the development of a prospective system for a given moment of time, specified by the condition of the problem, are obtained. The received characteristics can be input data for the subsequent design of the system itself.

2. Formulation of the problem

Predicting the development of the characteristic parameters of a complex information system, it is necessary to determine its characteristics and the depth of the level of forecasting in time through the errors of the first and second kind specified by the condition of the stated problem as functional dependences of the entropies. As you know, the degree of organization and orderliness of the structure of each system can be described using the concept of entropy [7-9]. Because of this, each period of development of the system can be determined in the form of an appraising potential, in the quality of which is entropy. Thus, the conceptual approach applied in this work is based on research [5-8], the mathematical basis and apparatus are based on the method proposed in [7, 9].

The formalization of the task is considered from the standpoint of the legitimate existence of a scheme for determining a complex preference based, in turn, on a set of time intervals for the development of the system, and is implemented based on the so-called method of committees. Under the committee for assessing the effectiveness of decisions taken to implement, we mean a certain finite set consisting of vectors of the form $R^n : K = (x_1, x_2, ..., x_n) \in R^n$. For a given set, the acceptance of the null hypothesis H_0 for a given value of the significance level, adopted based on the conditions of the task, provides a certain number of components of the vector X_m, which in its essence is the number of predicted characteristics of the system.

3. Information modeling of the task

In this paper, we consider an improved method for estimating the depth of forecasting the evolution of the characteristic parameters of an information system. The solution of the problem posed in the paper is connected with the definition of a vector X_m that quantitatively reflects the potential of the system, as well as an estimate of the magnitude of the criterion's power β_i^{eff} and the level α_0^{eff} of significance of the strategic forecast. In this case, the information display of the particular situation in question occurs according to the following outline.

Let there be comparable n periods of development of the system. Each period of existence corresponds to a set of m predicted characteristics that determine the potential of the system at a particular time. The forecasting situation is described by a prediction form, in which one input is formed by a set of its characteristics X_m, and the other input is formed by a set of time scale for the development of the system $\{A_i\}$. This situation is presented in table 1.

When forming the estimated potential, the weight of the parameters is characterized by a quantitative measure that exists in a situation of uncertainty. This is due to the distribution of probabilities p_{ji}. The characteristics of the system $\{X_{ji}\}$ can have different units of measurement and exist initially both in a single and in different physical scales. In the case of different scales to bring the characteristics of the system $\{X_{ji}\}$ to a single scale, it is necessary to carry out the normalization performed with respect to the extrema of the values of the vector $\{X_{ji}\}$ components in two possible variations:

1) without implementing a change of vector to the opposite with the mapping $x_{ji} \to r[0,1]$:

$$r_{ji} = \frac{x_{ji}}{x_{\max j}}; \tag{1}$$

2) with the implementation of the change of the vector to the opposite with the mapping $x_{ji} \to r[0,1]$:

$$r_{ji} = \frac{x_{\min j}}{x_{ji}}. \tag{2}$$

Table 1. Form of the prediction matrix.

Characteristics of the system	System development time scale				
	$\{A_1\}$	...	$\{A_i\}$	...	$\{A_n\}$
X_1	X_{11}	...	X_{1i}	...	X_{1n}
X_j	X_{j1}	...	X_{ji}	...	X_{jn}
...	...	...	...	...	...
X_m	X_{m1}	...	X_{mi}	...	X_{mn}

Dependencies (1) and (2) provide a mapping of the sampling space X_{ji} shown in table 1 to the space r_{ji} shown in table 2.

Table 2. The form of the decision matrix.

Characteristics of the system	System development time scale				
	$\{A_1\}$	...	$\{A_i\}$	...	$\{A_n\}$
X_1	r_{11}	...	r_{1i}	...	r_{1n}
X_j	r_{j1}	...	r_{ji}	...	r_{jn}
...	...	...	...	...	...
X_m	r_{m1}	...	r_{mi}	...	r_{mn}

Elements of space r_{ji}, reduced to a single scale, correspond to elementary events. The value r_{ji} determined by the normalization will correspond to the probability $p(r)$, which, in turn, corresponds to the concept of the integral potential for a given set of elementary events. In fact, the value r_{ji} is introduced to interpret the concept of probability. Probability is a combination of the degree of subjective confidence in the occurrence of an event, as well as measures of an objective possible realization of the event as a whole. To formalize the problem of choosing a solution $\{A_n\}$, let us set up a set of systems with an event space, and the set of attributes is identified by the set $\{X_{mn}\}$. Then the connection between the components forming the estimated potential for the realization of the situation by the components will be realized through a specified for them standardized value r_{ji}, identified with probability $p(A)$. The probability $p(A)$ distribution obtained in the simulation of the situation will provide an estimate of the null hypothesis H_0 of the power of the criterion β_0^{eff} and the level α_0^{eff} of significance of the predicted development of the system under given conditions of the problem posed.

Thus, in this method, the amount of information is determined through the probability at the expense of entropy, and the very concept of information is viewed as a certain reduced diminishing uncertainty. By virtue of this approach, it is necessary to conduct research on the laws by which the information from the Cartesian components, which are represented as the product of two sets (a set of characteristics and a set of systems), is converted into quantitative components of information for further construction of the methodology. For this, such concepts as conditional, a priori and a posteriori probabilities are introduced into the logic scheme. In addition, full-probability and Bayesian formulas are applied. Also introduced is the notion of conditional probability p of the performance j of a characteristic in the formation of the estimated potential, under the conditions under which those events forming the estimated potential were implemented.

In order to obtain the dependence of the definition of a quantity $p(r)$ that is a normalized quantity r_{ji}, we use the well-known principle that the notion of the estimated potential of a given set of elementary events is identified with the membership function assigned to each real number in the interval [0, 1]. Then, the dependence of the membership function we are looking for can be represented in the form:

$$p_{ji}(r) = \frac{r_{ji}}{\sum_{i=1}^{n} r_{ji}}. \tag{3}$$

One of the existing methods for calculating the probabilities of the performance j of a characteristic in the formation of the estimated potential for compared time intervals is based on the concept of the potential probability distribution introduced in [7]. In this case, the calculation of probabilistic estimates is based on the principle of maximum uncertainty, and the estimates themselves are obtained as a result of solving the problem of conditional extremum:

$$H(P) = -\sum_{j=1}^{m} p_j \log p_j \to \max ; \tag{4}$$

$$\sum_{j=1}^{m} p_j = 1; \tag{5}$$

$$\prod_{j=1}^{m} r^{p_j} = const , \tag{6}$$

where p_j – probability of realization j of the characteristic.

Formula (4) is the informational entropy of Boltzmann-Shannon, acting as a measure of uncertainty. Formula (5) is a normalization condition, and formula (6) defines the space of the mean geometric exponent:

IOP Conf. Series: Journal of Physics: Conf. Series **973** (2018) 012031 doi:10.1088/1742-6596/973/1/012031

$$r_{ji} = \sum_{i=1}^{n} r_{ji}^{p_j} . \tag{7}$$

Based on formulas (4)-(6) and the rules of the indeterminate Lagrange multiplier, we obtain the relation for estimating the vector of the potential probability distribution by means of which the extremum-maximum of the objective function is searched:

$$\widehat{p}_j(r) = \frac{\sum_{i=1}^{n} r_{ji}}{\sum_{j=1}^{m}\sum_{i=1}^{n} r_{ji}} . \tag{8}$$

The principle of the potential distribution of probabilities (8) is based on the idea of preferring a choice of those characteristics of the system whose properties contribute more to the total value of the estimated potential. A priori information on the behavior of characteristics for this principle is based on the approach of insufficient knowledge.

Obviously, the weight of the various features in the construction of the evaluation potential function is different for different characteristics. The derivation of the estimate of the a priori distribution p_j is dependent on the ratio of the order of preferences studied and described in [10]. In it, for a linear dependence of the order of estimating the a priori probabilities, a decreasing arithmetic progression is used:

$$\breve{p}_J = \frac{2(m-j+1)}{m(m+1)} . \tag{9}$$

By means of this input of the a priori probability value, it becomes possible, by specifying the various input data of the model, to take into account the different weights of the corresponding characteristics, and to build the function of the estimated potential of the system under study by these characteristics. Further, using the Bayes theorem and the principle of the potential distribution (8), the output of the simulation results in a posteriori conditional probabilities in the form:

$$p_j = \frac{\sum_{i=1}^{n} r_{ji}\breve{p}_J}{\sum_{j=1}^{m}\sum_{i=1}^{n} r_{ji}\breve{p}_J} . \tag{10}$$

The next stage in the development of the information model is the determination of the estimation of the likelihood of realizing j the characteristic i of the variant in order to obtain the value of the estimated potential. For this, Bayes theorem on the reversibility of the order of the assertions was used when finding the magnitude of the conditional probability. A search is made for the relationship between the quantities $p_{ji}(r)$ and p_j. The probability $p(A)$ is given by:

$$p_{ji}(A) = \frac{p_{ji}(r)p_j}{\sum_{i=1}^{n}\sum_{j=1}^{m} p_{ji}(r)p_j} . \tag{11}$$

Using the analogy of statistical concepts, namely the concept of error of the first kind and errors of the second kind, the construction of the solving algorithm is carried out. By the notion of error of the first kind, the deviation of the hypothesis H_0 under test is assumed when it is true. The probability of such an error is taken into account by introducing a level of significance α_0^{eff} that characterizes the risk when developing a new system. The value of this quantity, characteristic for the time period specified by the condition, is the higher, the smaller the degree of organization and ordering of the system under consideration. In this case, it corresponds to the probability to be at the level of the characteristics of the old system, that is, when the parameters of the new system slightly differ from the parameters of the existing one. This case is not the creation of a new system, but the modernization of the old one.

Under the error of the second kind is meant the acceptance of the hypothesis H_0 under test with its infidelity. The probability of making such an error when developing a new system is taken into account by introducing a power criterion β_1^{eff} characterizing the degree of risk of the consumer acting as a customer. The degree of risk in the consumer exists not only when the alternative option is inherent in a low degree of organization and ordering, but also when the proposed option has a low degree of adaptation to changes in external conditions. An error of the second kind characterizes the probability of realizing a case in which the predicted characteristic parameters do not have a continuity from the initial parameters of the initial system. That is, the error of the second kind shows the magnitude of the entropy increment of the successive characteristic parameters from the previous system. It shows the ability of the parameters under study to adapt to new conditions in the new system under investigation.

In the case of obtaining a large value (greater than the permissible value given by the conditions of the problem), it is concluded that there is no reason for further design, since a large value of the error indicates the existence of a large share of the uncertainty of the use of the system. This, in turn, emphasizes that the characteristic parameters of the predicted system will have a low degree of adaptation to existing environmental conditions, that is, they will not be related to reality. Measurement of the degree of organization of the system and the degree of its ordering is carried out using the amount of entropy. The significance level is expressed through the uncertainty function:

$$H_i(p) = -\sum p_{ji}(a)\log p_{ji}(a). \tag{12}$$

As a starting point in the calculation, it is advisable to take a predictable hypothetical system that has the best characteristic parameters in the given information situation. The level of significance in this case is calculated as

$$\alpha_i^{\text{eff}} = H_{\text{best}}(p) - H_i(p), \tag{13}$$

where $H_i(p)$ – the entropy value for a variety of the existing system; $H_{\text{best}}(p)$ – the entropy value for the predicted hypothetical system with the best characteristic parameters in the given information situation.

The amount of information I_i accumulated and stored in a set of systems corresponds to a quantitative decrease in the value of their entropy $\Delta H_i(p)$. The difference between the value of entropy $H_i(p)$ and the maximum value of entropy $H_{\max}$ in the information situation in question will correspond to the required amount of information I_i:

$$I_i = H_{\max} - H_i(p), \tag{14}$$

where $H_i(p)$ – the entropy value for the i version of the system, existing when assessing the degree of influence of each characteristic parameter on the formation of the value of the estimated potential; $H_{\max}$ – the maximum entropy of a particular situation under consideration.

In order that, in the process of evolutionary development of the system, it could not reach the limit of its "fitness" [9] (the limit upon which the system can exist only under certain strictly determined conditions), it is necessary that the system retain some degree of unpredictability, described a certain increment of entropy, the specific weight of which was found from the relation:

$$G_H^i = \frac{H_i(p)}{I_i}. \tag{15}$$

The next step is to solve the problem of minimizing the degree of uncertainty, which will allow us to determine the specific gravity of entropy.

In the models presented in [6-8], the existence of some optimal value of the magnitude of the increment G_H characterizing the value of the specific gravity of entropy was proved. In them, the optimal increment, representing the best combination of determinism and unpredictability, was $G_H^{\text{opt}} = 0.27$. Accordingly, the greater the difference between the calculated value G_H^i and the optimal value G_H^{opt}, the more likely the hypothesis will be accepted in the case when this hypothesis G_0 is not true. In the formular form, the error of the second kind, defined as the power of the criterion, is represented as follows:

$$\beta_i^{\text{eff}} = G_H^i - G_H^{\text{opt}}. \tag{16}$$

The values of the significance level α_0^{eff} and the power of the criterion β_0^{eff} characterize the system of arrangements in numerical form. With such a system, the side of the customer (customer) and the developer side determine a certain compromise numerical value of the errors of the first and second kind when the hypothesis H_0 is accepted. If the set of inequalities is fulfilled:

$$\begin{bmatrix} \alpha_i^{\text{eff}} \le \alpha_0^{\text{eff}} \\ \beta_i^{\text{eff}} \le \beta_0^{\text{eff}} \end{bmatrix}, \tag{17}$$

variant i of the system, that is, the null hypothesis H_0 is adopted and the process of forecasting the information model is completed.

4. Practical implementation of the method

The variant of application of the proposed method will be considered on the example of the forecast of the development of the situation of unauthorized access of internal violators to the information system.

Denote the following types of infringer security information system [12-14]: X_1, X_2, X_3, X_4, X_5, X_6. We compare each type of territory with the influence of the violator (1), the characteristics of the infringer of the information security system (2), the capabilities of the infringer of the information security system (3).

X_1 – (1) territory of telecommunication system, object; (2) employees who have the right of authorized access to the territory, but do not have the right to access premises and buildings; (3) the ability to intercept data over technical communication channels; facilitating unauthorized access to communication channels that go beyond the boundaries of the facility.

X_2 – (1) construction of telecommunications, object buildings; (2) employees who have authorized access to buildings, to the territory, but do not have access to office space; (3) the ability to intercept data over technical communication channels; facilitating unauthorized access to communication channels that go beyond the boundaries of the facility; possession of information on the video surveillance system, the deployment of a guard post.

X_3 – (1) communications, representation rooms; (2) employees who have access to special premises, but do not have access to office space; (3) the ability to intercept data over technical communication channels; facilitating unauthorized access to communication channels that go beyond the boundaries of the facility; possession of information on the video surveillance system, the deployment of a guard post.

X_4 – (1) offices of administration and users of IS; (2) registered users of IS, possessing the rights of the security administrator; registered users with system administrator rights; registered users with

IOP Conf. Series: Journal of Physics: Conf. Series **973** (2018) 012031 doi:10.1088/1742-6596/973/1/012031

system administrator rights for the segment; registered users with remote access to IS; Registered IS users with limited access to system resources; programmers, software developers (software); employees performing maintenance of IS; (3) has access to IS protection techniques; has full information about IS and has full access; has full information about IS; Has data on the topology of the segment, has access to network elements and segments; has access to pieces of sensitive data. Has incomplete information about the topology of IS. Possesses the ability to create firmware bookmarks; has software for processing data in the IS and data on algorithms; has access to make bookmarks in the technical means of IS.

X_5 – (1) server rooms, premises for confidential negotiations; (2) registered users who have the rights of a system administrator; registered users with security administrator rights; employees who have the right to enter premises of confidential negotiations; (3) have authorized access to the premises, has access to the configuration of the IS; has access to the configuration of the network segment; has access to the premises.

X_6 – (1) methods of information protection; (2) registered users with IS security administrator rights; (3) has access to all information about IS; has access to all premises.

As a characteristic (characteristics) of the system, the type of intruder IS.

According to the statistics of the criminal situation, for each type of threats on certain types of telecommunications objects in the region for several years, presented in table 3, the forecasting of the characteristic parameters of the situation of a security breach of the information system by the method of least squares by criterion

$$\min_i \sum_{i=0}^{n}\left[f(x_i)-P_3(x_i)\right]^2 , \tag{18}$$

where $P_3(x_i)$ – third-degree polynomial.

The choice of the third-degree polynomial is due to the smallest error in matching $f(x_i)$ with the statistical data in table 3 [6, 11].

Table 3. Statistics of the criminal situation at the IS site.

Designate violators of IS	System development time scale (year)				
	2011	2012	2013	2014	2015
X_1	2	3	2	2	3
X_2	4	5	4	5	6
X_3	7	6	8	7	8
X_4	2	3	1	4	3
X_5	3	2	3	4	4
X_6	2	1	3	1	2

Further, using the approximating function, the results of predicting the situation of a security breach of the information system were obtained using the above simulation method, according to formulas (1)-(3), (8)-(11), (14), (16), (18) the error of the first kind – the level of significance α_i^{eff} and the error of the second kind – the power of the criterion β_i^{eff}. The problem was solved iteratively. The process of the cycle was carried out until the errors reached the specified values. For the calculations, a mathematical package of applied mathematical modeling software Mathcad was used. The results of the calculations are presented in table 4. Fractional values in the results of forecasting are obtained as calculated data of the unknown probability of carrying out unauthorized actions by this group of employees and to improve accuracy are left without rounding.

IOP Conf. Series: Journal of Physics: Conf. Series **973** (2018) 012031 doi:10.1088/1742-6596/973/1/012031

Table 4. Results of forecasting the situation of IS security violation.

year	2011	2012	2013	2014	2015	2016	2017	2018	2019	2020	2021
X_1	2	3	2	2	3	2.8	2.9	3.0	3.1	3.2	3.3
X_2	4	5	4	5	6	6.4	6.8	7.2	7.6	8.0	3.4
X_3	7	6	8	7	8	8.4	8.7	9	9.3	9.5	3.5
X_4	2	3	1	4	3	3.8	4.1	4.4	4.7	5.0	5.3
X_5	3	2	3	4	4	4.8	5.2	5.6	6	6.4	5.8
X_6	2	1	3	1	2	1.8	1.8	1.8	1.8	1.8	1.8
H_i	0.701	0.756	0.703	0.775	0.870	0.942	0.967	0.99	1.012	1.034	1.055
α_0	0.169	0.114	0.167	0.095	0	0.038	0.071	0.101	0.13	0.159	0.185
β_0	0.093	0.079	0.093	0.074	0.048	0.027	0.020	0.013	0.007	$5\cdot10^{-5}$	0.006

The results of the calculations, presented in graphical form in fig. 1, show that the prediction of time for a criminal situation is appropriate to produce for a period of years. In this case, the magnitude of the error of the first kind, that is, the level of significance, is $\alpha_0 = 0.185$, and the magnitude of the error of the second kind – the power of the criterion – $\beta_0 = 0.006$.

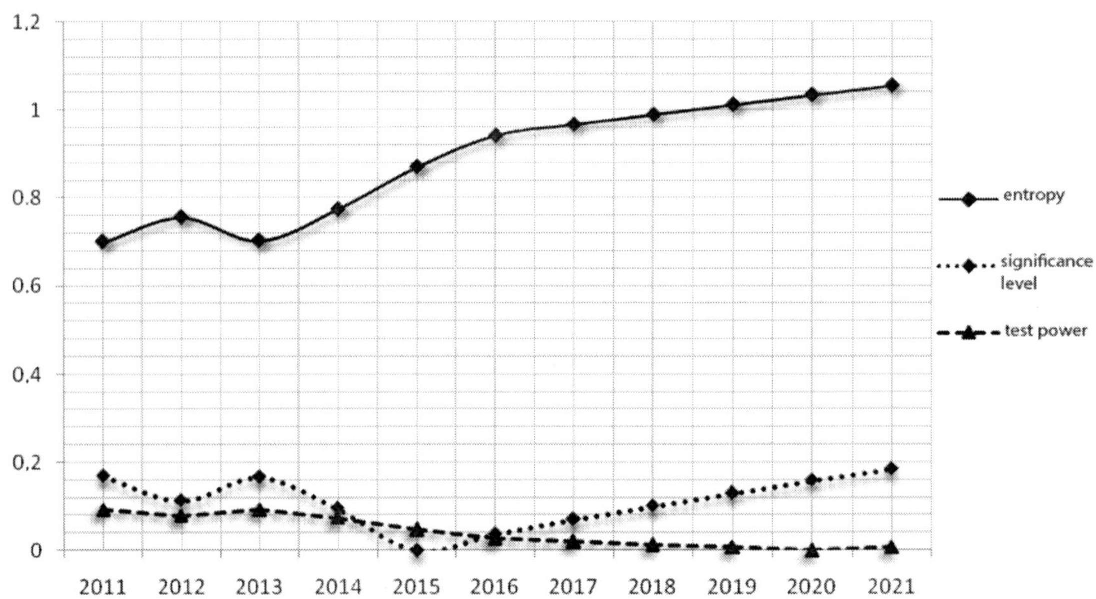

Figure 1. Graph of values of entropy, significance level, power of the criterion of the forecasted situation.

5. Conclusion

In order to reduce the degree of risk in the development and creation of information systems, they try to make decisions based on preliminary forecasts of evolutionary development of the system based on various forecasting methods, including probabilistic ones. Such methods include the improved method of predicting the evolution of the characteristic parameters of an information system, considered in this paper.

In this paper we show a concept for solving problems of estimating the effectiveness and forecasting of information systems, which creates the basis for the method of estimating the efficiency and forecasting of complex systems that differs from other existing methods considered in [1-5, 7, 10-11] by the wide application of the conceptual apparatus of the subject the field of information technology, as well as a qualitative justification for the probabilistic approach in the implementation of a comprehensive assessment of complex systems. Through the presented model it is possible to evaluate the development of the system before the design begins. In the course of the study, calculations were made to approbate the method of determining the prognostic values of the information system by examining the situation of a security breach of the information system. At the same time, the concepts of the significance level, the power of the criterion were used. Approbation of the method, carried out to assess the depth of the level of forecasting the development of the situation in order to form the characteristic parameters, has shown its effectiveness and the possibility of further use in forecasting the development of the characteristics of various complex systems. Introduction of the described method in practice is able to provide new opportunities for predicting new structures of systems with a complex hierarchy of characteristic parameters and characteristics.

References

[1] Bakhvalov N S, Zhidkov N P and Kobelkov G M 2004 *Numerical methods* (Moscow: BINOM) p 636

[2] Dushkin A V, Kravchenko A S, Novoseltsev V I, Smolentseva T E and Sumin V I 2014 *Mathematical models and information processes of complex object management* (Voronezh: Scientific Book) p 125

[3] Barkalov S A, Dushkin A V, Kolodyazhny S A, Sumin V I and Novoseltsev V I 2017 *Introduction to system analysis* (Moscow: Hot line – Telecom) p 234

[4] Libal G 1972 *Information Theory and the Living System* (N.Y.: Columbia University Press) p 174

[5] Volkenstein M 1986 *Entropy and information* (Moscow: Science) p 193

[6] Mushkov A Yu, Tikhomirov V A and Tikhomirov A V 2003 Models and methods of strategic management of complex economic and technological systems (Tver: Air Defense Force) p 244

[7] Sedov E A 1986 *Interrelation of information, energy and physical entropy in the processes of management and self-organization. Information and management* (Moscow: Science) p 214

[8] Kostin V N and Danshin D V 2015 A method for estimating the depth of the forecast (evolution) of characteristics of complex systems based on the entropy approach *Infocommunication technologies* **1** pp 62-67

[9] Fishburne P S 1978 *The utility theory for decision-making* (Moscow: Science) p 352

[10] Kostin V N and Shevchenko S N 2013 Methodology for the formation of requirements for a physical protection system based on a conceptual simulation model *Infocommunication technologies* **2** pp 91-98

[11] GOST R ISO / IEC 27033-3-2014 *Information technology, Methods and means of ensuring security, Network security* Part 3 Reference network scenarios: Threats, design methods and management issues

[12] Shangin V F 2012 *Protection of information in computer systems and networks* (Moscow: DMK Press) p 591

[13] Dushkin A V and Dubrovin A S 2016 *Methodological foundations for the construction of secure automated systems* (Voronezh: Scientific Book) p 76

[14] Varlataya S K and Kiryanenko A V 2015 Analysis of threats to the infringement of information security systems, existing models and methods of countering computer attacks *Actual problems of engineering in Russia and abroad: Collection of scientific papers on the results of an international scientific and practical conference* (Novosibirsk) p 162

IOP Conf. Series: Journal of Physics: Conf. Series **973** (2018) 012032 doi:10.1088/1742-6596/973/1/012032

Bottom friction models for shallow water equations: Manning's roughness coefficient and small-scale bottom heterogeneity

Tatyana Dyakonova, Alexander Khoperskov

Volgograd State University, Department of Information Systems and Computer Simulation, Volgograd, 400062, Russia

E-mail: khoperskov@volsu.ru

Abstract. The correct description of the surface water dynamics in the model of shallow water requires accounting for friction. To simulate a channel flow in the Chezy model the constant Manning roughness coefficient is frequently used. The Manning coefficient n_M is an integral parameter which accounts for a large number of physical factors determining the flow braking. We used computational simulations in a shallow water model to determine the relationship between the Manning coefficient and the parameters of small-scale perturbations of a bottom in a long channel. Comparing the transverse water velocity profiles in the channel obtained in the models with a perturbed bottom without bottom friction and with bottom friction on a smooth bottom, we constructed the dependence of n_M on the amplitude and spatial scale of perturbation of the bottom relief.

1. Introduction

The models of shallow water are successfully applied for calculation of a wide range of geophysical currents including the dynamics of surface and groundwater, landslides [1], torrential flows [2, 3], the pyroplastic and granular masses movement [4], the simulation of large-scale atmospheric, sea and ocean currents [5], and the technospheric security problems [6, 7]. To simulate the dynamics of channel and floodplain currents in the shallow water models, it is essential accounting for the hydraulic resistance [8, 9, 10, 11, 12]. The latter can be described using a number of developed phenomenological models [12]. The Chezy model, proposed as far back as the 18th century, is the most common [11]. There are several expressions determining the Chezy coefficient, but the Manning formula containing the roughness coefficient n_M [8] is the most often used in practice. The parameter n_M accounts for the features of both small-scale low heterogeneity and watercourse morphology on a large scale. The Manning roughness coefficient is a composite value generally taking into account the influence of vegetation, various obstacles, meandering, turbulence and other factors [13, 14, 15, 16]. Since the coefficient n_M widely varies for different watercourses [10] choosing its value in practice usually causes difficulties. In the Refs. [17, 18, 19] the estimates for n_M obtained from the Manning equation are given for various rivers. The coefficient n_M is assumed to be a function of coordinates and may depend on the water layer thickness, H, under the conditions of the nonstationary flow [20].

It should be noted, that in the engineering applications of the shallow water model the Manning coefficient (or similar parametr) should be determined as correctly as possible for the

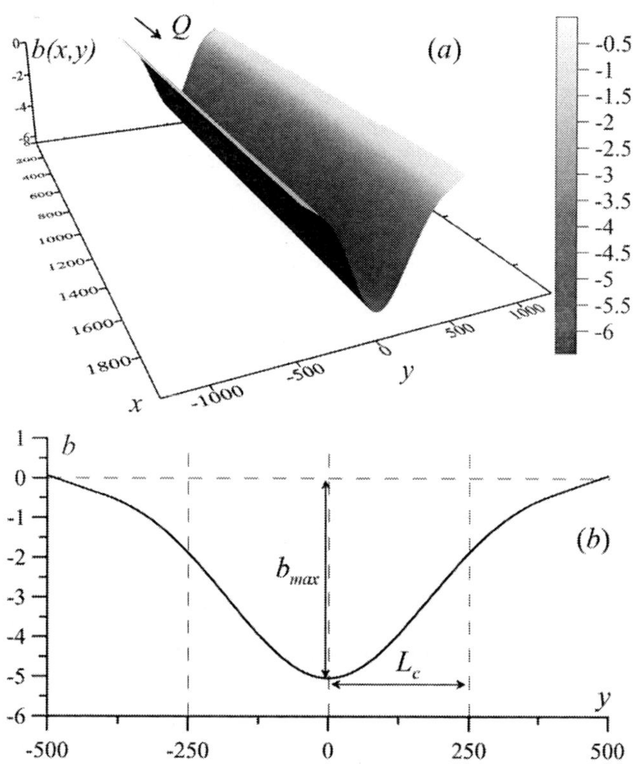

Figure 1. The geometry of the computed area with a watercourse (meters are used as the coordinate units). The fixed volume of water per time unit Q (m^3/ sec) is transferred to the watercourse input.

existing small-scale inhomogeneity of the bottom relief [2, 20, 21, 22]. The main aim of this article is the research of the Manning coefficient n_M dependence on the characteristic amplitude and spatial scales of the irregular structure of the channels relief. The estimations of the bottom friction coefficient for the Chezy formula have been obtained in a large series of the computer simulations results within the shallow water model. In particular, the relationship between the parameters of the small-scale bottom inhomogeneity in the channel and the Manning coefficient has been revealed.

2. Materials and Methods

2.1. Mathematical model

The hydrodynamic model of the surface water dynamics should account for the heterogeneity of the relief, external and internal forces acting on the water layer, evaporation from the water surface, and filtration into the soil [9]. Our model utilizes the following Saint-Venant equations:

$$\frac{\partial H}{\partial t} + \frac{\partial HV_x}{\partial x} + \frac{\partial HV_y}{\partial y} = \sigma, \tag{1}$$

$$\frac{\partial V_x}{\partial t} + V_x\frac{\partial V_x}{\partial x} + V_y\frac{\partial V_x}{\partial y} = -g\frac{\partial \eta}{\partial x} + F_x + \frac{\sigma}{H}(U_x - V_x), \tag{2}$$

$$\frac{\partial V_y}{\partial t} + V_x\frac{\partial V_y}{\partial x} + V_y\frac{\partial V_y}{\partial y} = -g\frac{\partial \eta}{\partial y} + F_y + \frac{\sigma}{H}(U_y - V_y), \tag{3}$$

where H is the water depth, V_x, V_y are the components of the water velocity vector $\vec{V}$ averaged over the vertical coordinate, σ is the surface density of water sources and sinks (m/sec), g is the gravitational acceleration, $\eta(x, y, t) = H(x, y, t) + b(x, y)$ is the vertical coordinate of the water surface, $b(x, y)$ is the relief function, U_x, U_y are the components of the water velocity vector in the source or drain $\vec{U}$, F_x, F_y are the components of the force vector $\vec{F}$.

Neglecting the internal viscosity and interaction with the surface layer of the atmosphere the expression for the force can be written as:

$$\vec{F} = \vec{F}^{fric} + \vec{F}^{cor}, \tag{4}$$

where $\vec{F}^{fric} = -\frac{\lambda}{2}\vec{V}|\vec{V}|$ is the bottom friction force, $\lambda = \frac{2gn_M^2}{H^{4/3}}$ is the coefficient of hydraulic resistance, n_M is the Manning's roughness coefficient (sec/m$^{1/3}$), $\vec{F}^{cor} = 2[\vec{V} \times \vec{\Omega}]$ is the Coriolis force, $\vec{\Omega}$ is the angular velocity of the Earth's rotation.

2.2. Numerical model

The numeric solution of the Saint-Venant equations has been obtained using the CSPH-TVD (Combined Smooth Particle Hydrodynamics – Total Variation Diminishing) scheme, which has the second order of accuracy in time and space, conservativeness, and well balance. Its advantage is the possibility of the through calculations of the unsteady "water – dry bottom" boundaries on a substantially inhomogeneous relief. For the boundary regions where the water leaves the calculation zone ($x = x_{\max}$, See Figure 1) we have specified the boundary conditions by the method described in Ref. [23]. The essence of the CSPH-TVD method has been explained in detail in Refs. [20, 24]. The calculation is carried out on a regular grid with cells $\Delta x = \Delta y$. The "liquid" particles are placed inside the cells. In accordance with the Lagrangian approach, the changes in the characteristics of the particles are calculated under the action of hydrodynamic and internal forces. At the Euler stage, the characteristics with respect to the flows through the cell's boundaries are refined. Since this algorithm is suitable for parallelization, we utilized the NVIDIA CUDA GPU technology in the developed software [25].

Figure 1 shows a watercourse situated along the x-coordinate which is specified by a function of bottom $b(x, y)$. The figure 1 *a* demonstrates only a part of the watercourse, because its total length amounts 20480 m according to our model. The shape of the watercourse's bottom is chosen to be an exponential function:

$$b_0(x, y) = -b_{\max} \cdot \exp(-y^2/L_c^2) \cdot (1 + Ix) \tag{5}$$

where $b_{max} = 5\,\text{m}$, and $L_c = 250\,\text{m}$. A small modification of Eq. (5) for $|y| > L_c$ provides a slow linear growth of $b_0(y)$ outside of the main part of the watercourse. The coefficient I (m/km) in Eq. (5) provides a weak slope of the watercourse's bottom along the x-coordinate. We name function $b_0(x, y)$ the smooth bottom profile, and in all our calculations it is assumed to be unchanged. We have chosen typical for the large plain rivers value of slope $I = 0.07\,\text{m/km}$ (for example, for the Volga, Don, Dnepr, Oka, Amazon, and Madeira the slope lies within $0.01 \leq I \leq 0.1$, for the Western Dvina, Neman, Kama, Seversky Donets, Mississippi, and Lower Vistula it is within $0.1 < I \leq 0.2$, and for the Thames and Congo it is within $0.2 < I \leq 0.4$).

The length of the numerical cell is $\Delta x = \Delta y = 10\,\text{m}$, thus for the computational domain with number of cells 2048×256 the total size of the computing area is $20480\ \text{m} \times 2560\ \text{m}$.

A fixed source of water is maintained at the entrance of the watercourse with the capacity $Q = 1500\,\text{m}^3/\text{sec}$, which roughly corresponds to the water discharge values for large plain rivers (for example, the Don River $Q = 680\,\text{m}^3/\text{sec}$, Oka River $Q = 1260\,\text{m}^3/\text{sec}$, Kama River $Q = 4000\,\text{m}^3/\text{sec}$, Rhine River $Q = 2200\,\text{m}^3/\text{sec}$, Vistula Rive $Q = 1080\,\text{m}^3/\text{sec}$). We have set the water volume $\Delta V^n = Q\Delta t^n$ for the area $S_Q = 50$ m^2 at the left boundary inside the watercourse ($x = 0$, See Figure 1 a) at each time step. The particular choice of S_Q does not influence on final results. The value $Q = 1500$ m^3/sec ensures the water depth in the fairway ($y = 0$) within $H_{\max} \sim 2 \div 4$ m depending on other factors. The calculation is conducted up to $t_{\max} = 144000$ sec, which certainly guarantees the establishment of a quasi-stationary solution.

Water does not reach boundaries $y = y_{\max}$ and $y = y_{\min}$ in our simulations, thus we have set the boundary conditions of the "water-dry bottom" type within the calculation area. At the boundary $x = 0$ we have put a solid wall and set the conditions of the waterfall type to $x = x_{\max}$ (see the details in [23]).

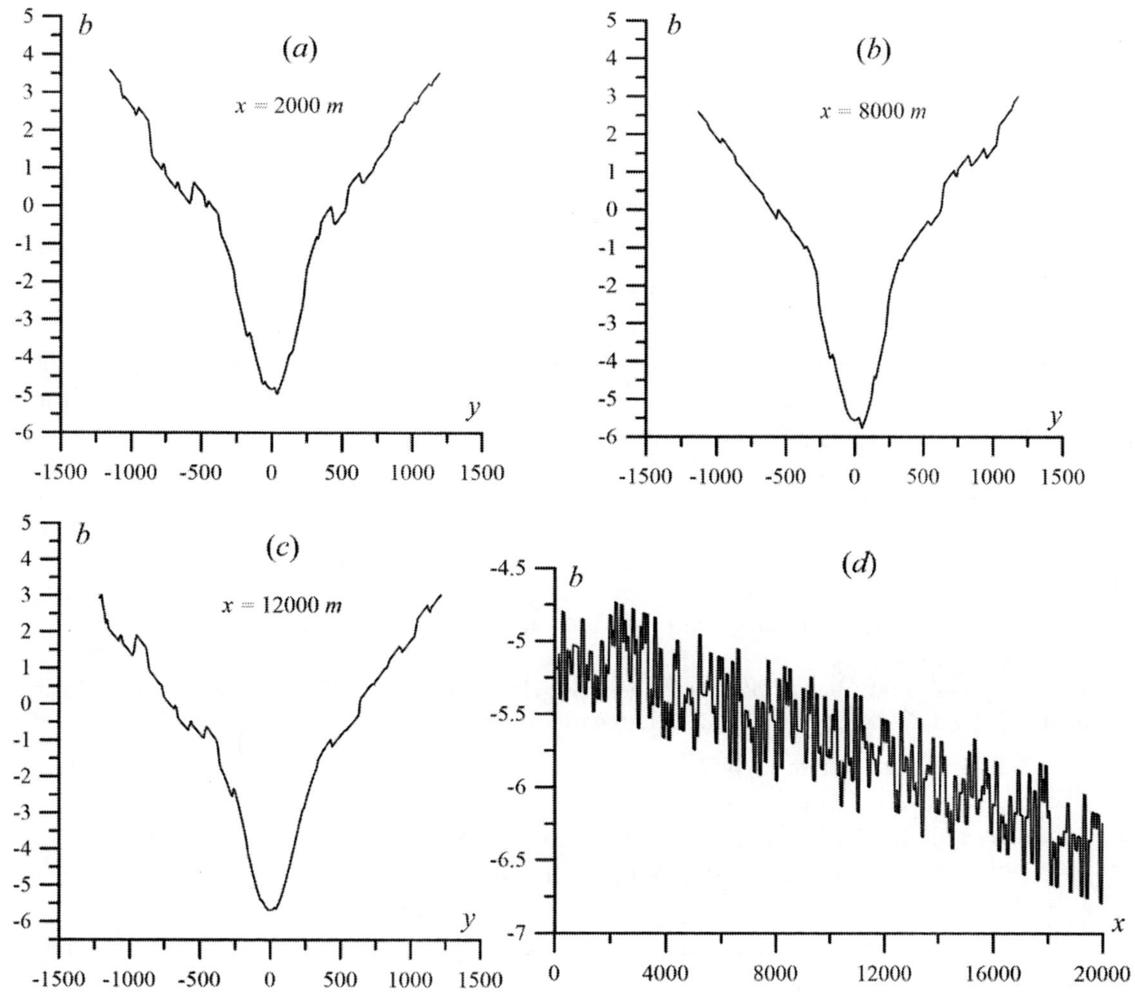

Figure 2. The profiles of the perturbed bottom across the channel $b(y)$ for the three values of the longitudinal coordinate: $x = 2000$ m (a), $x = 8000$ m (b), and $x = 12000$ m (c). The rest of the parameters are $\Delta b_{\max} = 0.4$ m, $\ell = 100$ m. The panel (d) demonstrates the dependence of $b(x)$ for fairway line $y = 0$.

IOP Conf. Series: Journal of Physics: Conf. Series **973** (2018) 012032 doi:10.1088/1742-6596/973/1/012032

2.3. The problem of determining the Manning coefficient in the Chezy model

The Chezy model takes into account the connection between the average flow velocity V_{av} and the hydraulic slope I, the hydraulic radius R, as well as the Chezy coefficient C which describes the frictional resistance:

$$V_{av} = C(IR)^{1/2} . \tag{6}$$

The Pavlovsky formula $C = R^r/n_M$ connects the parameter C with the roughness coefficient (or the Manning coefficient) n_M which leads to the Manning formula in the particular case $r = 1/6$:

$$V_{av} = \frac{R^{2/3} I^{1/2}}{n_M} . \tag{7}$$

The other kind of the Manning's formula for the discharge is

$$Q = \frac{S_n R^{2/3}}{n_M} \sqrt{I} , \tag{8}$$

where S_n is the area of the wetted watercourse cross-section.

The bottom roughness coefficient for the hydraulic resistance value calculations should be represented in the form [20, 13, 14, 15]:

$$n_M = \sum_{k=0}^{K} n_{Mk} , \tag{9}$$

accounting for K of various physical factors which characterize the underlying surface of the bottom and the morphology of the watercourse. These quantities, as a rule, are empirical parameters and there are some problems with their determination [8]. The parameter n_{M0} is defined by the properties of bottom itself on very small scales $\ell \ll H_{av}$ (H_{av} is the average depth along the transverse profile; for typical large rivers it lies within $2 \div 7$ m). In the case (9), we can consider the effect on real bottom roughness caused by small-scale irregular watercourse inhomogeneity (n_1), braking caused by vegetation (n_2), meandered channel (n_3), channel width changes (n_4), turbulence (n_5), impurity transfer and sediment movement (n_6), and etc. Experimental measurements of the parameter n_N for grass have given the result $n_M = 0.031 \div 0.035$ [26].

There is a large number of papers focused on the Manning coefficient estimation for various natural objects [27, 28, 29, 16, 18, 30]. Let us mention results of such research:

- $n_M \approx 0.043$ for the Acheron River, Mitta River, Tambo River; $n_M \approx 0.079$ for the Merrimans Creek [29];
- $n_M = 0.02$ and 0.035 for floodplains of the Sistan Basin [28];
- $n_M = 0.045$ during the wet period and $n_M = 0.101$ during the dry period for the Karun River [16];
- $n_M = 0.027 \div 0.042$ for Promaxonas, n_M from 0.134 to 0.942 for Peponia, n_M from 0.025 to 0.105 for Amfipoli [18];
- $n_M = 0.011$ for the Yellow River [30].

In the general case, the quantity n_M is inhomogeneous in space. The dependence of $n_M(x, y)$ on the coordinates is caused by the spatial changes in the properties of the underlying surface. This is especially evident in problems of flooding territories or for river streams with extended floodplain.

Let us discuss the factor n_{M1} in (9) resulting from the small-scale structure of the bottom heterogeneity [9]. We have considered the model problem of fluid flow in a channel to study the relationship between the structure of the bottom and the roughness coefficient.

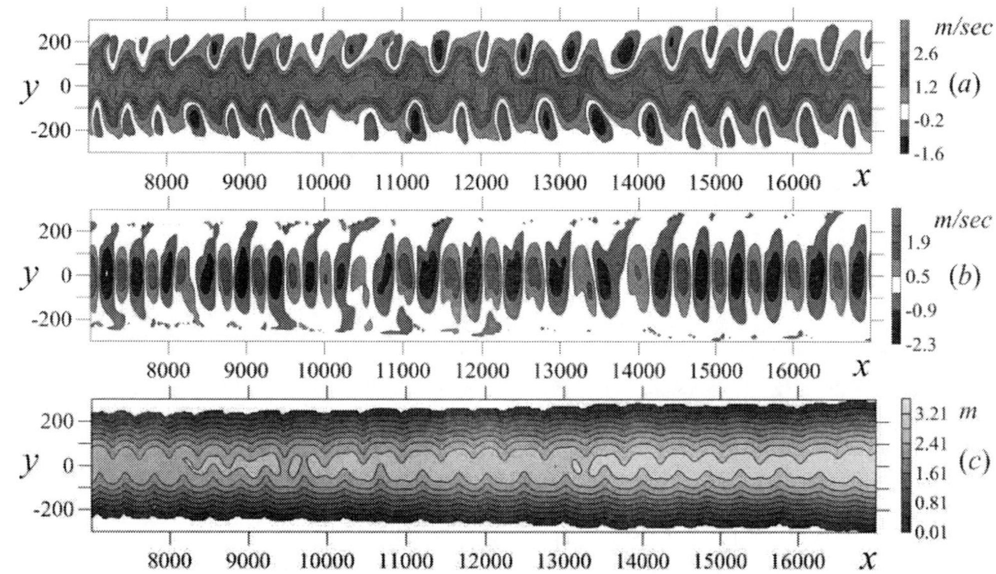

Figure 3. The spatial distributions of the velocity components $V_x(x, y)$ (a), $V_y(x, y)$ (b) and depth $H(x, y)$ (c) for the channel fragment $7\,\text{km} \leq x \leq 17$ km with $n_M = 0$.

The surface of bottom have been perturbated $b(x, y) = b_0(x, y) + \Delta b(x, y)$. The spatial scale ℓ and the maximum amplitude $\Delta b_{\max}$ are the primary characteristics of perturbation. These perturbations $\Delta b(x, y)$ have the form of local hills or valleys depending on the sign of the random variable Δb. To construct the perturbed relief we have chosen the uniformly distributed set of random variables $-\Delta b_{\max} \leq \Delta b \leq \Delta b_{\max}$ (Fig. 2). We performed a series of numerical experiments on the perturbed reliefs with $\Delta b_{\max} = 0.2$ m; 0.4 m; 0.6 m; 0.8 m for the maximum depth of the unperturbed channel $b_{\max} = 5$ m (See Figure 1).

3. Results and Discussion

Consider the limiting case $n_M = 0$. In such a case the water stream is unstable leading to hydrodynamic instability (Figure 3). Strong transverse motions arise with formation of the vortex structures. A sequence of non-stationary pulsations is formed along the entire flow (Figure 3).

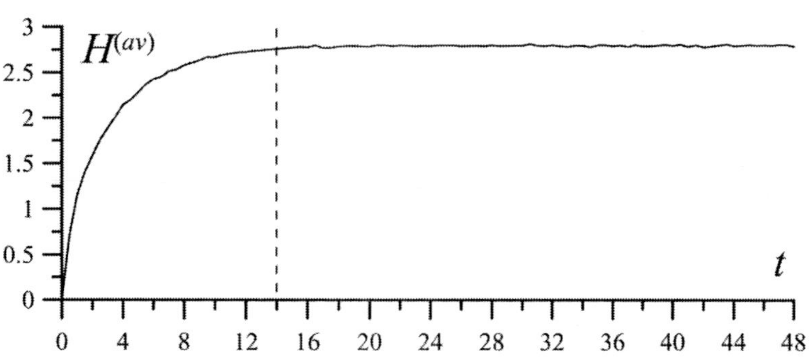

Figure 4. The time dependence of $H^{(av)}$ in the fairway $y = 0$ ($[H^{(av)}]$ =m, $[t]$ =hours)

At $n_M = 0$ the flow becomes substantially two-dimensional due to the appearance of the velocity transverse component $V_y(x, y, t)$ ($\max V_y \sim 0.5 \max V_x$). In small times and small spatial scales ($\sim L_c$), the flow is nonstationary, the velocity amplitude pulsation of which δV has the order of the flow velocity $V_{\max}$. These pulsations ensure the flow braking due to the internal numerical viscosity at small scales ($\sim \Delta x$), and as a result, a quasi-stationary state is established. The velocity profiles averaged over the x-coordinate for the cross-section of the channel can be determined as follows:

$$V_x^{(av)}(y) = \frac{1}{x_{\max}^{(av)} - x_{\min}^{(av)}} \int_{x_{\min}^{(av)}}^{x_{\max}^{(av)}} V_x(x, y)\, dx\,, \tag{10}$$

where $x_{\max}^{(av)} = 20000$ m and $x_{\min}^{(av)} = 50$ m in our models. The average water depth profile is defined similarly:

$$H^{(av)}(y) = \frac{1}{x_{\max}^{(av)} - x_{\min}^{(av)}} \int_{x_{\min}^{(av)}}^{x_{\max}^{(av)}} H(x, y)\, dx\,. \tag{11}$$

After $t > 14$ hrs, the profiles $H^{(av)}(y)$, $V_x^{(av)}(y)$ become almost stationary (Figure 4).

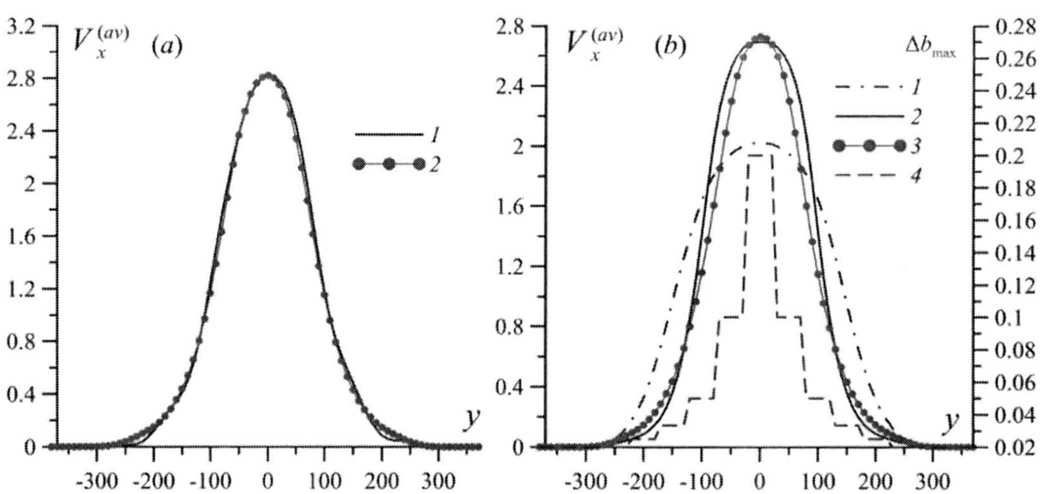

Figure 5. The result of concordance of the velocity profiles $V_x^{(av)}(y)$ for the model with perturbed bottom not accounting for bottom friction ($n_M = 0$) (curve *1*) and for the flow on the smooth bottom relief $b_0(x, y)$ with $n_M > 0$ (curve *2*) (*b*). The functions $V_x^{(av)}(y)$ (left axis): the curve *1* is a model with a constant distribution of the value $\Delta b_{\max}(y) = 0.2$ m in the section of the channel; the curve *2* corresponds to a model with a non-uniform profile $\Delta b_{\max}(y)$ (See curve *4*); the curve *3* has the best agreement in the case $n_M = 0.008$. Curve *4* is the distribution of $\Delta b_{\max}(y)$ in the riverbed of the channel (right axis) (*b*).

Random inhomogeneities with different amplitude and spatial scale resist to the water flow, and as a result, its average velocity V_{av} decreases. Thus, the transverse velocity profile $V_x(y)$

depends on the perturbed bottom parameters $\Delta b(x, y)$. We have conducted calculations with the Manning coefficient $n_M = 0$ for the perturbed bottom. Then we have performed a fitting of the function $V_x^{(av)}(y)$ with the velocity profiles obtained on the smooth (unperturbed) relief $b_0(x, y)$ by varying the parameter n_M. Figure 5 *a* shows the result of such procedure for perturbations of the bottom with parameters $\ell = 100$ m and $\Delta b_{\max} = 0.4$ m. Such velocity profiles on the perturbed bottom relief can be reproduced for flows on a smooth bottom with good accuracy for certain values of the Manning coefficient n_M. Current approach allows to connect the value of n_M with the heterogeneity parameters of the bottom within the chosen model $\Delta b(x, y)$.

For small values of $\ell \to \Delta x$ the profiles $V_x^{(av)}(y)$ for the models with perturbated bottom (with $n_M = 0$) and with smoothed bottom (with $n_M \neq 0$) are hardly consistent. Figure 5 *b* demonstrates stronger differences between the solutions for the most small-scale perturbations of the bottom with $\ell = \Delta x$, which are limitative in our numerical model of shallow water.

For this case, coincidence can be obtained either for speed along the fairway line, or for the value $\int\limits_{y_{\min}}^{y_{\max}} V_x^{(av)}(y)\,dy$. These methods give different estimates of n_M. Therefore, we have also considered the inhomogeneous form of $n_M(y)$ with a bell-shaped profile (see curve *4* in Figure 5 *b*). The latter approach allows the velocity profiles to be in agreement with each other in two different models that indicates the relation between n_M and H.

More than hundred computational experiments have been conducted for various ℓ, $\Delta b_{\max}$, n_M. The latter has allowed establishing a link between the heterogeneity parameters of the bottom and the Manning parameter. The maximum values of the longitudinal velocity profiles $V_x^{(av)}(y)$ decrease with increasing of n_M in the model on smooth (undisturbed) bottom relief, which almost always allows to choose the Manning coefficient for corresponding pertrubated bottom (See Figure 6).

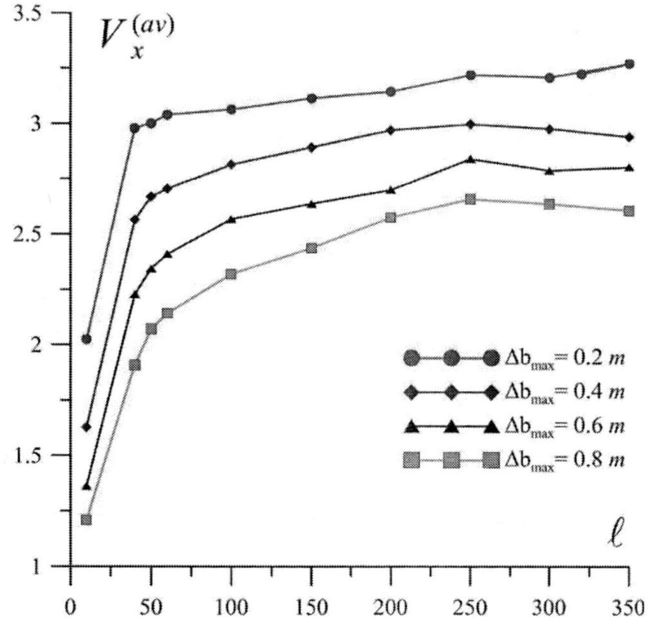

Figure 6. The dependence of the maximum velocity at the fairway line $V_x^{(av)}(y = 0)$ (m/sec) on the inhomogeneity spatial scale ℓ (m)

Increase of the maximum amplitude of perturbation of bottom $\Delta b_{\max}$ leading to the monotonically increase of the corresponding value of n_M. The water flow moves more slowly with increasing of $\Delta b_{\max}$ parameter. The dependence of the maximum flow velocity $V_x^{(max)}$ on scale ℓ has more complicated character (See Figure 6). The braking of flow is stronger for small-scale perturbations of the bottom which have an order of magnitude approximately $\ell < 50$ m. Our estimates of the Manning coefficient for various values of $\Delta b_{\max}$ and ℓ are presented in the Figure 7. For the range of parameters $\ell = 50 \div 350\,\text{m}$ and $\Delta b_{\max} = 0.2 \div 0.8\,\text{m}$ the Manning coefficient does not exceed 0.02.

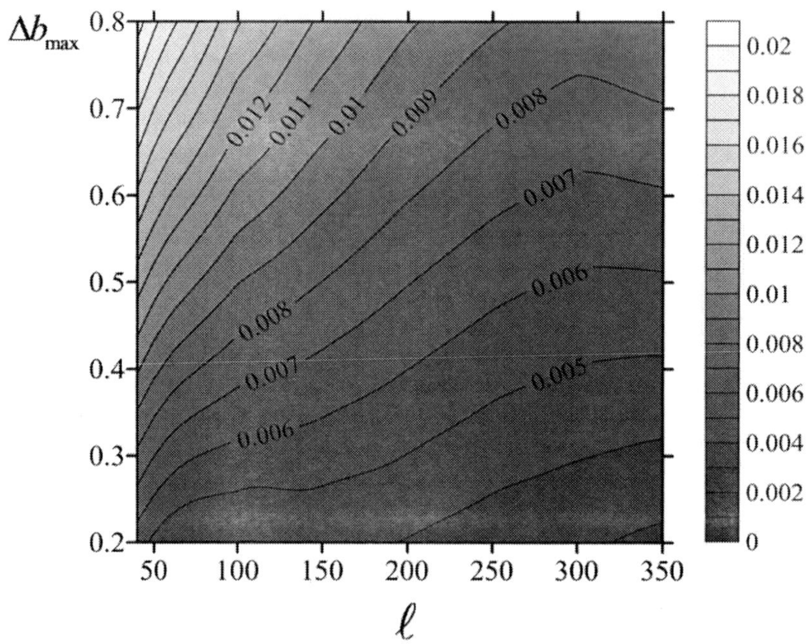

Figure 7. The distribution of the roughness coefficient n_M in the parameter plane ℓ (m) and $\Delta b_{\max}$ (m).

4. Conclusion

The numerical simulation of hydrodynamic currents of surface water on a realistic terrain relief usually containing a wide range of scales of bottom inhomogenieties has shown that these inhomogeneities exert an additional resistance to the flow. This factor should be taken into account when choosing the value of the Manning coefficient n_M. The estimates of n_M accounting for the inhomogeneity parameters of the bottom relief in the channel have been obtained from a series of numerical experiments in the shallow water model. It should be mentioned that these results largely depend on the adopted model of the surface perturbation of bottom relief which requires additional research.

The results obtained within the traditional shallow water model with a constant Manning coefficient ($n_M = \text{const}$) are in agreement with the results of our direct numerical experiments on the inhomogeneous bottom relief with scales $\ell \geq 50$ m. The n_M dependence on the coordinates should be accounted for better coincidence in the case of small-scale inhomogeneities ($\ell < 50$ m).

Acknowledgments

The authors are grateful to A.Yu. Klikunova, S.S. Khrapov, and A.V. Pisarev for help in the work and useful discussions. A. Khoperskov is thankful to the Ministry of Education and Science of the Russian Federation (government task No.2.852.2017/4.6). T. Dyakonova is thankful RFBR and Volgograd Region Administration (grant No. 15-45-02655), grant of young researchers of VolSU in 2017 also. The research is carried out using the equipment of the shared research facilities of HPC computing resources at Lomonosov Moscow State University.

References

[1] Shokina N and Aizinger V 2015 *Environmental Earth Sciences* **74** 7387–7405
[2] Agafonnikova E O, Klikunova A Y and Khoperskov A V 2017 *Bulletin of the South Ural State University, Series: Mathematical Modelling, Programming and Computer Software.* **10** 148–155
[3] Shokin Y I, Rychkov A D, Khakimzyanov G S and Chubarov L B 2016 *Russian Journal of Numerical Analysis and Mathematical Modelling* **31** 217–227
[4] Cui X and Gray J 2013 *Journal of Fluid Mechanics* **720** 314–337
[5] Majda A 2003 *Introduction to PDEs and Waves for the Atmosphere and Ocean* vol 9 (American Mathematical Soc.)
[6] Luo Z, Wu Q and Zhang L 2017 *Journal of Physics: Conference Series* vol 916 (IOP Publishing) p 012042
[7] Kulagin V, Moskvichev V, Makhutov N, Markovich D and Shokin Y I 2016 *Herald of the Russian Academy of Sciences* **86** 454–465
[8] Baryshnikov S O and Pagin A O 2010 *Vestnik gosudarstvennogo universiteta morskogo i rechnogo flota imeni admirala S.O. Makarova* **2 (6)** 90–93
[9] Dyakonova T A, Pisarev A V, Khoperskov A V and Khrapov S S 2014 *Science Journal of Volgograd State University. Mathematics. Physics* **20** 35–44
[10] Pisarev A V, Khrapov S S, Agafonnikova E O and Khoperskov A V 2013 *Bulletin of Udmurt University. Mathematics, Mechanics* **1** 114–130
[11] Jong-Seok L and Pierre Y J 2017 *Journal of flood engineering* **8 (2)** 55–75
[12] Habibi M, Namaee M R and Saneie M 2014 *KSCE Journal of Civil Engineering* **18** 1176–1184
[13] De Doncker L, Troch P, Verhoeven R, Bal K, Meire P and Quintelier J 2009 *Environmental fluid mechanics* **9** 549–567
[14] Hadiani M, Asl S J, Banafsheh M R and Dinpajouh Y 2013 *World Applied Sciences Journal* **22** 307–312
[15] Herget J and Meurs H 2010 *Global and Planetary Change* **70** 108–116
[16] Mohammadi S and Kashefipour S M 2014 *Water resources* **41** 412–420
[17] Hatzigiannakis E, Pantelakis D, Hatzispiroglou I, Arampatzis G, Ilias A and Panagopoulos A 2016 *Environmental Processes* **3** 263–275
[18] Hatzigiannakis E, Pantelakis D, Hatzispiroglou I, Arampatzis G, Ilias A and Panagopoulos A 2014 888–895
[19] Lyra G B, Cecilio R A, Zanetti S S and Lyra G B 2010 *Revista Brasileira de Engenharia Agricola e Ambiental-Agriambi* **14**
[20] Khrapov S S, Pisarev A V, Kobelev I A, Zhumaliev A G, Agafonnikova E O, Losev A G and Khoperskov A V *Advances in Mechanical Engineering* **5** 787016
[21] Nistoran D G, Ionescu C, Patru G, Armas I and Omrani S G 2017 *Energy Procedia* **112** 67–74
[22] Voronin A, Isaeva I, Khoperskov A and Grebenjuk S 2017 *Conference on Creativity in Intelligent Technologies and Data Science* (Springer) pp 419–429
[23] D'yakonova T, Khrapov S and Khoperskov A 2016 *Vestnik Udmurtskogo Universiteta: Matematika, Mekhanika, Komp'yuternye Nauki* **26** 401–417
[24] Khrapov S S, Khoperskov A V, Kuz'min N M, Pisarev A V and Kobelev I A 2011 *Vychisl. Metody Programm* **12** 282–297
[25] Dyakonova T, Khoperskov A and Khrapov S 2016 *Communications in Computer and Information Science* **687** 132–145
[26] Jakubis M 2000 *Journal of the International Commission on Irrigation and Drainage* **49** 41–54
[27] Lang S, Ladson A, Anderson B and Rutherford I 2004 *Stream roughness, Four case studies from Victoria, AJWR* **28**
[28] Motallebian M and Hassanpour F 2014 *Journal of Civil Engineering and Urbanism* **4 (5)** 540–545
[29] Lang S, Ladson T, Anderson B and Rutherford I 2004 *National Rivers Consortium: Canberra*
[30] Xia J, Lin B, Falconer R A and Wang Y 2012 *Proceedings of the Institution of Civil Engineers* **165** 377

A method for minimum risk portfolio optimization under hybrid uncertainty

Yu E Egorova and A V Yazenin

Department of Information Technology, Tver State University, 33 Zhelyabova Str., Tver 170000, Russian Federation

E-mail: j.e.egorova@gmail.com, yazenin.av@tversu.ru

Abstract. In this paper, we investigate a minimum risk portfolio model under hybrid uncertainty when the profitability of financial assets is described by fuzzy random variables. According to Feng, the variance of a portfolio is defined as a crisp value. To aggregate fuzzy information the weakest (drastic) t-norm is used. We construct an equivalent stochastic problem of the minimum risk portfolio model and specify the stochastic penalty method for solving it.

1. Introduction

Nowadays, various methods for solving portfolio optimization problems under hybrid uncertainty are being developed [1-3]. In the case of hybrid uncertainty of possibilistic-probabilistic type, portfolio optimization problems can be solved by techniques of possibilistic-probabilistic programming.

In this work we investigate a model of minimum risk portfolio where the risk of a portfolio is modeled by the variance. The variance of a portfolio can be defined in two different ways: as a crisp value or as a fuzzy variable. We use the former definition [4]. To aggregate fuzzy information the weakest (drastic) t-norm is used, that is, we consider that all fuzzy parameters are mutually T_W-related.

This mathematical model of a minimum risk portfolio has been studied in [3] where equivalent crisp problems have been constructed. However, obtaining equivalent crisp problems requires the calculation of the precise values of the expected value and the variance of a random function. In the case of the weakest t-norm, the expected value and the variance of a portfolio are of such a complex nature that the calculation of their precise values is almost impossible. Moreover, the results of this calculation are nonlinear non-convex functions, which lead to a non-convex optimization problem.

To avoid difficult calculations the methods of stochastic programming, such as the stochastic quasigradient methods, can be used [5]. In [6, 7] the stochastic quasigradient method is used to solve possibilistic-probabilistic optimization problems in the case of the weakest t-norm.

In this work we construct an equivalent stochastic problem of the minimum risk portfolio optimization problem. To solve it we specify the stochastic penalty method, which is a modification of the stochastic quasigradient method. Using of the stochastic penalty method is demonstrated on numerical examples and the solutions are compared with the results obtained in [3].

2. Materials and methods

2.1. Necessary concepts and notations

We introduce a number of definitions and concepts from the possibility theory following [8-10]. Let $(\Gamma, \mathfrak{P}(\Gamma), \tau)$ and $(\Omega, \mathcal{B}, \mathrm{P})$ be possibilities and probabilities spaces where Ω is a sample space with possible outcomes $\omega \in \Omega$, Γ is a pattern space with elements $\gamma \in \Gamma$, $\mathcal{B}$ is an σ-algebra of events, $\mathfrak{P}(\Gamma)$ is the discrete topology on Γ, $\tau \in \{\pi, \nu\}$, π and ν are measures of possibility and necessity respectively, P is a probability measure and E^1 is the real line.

Definition 1. *Fuzzy random variable Y is a real function $Y : \Omega \times \Gamma \to E^1$, which is σ-measurable for each fixed γ and*

$$\mu_Y(\omega,t) = \pi\{\gamma \in \Gamma : Y(\omega,\gamma) = t\}$$

is called its distribution function.

It follows from Definition 1 that the distribution function of a fuzzy random variable depends on a random parameter, that is, it is a random function.

Definition 2. *Let $Y(\omega,\gamma)$ be a fuzzy random variable. Its expected value* $\mathrm{E}[Y]$ *is a fuzzy variable with the possibility distribution function*

$$\mu_{\mathrm{E}[Y]}(t) = \pi\{\gamma \in \Gamma : \mathrm{E}Y(\omega,\gamma) = t\},$$

where E *is the mathematical expectation operator*

$$\mathrm{E}[Y(\omega,\gamma)] = \int_\Omega Y(\omega,\gamma)\mathrm{P}(\mathrm{d}\omega).$$

In this case, the distribution function of the expected value of a fuzzy random variable is no longer dependent on a random parameter and is therefore deterministic. We define second central moment following [4]. Let X and Y be fuzzy random variables.

Definition 3. *A covariance of fuzzy random variables X and Y is defined as*:

$$\mathrm{cov}(X,Y) = \frac{1}{2}\int_0^1 \left(\mathrm{cov}\left(X_\omega^-(\alpha), Y_\omega^-(\alpha)\right) + \mathrm{cov}\left(X_\omega^+(\alpha), Y_\omega^+(\alpha)\right)\right)\mathrm{d}\alpha,$$

where $X^-(\alpha)$, $Y^-(\alpha)$, $X^+(\alpha)$, $Y^+(\alpha)$ are left and right boundaries of α-level sets of fuzzy variables X_ω and Y_ω.

Definition 4. *A variance of a fuzzy random variable Y is*

$$\mathrm{D}[Y] = \mathrm{cov}(Y,Y).$$

The expected value, variance and covariance of fuzzy random variables determined in such a way inherit basic properties of the corresponding characteristics of real-valued random variables.

The most interesting representation of a fuzzy random variable is a shift-scale representation [10]:

$$Y(\omega,\gamma) = a(\omega) + \sigma(\omega)Z(\gamma),$$

where $a(\omega)$, $\sigma(\omega)$ are random variables defined on the probability space $(\Omega, \mathcal{B}, \mathrm{P})$ and $Z(\gamma)$ is a fuzzy variable defined on the possibility space $(\Gamma, \mathfrak{P}(\Gamma), \tau)$. Random components $a(\omega)$ and $\sigma(\omega)$ are called the shift and the scale respectively.

LR-type distributions are often used to model fuzzy numbers [11].

Definition 5. *$Z(\gamma)$ is called an LR-type fuzzy variable if its distribution function has the form*

$$\mu_Z(t)=\begin{cases} L\left(\dfrac{\underline{m}-t}{\underline{d}}\right), & t<\underline{m}, \\ 1, & \underline{m}\le t\le \overline{m}, \\ R\left(\dfrac{t-\overline{m}}{\overline{d}}\right), & t>\overline{m}, \end{cases}$$

where $L(t)$, $R(t)$ are shape functions [11].

In this case $Z(\gamma)$ is written in the form $Z=[\underline{m},\overline{m},\underline{d},\overline{d}]_{LR}$, where $\underline{m}\le\overline{m}$ are left and right limits of tolerance (modal) interval, $\underline{d}>0$, $\overline{d}>0$ are coefficients of fuzziness.

We use triangular norms and conorms (t-norms and t-conorms) as an instrument for aggregation of fuzzy information. They extend min and max operations on which actions on fuzzy sets and fuzzy variables are based [12].

In particular, in this work we consider two extreme t-norms

$$T_{\mathrm{M}}(x,y)=\min\{x,y\} \quad \text{and} \quad T_{\mathrm{W}}(x,y)=\begin{cases} \min\{x,y\}, & \text{if } \max\{x,y\}=1, \\ 0, & \text{otherwise.} \end{cases}$$

T_{M} and T_{W} are called the strongest and the weakest t-norm respectively, since for any arbitrary t-norm T and $\forall x,y\in[0,1]$ the following inequality holds [12]:

$$T_{\mathrm{W}}(x,y)\le T(x,y)\le T_{\mathrm{M}}(x,y).$$

One of the main properties of t-norms is their ability to control uncertainty ("fuzziness") growth, The growth of fuzziness can appear, for example, when performing arithmetic operations on fuzzy numbers: if two fuzzy numbers of LR-type are summed using the strongest t-norm T_{M}, corresponding coefficients of fuzziness are also summed, therefore uncertainty is growing. With the help of t-norms other than T_{M} we can slow the growth of fuzziness. The extreme triangular norms, which are considered in this work, give us boundaries for control of fuzziness in our minimum risk portfolio model.

Following [13] we introduce the notion of mutual t-relatedness of fuzzy sets and fuzzy variables. It is used as an instrument for constructing joint possibility distribution functions.

Definition 6. *Fuzzy sets $A_1,\dots,A_n\in\mathfrak{P}(\Gamma)$ are called mutually T-related, if for any index set $\{i_1,\dots i_k\}\subset\{1,\dots,n\}$, $k=\overline{1,n}$, we have*

$$\pi\left(A_{i1}\cap\ldots\cap A_{ik}\right)=T\left(\pi(A_{i1}),\ldots,\pi(A_{ik})\right),$$

where

$$T\left(\pi(A_{i1}),\ldots,\pi(A_{ik})\right)=T\Big(T\big(\ldots T\big(T\big(\pi\big(A_{i1},\pi(A_{i2})\big),\pi(A_{i3})\big),\ldots\big),\pi(A_{ik})\Big).$$

We can extend the notion of mutual T-relatedness of fuzzy sets to T-relatedness of fuzzy variables. Let $Z_1(\gamma),\dots,Z_n(\gamma)$ be fuzzy variables defined on the possibility space $(\Gamma,\mathfrak{P}(\Gamma),\pi)$.

Definition 7. *Fuzzy variables* $Z_1(\gamma), \ldots, Z_n(\gamma)$ *are called mutually T-related if for any index set,* $\{i_1,\ldots i_k\} \subset \{1,\ldots,n\}$, $k = \overline{1,n}$, *we have*

$$\mu_{Z_{i1},\ldots,Z_{ik}}(t_{i1},\ldots,t_{ik}) = \pi\{\gamma \in \Gamma : Z_{i1}(\gamma) = t_{i1},\ldots,Z_{ik}(\gamma) = t_{ik}\}$$
$$= \pi\{Z_{i1}^{-1}[t_{i1}] \cap \ldots \cap Z_{ik}^{-1}[t_{ik}]\} = T\left(\pi\left(Z_{i1}^{-1}[t_{i1}]\right),\ldots,\pi\left(Z_{ik}^{-1}[t_{ik}]\right)\right),\ t_{ij} \in E^1.$$

2.2. Minimum risk portfolio under conditions of hybrid uncertainty

In conditions of hybrid uncertainty of the possibilistic-probabilistic type the return of an investment portfolio can be represented as a fuzzy random function

$$R_P(x,\omega,\gamma) = \sum_{i=1}^{n} R_i(\omega,\gamma)x_i,$$

where fuzzy random variables $R_i(\omega,\gamma)$ model the profitability of financial assets and have a shift-scale representation [10]:

$$R_i(\omega,\gamma) = a_i(\omega) + \sigma_i(\omega)Z_i(\gamma).$$

Further in this work we assume that in this representation fuzzy variables $Z_i(\gamma) = [\underline{m}_i, \overline{m}_i, \underline{d}_i, \overline{d}_i]_{LR}$ are mutually T_W-related and random parameters $a_i(\omega)$ and $\sigma_i(\omega)$ are independent.

For a better intuitive understanding of such a representation of a fuzzy random variable imagine a situation where some financial expert is asked to estimate a return of a certain financial asset. Both the return and its estimation by the expert are uncertain quantities. We assume that the uncertainty determined by market conditions has a probabilistic nature. On the other hand, uncertainty of the estimation by the expert is described by some possibility distribution. This model seems to be quite plausible if we assume that the degree of fuzziness of the expert depends mainly on the scale of variation of the estimated variable and not on its true value [10].

In accordance with the classical Markowitz approach [14], and with the help of [4], we can construct a risk function for the portfolio and include its expected return in the system of constraints. Since the expected portfolio return in the case of fuzzy random data is fuzzy the uncertainty of possibilistic type can be removed by imposing requirements on the possibility/necessity of fulfilling an investor's constraints on an acceptable level of the expected portfolio return. In this case the model of feasible portfolios by Markowitz can be represented in the following form

$$\mathrm{D}[R_p(x,\omega,\gamma)] \to \min_{x \in G_p},$$

$$G_P = \begin{cases} \tau\{\mathrm{E}[R_P(x,\omega,\gamma)]\ \mathcal{R}\ m_d\} \geq \alpha, \\ \sum_{i=1}^{n} x_i = 1, \\ x \in E_+^n. \end{cases}$$

Here $\mathrm{D}[R_p(x,\omega,\gamma)]$ is the variance and $\mathrm{E}[R_P(x,\omega,\gamma)]$ is the expected return of a portfolio; $E_+^n = \{x \in E^n : x \geq 0\}$; τ is the measure of possibility π or necessity ν; $\mathcal{R}$ is a crisp relation, $\mathcal{R} \in \{\geq, =\}$; $\alpha \in (0,1]$ and m_d is an acceptable level of the expected return of a portfolio.

3. Results and discussion

3.1. Equivalent stochastic problem of the model of minimum risk portfolio

Let $a_i(\omega)$ and $\sigma_i(\omega)$ be independent random variables and fuzzy components $Z_i(\gamma)$ be represented by symmetrical fuzzy numbers of LR-type, i.e. $Z_i(\gamma)=[m_i,m_i,d_i,d_i]_{LR}$, $i=\overline{1,n}$ and $L(t)=R(t)$, $\forall t\geq 0$.

Then for each x the expected return of a portfolio is a fuzzy variable with the following possibility distribution [10]

$$E_P(x,\gamma)=[m_{\mathrm{ER}_P}(x),\ m_{\mathrm{ER}_P}(x),\ d_{\mathrm{ER}_P}(x),\ d_{\mathrm{ER}_P}(x)]_{LR},$$

and the risk of the portfolio takes the form [3]

$$D_P(x)=m_{\mathrm{DR}_P}(x)+\eta d_{\mathrm{DR}_P}(x).$$

Here

$$m_{\mathrm{ER}_P}(x)=\mathrm{E}[m_{R_P}(x,\omega)],\quad d_{\mathrm{ER}_P}(x)=\mathrm{E}[d_{R_P}(x,\omega)],$$

$$m_{\mathrm{DR}_P}(x)=\mathrm{D}[m_{R_P}(x,\omega)],\quad d_{\mathrm{DR}_P}(x)=\mathrm{D}[d_{R_P}(x,\omega)],\quad \eta=\int_0^1 (L(\alpha))^2\,d\alpha,$$

and

$$m_{R_P}(x,\omega)=\sum_{i=1}^{n}(a_i(\omega)+\sigma_i(\omega)m_i)x_i,\qquad d_{R_P}(x,\omega)=\max_{i=\overline{1,n}}\{|\sigma_i(\omega)|d_i x_i\}.$$

The constraint on the expected return can be replaced by [7]

$$E_P(x)=\lambda m_{\mathrm{ER}_P}(x)+\beta d_{\mathrm{ER}_P}(x)\leq \lambda m_d,$$

where $\lambda=\pm 1$, $\beta=\pm L^{-1}(\alpha_i)$ or $\pm L^{-1}(1-\alpha_i)$ depending on $\mathcal{R}$ and τ.

Let us write the constraints on assets of a portfolio in the form

$$X=\begin{cases}\sum_{i=1}^{n}x_i=1,\\ x\in E_+^n.\end{cases}$$

Thus, the model of a minimum risk portfolio under conditions of hybrid uncertainty of possibilistic-probabilistic type can be replaced by the following stochastic program

$$D_P(x)\to\min_{x\in F_P},\tag{1}$$

$$F_P=\begin{cases}E_P(x)\leq\lambda m_d,\\ x\in X.\end{cases}\tag{2}$$

Note that in general the model (1)-(2) is not convex. However, in [15] Nurminski studied one class of non-convex non-differentiable functions which he named weakly convex. For this class methods similar to the stochastic quasigradient methods [5, 15] can be specified.

Definition 8. *A function $F(x)$ is called weakly convex if for any x there exists a non-empty set of vectors $G(x)$ that for all z and $\hat{F}(x) \in G(x)$ the following inequality holds*

$$F(z) - F(x) \geq \langle \hat{F}(x),\ z - x \rangle + r(z,x), \tag{3}$$

where $r(x,y) \| x - y \|^{-1} \to 0$ for $y \to x$ uniformly in X. A vector $\hat{F}(x)$ that satisfies (3) is called quasi-gradient of a weakly convex function $F(x)$.

The term "weakly convex" has been suggested by analogy to "strongly convex" functions [15] that satisfy the inequality (3) with additional constraints on $r(x,y) = \delta(\| x - y \|) \leq 0$ and $\delta(t)/t \to 0$ for $t \to +0$.

While the class of strongly convex functions are a subset of the class of convex functions, the class of weakly convex functions includes convex functions as well as non-convex differentiable. For non-convex differentiable functions the inequality (3) transforms into

$$F(z) - F(x) = \langle F_x(x),\ z - x \rangle + r(z,x)$$

where $F_x(x)$ is the gradient of $F(x)$.

Theorem 1. *If the function $d_{ER_P}(x)$ is differentiable then $D_P(x)$ and $E_P(x)$ are weakly convex.*

Proof. Since $m_{ER_P}(x)$ is differentiable, then $E_P(x)$ is also differentiable as well as weakly convex. On the other hand the function

$$d_{DR_P}(x) = \mathrm{E}\left[\max_{i=1,\dots,n} \left\{ |\sigma_i(\omega)| d_i x_i \right\} \right]^2 - d_{ER_P}^2(x)$$

is weakly convex because it is the sum of a convex function and a differentiable function. Hence, $D_P(x)$ is weakly convex and this completes the proof.

3.2. The stochastic penalty method

The model (1)-(2) can be solved by deterministic techniques of mathematical programming. However, these techniques require the calculation of the precise values of the expected value and the variance of a random function. An alternative approach is to use the stochastic quasigradient (SQG) methods [5, 15], for example the stochastic penalty methods.

The SQG methods allow us to solve optimization problems with objective functions and constraints of such a complex nature that the calculation of the precise values of these functions (all the more of their derivatives) is impossible. The main idea of these methods consists of using statistical estimates for the values of the functions and of their derivatives instead of their precise values. The SQG methods generalize the well-known stochastic approximation methods for unconstrained optimization of the expectation of random functions.

Instead of the model (1)-(2) we consider the following minimization problem

$$F(x) = D_P(x) + c\big(E_P(x)\big) E_P(x) \to \min_{x \in X}, \tag{4}$$

where $c(\alpha)$ is a penalty function

$$c(\alpha) = \begin{cases} C, & \alpha > 0, \\ 0, & \alpha \leq 0, \end{cases}$$

C is the penalty coefficient. If Theorem 1 holds then it is easy to see that the function $F(x)$ is weakly convex. So, for solving (4) we can specify the stochastic quasigradient method with the averaging operation [5].

Let Y and Z be convex compact sets such that $d_{\mathrm{ER}_p}(x)\in Y$ and $E_p(x)\in Z$, $\forall x\in X$. We will define a sequence of points (x^s, y^s, z^s) by the following relations

$$x^{s+1}=\begin{cases} x^s-\rho_1^s\xi^s, & \|x^s\|\le 1,\\ b^{s+1}\in B, & \|x^s\|>1,\end{cases}\qquad y^{s+1}=\Pi_Y\left(y^s-\rho_2^s[y^s-\zeta^s]\right),\qquad z^{s+1}=\Pi_Z\left(z^s-\rho_3^s[z^s-\chi^s]\right),$$

where (x^0, y^0, z^0) is an initial point, $s=0,1,\ldots$, $B\in A=\{x \mid \|x\|\le 1\}$; the step sizes ρ_j^s, $j=1,2,3$, are non-negative and measurable with respect to a σ-algebra $\mathcal{B}_s$ generated by the sequence of points (x^s, y^s, z^s). Π_Y and Π_Z are projections on the sets Y and Z respectively; ξ^s is a random vector and ζ^s, χ^s are random variables such that

$$\xi^s=\delta^s_{\mathrm{DR}_p}+\eta\vartheta_2^s-2\eta y^s\vartheta_1^s+c(z^s)(\lambda\delta^s_{\mathrm{ER}_p}+\beta\vartheta_1^s),$$

$$\zeta^s=d_{R_p}(x^s,\omega^s),\qquad \chi^s=\lambda m_{R_p}(x^s,\omega^s)+\beta d_{R_p}(x^s,\omega^s).$$

If we consider $\hat{a}_i=\mathrm{D}[a_i(\omega)]$, $\hat{\sigma}_i=\mathrm{D}[\sigma_i(\omega)]$ and $\bar{a}_i=\mathrm{E}[a_i(\omega)]$, $\bar{\sigma}_i=\mathrm{E}[\sigma_i(\omega)]$ then

$$\delta^s_{\mathrm{ER}_p}=\left(\bar{a}_1+\bar{\sigma}_1 m_1,\ \ldots,\ \bar{a}_n+\bar{\sigma}_n m_n\right),\qquad \vartheta_1^s=(0,\ldots,|\sigma_{i^s}(\omega^s)|d_{i^s},\ldots,0),$$

$$\delta^s_{\mathrm{DR}_p}=\left(2(\hat{a}_i+\hat{\sigma}_1 m_1^2)x_1,\ \ldots,\ 2(\hat{a}_n+\hat{\sigma}_n m_n^2)x_n\right),\qquad \vartheta_2^s=(0,\ldots,2\sigma_{i^s}^2(\omega^s)d_{i^s}^2 x_{i^s},\ldots,0).$$

Here x^s is the value of x at the sth iteration, $\sigma_{i^s}(\omega^s)$ are the realization of a random variable $\sigma_{i^s}(\omega)$ at the sth iteration, i^s is such an index that for any $i=\overline{1,n}$

$$|\sigma_{i^s}(\omega^s)|d_{i^s}x^s_{i^s}\ge|\sigma_i(\omega^s)|d_i x_i^s.$$

We will show that under the assumptions mentioned above the vector ξ^s and the variables ζ^s, χ^s satisfy the requirements of convergence of the stochastic penalty method [5].

Theorem 2. *If $|y^s-d_{\mathrm{ER}_p}(x^s)|\to 0$ and $|z^s-E_P(x^s)|\to 0$ with probability 1 for $s\to\infty$, then for small ε and $\|x-x^s\|\le\varepsilon$ the following inequality holds*

$$\left|\mathrm{E}\left[\xi^s/(x^s,y^s,z^s)\right]-\hat{F}(x^s)\right|\to 0, \tag{5}$$

where $\hat{F}(x)$ is a quasi-gradient of the weakly convex function $F(x)$.

Proof. To proof this theorem it is sufficient to show that the following statements hold:

1. Vectors $\mathrm{E}[\delta^s_{\mathrm{DR}_p}/\mathcal{B}_s]$ and $\mathrm{E}[\delta^s_{\mathrm{ER}_p}/\mathcal{B}_s]$ are quasi-gradients of the functions $m_{\mathrm{ER}_p}(x)$ and $m_{\mathrm{DR}_p}(x)$ respectively.
2. Vectors $\mathrm{E}[\vartheta_1^s/\mathcal{B}_s]$ and $\mathrm{E}[\vartheta_2^s/\mathcal{B}_s]$ are quasi-gradients of the functions $d_{\mathrm{ER}_p}(x)$ and $\mathrm{E}d^2_{R_p}(x,\omega)$ respectively.
3. If $|y^s-d_{\mathrm{ER}_p}(x^s)|\to 0$ for $s\to\infty$ then a vector $\mathrm{E}[-2y^s\vartheta_1^s/\mathcal{B}_s]$ converges to a quasi-gradient of the function $d^2_{\mathrm{ER}_p}(x)$.

4. If $\left|z^s - E_P(x^s)\right| \to 0$ for $s \to \infty$ then a vector $\mathrm{E}\left[c(z^s)(\lambda\delta^s_{ER_P} + \beta\vartheta^s_1) / \mathcal{B}_s\right]$ converges to a quasi-gradient of the function $E_P(x)$.

It is obvious that the statement 1 holds. The proof that the vector $\mathrm{E}(\vartheta^s_1 / \mathcal{B}_s)$ is a quasi-gradient of the function $d_{ER_P}(x)$ is presented in [6]. Since $|\sigma_i(\omega)| d_i x_i \ge 0 \;\; \forall i = \overline{1,n}$, then

$$d^2_{ER_P}(x) = \mathrm{E}\left[\max_{i=\overline{1,n}}\left\{|\sigma_i(\omega)| d_i x_i\right\}\right]^2 = \mathrm{E}\left[\max_{i=\overline{1,n}}\left\{\sigma^2_i(\omega) d^2_i x^2_i\right\}\right].$$

Here $(0,\ldots,2\sigma^2_i(\omega)d^2_i x_i,\ldots,0)$ is a quasi-gradient of a function $\sigma^2_i(\omega)d^2_i x^2_i$. The following proof is similar to the proof for the vector $\mathrm{E}[\vartheta^s_1 / \mathcal{B}_s]$.

Since the vector $\mathrm{E}[\vartheta^s_1 / \mathcal{B}_s]$ is the quasi-gradient of $d_{ER_P}(x)$ then the vector

$$\mathrm{E}[-2y^s\vartheta^s_1 / \mathcal{B}_s] = -2y^s \cdot \mathrm{E}[\vartheta^s_1 / \mathcal{B}_s]$$

for $y^s \to d_{ER_P}(x^s)$ converges to a quasi-gradient of $d^2_{ER_P}(x)$.

Considering that $\mathrm{E}[\lambda\delta^s_{ER_P} + \beta\vartheta^s_1 / \mathcal{B}_s]$ is the quasi-gradient of $E_P(x)$ [6] and the form of the penalty function $c(\alpha)$, the vector $\mathrm{E}\left[c(z^s)(\lambda\delta^s_{ER_P} + \beta\vartheta^s_1)/\mathcal{B}_s\right]$ for $z^s \to E_P(x^s)$ converges to the quasi-gradient of the function $c\left(E_P(x)\right)E_P(x)$ and this completes the proof.

Theorem 3. *Assume that the condition (5) is fulfilled, the random variables* $a_i(\omega)$ *and* $\sigma_i(\omega)$ *are bounded and their expectation and variance are finite; the function* $d_{ER_P}(x)$, $E_P(x)$ *are continuous at* x; *the step sizes satisfy with probability 1 the conditions*

$$\rho^s_j \ge 0, \quad \frac{\rho^{s+1}_j}{\rho^s_j} \to 1, \quad \sum_{s=0}^{\infty}\rho^s_j = \infty,$$

$$\sum_{s=0}^{\infty}\mathrm{E}\left[\left(\rho^s_j\right)^2\right] < \infty, \quad j = 1,2,3.$$

Also assume that for $s \to \infty$

$$\frac{\rho^s_1}{\rho^s_2} \to 0 \quad \text{and} \quad \frac{\rho^s_1}{\rho^s_3} \to 0,$$

the functions $d_{ER_P}(x)$ *and* $E_P(x)$ *satisfy a local Lipschitz condition and*

$$\max_{x\in B} F(x) < \inf_{\|x\|>1} F(x).$$

Then $\lim_{s\to\infty} F(x^s) = F(x^*)$, $x^* \in X^*$ *with probability 1.*

Proof. Let us show that there exists a constant C that

$$\left\|\xi^s\right\| + \left|\zeta^s\right| + \left|\chi^s\right| + \left\|\hat{F}(x^s, y^s, z^s)\right\| \le C.$$

It is easy to see that there exist such constants C_ζ and C_χ that

$$\left|\zeta^s\right| = \left|d_{R_P}(x^s, \omega^s)\right| = \left|\sigma_{i^s}(\omega^s) d_{i^s} x^s_{i^s}\right| < C_\zeta$$

and

$$\left|\chi^s\right| = \left|\lambda m_{R_p}(x^s,\omega^s) + \beta d_{R_p}(x^s,\omega^s)\right| \le \left|m_{R_p}(x^s,\omega^s)\right| + \left|\beta\sigma_{i^s}(\omega^s)d_{i^s}x^s_{i^s}\right| < C_\chi .$$

The components of the vector ξ^s can be written as

$$\xi^s_{i^s} = 2(\hat{a}_{i^s} + \hat{\sigma}_{i^s} m^2_{i^s})x_{i^s} + 2\eta\sigma^2_{i^s}(\omega^s)d^2_{i^s}x^s_{i^s}$$
$$- 2\eta y^s \sigma_{i^s}(\omega^s)d_{i^s} + c(z^s)\cdot\left[\lambda\left(\overline{a}_{i^s} + \overline{\sigma}_{i^s} m_{i^s}\right) + \beta\sigma_{i^s}(\omega^s)d_{i^s}\right],$$
$$\xi^s_j = 2(\hat{a}_j + \hat{\sigma}_j m^2_j)x_j + c(z^s)\cdot\lambda\cdot\left(\overline{a}_j + \overline{\sigma}_j m_j\right), \quad j = \overline{1,n}, \ j \ne i^s .$$

It is easy to see that $\left|\xi^s_{i^s}\right|$ and $\left|\xi^s_j\right|$ are bounded. So there exists C_ξ that $\left\|\xi^s\right\| < C_\xi$. On the other hand

$$\left\|\hat{F}(x^s,y^s,z^s)\right\| = \left\|\mathrm{E}\xi^s\right\|,$$

which means that $\left\|\hat{F}(x^s,y^s,z^s)\right\| < C_\xi$.

The following steps are similar to the proof presented in [5].

3.3. Numerical examples

In this section, we will give numerical examples for $n = 2$ and compare the results of using the stochastic penalty method with the results that was obtained in [3]. We consider that the shift and scale coefficients $a_i(\omega)$ and $\sigma_i(\omega)$ are uniformly distributed on [0,1], all fuzzy variables have symmetrical triangular forms with $R(t) = L(t) = \max\{0,\ 1-t\}$, $t \ge 0$.

Let $Z_1 = [0.3, 0.3, 3.5, 3.5]_{LR}$ and $Z_2 = [2.8, 2.8, 1.5, 1.5]_{LR}$, $\alpha = 0.65$, the relation $\mathcal{R}$ is "greater than or equal" ($\ge$). Then the equivalent stochastic problem of the minimum risk portfolio (1)-(2) takes the following forms in possibility context

$$\frac{1}{12}\left(1.09x_1^2 + 8.84x_2^2\right) + d_{\mathrm{D}R_p}(x) \to \min_{x\in F_p}, \tag{6}$$

$$F_p = \begin{cases} -0.65x_1 - 1.9x_2 - 0.35\cdot d_{\mathrm{E}R_p}(x) \le -m_d, \\ x \in X, \end{cases} \tag{7}$$

and in necessity context

$$\frac{1}{12}\left(1.09x_1^2 + 8.84x_2^2\right) + d_{\mathrm{D}R_p}(x) \to \min_{x\in F_p}, \tag{8}$$

$$F_p = \begin{cases} -0.65x_1 - 1.9x_2 + 0.65\cdot d_{\mathrm{E}R_p}(x) \le -m_d, \\ x \in X, \end{cases} \tag{9}$$

where

$$d_{\mathrm{E}R_p}(x) = \mathrm{E}[d_{R_p}(x,\omega)], \qquad d_{\mathrm{D}R_p}(x) = \mathrm{D}[d_{R_p}(x,\omega)],$$

and

$$d_{R_p}(x,\omega) = \max\left\{3.5\cdot\sigma_1(\omega)x_1,\ 1.5\cdot\sigma_2(\omega)x_2\right\}.$$

In [16] it is shown that in the case of uniformly distributed random components the function $d_{ER_p}(x)$ is differentiable. Moreover, it is easy to show that all requirements of Theorem 3 are fulfilled, so we can solve these examples using the stochastic penalty method.

The figure 1 shows solutions of the problems (6)-(7) and (8)-(9) depending on the acceptable level of the expected portfolio return m_d. Curves (a) and (b) represent the dependence of the minimum risk of the portfolio on its expected return in necessity context (8)-(9) and in possibility context (6)-(7) respectively.

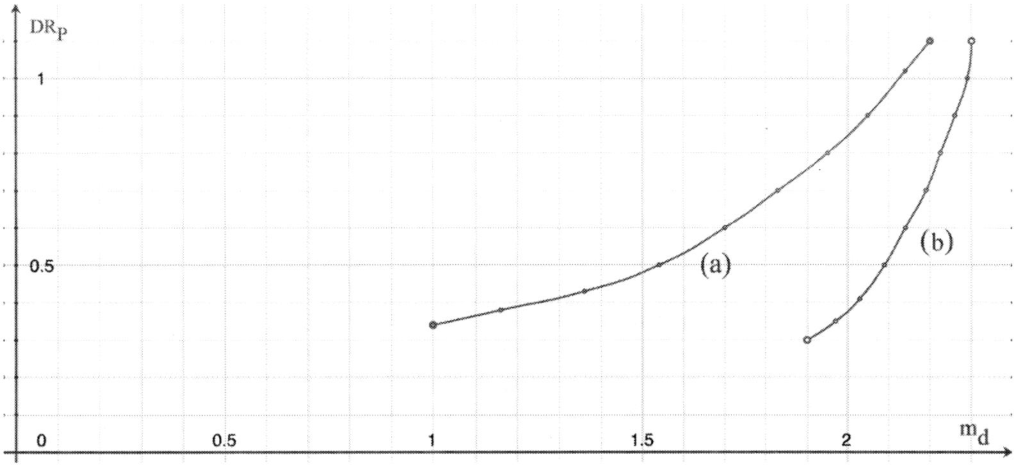

Figure 1. Dependence of the minimum risk of the portfolio on its expected return in necessity context (a) and possibility context (b).

The numerical examples confirm the theoretical results presented in this paper. Their solutions are similar to the results obtained in [3] where the problems (6)-(7) and (8)-(9) were replaced with their equivalent crisp problems and a genetic algorithm was used to solve them.

4. Conclusion

In the present paper, we have contributed to the research area of the portfolio optimization in the following aspects: 1) a mathematical model of the minimum risk portfolio has been described under the conditions of a hybrid uncertainty of possibilistic-probabilistic type for the case of T_W-related fuzzy components; 2) the stochastic equivalent problem for the minimum risk portfolio model has been obtained; 3) the stochastic penalty method has been specified for solving the equivalent stochastic program. The approach is demonstrated on numerical examples when its random parameters are uniformly distributed on [0,1] and fuzzy factors are symmetrical fuzzy numbers.

References

[1] Yazenin A V 2007 Possibilistic-probabilistic models and methods of portfolio optimization *Stud. Comp. Intel.* **36** 241–59

[2] Xu J and Zhou X 2011 Fuzzy-like multiple objective decision making *Stud. Fuzz. Soft Comp.* (Berlin: Springer//Heidelberg) **263**

[3] Yazenin A and Soldatenko I 2017 Portfolio of minimum risk in a hybrid uncertainty of a possibilistic-probabilistic type: comparative study *Advances in Intelligent Systems and Computing* **643** 551–63

[4] Feng Y, Hu L and Shu H 2001 The variance and covariance of fuzzy random variables and their applications *Fuzzy Sets Syst.* **120** 487–97

[5] Ermolev Yu M 1976 *Methods of Stochastic Programming* (Moscow: Nauka)
[6] Egorova Yu E and Yazenin A V 2014 Stochastic quasi-gradient method for solving possibilistic-probabilistic optimization task of one class *Vestnik TvGU. Seriya: Prikladnaya matematika (Herald of Tver State University. Series: Applied Mathematics)* **4** 57–70
[7] Egorova Yu E and Yazenin A V 2017 The problem of possibilistic-probabilistic optimization *J. Comput. Sys. Sci. Int.* **56** 652–67
[8] Nahmias S 1979 Fuzzy variables in a random environment *Advances in Fuzzy Sets Theory and Applications* 165–80
[9] Puri M L and Ralescu D A 1986 Fuzzy random variables *Journal of Mathematical Analysis and Applications* **114** 409–22
[10] Yazenin A V 2016 *Basic Concepts of Possibility Theory: A Mathematical Apparatus for Decision-making Under Hybrid Uncertainty Conditions* (Moscow: Fizmatlit Publ.)
[11] Dubois D and Prade H 1988 *Possibility Theory* (New York: Plenum Press)
[12] Nguyen H T and Walker E A 1997 *A First Course in Fuzzy Logic* (Boca Raton: CRC Press)
[13] Hong D H 2001 Parameter estimations of mutually T-related fuzzy variales *Fuzzy Sets Syst.* **123** 63-71
[14] Markowitz H M 1952 Portfolio selection *The Journal of Finance* **7** 77–91
[15] Nurminski E A 1980 An application of nondifferentiable optimization in optimal control *Numerical Optimization of Dynamic Systems* ed L Dixon and G P Szego (Amsterdam: North Holland Publishing Company) 137–57
[16] Yazenin A V and Egorova Yu E 2013 On solution methods for tasks of possibilistic-probabilistic optimization *Vestnik TvGU. Seriya: Prikladnaya matematika (Herald of Tver State University. Series: Applied Mathematics)* **4** 85–103

IOP Conf. Series: Journal of Physics: Conf. Series **973** (2018) 012034 doi:10.1088/1742-6596/973/1/012034

Models of multidimensional discrete distribution of probabilities of random variables in information systems

Yu Yu Gromov, Yu V Minin, O G Ivanova and O N Morozova

Department of Information Systems and Information Protection, Tambov State Technical University, Sovetskaya 106, Tambov, Russia

E-mail: gromovtambov@yandex.ru

Abstract. Multidimensional discrete distributions of probabilities of independent random values were received. Their one-dimensional distribution is widely used in probability theory. Producing functions of those multidimensional distributions were also received.

1. Introduction

Any model of information system is an abstract, formally described object. It describes the formalized process of information system functioning and it is capable to cover only its basic, specific characteristics, while leaving aside minor unnecessary factors.

The formalization of information process is preceded by the structure study, in this way it is developed a meaningful description of the process, i.e. the first attempt to get across the specific characteristics of investigated information process and to set a task. This gives information about the physical nature and quantitative characteristics of the process elementary phenomenon, the nature of their interaction, the place of each phenomenon in general process.

A meaningful description serves to construct a formalized scheme and a model of information process.

A formalized scheme of the process is elaborated in case when, due to the complexity of the process or the difficulties of formalizing some of its elements, a direct transition from a meaningful description to a model is impossible or inexpedient. To construct a formalized scheme, it is necessary to select the characteristics of the process; establish a system of parameters providing the process; to determine all dependencies between characteristics and parameters in view of factors considering when formalizing.

As a rule, the most difficult stage of simulation process is the conversion of the identified essential factors into the language of mathematical concepts and determination of interactions between these quantities. The fact is that the requirements of content and deductivity of the model are contradictory in their essence. To satisfy the requirement of content, it is necessary to consider in the model as many factors of information process as possible. Herewith, naturally, the model becomes more complex, which makes it difficult to study and obtain meaningful results. However, the desire to obtain a result in easier way leads to the need to simplify the model, thereby reducing its content. Thus, it is required to achieve a reasonable compromise, ensuring the possibility of obtaining non-trivial results and not missing the essence of the real process. In this case, an accurate set of all initial data, known parameters and initial conditions is attached.

A meaningful description may not provide the necessary information for constructing a formalized scheme, and then additional experiments and observations of the process under investigation are necessary. But in this case, when developing a formalized scheme, they should be fully used.

Further transformation of the formalized scheme into a model is fulfilled without inputting the additional information.

Depending on the complexity of the analytical description of the system, the following main methods of using mathematical models are distinguished: analytical research; qualitative research; research using numerical methods; simulation modelling.

In most cases, when modeling the information systems, there is no complete mathematical formulation of the problem, and even though there are analytical methods and approaches, but they are very complex and laborious, and besides evaluating certain parameters it is necessary to implement monitoring of the process during a certain period, and in some cases, it requires slowing down or accelerating the model time. In this context, simulation of information systems is simpler way of research than analytical.

For qualitative evaluation of a complex system, it is convenient to use the results of the theory of random processes. The experience of observing objects shows that they function under the conditions of many random factors, so the prediction of the behavior of complex system can be meaningful within the framework of probability categories. In other words, only the probabilities of their occurrence can be indicated for expected events, and in respect to some values it is necessary to confine to the laws of their distribution or other probability characteristics (for example, mean values, variances, etc.).

To study the process of functioning of each complex system, considering random factors, it is necessary to have a clear idea of the sources of random effects and very reliable data on their quantitative characteristics. Therefore, any calculation or theoretical analysis associated with the study of a complex system is preceded by an experimental accumulation of statistical material characterizing the behavior of individual elements and the whole system in real conditions. This material processing allows to obtain the initial data for calculation and analysis.

The main sources of random influences are the factors of the "external" environment and deviations from the normal operation modes (errors, noises, etc.) occurring within the system.

It follows from what has been said that a very serious attention must be given to the consideration of random factors while investigating the complex systems using random numbers with specified probabilistic characteristics. In this case, the results obtained with one-time modeling should be regarded only as an implementation of a random process. Each of these realizations separately cannot serve as an objective characteristic of system under study. The required values are usually determined by averaging and statistical processing of the data of many realizations, so this approach is also called the statistical modeling method.

2. Statement of the problem

Nowadays multidimensional discrete distributions are presented more often in tables, or matrixes [1,2], i.e. analytic expressions for them are used little in the theory of probability. Among them, we should mention first of all the polynomial probability distribution of dependent random variables (RV) $X_1,\ldots,X_k$, which take integer non-negative values $n_1,\ldots,n_k$ with probabilities [1-3]

$$P\{X_1 = n_1,\ldots,X_k = n_k\} = \frac{n!}{n_1!\ldots n_k!} p_1^{n_1}\ldots p_k^{n_k}, \tag{1}$$

where $0 < p_j < 1;\ p_1 + \ldots + p_k = 1;\ n_1 + \ldots + n_k = n;\ k \geq 2.$

Producing function of distribution (1) is defined by an expression

$$G(z_1,\ldots,z_k) = (p_1 z_1 + \ldots + p_k z_k)^n, \tag{2}$$

where z_j – is complex number $|z_j| \le 1$.

In paper [3] it is noted, that distribution (1) is (*k*-1)-dimensional, i.e. in dimension R^k it is degenerated. That's why for k -dimensional polynomial probability distribution of dependent RV $X_1, \ldots, X_k$ instead of expression (1) is offered to use ratio

$$P\{X_1 = n_1, \ldots, X_k = n_k\} = \frac{n!}{n_0! n_1! \ldots n_k!} p_0^{n_0} p_1^{n_1} \ldots p_k^{n_k}. \tag{3}$$

Distribution (3) can be considered as a modified polynomial distribution. Its producing function is defined by an expression.

$$G(z_1, \ldots, z_k) = (p_0 + p_1 z_1 + \ldots + p_k z_k)^n \tag{4}$$

Negative polynomial probability distribution of dependent RV $X_1, \ldots, X_k$, taking integer non-negative values $m_1, \ldots, m_k$ with probabilities [3] is used sometimes

$$P\{X_1 = m_1, \ldots, X_k = m_k\} = \frac{\Gamma(\alpha + m_1 + \ldots + m_k)}{\Gamma(\alpha) m_1! \ldots m_k!} p_0^{\alpha} p_1^{m_1} \ldots p_k^{m_k}, \tag{5}$$

where $\Gamma(z)$ – is gamma-function; $\alpha > 0$; $0 < p_j < 1$; $p_0 + \ldots + p_k = 1$; $k \ge 1$.

Producing function for distribution (5) is defined by an expression

$$G(z_1, \ldots, z_k) = p_0^{\alpha} (1 - p_1 z_1 - \ldots - p_k z_k)^{-\alpha}. \tag{6}$$

The main purpose of the work is to receive models of multidimensional discrete probability distributions of dependent RV, similar to probability models (3) and (5)

3. The solution of the problem

Table 1 shows the expressions of one-dimensional discrete distributions, widely used in probability theory [1-4]. Table 2 shows expressions for producing functions of these distributions.

Table 1. One-dimensional discrete distributions.

№ n/n	Type of distribution	Expression for distribution p(x)
1	Binomial	$p(x) = \frac{n!}{(n-x)!x!}(1-p)^{n-x} p^x$ $x = 0, 1, \ldots, n;\ 0 < p < 1,\ n \ge 1$
2	Negative binomial	$p(x) = \frac{\Gamma(\alpha + x)}{\Gamma(\alpha) x!}(1-q)^x q^{\alpha}$, $x = 0, 1, \ldots, \infty;\ 0 < q < 1,\ \alpha > 0$
3	Poisson's	$p(x) = \frac{\lambda^x}{x!}\exp(-\lambda),\ x = 0, 1, \ldots, \infty.$
4	Hypergeometric	$p(x) = \frac{n!}{(n-x)!x!} \frac{N_0! N_1! (N-n)!}{(N_0 - n + x)!(N_1 - x)! N!}$, $x = 0, 1, \ldots, n;\ N_0 \ge n,\ N_1 \ge n,\ n \ge 2;\ N_0 + N_1 = N$

5	Negative hypergeometric	$p(x)=\frac{n!}{(n-x)!x!}\frac{(b_1)_x(b_0)_{n-x}}{(b)_n}$, $x=0,1,\dots,n;\ b_0>0,\ b_1>0,\ n\geq 2;\ b_0+b_1=b$
6	Negative beta-binomial	$p(x)=\frac{\Gamma(a+b_0+1)\Gamma(b+1)(a)_x(b_1)_x}{\Gamma(b_0+1)\Gamma(a+b+1)(a+b+1)_x\, x!}$, $x=0,1,\dots,\infty;\ a>0,\ b_0>0;\ b_1\geq a;\ b_0+b_1=b$

Table 2. Producing functions of one-dimensional discrete distributions.

№ n/n	Type of distribution	Expression for producing function G(z)
1	Binomial	$G(z)=(q+pz)^n$ $0<p<1,\ q=1-p;\ n\geq 1$
2	Negative binomial	$G(z)=q^\alpha(1-pz)^{-\alpha}$ $0<p<1;\ q=1-p;\ \alpha>0$
3	Poisson's	$G(z)=\exp(\lambda z-\lambda);\ \lambda>0$
4	Hypergeometric	$G(z)=\frac{N_0!(N-n)!}{(N_0-n)!N!}\,{}_2F_1(-n,-N_1;1+N_0-n;z)$ $N_0\geq n,\ N_1\geq n,\ n\geq 2;\ N_0+N_1=N$
5	Negative hypergeometric	$G(z)=\frac{(b_0)_n}{(b_0+b_1)_n}\,{}_2F_1(-n,b_1;1-b_0-n;z)$ $b_0>0,\ b_1>0,\ n\geq 2;\ b_0+b_1=b$
6	Negative beta-binomial	$G(z)=\frac{\Gamma(a+b_0+1)\Gamma(b+1)}{\Gamma(b_0+1)\Gamma(a+b+1)}\,{}_2F_1(a,b_1;a+b+1;z)$ $a>0,\ b_0>0;\ b_1\geq a;\ b_0+b_1=b$

If we put $k=1$, $n_1=x$, $n_0=n-x$, $p_1=p$, $p_0=q=1-p$ in ratios (3) and (4), then they will transform respectively into one-dimensional binomial distribution and its producing function. Their expressions are shown in tables 1 and 2 (1 line). That's why, in the role of k-dimensional binomial probability distribution of dependent RV-s $X_1,\dots,X_k$ and its producing function we can use respectively k-dimensional modified polynomial distribution (3) and producing function (4).

Analogically we can make sure, that if in ratios (5) and (6) to accept $k=1$, $m_1=x$, $p_1=p$, $p_0=q=1-p$, then they will transform respectively in one-dimensional negative binomial distribution and its producing function. Their expressions are shown in tables 1 and 2 (line 2).

That's why, in the role of k-dimensional binomial probability distribution of dependent RV-s $X_1,\dots,X_k$ and its producing function we can use respectively k-dimensional modified polynomial distribution (5) and producing function (6).

It's known from probability theory, that the limiting case of one-dimensional binomial distribution, when $p\to 0$ and $n\to\infty$, and product np aims to some positive constant value λ ($np\to\lambda$), is one-dimensional Poisson distribution [1-3]. Expressions to Poisson distribution and its producing function are shown in the tables 1 and 2 (line 3). Respectively, the limiting case for k-dimensional binomial distribution (3) is k-dimensional Poisson's distribution [5]

$$P\{X_1 = n_1, \dots, X_k = n_k\} = \exp\left(-\sum_{i=1}^{k} \lambda_i\right) \prod_{i=1}^{k} \frac{\lambda_i^{n_i}}{n_i!}. \tag{7}$$

It follows from analysis of table 2 (line 4-6), that the basis of producing functions hypergeometric, negative hypergeometric and negative beta-binomial distributions is a hypergeometric Gaussian function ${}_2F_1(a, b; c; z)$ [6]. That's why for these distributions expression of producing function can be presented in general view

$$G(z) = K_н \; {}_2F_1(a, b; c; z) = K_н \sum_{x=0}^{\infty} \frac{(a)_x (b)_x}{(c)_x \, x!} z^x = \sum_{x=0}^{\infty} p(x) z^x . \tag{8}$$

where $(a)_k = a(a+1)\dots(a+k-1)$ – symbol of Poghammer; $K_н = \frac{\Gamma(c-a)\Gamma(c-b)}{\Gamma(c)\Gamma(c-a-b)}$ normalization coefficient.

From (8) it is easy to receive in general view the expression for one-dimensional probability distribution of shown distribution laws

$$p(x) = \frac{\Gamma(c-a)\Gamma(c-b)}{\Gamma(c)\Gamma(c-a-b)} \frac{(a)_x (b)_x}{(c)_x \, x!}. \tag{9}$$

It's one of comfortabilities of using producing functions.

In the analogy with one-dimensional case (8) producing function of multidimensional function must contain hypergeometric function of some dependent values. Such function can be Laurichell function $F_D^{(k)}(a, b_1 \dots b_k; c; z_1, \dots, z_k)$, which in the case of 1 variable (k=1) transforms in hypergeometric Gaussian function [6]. In this producing function either

$$G(z_1, \dots, z_k) = K_н \, F_D^{(k)}(a, b_1 \dots b_k; c; z_1, \dots, z_k) \tag{10}$$

or

$$G(z_1, \dots, z_k) = K_н \sum_{n_1, \dots, n_k = 0}^{\infty} \frac{(a)_{n_1 + \dots + n_k} (b_1)_{n_1}, \dots, (b_k)_{n_k}}{(c)_{n_1 + \dots + n_k}} \frac{z_1^{n_1} \dots z_k^{n_k}}{n_1! \dots n_k!}, \tag{11}$$

where $b = b_0 + \dots + b_k$; $K_н = 1/F_D^{(k)}(a, b_1 \dots b_k; c; 1)$.

From (11) follows in general case an expression for k-dimensional probability distribution of dependent RV-s $X_1, \dots, X_k$

$$p(n_1, \dots, n_k) = K_н \frac{(a)_{n_1 + \dots + n_k} (b_1)_{n_1}, \dots, (b_k)_{n_k}}{(c)_{n_1 + \dots + n_k} \, n_1! \dots n_k!}. \tag{12}$$

If $a = -n$, $b_1 = -N_1, \dots, b_k = -N_k$, $b = -N$, $c = 1 + N_0 - n$, then multidimensional distribution (12) transforms in k-dimensional hypergeometric distribution, defined by an expression

$$p(n_1, \dots, n_k) = \binom{N_0}{n_0}\binom{N_1}{n_1} \dots \binom{N_k}{n_k} \Big/ \binom{N}{n}, \tag{13}$$

IOP Conf. Series: Journal of Physics: Conf. Series **973** (2018) 012034 doi:10.1088/1742-6596/973/1/012034

where $n_0 = n - n_1 - \ldots - n_k$; $\binom{N}{n} = \frac{N!}{(N-n)!n!}$ – binomial coefficients.

If $a = -n$, $b = b_0 + \ldots + b_k$, $c = 1 - b_0 - n$, then multidimensional distribution (12) transforms in k-dimensional negative hypergeometric distribution, defined by an expression

$$p(n_1, \ldots, n_k) = \frac{n!}{(b)_n} \frac{(b_0)_{n_0}(b_1)_{n_1} \ldots (b_k)_{n_k}}{n_0!n_1!\ldots n_k!}. \tag{14}$$

If $b = b_0 + \ldots + b_k$, $c = 1 + a + b$, then multidimensional distribution (12) transforms in k - dimensional negative beta-binomial distribution, defined by an expression

$$p(n_1, \ldots, n_k) = \frac{\Gamma(b+1)\Gamma(a+b_0+1)}{\Gamma(a+b+1)\Gamma(b_0+1)} \frac{(a)_{n_1+\ldots+n_k}(b_1)_{n_1}, \ldots, (b_k)_{n_k}}{(a+b+1)_{n_1+\ldots+n_k} n_1!\ldots n_k!}. \tag{15}$$

In practice, among the multidimensional discrete distributions, two-dimensional distributions can most often be used. Table 3 shows the expressions for two-dimensional discrete distributions of dependent random variables one-dimensional distributions of which are presented in Table 1.

Table 4 shows expressions for producing functions of these distributions.

Table 3. Two-dimensional discrete distributions.

№ n/n	Type of distribution	Expression for distribution p(x,y)
1	Binomial	$p(x,y) = \frac{n!}{(n-x-y)!x!y!}(1-p_1-p_2)^{n-x-y} p_1^x p_2^y$ $x = 0, 1, \ldots, n;\ y = 0, 1, \ldots, n-x;\ 0 < p_i < 1,\ n \geq 1$
2	Negative binomial	$p(x,y) = \frac{\Gamma(\alpha + x + y)}{\Gamma(\alpha)\, x!y!} p_1^x p_2^y (1-p_1-p_2)^{\alpha},$ $x = 0, 1, \ldots, \infty;\ y = 0, 1, \ldots, \infty;\ 0 < p_i < 1,\ \alpha > 0$
3	Poisson's	$p(x,y) = \frac{\lambda_1^x \lambda_2^y}{x!y!} \exp(-\lambda_1 - \lambda_2),\ 0 \leq x < \infty;\ 0 \leq y < \infty.$
4	Hypergeometric	$p(x,y) = \frac{n!}{(n-x-y)!x!y!} \frac{N_0!N_1!N_2!(N-n)!}{(N_0-n+x+y)!(N_1-x)!(N_2-y)!N!},$ $N_0 \geq n,\ N_1 \geq n,\ N_2 \geq n,\ n \geq 2;\ N_0 + N_1 + N_2 = N$
5	Negative hypergeometric	$p(x,y) = \frac{n!}{(n-x-y)!x!y!} \frac{(b_1)_x (b_2)_y (b_0)_n}{(n+b_0-x-y)_{x+y} (b)_n},$ $b_0 > 0,\ b_1 > 0,\ b_2 > 0,\ n \geq 2;\ b_0 + b_1 + b_2 = b$
6	Negative beta-binomial	$p(x,y) = \frac{\Gamma(a+b_0+1)\Gamma(b+1)(a)_{x+y}(b_1)_x(b_2)_y}{\Gamma(b_0+1)\Gamma(a+b+1)(a+b+1)_{x+y}\, x!y!},$ $a > 0,\ b_0 > 0;\ b_1 \geq a;\ b_2 \geq a;\ b_0 + b_1 + b_2 = b$

Table 4. Producing functions of two-dimensional discrete distributions.

№ n/n	Type of distribution	Expression for distribution $G(z_1,z_2)$
1	Binomial	$G(z_1,z_2)=(q+p_1 z_1+p_2 z_2)^n$ $0<p_i<1,\ q=1-p_1-p_2;\ n\geq 1$
2	Negative binomial	$G(z_1,z_2)=q^{\alpha}(1-p_1 z_1-p_2 z_2)^{-\alpha}$ $0<p_i<1;\ q=1-p_1-p_2;\ \alpha>0$
3	Poisson's	$G(z_1,z_2)=\exp(\lambda_1 z_1+\lambda_2 z_2-\lambda_1-\lambda_2);\ \lambda_1>0;\ \lambda_2>0$
4	Hypergeometric	$G(z_1,z_2)=\frac{N_0!(N-n)!}{(N_0-n)!N!}F_D^{(2)}(-n,-N_1;-N_2;1+N_0-n;z_1,z_2)$ $N_0\geq n,\ N_1\geq n,\ N_2\geq n,\ n\geq 2;\ N_0+N_1+N_2=N$
5	Negative hypergeometric	$G(z_1,z_2)=\frac{(b_0)_n}{(b)_n}F_D^{(2)}(-n,b_1;b_2;1-b_0-n;z_1,z_2)$ $b_0>0,\ b_1>0,\ b_2>0,\ n\geq 2;\ b_0+b_1+b_2=b$
6	Negative beta-binomial	$G(z_1,z_2)=\frac{\Gamma(a+b_0+1)\Gamma(b+1)}{\Gamma(b_0+1)\Gamma(a+b+1)}F_D^{(2)}(a,b_1;b_2;a+b+1;z_1,z_2)$ $a>0,\ b_0>0;\ b_1\geq a;\ b_2\geq a;\ b_0+b_1+b_2=b$

Thus, the proposed approach makes it possible to obtain expressions for multidimensional discrete distributions of probabilities of independent random values, as well as their producing functions.

References

[1] Gnedenko B V 1988 *Course of the theory of probability* (Moscow: Science) p 448
[2] Fedotkin M A 2012 *Models in the theory of probability* (Moscow: FIZMATLIT) p 608
[3] *Probability and mathematical statistics: encyclopedia* ed Yu V Prokhorov 1999 (Moscow: The Great Russian Encyclopedia) p 910
[4] Johnson N L, Kots S and Kemp A 2010 *One-dimensional discrete distributions* (Moscow: BINOM) p 559
[5] Karpov I G 1998 The laws of distribution of discrete random variables and their identification *IV International Scientific and Technical Conference "Radiolocation, Navigation and Communication"* **1** pp 115-125
[6] Prudnikov A P, Brychkov Yu A and Marichev O I 1986 *Integrals and series. Additional chapters* (Moscow: Nauka) p 800

IOP Conf. Series: Journal of Physics: Conf. Series **973** (2018) 012035 doi:10.1088/1742-6596/973/1/012035

Algorithm for predicting the evolution of series of dynamics of complex systems in solving information problems

T I Kasatkina, A V Dushkin, V A Pavlov, R R Shatovkin

ET Department, Voronezh Institute Federal Penitentiary Service of Russia,
1a Irkutskaya Street, Voronezh, RU

E-mail: kasatkinatatian@googlemail.com

Abstract. In the development of information, systems and programming to predict the series of dynamics, neural network methods have recently been applied. They are more flexible, in comparison with existing analogues and are capable of taking into account the nonlinearities of the series. In this paper, we propose a modified algorithm for predicting the series of dynamics, which includes a method for training neural networks, an approach to describing and presenting input data, based on the prediction by the multilayer perceptron method. To construct a neural network, the values of a series of dynamics at the extremum points and time values corresponding to them, formed based on the sliding window method, are used as input data. The proposed algorithm can act as an independent approach to predicting the series of dynamics, and be one of the parts of the forecasting system. The efficiency of predicting the evolution of the dynamics series for a short-term one-step and long-term multi-step forecast by the classical multilayer perceptron method and a modified algorithm using synthetic and real data is compared. The result of this modification was the minimization of the magnitude of the iterative error that arises from the previously predicted inputs to the inputs to the neural network, as well as the increase in the accuracy of the iterative prediction of the neural network.

1. Introduction

In the development of information systems and programming in solving problems of forecasting the state of the system and, accordingly, to implement the possibility of influencing the development of the dynamic system, neural network methods have recently been applied. They are used where it is necessary to forecast parameters and parameters of the system for some pre-emptive period. Neural network methods are more flexible than existing analogues and are capable of accounting for the nonlinearities of series. Now, there is a wide variety of approaches to predicting the series of dynamics, for example, the use of statistical methods described in [1-10]. Of particular popularity recently acquired more flexible, in comparison with existing analogs, and capable of taking into account the nonlinear nature of the series of dynamics, neural network methods, for example, presented in [11-17]. When predicting the dynamics series, various neural network approaches, such as the use of recurrent neural networks, the use of generalized regression neural networks (the GRNN method), the use of RBF networks, the use of multilayer perceptrons [18-23] are also used. In order to obtain the maximum forecasting effect, several approaches and different types of neural networks are usually used at once. In this paper, we offer a unifying approach to predicting the series of dynamics, including the method of training neural networks, the method of describing and presenting input data, based on approaches to forecasting based on a multilayer perceptron. The proposed method can act both as an independent approach to predicting the series of dynamics, and be one of the parts of the forecasting system.

2. Statement of the problem with the traditional method of describing a number of dynamics

The predicted series of dynamics can be specified in the form of some sequence:

$$x_1, \quad x_2, \quad \dots \quad x_l\,. \tag{1}$$

It is required to define its continuation in the form:

$$x_{l+1}, \quad x_{l+2}, \quad \dots \tag{2}$$

In this case, the predicted nonlinear model can be defined as follows:

$$x_t = F\left(x_{t-1}, \quad x_{t-2}, \quad \dots \quad x_{t-k}\right), \tag{3}$$

where $t = \overline{k+1, \quad n}$; F – nonlinear functional dependence, constructed on the basis of artificially created neural network; k – the size of the sliding window, numerically equal to the number of neurons entering the network n.

To predict a number of dynamics for given values, a ***multilayer perceptron*** (MLP, Multilayer Perceptron) can be used as the main architecture. In [24] it was proved that an artificially created neural network could approximate with any desired accuracy any continuous function. Another no less important characteristic of a multilayer perceptron is the possibility of generalizing information that is in the set of training. Thus, the multilayer perceptron is a powerful tool for predicting the evolution of the series of dynamics of complex systems. In fig. 1 shows a multilayer perceptron used in the present work to predict the evolution of the series of dynamics.

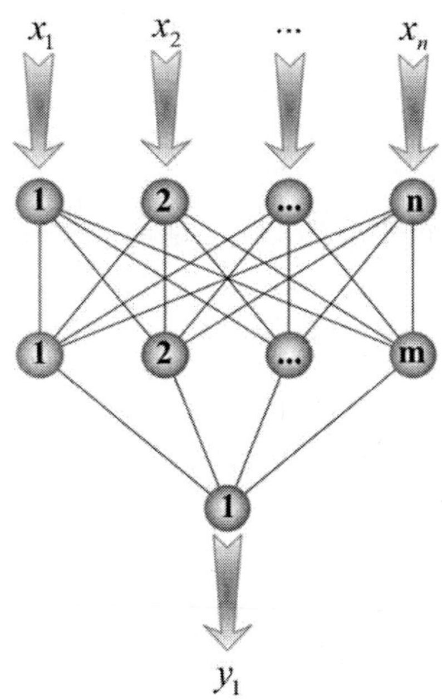

Figure 1. The multilayer perceptron.

When using the traditional method of describing the predicted series of dynamics, only the previous values (1) set in the sequence will be taken into account in predicting subsequent values. In the present work, the prediction was carried out on images that differed from the sample of images fed from the set of training. To ensure the correct functioning of the neural network, it is required to normalize the input data. There are many options for carrying out such an operation. Some of the best results are achieved by projecting the domain of the definition of the set of input data (1) into a segment $[0;1]$.

3. Statement of the problem with a modified method for describing a number of dynamics

This method consists in a special approach to learning the neural network in forecasting, as well as in a special way of specifying the input data set. At the first stage of the evolution prediction, the following transformation of a number of dynamics is performed.

1. In a number of dynamics only the extremum values are chosen for forecasting: the maximum points and the minimum points.
2. The value at the extremum point and the time interval that has elapsed since the appearance of the previous extremum value are time-value pairs, which are further used as inputs to the neural network. In fig. 2 shows a method for specifying time-value pairs.

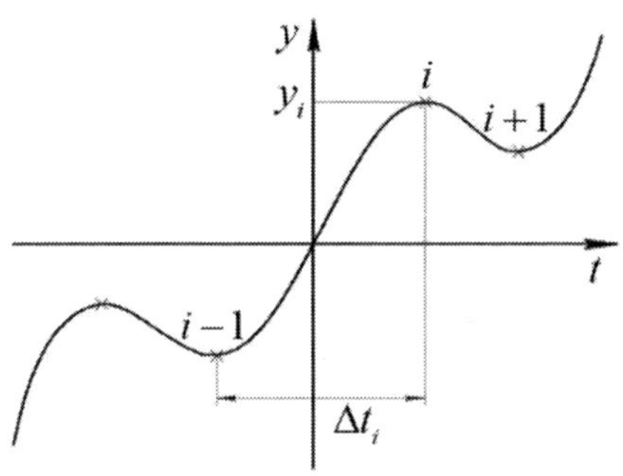

Figure 2. Method for specifying time-value pairs.

Thus, to construct a neural network as input data, it is expedient to use the values of a number of dynamics at the extremum points and time values corresponding to them and formed on the basis of the sliding window method. As a result, time-value pairs n are formed and fed to the network input, which are the values of a number of dynamics. Each such pair includes the value at the very extremum, and the time interval elapsed from the previous extremum to the moment of the present appearance. The output layer in the multilayer perceptron, respectively, contains both series of such values, and the output is as follows:

$$x_{n+1} = F(x_1,\ x_2,\ \ldots,\ x_n), \tag{4}$$

$$t_{n+1} = F(t_1,\ t_2,\ \ldots,\ t_n). \tag{5}$$

In fig. 3 is an image of the implemented neural network architecture.

The proposed modification of the approach to learning a neural network is that the classical traditional neural network learning process is first conducted on the basis of the accepted master sample data, and then the training is based on a sample of the predicted data. The result of this modification is the minimization of the error value resulting from the previously predicted inputs to the inputs to the neural network, the so-called iterative error. As a result, the learning process of a neural network is first performed on reference values, and then, once the iterative error reaches an acceptable level, the learning process of the neural network is started on the predicted data. The result of this approach is a reduction in the forecast error as a whole, as well as an increase in the accuracy of the iterative prediction of the neural network.

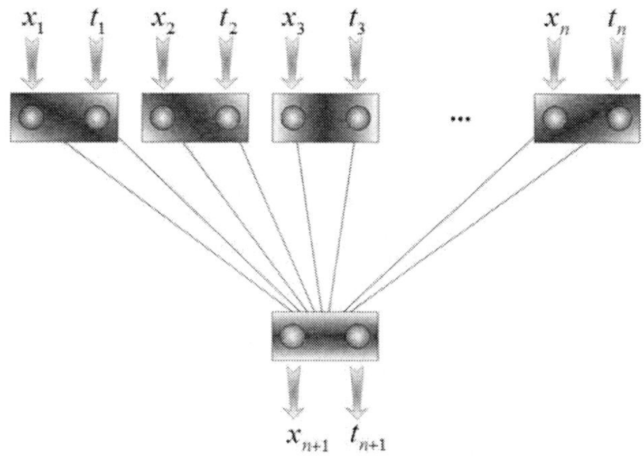

Figure 3. Architecture of neural network.

4. Results of the experimental study

In the course of the experimental study of the proposed approach, the evolution of chaotic series of dynamics based on the use of a multilayer perceptron with the use of synthetic data, as well as the use of real data, was predicted. As an explored chaotic system of synthetic data, the Enon attractor was used. To describe the Enon process, the following equations are applied:

$$\begin{cases} x_{n+1} = 1 - \alpha \cdot x_n^2 + y_n \\ y_{n+1} = \beta \cdot x_n \end{cases}, \quad (6)$$

where the coefficients of the chaotic behavior of the system are $\alpha = 1.4$, $\beta = 0.3$.

In fig. 4 shows the first 200 elements of the Enon dynamics series.

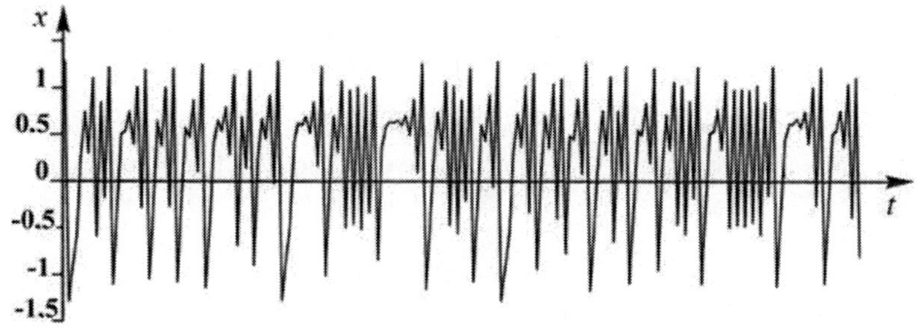

Figure 4. A number of dynamics of Enon.

The coordinate X of a number of dynamics of Enon is applied as a set of input data. When demonstrating the evolution of the series in the process of training the neural network, the first 400 elements of a series of dynamics and 300 elements other than the elements involved in training were used. The neural network was trained on the basis of the sliding window method. The forecasting process was performed on data that was different from the elements in the set of data used in training the network. This is necessary in order to make the results compare more clearly. The work used a multi-step and one-step forecast types. A one-step forecast is necessary for short-term forecasting of evolution, forecasting one step in time. When implementing this type of forecasting, only the reference actual data was used. Inputs from the neural network were fed from the data sample for forecasting. In fig. 5 is a diagram of a short-term one-step forecast of a number of dynamics.

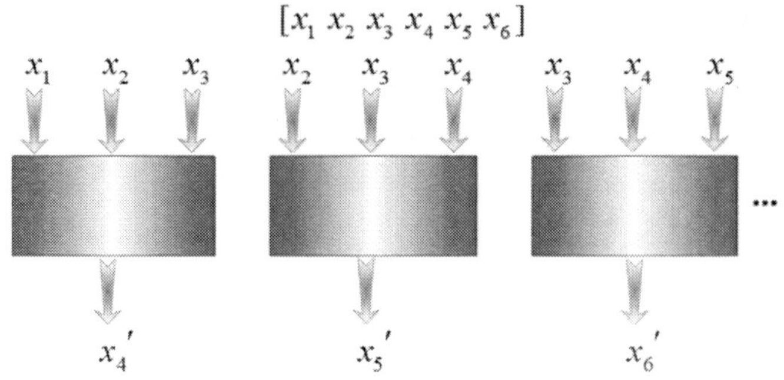

Figure 5. Scheme of short-term one-step forecasting of the evolution of a number of dynamics.

A multi-step view of predicting the evolution of the dynamics series, carried out by repeating the iterations, was carried out using a series of data obtained at the output from the previous stage of the forecasting process. In fig. 6 is a diagram of a long-term multi-step forecast of a number of dynamics. Estimation of the quality of forecasting the evolution of a number of dynamics was carried out using the MAP (Mean absolute percentage error) error and the MSE total root-mean-square error. The MAPE error is the average absolute error expressed in percentages and, as a rule, is used to estimate the accuracy of predicting the dynamics series for large volumes of statistical data. The calculation of the MARE error was carried out using the formula:

$$\mathrm{MAPE} = \frac{1}{n}\sum_{t=1}^{n}\left|\frac{A_t - F_t}{A_t}\right|, \tag{7}$$

where A_t – real value of the data set; F_t – the value obtained as a result of predicting the evolution of a number of dynamics.

Calculation of the mean square error MSE was carried out by the formula:

$$\mathrm{MSE} = \frac{1}{2}\sum_{t=1}^{n}(A_t - F)_t^2 . \tag{8}$$

When predicting the evolution of a number of dynamics of Enon with the help of the classical algorithm of a multilayer perceptron, the neural network showed the best results using architecture 28-12-1.

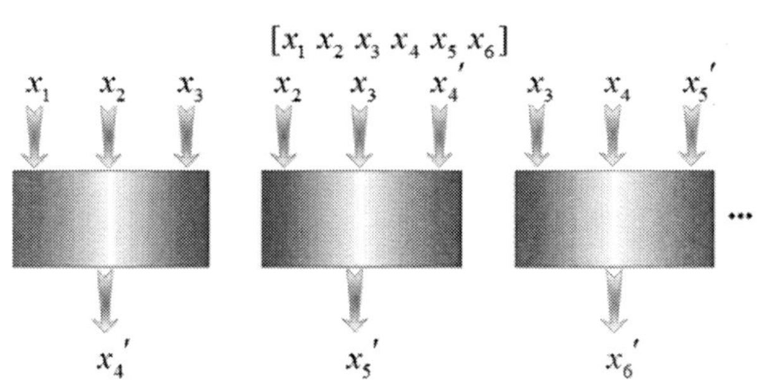

Figure 6. Scheme of long-term multi-step prediction of the evolution of a number of dynamics.

In single-step forecasting, the results of forecasting are presented in fig. 7, the average value of the MAPE error was 38.4%, and the average MSE error was 0.033. In fig. 7, in order to better visualize the calculated data, only the first 20 values obtained are shown.

In the multi-step prediction of the evolution of the dynamics series due to iteration repeatability under the classical algorithm of the multilayer perceptron, the average value of the MAPE error was 237.2%, and the average MSE error value was 0.96. The results of multi-step prediction of a number of Enon dynamics are shown in fig. 8. For the best visibility of the calculated data in fig. 8 only the first 20 predictions are shown.

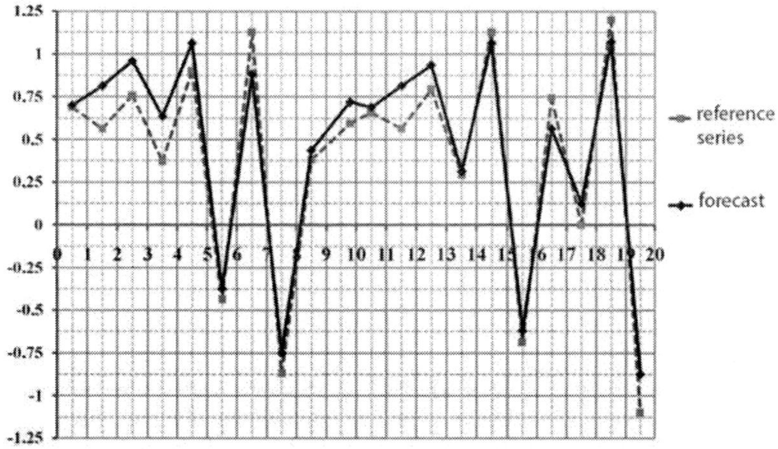

Figure 7. Results of a one-step prediction of a number of dynamics of Enon using the classical algorithm of a multilayer perceptron.

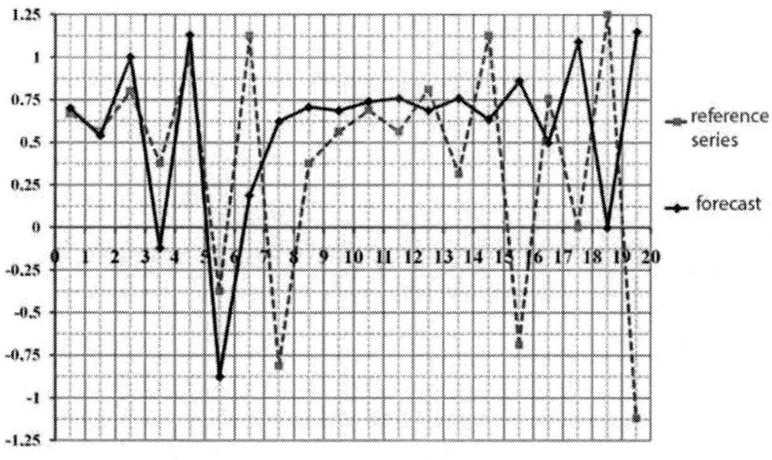

Figure 8. Results of multi-step prediction of a number of dynamics of Enon using the classical algorithm of a multilayer perceptron.

The results obtained for the short-term one-step and long-term multi-step prediction of the evolution of the dynamics series indicate the possibility of using the traditional classical multi-layer perceptron method for short-term one-step prediction of the evolution of a number of dynamics and the impossibility of using it for long-term forecasting, since by the fifth step of the iteration the MARE error was 125.9%.

The proposed modified forecasting method is implemented using the network architecture 30-13-2 and using the sigmoid function of nonlinear transformations in the output and hidden layers. In fig. 9 presents the results of a one-step short-term forecast. The average value of the MARE error was 37.1%, and the average value of the MSE error was 0.086. For the best visibility of the calculated data, only the first 20 values obtained as a result of the prediction are shown.

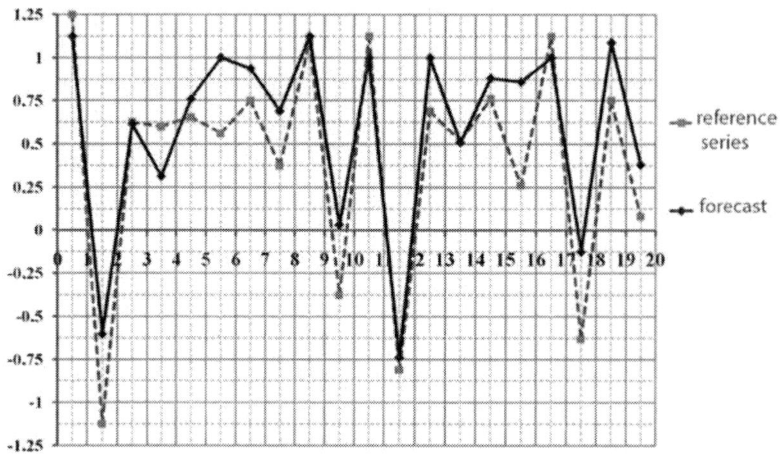

Figure 9. Results of a one-step short-term prediction of a number of dynamics of Enon using a modified algorithm of a multilayer perceptron.

As can be seen from comparing the values of the MARE error criterion for the classical and modified methods, the neural network performs a prediction with the modified algorithm more accurately than with the help of the classical algorithm. The MAPE error is 38.4%. Further, a multi-step long-term iterative forecasting study was conducted, since the main goal of the proposed modified training is to minimize the error of the iteration. The experimental results for the first 20 values are shown in fig. 10. The average value of the MAPE error was 47.7%, and the average MSE error value was 1.159. As can be seen from fig. 10, the forecast accuracy for 20 values is small, but there is no accumulation of the error value as a whole. That is, the received error value is kept constant throughout the entire duration of the prediction process. It does not increase.

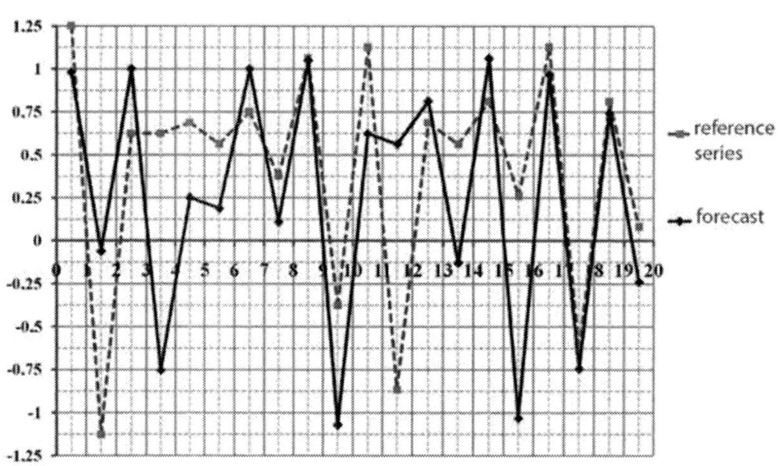

Figure 10. Results of multi-step prediction of a number of dynamics of Enon using a modified algorithm of a multilayer perceptron.

In order to confirm this conclusion, fig. 11 shows calculated and reference data for 120 consecutive iterations. Thus, an estimation of the prediction error by 120 steps forward is made. Based on the data presented, it can be concluded that the quality of the iterative training of the neural network is quite high. In addition to this statement, the average value of the MARE error, which is 47.6%, is indicative.

Therefore, we can assume that the modified algorithm with multi-step prediction shows much better results than the classical algorithm of the multilayer perceptron. For clarity of comparison, the results of the study are shown in table 1 and 2. Table 1 shows the values of the MARE error criterion, and in table 2 – the values of the MSE error criterion.

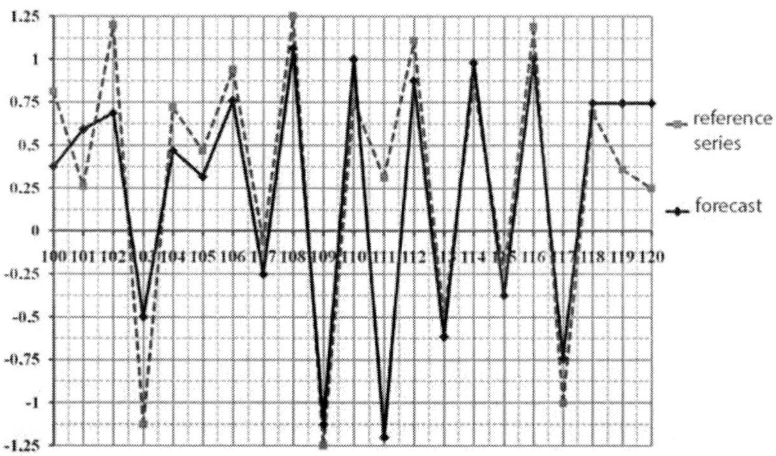

Figure 11. Estimation of the prediction error at 120 consecutive iterations of the modified multilayer perceptron algorithm.

Table 1. Values of the MARE error

Type of algorithm	Method of forecasting	
	one-step short-term forecast	multi-step long-term forecast
Multilayer Perceptron	38.4%	237.2%
Modified algorithm	37.1%	47.7%

Table 2. MSE error values

Type of algorithm	Method of forecasting	
	one-step short-term forecast	multi-step long-term forecast
Multilayer Perceptron	0.033	0.96
Modified algorithm	0.086	1.159

Verification of the operability and quality level of the proposed modified algorithm was carried out on the real experimental data proposed in [25]. For this purpose, a perceptron with architecture 28-13-1 with sigmoid function of nonlinear transformations, activating neurons in the output and hidden layers was considered. As training data, the first 400 elements of the proposed series of dynamics were taken, which was 60% of all available elements. In order to assess the quality of the forecast, the remaining 40% of the series was used. With a one-step short-term forecast, the average MAP error obtained was 63.4%, and the average MSE error was 296.823. The results of the forecast for the first 20 values are shown in fig. 12.

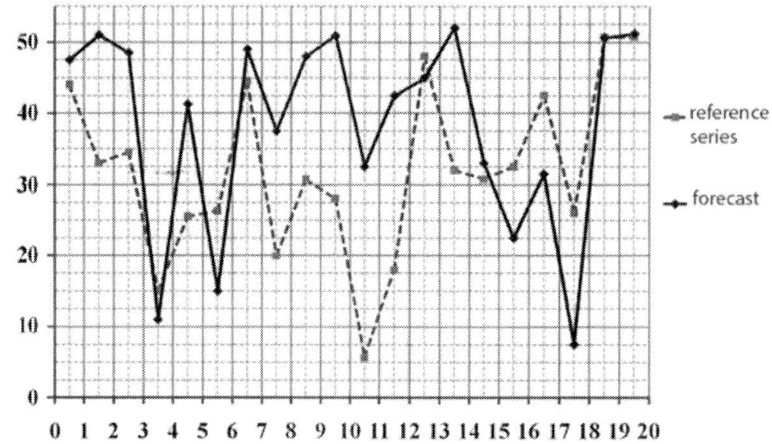

Figure 12. The results of a one-step prediction of a number of dynamics on the basis of real data using the classical algorithm of a multilayer perceptron.

Comparison of the shape of the graphs presented in fig. 12, as well as the average value of the MSE error, indicates that this architecture can not be applied to such forecasts due to a sufficiently large amount of error occurring at the first step of the prediction. Next, we will consider the application of the proposed modified multilayer perceptron algorithm for a one-step short-term prediction based on the 30-13-2 network architecture. Based on the results of the prediction of the evolution of a number of dynamics on the basis of real data, presented for 20 values in fig. 13, the average value of the MAPE error was 31.8%, and the average value of the MSE error was 72.04.

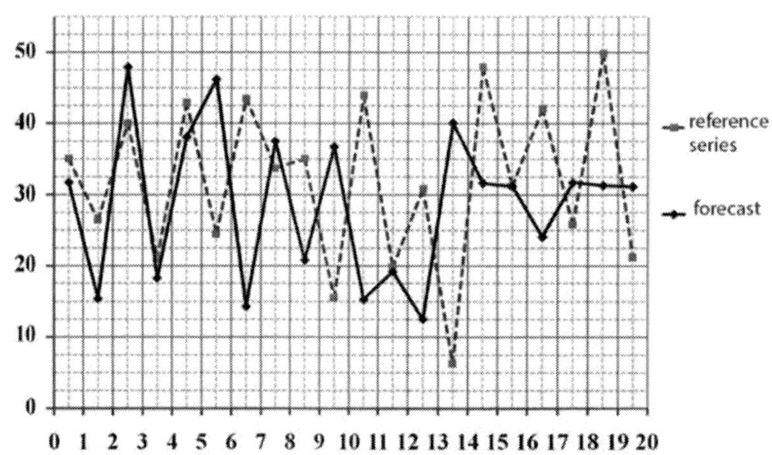

Figure 13. Results of a one-step short-term forecasting of a number of dynamics on the basis of real data using the modified algorithm of a multilayer perceptron.

The use of the modified forecasting algorithm gives better results, in favor of which low values of the MARE error criterion testify. Based on this, it can be concluded that this neural network architecture can be used for short-term one-step forecasting of the dynamics series.

The results of multi-step prediction of a number of dynamics using the modified multilayer perceptron algorithm for the first 20 values are shown in fig. 14. The average value of the MARE error criterion for this algorithm was 27.8%, and the mean value of the MSE error criterion was 34.02. Comparing these results with those for the classical algorithm (the average value of the MAPE error is 68.953%, the average value of the MSE error is 332.197), one can conclude that the proposed modified algorithm for predicting the evolution of the dynamics series in comparison with the classical algorithm is undeniable.

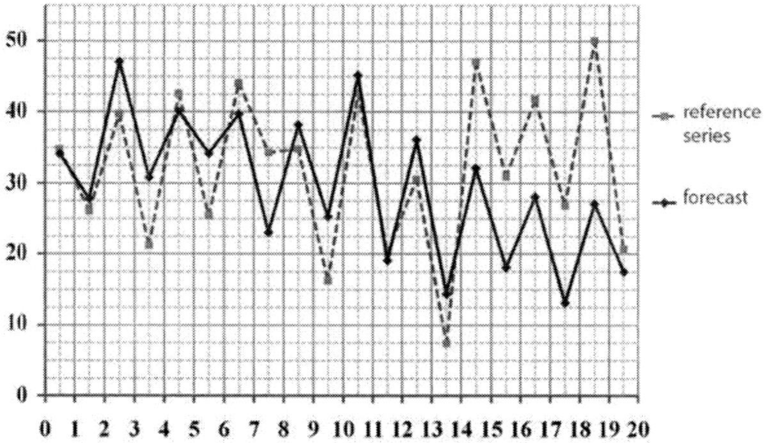

Figure 14. The results of multi-step prediction of a number of dynamics on the basis of real data using the modified algorithm of a multilayer perceptron.

5. Conclusion

The paper proposed a unifying approach to forecasting the series of complex system dynamics, which includes a method for training neural networks, a method for describing and presenting input data, based on approaches to forecasting based on a multilayer perceptron. In the course of the research it was shown that the modified neural network algorithm is able to compete with existing analogs presented in [18-26]. It is flexible enough compared to existing analogs [24-26] and has the ability to take into account the nonlinearity of a complex chaotic system. As inputs to the construction of a neural network, it is suggested to use the values of a number of dynamics at the extremum points, as well as the corresponding time values obtained by computation using the sliding window method. This algorithm can be used both as an independent approach to predicting the series of dynamics, and as part of the forecasting system. The efficiency of predicting the evolution of the dynamics series for a one-step and multi-step forecast based on the classical multilayer perceptron method and on the basis of the modified algorithm considered is compared. For comparison, synthetic chaotic data were used, as well as real data obtained from experiments [10, 14, 22].

Thus, the study showed that the modified algorithm of the multilayer perceptron is a powerful tool for predicting the evolution of chaotic series of the dynamics of complex systems. The result of this modification is to minimize the criterion of the iterative error that occurs when the previously predicted data is applied to the inputs of the neural network. A positive result is an increase in the accuracy of the iterative prediction of the neural network.

References

[1] Shah S, Jani H and Shetty S 2013 *International Journal of Computer Applications* **5** pp 17-23
[2] Tatyankin V M 2014 Algorithm for forming the optimal architecture of a multilayer neural network *A new word in science: development prospects: materials III international. scientific-practical Conf.* pp 187-188
[3] Golovko V A, Kroshchanka A, Rubanau U and Jankowski S 2014 Learning Technique for Deep Belief Neural Networks *Neural Networks and Artifical Intelligence* (Moscow: Communication in Computer and Information Science) pp 136-146

[4] Scholz M, Fraunholz M and Selbig J 2008 Nonlinear principal component analysis: neural network models and applications *Principal manifolds for data visualization and dimension reduction*. (Berlin: Springer Berlin Heidelberg) pp 44-67
[5] Erhan D, Bengio Y, Courville A, Manzagol P-A, Vincent P and Bengio S 2010 Why does unsupervised pre-training help deep learning? *Journal of Machine Learning Research* **11** pp 625-660
[6] Hinton G, Osindero S and Teh Y 2006 A fast learning algorithm for deep belief nets *Neural Computation* **18** pp 1527-1554
[7] Bengio Y, Lamblin P, Popovici D and Larochelle H 2007 Greedy layer-wise traning of deep networks *Advances in neural information processing systems* (MA: MIT Press, Cambridge) pp 153-160
[8] Hington G 2002 Training products of experts by minimizing contrastive divergence *Neural Computation* **14** pp 1771-1800
[9] Bengio Y 2009 Learning deep architectures for AI *Foundations and Trends in Machine Learning* **2** pp 1-127
[10] Barkalov S A, Dushkin A V, Kolodyazhny S A, Novoseltsev V I and Sumin V I 2017 *Introduction to the System Analysis* (Moscow: Hot Line –Telecom) p 234
[11] Dushkin A V, Kravchenko A S, Novoseltsev V I, Smolentseva T E and Sumin V I 2014 *Mathematical Models and Information Processes of Management of a Difficult Object* (Voronezh: Scientific book) p 125
[12] Mansourabad A, Beheshti M and Simab M 2013 A Hybrid PSO-Fuzzy-PID Controller for Gas Turbine Speed *International Journal of Control and Automation* **1** pp 48-56
[13] Fixations S, Fixations H and Riche N 2013 State-of-the-Art and Study of Comparison Metrics *IEEE International Conference on Computer Vision (ICCV): proceedings* (Sydney: IEEE) pp 1153-1160
[14] LeCun Y and Bengio Y 2015 *Deep Learning* (Hinton) **521** pp 436-444
[15] Golovko V, Kroishchanka A, Turchenko V, Stanislaw J and Treadwell D 2015 A New Technique for RESTRICTED Boltzmann Machine Learning *Proceedings of the 8 IEEE International Conference IDAACS-2015* (Warsaw) pp 182-186
[16] Jankowski S, Szymanski Z, Dziomin V, Golovko V and Barcz A 2016 Deep learning classifier based on NPCA and orthogonal feature selection *Proceedings of the International Conference on Photonics Applications in Astronomy, Communications, Industry, and High-Energy Physics Experiments* pp 5-9
[17] Borji A 2013 Analysis of scores, datasets and models in visual saliency prediction *IEEE International Conference on Computer Vision (ICCV): proceedings* (Sydney: IEEE) pp 921-928
[18] Ramik D M 2012 *Contribution to complex visual information processing and autonomous knowledge extraction: application to autonomous robotics*: Ph D thesis (Paris: Universite Paris-Est) p 176
[19] Hinton G 2010 *A practical guide to training restricted Boltzmann mashines* (Toronto: Machine Learing Group, University of Toronto) p 123
[20] Kachurka V 2015 Visual Saliency Based Approach to Object Detection in Computer Vision Systems: Real Life Applications *Intelligent Data Acquisition and Advanced Computing Systems (IDAACS): proceedings* (Warsaw: IEEE) **1** p 239-245
[21] Hinton G and Salakhutdinov R 2006 Reducing the dimensionality of data with neural networks *Science* 313 (5786) pp 504-507
[22] Golovko V 2016 Deep Neural Networks: A theory, application and new trends *Proccedings of the 13-th International Conference on Pattern recognition and Information Processing* (Minsk: publishing Center of BSU) pp 33-37
[23] Kachurka V 2014 A Statistical Approach to Human-Like Visual and Saliency Detection for Robot Vision: Application to Wildland Fires Detection *Communications in Computer and*

Information Science vol. 440: Neural Networks and Artificial Intelligence (Springer International Publishing) pp 124-135

[24] Tatyankin V M 2014 Modified algorithm for back propagation of error *Priority directions of development of science and education: materials III international scientific-practical Conf.* (Cheboksary: CNS Interactive Plus) pp 124-131

[25] Golovko V, Artsiomenka S, Kisten V and Evstigneev V 2015 Towards Automatic Epileptic Seizure Detection in EEGS based on Neural Networks and Largest Lyapunov Exponent *International Journal of Computing* **14** pp 36-47

[26] Dziomin U A, Kabysh A, Stetter R and Golovko V 2015 Multi-Agent Reinforcement Learning Approach for the Efficient Control of Mobile Robots *Advanced in Robotics and Collaborative Automation* (River Publishers) pp 123-146

IOP Conf. Series: Journal of Physics: Conf. Series **973** (2018) 012036 doi:10.1088/1742-6596/973/1/012036

The method of similar operators in the study of the spectra of the adjacency matrices of graphs

Serge Kozlukov
Department of Nonlinear Oscillations
Voronezh State University, 1 Universitetskaya Ploshad', Voronezh, RU 394036
E-mail: newkozlukov@gmail.com

Abstract. The method of similar operators [1, 2, 3] is used to investigate spectral properties of a certain class of matrices in the context of graphs [4, 5]. Specifically, we consider the adjacency matrix of an "almost-complete graph". Then we generalize the result to allow the matrices obtained as combinations of the Kronecker products [6, 7] and the small-norm perturbations. We derive the estimates of the spectra and the eigenvectors of such matrices.

1. Introduction

Consider a directed graph on N vertices. It is uniquely defined by its adjacency matrix $A = (a_{ij})$ of the size $N \times N$, in which a_{ij} is the number of edges from the vertex i to the vertex j. The *simple spectrum* of a graph is defined as the spectrum of its adjacency matrix. Different spectra of a graph are defined in terms of different matrices corresponding to this graph. It is customary to consider combinations of the matrices A, D (the out-degree matrix), E (the identity matrix) and J_N (the all-ones matrix). These matrices arise naturally in some stochastic models [8, p. 184]. Spectral properties of these matrices often play a vital role in such models. For instance, the Markovian random walk on a graph yields the notion of eigenvector centrality in a network [9, 10]. The score of a vertex i is defined as the i'th coordinate of a dominating (left) eigenvector of the transition matrix of such a process. This eigenvector defines the only stationary distribution of this random walk. The PageRank [9] algorithm originally used by Google to compute the eigenvector centrality relies on the Power-Method. Its speed of convergence depends on the ratio of the two largest absolute eigenvalues. Stability of a stationary distribution is determined [11] by the condition number which is bounded from below by the spectral gap — the distance between the two largest eigenvalues of the transition matrix of a process. The method of estimation of almost-invariant sets proposed in [12] also relies on spectral decompositions of such matrices. In the Susceptible-Infective-Susceptible model a viral spread in a network is modeled [13, 14] as a Markov process with 2^N states. An asymptotic (endemic or epidemic) behaviour of such a system is determined by the spectral radius (the largest absolute eigenvalue) of the adjacency matrix and the rates of curage and infection.

We should also note that the Kronecker products of the adjacency matrices are themselves of interest as they correspond to the adjacency matrices of the *non-complete extended p-sums* (NEPS) of graphs [8, p. 44].

For more details and comprehensive description of the graph spectra theory and its applications refer to [5, 8, 15].

2. The method of similar operators

In some applications it is infeasible to compute the exact eigenvalues. Then one needs to make reasonable estimates. We use the abstract method of similar operators to estimate the eigenvalues and eigenvectors for a certain class of matrices. This method originates from Friedrichs [16] and was later developed in abstract setting by Baskakov [1, 2, 3]. It relies on contraction mappings in Banach spaces and the Banach fixed-point theorem. This approach is often superior to usual methods of perturbation analysis that use series expansions. We are only concerned with finite-dimensional problems so we will only state the required notation and theorems in a simplified form.

Let $\mathbb{K} \in \{\mathbb{R}, \mathbb{C}\}$ be a field of either real or complex numbers. We consider the vector space $\mathbb{K}^n$, $n \in \mathbb{N}$ supplied with Euclidean structure:

$$(x, y)=\sum_{k=1}^{n} x_k \overline{y_k},\ x=(x_1, \dots, x_n),\ y = (y_1, \dots, y_n) \in \mathbb{K}^n$$

and the L_2-norm: $\|x\|_2^2=(x, x)$. We also consider the canonical basis $e_1, \dots, e_n$ in $\mathbb{K}^n$ given by $(e_i)_j = \delta_{ij}$, $i, j = \overline{1, n}$ (δ_{ij} is the Kronecker symbol). When V_1, V_2 are normed vector spaces we denote by $L(V_1, V_2)$ the space of bounded linear mappings from V_1 to V_2. An algebra of bounded linear endomorphisms from a Banach space V into itself is denoted by $L(V) = L(V, V)$. It is a Banach algebra with the operator norm:

$$\|A\|_{\text{op}} = \sup_{\substack{\|x\|=1, \\ x \in V}} \|Ax\|,\ A \in L(V).$$

Together with $L(\mathbb{K}^n, \mathbb{K}^m)$ we consider its isomorphic space $\mathbb{K}^{m\times n}$ of matrices of the size $m\times n$ with entries from the field $\mathbb{K}$. The space $\mathbb{K}^{n\times n} \sim L(\mathbb{K}^n)$ forms a Banach algebra when supplied with a submultiplicative norm $\|\cdot\|$, e.g.: $\|A\|_{\text{op}} = \sup_{\|x\|_2=1,\ x\in\mathbb{K}^n} \|Ax\|_2$, $\|A\|_{\text{F}} = \sqrt{\sum_{i,j} |a_{ij}|^2}$, for $A=(a_{ij}) \in \mathbb{K}^{n\times n}$. Finally we will also be dealing with the isomorphic spaces $L(L(\mathbb{K}^n))$ and $L(\mathbb{K}^{n\times n})$ with the operator norm. We will follow Krein and refer to elements of $L(\mathbb{K}^{n\times n})$ as "transformers".

The spectrum of a matrix A (the set of its eigenvalues) will be denoted as $\sigma(A)$. We call two matrices A_1, A_2 *similar* if there is an invertible matrix U (the similarity matrix) such that $A_1 U = U A_2$. Similar matrices share some spectral properties: they are isospectral ($\sigma(A_1) = \sigma(A_2)$) and U maps the eigenvectors of one to another's: $A_2 x = \lambda x \implies A_1 U x = \lambda U x$.

The most important notion in the abstract method of similar operators is that of an *admissible triple*. For our specific purposes it suffices to say that $(\mathbb{K}^{n\times n}, J, \Gamma)$ forms an *admissible triple* for a matrix $A \in \mathbb{K}^{n\times n}$ if the following conditions are met:

- $J, \Gamma \in L(\mathbb{K}^{n\times n})$ are transformers;
- J is a projection ($J^2 = J$);
- Γ satisfies the equations:

$$A\Gamma X - (\Gamma X)A = X - JX,$$

$$J\Gamma X = 0,\ X \in \mathbb{K}^{n\times n}.$$

The main theorem of the method may be now formulated as follows:

Theorem 1. *Consider a matrix $A - B$ with $A, B \in \mathbb{K}^{n\times n}$. Suppose $(\mathbb{K}^{n\times n}, J, \Gamma)$ is an admissible triple for the matrix A and suppose the following inequality holds:*

$$\|B\|\|\Gamma\|_{\text{op}} \leq \frac{1}{4}.$$

Then there exists such a matrix $X^o \in \mathbb{K}^{n\times n}$ that $A - B$ is similar to $A - JX^o$; the similarity matrix is $E + \Gamma X^o$; the following estimates are valid:

$$\|X^o - B\| \leq 3\,\|B\|\,,$$

$$\mathrm{spr}(X^o) \leq \|X^o\| \leq 4\,\|B\|\,,$$

where $\mathrm{spr}(X^o)$ is the spectral radius of X^o (the largest absolute eigenvalue). Such X^o can be found as the limit of a convergent sequence $\left(\Phi^k(0);\ k \in \mathbb{N}\right)$ in a Banach algebra $\mathbb{K}^{n\times n}$. Here Φ is a nonlinear contraction mapping defined on the ball $\{X \in \mathbb{K}^{m\times n};\ \|X - B\| \leq 3\|B\|\}$ and given by

$$\Phi(X) = B\Gamma X - (\Gamma X)J(B + B\Gamma X) + B$$

and $\Phi^k = \underbrace{\Phi \circ \cdots \circ \Phi}_{k\ copies}$ denotes the composition.

3. Results and discussion

Almost-complete graph example.

Consider a digraph defined by the following adjacency matrix:

$$A = J_N - B = \begin{pmatrix} 1 & \cdots & 1 \\ \vdots & \ddots & \vdots \\ 1 & \cdots & 1 \end{pmatrix} - B,$$

Here J_N is the all-ones matrix. The unity on the intersection of i'th row and j'th column of B corresponds to an edge from i to j being absent in the graph. This example has already been considered in [17] and here we will reproduce some steps to demonstrate the technique.

One can easily find the minimal annihilating polynomial of J_N to be $\lambda(\lambda - N)$. This comes from the fact that $J_N^2 = NJ_N$. Consequently the spectrum of J_N is

$$\sigma(J_N) = \{0, N\}\,.$$

The only non-zero eigenvalue N has the corresponding eigenvector

$$h_N = \frac{1}{\sqrt{N}}\,(1, \ldots, 1) \in \mathbb{R}^N.$$

The null-space of J_N is orthogonal to h_N and allows the orthonormal basis:

$$h_k = \frac{1}{\sqrt{k(k+1)}}\left(\underbrace{1, \ldots, 1}_{k\text{ copies}}, -k, 0, \ldots, 0\right),\quad k = \overline{1, N-1}.$$

We conclude that the adjacency matrix $A = J_N - B$ of this graph is similar to $\mathcal{A} - \mathcal{B}$ where $\mathcal{A}$ is a block matrix (subscripts denote block sizes; in what follows throughout this subsection block sizes are the same and will be omitted):

$$\mathcal{A} = \left(\begin{array}{c|c} N & \mathbf{0}_{1\times(N-1)} \\ \hline \mathbf{0}_{(N-1)\times 1} & \mathbf{0}_{(N-1)\times(N-1)} \end{array}\right) \in \mathbb{R}^{N\times N}$$

and $\mathcal{B}$ is obtained with a similarity transform: $\mathcal{B} = U^{-1}BU \in \mathbb{R}^{N\times N}$. The similarity matrix U is given by stacking the eigenvectors in columns:

$$U = \mathrm{columns}(h_N, h_1, \dots, h_{N-1}) = \begin{pmatrix} | & | & & | \\ h_N & h_1 & \dots & h_{N-1} \\ | & | & & | \end{pmatrix}.$$

Following the general method, we should first construct an admissible triple. Since $\mathcal{A}$ is block-diagonal it is natural to set J with the formula

$$JX = \left(\begin{array}{c|c} x_{11} & \mathbf{0} \\ \hline \mathbf{0} & X_{22} \end{array}\right)$$

for all block matrices

$$X = \left(\begin{array}{c|c} x_{11} & X_{12} \\ \hline X_{21} & X_{22} \end{array}\right) \in \mathbb{K}^{N\times N}.$$

Then $A - JX$ is block-diagonal for any X and its spectrum is the union of its diagonal blocks' spectra: $\sigma(A - JX) = \{N, \sigma(-X_{22})\}$.

Now we can find the corresponding Γ. Suppose it is defined by the formula

$$\Gamma X = \begin{pmatrix} \Gamma_{11}(X) & \Gamma_{12}(X) \\ \Gamma_{21}(X) & \Gamma_{22}(X) \end{pmatrix}.$$

Then the equations

$$\mathcal{A}\Gamma X - (\Gamma X)\mathcal{A} = N\begin{pmatrix} 0 & \Gamma_{12}(X) \\ -\Gamma_{21}(X) & 0 \end{pmatrix} = X - JX,$$

and $J\Gamma X = 0$ yield the result

$$\Gamma X = \frac{1}{N}\begin{pmatrix} 0 & X_{12} \\ -X_{21} & 0 \end{pmatrix}, \quad X \in \mathbb{K}^{N\times N}.$$

The last step is to estimate the norms and apply the theorem. One can easily check that $\|\Gamma\|_{\mathrm{op}} = \frac{1}{N}$. It is also apparent that $\|\mathcal{B}\|_{\mathrm{op}} = \|B\|_{\mathrm{op}} \leq \|B\|_{\mathrm{F}}$ since multiplication by the orthogonal matrix U is an isometry in Euclidean space. The Frobenius norm $\|B\|_{\mathrm{F}} = \sqrt{\sum_{ij} b_{ij}^2} = M$ of B reduces to the square root of the number of absent edges. This directly implies the following

Theorem 2. *Suppose the number of absent edges is*

$$M < \frac{1}{16}N^2.$$

Then the spectrum of the adjacency matrix $A = J_N - B$ *can be represented as disjoint union*

$$\sigma(A) = \{N - x_{11}^o\} \cup \sigma_2.$$

The dominating eigenvector of A *is*

$$\hat{h}_N = U(E + \Gamma X^o)e_1 = h_N - \frac{1}{N}(X_{21,(1)}^o h_1 + \dots + X_{21,(N-1)}^o h_{N-1}),$$

where $X_{21,(i)}^o$, $i = \overline{1, N-1}$ *are the coordinates of the vector* X_{21}^o. *Moreover* $\hat{h}_N \in \mathbb{R}^N$, $x_{11}^o \in \mathbb{R}$ *and* $\sigma_2 \subset \mathbb{C}$ *satisfy the following inequalities:*

$$\|\hat{h}_N - h_N\|_2 \leq 4\frac{\sqrt{M}}{N},$$

$$|x_{11}^o|, \ \sup_{\lambda\in\sigma_2}|\lambda| \leq 4\sqrt{M}.$$

IOP Conf. Series: Journal of Physics: Conf. Series **973** (2018) 012036 doi:10.1088/1742-6596/973/1/012036

A-tiled matrix example.
Now let $A \in \mathbb{K}^{M\times M}$ and consider the following ("A-tiled") block-matrix

$$\mathbb{A} = \begin{pmatrix} A & \cdots & A \\ \vdots & \ddots & \vdots \\ A & \cdots & A \end{pmatrix} \in \mathbb{K}^{MN\times MN}$$

and the perturbed matrix

$$\mathbb{A} - \mathbb{B},\ \mathbb{B} \in \mathbb{K}^{MN\times MN}.$$

Lemma 1. *Suppose A is an invertible self-adjoint matrix. Then it has M orthonormal eigenvectors $h_1, \ldots, h_M$ ($\|h_i\|_2 = 1,\ i=\overline{1,M}$) with the corresponding eigenvalues $\lambda_1, \ldots, \lambda_M \neq 0$. The spectrum of $\mathbb{A}$ is*

$$\sigma(\mathbb{A}) = \{0\} \cup N\sigma(A) = \{0\} \cup \{N\lambda;\ \lambda \in \sigma(A)\}.$$

Non-zero eigenvalues of $\mathbb{A}$ have corresponding block eigenvectors:

$$f_j = \frac{1}{\sqrt{N}}(h_j, \ldots, h_j) \in \mathbb{K}^{MN},\ j = \overline{1, M}.$$

The null-space of $\mathbb{A}$ has an orthonormal basis:

$$f_{j,k} = \frac{1}{\sqrt{k(k+1)}}(\underbrace{e_j, \ldots, e_j}_{k\ copies}, -ke_j, 0, \ldots, 0) \in \mathbb{K}^{MN\times MN}$$

The matrix $\mathbb{A}$ is similar to a block-diagonal matrix:

$$\mathcal{A} = \left(\begin{array}{c|c} \mathrm{diag}(N\lambda_1, \ldots, N\lambda_M) & \mathbf{0} \\ \hline \mathbf{0} & \mathbf{0} \end{array}\right) \in \mathbb{K}^{MN\times MN};$$

the similarity transform matrix is

$$U = \text{columns}\left(f_1, \ldots, f_M, f_{1,1}, \ldots, f_{1,N-1}, \ldots, f_{M,N-1}\right).$$

Throughout this subsection we will consider block matrices of the size $MN\times MN$ in the form

$$X = \left(\begin{array}{ccc|c} x_{11} & \cdots & x_{1M} & x_{1,M+1} \\ \vdots & \ddots & \vdots & \vdots \\ x_{M1} & \cdots & x_{MM} & x_{M,M+1} \\ \hline x_{M+1,1} & \cdots & x_{M+1,M} & X_{M+1,M+1} \end{array}\right),$$

where $X_{ij}{=}x_{ij}$, $X_{M+1,j}{=}x_{M+1,j}$, $X_{i,M+1}{=}x_{i,M+1} \in \mathbb{K}$ for $1 \leq i, j \leq M$, and $X_{M+1,M+1}$ is a block of the size $M(N{-}1){\times}M(N-1)$.

Just like before we are going to investigate the spectral behaviour of $\mathbb{A}$ under perturbations. We begin with constructing an admissible triple. A natural choice for J in this case is

$$JX = \left(\begin{array}{ccc|c} x_{11} & & 0 & 0 \\ & \ddots & & \vdots \\ 0 & & x_{MM} & 0 \\ \hline 0 & \cdots & 0 & X_{M+1,M+1} \end{array}\right).$$

Lemma 2. *Suppose A has a simple spectrum, i.e. its eigenvalues are pairwise distinct: $\lambda_i \neq \lambda_j$ for all $1 \leq i{\neq}j \leq M$.*

Then a tuple $(\mathbb{K}^{MN\times MN}, J, \Gamma)$ forms an admissible triple if we define

$$\Gamma X = \frac{1}{n}\left(\begin{array}{cccc|c} 0 & \gamma_{12}x_{12} & \cdots & \gamma_{1M}x_{1M} & \gamma_{1,M+1}x_{1,M+1} \\ \gamma_{21}x_{21} & 0 & \cdots & \gamma_{2M}x_{2M} & \gamma_{2,M+1}x_{2,M+1} \\ \vdots & \vdots & \ddots & \vdots & \vdots \\ \gamma_{M1}x_{M1} & \gamma_{M2}x_{M2} & \cdots & 0 & \gamma_{M,M+1}x_{M,M+1} \\ \hline \gamma_{M+1,1}x_{M+1,1} & \gamma_{M+1,2}x_{M+1,2} & \cdots & \gamma_{M+1,M}x_{M+1,M} & \mathbf{0} \end{array}\right),$$

$$\gamma_{ij} = \begin{cases} \dfrac{1}{\lambda_i - \lambda_j}, & 1 \leq i{\neq}j \leq M{+}1, \\ 0, & i = j \end{cases}$$

and we use the convention:

$$\lambda_{M+1} = 0.$$

The operator norm of Γ is:

$$\|\Gamma\|_{\mathrm{op}} = \frac{1}{N}\frac{1}{\min\limits_{1\leq i\neq j\leq M+1}|\lambda_i - \lambda_j|} =$$

$$= \frac{1}{N}\max\left\{\frac{1}{\min\limits_{1\leq i\neq j\leq M}|\lambda_i - \lambda_j|}, \frac{1}{\min\limits_{1\leq j\leq M}|\lambda_j|}\right\}$$

Theorem 3. *Suppose A has simple spectrum and the following inequality holds:*

$$\|\mathbb{B}\|_{\mathrm{op}} \leq \frac{N}{4}\min\left\{\min_{1\leq i\neq j\leq M}|\lambda_i - \lambda_j|, \min_{1\leq j\leq M}|\lambda_j|\right\}.$$

Then the spectrum of disturbed matrix $\mathbb{A} - \mathbb{B}$ is

$$\sigma(\mathbb{A}) = \{N\lambda_1 - x^o_{11}, \ldots, N\lambda_M - x^o_{MM}\} \cup \sigma_{M+1}.$$

The eigenvectors $\hat{f}_j$, $\hat{f}_{j,k}$, $j{=}\overline{1,M}$, $k{=}\overline{1,N{-}1}$ of the matrix $\mathbb{A}{-}\mathbb{B}$, the values x^o_{jj}, $j{=}\overline{1,M}$ and the set σ_{M+1} are in the following bounds:

$$|x^o_{jj}|, \max_{\lambda\in\sigma_{M+1}}|\lambda| \leq 4\|B\|,$$

$$\left\|\hat{f}_j - f_j\right\|_2, \left\|\hat{f}_{j,k} - f_{j,k}\right\|_2 \leq \frac{4}{N}\|B\|\max\left\{\frac{1}{\min\limits_{1\leq l\neq p\leq M}|\lambda_l - \lambda_p|}, \frac{1}{\min\limits_{1\leq l\leq M}|\lambda_l|}\right\}$$

for all $j{=}\overline{1,M}, k{=}\overline{1,N-1}$.

Kronecker products
Now we consider Kronecker product

$$A \otimes B = \begin{pmatrix} a_{11}B & \cdots & a_{1N}B \\ \vdots & \ddots & \vdots \\ a_{N1}B & \cdots & a_{NN}B \end{pmatrix} \in \mathbb{K}^{MN \times MN}$$

of squared matrices $A = (a_{ij}) \in \mathbb{K}^{N \times N}$, $B = (b_{ij}) \in \mathbb{K}^{M \times M}$. We will analyze its spectral properties under small-norm perturbations:

$$A \otimes B - F. \tag{1}$$

Kronecker product has several appealing properties [6].

- It is associative:
$$A \otimes (B \otimes C) = (A \otimes B) \otimes C.$$
- It is distributive with respect to addition:
$$(A + B) \otimes (C + D) = A \otimes C + A \otimes D + B \otimes C + B \otimes D.$$
- The Kronecker product of matrix products has the following property:
$$(AB) \otimes (CD) = (A \otimes C)(B \otimes D)$$
whenever products AB and CD make sense.
- The trace of $A \otimes B$ is
$$\operatorname{tr}(A \otimes B) = \operatorname{tr} A \operatorname{tr} B.$$
- If A and B are both symmetric matrices, then $A \otimes B$ is symmetric as well.

Note also that the "tiled" matrix from the last example can be represented as a Kronecker product:

$$\mathbb{A} = \begin{pmatrix} A & \cdots & A \\ \vdots & \ddots & \vdots \\ A & \cdots & A \end{pmatrix} = J_N \otimes A.$$

Lemma 3. *Suppose A and B have simple structure, i.e. A has N eigenvectors $f_1, \ldots, f_N$ whose corresponding eigenvalues are $\mu_1, \ldots, \mu_N$ and B has eigenvectors $h_1, \ldots, h_M$ with the eigenvalues $\lambda_1, \ldots, \lambda_M$. Then $A \otimes B$ also has simple structure; it has MN independent eigenvectors $f_i \otimes h_j$, $i = \overline{1, N}, j = \overline{1, M}$ and the corresponding eigenvalues are $\mu_i \lambda_j$.*

Now suppose that among these pairwise products $\mu_i \lambda_j$ there are only s distinct values $\nu_1, \ldots, \nu_s$. To each eigenvalue ν_k $(k = \overline{1, s})$ there corresponds an eigenspace

$$E_k = \operatorname{span}(f_i \otimes h_j;\ \mu_i \lambda_j = v_k,\ i = \overline{1, N},\ j = \overline{1, M}) \subset \mathbb{K}^{MN}.$$

These eigenspaces form a direct-sum decomposition of $\mathbb{K}^{MN}$:

$$\mathbb{K}^{MN} = E_1 \oplus \cdots \oplus E_s.$$

Any vector $x \in \mathbb{K}^{MN}$ can be uniquely represented in the form

$$x = x_1 + \cdots + x_s,\ x_k \in E_k,\ k = \overline{1, s}. \tag{2}$$

This direct-sum decomposition of $\mathbb{K}^{MN}$ corresponds to a decomposition of the identity matrix $E \in \mathbb{K}^{MN\times MN}$ (which defines an identity operator) into a sum of matrices of the spectral projections:

$$E = P_1 + \cdots + P_s.$$

The spectral projection $\mathcal{P}_k$ ($k{=}\overline{1,s}$) is given by the formula

$$\mathcal{P}_k x = x_k \in E_k \subset \mathbb{K}^{MN}$$

with respect to the decomposition (2) of x.

For any matrix $X \in \mathbb{K}^{MN\times MN}$ the following trivial equality holds:

$$X = \sum_{i,j=1}^{s} P_i X P_j.$$

The matrix $A \otimes B$ can be decomposed into

$$\mathcal{A} = \sum v_j P_j.$$

Now we should be able to reproduce the same steps as before to retrieve the estimates.

The natural way to define J is as follows:

$$JX = \sum_{j=1}^{s} P_j X P_j.$$

The system of equations

$$\begin{cases} \mathcal{A}\Gamma X - (\Gamma X)\mathcal{A} = X - JX, \\ J\Gamma X = 0, \ X \in \mathbb{K}^{MN\times MN} \end{cases}$$

has the unique solution:

$$\Gamma X = \sum_{1\le i\ne j\le s} \frac{1}{\nu_i - \nu_j} P_i X P_j.$$

The norm of Γ is:

$$\|\Gamma\|_{\text{op}} = \gamma = \frac{1}{\min_{1\le i\ne j\le s}|\nu_i - \nu_j|}$$

Theorem 4. *Consider the perturbed matrix* (1)

$$A{\otimes}B - F.$$

Let $A \in \mathbb{K}^{N\times N}$ and $B \in \mathbb{K}^{M\times M}$ be diagonalizable matrices. Let $f_1, \ldots, f_N$ be the eigenvectors of A corresponding to the eigenvalues $\mu_1, \ldots, \mu_N$ and let $h_1, \ldots, h_M$ be the eigenvectors of B corresponding to the eigenvalues $\lambda_1, \ldots, \lambda_M$. The spectrum of their Kronecker product $A{\otimes}B$ is composed of all the possible pairwise products $\mu_i\lambda_j$ and the corresponding eigenvectors are $f_i \otimes h_j$. Suppose that out of these MN eigenvalues only s are distinct: $\nu_1, \ldots \nu_s$.

Suppose

$$\|F\| \le \frac{1}{4}\gamma^{-1} = \frac{1}{4}\min_{1\le i\ne j\le s}|\nu_i - \nu_j|.$$

Then $A{\otimes}B - F$ is similar to

$$\sum_{k=1}^{s} \nu_k P_k - JX^o = \sum_{k=1}^{s}(\nu_k P_k - P_k X^o P_k)$$

for some $X^o \in \mathbb{K}^{MN\times MN}$, $\|X^o - F\| \le 3\|F\|$.

All the eigenvalues of $A\otimes B - F$ *are contained in the circles*

$$\Omega_k = \{\lambda \in \mathbb{C};\ |\lambda - \nu_k| \le 4\|F\|\},\ k=\overline{1,s}.$$

There is at least one eigenvalue in each of these circles.

Suppose the eigenvalue $\nu_k = \mu_{i_k}\lambda_{j_k}$ *of* $A\otimes B$ *has multiplicity* 1, *that is it has the only eigenvector* $v_k = f_{i_k}\otimes h_{j_k}$. *It is equivalent to the statement that the eigenvalue* μ_{i_k} *of* A *and the eigenvalue* λ_{j_k} *of* B *are both of multiplicity* 1. *Then* $A\otimes B - F$ *has eigenvalue in the circle* Ω_k *and the corresponding eigenvector* $\hat{v}_k$ *is within bounds*

$$\|\hat{v}_k - v_k\| \le 4\gamma\|F\|.$$

If ν_k *is well separated from all the other eigenvalues of* $A\otimes B$*:*

$$\min_{l\neq k}|\nu_k - \nu_l| \ge 4\|F\|,$$

then ν_k *is the only eigenvalue of* $A\otimes B - F$ *in that circle.*

For example, in the case of a "tiled" matrix

$$J_N\otimes B = \begin{pmatrix} B & \cdots & B \\ \vdots & \ddots & \vdots \\ B & \cdots & B \end{pmatrix}$$

we would have $\nu_1 = N$, $\nu_2 = 0$. Let $\lambda_1, \dots, \lambda_M$ be the eigenvalues of B. Spectrum of $J_N\otimes B$ is

$$\sigma(J_N\otimes B) = \{\mu_i\lambda_j;\ i=\overline{1,2},\ j=\overline{1,M}\} = \{0\} \cup N\sigma(B).$$

All of these eigenvalues except for 0 are of multiplicity 1 and well-separated for sufficiently large N. Then $\gamma = \frac{1}{N}$. This directly implies the theorem of the previous section.

These results might be refined with the use of the theorem on splitting an operator [18] which allows to consider each eigenvalue individually and obtain more precise estimates for the corresponding eigenvalue of the perturbed matrix.

Conclusion

We have derived bounds for perturbations of Kronecker products of matrices using the method of similar operators which is a powerful tool in perturbation theory. Developed in the context of Banach spaces this method turns out to be just as useful in finite-dimensional problems.

The problem of perturbation analysis for the Kronecker products despite being old still has a whole open field for future research. An example of more general yet quite similar problem is that of an analysis of tensor products of finite-dimensional operators (defined by matrices) and abstract operators in Banach spaces.

References

[1] Baskakov A G 1983 *Siberian Mathematical Journal* **24** 17–32
[2] Baskakov A G and Polyakov D M 2017 *Sb.: Mathematics* **208** 1
[3] Baskakov A G and Krishtal I A 2013 *Journal of Mathematical Analysis and Applications* **407** 157–178
[4] Van Dam E R and Haemers W H 2003 *Linear Algebra and its applications* **373** 241–272
[5] Cvetković D, Doob M and Sachs H 1980 Spectra of graph-theory and applications
[6] Bellman R 1997 *Introduction to matrix analysis* (SIAM)

IOP Conf. Series: Journal of Physics: Conf. Series **973** (2018) 012036 doi:10.1088/1742-6596/973/1/012036

[7] Xiang H, Diao H and Wei Y 2005 *Journal of Computational and Applied Mathematics* **183** 210 – 231 ISSN 0377-0427 URL `http://www.sciencedirect.com/science/article/pii/S0377042705000269`
[8] Cvetkovic D M, Rowlinson P and Simic S 2010 *An introduction to the theory of graph spectra* (Cambridge University Press Cambridge, UK)
[9] Page L, Brin S, Motwani R and Winograd T 1999 The pagerank citation ranking: Bringing order to the web. Technical Report 1999-66 Stanford InfoLab previous number = SIDL-WP-1999-0120 URL `http://ilpubs.stanford.edu:8090/422/`
[10] Bonacich P 1972 *Journal of Mathematical Sociology* **2** 113–120
[11] Meyer C D 1994 *SIAM Journal on Matrix Analysis and Applications* **15** 715–728
[12] Schwartz I B and Billings L 2006 Fluctuation induced almost invariant sets Tech. rep. NAVAL RESEARCH LAB WASHINGTON DC PLASMA PHYSICS DIV
[13] Wang Y, Chakrabarti D, Wang C and Faloutsos C 2003 Epidemic spreading in real networks: An eigenvalue viewpoint *Reliable Distributed Systems, 2003. Proceedings. 22nd International Symposium on* (IEEE) pp 25–34
[14] Chakrabarti D, Wang Y, Wang C, Leskovec J and Faloutsos C 2008 *ACM Transactions on Information and System Security (TISSEC)* **10** 1
[15] Godsil C and Royle G F 2013 *Algebraic graph theory* vol 207 (Springer Science & Business Media)
[16] Friedrichs K O 1965 *Advanced ordinary differential equations* (CRC Press)
[17] Kozlukov S *Mathematical Physics and Computer Simulation*
[18] Baskakov A G 1987 *Mathematics of the USSR-Izvestiya* **28** 421

IOP Conf. Series: Journal of Physics: Conf. Series **973** (2018) 012037 doi:10.1088/1742-6596/973/1/012037

Analysis of additive generators of fuzzy operations represented by rational functions

T M Ledeneva

Department of Applied Mathematics, Informatics and Mechanics, Voronezh State University, 1, University Square, Voronezh, 394018, Russia

E-mail: ledeneva-tm@yandex.ru

Abstract. This article presents an approach for determining additive generators of commutative and associative operations. Its applicability for finding generators of triangular norms and conorms is shown. Conditions for the parameters of increasing generators that generate triangular conorms in the class of rational functions are determined.

1. Introduction

A purposeful approach to the formation of various fuzzy operations becomes possible by the introduction of triangular norms and conorms. Triangular norms appeared as a natural generalization of the classical "triangle inequality" when considering the probability metric spaces and the corresponding axioms were proposed by Schweizer and Sclar in their studies [1]. The development of triangular norms and conorms, on the one hand, is based on the theory of functional equations. The foundations of this approach were laid in [2], and families of triangular norms and conorms that are obtained as solutions of functional equations (for example [3]) are widely known. The most important result is the characterization of certain classes of triangular norms by means of additive generators used to represent associative operations [2]. Another approach to the analysis of triangular norms and conorms is related to the theory of semigroups and allows them to be represented by means of ordinal sum. The main properties and representations of various families of triangular norms and conorms are presented in [4–6]. Among fuzzy operations, a large class consists of operations representable by the relation of two polynomials or, in a particular case, by a polynomial. The purpose of this article is to present a methodology for analyzing operations related to this class.

2. Materials and methods

2.1. Basic definitions [4, 7]

Definition 1. *A triangular norm (t-norm)* is a binary operation $T : [0,1] \times [0,1] \to [0,1]$ such that for all $x, y, z \in [0,1]$ we have the following properties:

a) $T(x,y) = T(y,x)$ (commutativity),

b) $T(x, T(y,z)) = T(T(x,y), z)$ (associativity),

c) $T(x,y) \leqslant T(x,z)$, if $y \leqslant z$ (monotonicity),

d) $T(x,1) = x$ (neutral element 1).

The system $\langle [0,1], T, \leqslant \rangle$ is an Abelian semigroup with neutral element 1 and order relation $\leqslant$. The triangular norm $T(x,y)$ models the multiplying operators (intersection of fuzzy sets, conjunction).

Definition 2. *A triangular conorm (t-conorm)* is a binary operation $S : [0,1] \times [0,1] \to [0,1]$ such that for all $x, y, z \in [0,1]$ the commutativity, associativity, monotonicity properties and property $S(x,0) = x$ are fulfilled.

The system $\langle [0,1], S, \leqslant \rangle$ is an Abelian semigroup with neutral element 0 and order relation $\leqslant$. The triangular conorm $S(x,y)$ models the adding operators (union of fuzzy sets, disjunction).

Comparing Definitions 1 and 2 we can see that they differ only in the boundary conditions. The monotonicity with respect to the second argument together with the commutativity property causes monotonicity in both arguments. An important additional requirement is the continuity property.

Definition 3. A continuous, strictly monotonic in both arguments, triangular norm T that satisfies the condition $T(x,x) < x$ (subidempotency) is called *Archimedean.*

Definition 4. A triangular conorm $S(x,y)$ is called the Archimedean conorm if it is continuous, strictly monotonic in both arguments, and satisfies condition $S(x,x) > x$ (superidempotency).

Definition 5. The pairs of triangular norms and conorms $T(x,y)$ and $S(x,y)$ are called *dual* ones if for them the equalities (de Morgan laws)

$$S(x,y) = N(T(N(x), N(y))), \quad T(x,y) = N(S(N(x), N(y))),$$

where $N(x)$ is a strong negation function, are satisfied for all $(x,y) \in [0,1]^2$.

Examples of dual triangular norms and conforms [4, 7] obtained by different authors are presented in Table 1. Most of them are rational functions. Classical operations T_M and S_M which have good characteristics from the mathematical point of view are most often used in applications. In particular, these are the only idempotent operations. Of the algebraic properties, only the laws of complementarity are not fulfilled for them. However, these operations are "strict", which affects the quality of fuzzy models, in particular, their sensitivity to the initial data. For the remaining pairs of fuzzy operations, the distributivity and idempotency laws are not fulfilled. Availability of parameters allows providing flexibility of fuzzy models and their adaptation to a specific task. However, as a rule, additional research is needed to determine the optimum values of the parameters.

2.2. Additive generators

We note that triangular norms and conorms are commutative and associative operations. Commutativity means that the result of the operation depends only on the values of the arguments and does not depend on their order. By associativity is understood the absence of hierarchy among the arguments. It is known that associative operations are represented by additive generators.

Definition 6 ([7])**.** A strictly decreasing continuous function $\varphi : [0,1] \to [0,\infty)$ such that $\varphi(1) = 0$ is called a *decreasing generator.*

A function $\varphi^{-1} : [0,\infty) \to [0,1]$ such that

$$\varphi^{-1}(x) = \begin{cases} \varphi^{-1}(x), & \text{if } x \in [0, \varphi(0)], \\ 0, & \text{if } x \in (\varphi(0), \infty), \end{cases}$$

is called pseudoinverse function for $\varphi(x)$.

IOP Conf. Series: Journal of Physics: Conf. Series **973** (2018) 012037 doi:10.1088/1742-6596/973/1/012037

Table 1. Some classes of fuzzy t-norms and t-conorms.

N	$T(x,y)$	$S(x,y)$
1	$T_M(x,y)=\min(x,y)$	$S_M(x,y)=\max(x,y)$
2	$T_P(x,y)=xy$	$S_P(x,y)=x+y-xy$
3	$T_L(x,y)=\max(0,x+y-1)$	$S_L(x,y)=\min(1,x+y)$
4	$T_D(x,y)=\begin{cases}0, \text{ if } (x,y)\in[0,1)^2\\ \min(x,y), \text{ otherwise}\end{cases}$	$S_D(x,y)=\begin{cases}1, \text{ if } (x,y)\in[0,1)^2\\ \max(x,y), \text{ otherwise}\end{cases}$
5	$T_0(x,y)=\dfrac{xy}{x+y-xy}$	$S_{-1}(x,y)=\dfrac{x+y-2xy}{1-xy}$
6	$T_\alpha(x,y)=\dfrac{xy}{\alpha+(1-\alpha)(x+y-xy)},\ \alpha>0$	$S_\beta(x,y)=\dfrac{(\beta-1)xy+x+y}{1+\beta xy},\ \beta>-1$
	$T_H(x,y)=\dfrac{xy}{h+(1-h)(x+y-xy)},\ h>0$	$S_H(x,y)=\dfrac{x+y+(h-2)xy}{h+(h-1)xy},\ h>0$
7	$T_\lambda(x,y)=\max(0,x+y-1-\lambda(1-x)(1-y)),$ $\lambda>-1$	$S_\lambda(x,y)=\min(1,x+y+\lambda xy),$ $\lambda>-1$
8	$T_w(x,y)=\max\left(0,\dfrac{x+y+wxy-1}{1+w}\right),$ $w>-1$	$S_w(x,y)=\min\left(1,x+y-\dfrac{w}{1-w}xy\right),$ $w>-1$

Definition 7 ([7]). A strictly increasing continuous function $\varphi:[0,1]\to[0,\infty)$ such that $\varphi(0)=0$ is called an *increasing generator.*

We note that in this case the pseudoinverse function looks like this:

$$\varphi^{-1}(x)=\begin{cases}\varphi^{-1}(x), & \text{if } x\in[0,\varphi(1)],\\ 1, & \text{if } x\in(\varphi(1),\infty),\end{cases}$$

In both cases for $[0,1]$ we have $\varphi^{-1}(\varphi(x))=x$.

It is known that a decreasing generator can be associated with an increasing generator and vice versa [4].

We have

Theorem 1 ([7], Characterization Theorem of t-norms). *The binary operation $T:[0,1]^2\to[0,1]$ is an Archimedean t-norm if and only if there exists a decreasing generator $t(x)$ such that for all $(x,y)\in[0,1]^2$ we have*

$$T(x,y)=t^{-1}(t(x)+t(y)). \tag{1}$$

There is an analogous representation for $(x,y)\in[0,1]^2$ for Archimedean triangular conorms

$$S(x,y)=s^{-1}(s(x)+s(y)), \tag{2}$$

where $s(x)$ is an increasing generator.

Functions $t(x)$ and $s(x)$ are also called *additive generators.* They are determined up to a positive multiplicative constant.

Characterization of the main classes of additive generators is one of the most important problems of the functional representation of fuzzy operations [8]. It is closely related to the well-known Kolmogorov's representation of associative operations [9].

IOP Conf. Series: Journal of Physics: Conf. Series **973** (2018) 012037 doi:10.1088/1742-6596/973/1/012037

2.3. The main classes of additive generators for commutative and associative operations representable by rational functions

We consider the class of binary fuzzy operations represented by rational functions — the relation of polynomials or, in a particular case, a polynomial where all variables are not to a higher power than the first, i. e.

$$\tilde{F}(x,y)=\frac{\alpha_0+\alpha_1 x+\alpha_2 y+\alpha_3 xy}{\beta_0+\beta_1 x+\beta_2 y+\beta_3 xy},\quad (x,y\in[0,1]).$$

Taking account of commutativity, from $\tilde{F}(x,y)$ we proceed to the following representation:

$$F(x,y)=\frac{a_0+a_1(x+y)+a_2 xy}{b_0+b_1(x+y)+b_2 xy}. \tag{3}$$

This class of functions will be called an *RF-class.*

Within the framework of the study, the following statements are proved.

Proposition 1 ([10]). *The function $F(x,y)$ is associative under the following conditions:*

$$\begin{cases} b_1a_0+a_1^2 &= a_1b_0+a_2b_0,\\ b_2a_1+b_1^2 &= b_0b_2+a_2b_1,\\ a_1b_1 &= a_0b_2. \end{cases}$$

Proposition 2 ([10]). *Let $F(x,y)=\dfrac{a_0+a_1(x+y)+a_2xy}{b_0+b_1(x+y)+b_2xy}$ be an associative and commutative operation on $[0,1]$, $b_0+b_1(x+y)+b_2xy\neq 0$ and simultaneously the following values $\alpha=a_1b_0-a_0b_1$, $\beta=a_2b_0-a_0b_2$, $\gamma=a_2b_1-a_1b_2$ are not equal to 0. Then there are the following classes K_i of additive generators (determined up to a positive multiplicative constant):*

1) K_1: if $\alpha=\beta=0$ and $\gamma\neq 0$, then $\varphi_1(x)=\pm\dfrac{1}{\gamma x}+C$;

2) K_2: if $\alpha=\gamma=0$, $\beta\neq 0$, then $\varphi_2(x)=\pm\dfrac{1}{\beta}\ln|x|+C$;

3) K_3: if $\alpha=0$, $\beta\neq 0$, $\gamma\neq 0$, then $\varphi_3(x)=\pm\dfrac{1}{\beta}\ln\left|\dfrac{\gamma x}{\beta+\gamma x}\right|+C$;

4) K_4: if $\alpha\neq 0$, $\beta=\gamma=0$, then $\varphi_4(x)=\pm\dfrac{1}{\alpha}x+C$;

5) K_5: if $\alpha\neq 0$, $\beta=0$, $\gamma\neq 0$, then $\varphi_5(x)=\begin{cases}\pm\operatorname{arctg}\left(x\sqrt{\dfrac{\gamma}{\alpha}}\right)+C, & \text{if } \dfrac{\alpha}{\gamma}>0,\\ \pm\ln\left|\dfrac{x+\sqrt{-\alpha/\gamma}}{x-\sqrt{-\alpha/\gamma}}\right|+C, & \text{if } \dfrac{\alpha}{\gamma}<0;\end{cases}$

6) K_6: if $\alpha\neq 0$, $\beta\neq 0$, $\gamma=0$, then $\varphi_6(x)=\pm\dfrac{1}{\beta}\ln|\alpha+\beta x|+C$;

7) K_7: let $D=\beta^2-4\alpha\gamma$, then

7.1) if $D>0$, then $\varphi_7(x)=\pm\ln\left|1-\dfrac{2\sqrt{D}}{2\gamma x+\beta+\sqrt{D}}\right|+C$;

7.2) if $D=0$, then $\varphi_7(x)=\pm\dfrac{1}{\gamma x+\beta/2}+C$;

7.3) if $D<0$, then $\varphi_7(x)=\pm\operatorname{arctg}\dfrac{2\gamma x+\beta}{\sqrt{-D}}+C$.

When using the above formulas, the constant C is found from the additional conditions: for a decreasing generator $\varphi(1) = 0$, for an increasing — $\varphi(0) = 0$.

Example 1. Consider a t-norm of the form $T_r(x,y) = \dfrac{xy}{r + (1-r)(x+y-xy)}$ [7]. Here

$$a_0 = 0,\ a_1 = 0,\ a_2 = 1,\ b_0 = r,\ b_1 = 1-r,\ b_2 = 1-r.$$

We compute $\alpha = 0$, $\beta = r$, $\gamma = 1-r$. Thus, we have a decreasing generator $\varphi(x) = = -\dfrac{1}{r}\ln\left|\dfrac{(1-r)x}{r+(1-r)x}\right| + C$ from the class K_3. Let us find the constant $C = \dfrac{1}{r}\ln|1-r|$ from the condition $\varphi(1) = 0$. Substituting C, we get $\varphi(x) = -\dfrac{1}{r}\ln\left|\dfrac{x}{r+(1-r)x}\right|$. We note that in [7] $r > 0$, then we finally obtain the generator $\varphi(x) = -\ln\left|\dfrac{x}{r+(1-r)x}\right|$.

3. Results and discussion

3.1. Definition of commutative and associative functions from the RF-class based on known additive generators

Analyzing the obtained classes we can identify the following types of additive generators that generate commutative and associative operations in the class of rational functions of two variables:

$$\varphi_1(x) = k\cdot\frac{ax+b}{cx+d} + C,\quad \varphi_2(x) = k\cdot\ln\left|\frac{ax+b}{cx+d}\right| + C,\quad \varphi_3(x) = k\cdot\operatorname{arctg}\frac{ax+b}{cx+d} + C.$$

where k is a certain constant whose sign is essential.

In solving the inverse problem of finding the coefficients of the function (3) on the basis of additive generators, the following results were obtained.

Proposition 3. *An additive generator $\varphi_1(x)$ generates a commutative and associative operation $F(x,y)$ of the form (3) with coefficients*

$$a_0 = bd^2,\quad a_1 = ad^2,\quad a_2 = 2adc - bc^2,\quad b_0 = ad^2 - 2bcd,\quad b_1 = -bc^2,\quad b_2 = -ac^2.$$

Proposition 4. *When $ad - bc \neq 0$, the additive generator $\varphi_2(x)$ generates a commutative and associative operation $F(x,y)$ of the form (3) with coefficients*

$$a_0 = \frac{db^2 - bd^2}{ad-bc},\quad a_1 = \frac{bd(a-c)}{ad-bc},\quad a_2 = \frac{da^2 - bc^2}{ad-bc},$$
$$b_0 = \frac{ad^2 - cb^2}{ad-bc},\quad b_1 = \frac{ac(d-b)}{ad-bc},\quad b_2 = \frac{ac^2 - ca^2}{ad-bc}.$$

Proposition 5. *An additive generator $\varphi_3(x)$ generates a commutative and associative operation $F(x,y)$ of the form (3) with coefficients*

$$a_0 = b(d^2+b^2),\quad a_1 = a(d^2+b^2),\quad a_2 = 2acd - b(c^2 - a^2),$$
$$b_0 = a(d^2-b^2) - 2bcd,\quad b_1 = -b(a^2+c^2),\quad b_2 = -a(a^2+c^2).$$

Example 2. Consider a generator $\varphi(x) = -\ln\left|\left(1 - {}^1\!/_2\, x\right)/(1+x)\right|$. In this case $a = -1$, $b = 2$, $c = 1$, $d = 1$. According to Proposition 4, we obtain the following coefficients:

$$a_0 = -2, \quad a_1 = 4, \quad a_2 = 1, \quad b_0 = 5, \quad b_1 = -1, \quad b_2 = 2.$$

The corresponding commutative and associative operation has the form

$$F(x, y) = \frac{-2 + 4(x+y) + xy}{5 - (x+y) + 2xy}.$$

We note that the generator $\varphi(x)$ is an increasing generator, $\varphi(0) = 0$, but the corresponding operation $F(x, y)$ is not a t-conorm. The reason for this fact is that the condition $F(0, x) = x$ is not considered.

Thus, there is the problem of characterizing such additive generators that generate precisely triangular norms and conorms.

3.2. Features of the representation of additive generators for triangular norms and conorms from the RF-class

When additional conditions provided by the definitions are fulfilled, the found functions $\varphi_1(x)$, $\varphi_2(x)$, $\varphi_3(x)$ become additive generators of triangular norms and conorms.

Let $\varphi(x)$ be an increasing generator. Then a condition $\varphi(0) = 0$ must be fulfilled and this condition allows us to find a constant C and go to the following functions:

1) $\tilde{\varphi}_1(x) = \dfrac{(ad-bc)x}{d(cx+d)} = \dfrac{tx}{rx+1}$, where $t = ad - bc$, $r = cd$, $l = d^2$;

2) $\tilde{\varphi}_2(x) = k \cdot \ln\left|\dfrac{(ax+b)\cdot d}{(cx+d)\cdot b}\right| = k \cdot \ln\left|\dfrac{1+mx}{1+nx}\right|$, where $m = {}^a\!/_b$, $n = {}^c\!/_d$;

3) $\tilde{\varphi}_3(x) = k \cdot \operatorname{arctg}\left(\dfrac{x(ad-bc)}{x(ab+dc)+(b^2+d^2)}\right) = k \cdot \operatorname{arctg}\left(\dfrac{ux}{vx+w}\right)$,

where $u = ad - bc$, $v = ab + cd$, $w = b^2 + d^2$.

We note that since the generators are determined up to a positive multiplicative constant, actually, it is the sign of the constant that makes sense in formulas 2) and 3).

We have the following assertions.

Proposition 6. *If $t > 0$ and $l \neq 0$, then $\tilde{\varphi}_1(x)$ is an increasing generator when the following conditions are fulfilled:*

$$a)\ r > 0 \text{ and } -{}^l\!/_r < 0; \quad b)\ r < 0 \text{ and } -{}^l\!/_r > 1.$$

There is no increasing generator for $k < 0$.

Proposition 7. *The function $\tilde{\varphi}_2(x)$ is an increasing generator only if the following conditions are fulfilled:*

$$a)\ k > 0,\ m > n,\ m > 0,\ n \in [-1, 0); \quad b)\ k < 0,\ m < n,\ n > 0,\ m \in [-1, 0).$$

Proposition 8. *The function $\tilde{\varphi}_3(x)$ where $w > 0$ is an increasing generator if $k \cdot u > 0$ and one of the conditions a) $-\dfrac{w}{v} < 0$ or b) $-\dfrac{w}{v} > 1$ is satisfied.*

Example 3. Let us consider the function $\varphi_r(x) = -\ln\left(\dfrac{1-x}{r+(1-r)(1-x)}\right)$ from [7] which is known as an increasing generator. Let us bring it to a form $\varphi_r(x) = \ln\left(\dfrac{1+(r-1)x}{1-x}\right)$. It is

IOP Conf. Series: Journal of Physics: Conf. Series **973** (2018) 012037 doi:10.1088/1742-6596/973/1/012037

easy to verify that $\varphi_r(x)$ satisfies Proposition 7. Here $a = r - 1$, $b = 1$, $c = -1$, $d = 1$. On the basis of Proposition 4, we find the coefficients and obtain

$$S_r(x, y) = \frac{x + y + (r - 2)xy}{1 + (r - 1)xy}. \tag{4}$$

We note that the t-conorm corresponding to $\varphi_r(x)$ in [7] is represented in the form

$$S(x, y) = \frac{x + y + (r - 2)xy}{r + (r - 1)xy},$$

but this formula is wrong because it does not satisfy the definition $\left(S(x, 0) = \frac{x}{r} \neq x\right)$. On the other hand, all the properties of the triangular conorm are satisfied for the function (4).

3.3. Additive generators for triangular conorms from the RF-class

Taking into account the property $S(x, 0) = x$, the t-conorms from the RF-class have the form

$$S(x, y) = \frac{x + y + a_2xy}{1 + b_2xy}.$$

Taking into account the notations $p = a_2$, $q = b_2$ we consider a t-conorm of the form

$$S(x, y) = \frac{x + y + pxy}{1 + qxy}. \tag{5}$$

We can see that the conditions for commutativity and associativity are satisfied.

An example of a t-conorm of the form (5) is, for example, the function (4) if we assume that $p = r - 2$, $q = r - 1$.

The study of the parameters included in (5) made it possible to obtain families of corresponding additive generators.

We have the following assertions.

Proposition 9. *If $p = 0$ and $q \neq 0$, then the t-conorm $S_q(x, y) = \frac{x + y}{1 + qxy}$ is generated by increasing generators: a) $s_q(x) = \text{arcg}(-qx)$ when $q < 0$, b) $s_q(x) = \ln\left|\frac{1 + x + \sqrt{q}}{1 - x\sqrt{q}}\right|$ when $q \in (0, 1)$.*

Proposition 10. *If $p \neq 0$ and $q = 0$, then when $p > -1$ there is an increasing generator $s_p(x) = \frac{1}{p}\ln|1 + px|$ that generates a t-conorm $S_p(x, y) = x + y + pxy$.*

Proposition 11. *Let $p \neq 0$ and $q \neq 0$, $D = p^2 + 4q$.*

If $D > 0$, $p \in \left(-\sqrt{D}, \sqrt{D}\right)$, $q \in \left(0, \frac{p + \sqrt{D}}{2}\right)$ then there exists an increasing generator

$$s_{p,q}(x) = \ln\left|\left(1 - \frac{2qx}{p - \sqrt{D}}\right)\Big/1 - \frac{2qx}{p + \sqrt{D}}\right|$$

which generates a t-conorm (5).

If $D = 0$, then there is no increasing generator.

If $D < 0$, then when $q < 0$, then the increasing generator has the form

$$s_{p,q}(x) = \text{arctg}\left(-\frac{Dx}{2 + px}\right),$$

which generates t-conorm (5).

Example 4. Let us consider the function $S_r(x,y) = \dfrac{x+y+(r-2)xy}{1+(r-1)xy}$ again. In accordance with the representation (5), here we have $p = r-2$, $q = r-1$. We notice that $q = p+1$. Since, in general, $p \neq 0$, $q \neq 0$, in accordance with Proposition 11, we define $D = p^2 + 4q = = p^2 + 4(p+1) = (p+2)^2 > 0$ and obtain

$$s_{p,p+1}(x) = \ln\left|\left(1 - \frac{2(p+1)x}{p-(p+2)}\right)\Big/ 1 - \frac{2(p+1)x}{p+(p+2)}\right| = \ln\left|(1+(p+1)x)/(1-x)\right|.$$

Substituting $p = r-2$ we obtain an increasing generator

$$s_{r-2,r-1}(x) = \ln\left|\frac{1+(r-1)x}{1-x}\right|$$

which coincides with the generator $\varphi_r(x)$ of Example 3 for $x \in [0,1]$. From Proposition 11 it also follows that for the parameter r we must have the inequality $r > 0$.

Similar results were obtained for triangular norms.

3.4. Analysis of the main classes of additive generators for triangular conorms

There are no increasing generators in classes K_1, K_2 and K_3.

In the class K_4 there is an increasing generator $s_L(x) = x$ which generates t-conorm $S_L(x,y)$.

In the class K_5 for $\alpha > 0$ and $\gamma > 0$ we have increasing generator $\varphi_1(x) = \operatorname{arctg}\left(x\sqrt{\gamma/\alpha}\right)$ that generates triangular conorm $S_{\alpha,\gamma}(x,y) = \dfrac{x+y}{1-\frac{\gamma}{\alpha}xy}$.

In the class K_6 we can identify the following situation $\left[\begin{array}{l}\alpha > 0,\ \ \beta > -\alpha \\ \alpha < 0,\ \ \beta > |\alpha|/x\end{array}\right.$, then the increasing generator has the form $\varphi_1(x) = -\dfrac{1}{\beta}\ln\left(1 + \dfrac{\beta}{\alpha}x\right)$ and the corresponding triangular conorm — $S_{\alpha,\beta}(x,y) = x + y + \dfrac{\beta}{\alpha}xy$, in this case, when $\left\{\begin{array}{l}\beta = -\alpha \\ \beta > 0\end{array}\right.$ we get $S_P(x,y)$, $s_P(x) = -\ln(1-x)$.

The class K_7 is represented by several types of additive generators. For their enumeration we denote $D = \beta^2 - 4\alpha\gamma$, $v_1 = \dfrac{-\beta-\sqrt{D}}{2\gamma}$, $v_2 = \dfrac{-\beta+\sqrt{D}}{2\gamma}$. The following situations are possible:

a) if $D > 0$, $\gamma > 0$, $\left\{\begin{array}{l}v_1 < 0 \\ v_2 < 0\end{array}\right.$ or $\left\{\begin{array}{l}v_1 > 1 \\ v_2 > 1\end{array}\right.$ or $\left[\begin{array}{l}v_1 \in [0,1] \\ v_2 \in [0,1]\end{array}\right.$, then the increasing generator has the form $\varphi_1(x) = \ln\left(\dfrac{1-x/v_2}{1-x/v_1}\right)$, and the corresponding triangular conorm — $S_{v_1,v_2}(x,y) = \dfrac{v_1v_2(x+y)-(v_1+v_2)xy}{v_1v_2 - xy}$;

b) if $D > 0$, $\gamma < 0$, $\left\{\begin{array}{l}v_1 > 1 \\ v_2 > 1\end{array}\right.$ or $\left[\begin{array}{l}v_1 \in [0,1] \\ v_2 \in [0,1]\end{array}\right.$, then the increasing generator has the form $\varphi_2(x) = -\varphi_1(x)$;

c) if $D = 0$ and $\gamma > 0$, then the increasing generator has the form $\varphi_3(x) = 2x\Big/\beta\left(x + \dfrac{\beta}{2\gamma}\right)$ and the corresponding conorm — $S_{\beta,\gamma}(x,y) = \dfrac{x+y+\left(2\beta/\gamma\right)xy}{1-\left(4\gamma^2/\beta^2\right)xy}$;

d) if $D = 0$ and $\gamma < 0$, then the increasing generator has the form $\varphi_4(x) = -\varphi_3(x)$.

In the class K_7 with an additional constraint $\alpha + \beta + \gamma = 0$, where $\alpha \neq \gamma$ from $\varphi_1(x)$, taking into account the change of notation $\beta = -\gamma/\alpha$, we get a generator $s_\beta(x) = \ln \dfrac{1+\beta x}{1-x}$ that generates a t-conorm $S_\beta(x, y)$. If for $\varphi_5(x)$ we have an additional constraint $2\gamma/\beta = -1$, then we obtain a generator $s_{-1}(x)$ for a triangular conorm $S_{-1}(x)$.

We note that for triangular norms and conorms in general there are several additive generators that generate them.

4. Conclusion

In this paper, we consider a family of triangular norms and conorms that are representable in the class of rational functions. For such operations, additive generators are obtained based on the gradual implementation of axioms. Most of the presented results concern triangular conorms. However, similar results were also obtained for triangular norms. Additive generators for commutative and associative operations are of interest by themselves. These properties are defining ones in the axiomatics of aggregation functions.

References

[1] Schweizer B and Sklar A 1983 *Probabilistic Metric Spaces* (New York: North-Holland)
[2] Aczel J 1966 *Lectures on Functional Equations and their Applications* (New York: Academic Press)
[3] Frank M J 1979 *Aequationes Math* **19** 194–226
[4] Klement E P, Mesiar R and Pap E 2004 *Fuzzy Set and Systems* **143** 5–26
[5] Klement E P, Mesiar R and Pap E 2004 *Fuzzy Set and Systems* **145** 439–454
[6] Klement E P, Mesiar R and Pap E 2004 *Fuzzy Set and Systems* **145** 411–438
[7] Klir G J and Yuan B 1995 *Fuzzy sets and fuzzy logic: theory and applications* (New Jersey: Prentice Hall PTR)
[8] Klement E P, Mesiar R and Pap E 2004 *Fuzzy Set and Systems* **145** 471–479
[9] Gini K 1970 *Average values* (Moscow: Statistics)
[10] Ledeneva T M 1997 *News of Higher Educational Institutions. Mathematics* 33–40

IOP Conf. Series: Journal of Physics: Conf. Series **973** (2018) 012038 doi:10.1088/1742-6596/973/1/012038

Fractal model of polarization switching kinetics in ferroelectrics under nonequilibrium conditions of electron irradiation

A G Maslovskaya and T K Barabash

Mathematics and Computer Science Department, Amur State University, 21 Ignatyevskoe Shosse, Blagoveshchensk, 675000, Russia

E-mail: maslovskayaag@mail.ru

Abstract. The paper presents the results of the fractal and multifractal analysis of polarization switching current in ferroelectrics under electron irradiation, which allows statistical memory effects to be estimated at dynamics of domain structure. The mathematical model of formation of electron beam-induced polarization current in ferroelectrics was suggested taking into account the fractal nature of domain structure dynamics. In order to realize the model the computational scheme was constructed using the numerical solution approximation of fractional differential equation. Evidences of electron beam-induced polarization switching process in ferroelectrics were specified at a variation of control model parameters.

1. Introduction

Presently the study of nonequilibrium processes in complex physical systems becomes one of the major scientific directions which widely use the methods of mathematical modelling and computer simulation. In particular, the well-known examples of simulated process in such systems include the nonequilibrium dynamics, phase transitions and domain kinetics in ferroelectric materials. Ferroelectrics are related to the promising subclass of polar dielectrics, which have a wide range of applications in microelectronics, optoelectronics, acoustics, radio engineering, piezo- and pyrotechnics due to polarization switching under external exposure. The numerous studies have been devoted to theoretical and experimental justifications of the domain boundaries movement, restructuring of domain configurations and kinetics of polarization switching in ferroelectrics. These findings have also indicated that domain structures reveal fractal properties and self-similar behavior [1-5].

In order to describe the quantitative characteristics of self-organization processes in complex physical systems the methods the fractal theory can be used [6]. The concept of a fractal is geometrical one and it sets the self-similar object which can be described by fractal dimension. However, not only the geometrical forms of objects can have a fractal structure but also dynamic characteristics of the processes and the phenomena also can find the self-similar behavior in materials with self-similar structure [7-8].

Previous works have shown that the self-similarity property of ferroelectric domain structure is not the only fractal characteristic of ferroelectrics. Dynamic characteristics of ferroelectrics demonstrate fractal properties likewise [9-11]. Indeed, the methods of fractal and multifractal analysis of time series can be effectively used to study the scaling characteristics of polarization switching of process in ferroelectrics.

Notice that the methods of fractal analysis allow us to identify only a posteriori estimates of physical process characteristics. So as to construct the mathematical model of physical system dynamics traditionally one can use the ordinary or partial differential equations. Nevertheless such approach cannot provide model adequacy for many objects and processes which possess self-similarity properties. The application of concepts of the fractional differential calculus theory is the relevant aspect of mathematical models modification to describe the behavior of fractal physical systems. In this case, to formalize the memory processes and phenomena one can employ a time fractional derivative. In addition, space fractional derivatives enable us to take into account the fractal properties of the material inhomogeneity [12-13].

The diagnostics of domain structure as well as modification of electrical properties of ferroelectrics can be performed using scanning electron microscope (SEM) techniques [14-17]. In particular, the application of the electron beam stimulated polarization current mode allows us to visualize SEM-contrast of ferroelectric domains and simultaneously to observe the process of polarization switching [17]. The main dynamic characteristic of polarization switching process is determined by the video signal which is electric response of ferroelectric sample. The polarization switching current is represented as time dependence which is attributed to intensity of the polarization processes in a crystal. Therefore the specific character of this problem requires developing of the theoretical justification. This allows us to formalize observed processes, to investigate scaling characteristics of polarization switching process in ferroelectrics and to predict behavior of system based on computing experiments [18]. The following study was undertaken to develop the mathematical model of electron beam stimulated polarization switching in ferroelectrics taking into account the results of fractal and multifractal analysis of polarization current registered in the injection mode.

2. Fractal and multifractal analysis of dynamic characteristics of polarization switching in ferroelectrics under electron irradiation

Fractal regularities of dynamic characteristics behavior are due stochasticity of nucleation process in ferroelectrics, nonlinear reorganization and dynamics of the domain structure, existence of defect structure as well as memory effects which, in particular, are shown by repeatability of the nucleation centers at cyclic of polarization switching. In this notation, nonlinear dynamics of domain structure is caused by lateral or through domain growth during polarization switching process and interaction of domain boundaries with defects notably pinning.

To be definite, assume that the object of fractal analysis is set to be time dependence of polarization switching currents in typical ferroelectric crystal TGS ($(NH_2CH_2COOH)_3 \cdot H_2SO_4$, triglycine sulfate) registered with the electron beam-stimulated polarization current mode [17]. The study was based on methods of fractal and multifractal analysis.

The Hurst method (so-called rescaled range analysis or R/S-analysis) and Fourier method were applied to estimate fractal dimension of polarization switching current represented as time series [6]. The Hurst method includes the following steps. First time series is splitting into non-overlapping segments I_a of size n. Then, the integrated term is calculated in each segment and the local trend is also estimated. Secondly the rescaled range is averaged over all segments with respect to n and the Hurst exponent H can be obtained using the following equation:

$$\left(\frac{R}{S}\right)_n = \frac{1}{A}\sum_{a=1}^{A}\frac{R_{Ia}}{S_{Ia}} \propto n^H, \tag{1}$$

where $R_{I_a} = \max\left(X_{k,a}\right) - \min\left(X_{k,a}\right)$ is maximum value of the range; $X_{k,a}$ is time series of accumulated deviations from average value for each period I_a; S_{I_a} is standard deviations calculated for each period I_a.

The angular coefficient of linear approximation of functional dependence (1) by least-squares procedure in log-log scale is determined by Hurst parameter H and fractal dimension $D = 2 - H$.

Generally three various dynamics depending on Hurst parameters are usually considered. While value $0 < H < 0.5$ indicates long-term anti-correlated behavior of the data; $H = 0.5$ corresponds to the random and uncorrelated events; $0.5 < H < 1$ is referred to as persistent series and the trending stability of behavior increases as long as H tends to the unit.

In addition, Fourier method was used to estimate the fractal parameters [6]. This method is based on Fourier transform. The procedure of Fourier decomposition of time series $y(t)$ permits us to obtain the values of amplitude c and phase φ for some fixed number of frequencies f:

$$y(t) = \sum_{f=f_{\min}}^{f_{\max}} c(f) \cdot \sin\left(2\pi \cdot f \cdot t + \varphi(f)\right). \qquad (2)$$

Therefore, the fractal dimension D of time series can be evaluated provided we calculate the power spectral density $c^2(f)$. Using the following relation $c^2(f) \propto f^{\beta}$ and linear approximation of $c^2(f)$ with double log scale we also can estimate the angular coefficient β related to fractal dimension according to the expression $D = (4 + \beta)/2$.

Moreover, the methods of multifractal analysis can be more informative techniques used to study scaling characteristics of dynamic data at diagnostics of physical systems. The multifractal methods are perspective tools used to research of complex systems behavior. This allows to determine a continuous spectrum of fractal dimensions $f(\alpha)$ and to localize time as well as scale properties of time series. In this study, the method of multifractal detrended fluctuation analysis (MDFA) [7] and the wavelet transform maxima modules method (WTMM) [8] were implemented.

The MDFA procedure is based on the following principles [7]. At the first step initial time series $y_k(t_k)$ of length N+1 is transformed into a new series using the equation $z_i(x_i) = \ln(y_{k+1} / y_k)$, $i = \overline{1,N}$. After that, a profile is specified as $x_j = \sum[z_i - \overline{z}]$, $i = \overline{1,j}$, $j = \overline{1,N}$, where $\overline{z}$ is the mean of series z. Then x_j is splitting into $N_s \equiv \mathrm{int}(N/s)$ nonoverlapping segments of equal length s. It follows that the local trend and variance $F^2(v,s)$ are determined for each of the $2N_s$ segments lengthwise the forward and reverse directions. Then the average over all segments is calculated to obtain the q-th order fluctuation function:

$$F_q(s) = \left[\frac{1}{2N_s} \sum_{v=1}^{2N_s} \left[F^2(v,s)\right]^{q/2}\right]^{1/q} \propto s^{h(q)}. \qquad (3)$$

Finally fluctuation function $F_q(s)$ is analyzed using logarithmic scales. Multifractal scaling exponents $\tau(q)$ are determined by formula $\tau(q) = qh(q) - 1$. Also the singularity spectrum $f(\alpha)$ and singularity strength α related to $\tau(q)$ characterize a multifractal series via Legendre transformation according to the following equations:

$$\alpha = \tau'(q), \quad f(\alpha) = q\alpha - \tau(q). \qquad (4)$$

Further, WTMM uses the wavelet transform with continuous basis soliton-like functions. Here the local extremum lines of a surface of wavelet coefficient $W(a,b)$ for each scale a are calculated. Then partial functions $Z(q,a)$ are evaluated as:

$$Z(q,a) = \sum_{l \in L(a)} \left(\sup_{a' \le a} \left|W(a', b_l(a'))\right|\right)^q \propto a^{\tau(q)}, \; W(a,b) = (1/a) \int_{-\infty}^{\infty} x(t)\psi\left((t-b)/a\right)dt, \qquad (5)$$

where $W(a,b)$ is the continuous wavelet transform of function $x(t)$; a is a scale, b is a time position, ψ is a mother wavelet (in particular, "Mexican hat" wavelet was used); the largest value of the wavelet transform coefficients modules is selected along each line on scales less then a, $L(a)$ is a set of all (l) local maxima lines of wavelet transform modules existing on scale a, $x_l(a)$ is the line maximum l position on the scale a. The scaling exponents $\tau(q)$ characterize the multifractal properties related to singularity spectrum $f(\alpha)$ obtained by Legendre transformation.

The program application realizing the methods of fractal and multifractal analysis was suggested to perform computing experiments. The software was designed using program application package Matlab. The structure of program application is shown in figure 1.

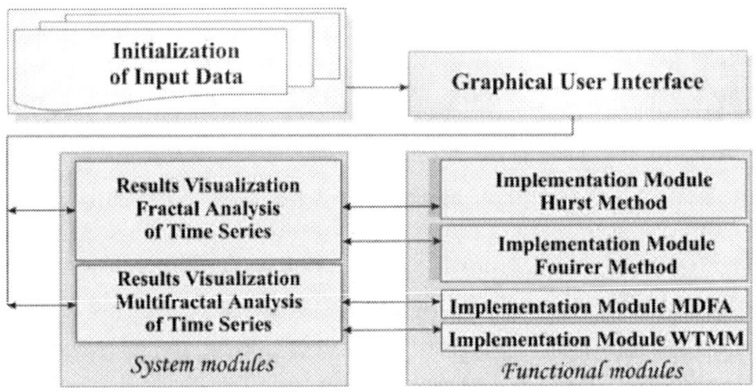

Figure 1. Program application structure for fractal and multifractal analysis of time series.

Program verification was performed using the generated time series with a priori specified fractal dimension. Hurst method allows us to obtain the result with accuracy order 0.001 while Fourier method is less exact and it has the order of accuracy of 0.01.

The Figure 2 (a) demonstrates the fractal characteristics calculated for time dependence for polarization switching current in TGS crystal registered with injection mode of SEM.

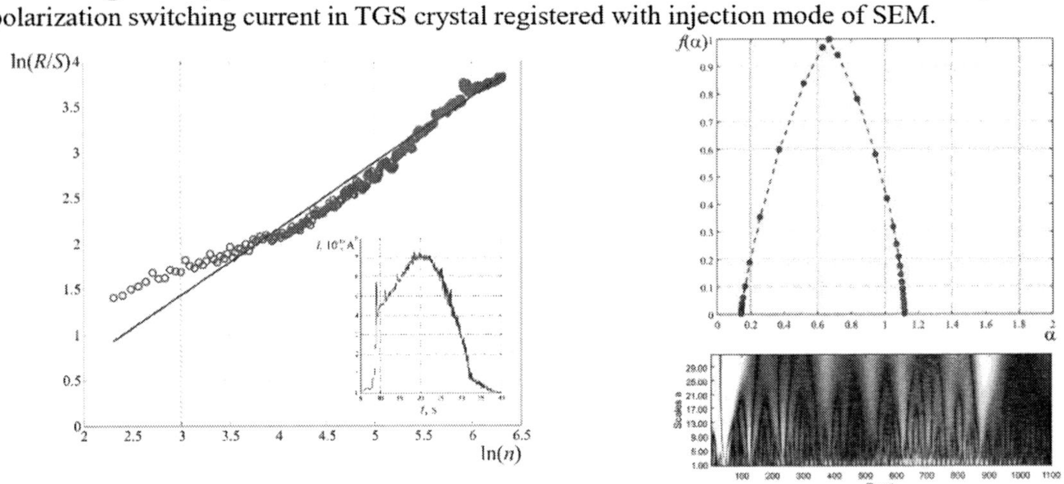

Figure 2. The result of *R*/*S*-analysis – (a); the singularity spectrum $f(\alpha)$ calculated by MDFA and the image of wavelet coefficient surface calculated by WTMM – (b) for polarization switching current in TGS (displayed in the insert figure).

The computed characteristics lead to the following results. Hurst parameter equals to $H = 0.73$ and the corresponding value of fractal dimension is going to be $D = 1.27$. This finding suggests the self-similarity organized movement of domain boundaries during the polarization reversal processes in ferroelectric crystals under external exposure.

Fractal behavior of domain structure dynamics can be characterized by persistent features due to effects of long-term memory. The polarization reorientation observed in ferroelectrics under electron beam irradiation is not actually a random process. This means that previous positions of domain boundaries impact the process of the formation of subsequent positions.

The results of multifractal analysis performed using multifractal detrended fluctuation analysis and the wavelet transform maxima modules method are presented in figure 2 (b). The image of wavelet coefficient surface demonstrates hierarchical arrangement of polarization current fluctuations referred to domain restructuring. This feature is caused by the fact that the observable signal has self-similarity properties. The calculated spectrum of fractal dimensions exhibits a variation of fractal dimension values over the range $0.9 \le D \le 1.9$, which confirms the assumption of the multifractal nature of polarization switching process in ferroelectrics exposed to electron irradiation. The maximum abscissa of a spectrum curve corresponds to Hurst parameter of fractal process. We may assume that the fractal characteristics specify characteristic dimension of kinetic process. Hence, scaling characteristics identified by fractal analysis methods can be used for mathematical model construction to describe the polarization switching process in ferroelectrics irradiated by electron beam.

3. Experimental aspects of polarization switching kinetics in ferroelectrics and conceptual problem definition

In order to formalize the electron beam-induced reorientation of domain structure in ferroelectrics we accepted a number of assumptions. Let us consider the evolution of 180° ferroelectric domains. To be defifnite also we assume that electron beam injection into sample on the depth of l results in accumulating charge with surface charge density σ in the irradiated layer and charges with surface charge densities σ_1 and σ_2 on the top and bottom electrodes respectively. By turn this induces the occurrence of the electric fields $\mathbf{E}_1$ in the irradiated layer and $\mathbf{E}_2$ in the main volume of material. The field intensity $\mathbf{E}_2$ can activate polarization switching processes in domains "tail-to-beam" orientated and, in the converse case, it can stabilize the domain structure with "head-to-beam" orientation of spontaneous polarization vector $\mathbf{P}_S$ as shown in figure 3.

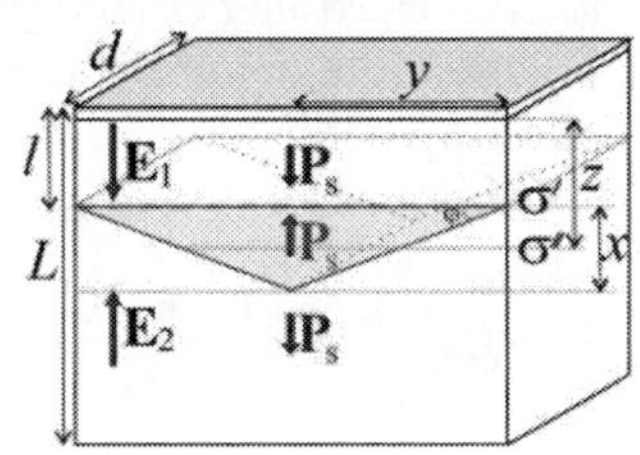

Figure 3. The diagram of domain growth (l is electron injection depth, x is instant position of domain boundary; z is corresponding average instant position of domain boundary).

To describe the main dynamic characteristics of process, which is given by polarization switching current, we use the general equation known as Kolmogorov-Avrami model [19-21]. In compliance with this approach the polarization current is expressed by a relation of the switched volume to the total volume of a crystal. In our case we suggest to modify this equation using the dependence of domain boundary velocity v, which is transformed to the formula:

$$I(t) = \frac{2P_S S}{L} \cdot \frac{dx}{dt}, \quad v = \frac{dx}{dt} = v_\infty \exp\left(-\frac{\delta}{E_2}\right), \tag{6}$$

IOP Conf. Series: Journal of Physics: Conf. Series **973** (2018) 012038 doi:10.1088/1742-6596/973/1/012038

where L is sample thickness in m; v_{∞} is maximum value of domain boundary velocity at $E \to \infty$ in m/s; δ is activation field in V/m; $S = 2yd$ is irradiated area in m^2.

In view of the importance of polarization switching process in the electron beam polarization current mode, let us assume the following mechanisms of domain growth. The polarization switching is realized by means of growth of domains with wedgewise form with the fixed width of the base $2y$ and the changing angle of domain boundary slope φ. The geometrical scheme of domains growth is shown in figure 4.

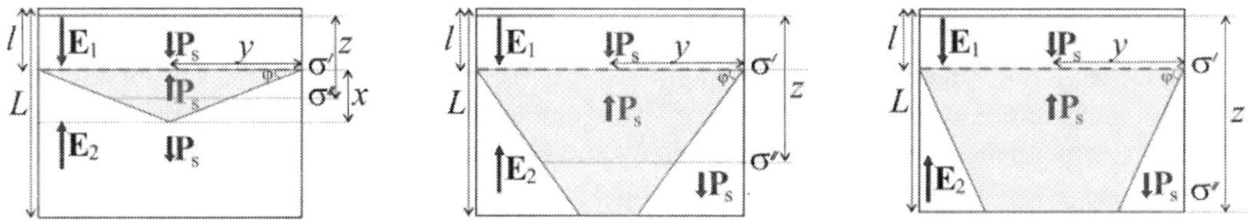

Figure 4. The geometrical scheme of sequential stages of growth of wedgewise domains.

The polarization switching process is going to be completed as long as the middle of wedge $z(t) = l + x(t)/2$ reaches the boundary of crystal L. The use of such mechanism permits us to specify falling of the polarization current at completion of polarization switching process.

4. Mathematical model: fractal approach

The evidences caused by fractal analysis of polarization current in ferroelectrics confirm that the reorganization of domain structure demonstrates a complex scaling behavior. In this terms polarization switching process can be characterized by long-term memory effect. Reorientation of domains polarization has nonrandom character provided the formation of the subsequent conditions based on the previous states of domain structure. Note also that the agreement of simulation results with experiment data can be observed if one uses the modified Kolmogorov-Avrami models with fractional values of kinetic process dimension [5], [10], [20], [21].

We applied fractional differential calculus to describe the domain boundary dynamics at switching polarization as fractal process with power-series memory. The fractal nature of the movement of domain boundaries can be under consideration using a fractional order derivative with respect to time. Therefore, we express the equation defining the polarization switching process (6) subjected to a dimensionless form with use of fractional differentiation operator. After some transformations we can rewrite the equation of domain dynamics in the following form:

$$\frac{d^{\alpha}s}{dw^{\alpha}} = \exp\left(-\tau_2\left[\tau_1 w + \frac{\tau_3}{2l}\left(1+\cos^2\varphi\right)\frac{Ls}{2}\right]^{-1}\right), \quad s(w_0)=0, \quad 0 \le s(w) \le 2(1-l/L), \qquad (7)$$

where $s = x/L$ and $w = t/\tau_1$ are dimensionless distance and time variables; α is the dynamic fractal dimension varied over the range of $(0.5, 0.99)$; $\cos^2\varphi = y^2/\left(x^2+y^2\right)$; $\tau_1 = L/v_{\infty}$ is the characteristic time of boundary growth through a crystal in s; $\tau_2 = \delta\varepsilon\varepsilon_0 L/(jl)$ is the characteristic time of accumulation of charge generating the field $E_2 = \delta$ at given current density j (in $\mathrm{A/m^2}$) in s; δ is activation field in V/m; $\tau_3 = 2P_S/j$ is the characteristic time period during which the charge density σ is defined to be $2P_S$ in s.

The expression for the field E_2 in unirradiated part of sample was used in general equation (7) in the following form:

$$E_2(w) = \frac{jl\tau_1 w}{\varepsilon\varepsilon_0 L} + \frac{P_S}{\varepsilon\varepsilon_0 L} \cdot \frac{Ls}{2}\left(1+\cos^2\varphi\right). \quad (8)$$

The formula (8) was determined taking into account the effect of boundary profile bending which is formed during domain boundary moving. Furthermore we can express the polarization switching current as follows:

$$I(w) = \begin{cases} \dfrac{4 \cdot P_S \cdot d \cdot y}{\tau_1} \cdot \dfrac{d^\alpha s}{dw^\alpha}, & if \quad 0 \le s \le 1 - \dfrac{l}{L}, \\ \dfrac{4 \cdot P_S \cdot d \cdot y}{s^2 \cdot \tau_1} \cdot L^2 \cdot \left[1 - \dfrac{k}{M}\right] \cdot \dfrac{d^\alpha s}{dw^\alpha}, & if \quad 1 - \dfrac{l}{L} < s \le 2\left(1 - \dfrac{l}{L}\right), \end{cases} \quad (9)$$

where wedge basis y is splitting into M number of discrete intervals, $k = \overline{0, M}$.

Indeed, the polarization switching process is attributed to growth of n wedge-shaped domains with an identical speed and width of the base $2y/n$. The polarization current is determined by superposition of deposits which are given by separate switching areas.

As a result, model of polarization switching current and domain dynamics includes the initial value problem for the fractional differential equation (7), expressions for the field intensity (8) and switching current (9). To obtain the dependences $I(t)$ and $E_2(t)$ one needs to execute the inverse change of variable $t = w \cdot \tau_1$.

5. Numerical scheme of model implementation

The problem (7) can not be solved by analytical methods and requires using numerical procedures. The numerical scheme was constructed to solve the fractional differential equation. The Grunwald-Letnikov definition provides a convenient finite difference discretization of fractional derivative operators [12-13]. To approximate the fractional derivative of p with respect to variable ξ the Grunvald-Letnikov formula was used [22]:

$$\frac{d^\alpha p(\xi)}{d\xi^\alpha} = \lim_{h \to 0} \frac{1}{\Gamma(-\alpha)} \cdot \frac{1}{h^\alpha} \cdot \sum_{i=0}^{k} \frac{\Gamma(i-\alpha)}{\Gamma(i+1)} p(\xi - ih), \quad (10)$$

where h is time step; $\Gamma(\alpha)$ is the gamma function.

The fractional derivative is nonlocal convolution type operator. This implies that the value of a fractional derivative with time depends on function values at all previous time moments. Thus, the value of function is defined by all previous states in each time moment. Sequential approximation of a fractional derivative for each element i on a grid $\Omega = \{w_i = w_0 + ih,\, i = \overline{0, N}\}$ can be expressed in the following form:

$$\frac{1}{\Gamma(-\alpha)} \frac{1}{h^\alpha}\left(\frac{\Gamma(-\alpha)}{\Gamma(1)} s_1 + \frac{\Gamma(1-\alpha)}{\Gamma(2)} s_0\right) = f(s_0, w_0),\ i = 1;$$

$$\frac{1}{\Gamma(-\alpha)} \frac{1}{h^\alpha}\left(\frac{\Gamma(-\alpha)}{\Gamma(1)} s_2 + \frac{\Gamma(1-\alpha)}{\Gamma(2)} s_1 + \frac{\Gamma(2-\alpha)}{\Gamma(3)} s_0\right) = f(s_1, w_1),\ i = 2;...$$

where $f(s, w)$ is right hand side of differential equation (7).

Hence, explicit numerical scheme can be writen in the form:

$$s_{i+1} = \Gamma(1)\left[h^{\alpha} f(s_i, w_i) - \sum_{j=1}^{i+1} \Lambda_j^{\alpha} s_{i-j+1} \right], \quad \Lambda_j^{\alpha} = \frac{\Gamma(j-\alpha)}{\Gamma(-\alpha)\cdot\Gamma(j+1)}. \tag{11}$$

The general scheme (11) demonstrates the main idea of fractional derivative used to simulate physical system with memory effect. This means that to simulate the self-similar domain boundary dynamics previous provisions time depended are taken into account.

6. Initialization of simulation parameters

To implement a computing experiment we have to specify a set of the parameters corresponding to physical experiment. Simulation of polarization reversal processes was performed with respect to ferroelectric crystal TGS which can be switched under the conditions of electron beam irradiation [17]. Geometrical parameters of a sample were set to be the following: $L = 10^{-3}$ m is thickness of a crystal; $S = 0.25 \cdot 10^{-6}$ m^2 is the irradiated area on the top electrode of a sample; the linear sizes of the irradiated area are correspond to $d = 2y = 0.5 \cdot 10^{-3}$ m (the base of switching area is splitting into $n = 20$ elements).

Physical parameters were also defined. Let us specify the dielectric permittivity $\varepsilon = 50$, electric constant $\varepsilon_0 = 8.85 \cdot 10^{-12}$ F/m. The spontaneous polarization is equal to $P_S = 2 \cdot 10^{-2}$ C/m^2 for TGS crystal. Time characteristic parameters were estimated to be $\tau_1 = 13.5$ s, $\tau_2 = 37$ s, $\tau_3 = 16.5$ s, which correspond to value of the activation field $\delta = 6 \cdot 10^5$ V/m and injection current density $j = 2.4 \cdot 10^{-3}$ A/m^2 at probe current $I = 6 \cdot 10^{-10}$ A.

Thickness of injection layer l was determined using the results of Monte Carlo simulation of electron transport in irradiated target [23-24]. As an example the result of computer simulation of electron trajectories for TGS crystal at start beam energy $E_0 = 15$ keV is shown in figure 5. Here we retraced $N = 10000$ stories of electrons.

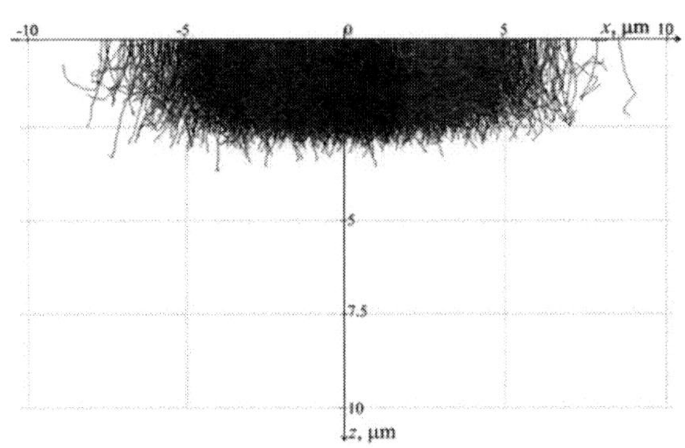

Figure 5. The result of Monte-Carlo simulation of electron trajectories for TGS crystal.

Note that Monte Carlo simulation of electron injection for ferroelectrics coated by metal (silver, aluminum or gold) electrodes at the same value of start beam energy leads to similar geometry of electron interaction zone with material. In this case, the thickness of an electrode should be smaller than 1 μm. In this way, the electrons can freely penetrate into the sample and accumulate in surface layer. This creates the necessary condition to start polarization switching process in ferroelectrics in injection mode of SEM. The trajectories visualized by dark color are correspond to transmitted electrons while emitted trajectories are visualized by light color. The depth of electron injection is approximately equal to $l \approx 3\ \mu\text{m}$.

7. Computing experiments and discussions

The model (7)-(9) was implemented using the numerical algorithm (11). Verification of computer simulation results was performed in two steps. First, simulation data was compared to analytical solutions for test problems which were presented in the form of initial problem for fractional differential equations. Secondly, we established a numerical convergence of calculation results obtained with the scheme (11) (at the fractal parameter $\alpha = 0.99$) and the results of determinate model calculated by predictor-corrector Adams methods of third-order accurate.

The results of fractal and multifractal analysis of polarization switching current in TGS allow us to conclude that Hurst parameter complies the fractal parameter α in the model of polarization current formation (7)-(11). Therefore, we can assume that fractal parameter α as a scaling parameter of domain kinetics can be calculated in terms of fractal and multifractal methods.

The figure 6 *a* represents the simulation result of switching current impulse compared to experimental curve (calculation is performed at given above parameters and fractal parameter is set to be 0.73). The curve of polarization switching current reproduces peculiarities experimentally observed (when signal is recorded with the bottom electrode). The following features can characterize the impulse of polarization switching current in TGS crystal. The curve of switching current has the "peaked" maximum, which is characteristic for a pulse mode of SEM. The polarization current impulse has the duration of tens of seconds and the total charge of polarization reversal corresponds to the relation $2P_S \cdot S$. The processing of data presented in figure 6 (a) gives the estimation of standard deviation of model from experimental curve equal to $5 \cdot 10^{-11}$.

The figure 6 (b) illustrates time dependence of domain boundary dynamics. The start time required to accumulate a charge and also to begin the switching process is estimated to be $t_{st} = 8\ \text{s}$. During the process observation ($42 - 45\ \text{s}$) the domain boundary reaches the bottom electrode when the switching process is finished and the main volume of a crystal will be characterized by inverse value of a vector $\mathbf{P}_S$.

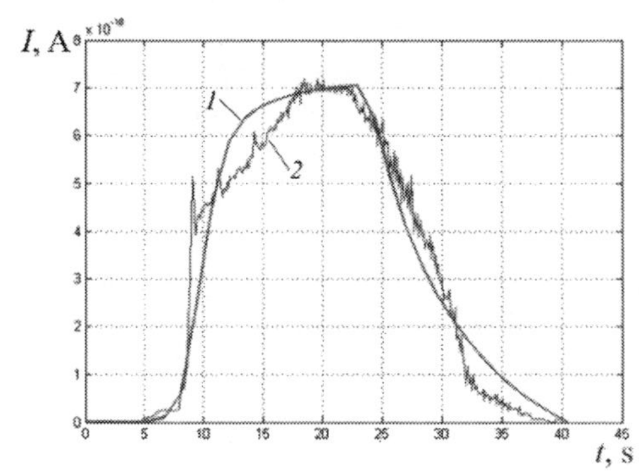

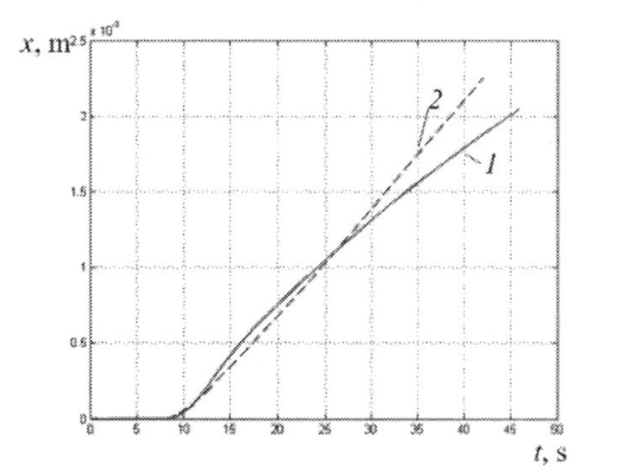

Figure 6. The model representation of polarization switching current in TGS (*1*) compared with experimental data [17] (2) – (a); time dependence of coordinate of domain boundary (the curve *1* is corresponded to fractal model, the curve *2* displays the result obtained with deterministic model) – (b).

The comparative analysis of the curves calculated by means of fractal and deterministic models indicates that domain dynamics has complex character. Domain boundary movement can be characterized by non-linear behavior compared with almost linear growth of $x(t)$ in of the deterministic model. This is caused by memory effect occurrence in physical system.

IOP Conf. Series: Journal of Physics: Conf. Series **973** (2018) 012038 doi:10.1088/1742-6596/973/1/012038

Variation of the control parameters allows modelling main characteristics of domain structure switching related to conditions of experimental observation. The start time t_{st} of polarization switching process in TGS as a function of injection current density j (change in parameters τ_3 and τ_2 depending on j) for seven realization of model is shown in figure 7.

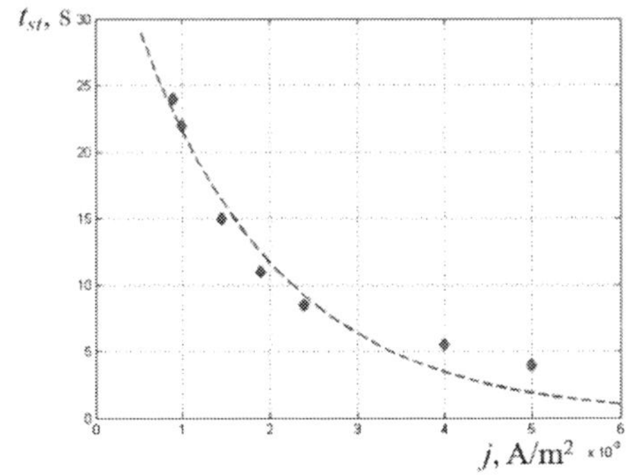

Figure 7. The start time of polarization switching as a function of injection current density for TGS (approximation by exponential relationship).

Also we can conclude that the character of model behavior corresponds to experimentally observed regularities of process. Here start time of polarization switching process decreases exponentially from 24 s to 4 s when injection current density increases from $0.9 \cdot 10^{-3}$ A/m^2 to $5 \cdot 10^{-3}$ A/m^2.

8. Conclusions

Thus, fractal and multifractal analysis of electron beam-induced polarization switching currents in ferroelectrics represented as time series was performed. The analysis was based on Hurst method, Fourier method as well as multifractal detrended fluctuation analysis and the wavelet transform maxima modules method. The fractal dimension, Hurst parameter, scaling exponents, singularity spectrum were calculated. The calculated spectrum gives a variation of fractal dimension in the wide range. The image of wavelet coefficient surface demonstrates hierarchical arrangement of polarization current fluctuations. These results suggest that domain structure reorganization demonstrates the complex self-similar nature. The dynamics of polarization switching process can be characterized by effect of statistical memory and formation of the subsequent conditions is based on previous states during domain reorientation process. Software application was developed to support fractal and multifractal analysis of complex signals.

The present study also demonstrates that characteristics specified by fractal analysis can be used to simulate the polarization switching process in ferroelectrics. The mathematical model of polarization switching in the injection mode was presented. The model permits us to simulate polarization current formation taking into account fractal nature of domain dynamics. The model is expressed by the initial problem for fractional differential equation describing the domain boundary movement as well as expression for calculation of polarization switching current.

The computational algorithm was proposed to realize this model. The numerical scheme was based on numerical approximation of fractional derivative by the Grunvald-Letnikov formula. Model calculation of polarization switching current in TGS crystal was implemented under the conditions corresponding experiment data in injection mode of SEM. The model reproduces correctly the main features of observed polarization switching currents in TGS. Computing experiment allows identifying characteristics of polarization switching kinetics depending on a variation of control scanning parameters in SEM, which are probe current and the accelerating voltages.

References

[1] Galiyarova N and Donysova L 1999 *Ferroelectrics* **222** 269
[2] Ozaki T, Fujii K and Ohgami J 1995 *J. Phys. Soc. Jpn.* **64** 2282
[3] Wu Z, Duan W, Wu J and Gu B-L 2007 *Nanotechnology* **18** 325703
[4] Roy M K, Paul J and Dattagupta S 2010 *IEEE Xplore* **109** 014108
[5] Scott J F 1998 *Fractal dimensions in switching kinetics of ferroelectrics* (Cambridge: University of Cambridge Press) p 9
[6] Bojokin S V and Parshin D A 2001 *Fractals and multifractals* (Moscow: R&C Dynamics)
[7] Kantelhardt J W *Fractal and multifractal time series* (Halle-Wittenberg: Institute of Physics, Martin-Luther-University) p 42
[8] Pavlov A N and Anishchenko V S 2007 *UFN* **177** 859
[9] Tadic B 2002 *Eur. Phys. J. B.* **28** 81
[10] Shur V Ya, Nikolaeva E V, Shishkin E I, Kozhevnikov V L and Chernykh A P 2002 *Physics of the Solid State* **44** 2151
[11] Maslovskaya A G and Barabash T K 2012 *Physics Procedia* **23** 81
[12] Kenneth S M and Bertram R 1993 *An Introduction to the fractional calculus and fractional differential equations* (New York: Wiley-Interscience) p 384
[13] Nakhushev A M 2003 *Fractional Calculus and Its Application* (Moscow: Fizmatlit) p 272
[14] He J, Tang S H, Qin Y Q, Dong P, Zhang H Z, Kang C H, Sun W X and Shen Z X 2003 *J. Appl. Phys.* **93** 9943
[15] Mateos L, Bausa L E and Ramırez M O 2013 *Applied Physics Letters* **102** 042910
[16] Mailis S, Sones C L and Eason R W 2014 Micro-structuring and ferroelectric domain engineering of single crystal lithium niobate *Ferroelectric crystals for photonic applications* (vol 91) ed P Ferraro, S Grilli et al (Berlin, Heidelberg: Springer Series in Materials Science) pp 3-19
[17] Sogr A, Kopylova I and Maslovskaya A 2005 *Proc. of SPIE* **5851** 246
[18] Pavelchuk A, Barabash T and Maslovskaya A 2016 *IOP Conf. Series: Materials Science and Engineering* **110** 012080
[19] Ishibashi Y 1990 *J. Phys. Soc. Jpn.* **59** 4158
[20] Meilanov R P and Sadykov S A 1999 *Technical Physics* **44** 595
[21] Zhang Y, Zhong X L, Zhang Z H, Wang J B and Zhou Y C 2012 *Proc. of SPIE* **8409** 84091Q1
[22] Scherera R, Kallab S L, Tangc Y and Huang J 2011 *Computers & Mathematics with Applications* **62** 902
[23] Joy D C 1995 *Monte-Carlo modeling for electron microscopy and microanalysis* (New York: Oxford University Press) p 216
[24] Demers H, Poirier-Demers N, Couture A R, Joly D, Guilmain M, de Jonge N and Drouin D 2011 *Scanning* **33** 135

IOP Conf. Series: Journal of Physics: Conf. Series **973** (2018) 012039 doi:10.1088/1742-6596/973/1/012039

A dual approach to solving a multi-objective assignment problem

O A Medvedeva and S N Medvedev

Department of Applied Mathematics, Informatics and Mechanics, Voronezh State University, 1 Universitetskaya pl., Voronezh, 394006, Russia

E-mail: s_n_medvedev@mail.ru

Abstract
Two approximation algorithms applied for solving the multi-objective assignment problem (MOAP) are proposed. The reduction to the single-objective problem is realized and dual Uzawa method is applied. The differences between algorithms are the Lagrangian construction technics, the application technics of Hungarian method or independent column minimization in each iteration. The results of the comparative computational experiment are presented.

1. Introduction

The assignment problem (AP) is one of the most-studied, well-known and important problem of discrete optimization. The classical linear AP has been well studied and many algorithms have been produced to solve it, e.g. Hungarian method.

The classical AP can be stated as: Let a number n of jobs be given that must be performed by n workers, where the costs depend on the specific assignments. Each job must be assigned to one and only one worker and each worker has to perform one and only one job. Let c_{ij} be the cost associated with worker i ($i = \overline{1,n}$) who has performed job j ($i = \overline{1,n}$). The problem is to find such an assignment that the total costs, i.e., the sum of the single assignments, become a minimum.

The mathematical model of the classical AP has the form:

$$L(X) = \sum_{i=1}^{n}\sum_{j=1}^{n} c_{ij}x_{ij} \to \min$$

$$\sum_{i=1}^{n} x_{ij} = 1, \quad j = \overline{1,n},$$

$$\sum_{j=1}^{n} x_{ij} = 1, \quad i = \overline{1,n},$$

$$x_{ij} = \{0,1\}, \quad i,j = \overline{1,n},$$

x_{ij} denotes that job j is to be assigned to the worker i.

Combinatorial optimization problems are usually considered in single-objective formulation. Mathematical optimization problems, having feasible solutions set corresponding to the classical AP and involving more than one objective function to be optimized simultaneously, are still not fully understood, but important for practical applications. MOAP [1, 2] is one of such problems.

MOAP can be formalized as classical AP in more than one factories simultaneously. The mathematical model of the MOAP has the form:

$$L_1(X_1) = \sum_{i=1}^{m}\sum_{j=1}^{n_1} c_{ij}^1 x_{ij}^1 \to \min \tag{1}$$

...

$$L_K(X_K) = \sum_{i=1}^{m}\sum_{j=1}^{n_K} c_{ij}^K x_{ij}^K \to \min \tag{2}$$

$$\sum_{i=1}^{m} x_{ij}^k = 1, \quad k = \overline{1,K}, \ j = \overline{1,n_k}, \tag{3}$$

$$\sum_{j=1}^{n_k} x_{ij}^k \le 1, \quad i = \overline{1,m}, \quad k = \overline{1,K}, \tag{4}$$

$$\sum_{k=1}^{K}\sum_{j=1}^{n_k} x_{ij}^k \le 1, \quad i = \overline{1,m}, \tag{5}$$

$$x_{ij}^k = \{0,1\}, \quad i = \overline{1,m}, \quad k = \overline{1,K}, \ j = \overline{1,n_k}, \tag{6}$$

where $C^k = \{c_{ij}^k\}_{m \times n_k}$, $k = \overline{1,K}$ are the cost matrices; value c_{ij}^k represents the cost associated with worker i who has performed job j in the factory k.

Multi-objective optimization methods for such problems can be conventionally classified in two categories. Methods of the first category transform the MOAP into a single-objective AP by using a convolution of the vector of individual objective functions into supercriterion (i. e., some generalized scalar objective function) [2–5]. The second category includes such methods as compromise method, step-by-step method, hierarchy analysis technique [5–7], etc.

In this paper, we propose the approach, which uses the convolution of the vector of objectives. The main results in this area are related to the studying the possibility of application of linear convolution algorithms to find the set of Pareto-optimal solutions (Pareto optimal set) or its subsets [8, 9].

In [10, 11] it has been proved that the MOAP is NP-hard problem if its objective vector contains at least two functions of the following type

$$L_\nu(X) = \sum_{i=1}^{n}\sum_{j=1}^{n} c_{ij}^\nu x_{ij} \to \min.$$

Works [12, 13] propose polynomial algorithms to solve the AP with two objective functions

$$L_1(X) = \sum_{i=1}^{n}\sum_{j=1}^{n} c_{ij} x_{ij} \to \min,$$

$$L_2(X) = \max_{i=\overline{1,n}} \max_{j=\overline{1,n}} c_{ij} x_{ij} \to \min.$$

Using the linear convolution of $L_1(X)$ and $L_2(X)$, this problem can be transformed to the single-objective AP, which are solvable by well-studied techniques, e.g. Hungarian method.

Paper [14] proposes the MOAP solving algorithm based on the Hungarian method and Hopcroft-Karp algorithm [15].

The algorithm for finding the Pareto optimal set presented in [16] differs from the previous one presented in [14]: in [16] the Kahichko algorithm [17] is used in some certain cases instead of the Hopcroft-Karp algorithm.

In [7] it has been shown that if the objective functions are positive, the following non-linear convolution can be used for solving the MOAP

$$L(X) = \max_{\nu \in \{1,\dots,N\}} \{\lambda_\nu L_\nu(X)\}.$$

Thus, in order to solve the MOAP, the main researches are related to the application of algorithm of linear convolution (ALC). However, this is not the unique approach, and use of other types of convolution is a promising direction. Further development of this research area is related not only to the application but also to the modification and addition of well-known vector optimization techniques, taking into account the problem specific character.

It is the purpose of this paper to develop MOAP-solving approximation algorithms based on transforming into a single-objective AP and further applying Uzawa method [18].

2. Materials and methods

Consider the mathematical model of MOAP (1)–(6).

Constraints (3) and (4) are standard MOAP restrictions for each fixed factory; the constraint (5) means that one and the same worker cannot simultaneously perform job in more than one factory. Let us remark here that (4) is a consequence of (5). However, constraint (4) is not removed from the mathematical model for the convenience of further algorithmization.

Statement 1. MOAP (1)–(6) has a solution on condition that $m \geq \sum_{k=1}^{K} n_k$.

Proof.

Summing up all inequalities in (5) with respect to $i = \overline{1,m}$ yields

$$\sum_{i=1}^{m}\sum_{k=1}^{K}\sum_{j=1}^{n_k} x_{ij}^k \leq m.$$

Summing up all equalities in (3) with respect to $j = \overline{1,n_k}$ for each $k = \overline{1,K}$ yields

$$\sum_{j=1}^{n_k}\sum_{i=1}^{m} x_{ij}^k = n_k, \quad k = \overline{1,K}.$$

Finally, summing up the latest equalities with respect to $k = \overline{1,K}$ yields $\sum_{k=1}^{K}\sum_{j=1}^{n_k}\sum_{i=1}^{m} x_{ij}^k = \sum_{k=1}^{K} n_k$.

Thus, $m \geq \sum_{k=1}^{K}\sum_{j=1}^{n_k}\sum_{i=1}^{m} x_{ij}^k = \sum_{k=1}^{K} n_k$.

The statement is proved.

We provide two MOAP-solving algorithms based on transforming into a single-objective AP and further applying Uzawa method [18].

One of these algorithms is generalization of two-objective AP solving method proposed in [19, 20], whereby two-objective AP transforms into the classical AP at every stage. In the paper under consideration we propose different way for construction the Lagrangian that simplifies very much the solving algorithm.

Consider an application of Uzawa method for MOAP (1)-(6).

As we aim to minimize each of the objective functions, consider the objective function of following type [20]

$$\min\{\max(L_1(X_1), L_2(X_2), ..., L_K(X_K))\}.$$

Let us set $\mu = \max(L_1(X), L_2(X), ..., L_K(X))$. Note that $L_i(X)$, $i=\overline{1,K}$ take on the nonnegative values ($c_{ij}^k \geq 0$, $x_{ij}^k \in \{0,1\}$, $i=\overline{1,m}$, $k=\overline{1,K}$, $j=\overline{1,n_k}$). Therefore, it can be argued that $\mu \geq 0$.

MOAP (1)–(6) can be restated in equivalent form:

$$\mu \to \min$$

$$L_1(X_1) - \mu \leq 0,$$

$$...$$

$$L_K(X_K) - \mu \leq 0,$$

$$\sum_{k=1}^{K}\sum_{j=1}^{n_k} x_{ij}^k \leq 1, \quad i=\overline{1,m},$$

$$x_{ij}^k \in S, \quad i=\overline{1,m}, \quad k=\overline{1,K}, \quad j=\overline{1,n_k},$$

$$\mu \geq 0.$$

The set S will be described further.

The Lagrangian for this problem can be written as

$$\Phi(x,\mu,u,w) = \mu + u_1\left(\sum_{i=1}^{m}\sum_{j=1}^{n_1} c_{ij}^1 x_{ij}^1 - \mu\right) + ... + u_K\left(\sum_{i=1}^{m}\sum_{j=1}^{n_K} c_{ij}^K x_{ij}^K - \mu\right)$$

$$+\sum_{i=1}^{m} w_i\left(\sum_{k=1}^{K}\sum_{j=1}^{n_k} x_{ij}^k - 1\right), \quad u,w \geq 0, \quad x \in S, \quad \mu \geq 0.$$

Using the Lagrangian, the MOAP (1)–(6) and its dual problem can be respectively written as

$$\min_{\substack{x\in S \\ \mu\geq 0}} \max_{u,w\geq 0} \Phi(x,\mu,u,w); \quad \max_{u,w\geq 0} \min_{\substack{x\in S \\ \mu\geq 0}} \Phi(x,\mu,u,w).$$

We call $\omega(u,w) = \min_{\substack{x\in S \\ \mu\geq 0}} \Phi(x,\mu,u,w)$ the dual function. The dual problem $\max_{u,w\geq 0} \omega(u,w)$ is more simply than the primal. The dual problem is continual (u and w take on any real values), whereas the primal problem is discrete.

Let us remark here that $\omega(u,w)$ is a convex function [18]; this makes it possible to maximize it using iterative subgradient method.

The values of the dual variables u and w are calculated in each iteration by formula

$$\begin{pmatrix} u \\ w \end{pmatrix}^{N+1} = \begin{pmatrix} u \\ w \end{pmatrix}^{N} + \begin{pmatrix} \alpha_N \\ \beta_N \end{pmatrix} \hat{\nabla}\omega(u^N, w^N),$$

where $\hat{\nabla}\omega$ is the corresponding value of subgradient of the dual function $\omega(\cdot)$; α_N and β_N are the step sizes.

IOP Conf. Series: Journal of Physics: Conf. Series **973** (2018) 012039 doi:10.1088/1742-6596/973/1/012039

Thus, the dual Uzawa method for the problem under consideration is summarized as follows:
Step 0. Initialization. Enter initial data $u^0 \geq 0,\ w^0 \geq 0,\ N = 0$.
Step 1. Find a solution of the problem

$$\Phi(x, \mu, u^N, w^N) \to \min_{x \in S, \mu \geq 0}.$$

Step 2. STOP-test. If the stopping condition is fulfilled, we obtain the optimal solution $X_k^* = X_k^{N+1},\ k = \overline{1,K}$. Otherwise, we have to calculate μ^{N+1}.

Step 3. Calculate the values of dual variables $\begin{pmatrix} u \\ w \end{pmatrix}^{N+1} \in R^{K+n}$. Assume $N = N+1$. Proceed to Step 1.

The main stopping criterion is fulfillment of constraints (5). The additional stopping conditions are: algorithm execution time, total number of iterations.

To implement the computing circuit, the Lagrangian is converted as follows:

$$\Phi(x, \mu, u, w) = \sum_{i=1}^{m}\sum_{j=1}^{n_1}\left(u_1 c_{ij}^1 + w_i\right)x_{ij}^1 + \ldots + \sum_{i=1}^{m}\sum_{j=1}^{n_K}\left(u_K c_{ij}^K + w_i\right)x_{ij}^K + \mu\left(1 - \sum_{k=1}^{K} u_k\right)$$

for $u, w \geq 0,\ x \in S,\ \mu \geq 0$.

As a result, Step 1 of the algorithm solves the following problems:

$$1. \quad \sum_{i=1}^{m}\sum_{j=1}^{n_k}(u_k^N c_{ij}^k + w_i^N)x_{ij}^k \to \min_{x \in S},\ k = \overline{1,K}, \tag{7}$$

$$2. \quad \mu(1 - \sum_{k=1}^{K} u_k) \to \min_{\mu \geq 0}. \tag{8}$$

Note that there are several alternatives of defining the structure of set S for solving the problem (7).

1. If S is defined as a set of $\{x_{ij}^k\}$ that satisfy the constraints (3), (4) and (6), the following APs will be solved in each iteration by Hungarian method holding dual variables $u_k,\ k = \overline{1,K}$ and $w_i,\ i = \overline{1,m}$ constant

$$\sum_{i=1}^{m}\sum_{j=1}^{n_k}\left(u_k^N c_{ij}^k + w_i^N\right)x_{ij}^k \to \min_{x \in S},\ k = \overline{1,K}. \tag{9}$$

2. If S is defined as a set of $\{x_{ij}^k\}$ that satisfy the constraints (3) and (6), the following problem will be solved in each iteration

$$\sum_{i=1}^{m}\sum_{j=1}^{n_k}\sum_{k=1}^{K}\left(u_k c_{ij}^k + w_i\right)x_{ij}^k \to \min \tag{10}$$

$$\sum_{i=1}^{m} x_{ij}^k = 1, \tag{11}$$

$$x_{ij}^k = \{0,1\},\ i = \overline{1,m},\ j = \overline{1,n},\ k = \overline{1,K}. \tag{12}$$

It is the problem of independent column minimization in the given cost matrix.

In view of definition of μ, the following restrictions are valid

$$0 \leq \mu \leq \max(L_1(X_1^N), \ldots, L_K(X_K^N)).$$

Then problem (9) is solved as follows:

$$\mu^N = \begin{cases} 0, \ \text{if} \sum_{k=1}^{K} u_k < 1, \\ \text{any}, \ \text{if} \ \sum_{k=1}^{K} u_k = 1, \\ \max\left(L_1\left(X_1^N\right),\dots,L_K\left(X_K^N\right)\right), \ \text{if} \ \sum_{k=1}^{K} u_k > 1. \end{cases}$$

The computational experiment has shown that in case of $\sum_{k=1}^{K} u_k = 1$, it is necessary to assume $\mu^N = \mu^{N-1}$.

Finally, the MOAP-solving algorithm takes the following form.

2.1. MOAP-solving algorithm

Step 1. Enter initial data $u^0 \geq 0, \ w^0 \geq 0, \ N = 0, \ \alpha^N, \beta^N, \ N_{\max}$.

Step 2. Solve K assignment problems (9) or the problem of independent column minimization (10)–(12).

Step 3. Check: $\sum_{k=1}^{K}\sum_{j=1}^{n_k} x_{ij}^k \leq 1, \ i = \overline{1,m}$? If **yes**, then the optimal solution $X_k^N, \ k = \overline{1,K}$ is obtained and algorithm is finished. If **no**, then proceed to step 4.

Step 4. Check: $N < N_{\max}$? If **yes**, then proceed to step 5. If **no**, then approximate solution $X_k^N, \ k = \overline{1,K}$ is obtained and algorithm is finished.

Step 5. Solve the problem (8) for μ.

Step 6. Recalculate the values of dual variables by formulas

$$u_k^{N+1} = \left[u_k^N + \alpha^N \left(\sum_{i=1}^{m}\sum_{j=1}^{n_k} c_{ij}^k x_{ij}^{kN} - \mu^N \right) \right]^+, \ k = \overline{1,K},$$

$$w_i^{N+1} = \left[w_i^N + \beta^N \left(\sum_{k=1}^{K}\sum_{j=1}^{n_k} x_{ij}^{kN} - 1 \right) \right]^+, \ i = \overline{1,m},$$

where $[x]^+ = \begin{cases} x, \ \text{if}\, x > 0, \\ 0, \ \text{otherwise}. \end{cases}$

Assume $N = N + 1$. Proceed to step 2.

If the inequality $N < N_{\max}$ is failed, it is possible to get an infeasible point. In this case, we have to correct the obtained solution by reducing the dimension of the cost matrices $C^k = \left\{c_{ij}^k\right\}, \ k = \overline{1,K}$. If the stopping condition $\sum_{k=1}^{K}\sum_{j=1}^{n_k} x_{i^*j}^k = 1$ is reached on the element $x_{i^*l}^h = 1$ for a given i^*, then row i^* is removed from all matrices and column l is removed from matrix C^h.

This leads to reducing the problem size. Thereafter, the MOAP-solving algorithm is restarted for the obtained matrices.

The general block scheme of the solving algorithms for the problem (1)–(6) is presented in Figure 1. The difference between algorithms is in the implementation of the unit highlighted in black (see the Figure 1). The implementation of this unit is determined by the method of defining the set S.

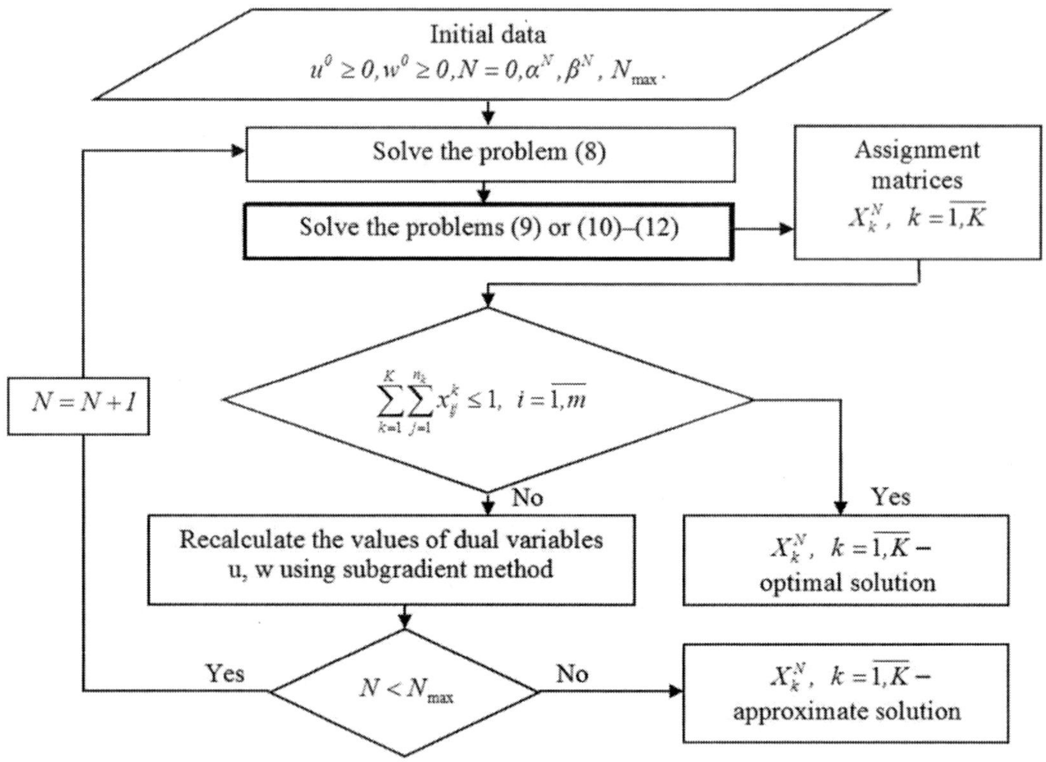

Figure 1. General block scheme of the MOAP-solving algorithms.

Statement 2 [21]. The computational complexity of single iteration of the algorithm using the Hungarian method is equal to $O(Km^3)$.

Statement 3 [21]. The computational complexity of single iteration of the algorithm using the independent column minimization is equal to $O(mn)$, where $n = \sum_{k=1}^{K} n_k$.

3. Results and discussion

A software complex to test and analyze the algorithms proposed in this paper has been developed in the Borland Delphi 7.0. Using this software we have conducted a computational experiment which made it possible to elaborate recommendations for using the algorithms according to problem sizes. The algorithms were tested on the input matrices of different dimension. The number of iterations needed to complete all stops and the algorithm execution time were evaluated.

Further, we introduce the notations:
Algorithm I – algorithm using the Hungarian method;
Algorithm II – algorithm using the independent column minimization.

Table 1 shows the results of the MOAP-solving Algorithm I in each iteration. 100 experiments were carried out for each problem size. The input matrices were filled with random elements between 1 and 100. Each table cell shows the number of these problems that have been solved at the corresponding number of iterations.

Table 1. Number of problems solved by Algorithm I.

Problem size, $K \times m \times n$	Number of iterations						Average execution time for one problem (sec)
	Less than 50	From 50 to 100	From 100 to 200	From 200 to 500	From 500 to 100	More than 1000	
$2\times6\times3$	83	9	3	3	2	0	0.01
$2\times10\times5$	72	10	8	6	2	2	0.05
$2\times20\times10$	55	12	11	7	9	6	0.20
$2\times30\times15$	20	5	13	16	15	31	0.55
$3\times9\times3$	79	11	7	2	1	0	0.02
$3\times15\times5$	45	12	20	8	5	10	0.10
$3\times30\times10$	23	6	13	9	13	36	1.30
$4\times12\times3$	36	14	25	15	3	7	0.07
$4\times20\times5$	27	2	3	16	23	29	0.25

Comment on the Table 1. For example, consider the second line corresponding to the problem size $2\times10\times5$: 72 problems (of hundred) are solved at less than 50 iterations, the average time for solving one problem is about 0.05 seconds.

As seen from Table 1, the algorithm has a great convergence rate for problems with $m<10$ in most cases. In addition, only 2 problems are not solved at 1000 iterations. If $10<m<20$ about 90% of problems are solved at 1000 iterations.

When parameter m increases, the number of problems unsolved at 1000 iterations increases.

The similar conclusions can be drawn for a parameter K.

Consider the results of the MOAP-solving Algorithm II in each iteration.

As seen from Table 2, the algorithm has a great convergence rate only for smaller sized problems in most cases. More than 10% of problems are not solved at 1000 iterations if $m=15$.

Table 2. Number of problems solved by Algorithm II.

Problem size, $K \times m \times n$	Number of iterations						Average execution time for one problem (sec)
	Less than 50	From 50 to 100	From 100 to 200	From 200 to 500	From 500 to 100	More than1000	
$2\times6\times3$	78	13	2	5	0	2	<0.001
$2\times10\times5$	47	12	8	14	5	4	0.005
$2\times20\times10$	2	33	5	19	18	23	0.015
$2\times26\times13$	0	23	14	12	21	30	0.02
$3\times9\times3$	57	2	5	2	0	6	<0.001
$3\times15\times5$	3	56	9	13	3	16	0.01
$3\times21\times7$	0	4	13	42	12	29	0.02
$4\times12\times3$	22	41	10	13	9	5	0.01
$4\times20\times5$	0	29	13	8	18	32	0.04

Evaluate now the performance of the both methods by solving the larger sized problems. As seen from Tables 1 and 2, the number of iterations and average execution time increase in this case.

In this connection, the stopping criterion is algorithm execution time or total number of iterations (depending on the solved problem). In this situation, we obtain interim solution for which the constraints (5) are violated.

The average number of unfulfilled constraints (5) in the first and last iterations for both algorithms is presented in Tables 3 and 4. The algorithm execution time is 60.0 seconds. 500 experiments were carried out for each problem size. The input matrices were filled with random elements between 1 and 100.

Table 3. The average number of unfulfilled constraints (5) for Algorithm I.

Problem size, $K\times m\times n$	Average number of constraints violated in the first iteration	Average number of constraints violated in the last iteration
$2\times 200\times 100$	51.7	23.9
$3\times 150\times 50$	39.9	22.0
$4\times 200\times 50$	51.2	32.5

Table 4. The average number of unfulfilled constraints (5) for Algorithm II.

Problem size, $K\times m\times n$	Average number of constraints violated in the first iteration	Average number of constraints violated in the last iteration
$2\times 200\times 100$	53.2	15.6
$3\times 150\times 50$	40.0	5.3
$4\times 200\times 50$	47.9	17.9

As seen from Tables 1 and 2, the Algorithm I performs more stops and has a greater convergence rate at 1000 iterations versus the Algorithm II. However, the computational complexity of one iteration of Algorithm I is higher than the computational complexity of one iteration of Algorithm II, and as consequence, the Algorithm I execution time is greater than Algorithm II running time. Therefore, Algorithm II performs a greater number of iterations and stops than the Algorithm I at the same time. Tables 3 and 4 confirm this fact.

Note that the convergence of the subgradient method is essentially dependent on the type of step-size rules for α^N and β^N. It is known [18] that the subgradient method converges if any step sizes satisfy:

$$\sum_{N=1}^{\infty}\alpha^N = +\infty ,$$

where $\alpha^N \to 0,\quad N\to\infty$.

The computational experiment showed that if $\alpha^N = 1/(N+1)$, the algorithms perform efficiently in any sized problems. As for β^N, step-size rule $\beta^N = \max\limits_{i,j,k} c_{ij}^k \big/ (N+1)$ gives great convergence rate of algorithms only if $m<20$.

Consider the different types of step-size rules for β^N for MOAP $3\times 30\times 10$. The results of the computational experiment are presented in Table 5. Each table cell shows the number of problems that have been solved at the corresponding number of iterations.

As seen from Table 5, it is more efficient to use the following step-size rule for considered problem:

$$\beta^N = \begin{cases} \max\limits_{i,j,k} c_{ij}^k \big/ (N+1), & N < \max\limits_{i,j,k} c_{ij}^k, \\ 1 - N/(N_{\max}+1), & \text{otherwise.} \end{cases}$$

Table 5. The number of solved problems according to the type of step-size rules for β^N.

β^N	Number of iterations: Less than 50	From 50 to 100	From 100 to 200	From 200 to 500	From 500 to 100	More than 1000	Average execution time for 100 iterations (sec)
$\beta_1 = \max_{i,j,k} c_{ij}^k / (N+1)$	23	6	13	9	13	36	1.3
$\beta_2 = 1/(N+1)$	2	1	2	2	3	90	0.8
$\beta_3 = \max_{i,j,k} c_{ij}^k \left(1 - N/(N_{\max}+1)\right)$	1	1	2	2	57	38	1.1
$\beta_4 = \begin{cases} \max_{i,j,k} c_{ij}^k / (N+1), & N < \max_{i,j,k} c_{ij}^k, \\ 1 - N/(N_{\max}+1), & \text{otherwise.} \end{cases}$	25	9	13	10	19	24	0.9

In Table 5 $N_{\max} = 1000$.

Thus, the results of computational experiment are summarized as follows:

1. Algorithms performs efficiently and can be used as alternative methods of MOAP-solving.
2. For smaller sized problems ($m < 15$) about 90% of problems are solved at 1000 iterations by both algorithms; the main stopping criterion in this case is constraints (5). Increasing the size problem leads to the decreasing the number of exactly solved problems; the stopping criterion in this case is number of iterations.
3. The Algorithm I is more efficient than the Algorithm II when stopping criterion is number of iterations.
4. The Algorithm II is more efficient than the Algorithm I when stopping criterion is algorithm execution time.
5. The presented algorithms are polynomial (see the Statements 2 and 3).

4. Conclusion

The considered multi-objective assignment problem is NP-hard and the pertinent question is to develop the heuristic solving algorithm for it. The approximate MOAP solution has been obtained using properties of the primal and dual problems. The dual Uzawa method has been applied.

The proposed algorithms are based on the dual problem solving with the help of Lagrangian relaxation of some constraints of the primal problem. The structure of the set formed by these constraints radically influences on the solving algorithms.

The computational experiment has made it possible to analyse the convergence rate of algorithms. The effective ways to specify algorithm parameters for the given problem sizes are obtained. The suggestions for opting the solving algorithm according to the problem size are developed.

References

[1] Larichev O I 2000 Theory and Methods of Decision-Making, and also the Chronicle of Events in the Enchanted Countries (Moscow: Logos) (in Russian)
[2] Nikonov O Ya, Podolyaka O A, Podolyaka A N and Skakalina E V 2011 *Bulletin of KhNAHU* **55** pp 103-112
[3] Baeva N B and Bondarenko Yu V 2003 Theoretical Framework Computational Schemes of Vector Optimization (Voronezh: Publishing House of VSU) (in Russian)
[4] Shvarc D T 2013 *Science and education*. (Moscow: BMSTU) (2013)003 (in Russian)
[5] Steuer R E 1986 Multiple Criteria Optimization: Theory, Computation, and Application (New York: Wiley)
[6] Kashirina I L and Semenov B A 2007 *Information Technology* **5** pp 62-68 (in Russian)

[7] Podinovskij V V and Nogin V D 2007 Pareto-Optimal Solutions of Multicriteria Problems (Moscow: Nauka) (in Russian)

[8] Volkonskij V A, Eganyan G K and Pomanskij A B 1983 *Siberian Mathematical Journal* **24**(2) pp 9-17

[9] Brucker P 1972 *Operat. Res.* **16**(5) pp 189-197

[10] Emelichev V A and Perepelica V A 1988 *Bulletin of the Georgian SSR Academy of Sciences* **131**(3) pp 501-504 (in Russian)

[11] Perepelica V A 1989 Multicriteria Problems of Graph Theory. Algorithmic Approach (Kiev: UMK VO) (in Russian)

[12] Sergienko I V and Perepelica V A 1987 *Cybernetics* **5** pp 85-93 (in Russian)

[13] Aneja P and Nair K R 1979 *Manag. Sci.* **25**(1) pp 73-78

[14] Emelichev V A and Perepelica V A 1989 *Computational Mathematics and Mathematical Physics* **29**(2) pp 171-183 (in Russian)

[15] Papadimitriou C H and Steiglitz K 1998 Combinatorial Optimization. Algorithms and Complexity (New York: Dover)

[16] Perepelitza V A 1989 *14-th IFIP Conf. on Syst. Mod. and Opt.* (Leipzig, GDR) pp 13-14

[17] Kakhichko A A 1986 *Methods for Solving Nonlinear Problems and Data Processing: Collection* (Dnipropetrovsk: Leningrad State University) pp 41-44 (in Russian)

[18] Minoux M 1990 Mathematical Programming: Theory and Methods (Moscow: Nauka) (in Russian)

[19] Medvedeva O A 2013 *Bulletin of St. Petersburg University: Applied Mathematics. Computer Science. Control Processes* **10**(1) pp 85-94 (in Russian)

[20] Medvedeva O A, Medvedev S N and Tchernyshova G D 2013 *Bulletin of Voronezh State University. Series: Systems analysis and information technologies* **2** pp 38-41 (in Russian)

[21] Medvedeva O A 2013 Models and algorithms of solving assignment problems with additional constraints: PhD dissertation (Mathematics) (Voronezh: Voronezh State University) p 138

IOP Conf. Series: Journal of Physics: Conf. Series **973** (2018) 012040 doi:10.1088/1742-6596/973/1/012040

Models of resource allocation optimization when solving the control problems in organizational systems

V Menshikh, A Samorokovskiy, O Avsentev

Voronezh Institute of the Ministry of Internal Affairs of Russia, Voronezh, Russia.

E-mail: samorokovskii_an@mail.ru

Abstract. The mathematical model of optimizing the allocation of resources to reduce the time for management decisions and algorithms to solve the general problem of resource allocation. The optimization problem of choice of resources in organizational systems in order to reduce the total execution time of a job is solved. This problem is a complex three-level combinatorial problem, for the solving of which it is necessary to implement the solution to several specific problems: to estimate the duration of performing each action, depending on the number of performers within the group that performs this action; to estimate the total execution time of all actions depending on the quantitative composition of groups of performers; to find such a distribution of the existing resource of performers in groups to minimize the total execution time of all actions. In addition, algorithms to solve the general problem of resource allocation are proposed.

1. Introduction

One of the control optimization methods in organizational systems is optimization of operation of the executive subsystem [1]. This subsystem, when accomplishing the control tasks, performs some set of actions in order to form control actions on the managed system. These actions are performed in a certain order; execution of separate consecutive actions may depend on previous actions. Separate actions can be performed in parallel.

The execution time of each action depends on the amount of the resource allocated (in general, nonlinearly), under which in the organizational systems, as a rule, decision-makers act. Taking into account this circumstance, we will further assume that the resource is of a discrete nature [2].

In order to reduce the performance time of all actions, various resource allocation options may be used. With this purpose, it is necessary to determine the optimal resource allocation choice before the beginning of performance of the set of actions. As a criterion for optimization is the a priori estimated time of implementation of the whole set of actions for the formation of the control action [3–6].

The purpose of the work is development of a mathematical model for resource allocation optimization in order to reduce the time for taking managerial decisions and algorithms for solving the general resource allocation problem.

2. Problem description and formalization of the resource allocation option selection optimization problem

Let us consider the resource allocation option optimization problem for reduction of the total time for performance of a certain task.

It will be considered that the task implies execution of some set of partially ordered acts by groups of performers. And each action is assigned to a certain group of performers. The action, in turn, is a set of partially ordered operations (subactions), each can be performed by any performer from the respective group. Hence, the action performance time depends on the number of performers.

By resource allocation we shall mean the increase of the number of performers in the groups. This may lead to a reduction of the time of action performance by the groups, to which performers are added, and hence the total performance time of the entire set of actions. The mentioned problem is a complex three-level combinatorial problem, whose evolving required solving of the following specific problems:

1) assessment of the duration of performance of each action depending on the number of performers in the group that performs this action;

2) assessment of the total performance time of all actions depending on the quantitative composition of the group of performers;

3) finding such allocation of available resource of performers by the groups that minimizes the total performance time of all actions [7–10].

Let us formalize this problem. Let us write $\Delta = \{D_1, D_2, \ldots, D_n\}$ — a set of actions (n — number of actions). On set Δ, we will determine partial order relation $E \in \Delta \times \Delta$ such that (D_i, D_j), if action D_i must precede action D_j. It will be considered that for each arbitrary action $D \in \Delta$ a group of performers is set $G(D) \in \Theta = \{G_1, ..., G_m\}$ (m — number of groups of performers, $m \leq n$) and action performance time $T(D)$.

Each action $D_i \in \Delta$ supposes performance of set of operations $D_i = \{d_i^1, ..., d_i^{ki}\}$ (k_i — number of operations when performing actions D_i). Each operation d_i^j, in turn, is characterized by its performance time t_i^j.

Assume that on set D_i is also set partial order relation $E_i \in D_i \times D_i$ such that $(d_i^j, d_i^k) \in E_i$, if operation d_i^j must precede operation d_i^k.

For each resource allocation option, each group G_j is characterized by available number of performers $N(G_j)$.

3. Assessment of duration of performance of each action.

Initially, let us consider the problems of assessment of performance duration of the entire aggregate of actions for the set numbers of performers in each group $N_1, N_2, ..., N_m$.

At the first stage it is necessary to assess duration T_i of performance of each action D_i. Obviously, T_i is determined by the values of duration of operations t_i^j, partial order relation E_i and number of performers N_i in the group that performs this action. Considering that t_i^j and E_i will not subsequently change, and the number of performers during resource allocation may vary, let us consider $T_i = f(N_i)$.

Let us address the problem of assessment of values $T_i = f(N_i)$. Let us represent the process of performance of operations of action D_i as a sequence of steps, such that each step is a performance of one operation by one performer. Let us write the performers of the group $G \in \Theta$, performing action D_i. Write

$$X_{js}^k = \begin{cases} 1, \text{ if at the } k\text{-th step, performer } g_s \text{ performs operation } d_i^j; \\ 0, \text{ if otherwise.} \end{cases}$$

Let us address the description of functional dependencies between the operations and the performers. Write A_i^j — the moment of beginning of performing operation d_i^j, then $A_i^j + t_i^j$ — the moment of its ending. The action ends after the last operations performance step. Hence,

$$T_i = \max(A_i^j(X) + t_i^j)X_{js}^k, \text{ where } i = 1, ..., n. \quad (1)$$

Let us set partial order relation E_i by matrix $B_i = (b_{st}^i), s, t = 1, ..., k_j$, where

$$b_{st}^i = \begin{cases} 1, & \text{if action } d_i^s \text{ is preceded by action } d_i^t; \\ 0, & \text{if otherwise.} \end{cases}$$

Let us describe the constraints for the sequence of performance of steps:

$$b_{st}^i(A_i^S(X) + t_i^S) \le A_i^t(X) \quad (2)$$

— a condition that operation d_i^S must end before the beginning of operation d_i^t;

$$A_i^S((X) + t_i^S)X_{sj}^{k-1}X_{tj}^k \le A_i^t X_{sj}^{k-1} X_{tj}^k \quad (3)$$

— a condition that if on step $k-1$ is performed operation d_i^S, and on the k-th step — operation d_i^t, then d_i^S must end before the beginning of d_i^t;

$$X_{sj}^k X_{tj}^k = 0, s \ne t \quad (4)$$

— a condition that at each step only one operation is performed.

$$\sum_{j=1}^{ki}\sum_{k=1}^{ki} X_{sj}^k = 1 \quad (5)$$

— a condition that each operation is performed at one step only.

The general problem of assessment of performance time for action D_i looks like a problem of nonlinear Boolean programming of finding X^*, minimizing (1) under constraints (2)–(5).

4. Assessment of the total time of performance of all actions

Value T is determined by the values of durations T_i of certain actions, defined at the previous stage, as well as by partial order relation E and distribution of the groups of performers by actions [15, 16, 17, 18].

Let us again assume that performance of the actions is a sequence of steps such that each step is performance of one action by a respective group of performers. Write

$$Y_i^{km} = \begin{cases} 1, & \text{if at } k \text{ step group of performers } G_m \text{ fulfills action } D_i, \\ 0, & \text{if otherwise.} \end{cases}$$

$$P_i^m = \begin{cases} 1, & \text{if action } D_i \text{ is fulfilled by group of performers } G_m, \\ 0, & \text{if otherwise.} \end{cases}$$

$A(Y)$ — moment of the beginning of performing action D_i.

The entire aggregate of actions ends when the last action ends, i.e.

$$T = \max(A_i(Y) + T_i), \text{where } i = 1, ..., n. \quad (6)$$

Let us set partial order relation E by matrix $B = (b_{st}), s, t, ..., k_i$, where

$$b_{st} = \begin{cases} 1, & \text{if action } D_s \text{ precedes action } D_t, \\ 0, & \text{if otherwise.} \end{cases}$$

IOP Conf. Series: Journal of Physics: Conf. Series **973** (2018) 012040 doi:10.1088/1742-6596/973/1/012040

By analogy with the previous stage, let us determine the sequence of actions

$$b_{st}(A_s(Y) + T_s) \leq A_t(Y) \tag{7}$$

$$(A(Y) + T_s)Y_s^{(k-1)m}Y_t^{km} \leq A_t(Y), \text{where } Y_s^{(k-1)m}Y_t^k = 0, s \neq t, \tag{8}$$

$$\sum_{k=1}^{p}\sum_{s=1}^{p} Y_t^{km} = 1. \tag{9}$$

Moreover, unambiguity of setting the group of performers for each action is determined by condition

$$\sum_{k=1}^{p} Y_t^{km} = 1. \tag{10}$$

5. Allocation of the resource of performers

In order to minimize the total performance time of all actions.

Write $Z = \{Z_1, ..., Z_m\}$ — an option of allocation of the resource of performers by groups Z_j, — resource allocated to group G_j.

Then, the performance time of action D_i, which is performed by group G_j is defined as $T_i = T_i(X, Z)$, where T_i — a function describing the solution of problem (1)–(6).

The total performance time for all actions is defined as $T = T(Y, T_i(X, Z))$, where T — a function describing the result of solving problem (1)–(6) at the second stage [19,20,21].

For values $Z = \{Z_1, ..., Z_m\}$, such that $Z_i \in \tilde{Z}$, where $\tilde{Z}$ — a set of integers, the following constraints shall be met

$$\sum_{i=1}^{m} Z_i \leq \hat{Z}, \tag{11}$$

$$Z_i \geq 0. \tag{12}$$

Thus, the general resource allocation problem looks like

Find

$$(X^*, Y^*, Z^*) = \underset{X,Y,Z}{Argmin}\, T(Y, T_i(X, Z))$$

under constraints (2)–(5), (7)–(12).

6. Algorithms of solving the general resource allocation problem

In order to construct an algorithm for finding the optimal option of resource allocation, it is necessary to determine groups G_i, which perform their actions D_i over maximum time $T(D)$. It is conditioned by the fact that the use of the resource with maximum action performance duration leads to a more considerable reduction of the action performance time. $t_i^j(M_i)$ — duration of operation when adding M number of performers. It is necessary to find the maximum number of performers $M_i^{\max}$, which is appropriate to be added to groups G_i, for maximum reduction of the performance time of all actions $T(D)$.

If the number of resource units is small, then the problem can be solved by complete enumeration. In other cases, only approximate solution can be found. Within this work, we shall present a general description of two algorithms without their detailing [1].

1) At the first stage, using an algorithm falling within the class of "greedy algorithms", we will find the approximate solution for using the allocated resource, which allows us to maximally use available resource $Z^{\max}$.

Assume that all groups of performers are ordered in descending order of the total duration of actions being performed.

Initially we allocate such a minimal part of the resource that ensures maximal reduction of the total time of performance of all actions.

Similarly, we allocate the resource for the second and subsequent groups of performers until it is exhausted. Graphically, it looks like as follows (Figure 1).

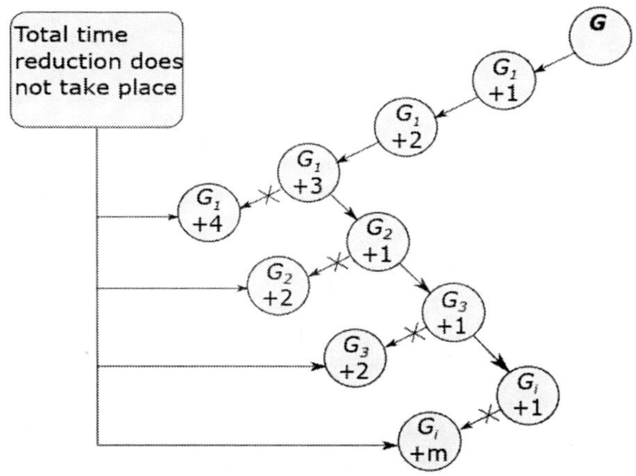

Figure 1. Resource allocation option based on the "greedy" algorithm

At subsequent stages, the obtained result is improved based on the selection of alternative options of resource allocation by groups.

First all rational, i.e. resulting in the total time reduction, options of allocation of the last performers out of the resource are considered. If in this case the total time is improved, then we will memorize the new option.

Then we consider the options of allocation of the two last performers etc. The process continues until all options are considered or the problem solving time limit is exhausted.

2) Another problem solving algorithm can be based on the ideas of "gradient descent" algorithm. Each time, one resource unit is allocated and an option that ensures maximum time reduction is selected (Figure 2). The algorithm continues until the entire resource is exhausted.

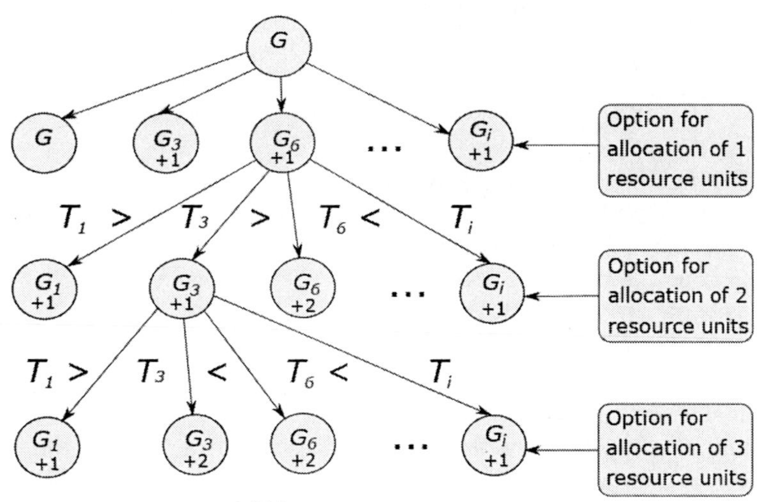

Figure 2. Resource allocation algorithm by the "gradient descent" method

A numerical example of a solution of a problem. Let us suppose that groups of performers g1, g2,. . . , g5 to form control actions must perform partially ordered actions $d_1, d_2, ...d_8$ (Figure 3).

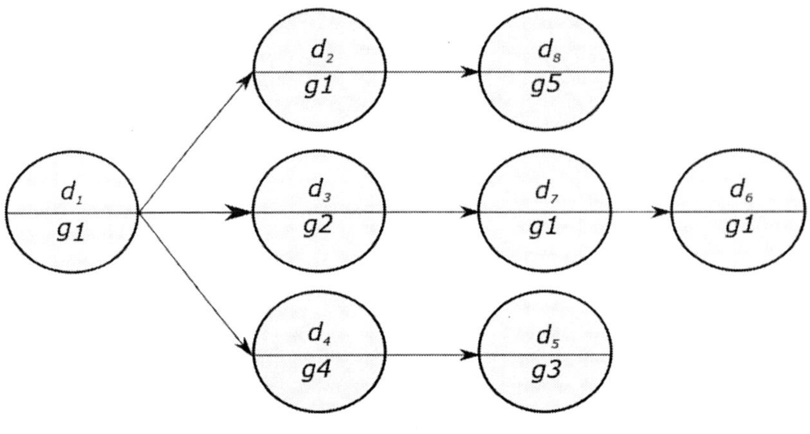

Figure 3. Graph of actions sequence execution.

On the basis of the graph of actions sequence the schedule of actions depending on the execution time of each action is constructed (Figure 4).

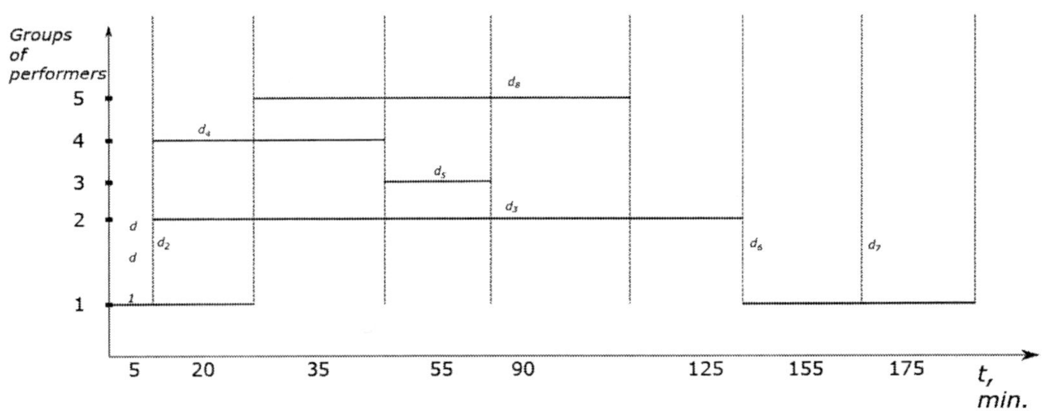

Figure 4. Schedule of actions depending on time.

The time for the implementation of individual actions can be reduced by attracting additional resources. Suppose that two more performers can be drawn into groups. The dependence of the reduction in the execution time of actions is shown in Table 1.

It can be seen from the table that the maximum reduction in the action time (40 minutes) occurs when two people are added to the group g2.

7. Conclusion

The developed model and algorithms can be used for optimizing the selection of the option of resource allocation in subsystems of the control element in order to reduce the total time of development of controlling actions in organizational systems of control of various purpose.

IOP Conf. Series: Journal of Physics: Conf. Series **973** (2018) 012040 doi:10.1088/1742-6596/973/1/012040

Table 1. Dependence of task execution time on resource usage.

		Performers							
		group g1				*group g2*	*group g3*	*group g4*	*group g5*
Actions		*d1*	*d2*	*d6*	*d7*	*d3*	*d4*	*d5*	*d8*
		Time, min.							
Attraction of a reserve	0	5	15	30	20	120	30	20	70
	1	5	13	25	17	100	30	20	70
	2	5	13	20	15	80	30	20	70
Time gain by action		0	2	5	3	20	0	0	0
Absolute time gain		15				40	0		

References

[1] Burkov V N and Novikov D A 2001 *Graph theory in managing organizational systems* (Moscow: SINTEG)

[2] Barkalov P S 2002 *Resource allocation problems in project management* (Moscow: IPU Russian Academy of Sciences)

[3] Korshunov Y M 1980 *Mathematical foundations of Cybernetics: textbook for universities* (Moscow: Energy)

[4] Menshikh V V 1993 *Automatic Control and Computer Sciences* **27** 3–9

[5] Menshikh V V 1996 *Automatic Control and Computer Sciences* **30** 51–54

[6] Menshikh V V 1996 *Automatic Control and Computer Sciences* **30** 35–40

[7] Menshikh V V and Samorokovskiy A F 2016 *Bulletin of Voronezh Institute of MIA of Russia* **1** 86–95

[8] Menshikh V V and Nikulina E Y 2011 *Optimization of temporal characteristics of information systems* (Voronezh: VI Ministry of internal Affairs of Russia)

[9] Menshikh V V and Sysoev V V 2002 *Structural adaptation of control systems* (Moscow: Radio Engineering)

[10] Burkov V N 1967 *The network model and the control problem* (Moscow: Soviet radio)

Mathematical model of information process of protection of the social sector

D A Novikov, E G Tsarkova, A S Dubrovin, A S Soloviev

ET Department, Voronezh Institute Federal Penitentiary Service of Russia, 1a Irkutskaya Street, Voronezh, RU

E-mail: university69@mail.ru

Abstract. In work the mathematical model of information protection of society against distribution of extremist moods by means of impact on mass consciousness of information placed in media is investigated. Internal and external channels on which there is a dissemination of information are designated. The problem of optimization consisting in search of the optimum strategy allowing to use most effectively media for dissemination of anti-terrorist information with the minimum financial expenses is solved. The algorithm of a numerical method of the solution of a problem of optimization is constructed and also the analysis of results of a computing experiment is carried out.

1. Introduction

In modern conditions terrorism became a problem of world international level. The global nature of modern terrorism is caused by development of means of communication, increase, due to various reasons, in migration flows, dense contact of the different communities differing in world outlook, religious and other traditions. Social and religious tension is the fertile field for replenishment of ranks of the terrorist organizations from among the people who haven't found the place in modern society and people with the deformed mentality. At the same time leaders of the terrorist organizations often are not the fanatics, but the people who have got a good education, pursuing the far-reaching aims. Modern terrorism has perfectly learned to use modern technical means of communications (for example, the Internet) for distribution of the ideology and recruitment of followers, has adapted to use the opportunities given by the democratic institutes providing the rights and personal freedom to express the views. The international nature of the terrorism which isn't knowing national borders demands combination of efforts of all countries for fight against him.

The leadership of the Russian Federation pays considerable attention to fight against terrorism. One of the directions of this fight is elimination of the social factors promoting growth of extremist moods, especially among youth further capable to bring them into ranks of terrorists. Other direction is a preventing to distribution of the ideas of the terrorist organizations, recruitment of new members. In the Concept of counteraction to terrorism in the Russian Federation approved as the President of the Russian Federation on October 5, 2009 distribution of the ideas of terrorism and extremism through the information and telecommunication Internet and mass media is called one of the major external factors promoting emergence and distribution of terrorism in the Russian Federation.

Often mass media involuntarily promote distribution of the ideas of terrorism, coming across tricks of the terrorist organizations as it is already noted above, perfectly learned to use freedom to express the opinion provided by institutes of democracy. For an opportunity to address wide audience terrorists manipulate public opinion, are covered with a mask of fighters against injustice, loosen moral and moral principles of society by substitution of concepts and ideals. Fight against similar terrorist activity demands the relevant activities, work with mass media in the direction of a point of their attention to need of increase in vigilance, professionalism and strengthening of attention to sources of providing information. The activity of terrorist promotion in mass media and the Internet needs to oppose effective counter-propaganda, namely, the purposeful work explaining to the population true tasks and the purposes of terrorism. Except professionals in the field of journalism and work with mass media it is necessary to attract widely the scientific capacity of the country for development of techniques of identification and blocking of distribution of sources of terrorist promotion. Development of methods of identification of information terrorist threats, their prevention and also assessment of a measure of their impact on the population is necessary.

In this work the model of distribution of counterterrorist promotion in society, including, on the Internet is investigated.

2. Theoretical part

2.1. Problem definition

We will consider some social group with a number N^* which members are subject to information influence during the anti-terrorist promotion which is carried out through media on information channels which are important for this group [1-2].

Let $x(t)$ – the number of the persons who have apprehended information distributed by a source by the time t of time, accepted the transferred ideas, norms, etc. We believe that at the initial moment $t = t_0$, $x(t_0) = x_0$.

We will assume that dissemination of information happens on two channels: external and internal.

The channel, external in relation to group, is characterized by size $u(t) \geq 0$ – number (intensity) of equivalent information acts of information transfer in unit of time.

The internal channel is defined by interpersonal communication of members of social group, at the same time for the characteristic of its intensity the parameter $\alpha \geq 0$ (a positive constant) is entered. As a result of such communication the members of the group who have already apprehended information, exerting impact on members yet not subject to information influence, make an own contribution to process of dissemination of information.

In this case the speed of change of number of the members of the group who have accepted the imparted norms is the sum of speeds of external and internal influences. This dependence is expressed by the equation [1]:

$$\dot{x}(t) = \left(u(t) + \alpha x(t)\right)\left(N^* - x(t)\right),\ t \in [0, T]. \tag{1}$$

We believe that the resource of the operating influence is limited on the set period $[0, T]$: $u(t) \leq u_{\max}$.

As criterion of optimality we consider maximizing number of the members of the group who have apprehended information at minimization of costs of these purposes on the fixed period. In this case criterion function will take a form:

$$J(u) = \int_0^T \left(x(t) - ku(t)\right)^2 dt \to \max, \tag{2}$$

where k – coefficient of cost of external influence (a payment for placement of information in media, financial and time expenditure).

As a result we come to a problem of optimum control:

IOP Conf. Series: Journal of Physics: Conf. Series **973** (2018) 012041 doi:10.1088/1742-6596/973/1/012041

$$J(u)=\int_0^T\left(x(t)-ku(t)\right)^2 dt\to\max,\tag{3}$$

at dynamic restriction:

$$\dot{x}(t)=\left(u(t)+\alpha x(t)\right)\left(N^*-x(t)\right),\quad t\in[0,T],\tag{4}$$

restriction for management:

$$0\le u(t)\le u_{\max},\ t\in[0,T],\tag{5}$$

entry condition:

$$x(t_0)=x_0.\tag{6}$$

2.2. Method of solution

We will apply the principle of a maximum of Pontryagin to the decision.

Pontryagin's function tasks (3)-(6) has an appearance:

$$H(t,x,u,p)=-\left(x(t)-ku(t)\right)^2+p(t)\left(\left(u(t)+\alpha x(t)\right)\left(N^*-x(t)\right)\right)\tag{7}$$

According to the principle of a maximum [3]-[4], the interfaced function $p(t)$ meets conditions:

$$\dot{p}(t)=-\frac{\partial H}{\partial x(t)}=\left(\alpha N^*-2x(t)\alpha-u(t)\right)p(t)-2k\alpha+2x(t),\quad t\in[0,T],\tag{8}$$

$$p(T)=-\frac{\partial H}{\partial x(T)}=0.\tag{9}$$

We will enter switching function:

$$\psi(t)=\frac{2x(t)k-p(t)\left(N^*-x(t)\right)}{2k^2}-\frac{u_{\max}}{2},\quad t\in[0,T].\tag{10}$$

Then optimum control $\bar{u}(t)$ meets a condition:

$$\bar{u}(t)=\begin{cases}u_{\max},\psi(t)\le 0,\\ 0,\psi(t)>0,\end{cases}\quad t\in[0,T].\tag{11}$$

The regional problem of the principle of a maximum has an appearance:

$$\begin{cases}\dot{x}(t)=\left(\bar{u}(t)+\alpha x(t)\right)\left(N^*-x(t)\right),\\ \dot{p}(t)=-\dfrac{\partial H}{\partial x(t)}=\left(\alpha N^*-2x(t)\alpha-\bar{u}(t)\right)p(t)-2k\alpha+2x(t),\\ x(t_0)=x_0,\\ p(T)=0,\end{cases}\tag{12}$$

где

$$\bar{u}(t)=\begin{cases}u_{\max},\psi(t)\le 0,\\ 0,\psi(t)>0,\end{cases}\quad t\in[0,T].\tag{13}$$

Thus, the solution of the task of optimum control (3)-(6) can be found as the solution of system of differential equations (12)-(13).

In some cases the system of a look (12)-(13) can be solved analytically. However, in the considered task the analytical decision is rather labor-consuming, besides even the most insignificant modification of model (for example, input of additional restrictions for a phase variable $x(t)$) does the decision only by an analytical method difficult. At the same time numerical methods of the decision (a method of a projection of a gradient, the interfaced directions, Newton, etc.) show high efficiency at the solution of the considered class of tasks by methods of nonlinear programming [3-19].

We approximate the considered task (3)-(6) problem of nonlinear programming and we will construct an algorithm of her numerical decision with use of a method of a projection of a gradient.

2.3. Discrete approximation

We will break a piece $[0,T]$, *q-1* point on q partial intervals of length $\Delta t = T/q$, we will put $t^i = i\Delta t$, $t^0 = 0$, $t^q = T$, $x(t^i) = x^i$, $u(t^i) = u^i$, $i = \overline{0, q-1}$.

We calculate target functionality with use of a formula of the left rectangles:

$$I(u) = \sum_{i=0}^{q-1} \left(x^i - ku^i\right)^2 \Delta t. \tag{14}$$

For approximation of derivatives we use Euler's formula of the 1st order of accuracy:

$$\dot{x}(t) \approx \frac{x^{i+1} - x^i}{\Delta t}, \quad x^0 = x_0. \tag{15}$$

The task approximating an initial problem of optimum control with an accuracy $O(\Delta t)$, has an appearance:

$$I(u) = \sum_{i=0}^{q-1} \left(x^i - ku^i\right)^2 \Delta t \to \max, \tag{16}$$

at restrictions:

$$x^{i+1} = x^i + \left(u^i + \alpha x^i\right)\left(N^* - x^i\right)\Delta t, \; x^0 = x_0, \tag{17}$$

$$0 \le u^i \le u_{\max}, \; i = \overline{0, q-1}. \tag{18}$$

We build Lagrange's function:

$$L(x, c, u, p) = -\sum_{i=0}^{q-1} \left(x^i - ku^i\right)^2 \Delta t + \sum_{i=0}^{q-1} p^{i+1}\left(x^{i+1} - x^i - \left(u^i + \alpha x^i\right)\left(N^* - x^i\right)\Delta t\right). \tag{19}$$

From stationarity conditions: $\frac{\partial L}{\partial x^i}(x, c, u, p) = 0$, $i = \overline{0, q}$, we receive recurrence relations for the interfaced variables p^i, $i = \overline{0, q-1}$:

$$p^i = p^{i+1} - \left(\left(\alpha N^* - 2x^i\alpha - u^i\right)p^i + 2k\alpha + 2x^i\right)\Delta t, \; i = \overline{0, q-1}, \; p_1^q = 0. \tag{20}$$

We calculate derivative functions of Lagrange in the direction u:

$$\frac{\partial L}{\partial u^i} = \frac{2x^i k\Delta t - p^i\left(N^* - x^i\right)\Delta t}{2k^2} - \frac{u_{\max}}{2}\Delta t, \; i = \overline{0, q-1}. \tag{21}$$

Taking into account the received formulas the algorithm of the numerical decision can be developed [3].

2.4. Gradient projection method algorithm

1. We set admissible approach of a vector of management $u^{(0)}$ on a zero step:

$$0 \le \left(u^{i}\right)^{(0)} \le u_{\max}, \ i = \overline{0, q-1}. \tag{22}$$

2. We count the initial trajectory corresponding to him: $\left(x^{i}\right)^{(0)}$, $i = \overline{0, q}$, on a formula:

$$\left(x^{i+1}\right)^{(0)} = \left(x^{i}\right)^{(0)} + \left(\left(u^{i}\right)^{(0)} + \alpha\left(x^{i}\right)^{(0)}\right)\left(N^{*} - \left(x^{i}\right)^{(0)}\right)\Delta t, i = \overline{0, q-1}, \left(x^{0}\right)^{(0)} = 0. \tag{23}$$

3. We calculate initial value of functionality:

$$I^{(0)} = -\sum_{i=0}^{q-1}\left(\left(x^{i}\right)^{(0)} - k\left(u^{i}\right)^{(0)}\right)^{2} \Delta t. \tag{24}$$

We define the interfaced variables p^{i}:

$$p^{i} = p^{i+1} - \left(\left(\alpha N^{*} - 2\left(x^{i}\right)^{(0)}\alpha - \left(u^{i}\right)^{(0)}\right)p^{i+1} + 2k\alpha - 2\left(x^{i}\right)^{(0)}\right)\Delta t, \ i = \overline{0, q-1}; \tag{25}$$

$$p^{q} = 0.$$

4. We set initial value of size of a step of gradient descent $\alpha^{(k)}$ also we will organize a cycle on steps.

5. We calculate management $(u)^{(k+1)}$, corresponding *(k+1)* iterations:

$$(u)^{(k+1)} = (u)^{(k)} - \alpha^{(k)} \frac{\partial L}{\partial u^{i}}\left(\left(u^{i}\right)^{(k)}\right), \ i = \overline{0, q-1}. \tag{26}$$

6. For new value $(u)^{(k+1)}$ vector of management we check an admissibility condition:

$$0 \le \left(u^{i}\right)^{(k+1)} \le u_{\max}, \ i = \overline{0, q-1}. \tag{27}$$

7. If the condition for some hub *i* isn't satisfied, then we carry out a projection of management to an admissible set.

8. We count the trajectory corresponding to this management $(x)^{(k+1)}$:

$$\left(x^{i+1}\right)^{(k+1)} = \left(x^{i}\right)^{(k+1)} + \left(\left(u^{i}\right)^{(k+1)} + \alpha\left(x^{i}\right)^{(k+1)}\right)\left(N^{*} - \left(x^{i}\right)^{(k+1)}\right)\Delta t, \tag{28}$$

$$\left(x^{0}\right)^{(k+1)} = 0.$$

9. We calculate the next approach of criterion function:

$$I^{(k+1)} = -\sum_{i=0}^{q-1}\left(\left(x^{i}\right)^{(k+1)} - k\left(u^{i}\right)^{(k+1)}\right)^{2} \Delta t. \tag{29}$$

10. We carry out check of a condition of monotony in a method of gradient descent. If the condition is satisfied: $I^{(k)} < I^{(k+1)}$, that we reduce a step of gradient descent also we pass to a step 5, otherwise we believe: $\alpha^{(k+1)} = \alpha^{(k)}$ also we pass to a step 10.

11. Iterative process continues until the set accuracy of calculations is reached ε.

12. If $\left|I^{(k+1)} - I^{(k)}\right| < \varepsilon$, we pass that to a step 12, otherwise we believe $(u)^{(k)} = (u)^{(k+1)}$, $(x)^{(k)} = (x)^{(k+1)}$ also we repeat all listed algorithm stages, since a step of the 5th algorithm where ε – the required accuracy of calculation of functionality.

13. $I^{(k+1)}$, $(x)^{(k+1)}$, $(u)^{(k+1)}$ – solution of a task.

3. Computing experiment

The considered algorithm is realized in the environment of *Microsoft Visual Studio* 2013 in the form of *Windows* – the application. The numerical solution of a task at the choice of parameters is constructed: $N^* = 100$, $u_{max} = 0.1$, $\alpha = 0.001$, $b = 0.8$, $k=1$, $T=10$, $q=1000$, $\varepsilon = 10^{-5}$. The optimal solution is presented in figures 1-2.

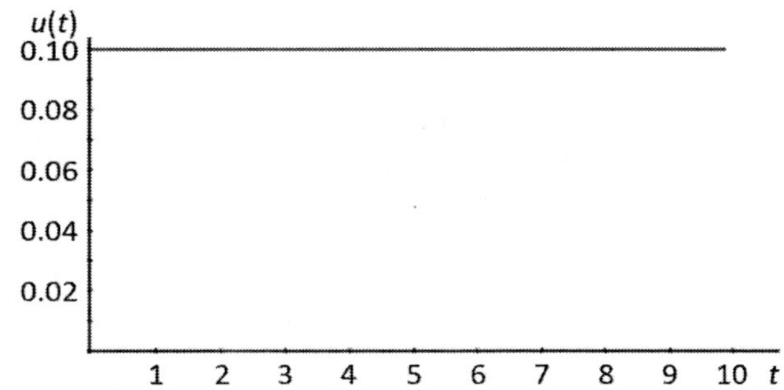

Figure 1. Schedule $\bar{u}(t)$.

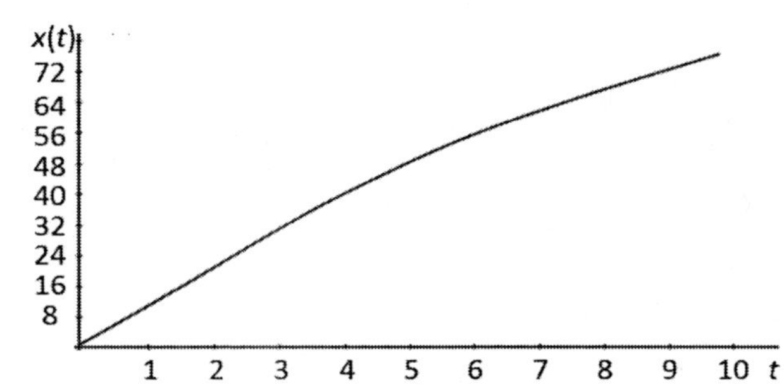

Figure 2. Schedule $\bar{x}(t)$.

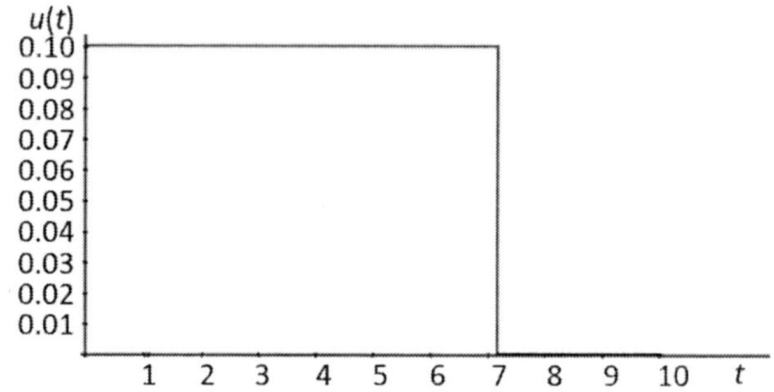

Figure 3. Schedule $\bar{u}(t)$.

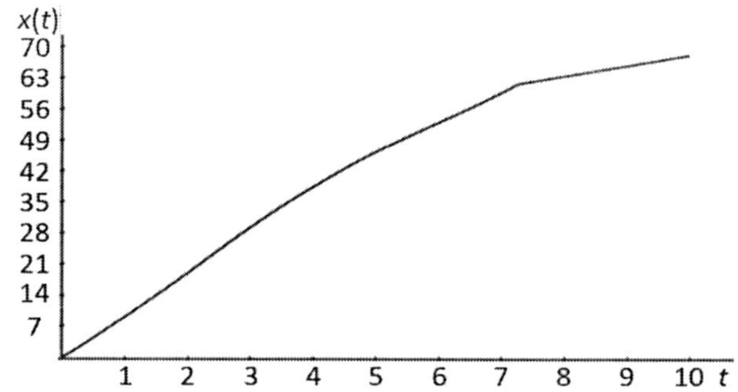

Figure 4. Schedule $\bar{x}(t)$.

At the choice of values of parameters: $N^* = 100$, $u_{\max} = 0.1$, $\alpha = 0.001$, $b = 0.8$, *k=100, T=10, q=1000,* $\varepsilon = 10^{-5}$ the optimal solution presented in figures 3-4 is received.

The solution of a task at values is provided on figures 5-6: $N^* = 100$, , $u_{\max} = 0.1$, $\alpha = 0.001$, $b = 0.8$, *k=500, T=10, q=1000,* $\varepsilon = 10^{-5}$.

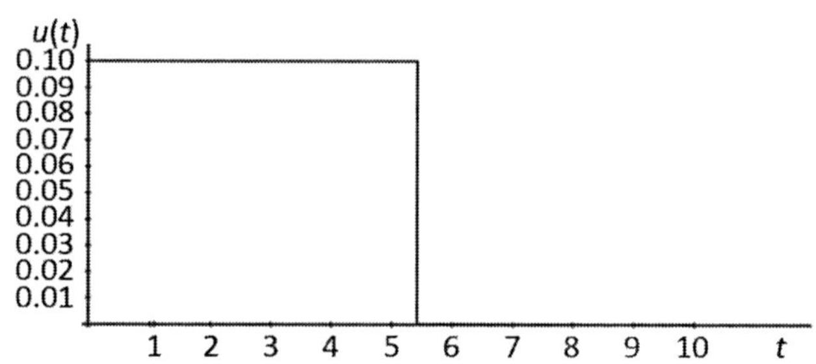

Figure 5. Schedule $\bar{u}(t)$.

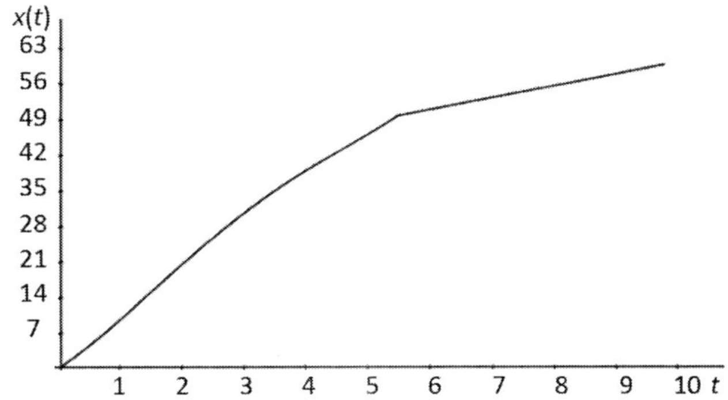

Figure 6. Schedule $\bar{x}(t)$.

From the analysis of schedules it is possible to draw a conclusion that at increase in a payment for placement of information in media in some timepoint the management which is responsible for distribution on the external channel is turned off and distribution continues only due to personal communications in group. Shutdown of the external channel happens the earlier, than cost is higher. Established facts will quite be coordinated with common sense that emphasizes adequacy of model of the considered situation.

4. Conclusion

Thus, at insignificant costs of placement of information in media strategy with the maximum intensity of external influence is optimum. At increase in cost of placement external influence initially as much as possible, and then is turned off. At the same time the activity of an internal channel of communication between members of social group remains. The results of the solution of a task received in number show high efficiency of the applied computing methods which keep the relevance in various modifications of model (for example, at input of the additional parameters expressing a oblivion of the transferred information, heterogeneity of social group, etc.) In this regard specification of model and also the analysis of productivity of numerical methods at the solution of the considered task is of special interest. The practical value of a research consists in development of the software which can be used at the organization of information protection of society against influence of ideology of extremism and also promotion of other socially important norms.

References

[1] Minayev V A, Ovchinsky A S, Skryl S V and Trostyansky S N 2013 *How to operate mass consciousness: modern models* (Moscow: RosNOU) p 202
[2] Balan V P, Dushkin A V, Novoseltsev V I and Sumin V I 2016 *Risk and reflection: monograph* (Moscow: Hot line – Telecom) p 136
[3] Anosov O L, Butkovsskii O Ya and Kravtsov Yu A 1996 *Predictability of Complex Dynamical Systems* (Berlin, Heidelberg: Springer-Verlag) p 234
[4] Minayev V A 2014 Modeling of developments of steady tourism *Service in Russia and abroad* **9** p 140-149
[5] Minayev V A and Sesyolkin A I 2014 Modeling of sustainable development of tourism in regions *Bulletin of the University* **21** pp 40-46
[6] Minayev V A and Faddeev A O 2009 *Assessment of geoecological risks: modeling the safety of tourist and recreational areas* (Moscow: Finance and statistics; INFRA-M) p 336
[7] Kara-Murza S G 2000 *Manipulation with consciousness* (Moscow: Algoritm) p 464
[8] Forester J 2003 *World dynamics* (Moscow: ACT) p 379
[9] Forrester J 1971 *Fundamentals of cybernetics of the enterprise* (*Industrial dynamics*) (Moscow: Progress) p 340

IOP Conf. Series: Journal of Physics: Conf. Series **973** (2018) 012041 doi:10.1088/1742-6596/973/1/012041

[10] Moiseyev N N 1982 *Man, environment, society: problem of formalization of the description* (Moscow: Science) p 240
[11] Moiseyev N N 1986 *Man, nature and the future of civilization: "Nuclear winter" and the problem of the "forbidden trait"* (Moscow: Novosti Press Agency) p 92
[12] Moiseyev N N, Alexandrov V V and Tarko A M 1985 *Man and the biosphere: Experience in system analysis and experiments with models* (Moscow: Science) p 271
[13] Dushkin A V, Barkalov S A, Kolodyazhny S A and Sumin V I 2017 *Introduction to the system analysis: monograph* (Moscow: Hot line – Telecom) p 234
[14] Dushkin A V, Kravchenko A S, Novoseltsev V I, Smolentseva T E and Sumin V I 2014 *Mathematical models and information processes of management of a difficult object: monograph* (Voronezh: Scientific book) p 125
[15] Dushkin A V, Kravchenko A S, Novoseltsev V I and Sumin V I 2014 *Models of management of difficult social objects in the system of training of employees of penal correction system: monograph* (Voronezh: Scientific book) p 156
[16] Mikhaylov A P and Klyusov N V 2002 About properties of the simplest mathematical model of distribution of information threat *Mathematical modeling of social processes* **4** p 115-123
[17] Samara A A and Mikhaylov A P 2001 *Mathematical modeling. Ideas, methods, examples* (Moscow: Science) p 164
[18] Andreyeva E A and Tsiruleva V M 2000 *Calculus of variations and methods of optimization* (Tver) p 143
[19] Andreyeva E A and Semykina N A 2010 *Optimum control. Manual* (Tver) p 167

IOP Conf. Series: Journal of Physics: Conf. Series **973** (2018) 012042 doi:10.1088/1742-6596/973/1/012042

Mathematical model and coordination algorithms for ensuring complex security of an organization

V I Novoseltsev, D E Orlova, A S Dubrovin, V P Irkhin

ET Department, Voronezh Institute Federal Penitentiary Service of Russia,
1a Irkutskaya Street, Voronezh, RU

E-mail: victor_novo@mail.ru

Abstract. The mathematical model of coordination when ensuring complex security of the organization is considered. On the basis of use of a method of casual search three types of algorithms of effective coordination adequate to mismatch level concerning security are developed: a coordination algorithm at domination of instructions of the coordinator; a coordination algorithm at domination of decisions of performers; a coordination algorithm at parity of interests of the coordinator and performers. Assessment of convergence of the algorithms considered above it was made by carrying out a computing experiment. The described algorithms of coordination have property of convergence in the sense stated above. And, the following regularity is revealed: than more simply in the structural relation the algorithm, for the smaller number of iterations is provided to those its convergence.

1. Introduction

Security of the organization represents the complex category including at least the following local components: providing mode and protection, fire security, information security, health and security of people. Need of coordination arises owing to the fact that the persons making private decisions on ensuring separate aspects of security are allocated with a certain independence. Not only the exception, but infringement of this independence reduces security as is followed by removal of responsibility from subordinates when performing of the functional duties by them. At the same time, the independence leads to the conflict of interests "private-part" and "private-general" which permission requires adoption of the complex coordinated decision considering both local interests, and interests at the level of all system in general [1-15].

2. Formulation of a problem

Let $S_{1234}^{1234}\left[\mathbb{R}(\vec{r}),\left(\mathbb{Q}_1(\vec{q}_1),\mathbb{Q}_2(\vec{q}_2),\mathbb{Q}_3(\vec{q}_3),\mathbb{Q}_4(\vec{q}_4)\right),t\right]$ – the tetravalent tensor describing security status of penal correction system institution; $\mathbb{R}(\vec{r})$ – space of indicators of complex security $(\vec{r})=(r_1,...,r_K)$; $\mathbb{Q}_1(\vec{q}_1),\mathbb{Q}_2(\vec{q}_2),\mathbb{Q}_3(\vec{q}_3),\mathbb{Q}_4(\vec{q}_4)$ – spaces of indicators of private aspects of security $(\vec{q}_1,...,\vec{q}_4)$, namely: mode and protection ($\vec{q}_1$), fire security ($\vec{q}_2$), information security ($\vec{q}_3$), health and security ($\vec{q}_4$), t – time. Let $S_{1234}^{1234}\left[\mathbb{R}'(\vec{r}'),t\right]$ – the tensor standard setting the required levels of complex security $(\vec{r}')$. Then the problem of ensuring complex security in her mathematical representation will consist in the decision relatively $C^{\mathbb{R}',\mathbb{Q}'_1,...,\mathbb{Q}'_4}_{\mathbb{R},\mathbb{Q}_1,...,\mathbb{Q}_4}$ и $C^{q'_i}_{q_i}(i=\overline{1,4})$ systems of the equations [1]:

$$\begin{cases} S^{1234}_{1234}\left[\mathbb{R}(\vec{r}),(\mathbb{Q}_1(\vec{q}_1),\mathbb{Q}_2(\vec{q}_2),\mathbb{Q}_3(\vec{q}_3),\mathbb{Q}_4(\vec{q}_4)),t\right]M(\vec{q}_1,...,\vec{q}_4)C^{\mathbb{R}',\mathbb{Q}'_1,...,\mathbb{Q}'_4}_{\mathbb{R},\mathbb{Q}_1,...,\mathbb{Q}_4}=S^{1234}_{1234}\left[\mathbb{R}'(\vec{r}'),t\right]; \\ \left.\begin{matrix} s^1(\vec{q}_1,t)K(\vec{q}_1,...,\vec{q}_4)C^{q'_1}_{q_1}=s^1(\vec{q}'_1,t); \\ ... \\ s^4(\vec{q}_4,t)K(\vec{q}_1,...,\vec{q}_4)C^{q'_4}_{q_4}=s^4(\vec{q}'_4,t) \end{matrix}\right\}, \end{cases} \quad (1)$$

where $C^{\mathbb{R}',\mathbb{Q}'_1...\mathbb{Q}'_4}_{\mathbb{R},\mathbb{Q}_1,...,\mathbb{Q}_4}$ and $C^{q'_i}_{q_i}(i=\overline{1,4})$ – the transformation tensors putting in compliance the flowing and demanded security levels; $\|\alpha_{ij}\|$– matrix of mismatch of the private aspects of security defined in coordinates $\vec{q}_1,...,\vec{q}_4$ spaces $\mathbb{Q}_1,\mathbb{Q}_2,\mathbb{Q}_3,\mathbb{Q}_4$, respectively; $K(\vec{q}_1,...,\vec{q}_4)$ – the coordination matrix set in spaces $\vec{q}_1,...,\vec{q}_4$; $s'^i(\vec{q}'_i,t);i=\overline{1,4}$ – the tensors describing conditions of private aspects of security; $s''^i(\vec{q}'_i,t);i=\overline{1,4}$ – the tensors standards setting the required levels of separate aspects of security $(\vec{q}'_1,...,\vec{q}'_4)$.

The obvious shortcoming (1) is that she has no constructive decisions as in neyaspekta of management and coordination are set implicitly. Directly to mark out the specified aspects we detail and we concretize (1). For this purpose, we will consider the control system of complex security of the organization represented on the scheme figure 1.

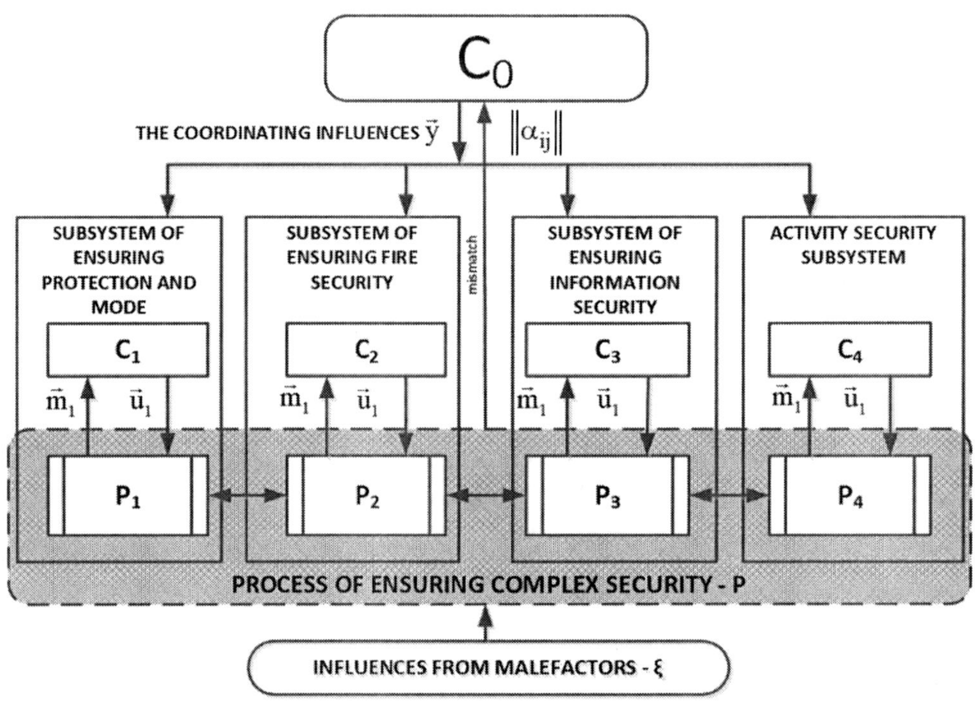

Figure 1. Control system of complex security of the organization.

On this scheme are designated by symbols: C_0 - the coordinator who is responsible for all system of security of the organization in general and coordinating work of performers; C_1, C_2, C_3, C_4 – performers, the exercising direct control of processes of ensuring local of the organization; P_1, P_2, P_3,

P_4 – the processes of providing the mode and protection (1), fire security (2), information security (3), health and security (4) making the general process of providing complex complex security of P; $\vec{y}_i(i=\overline{1,4})$ – vectors of the coordinating influences; $\vec{m}_i(i=\overline{1,4})$ – vectors of the operating influences; $\vec{u}_i(i=\overline{1,4})$ – vectors of information on a condition of private processes P_1, P_2, P_3, P_4; A = $\|\alpha_{ij}\|; i,j=\overline{1,4}; \alpha_{ii}=0$ – matrix of mismatch of processes P_1, P_2, P_3, P_4; ξ – matrix of mismatch of processes.

Functioning of such system is represented as follows. Coordinator C_0, obtaining information $\|\alpha_{ij}\|$ about the current mismatch of private processes of ensuring local security P_1, P_2, P_3, P_4, seeks to minimize a deviation of all process P from the required (reference) state that in an assumption of one-dimensionality of space $\mathbb{R}(\vec{r})$, that is when $(\vec{r})=r$, formally is expressed as the following criterion function:

$$r=\left[\int_0^T\sqrt{\frac{1}{4}\sum_{i=1}^{4}\mu_i\left(1-Q_i(t)\right)^2dt}\right]\Bigg|_{\|\alpha_{ij}\|}\to\min_{\vec{y}_i\in Y},\tag{2}$$

where $\mu_i\left(\sum_{i=1}^{4}\mu_i=1\right)$ – the expert coefficients reflecting the relative importance of private components of complex assessment of security; T – the considered time interval; $Q_1(\vec{m}_1,\vec{u}_1,\vec{y}_1,t)=\sqrt{\frac{1}{5}\sum_{i=1}^{5}\mu_{1i}\left(1-q_{1i}(\vec{m}_1,\vec{u}_1,\vec{y}_1,t)\right)^2}$; $Q_2(\vec{m}_2,\vec{u}_2,\vec{y}_2,t)=\sqrt{\frac{1}{7}\sum_{i=1}^{7}\mu_{2i}\left(1-q_{2i}(\vec{m}_2,\vec{u}_2,\vec{y}_2,t)\right)^2}$; $Q_3(\vec{m}_3,\vec{u}_3,\vec{y}_3,t)=\sqrt{\frac{1}{7}\sum_{i=1}^{7}\mu_{3i}\left(1-q_{3i}(\vec{m}_3,\vec{u}_3,\vec{y}_3,t)\right)^2}$; $Q_4(\vec{m}_4,\vec{u}_4,\vec{y}_4,t)=\sqrt{\frac{1}{5}\sum_{i=1}^{5}\mu_{4i}\left(1-q_{4i}(\vec{m}_4,\vec{u}_4,\vec{y}_4,t)\right)^2}$;

μ_{1i} - μ_{4i} – the rated expert coefficients reflecting the importance of indicators $\vec{q}_1,\dots,\vec{q}_4$ in the general structure of security of an object which contents reveals in table 1.

Performers of C_1, C_2, C_3, C_4, receiving from the coordinator of the instruction in a look $\vec{y}_i(i=\overline{1,4})$, aspire within these instructions to minimize deviations of the processes of P_1, P_2, P_3, P_4 from the required (reference) states that it is formally possible to express as the following system of criterion functions:

$$\left.\begin{aligned}
Q_1&=\left[\int_0^T Q_1(\vec{m}_1,\vec{u}_1,\vec{y}_1,t)dt\right]\Bigg|_{\vec{y}_1,\vec{u}_1,\xi}\to\min_{\vec{m}_1\in M_1}\\
Q_2&=\left[\int_0^T Q_2(\vec{m}_2,\vec{u}_2,\vec{y}_2,t)dt\right]\Bigg|_{\vec{y}_2,\vec{u}_2,\xi}\to\min_{\vec{m}_2\in M_2}\\
Q_3&=\left[\int_0^T Q_3(\vec{m}_3,\vec{u}_3,\vec{y}_3,t)dt\right]\Bigg|_{\vec{y}_3,\vec{u}_3,\xi}\to\min_{\vec{m}_3\in M_3}\\
Q_4&=\left[\int_0^T Q_4(\vec{m}_4,\vec{u}_4,\vec{y}_4,t)dt\right]\Bigg|_{\vec{y}_4,\vec{u}_4,\xi}\to\min_{\vec{m}_4\in M_4}
\end{aligned}\right\}.\tag{3}$$

Considering that indicators $\vec{q}_1,\dots,\vec{q}_4$ can be measured in different scales and have polar orientation, it is necessary before the decision (3) to carry out their normalization: $\hat{q}_{ij}=\left(q_{ij}-q_{ij}^{\min}\right)/\left(q_{ij}^{\max}-q_{ij}^{\min}\right)$ –

for the indicators focused on a maximum, that is the more the better and $\hat{q}_{ij} = 1 - (q_{ij} - q_{ij}^{min})/(q_{ij}^{max} - q_{ij}^{min})$ – for the indicators focused on a minimum that is the it is less, the better where q_{ij}^{max}, q_{ij}^{min} – minimum and maximum possible values of an indicator; $\hat{q}_{ij}$ – rated value of an indicator q_{ij}. Such normalization allows to reduce all indicators to the range of change from 0 to 1 and to identical orientation to a maximum. It gives the chance to construct a vector $\vec{q}_1,...,\vec{q}_4$ for a reference condition of P_1, P_2, P_3, P_4which components are in on a formula $q_j = \max_i q_{ij}$. Therefore, all components of a reference vector $\vec{q}_1,...,\vec{q}_4$, will be equal 1: $\vec{q}_1 = (1,1,1,1,1)$; $\vec{q}_2 = (1,1,1,1,1,1,1)$; $\vec{q}_3 = (1,1,1,1,1,1,1)$; $\vec{q}_4 = (1,1,1,1,1)$.

Table 1. Types and contents of indicators of private aspects of institution security.

Type	Maintenance of an indicator
Indicators of the mode and protection $(\vec{q}_1)$	q_{11} – security level when holding various actions;
	q_{12} – level of protection of buildings, rooms, equipment, production and technical means of ensuring production and other activity;
	q_{13} – level of personal security of the management and employees;
	q_{14} – timeliness of identification and elimination of the critical situations connected with actions of malefactors and terrorists;
	q_{15} – level of control of the mode of privacy and admission.
Indicators of fire security $(\vec{q}_2)$	q_{21} – level of completeness and quality of implementation of requirements of Fire security regulations and also other normative documents;
	q_{22} – level of security with fire extinguishing means and notifications;
	q_{23} – level of proficiency of staff to Fire security regulations and actions on a fire case;
	q_{24} – level of a fire-prevention condition of jobs and territories;
	q_{25} – level of the organization and ensuring fire security;
	q_{26} – level of potential danger of ignition of objects;
	q_{27} – timeliness of identification and elimination of the critical situations connected with the fires.
Information security indicators $(\vec{q}_3)$	q_{31} – reliability of providing information;
	q_{32} – completeness of output information;
	q_{33} – timeliness of providing information;
	q_{34} – reliability of output information;
	q_{35} – maintaining confidentiality of information;
	q_{36} – security level from unauthorized access and dangerous program and technical influences;
	q_{37} – timeliness of identification and elimination of the critical situations connected with information security.
Health and security indicators $(\vec{q}_4)$	q_{41} – levels of undesirable impacts on different person of streams of energy (mechanical, electromagnetic, ionizing);
	q_{42} – the doses received by the person during action on him the negative technogenic factors (ionizing, electromagnetic, etc.);
	q_{43} – concentration of the undesirable to the person toxic and polluting chemicals;
	q_{44} – volumes of emissions in the atmosphere and dumpings into the hydrosphere undesirable to the person toxic and (or) pollutants;
	q_{45} – timeliness of identification and elimination of the critical situations connected with danger of activity.

Follows from told that criterion functions (2) and (3) can be interpreted as follows: complex assessment of security is distance in 5, 7, 7 and 5 measured spaces $\mathbb{Q}_i(i=\overline{1,4})$ from a point with coordinates {(1,1,1,1,1), (1,1,1,1,1,1,1), (1,1,1,1,1,1,1), (1,1,1,1,1)}, соответствующей эталонному (требуемому) to condition of processes of P_1, P_2,P_3, P_4, to the point with coordinates $\vec{q}_1,...,\vec{q}_4$ corresponding to their current state.

Thus, the substantial party of mathematical model of coordination at adoption of the complex decision on security of the organization to be in that the coordinator on the basis of information on the level of mismatch of private decisions $\|\alpha_{ij}\|$ it developed and has brought such coordinating influences to subordinates $\vec{y}_i(i=\overline{1,4})$, which will help them to develop the managements of the security measures providing the greatest possible integrated effect at the level of the organization, but in too time keeping "optimality" of management of own processes from the point of view of local interests.

3. Assessment of mismatch of private processes of security of the organization

The theoretical base of such assessment is made by the typology of interference of private processes of ensuring local security of P_1, P_2, P_3, P_4 presented in table 2 [1, 5].

Table 2. Types of paired interaction between the private security processes of the institution and the corresponding level of inconsistency assessment.

Type interferences	Version	Interference manifestation form	Mismatch assessment (α_{ij})
Negative	Negative only	Achievement of the goal of one process excludes achievement of the goal of another	Arkh high (α_{ij} = - 1.0)
	Strictly negative	The greatest efficiency of one process is reached at the smallest efficiency of another	Very high (α_{ij} = - 0.8)
	Not strictly negative	Processes, exert negative impact at each other, but their purposes are achievable, though not fully	High (α_{ij} = - 0.5)
	Unilateral negative	One process exerts negative impact on another which doesn't exert any impact on the first	Not really high (α_{ij} = - 0.1)
Positive	Exclusively positive	The purposes of processes merge in one common goal	Zero (α_{ij} = 1.0)
	Strictly positive	Processes are united by unity of the purpose, but the contribution to her achievement at each of them is various	Very low (α_{ij} = 0.8)
	Not strictly positive	The purposes of processes are various, but are various on minor questions	Low (α_{ij} = 0.5)
	Unilateral positive	One process exerts positive impact on another which doesn't exert any impact on the first	Not really high (α_{ij} = 0.1)
Positive and negative	Normal	Despite counteraction, the purposes of processes are achievable	Very low (α_{ij} = 0.8)
	Abnormal	Despite assistance, the purposes of processes are unattainable	Very high (α_{ij} = - 0.8)
Neutral		Processes don't exert impact at each other	Zero (α_{ij} = 1.0)

Using data table 2, we leave a symmetric matrix $\|\alpha_{ij}\|; \alpha_{ii} = 0; i, j = 1,...,4$ by which we determine average degree of mismatch of private processes of security in the organization:

$$\alpha \approx 0,11\left[\sum_{i=1(i\neq j)}^{4}\left(\sum_{j=1(i\neq j)}^{4}\alpha_{ji}\right)\right]. \tag{4}$$

Size α defines type and the maintenance of an algorithm of coordination [3,4]. If $-1 < \alpha \leq 0.33$, then in system essential mismatches take place, and it is necessary for effective coordination that instructions of the coordinator dominated over decisions of performers. At $0.33 < \alpha \leq 1$ in system there are practically no mismatches concerning security, and effective coordination can be reached if the solution of problems is given not "payoff" to performers. If $-0.33 < \alpha \leq 0.33$, then in system insignificant mismatches on minor questions which can be eliminated on the basis of parity of interests of the coordinator and performers [5, 16-17] take place.

4. Coordination algorithm at domination of instructions of the coordinator

In this case the criterion of efficiency of coordination has to be based on comparison of sizes r and r', reflecting the flowing and demanded levels of security of all system in general. Taking into account it the algorithm of coordination comes down to the step-by-step iterative procedure of the following look.

Step 1. Using a program method of generation of random numbers, we set a starting vector of the coordinating influences $\vec{y}_i^{(0)} \in Y; (i = \overline{1,4})$. Random numbers have to get out of an interval from 0 to 1, at the same time their sequence has to submit to the uniform law of distribution.

Step 2. By method of calculus of variations, we solve problems (3). Also, we remember $Q_i^{(0)}\left(i = \overline{1,4}\right)$ the received values $\vec{m}_i^{*(0)}$.

Step 3. For the received values $Q_i^{(0)}\left(i = \overline{1,4}\right)$ we solve a problem (2) and we remember values $\vec{y}_i^{*(0)}(i = \overline{1,4})$ and $r^{(0)}$.

Step 4. We compare $r^{(0)}$ to the required level of complex security (r'). If $r^{(0)} \geq r'$ – the task is solved, the coordinating influences $\vec{y}_i^{*(0)}(i = \overline{1,4})$ are recognized as optimum. As $r^{(0)} < r'$ we continue the solution of a task.

Step 5. We choose a new casual vector $\vec{y}_i^{(1)} \in Y; (i = \overline{1,4})$.

Step 6. We solve problems (3), the received values $\vec{m}_i^{*(1)}\left(i = \overline{1,4}\right)$ and $Q_i^{(1)}\left(i = \overline{1,4}\right)$ remember.

Step 7. For the received values $Q_i^{(1)}\left(i = \overline{1,4}\right)$ we solve a problem (2) and we will remember values $\vec{y}_i^{*(1)}(i = \overline{1,4})$ and $r^{(1)}$.

Step 8. We compare $r^{(1)}$ to the required level of complex security (r'). If $r^{(1)} \geq r'$ – the task is solved, the coordinating influences $\vec{y}_i^{*(1)}(i = \overline{1,4})$ are recognized as optimum. As $r^{(0)} < r'$ we continue the solution of a task.

Step 9. We compare $r^{(1)}$ and $r^{(0)}$. If, $r^{(1)} < r^{(0)}$ that is the procedure repeats from a step 5. Otherwise $\left(r^{(1)} \geq r^{(0)}\right)$ the attempt is considered unsuccessful. Search stops after μ of unsuccessful steps is made: the problem of ensuring the required level of security of an object at the available basic data is considered impracticable.

5. Coordination algorithm at parity of interests of the coordinator and performers

Unlike two previous cases, the criterion of efficiency of coordination has to be based as on comparison of sizes r and r', reflecting the flowing and demanded levels of security of all system in general, and on comparison of sizes $Q_i\left(i=\overline{1,4}\right)$ and Q_i', reflecting the flowing and demanded levels of separate aspects of security. Taking into account it the coordination algorithm at parity of interests of the parties comes down to the step-by-step iterative procedure of the following look.

Step 1. Using a program method of generation of random numbers, we set a starting vector of the coordinating influences $\vec{y}_i^{(0)} \in Y;(i=\overline{1,4})$.

Step 2. One of methods of calculus of variations we solve problems (3). The received values $\vec{m}_i^{*(0)}\left(i=\overline{1,4}\right)$ и $Q_i^{(0)}\left(i=\overline{1,4}\right)$ remember.

Step 3. For the received values $Q_i^{(0)}\left(i=\overline{1,4}\right)$ we solve a problem (2) and we remember values $\vec{y}_i^{*(0)}(i=\overline{1,4})$ and $r^{(0)}$.

Step 4. Compare $r^{(0)}$ с r' и $Q_i^{(0)}\left(i=\overline{1,4}\right)$ с Q_i'. If $\left(r^{(0)} \geq r'\right)$ and $\underset{i=\overline{1,4}}{\forall}\left(Q_i^{(0)} \geq Q_i'\right)$ – the task is solved, the coordinating influences $\vec{y}_i^{*(0)}(i=\overline{1,4})$ are recognized as effective. If at least one their specified conditions aren't satisfied, then we continue the solution of a task.

Step 5. We choose a new casual vector $\vec{y}_i^{(1)} \in Y;(i=\overline{1,4})$.

*Step*6. We solve problems (3), the received values $\vec{m}_i^{*(1)}\left(i=\overline{1,4}\right)$ and $Q_i^{(1)}\left(i=\overline{1,4}\right)$ remember.

Step 7. For the received values $Q_i^{(1)}\left(i=\overline{1,4}\right)$ we solve a problem (2) and we remember values $\vec{y}_i^{*(1)}(i=\overline{1,4})$ and $r^{(1)}$.

Step 8. Compare $r^{(1)}$ с r' and $Q_i^{(1)}\left(i=\overline{1,4}\right)$ с Q_i'. If $\left(r^{(0)} \geq r'\right)$ и $\underset{i=\overline{1,4}}{\forall}\left(Q_i^{(1)} \geq Q_i'\right)$ – the task is solved, the coordinating influences $\vec{y}_i^{*(1)}(i=\overline{1,4})$ are recognized as effective. In a case $\left(r^{(1)} \geq r^{(0)}\right)$ the attempt μ is considered unsuccessful. Search stops after unsuccessful steps.

6. Conclusion

Assessment of convergence of the algorithms considered above it was made by carrying out a computing experiment. At the same time, it was necessary that the algorithm meets if for the number of iterative cycles (about 150-200) accepted for practice, it allows to receive established results, that is results having no more than 10% a deviation from average value at tenfold repetition of cycles. Results of a computing experiment are presented in table 3.

Table 3. The results of the computational experiment on the analysis of convergence of coordination algorithms.

Kind of an algorithm	Minimum of iterations for ensuring convergence	Deviation from an average at tenfold repetition of cycles
Coordination algorithm at domination of instructions of the coordinator	95-100	10-15%
Coordination algorithm at domination of decisions of performers	80-90	10-15%
Coordination algorithm at parity of interests of the coordinator and performers	120-160	10-15%

IOP Conf. Series: Journal of Physics: Conf. Series **973** (2018) 012042 doi:10.1088/1742-6596/973/1/012042

From the analysis of the obtained data it is visible that the described algorithms of coordination have property of convergence in the sense stated above. And, the following regularity is revealed: than more simply in the structural relation the algorithm, for the smaller number of iterations is provided to those its convergence. So, for example, for the simplest algorithm of coordination at domination of interests of performers the convergence is reached approximately for 80-90 cycles, in case of the most difficult a coordination algorithm at parity of interests of the coordinator and performers – for 120-160 cycles.

References

[1] Novoseltsev V I, Kochedykov S S and Orlova D E 2017 *Tensor Analysis Krone and his Applications* (Voronezh: Scientific book) p 220

[2] Aliyev R A and Liberzon M I 1987 *Methods and Algorithms of Coordination in Industrial Control Systems* (Moscow: Economy) p 168

[3] Mesarovich M, Mako D and Takahara I. 1973 *The Theory of Hierarchical and Multilevel Systems* (Moscow: World) p 344

[4] Sumin V I, Dushkin A V, Kravchenko A S and etc. 2014 *Modeling of Systems* (Voronezh: Scientific book) p 222

[5] Kinney R L and Rayfa X 1981 *Decision-Making at Many Criteria, Preferences and the Replacement* (Moscow: Radio communication) p 560

[6] 1999 *Common Criteria for Information Technology Security Evaluation* (*Electronic Materials*) (USA: National Institute of Standards and Technology & National Security Agency; Canada: Communication Security Establishment; United Kingdom: Communications-Electronics Security Group; Germany: Bundesamt fur Sichereit in der Informationstechnik; France: Service Central de la Securite des Systemes d'Information; Netherlands: National Communications Security Agency) ed 2.1

[7] FSTEC 2014 *Requirements to Ensuring Information Security in Automated Control Systems for Production and Technological Processes on Crucial Objects, Potentially Dangerous Objects and also the Objects Posing the Increased Hazard to Life and Human Health and for the Surrounding Environment* (Moscow: FSTEC RU 20140314)

[8] FSTEC 2007 *The General Requirements for Security of Information in the Key Systems of Information Infrastructure* (Moscow: FSTEC RU 20070518)

[9] FSTEC 2007 *Recommendations about Security of Information in the Key Systems of Information Infrastructure* (Moscow: FSTEC RU 20071119)

[10] FSTEC 2007 *Basic Model of Threats to Security of Information in the Key Systems of Information Infrastructure* (Moscow: FSTEC RU 20071119)

[11] FSTEC 2007 *A Technique of Definition of Relevant Threats to Security of Information in the Key Systems of Information Infrastructure* (Moscow: FSTEC RU 20071119)

[12] Balan V P, Dushkin A V, Novoseltsev V I and Sumin V I 2016 *Risk and Reflection* (Moscow: Hot Line – Telecom) p 136

[13] Balan V P, Dushkin A V and Novoseltsev V I 2015 *Management of the Conflicts: Manual for Higher Education Institutions* (Moscow: Hot Line – Telecom) p 144

[14] Novoseltsev V I and Tarasov B V 2011 *System Theory of the Conflict* (Moscow: Major) p 336

[15] Barkalov S A, Dushkin A V, Kolodyazhny S A, Novoseltsev V I and Sumin V I 2017 *Introduction to the System Analysis* (Moscow: Hot Line –Telecom) p 234

[16] Dushkin A V, Kravchenko A S, Novoseltsev V I, Smolentseva T E and Sumin V I 2014 *Mathematical Models and Information Processes of Management of a Difficult Object* (Voronezh: Scientific book) p 125

[17] Dushkin A V, Kravchenko A S, Novoseltsev V I and Sumin V I 2014 *Models of Management of Difficult Social Objects in the System of Training of Employees of Penal Correction System* (Voronezh: Scientific book) p 156

Assessment of efficiency of functioning the infocommunication systems a special purpose in the conditions of violation quality of relevance information

A V Parinov, L P Korotkikh, D B Desyatov, L V Stepanov

ET Department, Voronezh Institute Federal Penitentiary Service of Russia, 1a Irkutskaya Street, Voronezh, RU

E-mail: a_dushkin@mail.ru

Abstract. The uniqueness of information processing mechanisms in special-purpose infocommunication systems and the increased interest of intruders lead to an increase in the relevance of the problems associated with their protection. The paper considers the issues of building risk-models for the violation of the relevance and value of information in infocommunication systems for special purposes. Also, special attention is paid to the connection between the qualities of relevance and the value of information obtained as a result of the operation of infocommunication systems for special purposes. Analytical expressions for the risk and damage function in the time range in special-purpose infocommunication systems are obtained, which can serve as a mathematical basis for risk assessment. Further, an analytical expression is obtained to assess the chance of obtaining up-to-date information in the operation of infocommunication systems up to the time the information quality is violated. An analytical expression for estimating the chance can be used to calculate the effectiveness of a special-purpose infocommunication system.

1. Introduction

At present, the functioning of the Special-Purpose Infocommunication System (SPIS) plays a key role in ensuring security and making effective strategic decisions. At the same time, the uniqueness of information processing mechanisms in SPIS and the increased interest of malefactors lead to an increase in the urgency of problems related to their protection [1-7].

Attacks on special-purpose infocommunication systems are used to achieve the following objectives: obtaining information of varying degrees of confidentiality, as well as for reducing the usefulness of information received from SPIS after the attack. Problems of protection of SPIS were considered [8, 9]. In these works, the directions of their solution, the risks of special-purpose infocommunication systems [10-15] and the survivability of SPIS were analyzed to some extent [16-24]. However, these studies did not consider violations of the quality of the relevance and value of information, which is especially relevant in that the specificity of SPIS is related to the relevance of the data it produces.

The aim of the work is to develop a method for constructing a risk-model of the violation of the relevance and value of information in special-purpose infocommunication systems, as well as their interrelationships [25-32]. This is necessary to assess the effectiveness of such systems in the context of a violation of the quality of the relevance of information.

IOP Conf. Series: Journal of Physics: Conf. Series **973** (2018) 012043 doi:10.1088/1742-6596/973/1/012043

2. Risk assessment

Information has aging properties, since it is subject to the influence of time, therefore, information resources have a certain "life" period.

And the effect of aging ie. the decline in value over time and the artificial retention of information, depending on the importance of the decision taken, can lead to very harmful consequences. Risk analysis in this case is simply necessary.

A fundamental feature of the functioning of the SPIS is that its utility function has a non-linear nature of the buildup during the operation of the system. And, according to the law of marginal utility [4, 15], the first derivative in the dynamics of this function tends to the expression:

$$\frac{\partial \omega}{\partial x} = a\omega, \tag{1}$$

where ω – utility function, x – quantity of consumed goods, a – coefficient of value, depending on the specifics of the SPIS.

When operating an infocommunication system of a special purpose, its use is associated with the receipt of some information. Suppose that it comes with a relatively constant intensity λ_p. From here it is possible to write down $x = \lambda_p t$. Carrying out the normalization for the average duration of the successful operation of SPIS T_{cp}, we rewrite the last expression in the following form, $x = \bar{\lambda}_P \bar{t}$ where $\bar{t} = t/T_{av}$ and $\bar{\lambda}_P = \lambda_p T_{av}$.

This approach allows us to further operate with x as a dimensionless quantity.

Let us now return to expression (1) and carry out the corresponding substitution:

$$\frac{\partial \omega}{\partial \lambda_p t} = a\omega,$$

or

$$\frac{\partial \omega}{\partial t} = a\lambda_p \omega \Rightarrow \frac{\partial \omega(t)}{\partial t} = a\lambda_p \omega(t).$$

The solution of the last equation is obviously an exponential. Taking into account the start information b and the current $\lambda_y t$ loss of information during the operation of the SPIS, we have:

$$\omega(t) = \exp\left(a\lambda_p t + b - \lambda_y t\right),$$

or

$$\omega(t) = \exp\left[a\left(\lambda_p - \lambda_y\right)t + b\right],$$

or

$$\omega(t) = \exp\left[a\lambda t + b\right],$$

where $\lambda = \lambda_p - \lambda_y$, a and b – dimensionless (normalized) quantities.

In the normalized form, the last expression takes the following form:

$$\omega(\bar{t}) = \exp\left[\bar{a}\bar{\lambda}\bar{t} + \bar{b}\right]. \tag{2}$$

We now determine the chances and risks [3,11] of the violation of the quality of the information's relevance at a time $\overline{t_0}$. In this case, the probability density and the accumulated probability of violation of the quality of the information's topicality will be denoted by $f(t)$ and $F(t)$, respectively.

Usually, the risk is defined as the product of the amount of damage to the likelihood of its occurrence [4].

$$Risk(t) = f(t)\cdot u(t)\Delta t, \tag{3}$$

where $u(t)$ – amount of damage, $f(t)$ – the probability of occurrence of an event of a violation of the quality of the information's relevance, Δt – sampling step.

The damage function can be obtained with the utility function, the graph of this function is shown in figure 1.

The area of the shaded figure in the gap is the damage $[0;t_0]$ due to the current loss of information, and the area of the shaded figure on the gap $[t_0;T_{av}]$ represents a lost profit that a special-purpose info-communication system could have brought had it not lost its operability at time t_0.

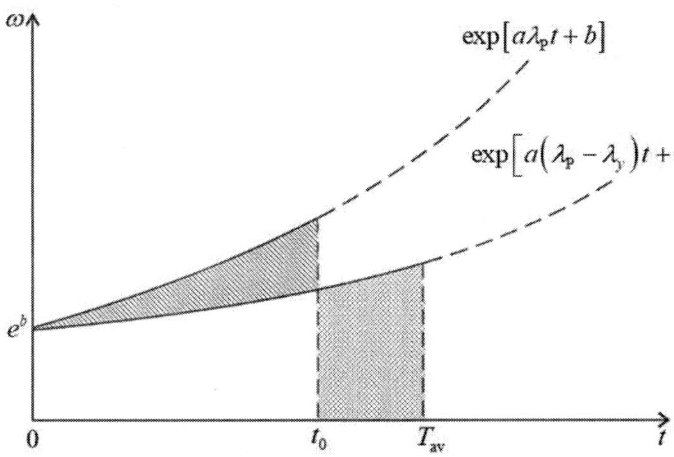

Figure 1. The graph of the utility function.

Hence the damage is:

$$u(t_0) = \int_0^{t_0} \exp[a\lambda_p t + b]dt - \int_0^{t_0} \exp\left[a\left(\lambda_p - \lambda_y\right)t + b\right]dt + \int_{t_0}^{T_{av}} \exp\left[a\left(\lambda_p - \lambda_y\right)t + b\right]dt.$$

Performing simple mathematical operations, we obtain the formula (4):

$$u(\mathrm{t}_0) = \int_0^{t_0} \exp\left[a\lambda_y t + b\right]\mathrm{d}t + \int_{t_0}^{T_{av}} \exp\left[a\left(\lambda_p - \lambda_y\right)t + b\right]\mathrm{d}t = $$
$$= \frac{\exp[b]\left(\exp\left[a\lambda_p T_{av} - a\lambda_y T_{av}\right] - \exp\left[a\lambda_p t_0 - a\lambda_y t_0\right]\right)}{a\left(\lambda_p - \lambda_y\right)} + \frac{\exp[b]\left(\exp\left[a\lambda_y t_0\right] - 1\right)}{a\lambda_y}. \tag{4}$$

The graph of the damage function is shown in the figure 2.

This chart is schematically similar to the utility function graph, which corresponds to its physical meaning, the ISSN brings little damage at the implementation stage, and since the loss of its efficiency, the value of the damage has increased dramatically.

Thus, the value of damage was found throughout the lifetime of the SPIS.

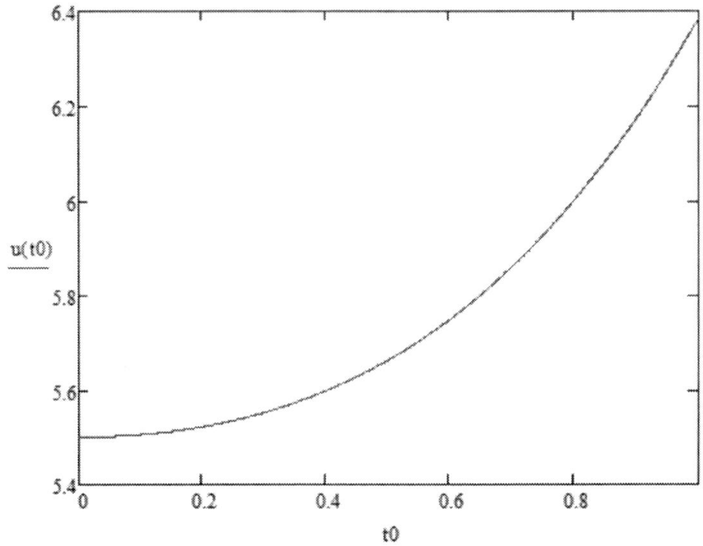

Figure 2. The schedule of the damage function.

According to [6, 12-14], cases of violation of the quality of the information's relevance correspond to the log-logistic distribution law [28-32]. The probability density function for the log-logistic distribution law is [7, 16-18]:

$$f(t) = \frac{(\beta/\alpha)(t/\alpha)^{\beta-1}}{\left[1+(t/\alpha)^{\beta}\right]^{2}}, \tag{5}$$

where $t > 0$, $\alpha > 0$, $\beta > 1$,

α – scale parameter,

β – form parameter.

The probability density function – mode [7] assumes the maximum value at the point:

$$t_{\max} = \alpha \sqrt[\beta]{\frac{\beta-1}{\beta+1}}. \tag{6}$$

Substituting $t_{\max}$ in (5) we obtain:

$$f_{\max} = \frac{(\beta-1)^{\frac{\beta-1}{\beta}} (\beta+1)^{\frac{\beta+1}{\beta}}}{4\alpha\beta}. \tag{7}$$

Thus, the step of sampling is:

$$\Delta t = \frac{4\alpha\beta}{(\beta-1)^{\frac{\beta-1}{\beta}} (\beta+1)^{\frac{\beta+1}{\beta}}}. \tag{8}$$

Taking into account formulas (4), (5) and (8), the analytic expression for the risk function is represented as follows:

$$Risk(t_0)=\left[\frac{\exp[b]\left(\exp\left[a\lambda_P T_{av}-a\lambda_y T_{av}\right]-\exp\left[a\lambda_P t_0-a\lambda_y t_0\right]\right)}{a\left(\lambda_P-\lambda_y\right)}+\frac{\exp[b]\left(\exp\left[a\lambda_y t_0\right]-1\right)}{a\lambda_y}\right]\times$$
$$\times\frac{(\beta/\alpha)(t_0/\alpha)^{\beta-1}}{\left[1+(t_0/\alpha)^\beta\right]^2}\times\frac{4\alpha\beta}{(\beta-1)^{\frac{\beta-1}{\beta}}(\beta+1)^{\frac{\beta+1}{\beta}}}. \quad (9)$$

In the subsequent calculations, to simplify the calculations, Δt one should take for a constant value $\Delta t = const$, and the $a\left(\lambda_{\Pi}-\lambda_y\right)=z,\ a\lambda_y=c$ value then:

$$Risk(t_0)=\left[\frac{e^b\left(e^{zT_{av}}-e^{zt_0}\right)}{z}+\frac{e^b\left(e^{ct_0}-1\right)}{c}\right]\times\frac{(\beta/\alpha)(t_0/\alpha)^{\beta-1}}{\left[1+(t_0/\alpha)^\beta\right]^2}\times\Delta t. \quad (10)$$

To solve the problems of preserving the quality of the actuality of information in general, it is necessary to determine the risk parameters. We calculate the risk parameters for the resulting damage expression and the log-logistic distribution law.

We now turn to the calculation of the mode and the peak of risk. Fashion is found by solving the task of investigating the risk function for extremes [19, 20].

Equate the resulting expression to zero:

$$\Delta t\frac{e^b(\beta/\alpha)(t_0/\alpha)^{\beta-1}}{cz\left[1+(t_0/\alpha)^\beta\right]^3}\times\left((t_0/\alpha)^\beta+1\right)\times\left(cze^{ct_0}-cze^{zt_0}+z\beta(e^{ct_0}-1)-\right.$$
$$\left.-z(e^{ct_0}-1)-c\beta\left(e^{zT_{av}}-e^{zt_0}\right)-c\left(e^{zT_{av}}-e^{zt_0}\right)\right)=0.$$

Critical points, functions $Risk(t_0)$ on the investigated interval $[0;T_{av}]$:

$$\frac{e^b(\beta/\alpha)(t_0/\alpha)^{\beta-1}}{cz\left[1+(t_0/\alpha)^\beta\right]^3}=0;$$

or

$$(t_0/\alpha)^\beta+1=0;$$

or

$$\left(cze^{ct_0}-cze^{zt_0}+z\beta(e^{ct_0}-1)-z(e^{ct_0}-1)-c\beta\left(e^{zT_{av}}-e^{zt_0}\right)-c\left(e^{zT_{av}}-e^{zt_0}\right)\right)=0.$$

The first equation has a solution $t_0=0$, that does not satisfy the condition $t_0>0$. The second equation has no positive solution.

We now turn to the solution of the last equation:

$$cze^{ct_0}-cze^{zt_0}+z\beta e^{ct_0}-z\beta-ze^{ct_0}+z-c\beta e^{zT_{av}}+c\beta e^{zt_0}-ce^{zT_{av}}+e^{zt_0}=0$$
$$e^{ct_0}\left(cz+z\beta-z\right)-e^{zt_0}\left(cz-c\beta-c\right)-c\beta e^{zT_{av}}-ce^{zT_{av}}-z\beta+z=0.$$

To simplify the notation, we introduce the notation:

$$cz+\beta z-z=N,$$

IOP Conf. Series: Journal of Physics: Conf. Series **973** (2018) 012043 doi:10.1088/1742-6596/973/1/012043

$$cz - c\beta - c = P,$$

$$c\beta e^{zT_{av}} - ce^{zT_{av}} - z\beta + z = Q.$$

The equation takes the form:

$$e^{ct_0} \cdot N - e^{zt_0} \cdot P - Q = 0.$$

We use the expansion of the exponential in a Taylor series, restricting ourselves to the first three terms. We have:

$$\left(1 + t_0 c + \frac{t_0{}^2 c^2}{2}\right) N + \left(1 + t_0 z + \frac{t_0{}^2 z^2}{2}\right) P - Q = 0;$$

$$N + t_0 cN + \frac{c^2 N}{2} t_0{}^2 + P + t_0 zP + \frac{z^2 P}{2} t_0{}^2 - Q = 0;$$

$$\left(\frac{c^2 N}{2} + \frac{z^2 P}{2}\right) t_0{}^2 + (cN + zP) t_0 + N + P - Q = 0.$$

We have a quadratic equation:

$$D = (cN + zP)^2 - 4(N + P - Q)\left(\frac{c^2 N}{2} + \frac{z^2 P}{2}\right);$$

$$t_{1,2} = \frac{-(cN + zP) \pm \sqrt{D}}{c^2 N + z^2 P}.$$

Presumably, the value of the second root is equal to the negative value, which contradicts the assertion that then the value $t > 0$, of the mode is equal to:

$$t_1 = \frac{-(cN + zP) + \sqrt{(cN + zP)^2 - 4(N + P - Q)\left(\frac{c^2 N}{2} + \frac{z^2 P}{2}\right)}}{c^2 N + z^2 P}. \quad (11)$$

Let the value of the mode $t_1 = t_n$, then the maximum damage will be equal to:

$$u(t_n) = \frac{e^b\left(e^{zT_{av}} - e^{zt_n}\right)}{z} + \frac{e^b\left(e^{ct_n} - 1\right)}{c}. \quad (12)$$

Consequently, the calculation formula for the risk peak has the following form:

$$Risk(t)_{\max} = Risk(t_n) \cong \left[\frac{e^b\left(e^{zT_{av}} - e^{zt_n}\right)}{z} + \frac{e^b\left(e^{ct_n} - 1\right)}{c}\right] \times \frac{(\beta/\alpha)(t_n/\alpha)^{\beta-1}}{\left[1 + (t_n/\alpha)^\beta\right]^2} \times \Delta t.$$

The obtained analytical expressions for risk can serve as a mathematical basis for assessing the risk of SPIS. In turn, the risk function will be used by us in the future when assessing the viability of these systems.

3. Evaluation of the chance of obtaining up-to-date information

If the risk function has the meaning of a probabilistic evaluation of the occurrence of damage at some point in time, then the chance function is something else like a probabilistic estimate of the receipt by some system of use at an arbitrary moment in time.

In general, the function of chance is as follows [3, 21-24]:

$$Chance(t_0) = v(t_0)\left[1 - F(t_0)\right], \tag{13}$$

where $v(t_0)$ – the benefits of using a special-purpose infocommunication system,
$(1 - F(t_0))$ – probability of failure-free operation of system components up to the moment t_0.

Thus, we have:

$$F(t_0) = \int_0^{t_0} f(t)\mathrm{d}t = \frac{(t_0/\alpha)^\beta}{1+(t_0/\alpha)^\beta}. \tag{14}$$

Then the probability of failure-free operation:

$$1 - F(t_0) = \frac{1}{1+(t_0/\alpha)^\beta}. \tag{15}$$

We will count the favor as a benefit obtained from the use of SPIS in a period of time $[0;t_0]$. To find it, we use the utility function, developed by us earlier.

The benefit received by the system at the supposed moment of violation of the quality of the information's relevance is determined from the utility function by integrating the latter by the parameter dt in the interval $[0;t_0]$, where t_0 is the alleged moment of violation of the quality of the information's relevance.

$$v(\mathrm{t}_0) = \int_0^{t_0} \exp\left[a(\lambda_{\mathrm{p}} - \lambda_y)t + b\right]\mathrm{d}t = \frac{e^b\left(e^{at_0\lambda_{\mathrm{p}} - at_0\lambda_y} - 1\right)}{a(\lambda_{\mathrm{p}} - \lambda_y)}. \tag{16}$$

Thus, we have an expression for assessing the chance of obtaining up-to-date information from the infocommunication system up to the time of the violation of the quality of the information's relevance t_0:

$$Chance(t_0) = \frac{e^b\left(e^{at_0\lambda_{\mathrm{p}} - at_0\lambda_y} - 1\right)}{a(\lambda_{\mathrm{p}} - \lambda_y)} \cdot \frac{1}{1+(t_0/\alpha)^\beta} = \frac{e^b\left(e^{at_0\lambda_{\mathrm{p}} - at_0\lambda_y} - 1\right)}{\left[a(\lambda_{\mathrm{p}} - \lambda_y)\right]\left[1+(t_0/\alpha)^\beta\right]}. \tag{17}$$

We use the substitution as in solving the risk, let $a(\lambda_{\mathrm{p}} - \lambda_y) = z$, then:

$$Chance(t_0) = \frac{e^b\left(e^{zt_0} - 1\right)}{z\left[1+(t_0/\alpha)^\beta\right]}. \tag{18}$$

This formula shows the dependence of the chance on time t_0, as well as $a, b, \lambda_y, \lambda_{\mathrm{p}}, \alpha, \beta$.

Thus, an analytical expression was obtained to assess the chance of obtaining up-to-date information in the operation of SPIS until the quality of the information is violated t_0. The analytical expression for the assessment of the chance will be used by us in the future to calculate the performance of the infocommunication system for special purposes.

4. Evaluation of the effectiveness of the infocommunication system

In the context of the risk analysis of the violation of the quality of the information's relevance, it is appropriate to solve the task of determining the effectiveness of the work of the SPIS [5, 9].

Traditionally, the efficiency of a system is understood to mean the ratio of the cost to total costs.

We estimate the efficiency of the system using the expression [4, 11]:

$$E = \frac{Chance(t_0) - Risk(t_0)}{Z(t_0)} - 1, \tag{19}$$

where $Chance(t_0)$ and $Risk(t_0)$ – risk assessment and the chance of SPIS at the moment (t_0);

$Z(t_0)$ – total costs of maintaining the life of the SPIS at the time (t_0).

Costs in this case are understood, in particular, the cost of deployment and commissioning of SPIS, as well as the costs of maintaining the system of protection of SPIS and the elimination of destructive influences.

Thus, we will define an expression for the evaluation of the effectiveness of SPIS.

$$E = \frac{Chance(t_0) - Risk(t_0)}{Z(t_0)} - 1 =$$

$$= \frac{\left[\frac{e^b\left(e^{zt_0}-1\right)}{z}\cdot\frac{1}{1+(t_0/\alpha)^\beta}\right] - \left[\left(\frac{e^b\left(e^{zT_{av}}-e^{zt_0}\right)}{z}+\frac{e^b\left(e^{ct_0}-1\right)}{c}\right)\times\frac{(\beta/\alpha)(t_0/\alpha)^{\beta-1}}{\left[1+(t_0/\alpha)^\beta\right]^2}\right]}{Z(t_0)} - 1 = \tag{20}$$

$$= \frac{\left[1+(t_0/\alpha)^\beta\right]\left(e^{zt_0}-1\right)ce^b - \left[ce^b\left(e^{zT_{av}}-e^{zt_0}\right)+ze^b\left(e^{ct_0}-1\right)\right](\beta/\alpha)(t_0/\alpha)^{\beta-1}}{Z(t_0)zc\left[1+(t_0/\alpha)^\beta\right]^2} - 1.$$

The expression of efficiency over the whole period of the functioning of the SPIS, taking into account the discretization of time, can be written in the form:

$$E_\Sigma = \frac{\sum_{k=1}^{n}\left[1+(k/n\alpha)^\beta\right]\left(e^{zk/n}-1\right)ce^b - \sum_{k=1}^{n}\left[ce^b\left(e^{zT_{av}}-e^{zk/n}\right)+ze^b\left(e^{ck/n}-1\right)\right](\beta/\alpha)(k/n\alpha)^{\beta-1}}{\sum_{k=1}^{n}Z(k/n)zc\left[1+(k/n\alpha)^\beta\right]^2} - 1. \tag{21}$$

5. Calculation of risk sensitivity coefficients

The dynamics of the risk movement will be estimated using the sensitivity coefficients. Consider the risk function obtained earlier:

$$Risk(t_0) = \left[\frac{e^b\left(e^{zT_{av}}-e^{zt_0}\right)}{z}+\frac{e^b\left(e^{ct_0}-1\right)}{c}\right]\times\frac{(\beta/\alpha)(t_0/\alpha)^{\beta-1}}{\left[1+(t_0/\alpha)^\beta\right]^2}\times\Delta t, \tag{22}$$

where: t_0 – the moment of violation of the quality of the relevance of the information, T_{av} – the average duration of use of SPIS, α, β – the parameters of the probability distribution of the occurrence of a violation of the quality of the information's relevance.

To find the coefficient of differential sensitivity with respect to the parameter β, we find the partial derivative of the risk function with respect to the shape parameter:

$$Risk'(\beta) = \Delta t\left[\left(\frac{e^{b}\left(e^{ct_0}-1\right)}{c}+\frac{e^{b}\left(e^{zT_w}-e^{zt_0}\right)}{z}\right)\times\frac{(\beta/\alpha)(t_0/\alpha)^{\beta-1}}{\left[1+(t_0/\alpha)^{\beta}\right]^{2}}\right]'. \tag{23}$$

Let us analyze the behavior of the sensitivity function for various values of α for a given value vector of the shape parameter (fig. 3-4).

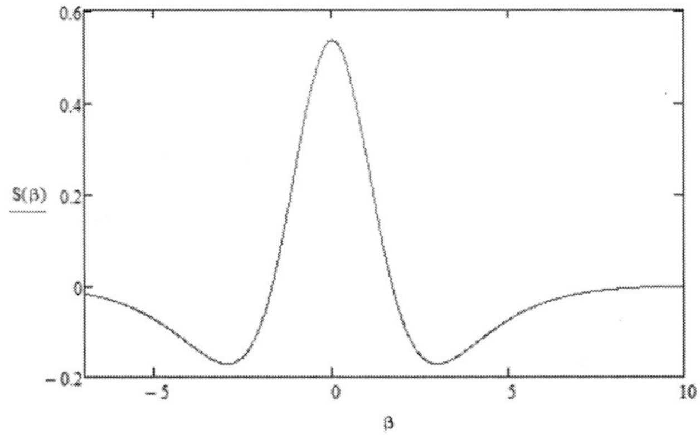

Figure 3. Graph of the differential sensitivity function depending on the form parameter when t_0=0.8, α =2.

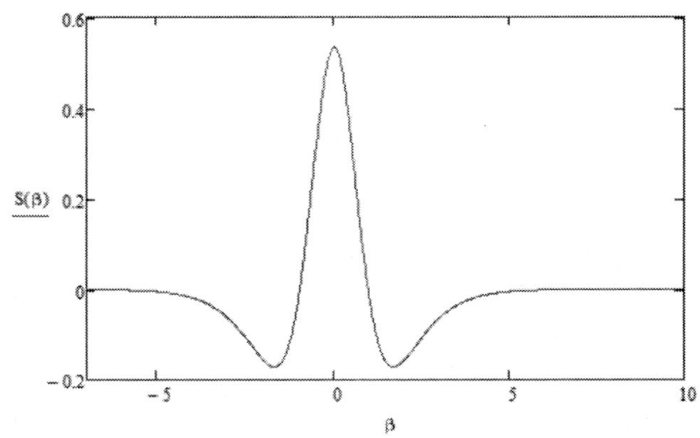

Figure 4. Graph of the differential sensitivity function depending on the form parameter when t_0=0.8, α =4.

Let us find the partial derivative with respect to the scale parameter:

$$\frac{\mathrm{d}Risk(t_0)}{\mathrm{d}\alpha} = \Delta t\left[\left(\frac{e^{b}\left(e^{ct_0}-1\right)}{c}+\frac{e^{b}\left(e^{zT_w}-e^{zt_0}\right)}{z}\right)\times\frac{(\beta/\alpha)(t_0/\alpha)^{\beta-1}}{\left[1+(t_0/\alpha)^{\beta}\right]^{2}}\right]'. \tag{24}$$

Let us analyze the behavior of the sensitivity function for various values of β for a given vector of parameter values (fig. 5-6).

Analysis of the graphs in figures 3-6 allows us to conclude that the risk function is sensitive to changing the shape and scale parameter over the entire time range, and it is advisable to take measures to regulate the risk of violating the quality of the information's topicality in the SPIS at any stage of its life cycle.

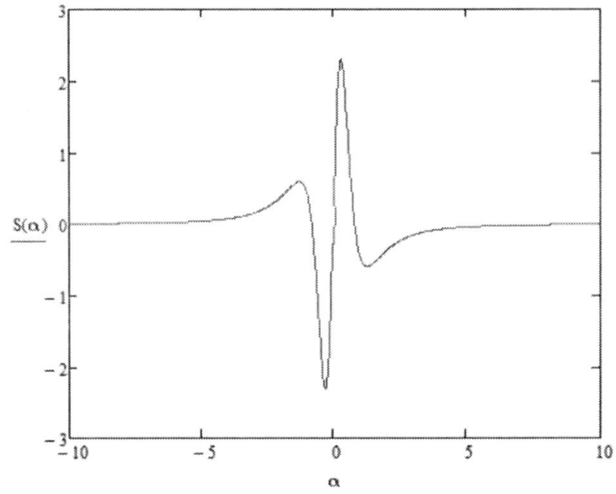

Figure 5. Graph of the differential sensitivity function depending on the scale factor when t_0=0.8, β =2.

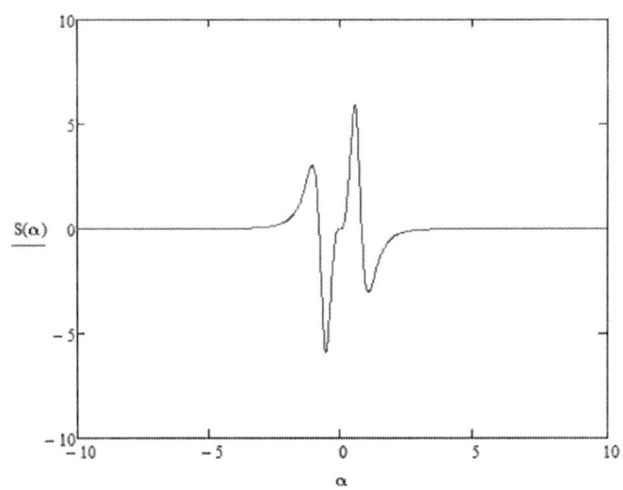

Figure 6. Graph of the differential sensitivity function depending on the scale factor when t_0=0.8, β =4.

6. Conclusion

Thus, in the work the questions of construction of risk-models of violation of the qualities of relevance and value of information in infocommunication systems of special purpose are considered. Particular attention is paid to the connection between the qualities of relevance and the value of information obtained as a result of the operation of infocommunication systems of a special purpose. Analytical expressions for the risk and damage function in the time range in special-purpose infocommunication systems are obtained, which can serve as a mathematical basis for risk assessment. An analytical expression is obtained for assessing the chance of obtaining up-to-date information in the operation of infocommunication systems up to the time when information quality is upsetting. An analytical expression for estimating the chance can be used to calculate the effectiveness of a special-purpose infocommunication system. The obtained analytical expressions can be useful for assessing the viability of infocommunication systems for special purposes.

References

1. Ostapenko A G, Ermilov E V and Kalashnikov A O 2013 Innovative development trends and information risks of IT sector development in the context of critical facilities provision *Information and Security* **3** pp 323-334
2. Ostapenko A G, Karpeev D O and Plotnikov D G 2009 Prospects for the development of the methodology of risk analysis of systems *Information and Security* **3** p 419-424
3. Ostapenko A G, Plotnikov D G, Makarov O Yu, Tikhomirov N M and Yurasov V G 2013 *Viability of attacked distributed systems: assessment of risks of fatal component failures* (Voronezh: Scientific book) p 160
4. Ostapenko A G, Ermilov E V and Kalashnikov A O 2013 Risks of damage, the chances of utility and resilience of the components of automated systems under the impact of information threats on them *Information and Security* **2** pp 215-218
5. Ostapenko A G, Kalashnikov A O, Ermilov E V and Korneeva N N 2014 Formalization of the risk management process in the information technology infrastructure of a critical facility *Information and security* **2** pp 164-179
6. Davydov A A 2010 Distribution of the rates of innovative development of the countries of the world *Official site of the ROS* p 11
7. Ostapenko A G, Plotnikov D G, Ostapenko O A and Maslikhov P A 2012 Loglogistic distribution of damage: the calculation of the risk of ITCS based on the risk parameters of its components *Information and Security* **3** pp 425-428
8. Ostapenko A G, Ermilov E V and Kalashnikov A O 2013 Risks of damage, the chances of utility and resilience of the components of automated systems under the impact of information threats on them *Information and Security* **2** pp 215-218
9. Ostapenko G A, Plotnikov D G and Rogozin A S 2013 The vitality of elements of critical information infrastructures: an analytical assessment with allowance for possible damages *Information and Security* **3** pp 353-364
10. Ostapenko G A, Plotnikov D G, Shcherbakova N Yu and Zarubin V S 2013 On the issue of assessing the damage and resilience of attacked distributed information systems: the development of methodological support *Information and Security* **1** pp 141-142
11. Ostapenko G A, Plotnikov D G and Meshkova E A 2010 Methodical and algorithmic support of calculation of risk parameters for components of distributed systems *Information and Security* **3** pp 335-350
12. Ostapenko G A, Plotnikov D G, Batishchev R V and Goncharov I V 2010 Distributed systems: methodologies for assessing effectiveness under attack conditions *Information and Security* **3** pp 359-366
13. Ostapenko A G, Plotnikov D G, Makarov O Yu, Tikhomirov N M and Yurasov V G 2013 Viability of attacked distributed systems: an assessment of the risks of fatal component failures (Voronezh: Scientific book) p 160
14. Ostapenko G A and Plotnikov D G 2012 On the issue of assessing the survival of information systems of an innovative nature *Information and Security* **3** pp 413-417
15. Parinov A V, Parinova L V and Yurasov V G 2013 Risk-assessment of mortality of innovative projects: scientific and methodological basis *Information and Security* **3** pp 423-426
16. Parinov A V Parinova L V and Zaryaev A V 2014 Risk management of innovative projects *Information and Security* **4** pp 602-605
17. Ostapenko A G 2010 Function of the possibility in assessing the risks, chances and effectiveness of systems *Information and Security* **2** pp 17-20

18. Ostapenko A G, Karpeev D O and Plotnikov D G 2009 Prospects for the development of the methodology of risk analysis of systems *Information and Security* **3** pp 419-424
19. Ostapenko G A, Plotnikov D G, Shcherbakova N Yu and Barannikov N I 2013 Models of survivability of the attacked distributed information system: risk-formalization with allowance for possible damage *Information and Security* **1** pp 63-68
20. Ostapenko G A, Plotnikov D G, Batishchev R V and Goncharov I V 2010 Distributed systems: methodologies for assessing effectiveness under attack conditions *Information and Security* **3** pp 359-366
21. Popov E A and Ostapenko G A 2014 Dos-attacks on innovative state distributed information systems: risk analysis in case of irregular distribution of damage *Information and Security* **3** pp 456-459
22. Rad'ko N M, Parinova L V, Pasternak Yu G, Razinkin K A and Tikhomirov N M 2013 Some assessments of risks, chances and survivability of networks in the context of information attacks of a viral nature *Information and Security* **4** pp 498-499
23. Ostapenko G A, Bursa M V and Tolstoy N N 2013 Risk-model of the innovative project, functioning in the face of threats of ddos-attacks *Information and Security* **3** pp 443-444
24. Ivankin E F, Mashin S V and Losev O A 2012 Virus attacks on the information resources of innovative projects: risk management *Information and Security* **3** pp 401-406
25. Dushkin A V, Novoseltsev V I and Sumin V I 2016 *Methodological bases of construction of the protected automated systems: the monography* (Voronezh: Scientific book) p 76
26. Balan V P, Dushkin A V, Novoseltsev V I and Sumin V I 2016 *Risk and reflection: monograph* (Moscow: Hot line – Telecom) p 136
27. Balan V P, Dushkin A V, Novoseltsev V I and Sumin V I 2015 *Conflict Management* (Moscow: Hot line – Telecom) p 144
28. Dushkin A V, Kravchenko A S, Novoseltsev V I, Smolentseva T E and Sumin V I 2014 *Mathematical models and information processes of complex object control: Monograph* (Voronezh: Scientific book) p 125
29. Barkalov S A, Dushkin A V, Kolodyazhny V I and Sumin V I 2017 *Introduction to system analysis: Monograph* (Moscow: Hot line – Telecom) p 234
30. Barkalov S A, Dushkin A V, Kolodyazhny S A and Sumin V I 2017 *Introduction to the system design of intellectual knowledge bases: monograph* (Moscow: Hot line – Telecom) p 108
31. Novoseltsev V I, Kochedykov S S and Orlova D E 2017 *Tensor analysis of Crohn and its applications: monograph* (Voronezh: Scientific book) p 220
32. Sumin V I, Dushkin A V and Kravchenko A S 2014 *Modeling of systems* (Voronezh: Scientific Book) p 222

W-algebra for solving problems with fuzzy parameters

A O Shevlyakov and M G Matveev

Department of Computer Sciences, Voronezh State University, Voronezh, 394006, Russia

E-mail: shevlyakov.a.o@mail.ru

Abstract. A method of solving the problems with fuzzy parameters by means of a special algebraic structure is proposed. The structure defines its operations through operations on real numbers, which simplifies its use. It avoids deficiencies limiting applicability of the other known structures. Examples for solution of a quadratic equation, a system of linear equations and a network planning problem are given.

1. Introduction

The fuzzy set theory introduced by Zadeh [1] has been successfully used to represent the uncertainty caused by the use of qualitative scales in the measurement of parameters and variables of the objects. In particular, this uncertainty is typical for many control and decision-making tasks. Mathematical models of such problems include the parameters and variables that are defined as fuzzy numbers or fuzzy variables. It allows to receive the numerical and analytic solutions for control and decision-making problems under uncertainty. However, in practice, these options are not used in full due to the problems with implementation of fuzzy information processing. Classical approaches for handling the fuzzy numbers are based on the Zadeh's extension principle [2]. Representation of fuzzy numbers in the form of LR-numbers with a convex membership functions allows to significantly simplify the computational process. On this basis a wide variety of algebraic approaches to the implementation of fuzzy calculations developed in [3], [4], [5], [6], [7], [8], [9]. Nevertheless, some characteristics of algebraic and computational operations with fuzzy numbers restrain their practical applications. We point out the following negative characteristics:

- operations with fuzzy numbers can be impractical, because they may lead to an unjustified extension of the result uncertainty;
- fuzzy results of solving problems aren't always able to be interpreted properly due to distortion of the form of a fuzzy numerical result or (and) the distortion of natural properties and relations of classical models (e.g., fuzzy parameters do not retain the identity of the equation after substituting the solution). This implies a decrease, and sometimes a loss of the models adequacy;
- there are a number of software tools to solve various mathematical tasks of control and decision-making, which designed to work with the real numbers. They cannot be used in the fuzzy data processing due to implementation of arithmetic operations with fuzzy numbers and problem of comparing of the fuzzy numbers.

Overcoming these problems is possible by constructing a suitable algebraic structure of fuzzy information processing, which is the purpose of this work. The article consists of eight sections. We look at the theoretical foundations of fuzzy theory and fuzzy numbers in Section 3. We construct the algebra of the single component numbers in section 4. Section 5 introduces *W*-algebra and shows some of its properties. Section 6 is devoted to the numerical examples showing some characteristics of compared approaches. Section 7 contains results and discussion of our work. Section 8 sets out the basic conclusions.

2. Materials and methods

Fuzzy set theory and the methods of algebraic structure theory used to develop and study the algebraic structure of fuzzy information processing that overcomes the previously mentioned characteristics. Linear programming methods used to find the critical path on a project's network graph in the numerical examples section. Microsoft Office Excel© used to solve equations and find optimal solutions.

3. Fuzzy numbers

A classical (crisp) set is a collection of elements or objects $.x \in X$. that can be finite, countable, or over-countable. Each single element can either belong to or not belong to a set A, . Such a classical set can be described using the characteristic function, in which 1 indicates membership and 0 non-membership. For a fuzzy set, the characteristic function allows various degrees of membership for the elements of a given set.

Definition 1. If X is a collection of objects denoted generically by x, then a fuzzy set A in X is a set of ordered pairs:

$$A = \{x, \mu_A(x) \mid x \in X\} \tag{1}$$

$\mu_A(x)$ is called the membership function or grade of membership (also degree of compatibility or degree of truth) of x in A that maps X to the membership space M.

Definition 2. The support of a fuzzy set A, $S(A)$, is the crisp set of all $x \in X$ such that $\mu_A(x) > 0$.

Definition 3. The (crisp) set of elements that belong to the fuzzy set A at least to the degree r is called the r-level set [10]:

$$A_r = \{x \in X \mid \mu_A(x) \geq r\}. \tag{2}$$

Definition 4. Fuzzy number A is a fuzzy set $\mu_A(x): X \to [0,1]$ which satisfies:

- μ is upper semicontinuous;
- $\mu(x) = 0$ outside some interval [c, d];
- there are real numbers a, b: $c \leq a \leq b \leq d$ for which $\mu(x)$ is monotonic increasing on [c, a], $\mu(x)$ is monotonic decreasing on [b, d], and $\mu(x) = 1$ at $a \leq x \leq b$.

Definition 5. A fuzzy number A in parametric form is a pair $(\underline{a}(r), \overline{a}(r))$, $.r \in [0,1]$., which satisfies the following requirements [11]:

$\underline{a}(r)$ is a bounded left continuous monotonic increasing function over $[0,1]$;

$\overline{a}(r)$ is a bounded left continuous monotonic decreasing function over $[0,1]$;

$a(r) \leq \overline{a}(r)$, $r \in [0,1]$.

Triangular fuzzy numbers are often used when there is not enough information to use more complex nonlinear membership functions. Also triangular fuzzy numbers help to reduce computational complexity and achieve sufficient accuracy.

Definition 6. Following [12] a triangular fuzzy number defined in parametric form can be presented as:

$$A(r) = (\underline{a}(r); \overline{a}(r)) = (\underline{a} + b_1 r; \overline{a} - b_2 r); \quad r \in [0,1], \tag{3}$$

where $\underline{a}(r)$ is the left boundary and . $\overline{a}(r)$. is the right boundary of the corresponding r - interval, b are non-negative coefficients of fuzziness.

The theoretical basis of arithmetic operations with fuzzy numbers is Zadeh's extension principle [2].

Definition 7. If $y = f(x_1, x_2, ..., x_n)$ is function of the fuzzy arguments where fuzzy numbers can be denoted as:

$$y = \bigcup_{r\in[0,1]} (\underline{y}_r, \overline{y}_r), x_i = \bigcup_{r\in[0,1]} (\underline{x}_{ir}, \overline{x}_{ir}), i = \overline{1,n} \quad (4)$$

then the function is calculated according to the expressions for any r -level:

$$\underline{y}_r = \inf(f(x_{1r}^*, x_{2r}^*, ..., x_{nr}^*)), \overline{y}_r = \sup(f(x_{1r}^*, x_{2r}^*, ..., x_{nr}^*)), \quad (5)$$

where $x_{ir}^* \in [\underline{x}_{ir}, \overline{x}_{ir}], i = \overline{1,n}.$

The work aims at overcoming these shortcomings. It contains an original approach to the implementation of operations on fuzzy numbers and can overcome negative shortcomings of the extension principle.

4. Algebra of single-component numbers (*X*-algebra)

The concept of a single component number is used in the construction of the W -algebra.

Definition 8. One-component number X (hereinafter the X -number) is called the structure represented in the form $X = x(r) = (a + br), r \in [0,1]$.

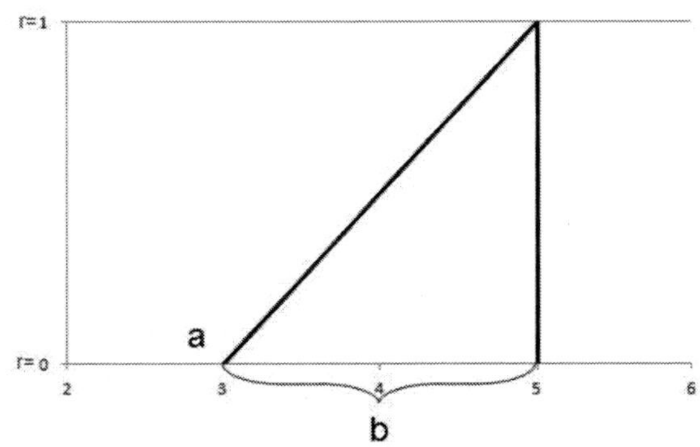

Figure 1. *X*-number.

The one-component number (3 + 2*r*) is shown in figure 1.

It is known that an algebraic structure consists of a carrier set K, collective operations (signatures) S over the set elements, and collective relations R over the set K. Let the set be defined as a set of fuzzy L- or R-numbers: $K = \{x(r)\}; x(r) = a + br; r \in [0;1]$.

If we want a substitution of a fuzzy solution in the equation with fuzzy parameters to guarantee identical equivalency of the left and right parts, then it is necessary to introduce a non-fuzzy equivalency relation R on the set K as follows:

Definition 9. Two fuzzy numbers $x_1(r) = a_1 + b_1 r$ and $x_2(r) = a_2 + b_2 r$ $x_2(r)$ are equivalent, i.e. $x_1(r) R x_2(r)$ if and only if $a_1 = a_2$ and $b_1 = b_2$. Here R is the equivalency relation.

Let us define a task of constructing an algebra $P = \langle K; +, * \rangle$ on the set $K = \{x(r)\}; x(r) = a + br; r \in [0;1]$ of fuzzy L- and R-numbers satisfying the following conditions:

- limitation of an unjustifiable extension of uncertainty of the calculation result;

- preservation of natural relations in the usual mathematical model of selection, including fuzzy parameters; for example, identity of an equation after substitution of its solution in the sense of exact equality of the corresponding inclusion functions;
- a guarantee that the carrier set K is closed in relation to the algebra operations.

Let us introduce a binary summation operation on the set K :

$$x_1(r)+x_2(r)=r_1(r)=a_1+a_2+(b_1+b_2)r\in K. \tag{6}$$

Let us define a single null element $O=(0+0r)\in K$, such that $\forall x(r)\in K: x(r)+O=a+br+0+0r=x(r)$.

Let us introduce a binary multiplication operation on the set K, such that it guarantees preservation of the result in the form of a triangular number of type L or R:

$$x_1(r)\cdot x_2(r)=r_2(r)=a_1a_2+(a_1b_2+a_2b_1+b_1b_2)r;\ r_2(r)\in K \tag{7}$$

Let us define a singular number $I=1+0r\in K$ such that $\forall x(r)\in K:\ I\cdot x(r)=(1+0r)(a+br)=x(r)$.

For all elements $x(r)=a+br\in K$ let us define a single reverse element $-x(r)=-a-br$, such that $x(r)+(-x(r))=a+br-a-br=O$.

Then, the subtraction operation becomes:

$$x_1(r)-x_2(r)=x_1(r)+(-x(r)). \tag{8}$$

In order to define the reverse element $x^{-1}(r)\in K$, such that the condition $x(r)\cdot x^{-1}(r)=1$ is fulfilled, the following form is used:

$$x^{-1}(r)=\frac{1}{a}-\frac{b}{a(a+b)}r,\quad a\neq 0,\ a+b\neq 0 \tag{9}$$

Then, the division operation becomes:

$$x_1(r)/x_2(r)=x_1(r)\cdot x_2^{-1}(r). \tag{10}$$

Limitations on the parameters of a fuzzy number in equation (9) should be accounted for during the formalization of an expert statement in the form of a fuzzy number.

Obtained reverse elements are used to define the subtraction and division operations of L and R-numbers as summation and multiplication operations with the corresponding opposite and reverse numbers.

Example 1. Addition of two *X*-numbers $A=(4+2r)$ and $B=(1-3r)$: $A+B=(4+1+(2-3)r)=(5-r)$.

Example 2. Subtraction of two *X*-numbers $A=(4+2r)$ and $B=(1-3r)$: $A-B=(4+2r)+(-1+3r)=(3+5r)$., wherein $A-A=O$.

Example 3. Multiplication of two *X*-numbers $A=(4+2r)$ and $B=(1-3r)$: $A\cdot B=(4\cdot 1+(4\cdot(-3)+1\cdot 2+2\cdot(-3))r)=(4+(-16)r)$

Example 4. Division of two *X*-numbers $A=(4+2r)$ and $B=(1-3r)$: $A\div B=(4+2r)\cdot(\frac{1}{1}-\frac{-3}{1\cdot(1-3)}r)=(4+2r)\cdot(1-1,5r)=(4-7r)$., wherein $A/A=I$.

It can easily be verified that the introduced algebra satisfies predefined conditions. However, implementation of operations (6,7,8,10) differs from the implementation of analogous operations on real numbers, which does not allow the use of standard computational software and justifies the need for construction of a more useful isomorphic algebra. Besides, known operations for comparison of fuzzy numbers dictate only a partial order, which is insufficient for implementation of many algorithms of selection.

Let us denote the obtained algebra as $P=\langle K_1;S_1\rangle$.

Definition 10. Isomorphism of the algebra $P=\langle K_1;S_1\rangle$ on to the algebra $G=\langle K_2;S_2\rangle$ is a mapping $\Gamma: K_1 \to K_2$ satisfying the following conditions:

$$\Gamma(S_1(k_1)) = S_2(\Gamma(k_1)), \quad k_1 \in K_1. \tag{11}$$

$$S_1(\Gamma^{-1}(k_2)) = \Gamma^{-1}S_2(k_2)), \quad k_2 \in K_2 \tag{12}$$

It is necessary to find isomorphism Γ, such that the implementation of algebra G uses operations on real numbers and allows the application of the standard software packages for solving problems with fuzzy numbers.

Let the elements of a set K_2 be represented as $x(r)=x(0)+(x(1)-x(0))r$, where $x(0), x(1)$ are real numbers. Signature S_2 is defined by

$$x_1(r) * x_2(r) = rx(1) + (1-r)x(0) = x(0) + (x(1) - x(0))r, \tag{13}$$

where $x(0)=x_1(0)*x_2(0)$; $x(1)=x_1(1)*x_2(1)$; $*\in\{+,-,\cdot,/\}$ and all operations satisfy the rules of operations with real numbers.

Theorem 1. Algebras P and G are isomorphic.

Proof. Consider elements of the sets K_2 and K_1. Obviously, any element of the set K_1 - $x(r)=a+br$ is obtained from the elements of K_2 by using a simple substitution $a=x(0)$, $b=(x(1)-x(0))$, in this case $K_1=K_2$, and the mapping Γ is a mapping onto itself. The set K_1 is closed with respect to operations (6,7,8,10), so to prove homomorphism (11) and the reverse homomorphism (12) it is sufficient to show correctness of the equality $S_1(k_1)=S_2(k_2)$. Since the result of operations in both algebras is a linear function of the parameter r, it is sufficient to show that these results are similar in two points, for instance at $r=0$ and $r=1$. Similarity of the results is easy to check by performing operations using rules (6,7,8,10) and (13) for two fuzzy numbers a_1+b_1r and a_2+b_2r [13].

5. Algebra of double-component numbers (*W*-algebra)

W-algebra is constructed using the *X*-numbers and *X*-algebra.

Definition 10. Two-component number W (hereinafter the *W*-number) is called the structure shown by the vector $(v^L(r);v^R(r))$, $v^L(r)\in X$, $v^R(r)\in X$, for which $v^L(1)=v^R(1)$.

Triangular fuzzy numbers can be represented as *W*-numbers, where $v^L(r)=a^L+b^Lr$, $a^L\geq 0$ and $v^R(r)=a^R+b^Rr$, $a^R\leq 0$.

Algebra of W-numbers (W-algebra) involves the independent performing of the operations on the left component with recording result to the left component of the final number and on the right component with the corresponding entry in the right component of the final number.

Definition 11. Let the left or right components of two *W*-numbers be represented as $x(r)=x(0)+(x(1)-x(0))r$, where $x(0),x(1)$ are crisp numbers. Then, as a result of the operation, the left or right component of the resulting *W*-number will be: $x_1(r)*x_2(r)=x(0)+(x(1)-x(0))r$, where $x(0)=x_1(0)*x_2(0), x(1)=x_1(1)*x_2(1), *\in\{+,-,\cdot,/\}$ and all operations are performed with crisp numbers.

Operations are linear with respect to r, the set of all *X*-numbers is closed for all operations, that is, $x_i(r), x_j(r)\in X$, then $x(r)=x_i(r)*x_j(r)\in X, *\in\{+,-,\cdot,/\}$.

Definition 12. Let $W_1 = (v_1^L(r); v_1^R(r))$ and $W_2 = (v_2^L(r); v_2^R(r))$ be two W-numbers. The operations between these numbers will be made in accordance with the following rule: $(v_1^L, v_1^R) * (v_2^L, v_2^R) = ([v_1^L * v_2^L]; [v_1^R * v_2^R]) = (v^L, v^R)$, where $* \in \{+, -, \cdot, /\}$.

Definition 13. Let $W_1 = (a_1, b_1, c_1)$ and $W_2 = (a_2, b_2, c_2)$ be two W-numbers. The operations between these numbers will be made in accordance with the following rule: $(a_1, b_1, c_1) * (a_2, b_2, c_2) = (a_1 * a_2, b_1 * b_2, c_1 * c_2)$, where $* \in \{+, -, \cdot, /\}$.

In this case, the W-algebra provides a complete overcoming of all these negative shortcomings of arithmetic, based on Zadeh's extension principle.

Theorem 2. W-number will be always the result of the operations on the W-numbers, i.e., closure condition always will be fulfilled.

Proof. We consider two W-numbers $W_1 = [a_1 + b_1 r, a_2 + b_2 r]$ and $W_2 = [c_1 + d_1 r, c_2 + d_2 r]$.

Since left and right components of W-numbers have the same value when r=1 (hereinafter modal value), we have:

$$a_1 + b_1 = a_2 + b_2 c_1 + d_1 \Rightarrow a_1 = a_2 - b_1 + b_2 c_1 = c_2 - d_1 + d_2 \tag{14}$$

Let us prove the equality of modal values for each of the operations (addition, subtraction, multiplication, division).

Addition:

$$\begin{cases} v^L = a_1 + b_1 r + c_1 + d_1 r \\ v^R = a_2 + b_2 r + c_2 + d_2 r \end{cases} \tag{15}$$

According to equation (14) for $r = 1$, we have:

$$\begin{cases} v^L = a_2 + b_2 + c_2 + d_2 \\ v^R = a_2 + b_2 + c_2 + d_2 \end{cases} \tag{16}$$

The proof is similar for the remaining operations.

Performing operations on W-numbers does not require special software, and as will be shown below, overcomes the shortcomings of the other arithmetic.

6. Numerical examples

We consider the results obtained by different approaches in the numerical examples.

Example 5. Subtraction of a triangular number from itself. Let us consider an arbitrary fuzzy number $A = (a, b, c)$.

$$\begin{cases} (A - A)^L = (r(b-a) + a) - (r(b-a) + a) = 0 \\ (A - A)^R = (c - r(c-b)) - (c - r(c-b)) = 0 \end{cases} \tag{17}$$

In general, it is shown that the operation's result of W-algebra equals zero.

Example 6. Division of a triangular number by itself. Let us consider an arbitrary fuzzy number $A = (a, b, c)$.

$$\begin{cases} (A / A)^L = r\left(\frac{b}{b} - \frac{a}{a}\right) + \frac{a}{a} = 1 \\ (A / A)^R = r\left(\frac{b}{b} - \frac{c}{c}\right) + \frac{c}{c} = 1 \end{cases} \tag{18}$$

In general, it is shown that the operation's result of W-algebra equals 1 (1,1,1).

Example 7. Solution of a linear equation. Let us consider the linear equation $Ax+2=0$, where A is a fuzzy number and $A=(3,5,6)$.

The solution of the linear equation:

$$\begin{cases} x^L = \frac{4r}{15} - \frac{2}{3} \\ x^R = -\frac{r}{15} - \frac{1}{3} \end{cases} \tag{19}$$

The result of substitution of the solutions in the equation:

$$\begin{cases} v^L = 0 \\ v^R = 0 \end{cases} \tag{20}$$

Example 8. Solution of a quadratic equation. Let us consider the quadratic equation $x^2+Bx+2=0$, where B is a fuzzy number and $B=(3,4,6)$.

The solution of the quadratic equation:

$$\begin{cases} \begin{cases} x_{1,L} = \left(\sqrt{2}-1\right)r-1 \\ x_{1,R} = \left(1+\sqrt{2}-\sqrt{7}\right)r+\sqrt{7}-3 \end{cases} \\ \begin{cases} x_{2,L} = -\sqrt{2}r-2 \\ x_{2,R} = \left(1-\sqrt{2}+\sqrt{7}\right)r-\sqrt{7}-3 \end{cases} \end{cases} \tag{21}$$

The result of substitution in the equation:

$$\begin{cases} \begin{cases} \left(\left(\sqrt{2}-1\right)r-1\right)^2+(r+3)\left(\left(\sqrt{2}-1\right)r-1\right)+2=0 \\ \left(\left(1+\sqrt{2}-\sqrt{7}\right)r+\sqrt{7}-3\right)^2+(6-2r)\left(\left(1+\sqrt{2}-\sqrt{7}\right)r+\sqrt{7}-3\right)+2=0 \end{cases} \\ \begin{cases} \left(-\sqrt{2}r-2\right)^2+(r+3)\left(-\sqrt{2}r-2\right)+2=0 \\ \left(\left(1-\sqrt{2}+\sqrt{7}\right)r-\sqrt{7}-3\right)^2+(6-2r)\left(\left(1-\sqrt{2}+\sqrt{7}\right)r-\sqrt{7}-3\right)+2=0 \end{cases} \end{cases} \tag{22}$$

Example 9. Solution of a system of linear equations.

Let us consider the system of linear equations:

$$\begin{cases} a_1x_1+b_1x_2=c_1, \\ a_2x_1+b_2x_2=c_2, \end{cases} \tag{23}$$

where $a_1=(1,2,4)$, $b_1=(-8,-5,-3)$, $c_1=(1,4,5)$, $a_2=(2,6,7)$, $b_2=(1,3,4)$, $c_2=(1,2,5)$ are the fuzzy numbers.

The solution of the system of linear equations:

$$\begin{cases} x_{1,L} = \frac{25r}{306} + \frac{9}{17}; \\ x_{1,R} = \frac{35}{37} - \frac{223r}{666}; \\ x_{2,L} = -\frac{76r}{153} - \frac{1}{17}; \\ x_{2,R} = -\frac{50r}{333} - \frac{15}{37}. \end{cases} \tag{24}$$

The result of substitution of $x_{1,L}$, $x_{2,L}$, $x_{1,R}$ and $x_{2,R}$ in the system of linear equations:

$$\begin{cases} -3r+\left(\frac{25r}{306}+\frac{9}{17}\right)(r+1)+\left(-\frac{76r}{153}-\frac{1}{17}\right)(3r-8)-1=0; \\ -(r+1)+\left(-\frac{76r}{153}-\frac{1}{17}\right)(2r+1)+\left(\frac{25r}{306}+\frac{9}{17}\right)(4r+2)=0; \\ -(5-r)+\left(-\frac{50r}{333}-\frac{15}{37}\right)(-2r-3)+\left(\frac{35}{37}-\frac{223r}{666}\right)(4-2r)=0; \\ \left(-\frac{50r}{333}-\frac{15}{37}\right)(4-r)+\left(\frac{35}{37}-\frac{223r}{666}\right)(7-r)-(5-3r)=0. \end{cases} \tag{25}$$

Example 10. Cauchy problem for ordinary differential equation.
Let us consider the problem:

$$\begin{cases} \frac{dy}{dt}=Ay; \\ y(t=0)=1; \end{cases} \tag{26}$$

where $A=(-2,0,1)$.

The solution is $y(\alpha)=(e^{-2t},1,e^{t})$. Substitution leads to the equality $(-2e^{-2t},0,e^{t})=(-2e^{-2t},0,e^{t})$.

Example 11. A network project planning problem under fuzzy initial conditions of work duration. Let us define a project in table 1:

Table 1. Project.

Operation	Preceding operations	Fuzzy operation duration, τ		
		τ^l	m	τ^r
A	-	1	2	5
B	-	2	4	5
C	A	3	7	9
D	B	4	6	9
E	B	9	10	12
F	C	4	5	6
G	D, E	1	5	6
H	F, G	2	4	7

Related network graph is shown in figure 2.

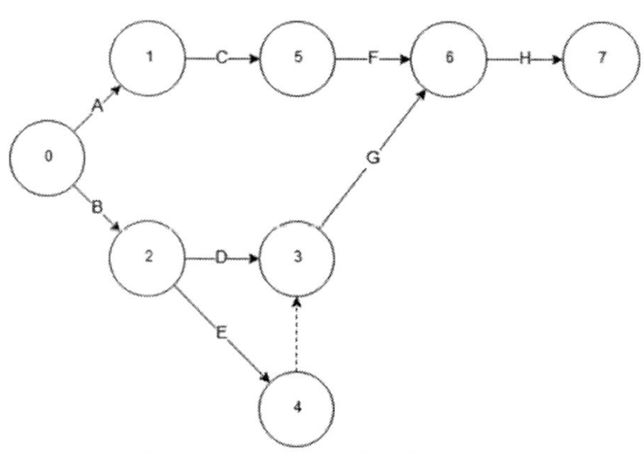

Figure 2. A network graph of the project.

It is required to determine the critical project operations and the overall fuzzy time for the project completion. Let us define a day as the unit of time measurement.

The necessary solution can be obtained by solving a linear programming problem with fuzzy parameters:

$$\begin{aligned} & t_n - t_0 \to \min; \\ & t_{js} - t_{is} \geq \tau_s, \quad \forall s = 1,...,8; \\ & \tau = \sum_{s1} \tau_{s1}; \end{aligned} \tag{27}$$

where t_0, t_n are the beginning and end times of the project; t_{is}, t_{js} are the beginning and end times of an s-operation; τ_s is the fuzzy duration of an .s.-operation; τ_{s1} is the fuzzy duration of the critical path, or in other words, those operations for which the solution of the problem satisfies the equality $t_{js1} - t_{is1} = \tau_{s1}$ (in the sense of equivalency relation R). Such constraints in linear programming problems are called deficit constraints; τ is the project duration.

The system of inequalities for (27) with respect to the data in table 1 takes the form:

$$\begin{cases} t_1 - t_0 \geq \tau_A; \\ t_2 - t_0 \geq \tau_B; \\ t_5 - t_1 \geq \tau_C; \\ t_3 - t_2 \geq \tau_D; \\ t_4 - t_2 \geq \tau_E; \\ t_6 - t_5 \geq \tau_F; \\ t_6 - t_4 \geq \tau_G; \\ t_7 - t_6 \geq \tau_H. \end{cases} \tag{28}$$

We point out that the proposed methodology eliminates the problem of comparing fuzzy numbers in the system of inequalities presented above, replacing it with a corresponding system of comparing non-fuzzy numbers.

Optimal operation start times are:

$$\begin{cases} t_0 = (0,0,0); \\ t_1 = (1,2,5); \\ t_2 = (2,4,5); \\ t_3 = (11,14,17); \\ t_4 = (11,14,17); \\ t_5 = (4,9,14); \\ t_6 = (12,19,23); \\ t_7 = (14,23,30). \end{cases} \tag{29}$$

The project's critical path contains operations B; E; G; H.

The results of solving the problem (27) have a natural interpretation:

- the sum of fuzzy critical path's operation durations equals (in the sense of equality of membership functions) the fuzzy time moment of the project's end, i.e., it is possible that the project is completed within the interval of $\tau = (14,23,30)$ days;
- the total time reserve for operations A, C, F , for example, is determined by the expression $(t_6 - t_0) - (\tau_A + \tau_C + \tau_F) = (12,19,23) - (8,14,20) = (4,5,3)$. Critical operations do not have a time reserve; for instance, the total time reserve for operation B equals to $(t_2 - t_0) - \tau_B = (2,4,5) - (2,4,5) = (0,0,0)$.

The fulfillment of such natural relations proves the obtained solution is trustworthy.

IOP Conf. Series: Journal of Physics: Conf. Series **973** (2018) 012044 doi:10.1088/1742-6596/973/1/012044

7. Results and discussion

We have constructed a methodology for solving problems with non-fuzzy mathematical models and fuzzy parameters. This approach preserves qualities of the initial non-fuzzy mathematical model.

Future development of the proposed approach, in our opinion, extends beyond just static problems. By introducing the notion of a fuzzy function and its derivative, we are planning to switch to solving problems of optimal control of dynamic objects with a fuzzy parametric uncertainty. Another development path of the proposed approach is to investigate the problem of stability of solutions under conditions of fuzzy initial data.

8. Conclusion

W-algebra overcomes the shortcomings of the arithmetic based on Zadeh's extension principle.

Natural relations $A-A=0$ and $A/A=1$ are always provided in the *W*-algebra and a crisp identity is provided by substitution of the solution in the equation. These relations are not provided by Zadeh's extension principle.

The application of the *W*-algebra allows us to solve problems with fuzzy parameters and variables and use all the existing methods, algorithms and software tools designed to work with crisp numbers. The solution of the problem with crisp variables at three points and the construction of the corresponding result in the form of *W*-numbers are sufficient to obtain a solution. Consequently, the problem of the fuzzy numbers' comparing does not occur in the process of solutions. Zadeh's extension principle uses special algorithms and software and has greater computational complexity than the *W*-algebra for solving problems with fuzzy parameters and variables.

References

[1] Zadeh L A 1965 Fuzzy Sets *Information and Control* **8** pp 338–53

[2] Zadeh L A 1975 The concept of a linguistic variable and its application to approximate reasoning: Parts 1,2,3 *Information Sciences* **8** 43–80,199–249,301–57

[3] Dubois D and Prade H 1978 Operations on fuzzy numbers *International Journal of Systems Science* **9(6)** pp 613–626

[4] Klir G 1997 Fuzzy arithmetic with requisite constraints *Fuzzy Sets and Systems* **91** pp 165–175

[5] Lodwick W 1999 Constrained interval arithmetic *CCM report 138* p 11

[6] Piegat A 2001 *Fuzzy Modeling and Control* (Heidelberg: Springer-Verlag) p 728

[7] Nagoor Gani A and Mohamed Assarudeen S N 2012 A new operation on triangular fuzzy number for solving fuzzy linear programming problem *Applied Mathematical Sciences* **11** pp 525–532

[8] Chalco-Cano Y, Lodwick W and Bede B 2014 Single level constraint interval arithmetic *Fuzzy Sets and Systems* **257** pp 146–168

[9] Vorontsov Y A and Matveev M G 2014 L-transform based algebraic operations on fuzzy LR-numbers *Software Engineering* **8** pp 23–30

[10] Zimmermann H J 2001 *Fuzzy Set Theory—and Its Applications* (Dordrecht: Kluwer Academic Publishers) p 514

[11] Ghazanfari B and Shakerami A 2012 Numerical solutions of fuzzy differential equations by extended Runge–Kutta-like formulae of order 4 *Fuzzy sets and systems* **189(1)** pp 74-91

[12] Goetschel Jr R and Voxman W 1986 Elementary fuzzy calculus *Fuzzy Sets and Systems* **18** pp 31–43

[13] Matveev M G, Shevlyakov A O, Semenov M E and Meleshenko P A 2017 Solution of selection problems with fuzzy parameters *Conf. proc. on 17th Int. Multidisciplinary Scientific Geoconf. SGEM 2017 (Bulgaria)* vol 7 Informatics, Geoinformatics and Remote Sensing issue 21 pp 595-602

IOP Conf. Series: Journal of Physics: Conf. Series **973** (2018) 012045 doi:10.1088/1742-6596/973/1/012045

Formation of the predicted training parameters in the form of a discrete information stream

T E Smolentseva, V I Sumin, V K Zolnikov, V V Lavlinsky

ET Department, Voronezh Institute Federal Penitentiary Service of Russia, 1a Irkutskaya Street, Voronezh, RU

E-mail: smoltan@bk.ru

Abstract. In work process of training in the form of a discrete information stream is considered. On each of stages of the considered process portions of the training information and quality of their assimilation are analysed. Individual characteristics and reaction trained for every portion of information on appropriate sections are defined. The control algorithm of training with the predicted number of control checks of the trainee who allows to define what operating influence is considered it is necessary to create for the trainee. On the basis of this algorithm the vector of probabilities of ignorance of elements of the training information is received. As a result of the conducted researches the algorithm on formation of the predicted training parameters is developed. In work the task of comparison of duration of training received experimentally with predicted on the basis of it is solved the conclusion is drawn on efficiency of formation of the predicted training parameters. The program complex on the basis of the values of individual parameters received as a result of experiments on each trainee who allows to calculate individual characteristics is developed, to form rating and to monitor process of change of parameters of training.

1. Introduction

During training process consideration as the system including all stages: knowledge acquisition, digestion of material and control check, it has been revealed that this process is carried out under the operating influence which is natural for presenting in the form of a discrete information stream.

At each stage of this process the trainee receives portions of information u_k, $k = 1, 2, \ldots, N$ which are in turn presented in the form of the renumbered elements on every portion of information:

$$u_k = \left(m_{k-1} + 1,\, m_{k-1} + 2, \ldots, m_k\right), \tag{1}$$

where m_k – the quantity of the elements of a portion of information given on k a training session, defined according to (1)

$$m_k = \max_{1 \le m^* \le m} \left\{ m^* : t_k \ge f \cdot \sum_{i \in m} p_i\left(t_i^k\right) \right\}, \tag{2}$$

where f – the average time of learning of an element of a portion of information at his first presentation to the trainee; m – total of elements of a portion of information on all this process.

For check of level of knowledge of the trainee testing in certain timepoints $t_j^k, j=1,2,\ldots,l$ is held provided that time of testing of each trainee doesn't exceed time allocated for studying of a portion of information: $t_j^k \le t^k$.

In this process of training the number of measurements s^* is set on the basis of the calculated dimension of a strange attractor. It is considered that studying of a portion of information in timepoint $t_j^k, j=1,2,\ldots$ on k a stage he acquires all volume of information and, therefore, the probability of ignorance is equal to zero.

$$p_i^k\left(t_j^k\right)=0. \tag{3}$$

The probability of ignorance P_k in timepoint $t_j^k, j=1,2,\ldots$ a stage is defined on k on the basis of exponential dependence [7-9].

$$P_k=\left(p_1^k, p_2^k,\ldots,p_{m_k}^k\right), \tag{4}$$

where

$$p_i^k=1-\exp\left[-\alpha_i^k t_i^k\right],\ i=1,2,\ldots,m_k,\ k=1,2,\ldots,N, \tag{5}$$

where

$$t_i^{k+1}=\begin{cases}\Delta t_k, i\in u_k;\\ t_i^k+\Delta t_k, i\notin u_k.\end{cases} \tag{6}$$

Considering the fact that over time information obtained by the trainee is forgotten over time, this process is formalized by the expression given below:

$$\alpha_i^{k+1}=\begin{cases}\alpha_i^k, i\notin\bigcup_{j=1}^{k}u_j;\\ \gamma_1\alpha_i^k, r_i^k=0\ u\ i\in\bigcup_{j=1}^{k}u_j;\\ \gamma_2\alpha_i^k, r_i^k=1\ u\ i\in\bigcup_{j=1}^{k}u_j,\end{cases} \tag{7}$$

where α^1, γ_1, γ_2 – individual characteristics of a condition of the trainee or parameters of adaptation, r_i^k – the reaction of k of a portion of information trained for influence defined according to a ratio (8) in the form of answers to control questions (testings):

$$r_i^k=\begin{cases}0, \text{ if the trainee has given the correct answer;}\\ 1, \text{ otherwise.}\end{cases} \tag{8}$$

Definition of reaction of the trainee, on every portion of information and also definition of individual characteristics will allow to improve all indicators on each trainee.

2. Theoretical part

Being based on above the given reasonings, the level of ignorance of a portion of information is defined by expression:

$$Q_j^k=\sum_{i=1}^{m_k}p_i^k q_i, \tag{9}$$

and

$$\alpha_i^{j+1} = \begin{cases} \gamma_1 \alpha_i^j, r_i^j = 0; \\ \gamma_2 \alpha_i^j, r_i^j = 1, i = 1, 2, \ldots, l. \end{cases} \quad (10)$$

$$t_i^{j+1} = \Delta t_j . \quad (11)$$

Naturally function of quality of training needs to be defined as average value of levels of ignorance on k a training session.

$$Q = \frac{1}{l} \sum_{j=1}^{l} Q_j^k . \quad (12)$$

In actual practice considering that individual abilities of each trainee aren't identical delivery of the portions of the training information (PTI), it can't be acquired by each trainee completely. In this regard an important role in the course of training will play the number of measurements of level of ignorance Q_j, information on which defines a condition of an object of management.

Considering that in our case process of training begins after giving of the first PTI, the vector of external influences of S is identical u_1. The control algorithm of training with the predicted number of control checks of the trainee defines what operating influence (a portion of information) needs to be created for the trainee.

We will in more detail consider a control algorithm of training with the predicted number of control checks of the trainee:

1. With use of testing the examination of the trainee of a portion of the training information u_k as a result of which the set $R_k = \left(r_1^k, r_2^k, \ldots, r_{m_k}^k\right)$ where are defined by a ratio r_i^k (8) is formed is carried out.

2. Using expressions in formulas (7), (10) adaptation of parameters of an object of management is made.

3. Based on a formula (5) with attraction of expressions (6)-(7), (10)-(11) the vector of probabilities of ignorance of EDI is calculated P_k.

4. Using formulas (11-12) values Q_j^k and function of quality of training of Q are defined.

5. Process of k of a grade level comes to an end if the criterion of quality of training $Q \le \sigma$ is carried out. Further to the trainee moves following PTI u_{k+1} according to points 1-5 of an algorithm.

If the criterion of quality of training, then is not executed if $j \le s^*$ (s^* – assessment of quantity of variables of function of quality received on the basis of a method of computation of correlative dimensionality on time series of observations) PTI u_k again are given to the trainee for a study according to points 1-5 of an algorithm.

Training achievement of the goal reflects a condition of an object with use of a control algorithm of training more precisely with the predicted number of control checks of the trainee which basis temporary ranks are doesn't use model of change of a condition of an object of management which because of variety and accident of factors can reflect a real condition of an object not absolutely precisely. Therefore, more exact assessment s^*, allows to adapt more effectively the synthesized system on the properties to optimum system.

We will consider in more detail process of management of training with the predicted number of control checks of the trainee [6, 9, 14]. When receiving PTI r_i^k the trainee, his knowledge will be characterized by some random variable $\lambda_i^k \in [0,1]$ having the following distribution:

$$P\left(\lambda_i^k = 1\right) = p_i^k ; \quad (13)$$

$$P\left(\lambda_i^k=0\right)=1-p_i^k,\quad i=1,2,\ldots,m_k. \tag{14}$$

Considering expression (5) it is possible to tell that the vector P_k unambiguously is defined by vectors:

$$\alpha^k=\left(\alpha_1^k,\alpha_2^k,\ldots,\alpha_{m_k}^k\right); \tag{15}$$

$$t^k=\left(t_1^k,t_2^k,\ldots,t_{m_k}^k\right). \tag{16}$$

Therefore, in $2m_k$ – measured Euclidean space vector function $f:R^{2m_k}\to R^{m_k}$, such is defined that

$$P_k=f(\alpha^k,t^k). \tag{17}$$

Based on (11) function determining the level of ignorance it is presented in the form of a scalar product: $Q_j=(P_k,q)$ where $q=\left(q_1,q_2,\ldots,q_{m_k}\right)\in R^{m_k}$, is a coordinate vector which are defined prior to training.

Proceeding from above stated, in $2m_k$ – measured Euclidean space vector function, such is defined $h:R^{2m_k}\to R^1$ that

$$Q_j=\left(f(\alpha^k,t^k),q\right)=h\left(\alpha^k,t^k\right). \tag{18}$$

Based on the analysis of expressions (15)-(18) we draw a conclusion that in timepoint t_j^k process of management of training can be described two casual m_k – measured vectors α^k and t^k.

3. Practical part. Obtaining experimental values of individual characteristics

Ratios (17)-(18) with use $\zeta_k=\left(\alpha^k,t^k\right)$ it is possible to write down in a look:

$$\begin{matrix}P_k=f(\zeta_k)\\ Q_j=h(\zeta_k)\end{matrix}. \tag{19}$$

For a further research of management of training we will consider a vector $\chi_k=\left(\chi_1^k,\chi_2^k,\ldots,\chi_{m_k}^k\right)$ with components:

$$\chi_i^k=\begin{cases}1,\ i\in\bigcup\limits_{j=1}^{k}u_j;\\ 0,\ i\notin\bigcup\limits_{j=1}^{k}u_j.\end{cases} \tag{20}$$

We will assume that $\omega=\left(\left(\alpha^1,t^1\right),\left(\alpha^2,t^2\right),\ldots,\left(\alpha^m,t^m\right)\right)$ – a training trajectory for the specific pupil from Ω – a set of all possible trajectories of century R^{2m}. Considering (7) we receive that random variables α_i^{k+1} have conditional distribution of probabilities:

$$\begin{gathered}P\left(\alpha_i^{k+1}=\alpha_i^k\left|\left(\alpha_i^k,t_i^k\right)\right.\right)=1-\chi_i^k;\\ P\left(\alpha_i^{k+1}=\gamma_1\alpha_i^k\left|\left(\alpha_i^k,t_i^k\right)\right.\right)=\left(1-p_i^k\right)\chi_i^k;\\ P\left(\alpha_i^{k+1}=\gamma_2\alpha_i^k\left|\left(\alpha_i^k,t_i^k\right)\right.\right)=p_i^k\chi_i^k,\quad i=1,2,\ldots,m.\end{gathered} \tag{21}$$

Expression (19) shows that on $(k+1)$ – the m a training session α^{k+1} completely is defined by a vector $\zeta_k = (\alpha^k, t^k)$ and doesn't depend on number of a session k. To define a vector we will consider that the period between sessions is constant and equal Δt. Therefore, expression (6) will take a form:

$$t_i^{k+1} = \begin{cases} \Delta t, i \in u_k \\ t_i^k + \Delta t, i \notin u_k \end{cases} = \begin{cases} \Delta t, \chi_i^k = 1; \\ t_i^k + \Delta t, \chi_i^k = 0 \end{cases}, \quad i = 1, 2, \ldots, m. \tag{22}$$

And expression (22) will take a form:

$$t_i^{k+1} = \Delta t + (1 - \chi_i^k) t_i^k = z(i, t^k). \tag{23}$$

Therefore, and on $(k+1)$ – the m a step, and on k-m a step, ζ_{k+1} is defined ζ_k, P_{k+1} and $Q^{k+1} = \frac{1}{l}\sum_{j=1}^{l} Q_j^{k+1}$ it means, as is defined by realization ζ_k and doesn't depend on k. It allows to draw the following conclusion. The probability of any state on k-m a step depends on a state previous, but doesn't depend on a way of transition from a state to a state. We will define conditional population means of parameters and functions of process of training $\zeta_k = (\alpha^k, t^k)$ at k a step of functions P_k and Q^k.

Using expressions (14) and (23) we will receive:

$$M(\alpha_i^{k+1} | \zeta_k) = \alpha_i^k (1 - \chi_i^k) + \gamma_1 \alpha_i^k (1 - p_i^k) \chi_i^k + \gamma_2 \alpha_i^k p_i^k \chi_i^k = \alpha_i^k (1 - \chi_i^k (1 - \gamma_1) - \chi_i^k (\gamma_1 - \gamma_2) p_i^k). \tag{24}$$

Considering that under a condition ζ_k time t_i^{k+1} is the determined size, we will receive:

$$M(\alpha_i^{k+1} t_i^{k+1} | \zeta_k) = \Delta \ t \alpha_i^k (1 - \chi_i^k (1 - \gamma_1) - \chi_i^k (\gamma_1 - \gamma_2) p_i^k) + (1 - \chi_i^k) \alpha_i^k t_i^k. \tag{25}$$

Considering (5) and (7) we will receive

$$\begin{aligned} M(p_i^{k+1} | \zeta_k) = M\left(1 - \exp\left[-\alpha_i^{k+1} t_i^{k+1}\right] | \zeta_k\right) = 1 - (1 - p_i^k) \exp\left[-t_i^{k+1}\left(\alpha_i^k (1 - \chi_i^k) + \chi_i^k \gamma_1\right)\right] \\ - p_i^k \exp\left[-t_i^{k+1}\left(\alpha_i^k (1 - \chi_i^k) + \chi_i^k \gamma_2\right)\right]. \end{aligned} \tag{26}$$

Considering (19) and (26), we receive

$$\begin{aligned} M(Q_j^{k+1} | \zeta_k) = \sum_{i=1}^{m_{k+1}} q_i M(p_i^{k+1} | \zeta_k) = \sum_{i=1}^{m_{k+1}} q_i - \sum_{i=1}^{m_{k+1}} q_i \Big[(1 - p_i^k) \exp\left[-t_i^{k+1}\left(\alpha_i^k (1 - \chi_i^k) + \chi_i^k \gamma_1\right)\right] \\ - p_i^k \exp\left[-t_i^{k+1}\left(\alpha_i^k (1 - \chi_i^k) + \chi_i^k \gamma_2\right)\right]\Big] = 1 - \sum_{i=1}^{m_{k+1}} q_i \Big[(1 - p_i^k) \exp\left[-t_i^{k+1}\left(\alpha_i^k (1 - \chi_i^k) + \chi_i^k \gamma_1\right)\right] \\ - p_i^k \exp\left[-t_i^{k+1}\left(\alpha_i^k (1 - \chi_i^k) + \chi_i^k \gamma_2\right)\right]\Big], \end{aligned} \tag{27}$$

where t_i^{k+1} is determined by a formula (25).

We define (27) for $\chi_i^k = 0$ and $\chi_i^k = 1$:

$$\begin{aligned} M(Q_j^{k+1} | \zeta_k) = \sum_{i=1}^{m_{k+1}} q_i M(p_i^{k+1} | \zeta_k) = 1 - \sum_{i \in \cup u_k} q_i \exp\left[-\Delta t \alpha_i^k \gamma_1\right] \\ + \sum_{i \in \cup u_k} q_i p_i^k \left(\exp\left[-\Delta t \alpha_i^k \gamma_1\right] - \exp\left[-\Delta t \alpha_i^k \gamma_2\right]\right) - \sum_{i \notin \cup u_k} q_i \exp\left[-(\Delta t + t_i^k) \alpha_i^k \gamma_1\right]. \end{aligned} \tag{28}$$

IOP Conf. Series: Journal of Physics: Conf. Series **973** (2018) 012045 doi:10.1088/1742-6596/973/1/012045

Considering (26) and (28) we find conditional population mean for function of quality of process of training

$$M(Q^{k+1}|\zeta_k)=\frac{1}{l}\sum_{j=1}^{l}M(Q_j^{k+1}|\zeta_k)=1-\frac{1}{l}\sum_{j=1}^{l}\left[\sum_{i\in\cup u_k}q_i\exp\left[-\Delta t\alpha_i^j\gamma_1\right]\right.$$
$$\left.+\sum_{i\in\cup u_k}q_ip_i^j\left(\exp\left[-\Delta t\alpha_i^j\gamma_1\right]-\exp\left[-\Delta t\alpha_i^j\gamma_2\right]\right)-\sum_{i\notin\cup u_k}q_i\exp\left[-\left(\Delta t+t_i^j\right)\alpha_i^j\gamma_1\right]\right]. \tag{29}$$

For determination of necessary level of proficiency, we will use function which defines number of the sessions of training necessary for achievement of level $Q\le\sigma$, depending from: m – volume of all information; m_k – volumes of portions of information; Δt_k – durations of intervals between training sessions; s^* – number of measurements of a state; Q – ignorance level; σ – ignorance borders.

This function has the appearance given in expression (30).

$$n(\omega)=F(m,\{m_k\}_{k=1,2,\ldots,N},\{\Delta t_k\}_{k=1,2,\ldots,N},s^*,\sigma,\left(Q^k(\omega)\right)_{k=1,2,\ldots,N}). \tag{30}$$

For definition of a type of function we will make the following assumptions:

1. On each session of training the identical quantity is given EDI $m_k=\dfrac{m}{N}=m$.
2. Time between sessions of training we are constant ($\Delta t_k=\Delta t$).
3. The number of measurements of s for every portion of information is constant.
4. Initial speeds of a forgetting are identical to all EDI.

Taking into account above stated (20) we will transform in a look:

$$\alpha_i^{k+1}=\gamma_i^k\alpha_i^k\chi_i^k+\alpha_i^k\left(1-\chi_i^k\right), \tag{31}$$

where

$$\gamma_i^k=\begin{cases}\gamma_1, & \chi_i^k=1 \text{ и } r_i^k=0;\\ \gamma_2, & \chi_i^k=1 \text{ и } r_i^k=1.\end{cases} \tag{32}$$

Considering (32) and (22), we will receive expression for ignorance level:

$$Q^{k+1}=\frac{1}{S}\sum_{j=1}^{s}\sum_{i=1}^{km}q_i\left(1-\exp\left[-\left[\gamma_i^j\alpha_i^j\Delta t\chi_i^j+\alpha_i^j\left(\Delta t+t_i^j\right)\left(1-\chi_i^j\right)\right]\right]\right). \tag{33}$$

Considering the assumption, and the number of repetitions, considering certain, one session of training contains s of cycles of training.

Therefore, PTI are described by vectors:

$$\underbrace{\chi^0-\left(\underbrace{1,\ldots,1}_{m},\underbrace{0,\ldots,0}_{m-m}\right)}_{S},\ \underbrace{\chi^1-\left(\underbrace{1,\ldots,1}_{m},\underbrace{1,\ldots,1}_{m},\underbrace{0,\ldots,0}_{m-2m}\right)}_{S},\ \ldots,$$
$$\underbrace{\chi^{r-1}=\left(\underbrace{1,\ldots,1}_{rm},\underbrace{0,\ldots,0}_{m-rm}\right)}_{S},\ \ldots,\ \underbrace{\chi^N=\left(\underbrace{1,\ldots,1}_{m-m},\underbrace{1,\ldots,1}_{m}\right)}_{S}. \tag{34}$$

Considering that on each session of m EDI s of times are given, and $0<\gamma_1\le\gamma_i^k\le\gamma_2<1$, and $(\gamma_1\alpha,\ldots,\gamma_1\alpha)\le\alpha^s\le(\gamma_2\alpha,\ldots,\gamma_2\alpha)$ through N sessions all elements will be given $r<Ns$ time, therefore

$$\alpha^{rs}\le\left((\gamma_2)^r\alpha,\ \ldots,\ (\gamma_2)^r\alpha\right);\ \alpha^{rs}\ge\left((\gamma_1)^r\alpha,\ \ldots,\ (\gamma_1)^r\alpha\right). \tag{35}$$

Transforming (34) taking into account (33) and (35), we receive assessment Q_{rs}:

$$\frac{1}{S}\sum_{j=1}^{s}\sum_{i=1}^{km}q_i\left(1-\exp\left[-(\gamma_1)^r\alpha N\Delta t\right]\right)\le Q^{rs}\le\frac{1}{S}\sum_{j=1}^{s}\sum_{i=1}^{km}q_i\left(1-\exp\left[-(\gamma_2)^r\alpha N\Delta t\right]\right), \tag{36}$$

as $\frac{1}{S}\sum_{j=1}^{s}\sum_{i=1}^{km}q_i=1$, that

$$Q^{rs}\le\left(1-\exp\left[-(\gamma_2)^r\alpha\Delta tN\right]\right)m,\ r=1,\ 2,\ \ldots \tag{37}$$

Using expression $Q\le\sigma$, we will receive the maximum number of issue of information to a portion necessary for achievement of level σ:

$$r_{\max}\le\frac{\ln\left(-\dfrac{\ln\left(\dfrac{m-\sigma}{m}\right)}{\alpha\,\Delta t\,N}\right)}{\ln\gamma_2}. \tag{38}$$

Therefore, duration of training $n=Nr$ is estimated by expression:

$$n=\left[N\frac{\ln\left(-\dfrac{\ln\dfrac{(m-\sigma)}{m}}{\alpha\,\Delta t\,N}\right)}{\ln\gamma_2}\right]+1. \tag{39}$$

Above stated allows to define process of formation of the predicted training parameters.

1. Determination of probability of ignorance of i EDI a session is defined on k as follows:

$$p_i^k=1-\exp\left[-\alpha_i^k t_i^k\right],i=1,2,\ldots,m,k=1,2,\ldots,N. \tag{40}$$

2. Definitions of all q_i by the teacher of material prior to training as importance of i of a concept.
3. Calculate

$$Q_k=\sum_{i=1}^{m_k}p_i^k q_i\,. \tag{41}$$

4. Definition of a threshold of quality of training:

$$Q = \frac{1}{l}\sum_{j=1}^{l} Q_j^k . \quad (42)$$

4. Threshold task σ and definition S^*.
5. We set the sequence of experimental values: $T(t): T_j = T(t_j)$.
6. We define also Q_j, $j > 0$ and calculate predictability interval size L.

Process of formation of the predicted parameters of training allows to evaluate on top training duration provided that portions of training and time slots identical among themselves. Results of experiments are given in table 1. As a result of experiments for each trainee personal characteristics i.e. fractal dimensionalities of assessment of quantity of the variables of function of quality necessary for execution of criterion of quality of training were defined and also the number of repeated passing of a portion of information was defined and comparing with predicted [5, 7, 15] was carried out. In table 1 are compared the duration of training received experimentally and predicted.

Table 1. The results of training received experimentally.

$N=8, \sigma=0.4, \Delta t=1$									
m_k	$\gamma_1=0.19, \gamma_2=0.28$			$\gamma_1=0.3, \gamma_2=0.49$			$\gamma_1=0.51, \gamma_2=0.68$		
	n_{rep}	s^*	Q	n_{rep}	s^*	Q	n_{rep}	s^*	Q
5	1	2	0.18	1	2	0.23	2	3	0.31
10	1	2	0.2	2	3	0.31	3	4	0.38
15	1	2	0.23	2	3	0.34	3	4	0.39
20	2	3	0.28	3	4	0.38	3	4	0.4

$N=12, \sigma=0.3, \Delta t=1$									
m_k	$\gamma_1=0.19, \gamma_2=0.28$			$\gamma_1=0.3, \gamma_2=0.49$			$\gamma_1=0.51, \gamma_2=0.68$		
	n_{rep}	s^*	Q	n_{rep}	s^*	Q	n_{rep}	s^*	Q
5	1	2	0.16	2	3	0.22	3	4	0.28
10	2	2	0.21	3	4	0.27	4	5	0.3
15	2	3	0.23	3	4	0.29	3	4	0.29
20	3	3	0.27	4	4	0.3	4	5	0.3

$N=20, \sigma=0.3, \Delta t=1$									
m_k	$\gamma_1=0.19, \gamma_2=0.28$			$\gamma_1=0.3, \gamma_2=0.49$			$\gamma_1=0.51, \gamma_2=0.68$		
	n_{rep}	s^*	Q	n_{rep}	s^*	Q	n_{rep}	s^*	Q
5	2	2	0.17	3	3	0.24	3	5	0.28
10	2	3	0.26	3	4	0.26	5	5	0.3
15	2	3	0.28	3	4	0.28	5	5	0.3
20	3	4	0.28	4	4	0.29	5	5	0.3

On the basis of values of table 1, values of table 2 are defined.

In table 2 the predicted values of duration of training also are compared to the values received on the basis of expression (31) and (32).

In table 2 the following designations are used:

n_{ex} – the average duration of training received experimentally;

n – the training duration received when using a formula (41).

$\frac{n}{n_{ex}}$ – shows in how many times assessment of duration of training is worse than the average duration of training of the studied process.

Table 2. Results of comparison of average duration of training and her assessment for different values of parameters γ_1 and γ_2.

$N=6, \sigma=0.4, \Delta t=1$									
	$\gamma_1=0.19, \gamma_2=0.28$			$\gamma_1=0.3, \gamma_2=0.49$			$\gamma_1=0.51, \gamma_2=0.68$		
m_k	n_{ex}	n	$\frac{n}{n_{ex}}$	n_{ex}	n	$\frac{n}{n_{ex}}$	n_{ex}	n	$\frac{n}{n_{ex}}$
5	14	20	1.43	14	34	2.43	14	66	4.71
10	14	22	1.57	14	40	2.86	14	77	5.5
15	14	23	1.64	14	42	3	14	80	5.71
20	20	25	1.25	20	44	2.2	20	84	4.2

$N=10, \sigma=0.3, \Delta t=1$									
	$\gamma_1=0.19, \gamma_2=0.28$			$\gamma_1=0.3, \gamma_2=0.49$			$\gamma_1=0.51, \gamma_2=0.68$		
m_k	n_{ex}	n	$\frac{n}{n_{ex}}$	n_{ex}	n	$\frac{n}{n_{ex}}$	n_{ex}	n	$\frac{n}{n_{ex}}$
5	20	37	1.85	30	69	2.3	40	130	3.25
10	20	43	2.15	40	92	2.3	50	149	2.98
15	20	45	2.25	40	94	2.35	50	152	3.04
20	30	46	1.53	40	110	2.75	50	160	3.2

$N=20, \sigma=0.3, \Delta t=1$									
	$\gamma_1=0.19, \gamma_2=0.28$			$\gamma_1=0.3, \gamma_2=0.49$			$\gamma_1=0.51, \gamma_2=0.68$		
m_k	n_{ex}	n	$\frac{n}{n_{ex}}$	n_{ex}	n	$\frac{n}{n_{ex}}$	n_{ex}	n	$\frac{n}{n_{ex}}$
5	40	88	2.2	60	160	2.67	60	298	4.97
10	60	99	1.65	60	179	2.99	100	326	3.26
15	60	103	1.72	60	186	3.1	100	338	3.38
20	60	106	1.78	80	191	2.39	100	356	3.56

We will consider in more detail dependence $\frac{n}{n_{ex}}$ which depends on EDI m number at various values γ_1 and γ_2. We will make approximation $\frac{n}{n_{ex}} = f(m)$ of expression function:

$$\frac{n}{n_{ex}} = f(m) = a \ln m, \quad m > 1, \tag{43}$$

where a – parameter.

At approximation we will use a method of the smallest squares. We will make assessment for various values γ_1 and γ_2. Therefore, considering expression (40) and (41), we will receive assessment of average duration of process of training in a look:

IOP Conf. Series: Journal of Physics: Conf. Series **973** (2018) 012045 doi:10.1088/1742-6596/973/1/012045

$$n = \frac{N \dfrac{\ln\left(-\dfrac{\ln \dfrac{(m-\sigma)}{m}}{\alpha \Delta t N} \right)}{\ln \gamma_2} + 1}{a \ln m}. \tag{44}$$

In spite of the fact that the received assessment is approximate it it is possible to use for forecasting of number of sessions of training for which the trainee with his individual parameters γ_1, γ_2, and α at set N, Δt, m will reach proficiency level σ. It is necessary to emphasize the fact that it is possible to solve also the return problem i.e. at the set number of sessions of training $N_{\max}$ it is possible to define quantities of EDI from m necessary trained at each session of training to reach level σ not later $N_{\max}$.

4. Conclusion

Development process of program and information components not only defines all necessary components when forming a program complex, but also sets the necessary sequence of actions (fig. 1-3).

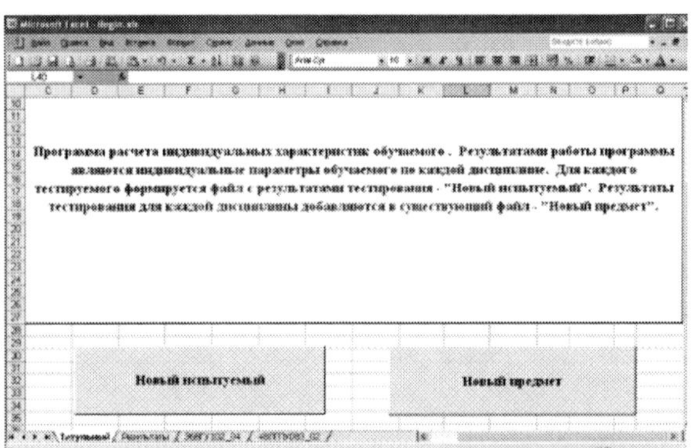

Figure 1. Individual parameters.

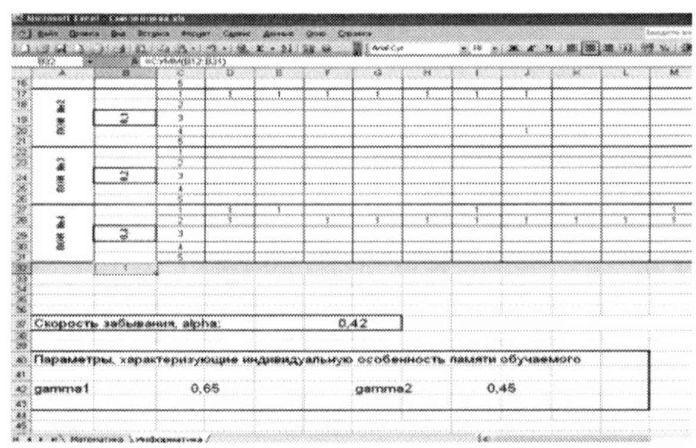

Figure 2. Indicators of parameters of training.

The sequence of creation has the following appearance:

- a choice of work benches as by development of both components work benches superimpose the requirements of process of their creation;
- creation of the DB physical model;
- creation of software modules.

The control of social systems described above meets certain difficulties on the translation of these difficult systems from start state in finite, a status according to a main goal of control because of impossibility of the formalized determination of initial conditions of such system [2, 13].

To monitor computation process and dynamics of change of figures of merit of assimilation of discipline, information systems allow, it is information – computer systems and databases which save information on each trainee throughout the entire period of training [1, 16-19].

The developed program complex on the basis of the values of personal parameters received as a result of experiments on each trainee allows to calculate personal characteristics, to create rating and to monitor process of change of parameters.

IOP Conf. Series: Journal of Physics: Conf. Series **973** (2018) 012045 doi:10.1088/1742-6596/973/1/012045

```
36102_04.txt    48083_02.txt
4 20

0,21 0,19 0,23 0,33 0,13 0,26 0,43 0,15 0,29 0,23 0,17 0,26
0,34 0,13 0,26 0,21 0,18 0,24 0,29 0,14 0,28 0,43 0,15 0,26
0,25 0,18 0,21 0,34 0,14 0,25 0,27 0,16 0,27 0,25 0,14 0,25
0,33 0,14 0,26 0,38 0,17 0,27 0,26 0,15 0,28 0,46 0,15 0,29
0,50 0,17 0,24 0,19 0,15 0,28 0,61 0,17 0,20 0,28 0,14 0,35
0,16 0,16 0,25 0,43 0,17 0,20 0,23 0,16 0,25 0,31 0,16 0,26
0,43 0,17 0,27 0,29 0,16 0,25 0,43 0,40 0,21 0,33 0,15 0,29
0,61 0,15 0,28 0,27 0,14 0,21 0,25 0,14 0,26 0,53 0,17 0,28
0,23 0,14 0,20 0,26 0,15 0,26 0,46 0,15 0,26 0,43 0,14 0,27
0,43 0,15 0,24 0,30 0,14 0,26 0,28 0,16 0,25 0,26 0,18 0,29
0,25 0,14 0,25 0,24 0,19 0,21 0,31 0,15 0,29 0,26 0,16 0,28
0,46 0,18 0,24 0,24 0,20 0,26 0,33 0,17 0,28 0,43 0,17 0,27
0,34 0,16 0,23 0,26 0,16 0,24 0,34 0,17 0,30 0,26 0,15 0,28
0,53 0,17 0,25 0,28 0,18 0,25 0,32 0,15 0,31 0,28 0,14 0,20
0,43 0,15 0,29 0,31 0,16 0,23 0,26 0,14 0,35 0,31 0,18 0,25
0,26 0,17 0,28 0,33 0,14 0,25 0,21 0,15 0,26 0,33 0,16 0,21
0,26 0,16 0,27 0,34 0,15 0,29 0,34 0,14 0,29 0,53 0,14 0,26
0,43 0,18 0,28 0,32 0,13 0,28 0,25 0,18 0,28 0,43 0,16 0,29
0,26 0,16 0,24 0,26 0,18 0,29 0,33 0,14 0,27 0,31 0,15 0,28
0,42 0,17 0,27 0,40 0,14 0,28 0,50 0,15 0,29 0,23 0,17 0,27

0,300   0,160   0,260
0,318   0,150   0,260
0,278   0,155   0,245
0,358   0,152   0,275
0,395   0,157   0,268
0,283   0,162   0,240
0,370   0,220   0,255
0,415   0,150   0,258
0,345   0,145   0,248
0,318   0,158   0,260
0,265   0,160   0,258
0,365   0,180   0,263
0,300   0,160   0,263
0,353   0,160   0,253
0,328   0,158   0,280
```

Figure 3. An example of the text file with total values.

In the developed information complex, perhaps to store, process results of researches, to look through total parameters and to form the rating of students [3, 18].

References

[1] Goldberge E 1990 Haos and fractals in physiology of the person *In a pattern of science* **4** pp 25-32
[2] Gorev A 1997 *DBMS Effective operation with the* (St Petersburg) p 704
[3] Grinchenko V T 2007 *Introduction to non-linear dynamics Chaos and fractals* (Moscow: LKI publishing house) p 264
[4] Degtyarev Y I 1976 *Fundamentals of cybernetics* (Moscow: High school) p 408
[5] Durant B and Odel P 1997 *Cluster analysis* (Moscow: Statistics) p 128
[6] Zhukovsky V I 1988 *Multicriteria decision-making in the conditions of uncertainty* (Moscow: MNIIPU) p 130
[7] Zeldovich Y B 1985 Fractals, similarity, the intermediate asymptotics *UFN* **3** pp 493-506
[8] Peters A A 2004 *Fractal analysis of the financial markets: Application of the chaos theory in investments and economy* (Moscow: Internet trading) p 304
[9] Savelyev A Ya 1986 *Preparation of information for automated learning systems* (Moscow: Higher school) p 176
[10] Sumin V I 2008 Formalization of the task of control of social systems *Theory of the conflict and its application* (Voronezh: Voronezh institute of high technologies) pp 133-137
[11] Sumin V I 2009 Konkretization of the training purpose in the trainee's model as object of control means *Mathematical Methods and Information Technology: All scientific and practical work conference collection of materials* (Krasnodar) pp 187-190

[12] Smolentseva T E 2011 Realization of function of quality of training *International scientific and practical conference Modern science: theory and practice* (Stavropol) pp 140-143
[13] Tarasenko V V 2000 Fractal geometry of the nature: sociocultural measurement *Synergetic paradigm The variety of searches and approaches* (Moscow: Progress-Tradition) pp 191-214
[14] Trakhtengerts E A 2001 Opportunities and realization of computer systems of support of decision-making *News of RAS Theory and control systems* **3** pp 86-113
[15] Dushkin A V 2017 *Introduction to the system analysis* (Moscow: Hot line – Telecom) p 234
[16] Dushkin A V 2014 *Mathematical models and information processes of management of a difficult object* (Voronezh: Scientific book) p 125
[17] Dushkin A V 2014 *Model of management of difficult social objects in the system of training of employees of penal correction system* (Voronezh: Scientific book) p 125
[18] Trubetskov D N 1998 Turbulentnost and the determined chaos *Sorovsky the educational magazine* **1** pp 77-83
[19] Feygenbaum M 1983 *Universalnost in behavior of nonlinear systems* (North Florid: UNF) pp 343-374

IOP Conf. Series: Journal of Physics: Conf. Series **973** (2018) 012046 doi:10.1088/1742-6596/973/1/012046

Mathematical modeling of the crack growth in linear elastic isotropic materials by conventional fracture mechanics approaches and by molecular dynamics method: crack propagation direction angle under mixed mode loading

Larisa Stepanova, Sergej Bronnikov

Samara National Research University, Moscovskoe shosse, 34 443011 Samara

Email: stepanovalv@samsu.ru

Abstract. The crack growth directional angles in the isotropic linear elastic plane with the central crack under mixed-mode loading conditions for the full range of the mixity parameter are found. Two fracture criteria of traditional linear fracture mechanics (maximum tangential stress and minimum strain energy density criteria) are used. Atomistic simulations of the central crack growth process in an infinite plane medium under mixed-mode loading using Large-scale Molecular Massively Parallel Simulator (LAMMPS), a classical molecular dynamics code, are performed. The inter-atomic potential used in this investigation is Embedded Atom Method (EAM) potential. The plane specimens with initial central crack were subjected to Mixed-Mode loadings. The simulation cell contains 400000 atoms. The crack propagation direction angles under different values of the mixity parameter in a wide range of values from pure tensile loading to pure shear loading in a wide diapason of temperatures (from 0.1 K to 800 K) are obtained and analyzed. It is shown that the crack propagation direction angles obtained by molecular dynamics method coincide with the crack propagation direction angles given by the multi-parameter fracture criteria based on the strain energy density and the multi-parameter description of the crack-tip fields.

1. Introduction

One of the fundamental ideas in the fracture assessment of brittle fracture is the local mode I concept [1]. The proposal of mode I dominance was suggested when dealing with cracked plates under plane loading and transverse shear, where the crack grows in the direction almost perpendicular to the maximum tangential stress (MTS) in radial direction from its tip [1]. This theory is one of the widely used theories for mixed mode crack growth [2-16]. In more detail the criterion states that the crack propagation starts along the direction on which the tangential stress becomes maximum. The fracture occurs when the maximum tangential stress reaches a critical value for the material equal to the fracture stress in uniaxial tension. For the calculation of the tangential stress a critical distance from the crack tip must be introduced. But the question is what one can use as a critical distance from the crack tip? To overcome this difficulty the concept of a core region surrounding the crack tip has been proposed by Sih [2]. The idea is the continuum mechanics solution, as well as experimental measurements, stop at some distance from the crack tip. The distance serves as a scale size of analysis at the continuum level. Together with the MTS criterion the strain energy density (SED) has been used to formulate failure criteria for materials exhibiting both ductile and brittle behavior. Dealing with the strain energy concept it is worth to note that it is necessary to introduce the radius of "core region" surrounding the crack tip [1, 2]. The key idea is that the continuum mechanics stops short at a distance from the crack tip. The strain energy density factor (S) was defined as a product of the strain energy density by a critical distance from the point of singularity. Therefore it is necessary to introduce the critical distance from the crack tip [3-16].

In [5,6] a discussion on the need to consider several more initial terms of the Williams series expansion during fracture behaviour assessment is presented. The paper [6] shows that the classical (one-parameter) MTS criterion gives significantly different crack propagation angle values in comparison to data calculated numerically by means of the finite element method. It is recommended that several more higher-order terms of the Williams expansion should be used for the description of the stress field as an input for the fracture criterion, especially when the criterion is applied at a large distance from the crack tip. In [10] it is noted that types of structural materials differ in the size of the whole nonlinear zone around the crack tip and also in the proportions of its individual subsets. Due to these effects it is clear that the stress field at a greater distance from the crack tip must be described more precisely. In [12] fracture behavior (crack deflection) of a crack in a silicate-based composite is studied. The three-point bending test is simulated numerically by means of the finite element method and the influence of existence of a stiff aggregate ahead of the crack tip is studied and discussed. The maximum tangential stress criterion is applied in order to calculate the initial crack propagation angle. Various geometrical configurations are investigated (the aggregate eccentricity and depth are varied) and the generalized MTS criterion based on the Williams expansion is tested considering various numbers of initial terms of the series. The authors [12] show that using the generalized MTS criterion is strongly limited by the elastic mismatch caused by the presence of the stiff aggregate near the crack tip. Therefore, as it is concluded in [12] additional extensive investigations are unavoidable in this field in order to find out how the stress field ahead of a crack tip behaves when a material interface exists at a small distance from it.

A new variant of the Element-Free Galerkin (EFG) method that combines the diffraction method, to characterize the crack tip solution, and the Heaviside enrichment function for representing discontinuity due to a crack, has been used in [13] to model crack propagation through non-homogenous materials. In the case of interface crack propagation, the kink angle is predicted by applying the maximum tangential principal stress (MTPS) criterion in conjunction with consideration of the energy release rate (ERR). The MTPS criterion is applied to the crack tip stress field described by both the stress intensity factor (SIF) and the T-stress, which are extracted using the interaction integral method. The proposed EFG method has been developed and applied for 2D case studies involving a crack in an orthotropic material, crack along an interface and a crack terminating at a bi-material interface, under mechanical or thermal loading; this is done to demonstrate the advantages and efficiency of the proposed methodology. The computed SIFs, T-stress and the predicted interface crack kink angles are compared with existing results in the literature and are found to be in good agreement. An example of crack growth through a particle-reinforced composite materials, which may involve crack meandering around the particle, is reported.

Nowadays one can estimate the crack propagation direction angle using molecular dynamics simulation [17-22]. Molecular dynamics (MD) is a powerful tool in characterizing the inception and evolution of plastic deformation and associated fracture mechanism at the atomic scale [17]. Applied to the problem of crack propagation and growth, the method has been used extensively to analyze the fundamental mechanisms of material's fracture in the past [17-22]. Thus, in [18] molecular dynamics (MD) simulations are performed to investigate a crack propagation behavior in ferrite iron with a bodycentered cubic (BCC) crystal structure containing a solute nanocluster under pure and mixed-mode loadings. In the MD simulations, coherent Cu and/or Ni nanoclusters are considered to be the solute nanoclusters formed in ferrite iron during neutron irradiation. A MD simulation to assess the effect of crack length on the ultimate tensile strength of infinitely large armchair and zigzag graphene sheets is presented in [19]. The authors of [19] note that the strength of graphene is inversely proportional to the square-root of crack length as in continuum fracture theories. Further comparison of the strength given by MD simulations with Griffith's energy balance criterion demonstrates a reasonable agreement. Armchair and zigzag graphene sheets with 2.5 nm long crack exhibit around 55% of the strength of pristine sheets. Investigation of the influence of temperature on the strength of grapheme performed in [19] indicates that sheets at higher temperatures fail at lower strengths, due to high kinetic energy of atoms. The main result of [19] is that strength obtained from molecular dynamics simulations shows inverse square-root proportionality with crack length as in continuum fracture mechanics theories.

In [20] crack propagation at the Cu/SiC interface under Mode I and mixed mode loading conditions is analyzed by MD simulations. The thickness of the Cu/SiC interface can be defined with monitoring the atomic energy change across the interface and crack growth is observed at strong Cu/SiC interface. For the mode I loading conditions, the crack propagates in an asymmetrical manner. The dislocations generated by lattice mismatch are believed to be responsible for the asymmetric crack propagation. The Rice and Thomson (R–T) model is adopted here as a criterion to identify the crack propagation style by comparing the energy for dislocation nucleation and the decohesion energy of the interface. The style of crack propagation predicted by R–T model is cleave and asymmetrical, which fits well with the results of MD simulations. For the mix mode loading conditions, six different loading angles are simulated by MD method. For the six mixed mode loading conditions, the style of the crack propagation is dependent on the loading angle. The MD simulations proved that cleave and blunt are two typical behaviors of the interfacial cracks, which are consistent with theoretical prediction by R–T model. Such experimental observations and numerical predictions represent a qualitative nanoscale understanding of the fracture processes, providing valuable insights into the crack tip plasticity and associated deformation mechanisms-dislocation nucleation and multiplications [17].

The present study is aimed at the determination of the crack propagation direction angle in a wide range of mixed-mode loading using 1) the multi-parameter maximum tangential stress criterion; 2) the multi-parameter strain energy density criterion; 3) molecular dynamics simulation of the crack propagation behavior under mixed-mode loading in a full range of the mixity parameter.

2. Methods of crack growth simulation

There are several methods of simulating crack behavior, including conventional fracture mechanics analysis [1-16], molecular mechanics, molecular dynamics [17-22], continuous mechanics, finite element method and methods based on DFT theory. Focus of this article is comparison of molecular dynamics and continuum fracture mechanic approach.

2.1. Continuum linear fracture mechanics approach

In the development of fracture mechanics Williams made a major breakthrough in the analysis of asymptotic stress field at the vicinity of the crack tip in isotropic linear elastic plane media. With the well-known eigenfunction expansion method it is possible to establish the separable variable nature of the solution and to obtain asymptotic expressions for the stress field in a plane medium with a traction-free crack submitted to mode I, mode II and mixed-mode (mode I and mode II) loading conditions:

$$\sigma_{ij}(r,\theta)=\sum_{m=1}^{2}\sum_{k=-\infty}^{\infty}a_k^m r^{k/2-1} f_{m,ij}^{(k)}(\theta), \tag{1}$$

with index m associated to the fracture mode; a_k^m amplitude coefficients related to the geometric configuration, load and mode; $f_{m,ij}^{(k)}(\theta)$ angular functions depending on stress component and mode. Analytical expressions for angular eigenfunctions $f_{m,ij}^{(k)}(\theta)$ are available [3,4]:

$$\begin{aligned} f_{1,11}^{(k)}(\theta)&=\frac{k}{2}\left[\left(2+k/2+(-1)^k\right)\cos(k/2-1)\theta-(k/2-1)\cos(k/2-3)\theta\right],\\ f_{1,22}^{(k)}(\theta)&=\frac{k}{2}\left[\left(2-k/2-(-1)^k\right)\cos(k/2-1)\theta+(k/2-1)\cos(k/2-3)\theta\right],\\ f_{1,12}^{(k)}(\theta)&=\frac{k}{2}\left[-\left(k/2+(-1)^k\right)\sin(k/2-1)\theta+(k/2-1)\sin(k/2-3)\theta\right], \end{aligned} \tag{2}$$

IOP Conf. Series: Journal of Physics: Conf. Series **973** (2018) 012046 doi:10.1088/1742-6596/973/1/012046

$$f_{2,11}^{(k)}(\theta)=-\frac{k}{2}\left[\left(2+k/2-(-1)^k\right)\sin(k/2-1)\theta-(k/2-1)\sin(k/2-3)\theta\right],$$
$$f_{2,22}^{(k)}(\theta)=-\frac{k}{2}\left[\left(2-k/2+(-1)^k\right)\sin(k/2-1)\theta+(k/2-1)\sin(k/2-3)\theta\right], \quad (3)$$
$$f_{2,12}^{(k)}(\theta)=\frac{k}{2}\left[-\left(k/2-(-1)^k\right)\cos(k/2-1)\theta+(k/2-1)\cos(k/2-3)\theta\right].$$

The multi-parameter fracture mechanics concept consists in the idea that the crack-tip stress field is described by means of the Williams expansion (1). In this work the central crack in an infinite plane medium is considered. Analytical determination of coefficients in crack-tip expansion for a finite crack in an infinite plane medium is given in [3]:

$$a_{2n+1}^{1}=(-1)^{n+1}\frac{(2n)!\sigma_{22}^{\infty}}{2^{3n+1/2}(n!)^2(2n-1)a^{n-1/2}}, \quad (4)$$

$$a_2^1=-\sigma_{22}^{\infty}/4, \quad a_{2k}^1=0, \quad (5)$$

for a Mode I crack;

$$a_{2n+1}^{2}=(-1)^{n+1}\frac{(2n)!\sigma_{12}^{\infty}}{2^{3n+1/2}(n!)^2(2n-1)a^{n-1/2}}, \quad (6)$$

$$a_{2k}^2=0, \quad (7)$$

for a Mode II crack.

Since all the coefficients in the Williams asymptotic series expansion for the geometry considered are known it is possible to estimate the crack propagation direction angle by means of the multi-parameter fracture mechanics concept. In this paper two fracture criteria were chosen for the estimation of the initial crack growth direction: maximum tangential stress (MTS) criterion and strain energy density (SED) criterion [2]. The most well-known criterion for the estimation of the crack propagation direction is MTS (maximum tangential stress). This criterion is strictly stress-based and its analysis does not depend on the conditions of plane stress and plane strain. It assumes that a crack extends in the direction of the maximum tangential stress. Mathematically the criterion of the maximum tangential stress is written as:

$$\frac{\partial\sigma_{\theta\theta}}{\partial\theta}=0, \quad \frac{\partial^2\sigma_{\theta\theta}}{\partial\theta^2}<0 \quad (8)$$

In this work, the multi-parameter form of the MTS criterion (8) is tested while considering various numbers of the initial terms of the Williams expansion (1). Therefore, the tangential stress has to be expressed via the power series and then its maximum is sought numerically. The results obtained by the MTS criterion are shown in Table 1 where N is the number of terms keeping in the Williams series expansion, $r_c=r/a$ is the dimensionless distance from the crack tip. The first column gives the crack propagation direction angles obtained by the MTS criterion when the leading term of the Williams asymptotic expansion is taken into account. While the following columns show the crack propagation direction angles given by the multi-parametric asymptotic expansion of the stress field in the vicinity of the crack tip. Here the multi-parameter crack tip expansion of the stress field contains 100 terms. One can see that the values of the crack propagation direction angle obtained by the one-parameter fracture criterion are close to the values of the second column especially for Mode II loading. Obviously it can be seen the influence of T-stresses at the crack tip. As the distance from the crack tip increases the difference between results given by the one-parameter fracture criterion and the multi-parameter fracture criterion enhances.

The criterion based on the minimum of the strain energy density deals with strain energy and therefore the conditions for predicting crack growth direction depend on whether or not the specimen is subjected to plane strain or plane stress conditions. Specifically, the minimum strain energy density criterion states that a crack will grow in the direction where the strain energy density reaches its minimum. This condition can be written as:

$$\frac{\partial S}{\partial \theta}=0, \quad \frac{\partial^2 S}{\partial \theta^2}>0, \quad S=\frac{1}{2\mu}\left[\frac{\kappa+1}{8}\left(\sigma_{rr}+\sigma_{\theta\theta}\right)^2-\sigma_{rr}\sigma_{\theta\theta}+\sigma_{r\theta}^2\right]. \tag{9}$$

The Williams series expansion (1) is used for approximation of the stress tensor components and the minimum is sought through numerical methods. A procedure for searching the extreme of the function is programmed in Waterloo Maple code. The relevant quantities are expressed by means of the Williams series expansion considering both various terms numbers in Williams series expansion and various distances from the crack tip. The results of the calculations are shown in Tables 2-5.

Table 1. Crack propagation direction angles obtained from the MTS fracture criterion at various radial distances from the crack tip.

$N=1$	$r_c=0.05$	$r_c=0.1$	$r_c=0.25$	$r_c=0.5$	$r_c=0.75$	$r_c=1.25$	$r_c=1.5$	$r_c=1.75$	M^e
-70.53	-66.21	-62.49	-55.08	-49.65	-47.39	-45.72	-45.40	-45.29	0
-67.53	-67.53	-62.20	-58.34	-51.34	-46.55	-44.63	-42.99	-42.70	0.1
-64.47	-58.01	-54.10	-47.60	-43.44	-41.83	-40.72	-40.52	-40.67	0.2
-62.86	-55.83	-51.91	-45.69	-41.84	-40.38	-39.39	-39.06	-39.06	0.25
-61.18	-53.53	-49.64	-43.72	-40.18	-38.87	-38.01	-37.87	-37.56	0.3
-57.48	-48.60	-44.83	-39.58	-36.65	-35.63	-35.01	-34.93	-34.14	0.4
-53.13	-43.03	-39.50	-35.00	-32.62	-31.95	-31.56	-31.52	-31.37	0.5
-47.72	-36.61	-33.46	-29.80	-28.11	-27.63	-27.44	-27.45	-27.35	0.6
-40.61	-29.12	-26.52	-23.77	-22.66	-22.42	-22.40	-22.44	-22.61	0.7
-36.12	-24.93	-22.68	-20.40	-19.54	-19.39	-19.44	-19.49	-19.47	0.75
-30.81	-20.42	-18.55	-16.75	-16.13	-16.06	-16.14	-16.20	-16.26	0.8
-17.19	-10.57	-9.59	-8.71	-8.60	-8.46	-8.55	-8.59	-8.53	0.9

Table 2. Crack propagation direction angles obtained from the SED fracture criterion at various radial distances from the crack tip (plane strain conditions), $\nu=0.3$.

$N=1$	$r_c=0.05$	$r_c=0.1$	$r_c=0.25$	$r_c=0.5$	$r_c=0.75$	$r_c=1.25$	$r_c=1.5$	$r_c=1.75$	M^e
-82.34	-82.36	-82.33	-81.74	-79.05	-75.27	-68.55	-66.02	-63.70	0
-76.19	-75.98	-75.84	-74.94	-72.04	-68.56	-62.68	-60.68	-57.95	0.1
-70.14	-69.52	-69.20	-67.94	-64.95	-61.81	-56.72	-54.81	-52.18	0.2
-67.14	-66.24	-65.81	-64.35	-61.37	-58.40	-53.69	-51.93	-51.36	0.25
-64.13	-62.90	-62.34	-60.72	-57.76	-54.97	-50.63	-49.00	-46.38	0.3
-58.10	-55.99	-55.17	-53.26	-50.45	-48.03	-44.38	-43.01	-40.55	0.4
-51.91	-48.65	-47.59	-45.52	-42.97	-40.93	-37.92	-36.81	-34.64	0.5
-45.35	-40.71	-39.47	-37.43	-35.24	-33.58	-31.20	-30.33	-24.63	0.6
-38.01	-31.95	-30.69	-28.88	-27.15	-25.90	-24.13	-23.48	-22.46	0.7
-33.82	-27.21	-26.01	-24.40	-22.93	-21.89	-20.42	-19.88	-20.53	0.75
-29.04	-22.20	-21.14	-19.77	-18.58	-17.74	-16.58	-16.15	-15.99	0.8
-16.75	-11.44	-10.82	-10.08	-9.48	-9.06	-8.49	-8.28	-8.56	0.9

Table 3. Crack propagation direction angles obtained from the SED fracture criterion at various radial distances from the crack tip (plane strain conditions), $\nu=0.5$.

$N=1$	$r_c=0.05$	$r_c=0.1$	$r_c=0.25$	$r_c=0.5$	$r_c=0.75$	$r_c=1.25$	$r_c=1.5$	$r_c=1.75$	M^e

-90.00	-89.28	-88.54	-86.07	-81.11	-76.21	-68.81	-66.17	-63.74	0
-83.99	-82.82	-81.85	-78.90	-73.92	-69.49	-63.00	-60.69	-58.01	0.1
-77.91	-76.14	-74.88	-71.50	-66.65	-62.70	-57.07	-55.05	-52.26	0.2
-74.82	-72.68	-71.26	-67.02	-62.97	-59.27	-54.03	-52.17	-51.41	0.25
-71.68	-69.12	-67.55	-63.86	-59.28	-55.88	-50.97	-49.24	-46.47	0.3
-65.19	-61.66	-59.80	-55.98	-51.78	-48.78	-44.68	-43.22	-40.65	0.4
-58.28	-53.60	-51.55	-47.80	-44.10	-41.57	-38.17	-36.97	-34.73	0.5
-50.68	-44.80	-42.70	-39.28	-36.16	-34.10	-31.38	-30.42	-28.69	0.6
-41.90	-35.06	-33.13	-30.28	-27.84	-26.27	-24.22	-23.50	-22.47	0.7
-36.84	-29.80	-28.05	-25.57	-23.50	-22.18	-20.47	-19.87	-20.27	0.75
-31.13	-24.27	-22.77	-20.70	-19.03	-17.97	-16.60	-16.11	-15.95	0.8
-17.21	-12.46	-11.63	-10.54	-9.70	-9.16	-8.47	-8.23	-8.38	0.9

Table 4. Crack propagation direction angles obtained from the SED fracture criterion at various radial distances from the crack tip (plane stress conditions), $\nu = 0.3$.

$N=1$	$r_c=0.05$	$r_c=0.1$	$r_c=0.25$	$r_c=0.5$	$r_c=0.75$	$r_c=1.25$	$r_c=1.5$	$r_c=1.75$	M^e
-79.66	-80.01	-80.26	-80.33	-78.36	-74.95	-68.48	-65.96	-63.70	0
-73.35	-73.59	-73.78	-73.60	-71.39	-68.24	-62.58	-60.40	-57.94	0.1
-67.19	-67.14	-67.20	-66.69	-64.35	-61.50	-56.60	-54.73	-52.16	0.2
-64.16	-63.88	-63.85	-63.16	-60.79	-59.09	-53.57	-51.84	-51.34	0.25
-61.14	-60.58	-60.44	-59.58	-57.21	-54.67	-50.51	-48.92	-46.36	0.3
-55.14	-53.80	-53.41	-52.25	-49.96	-47.76	-44.26	-42.94	-40.52	0.4
-49.09	-46.66	-46.03	-44.64	-42.54	-40.69	-37.83	-36.75	-34.62	0.5
-42.83	-39.00	-38.16	-36.70	-34.88	-33.39	-31.14	-30.30	-28.61	0.6
-36.00	-30.60	-29.67	-28.32	-26.88	-25.76	-24.10	-23.48	-22.45	0.7
-32.17	-26.06	-25.15	-23.93	-22.70	-21.77	-20.40	-19.89	-20.62	0.75
-27.82	-21.27	-30.44	-19.39	-18.40	-17.66	-16.57	-16.17	-16.02	0.8
-16.44	-10.97	-10.47	-9.89	-9.39	-9.02	-8.49	-8.30	-8.63	0.9

Table 5. Crack propagation direction angles obtained from the SED fracture criterion at various radial distances from the crack tip (plane stress conditions), $\nu = 0.5$.

$N=1$	$r_c=0.05$	$r_c=0.1$	$r_c=0.25$	$r_c=0.5$	$r_c=0.75$	$r_c=1.25$	$r_c=1.5$	$r_c=1.75$	M^e
-83.62	-83.50	-83.34	-82.44	-79.39	-75.43	-68.60	-66.04	-68.48	0
-77.53	-77.13	-76.84	-75.60	-72.36	-68.72	-62.74	-60.51	-61.59	0.1
-71.51	-70.66	-70.17	-68.54	-62.24	-61.96	-56.78	-54.86	-49.39	0.2
-68.51	-67.35	-66.74	-64.93	-61.65	-58.56	-53.75	-51.96	-42.16	0.25
-65.51	-63.98	-63.24	-61.26	-58.03	-55.12	-50.69	-49.04	-44.43	0.3
-59.42	-57.00	-55.99	-53.74	-50.69	-48.17	-44.43	-43.04	-35.74	0.4
-53.13	-49.56	-48.30	-45.93	-43.17	-41.04	-37.97	-36.84	-33.50	0.5
-46.41	-41.47	-40.07	-37.77	-35.40	-33.68	-31.24	-30.35	-29.23	0.6
-38.83	-32.55	-31.15	-29.14	-27.28	-25.97	-24.15	-23.49	-20.86	0.7
-34.47	-27.71	-26.40	-24.62	-23.04	-21.94	-20.43	-19.86	-17.90	0.75
-29.51	-22.61	-21.45	-19.95	-18.66	-17.79	-16.58	-16.15	-17.27	0.8
-16.86	-11.65	-10.98	-10.17	-9.52	-9.08	-8.49	-8.27	-9.30	0.9

The fracture criteria are applied at various distances from the crack tip (specifically at $r_c = 0.05, 0.1; 0.25; 0.5; 0.75; 1.25; 1.5; 1.75$). It is obvious that the fracture criteria in its classical one-parameter form are independent on the length parameter, i.e. on the radial distance. The following conclusions can be stated based on the results obtained from the MTS criterion: at larger distances from the crack tip, it is very useful to take into account more terms of the Williams series expansion; at small distances from the crack tip, the crack propagation direction is nearly independent of the number of Williams series expansion terms considered in the fracture criterion, which is caused by the strong singular stress field in the region. At larger distances from the crack tip, the initial crack propagation angle converges more slowly and it is necessary to keep high number of terms in the asymptotic expansion. In the present study 100 terms of the Williams series expansion are kept. Although the dependences in Tables 1-5 are not strictly uniform the qualitative trend is clear; with increasing distance from the crack tip a greater number of higher-order terms of the Williams series expansion is needed. As it is noted in [6] the very important question is how the choice of a right distance from the crack tip should be made. When non-singular terms of the stress expansion are taken into account it is impossible to avoid the dependence of the criterion on length parameter [6]. It is common that this value should be constant for each material and should be connected to materials properties. In order to consider this phenomenon in this study atomistic modeling of mixed-mode loading of the plane medium with the central crack is performed.

2.2. Atomistic modeling for mixed-mode loading of the plane medium with the central crack

Face-centered cubic (FCC) materials such as Cu are broadly used in various industries as structural materials [18-20]. Thus, in order to improve safety and reliability it is crucial to understand how such materials fail [20]. As noted in [20] even with the same FCC structure different materials exhibit diverse failure patterns due to their dissimilar material properties. The fracture of materials is intrinsically a multiscale phenomenon, which originates at the atomic scale as a result of the breaking of the bonds between atoms. The fracture behaviour of materials is, therefore, strongly dependent on the local atomic environment, such as the atomic structure, lattice orientation, and the discrete nature of matter distribution [18-20]. Thus, it is unsuitable to analyze fracture at the atomic scale by a conventional continuum mechanics-based approach. Atomistic scale modeling and simulation are required. Molecular dynamics is one of the most widely used numerical simulation techniques for investigating the fracture behavior of materials at the atomic scale [17-21]. In this paper to model copper plate under mixed loading we used large-scale atomic/molecular massively parallel simulator (LAMMPS) [21,22] in combination with embedded atom method (EAM) [17, 24]. EAM potentials are widely used in variety of different simulations, focusing mainly on mechanical properties [17, 24]. Potentials for copper were proposed by Foiles et al. [24] and shown decent results in elastic limit, therefore, all the strain considered in this work are elastic. Molecular dynamics simulations were performed with the plate with the central crack in Cu single crystal to investigate the governing deformation mechanisms at the crack tip under mixed-mode (Mode I and Mode II) loadings in a full range of the mixity parameter. Periodic boundary conditions were implemented in all three directions of the cell. To neglect the effect of neighbouring cells we choose the size of the central crack to be relatively small (1:10 ratio) to the size of the simulation cell. Furthermore, we added small non-interacting boundaries to the edge of the plane. Total number of atoms in the cell is 300000. Before mixed loading is applied, plate is optimized to the minimal energy with conjugated gradient method. When minimum energy state is achieved, we apply mixed strain. During all 50000 steps of simulation, we collect data of the state of all atoms in the cell. Results are shown in Figures 1-6, colour coding is obtained by OVITO [26, 27] tool. Brighter colours correspond to higher stress.

IOP Conf. Series: Journal of Physics: Conf. Series **973** (2018) 012046 doi:10.1088/1742-6596/973/1/012046

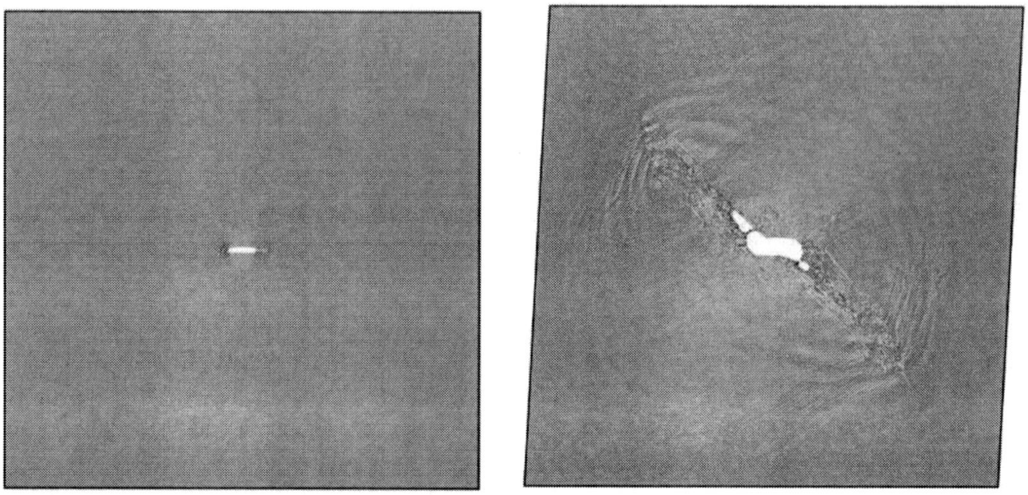

Figure 1. Copper plate just arter energy minimization and after 25 ps with stress tensor component σ_{11} shown.

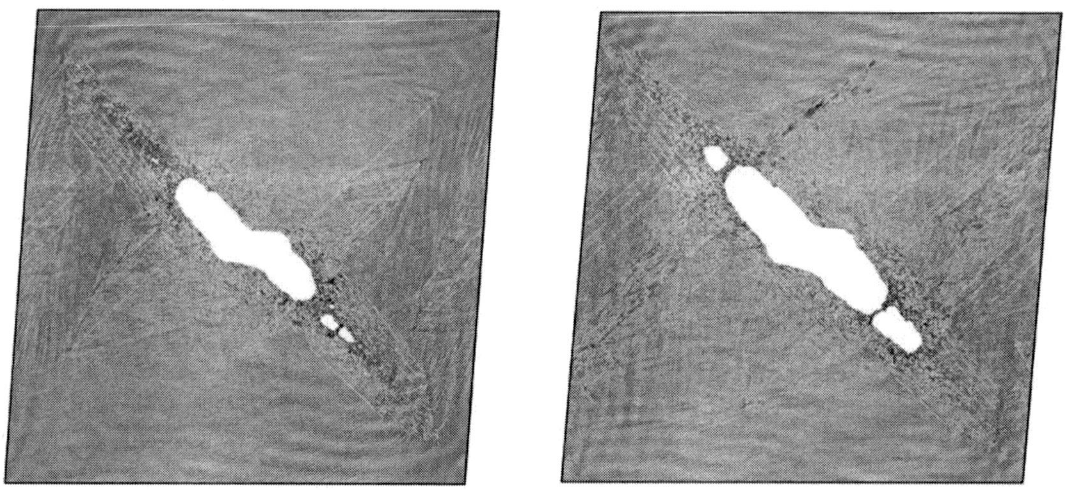

Figure 2. Copper plate after 35 ps and 45 ps.

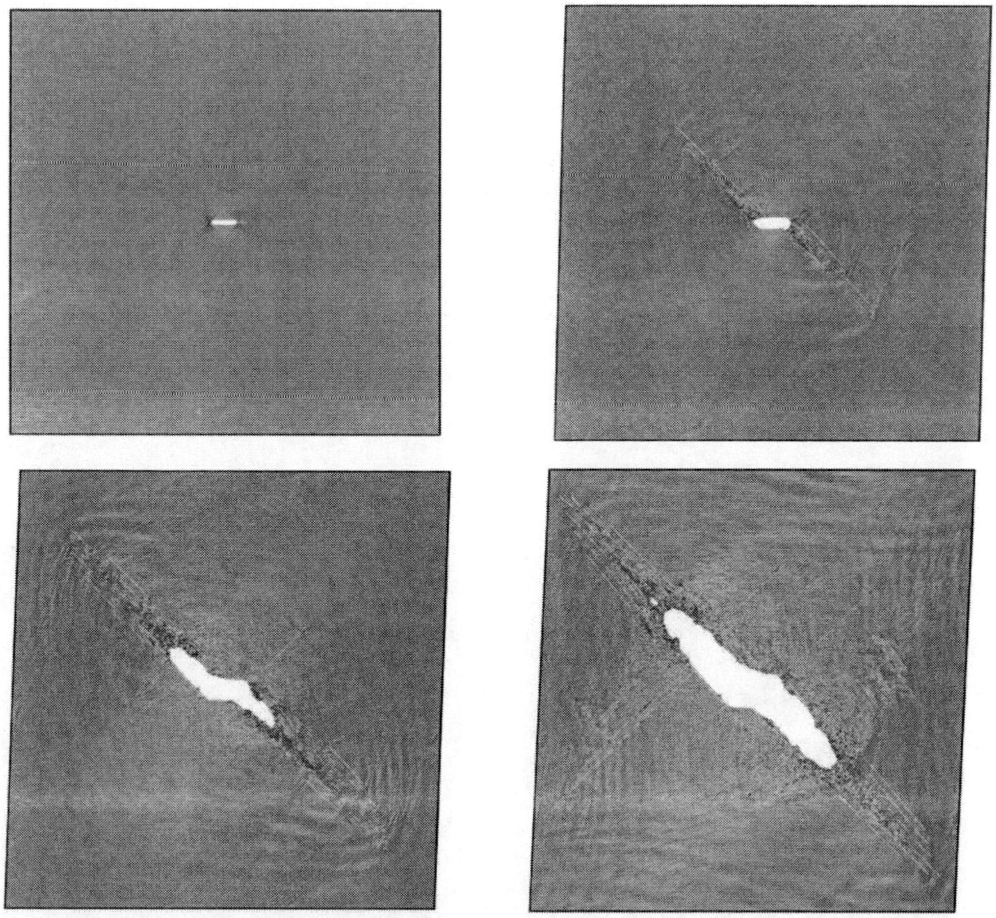

Figure 3. Plate for the mixity parameter $M^{p} = 0.5$ at 0 ps, 25 ps, 35 ps and 45 ps.

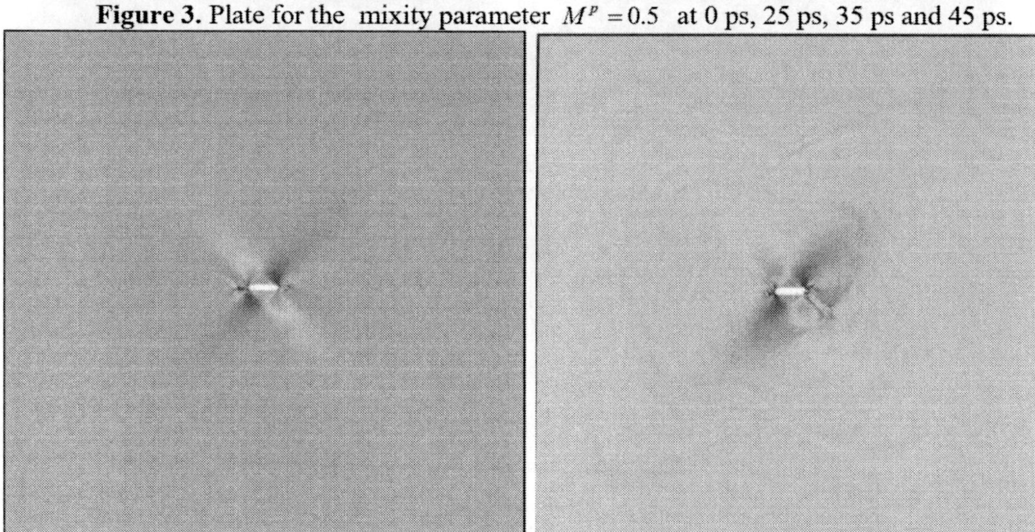

Figure 4. Copper plate just arter energy minimization and after 25000 steps with stress tensor component σ_{12} shown.

IOP Conf. Series: Journal of Physics: Conf. Series **973** (2018) 012046 doi:10.1088/1742-6596/973/1/012046

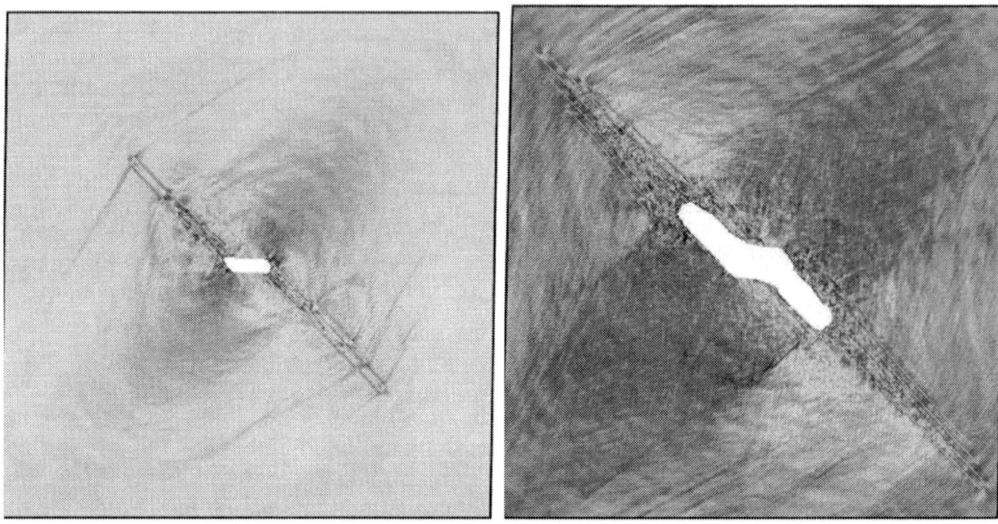

Figure 5. Copper plate for $M^p = 0.7$ after 35 ps and 45 ps steps with stress tensor component σ_{12} shown.

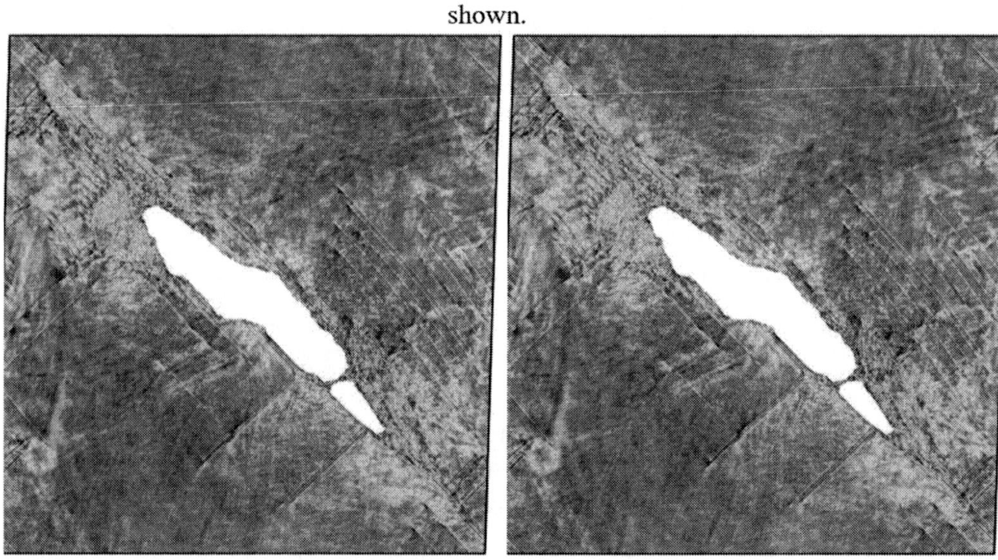

Figure 6. Plate with the growing for the mixity paramer $M^e = 0.7$ at 60 ps (left) and 75 ps (right) (stress tensor component σ_{12} is shown).

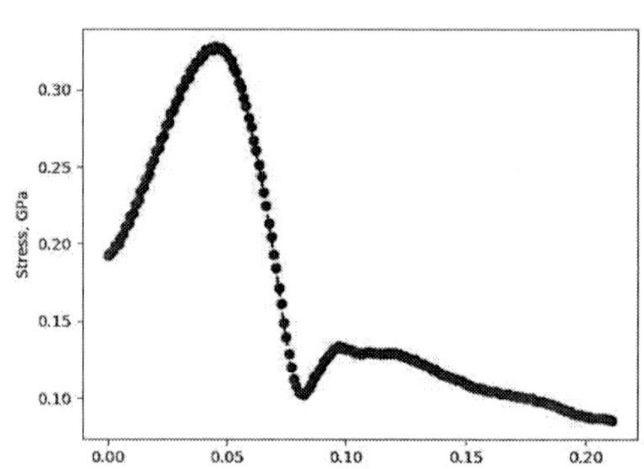

Figure 7. Stress – strain curve of Cu single crystal with the crack.

The stress-strain curve for Cu single crystal is shown in Figure 7. For this system material remains elastic till 5% strain. The subsequent plasticity caused by gliding of dislocations resulting in crack tip blunting (Figures 2, 3, 6). This phenomenon is in consistence with the previous numerical experiments [17]. Figures 1-3 show the deformation process and the crack growth for the mixity parameter $M^e = 0.5$ defined by

$$M^e = \frac{2}{\pi} arctan \frac{\varepsilon_{22}}{\varepsilon_{12}}. \tag{10}$$

Figures 4-6 show the deformation process and the crack growth for the mixity parameter $M^e = 0.7$. One can clearly see that in the vicinity of the macrocrack microcracks and vacancies appear, evolve and merge. The vacancies coalescence can be seen in Figure 8. The material shows elastic behavior till the first plastic effects begin to develop (Figure 8).

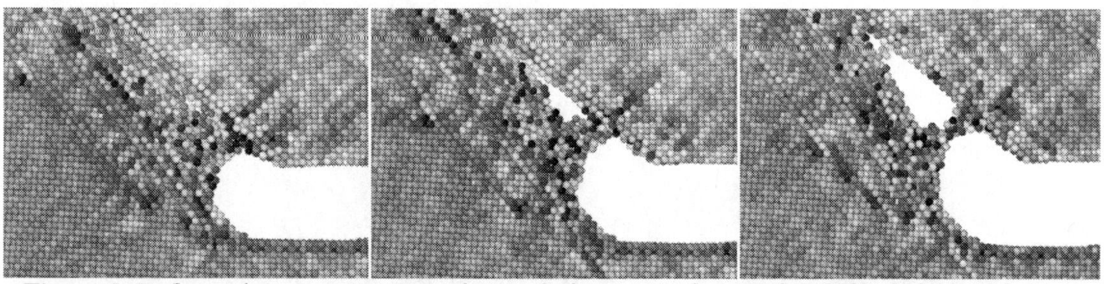

Figure 8. Deformation processes near the crack tip: onset of vacancies (left), their growth (center) and coalescence.

Figure 9 and 10 show the effects of temperatures on the crack propagation process. The crack propagation process at 0K is shown in Figure 9. The crack propagation process at 77K is shown in Figure 10. The mixity parameter for calculation illustrated by Figures 9 and 10 is equal to 0.5. It can be seen that the temperature influences on the crack propagation process and increasing the temperature leads to the blunting of the crack.

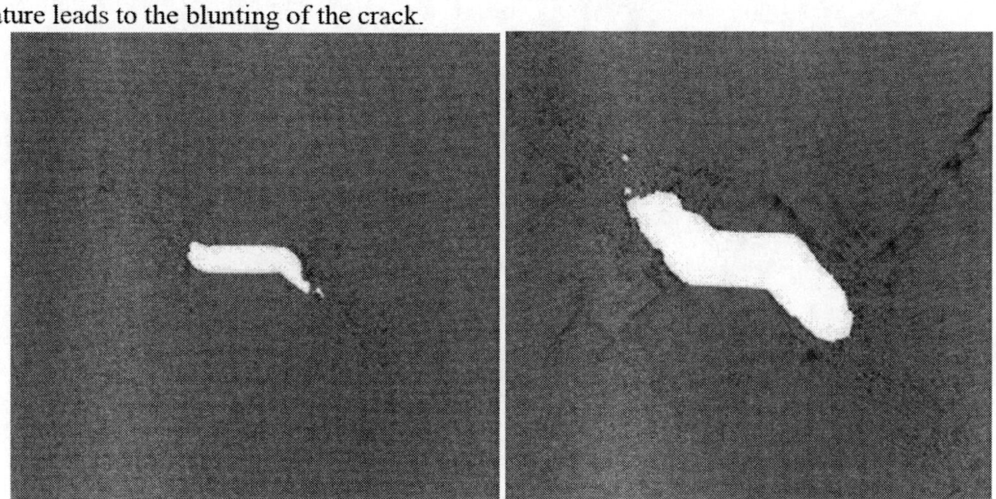

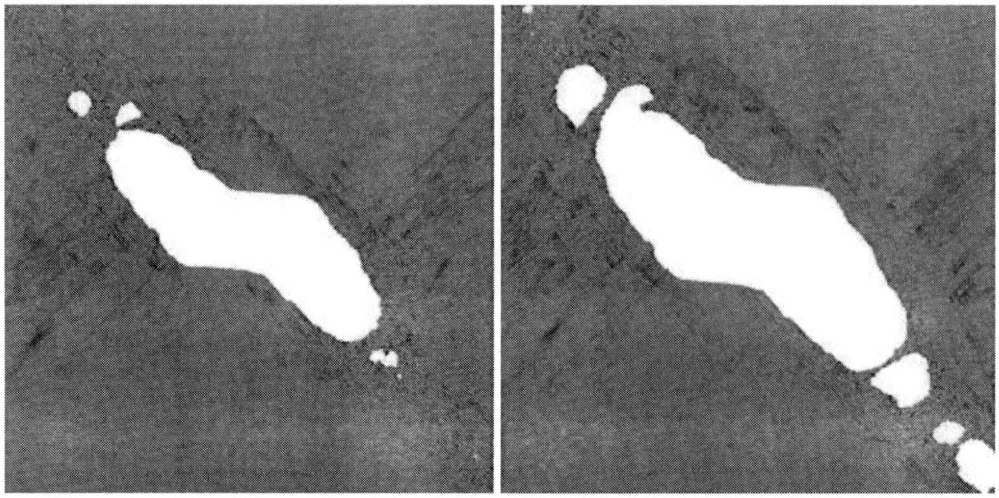

Figure 9. The crack propagation process at 0K.

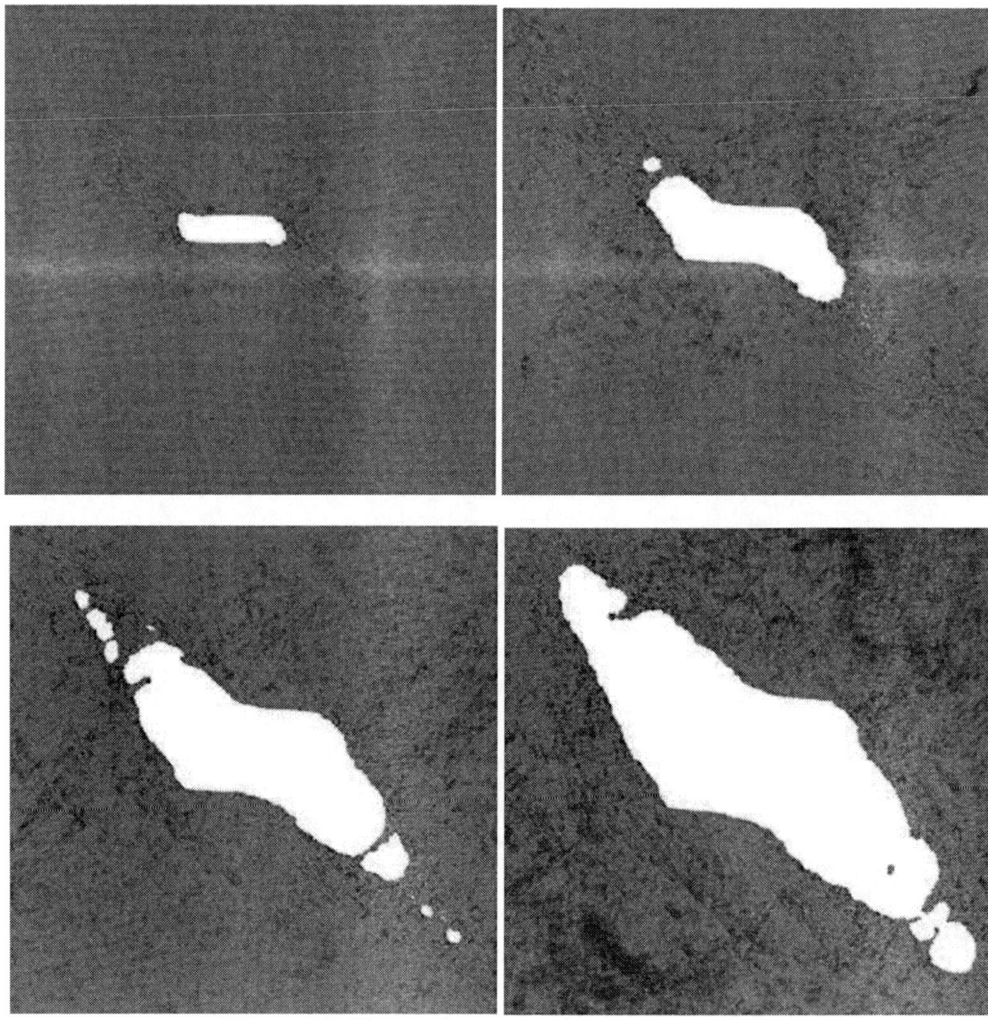

Figure 10. The crack propagation process at 77K.

3. Conclusions

The paper is focused on the application of the different approaches for the determination of the initial crack propagation angle. Copper plate with the central crack under complex mechanical stresses (Mode I and Mode II loading) is studied by extensive molecular dynamics simulations based on the EAM potential. On the other hand, the complete Williams expansion for the crack tip fields containing the higher-order terms is used. The crack propagation angle is obtained by 1) the multi-parameter fracture mechanics approach based on two fracture mechanics criteria, MTS and SED; 2) atomistic modeling for the mixed-mode loading of the plane medium with the central crack. From our simulations we can get crack propagation directions and crack angles. Calculations of MD method were run for three different values of M^e: 0.4, 0.5 and 0.6. Calculated values of crack angles were -51.5°, -46.6° and -42.2° accordingly. All the fracture criteria tested give similar values of the crack growth angle for different values of the mixity parameter. It is shown that the initial crack propagation angles given by the both approaches are very close especially for the case when the higher order terms of the Williams series expansion for the stress/displacement field description are taken into account (Tables 4, 5). Thus one can conclude that the criteria of classical continuum mechanics MTS and SED can give satisfactory predictions for crack initiation direction. The crack propagation direction angles given by the conventional fracture mechanics reasonably agree with the angles obtained from molecular dynamics simulations.

The paper is aimed at the application of the different approaches for the determination of the initial crack propagation angle. The crack propagation angle is obtained by 1) the multi-parameter fracture mechanics approach based on two fracture mechanics criteria, MTS and SED; 2) atomistic modeling for the mixed-mode loading of the plane medium with the central crack. It is shown that the initial crack propagation angles given by the both approaches are very close especially for the case when the higher order terms in the Williams series expansion are kept. One can conclude either that considerable efforts still remain while modeling the mechanical behavior of crystalline materials with empirical potential in the framework of MD simulations. One can conclude that it is crucial and natural to develop multiscale methods [28 -34]. For instance, in [28] the authors try to unify both the continuum mechanics and atomistic approaches and divide the simulation area into atomistic and continuum subdomains such that nanoscale defects are described at atomistic resolution while a continuum mechanics is employed elsewhere. Thus the combination of the continuum mechanics approach and the atomistic simulation allows us to get deeper understanding and realistic description of the crack growth behavior and deformation process under different complex loadings.

Acknowledgements

Financial support from the Russian Foundation of Basic Research (project No. 16-08-00571) is gratefully acknowledged.

References

[1] Berto F,Lazzarin P 2014 *Materials Science and Engineering R* **75** 1-48.
[2] Sih GC 1973 *Mechanics of Fracture (*Noordhoff: International Publishing Leiden) 21-45
[3] Hello G, Tahar M B, Roelandt and J-M 2012 *International Journal of Solids and Structures* **49** 556-566
[4] Stepanova L, Roslyakov P 2016 *International Journal of Solids and Structures* **100-101** 11-28
[5] Malikova L, Vesely V and Seitl S 2015 *Frattura ed Integrita Strutturale* **33** 25-32
[6] Malikova L 2015 *Engineering Fracture Mechanics* **143** 32-46
[7] Stepanova L V, Yakovleva E M 2014 *PNRPU Mechanics Bulletin* **3** 129-162
[8] Stepanova L V, Adylina E M 2014 *Journal of Applied Mechanics and Technical Physics* **55(5)** 885-895
[9] Stepanova L V, Igonin S A 2015 *Numerical Analysis and Applications* **8(2)** 168-181
[10] Stepanova LV, Fedina M E 2008 *Journal of Applied Mathematics and Mechanics* **72(3)** 360-368

[11] Vesely V, Sobek J, Frantik P and Seitl S 2016 *International Journal of Fatigue* **89** 20-35
[12] Stepanova L 2009 *Journal of Applied Mechanics and Technical Physics* **50(1)** 137-146
[13] Sajjadi S H, Salimi-Majd D and Ostad Ahmad Ghorabi M J 2016 *Engineering Fracture Mechanics* **155** 36-48
[14] Malikova L, Vesely V 2017 *Theoretical and Applied Fracture Mechanics*
[15] Stepanova L V, Roslyakov P S 2015 *PNRPU Mechanics Bulletin* **4** 188-225
[16] Muthu N, Maiti S K, Falzon B G and Yan W 2016 *Engineering Analysis with Boundary Elements* **72** 11-26
[17] Chandra S, Kumar N N, Samal M K, Chavan V M and Patel R J 2016 *Computational Materials Science* **117** 518-526
[18] Gao Y J, Deng Q Q, Huang L Ye, Wen Z C and Luo Zhi R 2017 *Computational Materials Science* **130** 64-75
[19] Al-Motasem A T, Mai N T, Choi S T and Posselt M 2016 *Journal of Nuclear Materials* **472** 20-27
[20] Dewapriya M A N, Rajapakse R K N D and Srikanta P A 2013 *13th International conference on Fracture. ICF 2013* **1** 847-852
[21] Yang Z, Zhou Y, Wang T, Liu Q and Lu Z 2014 *Computational Materials Science* **82** 17-25
[22] Cui C B, Beom HG 2014 *Materials Science and Engineering A* **609** 102-109
[23] Plimpton S 1995 *Journal of Computational Physics* **117** 1–19
[24] Daw M S, Baskes M I 1983 *Physical Review. Letters* **50** 1285–1288
[25] Foiles S M, Baskes M I and Daw M S 1986 *Physical Review B* **33** 7983-7991
[26] Stukowski A 2009 *Modelling and Simulation in Materials Science and Engineering* **18(1)** 015012
[27] Stukowski A. *Introduction to Atomistic Simulation Methods* Eds. Weinberger C R Tucker G J 2016 Springer. 317-336
[28] Xu S, Xiong L, Deng Q and McDowell D L 2016 International *Journal of Solids and Structure* **90** 144-152
[29] Li S, Urat S 2016 *Computational Methods of Applied Mechanical Engineering* **306** 452-478
[30] Stepanova L 2008 *Comptes Rendus – Mecanique* **336(1-2)** 232-237
[31] Chen H, Xu S, Li W, Ji W, Phan T and Xiong L 2018 *Computational Materials Science* **144** 1-10
[32] Chandra S, Samal MK, Chavan VM and Raghunathan S 2018 *International Journal of Plasticity* **101** 188 – 212
[33] Stepanova LV, Bronnikov SA and Belova ON 2017 *PNRPU Mechanics Bulletin* **4** 189-213
[34] Chen X, Diaz A, Xiong L, McDowell D and Chen Y 2018 *Journal of Computational Physics* **354** 393-402

IOP Conf. Series: Journal of Physics: Conf. Series **973** (2018) 012047 doi:10.1088/1742-6596/973/1/012047

Information model of trainee characteristics with definition of stochastic behavior of dynamic system

V I Sumin, T E Smolentseva, S V Belokurov, O V Lankin

ET Department, Voronezh Institute Federal Penitentiary Service of Russia, 1a Irkutskaya Street, Voronezh, RU

E-mail: smoltan@bk.ru

Abstract. In the work the process of formation of trainee characteristics with their subsequent change is analyzed and analyzed. Characteristics of trainees were obtained as a result of testing for each section of information on the chosen discipline. The results obtained during testing were input to the dynamic system. The area of control actions consisting of elements of the dynamic system is formed. The limit of deterministic predictability of element trajectories in dynamical systems based on local or global attractors is revealed. The dimension of the phase space of the dynamic system is determined, which allows estimating the parameters of the initial system. On the basis of time series of observations, it is possible to determine the predictability interval of all parameters, which make it possible to determine the behavior of the system discretely in time. Then the measure of predictability will be the sum of Lyapunov's positive indicators, which are a quantitative measure for all elements of the system. The components for the formation of an algorithm allowing to determine the correlation dimension of the attractor for known initial experimental values of the variables are revealed. The generated algorithm makes it possible to carry out an experimental study of the dynamics of changes in the trainee's parameters with initial uncertainty.

1. Introduction

In work process of formation of characteristics of the trainee with the subsequent their change is considered. Characteristics of trainees turn out as a result of testings according to each section of information on the chosen discipline. The results received in the course of testing are entrance data for dynamic system. The area of the operating influences is as a result formed.

2. Justification of a computing experiment

The initial data, which, when studying the dynamics of trainee characteristics, were the initial data in the dynamical system under consideration, were obtained by testing several training groups during one semester. Later, the processed values were presented graphically on the example of several students (figure 1).

Separation of the process of processing values to obtain experimental values allowed to achieve the necessary quality criterion.

Grade levels for several pupils are reflected in the figure 2.

The limited area of irregular solutions allows us to form the area of the control action and only its insignificant share of decisions can come out of this area.

For such decisions it was expedient to use the theory of the qualitative analysis of nonlinear systems.

The behavior of phase space trajectories in dynamical systems can depend on a number of characteristics, including on the instability [1-5].

We investigate a limit of the determined predictability of trajectories of dynamic systems in phase spaces with use of local or global attractors.

Change of behavior of a trajectory of phase space in systems can also be expressed the following expression which estimates the real change of a condition of system in general:

$$S_t x(0) = x(t), \tag{1}$$

where $x_0 = (x_1(0), \ldots, x_m(0))$ defines behavior of this system at the initial moment; $x(t)$ condition of system in discrete timepoints $t \in [t_0, t_n]$.

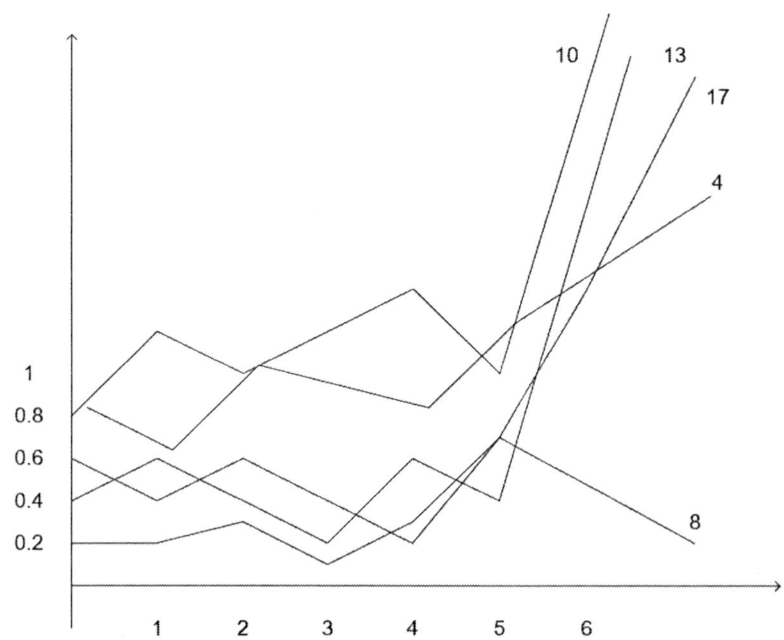

Figure 1. Conditions of trainees after each testing on appropriate section the training information.

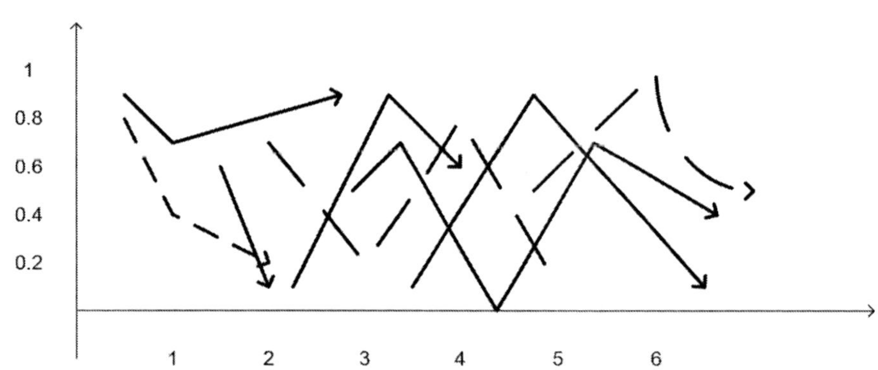

Figure 2. Characteristic grade levels of several people.

We will consider the distribution $P_t(C) = P_0(S^{-t}C)$ allowing us to define probabilities for all points $x(t), C \subset R^m$ in certain time intervals, i.e. $t \in [t_0, t_n]$ at $t > 0$ distribution don't change, and as a result the decision is considered steady, or otherwise the decision is unstable.

The theorem of existence and uniqueness of the decision assumes that for any initial $x_0 \in R^m$ data for all $t \geq 0$ [1, 3].

If, after all, to consider the result of a task (1) received on the basis of entrance true values $x(0) = x_0^*$ – not indignant movement of system, and the movement of system, corresponding to the available entrance conditions it is possible to carry to the indignant movement, then geometrical interpretation it is possible to present spheres of a certain radius in the form $\sqrt{\varepsilon}$.

If the movement is definitely steady, then for this sphere one more sphere with a radius $\sqrt{\delta}$, with the same set of properties has to be defined.

Coordinates of the points of phase space considered by us will allow to define deviations between parameters of not indignant and indignant movements.

In the course of the movement of these points of coordinate will correspond to internal area of the sphere ε , without reaching her surface.

At least one of these a trajectory of the representing point of M, from all available, eventually, in case of the unstable not indignant movement will cross the sphere ε from within outside at as much as close provision of a point M_0 and concerning the beginning of coordinates [2, 6-8].

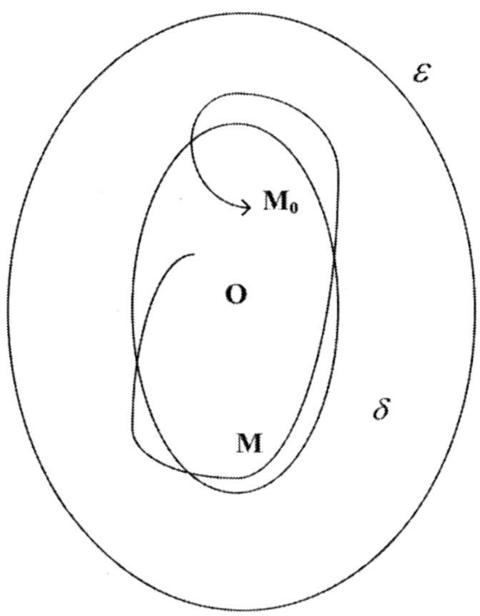

Figure 3. Stability across Lyapunov: the trajectories leaving from ε – close values remain $\delta(\varepsilon)$ relatives at all $t > t_0$.

We consider behavior of system steady, at the known initial values are known x_0 at an admissible error at $\|x_0\| < \delta_,$ [6, 8-16].

In phase space of the considered nonlinear dynamic equations strange attractors, i.e. the attracting areas arising in a combination of local instability of and global compression appear.

Let $y(t)$ – an observed signal, then autocorrelated function has an appearance

$$C(t)=\lim_{T\to\infty}\frac{1}{T}\int_0^T y(\tau)y(t+\tau)\mathrm{d}\tau, \tag{2}$$

where

$$y(\tau)=y(t)-\overline{y},\ \overline{y}=\lim_{T\to\infty}\frac{1}{T}\int_0^T y(t)\mathrm{d}t. \tag{3}$$

For discrete display $f^{i}(y_0)$ autocorrelated function has an appearance

$$C(m)=\lim_{N\to\infty}\frac{1}{N}\sum_{i=0}^{N}y_{m+i}y_i, \tag{4}$$

where

$$y_i=y_i-\overline{y},\ \overline{y}=\lim_{N\to\infty}\frac{1}{N}\sum_{i=0}^{N-1}f^{(i)}(y_0). \tag{5}$$

Correlative function defines a correlation measure for the variables differing in value of time slot τ, but in case of chaotic functioning the parameter value exponentially decreases.

For the discrete functioning the correlation tends to zero [3, 9, 13].

For determination of a range of power $P(\omega)=|y(\omega)|^2$ we will use Fourier transform $y(\omega)$, i.e. movement $y(t)$. If $y(t)$ has the irregular character, then the behavior of variables represents chaotic movement.

Other method of definition of the chaotic movement through Kolmogorov's *K*-entropy, allows to characterize the average speed of loss of information on behavior of dynamic system (if $K>0$, then it is sign of chaos in dynamic system), which can be presented as follows.

We will consider that R^n it is dynamic system with a strange attractor, will define then a trajectory $y(t),\ t>0$ of this system on an attractor.

We will break dynamic system into the appropriate cells of l of a certain length R^n. We will enter through variable joint probability $P_{i_0 i_1 \ldots i_n}$, then the point $y(0)$ will be in a cell with number i_0, and a point $y(\tau)$ in a cell with number i_1, etc.

Then the average speed of loss of the considered information on functioning of dynamic system can be determined on a formula of metric entropy of Renya:

$$K_s=-\lim_{\tau\to 0}\lim_{l\to 0}\lim_{N\to 0}\frac{1}{\tau N}\frac{1}{s-1}\ln\sum_{i_0 i_1\ldots i_n}P^s_{i_0 i_1\ldots i_n}. \tag{6}$$

In this case we will determine *K*-entropy by temporary series of observations using a ratio at $s=2$

$$K_2=-\lim_{\tau\to 0}\lim_{\varepsilon\to 0}\lim_{N\to 0}\frac{1}{\tau N}\ln\sum_{i_0 i_1\ldots i_n}P^2_{i_0 i_1\ldots i_n}. \tag{7}$$

Using Yorke [4] formulas we will define through correlation integral

$$K_2=\lim_{\tau\to 0}\lim_{\varepsilon\to 0}\lim_{n\to\infty}\frac{1}{\tau}\ln\frac{C_n(\varepsilon)}{C_{n+1}(\varepsilon)}, \tag{8}$$

where

$$C_n(\varepsilon) \approx \lim_{N\to\infty} \frac{1}{N^2} \sum_{i,j=1}^{N} \Theta\left(\varepsilon - \left(\sum_{k=0}^{n-1} \left(y_{i+k} - y_{j+k} \right)^2 \right)^{\frac{1}{2}} \right), \tag{9}$$

$\Theta(z)$ – Hevisayda function, i.e. $\Theta(z)=1$, if $z \geq 0$ и $\Theta(z)=0$, if.

If it is defined that, then $K \geq K_2$ the value K_2 can be used for value K assessment.

To define behavior of system in certain timepoints the state vector is applicable to system elements $x(k)$.

Then we will define changes in behavior of system in the form of the sequence $(Q_j)_{j>0}$, at the same time it is necessary to consider all possible decisions on an attractor. In this case for assessment K-entropy expressions (7)-(9) for the sequence of values are used $(Q_j)_{j>0}$ at $N=m$, $n=l$ – amount of the calculated values Q_j, $\varepsilon = \tau = \Delta t$.

Thus, it is natural to assume that phase space of dynamic system infinitely gradually whereas an attractor, and infinitely gradually phase space the dimension of an attractor is less than dimension of the most phase space [5].

A certain dimension of an attractor will allow to define dimension of space of the considered dynamic system which models it and by that will allow to estimate parameters of initial system.

To determine the parameter characterizing the dimension of the attractor, we take the set A, which is uniquely determined as a subset of the Banach space B.

We will cover a set A spheres A_j which radius ε_j doesn't surpass some value ε, i.e. $\varepsilon_j = \text{diam} A_j \leq \varepsilon$. We will define $\mu_d(A,\varepsilon)$:

$$\mu_{\substack{d \\ d>0}}(A,\varepsilon) = \inf \sum_j \left(\text{diam} A_j \right)^d = \inf \sum_j \varepsilon_j^d, \tag{10}$$

and radiuses of spheres it is less on all coverings of a set of A.

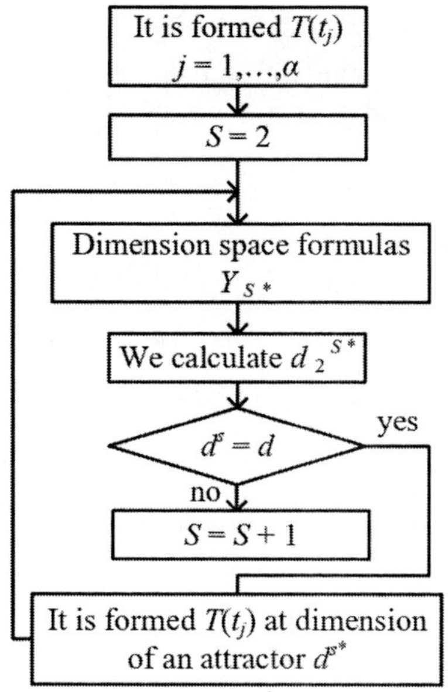

Figure 4. Algorithm of calculation of correlation dimension of an attractor.

Therefore, a hausdorfovy measure of dimension of d of a set of A is:

$$\lim_{\varepsilon\to 0}\mu_d\left(A,\varepsilon\right)=\mu_d(A). \tag{11}$$

For a set we define A dimension of parameter d_H in the form of a set:

$$d_H=\inf\left\{d:\mu_d(A)<\infty\right\}. \tag{12}$$

The attractor under consideration of a certain fractal dimension is:

$$d_F=-\lim_{\varepsilon\to 0}\frac{\log_2 M\left(\varepsilon\right)}{\log_2\frac{1}{\varepsilon}}, \tag{13}$$

where $M\left(\varepsilon\right)$ – number of spheres radius ε, covering sets A.

For dynamic system we will set a trajectory of functioning of her elements in a look: $y(t)=\left(y_1(t),y_2(t),...,y_n(t)\right)$ in timepoints $t=\tau_j,\ \tau>0,\ j=1,...,N$ on an attractor.

We will assume that the phase space is broken into fragments with the party l [12-14].

For each element with corresponding i number is defined a set of points $y(0),y(\tau),...,y(\tau N)$ (N_i – sets quantity of all points of this sequence) being in the fragment considered at the moment [15-17].

In this case the probability of hit of a point of an attractor with the set number can be presented to a fragment as follows:

$$P_i=\frac{N_i}{N}. \tag{14}$$

Substituting in (14) (13), we will receive

$$d_q=\lim_{l\to 0}\frac{1}{q-1}\frac{\ln\sum_{i=0}^{M(l)}P_i^q}{\ln l}, \tag{15}$$

where $M\left(l\right)$ – number of the cells covering an attractor at set l.

Having taken as a basis an algorithm of calculation of correlation dimension of an attractor, perhaps, to construct dimension spaces $s=2,3,4$ with the subsequent formation of the vectors determining system parameters.

The correlation dimension d_2 is defined at on $q=2$ a basis (14).

$$d_2=\lim_{l\to 0}\frac{\ln\sum_{i=0}^{M(l)}P_i^2}{\ln l}. \tag{16}$$

Usually correlation dimension d_2 uses a concept of correlation integral which defines correlation between values with a time interval τ.

$$C\left(l\right)=\lim_{N\to\infty}\frac{1}{N^2}\sum_{i,j=1}^{N}\Theta\left(l-\left|y\left(i\tau\right)-y\left(j\tau\right)\right|\right). \tag{17}$$

Considering that

$$\sum_{i=0}^{M(l)}P_i^2\approx C\left(l\right), \tag{18}$$

(13) will take a form:

$$d_2 = \lim_{l \to 0} \frac{\ln C(l)}{\ln l}. \tag{19}$$

We will read that in case of experiments process of movement of dynamic system the variables defining this movement are received, then correlative dimensionality is defined as follows.

We will select a vector component $x_j(t)$ из $x(t) = (x_1(t), x_2(t), \ldots, x_m(t))$, i.e. the vector which is the solution of some system of the nonlinear equations we will also make a vector $\xi(t)$ look which carries out an investment of the stream generated by system R^{2m+1} in at identical metric properties of space $\{x(t)\}$ и $\{\xi(t)\}$.

$$\xi(t) = (x_j(t),\ x_j(t+\tau), \ldots, x_j(t+2m\tau)),\ \tau > 0. \tag{20}$$

In process carrying out an experiment, as a rule, the dimension of an attractor is unknown, i.e. we accept values of this characteristic in the form of the following set of values $m = 2$, $m = 3$: etc. on the basis of the following expression:

$$d_2^m = \lim_{\varepsilon \to 0} \frac{\ln C_m(\varepsilon)}{\ln \varepsilon}, \tag{21}$$

where

$$C_m(\varepsilon) = \lim_{N \to \infty} \frac{1}{N^2} \sum_{i,j=1}^{N} \Theta\left(\varepsilon - \left|\xi^m(t_i) - \xi^m(t_j)\right|\right) \tag{22}$$

– the generalized correlation integral.

In expression (22) d_2^m increase of m is defined until d_2^m it is considerable will change around $d_2^{m^*}$ at some $m = m^*$.

At the first stage we build dimension spaces $s = 2, 3, 4$, etc. Further we form vectors

$$\begin{gathered} y_1 = \{T_1, T_2\};\quad y_2 = \{T_2, T_3\}, \ldots;\quad y_k = \{T_k, T_{k+1}\}, \ldots, \\ y_{S_2} = (T_{S_2}, T_{S_2+1}). \end{gathered} \tag{23}$$

For the sequence of two-dimensional vectors $\{y_1, y_2, \ldots, y_{S_2}\}$ in R^2 we will calculate correlation dimension of an attractor on formulas (18)-(20) at $s = 2$.

We form the sequence of three-dimensional vectors:

$$\begin{gathered} y_1 = \{T_1, T_2, T_3\};\ y_2 = \{T_2, T_3, T_4\}, \ldots, \\ y_{S_3} = (T_{S_3}, T_{S_3+1}, T_{S_3+2}). \end{gathered} \tag{24}$$

We calculate dimension $d_2^{(s=3)}$. Process continues until at some $s = s^*$ the value $d_2^{(s=s^*)}$ isn't stabilized. This value is also accepted to dimension of an attractor of some hypothetical dynamic system realizing the sequence $\{T(t_j)\}$, $j = 1, \ldots, L$.

On the basis of temporary series of observations it is possible to define an interval of predictability of all parameters.

Then the sum of positive indicators of Lyapunov which are a quantitative measure concerning all elements of system [5] will be a measure of predictability.

Let y_0 – initial value of a trajectory $f^N(y_0)$, where f – the function setting display, N – number of iteration, and $y_0+\delta$ - the indignant initial value, δ – an initial error or a mistake.

The point $y_0+\delta$ leaves a trajectory $f^N(y_0+\delta)$.

On each step of iteration differ at a certain size:

$$\left|f^N(y_0)-f^N(y_0+\delta)\right|. \tag{25}$$

Expression (20) can be simplified, believing that it will be equal to the work of an initial error on some coefficient [7]

$$\left|f^N(y_0)-f^N(y_0+\delta)\right|=\delta\cdot e^{N\lambda(y_0)}. \tag{26}$$

On the basis of expression (21) we will find $\lambda(y_0)$:

$$\lambda(y_0)=\lim_{N\to\infty}\frac{1}{N}\lim_{\delta\to 0}\ln\frac{\left|f^{(N)}(y_0)-f^{(N)}(y_0\cdot\delta)\right|}{\delta}. \tag{27}$$

The sum of L is a positive indicator of Lyapunov, i.e.

$$L=\sum_{k=1}^{p}\lambda_k^+,\quad \lambda_k^+>0,\ k=\overline{1,p}, \tag{28}$$

and the inverse value $T=\frac{1}{L}$ is the average time for which the volume of phase space will stretch in e time. In phase space R^m for one of observed phase coordinates $y(t)$ we will consider a vector:

$$y(t)=\left(y(t),y(t+\tau),...,y(t+(m-1)\tau)\right),\ \tau>0. \tag{29}$$

Applying an asymptotic formula [11, 13] we will receive

$$C_m(\delta)\approx\delta^{d_k}e^{-m\tau L},\delta\to 0,\ m\to\infty, \tag{30}$$

where $C_m(\delta)$ the generalized correlation integral (9).

Considering expression

$$\frac{C_{m+1}}{C_m}=e^{\tau L}, \tag{31}$$

get

$$L=\frac{1}{\tau}\ln\frac{C_{m+1}}{C_m}. \tag{32}$$

We will define $(Q_t)_{t>0}$ the sequence on the basis of the received experimental values of all variables from a set $T(t)$.

We will describe this process as follows: behind a control object we will watch under the influence of results on all to sections of learning information u_1 discretely through time intervals which in turn can be not uniform. We compare a parameter Δt_j value to value Q_j [15, 18-21].

We will calculate all values of ignorance received in case of similar values of personal characteristics of trainees α^1, γ_1, γ_2 on formulas (27)-(30) on the basis of the available selection.

On the basis of an algorithm of computation of correlative dimensionality we define the required sequences of vectors, using the generalized correlative integral and also, we will find parameter s^* values in case of which all values of this dimensionality are stabilized.

3. Results of a computing experiment

Further we will give values of the calculated parameters of *K*-entropy, correlation dimension and dimension of phase space for experimental values of function of quality of training:

The calculated values of correlation dimension have been based on entrance data on several educational institutions.

Graphical representation of the received parameters are given in figures 5 and 6.

The calculated values for Voronezh state pedagogical university (VSPU) are: $K = 1,23,\quad d_2 = 0,47,\quad s^* = 2$.

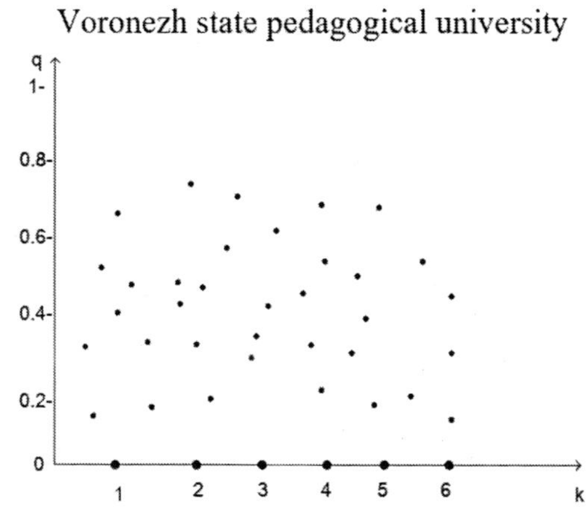

Figure 5. Values of *K*-entropy, correlation dimension and dimension of phase space for experimental values on an example VSPU

On the basis of expression (31) the size of an interval of predictability which corresponds to time necessary for the solution of initial system of the nonlinear equations is calculated, considering the limited sequence of measurements, the strange attractor allowing to enter the mode.

$$T = \frac{1}{L}. \tag{33}$$

For dynamic systems with instability stochastic properties are studied by numerical methods and don't present complexity in the computing plan.

Very often correlation dimension is defined on the basis of logarithming of an asymptotic ratio $C(l) = l^{\nu}, l \to 0$.

Calculated values for LSPU are: $K = 1.22,\quad d_2 = 0.71,\quad s^* = 2$.

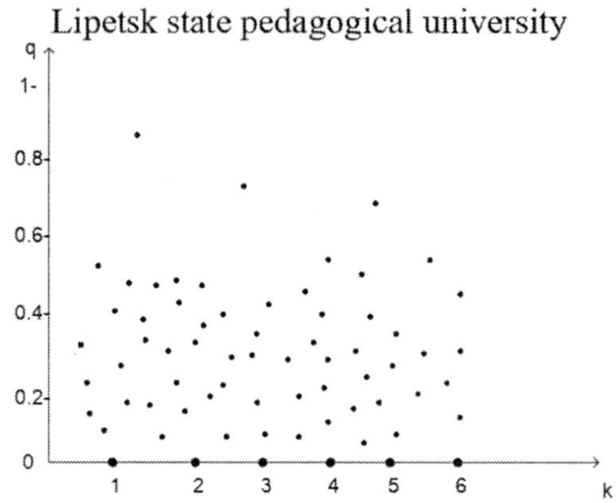

Figure 6. Values of *K*-entropy, correlation dimension and dimension of phase space for experimental values on an example LSPU

Considering the fact that expressions for determination of fractal dimensions contain limit transitions, these calculations for determination of correlation dimension make for dynamic systems from big N and small values l.

In dynamic systems measurement is understood as the smooth material function set on M $y:M \rightarrow R$.

4. Conclusion

Being based on above explained perhaps to create an algorithm of computation of correlative dimensionality of the attractor in case of the given sequence of the experimental values of variables.

The created algorithm allows to realize the pilot study of dynamics of change of parameters of the trainee with initial uncertainty.

Using the determined chaos the stochastic behavior of paths of dynamic system in phase space and possibility of the determined chaos owing to instability of dynamics is defined.

References

[1] Saati T V 1998 *Analytical planning: organization of systems* (Moscow: Radio and communication) p 224

[2] Vagin V N 2004 *A reliable and plausible conclusion in intellectual systems* (Moscow: Fizmatlit) p 231

[3] Zhilyakov E G 2011 A computer clustering of set of additive mathematical models of the interconnected processes *Radio electronics Issue* **1** pp 115-119

[4] Zhuravlev Y I 2006 *Recognition Mathematical methods Program system Practical* (Moscow: Phase) p 184

[5] Zamyatina O M 2009 *Modeling of systems* (Tomsk: TPU) p 204

[6] Mendel I D 1988 *Cluster analysis* (Moscow: Finance and statistics) p 176

[7] Stzheletski R I 2001 *Chaotic processes in the systems of power electronics* (Tomsk: Obverse) p 197

[8] Sumin V I 2014 *Modeling of training with use of temporary series of observations* (Voronezh: Scientific book) p 104

[9] Sumin V I 2015 Basic researches *A technique of grouping of a basic information for information processes of difficult systems* **2** pp 931-934

[10] Sumin V I 2010 About the algorithms and models given in solutions of problems of decision-making *Scientific sheets Belgorod state University Issue* **12** pp 120-128

[11] Sumin V I 2008 Description of functioning of information process of difficult systems *Modern problems of science and education* **1** pp 118-121
[12] Sumin V I 2008 Formalization of a task of management of social systems *All-Russian scientific-technical Voronezh institute of high technologies* (Voronezh: Voronezh institute of high technologies) pp 133-137
[13] Sumin V I 2009 A training purpose specification in the trainee's model as object of management *All-Russian scientific-technical Mathematical methods and information technical means* (Krasnodar) pp 187-190
[14] Tarasenko V V 2000 Fractal geometry of the nature: sociocultural measurement *Synergetic paradigm Variety of search and approaches* (Moscow: Progress-Tradition) pp 191-214
[15] Trakhtengerts E A 2001 Opportunities and realization of computer systems of support of decision-making *News of RAS Theory and control systems* **3** pp 86-113
[16] Dushkin A V 2014 *Mathematical models and information processes of management of a difficult object* (Voronezh: Scientific book) p 125
[17] Dushkin A V 2014 *Model of management of difficult social objects in the system of training of employees of penal correction system* (Voronezh: Scientific book) p 125
[18] Dushkin A V 2017 *Introduction to the system analysis* (Moscow: Hot line – Telecom) p 234
[19] Dushkin A V 2017 *Introduction to system design of intellectual knowledge bases* (Moscow: Hot line –Telecom) p 108
[20] Trubetskov D N 1998 Turbulence and the determined chaos *Sorovsky the educational magazine* **1** pp 77-83
[21] Feeder E. 1991 *Fractals* (Moscow: World) p 260

IOP Conf. Series: Journal of Physics: Conf. Series **973** (2018) 012048 doi:10.1088/1742-6596/973/1/012048

Determination of thermophysical characteristics of vulcanizable rubber products by the mathematical modeling method

S G Tikhomirov[1], Y V Pyatakov[1], O V Karmanova[2] and A A Maslov[1]

[1] Information and control systems department, Voronezh state university of engineering technologies, Revolution Av., 19 Voronezh, Russia
[2] Chemistry and chemical technology of organic compounds and polymers processing department, Voronezh state university of engineering technologies, Leninsky Av., 14 Voronezh, Russia

E-mail: tikhomirov_57@mail.ru, pyatakovjv@mail.ru, karolga@mail.ru, imasslove@mail.ru

Abstract. The studies of the vulcanization kinetics of elastomers were carried out using a Truck tyre tread rubber compound. The formal kinetic scheme of vulcanization of rubbers sulfur-accelerator curing system was used which generalizes the set of reactions occurring in the curing process. A mathematical model is developed for determining the thermal parameters vulcanizable mixture comprising algorithms for solving direct and inverse problems for system of equations of heat conduction and kinetics of the curing process. The performance of the model is confirmed by the results of numerical experiments on model examples.

1. Introduction

The intensification of the process of vulcanization is one of the main ways to reduce the cost of production of rubber products. Calculated methods for developing vulcanization regimes are the most effective and economically viable. They are based on modern theoretical concepts of heat transfer and kinetics of vulcanization. Also they allow to analyze and search the optimal mode of curing as existing in real production and design. This makes it possible at the design stage to evaluate the technological possibilities for implementing new designs and to plan the choice of materials and equipment.

The change in properties of vulcanization of rubber typically passes through the following four stages:

Stage I An induction period. There is no significant change in properties compared with raw rubber;

Stage II The vulcanization process. The rubber properties change monotonically, reaching the optimum properties of the vulcanized rubber;

Stage III The vulcanization plateau. The properties are maintained at the optimum level;

Stage IV Reversion or re-vulcanization. The properties of rubber deteriorate compared with the optimal.

In the presence of a vulcanization plateau, expensive thermal energy will be consumed in case of overheating. And in the case of underheating of a particular site the required set of properties will not be achieved and the product will be of poor quality. In connection with foregoing, it is very important to choose a vulcanization mode, which provides high quality products with the least energy costs, that is very important in the context of a continuous increase of energy prices.

For quantify the kinetics of vulcanization of rubbers under variable temperature conditions various methods have been proposed [1-3]. The most widely used method is reduced to calculating the effect of E and the equivalent time τ of vulcanization. To calculate the effect of vulcanization, the concept of intensity of vulcanization I is introduced as a quantity characterizing the rate of change of the acquired vulcanization properties of rubber depending only on changes in vulcanization temperature t. Taking the vulcanization intensity at temperature t_o as a unit, the intensity at temperature t in dimensionless units is expressed as follows [4]:

$$I = \frac{K \cdot (t - t_0)}{10}.$$

The effect of vulcanization E in the time interval from τ_1 to τ_2 for variable temperature and constant temperature coefficient of vulcanization K is represented in the following form:

$$E = \int_{\tau_1}^{\tau_2} K^{\frac{t(\tau)-t_0}{10}} d\tau.$$

To evaluate the effect of variable temperatures $t(\tau)$ it is necessary to know how the temperature t is changing in time τ and which values have temperature coefficient of vulcanization. Therefore, physical and mathematical modeling of temperature fields and degrees of vulcanization of rubbers in the product are used to optimize the vulcanization process of rubber products in the modern technology.

Equivalent time must be calculated and associated with vulcanization characteristics of rubbers for all layers of the product, especially for the internal, poorly heated and outer, highly heated layers where there is a danger of correspondingly under-vulcanization or re-vulcanization. The reaction rate of vulcanization and the degree of its perfection depend on the temperature of vulcanization, and therefore depend on the amount of heat supplied to the system [5] and released as a result of the reactions [3]. The amount of heat released during the reaction increases as the vulcanization temperature rises. The kinetics of heat release is correlated with the kinetics of addition of sulfur reaction.

It is established [7] that an increase in the vulcanization temperature practically does not affect the value of the coefficient of thermal conductivity, but reduces the value of thermal diffusivity of rubber compounds. Thus, thermophysical characteristics of rubbers significantly affect the process of heat transfer, however, they are not well studied for rubber compounds based on rubber regular structure.

2. Materials and methods

The kinetics of vulcanization was investigated using samples of the model rubber compounds based on butadiene-sterile rubber and industrial rubber compound based on a combination of stereoregular isoprene, butadiene, and styrene-butadiene rubbers intended for the production of tread of truck tires. Cylindrical samples of height 25 mm and diameter of 50 mm were vulcanized at various temperatures between the plates of the experimental setup equipped with sensors for temperature control in the sample at the surface of the plates and in the center.

The intensification of the process of vulcanization is one of the main ways to reduce the cost of production of rubber products. The calculated methods for developing vulcanization modes are the most effective and economically viable. They are based on modern theoretical concepts of heat transfer and kinetics of vulcanization. Also they allow to analyse and search the optimal mode of curing as existing in real production and design. This makes it possible at the design stage to evaluate the technological possibilities for implementing new designs and to plan the choice of materials and equipment.

Based on the analysis of literature data on the mechanism and kinetics of vulcanization of rubbers sulfur-accelerator vulcanizing systems formal kinetic scheme is proposed. This scheme, shown in Figure 1, generalizes a set of reactions occurring in the process of vulcanization.

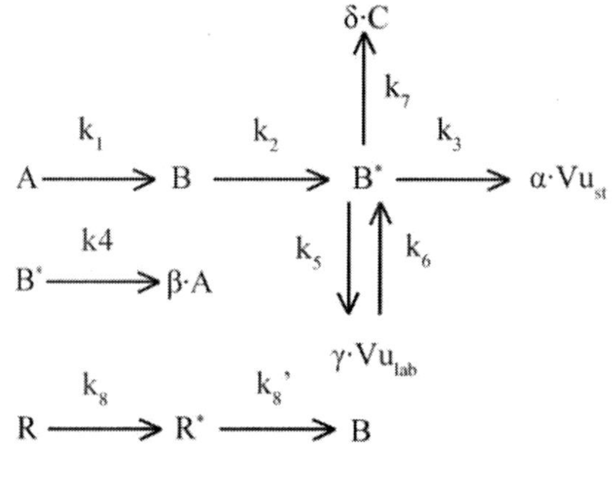

Figure 1. A kinetic scheme of vulcanization process.

A is the actual vulcanizing agent, *B* is the crosslinking precursor, *B** is its active form, *C* is the intramolecular bound sulfur, Vu_{St}, Vu_{Lab} are the stable and labile sites of the vulcanization grid, *R* is the rubber, *R** is the macroradical of the rubber, formed as a result of thermofluctuation decay; α, β, γ and δ are stoichiometric coefficients, k_1, k_2, k_3, ..., k_8, k_9 (k_8') are reaction rate constants related to the corresponding stages of the process.

3. Materials and methods

3.1. Formulation and solution of the direct problem for the system of equations of the kinetics of the process of vulcanization

The set of reactions occurring in the vulcanization process is described by a system of differential equations (DE):

$$\begin{cases} \dfrac{dC_A}{dt} = -k_1 \cdot C_A - k_4 \cdot C_A \cdot C_{B^*}, \\ \dfrac{dC_B}{dt} = k_1 \cdot C_A - k_2 \cdot C_B + \beta \cdot k_4 \cdot C_A \cdot C_{B^*} + k_9 \cdot C_{R^*}, \\ \dfrac{dC_{B^*}}{dt} = k_2 \cdot C_B - (k_3 + k_5 + k_7) \cdot C_{B^*} + k_6 \cdot C_{VuLab} - k_4 \cdot C_A \cdot C_{B^*}, \\ \dfrac{dC_{VuSt}}{dt} = \alpha \cdot k_3 \cdot C_{B^*}, \\ \dfrac{dC_{VuLab}}{dt} = \gamma \cdot k_5 \cdot C_B - k_6 \cdot C_{VuLab}, \\ \dfrac{dC_C}{dt} = \delta \cdot k_7 \cdot C_{B^*}, \\ \dfrac{dC_{R^*}}{dt} = k_8 \cdot C_R - k_9 \cdot C_{R^*}, \\ \dfrac{dC_R}{dt} = -k_8 \cdot C_R. \end{cases} \quad (1)$$

Initial concentration:

$$\begin{cases} C_A(0) = \varsigma \cdot [S_8] \cdot [A_C] \cdot [Akt]^{\theta} \cdot [C_R(0)]^{\eta}, \\ C_B(0) = 0, \\ C_{B^*}(0) = 0, \\ C_{VuSt}(0) = 0, \\ C_{VuLab}(0) = 0, \\ C_C(0) = 0, \\ C_{R^*}(0) = 0, \\ C_R(0) = C_R(0)^{\eta}. \end{cases} \quad (2)$$

The temperature dependence of the rate constants of the process is determined on the basis of generalized equation of Arrhenius [5]:

$$k_i(T) = k_{i0} \cdot e^{-\frac{E_i}{R \cdot T}}, \qquad i = 1,2,\ldots,9 \quad (3)$$

Further, the direct problem of kinetics (DPK) is the problem of finding the concentration of vulcanization nodes as a function of time.

When solving the DPK, the process is considered proceeding according to the proposed scheme (Figure 1). The purpose of the solution is to find the concentration of vulcanization nodes relative to the maximum cross-link density as a function of time.

$$X(t) = \frac{\left(C_{VuSt}(t) + C_{VuLab}(t)\right)}{C_{max}}, \qquad C_{max} = \max_{t \in [0, t_k]} [C_{VuSt}(t) + C_{VuLab}(t)]. \quad (4)$$

The problem reduces to the solution of the DE system (1-3), which is carried out by the Runge-Kutta method.

The expression for reaction rate constants (3) contains the temperature T, which is due to the need to determine the temperature fields in the products to be vulcanized.

At the present time, more and more widespread methods are found for calculating the temperature fields in the development of thermal vulcanization regimes, along with the experimental determination of temperatures. [1, 2]. The necessity for preliminary determination of thermophysical characteristics (TC) - heat capacity coefficients, thermal conductivity, etc., which depend on the temperature and composition of the cured mixture - is one of the complexities of using computational methods [1,2].

3.2. The formulation of the direct problem for the system of heat conduction equations

The methodology of calculation of thermophysical parameters of the mixture in the present work is based on the mathematical model represented by the heat equation:

$$C(T) \cdot \frac{\partial T(l,t)}{\partial t} = \lambda \cdot \frac{\partial^2 T(l,t)}{\partial l^2} + q(l,t), \qquad l \in (0, L), \qquad t \in (0, t_k). \quad (5)$$

$$T(l, 0) = T_0(l), \qquad l \in [0, L]. \quad (6)$$

$$T(0,t) = T_{top}(t), \qquad T(L,t) = T_{bottom}(t), \qquad t \in [0, t_k].$$

(7)

In equations (5)-(7) $T(l,t)$ are temperature; $q(l,t)$ is heat release density at a point at $l \in [0, L]$ at time t:

$$q(l,t) = q_0 \cdot m_s \cdot \frac{\partial X(l,t)}{\partial t}. \tag{8}$$

$X(l,t)$ is degree of completeness of the process of vulcanization; q_0 is the coefficient of heat release; m_s is mass fraction of sulfur in the mixture and composition of the rubber compound; L is the thickness of the test sample; t_k is the final time moment; $C(T)$ is the coefficient of the volume heat capacity and its dependence on temperature are approximated by a linear function of the form:

$$C(T) = C_0 + C_1 \cdot T, \tag{9}$$

λ is coefficient of thermal conductivity.

In the future, direct task (DPK) for the system of heat conduction equations will be called the task of determining the function $T(l, t)$ satisfying the equation (5), primary (6) and boundary (7) conditions for given values of thermophysical parameters C_0, C_1, λ.

4. Results and discussion

4.1. The solution of the direct problem for the system of heat conduction equations

We introduce a function $S(T)$ defined as follows:

$$S(T) = \int_0^T C(T')dT' = \int_0^T (C_0 + C_1 \cdot T') \cdot dT' = C_0 \cdot T + \frac{C_1}{2 \cdot T^2}, \tag{10}$$

In this case, equation (5) can be written in the form

$$\frac{\partial S(T(x,t))}{\partial t} = \lambda \cdot \frac{\partial^2 T(l,t)}{\partial l^2} + q(l,t), \qquad l \in (0, L), \qquad t \in (0, t_k). \tag{11}$$

The initial condition (6) will be:

$$\begin{cases} S(l,0) = S_0(l); \\ \Lambda(l,0) = \Lambda_0(l). \end{cases} \quad l \in [0, L] \tag{12}$$

Accordingly, the boundary conditions (7) will have the form

$$\begin{cases} S(0,t) = S_{top}(t); \\ S(L,t) = S_{bot.}(t); \\ \Lambda(0,t) = \Lambda_{top}(t); \\ \Lambda(L,t) = \Lambda_{bot.}(t). \end{cases} \quad t \in [0, t_k] \tag{13}$$

In (12), (13):

$$\begin{cases} S(l,0) = S(T(l,0)); \\ S_{top}(t) = S\left(T_{top}(t)\right); \\ \Lambda(l,0) = \Lambda(T(l,0)); \\ \Lambda_0(l) = \Lambda(T_0(l)); \\ S(0,t) = S(T(0,t)); \\ S(L,t) = S(T(L,t)); \\ \Lambda(0,t) = \Lambda(T(0,t)); \\ \Lambda_{top}(t) = \Lambda\left(T_{top}(t)\right). \end{cases} \tag{14}$$

We use explicit difference scheme to solve the differential equation (11) under conditions (12)-(14):

$$S_k^{n+1} = S_k^n + \frac{\Delta t \cdot \lambda \cdot (T_{k+1}^n - 2T_k^n + T_{k-1}^n)}{\Delta l^2} + q_0 \cdot \Delta t \cdot \frac{\partial X(l_k, t_n)}{\partial t}; \tag{15}$$

$$T_k^{n+1} = \frac{2 \cdot S_k^{n+1}}{\sqrt{C_0 \cdot C_1 + 2C_1 \cdot S_k^{n+1}} + C_0}; \; k = 1,2,\dots,K-1; n = 1,2,\dots,N-1. \tag{16}$$

$$S_k^n = S(T_k^n);\; L_k^n = L(T_k^n);\; T_k^n = T(l_k, t_n);\; l_k = \Delta l \cdot k;\; t_n = \Delta t \cdot n;\; \Delta l = \frac{L}{K}; \Delta t = \frac{t_k}{N}.$$

K is number of intervals over the spatial variable, N is number of sampling intervals by time. To ensure the stability of the solution (15), (16) the sampling intervals selected from the condition:

$$\Delta t \leq \frac{C_0 \cdot \Delta l^2}{2\lambda}. \tag{17}$$

The right side of equation (11) contains the value $q(l,t)$ that is determined by the relation (8) in terms of the quantity $X(l,t)$, which according to (4) is determined by solving the system of equations (1), (2). In turn, rate constants included in (1) on the basis of Arrhenius equation (3) are functions of the temperature.

Thus the solution of the direct problem of heat conduction and the direct problem of the kinetics is the simultaneous solution of equations (1), (2) and (5)-(7), which we carry out iteratively in the following sequence:

- to suppose as the initial approximation in (8) $X(l,t) = X^0(l,t) = 0$, calculate temperature $T(l,t)$ with $X(l,t) = X^{(0)}(l,t) = 0$ using the algorithm (16) and suppose $T^{(0)}(l,t) = T(l,t)$;
- to substitute the value $T^{(0)}(l,t)$ into the expression for the rate constants (3) and determine the value $X^{(1)}(l,t)$ by solving the system of equations of the kinetics (1)-(4);
- to calculate temperature $T^{(1)}(l,t)$ with $X(l,t) = X^{(1)}(l,t)$ using the algorithm (16).

As a condition of completion of the iterative process it is supposed the condition
$|T^{(m+1)}(l,t) - T^{(m)}| < \xi$ where ξ is a given accuracy of calculations.

4.2. Determination of thermophysical parameters

The inverse problem of heat conduction means the problem of determining the thermophysical parameters *C0, C1, λ, q0* using measured values of temperature of test specimens based on the reference data on thermophysical characteristics of the ingredients of the mixture.

T(l,t) is supposed function satisfying the system of equations (5)-(7). Then, the same function satisfies the system of equations of the form:

$$(1 + x_1 \cdot T) \cdot \frac{\partial T(l,t)}{\partial t} = x_2 \cdot \frac{\partial^2 T(l,t)}{\partial l^2} + x_3 \cdot m_s \cdot \frac{\partial X(l,t)}{\partial t}, \qquad l \in (0, L), \qquad t \in (0, t_k), \tag{18}$$

$$T(l, 0) = T_0(l), \qquad l \in [0, L] \tag{19}$$

$$T(0, t) = T_{top}(t), \qquad T(L, t) = T_{bottom}(t), \qquad t \in [0, t_k] \tag{20}$$

$$x_1 = \frac{C_1}{C_0}, \qquad x_2 = \frac{\lambda_1}{C_0}, \qquad x_3 = \frac{q_0}{C_0}.$$

On the other hand, if the function *T(l, t)* satisfies the system of equations (18)-(20), then it will satisfy the system of heat conduction equations (5)-(7) when the values of thermophysical parameters: *S0=C0, C1=C0• x1, λ = C0• x2, q0= C0• x3.*

The values of the parameters x1, x2, x3 are determined by minimizing the function

$$F(x) = \sum_{m=1}^{K^{samp.}} \sum_{i=1}^{N_m} [T_{i,m}^{meas.} - T_m^{calc.}(\boldsymbol{x}; l_0, t_i)]^2 \tag{21}$$

where $\boldsymbol{x}=(x_1, x_2, x_3)^T$, $T_{i,m}^{meas.}$ are values of the temperature of *m-th* testing sample in the middle part ($l_0=L/2$), measured in time t_i, the; $T_m^{calc.}(\boldsymbol{x}; l_0, t_i)$ are values of temperature for given values of the $\boldsymbol{x}$ calculated by solving the system of equations (18)-(20); $K^{samp.}$ - the number of samples.

The value of the parameter C_0 is determined depending on the composition according to the reference data. Indeed, the $m_1, m_2,...,m_M$ are supposed as mass fraction of the ingredients of the mixture; M is the number of ingredients; $c_1, c_2,...,c_M$ - specific heat capacity, $\rho_1, \rho_2,...,\rho_M$ - density of the ingredients of the mixture defined according to the reference data at a certain temperature t_0 (in the future, we assume $t_0 = 20°C$).

Then at the temperature t_0 the thermal conductivity of the mixture can be calculated by the formula [2]:

$$C(T_0) = \frac{\sum_{n=1}^{K} c_n \cdot m_n}{\sum_{n=1}^{K} \frac{m_n}{\rho_n}} \tag{22}$$

Substitute the value obtained in (9), and determine the value of C_0:

$$C_0 = \sum_{n=1}^{K} \frac{c_n \cdot m_n}{[(1 + x_1 \cdot T_0) \cdot \sum_{n=1}^{K} \frac{m_n}{\rho_n}]} \tag{23}$$

The minimization function *F(x)* in (21) will implement Newton's method. For this decompose function $T_m^{calc.}(\boldsymbol{x};l_0, l_i)$, included in the right hand side of (21) in a Taylor series in a neighbourhood of a point of the *k* approximation and limited to the linear part of the decomposition, we get:

$$T_m^{calc.}(x^{(k+1)}; l_0, t_i) = T_m^{calc.}(x^{(k)} + \Delta x^{(k)}; l_0, t_i) \approx$$
$$\approx T_m^{calc.}(x^{(k)}; l_0, t_i) + \nabla T_m^{calc.}(x^{(k)}; l_0, t_i) \cdot \Delta x^{(k)}$$

Substituting the resulting expression in (21), we obtain

$$F(x^{(k+1)}) = F(x^{(k)} + \Delta x^{(k)}) \approx$$
$$\approx \sum_{m=1}^{K^{samp.}} \sum_{i=1}^{N_m} \left[T_{i,m}^{meas.} - T_m^{calc.}(x^{(k)}; l_0, t_i) - \nabla T_m^{calc.}(x^{(k)}; l_0, t_i) \cdot \Delta x^{(k)}\right]^2 = \tag{24}$$
$$= \sum_{m=1}^{K^{samp.}} \sum_{i=1}^{N_m} \left[\Delta T_{i,m} - \nabla T_m^{calc.}(x^{(k)}; l_0, t_i) \cdot \Delta x^{(k)}\right]^2,$$

$$\Delta T_{i,m} = T_{i,m}^{meas.} - T_{i,m}^{calc.}(x^{(k)}),\ \Delta \boldsymbol{x}^{(k)} = \boldsymbol{x}^{(k+1)} - \boldsymbol{x}^{(k)}$$

K+1 approximation we find from the condition of minimum (24):

$$\frac{\partial F(\boldsymbol{x}^{(k)} + \Delta \boldsymbol{x}^{(k)})}{\partial \Delta x_j^{(k)}} = 0, j = 1,2,3; \tag{25}$$

where

$$\Delta \boldsymbol{x}^{(k)} = \boldsymbol{A}^{-1} \cdot \boldsymbol{b}, \tag{26}$$

where is $\boldsymbol{A} = \|a_{n,l}\|_{3\times3}$ the matrix with elements

$$a_{n,l} = \sum_{m=1}^{K^{samp.}} \sum_{i=1}^{N_m} \frac{\partial T_m^{calc.}(x^{(k)}; l_0, t_i)}{\partial x_n^{(k)}} \cdot \frac{\partial T_m^{calc.}(x^{(k)}; l_0, t_i)}{\partial x_l^{(k)}}, n, l = 1,2,3; \tag{27}$$

$$b = \sum_{m=1}^{K^{samp.}} \sum_{i=1}^{N_m} \Delta T_{i,m} \cdot \nabla T_m^{calc.}(x^{(k)}; l_0, t_i); \tag{28}$$

Thus, the value of point *k+1* approximation is determined by the formula

$$\boldsymbol{x}^{(k+1)} = \boldsymbol{x}^{(k)} + \Delta \boldsymbol{x}^{(k)}. \tag{29}$$

The conclusion of the iterative algorithm (29) is implemented with one of the conditions:
- $|\nabla F(\boldsymbol{x}^{(k+1)})| \le \varepsilon_1$, (the magnitude of the gradient $F(\boldsymbol{x}^{(k+1)})$);
- $|\Delta \boldsymbol{x}^{(k)}| \le \varepsilon_2$, (the absolute difference between values in adjacent iterations)
- $F(\boldsymbol{x}^{(k+1)}) > F(\boldsymbol{x}^{(k)})$ (in ascending order of the objective function)
- $k + 1 > K_{max}$ (the number of iterations)

where $\varepsilon_1, \varepsilon_2$ values.

4.3. Calculation of model examples of solution of direct problems of heat conduction equations
Example 1. To verify the performance of the algorithm for solving the direct problem in the example the calculation of the temperature field of the sample with thickness $L=2.5\times10^{-2}$ m was carried out. The thermophysical parameters of the sample correspond to parameters of rubber compounds used for the production of automobile tires:

$$C(T)=8000000+4900\cdot T\text{, J}\cdot\text{kg}^{-1}\cdot\text{deg}^{-1},\ \lambda=0.35\text{, W}\cdot\text{m}^{-1}\cdot\text{deg}^{-1}.$$

The boundary values of the temperature represented by the dependencies:

$$T_{top}(t)=T_{bot.}(t)=\begin{cases} 0{,}2t+30.0 & t\le 600; \\ 150.0 & 600\le t\le 2700; \\ 276.0-7/150t & t\ge 2700. \end{cases}$$

the initial value of the temperature $T_0(x)=30°\text{C}$
The sampling time $dt=1.0$ sec; $\varsigma=1.65$; $\eta=0$; $\theta=1.5$; $S_8=0.057456$ mole/kg;
$Ac=0.065664$ mole/kg; $Act=0.453974$ mole/kg; $R_1=0.003453$; $\alpha=1.0$; $\beta=2.0$; $\gamma=1.0$; $\delta=1.0$.
The activation energies of the stages of the process:
$E_1=E_2=E_3=E_4=E_5=E_6=E_7=E_8=E_9=1.081\times10^5$, J×mole^{-1};
Preexponential coefficients of reaction rates in stages:
$k_1=6.448\times10^{10}$, s^{-1}; $k_2=1.29\times10^{11}$, s^{-1}; $k_3=1.934\times10^{-8}$, s^{-1}; $k_4=1.29\times10^{11}$, s^{-1}×kg×mole^{-1};
$k_5=6.448\times10^{-9}$, s^{-1}; $k_6=2.579\times10^{-9}$, s^{-1}; $k_7=9.027\times10^{10}$, s^{-1}; $k_8=3.224\times10^{11}$, s^{-1};
$k_9=6.448\times10^{10}$, s^{-1}.
The coefficient q_0 is supposed to be equal to 4.0×10^8, J•kg^{-1}.
The calculation was performed for the time interval from 0 to 3600 sec. The results of calculations is shown in Figure 2.
The values of temperature field, which performed with different number of discretization points in the spatial coordinate, is shown in Table. 1. As can be seen from Table 1, the difference between the solutions obtained when K=12 and K=24 does not exceed 0.1 °C.

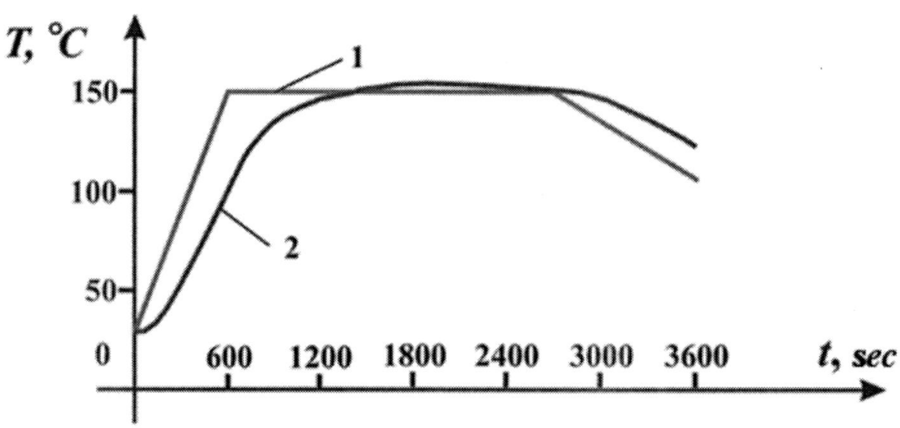

Figure 2. For example 1. 1 is a graph of the temperature on the upper and lower boundaries, 2 is a graph of the temperature in the center of the rod.

Table 1. Comparison of the results of calculations performed with different number points discretizing spatial coordinates.

	The temperature in the center of the rod , °C		
	The number of points of discretization the length of the rod		
A time value, sec	6	12	24
100	32.28	32.05	31.99
250	47.14	46.84	46.76
500	84.08	83.81	83.74
750	121.47	121.58	121.62
1000	139.15	139.43	139.50
2000	153.84	153.80	153.79
3000	146.69	146.76	146.78
3600	122.49	122.51	122.51

Example 2. In this example, we performed a computational experiment, whose goal was to test the performance of the algorithm for solving the inverse problem (IP) for the system of equations of heat conduction.

The testing was performed in the following sequence: at first, for given values of thermophysical parameters

(C_0= 8000000, j•kg^{-1}•deg^{-1}; C_1= 4900, j•kg^{-1}•deg^{-2}; λ=0.35, W·m^{-1}·deg^{-1}; q_0= 4.0×10^8, J·kg^{-1})

the calculation of the temperature $T(t)$ in the center of the rod was carried out using the algorithm for solving the direct problem (DP). Then, for the known temperature field $T(t)$ (calculated by algorithm of DP) the thermophysical parameters C_1, λ, of the sample q_0 were determined using the algorithm of IP. The other initial data are the same as in the previous example.

The values C_1=2450, J•kg^{-1}•deg^{-2}; λ = 0.175 W•m^{-1}•deg^{-2}; q_0 =0.0, J•kg^{-1} have been given as a priori (initial) values of determined parameters in the algorithm of solving the inverse problem.

The results of solving the inverse problem are presented in Table. 2.

Table 2. The results of solving the inverse problem.

Defined parameters	The results of solving the inverse problem by iteration of the optimization cycle				The exact solution
	1st iteration	2nd iteration	3rd iteration	4th iteration	
C_1	5057.76	4517.10	4890.53	4900.32	4900.00
L_0	0.297819	0.342972	0.354871	0.355009	0.35
q_0	874.89×10^5	3359.18×10^5	3990.81×10^5	4000.06×10^5	4000.00×10^5

5. Conclusion

As a result of the performed researches, a technique for determining the thermophysical parameters of the vulcanizable mixture was developed, including algorithms for solving direct and inverse problems for the system of heat conduction equations and the kinetics of the vulcanization process. The proposed solution of the direct problem is finding the concentration of vulcanization of nodes relative to the maximum density of cross-linkage, as a function of time. The solution of the direct problem for the system of equations of thermal conductivity was carried out using an explicit difference scheme. To solve the system of equations describing the kinetics of the process, the Runge-Kutta method was used. The proposed method for solving the inverse problem of the kinetics of vulcanization - the determination of the thermophysical parameters on the measured values of temperature of test specimens based on the reference data thermal characteristics of blend components. Minimization of functions of the residuals was carried out using the Newton method. The performance of the method is confirmed by the results of numerical experiments on model examples.

References

[1] Coran A 1995 Vulcanization: Conventional and Dynamic *Rubber Chem. Technol.* **68 (3)** pp 351-412

[2] Ghosh P, Katare S, Patkar P, Caruthers D and Venkatasubramanian V 2003 Sulfur vulcanization of natural rubber for benzothiazole accelerated formulations: from reaction mechanisms to a rational kinetic model *Rubber Chem.Technol.* **76** pp 592-693

[3] Dontsov A 1973 Mechanism of the formation of vulcanization structures in sulfur vulcanization *Polymer science Ser A* **15 (7)** p p 15455-51

[4] Gordeev V, Porotsky V, Saveliev V and Aldonina T 2000 Increasing the efficiency of tire vulcanization *Questions of practical tire manufacturing technology* **1** pp 85-98

[5] Markelov V and Soloviev M 2005 Approximation of the kinetic curves of vulcanization *Vestnik of YaSTU*, **5** pp 34-38

[6] Tikhomirov S, Pyatakov Y, Karmanova O and Maslov A 2015 Numerical algorithm for calculation of temperature fields of pneumatic tires during vulcanization *Vestnik VSUET* **2** pp 158-164

[7] Karmanova O, Molchanov V and Shashok Z 2012 Solution of the problem of vulcanization kinetics using numerical methods *Proceedings of BSTU* **4 (1)** pp 61-63

[8] Lukomskaya A, Badenkov P, Kepersha L 1978 *Calculations and prediction of modes of vulcanization of rubber products* (Moscow:Khimiya)

[9] Pyatakov Y, Tikhomirov S, Karmanova O and Maslov A 2016 Calculation of the temperature fields in the process of curing *Int. Symp. Problems of tires, rubbers and elastomeric composites* (Moscow: NPKTS VESKOM LLC) pp 405-413

[10] Ishchenko V and Shaptala M 2008 Features of calculations of vulcanization modes for pneumatic tires taking into account the three-dimensionality of the design *System technologies Regional intercollegiate collection of scientific papers* **2 (55)** pp 147-158

[11] Lukomskaya A, Badenkov P and Kepersha L 1972 *Thermal foundations of vulcanization of rubber products* (Moscow: Khimiya)

[12] Kuchanov S 1978 *Methods of kinetic calculations in the polymers chemistry* (Moscow:Mir)

[13] Khang T and Ariff Z 2012Vulcanization kinetics study of natural rubber compounds having different formulation variables *J Therm Anal Calorim* **109** pp 1545–53

[14] Isayev A and Deng J 1988 Non isothermal vulcanisation of rubber compounds *Rubber Chem Technol* **61(2)** p 340

[15] Marzocca A and Mansilla M 2006 Vulcanization kinetic of styrene–butadiene rubber by sulfur/TBBS *J Appl Polym Sci.* **101** pp 35–41

IOP Conf. Series: Journal of Physics: Conf. Series **973** (2018) 012049 doi:10.1088/1742-6596/973/1/012049

The analysis of factors of management of safety of critical information infrastructure with use of dynamic models

S N Trostyansky, A V Kalach, V V Lavlinsky, O V Lankin

ET Department, Voronezh Institute Federal Penitentiary Service of Russia, 1a Irkutskaya Street, Voronezh, RU

E-mail: avkalach@gmail.com

Abstract. Based on the analysis of the dynamic model of panel data by region, including fire statistics for surveillance sites and statistics of a set of regional socio-economic indicators, as well as the time of rapid response of the state fire service to fires, the probability of fires in the surveillance sites and the risk of human death in The result of such fires from the values of the corresponding indicators for the previous year, a set of regional social-economics factors, as well as regional indicators time rapid response of the state fire service in the fire. The results obtained are consistent with the results of the application to the fire risks of the model of a rational offender. Estimation of the economic equivalent of human life from data on surveillance objects for Russia, calculated on the basis of the analysis of the presented dynamic model of fire risks, correctly agrees with the known literary data. The results obtained on the basis of the econometric approach to fire risks allow us to forecast fire risks at the supervisory sites in the regions of Russia and to develop management solutions to minimize such risks.

1. Introduction

Not only ensuring collecting statistical data on the happening fires, but also their analysis by results of which adoption of the administrative decisions directed to minimization of fire risks to the Russian Federation is possible is relevant for elaboration of long-term strategy of management of the level of fire risks on objects of supervision in regions of the Russian Federation, for the purpose of their minimization. Results of a complex research of fire risks are presented in works [1-8]. In this work the leading role of social factors among the causes of the fires in Russia is noted.

It agrees [1], the reasons of the fires connected with a human factor and relating to prevented by the Public fire service (PFS) of Emercom of Russia make more than 65% of total of the fires. Believing linear dependence of number of such fires on total of economic and administrative objects of supervision with violations of requirements of fire safety, expression for probability of emergence of the fires on such objects in a certain interval of time (year), taking into account statistical determination of frequency of the fires, according to the model presented in [9, 10] it is possible to write down as

$$p = p_{\text{n}} + p_{\text{p}} = p_{\text{n}} + kC , \tag{1}$$

where k – regional coefficient of proportionality between probability of the fires on the objects of supervision caused by the prevented factors, and a share among objects which owners violate requirements of fire safety; p_{n}, p_{p} – probabilities of emergence of the fires in a year at the expense of respectively not prevented and prevented Public fire service of factors. Thus, probability of emergence of the fires at the expense of prevented Public fire service of factors is directly proportional to a multiplier which value is defined on the basis of economic model of the rational offender [9, 10]. The rationality of the offender means that violation happens only if the expected income from his commission exceeds possible, in case of fire and (or) loss punishments. When calculating level of violations on the basis of a hypothesis of the rational offender it is considered that the last, as the expected profit can consider economy on expenses on ensuring fire safety of objects, and as punishment can bear the following two types of losses: 1) losses at emergence of the fires on objects of supervision,

$$u = u_{\text{m}} + E_{\text{n}} \cdot R_{2\text{n}} / 10^2, \tag{2}$$

where u_{m} – loss of property; $R_{2\text{n}}$ – risk of death of the person at the fire on objects of supervision [1], where unit of his measurement has an appearance [the victim / 10^2 fires]; E_{n} – material equivalent of human life on economic and administrative objects [4]; 2) losses from penalties for violations of requirements of fire safety, at the probability expected them for a unit of time f. At the same time it is considered that the potential offender on the basis of the or others experience can estimate probability of emergence of the fire p in a year on objects of supervision.

Ignoring of requirements of fire safety in relation to the rational offender, taking into account possible penalties, happens only if the expected income of the offender meets a condition

$$(1-p)\cdot(b-fh) > pu. \tag{3}$$

At the same time it is considered that the potential offender on the basis of the or others experience can estimate probability of emergence of the fire of p in a year on objects of supervision of the region.

On the basis of a hypothesis of the rational offender [9, 10] in work [10] the formula for definition of an economic multiplier has been received C, the violation of requirements of fire safety reflecting economic idea of economic entities of expediency and defining a share of owners of objects, it is favorable to them to save means due to non-compliance with the called requirements. Expression for C, at non-stationary social and economic and administrative and legal factors, taking into account delay time τ between change of economic and administrative factors and change of C, takes a form

$$C_t = \int_0^\infty \rho_{\mu,\sigma_u}(u)_{t-\tau} \int_{\frac{p_{t-\tau}}{1-p_{t-\tau}}b+f_{t-\tau}H_{t-\tau}}^{\infty} \rho_{\eta,\sigma_b}(b)_{t-\tau}\, db du, \tag{4}$$

where

$$\rho_{\mu,\sigma_u}(u) = \frac{1}{\sqrt{2\pi}\cdot\sigma_u u}\exp\left[-\frac{[\ln(u)-\ln(\mu)]^2}{2\sigma_u^{\ 2}}\right],$$

$$\rho_{\eta,\sigma_b}(b) = \frac{1}{\sqrt{2\pi}\cdot\sigma_b b}\exp\left[-\frac{[\ln(b)-\ln(\eta)]^2}{2\sigma_b^{\ 2}}\right],$$

here $\rho_{\mu,\sigma_u}(u)$ – density lognormalno the distributed random variable of losses of owners of objects from one fire; μ – median value for the corresponding distribution of size of losses from one fire; $\sigma_u^{\ 2}$ – dispersion for normal distribution of a logarithm of size of losses from one fire $\ln(u)$; $\rho_{\eta,\sigma_b}(b)$ – density lognormalno the distributed random variable b profits of owners of objects in a year from economy on non-compliance with requirements of fire safety, with median value η for the corresponding distribution and dispersion σ_b^2 for normal distribution $\ln(b)$; f – probability of penalties for violators of requirements of fire safety in a year; H – average value of size of the penalties inflicted on violators of requirements of fire safety; indexes t and $t-\tau$ characterize time (for example, year) measurements of the corresponding characteristics.

At the same time regional coefficient of proportionality between probability and an economic multiplier C, and also probability p_{n} can depend on a number of regional social and economic and administrative and legal factors which influence production discipline of personnel of economic and administrative objects, power loading at operation and also degree of wear of the equipment on the objects, that is from factors influencing fire risks.

For the characteristic of fire risks on economic and administrative objects of supervision in regions of Russia, besides the probability of emergence of the fires on such objects in a year measured as the relation of the number of fires on objects of supervision in the region in a year to quantity of such objects, with dimension [the fire / (object year] it is possible to consider risk of death of the person, average on the region, from the fires on objects of supervision in a year which we will designate as R_{gn}, with dimension [victim / (object year)]. Integrated fire risk it is connected with probability of emergence of the fires on objects of supervision p and integrated fire risk of death of the person at the fire on object of supervision $R_{2\text{n}}$ [1] formula:

$$R_{\text{gn}} = p \cdot R_{2\text{n}} / 10^2 . \quad (5)$$

2. Analysis of panel data

Definition of determinants of fire risks on objects of supervision in regions of Russia, identification of their dependences on statistically registered socio-economic indexes and also from the indicators characterizing possibilities of the Ministry of Emergency Situations on suppression of the fires, such as indicators of time of rapid response to the fires is of practical interest. Use of the analysis of panel statistical data is relevant for the econometric analysis of the factors determining the level of fire risks on objects of supervision in regions of the Russian Federation. Panel data are observations of the same economic units which are carried out during the consecutive periods of time. In work [11], with use of statistical data on the Russian Federation, on the basis of the multidimensional regression analysis of dependence of number of the fires on a number of factors for which the coefficient of correlation has the highest value the dependence of the number of fires on 1 thousand people is received from number of the registered crimes per 100 thousand people of the population close to linear. Taking into account this result, assuming community of the socio-economic factors defining fire risks and crime rate in the Russian Federation, we will analyses empirical researches of influence of socio-economic factors on the different types of crime in regions of the Russian Federation executed on the basis of the analysis of panel statistical data in work [12]. In this work influence on crime of the following factors is revealed: average monetary per capita income of the population, Jeanie's coefficient (inequality measure in income), the level of abuse of alcohol among the population of the region, education level of the population of the region, average temperature of January in the region, the indicators connected with probability of criminal penalty for crimes. Then, carrying out the panel analysis of dependence of level of fire risks on objects of supervision in Russia from the socio-economic factors established in [12] as crime determinants, with replacement of the factors connected with probability and weight of criminal penalty for crimes by the factors connected with probability and weight of material and possible human losses is relevant for a research of the factors determining fire risks on objects of supervision. When carrying out the panel analysis, besides socio-economic factors, as the indicator characterizing possibilities of the Ministry of Emergency Situations on suppression of the fires also average time of arrival on the fire of the first fire divisions was considered.

For definition of influence of the considered factors on the level of fire risks on objects of supervision in regions of Russia the program has been used DPD98, written by Arellano and Bond [13]. It is necessary for asymptotic solvency of the received estimates that there was a sufficient number of the temporary periods ($T > 4$) and large number of objects ($N_{ob} > T$).

Information base for the analysis was made by panel data on 78 regions of the Russian Federation (autonomous areas as a part of areas, edges and also the regions which don't have a totality of data are excluded) for 2006-2016. Information on fire statistics and administrative and legal indicators has been received from these registered Public fire service of Emercom of Russia, in particular from [14], and social and economic indicators for regions and indicators of inflation have been taken from publications of Rosstat.

Believing that probability of the fires on objects of supervision in a year and risk of death of the person from the fires on objects of supervision in a year are linear functions from various factors, and aggregating on the population of the region, we will construct linear dynamic models for the description of fire risks, depending on a set of the set variables and also from a set of independent variables:

$$p_{it} = a_1 p_{i(t-1)} + a_2 R_{2nit} + a_3 U_{mit} + a_4 S_{it} + a_5 D_{it} + a_6 V_{it} + \\ + a_7 J_{it} + a_8 A_{it} + a_9 I_{it} + a_{10} G_{it} + a_{11} T^{\circ}_{it} + a_{12} t_{it} + C_1 , \quad (6)$$

$$R_{gnit} = b_1 R_{gni(t-1)} + b_2 U_{mit} + b_3 S_{it} + b_4 D_{it} + b_5 V_{it} + \\ + b_6 J_{it} + b_7 A_{it} + b_8 I_{it} + b_9 G_{it} + b_{10} T^{\circ}_{it} + b_{11} t_{it} + C_2 , \quad (7)$$

where lower indexes i and t designate the region and year respectively; dependent variables: in the equation (6) p_{it} – probability of emergence of the fire on objects of supervision in a year; in the equation (7) R_{gnit} – risk of death of the person, average on the region, from the fires on objects of supervision in a year; the set variable in the equation (6) R_{2nit} – risk of death of the person at the fire on object of supervision; the set variables in the equations (6,7): U_{mit} – taking into account inflation relatively 2006, accepting average material damage from one fire on objects of supervision in thousands of rubles for initial – average material damage in 2006; S_{it} – the average penalty in thousands of rubles which is ordered to pay for violation of requirements of fire safety on objects of supervision taking into account inflation relatively 2006, accepting for initial – an average penalty in 2006; independent variables in the equations (6,7): D_{it} – taking into account inflation relatively 2006, accepting average monetary income of the population in thousands of rubles for initial – average monetary income in thousands of rubles in 2006; V_{it} – per capita in thousands of rubles, taking into account inflation relatively 2006, accepting a gross regional product for initial – a gross regional product per capita in thousands of rubles in 2006; J_{it} – Jeanie's coefficient in regions (inequality measure in income); A_{it} – number of the patients with for the first time in life the established diagnosis of the psychotic frustration connected with alcohol intake and a syndrome of dependence on alcohol put under dispensary observation by psycho neurological and narcological institutions on 105 people of the population of the region (according to arguments from work [15], this indicator can characterize the level of abuse of alcohol among the population of the respective region); I_{it} – percent of students of educational institutions of higher education in the population of the region (this indicator can characterize education level in the region); G_{it} – percent of urban population in the region; T°_{it} – average temperature of January in degrees Celsius in the region; t_{it} – average time on the region (in minutes) arrivals on the fire of the first fire divisions; C_j – the constant including unaccounted factors.

The results of the regression analysis of panel data received with use of the program are presented in table 1 for models on the basis of the equations (6) and (7) DPD98 [9]. Also values of their standard mistakes (in brackets) and their importance are presented in table 1, except values of coefficients and constants, (z), which are defined as the attitude of value of coefficient or a constant towards value of the corresponding standard mistake and also levels of credibility of coefficients and constants are given.

Table 1. Results of the regression analysis of panel data from 2006 to 2016 on dependence of level of fire risks on objects of supervision in regions of the Russian Federation from socio-economic factors and an indicator of time of rapid respon.

Factors	Model 1	Model 2
	equation (6) for p_{it} – probabilities of emergence of the fire on objects of supervision in a year [fire / (object year)]	equation (7) for R_{gnit} – average on the region of risk of death of the person from the fire on object of supervision in a year [victim / (object year)]
$p_{i(t-1)}$ [fire / (object year)]	0.2607125 (0,0522585) z = 4.99	
$R_{gn(t-1)}$ [victim / (10^5 person year)]		0.0645238 (0,0599118) z = 1.08
R_{2nit} [victim /10^2 fire]	-0.000052974 (0.000085806) z = -0.62	
U_{mit} [thousand rubles/fire]	$-8.77*10^{-7}$ ** ($4.37*10^{-7}$) z = -2,01	$-2.83*10^{-8}$ ($3.44*10^{-8}$) z = -0.82
S_{it} [thousand rubles/penalty]	-0.0001907 *** (0.0000548) z = -3.48	$-7.86*10^{-6}$ ($4.33*10^{-6}$) z = -1.82
D_{it} [thousand rub]	-0.000275 (0.000282) z = -0.97	$-5.03*10^{-5}$ ** ($2.19*10^{-5}$) z = -2.29
V_{it} [thousand rub]	$1.23*10^{-8}$ ($7.66*10^{-8}$) = 0.16	$-2.99*10^{-9}$ ($5.93*10^{-9}$) z = -0.49
J_{it}	-0.1006677 ** (0.0439641) z = -2.29	0.0051511 (0.0036391) z = 1.42
A_{it} [patients /10^5 person]	0.0000247 * (0.0000128) z = 1.93	$-7.95*10^{-7}$ ($1.04*10^{-6}$) z = -0.76
I_{it} [%]	-0.0001333 (0.0007409) z = -0.18	0.0000422 (0.0000575) z = 0.73
G_{it} [%]	0.0005625 (0.000433) z = 1.30	0.0000162 (0.000037) z = 0.48
T°_{it} [degrees. C°]	0.0000458 (0.0000568) z = 0.81	$7.52*10^{-6}$ ($4.43*10^{-6}$) z = 1.69
t_{it} [minute]	0.0002154 (0.0001258) z = 1.71	$3.68*10^{-6}$ ($9.79*10^{-6}$) z= 0.38
C_j	0.0088513 (0.0376075) z = 0.24	-0.0022635 (0.0028878) z = -0.78

In the table 1 asterisks have designated levels of credibility: *** – 1%, ** – 5%, * – 10%.

IOP Conf. Series: Journal of Physics: Conf. Series **973** (2018) 012049 doi:10.1088/1742-6596/973/1/012049

Value of an economic equivalent of human life E_n for objects of supervision in Russia it is possible to estimate, having presented the equation (6) in the form:

$$p_{\mathrm{it}} = a_1 p_{\mathrm{i(t-1)}} + a_3(a_2 R_{2\mathrm{nit}} / a_3 + U_{\mathrm{mit}}) + a_4 D_{\mathrm{it}} + a_5 J_{\mathrm{it}} + a_6 A_{\mathrm{it}} + \\ + a_7 Z_{\mathrm{it}} + a_8 G_{\mathrm{it}} + a_9 S_{\mathrm{it}} + a_{10} T_{\mathrm{it}} + a_{11} t_{\mathrm{it}} + C_1 . \quad (8)$$

Taking into account a ratio (2), the equation (6) can also be written down in a look:

$$p_{\mathrm{it}} = a_1 p_{\mathrm{i(t-1)}} + a_3(E_{\mathrm{n}} R_{2\mathrm{nit}} / 10^2 + U_{\mathrm{mit}}) + a_4 D_{\mathrm{it}} + a_5 J_{\mathrm{it}} + a_6 A_{\mathrm{it}} + \\ + a_7 Z_{\mathrm{it}} + a_8 G_{\mathrm{it}} + a_9 S_{\mathrm{it}} + a_{10} T_{\mathrm{it}} + a_{11} t_{\mathrm{it}} + C_1 . \quad (9)$$

Follows (8) and (9) from comparison that value of an economic equivalent of human life E_{n} on objects of supervision in Russia is determined by a formula:

$$E_{\mathrm{n}} = 10^2 \cdot a_2 / a_3 . \quad (10)$$

Taking into account values and dimensions of coefficients and a_3 from the table 1: E_{n} =6040.36 [thousand rubles/victim], taking into account inflation relatively 2011.

The received assessment of an economic equivalent of human life on objects of supervision in Russia close will be coordinated with value of this indicator calculated for Russia by a technique of assessment of an economic equivalent of cost of the human life based that the economic equivalent of life of the average person is equal to the attitude of the average located monetary revenue towards average probability of death within a year [14-16]. From calculations of work [16], this value for Russia in 2013 was 12472 thousand rubles that there corresponded 9038 thousand rubles, taking into account official inflation relatively 2011. Such coordination of size E_{n}, received from model 1, with value of the economic equivalent of human life received by a technique [16] confirms correctness of model 1 in application to assessment of probability of emergence of the fires on objects of supervision in regions of Russia and also it will be agreed with a hypothesis of the rational offender.

The results of the regression analysis presented in table 1 for quantitative dependence on various socio-economic factors of probability of emergence of the fires on objects of supervision in a year and an average on the region of risk of death of the person from the fires on objects of supervision in a year will correctly be coordinated with the model of the rational offender given above. Really, according to model of the rational offender, on the basis of results of the work [17] performed for lognormalny distribution of income of owners of objects [9, 10] and losses from the fires on objects ratios follow:

$$\frac{\mathrm{d}p}{\mathrm{d}\ln(S)} < 0 ; \quad (11)$$

$$\frac{\mathrm{d}p}{\mathrm{d}\ln(U_\mu)} < 0 , \quad (12)$$

where S – average value of a penalty for violation of requirements of fire safety on objects of supervision in the region; U_μ – median value of size of losses from the fires on all objects of supervision of the region. Inverse relation follows from a ratio (11) p from , and from (12) inverse relation follows p from the factors determining the size of possible losses from the fires on objects of supervision: average values of loss of property U_{m} and $R_{2\mathrm{n}}$. It is also logical to assume that median value U_μ distributions of size of losses from the fires of owners of objects of supervision increases or decreases in the same party, as median value of distribution of size of legal income of owners of objects from which investments accumulate in the cost of objects and can be lost at the fire. It is possible to expect that the income of owners of objects of supervision in turn positively correlates with average value D monetary income of the population of the region. Then signs of derivatives $\frac{dp}{d\ln(U_{\mu})}$ and $\frac{dp}{d\ln(D)}$ have to coincide and be negative. Such negative dependence p_{it} from factors: S_{it}, U_{mit}, R_{2nit}, D_{it} it is received on the basis of the regression analysis of panel data from 2006 to 2016 on regions of the Russian Federation and presented in table 1.

The presented results of the regression analysis of fire risks for objects of supervision in regions of the Russian Federation will be coordinated with results of the works [9, 10] executed on the basis of model of the rational offender about existence of economic determination of fire risks.

Follows from the results presented in table 1 that for probability of emergence of the fires on objects of supervision p_{it} there is significant ($z = -2.29$) negative dependence on Jeanie's coefficient J_{it}, with high level of confidence (<1%); significant ($z = 1.30$) positive dependence from G_{it} – percent of urban population in the region; significant ($z = 1.71$) positive dependence from – average time on the region of arrival on the fire of the first fire divisions.

Also follows from results of table 1 that for R_{gnit} – average on the region to risk of death of the person from the fires on objects of supervision in a year, the significant negative dependence takes place ($z = -1.82$) from an average on the region of a penalty S_{it} for violation of requirements of fire safety on objects of supervision; significant negative dependence ($z = -2.29$) from D_{it} – average monetary income of the population in the region with high level of confidence (<5%); significant ($z = 1.42$) positive dependence on regional coefficient of Jeanie J_{it}; significant ($z = 1.69$) positive dependence from T°_{it} – the average January temperature in degrees Celsius in the region.

3. Conclusion

Results of the analysis of dynamic model of the panel data on regions of Russia from 2006 to 2016 including fire statistics on objects of supervision and statistics of set of regional socio-economic indexes and also indicators of time of rapid response of PFS Ministry of Emergency Situations to the fires allow to present quantitative dependences of probability of emergence of the fires in a year on objects of supervision in regions of the Russian Federation and an average on the region of risk of death of the person from the fires on objects of supervision in a year as linear functions from values of these indicators for the previous year, a set of regional social and economic and climatic indicators and also indicators of time of rapid response of Public fire service Ministry of Emergency Situations to the fires.

The received results correctly will be coordinated with results of application to fire risks of model of the rational offender. Assessment of an economic equivalent of human life, as of objects of supervision for Russia, calculated on the basis of the analysis of the presented dynamic model of fire risks, will correctly be coordinated with the values, known from literature, received on the basis of actuarial calculations. The results received on the basis of econometric approach to fire risks give information for the forecast of fire risks on objects of supervision in regions of Russia and administrative decisions on minimization of such risks.

References

[1] Brushlinsky N N, Glukhovenko Yu M, Korobko V B, Sokolov S V, Lupanov S A and Klepko E A 2004 *Fire risks Release Basic concepts* (Moscow: National Academy of Sciences of fire safety) p 47
[2] Brushlinsky N N and Sokolov S V 2012 A role of statistics of the fires in assessment of fire risks *Security concerns and emergency situations* **1** pp 112-124

[3] Brushlinsky N N, Sokolov S V and Grigorieva M P 2017 The analysis of the main fire risks in the countries of the world and in Russia *Fire and explosion safety* **2** pp 72-80
[4] Brushlinsky N N, Sokolov S V and Grigorieva M P 2016 About some regularities and features of the Russian fire statistics *Fire and explosion safety* **6** pp 33-38
[5] Ahrens M, Brushlinsky N N, Wagner P and Sokolov S V 2015 Situation with the fires in the world at the beginning of the 21st century *Fire and explosion safety* **10** pp 51-58
[6] Brushlinsky N N and Sokolov S V 2016 International fire statistics of the international association of rescue and fire fighting services *Bulletin of the Voronezh institute of GPS Emercom of Russia* **1** pp 71-103
[7] Brushlinsky N N and Sokolov S V 2013 About conceptual approaches to development of ensuring fire safety in modern society *Problems of the analysis of risk* **1** pp 34-39
[8] Brushlinsky N N, Sokolov S V, Klepko E A, Popkov S Yu and Belov V A 2013 Triad "DANGER-RISK-SAFETY" *Problems of the analysis of risk* **4** pp 42-49
[9] Trostyansky S N, Zenin Yu N, Skryl S V and Kalatch A V 2013 Fires and emergency situations: prevention, elimination *Fire and explosion safety* **4** pp 28-33
[10] Trostyansky S N, Zenin Yu N and Bakayeva G A 2014 Mathematical model and algorithms of calculation of level of regional fire risks on objects of supervision *Control systems and information technologies* **1** pp 204-208
[11] Dushkin A V, Novoseltsev V I and Sumin V I 2014 *Mathematical models and information processes of management of a difficult object*: *monograph* (Voronezh: Scientific book) p 125
[12] Dushkin A V, Novoseltsev V I and Sumin V I 2017 *Introduction to the system analysis*: *monograph* (Moscow: Hot line – Telecom) p 234
[13] Dushkin A V, Novoseltsev V I and Sumin V I 2017 *Introduction to system design of intellectual knowledge bases*: *monograph* (Moscow: Hot line – Telecom) p 108
[14] Dushkin A V, Novoseltsev V I and Sumin V I 2016 *Risk and reflection*: *monograph* (Moscow: Hot line – Telecom) p 136
[15] Sumin V I, Dushkin A V and Kravchenko A S 2014 *Modeling of systems* (Voronezh: Scientific book) p 222
[16] Novoseltsev V I, Dushkin A V and Sumin V I 2015 *Management of the conflicts: manual for higher education institutions* (Moscow: Hot line – Telecom) p 144

Finite element analysis of the biaxial cyclic tensile loading of the elastoplastic plate with the central hole: asymptotic regimes

Vera Turkova, Larisa Stepanova

Samara University, Samara, Russian Federation

E-mail: turkova.ver@yandex.ru , stepanovalv@samsu.ru

Abstract. For elastistoplastic structure elements under cyclic loading three types of asymptotic behavior are well known: shakedown, cyclic plasticity or ratcheting. In structure elements operating in real conditions ratcheting must always be excluded since it caused the incremental fracture of structure by means of the accumulation of plastic strains.

In the present study results of finite-element (FEM) calculations of the asymptotical behavior of an elastoplastic plate with the central circular and elliptic holes under the biaxial cyclic loading for three different materials are presented.

Incremental cyclic loading of the sample with stress concentrator (the central hole) is performed in the multifunctional finite-element package SIMULIA Abaqus. The ranges of loads found for shakedown, cyclic plasticity and ratcheting are presented. The results obtained are generalized and analyzed. Convenient normalization is suggested. The chosen normalization allows us to present all computed results, corresponding to separate materials, within one common curve with minimum scattering of the points. Convenience of the generalized diagram consists in a possibility to find an asymptotical behavior of an inelastic structure for materials for which computer calculations were not made.

1. Introduction

Asymptotic states of inelastic structures subjected to cyclic loading cause the particular interest in Solid Mechanics. Response of the inelastic structure under periodic mechanical loading is complicated and may contain inelastic strains. The reason of these difficulties is the need of calculations, which include the whole history of structure loading.

Elements of structures are often subjected to variable temperatures and loadings. If the body is elastic, then the durability is defined by the fatigue material data and fracture comes after large number of cycles. If the body is elastoplastic, the dangerous state can be reached after rather small number of cycles. It is necessary to distinguish two cases. The first one is when fracture comes owing to alternation of the plastic strains sign (e.g. after plastic stretching – plastic compression etc.). It is alternate plasticity (plastic or low-cyclic fatigue). The second one is when plastic strains increase with every cycle, which leads to its accumulation and then failure (progressing plastic deformation - ratcheting) [1]–[24].

In structural design shakedown is considered to be safe, ratcheting should be avoided. In this regard, requirement in knowledge on the early steps of configuration design of the asymptotic behavior of the inelastic structure after a large number of loading cycles becomes actual [2], [3].

Two classes of methods are developed in order to determine the asymptotic state: direct methods [2]–[7] and incremental methods [8]–[12].

Some works contain both analytical and incremental approaches, as in [8], where cyclic plasticity theory is used to study the three dimensional effects at the tip of rounded notches in plates of finite thickness to determine the actual stress and strain state arising. Different notch geometries and loading conditions are investigated.

As the problems of determining the asymptotic states of inelastic bodies cause considerable interest recently, in literature are proposed plenty solutions of applied durability problems (cyclic loading of rails, tools and machine parts) [9]–[25].

In recent work incremental step-by-step biaxial loading of the elastoplastic plate with the central hole is carried out. The applied load is cyclic. Three types of asymptotic behavior of the plate are revealed during the analysis of FEM calculations. Domains of loadings for three asymptotic regimes of an inelastic structure are found. Calculations are carried out for three different materials. The obtained ranges of shakedown, alternate plasticity and ratcheting were analyzed and general patterns of different regimes were revealed.

2. Computing experiments

Unfortunately, incremental methods are time-consuming and demand a considerable number of numerical experiments [1-9]. The aim of direct numerical methods is to overcome shortcomings of incremental analysis, however the developed theory is far from completion and practical use [2, 4, 9, 11, 13]. Owing to the specified reasons, in the recent work incremental loading of elastoplastic plate with central hole was carried out in order to identify different types of asymptotic behavior of inelastic structure and determine loads, which lead to regime changings.

Implementation and research of the three asymptotic regimes have been held out on the example of a square plate with central hole (elliptic and then circular). The geometry of the plate is shown in figure1.

The length of a side of the plate L is $16dm$. The plate's side length to its thickness ratio is 0.02. The semi major axis of the elliptical hole is $a=3dm$, and the semi minor axis is $b=2dm$. For the circular hole radius $r=2dm$.

Let's consider cases when the copper plate has different mechanical properties:

a) density is $\rho=5400kg/m^3$, Young's modulus is $E=4.3\times10^8 kg/m^2$, Poisson's ratio is $\nu=0.28$ [26];

b) density is $\rho=8920kg/m^3$, Young's modulus is $E=13\times10^9 kg/m^2$, Poisson's ratio is $\nu=0.28$ [27];

c) density is $\rho=8920kg/m^3$, Young's modulus is $E=1\times10^{10} kg/m^2$, Poisson's ratio is $\nu=0.3$ [27];

The plate is subjected to the biaxial loading P_1 and P_2 (figure1). The applied load $P_2(t)$ is cyclic. The applied loading is schematically shown in figure1.

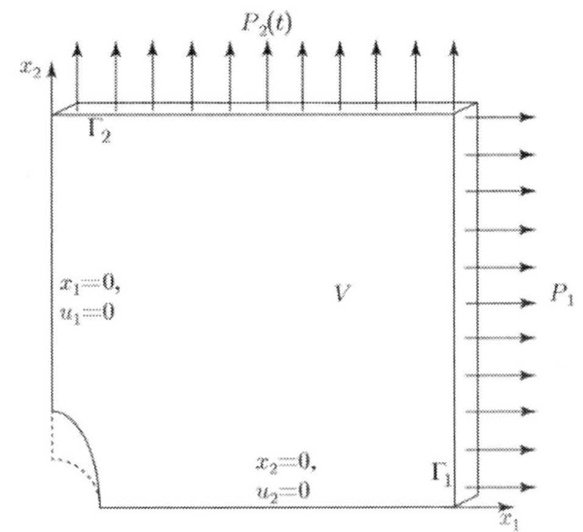

Figure 1. Geometry of the model: a quarter of a loaded plate.

2.1. Elasto-plastic analysis

In this section the finite-element method solution of the cyclic loaded elasto-plastic plate with central hole problem is given. The purpose of the present finite element analysis is to define load amplitudes (boundaries of shakedown, cyclic plasticity and ratcheting domains) when these regimes are realized.

Let's consider a body, occupying volume V (the plate with central hole, the quarter of which is given in figure 1). The boundary conditions are:
when $\vec{x} \in \Gamma_2$, $P_2(t) = k_2\sigma_T \sin(2\pi t / T)$, see figure 2. When $\vec{x} \in \Gamma_1$, $P_1 = k_1\sigma_T$, where σ_T is material yielding strength; k_1, k_2 are constants of proportionalities, which are used to vary the amplitude of the applied load.

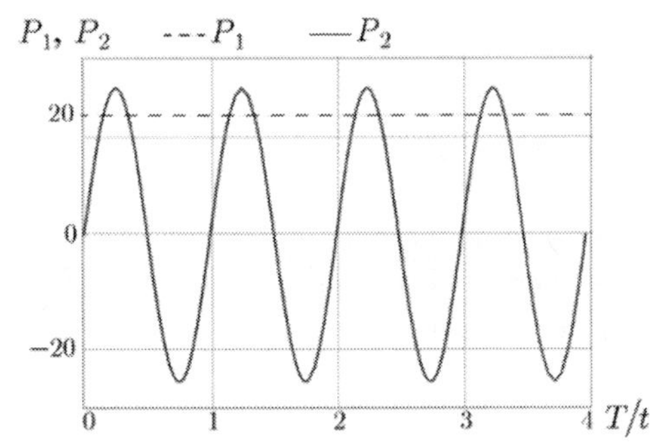

Figure 2. Periodic law for the load.

It is supposed that the plate is made of elasto-plastic material. The total strains consist of elastic and plastic components: $\varepsilon_{ij} = \varepsilon_{ij}^e + \varepsilon_{ij}^p$, where ε_{ij}^e is elastic strain and ε_{ij}^p is plastic strain. For the elastic part Hooke's law is valid: $\varepsilon_{ij}^e = C_{ijkl}\sigma_{kl}$. As physical ratios for plastic deformation the equations of the flow plasticity theory are applied. For the plastic environment in space of tension the flow surface $f(\sigma_{ij}) = s_{ij}s_{ij} - 2\sigma_T^2 / 3$ is introduced. Plastic strains are defined by the associate flow rule:

$$d\varepsilon_{ij}^p = d\lambda \frac{\partial f}{\partial \sigma_{ij}}.$$

It is necessary to determine the asymptotic state of the structure after a large number of loading cycles and define the possible regimes of the body behavior under the applied cyclic and static loadings. The mathematical problem is reduced to the system of equations:

equilibrium equation: $\sigma_{ij,j} = 0$ in V,

the constitutive law:

$$\dot{\varepsilon}_{ij} = \dot{\varepsilon}^{e}_{ij} + \dot{\varepsilon}^{p}_{ij} = \frac{1+\nu}{E}\dot{\sigma}_{ij} - \frac{\nu}{E}\dot{\sigma}_{kk}\delta_{ij} + \dot{\lambda}\frac{\partial f}{\partial \sigma_{ij}} \text{ in } V,$$

Cauchy's relations: $\varepsilon_{ij} = (u_{i,j} + u_{j,i})/2$ in V.

The boundary conditions are: periodic load $\sigma_{22} = P_2(t)$ on $\vec{x} \in \Gamma_2$ and static load $\sigma_{11} = P_1$ on $\vec{x} \in \Gamma_1$.

Boundary conditions on the symmetry planes of the quarter of a plate are shown in the figure 1.

2.2. *Finite element analysis*

The series of calculations was carried out in multifunctional finite-element package SIMULIA Abaqus [26].

Plastic material properties of the model are set by the table 1 (a, b, c). The first value in the first column is the material yielding strength, and the last one in the first column is the ultimate strength of material [26]–[27].

Table 1. Stress – plastic strain dependence for copper.

(a)		(b)		(c)	
σ, kg/m^2	ε^{pl}	σ, kg/m^2	ε^{pl}	σ, kg/m^2	ε^{pl}
54×10^4	0	210×10^6	0	60×10^6	0
58×10^4	0.0006	240×10^6	0.0055	100×10^6	0.09
63×10^4	0.0080	280×10^6	0.015	140×10^6	0.15
69×10^4	0.0013	300×10^6	0.02	170×10^6	0.23
74×10^4	0.0018	320×10^6	0.025	200×10^6	0.36
78×10^4	0.0023	360×10^6	0.03	220×10^6	0.45

The series of calculations of the structure asymptotic conditions have been carried out, where constants of proportionality (k_1, k_2) varied from 0 to 1.3. All the results in this section are given for the plate with elliptic hole. For the circular hole they are just similar.

Periodic nature of loading is seen in the figure of changes of equivalent stress during time for different types of structure behavior (figure 3).

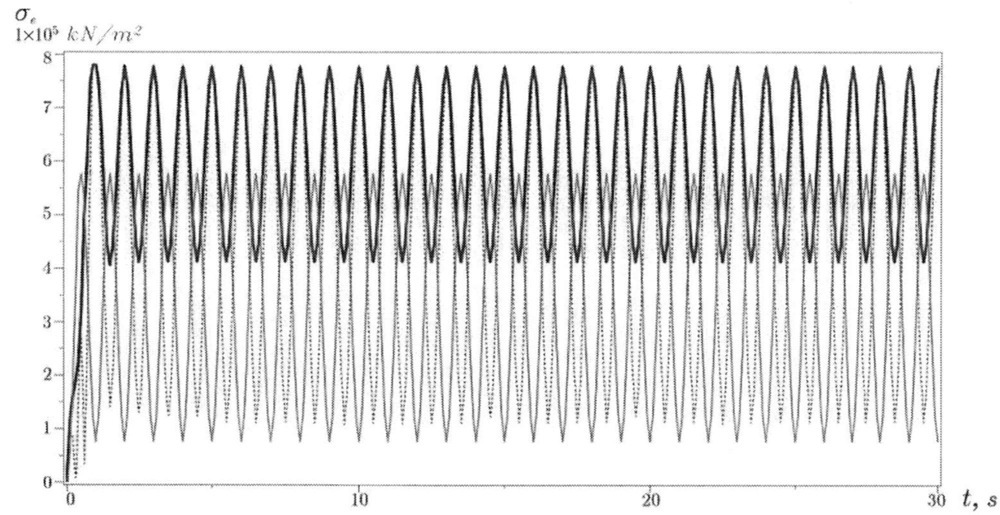

Figure 3. Cyclic character of the equivalent stress (blue curve corresponds to shakedown; green curve corresponds to cyclic plasticity, red curve shows ratcheting).

It is obvious that for different asymptotic regimes of the structure stresses always behave cyclically. This feature is the basis for the most part of direct methods which are used to determine the asymptotic state [2, 4, 9, 11, 13].

For the analyzed elasto-plastic plate all three regimes were obtained by changing the magnitude of the load (figures 4 – 6) and loadings that lead to the change of the regime were defined.

One can determine the asymptotic type of the structure after a number of loading cycles by the use of the plastic deformations character.

In the figure 4 plastic strain components of the copper plate (material b) for shakedown are shown:

1) ε_{11}^{p} (blue curve) and ε_{22}^{p} (green curve) for $P_1 = P_{2\max} = 100kN/m^2$;
2) ε_{11}^{p} (blue curve) and ε_{22}^{p} (green curve) for $P_1 = 100kN/m^2$; $P_{2\max} = 250kN/m^2$.

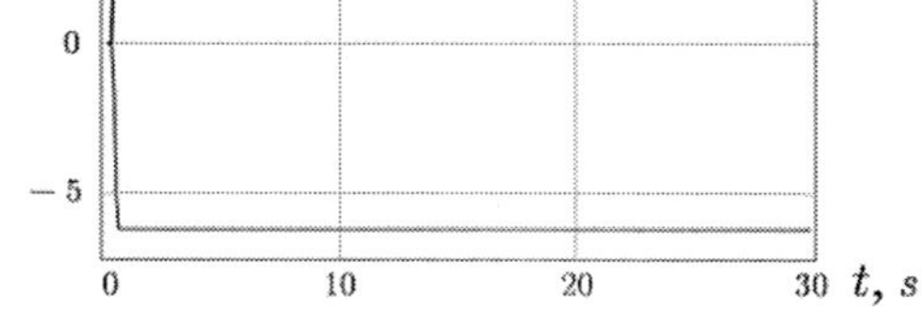

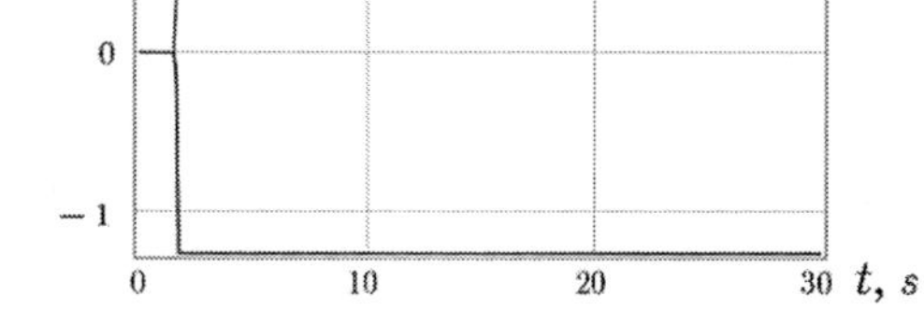

Figure 4. Plastic strains in the model for shakedown (case 1(left), 2(right)).

In the figure 5 plastic strain components of the copper plate (material a) for cyclic plasticity are shown:

1) ε_{11}^{p} (blue curve) and ε_{22}^{p} (green curve) for $P_1 = P_{2\max} = 380kN/m^2$;

2) ε_{11}^{p} (blue curve) and ε_{22}^{p} (green curve) for $P_1 = P_{2\max} = 400kN/m^2$.

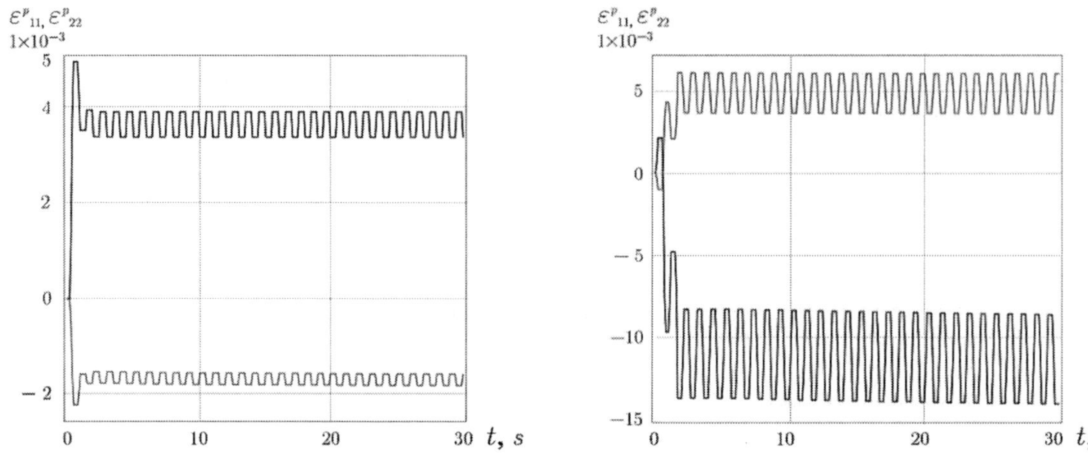

Figure 5. Plastic strains in the model for cyclic plasticity (case 1(left), 2(right)).

In the figure 6 plastic strain components of the copper plate (material b) for ratcheting are shown:

1) ε_{11}^{p} (blue curve) and ε_{22}^{p} (green curve) for $P_1 = 300kN/m^2$; $P_{2\max} = 50kN/m^2$;

2) ε_{11}^{p} (blue curve) and ε_{22}^{p} (green curve) for $P_1 = 100kN/m^2$; $P_{2\max} = 220kN/m^2$.

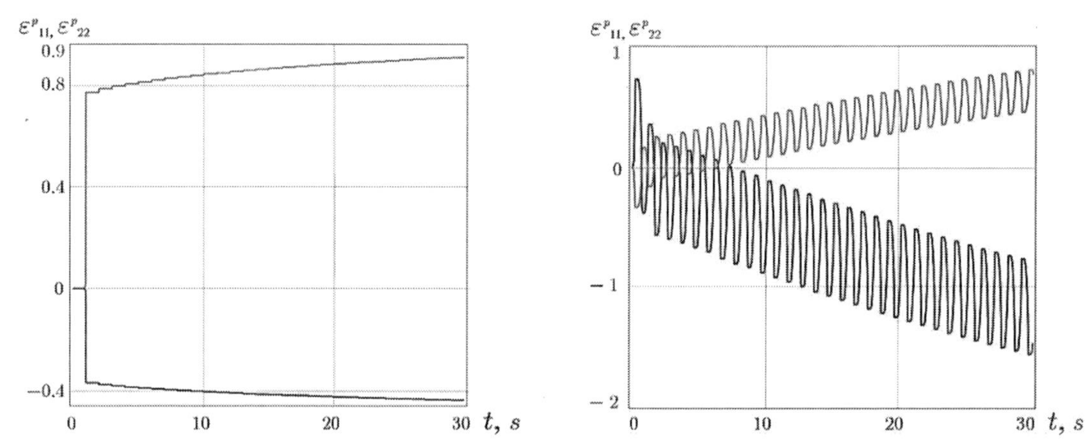

Figure 6. Plastic strains in the model for ratcheting (case 1(left), 2(right)).

Stress – strain curve also can brightly characterize the type of structure asymptotic behavior. In figure 7 dependence of ε_{11} and σ_{11} for shakedown is shown. Mathematically this relation for every moment $\tau = t/T$ (where t is the specific moment of time, T is the period of loading) inside the stabilized asymptotic cycle can be presented as: $\dot{\varepsilon}_{ij}^{pl,cs} = \lim_{n\to\infty} \dot{\varepsilon}_{ij}^{pl}(\tau) = 0$, where $\dot{\varepsilon}_{ij}^{pl,cs}$ is the plastic strain rate in the stabilized cycle.

In the figure 8 the strain ε_{11} and stress σ_{11} curve for cyclic plasticity is shown:

$\dot{\varepsilon}_{ij}^{pl,cs} = \lim_{n\to\infty} \dot{\varepsilon}_{ij}^{pl}(\tau) \neq 0$, $\Delta\varepsilon_{ij}^{pl,cs} = \int \dot{\varepsilon}_{ij}^{pl}(\tau) d\tau = 0$.

In the figure 9 stress – strain curve for ratcheting is shown [13]:

$$\dot{\varepsilon}_{ij}^{pl,cs} = \lim_{n\to\infty} \dot{\varepsilon}_{ij}^{pl}(\tau) \neq 0,\ \Delta\varepsilon_{ij}^{pl,cs} = \int \dot{\varepsilon}_{ij}^{pl}(\tau) d\tau \neq 0\,.$$

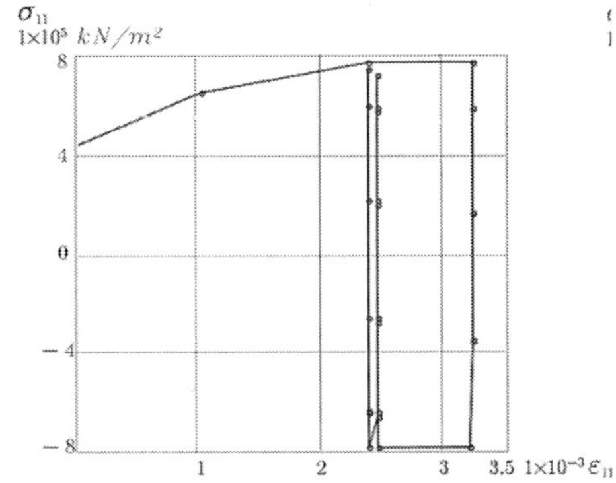

Figure 7. Stress – strain curve (30 loading cycles) for the model: shakedown.

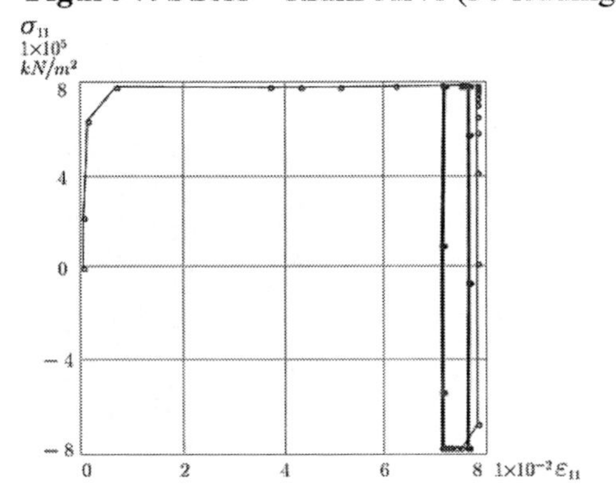

Figure 8. Stress – strain curve (30 loading cycles) for the model: cyclic loading.

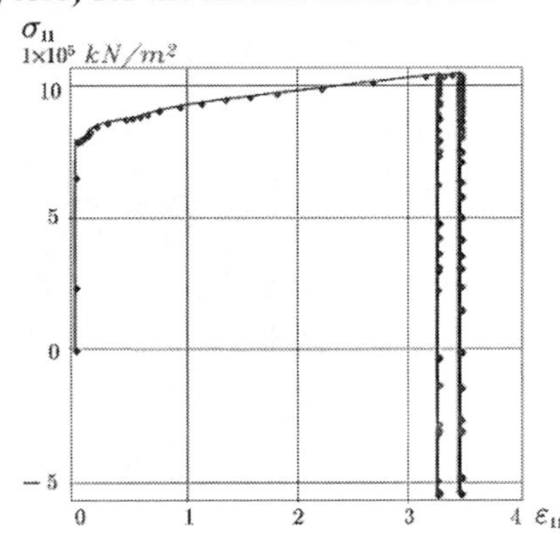

Figure 9. Stress – strain curve (30 loading cycles) for the model: ratcheting.

When modeling this regime one can definitely see the birth and progress of plastic flow in the plate (figure 10).

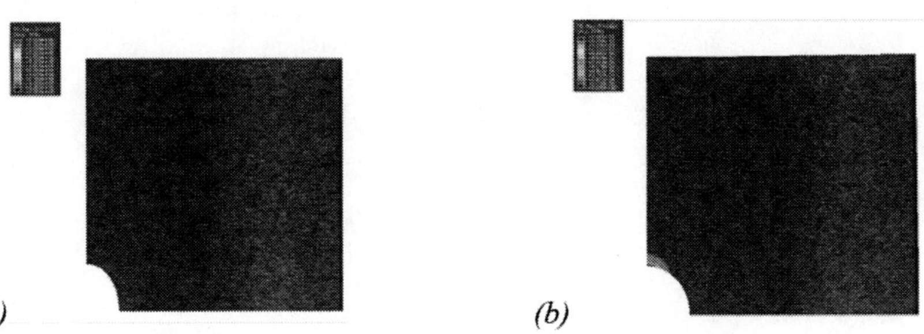

(a) *(b)*

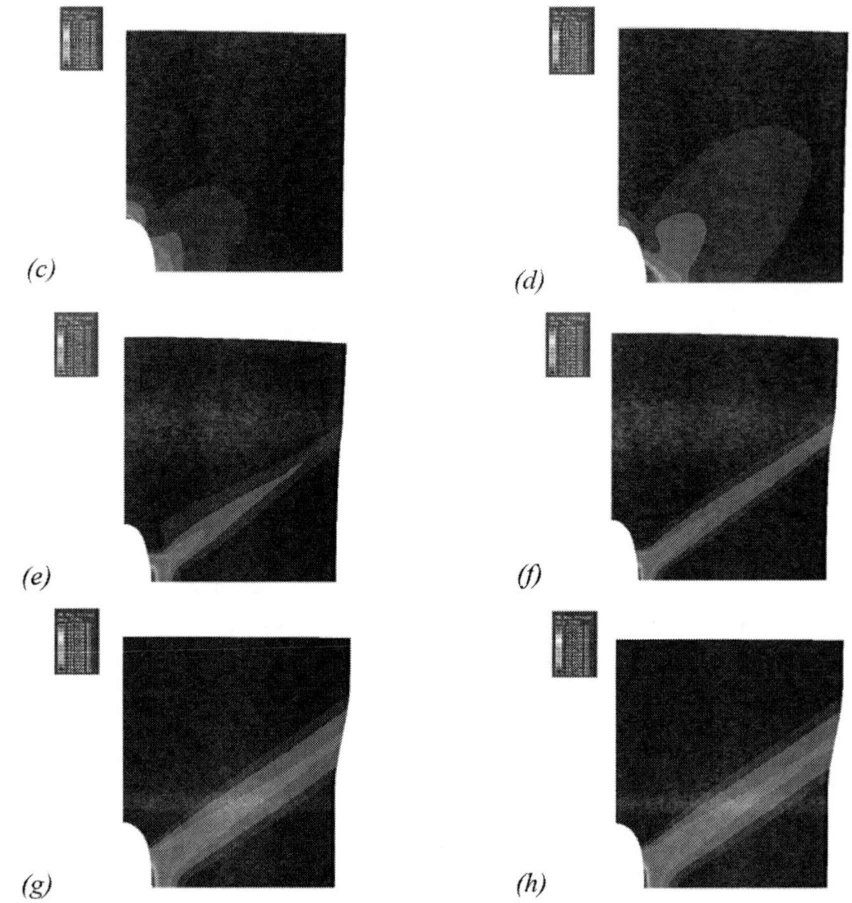

Figure 10. Plastic strains in the model.

The obtained data were generalized and represented in the form of the diagram for both of the plates (figure 11) in which the domains of loadings corresponding to each type of an asymptotical behavior of a plate are defined. The type of structure behavior was defined for the most deformed finite element of the plate. The location and number of this element varied according to the exact loading case.

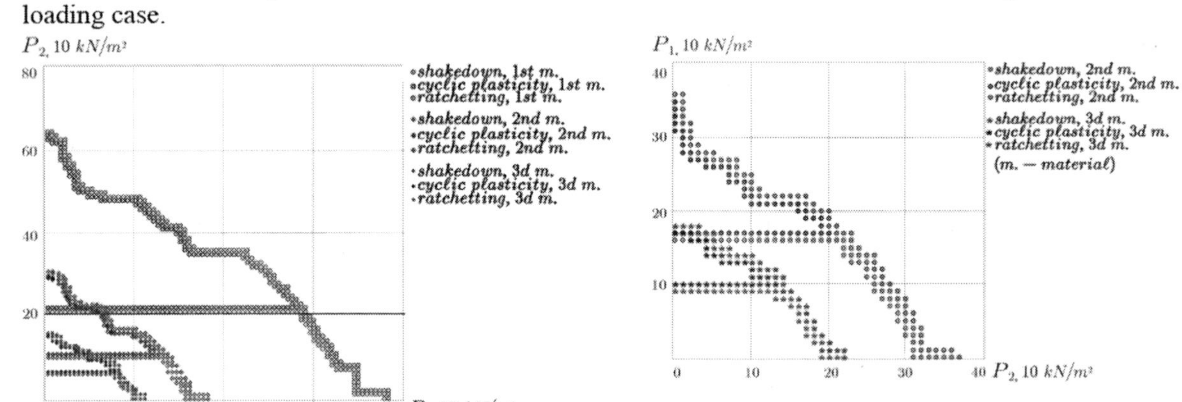

Figure 11. Results of numerical analysis: the diagram of loading domains for different asymptotic regimes of the plate: left – with elliptic hole, right – with circular hole.

Boundaries of the loading limits P_1 and $P_{2\max}$ for ratcheting are marked with red color, for cyclic plasticity – with blue, for shakedown – with green. From the diagram it is obvious that without cyclic loading ratcheting comes at the magnitude of the load which is equal to the material ultimate strength.

This diagram allows us to define safe magnitudes of loads for the structure and avoid dangerous (unsafe) regimes.

From figure 11 it is seen that for different materials limits of shakedown, cyclic plasticity and progressing plastic flow differ, but qualitative distribution of domains remains similar for all the materials.

Using this hypothesis, based on the supervision, we can introduce dimensionless parameters: $\pi_1 = P_1 / \sigma_B$, $\pi_2 = P_2 / \sigma_B$ and construct diagram P_1 and $P_{2\max}$ on the plane π_1, π_2 for all the three materials. Results are shown in the figure 12.

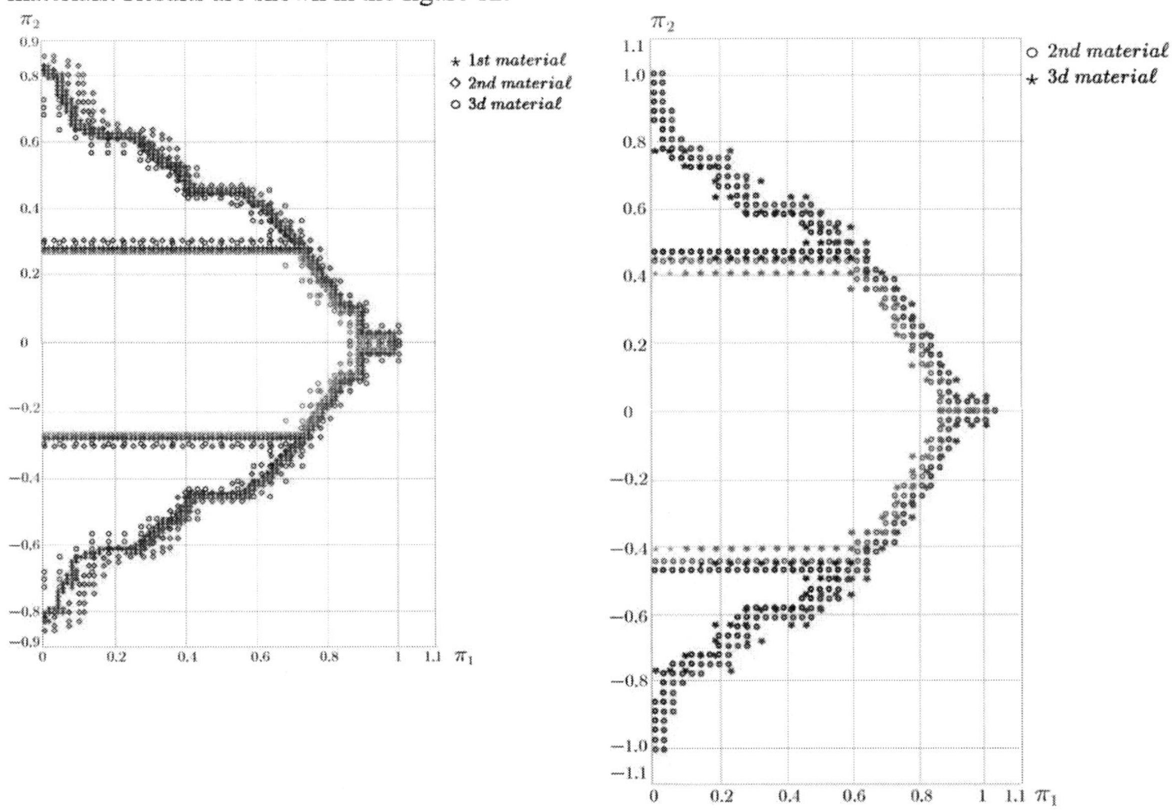

Figure 12. Results of numerical analysis: the generalized diagram of loading domains for different asymptotic regimes of the plate: left – with elliptic hole, right – with circular hole.

From figure 12 it is seen that all curves defining domain boundaries match and lay down on the uniform curve. On the plane π_1, π_2 all normalized calculated points from the numerical experiment lie on the uniform curve. The variables π_1, π_2 can be interpreted as similarity variables.

Thus, numerical analysis showed independence of characteristic domains from mechanical properties of materials. Therefore there is no need to calculate domains for every distinct material.

The chosen normalization allows us to present all computed charts for certain materials within one curve with the minimum point dispersion. The convenience of the generalized diagram is in

IOP Conf. Series: Journal of Physics: Conf. Series **973** (2018) 012050 doi:10.1088/1742-6596/973/1/012050

opportunity to know the asymptotic behavior of an inelastic structure for different materials without computing.

3. Conclusion

By means of varying k_1 and k_2 we can study the whole range of static and cyclic loads, applied to the structure. In the present paper large series of numerical computer modeling experiments for plate with central hole under cyclic loading (in the whole range values of coefficients k_1 and k_2) have carried out and interpreted. On the basis of the computer experiment ranges of loads for shakedown, cyclic plasticity and ratcheting have been defined.

Types of asymptotic behaviors of non-elastic structures are discovered. The typical diagrams shown in figure 12 are constructed; their main and intrinsic features are defined.

Borders of asymptotic behavior types of a certain structure subjected to identical structure loading, for different materials match and lay down on one curve.

For three materials with different mechanical features domains are similar. Thus a convenient normalization of computing results is offered. Experimental data are presented in a generalized diagram of structure's asymptotic behavior.

The obtained incremental analysis results may be useful for checking the data received with the direct methods. Direct methods give us an opportunity to decrease computing expenses while determining ranges of safe working regimes of test samples. This will allow us to organize design of structures in a new way.

Acknowledgements

Financial support from the Russian Foundation of Basic Research (project No. 16-08-00571) is gratefully acknowledged.

References

[1] Stolz C 2008 Optimal control approach in nonlinear mechanics *C. R. Mecanique* **1–2** 238–44

[2] Vetyukov Y 2014 *Nonlinear mechanics of thin-walled structures. Asymptotics, direct approach and numerical analysis* (Berlin: Springer)

[3] Peigney M 2014 Shakedown of elastic-perfectly plastic materials with temperature-dependent elastic moduli *J. of the Mechanics and Physics of Solids* **71** 112–31

[4] Spiliopoulos K V, Weichet D 2014 *Direct methods for limit states in structures and materials* (Berlin: Springer)

[5] Saxce G, Oueslati A, Charkaluk E, Tritsch J-B 2013 *Limit state of materials and structures* (Berlin: Springer)

[6] Fuschi P, Pisano A A, Weichert D 2015 *Direct Methods for Limit and Shakedown Analysis of Structures: Advanced Computational Algorithms and Material Modelling (Solid Mechanics and Its Applications)* (Berlin: Springer)

[7] Spiliopoulos K V, Panagiotou K D 2012 A direct method to predict cyclic steady state of elastoplastic structures *Comput. Methods Appl. Mech. Engrg.* **223–224** 186–98

[8] Campagnolo A, Berto F, Marangon C 2016 Cyclic plasticity in the three-dimensional notched components under in-phase multiaxial loading at R = 1*Theoretical and Appl. Fracture Mech.* **81** 76–88

[9] Benoit A, Maitoumam M H, Remy L, Oger F 2012 Cyclic behavior of structures under thermomechanical loadings: Application to exhaust manifolds *Int. J. of Fatigue* **38** 65–74

[10] Marques S P C, Creus G 2012 *J. Comp.Viscoelasticity* (Springer Briefs in Computational Mechanics)

[11] Turkova V A, Stepanova L V 2016 Finite element analysis of biaxial cyclic tensile loading of elasto-plastic plate with central elliptical hole *PNRPU Mechanics Bulletin* **3** 207–21

[12] Chen H F, Ponter A R S 2001 Shakedown and limit analyses for 3-D structures using the linear matching method *Int. J. of Pressure Vessels and Piping* **78** 443–51

[13] Ponter A R S, Chen H 2001 A minimum theorem for cyclic load in excess of shakedown, with application to the evaluation of a ratchet limit *Eur. J. Mech. A/Solids* **20** 539–53

[14] Du Z, Wang J, Fan X 2006 Direct cyclic method for solder joint reliability analysis *Proceedings of IMECE 2006: ASME International Mechanical Engineering Congress and Exposition* (USA, Chicago)

[15] Brassart L, Zhao K, Suo Zh 2013 Cyclic plasticity and shakedown in high-capacity electrodes of lithium-ion batteries *Int. J. of Solids and Structures* **50** 1120–29

[16] Spiliopoulos K V, Panagiotou K D 2014 A residual stress decomposition based method for the shakedown analysis of structures *Comput. Methods Appl. Mech. Engrg.* **276** 410–30

[17] Lee C-H, Do V N V, Chang K-H 2014 Analysis of uniaxial ratcheting behaviour and cyclic mean stress relaxation of a duplex stainless steel *Int. J. of Plasticity* **62** 17–33

[18] Mandal N K 2014 Ratcheting of railhead material of insulated rail joints (IRJs) with reference to endpost thickness *Engrg. Failure Analysis* **45** 347–62

[19] Tong J, Cornet C, Lin B, Lupton C, Li H-Y, Bowen P, Williams S, Hardy M 2016 Near-tip strain ratcheting and crack growth at elevated temperature *Int. J. of fatigue* **82** 514–20

[20] Da Costa Mattos H S, Peres J M A, Melo M A C 2015 Ratcheting behavior of elasto-plastic thin-walled pipes under internal pressure and subjected to cyclic axial loading *Thin-Walled Structures* **93** 102-11

[21] Halama R, Sedlak J, Fusek M, Poruba Z 2015 Uniaxial and biaxial ratcheting of ST52 steel under variable amplitude loading-experiments and modeling *Procedia Engrg* **101** 185–93

[22] Gustafson A, Moller M 2015 Experimental and numerical investigation of ratcheting in pressurized equipment *Procedia Engrg* **130** 1233–45

[23] Stepanova L V, Igonin S A 2014 Perturbation method for solving the nonlinear eigenvalue problem arising from fatigue crack growth problem in a damaged medium *Appl. Math. Modeling* **38** 3436–55

[24] Stepanova L V, Igonin S A 2015 Asymptotics of the near-crack-tip stress field of a growing fatigue crack in damaged materials: Numerical experiment and analytical solution *Numerical Analysis and Applications* **8(2)** 168–81

[25] Stepanova L, Roslyakov P 2016 Complete Williams asymptotic expansion of the stress field near the crack tip: Analytical solutions, interference-optic methods and numerical experiments *AIP Conf. Proceedings* **1785** 030029

[26] *ABAQUS Theory Manual* Information on the SIMULIA site: http://www.simulia.com/support/documentation.html

[27] Copper. Brass plate, bar copper, tape, wire. Available: http://polias.ru/index/0-16 .

[28] Bui H D 2006 *Fracture Mechanics: Inverse Problems and Solutions* (Berlin: Springer)

On flows of viscoelastic fluids under threshold-slip boundary conditions

E S Baranovskii

Department of Applied Mathematics, Informatics and Mechanics, Voronezh State University, Universitetskaya Pl. 1, 394018 Voronezh, Russia

E-mail: esbaranovskii@gmail.com

Abstract. We investigate a boundary-value problem for the steady isothermal flow of an incompressible viscoelastic fluid of Oldroyd type in a 3D bounded domain with impermeable walls. We use the Fujita threshold-slip boundary condition. This condition states that the fluid can slip along a solid surface when the shear stresses reach a certain critical value; otherwise the slipping velocity is zero. Assuming that the flow domain is not rotationally symmetric, we prove an existence theorem for the corresponding slip problem in the framework of weak solutions. The proof uses methods for solving variational inequalities with pseudo-monotone operators and convex functionals, the method of introduction of auxiliary viscosity, as well as a passage-to-limit procedure based on energy estimates of approximate solutions, Korn's inequality, and compactness arguments. Also, some properties and estimates of weak solutions are established.

1. Introduction

We consider the system of nonlinear partial differential equations governing steady-state flows of an incompressible viscoelastic fluid of Oldroyd type [1] in a bounded domain $\Omega \subset \mathbf{R}^3$ with sufficiently smooth boundary $\partial\Omega$ under the Fujita threshold-slip boundary condition [2]:

$$\rho(\mathbf{u} \cdot \nabla)\mathbf{u} - \operatorname{div} \mathbf{S} + \nabla \pi = \rho \mathbf{f} \quad \text{in } \Omega, \tag{1.1}$$

$$\mathbf{S} = \mathbf{E} + (1 - \omega)\mu \mathbf{D}(\mathbf{u}) \quad \text{in } \Omega, \tag{1.2}$$

$$\nabla \cdot \mathbf{u} = 0 \quad \text{in } \Omega, \tag{1.3}$$

$$\mathbf{E} + \lambda(\mathbf{u} \cdot \nabla)\mathbf{E} = \omega\mu \mathbf{D}(\mathbf{u}) \quad \text{in } \Omega, \tag{1.4}$$

$$\mathbf{u} \cdot \mathbf{n} = 0 \quad \text{on } \partial\Omega, \tag{1.5}$$

$$|(\mathbf{Sn})_{\tan}| \leq q \quad \text{on } \partial\Omega, \tag{1.6}$$

$$|(\mathbf{Sn})_{\tan}| < q \implies \mathbf{u}_{\tan} = \mathbf{0} \quad \text{on } \partial\Omega, \tag{1.7}$$

$$|(\mathbf{Sn})_{\tan}| = q \implies \mathbf{u}_{\tan} \upharpoonleft\downharpoonright (\mathbf{Sn})_{\tan} \quad \text{on } \partial\Omega, \tag{1.8}$$

where $\mathbf{u} = \mathbf{u}(\mathbf{x})$ is the velocity at the point $\mathbf{x} \in \Omega$, $\mathbf{S} = \mathbf{S}(\mathbf{x})$ is the extra-stress tensor, $\pi = \pi(\mathbf{x})$ is the pressure, $\mathbf{f} = \mathbf{f}(\mathbf{x})$ denotes the body force, $\mathbf{E} = \mathbf{E}(\mathbf{x})$ is the elastic part of the stress tensor, $\mathbf{D}(\mathbf{u})$ is the strain-rate tensor,

$$\mathbf{D}(\mathbf{u}) = \frac{1}{2}(\nabla \mathbf{u} + (\nabla \mathbf{u})^{\mathrm{T}}),$$

$\mu > 0$ is the viscosity coefficient, $\rho > 0$ is the density of the fluid, $\lambda > 0$ denotes the stress relaxation time, $\omega \in (0, 1)$ is a dimensionless parameter, $q = q(\mathbf{x}) \geq 0$ is the threshold of the tangential stress, $\mathbf{n} = \mathbf{n}(\mathbf{x})$ denotes the outward-directed unit normal vector to $\partial\Omega$, and $(\,\cdot\,)_{\tan}$ stands for the tangential component of a vector, i.e.,

$$\mathbf{u}_{\tan} = \mathbf{u} - (\mathbf{u} \cdot \mathbf{n})\mathbf{n}.$$

The symbol $\upharpoonleft\!\downharpoonright$ is used to denote oppositely directed vectors. In other words, for any $\mathbf{a}, \mathbf{b} \in \mathbf{R}^3$

$$\mathbf{a} \upharpoonleft\!\downharpoonright \mathbf{b} \iff \mathbf{a} \cdot \mathbf{b} + |\mathbf{a}||\mathbf{b}| = 0,$$

where $|\cdot|$ denotes the Euclidean norm.

The unknowns in system (1.1)–(1.8) are $\mathbf{u}$, $\mathbf{S}$, $\mathbf{E}$, and π, while all other quantities are assumed to be given.

In the present paper, we introduce the concept of weak solutions of problem (1.1)–(1.8) by the use of variational inequalities. Under appropriate conditions on the data, we prove a theorem on the existence and properties of weak solutions to (1.1)–(1.8).

The main novelty of our results is that the phenomenon of *threshold slip* at solid surfaces is taken into account. This study continues the series of articles written by the author [3–6], devoted to slip problems for flows of viscoelastic Oldroyd fluids.

Remark 1. Starting from pioneering works of M. Renardy [7] and C. Guillopé & J.-C. Saut [8], the equations of motion of Oldroyd-type fluids have been studied by many specialists (see, e.g., [9–19] and the references therein). A detailed analysis of mathematical results for the Oldroyd model and other similar non-Newtonian models can be found in the review article [20].

Remark 2. If we make $\lambda = \omega = 0$ formally, system (1.1)–(1.8) reduces to the classical Navier–Stokes equations with threshold-slip boundary conditions. For such type of slip problems, well-posedness results were obtained by H. Fujita [2]; in this regard, see also the papers [21, 22].

2. Notations

Denote by $\mathbf{R}^{3\times 3}_{\text{sym}}$ the space symmetric matrices of size 3×3 with the following scalar product:

$$\mathbf{A} : \mathbf{B} = \sum_{i,j=1}^{3} A_{ij} B_{ij}$$

for the matrices $\mathbf{A} = (A_{ij})$ and $\mathbf{B} = (B_{ij})$.

We use the standard notation

$$\mathbf{L}^p(\Omega, \mathbf{R}^d), \quad \mathbf{H}^m(\Omega, \mathbf{R}^d) = \mathbf{W}^{m,2}(\Omega, \mathbf{R}^d)$$

for the Lebesgue and Sobolev spaces of functions defined on Ω and with values in $\mathbf{R}^d$. For a detailed treatment of these spaces, consult, e.g., [23].

As is well known, it is possible to ascribe a value at the boundary (the trace) to vector functions from the space $\mathbf{H}^1(\Omega, \mathbf{R}^d)$. For that we use the trace operator [24]

$$\gamma_0 \colon \mathbf{H}^1(\Omega, \mathbf{R}^d) \to \mathbf{H}^{1/2}(\partial\Omega, \mathbf{R}^d)$$

such that $\gamma_0 \mathbf{w} = \mathbf{w}|_{\partial\Omega}$ when $\mathbf{w} : \bar{\Omega} \to \mathbf{R}^d$ is a smooth vector function. In the sequel, for $\mathbf{v} \in \mathbf{H}^1(\Omega, \mathbf{R}^d)$, we shall write simply $\mathbf{v}|_{\partial\Omega}$ instead of $\gamma_0 \mathbf{v}$.

Denote by $\mathbf{C}^\infty_0(\Omega, \mathbf{R}^{3\times 3}_{\text{sym}})$ the set of infinitely differentiable functions with support contained in Ω and with values in $\mathbf{R}^{3\times 3}_{\text{sym}}$.

Let us introduce the function spaces needed below:

$$\mathbf{H}_0^2(\Omega,\mathbf{R}^{3\times 3}_{\rm sym}) = \text{the closure of the set } \mathbf{C}_0^\infty(\Omega,\mathbf{R}^{3\times 3}_{\rm sym}) \text{ in the space } \mathbf{H}^2(\Omega,\mathbf{R}^{3\times 3}_{\rm sym}),$$

$$\mathbf{Q}(\Omega,\mathbf{R}^3) = \{\mathbf{v}\in \mathbf{C}^\infty(\bar\Omega,\mathbf{R}^3):\ \nabla\cdot\mathbf{v}=0,\ \ \mathbf{v}|_{\partial\Omega}\cdot\mathbf{n}=0\},$$

$$\mathbf{X}(\Omega,\mathbf{R}^3) = \text{the closure of the set } \mathbf{Q}(\Omega,\mathbf{R}^3) \text{ in the space } \mathbf{H}^1(\Omega,\mathbf{R}^3),$$

$$\mathbf{L}^2_+(\partial\Omega,\mathbf{R}) = \{v\in L^2(\partial\Omega,\mathbf{R}):\ \ v(\mathbf{x})\ge 0 \ \text{ for a.e. } \mathbf{x}\in\partial\Omega\}.$$

Assuming that $\Omega\subset\mathbf{R}^3$ is not a body of revolution, we define the norm in $\mathbf{X}(\Omega,\mathbf{R}^3)$ by the formula

$$\|\mathbf{v}\|^2_{\mathbf{X}(\Omega,\mathbf{R}^3)} = \int_\Omega |\mathbf{D}(\mathbf{v})|^2\,\mathbf{dx}.$$

It follows from Korn's inequality (see the Appendix, Proposition 2 (a)) that the norm $\|\cdot\|_{\mathbf{X}(\Omega,\mathbf{R}^3)}$ is equivalent to the norm induced from the Sobolev space $\mathbf{H}^1(\Omega,\mathbf{R}^3)$.

3. Weak (variational) formulation of problem (1.1)–(1.8)

Assume that the following conditions hold:

$$0<\lambda,\quad 0<\mu,\quad 0<\omega<1,\quad \mathbf{f}\in\mathbf{L}^2(\Omega,\mathbf{R}^3),\quad q\in\mathbf{L}^2_+(\partial\Omega,\mathbf{R}). \tag{3.1}$$

Definition. We say that a triplet

$$(\mathbf{u},\mathbf{S},\mathbf{E})\in\mathbf{X}(\Omega,\mathbf{R}^3)\times\mathbf{L}^2(\Omega,\mathbf{R}^{3\times 3}_{\rm sym})\times\mathbf{L}^2(\Omega,\mathbf{R}^{3\times 3}_{\rm sym})$$

is a *weak solution* of problem (1.1)–(1.8) if the condition (1.2) is valid and

$$\begin{aligned}
-\rho\sum_{i=1}^3\int_\Omega u_i\mathbf{u}\cdot\frac{\partial\mathbf{w}}{\partial x_i}\,\mathbf{dx} + \int_\Omega \mathbf{E}:\mathbf{D}(\mathbf{w})\,\mathbf{dx} - \frac{1}{\omega\mu}\int_\Omega|\mathbf{E}|^2\,\mathbf{dx}\\
+(1-\omega)\mu\int_\Omega \mathbf{D}(\mathbf{u}):\mathbf{D}(\mathbf{w}-\mathbf{u})\,\mathbf{dx} + \int_{\partial\Omega} q|\mathbf{w}|\,ds - \int_{\partial\Omega} q|\mathbf{u}|\,ds\\
\ge \rho\int_\Omega \mathbf{f}\cdot(\mathbf{w}-\mathbf{u})\,\mathbf{dx}\quad \forall\mathbf{w}\in\mathbf{X}(\Omega,\mathbf{R}^3),
\end{aligned}\tag{3.2}$$

$$\int_\Omega \mathbf{E}:\mathbf{F}\,\mathbf{dx} - \lambda\sum_{i=1}^3\int_\Omega u_i\mathbf{E}:\frac{\partial\mathbf{F}}{\partial x_i}\,\mathbf{dx} = \omega\mu\int_\Omega \mathbf{D}(\mathbf{u}):\mathbf{F}\,\mathbf{dx}\quad \forall\mathbf{F}\in\mathbf{H}_0^2(\Omega,\mathbf{R}^{3\times 3}_{\rm sym}). \tag{3.3}$$

Remark 3. Let us explain how relations (3.2) and (3.3) arise. Suppose that $(\mathbf{u},\mathbf{S},\mathbf{E},\pi)$ is a classical solution to problem (1.1)–(1.8) and $\mathbf{w}\in\mathbf{X}(\Omega,\mathbf{R}^3)$. If we take the scalar product of both sides of (1.1) by the vector function $\mathbf{w}-\mathbf{u}$ and integrate by parts over the domain Ω, then we obtain

$$-\rho\sum_{i=1}^3\int_\Omega u_i\mathbf{u}\cdot\frac{\partial\mathbf{w}}{\partial x_i}\,\mathbf{dx} + \int_\Omega \mathbf{S}:\mathbf{D}(\mathbf{w}-\mathbf{u})\,\mathbf{dx} - \int_{\partial\Omega}(\mathbf{Sn})\cdot(\mathbf{w}-\mathbf{u})\,ds = \rho\int_\Omega \mathbf{f}\cdot(\mathbf{w}-\mathbf{u})\,\mathbf{dx}.$$

Substituting (1.2) into the second term of the last equality, we get

$$-\rho\sum_{i=1}^{3}\int_{\Omega} u_i\mathbf{u}\cdot\frac{\partial\mathbf{w}}{\partial x_i}\,\mathbf{dx}+\int_{\Omega}\mathbf{E}:\mathbf{D}(\mathbf{w}-\mathbf{u})\,\mathbf{dx}+(1-\omega)\mu\int_{\Omega}\mathbf{D}(\mathbf{u}):\mathbf{D}(\mathbf{w}-\mathbf{u})\,\mathbf{dx}$$

$$-\int_{\partial\Omega}(\mathbf{Sn})\cdot(\mathbf{w}-\mathbf{u})\,ds=\rho\int_{\Omega}\mathbf{f}\cdot(\mathbf{w}-\mathbf{u})\,\mathbf{dx}. \tag{3.4}$$

Further, taking the scalar product of (1.4) with $\mathbf{E}$ and integrating over the domain Ω, we obtain

$$\int_{\Omega}|\mathbf{E}|^2\,\mathbf{dx}=\omega\mu\int_{\Omega}\mathbf{D}(\mathbf{u}):\mathbf{E}\,\mathbf{dx}, \tag{3.5}$$

where we used the following equality

$$\sum_{i=1}^{3}\int_{\Omega} u_i\frac{\partial\mathbf{E}}{\partial x_i}:\mathbf{E}\,\mathbf{dx}=0.$$

Combining (3.4) and (3.5), we find that

$$-\rho\sum_{i=1}^{3}\int_{\Omega} u_i\mathbf{u}\cdot\frac{\partial\mathbf{w}}{\partial x_i}\,\mathbf{dx}+\int_{\Omega}\mathbf{E}:\mathbf{D}(\mathbf{w})\,\mathbf{dx}-\frac{1}{\omega\mu}\int_{\Omega}|\mathbf{E}|^2\,\mathbf{dx}$$

$$+(1-\omega)\mu\int_{\Omega}\mathbf{D}(\mathbf{u}):\mathbf{D}(\mathbf{w}-\mathbf{u})\,\mathbf{dx}-\int_{\partial\Omega}(\mathbf{Sn})\cdot(\mathbf{w}-\mathbf{u})\,ds=\rho\int_{\Omega}\mathbf{f}\cdot(\mathbf{w}-\mathbf{u})\,\mathbf{dx}. \tag{3.6}$$

It can be shown that the system of boundary conditions (1.6)–(1.8) is equivalent to the following system:

$$|(\mathbf{Sn})_{\tan}|\le q \quad \text{on } \partial\Omega,$$

$$(\mathbf{Sn})_{\tan}\cdot\mathbf{u}_{\tan}+q|\mathbf{u}_{\tan}|=0 \quad \text{on } \partial\Omega.$$

Employing these relations and the Cauchy–Schwarz inequality, we deduce

$$-\int_{\partial\Omega}(\mathbf{Sn})\cdot(\mathbf{w}-\mathbf{u})\,ds=\int_{\partial\Omega}(\mathbf{Sn})_{\tan}\cdot(\mathbf{u}-\mathbf{w})_{\tan}\,ds$$

$$=-\int_{\partial\Omega}q|\mathbf{u}_{\tan}|+(\mathbf{Sn})_{\tan}\cdot\mathbf{w}_{\tan}\,ds\le-\int_{\partial\Omega}q|\mathbf{u}|\,ds+\int_{\partial\Omega}q|\mathbf{w}|\,ds.$$

If we combine this with (3.6), we obtain (3.2).

On taking the scalar product of both the left-hand and right-hand sides of (1.4) with a vector function $\mathbf{F}\in\mathbf{H}_0^2(\Omega,\mathbf{R}^{3\times3}_{\mathrm{sym}})$ and integrating over Ω, we get (3.3).

Finally, it should be also noted that if a weak solution $(\mathbf{u},\mathbf{S},\mathbf{E})$ of problem (1.1)–(1.8) is sufficiently regular, then there exists a function π such that $(\mathbf{u},\mathbf{S},\mathbf{E},\pi)$ is a classical solution to problem (1.1)–(1.8).

IOP Conf. Series: Journal of Physics: Conf. Series **973** (2018) 012051 doi:10.1088/1742-6596/973/1/012051

4. Main results

We are now in a position to state our main results.

Theorem. *Assume that* $\Omega \subset \mathbf{R}^3$ *is not a body of revolution and relations* (3.1) *hold. Under these conditions, we have*

(i) *boundary-value problem* (1.1)–(1.8) *has at least one weak solution* $(\mathbf{u}, \mathbf{S}, \mathbf{E})$ *such that*

$$(1-\omega)\mu \int_\Omega |\mathbf{D}(\mathbf{u})|^2 \,\mathbf{dx} + \frac{1}{\omega\mu}\int_\Omega |\mathbf{E}|^2 \,\mathbf{dx} + \int_{\partial\Omega} q|\mathbf{u}|\,ds \le \rho \int_\Omega \mathbf{f}\cdot\mathbf{u}\,\mathbf{dx}; \tag{4.1}$$

(ii) *the set of weak solutions to problem* (1.1)–(1.8) *is sequentially weakly closed in the space*

$$\mathbf{X}(\Omega, \mathbf{R}^3) \times \mathbf{L}^2(\Omega, \mathbf{R}^{3\times3}_{\mathrm{sym}}) \times \mathbf{L}^2(\Omega, \mathbf{R}^{3\times3}_{\mathrm{sym}});$$

(iii) *if* $(\mathbf{u}_0, \mathbf{S}_0, \mathbf{E}_0)$ *is a weak solution of problem* (1.1)–(1.8) *such that*

$$(\mathbf{u}_0, \mathbf{S}_0, \mathbf{E}_0) \in \mathbf{X}(\Omega, \mathbf{R}^3) \times \mathbf{L}^2(\Omega, \mathbf{R}^{3\times3}_{\mathrm{sym}}) \times \mathbf{H}^1(\Omega, \mathbf{R}^{3\times3}_{\mathrm{sym}}),$$

then

$$(1-\omega)\mu \int_\Omega |\mathbf{D}(\mathbf{u}_0)|^2 \,\mathbf{dx} + \frac{1}{\omega\mu}\int_\Omega |\mathbf{E}_0|^2 \,\mathbf{dx} + \int_{\partial\Omega} q|\mathbf{u}_0|\,ds = \rho \int_\Omega \mathbf{f}\cdot\mathbf{u}_0\,\mathbf{dx}; \tag{4.2}$$

(iv) *if there exists a potential* $h \in \mathbf{H}^1(\Omega, \mathbf{R})$ *such that* $\mathbf{f} = \nabla h$, *then* $(\mathbf{u}_0, \mathbf{S}_0, \mathbf{E}_0) = (\mathbf{0}, \mathbf{0}, \mathbf{0})$.

5. Sketch of the proof of the theorem

The proof of statement (i) is derived in four steps.

Step 1. For a fixed $\mathbf{u} \in \mathbf{X}(\Omega, \mathbf{R}^3)$, we consider the following linear problem depending on a positive parameter ε:

Find $\mathbf{E} \in \mathbf{H}^2_0(\Omega, \mathbf{R}^{3\times3}_{\mathrm{sym}})$ *such that*

$$\varepsilon \int_\Omega \Delta\mathbf{E} : \Delta\mathbf{F}\,\mathbf{dx} + \int_\Omega \mathbf{E} : \mathbf{F}\,\mathbf{dx} - \lambda \sum_{i=1}^{3} \int_\Omega u_i \mathbf{E} : \frac{\partial \mathbf{F}}{\partial x_i}\,\mathbf{dx}$$

$$= \omega\mu \int_\Omega \mathbf{D}(\mathbf{u}) : \mathbf{F}\,\mathbf{dx} \quad \forall \mathbf{F} \in \mathbf{H}^2_0(\Omega, \mathbf{R}^{3\times3}_{\mathrm{sym}}). \tag{5.1}$$

For any $\varepsilon > 0$ and $\mathbf{u} \in \mathbf{X}(\Omega, \mathbf{R}^3)$, problem (5.1) has a unique solution. Denote by $\mathbf{T}_\varepsilon(\mathbf{u})$ the solution of this problem. It can be proved that the operator

$$\mathbf{T}_\varepsilon \colon \mathbf{X}(\Omega, \mathbf{R}^3) \to \mathbf{H}^2_0(\Omega, \mathbf{R}^{3\times3}_{\mathrm{sym}})$$

is completely continuous.

Step 2. Let us consider one more auxiliary problem:

Find $\mathbf{u} \in \mathbf{X}(\Omega, \mathbf{R}^3)$ *such that*

$$-\rho \sum_{i=1}^{3} \int_\Omega u_i \mathbf{u} \cdot \frac{\partial \mathbf{w}}{\partial x_i}\,\mathbf{dx} + \int_\Omega \mathbf{T}_\varepsilon(\mathbf{u}) : \mathbf{D}(\mathbf{w}-\mathbf{u})\,\mathbf{dx} + (1-\omega)\mu \int_\Omega \mathbf{D}(\mathbf{u}) : \mathbf{D}(\mathbf{w}-\mathbf{u})\,\mathbf{dx}$$

$$+ \int_{\partial\Omega} q|\mathbf{w}|\,ds - \int_{\partial\Omega} q|\mathbf{u}|\,ds \ge \rho \int_\Omega \mathbf{f}\cdot(\mathbf{w}-\mathbf{u})\,\mathbf{dx} \quad \forall \mathbf{w} \in \mathbf{X}(\Omega, \mathbf{R}^3). \tag{5.2}$$

By $[\mathbf{X}(\Omega, \mathbf{R}^3)]^*$ denote the dual space of $\mathbf{X}(\Omega, \mathbf{R}^3)$ and introduce the following operators:

$$\mathbf{M}\colon \mathbf{X}(\Omega, \mathbf{R}^3) \to [\mathbf{X}(\Omega, \mathbf{R}^3)]^*,$$

$$\langle \mathbf{M}(\mathbf{u}), \mathbf{w} \rangle = (1-\omega)\mu \int_\Omega \mathbf{D}(\mathbf{u}) : \mathbf{D}(\mathbf{w})\, \mathbf{dx},$$

$$\mathbf{K}_\varepsilon\colon \mathbf{X}(\Omega, \mathbf{R}^3) \to [\mathbf{X}(\Omega, \mathbf{R}^3)]^*,$$

$$\langle \mathbf{K}_\varepsilon(\mathbf{u}), \mathbf{w} \rangle = -\rho \sum_{i=1}^{3} \int_\Omega u_i \mathbf{u} \cdot \frac{\partial \mathbf{w}}{\partial x_i}\, \mathbf{dx} + \int_\Omega \mathbf{T}_\varepsilon(\mathbf{u}) : \mathbf{D}(\mathbf{w})\, \mathbf{dx},$$

$$J\colon \mathbf{X}(\Omega, \mathbf{R}^3) \to \mathbf{R},$$

$$J(\mathbf{u}) = \int_{\partial\Omega} q|\mathbf{u}|\, ds.$$

In these notations, problem (5.2) can be written as the following variational inequality

$$\langle \mathbf{M}(\mathbf{u}) + \mathbf{K}_\varepsilon(\mathbf{u}) - \rho\mathbf{f}, \mathbf{w} - \mathbf{u} \rangle + J(\mathbf{w}) - J(\mathbf{u}) \geq 0 \quad \forall\, \mathbf{w} \in \mathbf{X}(\Omega, \mathbf{R}^3). \tag{5.3}$$

Observe that $\mathbf{M}$ is a monotone operator and

$$\langle \mathbf{M}(\mathbf{u}), \mathbf{u} \rangle = (1-\omega)\mu \|\mathbf{u}\|^2_{\mathbf{X}(\Omega,\mathbf{R}^3)}.$$

Moreover, it can be shown that the sum $\mathbf{M} + \mathbf{K}_\varepsilon$ is a pseudo-monotone operator and

$$\frac{\langle \mathbf{M}(\mathbf{u}) + \mathbf{K}_\varepsilon(\mathbf{u}), \mathbf{u} \rangle + J(\mathbf{u})}{\|\mathbf{u}\|_{\mathbf{X}(\Omega,\mathbf{R}^3)}} \to +\infty$$

as $\|\mathbf{u}\|_{\mathbf{X}(\Omega,\mathbf{R}^3)} \to +\infty$. Then from the existence results for variational inequalities with pseudo-monotone operators and convex functionals (see the Appendix, Proposition 1), it follows that for any $\varepsilon > 0$ inequality (5.3) has a solution $\mathbf{u}_\varepsilon \in \mathbf{X}(\Omega, \mathbf{R}^3)$.

Step 3. Let $\varepsilon_n > 0$ be a sequence such that $\varepsilon_n \to 0$ as $n \to \infty$. We denote by $\mathbf{u}_{\varepsilon_n}$ the solution of (5.3) corresponding to $\varepsilon = \varepsilon_n$. Set $\mathbf{E}_{\varepsilon_n} = \mathbf{T}_{\varepsilon_n}(\mathbf{u}_{\varepsilon_n})$. Obviously, we have

$$\varepsilon_n \int_\Omega \Delta\mathbf{E}_{\varepsilon_n} : \Delta\mathbf{F}\, \mathbf{dx} + \int_\Omega \mathbf{E}_{\varepsilon_n} : \mathbf{F}\, \mathbf{dx} - \lambda \sum_{i=1}^{3} \int_\Omega u_{\varepsilon_n i} \mathbf{E}_{\varepsilon_n} : \frac{\partial \mathbf{F}}{\partial x_i}\, \mathbf{dx}$$

$$= \omega\mu \int_\Omega \mathbf{D}(\mathbf{u}_{\varepsilon_n}) : \mathbf{F}\, \mathbf{dx} \quad \forall\, \mathbf{F} \in \mathbf{H}^2_0(\Omega, \mathbf{R}^{3\times 3}_{\mathrm{sym}}), \tag{5.4}$$

$$-\rho \sum_{i=1}^{3} \int_\Omega u_{\varepsilon_n i} \mathbf{u}_{\varepsilon_n} \cdot \frac{\partial \mathbf{w}}{\partial x_i}\, \mathbf{dx} + \int_\Omega \mathbf{E}_{\varepsilon_n} : \mathbf{D}(\mathbf{w} - \mathbf{u}_{\varepsilon_n})\, \mathbf{dx}$$

$$+(1-\omega)\mu \int_\Omega \mathbf{D}(\mathbf{u}_{\varepsilon_n}) : \mathbf{D}(\mathbf{w} - \mathbf{u}_{\varepsilon_n})\, \mathbf{dx} + \int_{\partial\Omega} q|\mathbf{w}|\, ds - \int_{\partial\Omega} q|\mathbf{u}_{\varepsilon_n}|\, ds$$

$$\geq \rho \int_\Omega \mathbf{f} \cdot (\mathbf{w} - \mathbf{u}_{\varepsilon_n})\, \mathbf{dx} \quad \forall\, \mathbf{w} \in \mathbf{X}(\Omega, \mathbf{R}^3). \tag{5.5}$$

Using (5.4) and (5.5), we establish estimates independent of ε_n:

$$\|\mathbf{u}_{\varepsilon_n}\|_{\mathbf{X}(\Omega,\mathbf{R}^3)} \leq \frac{\rho}{(1-\omega)\mu}\|\mathbf{f}\|_{[\mathbf{X}(\Omega,\mathbf{R}^3)]^*}, \tag{5.6}$$

$$\|\mathbf{E}_{\varepsilon_n}\|_{\mathbf{L}^2(\Omega,\mathbf{R}^{3\times 3}_{\mathrm{sym}})} \leq \rho\sqrt{\frac{\omega}{1-\omega}}\|\mathbf{f}\|_{[\mathbf{X}(\Omega,\mathbf{R}^3)]^*}. \tag{5.7}$$

Step 4. We want to pass to the limit $n \to \infty$ in (5.4) and (5.5). Taking into account the estimates (5.6) and (5.7), we can assume without loss of generality that

$$\lim_{n\to\infty} \mathbf{u}_{\varepsilon_n} = \mathbf{u}_* \quad \text{weakly in } \mathbf{X}(\Omega,\mathbf{R}^3), \tag{5.8}$$

$$\lim_{n\to\infty} \mathbf{E}_{\varepsilon_n} = \mathbf{E}_* \quad \text{weakly in } \mathbf{L}^2(\Omega,\mathbf{R}^{3\times 3}_{\mathrm{sym}}) \tag{5.9}$$

for some $\mathbf{u}_* \in \mathbf{X}(\Omega,\mathbf{R}^3)$ and $\mathbf{E}_* \in \mathbf{L}^2(\Omega,\mathbf{R}^{3\times 3}_{\mathrm{sym}})$.

The weak convergence (5.8) and the compactness of the embeddings

$$\mathbf{i}\colon \mathbf{X}(\Omega,\mathbf{R}^3) \to \mathbf{L}^4(\Omega,\mathbf{R}^3), \quad \gamma_0\colon \mathbf{X}(\Omega,\mathbf{R}^3) \to \mathbf{L}^2(\partial\Omega,\mathbf{R}^3)$$

imply that

$$\lim_{n\to\infty} \mathbf{u}_{\varepsilon_n} = \mathbf{u}_* \quad \text{strongly in } \mathbf{L}^4(\Omega,\mathbf{R}^3), \tag{5.10}$$

$$\lim_{n\to\infty} \mathbf{u}_{\varepsilon_n}|_{\partial\Omega} = \mathbf{u}_*|_{\partial\Omega} \quad \text{strongly in } \mathbf{L}^2(\partial\Omega,\mathbf{R}^3). \tag{5.11}$$

Putting $\mathbf{F} = \mathbf{E}_{\varepsilon_n}$ in (5.4), we see that

$$\int_\Omega \mathbf{D}(\mathbf{u}_{\varepsilon_n}) : \mathbf{E}_{\varepsilon_n}\,\mathbf{dx} \geq \frac{1}{\omega\mu}\int_\Omega |\mathbf{E}_{\varepsilon_n}|^2\,\mathbf{dx}.$$

If we combine this inequality with (5.5), we obtain

$$-\rho\sum_{i=1}^{3}\int_\Omega u_{\varepsilon_n i}\mathbf{u}_{\varepsilon_n}\cdot\frac{\partial\mathbf{w}}{\partial x_i}\,\mathbf{dx} + \int_\Omega \mathbf{E}_{\varepsilon_n} : \mathbf{D}(\mathbf{w})\,\mathbf{dx}$$
$$-\frac{1}{\omega\mu}\int_\Omega |\mathbf{E}_{\varepsilon_n}|^2\,\mathbf{dx} + (1-\omega)\mu\int_\Omega \mathbf{D}(\mathbf{u}_{\varepsilon_n}) : \mathbf{D}(\mathbf{w}-\mathbf{u}_{\varepsilon_n})\,\mathbf{dx}$$
$$+\int_{\partial\Omega} q|\mathbf{w}|\,ds - \int_{\partial\Omega} q|\mathbf{u}_{\varepsilon_n}|\,ds \geq \rho\int_\Omega \mathbf{f}\cdot(\mathbf{w}-\mathbf{u}_{\varepsilon_n})\,\mathbf{dx} \quad \forall\,\mathbf{w}\in\mathbf{X}(\Omega). \tag{5.12}$$

Using (5.8)–(5.11) and the following inequalities

$$\int_\Omega |\mathbf{D}(\mathbf{u}_*)|^2\,\mathbf{dx} \leq \liminf_{n\to\infty}\int_\Omega |\mathbf{D}(\mathbf{u}_{\varepsilon_n})|^2\,\mathbf{dx}, \tag{5.13}$$

$$\int_\Omega |\mathbf{E}_*|^2\,\mathbf{dx} \leq \liminf_{n\to\infty}\int_\Omega |\mathbf{E}_{\varepsilon_n}|^2\,\mathbf{dx}, \tag{5.14}$$

we pass to the lower limit in (5.12). The result is

$$-\rho\sum_{i=1}^{3}\int_\Omega u_{*i}\mathbf{u}_*\cdot\frac{\partial\mathbf{w}}{\partial x_i}\,\mathbf{dx} + \int_\Omega \mathbf{E}_* : \mathbf{D}(\mathbf{w})\,\mathbf{dx}$$

$$-\frac{1}{\omega\mu}\int_\Omega |\mathbf{E}_*|^2\,\mathbf{dx} + (1-\omega)\mu\int_\Omega \mathbf{D}(\mathbf{u}_*) : \mathbf{D}(\mathbf{w}-\mathbf{u}_*)\,\mathbf{dx}$$

$$+\int_{\partial\Omega} q|\mathbf{w}|\,ds - \int_{\partial\Omega} q|\mathbf{u}_*|\,ds \geq \rho\int_\Omega \mathbf{f}\cdot(\mathbf{w}-\mathbf{u}_*)\,\mathbf{dx} \quad \forall\,\mathbf{w}\in\mathbf{X}(\Omega,\mathbf{R}^3). \tag{5.15}$$

Further, fix an arbitrary vector function $\mathbf{\Phi}$ from the space $\mathbf{C}_0^\infty(\Omega,\mathbf{R}^{3\times 3}_{\mathrm{sym}})$. Putting $\mathbf{F}=\mathbf{\Phi}$ in (5.4) and integrating by parts the first term, we obtain

$$\varepsilon_n\int_\Omega \mathbf{E}_{\varepsilon_n} : \Delta(\Delta\mathbf{\Phi})\,\mathbf{dx} + \int_\Omega \mathbf{E}_{\varepsilon_n} : \mathbf{\Phi}\,\mathbf{dx} - \lambda\sum_{i=1}^{3}\int_\Omega u_{\varepsilon_n i}\mathbf{E}_{\varepsilon_n} : \frac{\partial\mathbf{\Phi}}{\partial x_i}\,\mathbf{dx} = \omega\mu\int_\Omega \mathbf{D}(\mathbf{u}_{\varepsilon_n}) : \mathbf{\Phi}\,\mathbf{dx}.$$

Taking into account (5.8)–(5.10), we pass to the limit $n\to\infty$ in the last equality and obtain

$$\int_\Omega \mathbf{E}_* : \mathbf{\Phi}\,\mathbf{dx} - \lambda\sum_{i=1}^{3}\int_\Omega u_{*i}\mathbf{E}_* : \frac{\partial\mathbf{\Phi}}{\partial x_i}\,\mathbf{dx} = \omega\mu\int_\Omega \mathbf{D}(\mathbf{u}_*) : \mathbf{\Phi}\,\mathbf{dx}.$$

Since the set $\mathbf{C}_0^\infty(\Omega,\mathbf{R}^{3\times 3}_{\mathrm{sym}})$ is dense in the space $\mathbf{H}_0^2(\Omega,\mathbf{R}^{3\times 3}_{\mathrm{sym}})$, the last equality remains valid if we replace $\mathbf{\Phi}$ with an arbitrary vector function $\mathbf{F}\in\mathbf{H}_0^2(\Omega,\mathbf{R}^{3\times 3}_{\mathrm{sym}})$. Thus, we have

$$\int_\Omega \mathbf{E}_* : \mathbf{F}\,\mathbf{dx} - \lambda\sum_{i=1}^{3}\int_\Omega u_{*i}\mathbf{E}_* : \frac{\partial\mathbf{F}}{\partial x_i}\,\mathbf{dx} = \omega\mu\int_\Omega \mathbf{D}(\mathbf{u}_*) : \mathbf{F}\,\mathbf{dx} \quad \forall\,\mathbf{F}\in\mathbf{H}_0^2(\Omega,\mathbf{R}^{3\times 3}_{\mathrm{sym}}). \tag{5.16}$$

Let us define $\mathbf{S}_*$ by the formula

$$\mathbf{S}_* = \mathbf{E}_* + (1-\omega)\mu\mathbf{D}(\mathbf{u}_*). \tag{5.17}$$

Then relations (5.15) and (5.16) together with (5.17) mean that the triplet $(\mathbf{u}_*,\mathbf{S}_*,\mathbf{E}_*)$ is a weak solution to problem (1.1)–(1.8).

In addition, setting $\mathbf{w}=\mathbf{0}$, from (5.12) we derive

$$(1-\omega)\mu\int_\Omega |\mathbf{D}(\mathbf{u}_{\varepsilon_n})|^2\,\mathbf{dx} + \frac{1}{\omega\mu}\int_\Omega |\mathbf{E}_{\varepsilon_n}|^2\,\mathbf{dx} + \int_{\partial\Omega} q|\mathbf{u}_{\varepsilon_n}|\,ds \leq \rho\int_\Omega \mathbf{f}\cdot\mathbf{u}_{\varepsilon_n}\,\mathbf{dx}.$$

Employing (5.10), (5.11), (5.13), and (5.14), we can pass to the lower limit in the last inequality as $n\to\infty$ and obtain the energy estimate (4.1) with $\mathbf{u}=\mathbf{u}_*$ and $\mathbf{E}=\mathbf{E}_*$. Thus, statement (i) is proved.

Further, using the passage-to-limit procedure as above, we establish that the set of weak solutions to problem (1.1)–(1.8) is sequentially weakly closed in the space

$$\mathbf{X}(\Omega,\mathbf{R}^3)\times\mathbf{L}^2(\Omega,\mathbf{R}^{3\times 3}_{\mathrm{sym}})\times\mathbf{L}^2(\Omega,\mathbf{R}^{3\times 3}_{\mathrm{sym}}).$$

Now we show that equality (4.2) holds. By definition, we have

$$\int_\Omega \mathbf{E}_0 : \mathbf{F}\,\mathbf{dx} - \lambda\sum_{i=1}^{3}\int_\Omega u_{0i}\mathbf{E}_0 : \frac{\partial\mathbf{F}}{\partial x_i}\,\mathbf{dx} = \omega\mu\int_\Omega \mathbf{D}(\mathbf{u}_0) : \mathbf{F}\,\mathbf{dx}$$

for any $\mathbf{F}\in\mathbf{H}_0^2(\Omega,\mathbf{R}^{3\times 3}_{\mathrm{sym}})$. After integrating the second term by parts, we get

$$\int_\Omega \mathbf{E}_0 : \mathbf{F}\,\mathbf{dx} + \lambda\sum_{i=1}^{3}\int_\Omega u_{0i}\frac{\partial\mathbf{E}_0}{\partial x_i} : \mathbf{F}\,\mathbf{dx} = \omega\mu\int_\Omega \mathbf{D}(\mathbf{u}_0) : \mathbf{F}\,\mathbf{dx}.$$

Since $\mathbf{H}_0^2(\Omega, \mathbf{R}_{\text{sym}}^{3\times3})$ is dense in the space $\mathbf{L}^2(\Omega, \mathbf{R}_{\text{sym}}^{3\times3})$, it follows that the last equality is valid for any vector function $\mathbf{F} \in \mathbf{L}^2(\Omega, \mathbf{R}_{\text{sym}}^{3\times3})$. Setting $\mathbf{F} = \mathbf{E}_0$, we find

$$\int_\Omega |\mathbf{E}_0|^2 \, \mathbf{dx} = \omega\mu \int_\Omega \mathbf{D}(\mathbf{u}_0) : \mathbf{E}_0 \, \mathbf{dx}. \tag{5.18}$$

Putting $\mathbf{u} = \mathbf{u}_0$, $\mathbf{E} = \mathbf{E}_0$, $\mathbf{w} = 2\mathbf{u}$ in the inequality (3.2) and taking into account (5.18), we derive the following inequality

$$(1-\omega)\mu \int_\Omega |\mathbf{D}(\mathbf{u}_0)|^2 \, \mathbf{dx} + \frac{1}{\omega\mu}\int_\Omega |\mathbf{E}_0|^2 \, \mathbf{dx} + \int_{\partial\Omega} q|\mathbf{u}_0| \, ds \geq \rho \int_\Omega \mathbf{f}\cdot\mathbf{u}_0 \, \mathbf{dx}. \tag{5.19}$$

On the other hand, substituting $\mathbf{u} = \mathbf{u}_0$, $\mathbf{E} = \mathbf{E}_0$, $\mathbf{w} = \mathbf{0}$ in (3.2), we obtain

$$(1-\omega)\mu \int_\Omega |\mathbf{D}(\mathbf{u}_0)|^2 \, \mathbf{dx} + \frac{1}{\omega\mu}\int_\Omega |\mathbf{E}_0|^2 \, \mathbf{dx} + \int_{\partial\Omega} q|\mathbf{u}_0| \, ds \leq \rho \int_\Omega \mathbf{f}\cdot\mathbf{u}_0 \, \mathbf{dx}.$$

This inequality together with (5.19) give (4.2).

Finally, if there exists a function $h \in \mathbf{H}^1(\Omega, \mathbf{R})$ such that $\mathbf{f} = \nabla h$, then

$$\int_\Omega \mathbf{f}\cdot\mathbf{u}_0 \, \mathbf{dx} = \int_\Omega \nabla h \cdot \mathbf{u}_0 \, \mathbf{dx} = \int_{\partial\Omega} h\,\mathbf{u}_0\cdot\mathbf{n}\, ds - \int_\Omega h \operatorname{div}\mathbf{u}_0 \, \mathbf{dx} = 0.$$

Combining this with (4.2) and applying Korn's inequality, we get $(\mathbf{u}_0, \mathbf{S}_0, \mathbf{E}_0) = (\mathbf{0}, \mathbf{0}, \mathbf{0})$.

The theorem is completely proved.

6. Concluding remarks

In this paper, we studied the threshold-slip problem for an incompressible viscoelastic fluid of Oldroyd type, assuming that the flow domain Ω is not rotationally symmetric. This condition plays an important role in the proof of the existence of weak solutions, because for such domains Korn's inequality ensures the coercivity of the relevant operators. Using Proposition 2 (b) (see the Appendix), one can obtain similar results for the case of a flow in a hollow body of revolution under the additional boundary condition $\mathbf{u}|_\Sigma = \mathbf{0}$, where Σ is a part of $\partial\Omega$.

7. Appendix

For the convenience of readers, we provide two statements used in Section 5.

Proposition 1 (see [25]). *Let* $\mathbf{V}$ *be a reflexive Banach space,* $\mathbf{V}^*$ *its the dual space,* $\mathbf{A}\colon \mathbf{V} \to \mathbf{V}^*$ *a pseudo-monotone operator, and* $J\colon \mathbf{V} \to \mathbf{R}$ *a lower semi-continuous convex functional. Suppose also that*

$$\frac{\langle \mathbf{A}(\mathbf{v}), \mathbf{v}\rangle + J(\mathbf{v})}{\|\mathbf{v}\|_\mathbf{V}} \to +\infty$$

as $\|\mathbf{v}\|_\mathbf{V} \to +\infty$. *Then, for an arbitrary* $\mathbf{z} \in \mathbf{V}^*$, *there exists an element* $\mathbf{u}_z \in \mathbf{V}$ *such that*

$$\langle \mathbf{A}(\mathbf{u}_z) - \mathbf{z}, \mathbf{w} - \mathbf{u}_z\rangle + J(\mathbf{w}) - J(\mathbf{u}_z) \geq 0 \qquad \forall \mathbf{w} \in \mathbf{V}.$$

Proposition 2 (Korn's inequalities; see, e.g., [26]). *Assume that* Ω *is a bounded domain in space* $\mathbf{R}^3$ *and* $\partial\Omega \in C^{0,1}$. *Under these conditions, we have*

(a) *if Ω is not a body of revolution, then there is a constant $C_1 = C_1(\Omega) > 0$ such that*

$$\|\mathbf{D}(\mathbf{v})\|^2_{\mathbf{L}^2(\Omega,\mathbf{R}^{3\times 3}_{\mathrm{sym}})} \geq C_1(\Omega)\|\mathbf{v}\|^2_{\mathbf{H}^1(\Omega,\mathbf{R}^3)} \quad \textit{for all} \quad \mathbf{v} \in \mathbf{H}^1(\Omega,\mathbf{R}^3) \quad \textit{such that} \quad \mathbf{v}|_{\partial\Omega} \cdot \mathbf{n} = 0;$$

(b) *if $\Sigma \subset \partial\Omega$ and the surface measure of Σ is positive, then there is a constant $C_2 = C_2(\Omega) > 0$ such that*

$$\|\mathbf{D}(\mathbf{v})\|^2_{\mathbf{L}^2(\Omega,\mathbf{R}^{3\times 3}_{\mathrm{sym}})} \geq C_2(\Omega)\|\mathbf{v}\|^2_{\mathbf{H}^1(\Omega,\mathbf{R}^3)} \quad \textit{for all} \quad \mathbf{v} \in \mathbf{H}^1(\Omega,\mathbf{R}^3) \quad \textit{such that} \quad \mathbf{v}|_{\Sigma} = \mathbf{0}.$$

Acknowledgments

This work was supported by the Russian Foundation for Basic Research, project no. 16–31–00182 mol_a.

References

[1] Oldroyd J G 1958 Non-Newtonian effects in steady motion of some idealized elastico-viscous liquids *Proceedings of the Royal Society* A **245** 278–297
[2] Fujita H 1994 A mathematical analysis of motions of viscous incompressible fluid under leak or slip boundary conditions *RIMS Kokyuroku* **888** 199–216
[3] Baranovskii E S 2014 On steady motion of viscoelastic fluid of Oldroyd type *Sbornik: Mathematics* **205** 763–776
[4] Baranovskii E S 2014 Optimal control for steady flows of the Jeffreys fluids with slip boundary condition *Journal of Applied and Industrial Mathematics* **8** 168–176
[5] Baranovskii E S 2017 On weak solutions to evolution equations of viscoelastic fluid flows *Journal of Applied and Industrial Mathematics* **11** 174–184
[6] Baranovskii E S and Artemov M A 2017 Global existence results for Oldroyd fluids with wall slip *Acta Applicandae Mathematicae* **147** 197–210
[7] Renardy M 1985 Existence of slow steady flows of viscoelastic fluids with differential constitutive equations *Zeitschrift für Angewandte Mathematik und Mechanik* **65** 449–451
[8] Guillopé C and Saut J-C 1990 Existence results for the flow of viscoelastic fluids with a differential constitutive law *Nonlinear Analysis* **15** 849–869
[9] Turganbaev E M 1995 Initial-boundary value problems for the equations of a viscoelastic fluid of Oldroyd type *Siberian Mathematical Journal* **36** 389–403
[10] Agranovich Yu Ya and Sobolevskii P E 1998 Motion of nonlinear visco-elastic fluid *Nonlinear Analysis* **32** 755–760
[11] Fernández-Cara E, Guillén F and Ortega R 1998 Some theoretical results concerning non Newtonian fluids of the Oldroyd kind *Annali della Scuola Normale Superiore di Pisa* **26** 1–29
[12] Lions P L and Masmoudi N 2000 Global solutions for some Oldroyd models of non-Newtonian flows *Chinese Annals of Mathematics, Series* B **21** 131–146.
[13] Vorotnikov D A 2004 On the existence of weak stationary solutions of a boundary value problem in the Jeffreys model of the motion of a viscoelastic medium *Russian Mathematics* **48** (2004) 10–14
[14] Le Roux C 2014 On flows of viscoelastic fluids of Oldroyd type with wall slip *Journal of Mathematical Fluid Mechanics* **16** 335–350
[15] Bejaoui O and Majdoub M 2013 Global weak solutions for some Oldroyd models *Journal of Differential Equations* **254** 660–685
[16] Elgindia T M and Liu J 2015 Global wellposedness to the generalized Oldroyd type models in R^3 *Journal of Differential Equations* **259** 1958–1966
[17] Fang D and Zi R 2016 Global solutions to the Oldroyd-B model with a class of large initial data *SIAM Journal on Mathematical Analysis* **48** 1054–1084
[18] Ye Z and Xu X 2016 Global regularity for the 2D Oldroyd-B model in the corotational case *Mathematical Methods in the Applied Sciences* **39** 3866–3879
[19] Maryani S 2016 Global well-posedness for free boundary problem of the Oldroyd-B model fluid flow *Mathematical Methods in the Applied Sciences* **39** 2202–2219
[20] Saut J-C 2013 Lectures on the mathematical theory of viscoelastic fluids *Lectures on the Analysis of Nonlinear Partial Differential Equations* MLM **3** 325–393
[21] Le Roux C and Tani A 2007 Steady solutions of the Navier–Stokes equations with threshold slip boundary conditions *Mathematical Methods in the Applied Sciences* **30** 595–624

IOP Conf. Series: Journal of Physics: Conf. Series **973** (2018) 012051 doi:10.1088/1742-6596/973/1/012051

[22] Kashiwabara T 2013 On a strong solution of the non-stationary Navier–Stokes equations under slip or leak boundary conditions of friction type *Journal of Differential Equations* **254** 756–778
[23] Adams R A and Fournier J J F 2003 *Sobolev Spaces* Vol. 40 of Pure and Applied Mathematics (Amsterdam: Academic Press)
[24] Lions J-L and Magenes E 1972 *Nonhomogeneous boundary value problems and applications* (Berlin, New York: Springer–Verlag)
[25] Brezis H 1968 Equations et inéquations non linéaires dans les espaces en dualité *Annale de l'Institut Fourier* **18** 115–175
[26] Oden J T and Kikuchi N 1988 *Contact Problems in Elasticity* SIAM Studies in Mathematics (Philadelphia: SIAM)

IOP Conf. Series: Journal of Physics: Conf. Series **973** (2018) 012052 doi:10.1088/1742-6596/973/1/012052

On some dynamical chameleon systems

I M Burkin and O I Kuznetsova

Department of Computational Mechanics and Mathematics, Tula State University, Tula, 92 Lenin Avenue, 300012 Tula, Russia

E-mail: i-burkin@yandex.ru

Abstract. It is now well known that dynamical systems can be categorized into systems with self-excited attractors and systems with hidden attractors. A self-excited attractor has a basin of attraction that is associated with an unstable equilibrium, while a hidden attractor has a basin of attraction that does not intersect with small neighborhoods of any equilibrium points. Hidden attractors play the important role in engineering applications because they allow unexpected and potentially disastrous responses to perturbations in a structure like a bridge or an airplane wing. In addition, complex behaviors of chaotic systems have been applied in various areas from image watermarking, audio encryption scheme, asymmetric color pathological image encryption, chaotic masking communication to random number generator. Recently, researchers have discovered the so-called "chameleon systems". These systems were so named because they demonstrate self-excited or hidden oscillations depending on the value of parameters. The present paper offers a simple algorithm of synthesizing one-parameter chameleon systems. The authors trace the evolution of Lyapunov exponents and the Kaplan-Yorke dimension of such systems which occur when parameters change.

1. Introduction

In general, an oscillation dynamical system can be easily localized numerically if the initial data from its open neighborhood in the phase space (with the exception of a minor set of points) lead to a long-term behavior that approaches the oscillation. Such oscillation (or a set of oscillations) is called an attractor, and its attracting set is called the basin of attraction (i.e., a set of initial data for which the trajectories tend to the attractor). At early stages of the study of real dynamic systems, their structure was, as a rule, so simple, that it was obvious that all trajectories could not tend to infinity and that they could only be excited from neighborhoods of unstable equilibria [1-3]. This meant that scientists of that time could compute such attractors by "starting" computational procedure from a small neighborhood of the equilibrium, observing how it is attracted and, thus, visualizes the attractor. Attractors, whose attracting set contains an arbitrarily small neighborhood of equilibrium, are called "self-excited" attractors. Nowadays, thousands of publications are devoted to the computation and analysis of self-excited chaotic oscillations of dynamical systems.

A further study showed that the self-excited periodic and chaotic oscillations did not give exhaustive information about the possible types of oscillations in dynamical systems. Oscillations of other type which in paper [4] have received the name "hidden oscillations" and "hidden attractors" have been found. This name was suggested for attractors whose basin of attraction does not intersect with small neighborhoods of equilibria. Numerical localization, computation, and analytical investigation of hidden attractors are much more challenging problems, since here there is no possibility to use information about equilibria for organization of similar transient processes in the standard computational procedure. Furthermore, in this case it is unlikely that the integration of trajectories with random initial data furnishes hidden attractor localization since a basin of attraction can be very small and the dimension of hidden attractor itself can be much less than the dimension of the considered system.

Systems containing hidden attractors are multistable and therefore it is a complex task to control their functioning [5, 6]. The latter circumstance has prompted many researchers to turn to the search for systems that have hidden attractors and to study the features of their dynamics. Studied were, for example, systems with no equilibrium points [7-9], with a single equilibrium point [10, 11] and with an infinite number of equilibria [12, 13]. It is necessary to note that researches [7-13], as well as many later studies, apply a "regular computer search" (i.e., as a matter of fact, a regular "scanning" of the phase space of the investigated system) for detection of hidden attractors. In research [4], another method, based on the use of the method of harmonic linearization, the method of a small parameter and the method of describing functions in combination with the idea of homotopy, is utilized to find hidden oscillations in multidimensional dynamical systems.

In a recent research [14], a hyper-chaotic four-dimensional system possessing simultaneously self-excited and hidden attractors has been found out. Motivated by this research, the authors of paper [15] introduced a new concept of "chameleon systems". These systems were so named because they demonstrate self-excited or hidden oscillations depending on the value of parameters. The system examined in this research is:

$$\dot{x} = y + b$$
$$\dot{y} = -7x + 4yz$$
$$\dot{z} = a - x^2 - y^2 + cxz$$

With the fixed values of parameters $b = 0.8$, $\mathrm{c} = 0.01$ and parameter $a > 0$ that changes in some range, this system has a self-excited attractor. With values of $b = 0$, $\mathrm{c} = 0.01$, the system also has an attractor, which is hidden, because the system has no equilibrium states with such values of parameters.

The goal of this paper is to describe a new method for synthesizing the simplest one-parameter chameleon systems. This method uses the idea of homotopy, as well as methods for finding hidden attractors in multidimensional systems, developed in [16, 17]. It will be demonstrated that it is not only the type of attractors but also Lyapunov exponents and the Kaplan-Yorke dimension that vary when the parameter in such systems get changed.

2. Models and methods

The main idea of using the homotopy method for constructing chameleon systems is as follows.

Let us consider the one-parameter family of systems:

$$\dot{x} = f(x, \varepsilon),\ \varepsilon \in [0,1],\ x \in R^n, \tag{1}$$

such that for small values of $\varepsilon > 0$, the system (1) has a ready-to-detect self-excited orbitally asymptotically stable cycle or self-excited chaotic attractor. The evolution of that cycle or attractor as ε grows up to 1 is traced. A bifurcation occurs at $\varepsilon \in (0,1)$, and the type of the attractor of the system changes. For example, some equilibrium points of the system become stable in the small at $\varepsilon = \varepsilon_0$ when their neighborhood ceases to excite an attractor. In this case, a bifurcation occurs. A self-excited cycle or a self-excited attractor disappears and a hidden cycle or a hidden chaotic attractor is born. At $\varepsilon = 1$, all the attractors of the system are hidden.

Obviously, the construction of the function $f(x, \varepsilon)$ with the above-mentioned properties is a key point in the represented algorithm. One of the methods of construction of a function with necessary properties for a class of systems of the form is:

$$\dot{y} = Ay + bf(\sigma, \varepsilon), \quad \sigma = c^T y, \tag{2}$$

where A is a real $n \times n$ matrix, b and c are n-vectors, and $f(\sigma, \varepsilon)$ is a scalar nonlinearity offered in [16, 17].

An important moment in the implementation of these methods is partition of the plane (σ, f) into sectors of linear stability and instability of some degree k of the system (2). For this purpose, we introduce the transfer function $\chi(p) = c^T (A - pI)^{-1} b$ of the system (2), where p is a complex variable. This function is assumed to be nondegenerate, i.e. it cannot be represented in the form of ratio of polynomials with degree of the denominator less than n.

Now we put in (2) $f(\sigma, \varepsilon) = \mu\sigma$ i.e. we consider the linear system

$$\dot{y} = (A + \mu bc^T)y. \tag{3}$$

We will suppose that number μ is chosen so that the system (3) has no periodic solution, i.e. the matrix $A + \mu bc^T$ has no pure imaginary eigenvalues. This assumption is equivalent to the requirement $1 + \mu\chi(i\omega) \neq 0$ for all $\omega \geq 0$. Since under continuous change of μ the spectrum of the matrix $A + \mu bc^T$ is continuously changing, then on the (σ, f)-plane we can construct a sector $S[\nu_1, \nu_2] = \left\{\sigma, f : \nu_1 \leq f/\sigma \leq \nu_2\right\}$ such that all linear systems with $f = \mu\sigma$, $\nu_1 \leq \mu \leq \nu_2$ have the same number of eigenvalues in the ring-hand half-plane and do not have them on the imaginary axis. Extending the sector $S[\nu_1, \nu_2]$, we arrive at values $\nu_1 = \mu_j$, $\nu_2 = \mu_{j+1}$ such that

$$1 + \mu_j \chi(i\omega_j) = 0, \quad 1 + \mu_{j+1}\chi(i\omega_{j+1}) = 0. \tag{4}$$

There can obviously be only finitely many numbers μ for which the condition (4) is fulfilled. We join to them the values $-\infty$ and ∞, and, in the case of degenerate matrix, also the number $\mu = 0$, and put

$$\mu_1 = -\infty < \mu_2 < \ldots < \mu_p < \mu_{p+1} = \infty. \tag{5}$$

It follows from the definition of the numbers (5) that the matrices of the linear systems for $\mu_h < \mu < \mu_{h+1}$ have the same number k of eigenvalues in the ring-hand half-plane and do not have them on the imaginary axis. Let us call the open sector $S(\mu_h, \mu_{h+1})$ a sector of linear stability for $k = 0$ or a sector of linear instability of degree k for $k > 0$. Thus, the (σ, f)- plane is divided by the straight lines $f = \mu_h\sigma$ into l sectors $S[\mu_h, \mu_{h+1}]$, $h = 1, 2, \ldots, l$.

In the theory of the oscillation of multidimensional nonlinear systems, the "Voronov conjecture" is well known, as formulated for systems of the form (2). The essence of this conjecture can be formulated as follows: "if for a certain value of the parameter ε the graph of a nonlinear function $f(\sigma, \varepsilon)$ in system (2) alternately visits the stability and instability sectors, then the system produces oscillation with rigid excitation". In terms of this paper, this means that the system under consideration can, in particular, contain hidden attractors. It is currently known that Voronov's conjecture is incorrect. However, the search for condition under which this hypothesis is valid [16] ultimately led [17] to the creation of analytical-numerical methods of finding hidden attractors in multidimensional dynamical systems of the form (2).

Analytical-numerical methods of finding hidden oscillation in systems of the form (2) are based on the following theorem, which is a special case of the assertion, proved in [17].

Theorem 1. Let at some value of parameter ε, the function $f(\sigma,\varepsilon)$ in system (2) is continuous and piecewise smooth and the following conditions are satisfied:

1) the graph of function $f(\sigma,\varepsilon)$ has a unique point of intersection $\sigma=0$ with the straight line $\sigma+\chi(0)f=0$ (or with the straight line $f=0$, if the matrix A is degenerate);
2) there are such numbers μ_1 and μ_2, that the conditions $\mu_1 \le f'(\sigma,\varepsilon) \le \mu_2$ are satisfied at all differentiability points of function $f(\sigma,\varepsilon)$;
3) for some $\lambda>0$ and all $\omega \ge 0$

$$1+(\mu_1+\mu_2)\operatorname{Re}\chi(i\omega-\lambda)+\mu_1\mu_2\left|\chi(i\omega-\lambda)\right|^2>0; \tag{6}$$

4) there is $f'(0,\varepsilon)$ and the matrix $A+f'(0,\varepsilon)bc^T$ has exactly two eigenvalues with positive real parts and have no one in the strip $-\lambda \le \operatorname{Re} p \le 0$;
5) for some $h\in(\mu_1,\mu_2)$ the matrix $A+hbc^T$ is Hurwitz matrix and $\left|f(\sigma,\varepsilon)-h\sigma\right|<\gamma<\infty$.

Then, the system (2) has at least one orbitally stable cycle whose attraction domain contains almost all points of a neighborhood of the equilibrium $y=0$.

Consider some system that can be represented in the form (2)

$$\dot{y}=Ay+b\varphi(\sigma),\quad \sigma=c^T y. \tag{7}$$

Suppose that it has a unique stable equilibrium state $y=0$. Let us perform a "linear analysis" of that system. Suppose that there exist numbers μ_1 and μ_2 such that relation (6) holds for some $\lambda>0$ and, in addition, some matrix $A+\mu bc^T$, where $\mu\in(\mu_1,\mu_2)$, has exactly two eigenvalues with positive real parts and have no one in the strip $-\lambda \le \operatorname{Re} p \le 0$. Let for some $h\in(\mu_1,\mu_2)$ the matrix $A+hbc^T$ is a Hurwitz matrix. Then, by using theorem 1, one can readily choose a nonlinearity $\psi(\sigma)$ so as to ensure that system (7) with such nonlinearity has at least one orbitally asymptotically stable cycle whose attraction domain contains almost all points in a small neighborhood of the equilibrium $y=0$ of the system. Let $y_0 \ne 0$ be some (arbitrary) point in a neighborhood of the equilibrium $y=0$. We numerically find a solution $y_0(t)$ of system (7) with nonlinearity $\psi(\sigma)$ on the interval $[0,T]$, where T is sufficiently large, and with the initial condition $y_0(0)=y_0$. The value $y_0(T)$ is sufficiently close to the cycle. Now consider a family of systems (7) with nonlinearities $\varepsilon_j\varphi(\sigma)+(1-\varepsilon_j)\psi(\sigma)$, where, for example, $\varepsilon_j=0.1j,\ j=1,2,\ldots,10$. The solutions of these systems are denoted by $y_j(t)$ for the numerical integration of each of systems of the family, as an initial condition $y_j(0)$, we take $y_{j-1}(T)$. If, for the integration of all systems of the family, we obtain an attractor, then for $j=10$, we find a hidden attractor of system (2) with the nonlinearity $\varphi(\sigma)$. But if the numerical integration does not detect an attractor for some value ε_j, then this implies a bifurcation and the vanishing of the attractor. Note if an attractor is not detected on some step of the implementation of the described algorithm, then this could imply that the next attractor has a very small attraction domain. In this case, it is reasonable to diminish the discretization increment with respect to ε and repeat the search procedure with the smaller increment. The process of searching for a hidden attractor is accompanied by visualization of the attractor obtained at each stage of the implementation of the described algorithm. Visualization can be performed through any mathematical package: Mathcad, Matlab, Maple.

If, as a result of the implementation of the described procedure, it proves to be possible to detect the hidden attractor of the system (7), this means that we have constructed a one-parameter chameleon system (2) with the function $f(\sigma,\varepsilon)=\varepsilon\varphi(\sigma)+(1-\varepsilon)\psi(\sigma)$.

The method of construction of one-parameter chameleon systems possessing an infinite number of equilibrium states will be demonstrated in Section 4.

3. Two examples

3.1. Fourth-order hyperjerk chameleon system

This section presents a four-order chameleon system, which we design, relying on the above algorithm. The system of the form $\frac{d^4x}{dt^4}=J\left(\frac{d^3x}{dt^3},\frac{d^2x}{dt^2},\frac{dx}{dt},x\right)$ is called "hyperjerk system" or "snap system" [18], where J is called as "jerk" which is the derivation of a single scalar variable x. We shall consider the system

$$\frac{d^4x}{dt^4}+\frac{d^3x}{dt^3}+3\frac{dx}{dt}+x=-f\left(\frac{d^2x}{dt^2},\varepsilon\right), \tag{8}$$

where $f\left(\frac{d^2x}{dt^2},\varepsilon\right)$ is a nonlinear function to be selected below. The system (8) can be written in the form (2), where

$$y=\begin{pmatrix}x\\ \dot{x}\\ \ddot{x}\\ \dddot{x}\end{pmatrix},\quad A=\begin{pmatrix}0&1&0&0\\0&0&1&0\\0&0&0&1\\-1&-3&0&-1\end{pmatrix},\quad \mathrm{b}=\begin{pmatrix}0\\0\\0\\-1\end{pmatrix},\quad c=\begin{pmatrix}0\\0\\1\\0\end{pmatrix},\quad \sigma=c^T\mathrm{y}.$$

For this system sector $\left(-\infty,\ {}^{10}\!/_{3}\right)$ is a sector of linear instability of degree 2 and sector $\left({}^{10}\!/_{3},\ \infty\right)$ is a sector of linear stability, $\chi(p)=p^2\left(p^4+p^3+3p+1\right)^{-1}$. The assumption 3) of the theorem 1 for such functions $\chi(p)$ is satisfied with $\mu_1=3,\ \mu_2=4.5,\ \lambda=0.3$.

Take

$$f(\sigma,\varepsilon)=2\varepsilon\cdot\operatorname{arctg}(4\sigma)\exp\left(\frac{\sigma}{3}\right)+(1-\varepsilon)\frac{4\left(\sigma^3+3\sigma\right)}{\sigma^2+4}. \tag{9}$$

As it is easy to check, for functions (9) at $\varepsilon=0$, assumptions 1), 2) and 4) of the theorem 1 are satisfied, together with assumption 5) at $h=4$. Therefore, according to the statement of this theorem, the system (8)-(9) has a self-excited attractor (cycle) if $\varepsilon=0$. Since for all $\varepsilon\in[0,1]$ the graph of the function $f(\sigma,\varepsilon)$ "alternately visits" sectors of linear stability and sectors of linear instability of degree 2 of system (8), so, according to results [16], for all $\varepsilon\in[0,1]$ system (8) with nonlinearity (9) may have an attractor. Since $f'(0,\varepsilon)=5\varepsilon+3$, this attractor will be self-excited for $\varepsilon\in\left[0,\ {}^{1}\!/_{15}\right)$ and hidden for $\varepsilon\in\left({}^{1}\!/_{15},\ 1\right]$.

Now we will use the method described above to numerically track the evolution of an existing one at $\varepsilon=0$ cycle of system (8)-(9) when ε varies from 0 up to 1. Figure 1 shows a projection of a self-excited cycle on the plane of (x_1, x_4) at $\varepsilon = 0$, obtained by integration of the system (8)-(9) with the initial conditions $(0.1, 0.2, -0.2, -0.2)$. The projection to the same plane of a hidden cycle at $\varepsilon = 0.2$ is presented in figure 2.

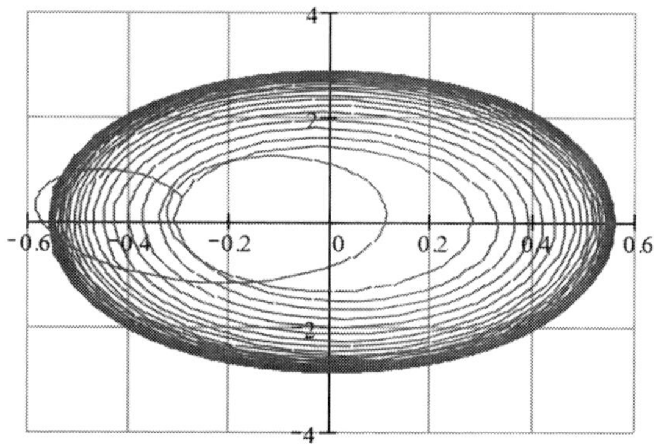

Figure 1. Self-excited cycle at $\varepsilon = 0$.

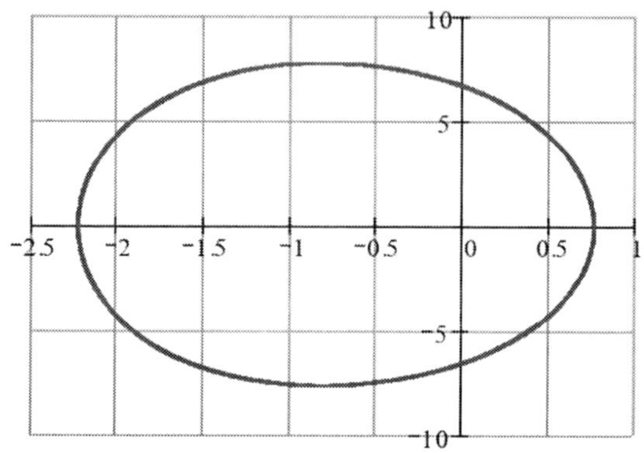

Figure 2. Hidden cycle at $\varepsilon = 0.2$.

When $\varepsilon \approx 0.637$ a doubling of hidden cycle is detected and, when $\varepsilon \approx 0.826$, the quadruple period is detected as well. In figures 3 the projection of hidden strange attractor of the system (8)-(9) with $\varepsilon = 0.88$ and $\varepsilon = 1$ on the plane (x_1, x_4) is presented.

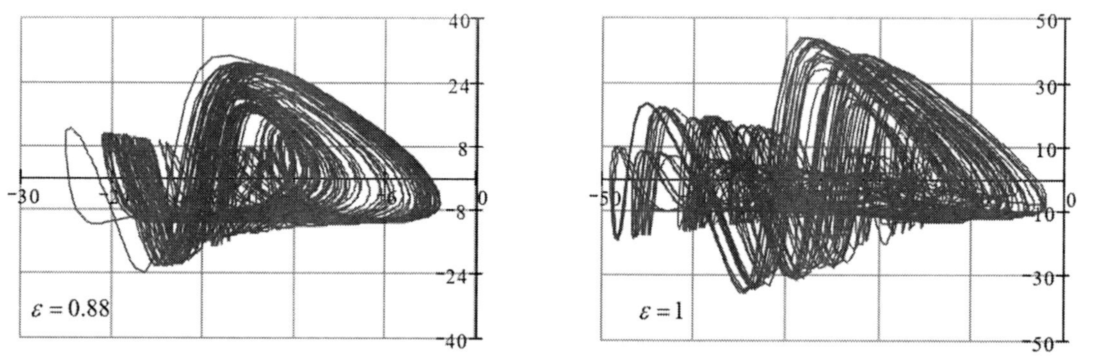

Figure 3. Projection of hidden strange attractor at $\varepsilon = 0.88$ and $\varepsilon = 1$ on the plane (x_1, x_4).

In order to verify that the attractors presented in figures 3 are chaotic, we calculate their Lyapunov exponents and the Kaplan-Yorke dimension. In a dynamical system, the existence of positive Lyapunov exponents confirms the chaotic behavior of the system. There are various methods of calculating Lyapunov exponents. In this paper we use the Jacobian method to calculate the LEs. We remember that in order to calculate the Kaplan-Yorke dimension of an attractor of a fourth-order system, it is necessary to calculate the Lyapunov exponents of the system $\lambda_1 \geq \lambda_2 \geq \lambda_3 \geq \lambda_4$. Also, the Kaplan–Yorke dimension of system is obtained as $D_{KY} = k + \left(\sum_{i=1}^{k} \lambda_i\right) |\lambda_{k+1}|^{-1}$, where $\sum_{i=1}^{k} \lambda_i \geq 0$ and $\sum_{i=1}^{k+1} \lambda_i < 0$. The results of calculations are given in table 1.

Table 1. The Lyapunov Exponents of the system (8)-(9).

ε	λ_1	λ_2	λ_3	λ_4	D_{KY}
0.88	0.104543	0.0	-0.222663	-0.860948	2.4695
1	0.061158	0.0	-0.132318	-0.915620	2.4623

3.2. Chameleon system of automatic control

It is well known that the equations of the kind (2) describe a wide class of automatic control systems [19]. Let us consider the system (2) with

$$A=\begin{pmatrix} 3.5 & 1 & -4.5 & -2 \\ -4 & -2 & 3.5 & 2.5 \\ 2.5 & 1 & -2.5 & -1 \\ -0.5 & -1 & -0.5 & 0 \end{pmatrix},\quad \mathrm{b}=\begin{pmatrix} -25.5 \\ 26.05 \\ -15.5 \\ -4.55 \end{pmatrix},\quad c=\begin{pmatrix} -9.2 \\ -20.2 \\ -21 \\ 10 \end{pmatrix},\quad \sigma=c^T\,\mathrm{y},\in$$

$$f(\sigma,\varepsilon)=(15-14\varepsilon)\frac{\exp(\sigma)-\exp(-\sigma)}{\exp(\sigma)+\exp(-\sigma)}+0.1(1-\varepsilon)\sigma,\quad \chi(p)=\frac{10.1p^3+10p^2}{(p^2+p+1)(\mathrm{p}^2+1)}.$$

One can readily see that, in this case, assumption 3) of the theorem 1 is satisfied if $\mu_1=0.1,\ \mu_2=20,\ \lambda=0.3$. For $\varepsilon=0$ the theorem guarantees that the considered system has a self-excited orbitally stable cycle. Evolution of this cycle when ε varies from 0 up to 1 is presented in figure 4. We note that at an ε of greater 5.2/14.1 the cycle is hidden.

The function $f(\sigma,1)$ satisfies the conditions $0<f'(\sigma,1)<2$, while $A+\mu bc^T$ is a Hurwitz matrix if $0<\mu<9.9$. Therefore, the considered system with nonlinearity $f(\sigma,1)$ satisfies all assumptions of the well-known Kalman conjecture [20].

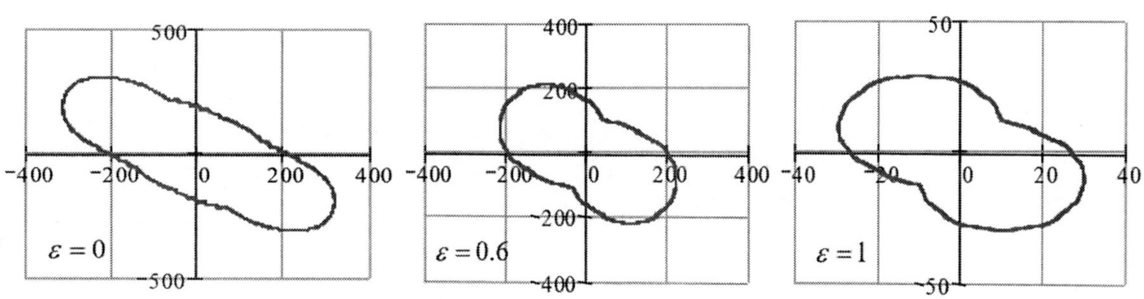

Figure 4. Evolution of the cycle of an automatic control system when ε changes from 0 to 1.

4. Chameleon systems with infinite number of equilibria

We will now consider a system of the following form

$$\begin{aligned} \dot{z}&=Pz+q\varphi(\sigma,\varepsilon),\\ \dot{\sigma}&=r^Tz+\beta\varphi(\sigma,\varepsilon), \end{aligned}\tag{10}$$

where P is the matrix of the size $(n-1)\times(n-1)$, q and r are $(n-1)$-dimensional vectors, β is a scalar, $\varphi(\sigma,\varepsilon)$ is a function 2π -periodic on a variable σ. Equations of the kind (10) describe a wide class of phase-locked loop (PLL). Nowadays various modification of PLL are widely used in radio systems, telecommunication systems, global positioning systems (GPS), computer architectures, and other domains.

If the function $\varphi(\sigma,\varepsilon)$ at some value of ε has zeros in period $[0.2\pi)$, the system (10) has an infinite number of equilibria.

Let $q^T(P-pI_{n-1})^{-1}r = m(p)[n(p)]^{-1}$ and $\chi(p)=[m(p)-\beta n(p)][pn(p)]^{-1}$ where $m(p)$ and $n(p)$ are polynomials. In order to search for hidden attractors in systems of the form (10) with an infinite number of equilibrium states, it is convenient to use the following theorem, proved in [21].

Theorem 2. Let at some value of parameter ε the function $\varphi(\sigma,\varepsilon)$ in system (10) is continuous and piecewise smooth, and the following conditions are satisfied:

1) the graph of the function $\varphi(\sigma,\varepsilon)$ has a unique point of intersection $\sigma=\tilde{\sigma}$ with the straight line $\varphi=0$;
2) there are such numbers μ_1 and μ_2, that the conditions $\mu_1 \le \varphi'(\sigma,\varepsilon) \le \mu_2$ are satisfied at all differentiability points of a function $\varphi(\sigma,\varepsilon)$;
3) for some $\lambda>0$ and all $\omega\ge 0$
$$1+(\mu_1+\mu_2)\operatorname{Re}\chi(i\omega-\lambda)+\mu_1\mu_2|\chi(i\omega-\lambda)|^2>0;$$
4) the polynomial $[m(p)-\beta n(p)]\varphi'(\tilde{\sigma},\varepsilon)+pn(p)$ has exactly two roots with positive real parts;
5) for some $h\in(\mu_1,\mu_2)$, all roots of the polynomial $[m(p)-\beta n(p)]h+pn(p)$ have negative real parts and $|\varphi(\sigma+\tilde{\sigma},\varepsilon)-h\sigma|<\gamma<\infty$.

Then, system (10) has at least one orbitally stable cycle whose attraction domain contains almost all points of a neighborhood of the equilibrium $z=0,\ \sigma=\tilde{\sigma}$.

Systems of the form (10), which have an infinite number of hidden attractors, were found in [21]. Based on the results of this paper the chameleon system, demonstrating, with different parameter values of ε, self-excited multi-scroll attractors and hidden attractors-twins, is proposed below.

Take

$$P=\begin{pmatrix}0 & 1\\ -1 & -1\end{pmatrix},\quad q=\begin{pmatrix}0\\ 1\end{pmatrix},\quad r=\begin{pmatrix}-15.8\\ -0.3\end{pmatrix},\quad \beta=-3.2,$$
$$\varphi(\sigma)=\frac{\pi}{6}\left[1.00052411\sin\sigma+\varepsilon(0.404720835\sin 2\sigma+0.1798759\sin 3\sigma+0.062532677\sin 4\sigma)\right]. \tag{11}$$

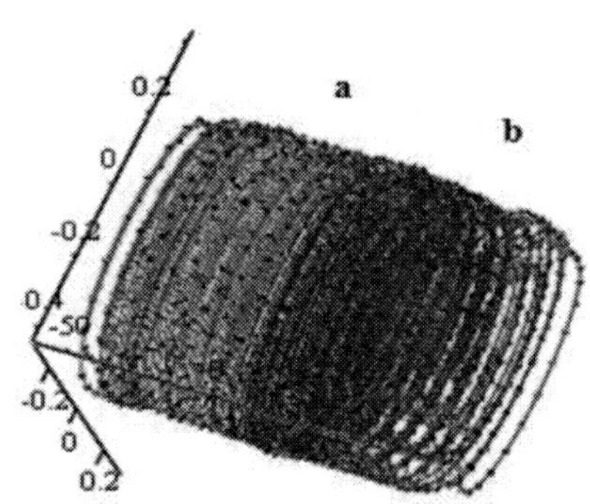

Figure 5. Multi-scroll attractor for $\varepsilon=0.2$.

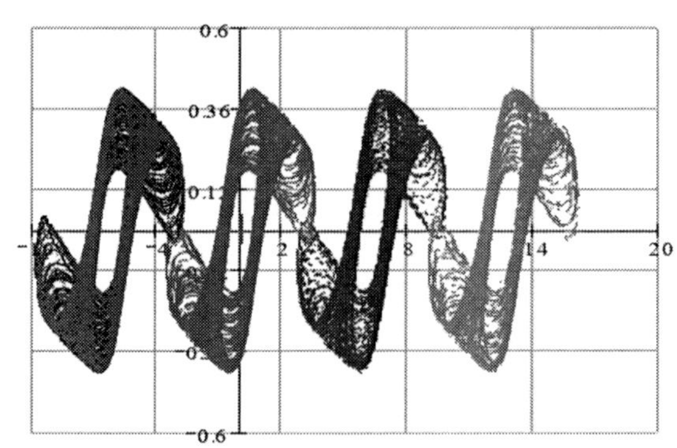

Figure 6. The attractor of the system for $\varepsilon=0.78$.

IOP Conf. Series: Journal of Physics: Conf. Series **973** (2018) 012052 doi:10.1088/1742-6596/973/1/012052

Attractors are self-excited from any point of neighborhood of all equilibria $z_1 = z_2 = 0,\ \sigma = \pi m,\ m \in Z$ of the considered system when $\varepsilon \in [0, 0.78)$. These attractors are multi-scroll. For $\varepsilon = 0.2$, a couple of such attractors excited from the neighborhood of the point $(0,0,0)$ (a) and the point $(0,0,\pi)$ (b) are presented in figure 5.

Equilibrium points $(0,0,2\pi m)$ become stable in the small when $\varepsilon \approx 0.78$, and attractors, shown in figure 6 (projection to the plane (z_2,σ)), excite from the neighborhood of unstable equilibrium points $(0,0,(2m+1)\pi)$.

The system under study has an infinite number of hidden attractors-twins if $\varepsilon = 1$. Each pair of these attractors is located in its own "band" $\pi(2k-1) < \sigma < \pi(2k+1),\ k = 0, \pm 1, \pm 2, \ldots$ These attractors are presented in figure 7. Lyapunov exponents and Kaplan-Yorke dimensions of attractors, represented in figures 5-7 are provided in table 2

Table 2. The Lyapunov Exponents of the System (11).

ε	λ_1	λ_2	λ_3	D_{KY}
0.2	0.222604	0.0	-2.705660	2.0827
0.78	0.423569	0.0	-1.266375	2.3345
1	0.666114	0.0	-1.034800	2.6437

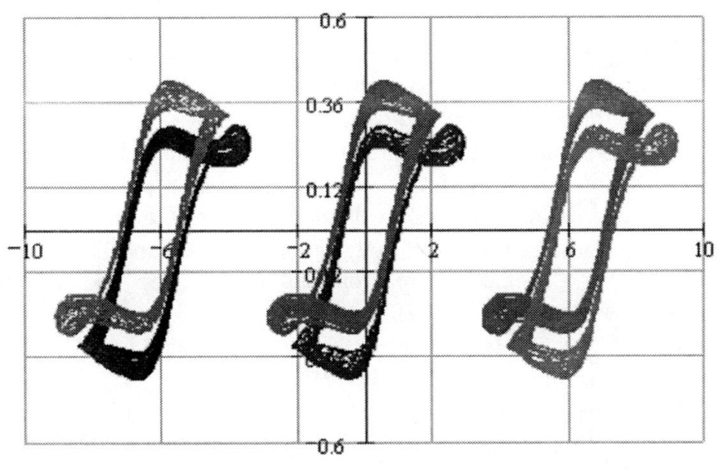

Figure 7. A family of hidden attractors (projection on the plane (z_2,σ), $\varepsilon = 1$).

5. Conclusion

This paper offers an algorithm for synthesizing one-parameter dynamic chameleon systems of the form (2). It demonstrates that the use of the algorithm to search for hidden attractors in the dynamical systems of the form (7) proposed in [17], always leads to the synthesis of a one-parameter chameleon system of the form (2). The authors of this paper know examples of using the homotopy method for synthesizing chameleon systems of a different kind. For example, in [22], a chameleon system was actually constructed, which, for some values of the parameter, has a self-excited attractor belonging to the generalized Lorentz system. When the parameter of this system changes in a certain range, the attractor of the generalized Lorentz system is transformed into a hidden attractor of the Glukhovsky-Dolzhansky system.

In the presence of a large number of currently discovered systems that have both self-excited and hidden attractors, the question remains open: can a hidden attractor of any system be localized by numerically tracing the evolution of the self-excited attractor of some auxiliary system when its parameters change in a certain range?

6. Acknowledgments

The results of the research project are published with the financial support of Tula State University within the framework of the scientific project № 2017-49 Publ.

7. References

[1] Lorenz E 1963 Deterministic nonperiodic flow *J. Atmos. Sci.* **20** pp 130–41
[2] Rössler O 1976 An Equation for Continuous Chaos *Physics Letters A.* **57** pp 397–8
[3] Chua L 1992 A zoo of Strange Attractors from the Canonical Chua's Circuits *Proc. IEEE 35th Midwest Symp. Circuits and Systems* (*Cat. No. 92CH3099-9*) **2** pp 916–26
[4] Leonov G, Kuznetsov N and Vagaitsev V 2011 Localization of hidden Chua's attractors *Phys. Lett. A.* **375** pp 2230-33
[5] Sharma P, Shrimali M, Prasad A, Kuznetsov N and Leonov G 2015 Control of multistability in hidden attractors *Eur Phys J Spec Top* **224** pp 1485–91
[6] Sharma P, Shrimali M, Prasad A, Kuznetsov N and Leonov G 2015 Controlling dynamics of hidden attractors *Int. J. Bifurcation and Chaos* **25** 1550061
[7] Pham V-T, Volos C, Jafari S, Wei Z and Wang X 2014 Constructing a novel no-equilibrium chaotic system *Int J Bifurcation and Chaos* **24** 1450073
[8] Tahir F, Jafari S, Pham V-T, Volos C and Wang X 2015 A novel no-equilibrium chaotic system with multiwing butterfly attractors *Int J Bifurcation and Chaos* **25** 1550056
[9] Jafari S, Pham V-T and Kapitaniak T 2016 Multiscroll chaotic sea obtained from a simple 3d system without equilibrium *Int J Bifurcation and Chaos* **26** 1650031
[10] Molaie M, Jafari S, Sprott J and Golpayegani SMRH 2013 Simple chaotic flows with one stable equilibrium *Int J Bifurcation and Chaos* **23** 350188
[11] Kingni S, Simo H and Woafo P 2014 Three-dimensional chaotic autonomous system with only one stable equilibrium: analysis, circuit design, parameter estimation, control, synchronization and its fractional-order form *Eur. Phys. J. Plus.* **129** pp 1–16
[12] Pham V-T, Jafari S, Volos C, Giakoumis A, Vaidyanathan S and Kapitaniak T. 2016 A chaotic system with equilibria located on the rounded square loop and its circuit implementation *IEEE Trans Circuits Syst II* **63** pp 878–82
[13] Pham V-T, Jafari S, Volos C 2017 A novel chaotic system with heart-shaped equilibrium and its circuital implementation *Optik* **131** pp 343–9
[14] Rajagopal K, Karthikeyan A and Duraisamy P 2017 Hyperchaotic chameleon: fractional order FPGA implementation *Complexity 2017* https://www.hindawi.com/journals/complexity/air/8979408/
[15] Rajagopal K, Akgul A, Jafari S, Karthikeyan A and Koyuncu I 2017 Chaotic chameleon: Dynamic analyses, circuit implementation FPGA design and fractional-order form with basic analyses *Chaos, Solitons and Fractals* **103** pp 476-87
[16] Burkin I 2002 The buffer phenomenon in multidimensional dynamical systems. *Diff. Equations* **38** pp 615-25
[17] Burkin I and Nguen NK 2014 Analytical-Numerical Methods of Finding Hidden Oscillations in Multidimensional Dynamical Systems *Diff. Equations* **50** pp 1695–717
[18] SprottJ and Fatma Y 2016 Simple Chaotic Hyperjerk System *Int. J. Bifurcation and Chaos* **26** 1650189
[19] Khalil H 2002 *Nonlinear Systems* 3-rd Edition Prentice Hall Upper Saddle River NJ 07458
[20] Kalman R 1957 Physical and Mathematical Mechanisms of Instability in Nonlinear Automatic Control Systems *Trans. ASME* **79.3** pp 553-66

[21] Burkin I 2017 Hidden attractors of some multistable systems with infinite number of equlibrium states *Chebyshevskiy sbornik* **18** pp 18-33 (in Russian).

[22] Leonov G, Kuznetsov N and Mokaev T 2015 Homoclinic orbits and self-excited and hidden attractors in a Lorenz-like system describing convective fluid motion *The European Physical Journal Special Topics, Multistability: Uncovering Hidden Attractors* **224** pp 1421–58

IOP Conf. Series: Journal of Physics: Conf. Series **973** (2018) 012053 doi:10.1088/1742-6596/973/1/012053

Spectral analysis of a difference operator with a growing potential

G V Garkavenko[1], **A R Zgolich**[2] **and N B Uskova**[3]

[1] Voronezh State Pedagogical University, Physico-Mathematical Faculty, 86 Lenina St, Voronezh, Russia

[2] Voronezh State University, Applied Mathematics, Informatics and Mechanics Faculty, 1 Universitetskaya Square, Voronezh, Russia

[3] Voronezh State Technical University, Information Technologies and Computer Security Faculty, 14 Moskovsky prospekt, Voronezh, Russia

E-mail: arsenij112@mail.ru, nat-uskova@mail.ru

Abstract. In this paper, we study the spectral properties of a second order difference operator with a growing potential. The operator acts in the complex Hilbert space $\ell_2(Z)$ of square summable complex sequences indexed by the integers. This operator is a discrete analogue of a second order differential operator with a growing complex potential. The study is based on a method of similar operators developed by A. G. Baskakov and his collaborators. This method allows us to reduce the study of the operator to one with a block-diagonal matrix. Asymptotic estimates of eigenvalues, eigenvectors, and spectral projections of a difference operator are obtained.

1. Introduction

In this paper, we explore the spectral properties of difference operators and the corresponding discrete Sturm-Liouville operators [1]. To this end, we employ the method of similar operators. The method has its origins in various similarity and perturbation techniques, such as the classical perturbation methods of celestial mechanics, Ljapunov's kinematic similarity method, Friedrichs' method of similar operators used in quantum mechanics, and Turner's method of similar operators (see [2–4]). The method has been extensively developed and used in the works of A. G. Baskakov and his collaborators (see [2–7] and references therein). It provides a foundation for finding estimates of eigenvalues and eigenvectors of difference and differential operators with a growing potential.

Let $\ell_2(Z)$ be the complex Hilbert space of square summable sequences indexed by the integers. The inner product and the norm in $\ell_2(Z)$ are defined by

$$(x, y) = \sum_{n \in Z} x(n)\overline{y(n)}$$

and

$$||x|| = \left(\sum_{n \in Z} |x(n)|^2 \right)^{1/2}, \quad x, y \in \ell_2(Z),$$

respectively.

We consider the closed linear difference operator $\mathcal{E} : D(\mathcal{E}) \subset \ell_2(Z) \to \ell_2(Z)$, generated by the expression

$$(\mathcal{E}x)(n) = a(n)x(n) + 2x(n) - x(n-1) - x(n+1),$$

with the domain $D(\mathcal{E}) \subset \ell_2(Z)$ defined by

$$D(\mathcal{E}) = \{x \in \ell_2(Z), \sum_{n \in Z} |x(n)|^2 |a(n)|^2 < \infty\}.$$

In the standard basis of the space $\ell_2(Z)$, the matrix of the operator $\mathcal{E}$ is given by

$$\begin{bmatrix} \cdots & \cdots & \cdots & \cdots & \cdots & \cdots & \cdots & \cdots \\ -1 & a(-2)+2 & 1 & 0 & \cdots & \cdots & \cdots & \cdots \\ 0 & -1 & a(-1)+2 & 1 & 0 & \cdots & \cdots & \cdots \\ \cdots & 0 & -1 & a(0)+2 & 1 & 0 & \cdots & \cdots \\ \cdots & \cdots & 0 & -1 & a(1)+2 & 1 & 0 & \cdots \\ \cdots & \cdots & \cdots & 0 & -1 & a(2)+2 & 1 & 0 \\ \cdots & \cdots & \cdots & \cdots & \cdots & \cdots & \cdots & \cdots \end{bmatrix}.$$

The sequence $a : Z \to C$ is such that $\lim\limits_{|n| \to \infty} |a(n)| = \infty$. We also assume that this sequence satisfies one of the following two groups of conditions:

1) $a(i) \neq a(j)$ for $i \neq j$ and $d_j = \inf_{i \neq j} |a(i) - a(j)| \to \infty$ as $|j| \to \infty$, $i, j \in Z$;

2) $a(i) = a(-i)$ for $i \in Z$ and $d_j \to \infty$ as $j \to \infty$, $i, j \in Z_+ = \{0\} \cup N$.

The first group of conditions corresponds to a growing potential in a general form, while the second one corresponds to an even growing potential.

In [1], a finite-dimensional analogue of the operator $\mathcal{E}$ was considered and its eigenvalue estimates were obtained by a variation method. In this paper, we provide estimates of eigenvalues and spectral projections of the operator $\mathcal{E}$ using the method of similar operators as in [2–4]. It is important to note that the method is commonly used to study the spectral properties for differential operators [5–7]. It has, however, also been used for difference operators (see [8–11]).

We write $\mathcal{E} = A - B$, where $A : D(A) = D(\mathcal{E}) \subset \ell_2(Z) \to \ell_2(Z)$ is defined by

$$(Ax)(n) = a(n)x(n) + 2x(n), n \in Z,$$

and B is a bounded operator given by

$$(Bx)(n) = x(n+1) + x(n-1), n \in Z.$$

We shall treat the operator $\mathcal{E}$ as the perturbation of A by B.

For a complex Hilbert space $\ell_2(Z)$ we denote by $End\,\ell_2$ the Banach algebra of all bounded linear operators in $\ell_2(Z)$. The operator B belongs to $End\,\ell_2$ and A is a normal closed linear operator. In the standard scheme of the method of similar operators, the unperturbed operator is a differential one and the perturbation is a multiplication operator by a potential function [2–7]. In this paper, the multiplication operator is chosen as the unperturbed one. This choice works because the matrix of the multiplication operator in the standard basis of $\ell_2(Z)$ is diagonal and, therefore, the spectral properties of this operator are easy to obtain.

The spectrum $\sigma(A)$ of the operator A can be written as

$$\sigma(A) = \bigcup_{l \in J} \sigma_l = \bigcup_{l \in J} \{\lambda_l\},$$

IOP Conf. Series: Journal of Physics: Conf. Series **973** (2018) 012053 doi:10.1088/1742-6596/973/1/012053

where $J \in \{Z, Z_+\}, Z_+ = N \cup \{0\}$. In the case of the first group of conditions, $\lambda_l = a(l)+2$, $l \in Z$, are simple eigenvalues. The corresponding eigenvectors are e_l, $l \in Z$, where $e_l(k) = \delta_{ek}, k \in Z$, and δ_{ek} is the standard Kronecker delta. The spectral projections $P_l = P(\sigma_l, A)$ are given by $P_l x = (x, e_l)e_l = x(l)e_l$, $l \in Z$, $x \in \ell_2(Z)$.

In the other case, $\lambda_l = a(l) + 2$, $l \in N$, are eigenvalues of multiplicity two, and $\lambda_0 = a(0) + 2$ is a simple eigenvalue. The corresponding eigenvectors are e_l, $l \in Z$. The spectral projections $P_l = P(\sigma_l, A)$ are given by $P_l x = (x, e_l)e_l + (x, e_{-l})e_{-l}$, $l \in N$, and $P_0 x = (x, e_l)e_0$, $x \in \ell_2(Z)$.

We will use the notation $H_k = ImP_k$, $k \in J$, $P_{(m)} = \sum_{|i| \leq m} P_i$, and $H_{(m)} = ImP_m$.

2. Materials and methods

Let $\mathcal{H}$ denote an abstract complex Hilbert space. We begin with the following definition.

Definition 1. Two linear operators $A_i : D(A_i) \subset \mathcal{H} \to \mathcal{H}, i = 1, 2$, are called *similar*, if there exists a continuously invertible operator $U \in End\mathcal{H}$ such that

$$A_1 U x = U A_2 x,\ x \in D(A_2), U D(A_2) = D(A_1). \tag{1}$$

The operator U is called the *similarity transform* of A_1 into A_2.

Directly from the definition (1), we have the following result about the spectral properties of similar operators.

Lemma 1. Let $A_i : D(A_i) \subset \mathcal{H} \to \mathcal{H}, i = 1, 2$, be two similar operators with the similarity transform U. Then the following properties hold.

(1) We have $\sigma(A_1) = \sigma(A_2), \sigma_p(A_1) = \sigma_p(A_2)$, and $\sigma_c(A_1) = \sigma_c(A_2)$, where σ_p denotes the point spectrum and σ_c denotes the continuous spectrum;

(2) Assume that the operator A_2 admits a decomposition $A_2 = A_{21} \oplus A_{22}$ with respect to a direct sum $\mathcal{H} = \mathcal{H}_1 \oplus \mathcal{H}_2$, where $A_{21} = A_2|\mathcal{H}_1$ and $A_{22} = A_2|\mathcal{H}_2$ are the restrictions of A_2 to the respective subspaces. Then the operator A_1 admits a decomposition $A_1 = A_{11} + A_{12}$ with respect to a direct sum $\mathcal{H} = \widetilde{\mathcal{H}_1} \oplus \widetilde{\mathcal{H}_2}$, where $A_{11} = A_1|\widetilde{\mathcal{H}}$ and $A_{12} = A_1|\widetilde{\mathcal{H}_2}$ are the restrictions of A_1 to the respective invariant subspaces. Moreover, if P is the projection onto $\mathcal{H}_1$ parallel to $\mathcal{H}_2$, then $\widetilde{P} = UPU^{-1}$ is the projection onto $\widetilde{\mathcal{H}_1}$ parallel to $\widetilde{\mathcal{H}_2}$.

(3) If λ_0 is an eigenvalue of the operator A_2 and x is a corresponding eigenvector, then $y = Ux$ is an eigenvector of the operator A_1 corresponding to the same eigenvalue λ_0.

We shall need to extend Property (2) in the Lemma 1 to the case of countable direct sums. To this end, we assume that the abstract Hilbert space $\mathcal{H}$ can be written as

$$\mathcal{H} = \bigoplus_{l \in J} \mathcal{H}_l,$$

where each $\mathcal{H}_l, l \in J$, is a closed nonzero subspace of $\mathcal{H}, \mathcal{H}_j$ is orthogonal to $\mathcal{H}_l$ for $l \neq j \in J$, and each $x \in \mathcal{H}$ satisfies $x = \sum_{l \in J} x_l$, where $x_l \in \mathcal{H}_l$ and $||x||^2 = \sum_{l \in J} ||x||_l^2$. In other words, we have a disjunctive resolution of the identity

$$\mathcal{P} = \{P_l, l \in J\}, \tag{2}$$

that is a system of idempotents with the following properties

1) $P_l^* = P_l, l \in J$;
2) $P_j P_l = \delta_{jl} P_l$, $j, l \in J$;
3) The series $\sum_{l \in J} P_l x$ converges unconditionally to $x \in \mathcal{H}$ and $||x||^2 = \sum_{l \in J} ||P_l x||^2$;
4) Equalities $P_l x = 0, l \in J$, imply $x = 0 \in \mathcal{H}$;
5) $\mathcal{H}_l = ImP_l, x_l = P_l x, l \in J$.

Definition 2. We say that a closed linear operator $A : D(A) \subset \mathcal{H} \to \mathcal{H}$ is represented as an *orthogonal direct sum* of bounded operators $A_l \in End\mathcal{H}_l, l \in J$, that is

$$A = \bigoplus_{l \in J} A_l,$$

if the following three properties hold.

1. $D(A) = \{x \in \mathcal{H} : \sum_{l \in J} ||A_l x_l||^2 < \infty, x_l = P_l x, l \in J\}$ and $\mathcal{H}_l \subset D(A)$ for all $l \in J$.

2. For each $l \in J$, the subspace $\mathcal{H}_l$ is an invariant subspace of the operator A and A_l is the restriction of A to $\mathcal{H}_l$. The operators $A_l, l \in J$, are called the *parts* of the operator A.

3. $Ax = \sum_{l \in J} A_l x_l, x \in D(A)$, where $x_l = P_l x, l \in J$, and the series converges unconditionally in $\mathcal{H}$.

Definition 3. Given a continuously invertible operator $U \in End\mathcal{H}$ and an orthogonal decomposition of $\mathcal{H}$, a *U-orthogonal decomposition* of $\mathcal{H}$ is the orthogonal direct sum

$$\mathcal{H} = \bigoplus_{l \in Z} U\mathcal{H}_l.$$

Definition 4. Given a continuously invertible operator $U \in End\mathcal{H}$, we say that a closed linear operator $A : D(A) \subset \mathcal{H} \to \mathcal{H}$ is a U-orthogonal direct sum of bounded linear operators $\widetilde{A}_l, l \in J$, if $\widetilde{A}_l = UA_lU^{-1}, l \in J$, and

$$A = \bigoplus_{l \in J} \widetilde{A}_l.$$

We remark that U-orthogonal decompositions and direct sums can be viewed as orthogonal with respect to the inner product

$$\langle x, y \rangle_U = \langle Ux, Uy \rangle, x, y \in \mathcal{H}.$$

An example of direct sums of operators is provided the operator A from the introduction. In particular, A is an orthogonal direct sum of operators $A_l = A|\mathcal{H}_l = (a(l) + 2)I_l$, where I_l is the identity operator on $\mathcal{H}_l = ImP_l, l \in J$. In other words, $A = \bigoplus_{l \in Z} (a(l) + 2)I_l$ in the first case, and $A = \bigoplus_{l \in Z_+} (a(l) + 2)I_l$ in the second case. This representation is with respect to the orthogonal decomposition of $\ell_2(Z)$ given by $\ell_2(Z) = \bigoplus_{l \in J} \mathcal{H}_l$.

Consider a new resolution of the identity

$$\mathcal{P}^{(m)} = \{P_{(m)}\} \cup \{P_l, |l| > m, l \in J\}.$$

Then the operator A may also be represented as an orthogonal direct sum

$$A = A_{(m)} \oplus \left(\bigoplus_{|l| > m, l \in J} A_l \right) = A_{(m)} \oplus \left(\bigoplus_{|l| > m, l \in J} (a(l) + 2)I_l \right), \tag{3}$$

$m \in Z_+, l \in J$, where $A_{(m)}$ is the restriction of A to $\mathcal{H}_{(m)} = ImP_{(m)}$. The representation (3) is with respect to the orthogonal decomposition $\ell_2(Z) = H_{(m)} \oplus \left(\bigoplus_{|l| > m, l \in J} \mathcal{H}_l \right)$. Observe that $A_{(m)} = \bigoplus_{|l| > m, l \in J} (a(l) + 2)I_l$ with respect to the decomposition $H_{(m)} = \bigoplus_{|l| < m, l \in J} H_j$.

The method of similar operators constructs a similarity transform for an operator $A - B : D(A) \subset \mathcal{H} \to \mathcal{H}$, where the spectrum of the operator A is known and has certain properties, and the operator B is A-bounded.

Definition 5 [2]. Let $A : D(A) \subset \mathcal{H} \to \mathcal{H}$ be a linear operator. A linear operator $B : D(B) \subset \mathcal{H} \to \mathcal{H}$ is A-bounded $D(B) \supseteq D(A)$ and $||B||_A = \inf\{c > 0 : ||Bx|| \leq c(||x|| + ||Ax||), x \in D(A)\} < \infty$.

The space $\mathcal{L}_A(\mathcal{H})$ of all A-bounded linear operators is a Banach space with respect to the norm $||\cdot||_A$. Moreover, given $\lambda_0 \in \rho(A)$, where $\rho(A) = C \setminus \sigma(A)$ is the resolvent set of A, we have $B \in \mathcal{L}_A(\mathcal{H})$ if and only if $B(\lambda_0 I - A)^{-1} \in End\mathcal{H}$ and $||B||_A = ||B(\lambda_0 I - A)^{-1}||_{End\mathcal{H}}$ defines an equivalent norm in $\mathcal{L}_A(\mathcal{H})$.

The method of similar operators uses the *commutator transform* $ad_A : D(ad_A) \subset End\mathcal{H} \to End\mathcal{H}$ defined by

$$ad_A X = AX - XA, \ X \in D(ad_A).$$

where the domain $D(ad_A)$ contains all $X \in End\mathcal{H}$ such that the following two properties hold:

1. $XD(A) \subseteq D(A)$;

2. The operator $ad_A X : D(A) \to \mathcal{H}$ (uniquely) extends to a bounded operator $Y \in End\mathcal{H}$; we then let $ad_A X = Y$.

The key notion of the method of similar operators is that of an admissible triplet.

Definition 6 [2]. Let $\mathcal{M}$ be a linear subspace of $\mathcal{L}_A(\mathcal{H})$, $J : \mathcal{M} \to \mathcal{M}$, and $G : \mathcal{M} \to B(\mathcal{H})$. The collection $(\mathcal{M}, J, G)$ is called an *admissible triplet* for the operator A, and the space $\mathcal{M}$ is the *space of admissible perturbations*, if the following six properties hold.

1. $\mathcal{M}$ is a Banach space that is continuously embedded in $\mathcal{L}_A(\mathcal{H})$, i.e., $\mathcal{M}$ has a norm $||\cdot||_*$ such that there is a constant $C > 0$ that yields $||X||_A \leq C||X||_*$ for any $X \in \mathcal{M}$.

2. J and G are bounded linear operators; moreover, J is an idempotent.

3. $(GX)D(A) \subset D(A)$ and

$$(ad_A GX)x = (X - JX)x, x \in D(A), X \in \mathcal{M};$$

moreover $GX \in End\mathcal{H}$ is the unique solution of the equation

$$ad_A Y = AY - YA = X - JX,$$

that satisfies $JY = 0$.

4. $XGY, (GX)Y \in \mathcal{M}$ for all $X, Y \in \mathcal{M}$, and there is a constant $\gamma > 0$ such that

$$||G|| \leq \gamma, \max\{||XGY||_*, ||(GX)Y||_*\} \leq \gamma ||X||_* ||Y||_*.$$

5. $J((GX)JY) = 0$ for all $X, Y \in \mathcal{M}$.

6. For every $X \in \mathcal{M}$ and $\varepsilon > 0$ there exists a number $\lambda_\varepsilon \in \rho(A)$, such that $||X(A - \lambda_\varepsilon I)^{-1}|| < \varepsilon$.

To formulate the main theorem of the method of similar operators [2] for an operator $A - B$, we use the function $\Phi : \mathcal{M} \to \mathcal{M}$ given by

$$\Phi(X) = BGX - (GX)(JB) - (GX)J(BGX) + B. \tag{4}$$

Theorem 1 [2]. Assume that $(\mathcal{M}, J, G)$ is an admissible triplet for an operator $A : D(A) \subset \mathcal{H} \to \mathcal{H}$ and $B \in \mathcal{M}$. Assume also that

$$4\gamma ||J|| ||B||_* < 1,$$

where γ comes from the Property 4 of Definition 6. Then the operator $A - B$ is similar to the operator $A - JX_*$, where $X_* \in \mathcal{M}$ is the (unique) fixed point of the function Φ given by equation (4), and the similarity transform of $A - B$ into $A - JX_*$ is given by $I + GX_* \in End\mathcal{H}$. Moreover, the map $\Phi : \mathcal{M} \to \mathcal{M}$ is a contraction in the ball $\{X \in \mathcal{M} : ||X - B||_* \leq 3||B||_*\}$, and the fixed point X_* can be found as a limit of simple iterations: $X_0 = 0, X_1 = \Phi(X_0) = B$, etc.

3. Results and discussions.

Theorem 2. There is a number $k \in Z_+$ and a continuously invertible operator $U \in End\,\ell_2$ such that operator $\mathcal{E} = A - B$ is similar to the operator $A - X_0$, where $X_0 \in End\,\ell_2, \mathcal{E}U = U(A - X_0)$ and the subspaces $\mathcal{H}_{(k)} = ImP_{(k)}, \mathcal{H}_l = ImP_l, |l| > k, l \in J$, are invariant for X_0. The operator $A - X_0$ is the orthogonal direct sum

$$A - X_0 = A - \left(X_{0_{(k)}} \oplus \left(\bigoplus_{|l|>k, l\in J} X_{0_l}\right)\right)$$

with respect to orthogonal decomposition $\mathcal{H} = \mathcal{H}_{(k)} \oplus \left(\bigoplus_{|l|>k,l\in J} \mathcal{H}_j\right)$ and the dimension of $\mathcal{H}_{(m)}$ is $2m + 1$. Moreover, the operator $\mathcal{E}$ is the U-orthogonal direct sum

$$\mathcal{E} = U\left(A - \left(X_{0_{(k)}} \oplus \left(\bigoplus_{|l|>k, l\in J} X_{0_{(l)}}\right)\right)\right)U^{-1}$$

with respect to the U-orthogonal decomposition $\mathcal{H} = U\mathcal{H}_{(k)} \oplus \left(\bigoplus_{|l|>k,l\in J} U\mathcal{H}_l\right)$.

Proof. Let $\mathcal{M} = End\ell_2 \subset \mathcal{L}_A(\ell_2)$ and

$$||Xx|| = ||XA^{-1}Ax|| \leq ||XA^{-1}||_\infty ||Ax|| < ||XA^{-1}||_\infty(||Ax|| + ||x||),$$
$$||X||_A \leq ||XA^{-1}||_\infty.$$

We deduce Property 1 of the Definition 6.

Given a resolution of identity $\mathcal{P}$ as in (2), it is often convenient to represent an operator $X \in End\ell_2$ in terms of its matrix. We write such matrices as $\widetilde{X} = (X_{jl})$, where $X_{jl} = P_j X P_l, j, l \in J$. In the case when the matrix of a linear operator is diagonal, this operator is the orthogonal direct sum (see definition 2).

Any operator $Z \in End\ell_2$ can be expressed (see [8]) as

$$Z = \lim_{n\to\infty} \sum_{|p|\leq n} (1 - \frac{|p|}{n})Z_p,\ Z_0 = \sum_{i-j=p} Z_{ij}.$$

The operator $Z_p, p \in Z$, is the p-th diagonal of the operator matrix of the operator Z.

We define a family of transformers $J_k X$ as follows:

$$J_k X = P_{(k)} X P_{(k)} + \sum_{|i|>k, i\in J} P_i X P_i,\ k \geq 0,$$

moreover,

$$JX = J_0 X = \sum_{i\in J} P_i X P_i,$$

where the series $\sum_{i\in J} P_i X P_i$ is convergent, because the operator X belongs to $End\ell_2$. Obviously, $||J_k|| = 1$, because

$$||J_k X|| \leq ||P_{(k)} X P_{(k)}||_2^2 + \sum_i ||P_i X P_i||_2^2 \leq ||X||_2^2.$$

If $X = \sum_i P_i X P_i$ (i.e., the matrix of the operator X is diagonal and X is an orthogonal direct sum of operators X_{ii}), then $J_k X = X$ and $||J_k X|| = ||X||$.

Let us proceed to the construction of the operator $G_0X : End\ell_2 \to End\ell_2$. First, we define it on the operator blocks $X_{ij} = P_iXP_j$, where $X \in End\ell_2$. For each $X_{ij}, i \neq j$, set $G_0X_{ij} = Y_{ij}$, where Y_{ij} is the solution of the equation

$$AY_{ij} - Y_{ij}A = X_{ij},\ i \neq j,\ i,j \in J,$$

and $Y_{ii} = 0$ for each $i \in J$. Note that the last equation can be represented in the form

$$A_iY_{ij} - Y_{ij}A_j = X_{ij}, \tag{5}$$

where $A_i = A|\mathcal{H}_i$, and $\mathcal{H}_i = RanP_i$. Since $\sigma(A_i) \cap \sigma(A_j) = \emptyset, i,j \in J$, it follows that equation (5) is solvable and the following inequality holds:

$$||Y_{ij}|| \leq \frac{c||X_{ij}||}{dist(\sigma_i, \sigma_j)} \leq \frac{c||X_{ij}||}{|a(i) - a(j)|} \leq c(min_{i \neq j}(|a(i) - a(j)|))^{-1}.$$

In addition, set $Y_{ii} = 0$ for each $i \in J$.

Now we form the operator G_0X from the operator blocks $Y_{ij} = (G_0X)_{ij}$ as follows: $G_0X = \sum_{i,j}(GX)_{ij}(see[2],[3])$.

The transformer G_0X belongs to the space $End\ell_2$ and $||G_0|| \leq d_0^{-1}$.

In addition, we define a family of transformers G_kX by setting

$$G_kX = G_0X - G_0(Q_kXQ_k) = G_0X - Q_k(G_0X)Q_k,\ k \geq 0.$$

It follows from the last relation that

$$||G_kX|| \leq cd_k^{-1}||X||, \gamma = \gamma_k = cd_k^{-1}.$$

If $X, Y \in End\ell_2$, then $X(G_kY), (G_kX)Y$ also belong to $End\ell_2$ and $||X(G_kY)|| \leq \gamma_k||X||||Y||$, $||(G_kX)Y|| \leq \gamma_k||X|| \cdot ||Y||$.

It follows from the preceding argument that condition 4 in Definition 6 holds with $\gamma = \gamma_k$ of the order of d_k^{-1}.

Operator J_0 is idempotent, because

$$J_0^2X = J_0(J_0X) = J_0(\sum_{i \in J} P_iXP) = \sum_{j \in J} P_j(\sum_{i \in J} P_iXP_i)P_j = \sum_{j \in J} P_jXP_j = J_0X.$$

It implies that condition 2 of Definition 6 holds true.

Let us check condition 3 of Definition 6. Let $Q_n = \sum_{|i|<n, i \in J} P_i$. We consider operator $AQ_nGX_0A^{-1}$ and represent it as

$$AQ_nG_0XA^{-1}x = Q_nG_0Xx + Q_n(X - J_0X)A^{-1}x,\ x \in \ell_2(Z).$$

Since $Q_nG_0X \to G_0Xx, Q_n(X - J_0X)A^{-1}x \to (X - J_0X)A^{-1}x$ as $n \to \infty$, then $AQ_nG_0XA^{-1}x \to y_0 \in \ell_2(Z)$. Let $Q_nG_0XA^{-1}x \to x_0 = G_0XA^{-1}x$, then due to the closedness of operator A we have $x_0 \in D(A)$ and $Ax_0 = y_0$, where $y_0 = \lim_{n \to \infty} y_n$.

Condition 6 is obvious since

$$||X(A - \lambda_\varepsilon I)^{-1}|| \leq ||X||||(A - \lambda_\varepsilon I)^{-1}||$$

at that, the first factor is finite and the other can be chosen arbitrarily small.

Thus, we have proven

Lemma 2. For each $k \geq 0$ the triple $(End\ell_2, J_k, G_k)$ is admissible for operator A.

Lemma 2, Theorem 1 and estimate (3) imply Theorem 2.

For the class of difference operators under consideration, one may also use as the admissible space of perturbations the space $End_1\ell_2 \subset End\,\ell_2$, which is defined as follows. An operator $Y \in End\,\ell_2$ belongs to the space $End_1\ell_2$ if the diagonals of the matrix of Y in the standard basis of ℓ_2 are summable. We note that the perturbation B belongs to $End_1\ell_2$ because the matrix of the operator B is tri-diagonal. The space $End_1\ell_2$ was chosen as the admissible space of perturbations in [10–11]. In this case, an analog of Theorem 2 may be proved along the same lines, but the operator X in the theorem must also belong to $End_1\ell_2$. If the sequence $d_i, i \in J$, is such that $\sum_{i\in J} d_i^{-2} < \infty$ then the admissible space of perturbations may be chosen as the ideal $\mathfrak{S}_2(\ell_2)$ of Hilbert-Schmidt operators. The operator X_0 in Theorem 2 will then also belong to the ideal $\mathfrak{S}_2(\ell_2)$. In general, however, we do not have $B \in \mathfrak{S}_2(\ell_2)$. This is why we have to start with a preliminary similarity transformation of the operator $A - B$ into the operator $A - \widetilde{B}$, where $\widetilde{B} \in \mathfrak{S}_2(\ell_2)$ [7].

Lemma 3. Under the assumptions of Theorem 2 we have

$$||P_i(X-B)P_i|| \le c_1 d_i^{-1},\ |i| > k, i \in J,$$
$$||P_i(X-B-BG_kB)P_i|| \le c_2 d_i^{-2},\ |i| > k, i \in J.$$

Proof. We have

$$X - B = BG_kX - G_kXJ_kB - G_kXJ_k(BG_kX);\ J_k(X-B) = J_k(BG_kX)$$

and

$$||P_i(X-B)P_i|| = ||P_i(BG_kX)P_i|| \le d_i^{-1}||B|| \cdot ||X|| \le 3d_i^{-1} \cdot ||B||^2 \le c_1 d_i^{-1},\ |i| > k, i \in J, c_1 > 0.$$

Analogically,

$$||P_i(X-B-BG_kB)P_i|| \le ||P_i(BG_k(X-B))P_i|| \le c_2 d_i^{-2}.$$

Assume that Theorem 2 holds true. Then the similarity of operators $A - B$ and $A - J_kX$ yields $\sigma(A-B) = \sigma(A - J_kX_0) = \sigma(A_{(k)}) \cup (\cup_{|i|>k} A_i)$, where $A_{(k)} = (A - Q_kX_0Q_k)|\mathcal{H}_{(k)}$. $\mathcal{H}_{(l)} = RanQ_l$ and $A_i = (P_iA - P_iX_0)|\mathcal{H}_i$. Since operator X_0 is unknown and we know only the first and the second approximation, then $A_i = (P_iA - P_iB - P_i(X_0 - B))|\mathcal{H}_i$ and the first two operators are known, while for the third operator we know only estimate from Lemma 3. Analogy, $A_i = (P_iA - P_iB - P_i(BG_k(X_0 - B))|\mathcal{H}_i$ and first three operators are known, while for the forth operator we know only estimate from Lemma 3.

The following thorem describes the spectral properties of the operator $\mathcal{E}$.

Theorem 3. Let the sequence $a : Z \to C$ satisfy conditions 1). There is a number $k \in Z_+$ such that the spectrum $\sigma(\mathcal{E})$ of the operator $\mathcal{E}$ satisfies

$$\sigma(\mathcal{E}) = \sigma_{(k)} \cup \left(\bigcup_{|l|>k} \sigma_l \right),$$

where $\sigma_{(k)}$ consists of no more than $2k+1$ eigenvalues. The sets $\sigma_l, |l| > k$, are one-point sets $\sigma_l = \{\mu_l\}$ and

$$\mu_l = a(l) + 2 + O(d_l^{-1}),$$
$$\mu_l = a(l) + 2 - \frac{a(l+1) - 2a(l) + a(l-1)}{(a(l+1) - a(l))(a(l-1) - a(l))} + O(d_l^{-3}),\ l \in Z.$$

The corresponding eigenvectors $\widetilde{e}_i, |i| > k$, satisfy the asymptotic estimates

$$||\widetilde{e}_i - \widetilde{y}_i|| = O(d_i^{-2}), |i| > k,$$

where $\widetilde{y}_i \in \ell_2(Z)$ and

$$\widetilde{y}_i(k) = \begin{cases} 1, & i = k; \\ (a(i \pm 1) - a(i))^{-1}, & k = i \pm 1; \\ 0, & \text{in other cases.} \end{cases}$$

The eigenvectors $\widetilde{e}_i, i \in Z$, constitute the Riesz basis in the space $\ell_2(Z)$.

Theorem 4. Let the sequence $a : Z \to C$ satisfy conditions 2). There is a number $k \in Z_+$ such that the spectrum $\sigma(\mathcal{E})$ satisfies

$$\sigma(\mathcal{E}) = \sigma_{(k)} \cup \left(\bigcup_{l>k} \sigma_l \right),$$

where $\sigma_{(k)}$ consists of no more than $2k + 1$ eigenvalues. The sets $\sigma_l, l > k$, satisfy $\sigma_l = \{\mu_l, \mu_{-l}\}$ and

$$\mu_{\pm l} = a(l) + 2 + O(d_l^{-1}),\ l > k,$$

$$\mu_{\pm l} = a(l) + 2 - \frac{a(l+1) - 2a(l) + a(l+1)}{(a(l+1) - a(l))(a(l-1) - a(l))} + O(d_l^{-3}),\ l > k.$$

Let $\widetilde{P_n}, n \in J$, be the spectral projections $\widetilde{P_n} = P(\sigma_n, \mathcal{E})$ corresponding to the sets $\sigma_n, n \in J$, described in Theorem 3 or 4.

Theorem 5. We have

$$\left\|\widetilde{P_n} - P_n\right\| = O(d_n^{-1}),$$

$$\left\|\sum_{n \geq m}^{N} \widetilde{P_n} - \sum_{n \geq m}^{N} P_n\right\| = O(d_m^{-1}),$$

where $m > k, N > k, m, N \in J$.

The spectral projections $\widetilde{P}_n$ also satisfy the following estimates of uniform unconditional equiconvergence of spectral expansions:

$$\left\|P(\sigma_{(k)}, \mathcal{E}) + \sum_{|i|>k}^{l} \widetilde{P}_i - \sum_{|i|<l} P_i\right\| = O(d_l^{-1}),\ \text{for } J = Z,$$

and

$$\left\|P(\sigma_{(k)}, \mathcal{E}) + \sum_{i>k}^{l} \widetilde{P}_i - \sum_{i=0}^{l} P_i\right\| = O(d_l^{-1}),\ \text{for } J = Z_+),$$

where $l \in Z_+, l > k$.

We need the following definition in the monograph [12].

Definition 7. Let $C : D(C) \subset \mathcal{H} \to \mathcal{H}$ be a linear operator in the space $\mathcal{H}$ whose spectrum can be represented as the union

$$\sigma(C) = \bigcup_{k \in J} \sigma_k \tag{6}$$

of pairwise disjoint sets $\sigma_k, k \in Z$, and let P_k be the Riesz projection corresponding to the spectral set σ_k. An operator C is said to be *spectral with respect to the expansion* (6) (or *generalized-spectral*) if the series $\sum_k P_k x$ is convergent for any vector $x \in \mathcal{H}$.

If $\sigma_k = \{\lambda_k\}, k \in J,$ – are one-point sets and $CP_k = \lambda_k P_k$ for all k, except for finitially many, then spectral with respect to the expansion (6) operator C is spectral operator. Operator A is the spectral operator of scalar type if $AP_k = \lambda_k P_k$ for all $k \in J$.

The following assertion is a straightforward consequence of Theorem 5.

Theorem 6. The operator L is spectral operator.

4. Conclusion

In this paper we studied the spectral properties of a second order difference operator with a growing potential. The operator acts in the complex Hilbert space $\ell_2(Z)$ of square summable complex sequences indexed by the integers. This operator is a dicrete analogue of a second order differential operator with a growing complex potential. Asymptotic estimates of eigenvalues, eigenvectors and the spectral projections were obtained. The main method of this exploration was the method of similar operators, which allowed us to reduce the study of the operator to one with a block-diagonal matrix. This method was thoroughly described in the paper and its basic definitions, such as admissible triplet and commutator transform, were given. Main results of the paper were formuled in five theorems and two lemmas, proofs of one theorem and two lemmas were also provided. The article is ten pages long and list of references consists of twelve issues.

Acknowledgments

The research was supported by the Russian Foundation for Basic Research (project no. 16-01-00197).

References

[1] Musilimov B and Otelbaev M 1981 *Comput. Math. Math. Phys.* **21** 5
[2] Baskakov A 1983 *Siberian Math J.* **24** 15
[3] Baskakov A 1995 *Izv. Math.* **45** 31
[4] Baskakov A, Derbushev A and Shcherbakov A 2011 *Izv. Math.* **75** 24
[5] Baskakov A and Krishtal I 2013 *J. Math. Anal. and Appl.* **407** 21
[6] Baskakov A, Krishtal I and Romanova E 2017 *J. Evol. Equat.* **17** 15
[7] Baskakov A and Polyakov D 2017 *Siberian Math. J.* **208** 43
[8] Baskakov A and Krishtal I 2014 *J. Funct. Anal.* **267** 54
[9] Garkavenko G and Uskova N 2016 *Proc. of Voronezh St. Univ.* **3** 10
[10] Garkavenko G, Uskova N and Zgolich A 2016 *Belgorod St. Univ. Scient. Bul. Math. Phys.* **43** 7
[11] Garkavenko G and Uskova N 2017 *Siberian Electron. Math. Izv.* **14** 16
[12] Dunford N and Schwartz J 1971 *Linear operators. Part III. Spectral operators* (New York: Wiley-Interscience)

On a numerical method for solving integro-differential equations with variable coefficients with applications in finance

O Kudryavtsev[1] **and V Rodochenko**[2]

[1] Department of Informatics and Information Customs Technologies, Russian Customs Academy Rostov branch, Budennovskiy 20, Rostov-on-Don 344002, Russia
[2] Department of Algebra and Discrete Mathematics, Institute for Mathematics, Mechanics, and Computer Science in the name of I.I. Vorovich, Southern Federal University, Milchakova str. 8a, Rostov-on-Don 344090, Russia

E-mail: [1] koe@donrta.ru, [2] vrodochenko@gmail.com

Abstract. We propose a new general numerical method aimed to solve integro-differential equations with variable coefficients. The problem under consideration arises in finance where in the context of pricing barrier options in a wide class of stochastic volatility models with jumps. To handle the effect of the correlation between the price and the variance, we use a suitable substitution for processes. Then we construct a Markov-chain approximation for the variation process on small time intervals and apply a maturity randomization technique. The result is a system of boundary problems for integro-differential equations with constant coefficients on the line in each vertex of the chain. We solve the arising problems using a numerical Wiener-Hopf factorization method. The approximate formulae for the factors are efficiently implemented by means of the Fast Fourier Transform. Finally, we use a recurrent procedure that moves backwards in time on the variance tree. We demonstrate the convergence of the method using Monte-Carlo simulations and compare our results with the results obtained by the Wiener-Hopf method with closed-form expressions of the factors.

1. Introduction

In the years since Dynkin, Feinman, and Katz published their fundamental works, it has been well known, that the solution to a diffusion equation in a given domain can be interpreted as an expected value of a function of Brownian motion at the first exit point from the domain. This idea was later generalized for the processes combining both diffusion and Lévy models, trajectories of which can be discontinuous. Nowadays Lévy processes are widely used for modeling various social end economical processes. Both researchers and financial institutes pay a lot of attention to the processes. From the point of view of financial mathematics, development of means for solving these equations allows to solve an essential applied problem — the problem of estimation of risk-neutral prices of derivative financial instruments in models with jumps, in particular — barrier options.

Thus, calculating functionals from Markov processes with jumps plays an important role in quantitative finance due to the fact that the presence of jumps even on liquid assets became more pronounced in recent decades. This task, nonetheless, is mathematically complex and comes down to solving complex partial integro-differential equations known as forward and backward

Kolmogorov equations with certain initial and boundary conditions. As a rule, analytical methods are useless for solving such equations, which brings a necessity to use cutting-edge methods of numerical analysis.

Let's consider a system of stochastic differential equations in Ito form that describes a dynamics of a position S_t of a particle on a line, that is connected with variance V_t driven by CIR process [1].

$$\begin{aligned} dS_t &= (r - \lambda_J \zeta) S_t dt + \sqrt{V_t} S_t dZ_t^S + (J-1) S_t dN_t, \\ dV_t &= \kappa_V(\theta_V - V_t)dt + \sigma_V \sqrt{V_t} dZ_t^V, \\ &< dZ_t^S, dZ_t^V >= \rho dt, \end{aligned} \tag{1}$$

where r is a non-negative parameter, Z_t^S and Z_t^V are Wiener processes with correlation coefficient ρ. Jumps are defined by Poisson process N_t with intensity λ_J, that counts a quantity of identically distributed jumps of size J to the moment t. The process N_t does not depend on processes Z_t^S and Z_t^V, and is independent of J as well. Parameter κ_V defines the speed of "regression" of variance process to a "long-run" variance average value θ_V, $\sigma_V > 0$ is referred to as a "volatility" of variance. The value of ζ is fixed by the distribution parameters of the jump size J so that $\exp(-rt)S_t$ is a martingale.

In financial applications S_t in (1) simulates dynamics of an underlying asset, whereas a parameter $r \geq 0$ denotes a risk-free interest rate. If jumps J are distributed log-normally, $J \sim \text{LogN}(\mu_J, \sigma_J^2)$, then $\zeta = e^{\mu_J + \frac{1}{2}\sigma_J^2} - 1$ and we obtain a well-known "Bates model" initially presented in paper [2]. By setting jump intensity to zero while keeping a variance process dynamics, we will obtain a system that represents a Heston model — one of the most widely-used models with stochastic volatility [3].

Replacing the process V_t with a positive constant V_0 will make the first equation of the system (1) describe a Lévy process which is referred to as "jump-diffusion" (see, for example, [4]). Particularly, if J is log-normally distributed, we get Merton model [5]. With the intensity of jumps set to zero the model finally comes down to the most deeply explored models in finance — the Black-Scholes model [6].

The goal of the research is to develop a numerical method to calculate the following conditional expectation

$$f(S, V, t) = E[e^{-r(T-t)} \mathbf{1}_{\underline{S}_T > H} G(S_T) | S_t = S, V_t = V], \tag{2}$$

for the class of models described by (1), where H is an absorbing barrier, $\underline{S}_T (= \inf_{0 \leq t \leq T} S_T)$ is infimum process for the process S_t.

Functionals like (2) arise in financial mathematics when solving a problem of pricing barrier options in models described by the system (1). A "barrier option" is a contract that pays a certain sum $G(S_T)$ on expiration time T, if and only if the underlying asset price S_T does not cross a certain barrier H from above (down-and-out barrier option) or from below H (up-and-out barrier option). For example, for an option that grant a holder a right to sell a certain asset for a certain price K ("put option"), $G(S) = \max\{0, K - S\}$, whereas for the option that grants a right to buy a certain asset for a certain price K ("call option"), $G(S) = \max\{0, S - K\}$.

Suppose $\tau = T - t$, then $F(S, V, \tau) (= f(S, V, T - \tau))$ satisfies the following (see, for example,

IOP Conf. Series: Journal of Physics: Conf. Series **973** (2018) 012054 doi:10.1088/1742-6596/973/1/012054

[7]) partial integro-differential equation with variable coefficients in the domain $S(\tau) > H$:

$$
\begin{aligned}
\frac{\partial F(S,V,\tau)}{\partial \tau} &= \frac{1}{2}VS^2\frac{\partial^2 F(S,V,\tau)}{\partial S^2} + \rho\sigma_V VS\frac{\partial^2 F(S,V,\tau)}{\partial S\partial V} + \\
\frac{1}{2}\sigma_V^2 V\frac{\partial^2 F(S,V,\tau)}{\partial V^2} &+ (r-\lambda_J\zeta)S\frac{\partial F(S,V,\tau)}{\partial S} + \kappa_V(\theta_V - V)\frac{\partial F(S,V,\tau)}{\partial V} - \\
(r+\lambda_J)F(S,V,\tau) &+ \lambda_J\int_0^\infty F(JS,V,\tau)f(J)dJ, \\
F(S,V,0) &= G(S),
\end{aligned}
$$

where $f(J)$ is a probability density function for jump amplitudes J. Because H is an absorbing barrier and S_t has an ability to cross the barrier, we need to include condition $F(S,V,\tau)=0, S(\tau)\leq H$.

Most common approaches to solve such equations are the Monte-Carlo method [8], finite-difference schemes (see, for example, [9]) and a hybrid approach that combines tree method and finite-difference schemes [10].

The most significant drawback of Monte-Carlo methods is low computational performance, because for the models with jumps and problems with barriers detailed trajectory simulation becomes inevitable. Finite-difference schemes approach highly depends on exact kind of Lévy model and does not provide much accuracy because a corresponding matrix of a system is dense (unlike diffusion models), even though many authors implicitly use only its tridiagonal part for its inversion. Some authors combine split method and finite-difference schemes while approximating an integral component as a function of a symbol of the corresponding pseudo-differential operator from finite differences [11], but for popular Lévy models it leads to computing matrix logarithms and exponents, which can lead to sufficient computational errors.

For the sake of dimension reduction some authors (see, for example, [12, 13]) suggest using Markov chain with continuous time for volatility process approximation. The simplest case of this approach is a binomial model [13]. As a result, we obtain Lévy model with regime-switching volatility. Markov chain is constructed in a way to save corresponding drift and volatility for transitional probabilities from every state.

In a series of articles (see, for example, [14]) a general method of approximate Wiener-Hopf factorization for a wide class of Lévy processes with a constant variation was used to solve problems with barriers for integro-differential equations, to which evaluation of prices for different options comes down. The numerical implementation of Wiener-Hopf operators was performed with the help of Fast Fourier Transform. This made the computational methods comparable to finite-difference schemes in simplicity of implementation, yet significantly faster and more accurate, as numerical experiments show [14].

One of the core ideas of a new numerical method proposed for solving a problem (2) is to use a new enhanced approximate Wiener-Hopf factorization, initially presented in [15]. The new formulae are more general-purposed in comparison to [14] and easier to use, as they do not necessarily require the separation of main part of a characteristic function under factorization. In order to solve a problem after time discretization (for example, Carr randomization) the coefficients of the integro-differential operator obtained are going to be approximated using a tree method which models a volatility process, leading therefore a problem in each state to a model with a constant volatility, and new explicit formulae for Wiener-Hopf factors are going to be used for a respective task. As a result, we obtain a method comparable in speed to a finite-difference scheme, but more general and accurate.

2. Materials and methods

2.1. *Process substitution*

One important feature of the model considered is an effect of correlation between Wiener processes in the equations for price and variation. To solve these equations one must take it into account while developing a numerical procedure. The method we use is similar to the one in [10], and aimed at constructing a suitable substitution for processes. Let us define

$$\hat{\rho} = \sqrt{1-\rho^2},$$
$$W = Z^V,$$
$$\rho W + \hat{\rho} Z = Z^S,$$

where W_t Z_t are independent Brownian motions. Let us first also introduce a substitution for the process S_t, setting

$$Y_t = \ln(\frac{S_t}{H}) - \frac{\rho}{\sigma_V} V_t.$$

In these terms,

$$S_t = H\exp(Y_t + \frac{\rho}{\sigma_V} V_t).$$

Being constructed in such way, the substitution allows to proceed a logarithmic scale and make a normalization with respect to a barrier at the same time, which makes computation more convenient. After making both substitutions, the system (1) will be as follows:

$$dY_t = (r - \tfrac{1}{2}V_t - \tfrac{\rho}{\sigma_V}\kappa_V(\theta_V - V_t) - \lambda_J\zeta)dt + \hat{\rho}\sqrt{V_t}dZ_t + \ln J dN_t,$$
$$dV_t = \kappa_V(\theta_V - V_t)dt + \sigma_V\sqrt{V_t}dW_t.$$

Let's introduce a simpler notation for the drifts we use later:

$$\mu_Y(V_t) = r - \tfrac{1}{2}V_t - \tfrac{\rho}{\sigma_V}\kappa_V(\theta_V - V_t) - \lambda_J\zeta,$$
$$\mu_V(V_t) = \kappa_V(\theta_V - V_t).$$

2.2. *Carr randomization*

In order to be able to operate on small timeframes we can use a procedure known as "Carr randomization" ("Canadization"), first introduced in [19] and generalized for a more general case of stochastic control problems in paper [17]. Let's denote

$$f_n(Y_{t_n}, V_{t_n}) = f(H\exp(Y_{t_n} + \frac{\rho}{\sigma_V}V_{t_n}), V_{t_n}, t_n)$$

a Carr approximation for the function $f(S, V, t)$ at the moment t_n, where $t_i = i\Delta\tau$ and $\Delta\tau = \frac{T}{N}$. Let $\{\tau_i\}_{i=1}^N$ — be a sequence of independent exponentially distributed variables with an average value of $\Delta\tau$. Let's denote $Z_\tau^n = Y_{t_n+\tau} + \frac{\rho}{\sigma_V}V_{t_n+\tau}$. Assuming $f_N(y, v) = G(H\exp(y + \frac{\rho}{\sigma_V}v))$, we have an opportunity to write down the following:

$$f_n(y,v) = E_{t_n}[e^{-r\tau_n}\mathbf{1}_{\underline{Z}^n_{\tau_n}>0} f_{n+1}(Y_{t_n+\tau_n}, V_{t_n+\tau_n})|Y_{t_n} = y, V_{t_n} = v], n = N-1, .., 0. \quad (3)$$

This form of expectation can be iteratively calculated for the case of Lévy model with fixed variance and comes down to a sequence of problems. However, the presence of stochastic variance in our case makes it necessary to use an approximation for the CIR process V, in order to compute it.

2.3. Approximation

The next goal we would like to achieve is to reduce the initial problem to a set of problems with lower dimension and obtain a Lévy model with a regime-switching volatility. For this purpose we construct approximation of a CIR process using a Markov chain following the procedure from [13]. Among the advantages of the scheme we can mention fast convergence towards the parent process and the ability to remain adequate in a wide range of parameters. In particular, keeping the Feller condition $2\kappa_V\theta_V > \sigma_V^2 r$ is not significant in this case for the procedure to remain correct.

Let us construct a binomial tree with "shared" vertices defined by formula:

$$V(n,k) = (\sqrt{V_0} + \frac{\sigma_V}{2}(2k-n)\sqrt{\Delta\tau})^2 \mathbf{1}_{(\sqrt{V_0}+\frac{\sigma_V}{2}(2k-n)\sqrt{\Delta\tau})>0}, n = 0,1,...,N,\ k = 0,1,...,n.$$

The idea of the approximation is that in every moment of time t_n a variation can take one of the states $V(n,k)$. In the moment t_{n+1} from the vertex (n,k) we can either ho "up" — to a vertex $(n+1,k_u)$, or "down" — to $(n+1,k_d)$. The vertices k_u and k_d are being calculated in accordance with the drift $\mu(V_{(n,k)})$, using the following rules:

$$k_u^{\Delta\tau}(n,k) = min\{k^* : k+1 \le k^* \le n+1, V(n,k) + \mu_V(V(n,k))\Delta\tau \le V(n+1,k^*)\},$$
$$k_d^{\Delta\tau}(n,k) = max\{k^* : 0 \le k^* \le k, V(n,k) + \mu_V(V(n,k)) \ge V(n+1,k^*)\}.$$

Let us define transitional probabilities as follows:

$$p^{\Delta\tau}_{k_u^{\Delta\tau}(n,k)} = \frac{\mu_V(V(n,k))\Delta\tau + V(n,k) - V(n+1,k_d^{\Delta\tau}(n,k))}{V(n+1,k_u^{\Delta\tau}(n,k)) - V(n+1,k_d^{\Delta\tau}(n,k))}.$$

In order to maintain correct work of the scheme for different parameters values we should introduce additional rules that make impossible for the negative probabilities to appear in some vertices:

$$p^{\Delta\tau}_{k_u^{\Delta\tau}(n,k)} := \begin{cases} 1, & p^{\Delta\tau}_{k_u^{\Delta\tau}(n,k)} > 1 \\ p^{\Delta\tau}_{k_u^{\Delta\tau}(n,k)}, & p^{\Delta\tau}_{k_u^{\Delta\tau}(n,k)} \in [0,1] \\ 0, & p^{\Delta\tau}_{k_u^{\Delta\tau}(n,k)} < 0 \end{cases}, \quad p^{\Delta\tau}_{k_d^{\Delta\tau}(n,k)} := 1 - p^{\Delta\tau}_{k_u^{\Delta\tau}(n,k)}.$$

2.4. Approximate factorization

The tree procedure described above can now be used to obtain problems with a fixed variation. With regard to the Markov chain, the expectation (3) for $n = N-1,..,0$ can be represented in the form:

$$\begin{aligned} f_n(y, V(n,k)) &= p^{\Delta\tau}_{k_d^{\Delta\tau}(n,k)} E_{t_n}[e^{-r\tau_n}\mathbf{1}_{\underline{Z}^n_{\tau_n}>0} f_{n+1}(Y_{t_n+\tau_n}, V(n+1,k_d))|Y_{t_n} = y] \\ &+ p^{\Delta\tau}_{k_u^{\Delta\tau}(n,k)} E_{t_n}[e^{-r\tau_n}\mathbf{1}_{\underline{Z}^n_{\tau_n}>0} f_{n+1}(Y_{t_n+\tau_n}, V(n+1,k_u))|Y_{t_n} = y]. \end{aligned}$$

Thus, for each of vertices $(n,k), n = N-1,...0$ two expectations to compute arise: one in the assumption that transition has gone up to a vertex $(n+1,k_d)$, another one in the assumption that transition has gone down to a vertex $(n+1,k_u)$.

Let us introduce a shorthand notation

$$f_n^k(y) = E_{t_n}[e^{-r\tau_n}\mathbf{1}_{\underline{Z}^n_{\tau_n}>0} f_{n+1}(Y_{t_n+\tau_n}, V(n+1,k))|Y_{t_n} = y].$$

It is well known (see e.g. [14]) that the expectation $f_n^k(y)$ is a solution to the following integro-differential equation:

$$\begin{cases} (q - L_{n+1,k})f_n^k(y) = \Delta t^{-1} f_{n+1}(y, V(n+1,k)) & y > -\frac{\rho}{\sigma}V(n,k) \\ f_n^k(y) = 0 & y \leq -\frac{\rho}{\sigma}V(n,k), \end{cases} \tag{4}$$

where $q = \Delta t^{-1} + r$ and

$$L_{n,k} := L_Y^{V(n,k)} = \mu_Y(V(n,k))\partial_y + \frac{1}{2}\hat{\rho}^2 V(n,k)\partial_{yy}^2.$$

The integro-differential operator $L_{n,k}$ can be interpreted as a pseudo-differential operator

$$L_{n,k}f(y) := L_Y^{V(n,k)}f(y) = \frac{1}{2\pi}\int_{-\infty}^{\infty} e^{iy\xi}\psi_{n,k}(\xi)\hat{f}(\xi)d\xi.$$

with the symbol $\psi_{n,k}(\xi)$ — a characteristic exponent of a process Y_t with $V_t = V(n,k)$:

$$\psi_{n,k}(\xi) = \hat{\rho}^2 \frac{V(n,k)}{2}\xi^2 - i\mu_Y(V(n,k))\xi + \phi(\xi),$$

where $\phi(\xi)$ is a characteristic exponent of a compound Poisson process. For example, for a Merton model it looks as follows: $\phi(\xi) = \lambda_J(1 - e^{\frac{\sigma_J^2}{2} + \mu_J})$.

To find each of $f_n^k(y)$, one should solve two problems:

$$\begin{cases} (q - L_{n+1,k_u})f^{k_u}(y) = \Delta t^{-1} f_{n+1}(y, V(n+1,k_u)), & y > -\frac{\rho}{\sigma}V(n,k), \\ f^{k_u}(y) = 0, & y \leq -\frac{\rho}{\sigma}V(n,k), \end{cases} \tag{5}$$

$$\begin{cases} (q - L_{n+1,k_d})f^{k_d}(y) = \Delta t^{-1} f_{n+1}(y, V(n+1,k_d)), & y > -\frac{\rho}{\sigma}V(n,k), \\ f^{k_d}(y) = 0, & y \leq -\frac{\rho}{\sigma}V(n,k), \end{cases} \tag{6}$$

where $q = \Delta t^{-1} + r$; $n = 0, 1, ..., N$; $k = 0, 1, ..., n$.

Solutions to each of the problems (4)-(5) mentioned can be written in terms of the Wiener-Hopf operators ε_q^+ and ε_q^- (see the paper [14]):

$$f_n^{k_d}(y) = (q\Delta\tau)^{-1}\, \varepsilon_q^- \mathbf{1}_{(-\frac{\rho}{\sigma_{V(n,k)}}V(n,k),+\infty)}(y)\, \varepsilon_q^+ f_{n+1}(y, V(n+1,k_d)),$$
$$f_n^{k_u}(y) = (q\Delta\tau)^{-1}\, \varepsilon_q^- \mathbf{1}_{(-\frac{\rho}{\sigma_{V(n,k)}}V(n,k),+\infty)}(y)\, \varepsilon_q^+ f_{n+1}(y, V(n+1,k_u)).$$

Further on, consequently calculating $f_n(y, V(n,k)) = p_{k_d^{\Delta t}(n,k)} f_n^{k_d}(y) + p_{k_u^{\Delta t}(n,k)} f_n^{k_u}(y)$ for $n = N-1, .., 0, k = 0, ..., n,$, we can get the approximate values for the sought-for functional (2) after returning to the initial notations.

The Wiener-Hopf operators can be represented in the form of pseudo-differential operators [14]:

$$\varepsilon_q^{\pm}u(x) = (2\pi)^{-1}\int_{-\infty}^{+\infty} e^{ix\xi}\phi_q^{\pm}(\xi)\hat{u}(\xi)d\xi.$$

Unlike a much simpler case of Heston model [18], the presence of jumps does not allow to use closed formulae for factors unknown at the moment.

IOP Conf. Series: Journal of Physics: Conf. Series **973** (2018) 012054 doi:10.1088/1742-6596/973/1/012054

Analytical formulae for the factors look as follows:

$$\phi_q^+(\xi) = \exp\left[(2\pi i)^{-1}\int_{-\infty+i\omega_-}^{+\infty+i\omega_-}\frac{\xi\ln(q+\psi(\eta))}{\eta(\xi-\eta)}d\eta\right],$$
$$\phi_q^-(\xi) = \exp\left[-(2\pi i)^{-1}\int_{-\infty+i\omega_+}^{+\infty+i\omega_+}\frac{\xi\ln(q+\psi(\eta))}{\eta(\xi-\eta)}d\eta\right].$$

The constants ω_+ and ω_- are so that $\omega_- < 0 < \omega_+$ play roles of parameters here and are being set to retain convergence of the respective integrals depending on Lévy process parameters. In the paper [15] a general and convenient for a numerical implementation representation of Wiener-Hopf factors was derived. The function $\phi_q^+(\xi)$ allows analytical continuation to a half-plane $\Im\xi > \omega_-$ and can be represented as follows:

$$\phi_q^+(\xi) = \exp\left[i\xi F^+(0) - \xi^2\hat{F}^+(\xi)\right],$$
$$F^+(x) = \mathbf{1}_{(-\infty,0]}(x)(2\pi)^{-1}\int_{-\infty+i\omega_-}^{+\infty+i\omega_-} e^{ix\eta}\frac{\ln(q+\psi(\eta))}{\eta^2}d\eta,$$
$$\hat{F}^+(\xi) = \int_{-\infty}^{+\infty} e^{-ix\xi}F^+(x)dx.$$

Analogously, $\phi_q^-(\xi)$ allows an analytical continuation to a halfplane $\Im\xi < \omega_+$ and can be represented as follows:

$$\phi_q^-(\xi) = \exp\left[-i\xi F^-(0) - \xi^2\hat{F}^-(\xi)\right],$$
$$F^-(x) = \mathbf{1}_{[0,+\infty)}(x)(2\pi)^{-1}\int_{-\infty+i\omega_+}^{+\infty+i\omega_+} e^{ix\eta}\frac{\ln(q+\psi(\eta))}{\eta^2}d\eta;$$
$$\hat{F}^-(\xi) = \int_{-\infty}^{+\infty} e^{-ix\xi}F^-(x)dx.$$

To calculate integrals we use an algorithm of Fast Fourier Transform.

3. Results and discussion

We provide the results of numerical experiments, that demonstrate the convergence and accuracy of the method proposed. We will refer to our method as M&WHFa, having in mind the use of a Markov Chain approximation and the approximation for Wiener-Hopf factors. To be able to compare the result with the existing methods, we use the European down-and-out put in Heston model with the parameters set as follows: $V_0 = 0.01$, $\kappa = 2$, $\theta = 0.01$, $\sigma = 0.2$, $\rho = 0.5$, $r = 0.095$. Option parameters are: $T = 1$, $K = 100$, $H = 90$. We also have closed formulae expressions for the Wiener-Hopf factors for the Heston model. So for comparison we will use the algorithm [18] that uses them for computation (which we will refer to as M&FWHFe).

As a benchmark for all methods we used a plain Monte-Carlo generator that uses a Mersenne Twister as a pseudo-random generator.

The advanced Alfonsi [8] MC results were obtained using the code implemented into the program platform Premia.

To perform the experiment we used a PC with a configuration: Intel CPU, 2.4 GHz, RAM 8 Gb, Windows 10. The program implementation was made using a scientific Python stack (numpy, scipy and pandas packages).

IOP Conf. Series: Journal of Physics: Conf. Series **973** (2018) 012054 doi:10.1088/1742-6596/973/1/012054

The algorithm parameters for the M&FWHe are as follows: N is the number of time steps, L is a scale coefficient, which controls the localization error of the FFT, and M is a number of points, meant to control its the discretization error. "Stock price" is a price of an underlying asset.

The algorithm parameters for the M&FWHa are as follows: N is the number of time steps, h is a distance between the points on a logarithmic scale, which controls the localization error of the FFT, and M is a number of points, meant to control its the discretization error. "Stock price" is a price of an underlying asset.

Table 1 illustrates the convergence of the method, Table 2 and Table 3 shows absolute and relative error values, respectively. It's therefore natural to assume that similar accuracy can be achieved in models where no closed formulae representations for factors is known.

Table 1. Barrier put option prices for the case of Heston model, calculated with Standard MC, Alfonsi MC, M&FWHFe and M&FWHa.
Heston model parameters: $v_0 = 0.01$, $\kappa = 2$, $\theta = 0.01$, $\sigma = 0.2$, $\rho = 0.5$, $r = 0.095$
Option parameters: $K = 100$, $H = 90$, $r = 0.072310$, $T = 1$.
Method parameters: h — space variable step, N — number of time steps (or an algorithm parameter for both WH methods), L — a scale multiplier parameter for M&FWHFe method, M — a number of points for WH methods, S — stock price.

Stock	standard MC	Alfonsi MC	M&FWHFe	M&WHFa	
price	(mean value) $h = 0.0001$ $N = 500000$	(mean value) $h = 0.0001$ $N = 600000$	$N = 300$ $M = 2^{18}$ $L = 3$	$N = 300$ $h = 5 \cdot 10^{-6}$ $M = 2^{20}$	$N = 300$ $h = 1 \cdot 10^{-6}$ $M = 2^{20}$
$S = 91$	0.21914	0.304862	0.218113	0.226777565	0.222095738
$S = 95$	0.57994	0.579616	0.587911	0.607220378	0.590497506
$S = 101$	0.26102	0.256442	0.267915	0.272406178	0.265426451
$S = 106$	0.08105	0.078067	0.082999	0.084105566	0.082232504
$S = 111$	0.02254	0.021498	0.022964	0.023224662	0.022804964
$S = 116$	0.00668	0.006012	0.006249	0.006828979	0.006737655

4. Conclusion

A general method for numerical solving of a special class of problems with barriers for partial integro-differential equations with variable coefficients arising in financial mathematics and other applications has been developed. The numerical method proposed is based on efficient formulae of approximate Wiener-Hopf factorization and allows fast calculation of values of functional of kind (2) for a wide class of models that can be represented as a diffusional process and a non-gaussian Lévy process and does not depend on the exact type of jump model.

Acknowledgments

The reported study was funded by Russian Foundation for Basic Research according to the project No. 18-01-00910 A.

Table 2. Absolute error values for barrier put option prices for the case of Heston model, calculated with Standard MC, Alfonsi MC, M&FWHFe and M&FWHa.
Heston model parameters: $v_0 = 0.01$, $\kappa = 2$, $\theta = 0.01$, $\sigma = 0.2$, $\rho = 0.5$, $r = 0.095$
Option parameters: $K = 100$, $H = 90$, $r = 0.072310$, $T = 1$.
Method parameters: h — space variable step, N — number of time steps (or an algorithm parameter for both WH methods), L — a scale multiplier parameter for M&FWHFe method, M — a number of points for WH methods, S — stock price.

Stock price	standard MC	Alfonsi MC	M&FWHFe	M&WHFa	
			$N = 300$	$N = 300$	$N = 300$
	$h = 0.0001$	$h = 0.0001$	$M = 2^{18}$	$h = 5 \cdot 10^{-6}$	$h = 1 \cdot 10^{-6}$
	$N = 500000$	$N = 600000$	$L = 3$	$M = 2^{20}$	$M = 2^{20}$
$S = 91$	0.00854646	0.085722	0.001027	0.007637565	0.002955738
$S = 95$	0.02783712	0.000324	0.007971	0.027280378	0.010557506
$S = 101$	0.01226794	0.004578	0.006895	0.011386178	0.004406451
$S = 106$	0.0037283	0.002983	0.001949	0.003055566	0.001182504
$S = 111$	0.00153272	0.001042	0.000424	0.000684662	0.000264964
$S = 116$	0.00044088	0.000668	0.000431	0.000148979	0.000057854

Table 3. Relative error values in % for barrier put option prices for the case of Heston model, calculated with Standard MC, Alfonsi MC, M&FWHFe and M&FWHa.
Heston model parameters: $v_0 = 0.01$, $\kappa = 2$, $\theta = 0.01$, $\sigma = 0.2$, $\rho = 0.5$, $r = 0.095$
Option parameters: $K = 100$, $H = 90$, $r = 0.072310$, $T = 1$.
Method parameters: h — space variable step, N — number of time steps (or an algorithm parameter for both WH methods), L — a scale multiplier parameter for M&FWHFe method, M — a number of points for WH methods, S — stock price.

Stock price	standard MC	Alfonsi MC	M&FWHFe	M&WHFa	
			$N = 300$	$N = 300$	$N = 300$
	$h = 0.0001$	$h = 0.0001$	$M = 2^{18}$	$h = 5 \cdot 10^{-6}$	$h = 1 \cdot 10^{-6}$
	$N = 500000$	$N = 600000$	$L = 3$	$M = 2^{20}$	$M = 2^{20}$
$S = 91$	3.9	39.1	0.5	3.5	1.3
$S = 95$	4.8	1.4	1.4	4.7	1.8
$S = 101$	4.7	2.6	2.6	4.4	1.7
$S = 106$	4.6	2.4	2.4	3.8	1.5
$S = 111$	6.8	1.9	1.9	3.0	1.2
$S = 116$	6.6	6.5	6.5	2.2	0.9

References

[1] Cox J C, Ingersoll J E and Ross S A 1985 A theory of the term structure of interest rates *Econometrica.* **53** 385–408
[2] Bates D S 1996 Jumps and stochastic volatility: exchange rate processes implicit in deutsche mark options *Review of Financial Studies.* **9** 69–107
[3] Heston L 1993 A closed-form solution for options with stochastic volatility with applications to bond and currency options. *Review of Financial Studies.* **6** 327–43
[4] Cont R and Tankov P 2004 *Financial Modelling with Jump Processes* (Boca Raton: Chapman & Hall/CRC

Financial Mathematics Series)
[5] Merton R 1976 Option pricing when underlying stock returns are discontinuous *Journal of Financial Economics* **3** 125–44
[6] Black F and Scholes M 1973 The pricing of options and corporate liabilities *Journal of Political Economy* **81**(3) 637–54
[7] Feng Y Q 2017 CVA under Bates model with stochastic default intensity *Journal of Mathematical Finance* **7** 682–98
[8] Alfonsi A 2010 High order discretization schemes for the CIR process: application to affine term structure and Heston models *Mathematics of Computation* **79** 209–37
[9] Chiarella C, Kang B and Meyer G H 2010 The evaluation of barrier option prices under stochastic volatility *Computers & Mathematics with Applications* bf 64 2034–48
[10] Briani D M, Caramellino L and Zanette A 2017 A hybrid approach for the implementation of the Heston model *IMA Journal of Management Mathematics* **28**(4) 467–500
[11] Itkin A 2017 *Pricing Derivatives Under Lévy Models* (Basel: Birkhauser)
[12] Chourdakis K 2005 Lévy processes driven by stochastic volatility *Asia-Pacific Financial Markets*(12) 333–52
[13] Apolloni E, Caramellino L and Zanette A 2015 A robust tree method for pricing American options with CIR stochastic interest rate *IMA Journal of Management Mathematics* **26**(4) 377–401
[14] Kudryavtsev O and Levendorskiĭ S 2009 Fast and accurate pricing of barrier options under Lévy processes *Finance and Stochastics* **13**(4) 531–62
[15] Kudryavtsev O 2016 Advantages of the Laplace transform approach in pricing first touch digital options in Lévy-driven models *Boletin de la Sociedad Matematica Mexicana* **22**(2) 711–31
[16] Carr P 1998 Randomization and the American put *Review of Financial Studies* **11** 597–626
[17] Bouchard B, Karoui N El and Touzi N 2005 Maturity randomization for stochastic control problems *Annals of Applied Probability* **15**(4) 2575–605
[18] Kudryavtsev O and Rodochenko V 2017 A Wiener-Hopf factorization approach for pricing barrier options in the Heston model *Applied Mathematical Sciences* **11**(2) 93–100

IOP Conf. Series: Journal of Physics: Conf. Series **973** (2018) 012055 doi:10.1088/1742-6596/973/1/012055

Stability integral manifold of the differential equations system in critical case

M I Kuptsov[1] and V A Minaev[2]

[1] Mathematics and Information Technology Department, Academy of Law and Management, 1, Sennaya str., Ryazan 390036, Russia
[2] Department of Informatics and Control Systems, Bauman Moscow State Technical University, 5, 2 Baumanskaya str., Moscow 105005, Russia

E-mail: kuptsov_michail@mail.ru, m1va@yandex.ru

Abstract. We consider the stability problem for non-zero integral manifolds of a non-linear finite-dimensional system of ordinary differential equations, where the right-hand side is a periodic vector function with respect to an independent variable and contains a parameter. It is assumed that the studied system has a trivial integral manifold for all values of the parameter, and the corresponding linear subsystem does not have the property of exponential dichotomy. The aim of the paper is to find sufficient conditions of existence in a neighborhood of the system equilibrium state for stable non-zero integral manifold to be lower dimension than the original phase space. For this purpose, based on the classical method of Lyapunov functions and the transforming matrix method operators are constructed, allowing solve the task by finding their fixed points. Due to the specific nature of the considered systems Lyapunov functions method is modified.

1. Introduction

Let the system of ordinary differential equations

$$\dot{y}=f(\nu,y,t) \tag{1.1}$$

for any ν, t have equilibrium state $y=0$ and in the field of Λ its solutions exist and are unique. Here and further f,ν,y – $(n+m)$–vectors, $\dot{y}=\frac{\mathrm{d}y}{\mathrm{d}t}$, $f(\nu,y,t+T)\equiv f(\nu,y,t)$, $\Lambda=\Lambda_1\times\Lambda_2$, $\Lambda_1=\{y:\|y\|\le\Delta_1\}\subset R^{n+m}$, $\Lambda_2=\{\nu:\|\nu\|\le\Delta_2\}\subset R^{n+m}$, Δ_1,Δ_2 – constants, R^d standard Euclidean space of dimension d и $\|\cdot\|$ – the Euclidean norm in R^d.

Suppose that the change of variables

$$y=\Gamma(\varepsilon,x,\phi,t),\ \nu=\xi(\varepsilon) \tag{1.2}$$

brings the system (1.1) to the form

$$\begin{cases}\dot{x}=X(\varepsilon,x,\phi,t)\cdot x,\\ \dot{\phi}=\Phi(\varepsilon,x,\phi,t),\end{cases} \tag{1.3}$$

where Γ,ξ,ε – $(n+m)$ – vectors, X – $(n\times n)$ – matrix, x – n–vector, ϕ,Φ – m – vectors, $\Gamma(\varepsilon,x,\phi,t)\neq 0$ at $x\neq 0$, $\Gamma(\varepsilon,0,\phi,t)\equiv 0$. In addition, we assume that the system

$$\begin{cases}\dot{x}=X(0,0,\phi,t)\cdot x,\\ \dot{\phi}=\Phi(0,0,\phi,t),\end{cases} \tag{1.4}$$

has m – parametric family of non-zero kT – periodic solutions $x=\bar{x}(\phi_0,t),\phi=\bar{\phi}(\phi_0,t)$.

We say that the system (1.1) has a n–dimensional nontrivial periodic integral manifold $\psi(\phi_0,t)$, if for all ϕ_0 exists such value $\nu_0=\xi(\varepsilon_0)$ of parameter ν, at which $\dot{\Gamma}(\varepsilon_0,\psi(\phi_0,t),\phi^{\psi}(\phi_0,t),t)\equiv f(\xi(\varepsilon_0),\Gamma(\varepsilon_0,\psi(\phi_0,t),\phi^{\psi}(\phi_0,t),t),t)$, moreover $\psi(\phi_0,t)$ does not become zero for any values ϕ_0 and t, being ω – periodic in components m – vector ϕ_0, kT – periodic in t, where k – natural number, $\omega=\text{colon}(\omega_1,\omega_2,...,\omega_m)$, $\phi=\phi^{\psi}(\phi_0,t)$ defines an integral curve on the manifold.

The problem of stable nontrivial periodic integral manifold existence of the system (1.1) near its equilibrium state is solved in this paper.

In applied research the invariant surface having a smaller dimension makes it possible to simplify these models, and also to study their typical properties and obtain additional information about the state of the studied systems. This is especially true for stable manifolds that "attract" trajectories over time. Therefore, the study of integral manifolds, the behavior of trajectories on them and near them, helps to solve many problems in the theory of nonlinear oscillations, dynamics, and problems of control theory, fluid mechanics etc. (see, for example [1, pp 6–9]).

However, the diversity of differential equations systems arising in applied research creates difficulties in obtaining general effective methods for investigation of integral manifolds (especially in critical cases), which determines the relevance of this work.

The most productive methods of integral manifolds study should include the method of a small parameter, point mapping, averaging, and the method of integral manifolds.

The method of integral manifolds developed by N. N. Bogolyubov, Yu. A. Mitropolsky and A. M. Samoilenko [2, 3, 4, 5], consists in constructing the Green's function and is successfully used for many systems of the form (1.3) (see, for example [6, 7, 8, 9]). However, in our case this approach cannot be implemented because the system (1.3) with all the parameter values has zero integral manifolds $x=0$ and the system (1.4) – m –parametric family of periodic solutions. These conditions can be overcome only by finding the solution to the auxiliary vector (bifurcation) equation and the transition in its neighborhood [10, 11].

The results presented in this paper are obtained on the basis of the transforming matrix method modification proposed in [1] and Lyapunov function method, application of which allowed us to obtain new sufficient conditions for the existence of a stable local integral manifolds of the system (1.1).

2. Basic symbols and definitions

Let $F(\phi,t)\in\Omega_1,\varepsilon(\phi)\in\Omega_2$ – vector-functions ω–periodic in components of the vector ϕ restricted respectively by numbers δ_{10} and δ_{20}, satisfying the Lipchitz condition:

$$\|F(\phi_1,t_1)-F(\phi_2,t_2)\|\leq\delta_{11}\|\phi_1-\phi_2\|+\delta_{12}|t_1-t_2|, \tag{2.1}$$

$$\|\varepsilon(\phi_1)-\varepsilon(\phi_2)\|\leq\delta_{21}\|\phi_1-\phi_2\|, \tag{2.2}$$

having, respectively, dimension n and l $(0<l\le n+m)$, $F(\phi,t)$ – kT–periodical in t, $\|F(\phi,t)\| = \left[\sum_{i=1}^{n} \sup_{\substack{\phi_i\in[0,\omega]\\ t\in[0,kT]}} |F_i(\phi,t)|\right]^{1/2}$, $\|\varepsilon(\phi)\| = \left[\sum_{i=1}^{l} \sup_{\phi_i\in[0,\omega]} |\varepsilon_i(\phi)|\right]^{1/2}$. Note that if we introduce the specified norm for sets then they become convex compacts sets [1, p 15].

To solve the differential equation

$$\dot{\phi} = \Phi\left(\varepsilon(\phi_0), F(\phi_0,t), \phi, t\right), \tag{2.3}$$

satisfying the initial data $\phi(0)=\phi_0$, we take the notation ϕ_t^F. Suppose, in addition, $Y_\varepsilon^F(\phi_0,t)$ – matrizant of the equation

$$\dot{x} = X\left(\varepsilon(\phi_0), F(\phi_0,t), \phi_t^F, t\right)\cdot x. \tag{2.4}$$

Here and further $F(\phi_0,t)\in\Omega_1$, $n+m-l$ values of vector components ε taken equal 0, but instead of remaining l values in the equations of the system (1.3) elements of the function are substituted $\varepsilon(\phi_0)\in\Omega_2$.

Definition 1. *Nonsingular functional $n\times n$–matrix $Q_\varepsilon^F(\phi_0)$ with constant determinant, continuous in all its variables and ω–periodic in components of the vector ϕ_0, will be called the transforming matrix of the system (1.3), if the matrix*

$$\left(Y_\varepsilon^F(\phi_0,kT) - I_n\right)\cdot Q_\varepsilon^F(\phi_0) \tag{2.5}$$

has the column $q_\varepsilon^F(\phi_0)$ that are not traded identically zero. Here I_n – is the unit $n\times n$–matrix.

Denote $\mathrm{X}=\{x:\|x\|\le\delta_1\}\subset R^n$, $\mathrm{E}=\{\varepsilon:\|\varepsilon\|\le\delta_2\}\subset R^{n+m}$, $\Theta=\{\phi:\phi_i\in[0,\omega_i]\}$ and let $\Gamma(\varepsilon,x,\phi,t)\to 0$ at $\|x\|\to 0$ uniformly with respect to variables ε,ϕ,t in the field $R^{m+1}\times\mathrm{X}\times\mathrm{E}$.

3. Theorems on the existence and stability of integral manifold

Here and below, we assume that the right-hand sides of the system (1.3) are ω–periodic in the components of the vector φ and T–periodic in the independent variable $t\in R$, in addition, they are continuous, assuring existence and uniqueness of the system (1.3) solutions in the field $R^{m+1}\times\mathrm{X}\times\mathrm{E}$ for sufficiently small δ_1 and δ_2. This, in particular, means that the substitution (1.2) not only retains the existence and uniqueness properties of the system (1.1) solutions, but also retains ω–periodicity in φ and T–periodicity in t.

Theorem 1. *Let transforming matrix of the system (1.3) be constructed so that there exist a number l $(0<l\le n+m)$ and such column $q_\varepsilon^F(\phi_0)$, for which in order to find the system solution*

$$\begin{cases} q_\varepsilon^F(\phi_0)=0, \\ \int_0^{kT} \Phi\left(\varepsilon(\phi_0), F(\phi_0,t), \phi_t^F, t\right)\mathrm{d}t = 0. \end{cases} \tag{3.1}$$

it is sufficient to find the solution to some, generally speaking, distinct from (3.1) system of l equations

$$S_\varepsilon^F(\phi_0)=0 \tag{3.2}$$

having for each function $F(\phi_0,t)\in\Omega_1$ *the unique solution* $\varepsilon^F(\phi_0)$ *from the set of* Ω_2. *In addition, suppose that for* $t\in[0;kT]$ *executed:*

$$\left\|Y_\varepsilon^F(\phi_0,t)\cdot Q_\varepsilon^F(\phi_0)\right\|\le r_0, \tag{3.3}$$

$$\left\|Y_\varepsilon^F(\phi_0^*,t^*)\cdot Q_\varepsilon^F(\phi_0^*)-Y_\varepsilon^F(\phi_0,t)\cdot Q_\varepsilon^F(\phi_0)\right\|\le r_1\left\|\phi_0^*-\phi_0\right\|+r_2\left|t^*-t\right|. \tag{3.4}$$

Then for any vector $\phi_0\in R^m$ *you can specify the value of the parameter* ν , *that the system (1.1) will have a nonzero integral manifold* $\psi(\phi_0,t)\in\Omega_1$ *in a neighborhood of the equilibrium state y=0.*

Proof. Since $\varepsilon=\varepsilon^F(\phi_0)$ is a solution to the system (3.2), then at $\varepsilon=\varepsilon^F(\phi_0)$ the expression (3.1) also becomes the identity. Consequently, the differential equation (2.4) for each function $F(\phi_0,t)\in\Omega_1$ has kT–periodic solution

$$x^F(\phi_0,t)=Y_\varepsilon^F(\phi_0,t)\cdot Q_\varepsilon^F(\phi_0)\cdot C, \tag{3.5}$$

where all elements of the constant n–vector C are zero, except for the element corresponding to the column number $q_\varepsilon^F(\phi_0)$, which is equal to an arbitrary constant c. Non-singularity of the transforming matrix ensures non-triviality $x^F(\phi_0,t)$.

In accordance with the conditions (3.3) and (3.4), $x^F(\phi_0,t)$ limited to $r_0\cdot|c|$, satisfies the Lipchitz condition with a constant $r_1\cdot|c|$ in the variable ϕ_0 and $r_2\cdot|c|$ at t. So by reducing c it is always possible to achieve $x^F(\phi_0,t)\in\Omega_1$. Thus, we have constructed the operator defined by equations (3.2) and (3.5), to which, because of the uniqueness values for $\varepsilon^F(\phi_0)$ each function $F(\phi_0,t)\in\Omega_1$, we can apply theorem [1, p 26]. Consequently, this operator has a fixed point

$$\Psi(\phi_0,t)=Y_\varepsilon^\Psi(\phi_0,t)\cdot Q_\varepsilon^\Psi(\phi_0)\cdot C.$$

It is clear that at $\varepsilon=\varepsilon^\Psi(\phi_0)$ function $\Psi(\phi_0,t)$, ϕ_t^Ψ a family of nonzero kT–periodic solutions to the system (1.3) is defined. Indeed, in order to verify this, it is sufficient to take into account appeal of equation (3.1) to the identity and to differentiate $\Psi(\phi_0,t)$, given that the function ϕ_t^Ψ satisfies the equation (2.3).

To complete the proof of the theorem we should return to the system (1.1) by means of the substitution (1.2). So, $\Psi(\phi_0,t)$ – the required n–dimensional nontrivial periodic integral manifold of the system (1.1).

The theorem is proved.

Note that the proof of Theorem 1 generally follows the proof of the theorem on the existence of integral manifolds in [12, p 79] for systems of differential equations, which are not solved with respect to derivatives.

The technique of constructing a transforming matrix demonstrates theorems, establishing the existence of integral manifolds by properties of the right-hand sides of system (1.3). The case when the problem of the system (1.1) local integral manifold existence is solved using linear terms of the matrix $X(\varepsilon,0,\phi,t)$ with respect to components of the vector ε is discussed in detail in [1, 13]. Therefore, we consider the case when the linear terms in components of the vector ε are either absent in the matrix $X(\varepsilon,0,\phi,t)$ (Theorem 2), or they do not allow us to solve the problem of the local integral manifold existence (Theorem 3).

Theorem 2. *Suppose that in the field* $R^{m+1}\times X\times E$ *at sufficiently small* δ_1 *and* δ_2 $X(\varepsilon,x,\phi,t),\Phi(\varepsilon,x,\phi,t)$ *– Lipchitz-continuous in all their variables,*

$$X(\varepsilon,x,\phi,t)=\mathrm{diag}\big(X_1(\varepsilon,x,\phi,t),X_2(\varepsilon,x,\phi,t)\big),\ \ \det\big(\exp\big(X_1(0,0,\phi,kT)-I_{n-p}\big)\big)\neq 0,$$

$$X_1(0,0,\phi,kT)\ -const,\ X_2(\varepsilon,x,\phi,t)=\mathrm{diag}(\alpha_1(\varepsilon,x,\phi,t),\alpha_2(\varepsilon,x,\phi,t),...,\ \alpha_p(\varepsilon,x,\phi,t)),$$

$$\Phi(\varepsilon,x,\phi,t)=\mathrm{colon}(\Phi_1(\varepsilon,x,\phi,t),\Phi_2(\varepsilon,x,\phi,t),...,\Phi_m(\varepsilon,x,\phi,t)),\ \ \Phi_i(\varepsilon,x,\phi,t)=\alpha_{p+i}(\varepsilon,x,\phi,t),$$

$$\alpha_i(\varepsilon,0,\phi,t)=\varepsilon_i^2-\overline{\alpha}_i(\varepsilon,0,\phi,t),$$

integrals $\int_0^{kT}\overline{\alpha}_i(\varepsilon,0,\phi,t)dt$ *– are positive definite forms at all* ϕ *with* ε *not lower than the fourth degree,* $\int_0^{kT}\overline{\alpha}_i(0,0,\phi,t)dt\equiv 0$, $\overline{\alpha}_i(\varepsilon,x,\phi,t)>0$ *at sufficiently small* $x\neq 0$, *where* $X=\{x:\|x\|\le\delta_1\}\subset R^n$, $E=\{\varepsilon:\|\varepsilon\|\le\delta_2\}\subset R^{n+m}$.

Here and further $X_1(\varepsilon,0,\phi,t),X_2(\varepsilon,0,\phi,t)$ *– respectively* $((n-p)\times(n-p))-$ *and* $(p\times p)-$ *matrix,* $\alpha_i(\varepsilon,x,\phi,t)$ *– scalar functions,* I_{n-p} *– unit* $((n-p)\times(n-p))$*–matrix.*

Then, for any vector $\phi_0\in R^m$ *one can specify the value of the parameter* ν *that the system (1.1) will have a nonzero integral manifold* $\psi(\phi_0,t)\in\Omega_1$ *in a neighborhood of the equilibrium state y=0.*

Proof. Based on the properties of the right-hand sides of the system (1.3), the solution ϕ_t^F to the equation (2.3) is restricted to satisfy the condition of Lipchitz in all its variables for $t\in[0,\mathrm{k}T]$ and $\omega-$ periodic in the initial data ϕ_0. And then matrizant $Y_\varepsilon^F(\phi_0,t)$ of the equation (2.4) has exactly the same properties (see, for example, [1, p 29]).

It can be shown (see, for example, [1, p 34]) that for the matrizant of equation (2.4) the representation $Y_\varepsilon^F(\phi_0,t)=\overline{Y}_\varepsilon^F(\phi_0,t)+\tilde{Y}_\varepsilon^F(\phi_0,t)$, where $\overline{Y}_\varepsilon^F(\phi_0,t)$ is a matrizant of the equations system $\dot{x}=X(\varepsilon(\phi_0),0,\phi_t^F,t)\cdot x$, and $\tilde{Y}_\varepsilon^F(\phi_0,t)$ is limited to satisfy Lipchitz condition with respect to all of its variables for $t\in[0,kT]$ with constant, which can be made arbitrarily small by decreasing the number of δ_{10} ($\|F(\phi_0,t)\|\le\delta_{10}$).

Since the matrix $X(\varepsilon,x,\phi,t)$ is quasi-diagonal, then $Y_\varepsilon^F(\phi_0,t)=\mathrm{diag}\big(Y_1^F(\phi_0,t),Y_2^F(\phi_0,t)\big)$, and $Y_2^F(\phi_0,t)=\mathrm{diag}\big(y_1^F(\phi_0,t),y_2^F(\phi_0,t),...,y_p^F(\phi_0,t)\big)$.

We'll define by the symbol J_p the following $(p\times p)-$ matrix

$$J_p=\begin{pmatrix}1&0&0&...&0&1\\0&1&0&...&0&1\\&&&...&&\\0&0&0&...&0&1\end{pmatrix}.$$

By choosing the transforming matrix of the system (1.3) in quasi-diagonal form $Q_\varepsilon^F(\phi_0)=\mathrm{diag}\big(I_{n-p},J_p\big)$ we'll fulfill (3.3) and (3.4).

In this case each nonzero component of the last column $q_\varepsilon^F(\phi_0)$ of the matrix (2.5) is a function $\varepsilon_i^2(\phi_0)\cdot kT - f_i^F(\varepsilon(\phi_0),\phi_0)$, in which by reducing constants δ_1 and δ_2, we can achieve the inequality $f_i^F(\varepsilon(\phi_0),\phi_0)>0$, and at $F(\phi_0,t)\equiv 0$ the function $f_i^F(\varepsilon(\phi_0),\phi_0)$ is a positive definite form at $\varepsilon(\phi_0)$ not lower than the fourth degree. The vector $\int_0^{kT}\Phi(\varepsilon(\phi_0),F(\phi_0,t),\phi_t^F,t)\mathrm{d}t$ has similar properties.

Therefore, any of $l=p+m$ equations of the system (3.1) can be represented as $\varepsilon_i(\phi_0)=\pm\frac{1}{kT}\sqrt{f_i^F(\varepsilon(\phi_0),\phi_0)}$.

Applying the operator (3.2) defined by equations $\varepsilon_i(\phi_0)=\frac{1}{kT}\sqrt{f_i^F(\varepsilon(\phi_0),\phi_0)}$, and, following the technique of [1], it is easy to prove that for sufficiently small δ_{ij} (see (2.1) and (2.2)), this operator is contractive and for each function $F(\phi_0,t)\in\Omega_1$ puts the space Ω_2 to Ω_2. This confirms that all conditions of the Theorem 1 are fulfilled, and, therefore, the local non-zero integral manifold of the system (1.1) exists.

The theorem is proved.

Theorem 3. *Suppose that in the field* $R^{m+1}\times \mathrm{X}\times \mathrm{E}$ *at sufficiently small* δ_1 *and* δ_2 $X(\varepsilon,x,\phi,t),\Phi(\varepsilon,x,\phi,t)$ - *Lipschitz-continuous in all their variables,* $X(\varepsilon,x,\phi,t)$ *has a quasitriangular form*

$$X(\varepsilon,x,\phi,t)=\begin{pmatrix} X_1(\varepsilon,x,\phi,t) & X_3(\varepsilon,x,\phi,t)\\ \Xi & X_2(\varepsilon,x,\phi,t)\end{pmatrix},$$

$$X(\varepsilon,0,\phi,t)=\mathrm{diag}(X_1(\varepsilon,0,\phi,t),X_2(\varepsilon,0,\phi,t)),\ \det(\exp(X_1(0,0,\phi,kT)-I_{n-p}))\neq 0,$$

$$X_1(0,0,\phi,kT)-const,\ X_2(\varepsilon,x,\phi,t)=\mathrm{diag}(\alpha_1(\varepsilon,x,\phi,t),\alpha_2(\varepsilon,x,\phi,t),\dots,\ \alpha_p(\varepsilon,x,\phi,t)),$$

$$\alpha_i(\varepsilon,0,\phi,t)\equiv\alpha_i(\varepsilon)\ at\ i\neq p,$$

$$\Phi(\varepsilon,x,\phi,t)=\mathrm{colon}(\Phi_1(\varepsilon,x,\phi,t),\Phi_2(\varepsilon,x,\phi,t),\dots,\Phi_m(\varepsilon,x,\phi,t)),\ \Phi_i(\varepsilon,0,\phi,t)=\alpha_{p+i}(\varepsilon),$$

$$at\ i\neq p\ done\ \alpha_i(\varepsilon)=a_{i1}\cdot\varepsilon_1+a_{i2}\cdot\varepsilon_2+\dots+a_{il}\cdot\varepsilon_l+\bar{\alpha}_i(\varepsilon),\ \bar{\alpha}_i(\varepsilon)=o(\|\varepsilon\|),$$

$$\alpha_p(\varepsilon,0,\phi,t)=\varepsilon_p^{\ 2}-\overline{\alpha_p}(\varepsilon,0,\phi,t),$$

integral $\int_0^{kT}\overline{\alpha_p}(\varepsilon,0,\phi,t)\mathrm{d}t$ *– at all* ϕ *is a positively determined form at* ε *no lower than the fourth degree,* $\int_0^{kT}\overline{\alpha_p}(0,0,\phi,t)\mathrm{d}t\equiv 0$, $\overline{\alpha_p}(\varepsilon,x,\phi,t)>0$ *at sufficiently small* $x\neq 0$, *where* $\mathrm{X}=\{x:\|x\|\le\delta_1\}\subset R^n, \mathrm{E}=\{\varepsilon:\|\varepsilon\|\le\delta_2\}\subset R^{n+m}$.

Here and further $X_1(\varepsilon,0,\phi,t),X_2(\varepsilon,0,\phi,t)$ *– respectively* $((n-p)\times(n-p))-$ *and* $(p\times p)-$ *matrix, elements of the constant* $(p\times(n-p))-$*matrix* Ξ *are zero,* $\alpha_i(\varepsilon,x,\phi,t)$ *– scalar functions,* I_{n-p} *– unit* $((n-p)\times(n-p))-$*matrix,* $l=p+m$, *determinant of* $((l-1)\times(l-1))-$*matrix*

$$A_0 = \begin{pmatrix} a_{11} & a_{12} & \dots & a_{1\,p-1} & a_{1\,p+1} & \dots & a_{1l} \\ a_{21} & a_{22} & \dots & a_{2\,p-1} & a_{2\,p+1} & \dots & a_{2l} \\ & & & \cdot & \cdot & \cdot & \\ a_{p-1\,1} & a_{p-1\,2} & \dots & a_{p-1\,p-1} & a_{p-1\,p+1} & \dots & a_{p-1\,l} \\ a_{p+1\,1} & a_{p+1\,2} & \dots & a_{p+1\,p-1} & a_{p+1\,p+1} & \dots & a_{p+1\,l} \\ & & & \cdot & \cdot & \cdot & \\ a_{l1} & a_{l2} & \dots & a_{l\,p-1} & a_{l\,p+1} & \dots & a_{ll} \end{pmatrix}$$

is different from zero.

Then, for any vector $\phi_0 \in R^m$ one can specify the value of the parameter v that the system (1.1) will have a nonzero integral manifold $\psi(\phi_0,t) \in \Omega_1$ in a neighborhood of the equilibrium state $y=0$.

Proof. Similar to the proof of the Theorem 2 we see that the solution ϕ_t^F and matrizant $Y_\varepsilon^F(\phi_0,t)$ are limited satisfying the Lipschitz condition in all of its variables for $t \in [0,kT]$ and ω–periodic in initial data ϕ_0. Additionally, $Y_\varepsilon^F(\phi_0,t) = \bar{Y}_\varepsilon^F(\phi_0,t) + \tilde{Y}_\varepsilon^F(\phi_0,t), \bar{Y}_\varepsilon^F(\phi_0,t)$ and $\tilde{Y}_\varepsilon^F(\phi_0,t)$ have the same properties as in theorem 2.

We set $Y_\varepsilon^F(\phi_0,kT) - I_n = \begin{pmatrix} Y_{11}(\varepsilon,F,\phi_0) & Y_{12}(\varepsilon,F,\phi_0) \\ Y_{21}(\varepsilon,F,\phi_0) & Y_{22}(\varepsilon,F,\phi_0) \end{pmatrix}$, where matrix $Y_{11}(\varepsilon,F,\phi_0)$, $Y_{12}(\varepsilon,F,\phi_0)$, $Y_{21}(\varepsilon,F,\phi_0)$, $Y_{22}(\varepsilon,F,\phi_0)$ has dimensions $((n-p)\times(n-p))$, $((n-p)\times p)$, $(p\times(n-p))$, $(p\times p)$, and $Y_{21}(\varepsilon,F,\phi_0) \equiv \Xi$ (as $X(\varepsilon,x,\phi,t)$ has a quasitriangular form). Then, due to execution of inequality $\det(\exp(X_1(0,0,\phi,kT) - I_{n-p})) \neq 0$, by reducing values of constants δ_1 and δ_2 we can achieve equality $\det Y_{11}(\varepsilon,F,\phi_0) = a_0 \neq 0$. This means that there is an inverse matrix $(Y_{11}(\varepsilon,F,\phi_0))^{-1}$ to matrix $Y_{11}(\varepsilon,F,\phi_0)$, which is also restricted, ω–periodic in ϕ_0 and satisfies the Lipschitz condition in ϕ_0.

Let the vector $q = \text{colon}(q_1,q_2,\dots,q_{n-p})$ determine by the equality $q = -(Y_{11}(\varepsilon,F,\phi_0))^{-1} \cdot Y_{12}(\varepsilon,F,\phi_0) \cdot J_1$, where elements of the constant $(p\times 1)$ – vector J_1 are equal to one.

Then the transforming matrix of the system (1.3) we will choose in the form $Q_\varepsilon^F(\phi_0) = \begin{pmatrix} I_{n-p} & \bar{Q} \\ \Xi & J_p \end{pmatrix}$.

Here among the elements of $((n-p)\times p)$ – matrix $\bar{Q}$ are only nonzero elements $q_1,q_2,\dots,q_{n-p}$ of the last column.

Since elements of the vector q are linear combinations of matrix elements $(Y_{11}(\varepsilon,F,\phi_0))^{-1}$ and $Y_{12}(\varepsilon,F,\phi_0)$, then $Q_\varepsilon^F(\phi_0)$ is bounded, ω–periodic in ϕ_0 and satisfies the Lipchitz condition with respect to ϕ_0 and hence, (3.3) and (3.4) will be executed.

Because of the transforming matrix choice $Q_\varepsilon^F(\phi_0)$ the last column $q_\varepsilon^F(\phi_0)$ of the matrix (2.5) will contain exactly p the non-zero component. Indeed, the first $(n-p)$ coordinate of the column $q_\varepsilon^F(\phi_0)$ will take the form of $Y_{11}(\varepsilon,F,\phi_0)\cdot q + Y_{12}(\varepsilon,F,\phi_0)\cdot J_1 \equiv 0$, and the rest of p coordinates will be written as follows $Y_{22}(\varepsilon,F,\phi_0)\cdot J_1$. The diagonal form of the matrix $X_2(\varepsilon,x,\phi,t)$ also provides diagonality of the matrix $Y_{22}(\varepsilon,F,\phi_0)$.

In this case, the p component of the last column $q_\varepsilon^F(\phi_0)$ of the matrix (2.5) is a function $\varepsilon_p^2(\phi_0)\cdot kT - f_p^F(\varepsilon(\phi_0),\phi_0)$, in which by reducing constants δ_1 and δ_2 it is possible to achieve inequality $f_p^F(\varepsilon(\phi_0),\phi_0)>0$, and with $F(\phi_0,t)\equiv 0$ function $f_p^F(\varepsilon(\phi_0),\phi_0)$ is a positively definite form at $\varepsilon(\phi_0)$ no lower than the fourth degree. Therefore, the equation number p of the system (3.1) can be represented as $\varepsilon_p(\phi_0)=\pm\frac{1}{kT}\sqrt{f_p^F(\varepsilon(\phi_0),\phi_0)}$.

Suppose that the operator $S_\varepsilon^F(\phi_0)$ is given by $l-1$ the non-zero equation of the system (3.1) (without the equation number p), and instead of equation number p of the system (3.1) to specify the operator $S_\varepsilon^F(\phi_0)$ we use equation $\varepsilon_p(\phi_0)=\frac{1}{kT}\sqrt{f_p^F(\varepsilon(\phi_0),\phi_0)}$. Hence, the system of equations (3.2) can be represented in the form $A\cdot\varepsilon(\phi_0)=y^F(\varepsilon(\phi_0),\phi_0)$, where for $p\neq i$ and for elements a_{pi} of $(l\times l)-$matrix $A=(a_{ij})$ is performed $a_{pi}=0$ and $a_{pp}=1$. In addition, the function $y^F(\varepsilon(\phi_0),\phi_0)$ is ω–periodic in ϕ_0 and limited, satisfying the Lipschitz condition in ϕ_0 with constants, which can be made arbitrarily close to zero by decreasing numbers δ_{ij}.

From nonsingularity of the matrix A_0 it follows that $\det A\neq 0$ and (3.1) is reduced to the form

$$\varepsilon(\phi_0)=A^{-1}\cdot y^F(\varepsilon(\phi_0),\phi_0). \tag{3.6}$$

By reducing δ_{ij} it is easy to verify that the operator defining the equation (3.6) is contractive and for each function $F(\phi_0,t)\in\Omega_1$ transfers the space Ω_2 into Ω_2. This means that all conditions of the Theorem 1 are fulfilled, and, hence, the local non-zero integral manifold of the system (1.1) exists.

The theorem is proved.

Suppose now that conditions of the Theorem 1 are fulfilled further and, therefore, the system (1.1) for certain values of the parameter $v_0=\xi(\varepsilon_0)$ has an integral manifold $\psi(\phi_0,t)\in\Omega_1$, $\psi(\phi_0,0)=\psi_0$. We denote by $x=x(x_0,\phi_0,t),\phi=\phi(x_0,\phi_0,t)$ – the solution to the system (1.3) satisfying the initial data $x(x_0,\phi_0,0)=x_0,\phi(x_0,\phi_0,0)=\phi_0$.

Lemma 1. *If $\varepsilon=\varepsilon_0$ for any number $\delta>0$ there exists a number $\Delta>0$, that the fulfillment of the inequality $\|x_0-\psi_0\|<\Delta$ for any $\phi_0\in\Theta$ implies $t\geq 0$ the inequality $\|x(x_0,\phi_0,t)-\psi(\phi_0,t)\|<\delta$ for all $\phi_0\in\Theta$, then with $v_0=\xi(\varepsilon_0)$ the integral manifold $\psi(\phi_0,t)$ of the system (1.1) is stable.*

Proof. The validity of the Lemma 1 follows from the identity $\Gamma(\varepsilon,0,\phi,t)\equiv 0$ and uniform convergence in the field $R^{m+1}\times \mathrm{X}\times \mathrm{E}$ of function $\Gamma(\varepsilon,x,\phi,t)$ to 0 with $\|x\|\to 0$.

Theorem 4. *If there exists a sign-definite function $V(x,t)$, for which function $\frac{dV}{dt}=\frac{\partial V(x(x_0,\phi_0,t)-\psi(\phi_0,t),t)}{\partial t}+\sum_{i=1}^{n}\frac{\partial V(x(x_0,\phi_0,t)-\psi(\phi_0,t),t)}{\partial x_i}$ has constant sign, opposite to $V(x,t)$, then with $v_0=\xi(\varepsilon_0)$ integral manifold $\psi(\phi_0,t)$ of the system (1.1) is stable.*

Proof. Let $\varepsilon=\varepsilon_0$ and $V(x,t)$ – be positive. Then for any number $\delta>0$ there exist such numbers $\Delta>0$ and $\bar{l}>0$, for which with $\|x\|=\Delta\leq\delta, t\geq 0$ will be done $V(x,t)\geq\bar{l}$, besides $\Delta\leq\delta_1$.

We now choose x_0 so that

$$V(x_0 - \psi_0, 0) < \bar{l} \tag{3.7}$$

and $\|x_0 - \psi_0\| \le \bar{\Delta} < \Delta$ for any $\phi_0 \in \Theta$. Due to the fact that

$$\frac{dV}{dt} \le 0, \tag{3.8}$$

inequality $\|x(x_0, \phi_0, t) - \psi(\phi_0, t)\| < \Delta \le \delta$ correct for all $\phi_0 \in \Theta$ with $t \ge 0$. Indeed, if this were not so, there would be such constant $\phi_0^* \in \Theta$ and $t^* \ge 0$ for which $\|x(x_0, \phi_0^*, t^*) - \psi(\phi_0^*, t^*)\| = \Delta$ was executed and so $V(x(x_0, \phi_0^*, t^*) - \psi(\phi_0^*, t^*), t^*) \ge \bar{l}$, which contradicts to the simultaneous execution of relations (3.7) and (3.8).

The proof of Theorem 4 is completed by applying Lemma 1.

The theorem is proved.

As an illustration of the use of Theorem 4, we consider the simplest systems of the form (1.3), to which none of the sufficient criteria for the existence of stable integral manifolds can be applied [2, 3, 5, 10].

Example 1. Let in the system of two differential equations

$$\begin{cases} \dot{x} = (\varepsilon_1 + \varepsilon_2 - x)^3 \cdot x, \\ \dot{\phi} = \varepsilon_2 + \cos 2t \end{cases} \tag{3.9}$$

all variables are scalar.

Obviously, when $\varepsilon_1 = \bar{\varepsilon} > 0, \varepsilon_2 = 0$ system (3.9) has a stable integral manifold $\psi(\phi_0, t) \equiv \bar{\varepsilon}$, and function $\phi = \phi_0 + 0{,}5\sin 2t$ determines a periodic solution. In particular, to prove this fact, one can apply Theorem 3 by choosing $V(x,t) \equiv V(x) = x^2$. Then $\frac{dV(x-\bar{\varepsilon})}{dt} = -2(x-\bar{\varepsilon})^4 \cdot x \le 0$ with $|x - \bar{\varepsilon}| < 0{,}25 \cdot \bar{\varepsilon}$.

Example 2. For a system of four scalar differential equations

$$\begin{cases} \dot{x}_1 = -x_2 + \varepsilon_1 x_1^2 - x_3^5 \cdot \sin\phi \\ \dot{x}_2 = x_1 + \varepsilon_2 x_2 - x_3^7 \cdot \sin^2\phi \\ \dot{x}_3 = \varepsilon_3 + \varepsilon_1^2 x_2 - x_3^3 \cdot (1 + \cos^2\phi) \\ \dot{\phi} = \varepsilon_4 + x_1 \cdot \sin\phi \end{cases}$$

it is also possible to construct a function $V(x) = x_1^2 + x_2^2 + x_3^2$, satisfying all conditions of the Theorem 3. Indeed, the system (3.1) for each function $F(\phi_0, t) \in \Omega_1$ has a solution $\varepsilon^F(\phi_0)$, where $\varepsilon_4(\phi_0) = -\frac{1}{2\pi}\int_0^{2\pi} F_1(\phi_0, t) \cdot \sin\phi_t^F \, dt$. Then the integral manifold $\psi(\phi_0, t)$ is given by $\psi(\phi_0, t) = \mathrm{colon}(c \cdot \cos t, c \cdot \sin t, 0)$, where $c \ne 0$, and function $\phi = \phi^\psi(\phi_0, t)$ determines on it a periodic curve. If now considered $x = x(x_0, \phi_0, t)$ as a solution close to $\psi(\phi_0, t)$ $(\|x(x_0, \phi_0, t) - \psi(\phi_0, t)\| < 0{,}5 \cdot |c|)$, then

IOP Conf. Series: Journal of Physics: Conf. Series **973** (2018) 012055 doi:10.1088/1742-6596/973/1/012055

$$\frac{\mathrm{d}V}{\mathrm{d}t} = -2x_3^4 \cdot \left(1+\cos^2\phi^{\psi}\left(\phi_0,t\right)\right) - 2x_1x_3^5 \cdot \sin\phi^{\psi}\left(\phi_0,t\right) - 2x_1x_3^7 \cdot \sin^2\phi^{\psi}\left(\phi_0,t\right) \le 0$$

at small c and $\varepsilon_1 = \varepsilon_2 = \varepsilon_3 = 0$.

4. Conclusions

New sufficient conditions of existence of nonzero local stable integral manifold of the system (1.1) are formulated in Theorem 1-4. The critical case (linear subsystem $\dot{x} = X\left(0,0,\phi_t,t\right)\cdot x$ has a zero of characteristic indicators) is considered. The results obtained on the basis of the Lyapunov functions method and the transforming matrix method, the application of which allowed to reduce the solution of this problem to searching of nonlinear operators fixed points. The problem of stability of integral manifold of the system (1.1) with respect to all variables (y) is solved, taking into account only characteristics a part of the variables (x).The examples of finding the transforming matrices and Lyapunov functions for the various systems of the form (1.3) are made in this research paper.

References

[1] Kuptsov M I 1997 *The existence of integral manifolds and periodic solution of system of ordinary differential equations: PhD dissertation (Mathematics)* (Izhevsk: Udmurt State University) p 133

[2] Bogolyubov N N 1945 *About some statistical methods in mathematical physics* (Lvov: Academy of Sciences of the Ukrainian Soviet Socialist Republic) p 139

[3] Mitropol'skii Yu A and Lykova O B 1973 *Integral manifolds in nonlinear mechanics* (Moscow: Nauka) p 512

[4] Samoilenko A M and Dvorak A V 1978 *Ukrainian Mathematical Journal* **29(4)** 427–431

[5] Samoilenko A M 1987 *The elements of mathematical theory of multi frequency vibrations. Invariant tori* (Moscow: Nauka) p 301

[6] Kurbanshoev S Z and Nusairiev M A 2014 *Reports of the Academy of Sciences of the Republic of Tajikistan* **57(11–12)** 807–812

[7] Samoilenko A M, Petryshyn R I and Sopronyuk T M 2003 *Ukrainian Mathematical Journal* **55(5)** 773–800

[8] Shchetinina E V 2010 *Vestnik of Samara State University* **6** 93–105

[9] Sobolev V 2016 *Journal of Physics: Conference Series* **727(1)** 012017

[10] Bibikov Yu N 1991 *Multi frequent nonlinear vibrations and their bifurcation* (Leningrad: Leningrad State University) p 142

[11] Volkov D Yu 1988 *Vestnik Leningrad University* **1(2)** 102–103

[12] Kuptsov M I, Terekhin M T and Tenyaev V V 2017 *Middle-Volga Mathematical Society Journal* **19(2)** 76–84

[13] Kuptsov M I 1998 *Differential equations* **34(7)** 1005–1007

IOP Conf. Series: Journal of Physics: Conf. Series **973** (2018) 012056 doi:10.1088/1742-6596/973/1/012056

Two-point mean value formulas for some elliptic equations spaces with constant curvature

A L Muglanov[1], **I P Polovinkin**[2] **and M V Polovinkina**[3]

[1,2]Department of mathematical and applied analysis, Voronezh State University, Universitetskaya pl.,1, Voronezh, Russian Federation

[3]Department of higher mathematics and information technologies, Voronezh State University of Engineering Technologies, Revolution Av., 19, Voronezh, Russian Federation

E-mail: polovinkin@yandex.ru

Abstract. An analogue of the consequence of Bitsadze & Nakhushev's mean value formula was obtained in the Lobachevskiy space and on a sphere.

1. Introduction

The terms "mean value theorem" and "mean value formula" often do not mean similar facts to some extent in various sections and applied problems. Nevertheless varied results for different types of equations have in common mean of function at some set of points, as a finite set of points of the phase plane in mean formula for the equation of a vibrating string, a sphere, as in mean formula for a harmonic function [1], pair of spheres as in the Asgeirsson theorem [2] or the set which is structured more complicatedly. Mean value theorems for elliptic equations are known most widely. The most famous of those is the mean value theorem (by Gauss) for the Laplace equation. This fact applies to general elliptic equations. Mean value theorem is carried over to ordinary differential equations. Mean value formulae are used while studying a proper spectral decomposition of elliptic operators. In this context see works by Ilyin V. A. and Moiseev E. I. (the papers [3]–[4]). Mean value theorem for Laplaces equation in circular sector is proved in the [5]. As well above mean value theorem generalizes to Riemannian manifolds (see the book [6]). We should point out the papers [7], [8], where the condition of converse mean value theorem for harmonic functions is weakened.

Among mean value theorems for hyperbolic equations the mentioned above Asgeirsson theorem is one of the best known. The A.V. Bitsadze and A.M. Nakhushev mean value theorem for the wave equation [9] is also well known. In this we derived a two-point mean value formula for elliptic equation in the Lobachevskiy space and on the sphere from the mean value formula for the wave equation in those spaces.

2. Materials and methods

The paper uses classical methods of mathematical analysis, the theory of equations of mathematical physics and differential geometry

3. The mean value theorem for the wave equation in Euclidean space

Consider in space R^{n+1} a pair of points $(\chi^{(j)}, \tau^{(j)})$, $j = 1, 2$, where $\chi^{(j)} = (\chi_1^{(j)}, \chi_2^{(j)}, \ldots, \chi_n^{(j)})$, $j = 1, 2$, that satisfy the condition

$$|\chi^{(1)} - \chi^{(2)}| < |\tau^{(1)} - \tau^{(2)}|. \tag{1}$$

We construct a matrix A by the following rule. We fix an index $i \in \{1, \ldots, n\}$ and set

$$a_{ij} = \frac{\chi_j^{(1)} - \chi_j^{(2)}}{|\chi^{(1)} - \chi^{(2)}|}, \quad j = 1, \ldots, n.$$

The remaining elements of this matrix can be constructed from the conditions

$$AA^T = I, \quad \det A = 1,$$

where A^T is the matrix, transposed to the matrix A, I is a unit matrix. Let A_i denote the matrix obtained from the matrix by replacing the i-th column by zeros. Following the paper [9], we introduce an averaging operator S_τ by the formula

$$S^n v = S_\tau^n v = \gamma(n) \int\limits_{|\xi|=\tau} v(\eta, \sigma)\, d\omega_\xi, \; |\chi^{(1)} - \chi^{(2)}| > 0, \tag{2}$$

where $\gamma(n) = \sqrt{\pi}^{1-n}$,

$$\eta = \frac{|\tau^{(1)} - \tau^{(2)}|\, \xi_i(\chi^{(1)} - \chi^{(2)})}{|\chi^{(1)} - \chi^{(2)}|\sqrt{(\tau^{(1)} - \tau^{(2)})^2 - |\chi^{(1)} - \chi^{(2)}|^2}} +$$

$$+ \frac{\chi^{(1)} + \chi^{(2)}}{2} + A_i \xi,$$

$$\sigma = \frac{\tau^{(1)} + \tau^{(2)}}{2} - \frac{|\chi^{(1)} - \chi^{(2)}|\xi_i}{\sqrt{(\tau^{(1)} - \tau^{(2)})^2 - |\chi^{(1)} - \chi^{(2)}|^2}},$$

$d\omega_\xi$ is a sphere $|\xi| = t$ surface area element in R^n. When $\chi^{(1)} = \chi^{(2)}$ we define the operator S_τ by the formula

$$S^n = S_\tau^n v = \gamma(n) \int\limits_{|\xi|=\tau} v\left(\chi^{(2)} + \xi, \frac{\tau^{(1)} + \tau^{(2)}}{2}\right) d\omega_\xi. \tag{3}$$

Next we introduce the operator B_τ by formulas

$$B^n v = B_\tau^n v = \tau \left(\frac{\partial}{2\tau\partial\tau}\right)^{\frac{n-1}{2}} \frac{1}{\tau} S_\tau v, \; n = 1(\mathrm{mod}2), \tag{4}$$

$$B^n v = B_\tau^n v = \frac{1}{\sqrt{\pi}} \left(\frac{\partial}{2\tau\partial\tau}\right)^{\frac{n}{2}} \int\limits_0^\tau \frac{S_\tau v\, d\theta}{\sqrt{\tau^2 - \theta^2}}, \; n = 0(\mathrm{mod}2), \tag{5}$$

Theorem 1. (the mean value theorem for the wave equation [9]). *If a function $v(\chi_1, \ldots, \chi_n, \tau)$ is a regular solution of the wave equation*

$$\frac{\partial^2 v}{\partial \tau^2} = \sum_{k=1}^n \frac{\partial^2 v}{\partial \chi_k^2}, \tag{6}$$

then for any pair of points $(\chi^{(j)}, \tau^{(j)})$, $j = 1, 2$, *satisfying condition (1), we have the equality*

$$v(\chi^{(1)}, \tau^{(1)}) + v(\chi^{(2)}, \tau^{(2)}) = B_\varrho v, \tag{7}$$

$\varrho = \sqrt{\Delta\tau^2 - |\Delta\chi|^2}/2$, $\Delta\tau = \tau^{(1)} - \tau^{(2)}$, $\Delta\chi = \chi^{(1)} - \chi^{(2)}$.

There is an assertion for sufficiently smooth functions that is inverse to theorem 3. Let

$$l = \max\left\{n-1, \ \frac{n+3}{2}\right\}, \ n = 1 (\mathrm{mod}\, 2), \tag{8}$$

$$l = \max\left\{n, \ \left[\frac{n+3}{2} + 1\right]\right\}, \ n = 0 (\mathrm{mod}\, 2). \tag{9}$$

Theorem 2. (the inverse mean value theorem for the wave equation [10]). *Let the function* $v(\chi, \tau) \in C^l(R^{n+1})$ *satisfies relation (7) for any pair of points* $(\chi^{(j)}, \tau^{(j)})$, $j = 1, 2$, *which satisfy condition (1). Then* $v(\chi, \tau)$ *is a regular solution of the wave equation (6).*

4. The wave equation in the Lobachevskiy space and transformations of P. Lax and R. Phillips

Let's consider the realization of the Lobachevskiy geometry on half-space

$$R^n_+ = \{x = (x_1, \ldots, x_{n-1}, x_n) = (x', x_n) : x_n > 0\}, n \geq 2,$$

with a metric

$$ds^2 = dx^2/x_n^2,$$

where $dx = (dx_1, \ldots, dx_{n-1}, dx_n) = (dx', dx_n), dx^2 = dx \cdot dx = dx_1^2 + \cdots + dx_n^2$. This model of Lobachevskiy geometry will be designated by Π. In other words, components of the metric (contra-variant) tensor and the inverse metric (covariant) tensor in our case, respectively, have the form

$$g_{ij} = \delta_{ij}/x_n^2, \ g^{ij} = \delta_{ij} x_n^2, \ i, j = 1, \ldots, n, \tag{10}$$

where δ_{ij} is the Kronecker symbol. As is well known (see for ex. the book [11]) the Laplace-Beltrami operator in the metric $\|g_{ij}\|$ is defined by the formula

$$\Delta_\omega = \frac{1}{\sqrt{|g|}} \sum_{i=1}^{n} \frac{\partial}{\partial x_i} \sum_{j=1}^{n} g^{ik} \sqrt{|g|} \frac{\partial}{\partial x_j},$$

where $g = \det \|g_{ij}\|$. The Laplace-Beltrami operator in metric (10) has the form

$$\Delta_\omega = x_n^2 \sum_{j=1}^{n-1} \frac{\partial^2}{\partial x_j^2} + x_n^n \frac{\partial}{\partial x_n} \left(x_n^{2-n} \frac{\partial}{\partial x_n} \right). \tag{11}$$

We consider the wave equation in the Lobachevskiy space, which has the form

$$\frac{\partial^2 u}{\partial t^2} = \Delta_\omega u + \left(\frac{n-1}{2}\right)^2 u. \tag{12}$$

This equation was considered by P. Lax and R. Phillips in the scattering theory [12]. It was also studied by I.A. Kipriyanov and L.A.Ivanov (see the papers [13], [14], [16]). The concept of the wave equation in a space with constant curvature is clarified by these authors. Following these authors we call equation (12) the wave equation in the Lobachevskiy space.

We introduce mapping $\Pi \times R \longrightarrow R^{n+1}$ by means of formulas

$$s = e^t, \tag{13}$$

$$\tau = \tau(x,t) = \frac{s}{2}\left(\frac{1}{x_n} + x_n + \frac{|x'|^2}{x_n}\right), \tag{14}$$

$$\chi_n = \chi_n(x,t) = \frac{s}{2}\left(\frac{1}{x_n} - x_n - \frac{|x'|^2}{x_n}\right), \tag{15}$$

$$\chi' = \chi'(x,t) = s\frac{x'}{x_n}, \tag{16}$$

where $|x'| = \sqrt{x_1^2 + \cdots + x_{n-1}^2}, \chi' = (\chi_1, \ldots, \chi_{n-1}), \chi = (\chi', \chi_n)$.

Let note that the inequality

$$|\chi|^2 < \tau^2 \tag{17}$$

in terms of (x,t) variables is equivalent to the inequality $x_n^2 > 0$, that holds for all points $(x,t) \in \Pi \times R$. Therefore, the mapping defined by means of equalities (13)–(16), realizes a one-to-one correspondence between the space $\Pi \times R$ and a cone in the space R^{n+1} where the cone is defined by condition (17).

We add (14) and (15). We get

$$\tau + \chi_n = \frac{s}{x_n}. \tag{18}$$

By subtraction (15) from (14), we get

$$\tau - \chi_n = \frac{s|x|^2}{x_n}. \tag{19}$$

From (15) we find

$$x' = \frac{x_n}{s}\chi',$$

from there, taking into account (18), we get

$$x' = \frac{\chi'}{\tau + \chi_n}. \tag{20}$$

We substitute (20) in (19). Then

$$\tau - \chi_n = \frac{s}{x_n}\left(\frac{x_n^2}{s^2}|\chi'|^2 + x_n^2\right),$$

whence

$$x_n = \frac{\sqrt{\tau^2 - |\chi|^2}}{\tau - \chi_n}. \tag{21}$$

Further, taking into account (18) we have

$$s = x_n(\tau + \chi_n),$$

and considering (21), we get

$$s = \sqrt{\tau^2 - |\chi|^2},$$

and taking into account (13), we obtain

$$t = \ln \sqrt{\tau^2 - |\chi|^2}\ . \tag{22}$$

Thus, the transformation defined by formulas (20), (21), (22), is the inverse of the transformation defined by formulas (13)–(16).

In addition to transformations of the variables (13)–(16) and their inverse (20), (21), (22), we also introduce a change of function

$$v(\chi, \tau) = (\tau^2 - |\chi|^2)^{(1-n)/4} u(x(\chi, \tau), t(\chi, \tau)). \tag{23}$$

Transformations (13)–(16), their inverse (20), (21), (22), and also (23) were introduced for $n = 2$ in the book [12]. Therefore, it is appropriate to call them the Lax-Phillips transforms.

Theorem 3. A function $u(x,t) \in C^2(\Pi \times R)$ is a regular solution of the wave equation (12) in the Lobachevskiy space if and only if the function (23), obtained from u by means of Lax-Phillips transforms is a regular solution of the wave equation (6) in the cone $|\chi| < |\tau|$ of space R^{n+1} of variables $\chi = (\chi_1, \dots, \chi_n),\ \tau$.

Proof. We can make a direct substitution into the equation, but we will carry out the proof according to the scheme of P. Lax and R. Phillips. In accordance with this scheme, we consider the wave operator

$$\Lambda = \frac{\partial^2}{\partial \tau^2} - \sum_{j=1}^{n} \frac{\partial^2}{\partial \chi_j^2}$$

as the Laplace-Beltrami operator corresponding to an indefinite Lorentz metric

$$d\sigma^2 = d\tau^2 - |d\chi|^2. \tag{24}$$

In metric (24) we pass to the coordinates (s, x). For the derivatives of τ we have:

$$\frac{\partial \tau}{\partial s} = \frac{1}{2}\left(\frac{1}{x_n} + x_n + \frac{|x_n'|^2}{x_n}\right) = \frac{1 + |x|^2}{2\,x_n},$$

$$\frac{\partial \tau}{\partial x_j} = \frac{s x_j}{x_n},\ j = 1, 2, \dots, n-1,$$

$$\frac{\partial \tau}{\partial x_n} = \frac{s(x_n^2 - |x_n'|^2 - 1)}{2x_n^2}\ .$$

Therefore

$$d\tau = \frac{1 + |x|^2}{2\,x_n}\, ds + \frac{s}{x_n} x' \cdot dx' + \frac{s(x_n^2 - |x_n'|^2 - 1)}{2\,x_n^2}\, dx_n, \tag{25}$$

where $dx' = (dx_1, \dots, dx_{n-1})$, $x' \cdot dx'$ denotes the scalar product of the vectors x' and dx' in the space R^{n-1}. Squaring equality (25), we get

$$d\tau^2 = \frac{(1 + |x|^2)^2}{4\,x_n^2}\, ds^2 + \frac{s^2}{x_n^2}(x' \cdot dx')^2 + \frac{s^2(x_n^2 - |x_n'|^2 - 1)^2}{4\,x_n^4}\, dx_n^2 +$$

$$+ \frac{s(1 + |x|^2)}{x_n}\, ds\, x' \cdot dx' + \frac{s^2(x_n^2 - |x_n'|^2 - 1)}{x_n^3}\, dx_n\, x' \cdot dx' +$$

$$+\frac{s(x_n^2-|x_n'|^2-1)(1+|x|^2)}{2\,x_n^3}\,ds\,dx_n. \tag{26}$$

Further, for the derivatives of χ,

$$\frac{\partial\chi_\kappa}{\partial s}=\frac{x_\kappa}{x_n},\ \kappa=1,\dots,n-1,$$

$$\frac{\partial\chi_\kappa}{\partial x_j}=\frac{s\,\delta_{\kappa j}}{x_n},\ \kappa=1,\dots,n-1,\ j=1,\dots,n-1,$$

$$\frac{\partial\chi_\kappa}{\partial x_n}=-\frac{s\,x_\kappa}{x_n^2},\ \kappa=1,\dots,n-1,$$

we get

$$d\chi_\kappa=\frac{s}{x_n}\,dx_\kappa+\frac{x_\kappa}{x_n}\,ds-\frac{s\,x_\kappa}{x_n^2}\,dx_n,\ \kappa=1,\dots,n-1. \tag{27}$$

Squaring equality (27), and then summing over the κ, we get

$$|d\chi'|^2=\frac{s}{x_n^2}|dx'|^2+\frac{|x'|^2}{x_n^2}ds^2+\frac{s\,|x'|^2}{x_n^4}dx_n^2+\frac{2s}{x_n^2}\,ds\,x'\cdot dx'-$$

$$-\frac{2s^2}{x_n^3}\,dx_n\,x'\cdot dx'-\frac{2s\,|x'|^2}{x_n^3}\,ds\,dx_n. \tag{28}$$

For the last coordinate we have

$$d\chi_n=-\frac{s}{x_n}\,x'\cdot dx'+\frac{1-|x|^2}{2\,x_n}\,ds+\frac{s}{2\,x_n^2}(|x'|^2-1-x_n^2)\,dx_n,$$

whence, after squaring, we get

$$d\chi_n^2=\frac{s^2}{x_n^2}\,(x'\cdot dx')^2+\frac{(1-|x|^2)^2}{4\,x_n^2}\,ds^2+$$

$$+\frac{s^2\,(|x'|^2-1-x_n^2)^2}{4\,x_n^4}\,dx_n^2-\frac{s\,(1-|x|^2)}{x_n^2}\,ds\,(x'\cdot dx')-$$

$$-\frac{s^2\,(|x'|^2-1-x_n^2)}{4\,x_n^3}\,dx_n\,(x'\cdot dx')+\frac{s(1-|x|^2)(|x'|^2-1-x_n^2)}{2\,x_n^3}\,ds\,dx_n. \tag{29}$$

Replacing (26), (28) and (29) in (24), after identical transformations we obtain

$$d\sigma^2=ds^2-s^2\,\frac{|dx|^2}{x_n^2}\,. \tag{30}$$

The transition from the coordinate s to the coordinate t by the formula $s=e^t$ leads expression (30) to the form

$$d\sigma^2=e^{2t}\left(dt^2-\frac{|dx|^2}{x_n^2}\right). \tag{31}$$

It follows that the Laplace-Beltrami operator corresponding to the indefinite metric (31), has the form

$$\Lambda = e^{-(n+1)t}\frac{\partial}{\partial t}\left(e^{(n-1)t}\frac{\partial}{\partial t}\right) - x_n^2\, e^{-2t}\,\triangle_{x'} - x_n^n\, e^{-2t}\frac{\partial}{\partial x_n}\left(x_n^{2-n}\frac{\partial}{\partial x_n}\right). \qquad (32)$$

Thus, the fact that the function $v(\chi,\tau)$ satisfies the Euclidean wave equation (6) in cone (17), is equivalent to the fact that the function $u(x,t) = v(\chi(x,t),\tau(x,t))$ satisfies the equation

$$e^{(1-n)t}\frac{\partial}{\partial t}\left(e^{(n-1)t}\frac{\partial v}{\partial t}\right) = \triangle_\omega v, \qquad (33)$$

where the operator $\triangle_\omega v$ is defined by formula (11).

Now we replace the function in equation (33) by formula (23). Taking into account (22), in the coordinates (x,t), we get

$$v = e^{(1-n)/2}\, u. \qquad (34)$$

The substitution (34) gives equation (33) to equation (12), which completes the proof of the theorem.

5. The mean value theorem for the wave equation in the Lobachevskiy space

Now we prove an analog of theorem 3 for the wave equation in the Lobachevskiy space, basing on the connection of the wave equations in the Lobachevskiy space and the Euclidean space, which is established by theorem 4.

Theorem 4. Let a function $u(x,t)$ be a regular solution of the wave equation (12) in the Lobachevskiy space. Let $(x^{(j)},t^{(j)}) = (x_1^{(j)},\dots,x_{n-1}^{(j)},x_n^{(j)},t^{(j)}) = (x'(j),x_n^{(j)},t^{(j)})$, $j=1,2$, be a pair of points in the space $\Pi\times R$ that is satisfying the condition

$$(x_n^{(2)}\, e^{t^{(1)}} - x_n^{(1)}\, e^{t^{(2)}})\,(x_n^{(1)}\, e^{t^{(1)}} - x_n^{(2)}\, e^{t^{(2)}}) - e^{t^{(1)}+t^{(2)}}|\triangle x'|^2 > 0. \qquad (35)$$

Then the following equality holds (the mean value formula)

$$e^{t^{(1)}(1-n)/2}u(x^{(1)},t^{(1)}) + e^{t^{(2)}(1-n)/2}u(x^{(2)},t^{(2)}) = B_\varrho^n(fu), \qquad (36)$$

where $f = (\tau^2 - |\chi|^2)^{(1-n)/4}$, $\chi^{(j)} = \chi(x^{(j)},t^{(j)})$, $\tau^{(j)} = \tau(x^{(j)},t^{(j)})$, $j=1,2$, $u = u(x,t) = u(x(\chi,\tau),t(\chi,\tau))$, the operator B_ϱ^n is defined by formulas (4) and (5).
Conversely, if the function $u(x,t) \in C^l(\Pi\times R)$, where l is defined by formulas (8) and (9) satisfies the mean value formula (36) then it is a regular solution of the wave equation (12) in the Lobachevskiy space.

Proof. Let $\triangle\tau = \tau^{(1)} - \tau^{(2)}$. Then we have

$$\triangle\tau = \tau^{(1)} - \tau^{(2)} = \frac{s^{(1)}}{2x_n^{(1)}}(1+|x^{(1)}|^2) - \frac{s^{(2)}}{2x_n^{(2)}}(1+|x^{(2)}|^2) =$$

$$= \frac{s^{(1)}x_n^{(2)} + s^{(1)}|x^{(1)}|^2x_n^{(2)} - s^{(2)}x_n^{(1)} - s^{(2)}|x^{(2)}|^2x_n^{(1)}}{2x_n^{(1)}x_n^{(2)}}.$$

The increment $\triangle\chi_n = \chi_n^{(1)} - \chi_n^{(2)}$ is expressed by the formula

$$\triangle\chi_n = \frac{s^{(1)}x_n^{(2)} - s^{(1)}|x^{(1)}|^2x_n^{(2)} - s^{(2)}x_n^{(1)} + s^{(2)}|x^{(2)}|^2x_n^{(1)}}{2x_n^{(1)}x_n^{(2)}}.$$

Hence we obtain

$$\Delta\tau^2 - \Delta\chi_n^2 = (\Delta\tau - \Delta\chi_n)(\Delta\tau + \Delta\chi_n) =$$

$$= \frac{(s^{(1)}x_n^{(2)} - s^{(2)}x_n^{(1)})(s^{(1)}|x^{(1)}|^2 x_n^{(2)} - s^{(2)}|x^{(2)}|^2 x_n^{(1)})}{\left(x_n^{(1)}\right)^2\left(x_n^{(2)}\right)^2} . \tag{37}$$

Further, since

$$\Delta\chi' = \chi'^{(1)} - \chi'^{(2)} = \frac{s^{(1)}x'^{(1)}x_n^{(2)} - s^{(2)}x'^{(2)}x_n^{(1)}}{x_n^{(1)}x_n^{(2)}} ,$$

scalar squaring, we obtain

$$|\Delta\chi'|^2 = \frac{\left(s^{(1)}\right)^2|x'^{(1)}|^2\left(x_n^{(2)}\right)^2 - \left(s^{(2)}\right)|x'^{(2)}|^2\left(x_n^{(1)}\right)^2}{\left(x_n^{(1)}\right)^2\left(x_n^{(2)}\right)^2} . \tag{38}$$

Taking into account an equality

$$|\Delta\chi|^2 = |\Delta\chi'|^2 + \Delta\chi_n^2 ,$$

from (37) and (38) we get

$$\Delta\tau^2 - |\Delta\chi|^2 = \frac{\left(s^{(1)}x_n^{(2)} - s^{(2)}x_n^{(1)}\right)\left(s^{(1)}x_n^{(1)} - s^{(2)}x_n^{(2)}\right)}{x_n^{(1)}x_n^{(2)}} - s^{(1)}s^{(2)}\left|\Delta x'\right|^2 .$$

Taking into account that $s^{(j)} = e^{t^{(j)}}$ and the fact that $x_n^{(j)} > 0,\ j = 1, 2$, we obtain that condition (35) is equivalent to condition $\Delta\tau^2 - |\Delta\chi|^2 > 0$. Therefore, the points $(\chi^{(j)}, \tau^{(j)}),\ j = 1, 2$, satisfy the condition of theorem 3. Let

$$v(\chi,\tau) = f(\chi,\tau)u(x(\chi,\tau), t(\chi,\tau)) = (\tau^2 - |\chi|^2)^{(1-n)/4}u(x(\chi,\tau), t(\chi,\tau)). \tag{39}$$

In accordance to Theorem 4 function (39) satisfies the wave equation (6) in Euclidean space, and hence it satisfies the mean value formula. (7). If we express the function $v(\chi,\tau)$ in the formula (7) in terms of the function $u(x(\chi,\tau), t(\chi,\tau))$ by formula (39), we obtain the statement of the theorem. The proof is complete.

6. The Hadamard descent method and the two-point mean value theorem for the Laplace equation in Euclidean space

A two-point formula of the mean for the Laplace equation was obtained in [17] by means of the Hadamard descent method.

Suppose now that the function $v(\chi)$ satisfies the Laplace equation

$$\frac{\partial^2 v}{\partial\chi_1^2} + \ldots + \frac{\partial^2 v}{\partial\chi_n^2} = 0 \tag{40}$$

in a neighborhood of the ball $|\chi| < r$.

We carry out the fictitious variable τ in accordance with the Hadamard descent method assuming

$$v(\chi,\tau) = v(\chi,0) = v(\chi),\ |\chi| < \tau,\ \tau \in R$$

Without loss of generality, we set

$$\tau_1 = 0, \ \tau_2 = r > 0.$$

The function $v(\chi, \tau)$ will satisfy equation (6), which means that the mean value formula (7) is valid for it. Taking into account our assumptions the operator S_τ will act as follows:

$$S_\tau v = S_\tau^n v = \gamma(n) \int\limits_{|\xi|=\tau} v\,(\eta)\; d\omega_\xi, \ |\chi^{(1)} - \chi^{(2)}| > 0, \tag{41}$$

$$S_\tau = S_\tau^n v = \gamma(n) \int\limits_{|\xi|=\tau} v\left(\chi^{(2)} + \xi\right)\, d\omega_\xi, \ \chi^{(1)} = \chi^{(2)}, \tag{42}$$

and formula (7) is transformed to the form

$$v(\chi^{(1)}) + v(\chi^{(2)}) = B_\varrho v. \tag{43}$$

Let us clarify the geometric meaning of the formula (43). Without loss of generality, we set

$$\chi^{(1)} = \alpha = (\alpha_1, 0, \ldots, 0), \ \chi^{(2)} = 0 = (0, 0, \ldots, 0).$$

In (41) integration is carried on the sphere $\xi_1^2 + \xi_2^2 + ... + \xi_n^2 = \tau^2$. We replace the first variable

$$\nu = \frac{\alpha_1}{2} + \frac{r\xi_1}{\sqrt{r^2 - \alpha_1^2}}. \tag{44}$$

Expressing the ξ_1 in terms of ν from formula (44), we obtain

$$\xi_1 = \frac{\sqrt{r^2 - \alpha_1^2}}{r}(\nu - \frac{\alpha_1}{2}). \tag{45}$$

Substituting expression (45) into the sphere equation $|\xi| = r$, we obtain in the new variables $\nu, \xi_2, ..., \xi_n$ the ellipsoid Φ equation:

$$\frac{4(\nu - \frac{\alpha_1}{2})}{r^2} + \frac{4(\xi_2^2 + \cdots + \xi_n^2)}{r^2 - \alpha_1^2} = 1. \tag{46}$$

Integration in formula (43) is carried out on the ellipsoid Φ, which is given by equation (46). Thus, the following assertion is proved.

Theorem 5. (two-point mean value theorem for a harmonic function). Let a domain $\Omega \subset R^n$, in which the function $v(\chi)$ is harmonic, contains the points $\chi^{(1)}$ and $\chi^{(2)}$, $|\chi^{(1)} - \chi^{(2)}| < r$. Let another domain bounded by the ellipsoid Φ defined (in suitable coordinate system) by equation (46), lies in the domain Ω together with its closure. Then the two-point mean value formula (43) is valid.

For $n = 3$ this theorem takes on a very transparent form.

Theorem 5'.(two-point mean value theorem for a harmonic function of three variables). For $n = 3$ in the conditions of the theorem 6 we have the formula

$$v(\alpha) + v(0) = \frac{1}{2\pi} \oiint\limits_{\eta_1^2+\eta_2^2+\eta_3^2=1} v(\frac{\alpha_1}{2} + \frac{r\eta_1}{2}, \frac{\sqrt{r^2 - \alpha_1^2}\eta_2}{2}, \frac{\sqrt{r^2 - \alpha_1^2}\eta_3}{2})\, d\omega_\eta +$$

$$+\frac{\sqrt{r^2-\alpha_1^2}}{4\pi}\oiint\limits_{\eta_1^2+\eta_2^2+\eta_3^2=1}\nabla u(\frac{\alpha_1}{2}+\frac{r\eta_1}{2},\frac{\sqrt{r^2-\alpha_1^2}\eta_2}{2},\frac{\sqrt{r^2-\alpha_1^2}\eta_3}{2})\cdot\mu\,d\omega_\eta,$$

where $\mu=(\frac{r\eta_1}{\sqrt{r^2-\alpha_1^2}},\eta_2,\eta_3)$, the sign $\cdot$ denotes the scalar product in R^3.

Theorem 6. For $\chi^{(1)}=\chi^{(2)}=\chi^{(0)}$ in the case $n=1(\mathrm{mod}2)$ the mean value formula (43) coincides with the well-known Gauss formula for the harmonic function for a sphere of radius $r/2$:

$$v(\chi^{(0)})=\frac{1}{|S_n|\delta^{n-1}}\oint\limits_{|\xi-\chi^{(0)}|=\delta}v(\xi)\,d\omega_\xi,\ \delta=r/2.$$

This theorem is proved by the same scheme as in [10].

7. The Hadamard descent method and the two-point formula of the mean value for an elliptic equation in the Lobachevskiy space

We now apply the Hadamard descent method to the mean value formula (36). Let the function $u=u(x)$ be a regular solution of an equation

$$\Delta_\omega u+\left(\frac{n-1}{2}\right)^2u=0. \tag{47}$$

In accordance to the Hadamard descent method we carry out a fictitious variable t, assuming

$$u(x,t)=u(x,0)=u(x),t\in R.$$

Then formally the function $u(x,t)$ satisfies the wave equation (12) in the Lobachevskiy space. Therefore, it also satisfies the mean value formula (36). In this formula $t^{(1)}$ and $t^{(2)}$ should already be perceived as two parameter values satisfying condition (35). In fact, we have proved the two-point formula of the mean for equation (47): for any two points $x^{(1)}$, $x^{(2)}\in\Pi$ and any two values $t^{(1)}$ and $t^{(2)}\in R$, satisfying condition

$$(x_n^{(2)}e^{t^{(1)}}-x_n^{(1)}e^{t^{(2)}})(x_n^{(1)}e^{t^{(1)}}-x_n^{(2)}e^{t^{(2)}})-e^{t^{(1)}+t^{(2)}}|\Delta x'|^2>0, \tag{48}$$

the regular solution $u(x)$ of equation (47) satisfies the mean value formula

$$e^{t^{(1)}(1-n)/2}u(x^{(1)},t^{(1)})+e^{t^{(2)}(1-n)/2}u(x^{(2)},t^{(2)})=B_\varrho^n(fu). \tag{49}$$

8. The Hadamard descent method and the two-point formula of the mean value for an elliptic equation on a sphere.

We consider the wave equation on a sphere, which has the following form

$$\partial^2u/\partial t^2=\Delta_\omega u-((n-1)/2)^2u. \tag{50}$$

This equation was considered by P. Lax and R. Phillips [15], by I.A. Kipriyanov and L.A. Ivanov (see [13], [14], [16]). These authors refined the concept of the wave equation in a space with constant curvature. Following these authors, we call equation (50) the wave equation on the sphere.

We introduce a map $S_n\times R\longrightarrow R^{n+1}$ by means of formulas

$$T=\tau_\pm(x,z,t)=\pm\sin t/(z\mp\cos t),\ \chi=\chi_\pm(x,z,t)=\pm x/(z\mp\cos t). \tag{51}$$

IOP Conf. Series: Journal of Physics: Conf. Series **973** (2018) 012056 doi:10.1088/1742-6596/973/1/012056

where $\chi = (\chi_1, \ldots, \chi_n)$. Along with the coordinate transformations (51) we introduce a substitution of function

$$u(x, z, t) = (z - \cos t)^{-(n-1)/2} v(\chi(x, z, t), \tau(x, z, t)). \tag{52}$$

Transforms (51)–(52) are equivalent to the transformations introduced in the paper [15] in spherical coordinates on the sphere S_n. Therefore, it is appropriate to call them the P. Lax and R. Phillips transforms.

Theorem 7. The function $v(\chi, \tau)$ is a regular solution of the wave equation (6) in the cone $|\chi| < |\tau|$ of space R^{n+1} of variables $\chi = (\chi_1, \ldots, \chi_n)$, τ if and only if the function (52), obtained from v by means of Lax-Phillips transformations (51)–(52), is a regular solution of the wave equation (50) in the part of the space $S_n \times R$, determined by the condition $|z| > |\cos t|$, $t \in (0, 2\pi]$.

It is possible to prove the analogue of Theorem 1 for the wave equation on the sphere basing on the connection between the wave equations on the sphere and in Euclidean space, which is established by theorem 7.

Theorem 8. *Let the function $u(x, t)$ be a regular solution of the spherical wave equation (50) in a part of the space $S_n \times R$, defined by the condition $|z| > |\cos t|$, $t \in (0, 2\pi]$. Let $(x^{(j)}, z^{(j)}, t^{(j)})$, $j = 1, 2$ be a pair of points in this part of the space $S_n \times \mathbf{R}$, satisfying the condition*

$$(z^{(1)} z^{(2)} + x^{(1)} x^{(2)} - \cos(t^{(1)} - t^{(2)}))/((z^{(1)} - \cos t^{(1)})(z^{(2)} - \cos t^{(2)})) > 0. \tag{53}$$

Then following equality (the mean value formula)

$$(z^{(1)} \mp \cos t^{(1)})^{(n-1)/2} u(x^{(1)}, z^{(1)}, t^{(1)}) + (z^{(2)} \mp \cos t^{(2)})^{(n-1)/2} u(x^{(2)}, z^{(2)}, t^{(2)}) =$$

$$= B^n_\varrho (f(\chi, \tau)\, u(x(\chi, \tau), z(\chi, \tau), t(\chi, \tau))), \tag{54}$$

occurs, where $f = (((\tau^2 + |\chi|^2 + 1)^2 - 4\tau^2 |\chi|^2)/2)^{(1-n)/4}$, $\varrho = \sqrt{\Delta\tau^2 - |\Delta\chi|^2}/2$, $\Delta\tau = \tau^{(1)} - \tau^{(2)}$, $\Delta\chi = \chi^{(1)} - \chi^{(2)}$, $\chi^{(j)} = \chi(x^{(j)}, t^{(j)})$, $\tau^{(j)} = \tau(x^{(j)}, t^{(j)})$, $j = 1, 2$, $u = u(x, t) = u(x(\chi, \tau), t(\chi, \tau))$, the operator B^n_ϱ is defined by formulas (4) and (5). Conversely, if the function $u(x, t) \in C^l(S_n \times R)$, where l is defined by (8), satisfies the mean value formula (54), then it is a regular solution of the spherical wave equation (50).

We now apply the Hadamard descent method to the mean value formula (54). Let a function $u = u(x)$ be a regular solution of the equation

$$\Delta_\omega u + \left(\frac{n-1}{2}\right)^2 u = 0. \tag{55}$$

In accordance to the Hadamard descent method, we introduce the dummy variable t assuming

$$t \in R.$$

Then formally the function $u(x, t)$ satisfies the wave equation (50) on the sphere. Therefore, it also satisfies the mean value formula (54). In this formula $t^{(1)}$ and $t^{(2)}$ should already be perceived as two parameter values that satisfy condition (53). Thus we proved the two-point mean value formula for equation (55): for any two points $(x^{(1)}, z^{(1)})$, $(x^{(2)}, z^{(2)}) \in S_n$ and any two values of $t^{(1)}$, $t^{(2)} \in (0, 2\pi]$ that satisfy condition (53) the regular solution $u(x)$ of equation (55) satisfies the mean value formula

$$(z^{(1)} \mp \cos t^{(1)})^{(n-1)/2} u(x^{(1)}, z^{(1)}, t^{(1)}) + (z^{(2)} \mp \cos t^{(2)})^{(n-1)/2} u(x^{(2)}, z^{(2)}, t^{(2)}) =$$

$$= B^n_\varrho (f(\chi, \tau)\, u(x(\chi, \tau), z(\chi, \tau), t(\chi, \tau))).$$

[1] Gilbarg D and Trudinger N S 1983 *Elliptic Partial Differential Equations of Second Order* (Berlin-Heidelberg-New York-Tokyo, Springer-Verlag XIII 513 S)
[2] John F 1955 *Plane Waves and Spherical means* (New York, London: Interscience Publishes)
[3] Ilyin V A 1969 On Fourier series in fundamental systems of functions of the Beltrami operator *Differential Equations (Differ. Uravn.)* **5** 11. 1940–78
[4] Ilyin V A and Moiseev E I 1981 The mean value formula for the associated functions of the Laplace operator *Differential Equations (Differ. Uravn.)* **17** 10 1908–10
[5] Moiseev 2010 The mean value formula for a harmonic function in the circular sector Proceedings of the Russian Academy of sciences/Doklady: Mathematics **432** 5 592–3
[6] Helgason S 1964 *Differential geometry and symmetric spaces* (Moscow: Mir)
[7] Zalcman L 1980 Offbeat integral geometry *Amer. Math. Monthly* **87** 3 161–75
[8] Volchkov V V 1996 The final version of the mean value theorem in the theory of harmonic functions *Mathematical notes* **59** 3 351–8.
[9] Bitsadze A V and Nakhushev A M 1974 To the theory of equations of mixed type in multidimensional domains *Differential Equations (Differ. Uravn.)* **10** 12 2184–91
[10] Polovinkin I P 1991 The theorem on the mean value for wave equation / *Differential Equations (Differ. Uravn.)* **27** 11 1987–90
[11] Ibragimov N Kh 1983 *Groups of transformations in mathematical physics* (oscow: Science)
[12] Lax P and Phillips R 1976 *The scattering theory for automorphic functions* (Princeton, New Jersey: Princeton University Press)
[13] Kipriyanov I A and Ivanov L A 1983 On the wave equation in a Riemannian space *General theory of boundary value problems* (Kiev: Naukova Dumka) 89–94
[14] Ivanov L A 1981 On Kirchhoff formulas in symmetric spaces of rank 1 *Reports of the Academy of Sciences of the USSR* **247** 4 783–5
[15] Lax P D 1978 An axample of Huygens' principle *Pure and Appl. Math.* **31** 4 415–21
[16] Ivanov L A 1982 On the wave equation in a Riemannian space with constant curvature *Boundary value problems for nonlinear equations* **II** (Novosibirsk: IM SB AS USSR) 137–9
[17] Meshkov V Z and Polovinkin I P 2009 Mean value properties of solutions of linear partial differential equations *Journal of Mathematical Sciences* **160** 1 45–52

IOP Conf. Series: Journal of Physics: Conf. Series **973** (2018) 012057 doi:10.1088/1742-6596/973/1/012057

A spectral method for approximate solving a second-order linear differential equation

M N Oreshina

Department of Applied Mathematics, Lipetsk State Technical University, 30, Moskovskaya Str., Lipetsk 398600, Russia

E-mail: m_oreshina@mail.ru

Abstract. We suggest an approximate method for the linear differential equation $x''(t) + 2\alpha x'(t) + Ax(t) = f(t)$, where A is an unbounded self-adjoint operator, α is a given scalar and the operator $A - \alpha^2\mathbf{1}$ is positive semidefinite. A priori estimates of the approximation error are obtained. The results of numerical experiments for the hyperbolic equation are presented. The suggested approach can be used in the remodeling problems for complicated objects and systems if the initial mathematical model is such an equation.

1. Introduction

We consider a linear differential equation of the second order

$$x''(t) + 2\alpha x'(t) + Ax(t) = f(t) \tag{1}$$

with an unbounded self-adjoint operator coefficient A. We assume that the operator $Q = A-\alpha^2\mathbf{1}$ is positive semidefinite. In this case, Equation (1) is a hyperbolic equation in the abstract form [1]. The exact solution of such an equation can be written in terms of cosine and sine operator functions [2] generated by the operator Q.

In this article we suggest a new approximate method for solving Equation (1). It is based on the approximation of sine operator function by some rational function of the operator Q. If the free term f in Equation (1) has the form $f(t) = bv(t)$, where b is a given vector and v is a scalar function, we suggest a special version of our method. It allows one to calculate the solution at a given point t without finding the solution at the previous points. This, in particular, distinguishes our method from the difference schemes [3, 4], where the calculations are carried out sequentially from the initial to the final point. For both versions of the method we obtain (Theorems 2 and 3) a priori estimates of the approximation error. The proof technique uses the spectral theorem for self-adjoint operators and a functional calculus theorem.

A similar approach is applied for solving an abstract parabolic equation in [5], and some close ideas for solving second-order differential equations with matrix coefficients are discussed in [6].

In Section 2, we refine the statement of the problem, describe the method and prove theorems on the estimation of the approximation error.

In Section 3, we present the results of numerical experiments.

2. Method

Let $\mathbf{H}$ be a Hilbert space (real or complex), and let $A: D(A) \subset \mathbf{H} \to \mathbf{H}$ be a linear unbounded operator. By $\mathbf{1}$ we denote the identity operator.

We consider a second order linear differential equation with zero initial conditions

$$\begin{aligned} x''(t) + 2\alpha x'(t) + Ax(t) &= f(t), \\ x(0) &= 0, \\ x'(0) &= 0. \end{aligned} \tag{2}$$

We assume that the function $f: [0, +\infty) \to \mathbf{H}$ is continuous, $\alpha \geq 0$, the operator A is self-adjoint, and the operator $Q = A - \alpha^2\mathbf{1}$ is positive semidefinite.

We recall that the spectrum $\sigma(Q)$ of the positive semidefinite operator Q is contained [7, 8] in $[0, +\infty)$. Since the operator Q is self-adjoint, there exists [7, 8, 9] a unique resolution of identity E and one can define a functional calculus $g \mapsto g(Q)$ which maps every Borel measurable function g to the densely defined closed operator

$$g(Q) = \int_0^{+\infty} g(\xi)\, dE(\xi).$$

If the function g is bounded on $\sigma(Q)$, then the operator $g(Q)$ is bounded, and the inequality

$$\|g(Q)\| \leq \sup_{\xi \in \sigma(Q)} |g(\xi)| \tag{3}$$

holds [7, 8].

The impulse response of problem (2) is defined as the operator function

$$U(t) = s_t(Q) = \int_0^{+\infty} s_t(\xi)\, dE(\xi),$$

where

$$s_t(\xi) = e^{-\alpha t}\frac{\sin\sqrt{\xi}t}{\sqrt{\xi}}.$$

The function

$$x(t) = \int_0^t U(t-\tau) f(\tau)\, d\tau \tag{4}$$

is called a (*generalized*) *solution* of problem (2).

Sometimes we discuss the version of problem (2) with the free term $f(t) = bv(t)$, where $b \in \mathbf{H}$ is a given vector and $v: [0, +\infty) \to \mathbb{R}$ is a given continuous function, i.e. we solve the problem

$$\begin{aligned} x''(t) + 2\alpha x'(t) + Ax(t) &= bv(t), \\ x(0) &= 0, \\ x'(0) &= 0. \end{aligned} \tag{5}$$

We define the auxiliary function

$$\theta_t(\xi) = \int_0^t s_{t-\tau}(\xi) v(\tau)\, d\tau = \int_0^t s_\tau(\xi) v(t-\tau)\, d\tau \tag{6}$$

or, in more detail,

$$\theta_t(\xi) = \frac{1}{\sqrt{\xi}}\int_0^t e^{-\alpha(t-\tau)}\sin\left(\sqrt{\xi}(t-\tau)\right) v(\tau)\, d\tau = \frac{1}{\sqrt{\xi}}\int_0^t e^{-\alpha(\tau)}\sin\left(\sqrt{\xi}\tau\right) v(t-\tau)\, d\tau.$$

IOP Conf. Series: Journal of Physics: Conf. Series **973** (2018) 012057 doi:10.1088/1742-6596/973/1/012057

We note that for any fixed $t > 0$ the functions s_t and θ_t are bounded on $\sigma(Q)$, therefore, the operators $s_t(Q)$ and $\theta_t(Q)$ are bounded.

Theorem 1. *The (generalized) solution of problem (5) can be represented in the form*

$$x(t) = \theta_t(Q)b.$$

Proof. From formula (4) for $f(t) = bu(t)$ we have

$$\begin{aligned} x(t) &= \int_0^t U(t-\tau)bv(\tau)\,d\tau = \int_0^t \int_0^{+\infty} s_{t-\tau}(\xi)\,dE(\xi)\,bv(\tau)\,d\tau = \\ &= \int_0^t \int_0^{+\infty} s_{t-\tau}(\xi)v(\tau)\,dE(\xi)\,d\tau\,b = \int_0^{+\infty} \int_0^t s_{t-\tau}(\xi)v(\tau)\,d\tau\,dE(\xi)\,b = \\ &= \int_0^{+\infty} \theta_t(\xi)\,dE(\xi)\,b = \theta_t(Q)b. \end{aligned}$$

The proof of the theorem is complete.

Let us construct an approximate impulse response of problem (2). We fix $t > 0$ and approximate on $\sigma(Q) \subset [0, +\infty)$ the function $\xi \mapsto s_t(\xi)$ by a rational function

$$r_t(\xi) = p_0 + \sum_{k=1}^{K_1} \frac{p_k}{\xi - q_k} + \sum_{k=1}^{K_2} \frac{\alpha_k + \beta_k \xi}{\xi^2 + \mu_k \xi + \nu_k} \tag{7}$$

where the roots of the denominators lie outside $\sigma(Q)$.

For the *approximate impulse response* of problem (2) we take the operator function

$$\widetilde{U}(t) = r_t(Q). \tag{8}$$

To calculate a rational function of the operator Q, it is convenient to use the formula [5]

$$r_t(Q) = p_0\mathbf{1} + \sum_{k=1}^{K_1} p_k(Q - q_k\mathbf{1})^{-1} + \sum_{k=1}^{K_2} (\alpha_k\mathbf{1} + \beta_k Q)(Q^2 + \mu_k Q + \nu_k\mathbf{1})^{-1}.$$

We define the *approximate solution* of problem (2) by the formula

$$\tilde{x}(t) = \int_0^t \widetilde{U}(t-\tau)f(\tau)\,d\tau = \int_0^t r_{t-\tau}(Q)f(\tau)\,d\tau. \tag{9}$$

In the case of the problem (5), we can immediately construct an approximate solution at a given point $t > 0$. To do this, we approximate on $\sigma(Q) \subset [0, +\infty)$ the function $\xi \mapsto \theta_t(\xi)$ by rational function (7), and for *approximate solution* of problem (5) we take the function

$$\tilde{x}(t) = r_t(Q)b = p_0 b + \sum_{k=1}^{K_1} p_k(Q - q_k\mathbf{1})^{-1}b + \sum_{k=1}^{K_2} (\alpha_k\mathbf{1} + \beta_k Q)(Q^2 + \mu_k Q + \nu_k\mathbf{1})^{-1}b. \tag{10}$$

We note that to find vectors of the form $w = (Q - q_k\mathbf{1})^{-1}b$ from formula (10) it is sufficient [5] to solve (exactly or approximately) equations

$$Qw - q_k w = b.$$

IOP Conf. Series: Journal of Physics: Conf. Series **973** (2018) 012057 doi:10.1088/1742-6596/973/1/012057

Similarly, to calculate vectors of the form $w = (Q^2 + \mu_k Q + \nu_k \mathbf{1})^{-1}b$ it is sufficient to solve equations

$$Q^2 w + \mu_k Q w + \nu_k w = b.$$

To estimate the absolute errors of approximate impulse response (8) and approximate solution (9) of problem (2) we present the following theorem.

Theorem 2. *Let* $\varepsilon\colon [0, +\infty) \to [0, +\infty)$ *be a given function, such that for each* $t > 0$ *the estimate*

$$|r_t(\xi) - s_t(\xi)| \le \varepsilon(t), \qquad \xi \in [0, +\infty) \tag{11}$$

holds. Then the approximate impulse response (8) of problem (2) satisfies the estimate

$$\|\widetilde{U}(t) - U(t)\| \le \varepsilon(t),$$

and the approximate solution (9) of problem (2) satisfies the estimate

$$\|\tilde{x}(t) - x(t)\| \le \sup_{\tau \in [0,t]} \varepsilon(\tau) \int_0^t \|f(\tau)\|\, d\tau.$$

Proof. Consider the function $g_t(\xi) = r_t(\xi) - s_t(\xi)$. Obviously, the function g_t is bounded on $\sigma(Q) \subset [0, +\infty)$. Using estimates (3) and (11) we obtain

$$\|\widetilde{U}(t) - U(t)\| = \|r_t(Q) - s_t(Q)\| = \|g_t(Q)\| \le \sup_{\xi \in \sigma(Q)} |g_t(\xi)| \le \varepsilon(t).$$

This implies

$$\begin{aligned} \|\tilde{x}(t) - x(t)\| &= \left\| \int_0^t (\widetilde{U}(t-\tau) - U(t-\tau)) f(\tau)\, d\tau \right\| \le \\ &\le \int_0^t \left\| \widetilde{U}(t-\tau) - U(t-\tau)) f(\tau) \right\| d\tau \le \int_0^t \varepsilon(t-\tau) \|f(\tau)\|\, d\tau \le \\ &\le \sup_{\tau \in [0,t]} \varepsilon(\tau) \int_0^t \|f(\tau)\|\, d\tau. \end{aligned}$$

The proof of the theorem is complete.

Now we estimate the absolute error of approximate solution (10) of problem (5).

Theorem 3. *We fix* $t > 0$. *Let the rational function* $\xi \mapsto r_t(\xi)$ *approximates the function* $\xi \mapsto \theta_t(\xi)$ *on* $[0, +\infty)$ *with the absolute error* $\varepsilon(t)$, *i.e.*

$$|r_t(\xi) - \theta_t(\xi)| \le \varepsilon(t), \qquad \xi \in [0, +\infty). \tag{12}$$

Then approximate solution (10) of problem (5) satisfies the estimate

$$\|\tilde{x}(t) - x(t)\| \le \varepsilon(t)\|b\|.$$

Proof. Consider the function $g_t(\xi) = r_t(\xi) - \theta_t(\xi)$. Obviously, the function g_t is bounded on $\sigma(Q) \subset [0, +\infty)$. Using estimates (3) and (12) we obtain

$$\|\tilde{x}(t) - x(t)\| = \|g_t(Q)b\| \le \|g_t(Q)\|\,\|b\| \le \sup_{\xi \in \sigma(Q)} |g_t(\xi)|\,\|b\| \le \varepsilon(t)\|b\|.$$

The proof of the theorem is complete.

3. Results and discussion

The main theoretical results are given above in Theorems 2 and 3, where the estimates of the approximation error are constructed. In this Section we discuss the results of the numerical experiments for the suggested method.

3.1. Input function

Let the input function v has the form

$$v(t) = \begin{cases} 0, & t < 0; \\ t/r, & 0 \le t < r; \\ 1, & r \le t < w + r; \\ -(t - (w + 2r))/r, & w + r \le t < w + 2r; \\ 0, & w + 2r \le t. \end{cases} \tag{13}$$

The graph of this function in the case $r = 0.1$ and $w = 1$ is shown in Figure 1.

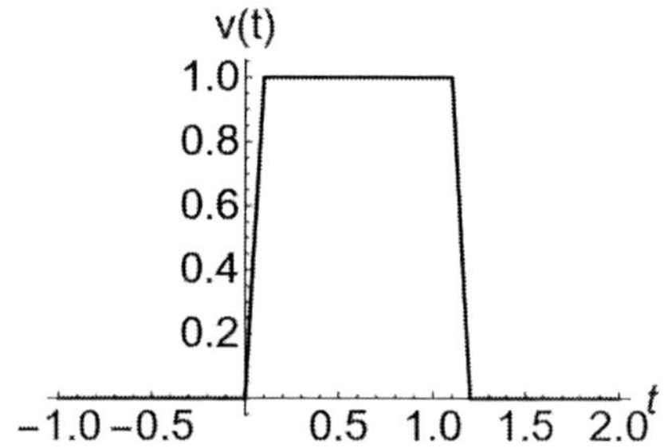

Figure 1. The graph of the input function v

Theorem 4. *In the case of the input function (13), the function (6) admits the representation*

$$\theta_t(\xi) = \frac{e^{-\alpha t}}{r\sqrt{\xi}(\alpha^2 + \xi)^2} \begin{cases} 0, & t < 0; \\ \theta_t^1(\xi), & 0 \le t < r; \\ \theta_t^2(\xi), & r \le t < w + r; \\ \theta_t^3(\xi), & w + r \le t < w + 2r; \\ \theta_t^4(\xi), & t \ge w + 2r. \end{cases}$$

where

$$\begin{aligned}
\theta_t^1(\xi) &= e^{\alpha t}\sqrt{\xi}\left(-2\alpha + \alpha^2 t + \xi t\right) + (\alpha^2 - \xi)\sin\left(\sqrt{\xi}t\right) + 2\alpha\sqrt{\xi}\cos\left(\sqrt{\xi}t\right), \\
\theta_t^2(\xi) &= (\alpha^2 + \xi)\sqrt{\xi}re^{\alpha t} + (\xi - \alpha^2)e^{\alpha r}\sin\left(\sqrt{\xi}(t - r)\right) - 2\alpha\sqrt{\xi}e^{\alpha r}\cos\left(\sqrt{\xi}(t - r)\right) + \\
&\quad +(\alpha^2 - \xi)\sin\left(\sqrt{\xi}t\right) + 2\alpha\sqrt{\xi}\cos\left(\sqrt{\xi}t\right), \\
\theta_t^3(\xi) &= 2(\alpha^2 + \xi)\sqrt{\xi}re^{\alpha t} + (\xi - \alpha^2)e^{\alpha r}\sin\left(\sqrt{\xi}(t - r)\right) - 2\alpha\sqrt{\xi}e^{\alpha r}\cos\left(\sqrt{\xi}(t - r)\right) + \\
&\quad +(\alpha^2 - \xi)e^{\alpha(r+w)}\sin\left(\sqrt{\xi}(r - t + w)\right) - 2\alpha\sqrt{\xi}e^{\alpha(r+w)}\cos\left(\sqrt{\xi}(r - t + w)\right) - \\
&\quad -(\alpha^2 + \xi)\sqrt{\xi}te^{\alpha t} + (\alpha^2 - \xi)\sin\left(\sqrt{\xi}t\right) + 2\alpha\sqrt{\xi}e^{\alpha t} + \\
&\quad +2\alpha\sqrt{\xi}\cos\left(\sqrt{\xi}t\right) + (\alpha^2 + \xi)\sqrt{\xi}we^{\alpha t},
\end{aligned}$$

$$\begin{aligned}\theta_t^4(\xi) &= (\xi-\alpha^2)e^{\alpha r}\sin\left(\sqrt{\xi}(t-r)\right) - 2\alpha\sqrt{\xi}e^{\alpha r}\cos\left(\sqrt{\xi}(t-r)\right) + \\ &\quad +(\alpha^2-\xi)e^{\alpha(2r+w)}\sin\left(\sqrt{\xi}(-2r+t-w)\right) + (\alpha^2-\xi)e^{\alpha(r+w)}\sin\left(\sqrt{\xi}(r-t+w)\right) + \\ &\quad +2\alpha\sqrt{\xi}e^{\alpha(2r+w)}\cos\left(\sqrt{\xi}(-2r+t-w)\right) - 2\alpha\sqrt{\xi}e^{\alpha(r+w)}\cos\left(\sqrt{\xi}(r-t+w)\right) + \\ &\quad +(\alpha^2-\xi)\sin\left(\sqrt{\xi}t\right) + 2\alpha\sqrt{\xi}\cos\left(\sqrt{\xi}t\right).\end{aligned}$$

In particular, for $r = 0$ we have

$$\theta_t(\xi) = \frac{e^{-\alpha t}}{\sqrt{\xi}(\alpha^2+\xi)}\begin{cases} 0, & t<0; \\ \sqrt{\xi}e^{\alpha t} - \alpha\sin\left(\sqrt{\xi}t\right) - \sqrt{\xi}\cos\left(\sqrt{\xi}t\right), & 0 \le t < w; \\ -\alpha\sin\left(\sqrt{\xi}t\right) - \sqrt{\xi}\cos\left(\sqrt{\xi}t\right) + \\ \quad +e^{\alpha w}\left(\alpha\sin\left(\sqrt{\xi}(t-w)\right) + \sqrt{\xi}\cos\left(\sqrt{\xi}(t-w)\right)\right), & t \ge w. \end{cases}$$

Proof. The proofs of these formulas reduce to direct calculations.

3.2. Example 1

We consider the problem of solving the hyperbolic equation

$$\frac{\partial^2}{\partial t^2}u(t,s) + \frac{\partial}{\partial t}u(t,s) - \frac{\partial^2}{\partial s^2}u(t,s) + u(t,s) = b(s)v(t), \qquad t \ge 0, \quad s \in [0,1], \tag{14}$$

with the boundary conditions of the third kind

$$u(t,0) - \frac{\partial u}{\partial s}(t,0) = 0, \quad u(t,1) + \frac{\partial u}{\partial s}(t,1) = 0, \tag{15}$$

and with the initial conditions

$$u(0,s) = 0, \quad \frac{\partial u}{\partial t}(0,s) = 0. \tag{16}$$

Here $b \in L_2[0,1]$ is a real function, and $v\colon [0,+\infty) \to \mathbb{R}$ is a continuous function.

In the Hilbert space $\mathbf{H} = L_2[0,1]$ we consider the operator $A = \mathbf{1} - \frac{d^2}{ds^2}$ with the domain

$$D(A) = \left\{z \in W_2^2[0,1] : z(0) - z'(0) = 0,\ z(1) + z'(1) = 0\right\}.$$

We set $\alpha = \frac{1}{2}$ $x(t)(s) = u(t,s)$. In this notation, Problem (14)–(16) has the form (5). It is easy to verify that the operator $Q = A - \alpha^2\mathbf{1} = \frac{3}{4}\mathbf{1} - \frac{d^2}{ds^2}$ is self-adjoint and positive semidefinite; therefore, we can use formula (10) to construct an approximate solution. Obviously, $D(Q) = D(A)$.

The vectors $w = (Q - q_k\mathbf{1})^{-1}b$ from formula (10) are given by the explicit formula

$$\begin{aligned}(Q-q_k\mathbf{1})^{-1}b(s) &= C_1(q_k,s)\int_0^s e^{\frac{\xi}{2}\sqrt{3-4q_k}}b(\xi)\,d\xi + C_2(q_k,s)\int_0^s e^{-\frac{\xi}{2}\sqrt{3-4q_k}}b(\xi)\,d\xi \\ &\quad + C_3(q_k,s)\int_s^1 e^{\frac{\xi}{2}\sqrt{3-4q_k}}b(\xi)\,d\xi + C_4(q_k,s)\int_s^1 e^{-\frac{\xi}{2}\sqrt{3-4q_k}}b(\xi)\,d\xi, \quad (17)\end{aligned}$$

where

$$\begin{aligned}
C_1(q_k,s) &= \frac{1}{C_0(q_k)}\left(e^{-\frac{1-s}{2}\sqrt{3-4q_k}}(1+4q_k) - e^{\frac{1-s}{2}\sqrt{3-4q_k}}(\sqrt{3-4q_k}+2)^2\right),\\
C_2(q_k,s) &= \frac{1}{C_0(q_k)}\left(-e^{-\frac{1-s}{2}\sqrt{3-4q_k}}(\sqrt{3-4q_k}-2)^2 + e^{\frac{1-s}{2}\sqrt{3-4q_k}}(1+4q_k)\right),\\
C_3(q_k,s) &= \frac{1}{C_0(q_k)}\left(e^{-\frac{1-s}{2}\sqrt{3-4q_k}}(1+4q_k) - e^{-\frac{1+s}{2}\sqrt{3-4q_k}}(\sqrt{3-4q_k}-2)^2\right),\\
C_4(q_k,s) &= \frac{1}{C_0(q_k)}\left(e^{\frac{1-s}{2}\sqrt{3-4q_k}}(1+4q_k) - e^{\frac{1+s}{2}\sqrt{3-4q_k}}(\sqrt{3-4q_k}+2)^2\right),\\
C_0(q_k) &= \sqrt{3-4q_k}(\sqrt{3-4q_k}-2)^2 e^{-\frac{1}{2}\sqrt{3-4q_k}} + e^{\frac{1}{2}\sqrt{3-4q_k}}\sqrt{3-4q_k}(2+\sqrt{3-4q_k})^2.
\end{aligned}$$

We note that to obtain formula (17) it suffices [5] to solve the boundary value problem

$$\begin{aligned}
-w'' + \left(\frac{3}{4} - q_k\right) w &= b,\\
w(0) - w'(0) &= 0,\\
w(1) + w'(1) &= 0.
\end{aligned}$$

In particular, for $b(s) = 1$, $s \in [0,1]$, formula (17) has the form

$$(Q - q_k\mathbf{1})^{-1}b(s) = \frac{8 - 8e^{\frac{1-s}{2}\sqrt{3-4q_k}} - 8e^{\frac{s}{2}\sqrt{3-4q_k}} + 4e^{\frac{1}{2}\sqrt{3-4q_k}}\left(\sqrt{3-4q_k}+2\right) - 4\sqrt{3-4q_k}}{\left(e^{\frac{1}{2}\sqrt{3-4q_k}}\left(\sqrt{3-4q_k}+2\right) - \sqrt{3-4q_k}+2\right)(3-4q_k)}.$$

Also there exists an explicit formula for the vectors $w = (\alpha_k\mathbf{1} + \beta_k Q)(Q^2 + \mu_k Q + \nu_k\mathbf{1})^{-1}b$ from formula (10), but it is very cumbersome, so we do not write it here. We note only that it follows from the formula for the the solution of the boundary value problem

$$\begin{aligned}
w^{IV}(s) - \left(\frac{3}{2} + \mu_k\right) w''(s) + \left(\frac{9}{16} + \frac{3}{4}\mu_k + \nu_k\right) w(s) &= b,\\
w(0) - w'(0) &= 0,\\
w(1) + w'(1) &= 0,\\
w''(0) - w'''(0) &= 0,\\
w''(1) + w'''(1) &= 0.
\end{aligned}$$

To construct an approximate solution $\tilde{x}$ as a rational function $\xi \mapsto r_t(\xi)$ we use the Padé approximation [10] of degree $(3,9)$ for the function $\xi \mapsto \theta_t(\xi)$ at zero. To compute $\tilde{x}(t)$, we expand $r_t(\xi)$ into the sum of the elementary fractions.

In Figure 2 we present the graph of the approximate solution $\tilde{x}$ of Problem (14)–(16) for $t \in [0,10]$, corresponding to the input function (13) with $r = 0$, $w = 1$ and $b(s) = 1$, $s \in [0,1]$. In this case we have

$$\max_{t\in[0,10]} \max_{\xi\in[0,+\infty)} |\theta_t - r_t(\xi)| \le \varepsilon = 0.0508.$$

From this estimate, by virtue of Theorem 3, we obtain the estimate for the error of the approximate solution

$$\|\tilde{x}(t) - x(t)\| \le \varepsilon\|b\| = 0.0508.$$

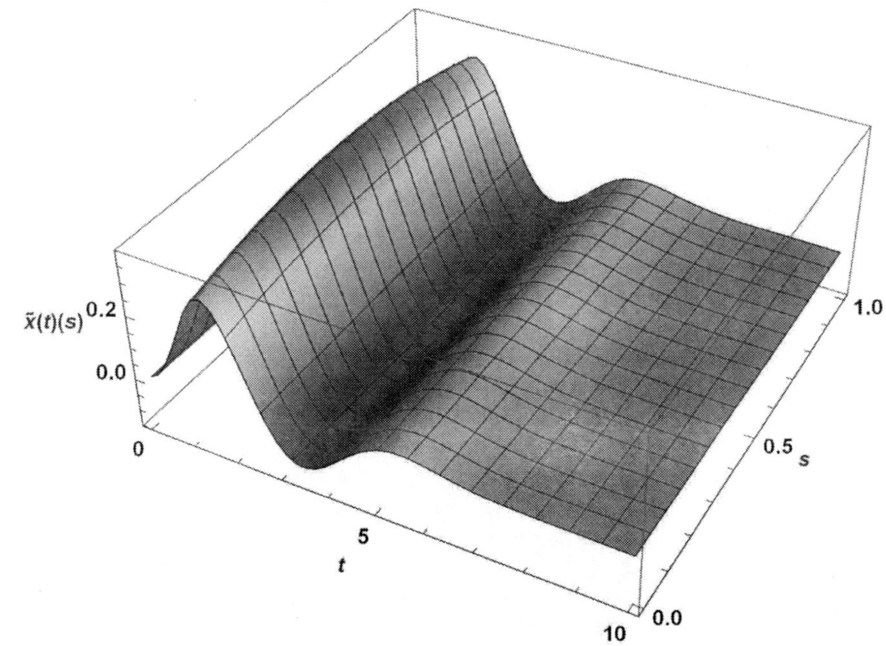

Figure 2. The graph of the approximate solution $\tilde{x}(t)(s) = \tilde{u}(t,s)$ of Problem (14)–(16)

3.3. Example 2

We consider the problem of solving the hyperbolic equation

$$\frac{\partial^2}{\partial t^2}u(t,s) + \frac{\partial}{\partial t}u(t,s) - \frac{\partial^2}{\partial s^2}u(t,s) + u(t,s) = b(s)v(t), \qquad t \geq 0, \quad s \in [0,1], \tag{18}$$

with the boundary conditions of the first kind

$$u(t,0) = 0, \quad u(t,1) = 0, \tag{19}$$

and with the initial conditions

$$u(0,s) = 0, \quad \frac{\partial u}{\partial t}(0,s) = 0. \tag{20}$$

Here $b \in L_2[0,1]$ is a real function, and $v\colon [0,+\infty) \to \mathbb{R}$ is a continuous function.

Similar to Example 1 we set $\alpha = \frac{1}{2}$, $x(t)(s) = u(t,s)$ and consider the operators $A = \mathbf{1} - \frac{d^2}{ds^2}$ and $Q = A - \alpha^2\mathbf{1} = \frac{3}{4}\mathbf{1} - \frac{d^2}{ds^2}$ with the domain

$$D(Q) = D(A) = \left\{z \in W_2^2[0,1] : z(0) = 0,\, z(1) = 0\right\}.$$

In this notation, problem (18)–(20) has the form (5).

To calculate the vectors $w = (Q - q_k\mathbf{1})^{-1}b$ from formula (10) it suffices to solve the boundary value problem

$$\begin{aligned} -w'' + \left(\frac{3}{4} - q_k\right) w &= b, \\ w(0) &= 0, \\ w(1) &= 0, \end{aligned}$$

IOP Conf. Series: Journal of Physics: Conf. Series **973** (2018) 012057 doi:10.1088/1742-6596/973/1/012057

and to obtain an explicit formula for the vectors $w = (\alpha_k \mathbf{1} + \beta_k Q)(Q^2 + \mu_k Q + \nu_k \mathbf{1})^{-1} b$ it suffices to solve the boundary value problem

$$\begin{aligned} w^{IV}(s) - \left(\frac{3}{2} + \mu_k\right) w''(s) + \left(\frac{9}{16} + \frac{3}{4}\mu_k + \nu_k\right) w(s) &= b, \\ w(0) &= 0, \\ w(1) &= 0, \\ w''(0) &= 0, \\ w''(1) &= 0. \end{aligned}$$

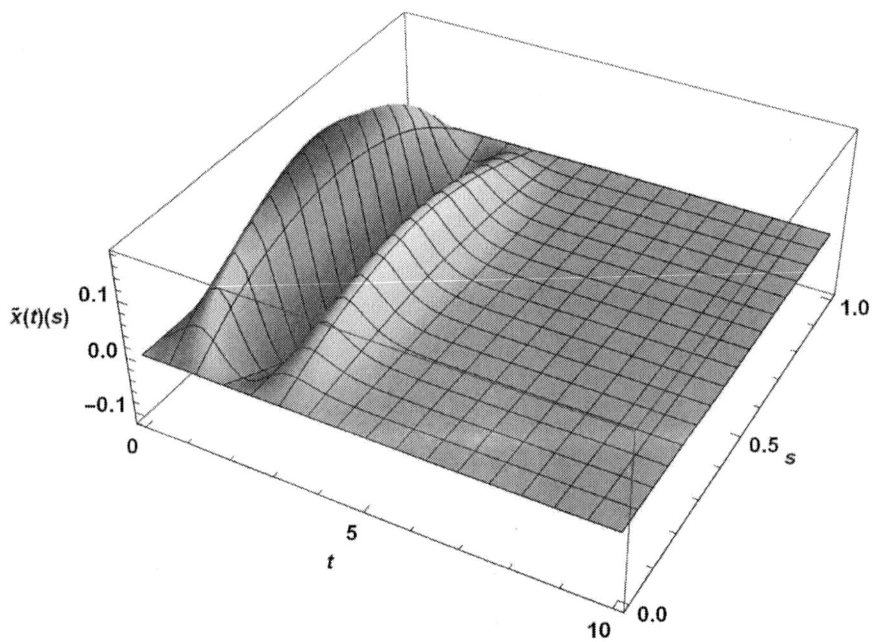

Figure 3. The graph of the approximate solution $\tilde{x}(t)(s) = \tilde{u}(t, s)$ of Problem (18)–(20)

In Figure 3 we present the graph of the approximate solution $\tilde{x}$ of Problem (18)–(20) for $t \in [0, 10]$, corresponding to the input function (13) with $r = 0$, $w = 1$ and $b(s) = 1$, $s \in [0, 1]$.

Since the input function is the same as in Example 1, we can use the same functions as scalar rational functions r_t which approximate θ_t on $[0, +\infty)$. Thus, due to Theorem 3, we have the same estimates of the approximate solution. However, the domains of the operator Q from Examples 1 and 2 do not coincide. Therefore, the results of substituting the operator Q into the function r_t are not equal and the corresponding solutions are different.

4. Conclusion

The suggested method allows one to construct an approximate solution by means of the rational approximation of the function s_t or θ_t. If one can write an explicit formula for the resolvent of operator Q or for the vectors $w = (Q - q_k \mathbf{1})^{-1} b$ the constructed solution is numerical-analytical. If the analytic form of the solution of the equation $Qw - q_k w = b$ is unknown, one can solve it approximately and apply the suggested approach. The numerical experiments show the effectiveness of the suggested method. To improve the accuracy of the approximation one can increase the degree of the Padé approximation or use the rational functions of the best approximation.

Acknowledgments
This work was funded by the Russian Foundation for Basic Research and Lipetsk region (research project No 17-47-480305-r_a).

References

[1] Krein S G 1967 *Linear Differential Equations in a Banach Space* (Moscow: Nauka)
[2] Vasil'ev V V, Krein S G and Piskarev S I 1990 Operator semigroups, cosine operator functions, and linear differential equations *Itogi Nauki i Tekhniki. Seriya "Matematicheskii Analiz"* **28** 87–202
[3] Samarskii A A and Gulin A V 2003 *Numerical Methods of Mathematical Physics* (Moscow: Scientific World)
[4] Ashyralyev A and Sobolevskii P E 2004 *New difference schemes for partial differential equations* (Basel-Boston-Berlin: Birkhäuser)
[5] Oreshina M N 2017 Approximate solution of a parabolic equation with the use of a rational approximation to the operator exponential *Differential Equations* **53(3)** 398–408.
[6] Kurbatov V G and Oreshina M N 2003 On approximate solution of second order linear differential equation *Vestnik Voronezhskogo gosudarstvennogo universiteta. Seriya: Fizika. Matematika* **2** 173–188
[7] Rudin W 1991 *Functional analysis* (New York: McGraw-Hill)
[8] Helemskii A Ya 2005 *Lectures and Exercises on Functional Analysis* (Providence: American Mathematical Society)
[9] Oreshina M N 2015 Spectral theorem for self-adjoint operators in a real Hilbert space *Vestnik Voronezhskogo gosudarstvennogo universiteta. Seriya: Fizika. Matematika* **3** 120–133
[10] Baker G A Jr and Graves-Morris P 1996 *Padé Approximants* (Cambridge: Cambridge Univ. Press)

IOP Conf. Series: Journal of Physics: Conf. Series **973** (2018) 012058 doi:10.1088/1742-6596/973/1/012058

Generalized Euler–Poisson–Darboux equation and singular Klein–Gordon equation

Elina L. Shishkina

Department of Applied Mathematics, Informatics and Mechanics, Voronezh State University, 394006, Universitetskaya pl., 1, Voronezh, Russia

E-mail: ilina_dico@mail.ru

Abstract. Solutions of the Cauchy problem for generalized Euler–Poisson–Darboux equation were obtained by Hankel transform method.

1. Introduction

The Euler–Poisson–Darboux equation is the one of the most extensively studied singular linear hyperbolic equation. This equation is closely related to the spherical mean transform, which integrates functions over spheres of arbitrary radius and centers located on a sphere and to the Riesz potential theory (see [1], [2]). In our article we study generalized Euler–Poisson–Darboux equation with both timelike and spacelike singularities and with a function multiplied by a constant on the right–hand side. In the particular case, such an equation is the singular Klein–Gordon equation and must be interesting for relativists ([3], [4]). We treat the problem with sufficient generality to develop generalized and regular solutions.

We apply Hankel transform method to solve the initial value problem

$$\left[\sum_{i=1}^{n}\left(\frac{\partial^2}{\partial x_i^2}+\frac{\gamma_i}{x_i}\frac{\partial}{\partial x_i}\right)-\left(\frac{\partial^2}{\partial t^2}+\frac{k}{t}\frac{\partial}{\partial t}\right)\right]u=c^2u, \tag{1}$$

$$u(x,0;k)=f(x),\quad u_t(x,0;k)=0,\quad u=u(x,t;k),\ \gamma_i>0,\ x_i>0,\ i=1,...,n,\ t>0. \tag{2}$$

We will call (1) the **generalized Euler–Poisson–Darboux equation**. We obtain the distributional solution of (1)–(2) in convenient space. Besides, we give formulas for regular solution of (1)–(2) in particular case of k and of Cauchy the the singular Klein–Gordon equation.

Equation

$$\frac{\partial^2 u}{\partial x^2}=\left(\frac{\partial^2 u}{\partial t^2}+\frac{k}{t}\frac{\partial}{\partial t}\right),\quad u=u(x,t)$$

appears in Leonard Euler's work (see [5] p. 227) and later was studied by Siméon Denis Poisson in [6], by Gaston Darboux in [7] and by Bernhard Riemann in [8].

Alexander Weinstein (see [9]) and then his students from "University of Maryland College Park" made significant progress in studies of (1) when $\gamma_i=0$, $i=1,...,n$ and $c=0$ in the classical sense. David Fox (one of the students of Weinstein) obtained the solution of (1)–(2) when $c=0$ and $k\neq-1,-3,-5,...$ in terms of the Lauricella function (see [10]). Carrol and Showalter (see [11]) applied method of transmutation operators based on the ideas of Delsarte to study Euler–Poisson–Darboux type equations. The benefits of the transmutation operators method see [12]. The solutions of the Euler–Poisson–Darboux type equations in the distributional sense were obtained by D. W. Bresters (see [13], [14]). Abstract Euler–Poisson–Darboux equation were studied in [15], [16], [17].

Let expand briefly on each of article sections. First, in the section 1, we give necessary definitions. Namely, we define space of weighted functionals, some weighted functionals connected with quadratic form, generalized convolution operator and Hankel transform. Next, in the section 2, we find explicit representation for Hankel transform of weighted functionals connected with quadratic form. In the section 3 we obtain an explicit form for the solution for arbitrary k (1)–(2). In the same section we show that for $c=0$ one obtains again the results of [18] and write the solution of a Cauchy for the singular Klein–Gordon equation. Finally, in the section 3 we show that for $k>n+|\gamma|$ distributional solution of (1)–(2) is regular and give some examples.

2. Definitions and propositions

We deal with the subset of the Euclidean space

$$R^n_+=\{x=(x_1,\dots,x_n)\in R^n,\ \ x_1>0,\dots,x_n>0\}.$$

Let denote $|x|=\sqrt{\sum\limits_{i=1}^n x_i^2}$ and Ω be finite or infinite open set in R^n symmetric with respect to each hyperplane $x_i=0$, $i=1,...,n$, $\Omega_+=\Omega\cap R^n_+$ and $\overline{\Omega}_+=\Omega\cap\overline{R}^n_+$ where

$$\overline{R}^n_+=\{x=(x_1,\dots,x_n)\in R^n,\ \ x_1\geq0,\dots,x_n\geq0\}.$$

The function f is said to be of class $C^\infty(\overline{\Omega}_+)$ if it has derivatives of all orders on $\overline{\Omega}_+$. Function $f\in C^\infty(\overline{\Omega}_+)$ we will call *even with respect to* x_i, $i=1,...,n$ if $\left.\frac{\partial^{2k+1}f}{\partial x_i^{2k+1}}\right|_{x=0}=0$ for all nonnegative integer k (see [19], p. 21). Class $C^\infty_{ev}(\overline{\Omega}_+)$ consists of functions from $C^\infty(\overline{\Omega}_+)$ even with respect to each variable x_i, $i=1,...,n$. Let $\overset{\circ}{C}{}^\infty_{ev}(\overline{\Omega}_+)$ be the space of all functions $f\in C^\infty_{ev}(\overline{\Omega}_+)$ with a compact support. Elements of $\overset{\circ}{C}{}^\infty_{ev}(\overline{\Omega}_+)$ we will call *test functions* and use the notation $\overset{\circ}{C}{}^\infty_{ev}(\overline{\Omega}_+)=\mathcal{D}_+(\overline{\Omega}_+)$. We will use notation $\mathcal{D}_+=\mathcal{D}_+(\overline{R}^n_+)$.

As the space of basic functions we will use the subspace of the space of rapidly decreasing functions:

$$S_{ev}(R^n_+)=\left\{f\in C^\infty_{ev}:\ \sup_{x\in R^n_+}\left|x^\alpha D^\beta f(x)\right|<\infty\quad\forall\alpha,\beta\in Z^n_+\right\},$$

where $\alpha=(\alpha_1,...,\alpha_n)$, $\beta=(\beta_1,...,\beta_n)$, $\alpha_1,...,\alpha_n,\beta_1,...,\beta_n$ are integer nonnegative numbers, $x^\alpha=x_1^{\alpha_1}x_2^{\alpha_2}\dots x_n^{\alpha_n}$, $D^\beta=D^{\beta_1}_{x_1}...D^{\beta_n}_{x_n}$, $D_{x_j}=\frac{\partial}{\partial x_j}$.

We deal with multi-index $\gamma=(\gamma_1,\dots,\gamma_n)$ consists of positive fixed reals $\gamma_i>0$, $i=1,...,n$, $|\gamma|=\gamma_1+\dots+\gamma_n$. Let $L^\gamma_p(\Omega_+)$, $1\leq p<\infty$, be the space of all measurable in Ω_+ functions even with respect to each variable x_i, $i=1,...,n$ such that

$$\int\limits_{\Omega_+}|f(x)|^p x^\gamma dx<\infty,$$

where and further

$$x^\gamma=\prod_{i=1}^n x_i^{\gamma_i}.$$

For a real number $p \geq 1$, the $L_p^\gamma(\Omega_+)$–norm of f is defined by

$$||f||_{L_p^\gamma(\Omega_+)} = \left(\int\limits_{\Omega_+} |f(x)|^p x^\gamma dx \right)^{1/p}.$$

Weighted measure of Ω_+ is denoted by $\text{mes}_\gamma(\Omega)$ and is defined by formula

$$\text{mes}_\gamma(\Omega_+) = \int\limits_{\Omega_+} x^\gamma dx.$$

For every measurable function $f(x)$ defined on R_+^n we consider

$$\mu_\gamma(f,t) = \text{mes}_\gamma\{x \in R_+^n : |f(x)| > t\} = \int\limits_{\{x \in R_+^n : |f(x)| > t\}} x^\gamma dx$$

We will call the function $\mu_\gamma = \mu_\gamma(f,t)$ a *weighted distribution function* $|f(x)|$.

A space $L_\infty^\gamma(\Omega_+)$ is defined as a set of measurable on Ω_+ and even with respect to each variable functions $f(x)$ such as

$$||f||_{L_\infty^\gamma(\Omega_+)} = \inf_{a \in \Omega_+} \{\mu_\gamma(f,a) = 0\} < \infty.$$

For $1 \leq p \leq \infty$ the $L_{p,loc}^\gamma(\Omega_+)$ is the set of functions $u(x)$ defined almost everywhere in Ω_+ such that $uf \in L_p^\gamma(\Omega_+)$ for any $f \in \overset{\circ}{C}{}_{ev}^\infty(\overline{\Omega}_+)$. Each function $u(x) \in L_{1,loc}^\gamma(\Omega_+)$ will be identified with the functional $u \in \mathcal{D}'_+(\overline{\Omega}_+)$ acting according to the formula

$$(u,f)_\gamma = \int\limits_{R_+^n} u(x)\, f(x)\, x^\gamma\, dx, \qquad f \in \overset{\circ}{C}{}_{ev}^\infty(\overline{R}{}_+^n). \tag{3}$$

Functionals $u \in \mathcal{D}'_+(\overline{\Omega}_+)$ acting by the formula 3 will be called *regular weighted functionals.* All other linear functionals $u \in \mathcal{D}'_+(\overline{\Omega}_+)$ will be called *singular weighted functionals.*

3. Singular Bessel differential operator, Laplace–Bessel operator, Bessel functions and Hankel transform

We will deal with the **singular Bessel differential operator** B_ν (see, for example, [19], p. 5):

$$(B_\nu)_t = \frac{\partial^2}{\partial t^2} + \frac{\nu}{t}\frac{\partial}{\partial t} = \frac{1}{t^\nu}\frac{\partial}{\partial t}t^\nu\frac{\partial}{\partial t}, \qquad t > 0.$$

and the elliptical singular operator or the Laplace–Bessel operator Δ_γ:

$$\Delta_\gamma = (\Delta_\gamma)_x = \sum_{i=1}^n (B_{\gamma_i})_{x_i} = \sum_{i=1}^n \left(\frac{\partial^2}{\partial x_i^2} + \frac{\gamma_i}{x_i}\frac{\partial}{\partial x} \right) = \sum_{i=1}^n \frac{1}{x_i^{\gamma_i}}\frac{\partial}{\partial x_i}x_i^{\gamma_i}\frac{\partial}{\partial x_i}. \tag{4}$$

The operator 4 belongs to the class of B–elliptic operators by I. A. Kipriyanovs' classification (see [19]).

The symbol j_ν is used for the normalized Bessel function:

$$j_\nu(t) = \frac{2^\nu\Gamma(\nu+1)}{t^\nu}J_\nu(t), \tag{5}$$

IOP Conf. Series: Journal of Physics: Conf. Series **973** (2018) 012058 doi:10.1088/1742-6596/973/1/012058

where $J_\nu(t)$ is the Bessel function of the first kind of order ν (see [20]):

$$J_\alpha(x) = \sum_{m=0}^{\infty} \frac{(-1)^m}{m!\,\Gamma(m+\alpha+1)} \left(\frac{x}{2}\right)^{2m+\alpha}.$$

Using formulas 9.1.27 from [21] we obtain that the function $j_\nu(t)$ is an eigenfunction of a linear operator B_ν:

$$(B_\nu)_t j_{\frac{\nu-1}{2}}(\tau t) = -\tau^2 j_{\frac{\nu-1}{2}}(\tau t). \tag{6}$$

We also will need some other Bessel functions (see [20]). Bessel functions of the second kind Y_α for non–integer α is related to J_α by:

$$Y_\alpha(x) = \frac{J_\alpha(x)\cos(\alpha\pi) - J_{-\alpha}(x)}{\sin(\alpha\pi)}.$$

In the case of integer order n, the function Y_n is defined by taking the limit as a non–integer α tends to n,

$$Y_n(x) = \lim_{\alpha\to n} Y_\alpha(x).$$

Hankel functions of the first and second kind $H_\alpha^{(1)}(x)$ and $H_\alpha^{(2)}(x)$, defined by:

$$H_\alpha^{(1)}(x) = J_\alpha(x) + iY_\alpha(x) \tag{7}$$

and

$$H_\alpha^{(2)}(x) = J_\alpha(x) - iY_\alpha(x). \tag{8}$$

Modified Bessel functions of the first and second kind $I_\alpha(x)$ and $K_\alpha(x)$ are defined by:

$$I_\alpha(x) = i^{-\alpha} J_\alpha(ix) = \sum_{m=0}^{\infty} \frac{1}{m!\,\Gamma(m+\alpha+1)} \left(\frac{x}{2}\right)^{2m+\alpha}, \tag{9}$$

$$K_\alpha(x) = \frac{\pi}{2} \frac{I_{-\alpha}(x) - I_\alpha(x)}{\sin(\alpha\pi)}, \tag{10}$$

when α is not an integer and when α is an integer, then the limit is used.

The **multidimensional Hankel (Fourier–Bessel) transform** of a function $f(x)$ is given by (see [22]):

$$\mathbf{F}_\gamma[f](\xi) = (\mathbf{F}_\gamma)_x[f(x)](\xi) = \widehat{f}(\xi) = \int\limits_{R_+^n} f(x)\,\mathbf{j}_\gamma(x;\xi) x^\gamma dx,$$

where

$$\mathbf{j}_\gamma(x;\xi) = \prod_{i=1}^{n} j_{\frac{\gamma_i-1}{2}}(x_i\xi_i), \qquad \gamma_1 > 0, ..., \gamma_n > 0.$$

For $f \in S_{ev}$ inverse multidimensional Hankel transform is defined by

$$\mathbf{F}_\gamma^{-1}[\widehat{f}(\xi)](x) = f(x) = \frac{2^{n-|\gamma|}}{\prod\limits_{j=1}^{n} \Gamma^2\left(\frac{\gamma_j+1}{2}\right)} \int\limits_{R_+^n} \mathbf{j}_\gamma(x,\xi)\widehat{f}(\xi)\xi^\gamma\, d\xi.$$

We will use the generalized convolution operator defined by the formula

$$(f * g)_\gamma = \int\limits_{R_+^n} f(y)({}^\gamma\mathbf{T}^y g)(x) y^\gamma dy,$$

where ${}^{\gamma}\mathbf{T}^{y}$ is multidimensional generalized translation

$$ {}^{\gamma}\mathbf{T}^{y} = {}^{\gamma_1}T^{y_1}_{x_1}...{}^{\gamma_n}T^{y_n}_{x_n}, $$

each one–dimensional operator ${}^{\gamma_i}T^{y_i}_{x_i}$, $i=1,...,n$ acts according to (see [23])

$$ {}^{\gamma_i}T^{y_i}_{x_i}f(x)=\frac{\Gamma\left(\frac{\gamma_i+1}{2}\right)}{\Gamma\left(\frac{\gamma_i}{2}\right)\Gamma\left(\frac{1}{2}\right)}\times $$

$$ \times\int\limits_0^{\pi} f(x_1,...,x_{i-1},\sqrt{x_i^2+y_i^2-2x_iy_i\cos\alpha_i},x_{i+1},...,x_n)\ \sin^{\gamma_i-1}\alpha_i\,d\alpha_i\,. $$

Based on the multidimensional generalized translation ${}^{\gamma}\mathbf{T}^{y}$ the weighted spherical mean $M_t^{\gamma}[f(x)]$ of a suitable function is defined by the formula

$$ M_t^{\gamma}[f(x)]=\frac{1}{|S_1^+(n)|_{\gamma}}\int\limits_{S_1^+(n)}{}^{\gamma}\mathbf{T}^{t\theta}_{x}f(x)\theta^{\gamma}dS, \tag{11} $$

where $\theta^{\gamma}{=}\prod\limits_{i=1}^{n}\theta_i^{\gamma_i}$, $S_1^+(n){=}\{\theta{:}|\theta|{=}1,\theta{\in}R_+^n\}$ and $|S_1^+(n)|_{\gamma}=\frac{\prod\limits_{i=1}^{n}\Gamma(\frac{\gamma_i+1}{2})}{2^{n-1}\Gamma(\frac{n+|\gamma|}{2})}$. It is easy to see that

$$ M_0^{\gamma}[f(x)]=f(x),\qquad \frac{\partial}{\partial t}M_t^{\gamma}[f(x)]\bigg|_{t=0}=0. \tag{12} $$

Lemma 1. Let $u\in S_{ev}$ then

$$ \mathbf{F}_{\gamma}[\Delta_{\gamma}f](\xi)=-|\xi|^2F_{\gamma}[f](\xi). \tag{13} $$

Proof. We have

$$ \mathbf{F}_{\gamma}[\Delta_{\gamma}f](\xi)=\int\limits_{R_+^n}[\Delta_{\gamma}f(x)]\,\mathbf{j}_{\gamma}(x;\xi)x^{\gamma}dx= $$

$$ =\sum_{i=1}^{n}\int\limits_{R_+^n}\left[\frac{1}{x_i^{\gamma_i}}\frac{\partial}{\partial x_i}x_i^{\gamma_i}\frac{\partial}{\partial x_i}f(x)\right]\,\mathbf{j}_{\gamma}(x;\xi)x^{\gamma}dx. $$

Integrating by parts by variable x_i and using formula (6), we obtain

$$ \mathbf{F}_{\gamma}[\Delta_{\gamma}f](\xi)=\sum_{i=1}^{n}\int\limits_{R_+^n}f(x)\left[\frac{1}{x_i^{\gamma_i}}\frac{\partial}{\partial x_i}x_i^{\gamma_i}\frac{\partial}{\partial x_i}\mathbf{j}_{\gamma}(x;\xi)\right]x^{\gamma}dx= $$

$$ =\sum_{i=1}^{n}(-\xi_i^2)\int\limits_{R_+^n}f(x)\,\mathbf{j}_{\gamma}(x;\xi)x^{\gamma}dx=-|\xi|^2\int\limits_{R_+^n}f(x)\,\mathbf{j}_{\gamma}(x;\xi)x^{\gamma}dx=-|\xi|^2\mathbf{F}_{\gamma}[f](\xi). $$

This completes the proof.

Lemma 2. The integral $\int\limits_{S_1^+(n)}\mathbf{j}_{\gamma}(r\theta,\xi)\theta^{\gamma}\,dS$ is calculated by the formula

$$ \int\limits_{S_1^+(n)}\mathbf{j}_{\gamma}(r\theta,\xi)\theta^{\gamma}\,dS=\frac{\prod\limits_{i=1}^{n}\Gamma\left(\frac{\gamma_i+1}{2}\right)}{2^{n-1}\Gamma\left(\frac{n+|\gamma|}{2}\right)}j_{\frac{n+|\gamma|}{2}-1}(r|\xi|). \tag{14} $$

Proof. It is well known that the function $\mathbf{j}_\gamma(x,\xi)$ is related to $e^{-i\langle x,\xi\rangle}$ by the formula (see [24], p. 32):

$$\mathbf{j}_\gamma(r\theta,\xi)=\mathcal{P}_\xi^\gamma\left(e^{-ir\langle\theta,\xi\rangle}\right)$$

therefore, one can use formula

$$\int\limits_{S_1^+(n)}\mathcal{P}_\xi^\gamma f(\langle\sigma,\xi\rangle)\sigma^\gamma dS_\sigma=\frac{\prod\limits_{i=1}^n\Gamma\left(\frac{\gamma_i+1}{2}\right)}{\sqrt{\pi}2^{n-1}\Gamma\left(\frac{|\gamma|+n-1}{2}\right)}\int\limits_{-1}^{1}f(|\xi|p)(1-p^2)^{\frac{n+|\gamma|-3}{2}}\,dp. \tag{15}$$

which is proved in [25]. We obtain

$$\int\limits_{S_1^+(n)}\mathbf{j}_\gamma(r\theta,\xi)\theta^\gamma\,dS=\frac{\prod\limits_{i=1}^n\Gamma\left(\frac{\gamma_i+1}{2}\right)}{\sqrt{\pi}2^{n-1}\Gamma\left(\frac{|\gamma|+n-1}{2}\right)}\int\limits_{-1}^{1}e^{-irp|\xi|}(1-p^2)^{\frac{n+|\gamma|-3}{2}}\,dp.$$

Replacing p by $-p$, we get

$$\int\limits_{-1}^{1}e^{-irp|\xi|}(1-p^2)^{\frac{n+|\gamma|-3}{2}}\,dp=\int\limits_{-1}^{1}e^{irp|\xi|}(1-p^2)^{\frac{n+|\gamma|-3}{2}}\,dp.$$

It is well known that the above-obtained integral satisfies the next formula (e.g., see [26], formula 2.3.5.3)

$$\int\limits_{-a}^{a}e^{itp}(a^2-p^2)^{\beta-1}\,dp=\frac{\sqrt{\pi}\,(2a)^{\beta-\frac{1}{2}}\Gamma(\beta)}{t^{\beta-\frac{1}{2}}}J_{\beta-\frac{1}{2}}(at).$$

Therefore

$$\int\limits_{S_1^+(n)}\mathbf{j}_\gamma(r\theta,\xi)\theta^\gamma\,dS=\frac{\prod\limits_{i=1}^n\Gamma\left(\frac{\gamma_i+1}{2}\right)}{\sqrt{\pi}2^{n-1}\Gamma\left(\frac{n+|\gamma|-1}{2}\right)}\frac{\sqrt{\pi}2^{\frac{n+|\gamma|}{2}-1}\Gamma\left(\frac{n+|\gamma|-1}{2}\right)}{(r|\xi|)^{\frac{n+|\gamma|}{2}-1}}J_{\frac{n+|\gamma|}{2}-1}(r|\xi|)=$$

$$=\frac{\prod\limits_{i=1}^n\Gamma\left(\frac{\gamma_i+1}{2}\right)}{2^{n-1}\Gamma\left(\frac{n+|\gamma|}{2}\right)}j_{\frac{n+|\gamma|}{2}-1}(r|\xi|).$$

That gives (14).

4. Weighted functionals connected with quadratic form and their Hankel transforms

We will use regular weighted functional $(w^2-|x|^2)_{+,\gamma}^\lambda$ defined by the formula

$$((w^2-|x|^2)_{+,\gamma}^\lambda,\varphi)_\gamma=\int\limits_{\{x\in R_+^n:|x|<w\}}(w^2-|x|^2)^\lambda\varphi(x)x^\gamma dx,\qquad \lambda\in C. \tag{16}$$

Here and further let $\varphi\in\mathcal{D}_+$.

Let $\lambda\in C$, $p,q\in N$, $n=p+q$ and

$$P=|x'|^2-|x''|^2=x_1^2+...+x_p^2-x_{p+1}^2-...-x_{p+q}^2,$$

where $x{=}(x_1,...,x_n){=}(x',x''){\in}R^n_+$, $x'{=}(x_1,...,x_p)$, $x''{=}(x_{p+1},...,x_{p+q})$. We define the weighted generalized functions $P^\lambda_{\gamma,+}$ by

$$(P^\lambda_{\gamma,+},\varphi)_\gamma = \int\limits_{\{P(x)>0\}^+} P^\lambda(x)\varphi(x)x^\gamma dx \qquad (17)$$

and

$$(P^\lambda_{\gamma,-},\varphi)_\gamma = \int\limits_{\{-P(x)>0\}^+} (-P(x))^\lambda\varphi(x)x^\gamma dx, \qquad (18)$$

where $\{\pm P(x)>0\}^+ = \{x\in R^n_+ : \pm P(x)>0\}$, $\lambda\in C$.

If quadratic form $\mathcal{P} = P_1 + iP_2$ has positive definite imaginary part P_2 then we can define a single–valued analytic function of λ

$$\mathcal{P}^\lambda = e^{\lambda(\ln|\mathcal{P}|+i\arg\mathcal{P})}, \qquad 0<\arg\mathcal{P}<\pi. \qquad (19)$$

To the function $\mathcal{P}^\lambda$ we compare the generalized function $\mathcal{P}^\lambda_\gamma = (P_1+iP_2)^\lambda_\gamma$ by formula

$$(\mathcal{P}^\lambda_\gamma,\varphi)_\gamma = \int\limits_{R^n_+} \mathcal{P}^\lambda(x)\varphi(x)x^\gamma dx.$$

Let us consider the non–degenerate quadratic form with real coefficients

$$P = \sum_{k=1}^n a_k x_k^2, \qquad a_k\in R.$$

Suppose that the quadratic form P has p positive summands and q negative, $p+q=n$. Let

$$\mathcal{P} = P + iP',$$

where P' is positive definite quadratic form with real coefficients. Without loss of generality, we assume

$$P' = \varepsilon(x_1^2+...+x_n^2), \qquad \varepsilon>0.$$

If $\mathcal{P}^\lambda_\gamma = (P+iP')^\lambda_\gamma$ then $(P+i0)^\lambda_\gamma$ and $(P-i0)^\lambda_\gamma$ when $\mathrm{Re}\,\lambda>0$ are defined by formulas

$$(P+i0)^\lambda_\gamma = \lim_{\varepsilon\to 0}(P+iP')^\lambda_\gamma,$$

$$(P-i0)^\lambda_\gamma = \lim_{\varepsilon\to 0}(P-iP')^\lambda_\gamma,$$

in which we pass to the limit under the integral sign $\int\limits_{R^n_+}\mathcal{P}^\lambda\varphi x^\gamma dx$. When $\mathrm{Re}\,\lambda<0$, $\lambda\neq -k$, $\lambda\neq-\frac{n+|\gamma|}{2}-k+1$, $k\in N$ for determining $(P+i0)^\lambda_\gamma$ and $(P-i0)^\lambda_\gamma$ we first apply

$$\mathcal{B}^k_{\gamma,a}(\partial)\mathcal{P}^{\lambda+k}_\gamma = 4^k(\lambda{+}1)...(\lambda{+}k)\left(\lambda{+}\frac{n{+}|\gamma|}{2}\right)...\left(\lambda{+}\frac{n{+}|\gamma|}{2}{+}k{-}1\right)\mathcal{P}^\lambda_\gamma,$$

where

$$\mathcal{B}_{\gamma,a} = \sum_{k=1}^n \frac{1}{a_k}\left(\frac{\partial^2}{\partial x_k^2} + \frac{\gamma_k}{x_k}\frac{\partial}{\partial x_k}\right)$$

and then we pass to the limit as $\varepsilon \to 0$ (see [27]).

In addition, we will consider similarly defined functions $(c^2+P+i0)^\lambda_\gamma$ and $(c^2+P-i0)^\lambda_\gamma$, where P is an arbitrary quadratic form and $c\in R$ does not depends on x.

Lemma 3. We have the following formula

$$(\mathbf{F}_\gamma)_x\left[\frac{(w^2-|x|^2)^\lambda_{+,\gamma}}{\Gamma(\lambda+1)}\right](\xi)=\frac{w^{n+|\gamma|+2\lambda}\prod\limits_{i=1}^{n}\Gamma\left(\frac{\gamma_i+1}{2}\right)}{2^n\Gamma\left(\frac{n+|\gamma|}{2}+\lambda+1\right)}\,j_{\frac{n+|\gamma|}{2}+\lambda}(w|x|), \tag{20}$$

where $(w^2-|x|^2)_{+,\gamma}^{\frac{k-n-|\gamma|-1}{2}}$ is defined by (16), $w>0$.

Proof. Let first $\operatorname{Re}\lambda>-1$. Passing to spherical coordinates in $\mathbf{F}_\gamma(w^2-|x|^2)^\lambda_{+,\gamma}$ and applying the formula (14) we obtain

$$(\mathbf{F}_\gamma)_x(w^2-|x|^2)^\lambda_{+,\gamma}=\int\limits_{B^+_w(n)}\mathbf{j}_\gamma(x,\xi)(w^2-|x|^2)^\lambda x^\gamma\,dx=\{x=r\theta, r=|x|\}=$$

$$=\int\limits_0^w(w^2-r^2)^\lambda r^{n+|\gamma|-1}dr\int\limits_{S^+_1(n)}\mathbf{j}_\gamma(r\theta,x)\theta^\gamma\,dS=$$

$$=\frac{\prod\limits_{i=1}^{n}\Gamma\left(\frac{\gamma_i+1}{2}\right)}{2^{n-1}\Gamma\left(\frac{n+|\gamma|}{2}\right)}\int\limits_0^w(w^2-r^2)^\lambda j_{\frac{n+|\gamma|}{2}-1}(r|x|)r^{n+|\gamma|-1}dr=$$

$$=|x|^{1-\frac{n+|\gamma|}{2}}2^{\frac{|\gamma|-n}{2}}\prod_{i=1}^{n}\Gamma\left(\frac{\gamma_i+1}{2}\right)\int\limits_0^w(w^2-r^2)^\lambda J_{\frac{n+|\gamma|}{2}-1}(r|x|)r^{\frac{n+|\gamma|}{2}}dr.$$

Using formula 2.12.4.6 from [28] which has the form

$$\int\limits_0^w r^{\nu+1}(w^2-r^2)^{\beta-1}J_\nu(\mu r)dr=\frac{2^{\beta-1}w^{\beta+\nu}\Gamma(\beta)}{\mu^\beta}J_{\beta+\nu}(\mu w),\ w>0,\ \operatorname{Re}\beta>0,\ \operatorname{Re}\nu>-1$$

we get

$$\int\limits_0^w(w^2-r^2)^\lambda J_{\frac{n+|\gamma|}{2}-1}(r|x|)r^{\frac{n+|\gamma|}{2}}dr=\frac{2^\lambda w^{\frac{n+|\gamma|}{2}+\lambda}\Gamma(\lambda+1)}{|x|^{\lambda+1}}J_{\frac{n+|\gamma|}{2}+\lambda}(|x|\omega)$$

for $\operatorname{Re}\lambda>-1$ and

$$(\mathbf{F}_\gamma)_x(w^2-|x|^2)^\lambda_{+,\gamma}=\frac{w^{n+|\gamma|+2\lambda}\Gamma(\lambda+1)\prod\limits_{i=1}^{n}\Gamma\left(\frac{\gamma_i+1}{2}\right)}{2^n\Gamma\left(\frac{n+|\gamma|}{2}+\lambda+1\right)}\,j_{\frac{n+|\gamma|}{2}+\lambda}(w|x|), \tag{21}$$

which coincides with (20). So we get (20) for $\operatorname{Re}\lambda>-1$. For other values of λ such that $\lambda\neq-1,-2,-3,...$ it can be analytically continued (see [29]).

The residue of $\frac{(w^2-|x|^2)^\lambda_{+,\gamma}}{\Gamma(\lambda+1)}$ at $\lambda=-m$, $m\in N$ is (see [27], [31])

$$\lim_{\lambda\to-m}\frac{(w^2-|x|^2)^\lambda_{+,\gamma}}{\Gamma(\lambda+1)}=\delta^{(m-1)}_\gamma(w^2-|x|^2)$$

and for $\lambda = -m$

$$(\mathbf{F}_\gamma)_x\left[\frac{(w^2-|x|^2)^\lambda_{+,\gamma}}{\Gamma(\lambda+1)}\right](\xi) = \int\limits_{R^n_+} \mathbf{j}_\gamma(x,\xi)\delta^{(m-1)}_\gamma(w^2-|x|^2)x^\gamma\,dx =$$

$$= \frac{w^{n+|\gamma|-2m}\prod\limits_{i=1}^n\Gamma\left(\frac{\gamma_i+1}{2}\right)}{2^n\Gamma\left(\frac{n+|\gamma|}{2}-m+1\right)}\, j_{\frac{n+|\gamma|}{2}-m}(w|x|).$$

This completes the proof.

Lemma 4. We have the following formulas

$$\mathbf{F}_\gamma(w^2+P+i0)^\lambda_\gamma =$$

$$= \frac{2^{\frac{|\gamma|-n}{2}+\lambda+1}e^{-\frac{1}{2}q\pi i}w^{\frac{n+|\gamma|}{2}+\lambda}\prod\limits_{i=1}^n\Gamma\left(\frac{\gamma_i+1}{2}\right)}{\Gamma(-\lambda)\sqrt{|\Delta|}}\left[\frac{K_{\frac{n+|\gamma|}{2}+\lambda}(wQ^{\frac{1}{2}}_{\gamma,+})}{Q_{\gamma,+}^{\frac{1}{2}(\frac{n+|\gamma|}{2}+\lambda)}} + \frac{i\pi}{2}\frac{H^{(1)}_{-\frac{n+|\gamma|}{2}-\lambda}(wQ^{\frac{1}{2}}_{\gamma,-})}{Q_{\gamma,-}^{\frac{1}{2}(\frac{n+|\gamma|}{2}+\lambda)}}\right] \quad (22)$$

and

$$\mathbf{F}_\gamma(w^2+P-i0)^\lambda_\gamma =$$

$$= \frac{2^{\frac{|\gamma|-n}{2}+\lambda+1}e^{\frac{1}{2}q\pi i}w^{\frac{n+|\gamma|}{2}+\lambda}\prod\limits_{i=1}^n\Gamma\left(\frac{\gamma_i+1}{2}\right)}{\Gamma(-\lambda)\sqrt{|\Delta|}}\left[\frac{K_{\frac{n+|\gamma|}{2}+\lambda}(wQ^{\frac{1}{2}}_{\gamma,+})}{Q_{\gamma,+}^{\frac{1}{2}(\frac{n+|\gamma|}{2}+\lambda)}} - \frac{i\pi}{2}\frac{H^{(2)}_{-\frac{n+|\gamma|}{2}-\lambda}(wQ^{\frac{1}{2}}_{\gamma,-})}{Q_{\gamma,-}^{\frac{1}{2}(\frac{n+|\gamma|}{2}+\lambda)}}\right], \quad (23)$$

where $Q = \sum\limits_{i=1}^n \frac{1}{a_i}\xi_i^2$ is the quadratic form dual to $P = \sum\limits_{i=1}^n a_i x_i^2$, Δ is the determinant of the coefficients of P, $H^{(1)}_\alpha$ and $H^{(2)}_\alpha$ are the Hankel functions of the first and second kind and K_α is modified Bessel function.

Proof. First we consider the Hankel transform of the generalized function $(c^2+P)^\lambda_\gamma$ for a positive definite quadratic form $P = |x|^2$ and $\operatorname{Re}\lambda < -\frac{n+|\gamma|}{2}$. Applying (14) we obtain

$$\mathbf{F}_\gamma[(w^2+P)^\lambda_\gamma](\xi) = \int\limits_{R^n_+}\mathbf{j}_\gamma(x,\xi)(c^2+|x|^2)^\lambda x^\gamma\,dx = \{x = r\theta, r = |x|\} =$$

$$= \int\limits_0^\infty (w^2+r^2)^\lambda r^{n+|\gamma|-1}dr\int\limits_{S_1^+(n)}\mathbf{j}_\gamma(r\theta,x)\theta^\gamma\,dS =$$

$$= \frac{\prod\limits_{i=1}^n\Gamma\left(\frac{\gamma_i+1}{2}\right)}{2^{n-1}\Gamma\left(\frac{n+|\gamma|}{2}\right)}\int\limits_0^\infty (w^2+r^2)^\lambda j_{\frac{n+|\gamma|-2}{2}}(r|\xi|)r^{n+|\gamma|-1}dr =$$

$$= |\xi|^{1-\frac{n+|\gamma|}{2}}2^{\frac{|\gamma|-n}{2}}\prod_{i=1}^n\Gamma\left(\frac{\gamma_i+1}{2}\right)\int\limits_0^\infty (w^2+r^2)^\lambda J_{\frac{n+|\gamma|}{2}-1}(r|\xi|)r^{\frac{n+|\gamma|}{2}}dr.$$

Using formula 2.12.4.28 from [28]

$$\int\limits_0^\infty x^{\nu+1}(x^2+z^2)^{-\rho}J_\nu(cx)dx = \frac{c^{\rho-1}z^{\nu-\rho+1}}{2^{\rho-1}\Gamma(\rho)}K_{\nu-\rho+1}(cz)$$

we obtain

$$\mathbf{F}_\gamma[(w^2+P)^\lambda_\gamma](\xi) = \frac{2^{\frac{|\gamma|-n}{2}+\lambda+1} w^{\frac{n+|\gamma|}{2}+\lambda} \prod\limits_{i=1}^{n} \Gamma\left(\frac{\gamma_i+1}{2}\right)}{|\xi|^{\frac{n+|\gamma|}{2}+\lambda}\Gamma(-\lambda)} K_{\frac{n+|\gamma|}{2}+\lambda}(w|\xi|), \tag{24}$$

where $\lambda < \frac{1-n-|\gamma|}{4}$. For other values of λ the Hankel transform $\mathbf{F}_\gamma(w^2+P)^\lambda_\gamma$ remains valid by analytic continuation in λ.

Now let P be any real quadratic form. We wish to consider the generalized functions $(w^2+P+i0)^\lambda_\gamma$ and $(w^2+P-i0)^\lambda_\gamma$ defined by

$$(w^2+P\pm i0)^\lambda_\gamma = \lim_{\varepsilon\to 0}(w^2+P\pm i\varepsilon P_1)^\lambda_\gamma,$$

where $\varepsilon>0$ and P_1 is a positive definite quadratic form. According to the uniqueness of analytic continuation, (24) implies that

$$\mathbf{F}_\gamma[(w^2+P\pm i0)^\lambda_\gamma](\xi) = \frac{2^{\frac{|\gamma|-n}{2}+\lambda+1} e^{\mp\frac{1}{2}q\pi} w^{\frac{n+|\gamma|}{2}+\lambda} \prod\limits_{i=1}^{n} \Gamma\left(\frac{\gamma_i+1}{2}\right)}{(Q\mp i0)^{\frac{1}{2}\left(\frac{n+|\gamma|}{2}+\lambda\right)}\Gamma(-\lambda)\sqrt{|\Delta|}} K_{\frac{n+|\gamma|}{2}+\lambda}(w(Q\mp i0)^{\frac{1}{2}}_\gamma), \tag{25}$$

where $Q=\sum\limits_{i=1}^{n}\frac{1}{a_i}\xi_i^2$ is the quadratic form dual to $P=\sum\limits_{i=1}^{n}a_i x_i^2$. Considering the definitions of modified Bessel functions of the first and second kind (9) and (10) we get

$$\frac{K_{\frac{n+|\gamma|}{2}+\lambda}(w(Q\mp i0)^{\frac{1}{2}}_\gamma)}{(Q\mp i0)^{\frac{1}{2}\left(\frac{n+|\gamma|}{2}+\lambda\right)}} = \frac{\pi}{2}\frac{I_{-\frac{n+|\gamma|}{2}-\lambda}(w(Q\mp i0)^{\frac{1}{2}}_\gamma) - I_{\frac{n+|\gamma|}{2}+\lambda}(w(Q\mp i0)^{\frac{1}{2}}_\gamma)}{\sin\left(\left(\frac{n+|\gamma|}{2}+\lambda\right)\pi\right)(Q\mp i0)^{\frac{1}{2}\left(\frac{n+|\gamma|}{2}+\lambda\right)}} =$$

$$= \frac{\pi}{2\sin\left(\left(\frac{n+|\gamma|}{2}+\lambda\right)\pi\right)}\sum_{m=0}^{\infty}\frac{1}{m!}\left(\frac{w^{2m-\frac{n+|\gamma|}{2}-\lambda}}{2^{2m-\frac{n+|\gamma|}{2}-\lambda}\Gamma\left(m-\frac{n+|\gamma|}{2}-\lambda+1\right)}(Q\mp i0)^{m-\lambda-\frac{n+|\gamma|}{2}}_\gamma - \right.$$

$$\left. - \frac{w^{2m+\frac{n+|\gamma|}{2}+\lambda}}{2^{2m+\frac{n+|\gamma|}{2}+\lambda}\Gamma\left(m+\frac{n+|\gamma|}{2}+\lambda+1\right)}(Q\mp i0)^m_\gamma\right).$$

Weighted generalized functions $(Q+i0)^\lambda_\gamma$ and $(Q-i0)^\lambda_\gamma$ can be expressed through weighted generalized functions $Q^\lambda_{\gamma,+}$ and $Q^\lambda_{\gamma,-}$ defined by (see [27]):

$$(Q\mp i0)^\mu_\gamma = Q^\mu_{\gamma,+} + e^{\mp\pi\mu i}Q^\mu_{\gamma,-}. \tag{26}$$

Then we get

$$\frac{K_{\frac{n+|\gamma|}{2}+\lambda}(w(Q\mp i0)^{\frac{1}{2}}_\gamma)}{(Q\mp i0)^{\frac{1}{2}\left(\frac{n+|\gamma|}{2}+\lambda\right)}} =$$

$$= \frac{\pi}{2\sin\left(\left(\frac{n+|\gamma|}{2}+\lambda\right)\pi\right)}\sum_{m=0}^{\infty}\frac{1}{m!}\left(\frac{w^{2m-\frac{n+|\gamma|}{2}-\lambda}}{2^{2m-\frac{n+|\gamma|}{2}-\lambda}\Gamma\left(m-\frac{n+|\gamma|}{2}-\lambda+1\right)}\times\right.$$

$$\times\left(Q^{m-\lambda-\frac{n+|\gamma|}{2}}_{\gamma,+} + e^{\mp\pi i\left(m-\lambda-\frac{n+|\gamma|}{2}\right)}Q^{m-\lambda-\frac{n+|\gamma|}{2}}_{\gamma,-}\right) -$$

$$-\frac{w^{2m+\frac{n+|\gamma|}{2}+\lambda}}{2^{2m+\frac{n+|\gamma|}{2}+\lambda}\Gamma\left(m+\frac{n+|\gamma|}{2}+\lambda+1\right)}\left(Q^m_{\gamma,+}+e^{\mp\pi i m}Q^m_{\gamma,-}\right)\Bigg)=$$

$$=\frac{\pi}{2}\frac{I_{-\lambda-\frac{n+|\gamma|}{2}}(wQ^{\frac{1}{2}}_{\gamma,+})-I_{\lambda+\frac{n+|\gamma|}{2}}(wQ^{\frac{1}{2}}_{\gamma,+})}{\sin\left(\left(\frac{n+|\gamma|}{2}+\lambda\right)\pi\right)(Q^{\frac{1}{2}}_{\gamma,+})^{\lambda+\frac{n+|\gamma|}{2}}}+$$

$$+\frac{i\pi}{2}\frac{J_{\lambda+\frac{n+|\gamma|}{2}}(wQ^{\frac{1}{2}}_{\gamma,-})-e^{\pm\left(\lambda+\frac{n+|\gamma|}{2}\right)\pi}J_{-\lambda-\frac{n+|\gamma|}{2}}(wQ^{\frac{1}{2}}_{\gamma,-})}{i\sin\left(-\left(\lambda+\frac{n+|\gamma|}{2}\right)\pi\right)(Q^{\frac{1}{2}}_{\gamma,-})^{\lambda+\frac{n+|\gamma|}{2}}}.$$

Noticing that

$$K_\alpha(x)=\frac{\pi}{2}\frac{I_{-\alpha}(x)-I_\alpha(x)}{\sin(\alpha\pi)},$$

$$H^{(1)}_\alpha(x)=\frac{J_{-\alpha}(x)-e^{-\alpha\pi i}J_\alpha(x)}{i\sin(\alpha\pi)},\qquad H^{(2)}_\alpha(x)=\frac{J_{-\alpha}(x)-e^{\alpha\pi i}J_\alpha(x)}{-i\sin(\alpha\pi)}.$$

we have

$$\frac{K_{\frac{n+|\gamma|}{2}+\lambda}(w(Q+i0)^{\frac{1}{2}}_\gamma)}{(Q\mp i0)^{\frac{1}{2}\left(\frac{n+|\gamma|}{2}+\lambda\right)}}=\frac{K_{\lambda+\frac{n+|\gamma|}{2}}(wQ^{\frac{1}{2}}_{\gamma,+})}{(Q^{\frac{1}{2}}_{\gamma,+})^{\lambda+\frac{n+|\gamma|}{2}}}+\frac{i\pi}{2}\frac{H^{(1)}_{-\lambda-\frac{n+|\gamma|}{2}}(wQ^{\frac{1}{2}}_{\gamma,-})}{(Q^{\frac{1}{2}}_{\gamma,-})^{\lambda+\frac{n+|\gamma|}{2}}},$$

$$\frac{K_{\frac{n+|\gamma|}{2}+\lambda}(w(Q-i0)^{\frac{1}{2}}_\gamma)}{(Q\mp i0)^{\frac{1}{2}\left(\frac{n+|\gamma|}{2}+\lambda\right)}}=\frac{K_{\lambda+\frac{n+|\gamma|}{2}}(wQ^{\frac{1}{2}}_{\gamma,+})}{(Q^{\frac{1}{2}}_{\gamma,+})^{\lambda+\frac{n+|\gamma|}{2}}}-\frac{i\pi}{2}\frac{H^{(2)}_{-\lambda-\frac{n+|\gamma|}{2}}(wQ^{\frac{1}{2}}_{\gamma,-})}{(Q^{\frac{1}{2}}_{\gamma,-})^{\lambda+\frac{n+|\gamma|}{2}}},$$

which, taking into account formula (25), gives (22) and (23)

Corollary. If P is positive definite then we have

$$\mathbf{F}_\gamma(w^2+P+i0)^\lambda_\gamma=\frac{2^{\frac{|\gamma|-n}{2}+\lambda+1}e^{-\frac{1}{2}q\pi i}w^{\frac{n+|\gamma|}{2}+\lambda}\prod_{i=1}^n\Gamma\left(\frac{\gamma_i+1}{2}\right)}{\Gamma(-\lambda)\sqrt{|\Delta|}Q^{\frac{1}{2}\left(\frac{n+|\gamma|}{2}+\lambda\right)}_{\gamma,+}}K_{\frac{n+|\gamma|}{2}+\lambda}(wQ^{\frac{1}{2}}_{\gamma,+})\tag{27}$$

and

$$\mathbf{F}_\gamma(w^2+P-i0)^\lambda_\gamma=\frac{2^{\frac{|\gamma|-n}{2}+\lambda+1}e^{\frac{1}{2}q\pi i}w^{\frac{n+|\gamma|}{2}+\lambda}\prod_{i=1}^n\Gamma\left(\frac{\gamma_i+1}{2}\right)}{\Gamma(-\lambda)\sqrt{|\Delta|}Q^{\frac{1}{2}\left(\frac{n+|\gamma|}{2}+\lambda\right)}_{\gamma,+}}K_{\frac{n+|\gamma|}{2}+\lambda}(wQ^{\frac{1}{2}}_{\gamma,+}),\tag{28}$$

while if P is negative definite

$$\mathbf{F}_\gamma(w^2+P+i0)^\lambda_\gamma=\frac{i\pi2^{\frac{|\gamma|-n}{2}+\lambda}e^{-\frac{1}{2}q\pi i}w^{\frac{n+|\gamma|}{2}+\lambda}\prod_{i=1}^n\Gamma\left(\frac{\gamma_i+1}{2}\right)}{\Gamma(-\lambda)\sqrt{|\Delta|}Q^{\frac{1}{2}\left(\frac{n+|\gamma|}{2}+\lambda\right)}_{\gamma,-}}H^{(1)}_{-\frac{n+|\gamma|}{2}-\lambda}(wQ^{\frac{1}{2}}_{\gamma,-}),\tag{29}$$

and

$$\mathbf{F}_\gamma(w^2+P-i0)^\lambda_\gamma=-\frac{i\pi2^{\frac{|\gamma|-n}{2}+\lambda}e^{\frac{1}{2}q\pi i}w^{\frac{n+|\gamma|}{2}+\lambda}\prod_{i=1}^n\Gamma\left(\frac{\gamma_i+1}{2}\right)}{\Gamma(-\lambda)\sqrt{|\Delta|}Q^{\frac{1}{2}\left(\frac{n+|\gamma|}{2}+\lambda\right)}_{\gamma,-}}H^{(2)}_{-\frac{n+|\gamma|}{2}-\lambda}(wQ^{\frac{1}{2}}_{\gamma,-}),\tag{30}$$

where $Q=\sum\limits_{i=1}^n\frac{1}{a_i}\xi_i^2$ is the quadratic form dual to $P=\sum\limits_{i=1}^n a_ix_i^2$, $H^{(1)}_\alpha$ and $H^{(2)}_\alpha$ are the Hankel functions of the first and second kind and K_α is modified Bessel function.

IOP Conf. Series: Journal of Physics: Conf. Series **973** (2018) 012058 doi:10.1088/1742-6596/973/1/012058

5. Solution of the Cauchy problem using the Hankel transform

The main tool used here consist of using the Hankel transform for solving Cauchy problem for generalized Euler–Poisson–Darboux equation. The greatest advantage of this method is that it gives solutions for all values of k. The case when $\gamma_i = 0$, $i = 1, ..., n$ was studied in [13].

We will be concerned with the solutions of the following initial value hyperbolic problem

$$[(\triangle_\gamma)_x - (B_k)_t]\, u = c^2 u, \quad k \in R, \quad c > 0, \quad u = u(x, t; k), \quad x \in R^n_+, \quad t > 0. \tag{31}$$

$$u(x, 0; k) = f(x), \qquad u_t(x, 0; k) = 0. \tag{32}$$

We will call (31) the **generalized Euler–Poisson–Darboux equation**. We are looking for the solution $u \in S'_{ev}(R^n_+) \times C^2(0, \infty)$ of (31)–(32). Notation $u \in S'_{ev}(R^n_+) \times C^2(0, \infty)$ means that $u(x, t; k)$ belongs to $S'_{ev}(R^n_+)$ by variable x and belongs to $C^2(0, \infty)$ by variable t.

Theorem 1. The solution $u \in S'_{ev}(R^n_+) \times C^2(0, \infty)$ of the (31)–(32) for $k \neq -1, -3, -5, ...$ is unique and defined by the formula

$$u(x, t; k) = C(n, \gamma, k) \left(t^{1-k} (t^2 - |x|^2)_+^{\frac{k-n-|\gamma|-1}{2}} j_{\frac{k-n-|\gamma|-1}{2}} \left((t^2 - |x|^2)_+^{\frac{1}{2}} \cdot c \right) * f(x) \right)_\gamma, \tag{33}$$

where

$$C(n, \gamma, k) = \frac{2^n \Gamma\left(\frac{k+1}{2}\right)}{\Gamma\left(\frac{k-n-|\gamma|+1}{2}\right) \prod\limits_{i=1}^{n} \Gamma\left(\frac{\gamma_i+1}{2}\right)}.$$

In the case when $k < 0$ of the (31)–(32) is not unique. When $k < 0$ and $k \neq -1, -3, -5, ...$ the difference between two arbitrary solutions is always of the form

$$A t^{1-k} u(t, x; 2-k), \qquad A = const, \tag{34}$$

where $u(t, x; 2-k)$ is solution of the Cauchy problem

$$[(\triangle_\gamma)_x - (B_{2-k})_t]\, u = c^2 u,$$

$$u(x, 0; 2-k) = \psi(x), \qquad u_t(x, 0; 2-k) = 0,$$

$\psi(x)$ is an arbitrary function or distribution belonging to S'_{ev}, When $k = -1, -3, -5, ...$ a nonunique solution of the Cauchy problem (31)–(32) will contain a terms (51) and

$$\frac{e^{\pm \frac{1}{2}\pi n i} \Gamma\left(\frac{n+|\gamma|-k+1}{2}\right)}{2^n \Gamma\left(\frac{1-k}{2}\right) \prod\limits_{i=1}^{n} \Gamma\left(\frac{\gamma_i+1}{2}\right)} t^{1-k} \left((t^2 - |x|^2 \pm i0)_\gamma^{\frac{k-n-|\gamma|-1}{2}} * f(x) \right)_\gamma.$$

Proof. Applying multidimensional Hankel transform to (31) with respect to the variables $x_1, ..., x_n$ only and using (13) we obtain

$$\left(|\xi|^2 + c^2 + \frac{\partial^2}{\partial t^2} + \frac{k}{t}\frac{\partial}{\partial t} \right) \widehat{u}(\xi, t; k) = 0, \tag{35}$$

$$\widehat{u}(\xi, 0; k) = \widehat{f}(\xi), \qquad \widehat{u}_t(\xi, 0; k) = 0, \tag{36}$$

where $\xi=(\xi_1, \xi_2, ..., \xi_n) \in R^n_+$ corresponds to $x=(x_1, ..., x_n) \in R^n_+$, $|\xi|^2 = \xi_1^2 + \xi_2^2 + ... + \xi_n^2$,

$$\widehat{u}(\xi, t; k) = (F_\gamma)_x [u(x, t; k)](\xi) = \int\limits_{R^n_+} u(x, t; k)\, \mathbf{j}_\gamma(x; \xi) x^\gamma dx$$

IOP Conf. Series: Journal of Physics: Conf. Series **973** (2018) 012058 doi:10.1088/1742-6596/973/1/012058

and $\widehat{f}(\xi) = \mathbf{F}_\gamma[f](\xi)$.

The solution solution $\widehat{G}^k(\xi,t)$ of the Cauchy problem

$$\left(|\xi|^2 + c^2 + \frac{\partial^2}{\partial t^2} + \frac{k}{t}\frac{\partial}{\partial t}\right)\widehat{G}^k(\xi,t) = 0,$$

$$\widehat{G}^k(\xi,0) = 1, \qquad \widehat{G}^k_t(\xi,0) = 0$$

was obtained in [13]. We have different solutions for nonnegative and negative values of k, specifically

(i) for $k \geq 0$

$$\widehat{G}^k(\xi,t) = j_{\frac{k-1}{2}}(\sqrt{|\xi|^2+c^2}\;t), \tag{37}$$

(ii) for $k < 0$, $k \neq -1,-3,-5,...$

$$\widetilde{G}^k(\xi,t) = At^{1-k}j_{\frac{1-k}{2}}(\sqrt{|\xi|^2+c^2}\;t) + j_{\frac{k-1}{2}}(\sqrt{|\xi|^2+c^2}\;t), \tag{38}$$

where A is arbitrary complex number which depend on ξ and c,

(iii) for $k = -1,-3,-5,...$

$$\widehat{G}^k(\xi,t) = Bt^{1-k}j_{\frac{1-k}{2}}(\sqrt{|\xi|^2+c^2}\;t) - \frac{\pi 2^{\frac{k-1}{2}}}{\Gamma\left(\frac{1-k}{2}\right)}\left[\sqrt{|\xi|^2+c^2}\,t\right]^{\frac{1-k}{2}} Y_{\frac{1-k}{2}}(\sqrt{|\xi|^2+c^2}\;t), \tag{39}$$

where B denotes an arbitrary complex number which depend on ξ and c.

From (37)–(39) we conclude that problem (31)–(32) has a unique solution for $k \geq 0$ only. Besides, we can see that the difference between two different solutions (38) is always of the form

$$At^{1-k}j_{\frac{1-k}{2}}(\sqrt{|\xi|^2+c^2}\;t). \tag{40}$$

Now let find $G^k(x,t) = \left((\mathbf{F}_\gamma^{-1})_\xi \widehat{G}^k(\xi,t)\right)(x)$. We call $G^k(x,t)$ the *fundamental solution* of problem (31)–(32). The inverse transform $\left((\mathbf{F}_\gamma^{-1})_\xi \widehat{G}^k(\xi,t)\right)(x)$ is most easily found by considering c as an additional independent variable. Setting $\xi' = (\xi_1,...,\xi_n,c)$ we can write $\widehat{G}^k(\xi,t) = j_{\frac{k-1}{2}}(|\xi'|\,t)$ for $k \geq 0$ and find an inverse Hankel transform of $j_{\frac{k-1}{2}}(|\xi'|\,t)$ by variable ξ' using (20). We obtain

$$\left((\mathbf{F}_{\gamma'}^{-1})_{\xi'} j_{\frac{k-1}{2}}(|\xi'|\,t)\right)(x') = \frac{2^{n+1}\Gamma\left(\frac{k+1}{2}\right)}{\Gamma\left(\frac{k-n-|\gamma'|}{2}\right)\prod\limits_{i=1}^{n+1}\Gamma\left(\frac{\gamma_i+1}{2}\right)} t^{1-k}(t^2-|x'|^2)_{+,\gamma'}^{\frac{k-n-|\gamma'|-2}{2}},$$

where $\gamma' = (\gamma_1,..,\gamma_n,\gamma_{n+1})$ and γ_{n+1} is an arbitrary positive number, $x' = (x,\sigma)$, $\sigma \in R_+$ is dual to c variable. Now in order to find $G^k(x,t)$ we need to apply a direct Hankel transform only on a one-dimensional variable σ. We have

$$G^k(x,t) = \frac{2^{n+1}\Gamma\left(\frac{k+1}{2}\right)}{\Gamma\left(\frac{k-n-|\gamma'|}{2}\right)\prod\limits_{i=1}^{n+1}\Gamma\left(\frac{\gamma_i+1}{2}\right)} t^{1-k}\left((F_{\gamma_{n+1}})_\sigma(t^2-x^2-\sigma^2)_{+,\gamma'}^{\frac{k-n-|\gamma'|-2}{2}}\right)(c) =$$

$$= \frac{2^n\Gamma\left(\frac{k+1}{2}\right)}{\Gamma\left(\frac{k-n-|\gamma|+1}{2}\right)\prod\limits_{i=1}^{n}\Gamma\left(\frac{\gamma_i+1}{2}\right)} t^{1-k}(t^2-x^2)_+^{\frac{k-n-|\gamma|-1}{2}} j_{\frac{k-n-|\gamma|-1}{2}}\left((t^2-|x|^2)_+^{\frac{1}{2}}\cdot c\right).$$

Therefore, the solution of (31)–(32) for $k \geq 0$ is then given by

$$u(x,t;k) = (G^k(x,t) * f(x))_\gamma. \tag{41}$$

It is easy to see that (41) still gives one of the possible solutions of (31)–(32) for $k < 0$, $k \neq -1, -3, -5, \ldots$. Moreover, in consideration of (40) the difference between two solutions for $k < 0$ has the form

$$(\mathbf{F}_\gamma^{-1})_\xi \left[A t^{1-k} j_{\frac{1-k}{2}}(\sqrt{|\xi|^2+c^2}\ t)\right](x). \tag{42}$$

Consequently, the difference between two arbitrary solutions for $k < 0$ is always of the form

$$A(t^{1-k}G^{2-k}(t,x) * \psi(x))_\gamma = A t^{1-k} u(t,x;2-k), \tag{43}$$

where $\psi(x)$ is an arbitrary function or distribution belonging to S'_{ev} , $u(t,x;2-k)$ is solution of the Cauchy problem

$$[(\triangle_\gamma)_x - (B_{2-k})_t\,]\, u = c^2 u,$$

$$u(x,0;2-k) = \psi(x), \qquad u_t(x,0;2-k) = 0.$$

and $G^{2-k}(t,x)$ is the corresponding fundamental solution.

Finally we consider the case $k = -1, -3, -5, \ldots$. In this case the solution will be of a different character than the solutions for other values of k and will always contain a term

$$G^k(t,x) = \frac{\pi 2^{\frac{k-1}{2}}}{\Gamma\left(\frac{1-k}{2}\right)} \mathbf{F}_\gamma^{-1}\left[\left(\sqrt{|\xi|^2+c^2}\,t\right)^{\frac{1-k}{2}} Y_{\frac{1-k}{2}}(\sqrt{|\xi|^2+c^2}\ t)\right].$$

It is clear that

$$G^k(t,x) = \frac{i\pi 2^{\frac{k-1}{2}}}{\Gamma\left(\frac{1-k}{2}\right)} \mathbf{F}_\gamma^{-1}\left[\left(\sqrt{|\xi|^2+c^2}\,t\right)^{\frac{1-k}{2}} H^{(1)}_{\frac{1-k}{2}}(\sqrt{|\xi|^2+c^2}\ t)\right]$$

and

$$G^k(t,x) = -\frac{i\pi 2^{\frac{k-1}{2}}}{\Gamma\left(\frac{1-k}{2}\right)} \mathbf{F}_\gamma^{-1}\left[\left(\sqrt{|\xi|^2+c^2}\,t\right)^{\frac{1-k}{2}} H^{(2)}_{\frac{1-k}{2}}(\sqrt{|\xi|^2+c^2}\ t)\right]$$

will also be a fundamental solution of our problem. Then using (29) and (30) we obtain

$$G^k(t,x) = \frac{e^{\pm\frac{1}{2}\pi n i}\Gamma\left(\frac{n+|\gamma|-k+1}{2}\right)}{2^n\Gamma\left(\frac{1-k}{2}\right)\prod\limits_{i=1}^{n}\Gamma\left(\frac{\gamma_i+1}{2}\right)} t^{1-k}(t^2-|x|^2 \pm i0)_\gamma^{\frac{k-n-|\gamma|-1}{2}}.$$

Since $(t^2-|x|^2)^\lambda_{+,\gamma}$ has its support in the interior of the part of the sphere $S_1^+(n)$ when $x_1 \geq 0, \ldots, x_n \geq 0$, we may conclude that in the case $k \neq -1, -3, -5, \ldots$ the generalized convolutions in exist for arbitrary $\varphi(x) \in S'_{ev}$. However, in the case $k = -1, -3, -5, \ldots$ the fundamental solution is no longer concentrated within the part of the sphere $S_1^+(n)$

Corollary 1.
The solution $u \in S'_{ev}(R^n_+) \times C^2(0,\infty)$ of the following initial value problem for the singular Klein–Gordon equation

$$\left[(\triangle_\gamma)_x - \frac{\partial^2}{\partial t^2}\right] v = c^2 v, \quad c > 0, \quad v = v(x,t), \quad x \in R^n_+, \quad t > 0. \tag{44}$$

$$v(x,0) = f(x), \qquad v_t(x,0) = 0, \qquad f(x) \in S'_{ev} \tag{45}$$

is

$$v(x,t;k) = \frac{2^n \sqrt{\pi}}{\Gamma\left(\frac{1-n-|\gamma|}{2}\right) \prod\limits_{i=1}^{n} \Gamma\left(\frac{\gamma_i+1}{2}\right)} t \left((t^2 - |x|^2)_{+,\gamma}^{\frac{-n-|\gamma|-1}{2}} * f(x)\right)_\gamma. \tag{46}$$

This solution was obtained by letting k tend to 0 in (50).
The Klein–Gordon equation

$$\left[\triangle_z - \frac{\partial^2}{\partial t^2}\right] v = c^2 v, \qquad v = v(z,t), \qquad z \in R^N \tag{47}$$

is the most frequently used wave equations for the description of particle dynamics in relativistic quantum mechanics. When we have that function v is radially symmetric by some groups of variables $z_1, ..., z_N$ in (47) we obtain (44) with a smaller number of spatial variables. In this case numbers γ_i, $i = 1, .., n$ in (44) will integer.
Corollary 2. $u \in S'_{ev}(R^n_+) \times C^2(0,\infty)$

$$[(\triangle_\gamma)_x - (B_k)_t]\, u = 0, \quad k \in R, \quad u = u(x,t;k), \quad x \in R^n_+, \quad t > 0. \tag{48}$$

$$u(x,0;k) = f(x), \qquad u_t(x,0;k) = 0. \tag{49}$$

for $k \neq -1, -3, -5, ...$ is defined by the formula

$$u(x,t;k) = \frac{2^n \Gamma\left(\frac{k+1}{2}\right)}{\Gamma\left(\frac{k-n-|\gamma|+1}{2}\right) \prod\limits_{i=1}^{n} \Gamma\left(\frac{\gamma_i+1}{2}\right)} t^{1-k} \left((t^2 - |x|^2)_{+,\gamma}^{\frac{k-n-|\gamma|-1}{2}} * f(x)\right)_\gamma. \tag{50}$$

Corollary 3. In the case $k > n + |\gamma| - 1$ the integral in (50) integral converges in the usual sense and we obtain unique classical solution of (31)–(32)

$$u(x,t;k) = A(n,\gamma,k)\, t^{1-k} \int\limits_0^t (t^2 - r^2)^{\frac{k-n-|\gamma|-1}{2}} j_{\frac{k-n-|\gamma|-1}{2}}\left(c\sqrt{t^2 - r^2}\right) r^{n+|\gamma|-1} M_r^\gamma[f(x)] dr, \tag{51}$$

$$A(n,\gamma,k) = \frac{2\Gamma\left(\frac{k+1}{2}\right)}{\Gamma\left(\frac{n+|\gamma|}{2}\right) \Gamma\left(\frac{k-n-|\gamma|+1}{2}\right)}.$$

Proof. For $k > n + |\gamma| - 1$ passing to spherical coordinates we obtain

$$u(x,t;k) = C(n,\gamma,k)\, t^{1-k} \int\limits_{B_t^+(n)} (t^2 - |y|^2)^{\frac{k-n-|\gamma|-1}{2}} j_{\frac{k-n-|\gamma|-1}{2}}\left((t^2 - |y|^2)^{\frac{1}{2}} \cdot c\right) {}^\gamma\mathbf{T}^y f(x)\, y^\gamma dy =$$

$$= C(n,\gamma,k) \int\limits_{B_1^+(n)} (1-|y|^2)^{\frac{k-n-|\gamma|-1}{2}} j_{\frac{k-n-|\gamma|-1}{2}} \left((1-|y|^2)^{\frac{1}{2}}\cdot tc\right) {}^\gamma\mathbf{T}^{ty} f(x)\, y^\gamma dy =$$

$$= C(n,\gamma,k) \int\limits_0^1 (1-r^2)^{\frac{k-n-|\gamma|-1}{2}} j_{\frac{k-n-|\gamma|-1}{2}} \left((1-r^2)^{\frac{1}{2}}\cdot tc\right) r^{n+|\gamma|-1} dr \int\limits_{S_1^+(n)} {}^\gamma\mathbf{T}^{tr\theta} f(x)\, \theta^\gamma dS =$$

$$= \frac{2\Gamma\left(\frac{k+1}{2}\right)}{\Gamma\left(\frac{n+|\gamma|}{2}\right)\Gamma\left(\frac{k-n-|\gamma|+1}{2}\right)} \int\limits_0^1 (1-r^2)^{\frac{k-n-|\gamma|-1}{2}} j_{\frac{k-n-|\gamma|-1}{2}} \left((1-r^2)^{\frac{1}{2}}\cdot tc\right) r^{n+|\gamma|-1} M_{rt}^\gamma[f(x)]dr =$$

$$= \frac{2\Gamma\left(\frac{k+1}{2}\right) t^{1-k}}{\Gamma\left(\frac{n+|\gamma|}{2}\right)\Gamma\left(\frac{k-n-|\gamma|+1}{2}\right)} \int\limits_0^t (t^2-r^2)^{\frac{k-n-|\gamma|-1}{2}} j_{\frac{k-n-|\gamma|-1}{2}} \left((t^2-r^2)^{\frac{1}{2}}\cdot c\right) r^{n+|\gamma|-1} M_r^\gamma[f(x)]dr.$$

Corollary 4. In the case $k > n + |\gamma| - 1$ the solution of

$$[(\triangle_\gamma)_x - (B_k)_t]\, u = 0, \quad k \in R, \quad u = u(x,t;k), \quad x \in R_+^n, \quad t > 0. \tag{52}$$

$$u(x,0;k) = f(x), \qquad u_t(x,0;k) = 0 \tag{53}$$

is unique and gives by the formula

$$u(x,t;k) = A(n,\gamma,k)\, t^{1-k} \int\limits_0^t (t^2-r^2)^{\frac{k-n-|\gamma|-1}{2}} r^{n+|\gamma|-1} M_r^\gamma[f(x)]dr, \tag{54}$$

$$A(n,\gamma,k) = \frac{2\Gamma\left(\frac{k+1}{2}\right)}{\Gamma\left(\frac{n+|\gamma|}{2}\right)\Gamma\left(\frac{k-n-|\gamma|+1}{2}\right)}.$$

Remark. Solution of the problem close to (52)–(53) was obtain in [10] (see also [11] p. 243) when $k \neq -1, -3, -5, \ldots$ in terms of the Lauricella function ([30], p. 33)

$$\mathbf{F}_\gamma^{(n)}(a_1,\ldots,a_n,b_1,\ldots,b_n;c;z_1,\ldots,z_n) = \sum_{m_1,\ldots,m_n}^{\infty} \frac{(a_1)_{m_1}\ldots(a_n)_{m_n}(b_1)_{m_1}\ldots(b_n)_{m_n}}{(c)_{m_1+\ldots+m_n}} \frac{z_1^{m_1}}{m_1!}\ldots\frac{z_n^{m_n}}{m_n!},$$

$$\max\{|z_1|,\ldots,|z_n|\} < 1.$$

More precisely, in [10] the solution of the problem

$$\frac{\partial^2 v}{\partial t^2} + \frac{k}{t}\frac{\partial v}{\partial t} - \sum_{i=1}^n \left(\frac{\partial^2 v}{\partial x_i^2} + \frac{\lambda_i}{x_i^2} v\right) = 0, \qquad v = v(t,x), \tag{55}$$

$$v(0,x) = T(x), \qquad \left.\frac{\partial v}{\partial t}\right|_{t=0} = 0 \tag{56}$$

has the form

$$v(t,x) = \frac{\Gamma\left(\frac{k+1}{2}\right)}{\pi^{\frac{n}{2}}\Gamma\left(\frac{k-n+1}{2}\right)} \int\limits_{|x-\xi|=|t|} |t|^{1-k}(t^2-|x-\xi|^2)^{\frac{k-n-1}{2}} T(\xi)\times$$

$$\times \mathbf{F}_\gamma^{(n)}\left(a_1,\ldots,a_n,b_1,\ldots,b_n;\frac{k-n+1}{2};z_1,\ldots,z_n\right) dS_\xi, \tag{57}$$

IOP Conf. Series: Journal of Physics: Conf. Series **973** (2018) 012058 doi:10.1088/1742-6596/973/1/012058

where

$$a_1 = \frac{1+\sqrt{1-4\lambda_1}}{2}, ..., a_n \frac{1+\sqrt{1-4\lambda_n}}{2}, \qquad b_1 = \frac{1-\sqrt{1-4\lambda_1}}{2}, ..., b_n = \frac{1-\sqrt{1-4\lambda_n}}{2},$$

$$z_1 = \frac{t^2-|x-\xi|^2}{2x_1\xi_1}, ..., z_n = \frac{t^2-|x-\xi|^2}{2x_n\xi_n}.$$

If $\lambda_k = \frac{\gamma_k}{2}\left(1-\frac{\gamma_k}{2}\right)$, $i = 1, ..., n$ and $u = x^{\frac{\gamma}{2}} v = x_1^{\frac{\gamma_1}{2}} ... x_n^{\frac{\gamma_n}{2}} v$ then we get our problem (52)–(53). As we see, the expression (54) gives a much more convenient formula for solving the problem (52)–(53). Also note that, in [18] in two different ways including $k = -1, -3, -5, ...$

6. Examples

Let consider the Cauchy problem for $k > n + |\gamma| - 1$

$$[(\triangle_\gamma)_x - (B_k)_t]\, u = c^2 u, \tag{58}$$

$$u(x, 0; k) = \mathbf{j}_\gamma(x;\xi), \qquad u_t(x, 0; k) = 0. \tag{59}$$

In this case the solution is unique and gives by the formula (51):

$$u(x,t;k) = A(n,\gamma,k)\, t^{1-k} \int\limits_0^t (t^2-r^2)^{\frac{k-n-|\gamma|-1}{2}} j_{\frac{k-n-|\gamma|-1}{2}}\left(c\sqrt{t^2-r^2}\right) r^{n+|\gamma|-1} M_r^\gamma \mathbf{j}_\gamma(x;\xi)dr,$$

$$A(n,\gamma,k) = \frac{2\Gamma\left(\frac{k+1}{2}\right)}{\Gamma\left(\frac{n+|\gamma|}{2}\right)\Gamma\left(\frac{k-n-|\gamma|+1}{2}\right)}.$$

For $M_r^\gamma \mathbf{j}_\gamma(x;\xi)$ we have the next formula (see [32])

$$M_r^\gamma \mathbf{j}_\gamma(x;\xi) = \mathbf{j}_\gamma(x;\xi) j_{\frac{n+|\gamma|}{2}-1}(r|\xi|). \tag{60}$$

Using (60) and (5) we get

$$u(x,t;k) = \frac{2\Gamma\left(\frac{k+1}{2}\right)}{\Gamma\left(\frac{n+|\gamma|}{2}\right)\Gamma\left(\frac{k-n-|\gamma|+1}{2}\right)} t^{1-k}\, \mathbf{j}_\gamma(x;\xi) \times$$

$$\times \int\limits_0^t (t^2-r^2)^{\frac{k-n-|\gamma|-1}{2}} r^{n+|\gamma|-1} j_{\frac{k-n-|\gamma|-1}{2}}\left(c\sqrt{t^2-r^2}\right) j_{\frac{n+|\gamma|}{2}-1}(r|\xi|)dr =$$

$$= \frac{2^{\frac{k-1}{2}}\Gamma\left(\frac{k+1}{2}\right)}{c^{\frac{k-n-|\gamma|-1}{2}} |\xi|^{\frac{n+|\gamma|}{2}-1}} t^{1-k}\, \mathbf{j}_\gamma(x;\xi) \times$$

$$\times \int\limits_0^t (t^2-r^2)^{\frac{k-n-|\gamma|-1}{4}} r^{\frac{n+|\gamma|}{2}} J_{\frac{k-n-|\gamma|-1}{2}}\left(c\sqrt{t^2-r^2}\right) J_{\frac{n+|\gamma|}{2}-1}(r|\xi|)dr.$$

Applying formula 2.12.35.2 from [28] of the form

$$\int\limits_0^t (t^2-x^2)^{m+\frac{\mu}{2}} x^{\nu+1+2l} J_\mu(c\sqrt{t^2-x^2}) J_\nu(hx)dx =$$

$$= t^{\mu+\nu-m-l+1} c^{\mu} h^{\nu} \left(\frac{\partial}{c\partial c}\right)^m \left(\frac{\partial}{h\partial h}\right)^l [(c^2+h^2)^{-\frac{\mu+\nu+m+l+1}{2}} J_{\mu+\nu+m+l+1}(t\sqrt{c^2+h^2})],$$
$$t > 0, \qquad \operatorname{Re}\nu > -l-1, \qquad \operatorname{Re}\mu > -m-1.$$

We have $k = m = 0$, $\nu = \frac{n+|\gamma|}{2} - 1$, $\mu = \frac{k-n-|\gamma|-1}{2}$, $h = |\xi|$ and

$$\int\limits_0^t (t^2-r^2)^{\frac{k-n-|\gamma|-1}{4}} r^{\frac{n+|\gamma|}{2}} J_{\frac{k-n-|\gamma|-1}{2}}\left(c\sqrt{t^2-r^2}\right) J_{\frac{n+|\gamma|}{2}-1}(r|\xi|)dr =$$

$$= \frac{t^{\frac{k-1}{2}} c^{\frac{k-n-|\gamma|-1}{2}} |\xi|^{\frac{n+|\gamma|}{2}-1}}{(\sqrt{c^2+|\xi|^2})^{\frac{k-1}{2}}} J_{\frac{k-1}{2}}(t\sqrt{c^2+|\xi|^2}).$$

Therefore

$$u(x,t;k) = \frac{2^{\frac{k-1}{2}}\Gamma\left(\frac{k+1}{2}\right)}{c^{\frac{k-n-|\gamma|-1}{2}} |\xi|^{\frac{n+|\gamma|}{2}-1}} t^{1-k} \mathbf{j}_\gamma(x;\xi) \frac{t^{\frac{k-1}{2}} c^{\frac{k-n-|\gamma|-1}{2}} |\xi|^{\frac{n+|\gamma|}{2}-1}}{(\sqrt{c^2+|\xi|^2})^{\frac{k-1}{2}}} J_{\frac{k-1}{2}}(t\sqrt{c^2+|\xi|^2}) =$$

$$= \mathbf{j}_\gamma(x;\xi) j_{\frac{k-1}{2}}(t\sqrt{c^2+|\xi|^2}).$$

The following three figures show the graphs of the solutions of the last problem for $n = 1$, $k = \frac{1}{3}$, $\gamma = \frac{1}{5}$ and for $c = 0$, $c = 1$, $c = 2$, $c = 4$, accordingly.

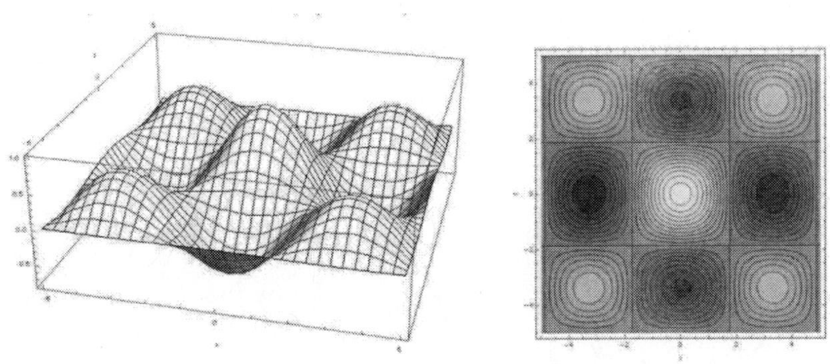

Figure 1. $u(x,t;k) = j_{-\frac{2}{5}}(x)\, j_{-\frac{1}{3}}\left(t\sqrt{c^2+1}\right), \quad c = 0.$

References

[1] John F 1981 *Plane Waves and Spherical Means* (New York: Springer Science & Business Media)
[2] Helgason S 2011 *Integral Geometry and Radon Transforms* (Springer)
[3] Davydov A S 1976 *Quantum Mechanics, 2nd Edition* (Pergamon Press)
[4] Weinberg S 2002 *The Quantum Theory of Fields* (Cambridge University Press)
[5] Euler L 1768 *Institutionum calculi integralis* (Lipsiae Et Berolini)
[6] Poisson S D 1823 *J. de L'École Polytechechnique* **6(19)** 215–248
[7] Darboux G 1888 *Leçons sur la théorie générale des surfaces et les applications géométriques du calcul infinitésimal* (Gauthier Villars Et Fils)
[8] Riemann B 1948 *On expansion of waves with finite amplitude* (Moscow–Leningrad: Gos. Izd-Tehn.-teoret. lit.)
[9] Weinstein A 1978 *Selecta* (Pitman Publishing)
[10] Fox D N 1959 *J. Math. Mech.* **8** 197–219.
[11] Carroll R W and Showalter R E 1976 *Singular and Degenerate Cauchy problems* (New York: Academic Press)
[12] Sitnik S M 2012 *Transmutations and applications: A survey* arXiv:1012.3741v1

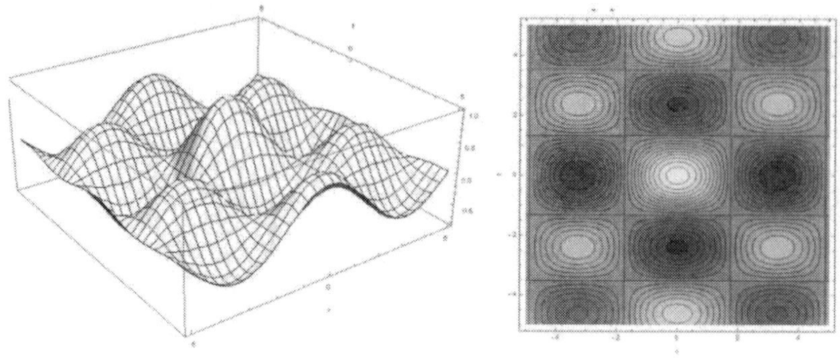

Figure 2. $u(x,t;k){=}j_{-\frac{2}{5}}(x)\,j_{-\frac{1}{3}}\left(t\sqrt{c^2+1}\right),\quad c=1.$

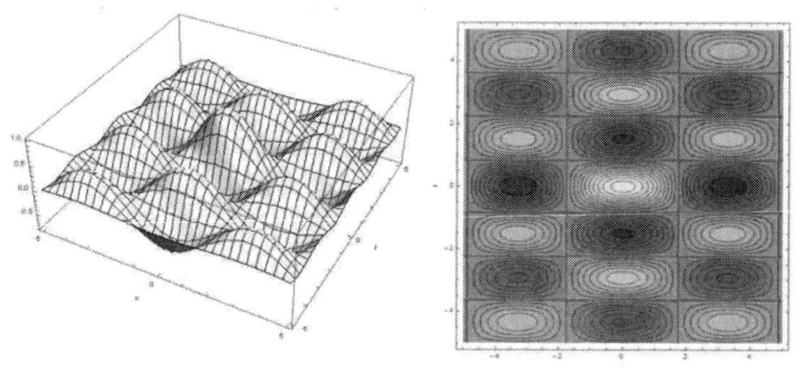

Figure 3. $u(x,t;k){=}j_{-\frac{2}{5}}(x)\,j_{-\frac{1}{3}}\left(t\sqrt{c^2+1}\right),\quad c=2.$

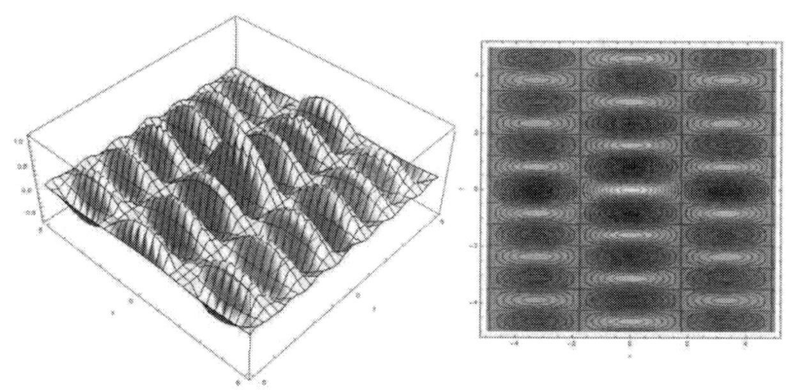

Figure 4. $u(x,t;k){=}j_{-\frac{2}{5}}(x)\,j_{-\frac{1}{3}}\left(t\sqrt{c^2+1}\right),\quad c=4.$

[13] Bresters D W 1973 *SIAM J. Math. Anal.* **4(1)** 31–41
[14] Bresters D W 1978 *SIAM J. Math. Anal.* **9(5)** 924–934
[15] Glushak A V and Pokruchin O A 2016 *Differential Equations* **52(1)** 39–57
[16] Glushak A V 2016 *Izvestiya VUZ. Matematika* **60(6)** 21–28
[17] Glushak A V and Popova V A 2008 *Journal of Mathematical Sciences* **149(4)** 1453–1468
[18] Shishkina E L and Sitnik S M 2017 *Journal of Inequalities and Special Functions* **8(1)** 49–67
[19] Kipriyanov I A 1997 *Singular Elliptic Boundary Value Problems* (Moscow: Nauka)

[20] Watson G N 1966 *A Treatise on the Theory of Bessel Functions* (Cambridge University Press)
[21] Abramowitz M and Stegun I A 1983 *Handbook of Mathematical Functions with Formulas, Graphs, and Mathematical Tables* (Applied Mathematics Series)
[22] Bateman H and Erdelyi A 1954 *Tables of Integral Transforms, Vol. 2* (New York: McGraw-Hill)
[23] Levitan B M 1951 *Uspekhi Mat. Nauk* **6, 2(42)** 102–143
[24] Lyakhov L N 2007 *B-Hypersingular Integrals and Their Applications to the Description of Kipriyanov Function Classes and to Integral Equations with B-Potentials Kernels* (Lipetsk: LGPU)
[25] Kipriyanov I A and Ivanov L A 1983 *Tr. Semin. im. S.L. Soboleva* **1** 55–77
[26] Prudnikov A P Brychkov Yu A and Marichev O I 1986 *Integrals and Series: Vol. 1: Elementary Functions* (New York: Gordon & Breach Sci. Publ.)
[27] Shishkina E L 2017 *Chelyabinsk Physical and Mathematical Journal* **2(1)** 88–98
[28] Prudnikov A P Brychkov Yu A and Marichev O I 1990 *Integrals and Series: Vol. 2: Special Functions* (New York: Gordon & Breach Sci. Publ.)
[29] Gel'fand I M and Shilov G E 1964 *Generalized functions. Vol. I: Properties and operations* (Boston: Academic Press)
[30] Srivastava H M and Karlsson P W 1985 *Multiple Gaussian Hypergeometric Series* (Chichester, England: Ellis Horwood)
[31] Shishkina E L 2016 *Issues Anal.* **5(23):2** 52–68
[32] Lyakhov L N Polovinkin I P and Shishkina E L 2014 *Differential Equations* **50(4)** 513–525

IOP Conf. Series: Journal of Physics: Conf. Series **973** (2018) 012059 doi:10.1088/1742-6596/973/1/012059

Models and Methods of Aggregating Linguistic Information in Multi-criteria Hierarchical Quality Assessment Systems

T V Azarnova[1], **I A Titova**[1] **and S A Barkalov**[2]

[1] Department of Applied Mathematics, Informatics and Mechanics, Voronezh State University, Voronezh, Russia

[2] Department of Economics, Management and IT, Voronezh State Technical University, Voronezh, Russia

E-mail: ivdas92@mail.ru, tia1404@yandex.ru

Abstract. The article presents an algorithm for obtaining an integral assessment of the quality of an organization from the perspective of customers, based on the method of aggregating linguistic information on a multilevel hierarchical system of quality assessment. The algorithm is of a constructive nature, it provides not only the possibility of obtaining an integral evaluation, but also the development of a quality improvement strategy based on the method of linguistic decomposition, which forms the minimum set of areas of work with clients whose quality change will allow obtaining the required level of integrated quality assessment.

1. Introduction

Client capital is an important factor in the competitiveness of modern companies working in the service sector. Manage client portfolio [1] and building long-term relationships with customers [2] are a priority in the work of each organization. To effectively manage customer capital, tools are needed to measure service quality from the perspective of customers, implementing feedback and assessing the degree to which the service meets the requirements and expectations of customers. The problem of assessing the quality of service is weakly structured, there are no established generally accepted approaches to its solution. Widely used methods are: secret client (Mystery Shopping); SCI (Customer Satisfaction Index), push-button surveys of customers and SERVQUAL [3]. The SERVQUAL method, on which the algorithms proposed in this article are based, is based on the assumption that the quality of service for any evaluation characteristic (service criterion) is measured by the discrepancy between the "expectation" and "perception" of the service level for a given characteristic. For the practical implementation of this approach to quality assessment, special tools are needed to measure the degree of discrepancy between "expectation" and "perception".

The article presents a hierarchical system of quality assessment, reflecting the process of aggregating evaluation information, structured by client groups. Depending on the specific activities of the organization's customers are divided into groups, each of which in turn is divided into sub-groups. The criteria for this segmentation can be such factors as socio-economic (income level, education level, etc.), geographical (region of residence, climate, etc.), demographic (sex, age, etc.), etc. Based on the proposed model and methodology, SERVQUAL developed a linguistic algorithm for assessing quality and a linguistic algorithm for shaping a quality improvement strategy.

2. Algorithms for the Formation of an Integral Linguistic Quality Assessment on a Multicriteria Hierarchical System

As the information base of the research, the results of the customer questioning for each selected segment are presented. The questionnaires correspond to the structure of the hierarchical model for each client category and provide a list of criteria and corresponding subcriteria reflecting the full range of services of the organization in question. For example, for a restaurant business [4], the criterion can be "Serving a waiter", and the sub criteria of this criterion is "Service speed", "Courtesy", etc. The evaluation criteria are divided according to their orientation (ordering of alternatives) into two groups, the criteria of the first group (symbol $\nearrow$) satisfy the condition — the higher the value of the evaluation, the higher the client's satisfaction, and the criteria of the second group (symbol $\swarrow$) to the condition — the lower the estimate value, the higher the customer satisfaction. In the process of questioning, the respondent for each of the selected items (subcriteria) should indicate: *an assessment of importance; assessment of quality expectations; assessment of the perception of quality.* The estimation is carried out in a linguistic scale [5] with the term set $S = \{S_1 = VL, S_2 = L, S_3 = M, S_4 = H, S_5 = VH\}$ (VL — very low, L — low, M — average, H — high, VH - very high).

The multicriteria hierarchical system for assessing the quality of the generalized (not tied to a particular problem) is shown in Figure 1. The above structure contains the following notation:

- K^V — final integrated customer service quality assessment;
- $K^{IV} = (k_1^{IV}, \ldots, k_p^{IV}, \ldots, k_{\overline{p}}^{IV})$ — vector integral group quality estimates;
- $p = \overline{1, \overline{p}}$ — group number, $\overline{p}$ — number of groups;
- $W^{IV} = (w_1^{IV}, \ldots, w_p^{IV}, \ldots, w_{\overline{p}}^{IV})$ — vector of assessments of the importance of the groups provided by the experts of the enterprise in question;
- $K^{III} = (k_{1p}^{III}, \ldots, k_{t_p p}^{III}, \ldots, k_{\overline{t_p} p}^{III})$ — vector of summary quality estimates of subgroups of the p-th group;
- $t_p = \overline{1, \overline{t_p}}$ — subgroup number of the p-th group, $\overline{t_p}$ — number of subgroups of the p-th group;
- $W^{III} = (w_{1p}^{III}, \ldots, w_{t_p p}^{III}, \ldots, w_{\overline{t_p} p}^{III})$ — vector of assessments of the importance of the subgroups provided by the experts of the enterprise in question;
- $K^{II} = (k_{1 t_p p}^{II}, \ldots, k_{m_{t_p} t_p p}^{II}, \ldots, k_{\overline{m_{t_p}} t_p p}^{II})$ — vector of integral quality estimates of the criteria of the p-th group of the t_p-th subgroup;
- $m_{t_p} = \overline{1, \overline{m_{t_p}}}$ — number of the criterion of the p-th group of the t_p-th subgroup, $\overline{m_{t_p}}$ — number of criteria for the p-th group of the t_p-th subgroup;
- $K^I = (k_{1 m_{t_p}}^I, \ldots, k_{n_{m_{t_p}} m_{t_p}}^I, \ldots, k_{\overline{n_{m_{t_p}}} m_{t_p}}^I)$ — vector of integral quality estimates of subcriteria of the m-th criterion of the p-th group of the t_p-th subgroup;
- $n_{m_{t_p}} = \overline{1, \overline{n_{m_{t_p}}}}$ — sub-criterion number of the m-th criterion of the p-th group of the t_p-th subgroup, $\overline{n_{m_{t_p}}}$ — number of subcriteria of the m-th criterion of the p-th group of the t_p-th subgroup;
- $K_{i=\overline{1,X}}^{is}$ — expectations of service quality, issued by the i-th client;
- $K_{i=\overline{1,X}}^{ex}$ — assessment of service quality perception, exposed by the i-th client;
- X — variable indicating the number of estimates for this sub-criterion;
- $\omega_{i=\overline{1,X}}$ — assessment of the importance of subcriteria, exposed by the i-th client;
- $\overline{K}_{i=\overline{1,X}}$ — stimates aggregated for each i-th client for each sub-criterion on the basis of estimates $K_{i=\overline{1,X}}^{is}$ and $K_{i=\overline{1,X}}^{ex}$.

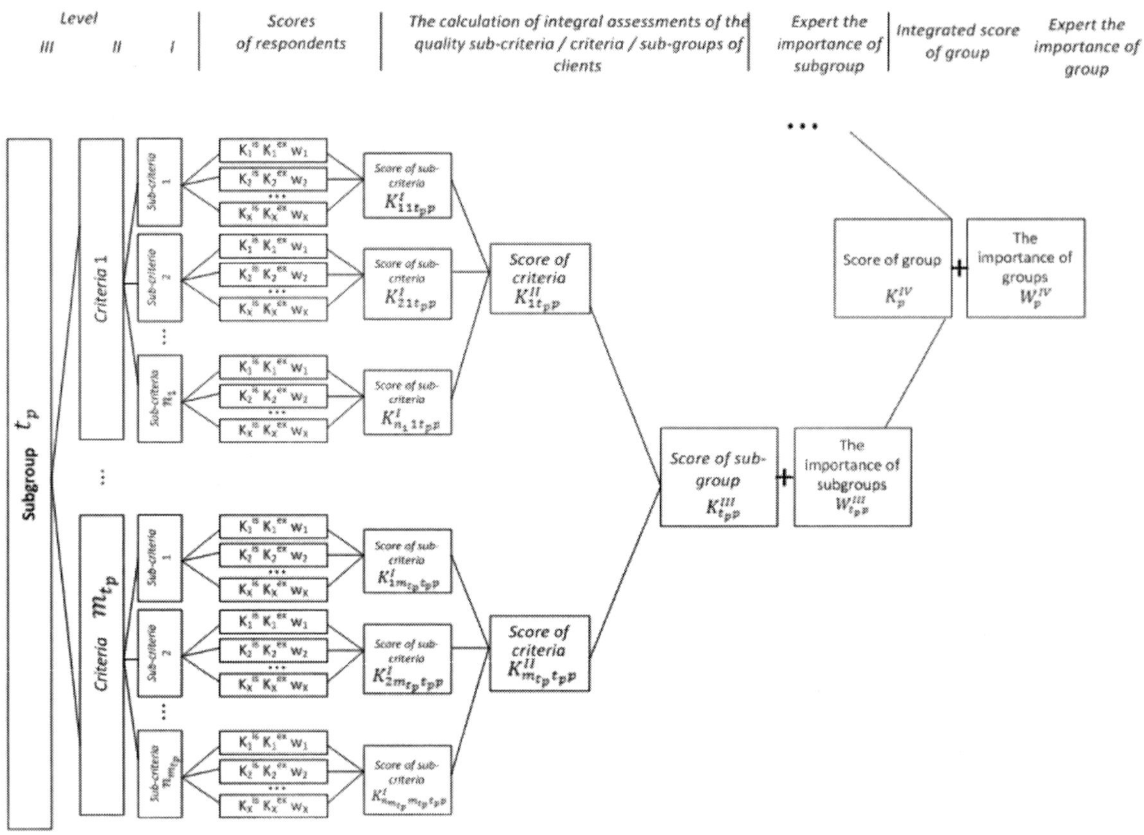

Figure 1. Scheme of calculation of integral quality assessment of customer service.

Algorithm 1. Computing an integral evaluation of the quality of customer service for each element of a multicriteria hierarchical system [6].

Step 1. The results of the questionnaire are structured in accordance with the main groups / subgroups of clients.

Step 2. For each respondent, linguistic assessments of quality are calculated based on the questionnaires that were filled in during questioning on the basis of linguistic assessments of expectations and perceptions of quality.

Let $K_m^{is} = S_i$ — the linguistic estimate of the expected quality from the m-th sub-criterion, $K_m^{ex} = S_j$ — the linguistic estimate of the perceived quality from the m-th sub-criterion, then the linguistic quality of service estimate for the m-th sub-criterion with orientation ($\nearrow$) will be calculated according to the formula:

$$\overline{K}_m = S_{\max\{\min\{2j-i;n\};1\}}, \tag{1}$$

where n — number of terms of the linguistic scale $S = \{S_i\}_{i=\overline{1,n}}$. For sub-criterions with orientation ($\swarrow$), the quality estimate is computed similarly after converting the estimates K_m^{is} and K_m^{ex} in to estimates of type ($\nearrow$) through the negation operation for linguistic variables.

For each subcriter, many pairs are formed $L = \{(w_l, k_l) : l = 1, \ldots, n\}$, where n — the number of respondents who rated the relevant item, w_l — assessment of the importance for l-th of respondent, k_l — quality assessment for l-th of respondent. The pairs are structured into two vectors: the importance vector $W = (w_1, \ldots, w_n)$ and quality vector $K = (k_1, \ldots, k_n)$.

The importance estimation vector $W = (w_1, \ldots, w_n)$ containing duplicate values generates a decomposition of the quality estimation vector into non-empty groups $K_1, \ldots, K_5$. Each nonempty group K_i, $i = 1, \ldots, 5$ is constructed from a certain value of importance and consists of the components of the vector $K = (k_1, \ldots, k_n)$ for which the corresponding components of the vector W are equal to the selected value of importance. Values of importance are viewed in ascending order $S_1 = VL$, $S_2 = L$, $S_3 = M$, $S_4 = H$, $S_5 = VH$. We denote by n_i the number of elements in the group K_j. For each of the groups $K_1, \ldots, K_s$, a generalized quality estimate is constructed by convolving their components with the help of the linguistic LOWA-operator $\Phi_U(K)$, which is defined as follows [7–9].

Let U – vector of weights, ($u_i \in [0,1]$, $\sum_{i=1}^{n} u_i = 1$), $K = \{k_1, k_2, \ldots, k_n\}$ — components of some non-empty group, then

$$\Phi_U(K) = C^n\{(u_k, b_k),\ k = \overline{1,n}\} = u_1 \otimes b_1 \oplus C^{n-1}\{(\lambda_k, b_k),\ k = \overline{2,n}\}, \tag{2}$$

where $B = (b_1, b_2, \ldots, b_n)$ — is the vector obtained from $K = \{k_1, k_2, \ldots, k_n\}$ ordering by non-increase of linguistic terms, $\lambda_k = \frac{\omega_k}{\sum_{m=2}^{n} \omega_m}$ $(k = \overline{1,n})$ — the normalized vector of weights obtained after removing the maximum weight u_1, $\sum_{m=2}^{n} \lambda_m = 1$; C^n, C^{n-1} — convex combinations, n and $(n-1)$ terms respectively.

For $n = 2$, the convex combination of linguistic terms $b_1 = S_j$ and $b_2 = S_i$ $(j \geqslant i)$ is determined by the rule:

$$C^2\{(u_1, b_1), (u_2, b_2)\} = u_1 \otimes b_1 \oplus u_2 \otimes b_2 = S_k, \tag{3}$$

where $k = \min\{T, i + round(u_1 \cdot (j - i)\}$, *round* corresponds to the rounding operation.

Vector of weights $U = (u_1, \ldots, u_n)$ for the aggregation operator is calculated through special functions of quantification $Q(x)$:

$$u_1 = Q\left(\frac{1}{n}\right)\ \forall i = \overline{2,n} \quad \left(u_i = Q\left(\frac{i}{n}\right) - Q\left(\frac{i-1}{n}\right)\right). \tag{4}$$

As a function of quantification $Q(x)$ the function is used $Q(x) = x^{\propto}$ with different values of α for different groups $K_1, \ldots, K_5$ with higher importance values, the aggregation strategy is closer to the conjunctive, with lower ones to disjunctive): for the group K_1 (importance $S_1 = VL$) — $Q_1(x) = x^{1/3}$; for the group K_2 (importance $S_2 = L$) — $Q_2(x) = x^{1/2}$; for the group K_3 (importance $S_3 = M$) — $Q_3(x) = x$; for the group K_4 (importance $S_4 = H$) — $Q_4(x) = x^2$; for the group K_5 (importance $S_5 = VH$) — $Q_5(x) = x^3$.

For the chosen quantification functions we calculate the weights $U = (u_1, \ldots, u_n)$ operator of aggregation and generalized estimates for groups $K_1, \ldots, K_s$.

As a result of the performed operations, three vectors are formed: the importance vector $W = (w_1, \ldots, w_n)$ the vector of generalized quality estimates $K = (k_1, \ldots, k_n)$, and the frequency vector $V = (v_1, \ldots, v_n)$, where $v_i = \dfrac{n_i}{n}$, n — the number of respondents evaluating the criterion in question.

Using the pairs (w_i, v_i), obtained, linguistic confidence coefficients q_i, are constructed that characterize the degree of confidence in the estimates of k_i. Confidence coefficients are calculated using linguistic scales, constructed individually for each value $s = 2, 3, 4, 5$ (the number of pairs (w_i, v_i)) and each valuation of VL, L, M, H, VH.

Algorithm 2. The construction of a scale for a specific value $w = S_{j^*}$.

Step 1. On the OX axis, the interval [0; 1] is the range of possible frequencies.

Step 2. At the point $\dfrac{1}{s}$, the membership function for a term equal to the value of the considered importance of S_{j^*} is assumed to be 1:

$$\mu_{S_{j^*}}\left(\frac{1}{n}\right) = 1.$$

Step 3. The remaining terms are completed as follows: for terms with indices $j = 1, \ldots, j^*$ the membership functions at the points $\dfrac{j-1}{n \cdot (j^*-1)}$ are assumed to be 1:

$$\mu_{S_j}\left(\frac{j-1}{n \cdot (j^*-1)}\right) = 1.$$

For terms with indices $j = j^*, \ldots, 5$, the membership functions at the points $\left(\dfrac{1}{n} + \dfrac{n-1}{n \cdot (5-j^*)} \cdot (j-j^*)\right)$ are assumed to be 1:

$$\mu_{S_j}\left(\frac{1}{n} + \frac{n-1}{n \cdot (5-j^*)} \cdot (j-j^*)\right) = 1.$$

The graphs mark the points obtained and complete the triangular numbers, connecting the right end of the i-th fuzzy number to the left end $(i+1)$-th in the marked values.

For $w = VL$, a fuzzy trapezoidal number with coordinates: $\left(0; \frac{1}{n}; 0; \dfrac{n-1}{4n}\right)$ corresponds to the term VL in the scale being formed. For $w = VH$, a fuzzy trapezoidal number with coordinates $\left(\dfrac{1}{n}; 1; \dfrac{1}{4n}; 0\right)$ corresponds to the VL term in the scale being formed.

Based on the vector of linguistic quality estimates $K = (k_1, k_2, \ldots, k_s)$ and the vector of linguistic confidence coefficients $Q = (q_1, q_2, \ldots, q_s)$ $(s \leqslant 5)$, a generalized quality estimate for each criterion $\overline{K}$ is constructed.

It is assumed that $s \geqslant 2$, the case $s = 1$ is trivial and does not require convolution construction. For $s = 2$, the number of possible different combinations (K, Q), where $Q = (q_1, q_2)$ and $K = (k_1, k_2)$ is 325, which allows us to consider and specify all possible rules for fuzzy inference:

– If $k_1 = VL$ and $q_1 = VL$, $k_2 = VL$ and $q_2 = VL$, then $K = VL$
– If $k_1 = L$ and $q_1 = VL$, $k_2 = VL$ and $q_2 = VL$, then $K = VL$
– If $k_1 = L$ and $q_1 = VL$, $k_2 = L$ and $q_2 = VL$, then $K = L$ ....

These rules can be presented in the form of a table, where for estimates (k_i, q_i) and (k_j, q_j) the generalized estimate $\overline{K}$ stands at the intersection of the i-th row and the j-th column.

For the three-dimensional case, the number of different combinations increases already to 7875 and the application of expert judgment on the values of generalized estimates becomes too time-consuming. Therefore, for $s \geqslant 3$, there is a need for a mechanism that would not involve experts. Let's describe a variant of such a mechanism.

Algorithm 3. Construction of generalized estimates $\overline{K}$ for the case $s \geqslant 3$.

Let there be 2 vectors: $K = (k_1, \ldots, k_s)$ and $Q = (q_1, \ldots, q_s)$. We denote by l the indices of the estimates in the vectors K and Q, $l = 1, \ldots, s$, and by i_l and j_l are the indices of the terms corresponding to these estimates to the scale S.

Step 1. Each pair of estimates $(k_l, q_l) = (S_{i_l}, S_{j_l})$ is translated into the numerical estimate $c_{i_l j_l}$ in accordance with Table 1.

Step 2. The values are calculated:

$$H_i = \prod_{l=1}^{s} c_{ij_l},\ i = \overline{1,5}, \qquad H^* = \prod_{l=1}^{s} c_{i_l j_l}. \tag{5}$$

Step 3. An output linguistic scale is constructed, containing 5 terms that correspond to the possible final estimates (VL, L, M, H, VH), as follows: on the OX axis the H_i points are marked.

Table 1. The mechanism of forming a numerical estimate.

$k \backslash q$	VL	L	M	H	VH
VL	55	45	20	13	4
L	66	55	40	27	9
M	79	67	60	42	30
H	90	85	80	65	64
VH	100	100	100	100	100

In these values, the membership functions of the corresponding evaluation terms will be equal to 1. Triangular numbers with the following coordinates are completed: (H_1, H_1, H_2) — for the term $S_1 = VL$; (H_{i-1}, H_i, H_{i+1}) — for the terms S_i, $i = 2, 3, 4$; (H_4, H_5, H_5) for the term $S_5 = VH$.

Step 4. The scale indicates the value of H^*. The value of the term S_i, for which the membership function for the given point takes the greatest value, is the final generalized estimate $\overline{K} = S_i$.

The algorithm described above allows us to form a vector of partial estimates of the subcriteries $\overline{K} = (\overline{k}_1, \overline{k}_2, \ldots, \overline{k}_m)$. For each term S_i from the scale S, the number of occurrences of this term in the vector of partial estimates K is counted. Let: r_1 — the number of estimates of VL in the vector $\overline{K}$; r_2 is the number of estimates of L in the vector $\overline{K}$; r_3 is the number of estimates of M in the vector $\overline{K}$; r_4 is the number of estimates of H in the vector $\overline{K}$; r_5 is the number of VH estimates in the vector $\overline{K}$. Denote the required integral estimate for a certain stage of aggregation by S_t, then the index t of the final estimate will be calculated by the formula:

$$t = round\left(\frac{1}{m}\sum_{i=1}^{5}(r_i * i)\right), \tag{6}$$

where *round* is the rounding operation. In this case, you can use 3 types of policies for calculating the final rating: hard, neutral, soft. With a *hard* policy, the rounding down method (the nearest integer not exceeding the value obtained) is used to round the result. When *neutral*, the index t is rounded in the traditional way. If you choose a *soft* policy when rounding, the round-up method is used (the nearest integer is larger than the value obtained).

3. Algorithm for forming a strategy for improving quality

The algorithm described in the previous section is of a constructive nature, its structure and the calculation technologies used allow us to develop tools for reverse decomposition and the development of a strategy for increasing the integral quality assessment [10].

Consider the algorithm for determining the minimum set of criteria, changing the integral linguistic quality assessment by which for some (if possible minimum) number of items up the linguistic scale, allow to increase the overall integral assessment of the quality of the organization as a whole.

Algorithm 4. The solution of the problem of reverse decomposition according to the multicriteria hierarchical system of quality assessment:

Step 1. Set the desired upper (fifth) level K^{V^*}.

Step 2. Determine all possible variants of the vector $K_i^{IV*} = (k_1^{IV*}, \ldots, k_p^{IV*})$, where K_i^{IV*} is the vector of quality grades by groups, vector K^{IV} and is the input vector for obtaining the desired estimate K^{V*}.

Step 3. Determine for each vector $K_i^{III*} = (k_1^{III*}, \ldots, k_p^{III*})$, where K_i^{III*} is the vector of quality estimates by subgroups, vector K^{III} and is the input vector for obtaining the desired estimates of the vectors K_i^{IV*}.

Step 4. For each vector K_i^{III*} find a minimum set of criteria, the assessment of which must be shifted one point up the linguistic scale.

Step 5. Construct a tree of options for the task of improving the quality assessment.

Step 6. Evaluate all the sets of criteria obtained in step 3, select expertly the best one.

Step 7. For the set of criteria selected in step 5, consider the sub-criteria, identify those that require increased attention from the enterprise.

Let us dwell in more detail on the implementation of steps 2, 3 of the algorithm.

The method of calculating the inverse problem will depend on the technique used in direct calculation, which in turn is determined by the number of given groups:

- for $p = 2$, the convolution was performed using a knowledge base represented in the form of a table,
- at $3 \leqslant p \leqslant 5$ — the convolution was performed using linguistic scales. Consider the formation of changes for both cases:

1. for $p = 2$ we modify the used table:

– at the intersection of rows and columns corresponding to the given estimates of the quality and importance of the groups, instead of the resulting estimate, we insert the value of the indexes of the first and second group quality ratings.

– the resulting estimates with the corresponding terms will be denoted by shading the areas in the table with the defined colors for each term as follows: VL — blue, L — light blue, M — yellow, H — light red, VH – dark red (Table 2).

Table 2. The inverse process table.

	ω_1^V	L	L	L	L	L
ω_2^V	k_1^V \ k_2^V	VL	L	M	H	VH
H	VL	2	3	4	5	6
H	L	3	4	5	6	7
H	M	4	5	6	7	8
H	H	5	6	7	8	9
H	VH	6	7	8	9	10

Algorithm 5. Formation of possible variants of changes for p = 2.

Step 1. Find the cell of the resulting estimate $K^V = S_j$ in the table and determine the tabular value corresponding to the given estimate. We denote it by z.

Step 2. Find the domain of the term for a given desired estimate $K^{V*} = S_{j^*}$.

Step 3. Find in the given area all the values of z^*, which are strictly right and/or lower than the value of z and for which the difference $(z^* - z)$ is minimal.

Step 4. Determine all possible quality estimation vectors $K^{IV*} = (k_1^{IV*}, k_2^{IV*})$, where k_1^{IV*} and k_2^{IV*} correspond to the values found z^* possible estimates of the groups.

If the number of groups p: $3 \leqslant p \leqslant 5$, then the quality estimates were no longer calculated using a table, but based on linguistic scales.

Algorithm 6. Formation of possible variants of changes for $3 \leqslant p \leqslant 5$.

In order to shift the estimate k_1^{IV} to the value k_1^{IV*}, it is necessary to shift the numerical value obtained on the basis of the table to the right according to the linguistic scale until the desired result is obtained. The minimum value on the linguistic scale at which the desired estimate k_1^{IV*} is obtained is the intersection of the scales of the terms of the estimate k_1^{IV*} and the estimate with the index is one less. The intersection point is the midpoint of the segment connecting the vertices of the scales of these terms.

We denote by l the indices of the estimates in the vectors k_1^{IV*}, W^{IV} and K^{IV*}, $l = 1, 2, 3$, and by i_l, j_l and i_l^* are the indices of the terms of the scale S.

Step 1. Translate each pair of estimates $(k_l^{IV}, w_l^{IV}) = (S_{i_l}, S_{j_l})$ into a numerical estimate of $c_{i_l j_l}$ in accordance with Table 1.

Step 2. Calculate the quantities:

$$H_i = \prod_{l=1}^{5} c_{ij_l}, \quad i = i_l^* - 1, i_l^*. \tag{7}$$

Step 3. Calculate the value of $\overline{H}$ — the middle of the segment $[H_{i_l^*-1}; H_{i_l^*}]$:

$$\overline{H} = \frac{H_{i_l^*-1} + H_{i_l^*}}{2}. \tag{8}$$

Step 4. Introduce the vector $K^{IV}_{a_1 \dots a_{\overline{p}}}$ with the coordinates $(S_{i_1+a_1}, \dots, S_{i_{\overline{p}}+a_{\overline{p}}})$ (a_i — integers that can take the values 0, 1, 2, 3 or 4, $\overline{p}$ is the number of groups, for example, for $\overline{p} = 3$, the vector $K^{IV}_{000} = K^{IV} = (k_1^{IV}, k_2^{IV}, k_3^{IV})$). Put k $k' = a_1 + \dots + a_{\overline{p}} = 1$, and for all a_i satisfying the above requirement, calculate the quantities:

$$H_{a_1 \dots a_{\overline{p}}} = \prod_{l=1}^{3} c_{(i_l+m_l)j_l} = c_{(i_1+m_1)j_1} * c_{(i_2+m_2)j_2} * \dots * c_{(i_{\overline{p}}+m_{\overline{p}})j_{\overline{p}}} \tag{9}$$

Step 5. If one of the quantities $H_{a_1 \dots a_{\overline{p}}} > \overline{H}$, then we put $K_i^{IV*} = (k_1^{IV*}, \dots, k_{\overline{p}}^{IV*}) =$ $= (S_{i_1+a_1}, \dots, S_{i_{\overline{p}}+a_{\overline{p}}})$, otherwise put $k' = k' + 1$ and go to Step 4.

Step 3 of Algorithm 4 assumes the presence of vectors of the third level and is performed in the same way as step 2, with the inverse problem being solved for the desired estimates of each group separately.

Let's consider in more detail step 4 of Algorithm 4.

All criteria are ranked by the value of the following value:

$$R_i = s_i * (u_i * d_i + v_i + g_i), \tag{10}$$

where

$$s_i = \begin{cases} 1, & if\ enough\ answers\ are\ received, \\ 0, & otherwise, \end{cases}$$

$$u_i = \begin{cases} 1, & if\ you\ can\ increase\ all\ the\ estimates\ f\ for\ the\ groups\ under\ consideration, \\ 0, & otherwise, \end{cases}$$

$$v_i = \begin{cases} 1, & if\ the\ evaluation\ of\ the\ criterion\ for\ a\ non\text{-}considered\ group\ changes, \\ 0, & otherwise, \end{cases}$$

$$g_i = \begin{cases} 1, & if\ in\ the\ groups\ under\ consideration\ there\ is\ an\ estimate\ VL, \\ 0.5, & if\ in\ the\ groups\ under\ consideration\ there\ is\ an\ estimate\ L, \\ 0, & otherwise \end{cases}$$

IOP Conf. Series: Journal of Physics: Conf. Series **973** (2018) 012059 doi:10.1088/1742-6596/973/1/012059

$$d_i = k * 5 - \sum j_i; \tag{11}$$

k — number of groups, $\sum j_i$ — sum of the indexes of the evaluation terms in these groups.

In the future, the selection criteria will be in accordance with the list obtained in the ranking.

Let $K^{III} = (k_1^{III}, k_2^{III}, k_3^{III})$, $k_j^{III} = S_j$ — the vector of subgroup quality assessments obtained in the direct aggregation process, $K^{III*} = (k_1^{III*}, k_2^{III*}, k_3^{III*})$, $k_j^{III*} = S_{j^*}$ — the vector of necessary estimates of subgroups found in the reverse process. Then for each of the subgroups we calculate the quantity:

$$\Delta = \begin{cases} \tilde{j} * m - \sum_{i=1}^{5}(r_i * i), & if\ k_j^{III*} > k_j^{III} \\ j^*, & if\ k_j^{III*} = k_j^{III}, \end{cases} \tag{12}$$

where

$$\tilde{j} = \begin{cases} (j^* - 0.95) & \textit{with a loyal policy of rounding} \\ (j^* - 0.5) & \textit{with a neutral rounding policy} \\ j^* & \textit{with a strict rounding policy} \end{cases}$$

The minimum number of criteria for which you need to shift the estimate to achieve the desired result will be calculated using the formula:

$$k = \max\{\Delta_1, \ldots, \Delta_{t_p}\}, \tag{13}$$

where $\Delta_1, \ldots, \Delta_{t_p}$ — rounded up values of Δ for each of the subgroups of clients.

In connection with the possibility of obtaining the same values of $rang_i$ in the ranking, it becomes possible to obtain more than one solution at the end of step 4. The previous steps (with the exception of the first one) are also characterized by many variants of the solution of the problem. In this connection, it becomes necessary to construct a tree of all possible solutions and to choose the optimal one among them. After determining the optimal set of criteria that should be analyzed by the organization in order to improve the evaluation of the quality of customer service, it is suggested to define as recommendations, those sub-criteria of the found criteria that have the greatest value and the lowest satisfaction rating from the point of view of consumers. To obtain an aggregated estimate of the importance of the subcriteria, a linguistic LOWA operator with a quantification function $Q(x) = x$ is used.

4. Conclusion

The algorithms proposed in the work allow to obtain an integral linguistic estimation of the quality of services from the perspective of clients, and also to implement the reverse process for finding weak positions in the service, the strengthening of which will increase the integral quality assessment. Algorithms are implemented in software, tested on test data and ready for practical use.

References

[1] Chinaryan R A 2013 *The clients and managing client portfolio* **2** 82–101

[2] Reicheld F and Stamps R 2013 *The Sincere loyalty. The key to winning customers for life* (Moscow: Mann, Ivanov and Ferber)

[3] Parasuraman A, Berry L and Zeithaml V 1988 *Journal of Retailing* **69** 197–199

[4] Mill R 2009 *To Manage a restaurant: a textbook* (Moscow: YUNITI-DATE)

[5] Borisov A N, Alekseyev A V and Krumberg O A 1982 *Models of Decision-making on the Basis of a Linguistic Variable* (Riga: Zinatne)

[6] Azarnova T V and Titova I A 2012 *Modern Economy: Problems and Solutions* **9** 151–157

[7] Ledeneva T M 2006 *Processing fuzzy information* (Voronezh: Voronezh State University)
[8] Beliakov G, Pradera A and Calvo T 2007 *Aggregation Functions: A Guide for Practitioners* (Berlin: Springer)
[9] Yager R R 1994 *Fuzzy Sets and Systems* **67** 129–145
[10] Proskurin D K, Asnina N G, Azarnova T V and Titova I A 2017 *Economics and management control systems* **2.1(24)** 172–182

IOP Conf. Series: Journal of Physics: Conf. Series **973** (2018) 012060 doi:10.1088/1742-6596/973/1/012060

System analysis of formation and perception processes of three-dimensional images in volumetric displays

Alexander Bolshakov[1], **Arthur Sgibnev**[2]

[1] Prof. Dr.-Ing., Peter the Great St. Petersburg Polytechnic University, Saint-Petersburg, RU
[2] PG student, Yuri Gagarin State Technical University of Saratov, Saratov, RU

E-mail: aabolshakov57@gmail.com

Abstract. One of the promising devices is currently a volumetric display. Volumetric displays capable to visualize complex three-dimensional information as nearly as possible to its natural – volume form without the use of special glasses. The invention and implementation of volumetric display technology will expand opportunities of information visualization in various spheres of human activity. The article attempts to structure and describe the interrelation of the essential characteristics of objects in the area of volumetric visualization. Also there is proposed a method of calculation of estimate total number of voxels perceived by observers during the 3D demonstration, generated using a volumetric display with a rotating screen. In the future, it is planned to expand the described technique and implement a system for estimation the quality of generated images, depending on the types of biplanes and their initial characteristics.

1. Introduction

One of the development trend of three-dimensional visualization is volumetric display systems. They create volumetric image by voxels (volumetric pixels) or vectors that fill the working space of the display. Working space is determined by the structural features of the device.

Important factors that significantly affect the widespread use of volumetric displays are: the cost, image quality, the possibility of simultaneous unlimited viewing without the using of special tools (various glasses, etc.). Due to the additional opportunities for physical realization that have appeared recently, the volumetric technology is on trend now. However, there are a number of problems need to be solved for the successful use of color-intensive dynamic visualization and eliminate existing contradictions.

The first, and the most significant of them is the contradiction between the required image quality and the data flow necessary to provide it. The color and size of the resulting volume image is substantially limited, in spite of use of high-technology projection devices and computing resources. It is still difficult or cost intensive to create real time volumetric image.

The second, there are the technical, mathematical and software development of tools for quality control of a generated three-dimensional visualization. The task is quite difficult and it includes an assessment of the prospects for the technical implementation of displays of a given type, as well as the rate of perception of the generated image.

Volumetric display technology is based on the principle of scattering or emission of a light from a certain point in the working space of the display. «Inertia» of a human vision creates an illusion of the object which volume fills by volumetric pixels, so-termed voxels [1–4]. The implementation of the working surface of the display can vary. There are layered arrangement of screens, rotating screen, screen with forward-backward motion, etc.). Thus, it is relevant to consider the process of forming volumetric images from the perspective of system analysis for evaluation the quality of observed images. It could help in developing of promising solutions in the design and use of volumetric displays. The work is carried out as part of the volumetric display development project [5, 6] with the use of the method described in [7].

2. Problem definition

There are a number of tasks that must be solved to construct volumetric display and also analyze quality of formed volumetric image. There are:

- Development of a semantic network, which describes relations between volumetric displays and tasks of volumetric dynamic visualization;
- Characteristics of variables describing the process of volumetric image formation;
- Determination of the observed resolving power of volumetric displays with a moving flat screen;

Now, we are move to the consideration of the results of solving above-described points.

3. The development of the semantic network for describing the object domain related with volumetric visualization

Analysis of various types of displays allowed to offer a semantic network describing the subject area associated with the formation of three-dimensional images by volumetric displays. Its fragment is shown in Fig.1.

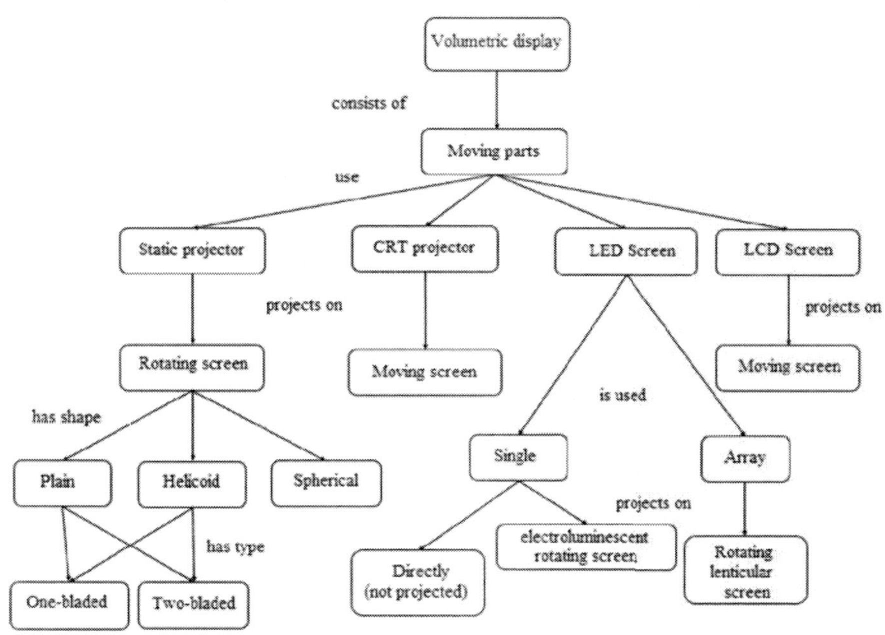

Figure 1. Fragment of semantic network

According to this scheme, the key concept is a 3D-display, which creates a three-dimensional image that has sets of properties. Among this set, we can split it into a subset of the monocular and binocular types. The monocular type has the following characteristics: parallax motion and object rotation. The binocular type, in turn, has the following characteristics: convergence and binocular disparity. If the image is formed from two-dimensional, then according to the full version of the semantic network, it has the subset, which is described by the following characteristics: color depth, resolution, size, brightness, contrast, etc.

In the future, the resulting semantic network is used to construct a mathematical model of the process of volumetric visualization, which makes it possible to automate the formation of a description of a specific version of the technical implementation of a volumetric display and the characteristics of its intermediate states obtained at various stages of the process.

As a result of the analysis, the main stages of the transformation of the source images into the final volumetric image were determined.

Source images X, sufficient to reproduce volumetric virtual object, are fed to the processing and the formation unit of primary images P. Source images include objects that are obtained using three-dimensional models, two-dimensional images from different angles or specially prepared "slices", and also video information for creating an animated visualization.

One of the most important stage in the structural sequence of the formation of the volumetric image P is the stage of creating the primary images, i.e. a cluster of two-dimensional images of the object, which forms a three-dimensional image (Fig.2).

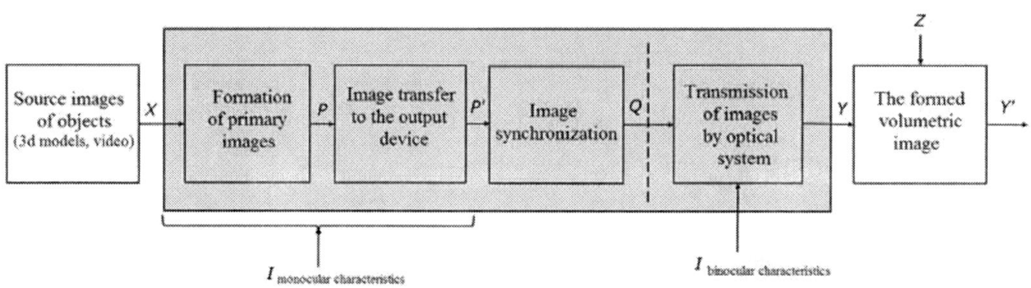

Figure 2. A generalized scheme of transformations of the three-dimensional image formation

As a result of the transfer of the primary images P using the output device, a composite image P is obtained, and the synchronization of the component parts of the images is required to correctly combine them. As a result, the image Q is synchronized in time and space, which is then transmitted to the optical system to generate the final volumetric image Y, which is directly perceived by the observer (user), taking into account his psychophysiological features Z. As the result we will get the final volumetric image Y.

4. Characteristics of variables describing the process of building 3D images

It is necessary to identify many characteristics of two-dimensional images that affect the final result. These characteristics determine the scales for measuring the values of the input and output variables of the system. It is necessary to take into account that the various technical means used impose certain restrictions on the values being formed. These restrictions also affect the final three-dimensional image and determine the quality of the perception of volume by a person.

Initial images are primary two-dimensional images, which are usually characterized by such parameters as color depth, resolution, size and brightness of the image.

The main ones are the characteristics of the images, which, in turn, depend on the parameters of the output devices: monitors, projectors or graphics cards of the primary image forming device. The quality of the output volumetric image directly depends on the degree of technical realizability of these parameters. The main criteria for the transfer of graphic information are speed, resolution, frequency, color rendition and viewing angle. In our case, these criteria are the input signals of the system.

The above characteristics affect the quality of flat images, therefore monocular and binocular characteristics should be considered as criteria for the perception of volume (Table 1). To estimate the degree of influence of image characteristics on the final volume perception, weights are introduced for image characteristics. Values of the coefficients are determined on the basis of expert estimates of the degree of influence of the system parameters on the final result. Playback with the specified values of the characteristics will allow to form a three-dimensional image to a certain extent.

We assign to the input variables the depth of the image, i.e. the ability of a person to perceive three-dimensionality of the image. Monocular attributes are taken into account when creating content. They are aimed at enhancing the perception of volume. Examples of similar monocular properties are: relative size of objects, texture gradient, relative positioning, linear and atmospheric perspectives, parallax motion; as well as binocular: binocular convergence, normal binocular disparity, occlusion.

The perception of the object is influenced by psychological factors and physiological characteristics of an individual. Therefore, one more aspect of the development of the method for the formation of a three-dimensional image is the need for its approbation at various groups of recipients.

Several output devices participate in the formation of the volume, which, when projecting the primary images onto the working plane, form a virtual volumetric object. When displaying a volumetric image through multiple devices, one of the main tasks is to synchronize them, while it is advisable to use the time of out-of-frame synchronization as the measured value.

The quality of the image formed by the autostereoscopic image system is estimated subjectively by the observer, the sensations of volume, clarity, color of the image are important at the same time. Therefore, at the first stage of the research, it is proposed to use a relatively rough ordinal scale with possible values: satisfactory, unsatisfactory, later on, more detailed estimates of the species may be used to improve accuracy: excellent, good, satisfactory, unsatisfactory.

As a unified mathematical apparatus for describing the proposed criteria and characteristics of the visualization process, it is proposed to use fuzzy models, which will allow to take into account the heterogeneity of the variables in different scales: from absolute to nominal.

5. The problem of the observed resolving power of a 3D display with a rotating flat screen

Using the above classification, it is planned to develop a mathematical model that will allow to connect the input variables of the volumetric displays with their output characteristics and find the most suitable ones. When constructing the model, a type of volumetric display with a rotating screen is selected for the subsequent analysis. It was widely used in various variations.

An example of such a display is Seelinder [8], which uses a rotating array of LEDs and a cylindrical parallax barrier. This achieves a viewing angle of 1 degree and a resolution of 1254×256 pixels.

Scientists from the University of Southern California report on the development of the Light Field Display [9], creating a three-dimensional structure (in a horizontal plane), which can be viewed from either side. The display includes a rapidly rotating mirror, with an anisotropic

Table 1. Characteristics and weights

Monocular factors	**Coefficient**
Motion parallax	0,05
Object rotating	0,05
Shadows	0,07
Perspective (direct, reverse, panoramic, spherical, air, perceptual)	0,1
Mutual overlapping of objects (occlusion)	0,03
Inhomogeneity of the shape	0,08
Gradient texture	0,07
Biniocular factors	
Convergence	0,06
Binocular disparity	0,15
2-Dimensional factors	
Size of the image area	0,07
Viewing Angle	0,05
Resolution	0,06
Lineage	0,04
Brightness	0,03
Contrast	0,03
Color bitness	0,04
Refresh rate	0,02
Total value	1

diffuser applied to it; microcontroller, engaged in decoding video signal, which is transmitted before the standard display; high-speed video projector, as well as a personal computer with the usual characteristics. The creators of the display argue that the resulting solution allows you to see a three-dimensional picture from anywhere around the display point and an unlimited number of viewers.

The resulting image has the following characteristics: refresh rate 15-20 Hz; the diameter of the image is 0.13 m; the rotation speed of the screen is 5400-7200 r/s; angular resolution - 1.25°; two colors are displayed, the resolution is 768×768 pixels. Getting a volumetric image provides a video card that creates more than 5000 slices of a three-dimensional object per second. Another similar display is RayModeler [10]. This is a full-color volumetric cylindrical display with a resolution of 96x128 pixels, developed by Sony Corporation and presented at the SIGGRAPH 2010 conference. It allows viewing a 3D image created using a rotating substrate with LEDs attached to it, from any angle and simultaneously from several points.

Consider a generalized scheme of displays of this type (Fig.2) and on its basis we will solve the problem of determining the number of human units perceived by a person depending on the radius of the working area, the resolution of the display and the distance to the observer.

Here it is necessary to make a number of remarks: from different distances and at different angles of vision, the person perceives the same image differently. The image on the display screen of 4K from a distance of 3 meters is practically the same as the image on the display of a standard notebook located at a distance that is optimal for perception of vision. If the display screen is turned azimuthally, the perceived horizontal resolution of the screen will decrease sharply. The screen of the volumetric 3D display rotates around its own vertical axis and, therefore, if the

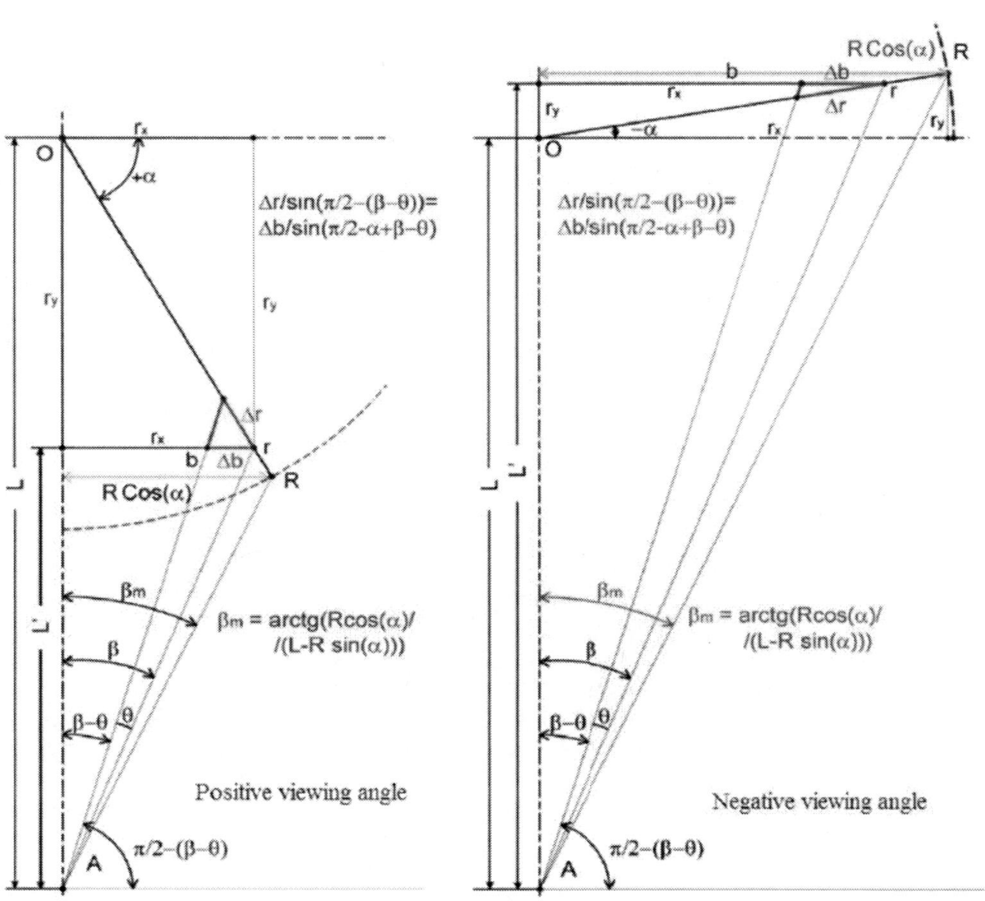

Figure 3. The scheme of viewing of volumetric display screen with a radius R from a distance L at a rotation angle α, an observation angle β, and a angle of image resolution unit θ

actual resolution of the supplied image is constant at different screen rotation angles, its perceived component will be substantially different, which must be taken into account when evaluating the quality of the 3D image being formed.

Proceeding from the geometric relations, we find Δr - the interval that is minimally discernable by the observer on the surface of the rotating screen of radius R. Let us assume that the angle at which the eye, at a distance L from the screen axis O, distinguishes two close vertical strokes is θ.

Define the interval that cuts this angle at the current radius r. We get:

$$\Delta r = \frac{2Lcos(\alpha)sin(\theta)}{cos(2\alpha - 2\beta + \theta) + cos(\theta)}, \tag{1}$$

where Δr is the visible interval of the image unit at the current angular position of the screen.

Let the number of pixels that are placed horizontally on the screen (horizontal resolution) is

Ng. Then the number of pixels placed horizontally per unit of screen width is: $ng = Ng/2R$, and the transverse interval per pixel is: $\Delta p = 2R/Ng$. The condition of the boundary of two zones (one perceptual zone directly pixels, and the other - the perceived single interval of the image): $\Delta p = \Delta r$, i.e.:

$$\frac{2R}{Ng} = \frac{2Lcos(\alpha)sin(\theta)}{cos(2\alpha - 2\beta + \theta) + cos(\theta)}. \quad (2)$$

Expressing the angle of observation β and solving this equation, we obtain three roots, which determine the zone where these units pass into each other, i.e., boundaries of calculated areas:

$$\beta_1 = \frac{4\alpha + 2\theta + 3\pi}{4} - \frac{\alpha sin(\dfrac{LNgsin(\alpha+\theta) - LNgsin(\alpha+\theta) + 2Rcos(\theta)}{2R})}{cos(2\alpha - 2\beta + \theta) + cos(\theta)},$$
$$\beta_2 = \frac{4\alpha + 2\theta - 3\pi}{4} - \frac{\alpha sin(\dfrac{LNgsin(\alpha+\theta) - LNgsin(\alpha+\theta) + 2Rcos(\theta)}{2R})}{cos(2\alpha - 2\beta + \theta) + cos(\theta)}, \quad (3)$$
$$\beta_3 = \frac{4\alpha + 2\theta + 3\pi}{4} + \frac{\alpha sin(\dfrac{LNgsin(\alpha+\theta) - LNgsin(\alpha+\theta) + 2Rcos(\theta)}{2R})}{cos(2\alpha - 2\beta + \theta) + cos(\theta)}.$$

Without reaching the boundary, at $\Delta p > \Delta r$, it is necessary to calculate the pixels directly perceived by the observer, i.e. The width of this part of the screen area should be divided by the width of the pixels. Passing through the boundary, with $\Delta p < \Delta r$, we should use the integral characteristic, i.e. determine the number of intervals Δr that fit into the remaining part of the screen.

Here is an example of the screen position in which are formed three different zones: two pixel and one integral. The behavior of the zone boundaries is depend on the angle α and angle β conduct based on the following data: $L = 250, R = 100, Ng = 800, \alpha = -60°, \theta = 1'$:

$$\frac{2 \cdot 100}{800} > \frac{2 \cdot 250cos(-60°)sin(1/60)}{cos(-2 \cdot 60 - 2\beta + 1/60) + cos(1/60)}. \quad (4)$$

Solving for the angle β, we obtain: $\beta_1 = 0.13246492, \beta_2 = 0.91502353, \beta_3 = -2.2265691$.

Thus, it can be argued that the condition $\Delta r > \Delta p$ is satisfied in the range from 0 to the first transition and from the second transition to $\pi/2$.

The length of the portion of the screen radius ΔR corresponding to a certain increment of the angle β is equal to:

$$\Delta R = \Delta R_2 - \Delta R_1 = \frac{Lsin(\beta + \Delta\beta)}{cos(\alpha - \beta - \Delta\beta)} - \frac{Lsin(\beta)}{cos(\alpha - \beta)}. \quad (5)$$

For ΔR can be written as:

$$\Delta R = \frac{Lcos(\beta)}{cos(\alpha)cos(\beta) + sin(\alpha)sin(\beta)} d\beta. \quad (6)$$

The number of visible image units within this interval is equal to the ratio: $\Delta n = \Delta R/\Delta r$. Substituting the previously obtained expression for Δr, we have:

$$\Delta n = \frac{cos(2\alpha - 2\beta + \theta) + cos(\theta)}{2Lcos(\alpha)sin(\theta)} \cdot \frac{Lcos(\beta)}{cos(\alpha - \beta)} d\beta. \quad (7)$$

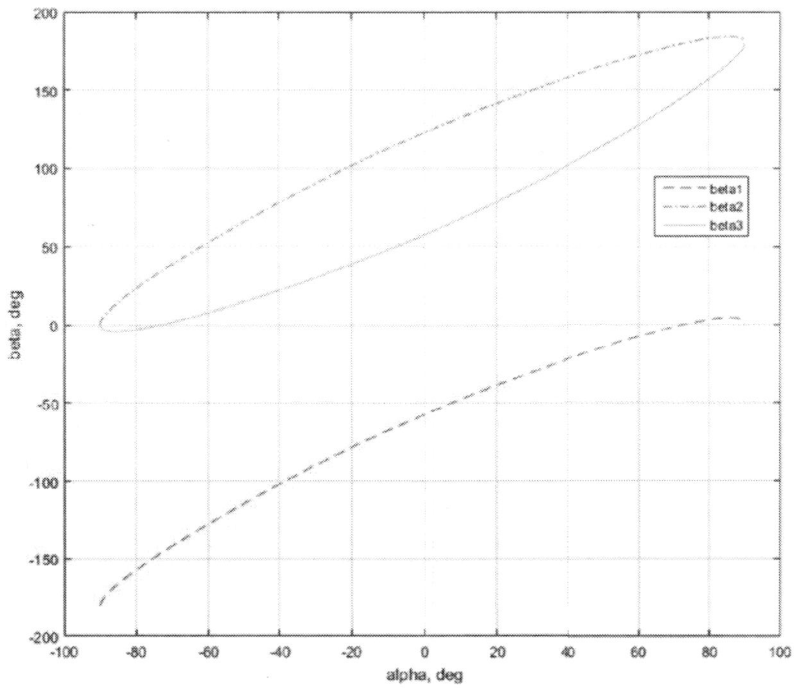

Figure 4. Graph of changing the edges of zones depending on the angle of the screen rotation

Expanding the expressions obtained into three separate relations and forming integrals of them, we obtain:

$$n_1 = \frac{cos(2\alpha+\theta)}{4cos(\alpha)sin(\theta)} \int_{\beta_3}^{\beta_2} \frac{4cos(\beta)^3 - 2cos(\beta)}{cos(\alpha-\beta)} d\beta,$$

$$n_2 = \frac{4sin(2\alpha+\theta)}{4cos(\alpha)sin(\theta)} \int_{\beta_3}^{\beta_2} \frac{sin(\beta)cos(\beta)^2}{cos(\alpha-\beta)} d\beta, \quad (8)$$

$$n_3 = \frac{cos(\theta)}{2cos(\alpha)sin(\theta)} \int_{\beta_3}^{\beta_2} \frac{cos(\beta)}{cos(\alpha-\beta)} d\beta.$$

The number of observed units (pixels) of the image along the horizontal direction of the screen with the given parameters of the system (display and observer) is equal to the sum of the solutions of the integral expressions after substituting the found limits:

$$\begin{aligned} n_1 &= \frac{cos(2\alpha+\theta)}{4cos(\alpha)sin(\theta)}((sin(3\alpha)-sin(\alpha))(ln(cos(\alpha-\beta_3))-ln(cos(\alpha-\beta_2)))+(\beta_3-\beta_2)\times \\ &\times(cos(3\alpha)-cos(\alpha))+sin(\alpha+2\beta_2)-sin(\alpha+2\beta_3)), \\ n_2 &= \frac{4sin(2\alpha+\theta)}{4cos(\alpha)sin(\theta)}((ln(cos(\alpha-\beta_2))-ln(cos(\alpha-\beta_3)))(cos(3\alpha)-cos(\alpha))+(\beta_3-\beta_2)\times \\ &\times(sin(3\alpha)-sin(\alpha))+(cos(\alpha+2\beta_3)-cos(\alpha+2\beta_2))), \\ n_3 &= \frac{cos(\theta)}{2cos(\alpha)sin(\theta)}(sin(\alpha)(ln(cos(\alpha)cos(\beta_2)+sin(\alpha)sin(\beta_2))-ln(cos(\alpha)cos(\beta_3)+ \\ &+\sin(\alpha)sin(\beta_3)))-(\beta_3-\beta_2)cos(\alpha)). \end{aligned} \tag{9}$$

Determine the number of image units, depending on the rotation angle of the screen $\alpha°$. For the initial values we take: $L = 250, R = 100, \alpha = 0° - 90°, \beta = 0° - 90°, \theta = 1', Ng = 800$. We define the behavior of the limits of the transition zones β_1, β_2, and β_3 and, by results, construct a graph of these functions (Fig.3).

We calculate the change in the total number of observed image units, including unmodified pixels, depending on the angle of rotation of the screen and the results of calculations, we construct the corresponding graph (Fig.4).

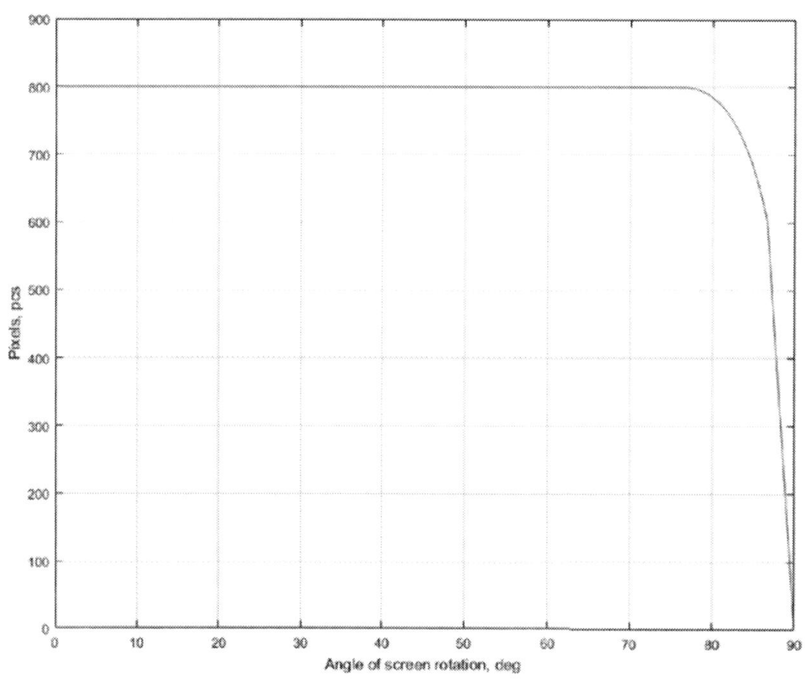

Figure 5. Graph of the dependence of the number of pixels on the angle of the screen rotation

Multiplying the resulting virtual horizontal screen resolution by its vertical component, which is determined in a similar way, taking into account the absence of azimuthal variations, we determine the screen resolution perceived by the observer for each of its azimuthal positions.

IOP Conf. Series: Journal of Physics: Conf. Series **973** (2018) 012060 doi:10.1088/1742-6596/973/1/012060

Knowing the screen update time, it is possible to calculate the actual number of voxels visible by the observer during the demonstration of the volumetric image, and also to estimate this characteristic for a particular display and determine the optimal values according to the conditions of formation, data transmission and observation.

Conclusion

Thus, the article attempts to structure and describe the interrelation of the essential characteristics of objects in the area of volumetric visualization. An example of an estimate of the total number of voxels perceived by observers during the demonstration of 3D images formed using a volumetric display with a rotating screen is given. The calculations can be useful for optimizing the operating parameters of volumetric volumetric displays, and in particular for determining the relationship between the quality of the generated image and the flow of generated and transmitted information.

The next step is to use developed method for a comprehensive assessment of the quality and process of perception of volumetric images displayed using autostereoscopic devices of other types.

Acknowledgment

This work is supported by a Grant (1995ГС1/26878) of Innovation Promotion Fund. ("Фонд содействия инновациям", *http://www.fasie.ru*

References

[1] Barry G Blundell and Adam J Schwarz 2013 *Volumetric three-dimensional display systems* (NY:John Wiley & Sons Inc) p 330

[2] Nagano K, Jones A 2013 *An Autostereoscopic Projector Array Optimized for 3D Facial Display* SIGGRAPH Emerging Technologies

[3] Nagano K, Jones A, Unger J et al. 2015 *An Automultiscopic Projector Array for Interactive Digital Humans* In SIGGRAPH ACM Press

[4] Annen T, Matusik W, H. Pfister, Seidel H, Zwicker M, 2006 *Distributed rendering for multiview parallax displays* Proc. SPIE 6055, Stereoscopic Displays and Virtual Reality Systems XIII

[5] Bolshakov A 2016 *Volumetric display testing unit for visualization and dispatching applications* (Springer: Lecture notes in computer science) vol 9812 pp 234-242

[6] Bolshakov A, Sgibnev A, Zhelezov M, Melnikov A, 2016 *Razrabotka ob'emnogo displeja dlja reshenija zadach vizualizacii i dispetcherizacii* Automation and Remote Control. vol 7 pp 31-34 (In Russian)

[7] Bolshakov A, Nikonov A *patent "Volumetric display and method for forming three-dimensional images* (http://www.freepatent.ru/patents/2526901, last accessed 27 November 2017)

[8] Endo(Yendo) T, Kajiki Y, Honda T and Sato M 2000 *Cylindrical 3-D Video Display Observable from All Directions* (Proc. Pacific Graphics)

[9] Jones A, McDowall I, Yamada H, Bolas M and Debevec P 2007 *Rendering for an interactive 360° light field display* (ACM Trans Graph 26(3) 40)

[10] Ito K, Kikuchi H, Sakurai H, Kobayashi I, Yasunaga H, Mori H, Tokuyama K, Ishikawa H, Hayasaka K 2010 *360-degree autostereoscopic display* (ACM SIGGRAPH Emerging Techncal)

IOP Conf. Series: Journal of Physics: Conf. Series **973** (2018) 012061 doi:10.1088/1742-6596/973/1/012061

Mathematical models and methods of assisting state subsidy distribution at the regional level

Yu V Bondarenko[1], **T V Azarnova**[1], **I L Kashirina**[1] **and I V Goroshko**[2]

[1] Department of Applied Mathematics and Mechanics, Voronezh State University, Voronezh, Universitetskaja Ploschad, 1, Russia

[2] The Academy of Management of Ministry Internal Affairs, Moscow, Ulitsa Zoi i Aleksandra Kosmodemjanskih, 8, Russia

E-mail: bond.julia@mail.ru

Abstract. One of the most common forms of state support in the world is subsidization. By providing direct financial support to businesses, local authorities get an opportunity to set certain performance targets. Successful achievement of such targets depends not only on the amount of the budgetary allocations, but also on the distribution mechanisms adopted by the regional authorities. Analysis of the existing mechanisms of subsidies distribution in Russian regions shows that in most cases the choice of subsidy calculation formula and its parameters depends on the experts' subjective opinion. The authors offer a new approach to assisting subsidy distribution at the regional level, which is based on mathematical models and methods, allowing to evaluate the influence of subsidy distribution on the region's social and economic development. The results of calculations were discussed with the regional administration representatives who confirmed their significance for decision-making in the sphere of state control.

1. Introduction

One of the essential strategic goals of any region is to provide a decent standard of living to the population. To achieve this goal, it is necessary to deal with a number of socio-economic issues existing in a certain region, the most urgent of which are high unemployment, low per capita income, slow growth of gross regional product etc.

We have studied the current situation in a number of Russian regions and the history of their development, and our analysis shows that there are not only social, but also economic reasons for most problems. These problems can be solved by way of creating effective mechanisms of cooperation between the authorities and the local enterprises. By providing financial support to businesses regional authorities can encourage enterprises' management to take an active part in tackling the most important socio-economic issues.

State support comprises various economic, legal and political measures aimed at promotion of entrepreneurship. Both, economic agents and public authorities, can benefit from that. One of the most common forms of state support in the world is subsidization [1–3].

By providing direct financial support to businesses, local authorities get an opportunity to set certain productivity and effectiveness goals and to control their achievement. As for companies' economic performance indicators that have special significance for most regions, we should mention profit, number of employees, output and average monthly nominal pay.

For regional authorities a successful result of economic support of business is the achievement of the set social and economic performance goals at the regional level. These goals are normally set for the following performance indicators: gross regional product, production index, average monthly nominal pay, number of profit-making enterprises, number of jobs, unemployment level etc. The target values for such performance indicators are reflected in the strategy and programs for social and economic development of the region.

In Russia subsidies are normally distributed at the federal, as well as regional level. At the federal level experts determine the general order of giving out subsidies, the industries to be supported and the amount of federal funds to be provided. Regional authorities, in their turn, provide for the order and concrete procedures of subsidies allocation. It must be stressed that the successful achievement of the set objectives depends not only on the amount of the budgetary allocations, but also on the distribution mechanisms adopted by the regional authorities.

The analysis of the existing mechanisms of subsidies distribution in Russian regions shows that in most cases the choice of subsidy calculation formula and its parameters depends on the experts' subjective opinion. It has been mentioned by researchers that the decisions taken are often not sufficiently grounded and highly subjective. This allows for information manipulation. Besides, little attention is paid to specific characteristics of the supported enterprises. As a result, budgetary funds are misspent at all levels [4, 5]. Therefore, it is crucial to develop scientifically grounded approaches to elaboration of subsidies allocation procedures in order to improve their objectivity and effectiveness.

The complexity of this task is associated with the specific aspects of the region as a sophisticated economic and social system. Among these aspects we should mention the variety of elements and connections, the unique character of each enterprise, the complexity of enterprises' behavior and the conflict between economic concerns of entrepreneurs and social concerns of the regional authorities, the lack of statistics available, high level on uncertainty etc. Mathematical models and methods are effective instruments helping to take decisions concerning such systems' management.

The aim of this research is to develop a subsidy distribution algorithm that will assist regional authorities in elaboration of the procedures of financial support provision. The algorithm will be based on mathematical models and methods allowing to:

– assess (on the basis of the statistical data available) the influence of the possible variants of the subsidy's distribution on the performance indicators of each enterprise having applied for financial support;

– assess the influence of the possible variants of the subsidy distribution on the regional development indicators;

– make well-grounded choice of the subsidy distribution rule with the aim of achieving the best possible social and economic results in the region's development;

– determine the target values of the enterprises' effectiveness indicators.

The theoretical part of our research was based on the scientific works dedicated to the problems of economic mechanisms theory (E. Maskin, R. Myerson and others [6, 7]); the research of market participants' motives and behavior (G. Stigler, J. Tirole, V. L. Tambovtsev, A. Y. Shastiko and others [8, 9]); works on development of the optimal mechanisms for Russian State regulation system (A. A. Yakovlev, F. T. Aleskerov, V. Z. Mazloyev and many others [4, 10, 11]).

The possibilities of mathematical methods application for state support optimization have been widely discussed and investigated from different points of view. In this respect we need to mention the works by V. N. Burkov, D. A. Novikov and K. A. Bagrinovsky dedicated to the development of resource planning and distribution mechanisms at various management levels [12, 13]; research by A. A. Yakushev concerning the problem of optimizing the mechanism for providing state financial support to small businesses by way of a probability classification model application [14]; works by G. A. Ougolnitsky, in which the author examines general hierarchical models for regional management [15] and many others. The main ideas and models of cooperation between the regional authorities and enterprises in the sphere of social and economic development indicators elaboration were discussed in the article by I. V. Goroshko and Y. V. Bondarenko [16].

The approach to the problem of subsidies distribution optimization proposed in this paper is characterized by special attention paid to the specificity of regional management and the peculiarities of the enterprises existing in the region. For each enterprise a separate optimization model for rational distribution of own funds and subsidies received is created, with due regard to the enterprise's individual characteristics. The models' parameters are determined on the basis of statistics available for the authorities of the region. The models allow regional authorities to calculate, with a certain level of accuracy, the anticipated values of economic performance indicators (output, profit, human resources, fixed assets cost) for each separate variant of subsidy distribution. To make the final decision concerning the procedure of subsidy distribution, the influence of the enterprises' performance on social and economic indicators of the region's development is evaluated. The main criterion of the optimal choice is maximization of the weighted total of the region's development indicators' standardized values.

2. Materials and methods

2.1. General description of the approach

To provide a formalized description of the problem of subsidies' distribution, we need to consider a subsystem of the region's socioeconomic system, which is comprised of the following elements:

1) regional enterprises performing their economic activities in the territory of the region;
2) regional government.

Each enterprise is associated with a governing body (administration of the enterprise), which is interested in getting maximum possible profit in the market of a certain region.

The regional government (regional authorities or regional administration) is responsible for handling strategic and tactical issues and determining the course of the region's development that will ensure decent living standards for the people of the region. The key indicators of social and economic development of the region are prescribed in the development strategy and programs as target indicators. The regional government sets target values for each target indicator.

The success in achieving each of the prescribed target values depends on a number of factors. We will further consider social and economic indicators that are determined by the results of certain economic entities' performance. Let us suppose that in order to achieve target values of performance indicators the authorities have taken a decision to provide state support in the form of a subsidy to certain entities.

For each subsidy, the regional authorities determine its recipient's performance indicators and set their target values. The subsidy is given to an enterprise at the beginning of the year, while the results achieved are analyzed at the end of the year. Let us introduce the following notation:

- B — the budget allocations for a subsidy provided in the current financial period (year) in the amount of B^f from the federal budget and B^r from the regional budget, $B = B^f + B^r$;
- M — the number of enterprises eligible for the subsidy allocation ($m = 1, \ldots, M$);
- $\overline{R}$ — the set of social and economic indicators of the region's development depending on the results of the enterprises' economic activities;
- J — the number of indicators of $\overline{R}$ set, ($j = 1, \ldots, J$);

- $\widetilde{R}_j$ — target values of the region's development indicators to be achieved during this period (year), $j = 1, \ldots, J$;
- N — the number of the enterprises' performance indicators determining the region's development indicators ($n = 1, \ldots, N$);
- $\widetilde{N}$ — the set of the enterprises' performance indicators used for effectiveness evaluation;
- $(s^m)^0 = \left((s_1^m)^0, (s_2^m)^0, \ldots, (s_N^m)^0\right)$ — indicators of enterprise m performance in the last accounting period, the values of which are taken by the authorities from the statistical reports and accounting statements ($m = 1, \ldots, M$).

Distribution of subsidies among the entities is performed in accordance with certain rules and calculation (distribution) formulas. We will consider each rule to be a function, the arguments of which are the enterprises' performance indicators, and the value is a nonnegative number — the amount of the subsidy provided. The formulas, as well as the parameters' values, may differ from rule to rule.

The simplest formula for subsidies' calculation is represented by a linear dependence:

$$F(s) = (\alpha_f + \alpha_r) \cdot s,$$

where s is the enterprise's performance indicator selected for the subsidy's calculation; α_f, α_r — are the rule's parameters (the subsidy amounts provided respectively by the federal and the regional budget per unit of measure of the enterprise's performance).

A rule can also have a more complex representation. For example, different parameters may be applied for different intervals of indicator s:

$$F(s) = \begin{cases} \alpha_1 \cdot s, & 0 \leqslant s \leqslant s_1, \\ \alpha_2 \cdot s, & s \geqslant s_1, \end{cases}$$

were α_1, α_2 are the rule's parameters, s_1 is the threshold value of the performance indicator.

As we have already noted, the arguments of the rule are the enterprises' performance indicators. Traditionally, in Russian regions these indicators are the amount of gross product sold and (or) intended for further processing by the recipient of the subsidy; cattle population; the amount of direct costs etc.

Naturally, not all the enterprises' indicators are considered while distributing subsidies in compliance with a certain rule, but for notational convenience we will assume that the indicators that are not mentioned in the rules are included in the formulas with zero coefficients.

In the general case, we will assume that the region's authorities have provided an H set of subsidy calculation rules that we will refer to as Γ:

$$\Gamma = \{F_1(s_1, \ldots, s_N), \ldots, F_H(s_1 \ldots, s_N)\}. \tag{1}$$

The amount of the financial support B_h^m provided in compliance with the rule F^h to enterprise m depends on the value of its performance indicators for the last accounting period:

$$B_h^m = F_h\left((s_1^m)^0, \ldots, (s_N^m)^0\right).$$

Naturally, each of the rules of subsidy distribution must be bearable for the budget. It means that the total amount of the subsidy must not exceed the amount of the budgetary provisions:

$$\sum_{m=1}^{M} B_h^m \leqslant B. \tag{2}$$

IOP Conf. Series: Journal of Physics: Conf. Series **973** (2018) 012061 doi:10.1088/1742-6596/973/1/012061

Different rules of subsidies' calculation have various effects on the performance of regional enterprises, and thus on the indicators of the region's social and economic development. The main objective of the regional authorities *is to work out such a rule for the subsidy's calculation (permissible for the budget)* $F^* \in \Gamma$, *that will allow for maximum approximation of the target values of the region's development indicators.*

We will call the above mentioned problem "*the problem of subsidies' distribution*". We propose to address this urgent management issue by way of following a seven-stage *algorithm of subsidy distribution among the enterprises of the region.*

Algorithm of subsidy distribution among the enterprises of the region

Stage 1. Specify the set of social and economic indicators depending on the results of the subsidy recipient's performance, $\overline{R}$ set generation. Determine the admissible deviations from the target values.

Stage 2. Determine the set of subsidy calculation rules (1) differentiated by formulas and parameters' values.

Stage 3. Evaluate the results of economic performance of the enterprises receiving state support in accordance with each rule of calculation.

Stage 4. For each variant of subsidy calculation, evaluate the changes in the selected indicators of the region's development program.

Stage 5. Determine an integrated index for comparing the rules and for selecting the best possible variant. Calculate the value of the integrated index for each distribution variant.

Stage 6. Choose the rule for financial support calculation, which provides for the maximum value of the integrated index.

Stage 7. Determine target performance values for each recipient of the subsidy.

Let us contemplate the mathematical tools used at different stages of the algorithm implementation.

At the first stage, the regional government analyzes the dependence between socioeconomic indicators of the region's development and the indicators of the subsidy recipient's performance. $\overline{R}$ set is specified. For each indicator the quantitative dependency between its value and the enterprises' performance indicators is determined:

$$R_j = R_j(s^1, s^2, \dots, s^M), \quad j = 1, \dots, J, \tag{3}$$

where $s^m = (s_1^m, \dots, s_N^m)$ — the set of performance indicators for enterprise m ($m = 1, \dots, M$).

We should also mention that the formulas (3) can either be used for calculations or reflect the dependency between the indicators in the form of regression equations.

For each R_j indicator the regional administration has determined the admissible deviation from the target value $\Delta_j \geqslant 0$. If an increase in R_j indicator value is considered advantageous for the region, the admissible range of values is $Z_j = \left[\widetilde{R}_j - \Delta_j, +\infty\right)$. Conversely, when decrease in the indicator value is considered advantageous, the admissible range of values is $Z_j = \left(-\infty, \widetilde{R}_j + \Delta_j\right]$.

At the second stage of the algorithm implementation the set of possible rules for the subsidy calculation is determined. The formulas for the rules are developed by the experts from the regional administration on the basis of general requirements and recommendations. We suggest that one of the following strategies is to be used for determination of the rules' parameters variants:

- experts' participation. In this case experts from the regional government determine the possible parameters for subsidy distribution rules on the basis of their own experience and expertise;
- parameters' generation. According to this strategy, each rule parameter is considered to be a random variable, the values of which are evenly distributed within the admissible variations interval. Different variants of random variable realization provide for different values of the parameters;
- mixed strategy combining the first two options.

The result of the second stage implementation must be a set of possible subsidy calculation rules:

$$\Gamma = \{F_1(s_1, \ldots, s_N), \ldots, F_H(s_1, \ldots, s_N)\}.$$

The third stage will be carried out with the assistance of a mathematical tool — *a model of rational finds distribution by the enterprise's management.* The mathematical model is created on the basis of the statistical data concerning the performance of each subsidy recipient, which is available to the regional administration. This model provides the regional authorities an opportunity to calculate the values of the enterprise's performance indicators which can be achieved through rational distribution of own funds and subsidies received. In this context rationality is understood as the natural urge to get maximum possible profit.

In order to provide a general representation of the model, let us consider the example of enterprise m. We will assume that at the beginning of the financial period (year) the enterprise's management has own funds in the amount of B_m and, besides, the enterprise receives a subsidy, in compliance with rule h, in the amount of B_m^h.

The funds available are distributed by the enterprise's management between current production and expansion (development) programs, which causes changes in the values of economic performance indicators. Each case of funds distribution can be represented as a manipulated value $u_m\ (B_m, B_m^h)$.

The model of rational funds distribution by the enterprise's management has to do with determining the management strategy that will provide for the maximum possible profit:

$$\pi^m\left(s^m, u_m(B_m, B_h^m)\right) \to \max.$$

On condition that

$$s_n^m = \varphi_n^m\left((s^m)^0, u_m(B_m, B_m^h)\right), \quad n1, \ldots, N, \tag{4}$$

where $\pi^m(\cdot)$ — the enterprise's profit, $s^m = (s_1^m, \ldots, s_n^m)$ — the set of economic performance indicators (variable models); φ_n^m — the rule of changing the value of indicator n for enterprise m.

The model described above (4) is an example of an optimization problem. The solution to this problem is represented by the optimal values of enterprise m performance indicators:

$$\widetilde{s}_h^m = \left((s_1^m)^h, \ldots, (s_n^m)^h\right).$$

After the third stage for each subsidy recipient the values of performance indicators that can be achieved through rational distribution of funds are determined with regard for each subsidy calculation rule F_h (where $h = 1, \ldots, H$):

$$\widetilde{s}_h = \left(\widetilde{s}_h^1, \widetilde{s}_h^2, \ldots, \widetilde{s}_h^M\right), \quad h = 1, \ldots, H. \tag{5}$$

We suggest that these values should be considered as the basis for the assessment of the influence that each of the subsidy distribution rules has over the enterprises' performance indicators.

The fourth stage has to do with the analysis and quantitative assessment of the changes in the region's development indicators resulting from the changes in the enterprises' performance indicators. The assessment is to be made separately for each variant of the subsidy's distribution:

$$R_j^h = R_j\left(\widetilde{s}_h^1, \widetilde{s}_h^2, \ldots, \widetilde{s}_h^M\right), \quad j = 1, \ldots, J \; h = 1, \ldots, H. \tag{6}$$

The subsidies' calculation rules where each R_j^h indicator value belongs to acceptable range Z_j are chosen for further consideration. The set of indices used in these rules will be denoted by H':

$$H' = \left\{h : R_j^h \in Z_j,\; j = 1, \ldots, J\right\}.$$

To make the final choice of the subsidy distribution rule you need to work out an integrated index allowing to evaluate the degree of conformity to the target values of the region development indicators:

$$\Phi(R_1, R_2, \ldots, R_J) = \sum_{j=1}^{J} \lambda_j \cdot R_j^0(s^1, s^2, \ldots, s^M), \tag{7}$$

where λ_j — is the priority index of the region development indicator j determined by the regional administration experts $\left(\lambda_j \geqslant 0,\; \sum_{j=1}^{J} \lambda_j = 1\right)$; $R_j^0\left(s^1, s^2, \ldots, s^M\right)$ — the standardized value of indicator j.

If an increase in j indicator value is considered advantageous for the region, its standardized value is calculated by the following formula: $R_j^0\left(s^1, s^2, \ldots, s^M\right) = R_j\left(s^1, s^2, \ldots, s^M\right)/\widetilde{R}_j$. Conversely, when decrease in the indicator value is considered advantageous, the standardized value is calculated as follows: $R_j^0\left(s^1, s^2, \ldots, s^M\right) = \widetilde{R}_j / R_j(s^1, s^2, \ldots, s^M)$.

The region's authorities have to choose such a subsidy distribution rule $F_{h^*} \in \Gamma$, where the integrated index (7) acquires the maximum possible value:

$$\Phi\left(R_1^{h^*}, \ldots, R_J^{h^*}\right) = \max_{h \in H'} \Phi\left(R_1^h, \ldots, R_J^h\right).$$

Now from the optimal values of the enterprises' performance indicators calculated for F_{h^*} rule in accordance with model (4) we need to choose those indicators' values that are used for effectiveness evaluation (included in $\widetilde{N}$ set) . We suggest that these values should be considered target values.

The results of the algorithm application are the following:

- subsidy distribution rule F_{h^*} provides for maximum approximation of the region's development indicators' values to the target levels;
- target performance levels of each subsidy recipient are calculated for F_{h^*} rule in compliance with the models (4).

It has to be noted that the results of the algorithm application are auxiliary tools used for decision making. The experts working for regional authorities can adjust them as needed.

IOP Conf. Series: Journal of Physics: Conf. Series **973** (2018) 012061 doi:10.1088/1742-6596/973/1/012061

2.2. Refinement of the model of rational funds distribution by the enterprise management
For practical realization of the algorithm of subsidy distribution among the regional enterprises, the third-stage models (4) have to be refined. Each model's restrictions are based on the results of research concerning the problems of Russian regions' development, which has allowed to determine the key enterprises' performance indicators.

Among these indicators we need to mention: profit, output, average monthly nominal pay, number of employees and fixed assets.

For convenience, when mentioning enterprise m, we are going to omit hereinafter designation m. Let us introduce the following notation for the selected performance indices:

- L — number of employees (human resources); ω — average annual nominal pay;
- K — fixed assets;
- π — the enterprise's profit.

The funds available are comprised of own and borrowed funds in the amount of B_0 as well as the subsidy B_h, calculated in compliance with F_h rule.

The parameters regulated by the management of the enterprise are:

- y — output;
- δ — the share of money intended for fixed assets acquisition;
- ν — the share of money intended for creation of new jobs.

The model of rational funds distribution by the enterprise's management will look like this:

$$\pi \to \max .$$

The following restrictions must be taken into account:

- the enterprise's manufacturing capability:

$$y \leqslant f(K, \omega \cdot L);$$

- allocation of own and donated capital for fixed assets acquisition:

$$K = (1-\gamma)K_0 + \delta(B_0 + B_h);$$

- allocation of own and donated capital for additional human resources:

$$L = L_0 + \nu(B_0 + B_h)/\omega';$$

- generation of current period profit:

$$\pi = y - (b \cdot y + \omega \cdot L/(1-\rho) + \gamma \cdot K_0);$$

- financial restrictions:

$$(\delta + \nu) \cdot (B_0 + B_h) + b \cdot y + \omega \cdot L/(1-\rho) + \gamma \cdot K \leqslant B_0 + B_h;$$

- restrictions on variables: δ, ν, $y \geqslant 0$.

For our model we use the following notation: $f(K, \omega \cdot L)$ — production function; γ — rate of fixed assets depreciation; ρ — payroll tax; b — manufacturing costs per one unit; K_0, L_0 — fixed assets and number of employees at the beginning of the year, respectively; ω' — cost of workplace creation.

The model provided has been worked out for an optimization problem with variables y, δ, ν. The target function is linear, while the system of restrictions contains an inequation providing for the enterprises' manufacturing capabilities, with a non-linear production function. The problem with a non-linear (and, in the general case, non-convex) production function should be solved by Sobol method that was described in [17].

For each subsidy distribution formula the ideal solution provides for such values $\widetilde{y}_h$, $\widetilde{\delta}_h$, $\widetilde{\nu}_h$ that will allow to calculate the enterprise's profit $\widetilde{\pi}_h$, the number of employees $\widetilde{L}_h$, fixed assets $\widetilde{K}_h$.

2.3. Results and discussion

To apply the subsidy distribution algorithm towards the companies of the region, a software package has been developed in C# environment. The package includes: a database containing information on a number of enterprises from Voronezh Region, a set of computation programs providing solutions for non-linear optimization problems and a procedure of generating parameters for subsidy calculation rules. We have tested this algorithm on training and real-life tasks, and verified the models' validity.

Let us consider, for example, a subsidy provided in Voronezh Region to agricultural companies (other than private farms) for partial compensation of costs associated with prevention of African swine fever epizooty.

The target indices of the development program and their target values are shown in Table 1.

Table 1. Targets indices, planned values and calculation formulas for the development program.

Targets of the development program	Target values (2017)	Calculation formulas	Indices
Pork production (live weight/meat on hoof)	130.5 ths. tons	$$R_1 = \sum_{m=1}^{M} y^m,$$ where y^m is pork production at enterprise m	Pork production
Creating new vacancies 100	100 staff positions	$$R_2 = \sum_{m=1}^{M} \Delta L^m,$$ where ΔL^m is the number of jobs created at enterprise m	The number of jobs created
Maintenance of index wages	19 867 rub/month	$$R_3 = \frac{1}{12 \cdot \mu} \sum_{m=1}^{M} \Phi^m,$$ where R_3 is average monthly nominal pay, Φ^m — annual payroll budget of enterprise m, μ — average number of employees per year at M enterprises	Average annual pay, number of employees

IOP Conf. Series: Journal of Physics: Conf. Series **973** (2018) 012061 doi:10.1088/1742-6596/973/1/012061

In the Government Decree of Voronezh Region No 132 dated 15 February 2017 the following order of subsidy calculation is provided for:

$$F = \alpha \cdot y^0, \tag{8}$$

where F is the subsidy amount paid to the recipient (rub); α — the subsidy rate; y^0 — the amount of pork sold by the subsidy recipient (in 2016).

The subsidy rate approved by the Voronezh region government in 2017 was $\alpha = 674$ rubles/ton.

Table 2 contains a fragment of a chart showing agricultural enterprises from Voronezh Region, the amounts of pork sold in 2016 and the subsidy amounts for 2017.

Table 2. Subsidy amounts for the enterprises of Voronezh Region (fragment).

Agricultural enterprises	Amount of pork sold by the subsidy recipient (thousands of tons) in the year 2016	Subsidy, ths. rub
"APK Agroeco"	21.2	14 288.8
"Agroresurce-Voronezh"	2.4	1 617.6
"SGC"	6.7	4 515.8
"Vishnevsky"	4.8	3 235.2
"Alfa"	3.8	2 561.2
"Agricultural Enterprise Moskovskoe"	3.6	2 426.4
"IX Pyatiletka"	2.5	1 685.0
"Kalacheevsky Meat Plant"	2.4	1 617.6
...		
Total		48 750

The actual amount of the budgetary provisions for subsidy in 2017 is 48 750 thousand rubles. All in all, 84 thousand tons of live weight pork was sold by the regional enterprises in 2016. The fugures show that the total amount of subsidy calculated by formula (8) must be 56 616 thousand rubles, but due to the lack of funds some of the enterprises were refused financial support. For organizational reasons, most of these enterprises are usually small companies whose role in the region's development is undoubted. Moreover, outbreaks of African swine fever are especially common in small farms.

In this calculation experiment we used type (8) rules with different rate values. We made calculations on the basis of the statistical data collected for 45 large- and medium-scale hog-raising farms of Voronezh region. Some of the calculations results are presented in Table 3.

Let us analyze the results of our calculations. It must be noted that the target value of the average monthly nominal pay was calculated with allowance for the models' restrictions.

The subsidy rate $\alpha = 750$ is not effective enough and cannot be recommended for further consideration.

The first variant offered (implying the current subsidy rate) provides for substantial support offered to important market players, which is likely to maximize the output growth. At the same time, the number of new jobs is not so impressive.

If the subsidy rate is lowered, more small to medium-scale businesses will be eligible for financial support, which will allow them to expand and to create new jobs.

IOP Conf. Series: Journal of Physics: Conf. Series **973** (2018) 012061 doi:10.1088/1742-6596/973/1/012061

Table 3. Some results of the calculation experiment.

Subsidy rate, rub/ton	Calculated values of target indicators	
	Pork production, tons (R_1)	Number of new vacancies, staff positions (R_2)
674	94 531.4	67
590	92 008.6	75
750	87 989.4	62
550	89 001.2	77

Let us develop an integrated criterion for choosing the subsidy distribution rule:

$$\Phi(R_1, R_2) = 0.3 \cdot R_1/130500 + 0.7 \cdot R_2/100.$$

The value of the integrated criterion for the rate $\alpha = 674$ is 0.686 points, while for the rate $\alpha = 590$ it is 0.736 points, while for $\alpha = 550$ it is 0.743 points. In accordance with the chosen integrated criterion, the following rule for subsidy calculation may be offered:

$$F = 550 \cdot y^0.$$

It must be noted that due to the importance of the problem under consideration, the results of the calculations must be verified and, if needed, amended by the government experts.

When we discussed this approach with the government representatives, the following advantages were mentioned:

1) substantial assistance in generating an expert opinion concerning subsidy distribution rules;

2) cutting down the time needed for taking a decision, and improving judgement at the same time;

3) the ability to carry out calculation experiments with different rules of subsidy distribution.

The main disadvantage of the method is associated with the necessity to keep a database. In future the method could be improved by way of developing instruments that would allow to consider various typical situations. We are also going to introduce our software product.

3. Conclusion

In this article we offer a new approach to assisting subsidy distribution at the regional level, which is based on mathematical models and methods. The algorithm proposed allows:

- to work out a rule for subsidy calculation, which will allow the region's development indices to come as close as possible to the target values;
- determine the target values of performance and effectiveness for each subsidy recipient.

The algorithm proposed is based upon a mathematical model of rational finds distribution by the enterprise's management. The restrictions of the models reflect the basic production dependencies, while the parameters are unique for each enterprise. The model provides an opportunity to calculate such values of performance indicators for each case of subsidy allocation, which can be reached on condition of rational funds distribution. The experiment with different variants of subsidy calculation rules provides an opportunity to make a justified choice of the rule and target values of the enterprises' performance indices. The use of statistics available for the region's authorities ensures practical implementation of the approach proposed.

The calculation experiments carried out prove that it is not only necessary, but also possible to make regional subsidy distribution mechanisms more objective and justified. Discussion of the approach and the experiment results with the regional administration shows its practical significance. The basic opportunities for further development of the method are now being considered.

References

[1] Ktzoev A B 2012 *SShA i Kanada: jekonomika, politika, kul'tura (USA and Canada: Economics, Politics, Culture)* **2** 105–109
[2] Goodwin B, Michra A and Ortalomagne F 2011 *The buck stops where? The distribution of agricultural subsidies* NBER Working Paper No. 16693 (Electronic Materials) (Cambridge: The National Bureau of Economic Research)
[3] Semina L A 2013 *Vestnik Altajskogo gosudarstvennogo universiteta (Bulletin of the Altai State University)* **11(109)** 117–122
[4] Mazloev V Z and Kzoev A B 2012 *Agroprodovol'stvennaja politika Rossii (Agro-food Policy of Russia)* **2** 61–67
[5] Gusalova A A and Plaeva A A 2014 *Izvestija gorskogo agrarnogo universiteta (News of the Mountain Agrarian University)* **3(51)** 241–244
[6] Maskin E S 2008 *American Economic Review* **98(3)** 567–576
[7] Myerson R 1986 *Econometrica* **54** 323–358
[8] Maskin E and Tirole J 1990 *Econometrica* **58** 379–409
[9] Tambovtsev V L 2014 *Economic theory of informal institutions* (Moscow: RG-Press)
[10] Yakovlev A A 2010 *Voprosy gosudarstvennogo i municipal'nogo upravleniya (State and Municipal Government Issues)* **2** 5–22
[11] Aleskerov F and Subochev A 2013 *Journal of Global Optimization* **56** 737–756
[12] Burkov V N and Burkova I V 2014 *Automation and Remote Control* **75(3)** 470–480
[13] Burkov V N, Novikov D A and Kuznetsov N A 2002 *Automation and Remote Control* **63(12)** 1947–65
[14] Yakushev A A 2016 *European Social Science Journal* **6** 89–101
[15] Yakushev A A 2016 *European Social Science Journal* **6** 89–101
[16] Goroshko I V and Bondarenko Y V 2015 *Problemy upravleniya (Control Science)* **1** 63–72
[17] Sobol I M 2006 *The choice of optimal parameters for multi-criteria problems* (Moscow: Drofa)

IOP Conf. Series: Journal of Physics: Conf. Series **973** (2018) 012062 doi:10.1088/1742-6596/973/1/012062

Application of cellular automatons and ant algorithms in avionics

A V Kuznetsov[1,4], N I Selvesiuk[2], G A Platoshin[2], E V Semenova[3]

[1]Voronezh State University, 1 Universitetskaja square, Voronezh, 394018, Russia
[2]State Research Institute of Aviation Systems, 7 Viktorenko street, Moscow, 125319, Russia
[3]Voronezh State Medical University, 10 Studencheskaja Street, Voronezh, 394036, Russia

[4]Corresponding author avkuz@bk.ru

Abstract. The paper considers two algorithms for searching quasi-optimal solutions of discrete optimization problems with regard to the tasks of avionics placing. The first one solves the problem of optimal placement of devices by installation locations, the second one is for the problem of finding the shortest route between devices. Solutions are constructed using a cellular automaton and the ant colony algorithm.

1. Introduction

Development of electronics and digital computer technics has led to quick increase in the number of placing aboard a plane electronics (so-called avionics). Multitude of electronic devices requires standards of its physical arrangement and data transmission buses. In 1990-thyearssetof heterogeneous united electronic devices was changed for system of computers and general purpose interfaces which look like plug-in architecture of personal electronic computer. Nowadays such systems of placing aboard a plane electronics are called integrated modular avionics (IMA).

IMA devices, called modules, fulfill several functions of on-board systems simultaneously using standardized software. Peripheral devices are connected with general system of input-output interfaces. Common broadband network with high bandwidth consolidates all IMA modules. Simultaneous realization of several functions of aircraft by one module enhances the capacities of modern processors. This makes it possible to reduce the number of aircraft electronic devices and at the same time to minimize mass and cost of avionic system, and also costs for developing, production and maintenance.

Nowadays systems of IMA devices are implemented in all modern civil and military aircrafts, business jets and helicopters. The second-generation systems of IMA devices, so-called distributed integrated modular avionics (DIMA), are implemented in aircrafts of the type Boeing 787 andA350 (see the [2]).

2. Objectives

The present paper seeks to achieve two objectives of DIMA: spatial distribution of devices and identifying of peripheral cables. Software of devices' provision identifies spatial distribution of DIMA in general structure of aircraft electronic devices. The devices are identified with locations which are limited by volume. The factor which limits process of displaying devices and its locations is set of resources required by devices and provided by locations. In addition, it is necessary to take into account the constraints of spatial distribution. This algorithm yields a scheme like represented in figure 1.

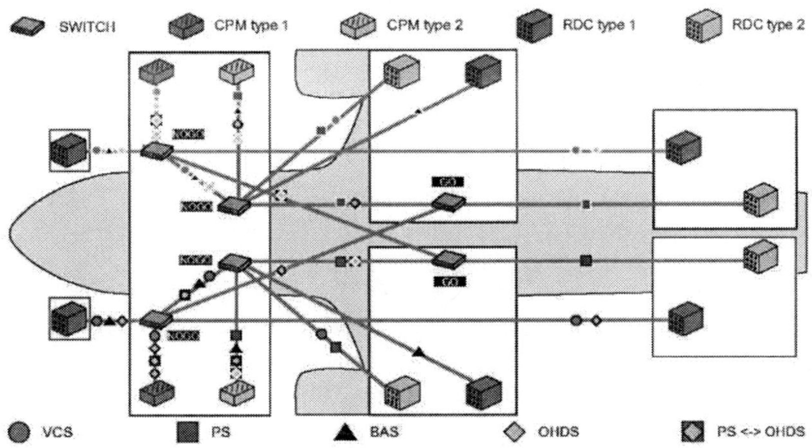

Figure 1. Sample designation of cable routes and locations

Algorithm of quasi-optimal spatial distribution of devices by location with the use of modified social potential is proposed as the same as the one adopted for modeling and simulation of swarm of agents in the [3]. The very objective of optimization is solved through the cellular automata model, this idea of using it for resource allocation was derived from the paper [4].

The peripheral cables connects peripherals to DIMA devices. This objective is being addressed under the limits on the number of cables and connections between the wires in cable routes. If two tasks which require peripherals should be segregated, it is estimated that peripheral cables also should be segregated, i.e. wires must not overlap anywhere in their routes. Identifying of peripheral cables yields segregated route for each peripheral cable. Such route is described together with all cable routes, connections and locations which it runs through. The full set of all routes must not exceed the amount of resources in each of cable routes, connections or locations, without any two routes disrupting the requirement of segregation. A search algorithm of quasi-optimal designation of cable routes is proposed as the one of ant colony seeking optimal way in graph of the same type as described in the[5].A significant characteristic of algorithm is simultaneous coexistence of several ant colonies.

3. An objective of minimization of a cable length

The idea of algorithm has been borrowed in its essence from the paper [6],where in some sense similar algorithm is applied for creation of random landscape with certain properties in order to verify a search algorithm of optimal way from the [7].

IOP Conf. Series: Journal of Physics: Conf. Series **973** (2018) 012062 doi:10.1088/1742-6596/973/1/012062

Suppose we are given an equal grid in $\mathbb{Z}^3$ and subset $U \subset \mathbb{Z}^3$. There are set of tools $V = \{v_i | i = \overline{1,n}\}$, which should be attributed to the grid cells U. We are given for each tool $v \in V$ list of tools $C(v) \subset V$, which should be connected to v directly. It is necessary to place tools through the grid in a such way that the distance between all connected tools would have been minimal.

Since $U \in \mathbb{Z}^3$ is, generally speaking, nonconvex and even disconnected one, we define distance ρ on U as geodesic distance, i.e. distance between cells x and y is equal to length of minimal lattice path in between, and if there is no such path, then distance is assumed to be infinite.

Let's introduce the concept of attractive-repulsive potential. Suppose for each tool $v_k \in V$ is given a function $\varphi_k: U \to \mathbb{R}_{\geq 0}$ and also a function $\varphi_0: U \to \mathbb{R}_{\geq 0}$. These functions are meeting the following requirements ($C_k \geq 0, \alpha_l > 0$) .

1. Suppose $d_0(i,j,l)$ is a distance from cell $(i,j,l) \in U$ to border ∂U (suppose, ∂U is not contained in U), then $\varphi_0(i,j,l)$ is increasing monotonically with the decreasing of $d_B(i,j,l)$. Let's suppose for ease that

$$\varphi_0(i,j,l) = -\frac{C_0}{d_0(i,j,l)^{\alpha_0}}.$$

2. Suppose there is tool v_k in the cell (i_1,j_1,l_1). Suppose $d(i_1,j_1,l_1;i_2,j_2,l_2) = d_k(i_2,j_2,l_2)$ is a distance between cells (i_1,j_1,l_1) and(i_2,j_2,l_2) . Then $\varphi_k(i,j,l)$ can be expressed as

$$\varphi_k(i,j,l) = \frac{C_{k1}}{d_k(i,j,l)^{\alpha_{k1}}} - \frac{C_{k2}}{d_k(i,j,l)^{\alpha_{k2}}}.$$

We can assume that tool $v_s \in V$ in a cell (i,j,l) is influenced by attractive-repulsive potential

$$\Phi_s(i,j,l) = \sum_{k=1}^{n} \delta_s(\varphi_k(i,j,l)),$$

where

$$\delta_s \circ \varphi_k = \begin{cases} 0, s = k, \\ 0, v_s \notin C(v_k), \\ 1, v_s \in C(v_k) \wedge s \neq k. \end{cases}$$

Let us set $P(i,j,l)$ as a tool (more exactly, as a tool number) in cell (i,j,l) and take $P(i,j,l) = 0$, if there is no tool in this cell. Let us set $\mathcal{D} = \{(i,j,l) | i,j,l = \overline{-1,1}\}$.

An algorithm is proposed as follows.

0.Tools from V are allocated arbitrarily in U.

1. For each cell $(i,j,l) \in U$, it is given the list $\{\varphi_k(i,j,l) | k = \overline{0,n}\}$.

2. Cellular automaton starts its operation with following local function of transition:

2.1.Tool v_s in cell (i,j,l) finds such possible direction $d = (d_1,d_2,d_3) \in \mathcal{D}$ that $\Phi_s(i + d_1, j + d_2, l + d_3) \to \max$,

2.2.If cell $(i + d_1, j + d_2, l + d_3)$ is free, then v_s moves to this cell,

2.3.If cell $(i + d_1, j + d_2, l + d_3)$ is not free, but its tool is planning to move in cell (i,j,l), then these tools change cells,

2.4. If cell $(i + d_1, j + d_2, l + d_3)$ is not free and its tool is not planning to move in cell (i,j,l), then we mark d as impossible and go to algorithm step 1.1.

3.We return to algorithm step 1.

4.If it's been more than largest possible number of cycles or automaton has become stable, algorithm is finished.

4. Designation of cable routes

To find optimal cable route we will use algorithm of the same type as an ant colony one. Suppose $V' \subseteq V$ is set of tools which have to be connected and hub devices. Also suppose we are given matrix A for tools from V', which need to be connected; if tools v_i and v_j have to be connected, $a_{ij} = 1$, otherwise $a_{ij} = 0$. Also suppose we are given matrix B; if tools v_i and v_j have to be hardware segregated, $b_{ij} = 1$, otherwise $b_{ij} = 0$.We give the list of possible communication equipment (switches) $L(v)$ for each tool $v \in V$.

Let's define a graph $\Gamma = (V, E, \psi)$, where E is set of connections between tools, $\psi: V \to U$ is one-to-one function of tool and location, which was linked on the previous steps, and its subgraph $\Gamma' = (V'', E'', \psi)$, where points are tools from V' and switches. Every ant $m \in K_i$ at every instant t has list $M_i(m,t)$ of the nodes Γ' already visited, in which it does not come back. Ants from K_i have "eyesight" in the sense that they tend to visit such points v_j, where $a_{ij} = 1$. Let us denote set of target points $m \in K_i$ as $T(t, v_i) = L(v_i) \backslash M_i(m,t)$.

Every $v_i \in V'$ corresponds to ant colony K_i, which marks edges Γ' by its own pheromone. The colonies can "be at enmity" with each other, so that pheromone of single ant colony is repellant for hostile one, what is necessary for segregation of the routes. To limit number of connections to one node let's assume that pheromone, the concentration of which is greater than a certain level, begins to act as a repellant. Consequently every edge (r, s) between the points v_r and v_s corresponds to ordinate vector of pheromone $\tau_{rs}(t) = (\tau_{rs}^1(t), \dots, \tau_{rs}^{|V'|}(t))$, $\tau_{rs}^j(t)$ is a pheromone level of the colony K_i at time t.

Let's define a function of pheromone detection for an ant from the colony number k

$$\theta_{rs}(k,t) = \begin{cases} \sum_{i=1}^{|V'|} \tau_{rs}^i(t), \forall(i) \quad b_{ik} = 0, 0 < \sum_{i=1}^{|V'|} \tau_{rs}^i(t) \leq m_{rs}, \\ \xi, \quad \sum_{i=1}^{|V'|} \tau_{rs}^i(t) = 0, \\ 0, \text{ otherwise.} \end{cases}$$

Here ξ is a small random number, $m_{rs} > 0$ is the limit number of connections via edge (r, s).

5. Conclusion

The report proposes algorithms which have made it possible to find quasi-optimal solutions of the objectives of spatial distribution of devices by location and hardware routing between these devices. In contrast to the standard algorithms of discrete optimization based on the total check, proposed methods offer suboptimal solutions which require much less computer time.

References

[1] P. Bieber, F. Boniol, M. Boyer, E. Noulard, and C. Pagetti. New challenges for future avionic architectures. AerospaceLab, 4(11), 2012.
[2] Buzdalov D.V., Zelenov S.V., Kornykhin E.V., Petrenko A.K., Strah A.V., Ugnenko A.A., Horoshilov A.V. Instrumentalniye sredstva proektirovaniya sistem integrirovannoy modulnoy avioniki[Development tools of integrated modular avionic systems]. In *Trudyi Instituta sistemnogo programmirovaniya RAN* [*Proc. of Institute for System Programming of the RAS*],2014, volume 26, issue 1, pages 201–230 (in Russian).
[3] M.P. Jones and D.D. Dudenhoeffer. A formation behavior for large-scale micro-robot force

deployment. In *Winter Simulation Conf.*, volume 01, pages 972–982, Los Alamitos, CA, USA, 2000. IEEE Computer Society.

[4] Epaminondas Sidiropoulos and Dimitrios Fotakis. Spatial optimization and resource allocation in a cellular automata framework. In *Cellular Automata-Simplicity Behind Complexity*, pages 67–86. InTech, 2011.

[5] Marco Dorigo, Gianni Di Caro, and Luca M. Gambardella. Ant algorithms for discrete optimization. *Artif. Life*, 5(2):137–172, April 1999. URL: http://dx.doi.org/10.1162/106454699568728

[6] Kuznetsov A.V. Random landscape generation with a specified configurational entropy and Total Edge. In *Vychislitelnyie tehnologii* [*Computational Technologies*], 2017,volume 22, № 4, pages 4–10.

[7] Kuznetsov A.V. A Model of the joint motion of agents with a three-level hierarchy based on a cellular automaton. In *Computational Mathematics and Mathematical Physics*, 2017, volume 57, № 2, pages 340–349 (in Russian).

[8] Björn Annighöfer. *Model-based Architecting and Optimization of Distributed Integrated Modular Avionics*. Aachen, ShakerVerlag, 2015. – 312 p.

Parameter meta-optimization of metaheuristics of solving specific NP-hard facility location problem

E S Skakov[1] and V N Malysh[2]
[1] Novolipetsk Steel, Metallurgov sq. 2, 398040, Lipetsk, Russia
[2] Department of Humanities and Natural Sciences, Lipetsk Branch of The Russian Presidential Academy of National Economy and Public Administration, Internatsional'naya str. 3, 398050, Lipetsk, Russia
E-mail: wallkirya@mail.ru

Abstract. The aim of the work is to create an evolutionary method for optimizing the values of the control parameters of metaheuristics of solving the NP-hard facility location problem. A system analysis of the tuning process of optimization algorithms parameters is carried out. The problem of finding the parameters of a metaheuristic algorithm is formulated as a meta-optimization problem. Evolutionary metaheuristic has been chosen to perform the task of meta-optimization. Thus, the approach proposed in this work can be called "meta-metaheuristic". Computational experiment proving the effectiveness of the procedure of tuning the control parameters of metaheuristics has been performed.

1. Introduction

1.1. The problem of placing elements of developing information systems

In this paper the tuning of the parameter values of the metaheuristics of solving the NP-hard facility location problem is carried out in the particular case of NP-hard facility location problem — the problem of placing elements of developing information systems (PPEDIS). PPEDIS is as follows: there is an information system (IS) of some configuration (let's call it the original IS). There is a need for modernization of the system. There are N_{cl} customers of information system. Upgraded IS should satisfy the needs of all users by connecting them to the elements (nodes) of information system. There is a predetermined set of potential sites (N_{ps}) where IS elements can be installed. There are N_{types} possible types of IS elements (with different characteristics) available for installation. The task is reduced to minimizing the cost of the transition from the initial state of the IS to the upgraded state with considering a number of restrictions. In [24], [25] and [26] metaheuristic algorithms for solving this problem are described: Evolutionary Algorithm (EA), Bee Colony Algorithm (BCO), Ant Colony Optimization (ACO), Multi-Start Algorithm (MS), Simulated Annealing (SA), Tabu Search (TS). A special case of PPEDIS is solved in this paper: the base stations location problem during planning and optimization of wireless data network.

1.2. Literature overview

The evaluation of the quality of the optimization algorithms is an important issue. Any solution found by some optimization algorithm can be estimated from the point of view of two characteristics:

IOP Conf. Series: Journal of Physics: Conf. Series **973** (2018) 012063 doi:10.1088/1742-6596/973/1/012063

- the time of finding the solution;
- the value of the objective function of the founded solution.

It is not possible to say exactly which combination of these characteristics is desirable. The ideal case is to find the exact optimal solution in extremely short time interval. However, for NP-hard problems it is impossible. The essence of NP-hardness implies the exponential time of finding the exact solution by the method of exhaustive search. Therefore, finding a solution using metaheuristics involves a compromise between the speed of finding and the accuracy of the obtained solution.

Parameters of metaheuristic algorithm that a decision-maker can specify are called control parameters or free parameters of the algorithm [27]. In [19] the vector (set) of values of free parameters of the algorithm is called the "strategy of the algorithm".

For each metaheuristic the following statement is true: different sets of control parameters of the algorithm provide solutions of different quality.

We can compare the quality of solutions obtained by the different metaheuristics. Also we can compare the solutions obtained by the same algorithm but with a different set of control parameters values.

By tuning of free parameters values of the optimization algorithm we mean a set of actions and procedures aimed at finding a set of values of free parameters in which the objective function (for the minimization task) reaches the smallest value. Theoretically, it is possible to use an exhaustive search to solve the problem of parameters tuning. However, if we have n parameters, each of which can take k values, we will need to perform n^k experiments. I.e. the task of tuning the free parameters has exponential time complexity.

There are two possible approaches to solving the problem of optimizing the parameters of metaheuristic algorithms:

- offline optimization;
- online optimization.

During offline optimization the values of the various parameters have been fixed before the metaheuristic is executed. The best set of parameter values is selected as the tuning result at the end of the series of experiments. The main disadvantage of offline optimization is high requirements for the time required for the procedure of tuning parameters. The advantage is universality: good method of offline optimization implies its application for several metaheuristic algorithms.

During online optimization the values of the control parameters are changing directly in the process of solving. Most algorithms that implement online optimization use dynamic approach or self-adaptive approach [27], [2]. The obvious lack of optimization directly in the search process is the non-universality of this approach. For example, in our case, it would be necessary to come up with modifications for each of the algorithms of PPEDIS solution. Strategies for self-adaptation of parameters applicable, for example, to evolutionary algorithms, are not suitable for a class of local search algorithms, etc. [27]. A classic example is the dynamic meta-optimization of the probability of a mutation for an evolutionary algorithm. It is based on the idea that in the early steps of the algorithm the probability must be large enough (for the latitude of the search space) gradually decreasing in the process of the algorithm functioning (to obtain a sufficiently accurate solution). Thus in order to carry out the dynamic adaptation correctly it is necessary to have at least an approximate idea of how each parameter should change during the functioning of the algorithm.

Evolutionary metaheuristic is a very young approach to solving optimization problems ([15], [16], [17]). Accordingly the idea of applying an evolutionary approach to parametric identification is even younger. Among the most significant works devoted to the identification of parameters of systems of various complexity we note the works of Demidova and Petrova [6],

Huang and Wang [18], Chang [5]. Studies of such scientists as Cachon and Vazquez [4], Ahmad et al. [1] are directed to the application of a genetic algorithm to optimizing the parameters of artificial neural networks.

The questions of tuning the parameters of metaheuristic algorithms by means of evolution metaheuristics are considred in the works of Gaertner and Clark [11], Wun et al. [31]. Real coding EA is used in both works. Note that in these works the algorithm of parameter adjustment is not given. In these articles the approach is simply indicated in words: the using of a genetic algorithm to optimize the parameters of the ant colony optimization [11] and bee colony algorithm [31]. Karpenko also considers the problem of meta-optimization ([19]) but does not consider the meta-metaheuristic approach as a way of solving this problem.

Thus this paper proposes the new numerical method based on an evolutionary approach which permits to optimize the control parameters for any metaheuristic algorithm aimed at solving PPEDIS.

2. Materials and methods

2.1. System analysis of the process of tuning control parameters for metaheuristic optimization algorithms

The problem of finding the parameter values of some metaheuristic algorithm can be formulated as an optimization problem. Since the metaheuristic algorithms are optimization algorithms, the offline optimization of metaheuristic parameters can be called "meta-optimization". Moreover, since meta-optimization can be performed using metaheuristics, the approach used in this paper can be called "meta-metaheuristic".

The scheme of meta-optimization is shown in Figure 1 [27]. Meta-optimization includes two levels: the meta-level and the base level. At the meta-level metaheuristic works with solutions that represent the parameters of the base-level metaheuristics, which need to be optimized. Solution x_i on the meta-level encodes specific parameter values optimized by the base-level metaheuristics (for example, the length of the tabu list, the probabilistic threshold, etc. in the tabu search). On the meta-level the objective function $F_{meta}(X)$ is associated with the best of the solutions found on the base level using metaheuristics with parameters encoded by the solutions x_i $(i = 1, 2, ..., n\_set)$. X is the set of all possible sets of parameters; n_set is the total number of different sets of parameters, the power of the X. Metaheuristics on the base level work with solutions that are direct solutions of the initial optimization problem (PPEDIS in our case). The objective function F_{base} is used by the base-level metaheuristics to designate the objective function of a particular solution with specific set of parameters x_i $(i = 1, 2, ..., n\_set)$. For the minimization problem we have the formula

$$F_{meta}(X) = \min_{i=1,2,...,n\_set} F_{base}(x_i). \tag{1}$$

The final solution to the meta-optimization task is the set of parameters x_{best_set}:

$$best\_set = arg \min_{i=1,2,...,n\_set} F_{base}(x_i). \tag{2}$$

2.2. Method of tuning control parameters for metaheuristic optimization algorithms

Further in this paper we will describe the evolutionary method of meta-optimizing the values of the parameters of some metaheuristics of solving the problem of placing elements of developing information systems (PPEDIS).

If the method of meta-optimization is considered as a system, then from the point of view of classification it can be called:

- abstract;

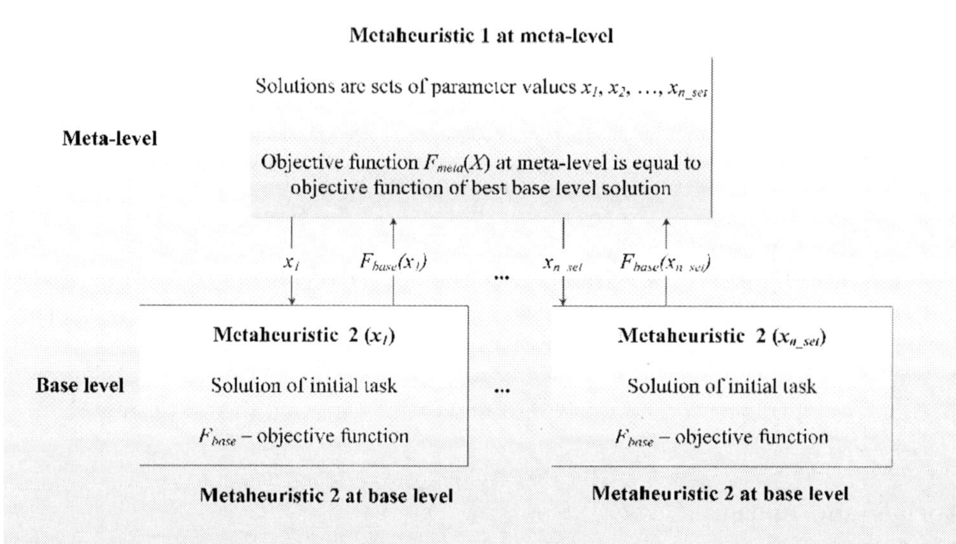

Figure 1. Meta-optimization using meta-metaheuristic approach

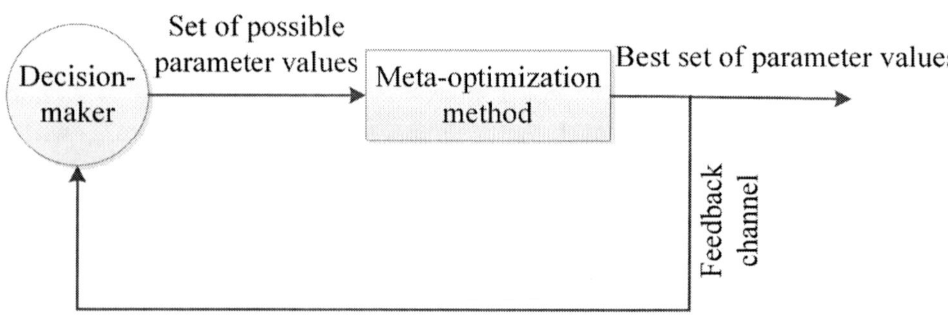

Figure 2. The decision-making process for meta-optimization

- mathematical;
- multi-element;
- heterogeneous;
- dynamic.

Figure 2 schematically shows the process of meta-optimization of metaheuristic algorithm. The decision-maker (DM) specifies a set of possible values of the free parameters of the original basic algorithm. The method of meta-optimization takes this set as an input and at the end of the calculations returns to the decision-maker the vector of values with which the base algorithm shows the best result. DM on the basis of the analysis of the meta-optimization results has the opportunity to start the calculations again, having transferred to the input another set of possible values of the control parameters.

The method of meta-optimization is a two-component system (see Figure 3). It includes a base-level algorithm for solving PPEDIS and a unit of evolutionary computations that actually produces meta-optimization. The evolutionary computation block passes the vector of parameter

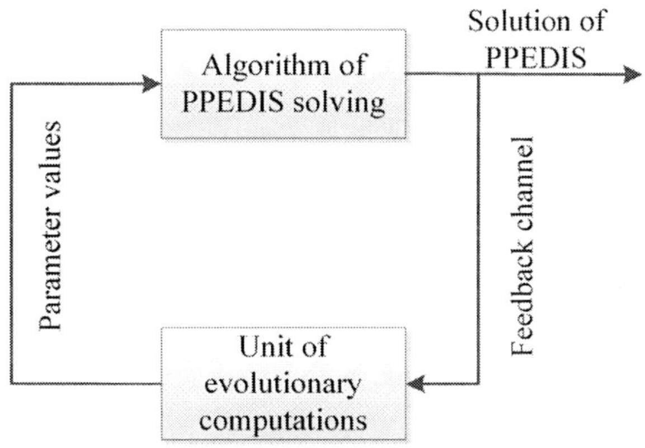

Figure 3. Method of meta-optimization

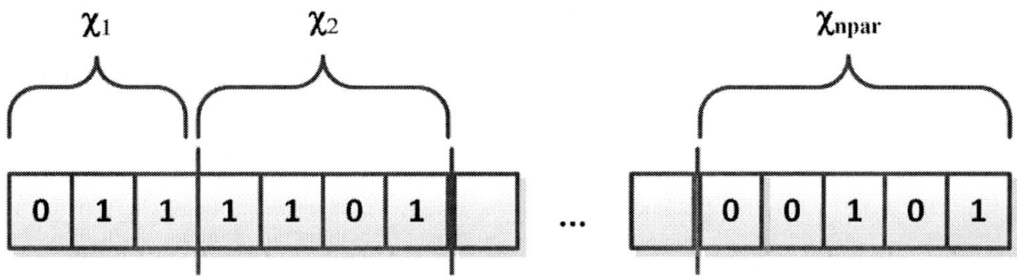

Figure 4. Representation of parameters of the algorithm in the form of a chromosome (example)

values to the input of the base-level algorithm. As an output the base-level algorithm gives the decision of PPEDIS (its objective function) obtained using the input parameter vector. The block of evolutionary computations is responsible for

- evaluating the solutions issued by the base-level algorithm;
- the organization of the population in each generation of the evolutionary algorithm;
- transferring sequentially all the individuals of the population to the base-level algorithm.

Let's pose the problem of meta-optimization as follows. We have PPEDIS of $N_{cl} \times N_{ps} \times N_{types}$ dimension. We are trying to solve it using some optimization algorithm. The functioning of the algorithm is determined by a set of *npar* variable control parameters: $\chi_1, \chi_2, ..., \chi_{npar}$. Each of the parameters is an element of some finite set: $\chi_1 \in A_1, \chi_2 \in A_2, ..., \chi_{npar} \in A_{npar}$. $A_1, A_2, ..., A_{npar}$ — the sets of possible parameter values from which the selection will occur, they are known in advance. Each set of parameters of a particular algorithm can be represented as a chromosome (see example in Figure 4).

Each "square" (bit) in Figure 4 can take the value 0 or 1. The chromosome consists of genes. Genes are not individual bits, but segments of chromosome (in Figure 4 they are separated by the vertical line). Each gene encodes one of the parameters of the algorithm. Each gene representing a sequence of bits is a binary number. We denote these numbers as follows: $z_1, z_2, ..., z_{npar}$. Because each of the sets $A_1, A_2, ..., A_{npar}$ is finite, we can regard any of these sets as a one-dimensional array (vector). Accordingly the genes $z_1, z_2, ..., z_{npar}$ encode the parameters $\chi_1, \chi_2, ..., \chi_{npar}$ position in arrays $A_1, A_2, ..., A_{npar}$.

IOP Conf. Series: Journal of Physics: Conf. Series **973** (2018) 012063 doi:10.1088/1742-6596/973/1/012063

Example in Figure 4: $z_1 = 011_2, z_2 = 1101_2, ..., z_{npar} = 00101_2$. We must translate these numbers into a decimal numeral system: $z_1 = 3, z_2 = 13, ..., z_{npar} = 5$. This means that the chromosome shown in Figure 4 represents the following: parameter χ_1 is the 4th element of array A_1, χ_2 is the 14th element of array A_2, χ_{npar} is the 6th element of array A_{npar} (elements of arrays will be numbered starting from 1).

The use of binary coding of parameters has the following advantages:

- Evolutionary algorithm with binary coding is much simpler than real-coding algorithm.
- Binary coding permits to combine the tuning of both integer and real parameters. Before starting the tuning for each of the parameters a decision-maker simply has to specify a set of possible values for this parameter.
- Binary coding permits to configure non-numerical (symbolic) parameters of metaheuristic algorithms. By non-numerical parameters we mean the parameters of the algorithm which are not numbers but describe "verbally" some of its properties. For example, there are two types of a bee colony algorithm: with and without global memory. According to the approach proposed in this work, we can code the parameter "a type of BCO from the point of view of memory usage" using a gene consisting of one bit, where 0 will correspond to the use of memory, and 1 — not to use (or vice versa, it is absolutely unimportant).

2.3. Pseudocode

Below there is a general scheme of the method for tuning control parameter values for a metaheuristic algorithm of the PPEDIS solution based on an evolutionary algorithm with binary coding.

(1) Formulate a test problem based on the solution of which the parameters will be tuned.

(2) Introduce the notation: n_{run} — the number of starts of the algorithm of finding the PPEDIS solution to determine the fitness of one individual from the proposed method; *time_alg* — time allocated to one launch of the algorithm for finding the PPEDIS solution; *time_overall* — time allocated to the free parameters tuning method.

(3) Initialize the parameters of the evolutionary algorithm of the control parameters tuning (population size N_{pop}, probability of mutation p_{mut}, etc.).

(4) Compose the list of control parameters: determine $npar$, sets $A_1, A_2, ..., A_{npar}$ (sets can be real, integer or non-numeric).

(5) For each integer i from 1 to $npar$: the gene encoding χ_i must consist of w_i bits, where w_i is the minimum of the numbers for which $|A_i| \leq 2^{w_i}$ is satisfied.

(6) The chromosome that represents an individual consists of W bits ($W = \sum_{i=1}^{npar} w_i$).

(7) Formation of the first generation. Create N_{pop} individuals as follows: each individual is a binary vector, each bit is 0 or 1 (generated randomly).

(8) Evaluate the fitness of the individuals of the population: launch the algorithm of solving PPEDIS *n_run* times for *time_alg* seconds each run and averaging the objective function of the results.

(9) Select N_{child} pairs of parents using tournament selection.

(10) Crossover. At the end of this step we should have N_{child} individuals of descendants.

(11) Apply the mutation operator to the children (the result of the mutation in EA with binary coding is the logical inversion of the bit).

(12) Formation of a new generation. It is carried out by selecting the N_{pop} of the best individuals from a population consisting of N_{pop} parents and N_{child} descendants.

(13) If the current running time is less than *time_overall*, then go to step 8.

(14) Return the set of parameter values presented by the best individual in the population as a final solution.

The fitness of each individual in the method of parameter optimization is calculated as follows. Each individual consists of a single chromosome that encodes the complete set of control parameter values of some PPEDIS solution algorithm. For each chromosome we have to start the PPEDIS solution algorithm n_run times ($time\_alg$ seconds per each run). After that we get n_run values of the objective function F_{base}. Denote them by F_base_i, where $i \in \{1, 2, ..., n\_run\}$. Denote the target function for the individual in the method of finding the control parameters as $Target\_func$. It is calculated by the formula:

$$Target\_func = \sum_{i=1}^{n\_run} \frac{F\_base_i}{n\_run}. \tag{3}$$

The fitness of the individual is inversely proportional to the value of $Target\_func$. The $Target\_func$ of the best individual in the last generation of the population is equal F_{meta}.

It should be noted that the method proposed in this paper is most effective in the situations where the number of control parameters (as well as the number of possible values of each of the parameters) is large. It is obvious that when the number of parameters is small, the simplest and the most effective strategy is to look over all possible combinations of control parameters.

We call the evolutionary algorithm with binary coding, presented above, the first phase of the method for optimizing the values of the control parameters of metaheuristic algorithms for solving PPEDIS. In case of availability of additional computer time it is possible to start the second phase of the parameter tuning method based on EA with real coding. During the second phase those genes that are integer and non-numerical will not be changed. There will be further optimization of only real parameters.

2.4. Parameter tuning for meta-algorithm

In this paper a metaheuristic is used to configure the control parameters of other metaheuristics. This fact raises the question of the necessity of tuning the control parameters directly for the evolutionary algorithm for tunning parameters (so-called meta-algorithm). It is necessary to determine the size of the population, the selection method, the probability of mutation (p_{mut}), etc. Theoretically the following infinite procedure is possible:

(1) Meta-algorithm is developed for tuning parameters of metaheuristics that solve PPEDIS.

(2) It is necessary to develop meta-meta-algorithm for tuning the control parameters for the evolutionary meta-algorithm.

(3) It is necessary to develop meta-meta-meta-algorithm for tuning the control parameters for the evolutionary meta-meta-algorithm.

...

And so on. Obviously, it is necessary to stop optimizing the parameters and hard-fix values of the algorithm parameters at some of the above steps. It is best to do this immediately in the first step. In this paper it is proposed to use an evolutionary meta-algorithm with the following parameters to tune the optimal values of the control parameters of metaheuristics of PPEDIS solution:

- $N_{pop} = 10$.
- $N_{child} = 10$.
- Selection method = "tournament".

IOP Conf. Series: Journal of Physics: Conf. Series **973** (2018) 012063 doi:10.1088/1742-6596/973/1/012063

- Method of forming a new population = "creation of an intermediate population of N_{pop} parents and N_{child} children with the transfer N_{pop} best individuals to the next generation".
- Number of crossover points (n_points) = 2;
- $p_{mut} = 0.05$.

2.5. "Default" values of the parameters

Below there are the "default" values of the control parameters of the metaheuristics that were applied to find PPEDIS solution. "Default" parameter values are the values of free parameters, which are recommended in the classic works on methaheuristics.

(1) Simulated Annealing ([20], [21], [30]):
  (a) cooling factor (α) = 0.99;
  (b) number of iterations at each temperature (N_iter) = 10
  (c) the probability of the worst solution acceptance at the early steps of the algorithm (p_0) = 0.9.
(2) Tabu Search ([13], [14], [12]):
  (a) tabu search length (l) = 100;
  (b) probabilistic threshold (p) = 0.05;
  (c) diversification usage = "yes".
(3) Multi-Start ([10], [23], [3]):
  (a) maximum number of iterations in a row without improving the objective function (N_iter_max) = 100;
  (b) additional local search usage = "no";
  (c) search intensification parameter (θ) = 1.25.
(4) Evolutionary Algorithm ([15], [16], [17]):
  (a) selection method = "tournament";
  (b) $N_{pop} = 100$;
  (c) $N_{child} = N_{pop}$;
  (d) $n\_points = 1$;
  (e) $p_{mut} = 0.05$;
  (f) method of forming a new population = "only children are transferred to the new generation
(5) Bee Colony Algorithm ([22], [28], [29]):
  (a) selection method = "roulette";
  (b) number of bees (B) = 5;
  (c) number of forward and backward passes (NC) = 15;
  (d) memory usage = "yes".
(6) Ant Colony Optimization ([7], [9], [8]):
  (a) algorithm type = "ACS";
  (b) number of ants (A) = N_{ps};
  (c) first colony relative importance of the pheromone versus the heuristic information (α_1) = 3;
  (d) second colony relative importance of the pheromone versus the heuristic information (α_2) = 3;
  (e) first colony evaporation rate (ρ_1) = 0.1;
  (f) second colony evaporation rate (ρ_2) = 0.1;
  (g) first colony pheromone decay coefficient (φ_1) = 0.1;
  (h) second colony pheromone decay coefficient (φ_2) = 0.1;

IOP Conf. Series: Journal of Physics: Conf. Series **973** (2018) 012063 doi:10.1088/1742-6596/973/1/012063

(i) first colony research factor (q_0^1) = 0.5;
(j) second colony research factor (q_0^2) = 0.5.

3. Results of the computational experiment

The developed algorithms have implemented as a software in the programming environment Embarcadero Delphi 2010. The simulation has performed on a computer with Intel Core i5-3470 processor.

Below there are the values of the control parameters of the metaheuristics of the PPEDIS solution ([26],[25] and [24]) found using the method given above.

(1) Simulated Annealing:
  (a) $\alpha = 0.97854$;
  (b) $N\_iter = 60$
  (c) $p_0 = 0.95712$.
(2) Tabu Search:
  (a) $l = N_{ps}$;
  (b) $p = 0.15615$;
  (c) diversification usage = “yes”.
(3) Multi-Start:
  (a) $N\_iter\_max = 50$;
  (b) additional local search usage = “yes”;
  (c) $\theta = 1.31773$.
(4) Evolutionary Algorithm:
  (a) selection method = “tournament”;
  (b) $N_{pop} = 40$;
  (c) $N_{child} = N_{pop} \times 2$;
  (d) $n\_points = 3$;
  (e) $p_{mut} = 0.03345$;
  (f) method of forming a new population = “creation of an intermediate population of N_{pop} parents and N_{child} children with the transfer N_{pop} best individuals to the next generation”.
(5) Bee Colony Algorithm:
  (a) selection method = “tournament”;
  (b) $B = 5$;
  (c) $NC = 12$;
  (d) memory usage = “yes”.
(6) Ant Colony Optimization:
  (a) algorithm type = “ACS”;
  (b) $A = N_{ps}$;
  (c) $\alpha_1 = 3$;
  (d) $\alpha_2 = 1$;
  (e) $\rho_1 = 0.09574$;
  (f) $\rho_2 = 0.11055$;
  (g) $\varphi_1 = 0.12560$;
  (h) $\varphi_2 = 0.10042$;
  (i) $q_0^1 = 0.14061$;
  (j) $q_0^2 = 0.22816$.

IOP Conf. Series: Journal of Physics: Conf. Series **973** (2018) 012063 doi:10.1088/1742-6596/973/1/012063

Table 1. The results of comparing different sets of control parameters for the developed metaheuristics (dimension 100×100×3).

	Parameters	SA	TS	MS	EA	ACO	BCO
δ	by method	5.850 %	5.402 %	4.498 %	5.585 %	3.496 %	2.233 %
	default	6.467 %	6.364 %	4.987 %	6.558 %	4.483 %	3.292 %
T	by method	9.37 sec	9.26 sec	8.96 sec	9.67 sec	7.07 sec	6.26 sec
	default	13.74 sec	10.91 sec	11.84 sec	12.77 sec	11.79 sec	11.48 sec
$succ$	by method	97.55 %	97.99 %	98.90 %	97.82 %	99.90 %	99.89 %
	default	96.93 %	97.04 %	98.41 %	96.84 %	98.92 %	98.83 %

Table 2. The results of comparing different sets of control parameters for the developed metaheuristics (dimension 500×500×3).

	Parameters	SA	TS	MS	EA	ACO	BCO
δ	by method	5.941 %	4.234 %	3.789 %	5.234 %	3.234 %	2.272 %
	default	6.412 %	4.859 %	4.206 %	6.260 %	3.842 %	2.614 %
T	by method	48.40 sec	50.27 sec	38.82 sec	51.53 sec	37.00 sec	36.49 sec
	default	67.72 sec	60.23 sec	49.39 sec	57.92 sec	49.22 sec	49.26 sec
$succ$	by method	97.46 %	99.17 %	99.61 %	98.17 %	99.83 %	99.21 %
	default	96.99 %	98.54 %	99.19 %	97.14 %	99.56 %	98.87 %

Further for the algorithms proposed in [26], [25] and [24] the efficiency of a set of "default" parameter values and a set of parameter values that found using the method proposed in this article has been compared.

Three criteria to evaluate the quality of algorithms have been used:

- Estimation of the average deviation (δ) of the solutions from the best known solution (record). The same algorithm running time is fixed for all algorithms.
- Estimation of the average time of algorithm convergence (T) — the average time measured from launching the algorithm to the moment when the record is reached with a specified accuracy.
- Estimation of the proportion of successful runnings ($succ$) — the ratio of the number of successful starts of the algorithm to the total number of starts. A successful launch means a situation when the algorithm succeeds in finding a solution of a predetermined quality within a fixed time.

Table 1 summarizes the results of a series of experiments that were aimed to evaluate and compare algorithms using three quality criteria for a (100×100×3)-dimension problem. Table 1 contains values of the corresponding quality criteria for the corresponding PPEDIS-solving algorithm with a set of control parameters found using the method proposed in this paper and with a set of the "default" control parameters. For correct comparison in the course of experiments all the algorithms were given the same conditions (running time, required accuracy, etc.).

Table 2 summarizes the results of a series of experiments that were aimed to evaluate and compare algorithms using three quality criteria for a (500×500×3)-dimension problem.

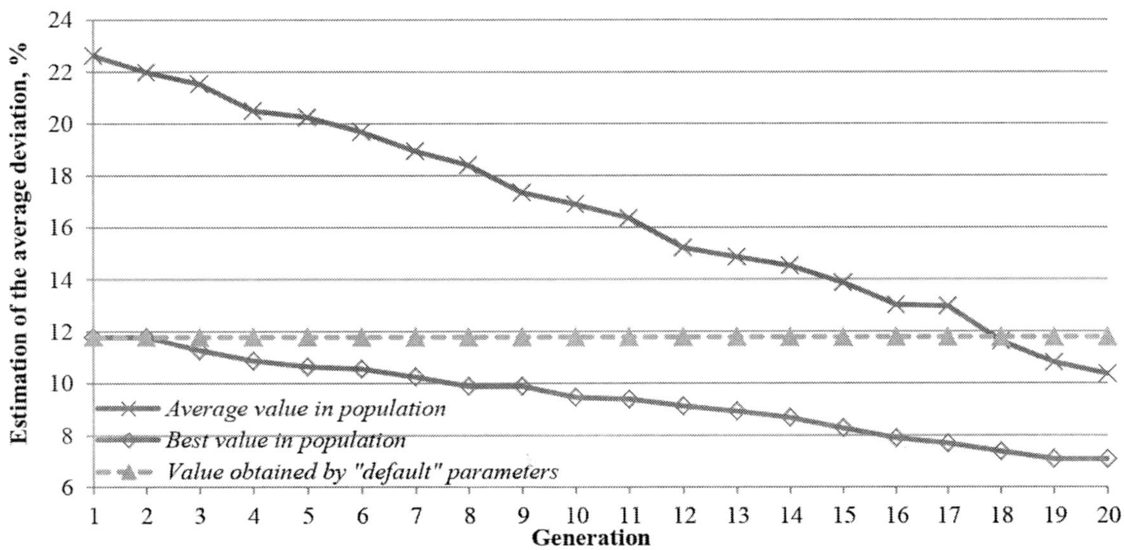

Figure 5. Dynamic of the average deviation during the meta-algorithm

We present Figure 5 to illustrate the work of the proposed evolutionary method. It shows for ACO the change of the value of δ with each EA generation relative to the static value of δ obtained by the "default" parameters. Note that $time\_alg = 10\ sec$ for data represented in Figure 5.

4. Conclusion

Thus the evolutionary method has been developed for tuning control parameters of metaheuristic algorithms for solving the problem of placing elements of developing information systems. The method implies the possibility of simultaneous selection of real, integer and symbolic parameters.

Solution to the problem of planning and optimization of information systems (for example, wireless data networks) using algorithms with free parameter values found by the method, proposed in this paper, can significantly improve the results compared with the "default" free parameter values. This improvement can reduce the cost of network modernization on average by a few percent which means that it increases the efficiency of planning and building wireless information systems.

Let's note possible further directions of researches on the topic of this article:

- Extension of the list of metaheuristics for which the developed method is applied. Adding algorithms such as particle swarm optimization, fish school search, bat search, etc.
- Using the concepts of parallel programming to improve the performance of metaheuristics for solving PPEDIS. Study of the effectiveness of the proposed method for tuning the parameters for parallel metaheuristics.
- Extension of the list of control parameters for each metaheuristic algorithm.

References

[1] Ahmad F, Isa N A M, Hussain Z, Osman M K and Sulaiman S N 2015 *Pattern Analysis and Applications* **18** 861
[2] Brest J, Greiner S, Boskovic B, Mernik M and Zumer V 2006 *IEEE Transactions on Evolutionary Computation* **10** 646

IOP Conf. Series: Journal of Physics: Conf. Series **973** (2018) 012063 doi:10.1088/1742-6596/973/1/012063

[3] Bronmo G, Christiansen M, Fagerholt K and Nygreen B 2007 *Computers and Operations Research* **34** 900
[4] Cachon A and Vazquez R A 2015 *Neurocomputing* **148** 187
[5] Chang W D 2007 *Applied Mathematical Modelling* **31** 541
[6] Demidova L A and Petrova N A 2013 *Ryazan State Radio Engineering University Bulletin* **45** 93
[7] Dorigo M, Birattari M and Stutzle T 2006 *IEEE Computational Intelligence Magazine* **4** 28
[8] Dorigo M and Blum C 2005 *Theoretical Computer Science* **344** 243
[9] Dorigo M and Gambardella L M 1997 *IEEE Transactions on Evolutionary Computation* **1** 53
[10] Dreo J, Petrowski A, Siarry P and Taillard E 2006 *Metaheuristics for Hard Optimization: Methods and Case Studies* (New York: Springer Science & Business Media)
[11] Gaertner D and Clark K 2005 *Proc. Int. Conf. on Artificial Intelligence (Las Vegas, USA)* vol 1 p 83
[12] Gendreau M, Hertz A and Laporte G 1994 *Management Science* **40** 1276
[13] Glover F 1989 *ORSA Journal on Computing* **1** 190
[14] Glover F 1990 *ORSA Journal on Computing* **2** 4
[15] Holland J H 1973 *SIAM Journal on Computing* **2** 88
[16] Holland J H 1975 *Adaptation in Natural and Artificial Systems* (Ann Arbor: University of Michigan Press)
[17] Holland J H and Goldberg D E 1988 *Machine Learning* **3** 95
[18] Huang C L and Wang C J 2006 *Expert Systems with Applications* **31** 231
[19] Karpenko A P 2014 *Modern Algorithms of Search Engine Optimization. Algorithms Inspired by Nature* (Moscow: MGTU Press)
[20] Kirkpatrick S, Gelatt C D and Vecchi M P 1983 *Science* **4598** 671
[21] Kirkpatrick S 1984 *Journal of Statistical Physics* **34** 975
[22] Lucic P and Teodorovic D 2003 *International Journal on Artificial Intelligence Tools* **12** 375
[23] Marti R, Resende M G C and Ribeiro C C 2013 *European Journal of Operational Research* **226** 1
[24] Skakov E S and Malysh V N 2015 *Journal of Information Technology and Applications* **5** 88
[25] Skakov E S and Malysh V N 2015 *Information and Control Systems* **76** 99
[26] Skakov E S and Malysh V N 2015 *Proc. Int. Conf. on Stability and Control Processes* 348
[27] Talbi E G 2009 *Metaheuristics: From Design to Implementation* (New York: Wiley-Interscience)
[28] Teodorovic D, Selmic M and Davidovic T 2015 *Yugoslav Journal of Operations Research* **25** 33
[29] Teodorovic D, Selmic M and Davidovic T 2015 *Yugoslav Journal of Operations Research* **25** 185
[30] Van Laarhoven P J M, Aarts E H L and Lenstra J K 1992 *Operations Research* **40** 113
[31] Wun M H, Wong L P, Khader A T and Tan T P 2014 *Proc. 4th World Congress on Information and Communication Technologies (Malacca, Malaysia)* vol 1 p 314

IOP Conf. Series: Journal of Physics: Conf. Series **973** (2018) 012064 doi:10.1088/1742-6596/973/1/012064

A project optimization for small watercourses restoration in the northern part of the Volga-Akhtuba floodplain by the geoinformation and hydrodynamic modeling

Alexander Voronin, Ann Vasilchenko, Alexander Khoperskov*

Mathematics and Information Technology Institute, Volgograd State University, Volgograd, 400062, Russia

E-mail: *khoperskov@volsu.ru

Abstract. The project of small watercourses restoration in the northern part of the Volga-Akhtuba floodplain is considered together with the aim of increasing the watering of the territory during small and medium floods. The topography irregularity, the complex structure of the floodplain valley consisting of large number of small watercourses, the presence of urbanized and agricultural areas require careful preliminary analysis of the hydrological safety and efficiency of geographically distributed project activities. Using the digital terrain and watercourses structure models of the floodplain, the hydrodynamic flood model, the analysis of the hydrological safety and efficiency of several project implementation strategies has been conducted. The objective function values have been obtained from the hydrodynamic calculations of the floodplain territory flooding for virtual digital terrain models simulating alternatives for the geographically distributed project activities. The comparative efficiency of several empirical strategies for the geographically distributed project activities, as well as a two-stage exact solution method for the optimization problem has been studied.

1. Introduction

The hydrographic network of the Volga-Akhtuba floodplain (VAF) is an extremely complex system, and its properties are largely determined by seasonal flooding of the area [1, 2]. The aim of our study is the research of northern part of the Volga-Akhtuba Floodplain with an area of approximately 900 km^2. Volzhskaya HPP fully controls the hydrological regime of the Volga River following the primary hydropower interests [3]. The latter leads to significant decrease in the spring floods volume in comparison with the natural volume in absence of the large dams on the Volga River.

The limitations of flood peaks due to the hydrological safety requirements for expanding agricultural and urbanized areas, the anthropogenic degradation of riverbeds of the Volga and Akhtuba Rivers, as well as numerous small channels, are the reasons of the aridization of the interfluve territory [3]. The interfluve flooding does not occur in the case of low spring water on the Volga River, that leads to an ecological catastrophe [4]. The most recent examples of the such years are 2006 and 2015 when in the absence of spring flooding the wetland landscapes transformed into the arid areas without vegetation and reproduction of fish stocks

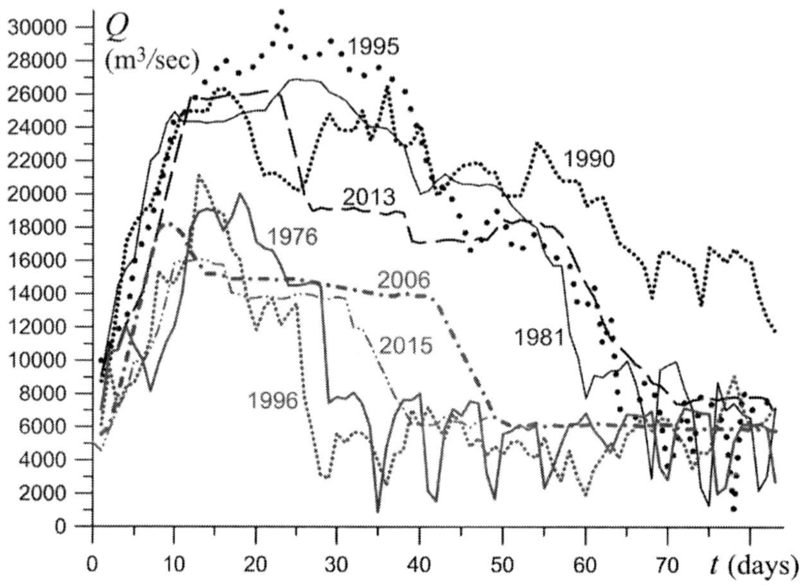

Figure 1. The examples of the Volga River HPP dam hydrographs for low (1976, 1996, 2006, 2015) and medium (1981, 1990, 1995, 2013) peaks of water discharge

[2, 5] (Figure 1).

One of the most important factors is the natural clean up of the interfluve territory due to very high water levels (with the maximum peak of water discharge $Q_{\max} > 32000\,\mathrm{m}^3/\mathrm{sec}$) and long-lasting (more 40 days) flood peak which occurs every few years. Such spring super floods have not been observed during the last 25 years. The implementation of hydrotechnical projects increasing the territory watering level in the case of small and medium floods is perhaps the only way to preserve the ecosystem of the VAF. One of the such projects is the restoration of small watercourses, about 884 km long, in the northern part of the VAF. The branching of the hydrographic network together with the irregularity of the terrain leads to a multidirectional dependence of the flooded area on local changes in the watercourses depths [6]. Therefore, an urgent task aimed at improving the hydrological state and the ecological situation as a whole is the search for the optimal geographically distributed types of activities for the restoration (deepening and cleaning) of small watercourses in the interfluve.

2. Geoinformation and hydrodynamic modeling of floods

The topography irregularity, the complex structure of the floodplain valley consisting of a large number of small watercourses, the presence of urbanized and agricultural areas require a careful preliminary analysis of the hydrological safety and efficiency of geographically distributed project activities. On the basis of digital terrain models and watercourses structure, a discretization of the projects optimization task in the geographically distributed project activities conditions has been carried out. Using the digital terrain and watercourses structure models of the floodplain, and the hydrodynamic flood model, the analysis of the hydrological safety and efficiency of the several project implementation strategies has been conducted [7].

Our objective function has shown the improvement of the interfluve hydrological regime due to clearing of small natural watercourses in this territory. The objective function values have been obtained from the hydrodynamic calculations of the floodplain territory flooding for virtual digital terrain models for simulation of the alternatives for the geographically distributed project

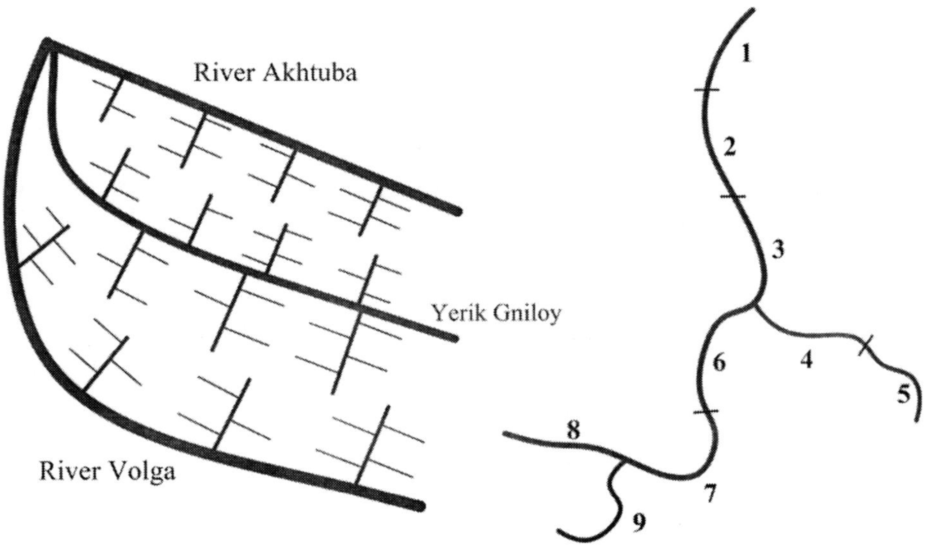

Figure 2. The selected zones and main watercourses of the VAF. The insert on the right shows the watercourses of the 37th zone.

activities. The comparative efficiency of the several empirical strategies for the geographically distributed project activities, as well as a two-stage exact solution method for the optimization problem, has been explored.

The satellite data from ASTER GDEM [8], SRTM [9], Landsat and Sentinel are utilized in our Digital Elevation Model for the northern part of the VAF $b(x_i, y_j)$ $(i = 1, ..., N_x;\ j = 1, ..., N_y)$. Using approximation and excluding artifacts in the original data, in the geoinformation system Professional GIS "Panorama" we construct basic grids with a cell size varying from 25 to 10 m in the SK-42 reference system (Krasovsky ellipsoid). The digital map of the riverbeds and river bottoms of the Volga, Akhtuba Rivers and small watercourses in the floodplain valley has been built combined with vectorization of space images, topographic maps, and pilot charts coupled with the results obtained from our field measurements [7, 10].

An important feature of the VAF hydrological regime is the submergence of about $25 - 40\,\%$ of the interfluve area in spring and drying up of reservoirs in summer. Monitoring the changes in the coastlines of such seasonal reservoirs allows us building up a system of topographical isolines by the actual digital terrain model for the most key-zones of floodplain [10, 11]. Current VAF vector map includes the layer of hydraulic system consisting of 1542 small riverbed objects of 1–3 m depth, the infrastructure layer including 118 settlements and the relief layer comprising more than 15000 relief objects.

To solve the main aim we conducted a hydrological structuring of the territory (Figure 2). The boundaries of 58 zones formed by the local watercourses systems are shown in Figure 3. For example the watercourses of zone 37 are also depicted there. We identified 58 of such zones $(i = 1, 2, \dots, 58)$ using geoinformation analysis tools for our topography.

The edges of a hierarchical graph correspond to the nodes of the digital terrain model grid denoting the watercourses. Three main watercourses (Volga River, Akhtuba River and Erik Gniloy) form the first (highest) level of the hierarchical graph (See Figure 2). The second level of the graph contains 58 main watercourses about 2–3 m deep for each i-th zone. The third level is formed by watercourses of 1 to 2 m deep connected to the riverbeds of the second-level watercourses.

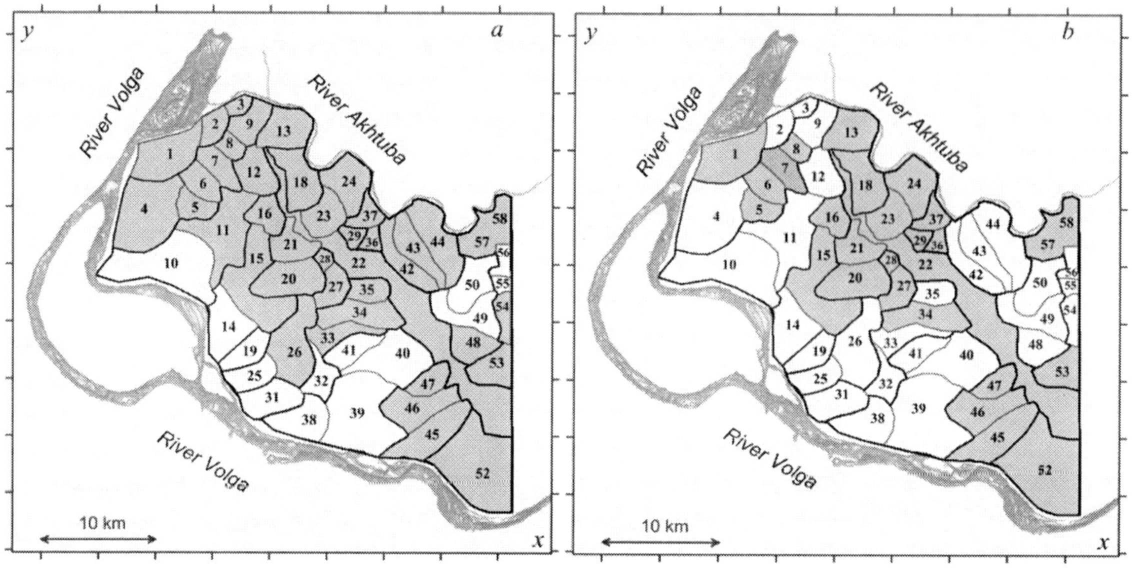

Figure 3. The zones with inessential effect of inter-zonal water flows for the domains B (panel a) and C (panel b) hydrograph parameters

We assume a numerical 2D shallow water dynamics model taking into the account all the principal aspects of the flooding: the inhomogeneous relief of the terrain $b(x, y)$, the bottom friction force $\mathbf{f}^{fric}$ for the inhomogeneous Manning coefficient $n_M(x, y)$, the evaporation σ^e, the infiltration σ^{in}, the Coriolis force $\mathbf{f}^{Cor}$, and the viscous force $\mathbf{f}^{visc}$ (this is described in detail in Refs. [12, 13, 11] for Combined Smoothed Particle Hydrodynamics — Total Variation Diminishing method). Our parallel OpenMP-CUDA software has been described in the work [14].

Calculation error estimates of the VAF flooded zones' areas appearing due to the inaccuracy of the digital terrain model have been evaluated by comparison of calculation results with GPS measurement data with special imitation hydrodynamic flood simulations accounting for a random height change at the grid nodes. The safety control has been carried out by the union, while the calculation of the objective function (the area of flooded zones) has been fulfilled by the intersection of the digital flood maps obtained from hydrodynamic simulations. The calculation of the objective function determined by the increment of the area of the flooded territory, has been conducted matching the intersection of these maps. The largest relative error in the VAF zones objective function calculation, ϵ, defined as the area difference in the union and intersection of the digital flood maps, amounts 5%. The total area of zones 19, 25, 31, and 38 excluded from the general analysis due to the excess of the ϵ value is 26 square kilometers. These territories are highlighted by a dark tint in Figure 3.

The analysis of space images and results of hydrodynamic simulations shows that the area of flooding territory S_F is determined by the flow rate constant Q^{max} of the VHPP hydrograph forming the first phase of the flood and flood duration t. The distributions of the parameters Q^{max} and t since 1990 and the approximate boundaries of small (A, B), mean (C, D) and large (E) floods are represented in Figure 4.

IOP Conf. Series: Journal of Physics: Conf. Series **973** (2018) 012064 doi:10.1088/1742-6596/973/1/012064

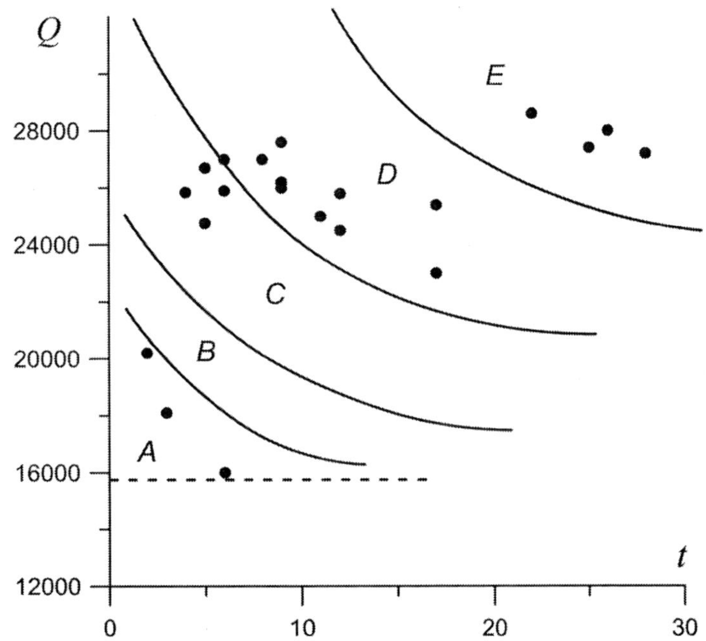

Figure 4. The peak values of water discharge $Q^{\max}$ versus the duration of maximum discharge (t) for 1995 – 2016 are represented by points. The boundaries of small (A, B), mean (C, D) and large (E) flood areas are shown by lines.

3. The optimization problem

Searching for a set of sections of each of the 1542 watercourses maximizing the safe flooding area, $\Delta S(L, Q, t)$ provides a parametric optimization of the project ($L \leq L^{\max}$, where L and $L^{\max}$ are the actual and limiting total lengths of the restored watercourses, respectively). If the solutions depend on these parameters significantly, the choice of the restored watercourses final sections should be approved by the decision-making experts [15, 16]. The latter requires attraction of an additional information.

The two-stage method for solving a discrete version of the optimization problem is described below. Its accuracy for the territories with an insignificant inter-zone water flows effect is limited by the accuracy of the geoinformation and hydrodynamic models. Let us consider a two-stage algorithm for solving the discrete parametric project optimization problem in the territory with a small inter-zone water flows effect. The project optimization for each zone is the first stage. Alternatives to the first stage of optimization in each of the zones are continuous fragments of the local watercourse systems connected to the main watercourses. The edges of the watercourses graph of the i-th zone ($i = 1, ..., N$, where $N = 42$ for the domain B, and $N = 28$ for the domain C) are numbered sequentially in the direction of increasing of the hierarchical level from 1 to k_i.

The discrete alternatives are continuous sequences of edges of the watercourses graph with monotonically increasing numbers. Thus, each alternative a_{ij} ($j = 1, ..., n_i$) of the i-th zone contains from 1 to k_i edges for which the sum of the lengths is L_{ij}. To reduce the discretization error, the long edges of the graphs can be replaced by a sequence of edges which length does not exceed a given value ℓ. This parameter determining the discretization error and simultaneously the number of necessary hydrodynamic simulations is limited by the available computational resource. Thus, formed alternatives constitute the vertices of the hierarchical graphs the edges

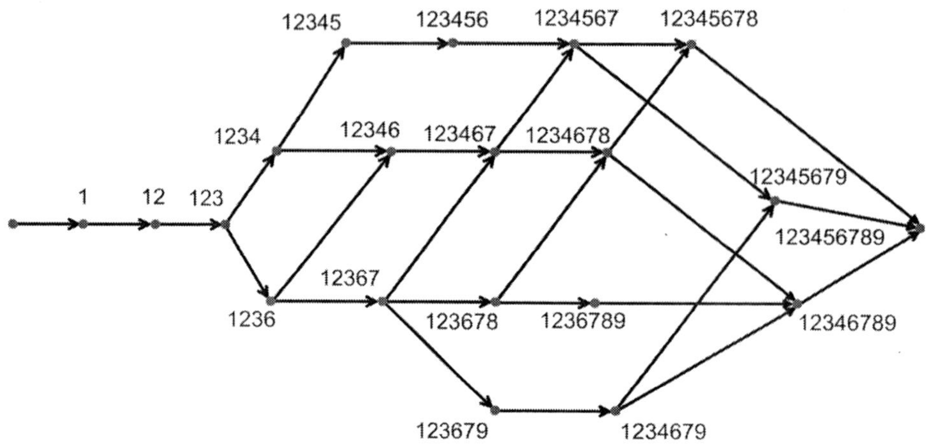

Figure 5. The alternatives graph $a_{37,j}$ (j=1,...,20) for the watercourses graph of 37-th zone for $\ell = 1$ km.

of which are directed from each alternative vertex (excluding the last) to the other vertex-alternative obtained from it by adding an edge. For example, Figure 5 shows the 9-level graph of hierarchical alternatives of the watercourses graph for the 37th zone with $\ell = 1$ km, including 20 vertices. The numbers of edges are indicated at Figure 2.

In the case of $\ell = 0.5$ km or less, the graph becomes rather complicated and inconvenient for visual analysis. The lengths of alternatives $L_{37,j}$ (km) are equal to the number of the corresponding hierarchical level of the graph G_{37}.

Thus, the first stage task is a conditional optimization of the project in each of the zones where the common lengths of the reconstructed watercourses in the zones (L_i) are used as the parameters. It can be written as follows:

$$\Delta S_i(a_{ij}) \longrightarrow \max_{a_{ij}}, \qquad L_{ij} = L_i\,, \quad 0 \le L_i \le L_i^{\max}\,,$$
$$(i = 1, ..., N;\ \ j = 1, ..., n_i)\,, \tag{1}$$

where $L_i^{\max}$ is the total length of the watercourses in the zone. To solve the problem in the ith zone of the parameter plane $(\Delta S_{ij}, L_{ij})$, one can construct a metrized graph G_i which vertices (a_{ij}) are located at points $(\Delta S_{ij}, L_{ij})$, while the edges are linear interpolations of the grid functions $\Delta S_{ij}(a_{ij})$ $(j = 1, ..., n_i)$. Increasing non-negative part of the upper boundary of the graph edges corresponds to the parametric dependence $\Delta S_i^*(L_i)$ for the solution of problem (1) in the i-th zone.

The result of problem solution (1) is the numbers of the optimal alternatives L_i for the zones $j_i^*(L_i)$, as well as the values of the grid function:

$$\Delta S_i^*(L_i) \quad \left(L_i = k_i\ell;\ \ k_i = 1, \ldots, \frac{L_i^{\max}}{\ell};\ \ i = 1, ..., N\right). \tag{2}$$

The values of the function between the grid nodes can be evaluated by the linear or cubic spline-interpolation.

At the second stage, the problem of conditional optimization of the flooding area total increment with the total length of the restored watercourses as a parameter is solved using the dependencies obtained at previous stage:

$$\Delta S = \sum_{j=1}^{n_i} \Delta S(L_i) \longrightarrow \max_{L_i},$$

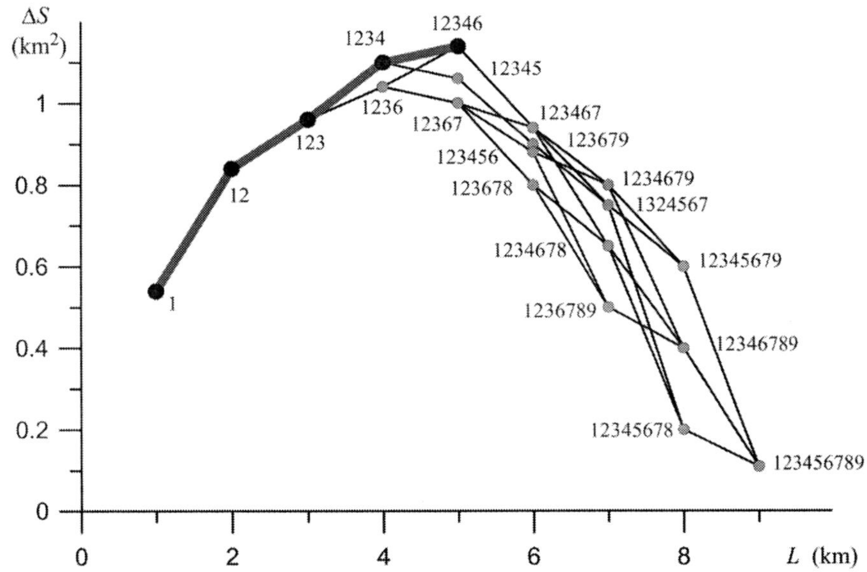

Figure 6. The metrized graph. The thick red line shows the solution of the problem (1) for the considered zone.

$$\sum_{i=1}^{N} L_i = L \quad (i = 1, ..., N), \; L \le L^{\max}, \tag{3}$$

where L is total length of restored watercourses, $L^{\max}$ is the limiting value of the total length L. A multi-step dynamic programming method is used for the problem (3) solution. The corresponding Bellman equation can be represented in the following form:

$$\begin{aligned} \int_{-\infty}^{\infty} W_k \;&=\; \max_{0 \le L_k \le \min(L_k^{\max}, C_k)} \left(\Delta S_k(L_k) + W_{k+1}(C_k - L_k)\right) \\ & (k = N-1, ..., 0), \quad 0 \le C_k \le L, \\ & 0 \le L \le L^{\max}, \quad W_N = \Delta S_N, \quad W_0 = \Delta S. \end{aligned} \tag{4}$$

The numerical solution has been obtained with steps $\Delta L_k = 0.1\ell$, $\Delta C_k = \ell$.

The results of numerical solution of the problem (1)—(3) has been obtained in the territory with a negligible effect of inter-zonal water flows for the parameters $Q = 22\,000\,\mathrm{m}^3/\mathrm{sec}$, $t =$ 5 days (See domain B in Figure 4), $\ell = 1\,\mathrm{km}$. The values of $\Delta S_i(a_{ij})$ $(i = 1, ..., N;\ j = 1, ..., n_i)$ for all the zones has been calculated simultaneously by the numerical hydrodynamic simulations. Hence, the total number of calculations has not exceeded the value of $\max_i \{n_i\}$ $(i = 1, ..., N)$. The actual number of calculations has been even less than this value, since the alternatives corresponding to the negative effect of the deepening of the first sections of the long watercourses have been excluded from consideration. Figure 6 shows a metrized graph (See Figure 5 for comparison), where the solution of problem (1) in this zone coincides with selected part of the graph. Our problem is more complex than simple distributed hydrological systems studied in the works [17, 18, 19, 20].

The equation (4) has been solved numerically with steps $\Delta L_k = 0.1\,\ell$, $\Delta C_k = \ell$. The functions $\Delta S^*(L)$ corresponding to the solution of problem (1), (3) are shown in Figure 7.

The efficiency of the specified algorithm (strategy I) is essentially limited by the condition of smallness of the interband flooding effect. In addition a study of the project limited optimization

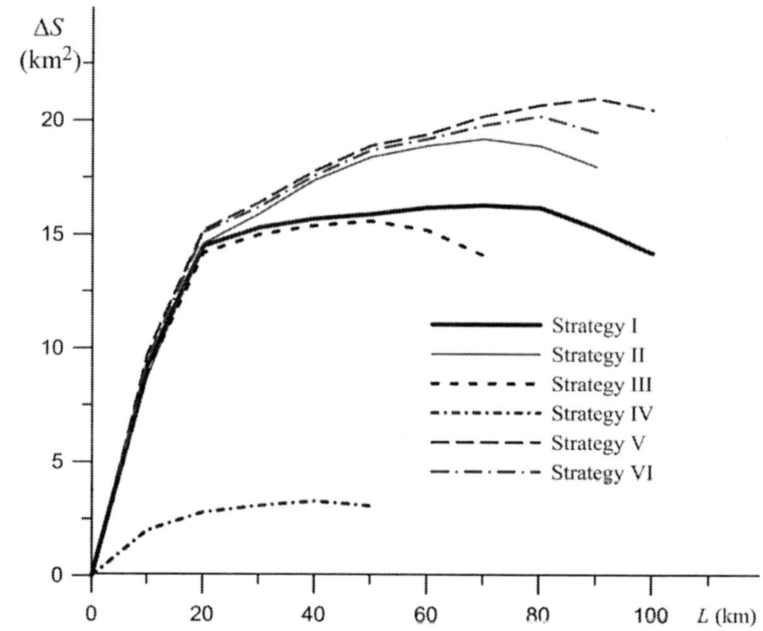

Figure 7. The parametric dependencies of $\Delta S(L)$ for the problem solutions (1), (3) (curve I) of the optimal realizations of empirical strategies II-IV and composite strategies V, VI ($Q = 22000\,\mathrm{m}^3/\mathrm{sec}$, $t = 5$ days).

algorithm with several empirical strategies of project activities in the VAF zones has been carried out: the restoration of only the second level watercourses (strategy II); the restoration of only sections of watercourses of the second level with a length of 1 km (strategy III); the restoration of all watercourses of each zone at a distance of 1 km from the beginning of the second-level watercourses (strategy IV).

The first stage for each strategy has been implemented on the whole VAF territory by parallel calculation of the project effect values in each zone. At the second stage, the zones have been ordered by the effect magnitude, and the integral effect $\Delta S(L)$ parametric curve has been built with the cumulative total number. The verification of the results by direct hydrodynamic calculations has showed that the error of the parallel calculation of the zonal effects does not exceed the simulation error.

Utilizing the obtained data, two composite strategies for two-stage optimization have been proposed combining, at the first stage, the results of the problem (1) solution in the territories with insignificant effect of interband flooding and strategies II and III in the territories with significant effect of interband flooding. Using obtained data two composite strategies for two-stage optimization combining the results of problem (1) solution at the first stage for the territories with insignificant effect inter-zonal flooding with strategies II and III for the territories with significant effect inter-zonal flooding have been suggested. At the second stage the problem (3) has been solved for the whole floodplain territory. The dependencies $\Delta S^*(L)$ for the solutions of these problems (curves V and VI, respectively) are constructed.

In general, as calculations have shown, the largest effect of the project is achieved in the area of small floods (B and C in the Figure 4). Its efficiency decreases sharply in zones A and D. The latter can be explained by the fact that during small floods the localization of flood waters in the deepened watercourses forms a significant fraction in their total volume, while during large floods small watercourses do not significantly affect the size of the flooded area.

4. Conclusion

We have considered a new approach to the hydrotechnical project optimization for small watercourses deepening in the interfluve to improve the efficiency of flood water use in the flooded area. A distinctive feature of current project is the project parameters spatial distribution and its optimal implementation at the inhomogeneous territory. The presence of global external parameters such as peak and duration of floods, the amount of project financing and etc. make our problem complicated and we somehow should overcome this uncertainty. Such projects may be implemented in valleys of large rivers to improve the conditions of agriculture, and mitigate the negative environmental consequences of the hydroelectric power stations construction. A discrete optimization combining precise and heuristic methods for the territory where the preliminary zoning is carried out underlies at the heart of our approach.

The possibility of application of exact two-stage optimization method is caused by smallness of the effect of the inter-zone (transboundary) flooding of the territory. The discretization error for small watercourses restricts this method due to the limitations of the computational complexity of the problem, as well as the digital elevation model (DEM) error. We include these errors in the algorithm of problem solution. The zoning procedure allows selecting areas for each of which an independent mini-project is considered. It enables to calculate their effectiveness simultaneously within the framework of one hydrodynamic calculation. At the second stage, the overall optimization of the whole project has been carried out. In addition to the exact method, we have proposed several empirical and composite optimization strategies useing the exact method for zones set with small transboundary flooding and the empirical strategies for rest of the territory.

The efficiency and specific features of the approach application have studied using the VAF Northern part of about 900 sq km area as an example. Our analysis for VAF has revealed that the efficiency is the best for the mean values of the flood peaks. In the case of small and large flood peaks the efficiency of the entire project has dramatically decreased. However, the project efficiency for individual zones can be very high, achieving 100 %. In the current paper we have limited the objective function by the flooding area. For more accurate estimates of the objective function it is necessary to take into account the data on the ecological and economic significance of different zones, as well as the control mechanisms based on regional co-financing [21, 22].

Acknowledgments

The authors are grateful to A.Yu. Klikunova, S.S. Khrapov, A.V. Pisarev for help in the work and useful discussions. A. Khoperskov is thankful to the Ministry of Education and Science of the Russian Federation (government task No. 2.852.2017/4.6). A. Voronin and A. Vasilchenko have been supported by the RFBR grant and Volgograd Region Administration (No. 16-48-340147_povolzhye, 15-45-02655_povolzhye). The research is carried out using the equipment of the shared research facilities of HPC computing resources at Lomonosov Moscow State University.

References

[1] Gorski K, Bosch L, Wolfshaar K, Middelkoop H, Nagelkerke L, Filippov O, Zolotarev D, Yakovlev S, Minin A, Winter H *et al.* 2012 *River Research and Applications* **28** 1121–1134
[2] Gorski K, Buijse A, Winter H, De Leeuw J, Compton T, Vekhov D, Zolotarev D, Verreth J and Nagelkerke L 2013 *River Research and Applications* **29** 1226–1236
[3] Kozlov A, Kozlova M and Skorik N 2016 *Remote Sensing* **8** 762
[4] Kuzmina Z V, Treshkin S and Karimova T Y 2015 *Arid Ecosystems* **5** 230–242
[5] Gorski K, De Leeuw J, Winter H, Khoruzhaya V, Boldyrev V, Vekhov D and Nagelkerke L 2016 *Inland Waters* **6** 105–110

[6] Vasilchenko A, Voronin A, Svetlov A and Antonyan N 2016 *International Journal of Pure and Applied Mathematics* **110** 183–192
[7] Voronin A A, Vasilchenko A, Pisarev A V, Khrapov S S and Radchenko Y E 2016 *Science Journal of Volgograd State University. Mathematics. Physics* **1 (32)** 24–37
[8] Suwandana E, Kawamura K, Sakuno Y and Kustiyanto E 2012 *Remote Sensing Letters* **3** 423–432
[9] Rabus B, Eineder M, Roth A and Bamler R 2003 *ISPRS Journal of Photogrammetry and Remote Sensing* **57** 241–262
[10] Presnyakova A N, Pisarev A V and Khrapov S S 2017 *Science Journal of Volgograd State University. Mathematics. Physics* **38** 66–74
[11] Khrapov S S, Pisarev A V, Kobelev I A, Zhumaliev A G, Agafonnikova E O, Losev A G and Khoperskov A V 2013 *Advances in Mechanical Engineering* **5** 787016
[12] Khoperskov A and Khrapov S 2017 *Numerical Simulations, InTech* 20
[13] Khrapov S S, Kuz'min N M and Butenko M A 2016 *Science Journal of Volgograd State University. Mathematics. Physics* **6 (37)** 166–173
[14] Dyakonova T, Khoperskov A and Khrapov S 2016 *Communications in Computer and Information Science* **687** 132–145
[15] Voronin A, Isaeva I, Khoperskov A and Grebenjuk S 2017 *Communications in Computer and Information Science* **754** 419–429
[16] Ahmad S and Simonovic S P 2006 *Water Resources Management* **20** 391–410
[17] Gupta V, Puig V and Blesa J 2017 A methodology for distributed fault diagnosis *Journal of Physics: Conference Series* vol 783 (IOP Publishing) p 012005 URL `http://stacks.iop.org/1742-6596/783/i=1/a=012005`
[18] Hartanto I M, van der Kwast J, Alexandridis T K, Almeida W, Song Y, van Andel S and Solomatine D 2017 *International Journal of Applied Earth Observation and Geoinformation* **57** 123–135
[19] Haahti K, Warsta L, Kokkonen T, Younis B A and Koivusalo H 2016 *Water Resources Research* **52** 246–263
[20] Khalid K, Ali M, Rahman N A, Mispan M, Rasid M, Haron S and Mohd M 2015 *ARPN Journal of Engineering and Applied Sciences* **10** 6628–6633
[21] Burkov V N, Goubko M, Korgin N and Novikov D 2017 *Introduction to theory of control in organizations* (CRC Press)
[22] Ougolnitsky G A, Rokhlin D B and Usov A B 2017 *Far East Journal of Mathematical Sciences* **102** 1319–1570

IOP Conf. Series: Journal of Physics: Conf. Series **973** (2018) 012065 doi:10.1088/1742-6596/973/1/012065

Implementing and analyzing the multi-threaded LP-inference

S Yu Bolotova[1], E V Trofimenko[1] and M V Leschinskaya[3]

[1] Associate professor, The Department of Applied and System Software, Voronezh State University, Voronezh, Russia

[2] Master, The Department of Applied and System Software, Voronezh State University, Voronezh, Russia

E-mail: Bolotova.svetlana@gmail.com

Abstract. The logical production equations provide new possibilities for the backward inference optimization in intelligent production-type systems. The strategy of a relevant backward inference is aimed at minimization of a number of queries to external information source (either to a database or an interactive user). The idea of the method is based on the computing of initial preimages set and searching for the true preimage. The execution of each stage can be organized independently and in parallel and the actual work at a given stage can also be distributed between parallel computers. This paper is devoted to the parallel algorithms of the relevant inference based on the advanced scheme of the parallel computations "pipeline" which allows to increase the degree of parallelism. The author also provides some details of the LP-structures implementation.

1. Introduction

Production systems became an essential part of various applications in a number of areas such as engineering, medicine, IT and many others [1]. The development of the expert systems as well as the theory of learning and problem solving solutions is dependent to a large extent on the production systems. Despite the fact that they're well known in the areas of computer science and mathematical research, they still attract significant interest which is proved by many recent papers [2–5].

At the same time production systems might run very slowly due to the high computational requirements. It's quite common for them to perform an intensive exchange with external memory and exponential number of computational operations. Optimizing the performance of the production systems can make them more reliable and bring attention of those experts who currently avoid working with such systems due to the serious requirements they demand [1].

The purpose of this paper is to focus on the possible ways of acceleration of the production system execution. Though there might be various ways to do so the authors chose an approach based on a parallel computing. As part of the practical implementation the paper presents the architecture of a program solution for the extended production systems and describes an object-oriented class LPStructure containing all the properties and methods that are necessary for the implementation. It heavily uses the multi-threading at its core part for searching logical reductions and solutions of the production logical equations [6]. The class was developed using C++ language and STL library and built as a dynamic C-library, called LPStructure.dll.

This article suppliments the previous paper [7] by adding the practical aspects of the program implementation of the parallel LP-structures.

IOP Conf. Series: Journal of Physics: Conf. Series **973** (2018) 012065 doi:10.1088/1742-6596/973/1/012065

2. A general description of the algorithm

The considered software system implements the relevant parallel backward chaining based on the solution of the productional logical equations and for this purpose it uses mathematical algorithms for finding the truth preimage and accelerating backward inference [6]. Hereafter we use the notations and definitions from the mentioned article.

LP-structure (lattice-production structure) — is the lattice with an additional binary relation set on it, which has a number of production-logical properties [6]. The relevant backward inference strategy is aimed at minimization of a query number to the external information source (either to a database or an interactive user). The inference based on equation solving starts with the creation of all minimal initial preimages in the LP-structure for the atoms that correspond to the values of an examined object. Using this constructed set it is enough to find the preimage that contains only true facts. If founded it makes possible to make a conclusion about the corresponding value of the object. The effective way to achieve that, is to prioritize viewing of the preimages containing the values of the most relevant objects [6]. These are first of all the objects, whose values are present in a maximal number of the constructed preimages. A negative answer to a unique query eliminates all subsequent queries about the elements of a facts subset. Along with a significantly reduced number of queries, when using LP-inference, the preference is given to testing the sets of facts of a minimal cardinality.

However, experiments have shown that the process of simultaneous construction of the minimal initial preimages for large knowledge bases and their "deep" structure may require an excessive amount of computational resources of the computer. Considering this relevant LP-inference method was modified to use parallel computing algorithms. Parallel relevant LP-inference allows speeding up the process of constructing sets of facts that are required in the inference, and their further processing [8].

As an illustration we shall give simplified descriptions of algorithms of a relevant LP-inference and a parallel LP-inference.

Input:
initial facts — true (set T), false (set F);
set of rules R;
hypothesis b.
Output:
true initial preimage $X^0 = \{x_1^0,\dots,x_n^0\}$ for b
(or X^0 = null).

```
// Relevant LP-inference
X^0 = null
{X} = getPreImages (b)
while X^0 = null and {X} ≠ ∅ do
  k = getRelevantIndex ({X}, T)
  Ask (x_k)
  foreach X_j ∈ {X} do
    if X_j ⊆ T then
      X^0 = X_j
      break
    end
    if X_j ∩ F ≠ ∅ then {X} = {X} \ X_j
  end
end
```

IOP Conf. Series: Journal of Physics: Conf. Series **973** (2018) 012065 doi:10.1088/1742-6596/973/1/012065

The function $Ask(x)$ asks a user (or an external database) about the truth of an atomic fact x and according to the answer modifies sets T and F. Function $getPreImages(b)$ solves the equation with a right-hand side b, i.e. it constructs set $\{X\}$ of all minimal initial preimages for atom b.

The equation solving is carried out by splitting relation R into "layers" $R_t, t \in T$, each containing no more than one solution [6]. The layer has the maximum possible set of relationship pairs with the unique right-hand sides, and two layers differ in one pair at least. In each separate layer the process of the equation solving is reduced to the problem of finding the set of initial nodes of the graph $G_{R_t,b}$, corresponding to a layer.

The work for different layers is organizing independently and in parallel. The primary application thread creates secondary threads (which are limited by the *MaxThreads* – the maximum number of threads), passing them a block of data. Hereinafter the "threads" are threads of execution [8]. Created thread searches the solution of production logic equation in a separate layer. When the sufficient number of preimages is calculated (or all preimages), the LP-inference program module immediately checks its "truth", sending queries to the external information source, if necessary. If one of the preimages is not true, the program stores the information about false facts (it reduces the possible number of rules, required for the next equation solution) and calculates the next preimage cluster which also significantly speeds up the process. After this calculation the thread exits.

In the case of large amount of running threads, in accordance with the Amdahl's low, the efficiency of parallel computing is declined. Frequent creation and termination of threads with a short lifecycle and switching their contexts increase the amount of resources. Therefore, in implementing a thread pool mechanism is used [8]. The maximum pool size is limited by parameter. Its value must be greater than the number of processors in order to achieve the maximum simultaneity.

The main thread takes a working thread from the pool and passes it the necessary data for processing. When the number of active threads reaches its maximum, the request is queued. If the number of active threads is less than the maximum, a new thread is created. When the number of active threads is maximized, request is put to the queue and starts waiting for the release of one of the threads.

For higher efficiency the processes of preimages calculation and their further analysis are performed asynchronously. Until a solution is found, an another preimages cluster is stored in the dynamic queue Q. The main thread takes the preimages clusters from the queue and (if one of the secondary threads is free) initiates the parallel analysis of their items relevance. The most relevant elements are placed in the priority queue P, based on a binary heap. The maximum queue size is specified by the maxRelevanceQueueSize parameter. Such data organization is used to sort investigated elements according to their relevance, to change their order by adding new elements, if necessary, and to access the element with maximum value in the constant time. The process of identifying the relevant objects of constructed clusters continues until a solution is found.

The general structure of the parallel algorithm of the relevant inference can be represented by the following scheme (figure 1).

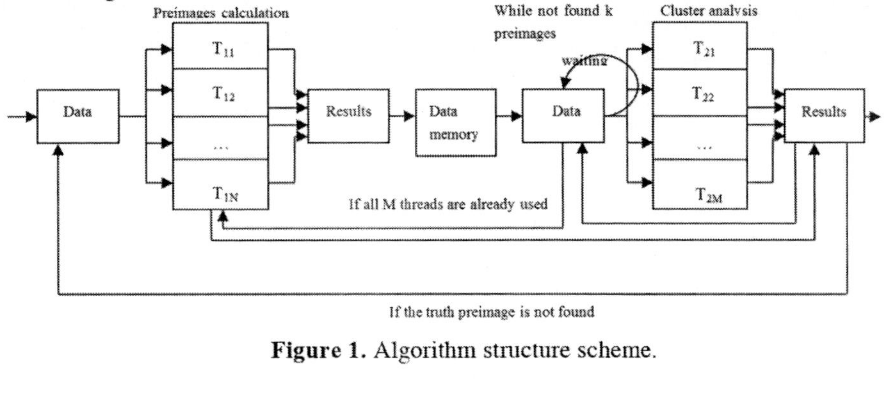

Figure 1. Algorithm structure scheme.

IOP Conf. Series: Journal of Physics: Conf. Series **973** (2018) 012065 doi:10.1088/1742-6596/973/1/012065

The parallel algorithm of the relevant inference is presented below.

```
//The parallel relevant inference
X^0 = null
while X^0 = null do
  asynchronous call {X}=getPreImagesCluster(b)
  if {X} ≠ ∅ then
     // Add the cluster to queue Q
     addClusterToQueue(Q, {X})
  end
  // If the queue is not empty
  while not clustersQueueIsEmpty(Q) do in parallel
   // Extract the cluster from queue
    {Y} = extractClusterFromQueue(Q)
    foreach ( Y_j ∈ {Y} ) do
      if Y_j ∩ F ≠ ∅ then {Y}={Y}\Y_j
    end
    while X^0 = null do
    // If P has the free place
    asynchronous if ( relevanceQueueSize(P)<maxRelevanceQueueSize) then in parallel
    // The relevant object index k ant its relevance
    foreach ( Y_j ∈ {Y} ) do
        if Y_j ∩ F ≠ ∅ then {Y}={Y}\Y_j
    end
    k=getRelevantIndex ({Y}, T,relevance=0)
     // Insert object to the priority queue
      InsertToQueue(P,k,relevance)
    else
       Waiting() // Wait until the queue is freed
    end
    asynchronous while (relevanceQueueSize(P)!=0) do
    // Element with a maximum relevance
       k=ExtractMaxFromQueue(P)
       Ask (y_k) // Check the fact truth
       foreach  Y_j ∈ {Y} do
         if  Y_j ⊆ T then
            X^0=Y_j
           break
         end
         if Y_j ∩ F ≠ ∅ then
            foreach ( y_m ∈ {Y_j} ) do
              //Element in P -lower relevance
              if elemIsMemberOfQueue(y_m,P) then
```

IOP Conf. Series: Journal of Physics: Conf. Series **973** (2018) 012065 doi:10.1088/1742-6596/973/1/012065

```
                changeRelevance(y_m,P)
              end
            end
          end
        end
      end
    end
end
```

The function getPreImagesCluster(b) launches the process of computing the fixed number of minimal initial preimages for the atom b. As already mentioned, this process is carried out in separate threads to provide the greatest efficiency.

Function $getRelevantIndex\ (\{Y\}, T, relevance = 0)$ finds the relevant element and it's relevance $relevance$ for each cluster in a separate thread. Then the element and it's relevance are placed in the priority queue. If this element is already there, the total relevance is calculated and the queue is rebuilt.

At the same time the $Ask\ (y_k)$ function asks the external source about the truth of the most relevant fact, extracted from the priority queue. If the negative request answer is obtained the preimages, which contain an y_k element, excluded from consideration, and already investigated elements relevances are modified.

As an illustration we shall give the computing model as an acyclic graph $G = (V, R)$, where the set of operations is represented as a set of vertices $V = \{1,...,|V|\}$, and the information dependencies between operations — as a set of arcs R [9]. The arc $r = (i, j)$ means that the operation j uses the result of another operation i. The operations of the algorithm that are not connected by an arc can be parallelized.

A possible way to describe the parallel execution of the algorithm solution search in a separate layer is shown in figure 2.

Hereafter is an algorithm of getPreImagesCluster(b) function, where R' is a set of layers $\{R_i\}$. As already mentioned, the solution in each layer is calculated in a separate thread. The maximum number of threads in the pool is limited by *MaxThreads* parameter. The number of active threads is stored in the *countUsedThreads* variable. If the pool has a free thread, then it schedules this thread for execution by passing it a pointer to a $FindEquationDesicion(R_i)$ function, that finds a solution in the next layer R_i.

The algorithm of $getPreImagesCluster(b)$ function is presented below. Denote the set of layers as $\{R_i\}$. As already stated, the solution of each layer is calculated in a separate thread. The number of currently active threads is stored in the *countUsedThreads* variable. If a pool has a free thread, the new thread is started and it calls the $FindEquationSolution(R_i)$ function, which finds a solution of the equation in another layer R_i.

```
// Finding the solutions of the equation for each layer in a separate thread
foreach R_i ∈ R' do in parallel
  if countUsedThreads < MaxThreads then
    BeginThread() // Start the thread
    countUsedThreads++
    FindEquationSolution(R_i)
    ExitThread() // End the thread
    countUsedThreads--
  else
    Waiting() // Wait until the thread is freed
  end
end
```

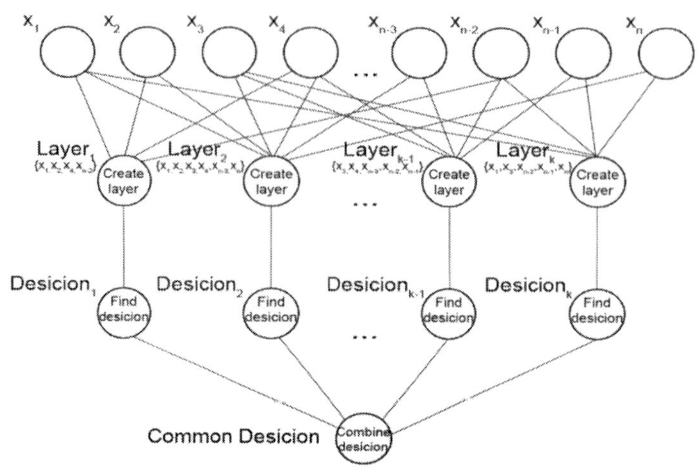

Figure 2. Count computing.

Function *getRelevantIndex* $(\{Y\}, T, relevance)$ finds the index k of the any of the most relevant and not checked for truth facts contained in the current cluster elements $\{Y\}$ and its *relevance* . The function uses two mentioned above indices of relevance to minimize the number of calls of the function $Ask\ (x_k)$. The process of identifying relevant objects is very expensive, so it is also parallelized. For each number of facts the system calculates the number of preimages, containing these facts using multiple threads (number of threads is limited by *MaxThreads* parameter). The number of current active threads is stored in the *usedThreadsNumber* variable. The fact, that belongs to the maximum number of preimages, increments its relevance score by 1. When accessing shared resources threads are synchronized using the critical sections mechanism. We also present computing graph for this algorithm (figure 3).

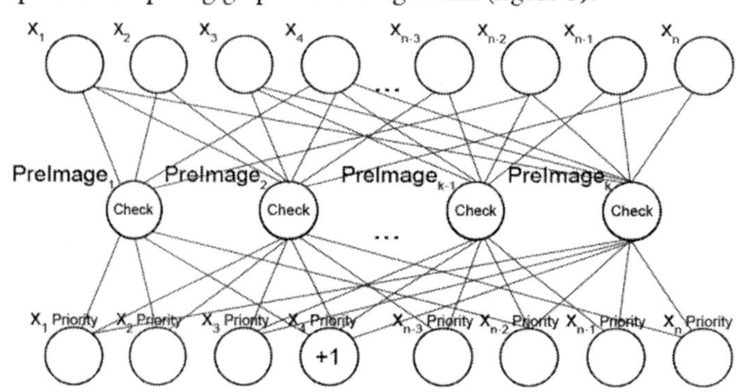

Figure 3. Computing graph.

A simple variant of the function *getRelevantIndex* is given below.

```
//Parallel relevance counting
int getRelevantIndex ({Y}, T,relevance)
if ( usedThreadsNumber < MaxThreadsCount ) or
  ( countUsedThreads < MaxThreads ) then
   BeginThread() // Start the thread
```

```
      if ( usedThreadsNumber < MaxThreadsCount ) then
        usedThreadsNumber++
        usedSelfThreads = true
      else
        countUsedThreads++
        usedSelfThreads = false
      end
       k=MostRelevantFind(relevance) // The relevant atom index
      ExitThread() // End the thread
      if usedSelfThreads then
         usedThreadsNumber--
      else
         countUsedThreads--
      end
     return k
    else
      Waiting() // Wait until the thread is freed
    end
```

The *MostRelevantFind*(*relevance*) function finds the index k of the most relevant object and its relevance.

The benefits of parallel LP-inference are confirmed experimentally. When processing the large knowledge bases the use of parallel algorithms speeds up LP-inference by up to 30%.

3. Implementation

The proposed algorithms were implemented as a dynamic C-library called Parallel LPStructure.dll. Below are some practical considerations for the LP structures. In accordance with [2], it is proposed to use a flexible bit vector which length can be adjusted by a "client" program to store lattice elements. The dimension of such a vector is specified by two parameters — the length in bytes of one cluster (eItemSize) and the number of these clusters (eLengthItems).

The value eItemSize determines the size of the cluster — a part of the vector that is processed by a single logical operation of the C ++ language. This value is stored as a static class constant and can be changed with recompiling. By setting eItemSize to 1, it is possible to restructure a bit vector to a sequence of bytes, which results in memory saving and makes the vector structure more transparent in terms of understanding the details of the algorithms. Setting eItemSize to 8 can significantly improve the speed of operations on vectors, since many cycles in the program will be 8 times shorter, and 64-bit operations will be more efficient for computers with the appropriate memory capacity.

The number of clusters for a bit vector is stored in the eLengthItems integer field. The constructor of the LPStructure class receives a corresponding integer value for setting this field as a parameter. This fact allows the "client" program to dynamically adjust the length of the bit vectors when creating an LP-structure, based on the lattice size and thus it allows to adjust the amount of the available knowledge base accordingly.

In the ParallelLPStructure class Elem and Pair types are defined through renaming the types of single and a tuple of integers. The type Elem assumes storing the address of the lattice element, namely, the bit vector corresponding to the element. The type Pair contains a contiguous pair of Elem values, so that it is intended to store a pair of elements (more precisely, their addresses) from some binary relation on the lattice. Of course, types Elem and Pair could also be implemented in a form of classes with overloaded operators (|, &, <=), corresponding to operations on the lattice. However, the decision to use the simplest types turns out to be more consistent in terms of the adopted strategy of saving memory.

When using STL libraries to implement LP structures, it's possible to use effectively parameterized vector and set container classes, as well as their combination. It is especially suitable to use set<Elem>

IOP Conf. Series: Journal of Physics: Conf. Series **973** (2018) 012065 doi:10.1088/1742-6596/973/1/012065

defined as ElemContainer for the representation of sets of lattice elements. To store the generic binary relations defined on the lattice, it's convenient to define the Pair-Container type as set <Pair>. Some algorithms use paired vectors — PairVector (vector<Pair>) to store some intermediate data. Taking into account all the previously mentioned types, the canonical binary relation might be described by the vector of sets vector <ElemContainer> named ElemContainerVector.

The value of aMaxThread specifies the maximum number of worker threads during execution. It is stored as a static constant field of the object-oriented class ParallelLPStructure.

Hereby we've listed the main types used in the ParallelLPStructure class.

4. Experiments results

To illustrate the parallel relevant LP-inference capabilities we created about five hundred knowledge bases that differed significantly in the scope of the rules sets. As a result of testing we obtained the timing indicators of the normal and parallel relevant backward inference. The maximum number of threads was also limited in the range from 1 to 20 in increments of 3. The results of the examinations were processed in the package Statistica 6.

In this package the one-factor and two-factor analysis of variance were carried out. The homogeneity of group dispersions is verified using the Leven test, the results of which are shown in figure 4.

	Levene's Test for Homogeneity of Variances (Spreadsheet7.sta) Effect: Threads Degrees of freedom for all F's: 5, 294			
	MS Effect	MS Error	F	P
Time	3,228410	0,609062	5,300625	0,000111

Figure 4. The Levene test results.

The results show that the variances do not differ (P << 0.05). Consequently, one of the conditions for applying the parametric variant of the variance analysis is not fulfilled. However, this method is very stable in case of deviation from the specified requirement, therefore it is possible to continue the study [10].

To verify the normal distribution of the analyzed data, the graphs of the normal probabilities and the histograms shown in figures 5 and 6 were plotted.

The figures show that the distribution is close to normal. In addition, the sample sizes allow a further research. The result of the variance analysis is shown in figure 7.

Since the error value for the null hypothesis shows that there is no connection between the number of threads and the execution time is P << 0.05, we can conclude that the time is statistically significantly different depending on the number of threads.

A posteriori analysis using the Tukey method (figure 8) showed that a statistically significant difference in the execution time exists between pairs of parallel relevant inference with 5 threads and 15 threads, as well as parallel relevant inference with 8 and 15 threads.

Two-factor analysis of variance (figure 9) revealed a significant impact of the number of threads and the number of facts, as well as their mutual impact on the duration of implementation (P << 0.05).

Nonparametric one-factor variance analysis of Kraskel-Wallis (figure 10) demonstrated the statistically significant differences between the compared groups (P << 0.05).

For clarity, we also constructed a sweep diagram (figure 11) and a dependence graph of the algorithm execution time and the threads number (figure 12).

The diagram shows the differences between the averages. It can be concluded that the use of 8 threads yields the highest time efficiency. Further increase in the number of threads reduces efficiency, increasing execution time.

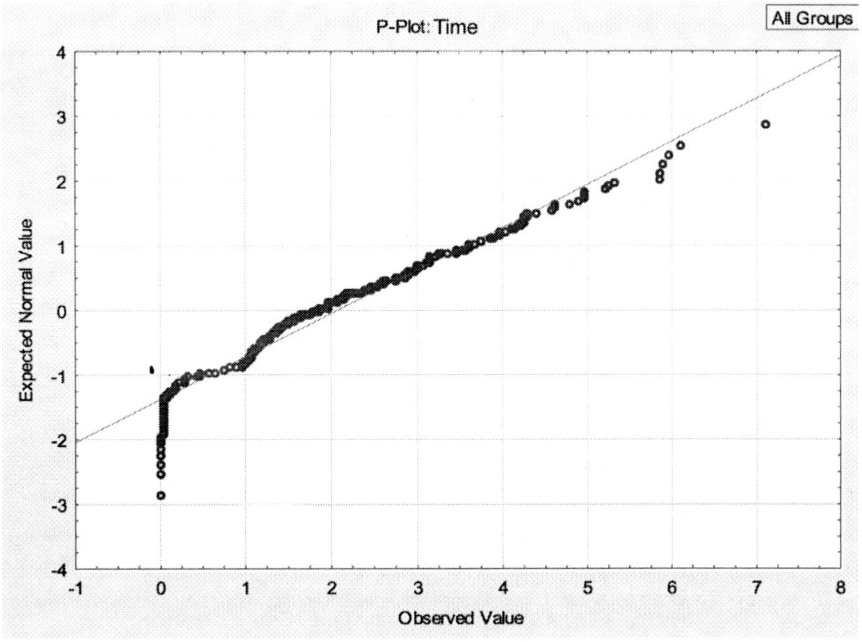

Figure 5. Normal probability graphs.

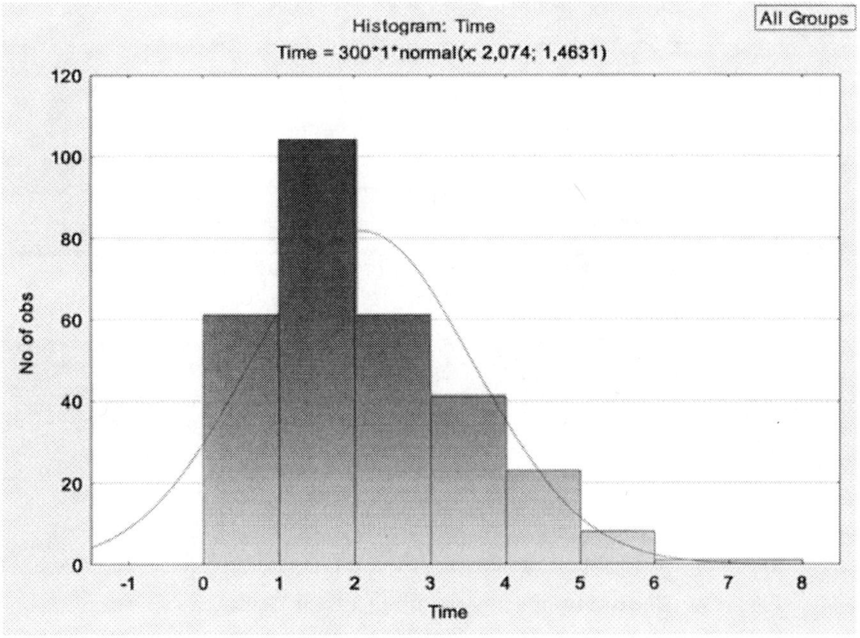

Figure 6. The histogram.

IOP Conf. Series: Journal of Physics: Conf. Series **973** (2018) 012065 doi:10.1088/1742-6596/973/1/012065

Effect	Univariate Tests of Significance for Time (Spreadsheet7.sta) Sigma-restricted parameterization Effective hypothesis decomposition SS	Degr. Of Freedom	MS	F	p
Intercept	1290,503	1	1290,503	632,6348	0,000000
Threads	40,327	5	8,065	3,9538	0,001742
Error	599,727	294	2,04		

Figure 7. The result of variance analysis.

Cell No.	Tukey HSD test; variable Time (Spreadsheet7.sta) Approximate Probabilities for Post Hoc Tests Error: Between MS = 2,0399, df = 294,00 Threads	{1} 2,2824	{2} 2,0739	{3} 1,8466	{4} 1,4901	{5} 2,0687	{6} 2,6826
1	1		0,978298	0,647656	0,061704	0,975810	0,726543
2	2	0,978298		0,968346	0,317385	1,000000	0,271218
3	5	0,647656	0,968346		0,812921	0,971366	0,040124
4	8	0,061704	0,317385	0,812921		0,327529	0,000440
5	10	0,975810	1,000000	0,971366	0,327529		0,262113
6	15	0,726543	0,271218	0,040124	0,000440	0,262113	

Figure 8. The posteriori analysis using Tukey method results.

Effect	Univariate Tests of Significance for Time (Spreadsheet7.sta) Sigma-restricted parameterization Effective hypothesis decomposition SS	Degr. Of Freedom	MS	F	p
Intercept	1290,503	1	1290,503	12592,53	0,00
Facts	550,367	9	61,152	396,71	0,00
Threads	40,327	5	8,065	78,70	0,00
Facts*Thread	24,764	45	0,550	5,37	0,00
Error	24,596	240	0,102		

Figure 9. The results of two-factor analysis of variance.

Depend.: Time	Kruskal-Wallis ANOVA by Ranks; Time (Spreadsheet7.sta) Independent (grouping) variable: Threads Kruskal-Wallis test: H (5, N=300)=14,42252 p=,0131 Code	Valid N	Sum of Ranks	Mean Rank
1	1	50	8124,500	162,4900
2	2	50	7609,000	152,1800
5	5	50	6967,000	139,3400
8	8	50	5860,500	117,2100
10	10	50	7664,500	153,2900
15	15	50	8924,500	178,4900

Figure 10. The results of nonparametric Kruskal-Wallis one-way analysis of variance.

A similar conclusion can also be drawn from the graph of the algorithm execution time and threads number dependence. On the X axis the number of rules in the knowledge bases is marked, on the Y axis — the execution time in seconds, and the colors of the curves reflect the number of threads with parallel algorithm implementation. For the case of a parallel relevant inference with 8 threads and the relevant inference, a t-criterion is used (figure 13). Its results indicate the presence of statistically significant differences between the mediums ($P << 0.05$).

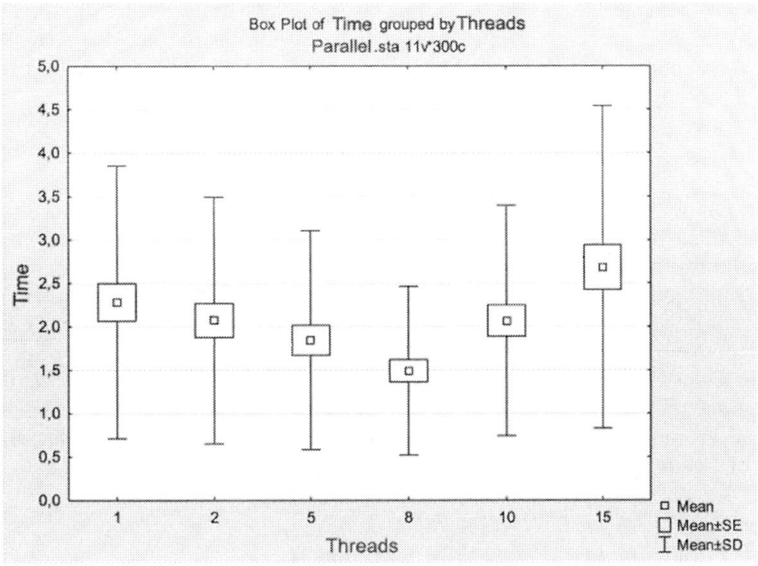

Figure 11. The span diagram.

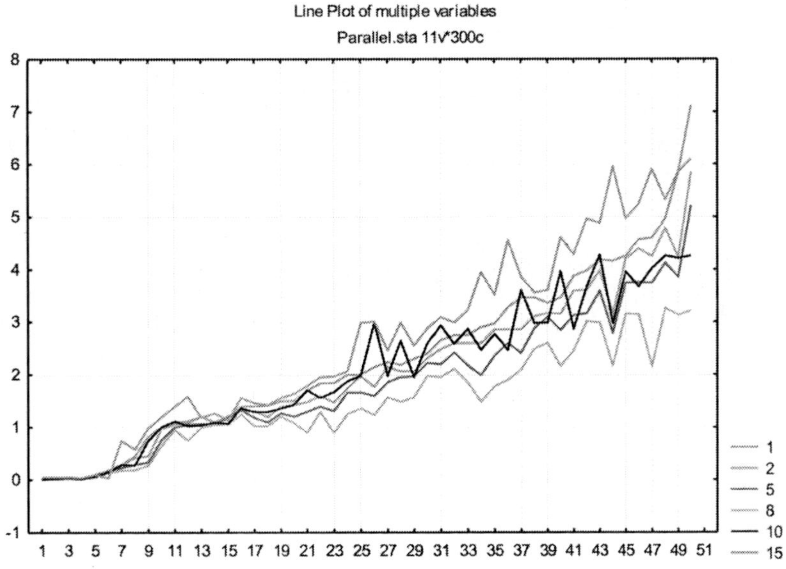

Figure 12. The graph of the dependence of the execution time of the algorithm from the number of threads.

IOP Conf. Series: Journal of Physics: Conf. Series **973** (2018) 012065 doi:10.1088/1742-6596/973/1/012065

T-tests; Grouping: Threads (Parallel.sta)
Group 1: 1
Group 2: 8

Variable	Mean 1	Mean 8	t-value	df	p	Valid N 1	Valid N 8	Std.Dev. 1	Std.Dev. 8	F-ratio Variances	p Variances
Time	2,282412	1,490060	3,034365	98,000000	0,003086	50	50	1,570445	0,971105	2,615248	0,001014

Figure 13. The t-criterion results.

All the demonstrated results indicate a decrease in the solution finding time in case of using multithreading by 20–25% on average. In this situation, we are talking about a statistical indicator, since even with normal output there is some probability of accidental achievement of the best result "from the first time."

5. Conclusion

It's quite common for the algorithms that process large knowledge bases to consume significant computational resources which makes it a big challenge for the researches to find a way to improve the efficiency of the mathematical algorithms and program implementations. The described library uses the theory of LP-structures to speed up the backward inference production systems along with implementing this theory using mutli-threading approaches. The effectiveness of parallel LP-inference is proved by experimental results: the execution time of the considered processes decrease on average by 30%.

Application of the methods described is demonstrated on the simple architecture of production systems in order to focus on the methods proper. A similar approach can also be applied to production systems having a more complex structure; many models in informatics have production nature, and the structures of information representation are often hierarchical. Besides, the methodology presented can be applied to logical systems of a more general form, including non-classical.

References

[1] Gupta A 1987 *Parallelism in production systems* (Pitman)
[2] Liberatore P 2008 *Artificial Intelligence* **172(23)** 265–299
[3] Maciol A 2008 *Expert Systems with Applications* **34(3)** 1825–1836
[4] Poli R and Langdon W B 2006 *Artificial Intelligence* **170(11)** 953–982
[5] Katerinenko R S and Bessmertnyi I A 2009 *Programming and Computer Software* **37(4)** 197–199
[6] Makhortov S D 2009 *Mathematical Foundations of Artificial Intelligence: Theory LP-structures for the construction and studying of knowledge models of production type* ed Vasenin V A (Moscow: Publ. MCCME)
[7] Bolotova S Yu and Makhortov S D 2015 *Modern Informatization Problems in the Technological and Telecommunication Systems Analysis and Synthesis: Proceedings of the XX-th International Open Science Conference* ed Kravets O Ja (Yelm: Science Book Publishing House LLC) pp 339–347
[8] Richter J 1999 *Programming applications for Microsoft Windows* (Microsoft Press)
[9] Voevodin V V and Voevodin Vl V 2002 *Parallel computing* (St. Petersburg : BHV-Petersburg)
[10] Vukolov E A 2008 *Bases of the statistical analysis. Workshop on statistical methods and operations research using the packages STATISTICA and EXCEL* (Moscow: Forum)

IOP Conf. Series: Journal of Physics: Conf. Series **973** (2018) 012066 doi:10.1088/1742-6596/973/1/012066

An Algebraic Approach to the Study and Optimization of the Set of Rules of a Conditional Rewrite System

S D Makhortov

Voronezh State University, Department of Applied and System Software, Universitetskaya pl. 1, 394018 Voronezh, Russia

E-mail: msd_exp@outlook.com

Abstract. An algebraic system containing the semantics of a set of rules of the conditional equational theory (or the conditional term rewriting system) is introduced. The following basic questions are considered for the given model: existence of logical closure, structure of logical closure, possibility of equivalent transformations, and construction of logical reduction. The obtained results can be applied to the analysis and automatic optimization of the corresponding set of rules. The basis for the given research is the theory of lattices and binary relations.

1. Introduction

The theory of term rewriting systems (TRS) is an effective tool for solving a number of computer algebra and artificial intelligence problems. It finds applications in such known areas as symbolic simplification of algebraic expressions, automated theorem proving, etc. Equivalent transformations and minimization of the sets of rules are important problems for the term rewrite systems. While for usual TRS similar problems have been solved in a number of works [1-2], for conditional TRS [3] they, apparently, are still open [4]. This fact can be explained by a more complicated structure of the rules for conditional TRS. For usual systems, the problem of minimization of a set of rules is eventually reduced to a transitive reduction of some binary relation ("elimination of transitivity"). For conditional systems, it is possible to speak about a more complicated problem of finding a logical reduction.

The starting point in defining a rewrite system is usually the equational theory, whose set of rules consists of equalities. The TRS rules are obtained by "orientation" of equalities and, probably, by completion of the rules for attaining the confluence property [2]. A similar approach is also applied to conditional TRS [3]. Since usually the equational theory is the criterion of equivalence of rewrite systems, the study of conditional TRS in this aspect can be started with reviewing the equivalence of conditional equational theories.

Let us assume that the equational theory (similar to [3]) contains a set of positive-conditional equalities of the form $s_1 = t_1, \dots, s_n = t_n : s = t$. The meaning of the set is as follows: if all the equalities $s_i = t_i, i = 1, \dots, n$ take place, then $s = t$ is also fulfilled. If instead of the terms one considers independent elements, then the problem of simplification of such a set of conditional equalities can be reduced to the minimization of Horn conjunctive normal forms studied in [5]. The same method can also be applied to the equalities between the terms, but the optimization can appear to be only partial.

Let us consider an example obtained by modification of the example taken from [3]:

1) $x + y = z : s(x) + y = s(z)$;

IOP Conf. Series: Journal of Physics: Conf. Series **973** (2018) 012066 doi:10.1088/1742-6596/973/1/012066

2) $x+y=z:g(x)+y=g(z)$;
3) $s(x)+y=s(z), g(x)+y=g(z):f(x)=f(z)$;
4) $x+y=z:f(x)=f(z)$.

Here the latter equality is "redundant", since it is deduced from the first three ones. In fact, let $x+y=z$. Then from 1)-2) we have $s(x)+y=s(z)$, $g(x)+y=g(z)$, whence by means of 3) we get $f(x)=f(z)$. Thus, at $x+y=z$ holds $f(x)=f(z)$, i.e. the conditional equality 4) is implicitly contained in 1)-3).

This fact can be revealed by the methods used in [5], since for this it is not required to take into account the peculiarities of the terms. We will now replace the second equality by the following one: 2) $h(x+y)=h(z):g(x)+y=g(z)$.

Now a formal application of the production inference will not help to find a "redundant" conditional equality 4), if one does not consider one more relation $x+y=z:h(x+y)=h(z)$, that follows from the properties of the terms.

In the present paper, the lattice-based algebraic model of the conditional equational theory is introduced. This model is an extension of LP structures theory models [6]. It takes into account possible connections between the terms, caused by the applications of functions and substitutions. The initial theory consists of conditional relations of the form $s_1=t_1,...,s_n=t_n:u_1=v_1,...,u_m=v_m$ ("if the equalities of the terms $s_i=t_i$, $i=1,...,n$ take place, then all $u_j=v_j$, $j=1,...,m$ are fulfilled"). We will call such relations (conditional) equational rules, or simply "rules", wherever this word will not cause any misunderstanding. The left side ($s_1=t_1,...,s_n=t_n$) will be called the condition, the right side ($u_1=v_1,...,u_m=v_m$) the conclusion of a rule. We interpret the equalities between the terms by the standard method of the equational theory: $s=t$, if the given equality can be obtained from the available set of equalities by means of the equational deduction under consideration. We will define the corresponding axioms and inference rules in section 2. They in a natural manner expand the set of axioms and inference rules described in [7] for conditional equalities of the form $s_1=t_1,...,s_n=t_n:s=t$.

In the proposed algebraic model, conditional equational rules are realized by a binary relation R on a lattice generated by the sets of equalities $\{s_i=t_i\}$. Note that R does not connect separate terms (as in the standard equational logic); it connects the sets of equalities between the terms: each rule $s_1=t_1,...,s_n=t_n:u_1=v_1,...,u_m=v_m$ is matched by the pair $(a,b)\in R$, where $a=\{s_1=t_1,...,s_n=t_n\}$ and $b=\{u_1=v_1,...,u_m=v_m\}$. Our model contains the logic of a production inference. We will call such inference a meta-inference, since it is applied not to the terms but to the subsets of equalities between the terms.

Apart from the model, the basic results of the paper (section 3) are as follows. For the given model the statement about the existence of a logical closure (theorem 3.1) is introduced, which makes it possible to formulate in the applications the notion of the equivalent equational theory. The possibility of local equivalent transformations of the initial relation (theorem 3.2) is grounded, as well as the possibility of transformation of a set of rules. The structure of a logical closure (theorem 3.3) is investigated, which allows one to use fast algorithms in its construction. The problems of existence and construction of a logical reduction of a binary relation (theorem 3.4) are studied. This provides a theoretical basis for the minimization of conditional equational theory. In our understanding, minimization is the obtaining of such equivalent system, from which it is impossible to remove any rule without violation of equivalence. The theorems are given without proofs because of the length limits for the paper.

In section 4, the conclusions are drawn and prospects for further investigations are specified.

2. Method: basic notions and notations

A binary relation R on some set F is called reflexive, if for any $a \in F$ holds $(a,a) \in R$; it is called transitive, if for any $a,b,c \in F$ from $(a,b),(b,c) \in R$ follows $(a,c) \in R$.

It is known that there exists a closure R^* of an arbitrary relation R with respect to the properties of reflexivity and transitivity (reflexive-transitive closure). There exists a notion of a transitive reduction of a binary relation. Its construction is an inverse problem with respect to the closure construction. For the given relation R we seek the minimum relation R', such that its transitive closure coincides with the transitive closure of R. As usual, for partially ordered sets we will discriminate between the notions of a minimal element (there is no smaller element for it) and the least element (it is the smallest one). In [8] an algorithm for the construction of a transitive reduction of the oriented graphs is given; it is shown that computationally this problem is equivalent to the problem of the construction of a transitive closure, and the uniqueness of a transitive reduction of an acyclic graph is proved.

A lattice is a partially ordered set $\mathbb{F}$, in which along with the relation $\le$ ("not greater than", "is contained") two two-place operations $\wedge$ ("meet") and $\vee$ ("join") calculating, respectively, the greatest lower and least upper bounds for any $a,b \in \mathbb{F}$ are introduced.

As is known, a set of all finite subsets $\lambda(U)$ of some set U is a lattice. In this paper, such form of lattices is considered. To underline this fact, instead of the symbols $\le$, $\ge$, $\wedge$ and $\vee$, we will use the signs of set-theoretic operations $\subseteq$, $\supseteq$, $\bigcap$ and $\bigcup$. However, we preserve the use of the term "lattice", since later on our results can also be extended to other forms of lattices [9].

Let us give some basic definitions connected with the terms [7].

Let Σ be an alphabet representing a union of the following non-intersecting sets: V is a set of variables; Σ_n, $n = 0,1,...$ are sets of n-arity functions (functional symbols); 0-arity functions are also called constants.

The set of terms $T(\Sigma)$ is defined recursively:

1) $V \subset T(\Sigma)$;
2) $\Sigma_0 \subset T(\Sigma)$;
3) if $f \in \Sigma_n$ and $t_1,...,t_n \in T(\Sigma)$, then $f(t_1,...,t_n) \in T(\Sigma)$.

The map $\sigma : V \to T(\Sigma)$ is called substitution. This notion is extended to all $t \in T(\Sigma)$ as follows:

1) if $t = x \in V$, then $\sigma(t) = \sigma(x)$;
2) if $t = f \in \Sigma_0$, then $\sigma(t) = f$;
3) if $f \in \Sigma_n$, $t_1,...,t_n \in T(\Sigma)$ and $t = f(t_1,...,t_n)$, then $\sigma(t) = f(\sigma(t_1),...,\sigma(t_n))$.

The pair (Σ, E) is called the equational theory, where Σ is the alphabet, consisting of a denumerable set of variables and a nonempty set of functional symbols (signature), and $E \subseteq T(\Sigma) \times T(\Sigma)$ is a set of equalities of the form $s = t$ $(s,t \in T(\Sigma))$. The notion of deducibility of the equality $s = t$ from E ($(\Sigma, E) \vdash s = t$) is defined:

1) $(\Sigma, E) \vdash t = t$;
2) if $s = t \in E$, then $(\Sigma, E) \vdash s = t$;
3) if $(\Sigma, E) \vdash s_1 = t_1,...,s_n = t_n$, then $(\Sigma, E) \vdash f(s_1,...,s_n) = f(t_1,...,t_n)$ $(f \in \Sigma_n)$;
4) if $(\Sigma, E) \vdash s = t$, then $(\Sigma, E) \vdash \sigma(s) = \sigma(t)$ for any substitution σ;
5) if $(\Sigma, E) \vdash t_1 = t_2, t_2 = t_3$, then $(\Sigma, E) \vdash t_1 = t_3$;
6) if $(\Sigma, E) \vdash s = t$, then $(\Sigma, E) \vdash t = s$.

For the given set of equalities E we will consider a set of finite subsets $\lambda(E)$. In it the relations of inclusion $\subseteq$, $\supseteq$, as well as "lattice" operations $\bigcap$ and $\bigcup$ (set-theoretic intersection and union) are given. Besides, we will need two more groups of operations connected with functions and substitutions of the terms, respectively:

1) if $a = \{s_i = t_i \mid i = 1,...,n\}$, $f \in \Sigma_n$, then $f(a) = \{f(s_1,...,s_n) = f(t_1,...,t_n)\}$;
2) if $a = \{s_j = t_j \mid j = 1,...,m\}$, then $\sigma(a) = \{\sigma(s_j) = \sigma(t_j) \mid j = 1,...,m\}$ for any substitution σ.

Definition 2.1. Let an equational theory .. be set. The lattice obtained by the completion of $\lambda(E)$ with respect to the operations 1)-2) defined above will be called an equational lattice $\mathbb{F}$.

As was already mentioned in section 1, we consider the conditional equational theory containing conditional rules of the form $s_1 = t_1,...,s_n = t_n : u_1 = v_1,...,u_m = v_m$. Thus, the condition and conclusion of a rule are elements of $\mathbb{F}$.

By analogy with [7], we will introduce axioms and rules of our conditional equational deduction. The axioms are generated by the production rules of the equalities specified above. The production rule 2) means the presence of a conditional rule (axiom) $a : f(a)$ for any $a = \{s_1 = t_1,...,s_n = t_n\}$ and $f \in \Sigma_n$; from 3) follows the axiom $a : \sigma(a)$ for any $a \in \mathbb{F}$ and substitution σ. In our logic, it is also possible to call such conditional rules tautologies. One more obvious tautology is the rule $a : b,\ a, b \in \mathbb{F}$ for $a \supseteq b$. We set these three axioms in our logic.

The inference rules in conditional equational logic are as follows:

1) $a : b \vdash \sigma(a) : \sigma(b)\ (a, b \in \mathbb{F})$ for any substitution σ (see a similar rule in [7] with initial notations);
2) $a : b,\ a : c \vdash a : b \cup c\ (a, b, c) \in \mathbb{F}$ (possible inference by parts);
3) $a : b,\ b : c \vdash a : c\ (a, b, c) \in \mathbb{F}$ (transitivity).

3. Results and discussion

In this section, we consider binary relations on the equational lattice. Below we will introduce the notion of a logical relation, which corresponds to a set of rules of the conditional equational theory. Properties of such a relation should reflect the axioms and rules of the conditional equational deduction formulated in section 2.

First, a logical relation R should contain all tautologies. For them we will introduce the general notation: $a \succcurlyeq b$, if $a \supseteq b$, $b = \sigma(a)$ or $b = f(a)$ Thus, for the logical relation R holds $\succcurlyeq \subseteq R$. Other properties of the logical relation follow from the deduction rules.

Definition 3.1. Let us call the relation R applicable, if for any substitution σ from $(a,b) \in R$ follows $(\sigma(a), \sigma(b)) \in R$.

Definition 3.2. Let us call the relation R $\cup$-distributive, if for any $(a,b_1),(a,b_2) \in R$ holds $(a, b_1 \cup b_2) \in R$.

The following definition summarizes the properties considered above.

Definition 3.3. A binary relation on the equational lattice is called logical, if it contains tautologies, and is applicable, $\cup$-distributive and transitive. The least logical relation containing R is called a logical closure of an arbitrary relation R.

Two relations R_1 and R_2, defined on a common equational lattice, are called equivalent, if their logical closures coincide. We denote this fact by $R_1 \sim R_2$. A minimal relation R_0 equivalent to it is called a logical reduction of the given relation R.

From the definition, no existence of a logical closure or reduction for an arbitrary binary relation follows. Below we will consider these problems.

Definition 3.4. Let some relation R be set on the equational lattice $\mathbb{F}$. We say that the relation R logically connects the ordered pair $a, b \in \mathbb{F}$ (we denote this fact by $a \xrightarrow{R} b$), if one of the following conditions is fulfilled:

1) $(a,b) \in R$;

2) $a \succcurlyeq b$:

2.1) $a \supseteq b$ or 2.2) $b = \sigma(a)$ or 2.3) $b = f(a)$;

3) there exist such $a_1, b_1 \in \mathbb{F}$ and substitution σ that $a = \sigma(a_1), b = \sigma(b_1)$, and $a_1 \xrightarrow{R} b_1$;

4) there exist such $b_1, b_2 \in \mathbb{F}$ that $b_1 \cup b_2 = b$, and $a \xrightarrow{R} b_1, a \xrightarrow{R} b_2$;

5) there exists an element $c \in \mathbb{F}$ such that $a \xrightarrow{R} c$ and $c \xrightarrow{R} b$.

Theorem 3.1. For an arbitrary relation R on the equational lattice a logical closure exists and coincides with a set $\xrightarrow{R}$ of all ordered pairs logically connected by the relation R.

Corollary 3.1. Let R be a binary relation on the equational lattice and $a_t \xrightarrow{R} b_t, \forall t \in T$. Then the relation $R' = R \cup \{(a_t, b_t) \mid t \in T\}$ is equivalent to R.

Let an arbitrary binary relation R on the equational lattice is given. Its equivalent transformation is such a replacement of the whole set of the ordered pairs R or a part of it that the new relation P obtained as a result is logically equivalent to R, i.e. $P \sim R$.

Theorem 3.2. Let R_1, R_2, R_3, R_4 be relations on the common equational lattice. If $R_1 \sim R_2$ and $R_3 \sim R_4$, then $R_1 \cup R_3 \sim R_2 \cup R_4$.

Corollary 3.2. Let R_1, R_2, R be relations on the common equational lattice. If $R_1 \sim R_2$, then $R_1 \cup R \sim R_2 \cup R$.

Corollaries 3.1 and 3.2 justify the principle of locality of equivalent transformations of logical relations.

Let us try to clarify the question, whether it is possible to select the stage, corresponding to the transitive closure in the general process of the construction of a logical closure. This will allow us to reduce the study of some important problems, regarding logical relations, to the corresponding problems of transitive relations. In particular, the construction of a logical closure or a reduction can be realized by means of fast algorithms (similar to Worshall's algorithm) [8]).

For an arbitrary relation R on the equational lattice we will consider a relation $\tilde{R}$, constructed with respect to the given R by a consecutive performance of the following steps:

1) add to R all pairs (a,b), for which $b = \sigma(a)$, or $b = f(a)$, and denote a new relation R_1;
2) add to R_1 all possible pairs of the form $(\sigma(a), \sigma(b))$, for which, $(a,b) \in R_1$, and denote a new relation R_2;
3) add to R_2 all possible pairs of the form $(a_1 \cup a_2 \cup \ldots \cup a_m, b_1 \cup b_2 \cup \ldots \cup b_m)$, where $(a_j, b_j) \in R_2$, $j = 1, \ldots, m$, and denote a new relation R_3;
4) combine R_3 with the relation $\supset$.

Note that by virtue of the infinity of the set $\mathbb{F}$, the described process of the construction of $\tilde{R}$ has a theoretical aspect. In applications, it is possible to take as $\mathbb{F}$ a final subset of the equational lattice, constructed under the boundedness of a maximal level of enclosure of the terms.

Theorem 3.3. A logical closure of the relation R on the equational lattice coincides with a transitive closure $\tilde{R}^*$ of the corresponding relation $\tilde{R}$.

Let us consider further a problem of existence and construction of a logical reduction of binary relations.

For the relation R on the equational lattice we will consider relation $\underset{\sim}{R}$ constructed by the consecutive performance of the steps given by R, inverse to the construction of $\tilde{R}$, namely:

1) exclude from R all pairs (a,b), for which $a \supset b$, and denote a new relation by R_{-1};
2) exclude from R_{-1} all pairs (a,b) of the form $(a_1 \cup a_2 \cup \ldots \cup a_m, b_1 \cup b_2 \cup \ldots \cup b_m)$, where $(a_j, b_j) \in R_{-1}$, $j = 1, \ldots, m$, and (a,b) coincides with no pair (a_j, b_j), and denote a new relation by R_{-2};

3) exclude from R_{-2} all possible pairs of the form $(\sigma(a),\sigma(b))$, for which $(a,b)\in R_{-2}$, and (a,b) does not coincide with the pair $(\sigma(a),\sigma(b))$, and denote a new relation by R_{-3};
4) exclude from R_{-3} all pairs (a,b), for which $b=\sigma(a)$, or $b=f(a)$.

The following theorem specifies a sufficient condition of existence and a way of construction of a logical reduction of the given relation.

Theorem 3.4. Let for the relation R on the equational lattice $\mathbb{F}$ the corresponding relation $\tilde{R}$ be constructed. Then, if for $\tilde{R}$ there exists a transitive reduction R^0, then the relation $\underset{\sim}{R}^0$ corresponding to it represents a logical reduction of the initial relation R.

Note that the requirement of existence of a transitive reduction of the relation $\tilde{R}$ under certain conditions can appear to be superfluous for the existence of a logical reduction of the initial relation R. It is obvious that when the set R is finite, it is always possible to remove sequentially "superfluous" pairs from it and to obtain, as a result, a logical reduction. However, if the lattice $\mathbb{F}$ is infinite and has the corresponding structure, then by combining R with the relation $\supseteq$ it is possible to construct relation $\tilde{R}$ having no transitive reduction. Thus, the question of the strengthening of theorem 3.4 arises. One of the possible ways is to use in the above constructions instead of the relation $\supseteq$ the relation $\supseteq_R$ containing only a subset of $\supseteq$ necessary for deriving a logical reduction of R. It is also possible to take a finite subset of the equational lattice constructed on the basis of the restriction of the level of the terms enclosure. In this case, the statements and proofs in the present work will become more cumbersome, but will not yield any essential innovations. Therefore, we will keep these ideas only as recommendations for concrete applications.

Let us briefly illustrate the process of minimization of the set of rules by the modified example from section 1:

1) $x+y=z:s(x)+y=s(z)$;
2) $h(x+y)=h(z):g(x)+y=g(z)$;
3) $s(x)+y=s(z),g(x)+y=g(z):f(x)=f(z)$;
4) $x+y=z:f(x)=f(z)$.

Let us introduce the elements of the lattice $\mathbb{F}$:

$$A\sim\{x+y=z\};\ B\sim\{s(x)+y=s(z)\};\ C\sim\{h(x+y)=h(z)\};$$
$$D\sim\{g(x)+y=g(z)\};\ E\sim\{f(x)=f(z)\}.$$

We denoted them by capital letters to avoid duplication of the symbols of the terms. The initial binary relation R consists of the pairs $(A,B),(C,D),(B\cup D,E),(A,E)$, corresponding to conditional rules 1)-4). We will not describe here a full process of construction of the relation $\tilde{R}$. For brevity, we will generate its subset, which will be enough for the problem solving. So, by virtue of the obvious equality $C=h(A)$, we have the tautology (A,C), which at construction of $\tilde{R}$ should be added at step 1). At the same step we will add one more tautology (B,B). Further, according to step 3) of the construction process of $\tilde{R}$ in this set it is necessary to include the pair $(A,B\cup C)$ (on the basis of the available pairs $(A,B),(A,C)$) and the pair $(B\cup C,B\cup D)$ (on the basis of $(B,B),(C,D)$). Thus, we have

$$\{(A,B),(C,D),(B\cup D,E),(A,E),(A,C),(B,B),(A,B\cup C),(B\cup C,B\cup D)\}\in\tilde{R}.$$

Note that in practice at the construction of $\tilde{R}$ it is not actually necessary to add the tautologies. It is enough to realize an effective mechanism of checking whether the given pair is a tautology.

Making a transitive reduction of the obtained set, we notice the presence of a chain of the pairs $(A,B\cup C),(B\cup C,B\cup D),(B\cup D,E)$ in it. Hence follows the transitivity in $\tilde{R}$ of the pair (A,E). Thus, the pair (A,E) will be excluded in the process of finding a transitive reduction R^0 of the relation $\tilde{R}$. Further operations correspond to the construction for R^0 of the relation $\underset{\sim}{R}^0$. At this stage, the pairs $(B\cup C, B\cup D)$, $(A, B\cup C)$ added earlier and the tautologies (B,B), (A,C) will be excluded.

As a result, we obtain the relation $\underset{\sim}{R}^0 = \{(A,B),(C,D),(B\cup D,E)\}$, which is a logical reduction of the initial relation R and corresponds to the first three conditional equational rules of the example.

4. Conclusions

In the present work, an algebraic lattice-based model of the conditional equational theory is proposed, and a theoretical study of this model is carried out. The results obtained justify local equivalent transformations of a set of conditional rules, as well as its optimization by deriving an equivalent system with a minimal set of rules.

Later on, it will be possible to consider a more general algebraic model, using as its basis the Lindenbaum-Tarsky lattice [10] instead of $\lambda(E)$. Then the conditional rules can, as premises and inferences, contain formulas of propositional calculus, whereas the general methods of research will remain just the same.

Another possible direction is connected with a deeper reviewing of the structure of conditional rules. In section 1, we pointed out that by replacing the terms by independent elements of some set it is possible to optimize the system of rules on the basis of the methods described in [5]. This is the first level of the research. In the present work, we propose a model of the second level, taking into account the connections between the equalities on the basis of functions and substitutions. A deeper third level could consider the structure of separate terms in equalities. Let us consider an example:

1) $x+y=z : s(x)+y=s(z)$;
2) $x+y=z : s(z)=g(x+z)$;
3) $x+y=z : s(x)+y=g(x+z)$.

Here conditional rule 3) is "superfluous": at $x+y=z$ by rules 1)-2) from $s(x)+y=s(z)$ and $s(z)=g(x+z)$ automatically follows $s(x)+y=g(x+z)$. However, this fact can be observed only at the third level of the research.

At the same level, the solutions are found to other problems, which depend on the structure of equalities and terms. For example, the problem of the influence of extra variables [11] on the possibility of a logical reduction (such variables are present in rules 1)-3)). The results obtained at the level of the research of the given paper do not depend on extra variables.

As far as the estimates of the complexity of algorithms of the construction of a logical reduction are concerned, here the forecasts are not always optimistic. As was shown in [5], the problem of minimization of Horn functions is NP-complete. This already corresponds to the complexity of minimization of the set of conditional rules at the upper level. In our work, we have described the construction not of the least, but of the minimal set of rules. The use of a fast algorithm for the construction of a transitive reduction [8] yields complexity $O(N^3)$, if it is possible to realize effectively the construction of relations $\tilde{R}$ and $\underset{\sim}{R}$. Note also that the number of conditional rules N can appear comparable to the cardinality of the power set $2^{n\times n}$, where n is the number of terms under consideration.

It would be interesting to consider the model of the oriented rules and clarify the question of how the logical equivalent transformations of a set of rules influence the basic properties of the initial conditional TRS (noetherian property and confluence [3, 12]). It is possible to assume that by virtue of the equivalence of transformations the basic properties of TRS will be preserved, however, this problem requires formal investigations.

References

[1] Metivier Y 1983 About the Rewriting Systems Produced by the Knuth-Bendix Completion

Algorithm *Inf. Process. Lett.* **16(1)** pp 31–4
[2] Toyama Y 1986 On Equivalence Transformations for Term Rewriting Systems *Proc. 1983 and 1984 RIMS Symp. on Software Science and Engineering II* vol 220, ed E Goto and K Araki et al (London: Lecture Notes In Computer Science, Springer-Verlag) pp 44–61
[3] Dershowitz N, Okada M and Sivakumar G 1988 Canonical Conditional Rewrite Systems *Proc. 9th Int. Conf. on Automated Deduction (May 23 - 26, 1988)* vol 310, ed E L Lusk and R A Overbeek (London: Lecture Notes In Computer Science, Springer-Verlag) pp 538–49
[4] Gmeiner K and Nishida N 2014 Notes on Structure-Preserving Transformations of Conditional Term Rewrite Systems *Proc. 1st Int. Workshop on Rewriting Techniques for Program Transformations and Evaluation (13th July 2014, Vienna, Austria)* ed M Schmidt-Schau and Masahiko Sakai et al pp 3–14
[5] Hammer P L and Kogan A 1993 Optimal compression of propositional Horn knowledge bases: complexity and approximation *Artif. Intel.* **64(1)** pp 131–45
[6] Makhortov S D 2012 Production-logic Relations on Complete Lattices *Autom. Remote Control* **73** pp 1937–43
[7] Klop J W 1992 Term rewriting systems *Handbook of Logic in Computer Science Vol. 2: Background: Computational Structures*, ed S Abramsky and D M Gabbay et al (New York: Oxford University Press) pp 1–116
[8] Aho A V, Garey M R and Ulman J D 1972 The transitive reduction of a directed graph *SIAM J. Computing* **1(2)** pp 131–7
[9] Makhortov S D 2009 LP Structures on Type Lattices and Some Refactoring Problems *Program. Comput. Soft.* **35** pp 183–9
[10] Rasiowa H and Sikorski R 1963 *The Mathematics of Metamathematics* (Warszawa: PWN) 519
[11] Ohlebusch E 1999 Transforming Conditional Rewrite Systems with Extra Variables into Unconditional Systems *Proc. 6th Int. Conf. on Logic Programming and Automated Reasoning (September 06-10 1999)* vol 1705, ed H Ganzinger and D A McAllester et al (London: Lecture Notes In Computer Science, Springer-Verlag) pp 111–30
[12] Gmeiner K 2017 Confluence of Conditional Term Rewrite Systems via Transformations *Proc. 3rd Int. Workshop on Rewriting Techniques for Program Transformations and Evaluation (23rd June 2016, Porto, Portugal)* ed H Cirstea and S Escobar pp 32–45

IOP Conf. Series: Journal of Physics: Conf. Series **973** (2018) 012067 doi:10.1088/1742-6596/973/1/012067

A solver for General Unilateral Polynomial Matrix Equation with Second-Order Matrices Over Prime Finite Fields

Filipp Burtyka

Junior Research Fellow, Institute of Computer Technologies and Information Security of Southern Federal University, 2 Chekhova St., Taganrog, Russia

E-mail: bbfilipp@yandex.ru

Abstract. The paper *firstly* considers the problem of finding solvents for *arbitrary* unilateral polynomial matrix equations with second-order matrices *over prime finite fields* from the practical point of view: we implement the solver for this problem. The solver's algorithm has two step: the first is finding solvents, having Jordan Normal Form (JNF), the second is finding solvents among the rest matrices. The first step reduces to the finding roots of usual polynomials over finite fields, the second is essentially exhaustive search. The first step's algorithms essentially use the polynomial matrices theory. We estimate the practical duration of computations using our software implementation (for example that one can't construct unilateral matrix polynomial over finite field, having any predefined number of solvents) and answer some theoretically-valued questions.

1. Introduction

Let $\mathbb{K}$ be a field and let $\mathsf{M}_n(\mathbb{K})$ denote the ring of all $n \times n$ matrices with entries in $\mathbb{K}$. The field $\mathbb{K}$ is meant to be $\mathbb{C}$ or $\mathbb{F}_p$, for some $p \in \mathbb{N}$ (when p is a prime number, then field is called *prime*).

The *unilateral matrix polynomial (UMP)* of n-th order and d-th degree over $\mathbb{K}$ is the expression of the form

$$\mathcal{F}(X) = X^d + \mathbf{F}_{d-1} \cdot X^{d-1} + \cdots + \mathbf{F}_2 \cdot X^2 + \mathbf{F}_1 \cdot X + \mathbf{F}_0, \tag{1}$$

where $\mathbf{F}_i \in \mathsf{M}_n(\mathbb{K}), i = 0, ..., d-1$ are coefficients and $X \in \mathsf{M}_n(\mathbb{K})$ is (unknown) variable.

The set of all such UMP for given n and $\mathbb{K}$ we denote by $(\mathsf{M}_n(\mathbb{K}))[X]$. The *root* (or *solvent*) of UMP over $\mathbb{K}$ is a matrix $\mathbf{S} \in \mathsf{M}_n(\mathbb{K})$, such that $\mathcal{F}(\mathbf{S}) = \mathbf{0}$, where $\mathbf{0}$ is zero n-th order matrix. The set of all UMP's roots we denote $\mathsf{Ker}(\mathcal{F}(X))$.

Remark 1. *This paper considers only left UMPs, while right unilateral matrix polynomials exist looking like* $\mathcal{F}(X) = X^d + X^{d-1} \cdot \mathbf{F}_{d-1} + \cdots + X^2 \cdot \mathbf{F}_2 + X \cdot \mathbf{F}_1 + \mathbf{F}_0$. *Our below discussion without loss of generality can be extended to them.*

The most important problems related to UMP $\mathcal{F}(X)$ are *(i)* whether $\mathcal{F}(X)$ has solvents and *(ii)* how to find these solvents. The main challenges are non-commutative nature of UMPs, the presence of zero-divisors in $\mathsf{M}_n(\mathbb{K})$ and as a consequense anomalous (in contrast to usual scalar polynomials) total number of these solutions. Consider two examples, where $\mathbb{K}$ is $\mathbb{C}$.

IOP Conf. Series: Journal of Physics: Conf. Series **973** (2018) 012067 doi:10.1088/1742-6596/973/1/012067

Example 1 (The unresolvable UMP).

$$\mathcal{F}(X) = X^2 - \begin{pmatrix} 0 & 1 \\ 0 & 0 \end{pmatrix} \in (\mathsf{M}_2(\mathbb{C}))[X]$$

The root of $\mathcal{F}(X)$ can be only the matrix, having exclusively zero eigenvalues (because the square of this matrix has only zero eigenvalues), i.e. nilpotent matrix. But the square of any nilpotent 2×2 matrix is the zero matrix.

Example 2 (The UMP having infinity roots).

$$\mathcal{F}(X) = X^2 - \begin{pmatrix} 0 & 0 \\ 0 & 0 \end{pmatrix} \in (\mathsf{M}_2(\mathbb{C}))[X]$$

The roots are all nilpotent 2×2 matrices.

The following theorem summarizes the general results about the total number of solutions for $\mathbb{K}$ is $\mathbb{C}$ from a plenty of works devoted to analysis of UMP over $\mathbb{C}$.

Theorem 1. *(i) There exist unresolvable UMPs $\mathcal{F}(X) \in (\mathsf{M}_n(\mathbb{C}))[X]$.*

(ii) There exist UMPs $\mathcal{F}(X) \in (\mathsf{M}_n(\mathbb{C}))[X]$, having infinity number of solvents.

(iii) If $\mathcal{F}(X) \in (\mathsf{M}_n(\mathbb{C}))[X]$ has finite number of solvents, then $|\mathsf{Ker}(\mathcal{F}(X))| \leq C^n_{n \cdot d}$.

An important question is: what can we say about the total number of solvents for given UMP $\mathcal{F}(X) \in (\mathsf{M}_n(\mathbb{F}_p))[X]$, where p is prime number? This problem hasn't been studied yet in literature for the case of arbitrary $\mathcal{F}(X)$. May be this was caused by the fact that until 2014 UMP-related applications consider only $\mathbb{C}$ case (it was applications from mechanics and control theory, see, for example [34, 35]). But today it has important practical applications in homomorphic cryptography [17, 18], because finding solvents of UMP over $\mathbb{F}_p$ in fact simulates cryptanalitic attack on some homomorphic encryption. In particular, we interested in problem: to what complexity class belongs the problem of finding UMP's solvents? For investigation this problem the first step is to determine, if for any number $t \in [0, p^{n^2}] \cap \mathbb{N}$ there exist UMP, having exactly t solvents? Another aim is to simulate *in practice* the process of solving UMP problem by attacker.

In this work we partially solve both tasks. We discuss algorithms for computing all solvents of $\mathcal{F}(X) \in (\mathsf{M}_2(\mathbb{F}_p))[X]$ belonging to the class of matrices $\in \mathsf{M}_2(\mathbb{F}_p)$ with all eigenvalues in $\mathbb{F}_p$ (the last class of matrices denoted by $Eig_2(\mathbb{F}_p)$), and estimate the complexity of these algorithms.

Our results are essentially based on the thorough study of computing the roots and estimating their number for second-degree UMP $\mathcal{F}(X) \in (\mathsf{M}_2(\mathbb{C}))[X]$ presented in [4]. We make an adaptation of methods (informally) described in [4] for $\mathbb{C}$ to $\mathbb{F}_p$.

The paper is organized as follows: Section 2 reviews previous works devoted to finding UMP's solvents, Section 3 gives necessary theoretical background concerning polynomial matrices, Sections 4 and 5 discuss algorithms for finding UMP's solvents, Section 6 shows experimental results and, finally Section 7 is conclusion.

2. Related works

2.1. The works concerning the UMPs over $\mathbb{C}$

The major part of works dealing with matrix polynomials is devoted to calculating solvents of UMP $\mathcal{F}(X) \in (\mathsf{M}_n(\mathbb{C}))[X]$. Several researchers [1, 2, 5, 10, 7] presented the result formalized in Theorem 1. This result was obtained using polynomial matrix theory [8] (see section 3). The detailed case study for second-order matrices was done in [4, 3].

IOP Conf. Series: Journal of Physics: Conf. Series **973** (2018) 012067 doi:10.1088/1742-6596/973/1/012067

In [10] the author thoroughly studies the properties of block-companion matrix. It can be composed by coefficients of UMP $\mathcal{F}(X) \in (\mathsf{M}_n(\mathbb{C}))[X]$ and gives an insight how matrix polynomials solvents are built inside. Also the conditions for the number of solvents to be infinite are stated and some cases of finite (positive) solvents number are considered.

Theorem 1 states if $\mathcal{F}(X) \in (\mathsf{M}_n(\mathbb{C}))[X]$ has the finite number of solvents t, then $t \leq C^n_{n \cdot d}$. But for arbitrary n it's unknown if $\mathcal{F}(X) \in (\mathsf{M}_n(\mathbb{C}))[X]$ exists having exactly t roots. In [5] the author solves this problem for $n = 2$. He shows for $n = 2$ and $\forall d, t \leq C^2_{2 \cdot d}$ d-th degree UMP exists with exactly t solvents. To achieve this result the author builds UMP of special construction. Such UMP has only diagonalizable (recall that a square matrix $\mathbf{A}$ is called *diagonalizable* if it is similar to a diagonal matrix, i.e., if there exists an invertible matrix $\mathbf{P}$ such that $\mathbf{P}^1\mathbf{AP}$ is a diagonal matrix) solvents, and their number could be exactly t.

Finally there is a plenty of works proposing numerical algorithms for computing solvents of UMP $\mathcal{F}(X) \in \mathsf{M}_n(\mathbb{C})[X]$. For example, in [1, 22, 23, 24, 25] the authors propose how to adapt numerical methods of computing matrix polynomials solvents.

2.2. The works concerning the UMPs over $\mathbb{F}_p$

To the best of our knowledges the problem of computing and estimating the number of roots for arbitrary given $\mathcal{F}(X) \in \mathsf{M}_n(\mathbb{F}_p)[X]$ is not studied in literature. But anyway some case studies have been done.

Equation $X^n = 0$ over $\mathsf{M}_n(\mathbb{F}_p)$ was investigated in the following works [21, 11, 12]. It's well known that solutions of this equations are nilpotent $n \times n$ matrices. It's proved in [21] the total number of $n \times n$ nilpotent matrices is p^{n^2-n}.

The study of solving $X^n = \mathbf{B}$ over $\mathsf{M}_n(\mathbb{F}_p)$ was done in [13, 14] , where $\mathbf{B} \in \mathsf{M}_n(\mathbb{F}_p)$ – some arbitrary matrix.

Also there are works researching the question how to find matrices $\mathbf{S} \in \mathsf{M}_n(\mathbb{F}_p)$ being solutions of given scalar polynomial $f(x) \in \mathbb{F}_p[x]$ (see, for example, [15]).

3. Theoretical background on polynomial matrices and solvents of matrix polynomials

Each UMP can be associated with a polynomial matrix (called λ-*matrix*) $\mathcal{F}(\lambda)$ which is obtained from $\mathcal{F}(X)$ through replacing matrix-variable $X \in \mathsf{M}_n(\mathbb{K})$ by scalar $\lambda \in \mathbb{K}$, i.e. $\mathcal{F}(\lambda) = \lambda^d + \mathbf{F}_{d-1} \cdot \lambda^{d-1} + \cdots + \mathbf{F}_0 \in (\mathsf{M}_n(\mathbb{K}[\lambda]))$.

Below definitions has been formulated in [10] for field $\mathbb{C}$, but here we rewrite them for arbitrary field $\mathbb{K}$.

Definition 1 ([10]). *The latent root (or eigenvalue) of λ-matrix $\mathcal{F}(\lambda)$ is $\alpha \in \mathbb{K}$ such that* $\det(\mathcal{F}(\alpha)) = 0$.

Definition 2 ([10]). *The latent vector $\vec{v} \in \mathbb{K}^n$ (or eigenvector) of $\mathcal{F}(\lambda)$, corresponding to eigenvalue $\alpha \in \mathbb{K}$, is a vector such that $\mathcal{F}(\alpha) \cdot \vec{v} = \vec{0}$.*

Definition 3 ([10]). *Given $\mathcal{F}(\lambda)$ of second degree and its eigenvalue $\alpha \in \mathbb{K}$, let $\mathcal{F}'$ denote the derivative of $\mathcal{F}$. If vectors $\vec{v_1}, \vec{v_2} \in \mathbb{K}^n$, with $\vec{v_1} \neq \vec{0}$, not necessary distinct or linearly independent satisfy*

$$\mathcal{F}(\alpha)\vec{v_2} + \mathcal{F}'(\alpha)\vec{v_1} = \vec{0}$$

we say $\vec{v_1}, \vec{v_2}$ is a Jordan chain (or generalized eigenvectors) of length 2 of $\mathcal{F}(\lambda)$ corresponding to the eigenvalue α.

Definition 4 ([10]). *Let there is $\mathcal{F}(\lambda)$ and the matrix $\mathbf{V}$ over $\mathbb{K}$ of dimension $n \times k$, $\mathbf{V} = [\vec{v_1} \ \dots \ \vec{v_k}]$, where $\vec{v_1}, \dots, \vec{v_k}$ is a Jordan chain of $\mathcal{F}(\lambda)$ corresponding to eigenvalue α. The pair $(\mathbf{V}, \mathbf{J})$, where $\mathbf{J}$ – Jordan block of size k with α in the main diagonal, is an eigenpair of $\mathcal{F}(\lambda)$.*

Theorem 2 ([10]). *If the matrix* $\mathbf{T} = [\mathbf{V}_1 \ \dots \ \mathbf{V}_t]$ *is a nonsingular matrix of order* n *and* $\mathbf{J}_0 = \mathsf{diag}(\mathbf{J}_1, \dots, \mathbf{J}_t)$ *is also matrix of order* n*, then* $(\mathbf{V}_i, \mathbf{J}_i), i = \overline{1,t}$ *are eigenpairs of* $\mathcal{F}(\lambda)$ *if and only if* $\mathbf{S} = \mathbf{T} \cdot \mathbf{J}_0 \cdot \mathbf{T}^{-1}$ *is a root of UMP* $\mathcal{F}(X) \in (\mathsf{M}_n(\mathbb{K}))[X]$.

Theorem 2 shows the internal structure of solvents of given matrix polynomial $\mathcal{F}(X) \in (\mathsf{M}_n(\mathbb{K}))[X]$. It states that any solvent of $\mathcal{F}(X)$ can be composed by some subsets of $\mathcal{F}(\lambda)$ eigenpairs. Of course if the basic field $\mathbb{K}$ is not algebraically closed, than some part of eigenpairs are matrices over extension of $\mathbb{K}$. The simple corollary of Theorem 2 is if $\mathbf{S}$ is a root of $\mathcal{F}(X)$, then all its eigenvalues and eigenvectors are eigenvalues and eigenvectors of $\mathcal{F}(\lambda)$.

Theorem 2 says how to compute all $\mathcal{F}(X)$ solvents: *find all eigenpairs and try to combine them in a different ways to get all different solvents of* $\mathcal{F}(X)$.

The next theorem was proved in [2], but it can be viewed as a direct consequence of Theorem 2. It shows how to get all diagonalizable solvents of $\mathcal{F}(X)$ over $\mathbb{K}$.

Theorem 3 ([2]). *Let* $\mathcal{F}(\lambda) \in (\mathsf{M}_n(\mathbb{K}))[\lambda]$ *has eigenvalues* $\lambda_1, \dots, \lambda_n \in \mathbb{K}$ *and their corresponding eigenvectors form linear independent system* $\{\vec{v}_1, \dots, \vec{v}_n\}$*. Then*

$$\mathbf{S} = \mathbf{V} \cdot \mathsf{diag}(\lambda_1, \dots, \lambda_n) \cdot \mathbf{V}^{-1} \in \mathsf{M}_n(\mathbb{K}) \tag{2}$$

is root of $\mathcal{F}(X)$*, where i-th column of* $\mathbf{V} \in \mathsf{M}_n(\mathbb{K})$ *is* $\vec{v}_i$*. Any diagonalizable root of* $\mathcal{F}(X)$ *can be built in this way.*

Denote as $\mathcal{D}_2(\mathbb{K})$ the set of all *diagonalizable* 2×2 matrices with entries from $\mathbb{K}$.

Since in this work we are going to deal only with 2×2 matrices, let's see how the roots of $\mathcal{F}(X) \in (\mathsf{M}_2(\mathbb{F}_p))[X]$ belonging to $Eig_2(\mathbb{F}_p)$ will look like. We have two types of solvents:

- Diagonalizable solvent is $\mathbf{S} = \mathbf{V} \cdot \begin{pmatrix} \alpha_1 & 0 \\ 0 & \alpha_2 \end{pmatrix} \cdot \mathbf{V}^{-1}$, where $\mathbf{V} = [\vec{v}_1 \ \vec{v}_2] \in \mathsf{M}_2(\mathbb{F}_p)$, $\alpha_i \in \mathbb{F}_p$, $(\alpha_i, \vec{v}_i), i = 1, 2$ are two eigenpairs of $\mathcal{F}(\lambda)$, $\vec{v}_1, \vec{v}_2$ are linear independent.
- Non-diagonalizable solvent is $\mathbf{S} = \mathbf{V} \cdot \begin{pmatrix} \alpha & 1 \\ 0 & \alpha \end{pmatrix} \cdot \mathbf{V}^{-1}$, where $\mathbf{V} = [\vec{v}_1 \ \vec{v}_2] \in \mathsf{M}_2(\mathbb{F}_p)$ is a Jordan chain of $\mathcal{F}(\lambda)$ corresponding to eigenvalue $\alpha \in \mathbb{F}_p$, $\vec{v}_1, \vec{v}_2$ are linear independent (latent) vectors, α has algebraic multiplicity ≥ 2. Recalling theorem 2 we get $\vec{v}_1, \vec{v}_2$ are solutions of the following equations system:

$$\begin{cases} \mathcal{F}(\alpha)\,\vec{v}_1 = \vec{0} \\ \mathcal{F}(\alpha)\,\vec{v}_2 + \mathcal{F}'(\alpha)\,\vec{v}_1 = \vec{0}. \end{cases} \tag{3}$$

4. Finding solvents by exhaustive search and by solving corresponding system of scalar polynomial equations

Since the number of solvents for UMP over $\mathbb{F}_p$ is always finite, one can find all of them using exhaustive search. The complexity of evaluating d-th degree UMP $\mathcal{F}(X) \in (\mathsf{M}_2(\mathbb{F}_p))[X]$ at some point $\mathbf{S} \in \mathsf{M}_2(\mathbb{F}_p)$ is $O(d)$, if one uses Horner's rule. And since the total number of 2×2 matrices is p^4, the complexity of brute force search is $O(d \cdot p^4)$ arithmetic operations in $\mathbb{F}_p$.

Another technique is just write in symbolic form the unknown solvent's entries and construct a system of usual polynomial equations that then solve using such techniques as Grobner basis, linearisation, e t.c. This was done in [31], where one can find tables for number of monic UMP's, having various numbers of solvents. In particular, in Table 1 from [31] one can see that for every $t \in [0, 16] \cap \mathbb{N}$. Such experiments was carried out for all $\mathbb{F}_2, \dots, \mathbb{F}_{31}$ and for matrix dimensions $2, \dots, 7$ and degrees 2..7 (for larger fields, dimensions and degrees our computer failed to finish). This allows to propose the following hypothesis:

Hypothesis 1. *For any prime field* $\mathbb{F}_p$ *and any* $t \in [0, p^n] \cap \mathbb{N}$ *there exist UMP* $\mathcal{F}(X) \in (\mathsf{M}_n(\mathbb{F}_p))[X]$ *having exactly* t *solvents.*

IOP Conf. Series: Journal of Physics: Conf. Series **973** (2018) 012067 doi:10.1088/1742-6596/973/1/012067

5. Finding solvents using λ-matrixes

Now let's proceed to algorithms for finding solvents using polynomial matrices theory. Almost immediately from the definition of λ-matrices and latent roots follows Algorithm 1.

5.1. *The overview of algorithm for finding all solvents*

We recall there are two types of solvents – diagonalizable and non-diagonalizable ones (see Section 3). Following [4] solvents of each of these types are search in separately ways. Because our algorithm for $\mathbb{F}_p$ is adaptation of algorithm from [4] for $\mathbb{C}$ we also will adhere to this strategy: algorithm FindDiagonalizableSolvents$_2(\mathcal{F}(X))$ described below finds actually $\mathsf{Ker}(\mathcal{F}(X)) \cap \mathcal{D}_2(\mathbb{F}_p)$ and algorithm FindNonDiagonalizableSolventsHavingJNF$_2(\mathcal{F}(X))$ also described further finds actually $\mathsf{Ker}(\mathcal{F}(X)) \cap (Eig_2(\mathbb{F}_p) \setminus \mathcal{D}_2(\mathbb{F}_p))$.

In the case, when solvents finding problem is considered over $\mathbb{C}$ every matrix has JNF, therefore we can find non-diagonalizable solvents exploiting this fact. While all matrixes from $\mathcal{D}_2(\mathbb{F}_p)$ still have JNF, the remaining matrices are divided into two sets: the set of matrices, having JNF ($Eig_2(\mathbb{F}_p)$) and the set of matrices, not having JNF.

$$\mathsf{Ker}(\mathcal{F}(X)) = \Big(\mathsf{Ker}(\mathcal{F}(X)) \cap \mathcal{D}_2(\mathbb{F}_p)\Big) \cup \Big(\mathsf{Ker}(\mathcal{F}(X)) \cap (Eig_2(\mathbb{F}_p) \setminus \mathcal{D}_2(\mathbb{F}_p))\Big) \cup \Big(\mathsf{Ker}(\mathcal{F}(X)) \cap (\mathsf{M}_2(\mathbb{F}_p) \setminus Eig_2(\mathbb{F}_p))\Big) \quad (4)$$

But since unlike $\mathsf{M}_2(\mathbb{C})$ eigenvalues of matrix from $\mathsf{M}_2(\mathbb{F}_p)$ may not be in $\mathbb{F}_p$, for now we are forced to work with $Eig_2(\mathbb{F}_p)$ instead of entire $\mathsf{M}_2(\mathbb{F}_p)$. We recall $Eig_2(\mathbb{F}_p)$ is a big class of matrices, according to [19]:

$$|Eig_2(\mathbb{F}_p)| = \frac{1}{2} \cdot p^4 + p^3 - \frac{1}{2} \cdot p^2.$$

Keeping in mind $|\mathsf{M}_2(\mathbb{F}_p)| = p^4$ we get fraction $\frac{|Eig_2(\mathbb{F}_p)|}{|\mathsf{M}_2(\mathbb{F}_p)|}$ decreases to a limiting value $\frac{1}{2}$ as p approaches infinity. And this means we take into account more than half of matrices from $\mathsf{M}_2(\mathbb{F}_p)$ in our current work.

So, the overall algorithm is Algorithm 1.

Algorithm 1: FindRootsByPolynomialMatrix$_2(\mathcal{F}(X))$

Input: UMP $\mathcal{F}(X) \in \mathsf{M}_2(\mathbb{F}_p[\lambda])$
Output: Set of all solvents $\mathcal{F}(X)$

1 $D :=$ FindDiagonalizableSolvents$_2(\mathcal{F}(X))$
2 $J :=$ FindNonDiagonalizableSolventsHavingJNF$_2(\mathcal{F}(X))$
3 $N :=$ FindSolventsNotHavingJNF$_2(\mathcal{F}(X))$
4 Return $D \cup J \cup N$

5.2. *Diagonalizable and Non-diagonalizable solvents having JNF*

We search such solvents using techniques from [4] and summarize this in [38].

5.3. *Handling matrixes not having JNF*

Algorithm FindSolventsNotHavingJNF$_2(\mathcal{F}(X))$ mentioned in Algorithm 1 is just subsequent substitution in $\mathcal{F}(X)$ of all matrixes not having JNF. The set $Eig_2(F_p)$ can be prepared in advance, before running FindSolventsNotHavingJNF$_2(\mathcal{F}(X))$. It can be done through exhaustive search, but we propose slightly more computationally effective algorithm. Let's see at Algorithm 2.

Algorithm 2: GetMatricesNotHavingJNF$_2$($deg2irreducibles(F_p)$)

Input: list $deg2irreducibles(\mathbb{F}_p)$ of all monic irreducible polynomials over $\mathbb{F}_p$
Output: Set of matrixes not having Jordan Normal Form over $\mathbb{F}_p$

```
N := ∅
for each irreducible polynomial x^2 + p1 x + p0 ∈ deg2irreducibles(F_p) do
    for d = 0 to p − 1 do
        a := −(p1 + d)
        for b = 0 to p − 1 do
            if b ≠ 0 then
                c := −(d^2 + p1 d + p0) · b^(−1)
                N := N ∪ {(a b; c d)}
            end
            else
                for c = 0 to p − 1 do
                    N := N ∪ {(a b; c d)}
                end
            end
        end
    end
end
Return N
```

Algorithm 2 uses, in turn list of irreducible polynomials of secon degree over $\mathbb{F}_p$. Such list of polynomials can be found, for example, at [31] or [33] for tables with polynomials or approaches for fast generating.

6. Experimental results

Using NTL library by Victor Shoup [30] we implement both exhaustive search and algorithms for finding solvents using polynomial matrices theory.

From our experimens we found that for $\mathbb{F}_p$, $p = 2, 3, ..., 104729$ (first 10000 prime numbers) there exist UMP's having *any* number of solvents from $[0, p^4] \cap \mathbb{N}$ which acknowledge Hypothesis 1 for these p. Also, were carried out experiments for these p in order to found UMPs having exactly t diagonalizable solvents for each $t \in [0, p^4] \cap \mathbb{N}$. The experiments results that for each p there exist such t that *there is no UMPs* having t diagonalizable solvents. For example, Table 1 presents experimental data about the number of UMPs from $(\mathsf{M}_2(\mathbb{F}_2))[X]$ with each number of solvents in $[0, 8] \cup \mathbb{N}$ ($\mathsf{M}_2(\mathbb{F}_2)$ has only 8 diagonalizable matrices).

Another observation is that for sufficiently large degrees of UMPs (different for each $\mathbb{F}_p$) pure exhaustive search algorithm finished faster then algorithm based on λ-matrices, while for smaller degrees algorithm based on λ-matrices beats pure exhaustive search. Below we give experimental results for our implementation that runs on laptop with CPU AMD Phenom(tm) II P960 Quad-Core 1.80 GHz and 4.00 Gb RAM.

IOP Conf. Series: Journal of Physics: Conf. Series **973** (2018) 012067 doi:10.1088/1742-6596/973/1/012067

Table 1. Number of UMPs from $(\mathsf{M}_2(\mathbb{F}_2))[X]$ having exact numbers of diagonalizable solvent.

	deg = 2	**deg = 3**	**deg = 4**	**deg = 5**
without solvents	171	2736	43776	700416
with 1 solvent	66	1056	16896	270336
with 2 solvents	0	0	0	0
with 3 solvents	18	288	4608	73728
with 4 solvents	0	0	0	0
with 5 solvents	0	0	0	0
with 6 solvents	0	0	0	0
with 7 solvents	0	0	0	0
with 8 solvents	1	16	256	4096

Table 2. Ranges of time it takes to get the solvents of UMPs with various parameters using exhausive search.

	p = 2	**p = 3**	**p = 5**	**p = 7**
deg = 2	0..1 ms	0..5 ms	10..52 ms	38..230 ms
deg = 3	1..14 ms	1..16 ms	17..88 ms	
deg = 4	1..25 ms	3.. 39 ms		
deg = 5	2..43 ms			

Table 3. Ranges of time it takes to get the solvents of UMPs with various parameters using Lambda-matrices based solver.

	p = 2	**p = 3**	**p = 5**	**p = 7**
deg = 2	4..14 ms	4..30 ms	6..369 ms	8..741 ms
deg = 3	2..150 ms	5..190 ms		
deg = 4	2..251 ms			
deg = 5	3..328 ms			

7. Conclusion

The solver for UMP problem over $\mathbb{F}_p$ was proposed and implemented. The solver's algorithms based on λ-matrices theory and beat the pure exhaustive search algorithm for sufficiently small degrees and relatively big p (for example for $p > 19$ and deg of UMP less than ten).

Using the solver the Hypothesis 1 was checked for $\mathbb{F}_2, ..., \mathbb{F}_{104729}$, $n = 2$ and was confirmed for all these parameter values.

The technique using in [5] for proof that for each number t from 0 to C_n^k there exist UMP over $\mathbb{C}$ having exactly t solvents and that is the construction of UMP having *only* diagonalizable solvents *now is obviously not applicable to the case of* $\mathbb{F}_p$ while the statement itself seems to be valid.

Acknowledgments
The work was supported by the RFBR grant no. 16-37-00125 mol_a.

References
[1] Dennis J E, Traub J F and Weber R P 1976 The algebraic theory of matrix polynomials. *SIAM Journal on Numerical Analysis* **13**(6) pp 831–845.
[2] Dennis J E, Traub J F and Weber R P 1978 Algorithms for solvents of matrix polynomials *SIAM Journal on Numerical Analysis* **15** (3) pp 523-533
[3] Gelfand S I 2004 O chisle reshenij kvadratnogo uravnenija [On the number of quadratic equation's solutions] *Obshhematematicheskij seminar "Globus". [All-mathematical workshop "Globus"].* **1** pp 124–133 (In Russian)
[4] Wilson R L 1999 Polynomial equations over matrices. *Rutgers University preprint.*
[5] Slusky M 2010 Zeros of 2 × 2 matrix polynomials.*Communications in Algebra.* **38**(11) 4212–4223.
[6] Hellenbrand K 2009 Solutions of Polynomial Equations over Matrices.
[7] Shavarovskii B Z 2007 Finding a complete set of solutions or proving unsolvability for certain classes of matrix polynomial equations with commuting coefficients. *Computational Mathematics and Mathematical Physics.*, **47**(12) pp 1902–1911.
[8] Gantmakher F 1959 *The theory of matrices – Volume 1* (Chelsea, New York)
[9] Fuchs D and Schwarz A 1996 A Matrix Vieta Theorem. *E.B. Dynkin Seminar, Amer. Math. Soc. Transl. Ser. 2* . p **196**.
[10] Pereira E 2003 On solvents of matrix polynomials. Applied numerical mathematics. **47** pp 197–208.
[11] Kirillov A A 1995 O chisle reshenij uravnenija $X^2 = 0$ v treugol'nyh matricah nad konechnym polem [On the number of solutions to the equation $X^2 = 0$ in triangular matrices over a finite field]. *Functional Analysis and Its Applications.* **29**(1) pp 64–68. (In Russian)
[12] Ekhad S B and Zeilberger D 1996 The number of solutions of $X^2 = 0$ in triangular matrices over *GF (q)*. *The Electronic Journal of Combinatorics 3*, #R2.
[13] Hodges J H 1958 The matrix equation $X^2 - I = 0$ over a finite field. *American Mathematical Monthly.* **65**(7) pp 518–520.
[14] Acosta-De-Orozco M T and Gomez-Calderon J 1993 On the matrix equation $X^n = B$ over finite fields. *International Journal of Mathematics and Mathematical Sciences.* **16**(3) pp 539–544.
[15] Hodges J H 1958 Scalar polynomial equations for matrices over a finite field. *Duke Math. Journal* **25**(3) 291–296
[16] Hua J 1999 Polynomial equations for matrices over finite fields. *Bulletin of the Australian Mathematical Society.* **59**(1) 59–64.
[17] Burtyka F B 2014 Paketnoe simmetrichnoe polnost'ju gomomorfnoe shifrovanie na osnove matrichyh polinomov [Batch symmetric fully homomorphic encryption using matrix polynomials]. *Trudy Instituta sistemnogo programmirovanija RAN [The Proceedings of Institute for System Programming RAS]* **26**(5) pp 99–115.
[18] Burtyka F B and Makarevich O B 2014 Symmetric fully homomorphic encryption using decidable matrix equations. *Proceedings of the 7th International Conference on Security of Information and Networks.* ACM.
[19] Kaylor L and Offner D 2014 Counting matrices over a finite field with all eigenvalues in the field. *Involve, a Journal of Mathematics.* **7**(5) pp 627–645.
[20] Olsavsky G 2003 The Number of 2 by 2 Matrices over $\mathbb{Z}/p\mathbb{Z}$ with Eigenvalues in the Same Field. *Mathematics magazine.* **76**(4) pp 314–317
[21] Gerstenhaber M 1961 On the number of nilpotent matrices with coefficients in a finite field *Illinois Journal of Mathematics.* **5** pp 330-333
[22] Higham N J and Kim H M 2001 Solving a quadratic matrix equation by Newton's method with exact line searches. *SIAM Journal on Matrix Analysis and Applications.* **23**(2) pp 303-316.
[23] Peichang G 2014 The Newton-Shamanskii method for solving a quadratic matrix equation arising in quasi-birth-death problems *arXiv preprint arXiv:1406.1075*
[24] Fernando M and Pereira E 2010 A fixed point method to compute solvents of matrix polynomials *Mathematica Bohemica* **135**(4) pp 355-362
[25] Seo J-H and Kim H M 2008 Solving matrix polynomials by Newton's method with exact line searches *Journal of the Korea Society for Industrial and Applied Mathematics* **12**(2) pp 55–68
[26] Von Zur Gathen J and Panario D 2001 Factoring polynomials over finite fields: A survey *Journal of Symbolic Computation* 31.1-2: pp 3–17.
[27] Aho A V and Hopcroft J E 1974 The design and analysis of computer algorithms. *Pearson Education India*
[28] Knuth D E 1975 Son of seminumerical algorithms. *ACM SIGSAM Bulletin* **9**(4) pp 10–11

[29] Hart W and Novocin A 2011 Practical divide-and-conquer algorithms for polynomial arithmetic *Proceedings of International Workshop on Computer Algebra in Scientific Computing, Springer Berlin Heidelberg*

[30] Shoup V 2010 Number theory library (NTL) for C++ *Available at Shoups homepage* `http://shoup. net/ntl`

[31] Burtyka F B 2015 On complexity of Boolean matrix polynomials solving *Matem. Mod.* **27**:7 pp 25–30 (In Russian)

[32] Lidl R and Niederreiter H 1997 Finite fields (Vol. 20). Cambridge university press.

[33] Lopez-Ruiz R 2004 An approach to the problem of generating irreducible polynomials over the finite field GF(2) and its relationship with the problem of periodicity on the space of binary sequences. arXiv preprint nlin/0401011.`https://arxiv.org/pdf/nlin/0401011`

[34] Gohberg I, Lancaster P and Rodman L 2005 Matrix polynomials (pp. 237–266). Birkhuser Basel.

[35] Kratz W and Stickel E 1987 Numerical solution of matrix polynomial equations by Newton's method.*IMA journal of numerical analysis*, **7**(3) pp 355-369.

[36] Higham N J, Kim H M 2000 Numerical analysis of a quadratic matrix equation. IMA Journal of Numerical Analysis, **20**(4), pp 499–519.

[37] Ugalde E S 2015 Computation of invariant pairs and matrix solvents (Doctoral dissertation, Universit de Limoges).

[38] Burtyka F B 2017 Algorithms for computing solvents of unilateral second-order matrix polynomials over prime finite fields using lambda-matrices. *Proceedings International conference on Computer Simulation in Physics and beyond (CSP2017)*

Comprehensive efficiency analysis of supercomputer resource usage based on system monitoring data

A A Mamaeva[1], D I Shaykhislamov[1,2], Vad V Voevodin[2] and S A Zhumatiy[2]

[1] Faculty of Computational Mathematics and Cybernetics, Lomonosov Moscow State University, 119991 Russia, Moscow, Leninskie gory, 1, bld. 52

[2] Research computing center, Lomonosov Moscow State University, 119234 Russia, Moscow, Leninskie gory, 1, bld. 4

E-mail: vadim@parallel.ru

Abstract. One of the main problems of modern supercomputers is the low efficiency of their usage, which leads to the significant idle time of computational resources, and, in turn, to the decrease in speed of scientific research. This paper presents three approaches to study the efficiency of supercomputer resource usage based on monitoring data analysis. The first approach performs an analysis of computing resource utilization statistics, which allows to identify different typical classes of programs, to explore the structure of the supercomputer job flow and to track overall trends in the supercomputer behavior. The second approach is aimed specifically at analyzing off-the-shelf software packages and libraries installed on the supercomputer, since efficiency of their usage is becoming an increasingly important factor for the efficient functioning of the entire supercomputer. Within the third approach, abnormal jobs – jobs with abnormally inefficient behavior that differs significantly from the standard behavior of the overall supercomputer job flow – are being detected. For each approach, the results obtained in practice in the Supercomputer Center of Moscow State University are demonstrated.

1. Introduction

Supercomputer technologies are used very widely in the modern world. In different fields of science, such as medicine, astronomy, quantum physics, hydrodynamics, molecular biology and many others supercomputers are used on an ongoing basis to conduct simulations and perform numerical calculations. Thus, the speed of scientific research in these areas depends to a large extent on the efficiency of supercomputer usage. However, it often turns out that this efficiency is very low. There are several reasons for this. The first reason is an incredible complexity of the supercomputer architecture, which makes it almost impossible to take into account all the features and capabilities of the hardware platform when writing a parallel program. Another reason is an increase of large-scale use of supercomputing technologies – more and more people tend to use supercomputers, and in most cases new users are excellent experts in their subject areas, but they are not so experienced in the field of parallel computing.

This leads to the fact that the efficiency of supercomputer resource usage can be very low [1]. To solve this problem, it is necessary to study the behavioral features of both the supercomputer as a whole and the individual applications being executed on it. There are a lot of ready-to-use tools for analyzing the performance of individual parallel applications (trace tools, profilers, debuggers, etc), but, in our opinion, the problem of studying the performance of a supercomputer as a whole has not been studied in many respects. One of the possible research directions in this area is to study the entire flow of supercomputer applications based on system monitoring data, which is the purpose of this research. Solving this issue will allow the management and system administrators of supercomputer centers to increase the performance of their computing systems, which, in turn, will allow users to solve scientific problems faster and more efficiently.

In this article, we will describe three different approaches for conducting a comprehensive study of the resource utilization statistics, the performance of application packages, and anomalies in the application flow, accordingly. All these approaches are being actively developed in Research computing center of Lomonosov Moscow state university (RCC MSU).

The results in this paper are obtained in the MSU Supercomputer Center, i.e. allow to analyze its efficiency. But these results can be applied outside this center as well, since they show how similar analysis can be carried out on other systems and what useful conclusions it allows to make. Moreover, all these approaches are being developed with their portability in mind. Also it should be noted that these studies are a part of a general work carried out in RCC MSU aimed at analyzing the most diverse efficiency aspects of both individual parallel applications and supercomputer center as a whole [1].

Next, the description of these approaches will be given, as well as several interesting results obtained using these approaches in practice.

2. A description of the approaches to analyze supercomputer resource utilization efficiency

All the approaches described in this article are based on using system monitoring data. This allows to obtain all of the needed information on the execution of various jobs, at the right level of detail. In this case, studying supercomputer utilization efficiency requires examining the flow of parallel applications executed on the supercomputer. For that purpose, various dynamic characteristics are measured for each job executed on the supercomputer, allowing to describe its behavior. Since we are primarily interested in the efficiency of job execution, we will analyze the following characteristics: CPU utilization (CPU user load on the processing cores, system load (loadavg)), interaction with the memory subsystem (cache misses per second, memory reads/writes per second, etc.), InfiniBand or Ethernet network usage intensity (bytes and packets sent and received per second), file system usage intensity, etc.

Various existing monitoring tools can be used to obtain this data, such as Nagios [2], Zabbix, Zenoss, etc. MSU supercomputers currently use a set of custom-built solutions, but these will be replaced in the near future with DiMMon [3], a system being developed at the RCC MSU, designed specifically to run on extremely large-scale supercomputers and therefore offering good scalability, flexibility, and high performance. Since in this study we are interested in the overall supercomputer efficiency, not the efficiency of each particular application, in most cases only integral values need to be considered for each measurement of a specific job – for example, average user load or average number of L1 cache misses per second during the entire job runtime. This is not enough to conduct a detailed analysis of individual applications, but quite sufficient for studying the behavior of the supercomputer in general. In this case, a reasonable amount of data that does not require specific methods for big data processing is produced, which is also important for this kind of research.

General information on each job should also be collected, such as the job ID and user name; start and finish times; the name of used partition of the supercomputer; the number of cores and nodes used, etc. This data can easily be obtained from the resource manager, which is responsible for distributing and queuing supercomputer applications. The Slurm resource manager is used in MSU supercomputers.

Next, we will discuss three individual approaches in more detail.

2.1. Analysis of computing resource utilization statistics

The first approach is aimed at collecting and analyzing usage statistics for the supercomputer's computing resources. It is based on the analysis of the job flow structure – information is collected on how each job utilizes the available resources [4]. This study helps answer the following questions, among others:

- Are supercomputer resources – compute cores, nodes, sections, etc. – efficiently loaded?
- Are policies and quotas configured appropriately, is there any imbalance between users' activity or partitions' usage?
- Are queue planning strategies working properly?

There aren't many studies aimed at analyzing computing resource utilization statistics based both on resource manager data and information from system monitoring tools. The works that do exist are mainly intended for use on a specific supercomputer only (e.g., [5, 6]), and cannot be applied in our case. However, studying these works helped us to determine needed research directions for this study.

As previously mentioned, this analysis was performed based on the integral values for all available dynamic characteristics that describe the supercomputer resource utilization efficiency. This information can be taken directly from the database of the monitoring system, so obtaining input data is not an issue in this case.

It is much more of a challenge to figure out what useful conclusions can be derived from this information. Described approach is intended to answer this question. A number of methods were suggested that help obtain the most relevant statistics and evaluate resource utilization from different perspectives. A description of the three methods and their results is presented below. All of the methods are described in the article [4].

2.1.1. Identifying different classes of jobs. We have identified several classes of jobs with specific peculiarities that help us to evaluate the structure of the supercomputer job flow. For example, one class is "ideal-for-supercomputer" programs that actively utilize both compute nodes and communication network. These programs are very well suited to run on supercomputers, so one should try to increase the share of such jobs running on the machine. The statistics on the number of jobs in this class on the Lomonosov-2 supercomputer, along with some other classes, are shown in Table 1.

We can see that "ideal" programs are unfortunately not very common – only 5.5% of the total number of jobs launched since the beginning of the year. However, one must keep in mind that by far not all algorithms, let alone their specific implementations, can generally exhibit the efficient behavior required to be included in this class. Other classes, such as "suspicious" (very low CPU load and loadavg) or "non-communicative" (a job that occupies many compute nodes but almost does not use the communication network) are, on the contrary, less efficient in occupying supercomputer resources, so the share of these jobs should be decreased if possible. As can be seen from Table 1, these jobs are not numerous – less than 2% in sum on the Lomonosov-2 supercomputer since the beginning of the year.

Table 1. Statistics on the number of jobs in different classes for Lomonosov-2 supercomputer (on October 14, 2017).

	Last day	Last week	Last month	Since the beginning of the year
Overall number of jobs	388	5 317	16 676	298 676
«Ideal»	17	187	725	16 254
«Suspicious»	9	64	171	450
«Non-communicative»	10	130	735	5 173
«Serial»	3	10	39	1 304
«Single-node»	83	1143	3 612	28 348

The last two classes in Table 1 refer to jobs that may use the resources quite efficiently but require very few of them – one node ("single-node" jobs) or even one compute core ("serial"). The supercomputer is probably not needed for these jobs; they can be executed just fine on a standard

IOP Conf. Series: Journal of Physics: Conf. Series **973** (2018) 012068 doi:10.1088/1742-6596/973/1/012068

server or a workstation. So, ideally the number of these tasks should be kept to a minimum as well. It can be seen that very few of the tasks running on the Lomonosov-2 supercomputer belong to the "serial" class – less than 0.5%. However, single-node tasks are fairly common; unfortunately, this situation is similar at many supercomputer centers.

2.1.2. Studying the dynamics of supercomputer behavior. Another useful way of analyzing resource utilization statistics is to study how different aspects of supercomputer behavior change over time. How did the overall resource utilization change over the last months? Did some policy/quota adjustments help to increase the overall efficiency? Is there any periodicity in the usage of the supercomputer during a week/month/year? Answering these questions helps administrators of a supercomputer center evaluate the overall trends.

One of such interesting trends that are useful to analyze is the behavior of "big" jobs – jobs that require a lot of compute cores. We have analyzed and compared the data from Lomonosov-2 supercomputer within this year for 2 classes: "big" jobs with more than 256 cores, and all others ("small" jobs). The results have shown that the average values of both CPU load and loadavg for these classes are very similar. This means that average CPU utilization does not differ much for computationally-intensive and more light-weight jobs in our case. This is also true in terms of memory usage intensity – the number of load/store operations per second tends to be quite the same.

But the situation is different for communication network intensity. Figure 1 shows the distribution by month of an average number of bytes sent per second for two classes (the results for January are not available due to the lack of system monitoring data). Here we can see that in general "big" jobs show more intensive network usage. But it should be noted that "small" class includes jobs running on one node that does not use network at all, so this kind of behavior seems like an expected one. Taking this into account, the low values for "big" class in February as well as in October could mean that "big" jobs may not have used communication network efficiently. Of course, this could happen for many different reasons and does not necessarily indicate a performance issue, but in any case system administrator should conduct a more detailed analysis to find root causes of such behavior.

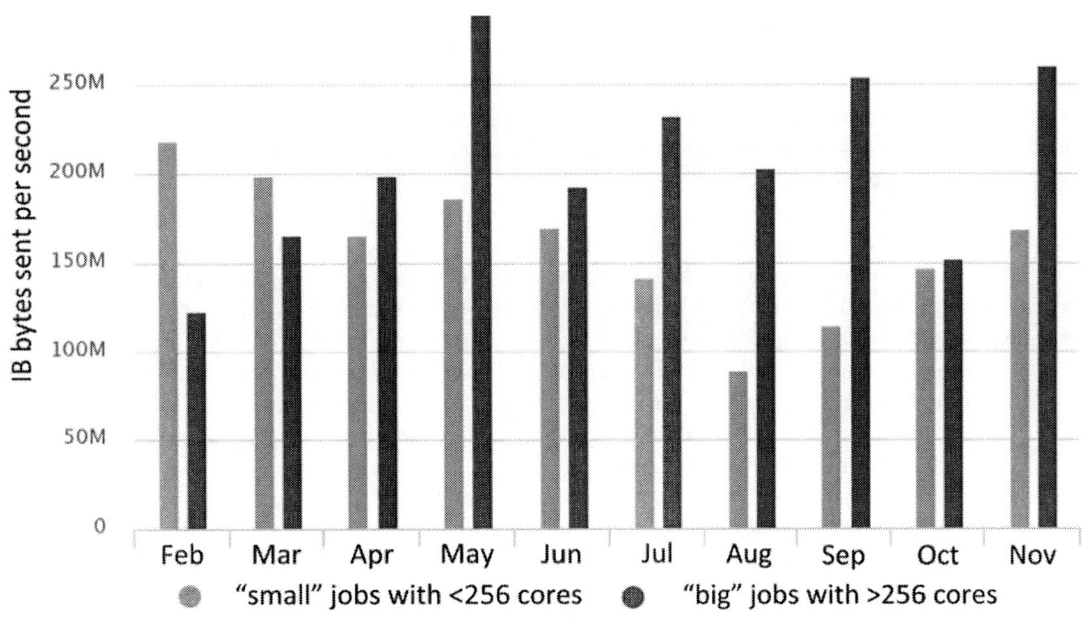

Figure 1. A number of bytes sent per second using Infiniband network, distribution by month for "small" and "big" classes.

IOP Conf. Series: Journal of Physics: Conf. Series **973** (2018) 012068 doi:10.1088/1742-6596/973/1/012068

2.1.3. Studying the distribution of values by characteristics. In this method, an independent detailed study of a certain characteristic is performed across the entire task flow. This helps identify specific patterns and assess certain performance features for a given supercomputer. Let's consider the following example we came across during our practical study. We made a list of users on the Lomonosov supercomputer with the lowest loadavg values (the average value was calculated among all the jobs launched by each user). Only those users whose jobs occupied more than 100,000 core-hours in total were included in the list. So, the list included users that: 1) consumed a substantial amount of computing resources; 2) used them inefficiently (using loadavg as a metric). One user stood out from the list, as his tasks had a very low average loadavg value of 0.07, which in general indicates totally abnormal behavior. Further study showed that most of his tasks were executed normally (with loadavg > 1), but more than 25 launches had a loadavg less than 0.1, which probably meant that the jobs were stalled. The appearance of such failed launches stopped after a certain point in time; apparently, the user was able to find the problem and fixed the situation. However, a lot of computing resources had been wasted up to that point. This is why this kind of information is very important for system administrators – situations like this need to be detected promptly and their root causes identified and excluded.

2.2. Studying the efficiency of application package usage

While the first approach is aimed at studying all jobs across the supercomputer job flow, the second approach is focused particularly on using off-the-shelf software packages. More and more users from all subject areas are using supercomputers these days; in this respect, the number of off-the-shelf application packages installed and used in practice is also increasing. This means the efficiency of these application packages will have an increasingly powerful effect on the overall supercomputer utilization efficiency. We are primarily interested in applied packages that are designed for solving particular tasks in different application areas – such as the Gromacs package [7] for modeling the molecular dynamics for physical and chemical processes, or the OpenFOAM package for the numeric modeling in continuous mechanics. However, it is also important to study the usage of system software packages – compilers, MPI libraries, etc. This study is aimed at addressing all these issues.

First of all, the study could be of interest to administrators at other supercomputer centers, as the software package study method described here can easily be implemented on most modern supercomputers. However, ordinary supercomputer users can also benefit from this information, as it helps evaluate and compare the efficiency of various packages on a given supercomputer.

Two tasks should be addressed before we can obtain the data needed to assess the efficiency of off-the-shelf software packages. The first task is to obtain execution efficiency data for all jobs; this can easily be done using monitoring systems, so we are not going to discuss this in detail.

The second task is to determine whether a certain job is using one software package or another. That would require obtaining additional information for each job, such as determining the list of linked static and dynamic libraries, the modules used, and also collecting data on the used compilers and MPI libraries as well as their launch options. In this study, we used XALT [8], a tool developed by the Texas Advanced Computing Center (TACC), to collect this kind of information. This tool was chosen because it is an open-source software that can be easily installed on most modern supercomputers and provides all the functionality needed for our study. This tool works as follows:

1. XALT replaces the standard link and launch scripts (particularly, ld and mpirun) used on supercomputers with its own wrappers.
2. Each time any user links or launches his program, XALT obtains all the information needed (libraries used, full file names, environment variables, input parameters used, etc.) and stores it in the user's home directory.
3. Obtained data is periodically stored in a single MySQL database. This database aggregates all of the "raw" data from the supercomputer and is in fact the entry point for analysis.
4. Using SQL queries, the administrator can obtain all of the relevant statistics on the utilization of software packages or libraries.

Using XALT on MSU supercomputers helped us analyze the application package behavior from various perspectives. In this article we will consider the results obtained on the Lomonosov-2 supercomputer for one specific application package – NAMD [9], which is frequently used in molecular dynamics.

Studying just the integral values for the package already provides interesting information. For example, it turned out that NAMD was the leading package in terms of job execution duration – the average execution time of a single job was more than 1 day. An average NAMD job requires about 80 compute cores, which is quite a lot for application packages. This means that the efficiency of each launch is particularly important, as it has a substantial impact on the efficiency of the entire supercomputer functioning. It is also interesting to note that NAMD uses the communication network very actively, with sustained intensity of more than 150 MB sent per second; the only off-the-shelf package showing higher network load is CP2K, which is also used in molecular dynamics.

Naturally, average values alone are not enough. For example, the average CPU load and loadavg values are 38.74% and 12.36, respectively. These may look like good values, but what is hiding behind these averages? Let's look at the distribution of these values by individual users (Figure 2).

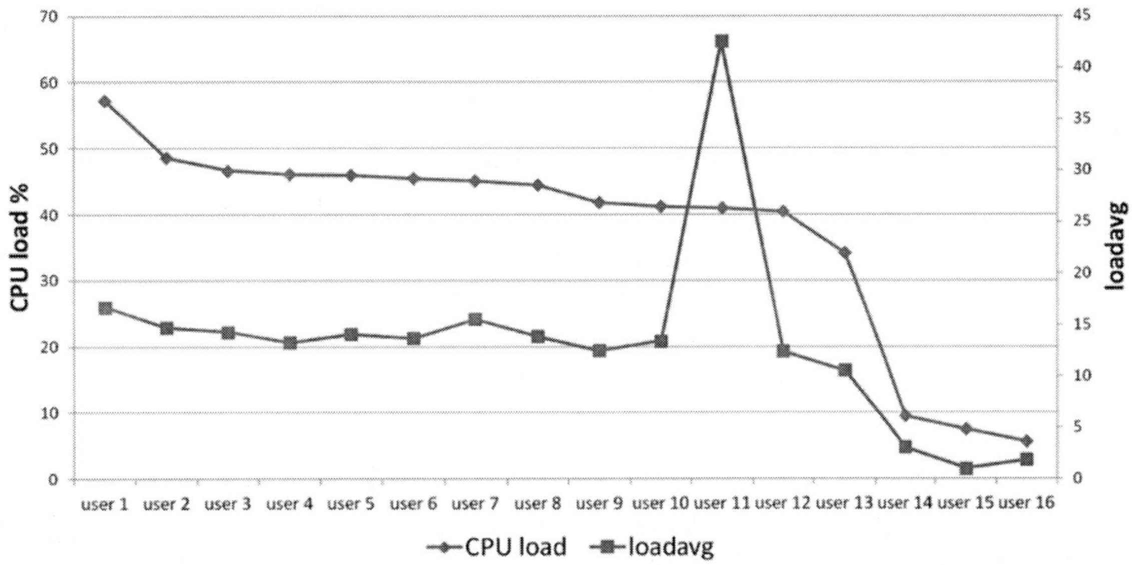

Figure 2. Average CPU load and loadavg values for NAMD package launches, distribution by individual users (sorted by the CPU load value).

You can see that most users use NAMD in a very similar manner, except for just 4 users. The most notable is user 11, as his loadavg value is clearly an outlier. A more detailed study of individual launches showed that the first three launches of this package by that user had a loadavg value of more than 100, which is definitely an abnormal behavior (normal loadavg value for Lomonosov-2 supercomputer is less than 30 active processes per node). But after that it returned to normal. Apparently the user was initially learning how to use the package (or maybe tuning package options for this particular supercomputer), after which he started utilizing the software in a normal fashion.

Users 14-16 are also outliers, with much lower average values than the other users. An analysis of individual launches for user 16 (Figure 3) shows that inefficient use of the NAMD package is quite common and continues to date – CPU load is always less than 10%, while a common value is above 40%. In this case we have a very clear suspicion that application package is used inefficiently on a regular basis. A direct contact with this user is needed to figure out if it is possible for him to use more efficient implementations within this package.

IOP Conf. Series: Journal of Physics: Conf. Series **973** (2018) 012068 doi:10.1088/1742-6596/973/1/012068

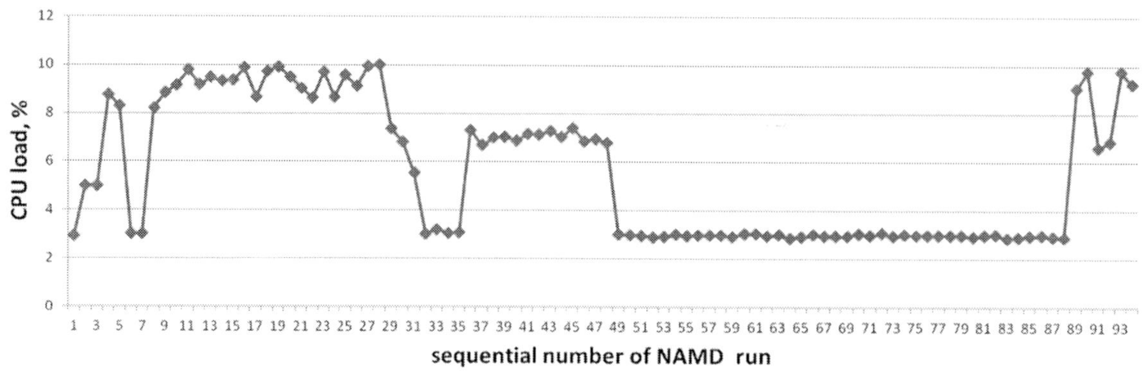

Figure 3. Average CPU load for each NAMD run by user 16, sorted by program start time.

One of the possible reasons why package usage efficiency can be quite low is that the particular version or the configuration of the package installed on the supercomputer does not suit the needs of the user. In this case – as well as in other cases when a user is unsatisfied with generally available package – the user may build by himself a local, more optimal for him version of this package. Administrators of supercomputer centers should also track this kind of activity: if a lot of people use other package versions installed locally, maybe it means that the generally available package installed by administrators should be updated or changed somehow. From the other point of view, maybe the user could not figure how to properly use the generally available package, in this case an administrator should help him to solve the issue.

That is why it is also important to track what common users install locally. We have collected the data about the usage of local libraries on Lomonosov-2 supercomputer since June this year. Top 15 most launched local libraries are shown in Table 2. Since we are interested in locally installed packages, and each package usually contains a lot of libraries that should be included in the list which makes it hard to analyze, only one mostly launched library is shown for each user.

Table 2. Top most used locally installed libraries on Lomonosov-2 supercomputer.

Library path	Number of launches
<USER1>/PROJECT/ergo_project/repos/chunks_and_tasks/source/libcht.a	731
<USER2>/soft/netcdf4-HDF5-serial/add/lib/libz.so.1.2.8	576
<USER3>/OpenFOAM/OpenFOAM-5.0/platforms/linux64GccDPInt32Opt/lib/libOpenFOAM.so	495
<USER4>/ww3/external/lib/libz.a	422
<USER5>/gromacs/src/gmxlib/.libs/libgmx_d.so.6.0.0	398
<USER6>/gcc-5.4/lib/gcc/x86_64-unknown-linux-gnu/5.4.0/crtbegin.o	291
<USER7>/progs/babel/build/lib/libopenbabel.so.5.0.0	284
<USER8>/opt_debug/lib/libnetcdf.so.7.2.0	225
<USER9>/gromacs-5.1.4/build/lib/libgromacs_mpi.so.1.4.0	209
<USER10>/Soft/third_party_libs/clapack/LIB/libblas.a	173
<USER11>/software/plumed-2.3.2/conftest-main.o	150
<USER12>/FATHOM_DIR_OMPI/zlib/zlib-1.2.8/build/lib/libz.so.1.2.8	131
<USER13>/boost_1_65_1/stage/lib/libboost_mpi.so.1.65.1	125
<USER14>/NAMD_2.10_Source/charm-6.6.1/mpi-linux-x86_64/tmp/test.o	111
<USER15>/gromacs-2016.4/build-double/lib/libgmock.a	84

IOP Conf. Series: Journal of Physics: Conf. Series **973** (2018) 012068 doi:10.1088/1742-6596/973/1/012068

It can be seen from this table that 3 different versions of GROMACS package (which is one of the most used packages on Lomonosov-2 supercomputer) are installed locally. Generally available GROMACS version is older – it seems like we should update it or install several versions for users to choose from. Also, there are 2 users that use NetCDF library which is currently not generally available on our supercomputer. Similar reasoning is also suitable for other packages and libraries on the list like NAMD, OpenFOAM, Boost, etc. All this information can be very helpful for administrators since it helps to understand what supercomputer users need in practice in terms of software packages.

2.3. Finding anomalies in the supercomputer job flow

The last approach stands somewhat aside from the first two: in this case we also study the entire range of jobs executed on the supercomputer, but the main goal is not to analyze the overall job flow statistics, or statistics for individual software packages, but to find individual applications with certain properties. Specifically, the purpose of this approach is to find individual launches with abnormally low efficiency in the overall job flow. In this case, an "anomaly" is understood not in absolute terms, but relative to the overall structure of the job flow. This way we can detect "complex" cases of inefficient behavior – low-efficiency launches that are difficult or impossible to trace using absolute threshold values. This approach uses the same system monitoring data for input as the other approaches previously described.

Since the criteria for abnormal behavior cannot be defined in advance (specifically because they will be different for job flows on other supercomputers), it was decided to use machine learning methods for this approach, namely classification methods. Each job from the overall job flow is analyzed and classified as "normal" (no efficiency issues found), "abnormal" (clearly exhibiting efficiency issues), or "suspicious" (no clear understanding of whether the job is behaving abnormally, a more detailed study is needed). It should be noted that supervised learning methods are used – the original training set is prepared and classified by hand, which is then used to train the classification method to identify the properties of a given job flow.

We considered several options for addressing this task. Initially only the integral values were considered for each property and each job [10] – just as in the two previous approaches described above. This method demonstrated a relatively high accuracy of identifying abnormal and suspicious launches, but it did have one serious drawback: it was impossible to obtain any additional information on abnormal behavior detected – when it started during the execution, its potential root causes, etc.

That is why we decided to try a different option. Execution time of each job is divided into smaller intervals. To identify intervals, we use an intellectual method that tries to detect substantial changes in the dynamic behavior of the program. Isolating such changes into separate intervals, we can accurately describe each of them using integral characteristics. Every interval is then classified separately, based on the monitoring data falling within the given interval. After each interval in the job is classified (as normal, abnormal or suspicious), the entire job is classified based on the interval information. The classification process is described in more detail in [10].

This classifier is currently working on the Lomonosov-2 supercomputer on a permanent basis, collecting statistics and notifying system administrators daily of any abnormal tasks. A brief description of the statistics collected by the developed classifier since it was launched into fully operational mode on the supercomputer (May 2017) is shown below.

A total of 251 abnormal and 5,882 suspicious jobs were detected during this period, which consumed 0.5 and 9.0 million core-hours respectively. For comparison – one day on the Lomonosov-2 supercomputer is 0.5 million core-hours; that is, all abnormal jobs have taken up about one day of supercomputer operation during the entire observation period. It should be noted that suspicious jobs have consumed a lot of computing time – about 18 days of the entire supercomputer operation; however, this suspicious behavior can be caused by a wide range of reasons. In particular, it is quite likely that the used algorithm simply cannot be implemented efficiently on the given supercomputer, so suspicious behavior may actually be normal, even though our detection of inefficient execution was correct.

Abnormal jobs clearly represent the most relevant area for analyzing supercomputer utilization efficiency – they waste computing resources, so their number should ideally be kept to a minimum. Figure 4 shows the total core-hours spent on abnormal jobs, with a breakdown by users. A total of 31 users executed abnormal jobs (and 72 had suspicious jobs). It can be seen that the first 4 users stand out from others. Even though their absolute values of consumed core-hours are not that high, this analysis warrants a discussion of the results with the users in question, to avoid wasting computing resources in the future.

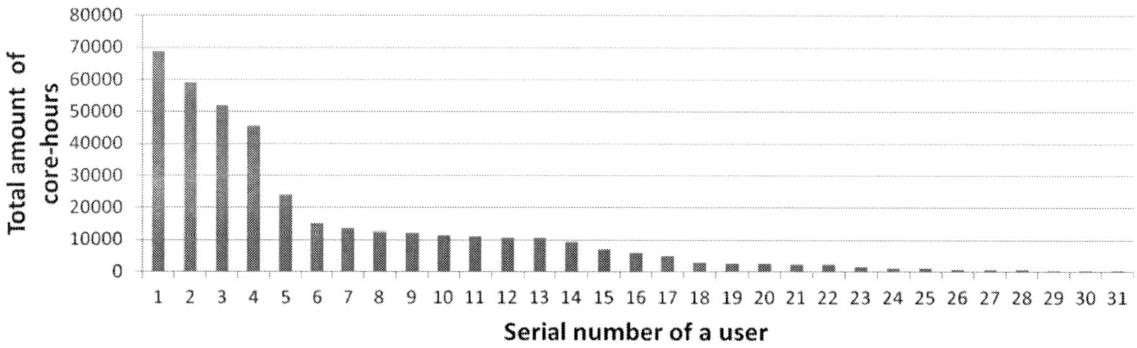

Figure 4. List of top users by the amount of core-hours spent on abnormal jobs.

3. Conclusion

In this article, we described three different approaches to analyzing supercomputer resource utilization efficiency based on system monitoring data, which are being developed at the MSU Research Computing Center.

The first approach is based on collecting and studying statistics across the entire job flow, which helps analyze the overall supercomputer resource utilization efficiency. In particular, this helps identify and analyze specific classes of jobs and the distribution of values for various characteristics to study the job flow structure.

The second approach focuses on studying more detailed statistics on the usage of application packages. Since these packages are becoming increasingly common in practice, we need to analyze and control their utilization efficiency. We addressed this task using XALT, a tool that helps track what application and system software packages were used in each job launch.

The third approach is based on detecting individual abnormal launches in the overall job flow. Its goal is to find jobs that substantially deviate from the standard behavior on a given supercomputer. Namely, this approach helps detect jobs with abnormally low execution efficiency. This task is accomplished by analyzing the dynamic characteristics for every job using machine learning techniques. Each job is divided into a set of time intervals; each interval is classified as normal, suspicious or abnormal independently. Then a classification of the whole job is performed based on interval classification results.

Each of these approaches enables us to comprehensively analyze supercomputer resource utilization efficiency from different perspectives. In the future, we plan to further explore each of these approaches, eventually merging the developed methods into a single tool aimed at supercomputer utilization efficiency analysis. One of the other main goals is to provide supercomputer users with the results of conducted efficiency analysis for their jobs, so they can also carry out the study by themselves if needed and provide us with valuable feedback on the correctness of the obtained results.

Acknowledgments
The results described in all sections except Subsection 2.3 were obtained in the Lomonosov Moscow State University with the financial support of the Russian Science Foundation (agreement N 17-71-20114). The research presented in Subsection 2.3 was supported by the Russian Foundation for Basic Research (No. 16-07-00972).

References

[1] Voevodin V and Voevodin V 2016 Efficiency of Exascale Supercomputer Centers and Supercomputing Education *High Performance Computer Applications: Proceedings of the 6th International Supercomputing Conference in Mexico (ISUM 2015)* (Springer, Cham) pp 14–23
[2] Infrastructure monitoring system Nagios URL: https://www.nagios.org/ (accessed on 22.11.2017)
[3] Stefanov K, Voevodin V, Zhumatiy S and Voevodin V 2015 Dynamically Reconfigurable Distributed Modular Monitoring System for Supercomputers (DiMMon) *Procedia Comput. Sci.* **66** pp 625–34
[4] (in Russian) Mamaeva A and Voevodin V 2017 Methods for statistical analysis of large supercomputer job flow *Supercomput. Third Russ. Supercomput. Days, RuSCDays 2017, Moscow, Russ. Sept. 25–26, 2017, Revis. Sel. Pap.* (MSU Publisher) pp 788–99
[5] Jones M D *et al.* 2017 Workload Analysis of Blue Waters *Preprint arXiv:1703.00924*
[6] Evans T, Barth W L, Browne J C, DeLeon R L, Furlani T R, Gallo S M, Jones M D and Patra A K 2014 Comprehensive Resource Use Monitoring for HPC Systems with TACC Stats *2014 First International Workshop on HPC User Support Tools* (IEEE) pp 13–21
[7] Abraham M J, Murtola T, Schulz R, Páll S, Smith J C, Hess B and Lindahl E 2015 GROMACS: High performance molecular simulations through multi-level parallelism from laptops to supercomputers *SoftwareX* **1–2** pp 19–25
[8] Agrawal K, Fahey M R, McLay R and James D 2014 User Environment Tracking and Problem Detection with XALT *2014 First International Workshop on HPC User Support Tools* (IEEE) pp 32–40
[9] Phillips J C, Braun R, Wang W, Gumbart J, Tajkhorshid E, Villa E, Chipot C, Skeel R D, Kalé L and Schulten K 2005 Scalable molecular dynamics with NAMD *J. Comput. Chem.* **26** pp 1781–802
[10] Shaykhislamov D 2016 Using Machine Learning Methods to Detect Applications with Abnormal Efficiency *Supercomputing. Second Russian Supercomputing Days, RuSCDays 2016, Moscow, Russia, September 26–27, 2016, Revised Selected Papers* (Springer, Cham) pp 345–55

IOP Conf. Series: Journal of Physics: Conf. Series **973** (2018) 012069 doi:10.1088/1742-6596/973/1/012069

Organization of functional interaction of corporate information systems

V V Safronov, V F Barabanov, S L Podvalniy and A M Nuzhnyy

Department of automated and computing systems, Voronezh State Technical University, VSTU, 396066, Russia, Voronezh, Moscow avenue, 179

E-mail: safronov.vitaliy@mail.ru

Abstract. In this article the methods of specialized software systems integration are analyzed and the concept of seamless integration of production decisions is offered. In view of this concept developed structural and functional schemes of the specialized software are shown. The proposed schemes and models are improved for a machine-building enterprise.

1. Introduction

Quite often it happens when that a lot of different software products are used by an industrial plant. This may be caused by integration of companies using different software, or simply by historical factors of development.

There are several reasons for the simultaneous application of a variety of specialized software systems:

- high complexity of today's products;
- manufacturers are transformed into transnational corporations, and for the organization of their operation data replication is required;
- assimilation of existing software infrastructures to maintain data integrity in mergers and take-over.

In this regard, analysis and integration of data from different software systems, and creation of joint documents are not only difficult, but also costly for the company.

It is important that each organizational unit operates and processes its information in its own manner. That is why in the course of corporate systems implementation special corporate standards are introduced for data exchange formats. Their integration is ensured by means of data conversion from one format to another, which often causes errors and degrades the information quality.

Among the shortcomings of the existing information systems developed by most design companies are:

- Narrow specialization of design and calculation causing incorrect problem statement and insufficient complete analysis of the results. At the same time there is no possibility for designers to carry out their own preliminary calculations of developed product, which results in complexity and timing increase in the project work as a whole.
- A large number of paper documents which has the following disadvantages: slow retrieval of documents; difficulty in document tracking at all stages of its life cycle; the duration of the

timing and coordination of documents; the greater likelihood of errors in the processing and transmission of information collection; increase in the information processing terms.
- A set of specialized and poorly integrated techniques implemented in the Microsoft Excel environment is usually used in calculations [1-2].
- Considerable length of the calculations caused by the need to perform time-consuming procedures of phased iterative calculation and parameters associating.

2. Materials and methods

2.1. Justification of the need to improve engineering enterprise's information system
Information systems of most engineering enterprises as a rule are not optimal and do not provide the possibility of carrying out all the necessary calculations for the company's specialists. It requires the involvement of third parties.

The computerization of traditional methods to solve engineering design problems at the machine-building enterprises is not systematized. This is mainly carried out by means of in-house design applications providing only particular design problems solutions [3].

Currently a full design study of the object characteristics with the required parameters of accuracy and turnaround time can be achieved through the creation of automated software and hardware systems that are based on a computer analysis using physical and mathematical models that describe the hydrodynamic, thermal, and other processes occurring in created product.

The method of defect elimination and improving the existing engineering techniques is the development of comprehensive automated software and hardware modules to be integrated into a single computational environment for managing and sharing data between software applications within a single company's information environment.

2.2. Methods of corporate software support systems for product life cycle maintenance integrating
Integration provides a solution to data mismatch problems in two or more systems used in the design organization and the construction of the organization's IT infrastructure.

Technical problems to be solved in the course of work on the systems integration include [4-5]:
- semantic data reconciliation - bringing data in different systems for "total solution";
- construction of single classifiers and directories - building a one-one correspondence between the elements of directories in different systems and fixing in the additional structures which exercise the "translation" functions;
- creation of software interfaces of integrable systems for data transfer and call system functions on external events;
- development of converters for data transmission from one system to another and output data formats for transmission including the realtime;
- logical systems binding - building algorithm which enables to display one system "events" to other systems;
- designing mechanisms for remote synchronization (replication) of data and their distributed development;
- designing interfaces that enable one to control the data flow, logic transformation and structures, define a single access rights and mechanisms for working together with data, etc;
- development of additional means of access, analysis, and collaboration in data processing.

An information system comprises the following components:
- The platform on which the other system components, including the hardware and system software, operate.
- The data that the system works with consists of DBMS and database [6].

- The applications that implement the business logic of the data system. They consist of the components of the business logic, user interface, auxiliary components and the application server which provides storage and access to application components.
- The business processes which are the scenario of the user interaction with the system [7].

It is believed that the integration of information systems is the integration of one or more components of integrable information systems.

The objectives of the platforms integration are:

- Ensuring interoperability between applications running on different hardware and software platforms.
- Enabling applications developed for a specified software and hardware platform to operate on platforms.

There are several ways to achieve these goals. Within each approach, there are different technologies:

- Remote Procedure Call.
- Middleware.
- Virtualization.

RPC (Remote Procedure Call) technology enables you to publish the procedure and to call it for applications running on other platforms. The elements of such technologies are common to all platforms description interface procedures language (IDL, WSDL), procedure's "adapter" which translates external calls into internal ones and transmits the results back, and managers responsible for the delivery of enqueries and results between platforms in the network.

The middleware ideology is to develop application software without using particular operating system services by means of middleware services.

"Virtualization" is the newest concept of platforms integration, as it greatly simplifies the use of different platforms and, accordingly, the use of systems demanding the presence of specific platforms for their functioning.

An information system works with data and is composed of a database consists of a database and a database management system. Integration at the data level suggests the data sharing from different systems. Often it turns out to be easier than the applications integration, as industrial databases which store data information systems have advanced programmatic access capabilities to stored data from other applications.

Approaches to Data Integration:

- Universal access to data.
- The data warehouse.

The universal data access technology provides uniform access to data of different DBMS through a dedicated driver [8-9].

The concept of data warehousing is to create a corporate data warehouse. A data warehouse is a database that stores the data collected from various information systems databases for further analysis. OLAP technology is used to create a data warehouse. Approaches to the creation and filling of data warehouses are reflected in the ETL paradigm (extraction, transformation, loading) [10-12].

The application level integration is put into practise through the use of ready-made application functions by other applications.

Existing approaches to application integration are:

- Applied Programming Interfaces.
- Messaging.
- Service-Oriented Architecture.
- User interfaces Integration.

Application Programming Interface of a particular system is a "declared" system's functionality which can be used outside. The functional is published as a set of functions or an object model.

Service-oriented architecture (SOA) is a modern and popular paradigm. It is a logical continuation of the Web-services concept which is to publish the functional blocks of an application form permitting other applications to get access to them by the Web. Web-service is a small add-on software application functionality that converts calls obtainable via the Web into the internal application function calls and returns the results back.

Integration at the level of enterprise applications (EAI, Enterprise Application Integration) means the use of shared code, not the internal application data. Programs are divided into components that are integrated via standardized programming interfaces and special communication software. This approach of to these components means to create a universal software kernel, which is used by all applications.

Integration at the level of user interface enables applications interconnection by means of special user interface tools.

The most comprehensive systems integration is the integration at the level of business processes. It provides application integration and data integration. Business Integration is "natural" for companies, since their work is based on business processes rather than applications, databases and platforms.

Corporate systems are complex software solutions. So it is impossible to use a single method for their integration. In order to provide specialized software integration, we suggest an approach structurally based on the individual functional features of the following methods of integration:

- platforms integration (remote procedure call, the use of middleware);
- data integration (the formation of a unified database as a part of enterprise information systems);
- integration of applications (using the API, work with a service-oriented architecture);
- integration at the level of enterprise applications;
- integration at the user interface level (creating cross-platform and inter-system interfaces, interaction of software systems);
- integration of business processes.

It is significant that the joint use of the various methods individual elements is not contrary to the basic requirements for the enterprise software interaction and operation organization. This solution creates a single solutions database for support, maintenance, and planning of the product life cycle, and for ensuring full interoperability between them. Some aspects of integration will ensure the software interaction without user's direct participation which is not typical for any of the original methods. Thus, the use of the composite system integration will allow seamless integration with the formation of a single, integrated database.

Seamless integration is ensuring the interaction of two or more software systems with a "simplification" of the user's influence on the data exchange between systems, due to the formation of a structured shared database; "embedding" translationing devices, conversion and transmission of data into the original software solution while maintaining its integrity and stability; creation and use of inter-module interfaces.

3. Results and discussion

3.1. Structural and functional schemes of the seamless enterprise software systems integration

Throughout the product lifecycle the same information is treated by different life cycle support software systems, but each system operates with data as generated and stored in an electronic database and unique information generated by this process.

Paperless technology strategy is to create a single information space for all participants throughout the product lifecycle with the creation of the EPD (Electronic Product Definition).

Software system in a single information space (SIS) provides:

- joint development of an interactive environment;
- structured electronic product description;
- data protection and access to information about the product;
- change management in an integrated database.

Without the formation of the SIS it is not possible to provide a functional, technological, information and logical compatibility and harmonization of the automated system design and technological complex with other software systems it interactes with. This means that the common principles and general rules for the formation and maintenance of information resources and information and telecommunication systems installed in the SIS should be respected by all actors of information relations, carrying out their activities and exchange of information within the framework of the company's SIS.

The main elements of the SIS are:

- information resources recorded on data carriers;
- organisational structures for the operation and development of the SIS;
- means of information interaction of subjects of information relations in the SIS to ensure regulated access to information resources on the basis of appropriate information technologies, including software, hardware, and legal documents.

The software as a part of corporate solutions is a set of highly specialized software products, what leads to a number of problems, such as their interaction and integration of software from different manufacturers. To resolve these problems, in accordance with the concept of seamless integration based on the structure of the interaction of an integrated information system in accordance with the paperless ideology we formed the scheme of functional interaction with support systems, planning and support of the product life cycle shown in Figure 1. Interaction with solutions support planning and tracking the life cycle involves the use of structural typed datasets, the introduction of the unification of data on the interaction of elements for all kinds of descriptions of Digital Prototyping allows us to develop cross-cutting process which provides a seamless integration with external software.

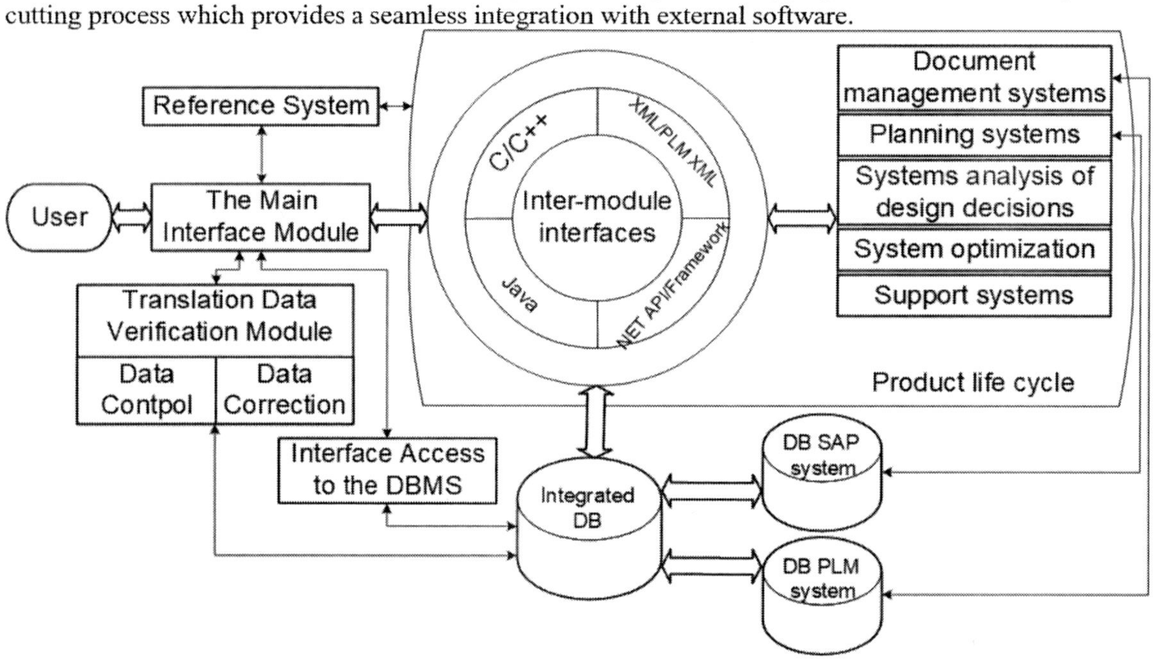

Figure 1. The general scheme of interoperability with the systems of planning, maintenance and life cycle support.

The Feature of seamless integration software solution is the organization of interaction with support creation, maintenance and life cycle planning with a single information systems integrated database by means of program interfaces inter-module integration (Figure 2).

On the basis of structural and functional scheme of software solutions a model of inter-module interfaces integration based on project data management with external specialized software was synthesized, Figure 3. The advantage of this is the formation of a seamless integration and integration of used software solutions in a single information space with an integrated data management system project.

The interaction with the database enables us to solve the problem of modules consistency and create a common information data space. The structure of the information system includes an external plug-in program graphics system, external information life cycle supporting environment, built-in reference system, external information design support environment, built-in projects integration environment, built-in information recovery data model that includes modules for interfacing with PLM, PDM, and ERP systems, modules of recovery data models, business model verification and correction modules. The structure and composition of the proposed software is not limited to the following list of modules, and it may be changed depending on the task.

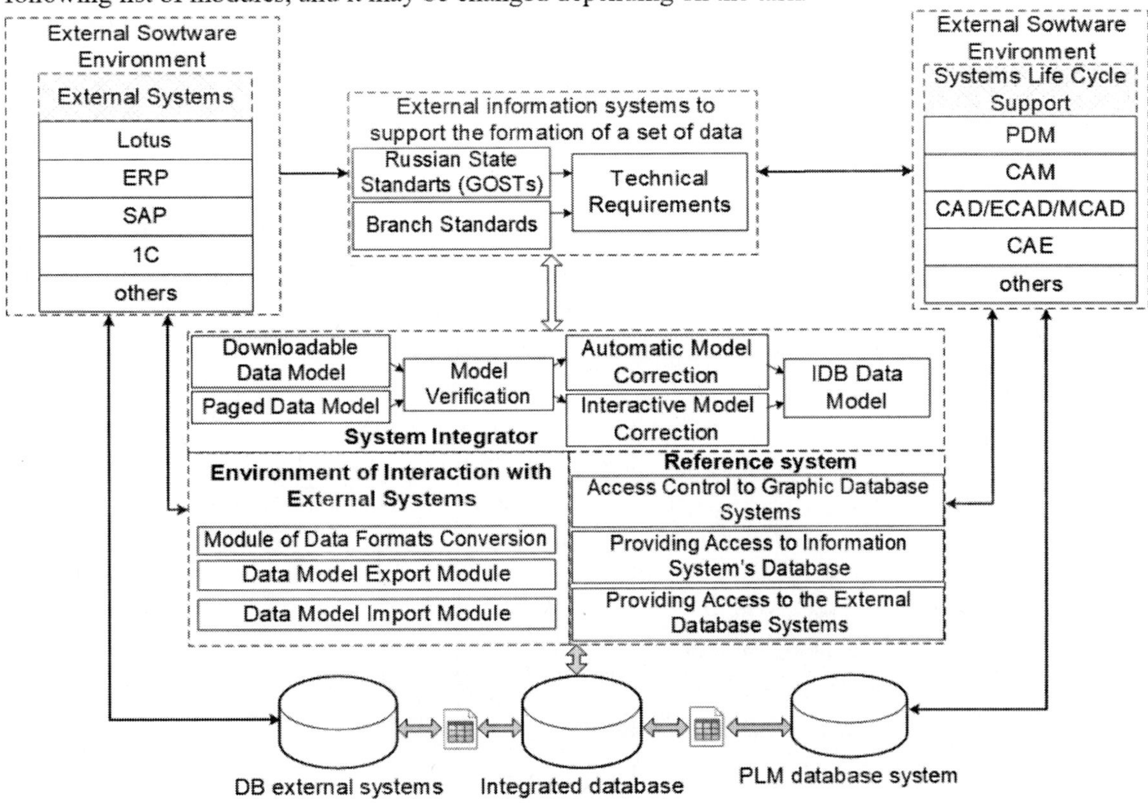

Figure 2. The Block diagram of seamless integration information system with an integrated database.

Thus, the proposed model of inter-module interfaces integration of enterprise software, modular structure of the information system and integrated graphic elements database provide a seamless integration of software systems and a complete solution to a wide range of tasks, taking into account the specificity of the subject area.

3.2. Decisions on system and subsystems structure

As an example, the realization of the automated system of a design and technological complex (AS of DTC) consisting of a set of the firmware modules (FM) is considered.

The AS of DTC has the flexible organization and possibility of adaptation under the changing external factors (such as changes in the organization of business processes, the current legislation, etc.) and provides:

- scalability by number of users and the processed information;

- archival storage of information – according to the legislation of the Russian Federation and the existing nomenclature of the engineering company affairs.

Each FM provides carrying out all necessary types of calculations and has a possibility of adaptation under the changing external factors. The offered subsystems allow capturing full activity of all projecting divisions and technological preparation of production divisions, executing integration of firmware modules into common information space, adjusting electronic document circulation and creating archive of electronic documentation.

The modules, making AS of DTC part are also intended for ensuring seamless input and output data integration. The data exchange module of electronic manufacturing techniques of products (EMTP) and standard reference information are intended for data exchange with external systems.

In PDM system, seamless integration of the input and output data of design and technological documentation preparation modules is provided by a set of the means allowing to make the automated formation of EMTP descriptions on the basis of data from the electronic structure of the product (ESP).

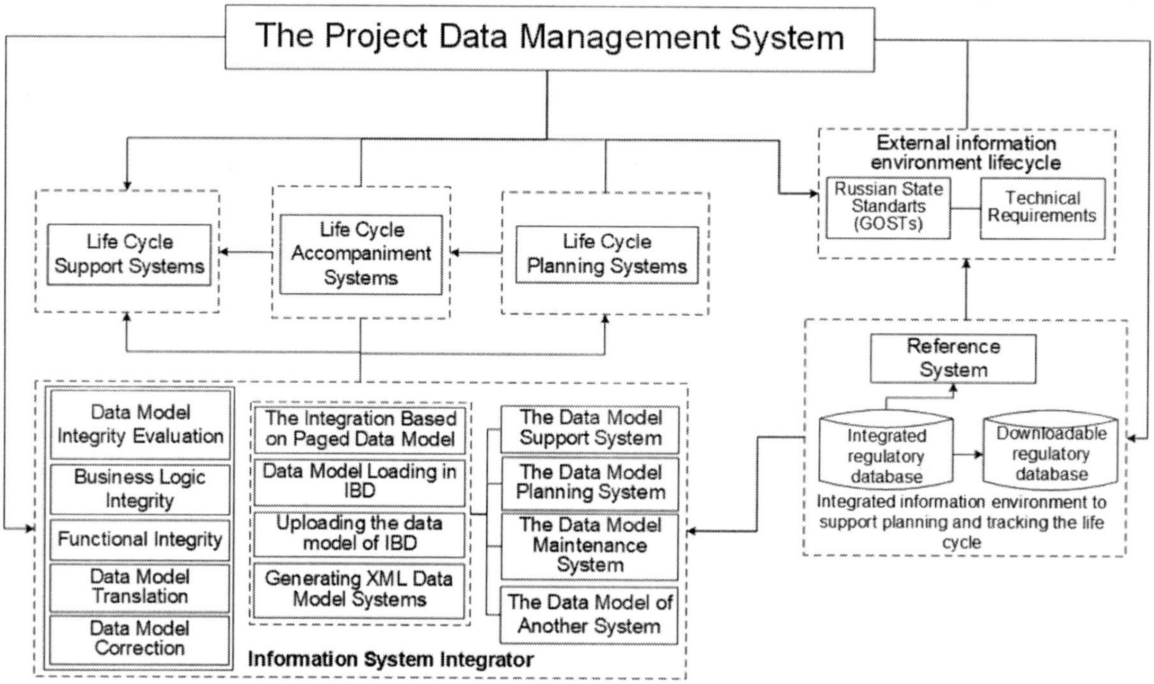

Figure 3. The Model of intermodule interfaces integration based on the project data system management.

All means of information exchange can be divided into two parts:

- means of the internal information flows organization;
- means of interaction with external information systems.

The AS of DTC provides the centralized storage, processing and control of data received from various design systems in the central database realized by means of PDM. For the solution of this task the following functional subsystems are allocated:

- design data formation subsystem;
- subsystem of the automated control and design data updating;
- subsystem of graphic design data translation;
- external information environment of design support;
- life cycle support maintenance subsystem;
- business management subsystem;
- maintenance and support subsystem.

The version of the system's block diagram is shown in Figure 4.

The Basic purpose of FM "EIS" in the AS of DTC structure is creation of a single information space taking into account maintaining electronic structure of the product, the organization of electronic document circulation of all types of the design documents prepared in AS of DTC subsystems and the solution of several tasks on the organization of the top level developed design system management functions.

The main information structure providing creation of SIS is ESP. Formation of ESP can be made in the following modes:

- the automatic mode – formation of the structural description of a product is made automatically, based on a product 3D model with use of the special modules providing the analysis of an object model of a graphic system;
- the automated mode – ESP is created based on specially prepared electronic spreadsheets (specifications) storing structure declaration and composition of a product;
- the manual mode – electronic structure of a product is created manually by ESP controls provided by PDM system.

Storage of all data as a part of ESP provides the integral description of a product used by all SIS components at all stages of life cycle.

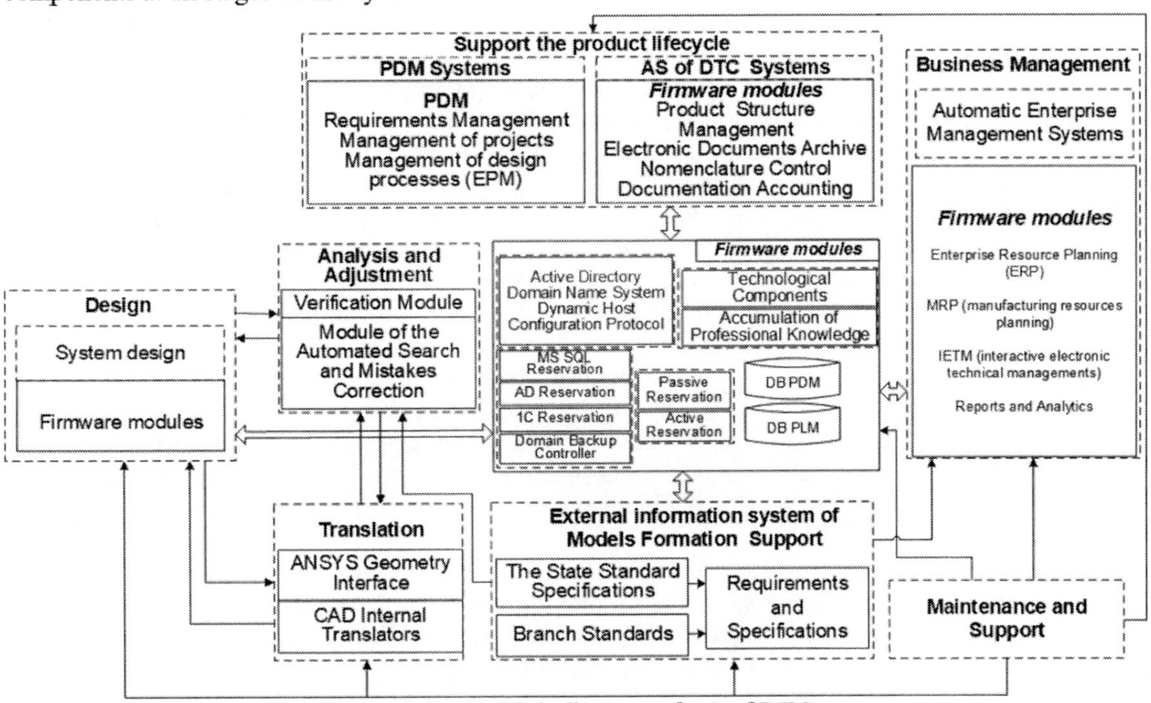

Figure 4. The blok diagram of AS of DTC.

One of the functions of FM "EIS" in AS of DTC structure is giving program and information means for design-engineering preparation of the production.

Today PDM-systems are widely used as an integration tool of the automated design and technological preparation of production systems. Also they are used for the electronic document flow organization, creation of the common information space at the enterprises engaged in design and production of difficult technical products; the organizational and technical systems providing management of all product information.

For firmware modules integration in the AS of DTC structure, for engineering data management solution, management of product information, specifications management, maintaining electronic archive of documentation it is possible to use system "1C:PDM. Engineering Data Management".

IOP Conf. Series: Journal of Physics: Conf. Series **973** (2018) 012069 doi:10.1088/1742-6596/973/1/012069

3.3. Means and ways of interaction for information exchange between system components

All means of information exchange can be divided into two parts:

- means of the internal data streams organization;
- means of interaction with external information system.

Taking into account the requirements and procedural restrictions, the decision is made:

- AS of DTC should be desighned with formation of a single information space on the basis of local area network (LAN) with newly formed FM usage.
- The expedient type of LAN organization is the multiuser client - server architecture constructed on "multicascade star" topology in the form of an independent information cluster.
- The structure of a cluster has to include client FM, servers, calculators, switching equipment, peripheral equipment.
- It is expedient to arrange cluster compactly.
- It is expedient to license the specialized software used on client AM on client-server architecture.

3.4. The basic methodological provisions

A single information space is the integrated set of the spatially distributed bases and databanks of the systems, technologies of their maintaining and use, information and telecommunication systems and networks functioning on the basis of the uniform principles and common rules, providing information exchange of all participants.

AS of DTC has hierarchical and multilevel structure of information resources.

The 1st level is the database of primary, most detailed, reliable and actual information about objects stored and used at the shop level in the system of integrated management.

The 2nd level includes the databases of the aggregated indicators characterizing condition of the subjects, objects and processes controlled by governing bodies and the enterprise.

The corporate network of data transmission for providing information exchange of AS of DTC with remote sources of information needs to be developed according to the following principles:

- possibility of further integration and use of the existing infrastructure of communication and telecommunication;
- priority investment into such infrastructure elements which will allow to achieve solving problems optimum expenses and in the shortest possible time, also will provide possibility of long operation and networks modernization without their essential reconstruction;
- choice of the specific telecommunications operator in accordance with the established procedure for carrying out management and administrative functions on the organization of users work in the corporate network;
- specific technological decisions and organizational forms of cooperation with the telecommunications operator striving for providing optimum operational costs and expenses.

Functioning of the software and hardware complex of AS of DTC and its interfacing to territorially remote local area networks is realized within a uniform corporate area network of machine-building enterprise.

4. Conclusion

The offered block and function schemes of the specialized software are developed on the basis of the concept of seamless integration with support, planning and life cycle maintenance systems. They are oriented to preservation of functional client-server model integrity.

As the integration platform management system is used for engineering data, providing structured storage of all product data in the electronic structure of the product. This approach makes it possible to construct a complete description of the product that can be effectively used components of the information structure of the enterprise at all stages of the software development life cycle.

The main requirement for support of a seamless integration of enterprise information systems is the organization of the centralized storage, control and uses of the descriptions of production executed using uniform qualifiers and reference manuals.

The software of a seamless integration is realized in the form of a series of the specialized program intermodular interfaces expanding possibilities of PDM system.

The offered means are aimed at providing interaction of planning, maintenance, support systems and problem-solving in the sphere of production life cycle management.

References

[1] Atakischev O I, Atakischeva I V and Atakischev A O 2016 *Izvestiya Yugo-Zapadnogo Gosudarstvennogo Universiteta* **20** pp 7-13
[2] Hisamutdinov M R 2015 *Scientific and technical herald in volga region* **6** pp 208-12
[3] Ratmanova I D and Pavlov M N 2006 *Information technology* **6** pp 2-11
[4] Velizhanin A S and Revnivykh A V 2012 *Bulletin of the Novosibirsk state university* **1** pp 94-103
[5] Dvoiehlazova M and Skiter I 2013 *Technical sciences and technologies* **65** pp 137-41
[6] Podvalny S L, Kravets O Ya and Barabanov V F 2014 *Automation and Remote Control* **12** pp 2225-30
[7] Podvalny S L, Vasiljev E V and Barabanov V F 2014 *Automation and Remote Control* **10** pp 1886-91
[8] Kenin S L, Barabanov V F, Nuzhny A M and Grebennikova N I 2013 *The Bulletin of Voronezh State Technical University* **3** (1) pp 4-8
[9] Nuzhny A M, Safronov V V, Barabanov A V and Gaganov A V 2013 *The Bulletin of Voronezh State Technical University* **6** (1) pp 23-7
[10] Nuzhny A M, Grebennikova N I, Barabanov A V and Povalyaev A V 2013 *The Bulletin of Voronezh State Technical University* **6** (2) pp 25-31
[11] Safronov V V, Barabanov V F, Kenin S L and Pitolin V M 2013 *Control Systems and Information Technology* **53** pp 95-9
[12] Ivaschenko A V, Barabanov V F and Podvalny E S 2015 *Automation and Remote Control* **6** pp 1081-87

IOP Conf. Series: Journal of Physics: Conf. Series **973** (2018) 012070 doi:10.1088/1742-6596/973/1/012070

Supercomputer modelling of an electronic structure for KCl nanocrystal with edge dislocation with the use of semiempirical and nonempirical models

Yu K Timoshenko[1], **V A Shunina**[2] **and A I Shashkin**[1]

[1] Voronezh State University, Universitetskaya pl., 1, Voronezh, 394006, Russia
[2] Voronezh State Technical University, Moscow Avenue, 14, Voronezh, 394026, Russia

E-mail: `yutim@phys.vsu.ru`

Abstract. In the present work we used semiempirical and non-empirical models for electronic states of KCl nanocrystal containing edge dislocation for comparison of the obtained results. Electronic levels and local densities of states were calculated. As a result we found a reasonable qualitative correlation of semiempirical and non-empirical results. Using the results of computer modelling we discuss the problem of localization of electronic states near the line of edge dislocation.

1. Introduction

Apparently the first calculations of equilibrium positions of atoms nearby from edge dislocation core made in 1960s [1]. Over the years many calculations of atomic and electronic structure of dislocations have been carried out. Principally were investigated the metals and the covalent semiconductors. However very few of such calculations are made for the ionic systems in view of the creation difficulties of a suitable mathematical model taking into account the polarization effects. However the situation becomes simpler if we consider a dislocation in a nanocrystal. The dislocations in nanosystems were studied for example in papers [2, 3].

2. Models and methods

In our work a comparison of the calculations in the framework of semiempirical and non-empirical mathematical models of electronic structure of the KCl nanocrystal with an edge dislocation are presented. The edge dislocation had Burgers vector $\langle 110 \rangle$ and was located in the glide plane $\{1\bar{1}0\}$ as well as in papers [4, 5]. The $K_{172}Cl_{172}$ nanocrystal was under consideration. The coordinates of ions of this nanocrystal with edge dislocation were the same as well as ones in paper [4]. In this paper the relaxation of the nanocrystal was considered in terms of the model of pointlike polarizable ions. In this approximation, the potential energy of interaction of the ions of the system is

$$W = W_C + W_R + W_{VdW} + W_{pol},$$

where W_C is the energy of the Coulomb interaction of the ions; W_R is the energy of repulsion of electron shells; W_{VdW} is the van der Waals energy with allowance for the dipole–dipole interaction; W_{pol} is the polarization term (see details in [4]). In the model under consideration,

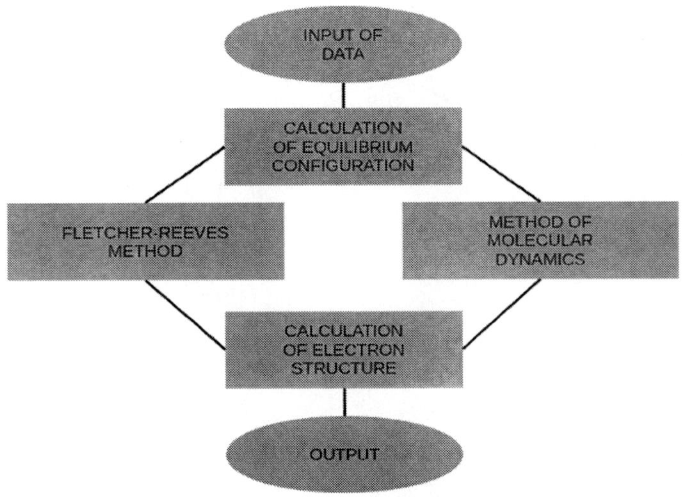

Figure 1. Scheme of semiempirical calculations of equilibrium configuration of nanocrystal and its electron structure.

each ion represents a pointlike quasi-elastic dipole. In the zero approximation, all moments of quasi-elastic dipoles were equal to 0. Then, in the process of the minimization of the potential energy of the system with respect to ion coordinates, which was performed by the Fletcher-Reeves method [6], the dipole moments were calculated in a self-consistent way. After the calculation of the equilibrium configuration of the nanocrystal, its electron structure was considered. We used the Slater-Koster variant of the tight-binding method modified by one of us (Yu. K. T.), in which the self-consistency was performed with respect to the effective charges and dipole moments of the ions. The results of semiempirical calculations can be seen in [4]. Thus the calculations of ions' coordinates of equilibrium configuration of nanocrystal and the calculations of its electron structure are two separate steps (see figure 1).

For checkout of a correctness of semiempirical results we have executed non-empirical calculations also. In principle, this geometry could be defined with the use of nonempirical mathematical model. However geometry optimization of a nanocrystal being composed of 344 ions requires too much computational time. The calculations of one-electronic energy spectrum and partial local density of states (PLDOS) in the framework of nonempirical mathematical model were carried out only. The Hartree-Fock-Rutaan approximation and pseudopotential [7] were used. For comparison the electronic structure of nanocrystal $K_{196}Cl_{196}$ with parallelepiped shape (in a unit of interionic distance) without a dislocation was calculated also. The equilibrium spatial arrangement of ions in this nanocrystal was calculated also with using of semiempirical model of pointlike polarizable ions.

3. Results

The one-electronic levels of these nanocrystals calculated in framework of semiempirical and non-empirical approaches are presented on figure 2. The energy levels of the highest occupied nanocrystal orbitals (HONO) for $K_{196}Cl_{196}$ in semiempirical and nonempirical calculations were assigned zero for convenience of comparison. Symbol "$\lhd$" marks the energy levels of the lowest unoccupied nanocrystal orbitals (LUNO). According to figure 2 the HONO and LUNO levels

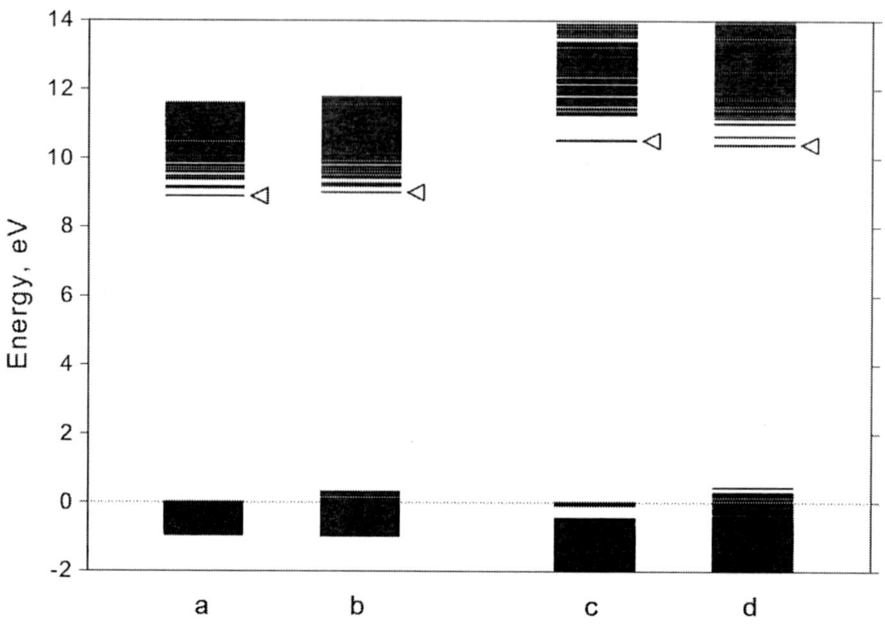

Figure 2. a, b: $K_{196}Cl_{196}$ and $K_{172}Cl_{172}$ (semiempirical model); c, d: $K_{196}Cl_{196}$ and $K_{172}Cl_{172}$ (non-empirical model)

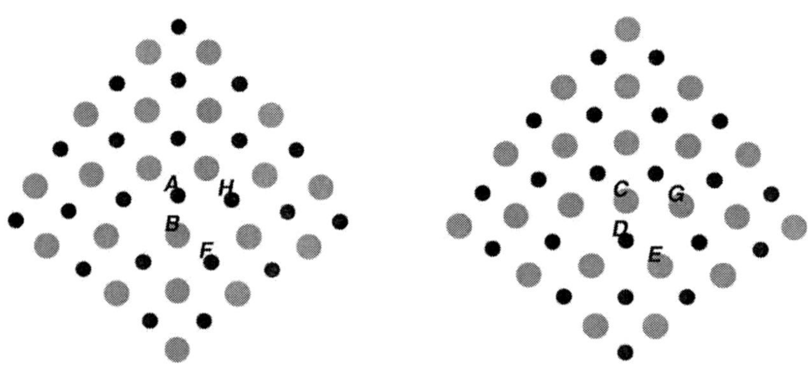

Figure 3. The central layers of nanocrystal $K_{172}Cl_{172}$. Black disks – cations; gray disks – anions.

are more widely separated from neighbouring levels in nonempirical calculations than in the semiempirical ones. It is remarkable that in semiempirical and nonempirical calculations the edge dislocation affects LUNO energy levels to only a small extent, but the HONO energy levels increase rather appreciably. The analysis of PLDOS has shown that HONO and LUNO levels of $K_{196}Cl_{196}$ (figure 2(c)) correspond to the one-electron quantum states localised near ions of nanocrystal corners. In case of $K_{172}Cl_{172}$ nanocrystal these levels are associated with the states localised on the surface ions, including dislocation line intersections with the surfaces of nanocrystal. Special interest causes the research of localisation conditions near dislocation core in central region of a nanocrystal where surface influence is weakened. Two central layers of a

IOP Conf. Series: Journal of Physics: Conf. Series **973** (2018) 012070 doi:10.1088/1742-6596/973/1/012070

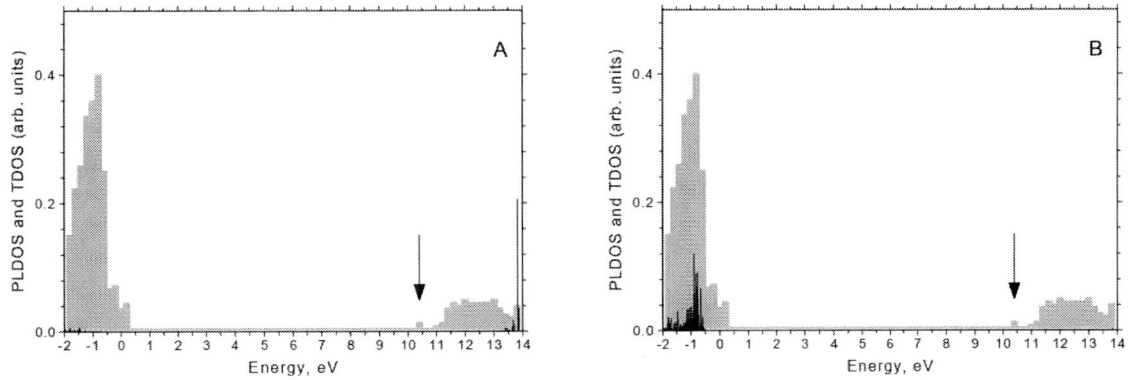

Figure 4. 3p PLDOS for ions A and B (see figure 3) and histogram of TDOS in arbitrary units. The arrow matches the E_{LUNO}.

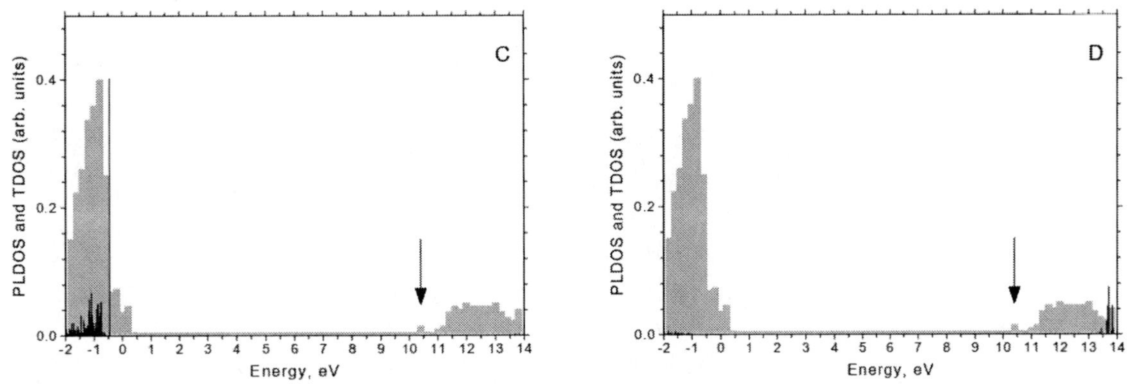

Figure 5. 3p PLDOS for ions C and D (see figure 3) and histogram of TDOS in arbitrary units. The arrow matches the E_{LUNO}.

nanocrystal are presented on figure 3, the letters mark the ions for which were calculated the partial local densities of states (PLDOS). The PLDOS for these ions are shown on figure 4, 5, 6. It was found that the occupied states are localised to a greater degree near anions C and G. Let's notice that the highest peaks in these PLDOS are located on 0.5 eV less than HONO energy. PLDOS for cations near with dislocation core in the central layers of nanocrystal do not contain the peaks testifying against some essential localisation of unoccupied states, related to edge dislocation presence. The results of non-empirical calculations are in reasonably good agreement with the results of semiempirical calculations [4, 5].

Non-empirical quantum-chemical calculations were made with the use of the program "Gaussian 03" [8] which was installed on the parallel computer cluster of the Digital Technologies Department of Computer Sciences Faculty, Voronezh State University, and with the use of the program "Firefly" [9, 10] on the supercomputer "Chebyshev" of Moscow State University.

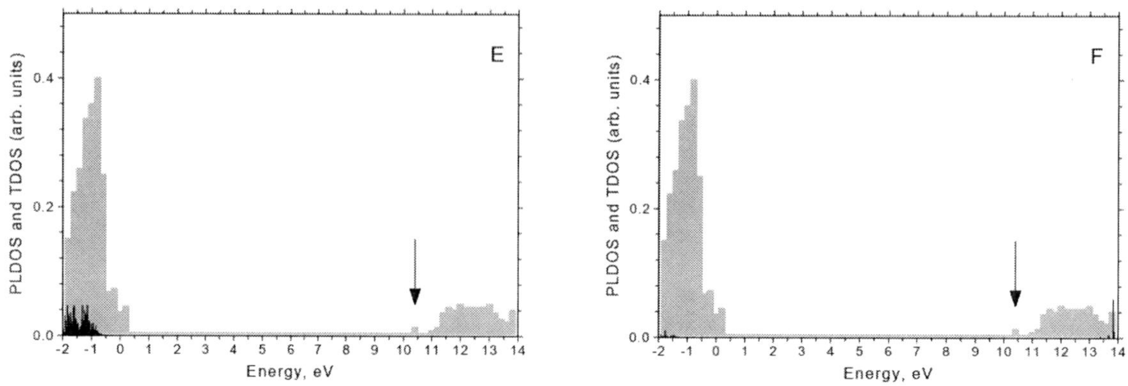

Figure 6. 3p PLDOS for ions E and F (see figure 3) and histogram of TDOS in arbitrary units. The arrow matches the E_{LUNO}.

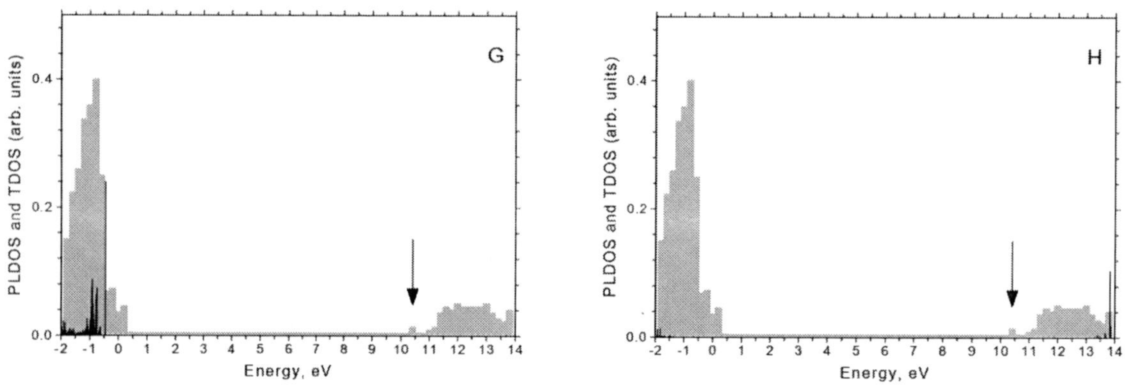

Figure 7. 3p PLDOS for ions G and H (see figure 3) and histogram of TDOS in arbitrary units. The arrow matches the E_{LUNO}.

4. Conclusions

Thus, by using of the obtained results it seems reasonable to say that the representations about strong localization of electron states on an edge dislocation line in potassium chloride crystal with positions from edges of bands on > 1.5 - 2 eV (see introduction of paper [4]) are rather doubtful.

Acknowledgments

The research is carried out using the equipment of the shared research facilities of HPC computing resources at Lomonosov Moscow State University. Many thanks to Professor S. D. Kurgalin for the possibility of using the computer cluster of Voronezh State University.

References

[1] Beeler J R 1970 *Advanced in Materials Research* vol 4 ed Herman H (New York: Interscience) p 295
[2] Masuda-Jindo K, Hung V V and Menon M 2005 *Phys. Stat. Sol. (c)* **2** 1781
[3] Omar M S and Taha H T 2009 *Physica B* **404** 5203

IOP Conf. Series: Journal of Physics: Conf. Series **973** (2018) 012070 doi:10.1088/1742-6596/973/1/012070

[4] Timoshenko Yu K and Shunina V A 2002 *Physics of the Solid State* **44** 888
[5] Timoshenko Yu K and Shunina V A 2005 *Phys. Stat. Sol. (c)* **2** 1788
[6] Bunday B D 1984 *Basic Optimisation Methods* (London: E. Arnold)
[7] Hay P J and Wadt W R 1985 *J. Chem. Phys.* **82** 270
[8] Frisch M J, Trucks G W, Schlegel H B, Scuseria G E, Robb M A, Cheeseman J R, J A Montgomery J, Vreven T, Kudin K N, Burant J C, Millam J M, Iyengar S S, Tomasi J, Barone V, Mennucci B, Cossi M, Scalmani G, Rega N, Petersson G A, Nakatsuji H, Hada M, Ehara M, Toyota K, Fukuda R, Hasegawa J, Ishida M, Nakajima T, Honda Y, Kitao O, Nakai H, Klene M, Li X, Knox J E, Hratchian H P, Cross J B, Adamo C, Jaramillo J, Gomperts R, Stratmann R E, Yazyev O, Austin A J, Cammi R, Pomelli C, Ochterski J W, Ayala P Y, Morokuma K, Voth G A, Salvador P, Dannenberg J J, Zakrzewski V G, Dapprich S, Daniels A D, Strain M C, Farkas O, Malick D K, Rabuck A D, Raghavachari K, Foresman J B, Ortiz J V, Cui Q, Baboul A G, Clifford S, Cioslowski J, Stefanov B B, Liu G, Liashenko A, Piskorz P, Komaromi I, Martin R L, Fox D J, Keith T, Al-Laham M A, Peng C Y, Nanayakkara A, Challacombe M, Gill P M W, Johnson B, Chen W, Wong M W, Gonzalez C and Pople J A 2004 *Gaussian 03, Revision C.02* (Wallingford CT: Gaussian, Inc.)
[9] Granovsky A A Firefly version 7.1.g http://classic.chem.msu.su/gran/firefly/index.html
[10] Schmidt M W, Baldridge K K, Boatz J A, Elbert S T, Gordon M S, Jensen J H, Koseki S, Matsunaga N, Nguyen K A, Su S, Windus T L, Dupuis M and Montgomery J A 1993 *J. Comput. Chem.* **14** 1347

Institute of Physics
Dirac House, Temple Back
Bristol BS1 6BE UK

ISSN: 1742-6588
ISBN 978-1-5108-6041-4

9 781510 860414

CURRAN ASSOCIATES INC.
proceedings.com